Organic Chemistry

Maitland Jones, Jr.

PRINCETON UNIVERSITY

W. W. NORTON & COMPANY
NEW YORK • LONDON

Organic Chemistry

THIRD EDITION

W.W. Norton & Company has been independent since its founding in 1923, when William Warder Norton and Mary D. Herter Norton first published lectures delivered at the People's Institute, the adult education division of New York City's Cooper Union. The Nortons soon expanded their program beyond the Institute, publishing books by celebrated academics from America and abroad. By mid-century, the two major pillars of Norton's publishing program—trade books and college texts—were firmly established. In the 1950s, the Norton family transferred control of the company to its employees, and today—with a staff of four hundred and a comparable number of trade, college, and professional titles published each year—W.W. Norton & Company stands as the largest and oldest publishing house owned wholly by its employees.

Composition: TSI Graphics

Illustrations: J/B Woolsey

Manufacturing by R. R. Donnelley

Book design by Rubina Yen

Cover design by Maitland Jones

Director of Manufacturing/College: Roy Tedoff

Editor: Vanessa Drake-Johnson

Copy Editor: Jeannette Stiefel

Production Editors: Chris Granville, Sarah Chamberlin

Editorial Assistant: Erin O'Brien

Library of Congress Cataloging-in Publication Data

Jones, Jr., Maitland, 1937–
    Organic chemistry/Maitland Jones, Jr.—3rd ed.
      p. cm.
    Includes bibliographical references and index.
    ISBN 0-393-92408-4
    1. Chemistry, Organic.   I. Title

QD253.2.J66 2003

547—dc22   2003061280

ISBN 0-393-92408-4

W. W. Norton & Company, Inc., 500 Fifth Avenue, New York, NY 10110

www.wwnorton.com

W. W. Norton & Company Ltd., Castle House, 75/76 Wells Street, London W1T 3QT

1 2 3 4 5 6 7 8 9 0

PERIODIC TABLE OF THE ELEMENTS

Atomic masses are based on ^{12}C.

Atomic masses in parentheses are for the most stable isotope.

Groups

Period	1A	2A	3B	4B	5B	6B	7B	8B	8B	8B	1B	2B	3A	4A	5A	6A	7A	8A
1	1 **H** 1.008																	2 **He** 4.003
2	3 **Li** 6.941	4 **Be** 9.012											5 **B** 10.811	6 **C** 12.011	7 **N** 14.007	8 **O** 15.999	9 **F** 18.998	10 **Ne** 20.180
3	11 **Na** 22.990	12 **Mg** 24.305											13 **Al** 26.982	14 **Si** 28.086	15 **P** 30.914	16 **S** 32.066	17 **Cl** 35.453	18 **Ar** 39.948
4	19 **K** 39.098	20 **Ca** 40.078	21 **Sc** 44.956	22 **Ti** 47.867	23 **V** 50.942	24 **Cr** 51.996	25 **Mn** 54.938	26 **Fe** 55.845	27 **Co** 58.933	28 **Ni** 58.693	29 **Cu** 63.546	30 **Zn** 65.39	31 **Ga** 69.723	32 **Ge** 72.61	33 **As** 74.922	34 **Se** 78.96	35 **Br** 79.904	36 **Kr** 83.30
5	37 **Rb** 85.468	38 **Sr** 87.62	39 **Y** 88.906	40 **Zr** 91.224	41 **Nb** 92.906	42 **Mo** 95.94	43 **Tc** (98)	44 **Ru** 101.07	45 **Rh** 102.906	46 **Pd** 106.42	47 **Ag** 107.868	48 **Cd** 112.411	49 **In** 114.818	50 **Sn** 118.710	51 **Sb** 121.760	52 **Te** 177.60	53 **I** 126.904	54 **Xe** 131.29
6	55 **Cs** 132.905	56 **Ba** 137.327	71 **Lu** 174.967	72 **Hf** 178.49	73 **Ta** 180.948	74 **W** 183.84	75 **Re** 186.207	76 **Os** 190.23	77 **Ir** 192.217	78 **Pt** 195.078	79 **Au** 196.967	80 **Hg** 200.59	81 **Tl** 204.383	82 **Pb** 207.2	83 **Bi** 208.980	84 **Po** (209)	85 **At** (210)	86 **Rn** (222)
7	87 **Fr** (223)	88 **Ra** 226.025	103 **Lr** (262)	104 **Rf** (261)	105 **Ha** (262)	106 **Sg** (263)	107 **Ns** (262)	108 **Hs** (265)	109 **Mt** (266)									

*Lanthanide Series

57 **La** 138.906	58 **Ce** 140.116	59 **Pr** 140.908	60 **Nd** 144.908	61 **Pm** (145)	62 **Sm** 150.4	63 **Eu** 151.96	64 **Gd** 157.25	65 **Tb** 158.925	66 **Dy** 162.50	67 **Ho** 164.930	68 **Er** 167.26	69 **Tm** 168.934	70 **Yb** 173.04

†Actinides

89 **Ac** 227.028	90 **Th** 232.038	91 **Pa** 231.036	92 **U** 238.029	93 **Np** 237.048	94 **Pu** (244)	95 **Am** (243)	96 **Cm** (247)	97 **Bk** (247)	98 **Cf** (251)	99 **Es** (252)	100 **Fm** (257)	101 **Md** (258)	102 **No** (259)

Key:
- Atomic number
- Symbol for element
- Atomic mass

Legend: Metal · Nonmetal · Metalloid

SUBSTITUTED HYDROCARBONS (continued)

Ethers

$CH_3CH_2OCH_2CH_3$
Diethyl ether

Ethyl phenyl ether

Sulfides (thioethers)

$CH_3CH_2SCH_2CH_3$
Diethyl sulfide

Nitro compounds	Nitroso compounds	Sulfoxides	Sulfones	Sulfonic acids
Nitromethane	Nitrosomethane	Dimethyl sulfoxide	Dimethyl sulfone	Methanesulfonic acid

CARBONYL (C=O) COMPOUNDS

Aldehydes

Benzaldehyde

Acetaldehyde

Ketones

Acetone
(dimethyl ketone)

Carboxylic acids

Benzoic acid

Acetic acid

DERIVATIVES OF CABOXYLIC ACIDS

Esters	Amides	Acid chlorides	Anhydrides	Nitriles (cyanides)
Ethyl acetate	Acetamide	Acetyl chloride	Acetic anhydride	CH_3CH_2—C≡N Propionitrile (Ethyl cyanide)
Methyl benzoate	Benzamide	Benzoyl chloride		

This book is dedicated to those who oppose tyrannies, large and small, everywhere. For personal reasons I want to single out the intrepid people of Tibet, who continue to resist the cruel occupation of their country, and those who were punished for their support for democracy at Tiananmen in June, 1989.

Brief Contents

Contents

20   Special Topic: Aromatic Transition States; Orbital Symmetry   1119

21   Special Topic: Intramolecular Reactions and Neighboring Group Participation   1175

Preface to the Third Edition

The third edition of this book, like the first and the second, is aimed not just at the chemistry major, but instead at anyone who wants a broad, yet modern introduction to the subject. General principles are stressed, because it is impossible to memorize all the details of this vast subject these days. One can, however, learn to make connections and to apply widely a set of organizing general principles, and that helps to render the prolific detail more manageable.

WHAT'S NEW?

In general, the changes in this Third Edition reflect suggestions and comments made by users. There has been a general, and vigorous call for "reactivity early" for example, and I agree with that notion. We responded by pushing simple HX addition reactions right up front in Chapter 3. Many users preferred not to separate the two carbonyl chapters. We discovered that by reuniting the two chapters, there was a way to do a substantial streamlining, and to put in a needed section on the properties of solvents.

What's not new? The voice of the book is the same; as before it is personal, and talks directly to the student. There are hints on how to do things everywhere, and we have added the features summarized below that try to make this "conversation" between student and book even more helpful. On occasion, the book will try to help students through a tough part of the subject by pointing out that it *is* tough and then suggesting ways to deal with it. Everyone who has taught this subject knows there are such places—when we talk to students ourselves, we try to use our experience to help students prevail, even when we know the going is likely to be difficult for some readers—the book tries to do the same thing.

Reactivity has been brought forward and now appears very early in the book. In Chapter two (Alkanes), there is a new treatment of Lewis acids and Lewis bases, done in the context of reactions with alkyl cations. This introduction leads to a longer set of six new sections in Chapter 3 (Alkenes and Alkynes) on addition reactions. Simple uncatalyzed and catalyzed HX additions to alkenes are described and regiochemistry (but not stereochemistry, which must wait for Chapter 4) is discussed. This material allows an introduction of selectivity and a beginning of synthesis at a very early point in the year.

In the First and Second Editions, the two chapters on carbonyl chemistry (addition reactions, and enol chemistry) were separated by a chapter collating, summarizing, and extending alcohol chemistry. In the new Third Edition, that intervening chapter is gone, as is a late chapter on amines. In their stead, there is a new Chapter 6 in which halides, alcohols, and amines are introduced and discussed. In addition, there is a new section in Chapter 6 on the properties of solvents. All of this introductory material leads directly into Chapter 7, which deals with the fundamental S_N1, S_N2, E1, and E2 reactions. Duplication is eliminated, carbonyl chemistry united, and an early exposure to the molecules extensively featured in Chapter 7 is gained. Chapter 7 also includes a new section on how to do synthesis.

Every chapter now begins with a Preview section in which the coming chapter is outlined. At the end of the Preview, there are small sections on the Essential Skills and Important Details coming in the chapter. At exam time, the student can use these sections as pointers for study and review.

New Summary sections have been incorporated into every chapter. Thus, at five or six points in every chapter the narrative is broken and the reader is brought up to date on the important points of the previous topic. Key concepts and points are highlighted. Of course, these summaries will serve as excellent "reminder and review" sections when a student is studying for an exam.

The end-of-chapter Summary sections of the previous editions have been maintained—The New Concepts, New Reactions and Mechanisms, Syntheses and Common Errors sections are all retained.

A BRIEF TOUR OF THE THIRD EDITION

To understand atoms and molecules one must first think sensibly about electrons, and for that we need to explore a bit of what quantum mechanics tells us. That does *not* mean we will all have to become mathematicians. Far from it. Our discussion here will be purely qualitative, since we need only grasp qualitatively what the mathematicians have to say to us. Qualitative molecular orbital theory is not too complicated a subject for students, and

requires no mathematics. Yet, this simple theory is amazingly powerful in its ability to rationalize and, especially, to predict structure and reactivity. The abbreviated tutorial in Chapter 1 on qualitative applications of molecular orbital theory is likely to be new to students. This material is very important, as it enables us to emphasize explanations throughout the rest of the book. Not only are traditional subjects such as aromaticity and conjugation (Chapters 12–14) more accessible with the background of Chapter 1, but explanations for the essential, building-block reactions of organic chemistry (Chapters 7 and 9, for example) become possible. There is, after all, no essential difference between the classic statement "Lewis acids react with Lewis bases" and the idea that the interaction of a filled and empty orbital is stabilizing. The latter formulation allows all sorts of seemingly disparate reactions to be gathered together—unified—in a very useful way. (For example, a hydride shift and the S_N1 reaction become partners in a unified theory rather than two wildly different reactions that must be memorized in all their detail.)

We also need a modicum of mathematics because of a problem of scale. We live in a world of certain sizes, and our intuitions are honed by everyday experience. We know how baseballs and basketballs behave. Unfortunately, our experience in the world of such objects poorly equips us to anticipate matters in the large-scale universe—pulsars, quasars, black holes, and other bizarre inhabitants of the macro-universe. What is generally less well appreciated by students is that we operate intuitively just as poorly on the micro scale—the world of atoms and electrons.

Atoms and molecules are really quite strange things; they are not like the objects that populate our daily experience. For one thing, molecular events take place on a time scale we cannot easily appreciate. Under typical conditions in the gas phase a single molecule undergoes 10^{10}–10^{11} collisions per second. Let's call one of those collisions an ordinary event in a molecule's "life," analogous to an ordinary event in a human life—a heartbeat, for example. If your heart beats 60 times a minute, you will experience more than 2.5 billion (2.5×10^9) beats over an 80-year lifespan. So, in one second, a molecule undergoes more ordinary events than your heart will beat in your entire life. We do not appreciate such differences easily!

The language that makes both the macro scale and the micro scale accessible to us is mathematics. Although we need not do the mathematical operations ourselves, we do need to appreciate some of the things that quantum mechanics has to say to us. Chapter 1 also focuses strongly on Lewis structures—pictorial representations of atoms and ions. The ability to write good Lewis structures easily and to determine the locations of charges in molecules with ease is an essential skill. This skill is part of the language of chemistry and will be as important in Chapter 23 as it is in Chapter 1.

After the introductory chapter comes a sequence of four chapters devoted largely to aspects of structure (2–5). Here the details of the archetypal structures of organic chemistry are introduced. Hybridization appears, and the wonderful three-dimensionality of the subject begins to grow in. Some functional groups are introduced, and stereochemistry is dealt with in depth. A particularly vexing and fundamental question concerns what makes two atoms or molecules the same or different? The third edition includes a preview of NMR spectroscopy in Chapter 2. This section is not detailed—it is only an introduction, but it allows a real discussion of that elusive question of "difference." It also allows reinforcement through a series of new problems introduced throughout the first half of the book.

As mentioned earlier, there is a new treatment of reactivity in Chapter 3. Addition of "HX" molecules to alkenes allows an introduction to synthesis, as well as a discussion of selectivity, catalysis, and reaction mechanism in general.

After the series of "structure" chapters comes another new feature, Chapter 6 on alcohols, amines, halides, and the properties of solvents. This new chapter functions as a lead in to a discussion of several building block reactions, the S_N2, S_N1, E2, and E1 reactions.

Chapter 7 is one of the key chapters, and the reactions discussed here—substitution and elimination—serve as reference points throughout the book; they are fundamental reactions to which we return over and over in later chapters in order to make analogies.

Once we have these basic, reference reactions under control, a general discussion of the role of energetics—kinetics and thermodynamics—becomes appropriate (Chapter 8). Chapters 9 and 10 introduce other building-block reactions, and other functional groups, in the context of an expansion of the earlier discussion of addition reactions in Chapter 3. Even at these early stages we introduce the biological applications of organic chemistry. For example, in the chapters devoted to the structure of alkanes, and, especially, alkenes, biorelevant examples appear. These do not obscure the essential information of the chapters, however. They are kept as examples, potential extensions and applications of what we have learned at this point. Later on in the book their role is expanded, with whole chapters (22 and 23) devoted to biological topics.

The basic reactions of the middle chapters provide a foundation for the chemistry of carbonyl compounds, the subject of a series of chapters in the second half of the book (16–19).

Chapter 15, set in the midst of this run of chapters, is devoted to spectroscopy. Sections of earlier chapters have already dealt with parts of this subject in an introductory way (Chapter 2 for NMR, Chapter 12 for UV/vis). As there is no consensus on where to put this material, I have tried to make this chapter free-standing.

To make an analogy to the study of a language, in the first sequence of reaction chapters (7, 9, and 10) we write sentences constructed from the vocabulary and grammar developed in the early, structural chapters. We will go on in later chapters in the book to more complicated mechanisms and molecules, and to write whole paragraphs and even short essays in organic chemistry. Some of those essays are contained in the "special topics" chapters toward the end of the book (20–23) in which biological chemistry and physical-organic chemistry are further explored with the material of the early chapters as a foundation.

SPECIAL FEATURES OF THIS BOOK

VOICE. A feature carried forward from the first edition is that this book talks directly to the reader, not only about the material at hand, but also about the "how and why" of organic chemistry. It tries to do so in a voice that is personal, not distant. I think it is much easier to enjoy, and learn, organic chemistry if a strong focus on "Where are we and why are we here?" and "What is the best way to do this?" is maintained. In this book, I often stop to discuss problem solving, and the number of these sections has increased in the third edition. Of course, there is no substitute for a thorough understanding of the material, and advice cannot substitute for that knowledge. But our subject is an experience-intensive one, and by definition students are short on that commodity. It can be a great help to be shown profitable (and unprofitable) approaches. The main arena for such talk must be the lecture, and no book can hope to substitute for the direction and inspiration given by the lecturer. At the same time, it is important for students to hear different voices, to see the path we are all on in different ways. This approach provides at least two views of the material, two different routes to understanding. The sum of these perspectives is likely to be greater than either of them alone.

There is a growing feeling among teachers of organic chemistry that the traditional lecture course is, by itself, not an efficient way to teach the skills necessary to become facile in the subject. Increasingly, we are moving to hybrid courses, in which small-group problem-solving sessions augment, or even replace, traditional lectures. The personal, "talk directly to the student" voice of this book is specifically designed to help in this endeavor. When students have to pick up material on their own, they prefer a voice that talks to them, and explains what's important and why, to a straightforward "list of facts" approach.

FLEXIBILITY. There is no consensus on the precise order in which to take up many subjects in organic chemistry. This book makes different decisions possible. For example, as already pointed out, the spectroscopy chapter is largely free-standing. Traditionally, it comes where it is here, roughly at the midpoint of the book. But cogent arguments can be made that spectroscopy should be introduced earlier, and that is possible, if one is willing to pay the price of delaying the serious discussion of chemical reactions one more chapter.

The last few chapters now explicitly constitute a series of special topics. No one really hopes to finish everything in an organic textbook in one year, and this book provides a number of choices. One might emphasize biological aspects of our science, for example, and Chapters 22 and 23 provide an opportunity to do this. Alternatively, a more physical approach would see the exciting chemistry of Chapters 20 and 21 as more appropriate.

EXTENSIVE USE OF FIGURES AND CONSTRUCTIVE USE OF COLOR. Organic chemistry is a highly visual subject. Chemists think by constructing mental pictures of molecules, and we talk to each other by drawing pictures. This book favors series of figures over long discussions in the text. The text serves to point out the changes in successive figures. Color is used to highlight change, and often to track the fates of atoms and groups in reactions. The use of color is judicious, however. There are times when "more is less," and the use of color in organic textbooks can easily fall into this trap. Color is for emphasis. Too much color leads to visual clutter, in which case the emphasis intended by color can be lost.

CROSS-REFERENCING. There is extensive cross-referencing in this book, which is facilitated by the early use of prototypal reaction examples to which analogy is repeatedly made. The reader is consistently reminded that the current material is "not really new—you have seen something like it before, and it's on page xxx." The hope is to supply a unifying treatment in which seemingly different topics are brought together. In turn, this reduces the need to memorize.

PROBLEMS. This book incorporates problems in two ways. First, there are many unsolved problems scattered throughout the text, and more, of all degrees of difficulty, are found at the end of each chapter. These range from drill exercises and simple examples designed to emphasize important skills and illustrate techniques, to sophisticated, challenging problems. In the latter cases, we are careful to provide hints and references to material useful for the solution. The third edition contains many new problems, most devoted to synthetically useful reactions. As before, answers to all unsolved problems can be found in the Study Guide. The Study Guide tries to help students learn how to approach problems, and gives much more than bare-bones answers to most problems. Second, and most critical, is the use of problems as an integral part of the discussion in the text. There are times when a skill must be mastered or a point well understood before further discussion can take place effectively. Rather than simply emphasizing this point in the discussion, a solved, illustrative problem is almost always provided. It is important that these problems be worked. The answer is not there just to be read. These problems are vital to effective progress through the text and must be "read with a pencil."

IMPORTANT DETAILS

AUTHENTICITY AND REACTION DESCRIPTIONS.
As in the earlier editions, the new reactions that have been added have been pursued back to the original literature in most cases, and at least to the review literature in all cases. When reactions are discussed, both the general case and at least one specific example are shown. Generality is emphasized, but the particular is not ignored.

CONVENTION ALERTS.
One factor that can make organic chemistry difficult is that a language must be learned. Organic chemists talk to each other using many different conventions, and at least some of that language must be learned or communication is impossible. In a real sense, this is analogous to learning the vocabulary and grammar of a language. In addition to general treatments of nomenclature at the beginnings of many chapters, we have incorporated numerous "Convention Alerts" in which aspects of the language chemists use are highlighted.

END-OF-CHAPTER SUMMARIES.
Each chapter ends with a summary of the important concepts, reaction mechanisms, and synthetic procedures discussed therein. These recapitulate and reinforce the material of the chapter. These summaries are further extended with problems and detailed solutions in the Study Guide. In each summary, there is also a section on common errors. What kinds of traps can one easily fall into? How can they be avoided?

IN-CHAPTER SUMMARY SECTIONS.
There are many moments in any course on organic chemistry when it is important to take stock of where we are. What have we done, why did we do it, and where are we going? The Preview sections in each chapter do some of this, but there are now many points in the book where we overtly pause

to take stock and summarize. Sometimes these are whole sections, sometimes just a sentence or two. These sections allow us to focus and keep the direction in which we are going straight in our heads. If you read them and it all makes sense, the proper bells ring, then you are probably in good shape. If something makes no sense, it is probably time to do some reading and problem working.

KEY TERMS.
In each chapter, important terms are printed in boldface at their first mention. These terms are then collected in a section at the end of the chapter and defined again in an overall glossary at the back of the book.

BIOLOGICAL EXAMPLES.
Throughout the book, reference is made to the connection between organic chemistry and the world of biology. Although many biological reactions are given, these are framed as examples and do not preempt discussions of simple, prototypal reactions. We all have to walk before we can run. Almost every chapter now has a section devoted to the biological relevance of the new reactions discussed.

ORGANIC REACTION ANIMATIONS.
Since its second year, this book has been graced by its association with the "Organic Reaction Animations" software created by Steven Fleming, Paul Savage, and Greg Hart of Brigham Young University. There are now over 50 different reactions in this splendid collection, and these have been fully integrated into the text of the third edition, with marginal icons like the one shown alongside identifying each reaction and its place in the book. All versions of ORA from Version 2.1 on also include tutorials on the reactions themselves.

Name of reaction

ORGANIC REACTION ANIMATIONS *INTERACTIVE*,
a new, interactive version of the animations software, allows the user to rotate and zoom within the reactions as they occur and includes optional voiceover narration. A new feature will allow instructors to expand the reactions to the full screen for use in lecture.

WEB STRUCTURES.
One of the most difficult things for students at the beginning of their study of organic chemistry is the three-dimensionality of the subject.

There is just no avoiding the fact that for many people learning to see in 3-D is not easy. To help in this effort, we have built a web site with over 400 compounds. The structures were made in Chem-3D and can be manipulated in space and viewed in several ways (ball and stick, space-filling, etc.). In addition there is a short writeup and usually a few questions (and answers!) for most of the molecules.

INSTRUCTOR RESOURCES

Instructor's Manual/Supplemental Problems Set, by Maitland Jones, Jr., *Princeton University,* and Timo Ovaska, *Connecticut College*
Power Point slides, available at wwnorton.com/nrl
Transparency set, with approximately 150 key figures from the text

STUDENT RESOURCES

Study Guide/Solutions Manual, by Maitland Jones, Jr., and Henry L. Gingrich, *Princeton University*
Supplementary Problems Set, Student Version, by Timo Ovaska, *Connecticut College*

ACKNOWLEDGMENTS. When I was 13, desperate to spend a summer or two on the tennis circuit, my parents had the wit to see that I was no future Laver or McEnroe, and to insist, over my strenuous objections, that I get a job. That "job" as a volunteer bottle washer and gopher began my lifelong association with William Doering. I went on as an undergraduate to learn organic chemistry from Professor Doering in what was the most exciting (and demanding) course I ever took, and I remained at Yale to do my graduate work with him. I remain his student today, and there is no easy way to acknowledge adequately my debt to him. I can only hope that this book helps others to get off to a similar start, and that they will encounter someone of Doering's intellectual honesty and power along their way. Some of Doering's work appears on the cover of this Third Edition.

After serving my sentence as a bottle washer for Doering I began real laboratory work, and was fortunate to fall under the influence of two masters, Lawrence H. Knox and Gunther Laber. Much of the pleasure of organic chemistry comes from the design and execution of a well-done experiment. From these two patient teachers I learned the great pleasure in carrying out the manipulations of organic chemistry: preparing the exquisite glassware, setting up the apparatus, and finally handling the wonderful liquids and powders that were to be combined to make something new. The cover of the second edition featured some of Larry Knox's research. This too-little-known master deserves far greater recognition.

I have been lucky enough to have studied with some wonderful teachers both before and after my days as a university student. Stanley Feret and Margaret Frankel taught me in elementary school, and were surely greatly responsible for setting me on my way. Arthur B. Darling, who taught me American History in high school, was particularly inspiring. One could ask for no better example of how to communicate a love for a subject than his. And 40 years ago I spent a postdoctoral year with Jerome Berson, a man who wrote complex papers on carbonium ions that I couldn't yet read sensibly. I learned to read those papers during my year with Jerry, but more important, I discovered an intellectual atmosphere that affected me forever.

Chemists have families, too. Some even have lives. If I have had one during the construction and revisions of this book, it is largely the result of the efforts of my family. Adequate thanks to all of them would be far too long to fit here, but their names should be mentioned, and they should know that I understand some of what they put up with. So, thanks to Stephanie and Matt for cheerful, if often crazed chatter by e-mail; to Mait and Perla for delightful dinners in Brooklyn; and to Hilary and Chad for being so splendidly all-Galaxy. It also seems that as time goes on, one acquires the wonderful creatures called grandchildren. There is not much in the world that can be described as pure pleasure—but these small people certainly seem to fill that role, at least for the generation once-removed. So, many thanks to the aforementioned kids for producing Io, Ray, Charlie, Helen, Adda, and Jane. My parents really did extract me from the tennis court and get me into a lab for the first time. They had the intelligence to see that a kid who was devoted to picking up frogs might find science a passion. I'm sorry that my father never got to see this book, but he surely had a lot to do with it. I'm glad that my mother did see it, and I hope she thought it was worthwhile driving that 13-year-old to and from the lab all those years ago. My wife Susan bore much of the brunt of the writing of this book. Too many absences, too much preoccupation, all coped with. Love and great thanks to her.

Books don't get written by setting an author on his or her way and then waiting for the manuscript to appear. There is a great deal of work to be done; misguided authors need to be straightened out, bribes and threats need to be administered, and commas inserted and deleted. In general, it is an editor's job to make it possible for the author to do the best of which he is capable. Don Fusting, Joe Wisnovsky and Vanessa Drake-Johnson at W. W. Norton were exemplary in their execution of those duties. Jeannette Stiefel was copy editor for all editions and has not only worked heroically but also taught me much. Kate Barry and Christopher Granville, as project editors at Norton, were always helpful in keeping me on track and in finding those errors and inconsistencies I missed. Joel Dubin and the rest of the crew at J/B Woolsey Associates did a superb job of producing the art for the first edition, and TSI Graphics kept up the high standard for the second edition; all coped with great skill and reasonable cheer with the many corrections.

This book also profited vastly from the comments and advice of an army of reviewers, and I am very much in their debt. Their names and affiliations follow this Preface. Two special reviewers, Henry L. Gingrich of Princeton and Ronald M. Magid of the University of Tennessee, read the work line by line, word by word, comma by missing comma. Their comments, pungent at times, but helpful always, were all too accurate in uncovering both the gross errors and lurking oversimplifications in the early versions of this work. Jerry Berson, Wes Borden, and Josef Michl all made special comments on parts of the first edition for which I am very grateful. The book is much improved by the efforts of all the reviewers.

Ron Magid, Peter Gaspar, and Bob Willcott remain steadfast friends from our days together at Yale, many years ago, and are still people with whom I can talk and learn about obscure chemical (and other) subjects. Many other colleagues over the years have also been especially helpful to me; Bob Moss at Rutgers and Matt Platz at Ohio State, as well as Jeff Schwartz and Bob Pascal at Princeton are prime examples.

Despite all the efforts of editors and reviewers, errors will persist. These are my fault only. When you find them, let me know.

Princeton,
November, 2003

FIRST EDITION REVIEWERS: William F. Bailey, University of Connecticut; Ronald J. Baumgarten, University of Illinois at Chicago; John I. Brauman, Stanford University; Donald B. Denney, Rutgers University; John C. Gilbert, University of Texas at Austin; Henry L. Gingrich, Princeton University; Richard K. Hill, University of Georgia; A. William Johnson, University of Massachusetts; Guilford Jones, II, Boston University; Joseph B. Lambert, Northwestern University; Steven V. Ley, Imperial College of Science, Technology and Medicine; Ronald M. Magid, University of Tennessee-Knoxville; Eugene A. Mash, Jr., University of Arizona; Robert J. McMahon, University of Wisconsin-Madison; Andrew F. Montana, California State University-Fullerton; Roger K. Murray, Jr., University of Delaware; Thomas W. Nalli, State University of New York at Purchase; Patrick Perlmutter, Monash University; R. M. Paton, University of Edinburgh; Matthew S. Platz, Ohio State University; Lawrence M. Principe, Johns Hopkins University; Charles B. Rose, University of Nevada-Reno; John F. Sebastian, Miami University; Carl H. Schiesser, Deakin University; Martin A. Schwartz, Florida State University; Jonathan L. Sessler, University of Texas at Austin; Robert S. Sheridan, University of Nevada-Reno; Philip B. Shevlin, Auburn University; Harry H. Wasserman, Yale University; Craig Wilcox, University of Pittsburgh; David R. Williams, Indiana University.

SECOND EDITION REVIEWERS: John Barbaro, University of Georgia; David M. Birney, Texas Tech University; Peter Buist, Carleton University; David Goldsmith, Emory University; Nancy S. Goroff, State University of New York, Stony Brook; David N. Harpp, McGill University; S. Bruce King, Wake Forest University; Grant Krow, Temple University; Lydia McKinstry, University of Nevada, Las Vegas; Kathleen S. Richardson, Capital University; Christian Rojas, Barnard College; Alan M.

Rosan, Drew University; Valerie V. Sheares, Iowa State University; Edward Turos, University of South Florida.

THIRD EDITION REVIEWERS: Mark Arant, University of Louisiana at Monroe; Arthur Ashe, University of Michigan; Michael Biewer, University of Texas at Dallas; David Birney, Texas Tech University; Jeffrey Charonnat, California State University, Northridge; Marc d'Arlacao, Tufts University; Robert Flowers, Lehigh University; David C. Forbes, University of South Alabama; B. Lawrence Fox, University of Dayton; Ian Hunt, University of Calgary; Richard Keil, University of Washington; Philip Le Quesne, Northeastern University; Robert Loeschen, California State University, Long Beach; Carl Lovely, University of Texas at Arlington; John McClusky, University of Texas at San Antonio; Keith Mead, Mississippi State University; Kathleen Morgan, Xavier University of Louisiana; Alan Rosan, Drew University; Matthew Sigman, University of Utah; William Tam, University of Guelph; David Wiedenfeld, New Mexico Highlands University.

Introduction

These days, a knowledge of science must be part of the intellectual equipment of any educated person. Of course, that may always have been true, but, I think there can be no arguing that an ability to confront the problems of concern to scientists is especially important today. Our world is increasingly technological, and many of our problems, and the answers to those problems, have a scientific or technological basis. Anyone who hopes to understand the world we live in, to evaluate many of the pressing questions of the present and the future—and vote sensibly on them—must be scientifically literate.

The study of chemistry is an ideal way to acquire at least part of that literacy. Chemistry is a central science in the sense that it bridges such disparate areas as physics and biology, and connects those long-established sciences to the emerging disciplines of molecular biology and materials science. Similarly, as this book shows, organic chemistry sits at the center of chemistry, where it acts as a kind of intellectual glue, providing connections between all areas of chemistry. One does not have to be a chemist, or even a scientist, to profit from the study of organic chemistry.

The power of organic chemistry comes from its ability to give insight into so many parts of our lives. How does penicillin work? Why is Teflon nonstick? Why does drinking a cup of coffee help me stay awake? How do plants defend themselves against herbivores? Why is ethyl alcohol a depressant? All these questions have answers based in organic chemistry. And the future will be filled with more organic chemistry—and more questions. What's a buckyball or a nanotube and how might it be important to my life? How might an organic superconductor be constructed? Why is something called the Michael reaction important in a potential cancer therapy? Read on, because this book will help you to deal with questions such as these, and many more we can't even think up yet.

WHAT IS ORGANIC CHEMISTRY? Organic chemistry is traditionally described as the chemistry of carbon-containing compounds. In the last century, it was thought that organic molecules were related in an immutable way to living things, hence the term "organic." The idea that organic compounds could be made only from molecules derived from living things was widespread, and gave rise to the notion of a vital force being present in carbon-containing molecules. In 1828 Friedrich Wöhler (1800–1882) synthesized urea, a certified organic substance, from heating ammonium cyanate, a compound considered to be inorganic.* Wöhler's experiment really did not speak to the question of vital force, and he knew this. The problem was that at the time there were no sources of ammonium cyanate that did not involve such savory starting materials as horns and blood—surely "vital" materials. The real *coup de grâce* for vitalism came some years later when Adolph Wilhelm Hermann Kolbe (1818–1884) synthesized acetic acid from elemental carbon and inorganic materials in 1843–1844 (see structures below).

Acetic Acid   **Urea**

Despite the demise of the vital-force idea, carbon-containing molecules certainly do have a strong connection to living things, including ourselves. Indeed, carbon provides the backbone for all the molecules that make up the soft tissues of our bodies. Our ability to function as living, sentient creatures depends on the properties of carbon-containing organic molecules, and we are about to embark on a study of their structures and transformations.

Organic chemistry has come far from the days when chemists were simply collectors of observations. In the beginning, chemistry was largely empirical, and the questions raised were, more or less, along the lines of, What's going to happen if I mix this stuff with that stuff?, or, I wonder how many different things I can isolate from the sap of this tree? Later, it became possible to collate knowledge and to begin to rationalize the large numbers of collected observations. Questions now could be expanded to deal with finding similarities in different reactions, and chemists began to have the ability to make predictions. Chemists began the transformation from the hunter--gatherer stage to modern times, in which we routinely seek to use what we know to generate new knowledge.

Many advances have been critical to that transformation; chief among them is our increased analytical ability. Nowadays the structure of a new compound, be it isolated from tree sap or produced in a laboratory, cannot remain a mystery for long. Today, the former work of years can often be accomplished in hours. This expertise has enabled chemists to peer more closely at the "why" questions, to think more deeply about reactivity of molecules. This point is important because the emergence of unifying principles has allowed us to teach organic chemistry in a different way; to teach in a fashion that largely frees students from the necessity to memorize organic chemistry. That is what this book tries to do: to teach concepts and tools, not vast compendia of facts. The aim of this book is to provide frameworks for generalizations, and the discussions of topics are all designed with this aim in mind.

We will see organic molecules of all types in this book. Organic compounds range in size from hydrogen (H_2)—a kind of honorary organic molecule even though it doesn't contain carbon—to the enormously complex collections of atoms in biomolecules, which typically contain thousands of atoms and have molecular weights in the hundreds of thousands. Despite this diversity, and the apparent differences between small and big molecules, the study of all molecular properties always begins the same way, with structure. Structure determines reactivity, which provides a vehicle for navigating from the reactions of one kind of molecule to another and back again. So, early on, this book deals extensively with structure.

WHAT DO ORGANIC CHEMISTS DO? Structure determination has traditionally been one of the things that practicing organic chemists do with their lives. Early on, such activity took the form of uncovering the gross connectivity of the atoms in the molecule in question: What was attached to what? Exactly what are those molecules isolated from the Borneo tree or made in a reaction in the lab? Such questions are quickly answered by application of today's powerful spectroscopic techniques, or, in the case of solids, by X-ray diffraction crystallography. And small details of structure lead to enormous differences in properties: morphine, a pain-killing agent in wide current use, and heroin, a powerfully addictive narcotic, differ only by the presence of

---

* Wöhler's urea is an end product of the metabolism of proteins in mammals, and is a major component of human urine. An adult human excretes about 25 g (6–8 level teaspoons) of urea each day. The formation of urea is our way of getting rid of the detritus of protein breakdown through a series of enzymatic reactions. If you are missing one of the enzymes necessary to produce urea, it's very bad news indeed, as coma and rapid death result.

two acetyl groups (CH$_3$CO units), a tiny difference in their large and complex structures.

Today, much more subtle questions are being asked about molecular structure. How long can a bond between atoms be stretched before it goes "boing" in its quiet, molecular voice, and the atoms are no longer attached? How much can a bond be squeezed? How much can a bond be twisted? These are structural questions, and reveal much about the properties of atoms and molecules—in other words, about the constituents of us and the world around us.

Many chemists are more concerned with how reactions take place, with the study of "reaction mechanisms." Of course, these people depend on those who study structure; one can hardly think about how reactions occur if one doesn't know the detailed structures—connectivity of atoms, three-dimensional shape—of the molecules involved. In a sense, every chemist must be a structural chemist. The study of reaction mechanisms is an enormously broad subject. It includes people who look at the energy changes involved when two atoms form a molecule, or, conversely, when a molecule is forced to come apart to its constituent atoms, as well as those who study the reactions of the huge biomolecules of our bodies—proteins and polynucleotides. How much energy is required to make a certain reaction happen? Or, how much energy is given off when it happens? You are familiar with both kinds of process. For example, burning is clearly a process in which energy is given off as both heat and light.

Chemists also want to know the details of how molecules come together to make other molecules. Must they approach each other in a certain direction? Are there catalysts—molecules not changed by the reaction—that are necessary? There are many such questions. A full analysis of reaction mechanism requires a knowledge of the structures and energies of all molecules involved in the process, including species called intermediates, molecules of fleeting existence that cannot usually be isolated because they go on quickly to other species. One also must have an idea of the structure and energy of the highest energy point in a reaction, called the transition state. Such species cannot be isolated—they are energy maxima, not energy minima—but they can be studied nonetheless. We will see how.

Still other chemists focus on synthesis. The goal in such work is the construction of a target molecule from smaller, available molecules. In earlier times the reason for such work was sometimes structure determination. One set out to make a molecule one suspected of being the product of some reaction of interest. Now, determination of structure is not usually the goal. And it must be admitted that Nature is still a *much* better synthetic chemist than any human. There is simply no contest; evolution has generated systems exquisitely designed to make breathtakingly complicated molecules with spectacular efficiency. We cannot hope to compete. Why, then, even try? The reason is that there is a cost to the evolutionary development of synthesis, and that is specificity. Nature can make a certain molecule in an extraordinarily competent way, but Nature can't make changes on request. The much less efficient syntheses devised by humans are far more flexible than Nature's, and one reason for the chemist's interest in synthesis is the possibility of generating molecules of Nature in systematically modified forms. We hope to make small changes in the structures and to study the influence on biological properties induced by those changes. In that way it could be possible to find therapeutic agents of greatly increased efficiency, for example, or to stay ahead of microbes that become resistant to certain drugs. Nature can't quickly change the machinery for making an antibiotic molecule to which the microbes have become resistant, but humans can.

WHAT'S HAPPENING NOW? IT'S NOT ALL DONE. In every age, some people have felt that there is little left to be done. All the really great stuff is behind us, and all we can hope for is to mop up some details; we won't be able to break really new ground. And every age has been dead wrong in this notion. By contrast, the slope of scientific discovery continues to increase. We learn more every year, and not just details. Right now the frontiers of molecular biology—a kind of organic chemistry of giant molecules, I would claim—are the most visibly expanding areas, but there is much more going on.

In structure determination, completely new kinds of molecules are appearing. For example, just a few years ago a new form of carbon, the soccer-ball-shaped C$_{60}$, was synthesized in bulk by the simple method of vaporizing a carbon rod and collecting the products on a cold surface. Even more recently it has been possible to capture atoms of helium and argon inside the soccer ball. These are the first neutral compounds of helium ever made. Molecules connected as linked chains or as knotted structures are now known. What properties these new kinds of molecules will have no one knows. Some will certainly turn out to be mere curiosities, but others will influence your lives in new and unexpected ways.

The field of organic reaction mechanisms continues to expand as we become better able to look at detail. For example, events on a molecular time scale are becoming visible to us as our spectrometers become able to look at ever smaller time periods. Molecules that exist for what seems a spectacularly short time—microseconds or nanoseconds—are quite long lived if one can examine them on the femtosecond time scale. Indeed, the Nobel prize for chemistry in 1999 was given to Ahmed H. Zewail (b. 1946) of Caltech for just such work. Nowadays we are moving ever further into the strange realm of the attosecond time regime. We are sure to learn much more about the details of the early stages of chemical reactions in the next few years. At the moment, we are still defining the coarse picture of chemical reactions. Our resolution is increasing, and we will soon see micro details we cannot even imagine at the moment. It is a very exciting time. What can we do with such knowledge? We can't answer that question yet, but chemists are confident that with more detailed knowledge will come an ability to take finer control of the reactions of molecules. At the other end of the spectrum, we are learning how macromolecules react, how they coil and uncoil, arranging themselves in space so as to bring two reactive molecules to just the proper orientation for reaction. Here we are seeing the bigger picture of how much of Nature's architecture is designed to facilitate positioning and transportation of molecules to reactive positions. We are learning how to co-opt Nature's methods by modifying the molecular machinery so as to bring about new results.

We can't match Nature's ability to be specific and efficient. Over evolutionary time, Nature has just had too long to develop methods of doing exactly the right thing. But we are learning how to make changes in Nature's machinery—biomolecules—that lead to changes in the compounds synthesized. It is likely that we will be able to coopt Nature's methods, deliberately modified in specific ways, to retain the specificity but change the resulting products. This is one frontier of synthetic chemistry.

The social consequences of this work are surely enormous. We are soon going to be able to tinker in a controlled way with much of Nature's machinery. How does humankind control itself? How does it avoid doing bad things with this power? Those are not easy questions, but there is no hiding from them. We are soon going to be faced with the most difficult social questions of human history, and how we deal with them will determine the quality of the lives of us and our children. That's one big reason that education in science is so important today. It is not that we will need more scientists; rather it is that we must have a scientifically educated population in order to deal sensibly with the knowledge and powers that are to come. So, this book is not specifically aimed at the dedicated chemist-to-be. That person can use this book, but so can anyone who will need to have an appreciation of organic chemistry in his or her future—and that's nearly everyone these days.

HOW TO STUDY ORGANIC CHEMISTRY

WORK WITH A PENCIL. My old professor told me, "Organic chemistry must be read with a pencil." Truer words were never spoken. You can't read this book, or any chemistry book, in the way you can read books in other subjects. You *must* write things as you go along. There is a real connection between the hand and the brain in this business, it seems. When you come to the description of a reaction, especially where the text tells you that it is an important reaction, by all means take the time to draw out the steps yourself. It is not enough to read the text and look at the drawings; it is not sufficient to highlight. Neither of these procedures is reading with a pencil. Highlighting does not reinforce the way working out the steps of the synthesis or chemical reaction at hand does. You might even make a collection of file cards labeled "reaction descriptions" on which you force yourself to write out the steps of the reaction. Another set of file cards should be used to keep track of the various ways you will accumulate to make molecules. At first, these cards will be few in number, and sparsely filled, but as we reach the middle of the course there will be an explosion in the number of synthetic methods available. This subject can sneak up on you and keeping a catalog will help you to stay on top of this part of the subject. We will try to help you to work in this interactive way by interrupting the text with problems, with solutions that follow immediately when we think it is time to stop, take stock, and reinforce a point before going on. These problems are important. You can read right by them of course, or read the answer without stopping to do the problem, but to do so will be to cheat yourself and make it harder to learn the subject. Doing these in-chapter problems is a part of reading with a pencil and should be very helpful in getting the material under control. There is no more important point to be made than this one. Ignore it at your peril!

DON'T MEMORIZE. In the old days, courses in organic chemistry rewarded people who could memorize. Indeed, the notorious dependence of medical school admission committees on the grade in organic chemistry may have stemmed from the need to memorize in medical school. If you could show that you could do it in organic, you could be relied upon to be able to memorize that the shin bone was connected to the foot bone, or whatever. Nowadays, memorization is the road to disaster; there is just too much material. Those who teach this subject have come to see an all too familiar pattern. There is a group of people who do very well early and then crash sometime around the middle of the first semester. These folks didn't suddenly become stupid or lazy; they were relying on memorization and simply ran out of memory. Success these days requires generalization, understanding of principles that unify seemingly disparate reactions or collections of data. Medical schools still regard the grade in organic as important, but it is no longer because they look for people who can memorize. Medicine, too, has outgrown the old days. Now medical schools seek people who have shown that they can understand a complex subject, people who can generalize.

WORK IN GROUPS. Many studies have shown that a most effective way to learn is to work in small groups. Form a group of your roommates or friends, and solve problems for each other. Assign each person one or two problems to be solved for the group. Afterward, work through the solution found in the chapter or Study Guide. You will find that the exercise of explaining the problem to others will be enormously useful. You will learn much more from "your" problems than from the problems solved by others. When I teach organic chemistry at Princeton, I increasingly replace lecture with small-group problem solving. In the second semester, there is no lecture at all. Students do at least as well under this regime as in the traditional class.

WORK THE PROBLEMS. As noted above, becoming good at organic chemistry is an interactive process; you can't just read the material and hope to become expert. Expertise in organic chemistry requires experience, a commodity that by definition you are very low on at the start of your study. Doing the problems is vital to gaining the necessary experience. Resist the temptation to look at the answer before you have tried to do the problem. Disaster awaits you if you succumb to this temptation, for you cannot learn effectively that way, and there will be no answers available on the examinations until it is too late. That is not to say that you must be able to solve all the problems straight away. There are problems of all difficulty in each chapter, and some of them are very challenging indeed. Even though the problem is hard or very hard, give it a try. When you are truly stuck, that is the time to gather a group to work on it. Only as a last resort should you take a peek at the Study Guide. There you will find not just a bare bones answer, but, often, advice on how to do the problem as well. Giving hard problems is risky, because there is the potential for discouraging people. Please don't worry if some problems, especially hard ones, do not come easily or do not come at all. Each of us in this business has favorite problems that we still can't solve. Some of these form the basis of our research efforts, and may not yield, even to determined efforts, for years. A lot of the pleasure in organic chemistry is working challenging problems, and it would not be fair to deprive you of such fun.

USE ALL THE RESOURCES AVAILABLE TO YOU. You are not alone. Moreover, everyone will have difficulty at one time or another. The important thing is to get help when you need it. Of course the details will differ at each college or university but there are very likely to be extensive systems set up to help you. Professors have office hours, there are probably teaching assistants with office hours, and there will likely be help, review, or question sessions at various times. Professors are there to help you, and will not be upset if you show enough interest to ask questions about a subject they love. "Dumb questions" do not exist! You are not expected to be an instant genius in this subject, and many students are too shy to ask perfectly reasonable questions. Don't be one of those people!

If you feel uncertain about a concept or problem in the book—or lecture—get help soon! This subject is highly cumulative, and ignored difficulties will come back to haunt you. I know that many teachers tell you that it is impossible to skip material and survive, but this time it is true. What happens in December or April de-pends on September, and you can't wait and wait, only to "turn it on" at the end of the semester or year. Almost no one can cram organic chemistry. Careful, attentive, daily work is the route to success, and getting help with a difficult concept or a vexing problem is best done immediately.

Over the life of the first and second editions of this book, I have interacted with many of you by e-mail, much to my pleasure. Of course, I can't begin to replace local sources of help, and I can't be relied upon in an emergency, as I might be out of touch with my e-mail, but I can usually be reached at mjjr@princeton.edu, and I look forward to your comments and questions.

Organic
Chemistry

Atoms and Molecules; Orbitals and Bonding

1

When it comes to atoms, language can be used only as in poetry. The poet too, is not nearly so concerned with describing facts as with creating images.

—Niels Bohr to Werner Heisenberg[*]

1.1   PREVIEW

The picture of atoms that all scientists had in their collective mind's eye at the beginning of the last century was little different from that of the ancient Greek philosopher Democritus (~460–370 B.C.), who envisioned small, indivisible particles as the constituents of matter. These particles were called **atoms**, from the Greek word for indivisible. The British chemist John Dalton (1766–1844) had the idea that atoms might have different characteristic weights, but he did not abandon the notion of a solid, uniform atom. That picture did not begin to change dramatically until 1897, when the English physicist J. J. Thomson (1856–1940) discovered the **electron**. Thomson postulated a sponge-like atom with the negatively charged electrons embedded within a positively charged material, rather like raisins in a pudding. Ernest Rutherford's (1871–1937) discovery, in 1909, that an atom was mostly empty space demolished the "pudding" picture and led to his celebrated "planetary" model, in which electrons were seen as orbiting a compact, positively charged nucleus.

It was Niels Bohr who made perhaps the most important modification of the planetary model. He made the brilliant and largely intuitive[†] suggestion that electrons were required to occupy only certain orbits. As an electron's energy depends on the distance of its orbit from the positively charged nucleus, Bohr's suggestion

---

[*]Niels Bohr (1885–1962) and Werner Heisenberg (1901–1976) were pioneers in the development of quantum theory, the foundation of our current understanding of chemical bonding.

[†]Some people's intuitions are better able than others' to cope with the unknown!

amounted to saying that electrons can have only certain energies. Thus, the energy states of the electron were (and still are) said to be quantized. Although this notion may seem strange, there are similar phenomena in the everyday world. One cannot create just any tone by blowing across the mouth of a bottle, for example. The tone you hear depends on the size and shape of the particular bottle. If one blows harder, the frequency does not change smoothly; instead a "quantum jump" to an overtone of the original sound occurs.

In the 1920s and 1930s, a number of mathematical descriptions emerged from the need to understand Bohr's quantum model. It became clear that one must take a probabilistic view of the subatomic world. Werner Heisenberg discovered that it was not possible to determine simultaneously both the position in space and momentum (mass times speed) of an electron.* Thus, one can determine where an electron is at any given time only in terms of probability. One *can* say, for example, that there is a 90% probability of finding the electron in a certain volume of space, but one *cannot* say that at a given instant the electron is at a particular point in space.

The further elaboration of this picture of the atom has given us the conceptual basis for all modern chemistry: the idea of the orbital. Loosely speaking, an orbital describes the region of space occupied by an electron or a pair of electrons of a certain energy. Both the combining of atoms to form molecules and the diverse chemical reactions these molecules undergo involve, at a fundamental level, the interactions of electrons in orbitals. This notion will appear throughout this book; it is the most important unifying principle of organic chemistry. In atoms, we deal with **atomic orbitals**, and in molecules, with **molecular orbitals**. Various graphic conventions are used in this book to represent atoms and molecules—letters for atoms, dots for electrons not involved in bonding, and lines for electrons in bonds—but it is important to keep in mind from the outset that the model that most closely approximates our current understanding of reality at the atomic and molecular level is the cloudy, indeterminate—one might even say poetic—image of the orbital.†

The following section, a version of which will appear in every chapter, is designed to alert you to the important parts of the forthcoming chapter, and, especially, to aid you in reviewing. After you finish the chapter, or before an examination, it is a good idea to return to this section and make sure you are clear on both the Essential Skills and the Important Details listed below.

---

*This idea is extraordinarily profound—and troubling. The **Heisenberg uncertainty principle** (which states that the product of the uncertainty in position times the uncertainty in momentum is a constant) seems to limit fundamentally our access to knowledge. For an exquisite exposition of the human consequences of the uncertainty principle, see Jacob Bronowski, *The Ascent of Man,* Chapter 11 (Little Brown, New York, 1973).

†In the wonderful quote that opens this chapter, Niels Bohr points out that once we transcend the visible world, all that is possible is modeling or image making. To me, what is even more marvelous about the quote is the simple word "too": It was obvious to Bohr that scientists speak in images, and he was pointing out to Heisenberg that there was another group of people out there in the world who did the same thing—poets.

ESSENTIAL SKILLS

1. Writing structures of atoms, ions (charged atoms and molecules), and neutral molecules; showing the atomic orbitals the electrons occupy; and the drawing of Lewis "dot" structures for atoms and molecules.

2. Writing the resonance forms (different electronic structures) that, taken together, give a more accurate picture of molecules than any single structure does.

IMPORTANT DETAILS

1. Remembering the sign convention for exothermic ($\Delta H°$ is negative) and endothermic ($\Delta H°$ is positive) reactions.

2. Determining the charge of an atom, especially an atom in a molecule.

3. Using the curved arrow formalism to "push" pairs of electrons in writing resonance forms and in sketching electron flow in chemical reactions.

1.2   ATOMS AND ATOMIC ORBITALS

In a neutral atom, the **nucleus**, or core of positively charged protons and neutral neutrons, is surrounded by a number of negatively charged electrons equal to the number of protons. If the number of electrons and protons is not equal, the atom or molecule containing the atom must be charged and is called an **ion**. A negatively charged atom or molecule is called an **anion**; a positively charged species is a **cation**. (One of the tests of whether or not you know how to "talk organic chemistry" is the pronunciation of the word cation. It is kat-eye-on, not kay-shun.)

The energy required to remove an electron from an atom to form a cation is called the **ionization potential**. In general, the farther away an electron is from the nucleus, the easier it is to remove, and the lower the ionization potential.

Single atoms (in this case, carbon) are often written $^{W}_{Z}C$, where W is the mass number and Z is the atomic number, the number of protons in the nucleus, which equals the nuclear charge. In this book, the superscript W is omitted.

The electrons surrounding the helium (He) nucleus completely fill the first energy level, and it is most difficult to remove an electron from He. Helium has an especially high ionization potential, 24.6 eV/atom = 567 kcal/mol.* Likewise, the chemical inertness of the other noble gases, which also have high ionization potentials, is due to the stability of their filled outer subshells.

Much of the chemistry of atoms is dominated by gaining or losing electrons in order to achieve the electronic configuration of one of the noble gases (He, Ne, Ar, Kr, Xe, or Rn). The noble gases have especially stable filled shells of electrons: 2 for He, 10 for Ne ($2 + 8$), 18 for Ar ($2 + 8 + 8$), and so on. The idea that filling certain

CONVENTION ALERT

---

*There are several units of energy in use. Organic chemists commonly use kilocalories per mole (kcal/mol); physicists use the electron volt (eV). One electron volt/molecule translates into about 23 kcal/mol. Recently, the International Committee on Weights and Measures has suggested that still another unit, the kilojoule (kJ), be substituted for kilocalorie. So far, organic chemists in some countries, including the United States, seem to have resisted this suggestion (1 kcal is equal to 4.184 kJ).

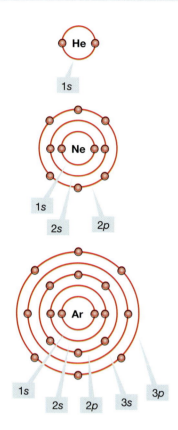

FIGURE 1.1 Highly schematic representations of He, Ne, and Ar.

shells creates especially stable configurations is known as the **octet rule**. The first shell, called 1s, is an exception, as it can hold only two electrons. The second shell can hold eight in two subshells called 2s and 2p, which can hold two and six electrons, respectively. We will explain these numbers and names shortly, but we need the labels first. Figure 1.1 shows highly schematic pictures of some noble gases, and uses this terminology. These pictures do not give good three-dimensional representations of these species, but they do show orbital occupancy and electron counting. Better pictures are forthcoming.

Electrons can be added to atoms as well as removed. The energy that is released by adding an electron to an atom to form a negatively charged anion is called the atom's **electron affinity**, measured in electron volts. The noble gases have very low electron affinities. Conversely, atoms to which adding an electron would complete a noble gas configuration have high electron affinities. The classic example is fluorine: the addition of a single electron yields a fluoride ion, F^-, which has the electronic configuration of Ne. Both F^- and Ne are 10-electron species (Fig. 1.2).

FIGURE 1.2 Addition of an electron to fluorine gives fluoride.

Table 1.1 shows ionization potentials and electron affinities of some elements arranged as in the periodic table. Notice that atoms with low ionization potentials, atoms that have electrons that are easy to remove to give a positive ion, cluster on the left side of the periodic table; atoms with high electron affinities, which accept electrons easily, are on the right side (excluding the noble gases).

TABLE 1.1   Some Ionization Potentials (black) and Electron Affinities (red)

H							He
13.60							24.59
0.75							~0
Li	Be		B	C	N	O	F	Ne
5.39	9.32		8.30	11.26	14.53	13.62	17.42	21.56
0.62	~0		0.24	1.27	~0	1.47	3.34	~0
Na	Mg		Al	Si	P	S	Cl	Ar
5.14	7.65		5.99	8.15	10.49	10.36	12.97	15.75
0.55	~0		0.46	1.24	0.77	2.08	3.61	~0

Atoms having low ionization potentials often form **ionic bonds** with atoms having high electron affinities by transferring an electron. In sodium fluoride (Na^+F^-), both Na^+ and F^- have filled second shells, and have achieved the stable electronic configuration of the noble gas Ne. In potassium chloride (K^+Cl^-), both ions have the electronic configuration of Ar, another noble gas (Fig. 1.3).

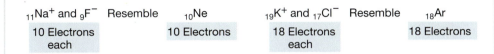

FIGURE 1.3 In the ionic compounds NaF and KCl, each atom can achieve a noble gas electronic configuration with a filled octet of electrons.

Compounds formed by ionic bonding are held together by the electrostatic attraction of opposite charges and are traditionally the province of inorganic chemistry. Nearly all of the compounds of organic chemistry are bound not by the electrostatic attractions present in ionic compounds but by the sharing of electrons in covalent bonds.

We have just developed pictures of some atoms and ions. It is time now to elaborate a bit to provide a much fuller picture of atomic orbitals. Your reward for bearing with an increase in complexity will be a much-increased ability to think about structure and reactivity. The most useful models of Nature, those that are best at generalizing, center on the qualitative nature of atomic (and molecular) orbitals. So, it is time we learn more about them.

The electrons in atoms do not occupy simple circular or spherical orbits. Erwin Schrödinger (1887–1961) described a negatively charged electron in the vicinity of a positive nucleus with a wave equation, which recognized that electrons could have the properties of both particles and waves. The solutions to Schrödinger's wave equation are called **wave functions**, written ψ (pronounced "sigh"). They have many of the properties of waves. They can be positive in one region (wave), negative in another (trough), or zero (in between).

A one-electron wave function is called an orbital, thus recalling earlier descriptions of atoms. It is the square of the wave function, ψ^2, that is proportional to the electron density. Although there are regions of space in which ψ and $\psi^2 = 0$ (zero electron density = zero probability of finding an electron), ψ does not vanish at a large distance from the nucleus, but maintains a finite value, even if very small.

Each wave function is determined by a set of **quantum numbers**. The first quantum number is n, the principal quantum number, which may have the integral values $n = 1, 2, 3, 4, \ldots$. It is related to the distance of the electron from the nucleus and hence to the energy of the electron. The higher n, the greater the average distance of the electron from the nucleus, and the greater its energy. The principal quantum number of the highest energy electron of an atom also determines the row occupied by the atom in the periodic table. Hydrogen and helium are in the first row, Li, Be, B, C, N, O, F, and Ne in the second, Na, Mg, Al, Si, P, S, Cl, and Ar in the third, and so on (Table 1.2).

TABLE 1.2
Principal Quantum Number (n) of the Highest Energy Electron

Atom	n
H, He	1
Li, Be, B, C, N, O, F, Ne	2
Na, Mg, Al, Si, P, S, Cl, Ar	3

TABLE 1.3
Relationship between n and l

n	l	Orbital Designation
1	0	$1s$
2	0	$2s$
2	1	$2p$
3	0	$3s$
3	1	$3p$
3	2	$3d$

TABLE 1.4
Relationship between n, l, and m_l

n	l	m_l	Orbital Designation
1	0	0	$1s$
2	0	0	$2s$
2	1	−1	$2p$
2	1	0	$2p$
2	1	+1	$2p$
3	0	0	$3s$
3	1	−1	$3p$
3	1	0	$3p$
3	1	+1	$3p$
3	2	−2	$3d$
3	2	−1	$3d$
3	2	0	$3d$
3	2	+1	$3d$
3	2	+2	$3d$

The second quantum number, l, is related to the shape of the orbital and depends on the value of n. It may have only the integral values 0, 1, 2, 3, . . . , $(n-1)$ [0 up to $(n-1)$]. So, for an orbital for which $n = 1$, l must be zero; for $n = 2$, l can be 0 or 1; and for $n = 3$, the three possible values of l are 0, 1, and 2. The quantum number l is always referred to by a letter, s for $l = 0$, p for $l = 1$, d for $l = 2$, and f for $l = 3$, which leads to the common designations of orbitals shown in Table 1.3.

The third quantum number m_l, depends on l as did l on n. It may have the integral values 0, ±1, ±2, ±3, . . . , ±l, and is related to the orientation of the orbital in space. Table 1.4 presents the possible values of n, l, and m_l.

Finally, there is s, the spin quantum number, which may have only the two values ±½. Table 1.5 gives a complete listing of the possible combinations of quantum numbers through $n = 3$.

As shown in Tables 1.3–1.5, orbitals are designated with a number and a letter in the following way:

$$1s\,(n=1, l=0) \quad 2s\,(n=2, l=0) \quad 2p\,(n=2, l=1)$$

For $n = 2$, $l = 1$ (a $2p$ orbital), m_l may take the values −1, 0, +1; thus, there are *three* $2p$ orbitals, one for each value of m_l. These equi-energetic orbitals are differentiated by arbitrarily designating them as $2p_x$, $2p_y$, or $2p_z$. A little later we will see that the "x, y, z" notation indicates the relative orientation of the $2p$ orbitals in space.

For $n = 3$, we have the following possibilities:

$$3s\,(n=3, l=0) \quad 3p\,(n=3, l=1)$$

Just as for $n = 2$, m_l may now take the values −1, 0, +1. The three $3p$ orbitals are designated as $3p_x$, $3p_y$, and $3p_z$.

$$3d\,(n=3, l=2)$$

As $l = 2$, m_l may now take the values: −2, −1, 0, +1, and +2; thus there are five $3d$ orbitals. These turn out to have the complicated designations $3d_{x^2-y^2}$, $3d_{z^2}$, $3d_{xy}$, $3d_{yz}$, and $3d_{xz}$. Mercifully, in organic chemistry we only very rarely have to deal with $3d$ orbitals and need not consider the f orbitals, for which $n = 4$ and which are even more complicated.

TABLE 1.5 Possible Combinations of Quantum Numbers for $n = 1$, 2, and 3

n	l	m_l	s	Orbital Designation
1	0	0	±½	$1s$
2	0	0	±½	$2s$
2	1	−1, 0, +1	±½ each value of m_l	$2p$
3	0	0	±½	$3s$
3	1	−1, 0, +1	±½ each value of m_l	$3p$
3	2	−2, −1, 0, +1, +2	±½ each value of m_l	$3d$

For each of these orbitals, the spin quantum number s may be either $+\frac{1}{2}$ or $-\frac{1}{2}$. We may now designate an electron occupying the lowest energy orbital, $1s$, as either $1s$, with $s = +\frac{1}{2}$, or $1s$, with $s = -\frac{1}{2}$, *but nothing else*. Similarly, there are only two possibilities for electrons in the $2p_x$ orbital: $2p_x$, with $s = +\frac{1}{2}$, or $2p_x$, with $s = -\frac{1}{2}$. That is why it is impossible for more than two electrons to occupy any orbital! When we write $1s^2$ we mean that the $1s$ orbital is occupied by two electrons, which have opposite (paired) spin quantum numbers ($s = \pm\frac{1}{2}$). The designation $1s^3$ is meaningless, as there is no way to put a *different* third electron in any orbital. This rule is called the **Pauli principle**, after Wolfgang Pauli (1900–1958), who first articulated it in 1925. These ideas are summarized in Figure 1.4.

	Electron No.1	Electron No. 2
$1s^2$ means the $1s$ orbital contains two electrons:	$n = 1, l = 0, m_l = 0, s = +\frac{1}{2}$	$n = 1, l = 0, m_l = 0, s = -\frac{1}{2}$

FIGURE 1.4 Two electrons in the same orbital must have different (paired) spins.

We can now write electronic descriptions for atoms, following what is known as the **aufbau principle** (*aufbau* is German for building up or construction), which simply makes the reasonable assumption that we should fill the available orbitals in order of their energies, starting with the lowest energy orbital. To form these descriptions of neutral atoms, we add electrons until they are equal to the number of positive charges (thus the number of protons) in the nucleus. Table 1.6 gives the electronic descriptions of H, He, Li, Be, and B.

For carbon, the next atom after boron, there is a choice to be made in adding the last electron. The first five electrons are placed as in boron, but where does the sixth and last electron go? One possibility would be to put it in the same orbital as the fifth electron to produce $_6C = 1s^2 2s^2 2p_x^2$ [Fig. 1.5(a)]. In such an atom, the spins of the two electrons in the $2p_x$ orbital must be paired (opposite spins).

TABLE 1.6
Electronic Descriptions of Some Atoms

Atom	Electronic Configuration
$_1$H	$1s$
$_2$He	$1s^2$
$_3$Li	$1s^2 2s$
$_4$Be	$1s^2 2s^2$
$_5$B	$1s^2 2s^2 2p_x$

(a)

(b)

(c) $1s^2 2s^2 2p_x 2p_y$

Lowest energy

FIGURE 1.5 An application of Hund's rule. The electronic configuration with the largest number of parallel spins is lowest in energy. Note the use of the arrow convention to show electron spin.

WORKED PROBLEM 1.1 Explain why the two electrons in the $2p_x$ orbital of carbon $(_6C = 1s^2 2s^2 2p_x^2)$ must have opposite (paired) spins.

ANSWER Because they are in the same orbital $(2p_x)$, both electrons have the same values for the three quantum numbers n, l, and m_l ($n = 2$, $l = 1$, m_l = either +1, 0, or −1). If the spin quantum numbers were not opposite, one +½, the other −½, the two electrons would not be different! The Pauli principle (No two electrons may have the same values of the four quantum numbers) ensures that two electrons in the same orbital must have different spin quantum numbers.

Alternatively, the sixth electron could be placed in another $2p$ orbital to produce $_6C = 1s^2 2s^2 2p_x 2p_y$ [Fig. 1.5(b)]. The only difference is the presence of two electrons in a *single* $2p$ orbital $(_6C = 1s^2 2s^2 2p_x^2)$, in Figure 1.5(a), versus one electron in each of two *different* $2p$ orbitals $(_6C = 1s^2 2s^2 2p_x 2p_y)$ in Figure 1.5(b). The three $2p$ orbitals are of equal energy, so how is one to make this choice? Electron–electron repulsion would seem to make the second arrangement the better one, but there is still another consideration. When two electrons occupy different, but equi-energetic orbitals, their spins can either be paired (↑↓) or unpaired (↑↑). **Hund's rule** (Friedrich Hund, 1896–1997) states that for a given electron configuration, the state with the greatest number of parallel spins has the lowest energy. So the third possibility [Fig. 1.5(c)] is actually the best one in this case. Essentially, giving the two electrons the same spin ensures that they cannot occupy the same orbital, and tends to keep them apart. So, Hund's rule tells us that the third configuration, $_6C = 1s^2 2s^2 2p_x 2p_y$, with the spins of the two electrons in the $2p$ orbitals the same (parallel), is the lowest energy state for a carbon atom. Figure 1.5 shows these possible electronic configurations for carbon.

CONVENTION ALERT

The convention used to designate spin shows the electrons as little arrows pointing either up ↑ or down ↓. Two arrows in opposite directions (↑↓) show paired spins and two arrows in the same direction (↑↑) show parallel, or unpaired, spins.

PROBLEM 1.2 Explain why the fifth and sixth electrons in a carbon atom may not occupy the same orbital as long as they have the same (parallel) spins.

Now we can write electronic descriptions for the rest of the atoms in the second row of the periodic table (Table 1.7). Notice that for nitrogen we face the same kind of choice just described for carbon. Does the seventh electron go into an already occupied $2p$ orbital, or into the remaining empty $2p_z$ orbital? The configuration in which the $2p_x$, $2p_y$, and $2p_z$ orbitals are all singly occupied by electrons with the same spin is lower in energy than any configuration in which a $2p$ orbital is doubly occupied. Again, Hund rules!

TABLE 1.7
Electronic Descriptions of the Remaining Atoms in the Second Row

Atom	Electronic Configuration
$_6C$	$1s^2 2s^2 2p_x 2p_y$
$_7N$	$1s^2 2s^2 2p_x 2p_y 2p_z$
$_8O$	$1s^2 2s^2 2p_x^2 2p_y 2p_z$
$_9F$	$1s^2 2s^2 2p_x^2 2p_y^2 2p_z$
$_{10}Ne$	$1s^2 2s^2 2p_x^2 2p_y^2 2p_z^2$

PROBLEM 1.3 Write the electronic configurations for the next row in the periodic table, $_{11}Na$ through $_{18}Ar$.

Wave functions, or orbitals (solutions to Schrödinger's equation), can be graphed to show the value of ψ as a function of r, the distance from the nucleus. As it is ψ^2 that is related to electron density, it is usually this parameter that is plotted against r. Figure 1.6 shows this plot for the lowest energy solution to the Schrödinger equation, the 1s orbital.

Figure 1.6 shows that the probability of finding an electron falls off rather sharply in all directions as we move out from the nucleus. There is no directionality—the 1s orbital is symmetrical in all directions. It is the quantum number l that is associated with orbital shape, and all s orbitals, for which l always equals zero, will be spherically symmetric. Note also that the electron density never goes to zero—there is a finite, though very, very small probability of finding an electron at a distance of several angstroms from the nucleus. Figure 1.7 translates this two-dimensional graph into a three-dimensional picture of the 1s orbital, a spherical cloud of electron density, at its maximum near the nucleus. Although this cloud of electron density is arbitrarily bound by the spherical surface at some distance from the nucleus, it does not really terminate sharply there. This surface simply indicates the volume within which we have a high chance of finding the electron. We can choose to put the boundary at any percentage we like—the 95% confidence level, the 99% confidence level, or any other value. It is this picture that we have been approaching throughout these pages. This representation of the probability function is the one you will remember best (probably) and the one that will be most useful in our study of chemical reactions. This is the region of space occupied by an electron in the 1s orbital.

Figure 1.6 correctly gives the probability of finding an electron at a particular point a distance r from the nucleus, but it doesn't recognize that as r increases, a given change in r (Δr) produces a greater volume of space and thus a greater number of points.

A different graph gives a picture of the probability of finding an electron at *all* points at a distance r from the nucleus (Fig. 1.8). This new picture takes account of the increasing volume of spherical shells Δr thick as the distance from the nucleus r increases.

FIGURE 1.6 A plot of $\psi^2(1s)$ versus distance r for hydrogen (1s). Note that the value of ψ^2 does not go to zero, even at very large r.

FIGURE 1.7 A three-dimensional picture of the 1s orbital. The surface of the sphere denotes an arbitrary cutoff.

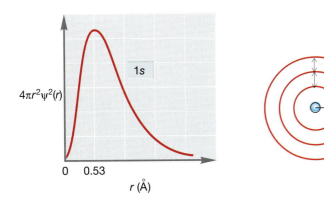

FIGURE 1.8 A slice through the three-dimensional 1s orbital. At relatively low values of r (short distance from the nucleus) Δr will contain a smaller volume than at higher values of r (larger distance from the nucleus). The outer spherical shell contains greater volume than the middle shell even though they are both Δr wide. This plot is weighted for the increasing influence of Δr as r gets large.

SUMMARY

1. We can do no better than to say that within some degree of certainty the electron is within a certain volume of space.
2. The cutoff is arbitrary—the electron density does not become zero at long distance from the nucleus.
3. An orbital can contain only two electrons, which will have what are called "paired" or "opposite" spin quantum numbers, $s = +\frac{1}{2}$ and $-\frac{1}{2}$.
4. It is the quantum number l that is related to the shape of the orbital, and all s orbitals have spherical symmetry.

The wave function ψ has many of the properties of other kinds of waves. There are, for example, the peaks and valleys evident in waves in liquids and in a vibrating string. Peaks correspond to regions for which the mathematical sign of ψ is positive and valleys correspond to negative regions. In a vibrating string, when we pass from a length of string in which the amplitude is positive to one where it is negative, we find a stationary point called a **node**. Nodes also appear in wave functions at the points at which they change sign; where $\psi = 0$. Of course, if $\psi = 0$, then $\psi^2 = 0$, and the probability of finding an electron at a node is also zero. Thus, a node is a region of space in which the electron density is zero. The number of nodes in an orbital is always one less than n, the principal quantum number. Thus, the $n = 1$ orbital (1s) has no nodes, all $n = 2$ orbitals have a single node, $n = 3$ orbitals have two nodes, and so on. As we turn our attention from the 1s orbital to higher energy orbitals, we will have to pay attention to the presence of nodes.

The next higher energy orbital is the 2s ($n = 2$, $l = 0$). Because $n = 2$, there must be a single node in this orbital, and we expect spherical symmetry, as with all s orbitals. Figure 1.9 shows a plot of $\psi^2(r)$ versus r and a three-dimensional representation of the 2s orbital. This figure shows a spherical node, the region at which the sign of ψ and ψ^2 is zero.

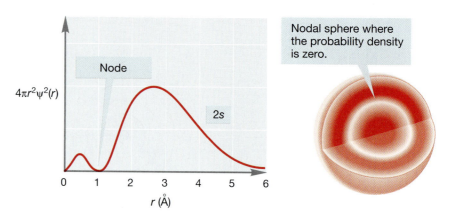

FIGURE 1.9 A plot of $\psi^2(r)$ versus r for a 2s orbital and a cross section through the orbital.

When $n = 2$, l can equal 1 as well as zero. This combination of quantum numbers gives the $2p_x$, $2p_y$, and $2p_z$ orbitals, each of which must possess a single node. Because l is not zero, the $2p$ orbitals are not spherically symmetrical. As you may have guessed from the x, y, z designations, the $2p$ orbitals are directed along the x, y, and z axes, which intersect at 90° angles. The lobes of $2p$ orbitals are slightly flattened spheres, and the orbital as a whole is shaped roughly like a dumbbell (Fig. 1.10). The node is the plane separating the two halves of the dumbbell. In one lobe of the orbital the sign of the wave function is positive; in the other it is negative. These signs are usually indicated by a change of color rather than (+) and (−) to avoid confusion with electrical charge. The three $2p$ orbitals together are shown in Figure 1.11. As you can see, the stylized dumbbells used to represent $2p$ orbitals are easier to draw than the more accurate picture shown in Figure 1.10.

CONVENTION ALERT

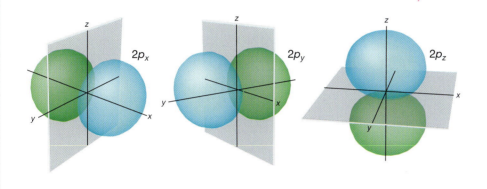

FIGURE 1.10 An accurate three-dimensional representation of three $2p$ orbitals. Note the nodal planes separating the lobes where the sign of ψ differs.

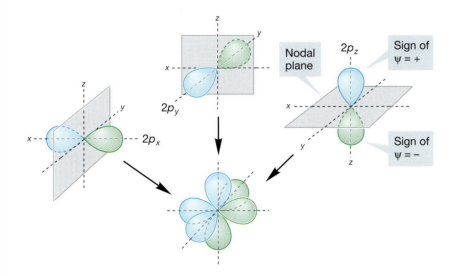

FIGURE 1.11 A schematic three-dimensional representation of three $2p$ orbitals, shown separately and in combination.

For the first time, we begin to get hints of the causes of the complicated three-dimensional structures of molecules: Electrons are the "glue" that holds the atoms of molecules together, and electrons are confined to regions of space that are by no means always spherically symmetrical.

We can now make some quite detailed pictures of atoms using the shapes described here. In the drawings in Figure 1.12, the $1s$ electrons have been omitted in the atoms past He. The other electrons are shown as dots.

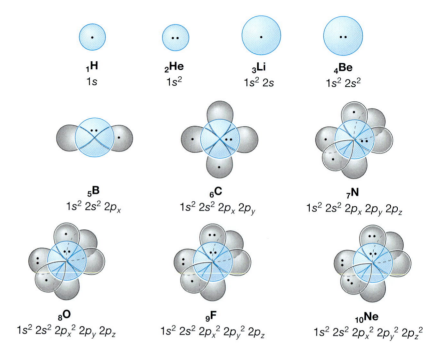

$_1$**H**
$1s$

$_2$**He**
$1s^2$

$_3$**Li**
$1s^2\ 2s$

$_4$**Be**
$1s^2\ 2s^2$

$_5$**B**
$1s^2\ 2s^2\ 2p_x$

$_6$**C**
$1s^2\ 2s^2\ 2p_x\ 2p_y$

$_7$**N**
$1s^2\ 2s^2\ 2p_x\ 2p_y\ 2p_z$

$_8$**O**
$1s^2\ 2s^2\ 2p_x^{\,2}\ 2p_y\ 2p_z$

$_9$**F**
$1s^2\ 2s^2\ 2p_x^{\,2}\ 2p_y^{\,2}\ 2p_z$

$_{10}$**Ne**
$1s^2\ 2s^2\ 2p_x^{\,2}\ 2p_y^{\,2}\ 2p_z^{\,2}$

FIGURE 1.12 Some schematic pictures of atoms.

SUMMARY

We now know that electrons on atoms are confined to certain volumes of space and that these volumes are not all alike in shape. A dumbbell-shaped $2p$ orbital is quite different from a spherical $2s$ orbital, for example. Electrons are filled into the available orbitals starting with the lowest energy orbital (the aufbau principle). In doing this filling, remember that an orbital can hold a maximum of two electrons (the Pauli principle). It makes no sense to think of an orbital holding more than two electrons—there is no quantum description available for a third electron in an orbital.

1.3   COVALENT BONDS AND LEWIS STRUCTURES

The formation of molecules through covalent bonding will be the subject of much of the remainder of this chapter. The bond formed by sharing a pair of electrons is called a **covalent bond**. A covalent bond is shown as a pair of dots between the two atoms or as a line joining the two. These two-electron bonds will be prominent in almost all of the molecules shown in this book. For many simple molecules, even the line between the atoms is left out, and the compound is written simply as A_2 or AB. You are left to supply mentally the missing two electrons (Fig. 1.13).

FIGURE 1.13 Bonding through the sharing of electrons is not ionic, but covalent. Covalent bonding can also result in stable electronic configurations.

The idea of covalent bonding was largely the creation of the American chemist Gilbert Newton Lewis (1875–1946) and the molecules formed by covalent bonds are usually written as what are called **Lewis structures**. In almost all covalently bonded molecules, every atom can achieve a noble gas electronic configuration by sharing electrons. In the H_2 molecule, each hydrogen atom shares two electrons and thus resembles helium. In F_2, each fluorine has a pair of $1s$ electrons in the lowest energy shell and is surrounded by eight other electrons, six unshared (**nonbonding** or **lone-pair** electrons) and two shared in the fluorine–fluorine bond. Thus, each fluorine in F_2 has a filled octet and, by virtue of sharing electrons, resembles neon. With the exception of H, electrons in $1s$ orbitals are not involved in bonding, and are not shown in Lewis structures. This omission can make the construction and interpretation of Lewis structures complicated. Always remember to account for the unshown $1s$ electrons. In HF, the hydrogen is helium-like and the fluorine is neon-like (Fig. 1.14).

FIGURE 1.14 Lewis dot structures for H_2, F_2, and HF. Every atom has a noble gas electronic configuration.

Hydrogen fluoride is different from any of the molecules mentioned so far in that the two electrons in the bond are shared unequally between the two atoms. In any diatomic molecule made from two different atoms there cannot be equal sharing of the electrons, because one atom of the two must attract the electrons more strongly than the other one does. The extreme case for this unequal electron sharing is the ionic bonding seen in Section 1.2 in Na^+F^- or K^+Cl^-. Molecules in which electrons are unequally shared contain **polar covalent bonds**. Often a small arrow pointing from positive to negative is used to indicate the direction of the dipole, or δ^+ (partial positive charge) and δ^- (partial negative charge) signs are placed on the appropriate atoms. We speak of such molecules as having **dipole moments**. The dipole moment is measured in debye units after the Dutch physicist and chemist Pieter Debye (1884–1966), who spent much of his working life at Cornell University and won the Nobel prize in 1936 (Fig. 1.15).

A—A	Two identical atoms share the electrons in a covalent bond equally: examples are H–H and F–F.
A—B	Two different atoms cannot share the electrons in a covalent bond equally. One atom will attract the electrons more strongly than the other.
δ⁺ δ⁻ A—B	Here B attracts the electrons more strongly than A. The direction of the dipole is shown with a crossed arrow, with the symbols δ⁻ and δ⁺ added to show the partial charges.

FIGURE 1.15 A polar covalent bond between atoms A and B. There is a dipole moment in this molecule.

The tendency of an atom to attract electrons is called **electronegativity**. Atoms with high electron affinities are the most electronegative and are found at the right of the periodic table. Atoms with low ionization potentials are called electropositive and lie at the left of the table. Table 1.8 gives some electronegativity values.

TABLE 1.8 Some Electronegativities

1 (IA)	2 (IIA)			3 (IIIA)	4 (IVA)	5 (VA)	6 (VIA)	7 (VIIA)
H 2.3								
Li 0.9	Be 1.6			B 2.1	C 2.5	N 3.1	O 3.6	F 4.2
Na 0.9	Mg 1.3			Al 1.6	Si 1.9	P 2.3	S 2.6	Cl 2.9
K 0.7								Br 2.7
Rb 0.7								I 2.4

PROBLEM 1.4 Indicate the direction of the dipole in the indicated bonds of the following molecules:

H–Cl, H–F, Li–F, H_3C–Cl, HO–Br, Br–Cl

Not all molecules containing polar bonds have dipole moments. Many do, including all those shown in Problem 1.4, but consider carbon dioxide (CO_2), a linear molecule of the structure O=C=O (the double lines represent double bonds in which the carbon atom shares two pairs of electrons with each oxygen atom). Oxygen is more electronegative than carbon (see Table 1.8), so there will be two polar bonds as shown in Figure 1.16. Even though carbon dioxide contains two polar bonds, there is no net dipole moment in the molecule. The bond dipoles cancel. The dipole moment of a molecule is the vector sum of all dipoles within the molecule. In O=C=O the two dipoles are equal in magnitude and pointed in exactly opposite directions; they cancel. There are two polar carbon–oxygen bonds in carbon dioxide, but no dipole moment.

In representing three-dimensional structures on a two-dimensional surface such as a blackboard or book page, it is necessary to devise some scheme for showing bonds directed away from the surface, either toward the front (up) or toward the rear (down). In the following figures, the solid wedges (➤) are coming out toward you and the dashed wedges (⫶⫶⫶) are retreating into the page. The blue dashed lines in the figure in Problem 1.5 show the outlines of the tetrahedra. If visualizing these molecules is hard for you at first, by all means make models.

Carbon dioxide

CONVENTION ALERT

PROBLEM 1.5 Which of the following two molecules has a dipole moment and which does not? Look carefully at the shapes of the two tetrahedral molecules.

Carbon tetrachloride
(CCl_4)

Chloroform
($CHCl_3$)

In order to write Lewis dot structures for molecules, we need to know the number of electrons in an atom. For atoms in the second row of the periodic table, one can tell immediately how many electrons are available by knowing the atomic number of the atom. The number of available electrons is equal to the total number of electrons in the atom, less the two $1s$ electrons. The electrons in the outermost subshell, the least tightly held electrons, are called the **valence electrons**. Electronic configurations are shown in shorthand by writing the name of the orbital and appending the number of electrons as a superscript. For example, the electronic configuration of helium ($_2$He) would be written $1s^2$ and that of beryllium ($_4$Be) as $1s^2 2s^2$.

For atoms in the third row, neither the $1s$ pair nor the eight electrons in the second shell ($2s^2 2p^6$) are shown in the Lewis structure. Instead, only the valence electrons in the third shell are indicated. In complete Lewis structures, every valence electron is written as a dot. In a slightly more abstract form of a Lewis structure, pairs of electrons in bonds are shown as lines and only the nonbonding valence electrons remaining on atoms appear as dots. Some examples of Lewis structures are shown in Figure 1.17.

H₂   H:H   or   H—H   HF   H:$\ddot{\ddot{F}}$:   or   H—$\ddot{\ddot{F}}$:

F₂   :$\ddot{\ddot{F}}$:$\ddot{\ddot{F}}$:   or   :$\ddot{\ddot{F}}$—$\ddot{\ddot{F}}$:   H₂O   H:$\ddot{\ddot{O}}$:H   or   H—$\ddot{\ddot{O}}$—H

CH₄   H:$\overset{H}{\underset{H}{\ddot{C}}}$:H   or   H—$\overset{\overset{H}{|}}{\underset{\underset{H}{|}}{C}}$—H

FIGURE 1.17
A few Lewis dot structures.

BH₃   H:$\overset{H}{\ddot{B}}$:H   or   H—$\overset{\overset{H}{|}}{B}$—H

NH₃   H:$\overset{H}{\ddot{N}}$:H   or   H—$\overset{\overset{H}{|}}{\underset{..}{N}}$—H

Let's draw Lewis structures for some of these more complicated molecules. To construct the Lewis dot structure of methane (CH₄), first determine that carbon (₆C) can contribute four electrons to bonding (six less the two inner-shell 1s electrons) and that each hydrogen contributes an electron. Thus four two-electron covalent bonds are formed, with no electrons left over (Fig. 1.18). For ammonia: (:NH₃), nitrogen (₇N) contributes five electrons (seven minus the 1s pair), and the three hydrogens contribute one electron each. Therefore three nitrogen–hydrogen bonds can be formed with two electrons, with a nonbonding pair left over (Fig. 1.18).

CH₄   ₆C contributes four electrons to bonding; each ₁H contributes one electron

NH₃   ₇N contributes five electrons to bonding; each ₁H contributes one electron

FIGURE 1.18   Construction of a Lewis structure for methane, CH₄, and ammonia, :NH₃.

WORKED PROBLEM 1.6   Construct Lewis structures for the following neutral molecules:

(a) BF₃   (b) H₂Be   (c) BH₃   (d) ClCH₃   (e) HOCH₃   (f) H₂N—NH₂

ANSWER   (a) BF₃   The first task is to determine the gross structure of BF₃. Is it B—F—F—F or some other structure with F—F bonds? Start by working out the number of electrons available for bonding in F and B. Fluorine (₉F) has seven electrons available for bonding (9 – 2 1s) and thus has a single "odd" electron. Boron (₅B) has three electrons (5 – 2 1s).

₉F
$1s^2\, 2s^2\, 2p_x^{\,2}\, 2p_y^{\,2}\, 2p_z$

₅B
$1s^2\, 2s^2\, 2p$

There are seven electrons available for bonding

There are three electrons available for bonding

:$\ddot{F}$·

·B·

(continued)

There is no easy way to form a molecule with both boron–fluorine and fluorine–fluorine bonds. Moreover, B has three electrons available for bonding, and a structure with three boron–fluorine bonds can be nicely accommodated through covalent bonds in which the "odd" electron on each fluorine is shared with one of boron's available three electrons:

Atoms of elements below the first row of the periodic table can participate in multiple bonding—there can be more than one bond to most other atoms. Usually, the structural formula will give you a clue when this is necessary. Ethane (H_3C-CH_3) requires no multiple bonds; indeed it permits none, as can be seen from construction of the Lewis structure. Each carbon is attached to four other atoms—all four of the electrons available for bonding to carbon are shared in single bonds to the three hydrogen atoms and the lone other carbon. Therefore, all electrons can be accounted for in single bonds between the atoms (Fig. 1.19).

FIGURE 1.19 Construction of a Lewis structure for ethane.

Carbon can also form multiple bonds, as in ethylene (H_2CCH_2) and acetylene (HCCH). In H_2CCH_2, only three of carbon's four bonding electrons are used up in forming single covalent bonds to one pair of hydrogens and the other carbon. There is an electron remaining on each carbon, and these two electrons are shared in a second carbon–carbon bond. Thus, ethylene contains a carbon–carbon double bond (Fig. 1.20).

FIGURE 1.20 Construction of a Lewis structure for ethylene.

In acetylene (HCCH), for each carbon, two of the four bonding electrons are used in forming single bonds to one hydrogen and the other carbon. Two electrons are left over on each carbon, and these are shared to create a triple bond between the carbons (Fig. 1.21).

FIGURE 1.21 Construction of a Lewis structure for acetylene.

As we examine the nature of the bonding involved in the various structural types of organic chemistry, it is vital to be able to draw Lewis structures quickly and easily.

WORKED PROBLEM 1.7 Draw Lewis structures for the following neutral atom and compounds. Use lines to indicate electrons in bonds and dots to indicate nonbonding electrons.

(a) CH_3 (c) Br (e) NH_2

(b) CH_2 (d) OH (f) H_3C-N

ANSWER In this problem, you must again determine the number of bonding electrons on each atom. Then make the possible bonds and see how many electrons are left over. As each compound is neutral, the number of bonding electrons will be equal to the atomic number less the $1s$ electrons. For hydrogen, there will always be only the single $1s$ electron.

(a) CH_3 Carbon has four bonding electrons and forms normal covalent bonds with three hydrogens, each of which contributes a single electron. There is an odd electron left over on carbon, which is shown as the usual Lewis dot.

(b) CH_2 Now carbon can form only two covalent bonds because only two hydrogen atoms are available. Two nonbonding electrons are left over.

WORKED PROBLEM 1.8 In each of the following compounds, there is at least one multiple bond. Draw a Lewis structure for each molecule. Use lines to indicate electrons in bonds and dots to indicate nonbonding electrons.

(a) F_2CCF_2 (c) H_2CO (e) $H_2CCHCHCH_2$ (g) H_3COCOH (with O above)

(b) H_3CCN (d) H_2CCO (f) H_3CNO

(continued)

ANSWER (b) H₃CCN The methyl (CH₃) group should be familiar by now. Its three covalent bonds are constructed by sharing three of carbon's four available electrons with the single electrons of the three hydrogens.

The other atoms, the carbon and the nitrogen, bring four and five electrons, respectively ($_6$C, 6 – 2 1s electrons = 4; $_7$N, 7 – 2 1s electrons = 5). Carbon–carbon and carbon–nitrogen covalent bonds can be formed, which leaves two unshared electrons on carbon and four on nitrogen.

Formation of a triple bond between carbon and nitrogen completes the picture, and nitrogen is left with a nonbonding pair of electrons.

It is also extremely important to be able to determine the charge on atoms in different structures. For single atoms, this process is easy as long as you remember to count the two 1s electrons, which are never shown except for H and He. The total number of electrons is compared with the nuclear charge, the atomic number. Figure 1.22 gives Lewis structures and charge determinations for some neutral atoms and charged species (ions).

H:⁻

$_1$H 1 Proton   = 1 Positive charge
2 Electrons = 2 Negative charges
Net 1– = H:⁻

:F̈:⁻

$_9$F 9 Protons   = 9 Positive charges
2 1s Electrons
8 Nonbonding
electrons   } 10 Negative charges
Net 1– = :F̈:⁻

H·

$_1$H 1 Proton   = 1 Positive charge
1 Electron = 1 Negative charge
Neutral = H·

Li·

$_3$Li 3 Protons   = 3 Positive charges
2 1s Electrons
1 Nonbonding
electron   } 3 Negative charges
Neutral = Li·

H⁺

$_1$H 1 Proton   = 1 Positive charge
0 Electrons = No negative charge
Net 1+ = H⁺

·C̈⁺

$_6$C 6 Protons   = 6 Positive charges
2 1s Electrons
3 Nonbonding
electrons   } 5 Negative charges
Net 1+ = ·C̈⁺

FIGURE 1.22 A few examples of charge determination.

In molecules, the determination of charge is a bit harder, because you must take care to account for shared electrons in the right way. As usual, remember that every atom except hydrogen and helium has a pair of unshown nonvalence $1s$ electrons. Count all nonbonding valence electrons as "owned" by an atom, as well as one electron from each two-electron bond to it. In H_2, for example, each hydrogen has only a share in the pair of electrons binding the two nuclei, and therefore each hydrogen is neutral. In F_2 each fluorine has two $1s$ electrons, which are not shown, six nonbonding (lone pair) electrons, and a share in the single covalent bond between the two fluorine atoms. This count gives a total of nine electrons, exactly balancing the nine positive nuclear charges. In $H_2C{=}CH_2$, each carbon has a pair of $1s$ electrons and a share in four covalent bonds for a total of six. As each carbon has a nuclear charge of +6, both are neutral. Figure 1.23 shows these examples.

FIGURE 1.23

Electron counting in some simple molecules.

$$H{:}H = H{-}H$$

$_1$H		
1 Proton	=	1 Positive charge
1 Shared electron	=	1 Negative charge
		Neutral H

$$:\!\ddot{F}\!:\!\ddot{F}\!: \ = \ :\!\ddot{F}\!-\!\ddot{F}\!:$$

$_9$F		
9 Protons	=	9 Positive charges
2 1s Electrons		
6 Nonbonding electrons		
1 Shared electron	}	9 Negative charges
		Neutral F

$_6$C		
6 Protons	=	6 Positive charges
2 1s Electrons		
4 Shared electrons	}	6 Negative charges
		Neutral C

Let's see why the carbon of the methyl anion ($^-{:}CH_3$) is negatively charged and why the nitrogen of the ammonium ion ($^+NH_4$) is positively charged. In the methyl anion, carbon has a pair of $1s$ electrons, two nonbonding electrons, and a share in three covalent bonds, for a total of seven electrons. The nuclear charge is only +6, and therefore the carbon must be negatively charged. The nitrogen atom of the ammonium ion has two $1s$ electrons and four shared electrons for a total of six. The nuclear charge is +7, and therefore the nitrogen is positively charged (Fig. 1.24).

FIGURE 1.24

Examples of the calculation of the charge on an atom in a molecule.

The methyl anion

$_6$C		
6 Protons	=	6 Positive charges
2 1s Electrons		
2 Nonbonding electrons		
3 Shared electrons	}	7 Negative charges
		Net 1−

The ammonium ion

$_7$N		
7 Protons	=	7 Positive charges
2 1s Electrons		
4 Shared electrons	}	6 Negative charges
		Net 1+

Figure 1.25 shows two more ions and works through the calculations of charge.

$_5$B
5 Protons	= 5 Positive charges
2 1s Electrons	
4 Shared electrons	6 Negative charges
	Net 1−

$_6$C
6 Protons	= 6 Positive charges
2 1s Electrons	
3 Shared electrons	5 Negative charges
	Net 1+

FIGURE 1.25
More calculations of charge in molecules.

PROBLEM 1.9 Draw Lewis structures for the following charged species. The charge is shown closest to the charged atom. For example, in $^+NO_2$, it is the nitrogen that is positive, not oxygen.

(a) ^-OH (d) ^-Cl (g) $^+NO_2$ (ON^+O)

(b) $^-BH_4$ (e) $^+CH_3$

(c) $^+NH_4$ (f) $^+OH_3$

PROBLEM 1.10 Add charges to the following compounds wherever necessary:

(a)   (b)   (c)   (d)   (e)   (f)   (g)   (h)
:CH$_2$ ·CH$_3$:CH   :ÖH   :OH$_3$ H$_2$C=Ö: :NH$_2$ HN=C=NH

PROBLEM 1.11 Add electrons to complete the following Lewis structures. The charge is placed as close as possible to the charged atom.

(a)   (b)   (c)   (d)   (e)   (f)   (g)   (h)
$^+CH_2$ $^-CH_2CH_3$ HC=CH$_2$ $^+OH_3$ ^-OH $^+NH_2$ $^-NH_2$ CH$_3$-C≡N$^+$-H

1.4   INTRODUCTION TO RESONANCE FORMS

Often, there will be more than one possible Lewis structure for a molecule. This statement is especially true for charged molecules, but it applies to many neutral species as well. (You may already have encountered this phenomenon in comparing your answers to Problems 1.9 and 1.10 with mine.) How do we decide which Lewis structure is the correct one? The answer almost always is that neither Lewis structure is complete all by itself, and that the molecule is best described as a combination of all reasonable Lewis structures, that is, a combination of **resonance forms**. Resonance forms are different *electronic* representations of the molecule. There can be no change in the position of the *atoms* in resonance forms. The electrons, but not the atoms, are distributed differently in different resonance forms. We shall return to this point many times, and in Chapter 9 resonance forms will be treated much more extensively.

Let's use nitromethane, H_3CNO_2, as an example. The nitro group, NO_2, is a **functional group**, a collection of atoms that behaves more or less the same way, wherever it appears in a molecule. As we work through the various structural types of organic chemistry, you will become familiar with many functional groups, but right now, at this early stage, almost everyone needs to look up a structure now and then. The problem is that the condensed formula "NO_2" doesn't really contain structural information about what is connected to what. For that matter, neither does "CH_3." Over time, you will become familiar with many of the functional groups and will be able to write the appropriate structures without even thinking about it. For now, you can consult the collection of functional groups and their structure inside the front cover.

Figure 1.26 gives one Lewis structure for nitromethane.

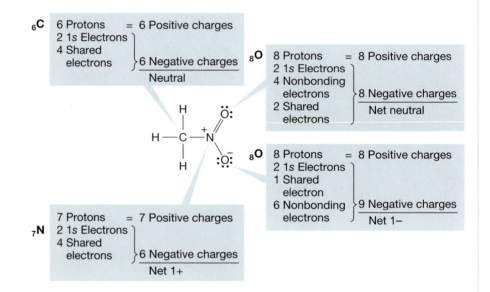

FIGURE 1.26 One electronic representation of nitromethane.

Even though nitromethane is a neutral molecule, there is no good way to draw it without separated charges. The nitrogen is positive and one of the oxygens is negative. But this structure is not the only Lewis form possible! We can draw the molecule so the other oxygen bears the negative charge. In the real nitromethane, the two oxygens share the negative charge equally. Thus, there are two equivalent electronic representations for nitromethane. These are called resonance forms. Notice the special, double-headed arrow between the two forms and the curved arrows used to convert one resonance form into the other (Fig. 1.27).

FIGURE 1.27 Two equivalent electronic representations for nitromethane.

Resonance arrow

Nitromethane

PROBLEM 1.12 Draw a Lewis structure for nitric acid (HO—NO$_2$), and verify that the nitrogen is positive and one of the oxygens is negative (see Fig. 1.26).

In Figure 1.27, curved arrows are used to interconvert the two Lewis structures, or resonance forms. Two things must be stressed here. First, although the curved arrows in Figure 1.27 have no physical significance, they do constitute an extraordinarily important bookkeeping device. We will use curved arrows throughout this book, and all organic chemists use them to keep track of electron movement, not only in drawing resonance forms, but in chemical reactions as well. Being able to draw resonance forms quickly and accurately is an essential skill for any organic chemist. Second, notice that the bookkeeping is accomplished by moving or "pushing" *pairs* of electrons. Be careful when doing this sort of thing not to violate the rules of valence, not to make more bonds than are possible, for this mistake is easy to make. These curved arrows are two-electron arrows. Figure 1.28 gives some examples of this kind of electron-pair pushing in molecules best represented as combinations of resonance forms. This device, the arrow formalism, will be discussed extensively in Chapter 7.

> CONVENTION ALERT

FIGURE 1.28 Resonance forms.

PROBLEM 1.13 Use the arrows to convert your Lewis structure for nitric acid (HO—NO$_2$, Problem 1.12) into a resonance form.

It is also important to be extremely clear on another point. Nitromethane does not spend half its time as one resonance form and half as the other. *There is no equilibration between the resonance forms.* Resonance involves only the movement of electrons, never atoms. Nitromethane is best described as the combination of the two forms representing different electronic structures.

Here's an analogy: Frankenstein was always Frankenstein. One might describe that poor constructed creature as part monster and part human, but he was always that combination—he did not oscillate between the two. In chemical terms, one would say that Frankenstein was a resonance hybrid of monster and human. On the other hand, Dr. Jekyll and Mr. Hyde were in equilibrium. When the good Dr. Jekyll drank the potion he became the monstrous Hyde. Later, when the potion wore off he reverted to Jekyll. Part of the time he was Jekyll and part of the time he was Hyde. The two were in equilibrium.

CONVENTION ALERT

The special double-headed arrow (◄──►) used in Figures 1.27 and 1.28 is re-served for resonance phenomena and is never used for anything but resonance. A pair of arrows (⇄) indicates equilibrium, the interconversion of two chemically dis-tinct species, and is never used for resonance (Fig. 1.29). This point is most impor-tant in learning the language of organic chemistry. It is difficult because it is arbitrary. There is no way to reason out the use of the different kinds of arrows; they simply must be learned.

Two chemically distinct species

$$A \rightleftharpoons B$$

One species, E, with two Lewis descriptions, C and D

$$\left[C \longleftrightarrow D \right] = E$$

FIGURE 1.29 The difference between equilibrium (two different species, A and B; two arrows) and resonance (different electronic representations, C and D, for the same molecule, E; double-headed arrow).

WORKED PROBLEM 1.14 Draw another structure for nitromethane in which every atom is neutral. *Hint*: There are only single bonds in this structure.

ANSWER You can arrive at the answer by using arrows to push electron pairs in the following way:

Note that, as a new bond is formed between the two oxygen atoms, the two charges orig-inally on oxygen and nitrogen are canceled. Notice also the long oxygen–oxygen bond. *Remember*: In drawing resonance forms, you move only electrons, not atoms. The oxy-gen–oxygen distance must be the same in each resonance form.

Is this cyclic form an important resonance form? Is it a good representation of the mol-ecule? The problem is the long bond between the oxygens. This bond must be weak just because it is so long. In the language of organic chemistry, we would say that the cyclic resonance form contributes little to the structure of nitromethane.

Is the cyclic structure below a resonance form? Resonance involves different electronic structures that may not differ in the positions of atoms. If atoms have been moved, as in the figure below in which there is a normal oxygen–oxygen bond, the two structures are in equilibrium and are not resonance forms.

PROBLEM 1.15 Verify that the C, N, and O atoms of your new structure are neutral.

The carbon–oxygen double bond gives us another opportunity to write resonance forms. As an example, take the simple compound formaldehyde ($H_2C\!=\!O$, Fig. 1.30).

FIGURE 1.30 Formaldehyde.   www   Resonance forms for formaldehyde

Carbon ($_6$C) has a pair of 1s electrons and shares in four covalent bonds; therefore it is neutral. Oxygen ($_8$O) has a pair of 1s electrons, four nonbonding electrons, and a share in two covalent bonds, for a total of eight electrons. Oxygen is also neutral. However, we can push electrons to generate a second Lewis structure in which the carbon is positive and the oxygen negative. The real formaldehyde is a combination, *not a mixture*, of these two resonance forms. Charge separation is energetically unfavorable so these two resonance forms will not contribute equally to the structure of the formaldehyde molecule. Still, neither by itself is a perfect representation of the molecule, and in order to represent formaldehyde well, both electronic descriptions must be considered.

PROBLEM 1.16 Verify that the carbon of the polar resonance form of formaldehyde in Figure 1.30 is positive and the oxygen is negative.

PROBLEM 1.17 There is a third resonance form for formaldehyde, but it contributes very little to the structure and is usually ignored. Can you find it and explain why it is relatively unimportant?

PROBLEM 1.18 Use the arrow formalism to convert each of the following Lewis structures into another resonance form. Notice that the last part of this question asks you to do something new—to move electrons one at a time in writing Lewis forms.

WORKED PROBLEM 1.19 Use the arrow formalism to write resonance forms that contribute to the structures of the following molecules:

(continued)

ANSWER (c) As we have seen, it is not only electron pairs (nonbonding electrons) that can be redistributed (pushed) in writing resonance forms, but bonding electrons as well. In this case, a pair of electrons in a carbon–carbon double bond is moved.

(c)

The positive charge on the right-hand carbon disappears, but reappears on the left-hand carbon. In the real structure, the two carbons share the positive charge equally, each bearing one-half the charge. The structure is sometimes written with dashed bonds to show this sharing:

Summary structure

1.5   THE COVALENT BOND: HYDROGEN (H₂)

In the first part of this chapter, we examined atomic structure and began a study of the collections of atoms called molecules. Now we will enlarge the discussion of atomic orbitals to include molecular orbitals, the regions of space occupied by electrons in molecules. Covalent bonding between atoms involves the sharing of electrons. This sharing takes place through overlap of atomic orbitals with other atomic or molecular orbitals.

Chemistry is largely the study of the structures and reactivities of molecules. A century ago, when chemistry was a young science, there seemed to be little time to worry too much about the "how and why" of the science; there was too much discovering going on, too much information to be collected.

Here is what Friedrich Wöhler (1800–1882), a great early unifier of organic chemistry, had to say on the subject: "Organic chemistry just now is enough to drive one mad. It gives one the impression of a primeval, tropical forest full of the most remarkable things, a monstrous and boundless thicket, with no way of escape, into which one may well dread to enter." Nowadays, no longer is some vast jungle being explored by brute force; chemists are now aiming more and more for new discoveries. The more we understand, and the better our models of Nature are, the more likely it is that the current transformation from trial-and-error methods to intellectually driven efforts will be successful.

Further, directed progress in chemistry depends on understanding how molecules react and the application of that understanding in creative ways. These days, the idea that both the structures of molecules and the reactions they undergo can be understood by examining the shapes and interactions of atomic and molecular orbitals has gained widespread acceptance in the chemical community, and an important branch of theoretical chemistry embraces these ideas. Although the study of molecular orbitals can be approached in a highly mathematical way, we will not follow that path. One of the great benefits of this approach is that it can be appreciated quite readily by nonmathematicians, and even very "low level," highly qualitative molecular orbital theory can provide striking insights into structure and reactivity.

In Section 1.3, we constructed Lewis dot structures for molecules. Our next task is to elaborate on this theme to produce better pictures of the bonds that hold atoms together in molecules. We'll need to consider how electrons act to bind nuclei together, and we'll take as our initial example hydrogen (H₂), the second simplest molecule.

PROBLEM 1.20 What is the simplest molecule? If H_2 is the second simplest molecule, the answer to this question must be "H_2 minus something." What might the "something" be? The answer will appear further along in the text, so think about this question for a while now.

We start with two hydrogen atoms, each consisting of a single proton, the nucleus, surrounded by an electron in the spherically symmetrical $1s$ orbital. In principle, the situation is this simple only as long as the two hydrogen atoms are infinitely far apart. As soon as they come closer than this infinite distance they begin to "feel" each other. *Remember*: Wave functions do not vanish as the distance from the nucleus increases. In practice, we can ignore the influence of one hydrogen atom on the other until they come quite close together, but as soon as they do, the energy of the system changes greatly. As one hydrogen atom approaches the other, the energy decreases until the two hydrogens are 0.74 angstrom ($Å$, where $1\ Å = 10^{-8}$cm) apart, from which point the energy of the system rises sharply, asymptotically approaching infinity as the distance between the two hydrogens declines toward zero (Fig. 1.31).

(a)
0.25 Å
Hydrogens too close

(b)
0.74 Å
Optimum distance

(c)
1 Å
Too far apart

(d)
2 Å
Much too far apart

FIGURE 1.31 A plot of the energy for the hydrogen molecule (H₂) as a function of the distance between the two hydrogen nuclei. The point of minimum energy corresponds to the equilibrium internuclear separation, the bond distance.

In the hydrogen molecule, two $1s$ electrons serve to bind the nuclei together—the system H—H is *more stable than two separated hydrogen atoms.* The two negatively charged electrons attract both nuclei and hold the molecule together. Finally, when the atoms approach too closely, the positively charged nuclei begin to repel each other and the energy goes up sharply.

We begin our description of bonding by combining the two $1s$ atomic orbitals (ψ's) of the two hydrogen atoms to produce two new orbitals called molecular orbitals. The first molecular orbital (Φ) results from a simple addition of the constituent atomic orbitals, and it can be written as $\psi(H_a, 1s) + \psi(H_b, 1s) = \Phi_{bonding} = \Phi_B$. This **bonding molecular orbital** is drawn in Figure 1.32, which shows that Φ_B looks much like what one would expect from a simple addition of two spherical $1s$ orbitals.

FIGURE 1.32 The bonding molecular orbital of H_2.

$$\psi(H_a) = \psi(H_a, 1s) \qquad \psi(H_b) = \psi(H_b, 1s) \qquad \Phi_B(H_2) = \psi(H_a) + \psi(H_b)$$

CONVENTION ALERT

Orbitals are customarily designated by Greek letters. We have already used ψ for atomic orbitals. Many people continue to use ψ for molecular orbitals as well. We will switch and use Φ (pronounced "fy"). Be careful in your other reading, as you will encounter situations in which ψ is used for all orbitals and even places in which our use of Φ and ψ is reversed.

As the bonding orbital Φ_B is concentrated between the two nuclei, two electrons in it can interact strongly with both nuclei. This attraction explains why Φ_B, in which the electron density is highest between the nuclei, is strongly bonding.

Quantum mechanics tells us another important thing. *The number of wave functions (orbitals) resulting from the combining process must equal the number of wave functions (orbitals) going into the calculation.* There must be another molecular orbital that results from our combination of the *two* hydrogen atomic $1s$ orbitals. It corresponds to $\psi(H_a, 1s) - \psi(H_b, 1s)$ and is called the **antibonding molecular orbital**, Φ_A. It is shown in Figure 1.33. Notice the node—a plane separating the two lobes of opposite sign.

FIGURE 1.33 The antibonding molecular orbital of H—H.

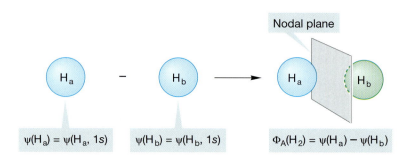

$$\psi(H_a) = \psi(H_a, 1s) \qquad \psi(H_b) = \psi(H_b, 1s) \qquad \Phi_A(H_2) = \psi(H_a) - \psi(H_b)$$

Be careful, it is easy to confuse the plus and minus signs meant to designate the sign of the wave function with electrical charges or other symbols in the figures.

Electrons can occupy both bonding and antibonding molecular orbitals. An electron in a bonding orbital acts to hold the two nuclei together. As the name "antibonding" suggests, an electron in an antibonding orbital does just the opposite—it contributes to the dissociation of the two nuclei. There is a third kind of molecular orbital in which an electron is neutral in its effect on the two nuclei—such an orbital is called a **nonbonding orbital.** We will see some nonbonding orbitals when we come to molecules larger than H_2. In Φ_A, unlike Φ_B, there is a nodal plane between the two hydrogen nuclei. Recall that a node is a region in which the sign of the wave function is zero. As we pass from a region where the sign of the wave function is positive to a region in which it is negative, the sign of Φ must go through zero. As noted earlier, the square of the wave function corresponds to the electron density, and if Φ is equal to zero, then Φ^2 is also zero. Therefore the probability of finding an electron at a node must be zero as well.

In contrast to Φ_B, in the antibonding molecular orbital (Φ_A), the electron density is zero between the nuclei, because that is where there is a node. Accordingly, when a pair of electrons occupies this orbital, the nuclei are poorly shielded from each other, and electrostatic repulsion forces the two positively charged nuclei apart.

Thus, the combination of two hydrogen 1s atomic orbitals yields two new, molecular orbitals, Φ_B and Φ_A. This idea seems intuitively reasonable: There are, after all, only two ways in which two 1s atomic orbitals can be combined. The signs of the wave functions can be either the same, as in the bonding orbital Φ_B, or opposite, as in the antibonding orbital Φ_A. Figure 1.34 shows a very simple, and most useful graphic device for summing up these relations.

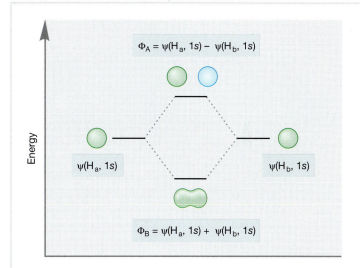

$$\Phi_A = \psi(H_a, 1s) - \psi(H_b, 1s)$$

Energy

$$\psi(H_a, 1s) \qquad \psi(H_b, 1s)$$

$$\Phi_B = \psi(H_a, 1s) + \psi(H_b, 1s)$$

FIGURE 1.34   A graphical representation of the combination of two atomic 1s orbitals to form a new bonding and antibonding pair of molecular orbitals, Φ_B and Φ_A.

The atomic orbitals (ψ's) going into the calculation are shown at the left and right sides of the figure. For molecular hydrogen, these orbitals are the two equivalent hydrogen 1s atomic orbitals, which must be of the same energy. They combine in a constructive, bonding way ($1s + 1s$) to give the lower energy Φ_B (bottom), and in a destructive, antibonding way ($1s - 1s$) to give the higher energy Φ_A (top).

CONVENTION ALERT

PROBLEM 1.21   Sketch the orbitals produced through the interaction of two carbon 2s atomic orbitals.

WORKED PROBLEM 1.22 Sketch the orbitals produced through the interaction of a carbon $2s$ atomic orbital overlapping end-on with a carbon $2p$ atomic orbital.

ANSWER The two orbitals can interact in a bonding way $(2s + 2p)$ or an antibonding way $(2s - 2p)$:

$$2s \quad - \quad 2p \qquad\qquad \text{Antibonding orbital — note the new node (bar)}$$

$$2s \quad + \quad 2p \qquad\qquad \text{Bonding orbital}$$

Note that the graphical construction of molecular orbitals takes place without reference to electrons (Fig. 1.34). Only after the diagram has been constructed do we have to worry about electrons. But now let's count them up and put them in. In the construction of H_2, each hydrogen atom brings one electron. In Figure 1.35, these are placed in the appropriate $1s$ orbitals ψ (H_a) and ψ (H_b). In the new molecule, H_2, the two electrons will go into the lower energy Φ_B. Because these two electrons are in the same orbital their spin quantum numbers must be different. The Pauli principle made this point earlier (p. 7). The antibonding molecular orbital (Φ_A) is empty.

We represent electron spin with two small arrows as in Figure 1.35. The arrows point in opposite directions to indicate that the spin quantum numbers are opposite ($s = \pm \frac{1}{2}$), or in the same direction to show two electrons with the same spin.

Not all combinations of orbitals are productive. If two orbitals approach each other in such a way that the new bonding interactions are exactly balanced by antibonds, there is no net interaction between the two. Such orbitals are called **orthogonal** orbitals. The way in which orbitals approach each other in space is very important. Figure 1.36 shows two such "net-zero" interactions. [In each case, the bonding interactions (blue–blue) are exactly canceled by antibonding interactions (blue–green).]

There is a connection between stability and energy. The lower the energy of an orbital, the more stable an electron in it is. A consequence of this stability is that the strongest bonds in molecules are formed by electrons occupying the lowest energy molecular orbitals. Throughout this chapter we will be taking note of what factors lead to low-energy orbitals and, thus, to strong bonding between atoms.

CONVENTION ALERT

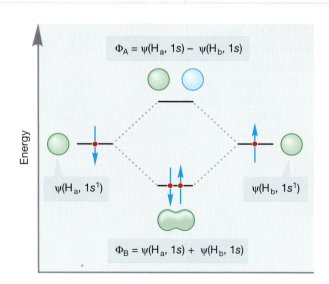

$$\Phi_A = \psi(H_a, 1s) - \psi(H_b, 1s)$$

$$\psi(H_a, 1s^1) \qquad\qquad \psi(H_b, 1s^1)$$

$$\Phi_B = \psi(H_a, 1s) + \psi(H_b, 1s)$$

Energy

FIGURE 1.35 The electronic occupancy for the molecule H_2.

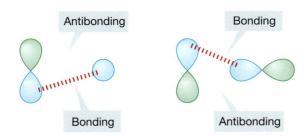

FIGURE 1.36 Two examples of noninteracting (orthogonal) orbitals.

WORKED PROBLEM 1.23 Contrast the interactions between two $2p$ orbitals approaching in the two different ways shown below.

(a)   (b)

Parallel   Perpendicular

ANSWER A pair of $2p$ orbitals interacting side by side produces two new orbitals (a), one bonding ($2p + 2p$), the other antibonding ($2p - 2p$). On the other hand, interaction end-on involves no net bonding or antibonding. The two exactly cancel, producing no net interaction (b). The two orbitals are orthogonal.

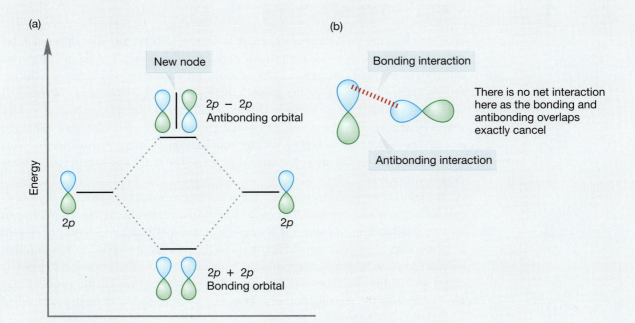

(a)

New node

$2p - 2p$
Antibonding orbital

Energy

$2p$

$2p$

$2p + 2p$
Bonding orbital

(b)

Bonding interaction

There is no net interaction here as the bonding and antibonding overlaps exactly cancel

Antibonding interaction

SUMMARY

Overlap of two orbitals produces two new molecular orbitals, one bonding, the other antibonding. Two electrons can be accommodated in the lower energy bonding molecular orbital. In this way, two atoms (or groups of atoms) can be bound through the sharing of electrons in what is called covalent bonding.

Some "rules" for orbital construction follow:

1. The number of orbitals produced at the end must equal the number at the beginning. If you start with n orbitals, you must produce n new orbitals. Here is a way to check your work as it proceeds.

2. Keep the process as simple as you can. Use what you know already, and combine orbitals in as symmetrical a fashion as you can.

3. The closer in energy two orbitals are, the more strongly they interact. At this primitive (but useful) level of theory, you need only interact the pairs of orbitals closest in energy to each other.

4. When orbitals interact in a bonding way (same sign of the wave function), the energy of the resulting orbital is lowered; when they interact in an antibonding way (different signs of the wave function), the energy of the resulting orbital is raised.

5. To order the product orbitals in energy, count the nodes; the more nodes in an orbital, the higher it is in energy.

1.6   BOND STRENGTH

How much is the stabilization apparent in Figure 1.35 worth? The difference between the energies of the two electrons in Φ_B and the energies of the two separate electrons in atomic $1s$ orbitals is 104 kcal/mol. This number is very large, and the hydrogen molecule is held together (bound) by this substantial amount of energy. That the hydrogen molecule is bound by 104 kcal/mol means that this amount of energy would be liberated by the creation of a molecule of H—H from separated hydrogen atoms. The 104 kcal/mol represents the stabilization of two electrons in Φ_B. This same amount of energy must be applied to break the bond and regenerate two hydrogen atoms (Fig. 1.37).

Actually, if we could measure it, the temperature would rise ever so slightly as two hydrogen atoms combined to make a hydrogen molecule. As the two hydrogen atoms came together to make a hydrogen molecule, 104 kcal/mol would be released as heat energy. The vessel would warm up as the process proceeded. A reaction that liberates heat, in which the products are more stable than the starting materials, is called an **exothermic reaction.** The opposite situation, in which the products are less stable than the starting material, requires the application of heat and is called an **endothermic reaction.**

Two separate hydrogen atoms ⇌ molecular H–H, or H_2

H• + H• ⟶ H—H
$\Delta H° = -104$ kcal/mol
This reaction is exothermic by 104 kcal/mol

H• + H• ⟵ H—H
$\Delta H° = +104$ kcal/mol
This reaction is endothermic by 104 kcal/mol

FIGURE 1.37 Molecular hydrogen (H_2) is more stable than two isolated hydrogen atoms by 104 kcal/mol.

The sometimes troublesome sign convention shows $\Delta H°$, the enthalpy (bond energy) change for an exothermic reaction, as a negative quantity: for A + B → C, $\Delta H° = -x$ kcal/mol. For an endothermic reaction, $\Delta H°$ is positive: for A + B → C, $\Delta H° = +x$ kcal/mol. In Figure 1.37, H• + H• → H—H, $\Delta H° = -104$ kcal/mol, is an exothermic reaction, and H—H → H• + H•, $\Delta H° = +104$ kcal/mol, is an endothermic reaction.

In graphs such as Figure 1.38, energy is plotted against a quality known as reaction progress or reaction coordinate. Reaction coordinate simply means "progress of the reaction," and monitors some geometrical change taking place as the reaction proceeds. In this reaction, the formation of molecular H_2 from two hydrogen atoms, a suitable reaction coordinate might be the distance between the two hydrogen atoms. It is a fair approximation at this point to take "reaction coordinate" as indicative of the progress of the overall reaction. We will use the term "reaction progress" from now on. Figure 1.38 shows the release of energy, in this case, 104 kcal/mol, as H—H is formed. If we read the figure from right to left, we also see the endothermic formation of two hydrogen atoms from the hydrogen molecule.

CONVENTION ALERT

FIGURE 1.38 Another schematic picture of the formation of molecular hydrogen from two hydrogen atoms.

This amount of energy, 104 kcal/mol for hydrogen, is called the **bond dissociation energy (BDE)**, and is the amount of energy that must be applied for **homolytic bond cleavage** (i.e., the energy needed to break the bond into a pair of neutral species). We could also imagine breaking a typical bond in another way, **heterolytic bond cleavage**, in which an anion–cation pair would be formed. Normally, this is not the path followed, as it costs even more energy to develop separated charges than to make neutral species.

Remember the arrow convention that keeps track of these two different kinds of bond breaking. Two-electron arrows have standard, double-barbed arrows, whereas single electron (homolytic) bond breakings are shown using pairs of single-barbed arrows (Fig. 1.39).

CONVENTION ALERT

FIGURE 1.39 Heterolytic cleavage produces a pair of ions. In the other kind of bond cleavage, the pair of binding electrons is split evenly between the atoms, giving a pair of neutral species. This much easier process is called homolytic cleavage.

WORKED PROBLEM 1.24 Sketch the profile of an endothermic reaction. See Figure 1.38 for the sketch of an exothermic reaction.

(continued)

ANSWER Aha! A trick question. If you have written an exothermic reaction, you have already written an endothermic reaction as well. You need only read your answer backward. We are psychological prisoners of our tendency to read from left to right. Nature has no such hangups! The formation of molecular hydrogen from two hydrogen atoms (Fig. 1.38, left to right) is exothermic (104 kcal/mol of energy is given off as heat), and the formation of two hydrogen atoms from a single hydrogen molecule is endothermic by the same amount (104 kcal/mol of heat energy must be applied).

In practice, a reaction that requires about 15–20 kcal/mol of thermal energy procedes quite reasonably at room temperature, about 25 °C. It is important to begin to develop a feeling for which bonds are strong and which are weak, to start to build up a knowledge of approximate bond strengths.* The bond in the hydrogen molecule is a strong one. As we continue our discussion of structure we'll make a point of noting bond energies as we go along. Unfortunately, there is no way to acquire this knowledge except by learning some bond strengths. Fortunately, there are not too many numbers to remember. Table 1.9 gives a few important bond dissociation energies. These numbers do not have to be known precisely, but it is important to have a rough idea of the bond strengths of common covalent bonds.

*To describe a bond as strong or weak is an arbitrary function of human experience. A 100-kcal/mol bond is "strong" in a world where room temperature supplies much less thermal energy. It would not be strong on the sunlit side of Mercury (where it's *hot*: the average temperature is about 377 °C, or 710 °F). Similarly, if you were a life form that evolved on Pluto (where it's *cold*: the average temperature is about − 220 °C, or − 361 °F), you would regard as strong all sorts of "weak" (on Earth) interactions, and your study of chemistry would be very different indeed. One area of research in organic chemistry focuses on extremely unstable molecules held together by weak bonds. The idea is that by understanding extreme forms of weak bonding we can learn more about the forces that hold together more conventional molecules. Chemists who work in this area deliberately devise conditions under which species that are normally most unstable can be isolated. They create very low temperature "worlds" in which other reactive (predatory) molecules are absent. In such a world, exotic species may be stable, as they are insulated from both the ravages of heat (thermodynamic stability) and the predations of other molecules (kinetic stability).

This table gives averages over a range of different compounds. A few compounds may lie outside these values. In later chapters more precise specific values will appear.

TABLE 1.9   Some Average Bond Dissociation Energies

Bond	BDE (kcal/mol)	Bond	BDE (kcal/mol)
C–H	96–105	C–I	55–57
N–H	93–107	C=C	~175
O–H	110–119	C=N	~143
S–H	82–87	C=O	173–181
C–C	83–90	C≡C	~230
C–O	85–96	C≡N	~204
C–N	69–75	H–I	71
C–F	105–115	H–Br	88
C–Cl	83–85	H–Cl	103
C–Br	72–74	H–F	136

Even the simplest molecule, H_2^+, is bound quite strongly.* The small amount of information already in hand—a molecular orbital picture of H_2 and the bond strength of the H–H bond (104 kcal/mol)—enables us to construct a picture of this exotic molecule and to estimate its bond strength. We can imagine making H_2^+ by allowing a hydrogen atom (H•) to combine with H^+, a bare proton. All the work necessary to create a molecular orbital diagram has already been done in the construction of Figure 1.34. We are still looking at the combination of a pair of hydrogen $1s$ atomic orbitals, so the building of a diagram for H_2^+ begins exactly as did the construction of the diagrams for H_2. Indeed, we come to precisely the same molecular orbital diagram, reproduced in Figure 1.40. The only difference comes when we put in the electrons. Instead of having *two* electrons, as does H_2, with one coming from each hydrogen atom, H_2^+ has only *one*. Naturally, it goes into the lower energy, bonding molecular orbital, Φ_B. What would we guess about the bond energy of this molecule? If *two* electrons in this orbital result in a

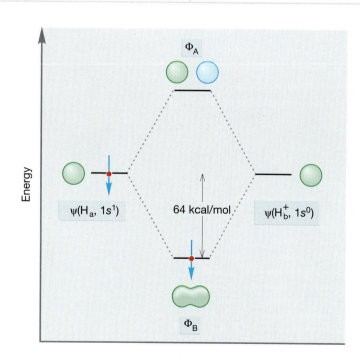

Φ_A

$\psi(H_a, 1s^1)$ 64 kcal/mol $\psi(H_b^+, 1s^0)$

Φ_B

FIGURE 1.40   An orbital interaction diagram for H_2^+.

*Here is the answer to Problem 1.20. The simplest molecule is hydrogen (H_2) minus one electron. Another electron cannot be removed to give something even simpler because H_2^{2+} is not a molecule—there are no electrons to bind the two nuclei.

bond energy of 104 kcal/mol, it seems reasonable to guess first that stabilization of a *single* electron would be worth about 52 kcal/mol. And we are amazingly close to being correct! The H_2^+ molecule is bound by 64 kcal/mol.

PROBLEM 1.25 Although our guess is quite close to reality, it is a little bit low. The H_2^+ molecule is more stable (lower in energy) than we thought. Why is our estimate of bond strength low? Why might the stabilization of two electrons in an orbital be less than twice the stabilization of one electron? *Hint*: Consider electron–electron repulsion.

It may seem somewhat counterintuitive that a strong (low energy) bond is associated with a large number. The lower the energy, the higher the number representing bond strength! The H_2 molecule has a bond dissociation energy of 104 kcal/mol, for example, whereas the more weakly bound H_2^+ is held together by only 64 kcal/mol. A strong bond means a low-energy species. The diagram in Figure 1.38 should help keep this point straight. A strong bond does not mean high energy; it means low energy. Remember also the sign convention: $2 \, H\cdot \rightarrow H-H$, $\Delta H° = -104$ kcal/mol.

As in H_2, the antibonding molecular orbital (Φ_A) is empty in H_2^+. Why the emphasis on Φ_A? What is an empty orbital, anyway? The easiest, nonmathematical way to think of Φ_A in H_2 is as the place the next electron would go. If the bonding molecular orbital Φ_B is filled, as it is in H_2, for example (Fig. 1.37), another electron cannot occupy Φ_B, and must go instead into the antibonding orbital, Φ_A, to create H_2^- ($H_2 + e^- \rightarrow H_2^-$). Dihelium, He_2, is an example of a molecule in which electrons must occupy antibonding orbitals. Each He, like each H in H_2, brings only a $1s$ orbital into the calculation. Just as in the molecules H_2 and H_2^+, the molecular orbital system is created by the combination of two $1s$ atomic orbitals. We can reuse our orbital interaction diagrams of Figures 1.34 and 1.40 to start to draw Figure 1.41.

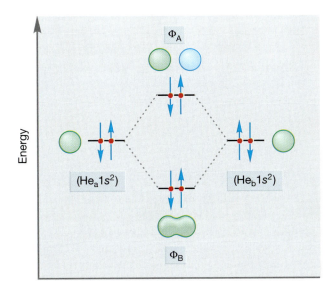

Φ_A

$(He_a 1s^2)$ $(He_b 1s^2)$

Φ_B

Energy

FIGURE 1.41 An orbital interaction diagram for He_2. There is no net bonding because the stabilization owing to the pair of electrons in the bonding molecular orbital is offset by the destabilization from the two electrons in the antibonding orbital.

Each helium atom brings two electrons, though, so the electronic occupancy of the orbitals will be different from what it is for H_2 or H_2^+. The bonding orbital (Φ_B) can only hold two electrons (Pauli principle, p. 7) so the next two must occupy Φ_A. The two electrons in this antibonding orbital are destabilizing to the molecule just as the two electrons in Φ_B are stabilizing, so there is no net binding in the hypothetical molecule He_2. Molecular helium (He_2) is, in fact, unknown.

SUMMARY

The energy of a bonding molecular orbital generated through overlap of two orbitals depends primarily on the amount of overlap between the two orbitals combining. Extensive overlap leads to a low energy bonding molecular orbital (and a high energy antibonding molecular orbital). Thus, one or two electrons in such an orbital will be well stabilized. The bond joining the two atoms is strong.

PROBLEM 1.26 Derive the molecular orbital diagram for He_2^+.

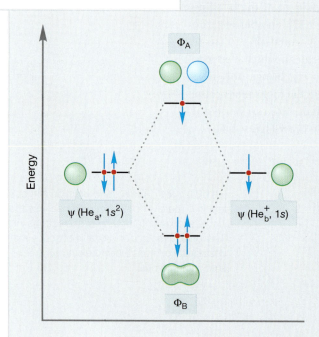

WORKED PROBLEM 1.27 Estimate the binding energy for He_2^+. How did you arrive at your estimate?

ANSWER The molecular orbital diagram for this molecule can be easily derived from Figure 1.41 by removing one electron. The He_2^+ molecule can be constructed from $He + He^+$. This molecule will have only three electrons.

Assuming that an electron in the antibonding orbital Φ_A is destabilized about as much as one in the bonding orbital Φ_B is stabilized, there is net one bonding electron in this molecule. The He_2^+ molecule should be stabilized by about the same amount as H_2^+, another molecule with a single electron in the bonding molecular orbital. This estimate turns out to be correct, and He_2^+ is bound by a little more than 60 kcal/mol.

In the figure (reading from the diagram): Φ_A, $\psi\,(He_a, 1s^2)$, Energy, $\psi\,(He_b^+, 1s)$, Φ_B

PROBLEM 1.28 We can use the answer to Problem 1.25 to work out the answer to a more subtle question. In He_2, both the bonding and antibonding orbitals are filled with two electrons. Consider electron–electron repulsion to explain why the stabilization of the two electrons in Φ_B is less than the destabilization of the electrons in Φ_A.

1.7   AN INTRODUCTION TO REACTIVITY: ACIDS AND BASES

Even at this early point we can begin to consider what many see as the real business of chemistry—reactivity. The orbital interaction diagrams we have just learned to draw lead us to powerful unifying generalizations. Remember that we can fit no more than two electrons into any orbital (p. 7). There are two ways to combine orbitals so as to produce a stabilization of two electrons. Figure 1.42(a) shows a pair of singly occupied orbitals combining to produce a bond in which two electrons are stabilized. Alternatively, we could imagine the situation outlined in Figure 1.42(b) in which a filled orbital (**Lewis base**) overlaps with an empty orbital (**Lewis acid**) to produce a similar stabilization.

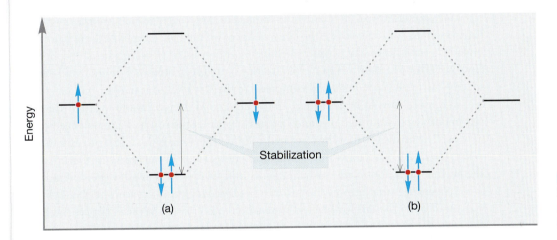

FIGURE 1.42 Two interactions that lead to stabilization of a pair of electrons.

Both these bond-forming scenarios are common, but the latter is especially important. We have already seen an example of the former case, when we considered a pair of hydrogen atoms combining to form H_2 (p. 26). We have not yet seen any real examples of the latter situation, but we will encounter hundreds as we work through this book, as our Figure 1.42(b) describes what is called acid–base chemistry. You are probably familiar with Brønsted acids and bases (proton donors and acceptors) but the situation described by Figure 1.42(b) is much more general. It would seem that a stabilizing, bond-forming interaction can be found whenever a filled orbital overlaps with an empty orbital. We will give this important idea the careful treatment it deserves in Chapters 3 and 7, but it is worth thinking about it a bit right now at this very early stage. In general, a base, called a Lewis base after G. N. Lewis (p. 13), will bear a reactive pair of electrons and a Lewis acid will have an empty orbital ready to receive those electrons. The combination of a Lewis base and a Lewis acid, conceptualized in the simple orbital interaction diagram of Figure 1.42(b), leads to a bond. The concepts that "Lewis bases react with Lewis acids" or "the interaction of a filled and empty orbital leads to bond making and stabilization of two electrons" run throughout organic (and other) chemistry. They allow us to generalize, and not to be swamped by all the prolific detail soon to come.

PROBLEM 1.29 Identify the Lewis bases and Lewis acids in the following possible reactions. Write the molecules formed by their reaction.

(a) H^+ $H{:}^-$ $\longrightarrow$?   (c) H_3C^+ $^-{:}CH_3$ $\longrightarrow$?

(b) H^+ $^-{:}\ddot{O}H$ $\longrightarrow$?   (d) H_3C^+ ${:}NH_3$ $\longrightarrow$?

1.8   SPECIAL TOPIC: QUANTUM MECHANICS AND BABIES

One might imagine that quantum mechanics, so critical in the microscopic world, would have little real influence on our day-to-day lives. We really do live in a world of baseballs, not electrons. Or so it would seem. Yet it has been argued* that "a new-born baby becomes conscious of . . . the consequences [of the Pauli principle] long before he finds it necessary to take account of the consequences of Newton's laws of motion." How so? Ernest Rutherford (p. 1) told us long ago that an atom is essentially all empty space—the nucleus is very small compared to the size of a typical atom, and electrons are even more tiny. Why is it then, that when "solid" objects come together they do not smoothly pass through one another, as they sometimes do in 3 a.m. science fiction movies? The answer is that the apparent solidity of matter is the result of Pauli forces. When your hand encounters the table top, the electrons in the atoms of those hands, and in that table top, have the same spins and nearly the same energies. The Pauli principle ensures that they cannot occupy the same regions of space, and the table top feels solid, even though Rutherford showed us that it is not. Were it not for the Pauli principle our apprehension of the world would be vastly different.

*By Walter Kauzmann (b. 1916) in an early book, *Quantum Chemistry*, Academic, 1957.

1.9   SUMMARY

New Concepts

Orbitals (wave functions) are defined by four quantum numbers, n, l, m_l, and s. Electrons may have only certain energies determined by these quantum numbers. Orbitals have different, well-defined shapes: s orbitals are spherically symmetric, p orbitals are roughly dumbbell shaped, and d and f orbitals are more complicated. Orbitals may contain a maximum of two electrons.

Some molecules cannot be well described by a single Lewis structure, but are better represented by two or more different electronic descriptions. This phenomenon is called "resonance." Be sure you are clear on the difference between an equilibrium between two different molecules, and the description of a single molecule using a number of different resonance forms.

When two orbitals overlap, two new orbitals are formed: one lower in energy than the starting orbitals, the other higher. Two electrons can be stabilized in the new, lower energy orbital. The amount of stabilization depends on the amount of overlap of the two orbitals and on the relative energy of the overlapping orbitals. The closer the two are in energy, the greater the stabilization. The overlap of orbitals and the attendant stabilization of electrons compared with those in separated atoms are the bases of covalent bonding.

This orbital-forming process can be generalized: n orbitals into the calculation requires n new orbitals (solutions) out.

Key Terms

anion (p. 3)	exothermic reaction (p. 32)	node (p. 10)
antibonding molecular orbital (p. 28)	functional group (p. 22)	nonbonding electrons (p. 13)
atomic orbitals (p. 2)	Heisenberg uncertainty principle (p. 2)	nonbonding orbital (p. 29)
atoms (p. 1)	heterolytic bond cleavage (p. 34)	nucleus (p. 3)
aufbau principle (p. 7)	homolytic bond cleavage (p. 34)	octet rule (p. 4)
bond dissociation energy (BDE) (p. 34)	Hund's rule (p. 8)	orthogonal (p. 30)
bonding molecular orbital (p. 28)	ion (p. 3)	Pauli principle (p. 7)
cation (p. 3)	ionic bonds (p. 5)	polar covalent bonds (p. 13)
covalent bond (p. 12)	ionization potential (p. 3)	quantum numbers (p. 5)
dipole moments (p. 13)	Lewis acid (p. 39)	resonance forms (p. 21)
electron (p. 1)	Lewis base (p. 39)	valence electrons (p. 15)
electron affinity (p. 4)	Lewis structures (p. 13)	wave functions (p. 5)
electronegativity (p. 14)	lone-pair electrons (p. 13)	
endothermic reaction (p. 32)	molecular orbitals (p. 2)	

Reactions, Mechanisms, and Tools

There are no reactions to speak of yet, but we have developed a number of tools in this first chapter.

Lewis structures are drawn by representing each electron by a dot. These dots are transformed into arrows if it is necessary to show electron spin. Exceptions are the 1s electrons, which are held too tightly to be importantly involved in bonding except in hydrogen. These electrons are not shown in Lewis structures except in H and He. Somewhat more abstract representations of molecules are made by showing electron pairs in bonding orbitals as lines—the familiar bonds between atoms.

The double-headed resonance arrow is introduced to cope with those molecules that cannot be adequately represented by a single Lewis structure. Different *electronic* representations for the molecule are written. The molecule is better represented by a combination of these so-called resonance forms than by any one form. Be very careful to distinguish the resonance phenomenon from chemical equilibrium. Resonance forms give multiple descriptions of a single species. Equilibrium describes two (or more) different molecules.

The curved arrow formalism is introduced to show the flow of electron pairs. This critically important device is used both in drawing resonance forms and in sketching electron flow in reactions throughout this book.

The formation of a bonding (lower in energy) and an antibonding (higher in energy) molecular orbital from the overlap of two atomic orbitals can be shown graphically. The overlap of two atomic orbitals produces two new molecular orbitals. One is lower in energy and bonding, the other higher in energy and antibonding (Fig. 1.43). Electrons are represented by vertical lines, which are transformed into arrows if it is necessary to show electron spin. Remember that only two electrons can be stabilized in the bonding orbital and that two electrons in the same orbital must have opposite spins.

FIGURE 1.43 The overlap of two atomic orbitals produces two new molecular orbitals.

Common Errors

Here, and in similar sections throughout the book, we will take stock of some typical errors made by those who attempt to come to grips with organic chemistry.

Electrons are not baseballs. Nothing is harder for most students to grasp than the consequences of this observation. Electrons behave in ways that the moving objects in our ordinary lives do not. No one who has ever kicked a soccer ball or caught a fly ball can doubt that on a practical level it is possible to determine both the position and speed of such an object at the same time. However, Heisenberg demonstrated that this is *not* true for an electron. We cannot know both the position and speed of an electron at the same time. Baseballs move at a variety of speeds (energies) and a baseball's energy depends only on how hard we throw it. Electrons are restricted to certain energies (orbitals) determined by the values of quantum numbers. Electrons behave in other strange and counterintuitive ways. For example, we have seen that a node is a region of space of zero electron density. Yet, an electron occupies an entire $2p$ orbital in which the two halves are separated by a nodal plane. A favorite question, How does the electron move from one lobe of the orbital to the other? simply has no meaning. The electron is *not* restricted to one lobe or the other, but occupies the *whole orbital* (Fig. 1.44). Mathematics makes these properties seem inevitable; intuition, derived from our experience in the macroscopic world, makes them very strange.

FIGURE 1.44 An electron occupies an entire $2p$ (or other) orbital—it is not restricted to only one lobe.

It is easy to confuse resonance with equilibrium. On a mundane, but nonetheless important level, this confusion appears as a misuse of the arrow convention. Two arrows separate two entirely different molecules (A and B), each of which might be described by several resonance forms. The amount of A and B present at equilibrium depends on the equilibrium constant. The double-headed resonance arrow separates two different electronic descriptions (C and D) of the same species, E (Fig. 1.29).

Constructing molecular orbitals through combinations of atomic orbitals (or other molecular orbitals) can be daunting, at least in the beginning. Remember these hints.

1. The number of orbitals produced at the end must equal the number at the beginning. If you start with n orbitals, you must produce n new orbitals. Here is a way to check your work as it proceeds.

2. Keep the process as simple as you can. Use what you know already, and combine orbitals in as symmetrical a fashion as you can.

3. The closer in energy two orbitals are, the more strongly they interact. At this level of theory, you need only interact the pairs of orbitals closest in energy to each other.

4. When orbitals interact in a bonding way (same sign of the wave function), the energy of the resulting orbital is lowered; when they interact in an antibonding way (different signs of the wave function), the energy of the resulting orbital is raised.

5. To order the product orbitals in energy, count the nodes; the more nodes in an orbital, the higher it is in energy.

These methods will become much easier than they seem at first, and they will provide a remarkable number of insights into structure and reactivity not easily available in other ways.

The sign convention for exothermic and endothermic reactions seems to be a perennial problem. In an exothermic reaction, the products are more stable than the starting materials. Heat is given off in the reaction and $\Delta H°$ is given a minus sign. Thus, for the reaction A + B → C, $\Delta H° = -x$ kcal/mol. An endothermic reaction is just the opposite. Energy must be applied to form the less stable products from the more stable starting material. The sign convention gives $\Delta H°$ a plus sign. Thus, A + B → C, $\Delta H° = +x$ kcal/mol.

1.10 ADDITIONAL PROBLEMS

PROBLEM 1.30   Use the arrow formalism to write structures for the resonance forms contributing to the structures of the following ions:

(a)

Carbonate ion

(b)

Sulfate ion

(c)

Nitrate ion

(d)

Guanidinium ion

(e)

A vinyl ammonium ion

PROBLEM 1.31   Use the arrow formalism to write resonance forms contributing to the structures for the following molecules:

(a)

(b)

(c)

(d)

(e)

(f)

PROBLEM 1.32   Draw resonance forms for the following cyclic molecules:

(a)

(b)

(c)

PROBLEM 1.33   Draw resonance forms for the following acyclic molecules:

(a)

(b)

PROBLEM 1.34   Ozone (O_3) is another molecule resembling those in Problem 1.31. These days it has a rather bad press, as it is present in too small an amount in the stratosphere and too great an amount in cities. Write a Lewis "dot" structure for ozone and sketch out contributing resonance forms. Write one neutral resonance form. Be careful with this last part; the answer is tricky.

PROBLEM 1.35   Add charges to the following molecules where necessary:

(a)

(b)

(c)

(d)

PROBLEM 1.36   Add charges to the following molecules where necessary:

(a)

(b)

(c)

(d)

(e)

(f)

(g)

(h)

(i)

(j)

(k)

(l)

PROBLEM 1.37   Write Lewis dot structures for the following neutral diatomic molecules. In F_2, there is a single bond between the two atoms, but in N_2 there is a triple bond between the two atoms.

$$N_2 \quad F_2$$

PROBLEM 1.38   Atomic carbon can exist in several electronic states, one of which is (of course) lowest in energy and is called the "ground state." Write the electronic description for the ground state and at least two higher energy, "excited states."

PROBLEM 1.39   Write the electronic configurations for the following ions:

(a) Na^+ (b) F^- (c) Ca^{2+}

PROBLEM 1.40   Write the electronic configurations for the atoms in the fourth row of the periodic table, $_{19}K$ through $_{36}Kr$. (*Hint:* The energy of the $3d$ orbitals falls between that of the $4s$ and $4p$ orbitals. Don't worry about the m_l designations of the $3d$ orbitals.)

PROBLEM 1.41   Write electronic configurations for $_{14}Si$, $_{15}P$, and $_{16}S$. Indicate the spins of the electrons in the $3p$ orbitals with a small up or down arrow.

PROBLEM 1.42   There is an instrument, called an electron spin resonance (ESR) spectrometer, that can detect "unpaired spin." In which of the following species would the ESR machine find unpaired spin? Explain.

$$O \quad O^+ \quad O^{2-} \quad Ne^+ \quad F^-$$

PROBLEM 1.43   For the Lewis structure of carbon monoxide shown below, first verify that both the carbon and the oxygen atoms are neutral.

Second, indicate the direction of the dipole moment in this Lewis structure.

$$:C{=}\ddot{O}:$$

As you have just shown, based on this Lewis structure, carbon monoxide should have a substantial dipole moment. In fact, the experimentally determined dipole moment is very small, 0.11 D. Draw a second resonance structure for carbon monoxide, verify the presence of any charges, and indicate the direction of any dipole in this second resonance form. Finally, rationalize the observation of only a very small dipole moment in carbon monoxide.

PROBLEM 1.44   Would you expect formaldehyde, shown below, to have a greater dipole moment than carbon monoxide? Why or why not?

PROBLEM 1.45   Consider three possible structures for methylene fluoride (CH_2F_2), one tetrahedral (structure **A**), the others flat (structures **B** and **C**). Does the observation of a dipole moment in CH_2F_2 allow you to decide between structures **A** and **B**? What about structures **A** and **C**?

A   **B**   **C**

PROBLEM 1.46   While wandering in an alternative universe you find yourself in a chemistry class and, quite naturally, you glance at the periodic table on the wall. It looks (in part) like this:

$_1$H	$_2$He							
$_3$Li	$_4$Be							
$_5$B	$_6$C	$_7$N	$_8$O	$_9$F	$_{10}$Ne	$_{11}$Na	$_{12}$Mg	
$_{13}$Al	$_{14}$Si	$_{15}$P	$_{16}$S	$_{17}$Cl	$_{18}$Ar	$_{19}$K	$_{20}$Ca	

Deduce the allowed values for the four quantum numbers in the alternative universe.

PROBLEM 1.47   In a different alternative universe, the following restrictions on quantum numbers apply:

$n = 1, 2, 3, \ldots$
$l = n - 1, n - 2, n - 3, \ldots, 0$
$m_l = l + 1, l, \ldots, 0, \ldots, -l - 1$
$s = \pm \frac{1}{2}$

Call the elements in this universe **1, 2, 3**, . . . and assume that Hund's rule and the Pauli principle still apply.
(a) For $n = 1, 2,$ and 3 show what orbital subshells are available and label them as *s, p, d,* and so on.
(b) How many electrons can be accommodated in the first three shells ($n = 1, 2,$ and 3) in this universe?
(c) Provide electronic descriptions for elements **1–14** in this universe. (In *our* universe Li would be written $1s^2 2s$.)
(d) Construct a periodic table for elements **1–23** in this universe.

PROBLEM 1.48   In Problem 1.23, you looked at two possible interactions of a pair of $2p$ orbitals, side by side and perpendicular, or end-on. Now let a pair of $2p$ orbitals interact end to end, and draw the two new molecular orbitals produced.

PROBLEM 1.49   Indicate whether the following reactions are exothermic or endothermic. Estimate by how much. Use Table 1.9 (p. 36).

(a) $H_3C\cdot + \cdot\ddot{B}\ddot{r}: \rightleftharpoons H_3C-\ddot{B}\ddot{r}:$

(b) $H_3C-\ddot{C}\ddot{l}: \rightleftharpoons H_3C\cdot + \cdot\ddot{C}\ddot{l}:$

(c) $H_2C{=}CH_2 + H-H \rightleftharpoons H_2C-CH_2$ with H, H below

(d) $H_2C{=}CH_2 + H-\ddot{C}\ddot{l}: \rightleftharpoons H_2C-CH_2$ with H, :$\ddot{C}\ddot{l}$: below

PROBLEM 1.50   Let's extend our discussion of H—H a little bit to make the molecule linear HHH. Use the molecular orbitals for H_2 and the $1s$ atomic orbital of H. Place the new H in between the two hydrogen atoms of H—H. Watch out for net-zero (orthogonal) interactions!

H—H ± H ⟶ H—H—H

H_2, Φ_B, Φ_A H, $1s$

H_2 Molecular ± H Atomic ⟶ Molecular orbitals
  orbitals   orbital   for HHH

(a) How many orbitals will there be in linear HHH?
(b) Use the molecular orbitals of H_2 and the $1s$ orbital of hydrogen to produce the molecular orbitals of linear HHH. Sketch the new orbitals.
(c) Order the new orbitals in energy (count the nodes).

The following molecular orbital problems are more challenging than the earlier ones, and may be fairly regarded as "special topics." Nonetheless, they do provide some remarkable insights, and for those who like orbital manipulation, they will actually be fun. Just remember the "rules" outlined on p. 32.

PROBLEM 1.51 Now make a set of molecular orbitals very much like the ones you made for linear HHH in Problem 1.50, but this time use $2p$ orbitals, not $1s$ orbitals. Place one $2p$ orbital in between the other two. Use the molecular orbitals $2p + 2p$ and $2p - 2p$ that you constructed in Problem 1.23.

Order the new orbitals in energy.

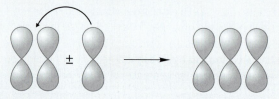

PROBLEM 1.52 One can generate the molecular orbitals for triangular H_3 simply by bending the orbitals generated in Problem 1.50 to transform the linear molecule into the bent one.
(a) Bend the three molecular orbitals for HHH to make the orbitals for the triangle. Make careful drawings. Note that as the old H(1) and H(3) come closer together in the triangle they will create a new bonding or antibonding interaction.
(b) Order the new molecular orbitals in energy by counting the nodes.

PROBLEM 1.53 Which will be lower in energy, linear or triangular $H_3{}^+$? What a sophisticated question! Yet, given the answer to Problems 1.50 and 1.52, it is easy.

PROBLEM 1.54 In the chapter, we made the two molecular orbitals for H_2 (p. 26), and in Problem 1.50 we used the two molecular orbitals of H_2 and a $1s$ orbital to make HHH. This time, generate the molecular orbitals for HHHH, linear H_4, from the molecular orbitals of two H_2 molecules placed end to end. *Remember*: At this level of "theory," you need only interact orbitals closest in energy.

Order the new molecular orbitals in energy by counting nodes, and place the proper number of electrons in the orbitals.

PROBLEM 1.55 Make the molecular orbitals for square H_4 by allowing the molecular orbitals of H_2 to interact as shown below:

$$\begin{array}{c} H-H \\ \pm \\ H-H \end{array} \longrightarrow \begin{array}{c} H-H \\ |\quad| \\ H-H \end{array}$$

Order the new molecular orbitals by counting nodes, and add the proper number of electrons. You might check your answer by deriving the same orbitals. Do this by bending the molecular orbitals for linear HHHH developed in Problem 1.54.

PROBLEM 1.56 Generate the molecular orbitals for planar ammonia, NH_3. Do this by taking combinations of the molecular orbitals for triangular H_3 (see Problem 1.52 for these orbitals) and the atomic orbitals of nitrogen.
(a) Show clear pictures of the molecular and atomic orbitals you are using to make the molecular orbitals of planar ammonia.
(b) How many molecular orbitals will ammonia have?
(c) Draw pictures of the molecular orbitals for planar ammonia. Show clearly how these are generated.
(d) Order the bonding and nonbonding molecular orbitals in terms of energy. Place them on a scale relative to the energy of a lone, nonbonding $2p$ orbital. You do not have to order the antibonding orbitals.
(e) Place the appropriate number of electrons in the orbitals. Be careful to indicate the spin quantum number for each electron (use a little ↑ or ↓ arrow to show spin).

PROBLEM 1.57 Generate the molecular orbitals for linear methylene, H—C—H, by combining the atomic orbitals of carbon with the molecular orbitals of hydrogen, H_2.
(a) Show clear pictures of the molecular and atomic orbitals you are using.
(b) How many molecular orbitals will linear methylene have?
(c) Draw pictures of the molecular orbitals for linear methylene.
(d) Order these molecular orbitals in terms of energy. Place them on a scale relative to the energy of a lone, nonbonding carbon $2p$ orbital.
(e) Place the appropriate number of electrons in the orbitals, being careful to indicate the spin quantum number for each electron (use a little ↑ or ↓ arrow to show spin).

Alkanes

2

God never saw an orbital.

—Walter Kauzmann*

February, 1964

2.1   PREVIEW

The chemical reactions of a given compound are a function of its detailed structure. No compound reacts exactly like another, no matter how similar their structures. Yet there are gross generalizations that can be made, and some order can be carved out of the chaos that would result if compounds of similar structure did not at least react in somewhat the same fashion. To a first approximation, the reactivity of a molecule depends on the presence of specific functional groups, commonly encountered groups of atoms that generally react in the same way in all molecules.

The inside front cover of this book presents an overview of some of the functional groups that are important in organic chemistry. This material is not to be memorized, and it is certainly not complete. You will want to work through it, however, being certain you can write good Lewis structures for all the different kinds of molecules, and begin to become familiar with the variety of functional groups important in organic chemistry. These different kinds of molecules will all be discussed in detail in later chapters. We start in this chapter with **alkanes**, the simplest members of the family of **hydrocarbons**—compounds composed entirely of carbon and hydrogen.

Alkanes all have the molecular formula $C_nH_{2n + 2}$. The simplest of the alkanes is **methane**, CH_4. Vast numbers of related molecules can be constructed from methane by replacing one or more of its hydrogens with other carbons and their attendant hydrogens. Linear arrays can be made, as well as branched structures and even rings (ring compounds, called cycloalkanes, have a slightly different formula: C_nH_{2n}).

*Walter Kauzmann (p. 40) was chairman of the Princeton University chemistry department in 1964 when I interviewed for a job. He made this comment in response to my typically convoluted molecular orbital answer to some simple question.

Figure 2.1 shows a few schematic representations. Notice, but don't worry about, the extremely schematic representations for these molecules. Before we can go farther we need to make sense of these schematics, and as usual, we must start with structure. The first task is to describe the shape of methane. Once we have determined this, we can go on to investigate other members of the alkane family.

FIGURE 2.1 Several different ways of drawing alkane structures.

Methane (CH₄)   Propane (C₃H₈)   Cyclopropane (C₃H₆)

PROBLEM 2.1 Draw Lewis structures for the following two linear alkanes (no branching—all carbons in a row): butane (C_4H_{10}) and pentane (C_5H_{12}).

ESSENTIAL SKILLS

1. Above all, it is important to master the structural model introduced in this chapter, hybridization. This new approximation does a good job of allowing us to predict the general structures of compounds, and is very nicely suited for use in outlining and following electron flow in chemical reactions. Thus, it fits in well with the arrow formalism.

2. It is important to be able to use the various structural formulas, which range from richly detailed three-dimensional representations to the ambiguous condensed formulas.

3. The concept of difference is essential in organic chemistry. When are two atoms the same (in exactly the same environment) or different (not in the same environment)? This question gets to the heart of structure and is much tougher to answer than it seems. In this chapter, we will introduce this subject and will return to it in Chapter 4.

IMPORTANT DETAILS

1. The naming convention for alkanes. In truth, it is not necessary to know every nuanced detail, but you do need to know some nomenclature.

2. The cis/trans convention. Cyclic molecules (rings) have sides, and attached groups can be on the same or opposite sides of the ring.

3. The Newman projection. Drawing and "seeing" Newman projections is a *critical* skill.

4. Bond strengths. You surely do not have to memorize all possible bond strengths, but by the end of this chapter you should have a good idea of some of the important ones, such as C—C (alkane), C—H, and H—H. More will appear in future chapters.

2.2 HYBRID ORBITALS: MAKING A MODEL FOR METHANE

2.2a Hybridization. Our task is to devise a bonding model for the structure of the simplest alkane, methane, CH_4. Physical chemists tell us the structure: methane is a tetrahedron, with all carbon–hydrogen bonds the same length (Fig. 2.2). We had an earlier look at a tetrahedron in Problem 1.5 (p. 15), where the molecule carbon tetrachloride appeared. Our job now is to work out a bonding scheme that leads to the experimentally determined tetrahedral structure of methane.

The method we use to generate our model of methane goes by the name of "**hybridization.**" Remember that we are working out a *model*—no matter how useful it is, what we produce here is bound to be flawed in some respects—it is not "ultimate truth," by any means! Our strategy will be to combine the four atomic orbitals of carbon ($2s$, $2p_x$, $2p_y$, $2p_z$) to produce four new orbitals, called hybrid orbitals, that can overlap with the $1s$ orbitals of four hydrogens to produce methane. Despite the fancy name "hybridization," we are really doing something quite similar to the combining of orbitals we have seen so many times before. Remember that we already know that a combination of *four* atomic orbitals must produce *four* new orbitals (p. 32), in this case four new hybrid orbitals.

It will greatly help our understanding of the hybridization process if we back up just a bit and look at two simpler hybridization models before we return to methane, so that is how we will proceed.

2.2b sp Hybridization. Let's first make a combination of a $2s$ and $2p_x$ orbital, as shown in Figure 2.3. The two can be combined in a constructive ($2p + 2s$) or a destructive ($2p - 2s$) way. The result is a pair of distorted, hybrid orbitals. No longer are there equal-sized lobes, as in a $2p$ orbital. In the first case ($2p + 2s$), we get an expansion of one of the original lobes (the one with the same sign as the $2s$ orbital) and a shrinking of the other (the one with the opposite sign) to produce a new orbital that looks very much like a lopsided $2p$ orbital. The second case ($2p - 2s$) is also lopsided, but in the opposite sense. These two new orbitals are called sp **hybrid** orbitals. The designation sp means that the orbitals are composed of 50% s orbital (50% s character) and 50% p orbital (50% p character).

Figure 2.3 also shows the real shape of one of these orbitals. It is traditional to use the schematic form in drawings, rather than attempt to reproduce the detailed structure of the orbital.

Note especially that these new sp hybrids are unsymmetrical, as they have a fat lobe and thin lobe. Overlap with another orbital will be especially good if

A tetrahedron

All angles are 109.5°

Methane, a tetrahedral molecule

FIGURE 2.2 A tetrahedron and a tetrahedral molecule, methane, showing the arrangement of the four hydrogens about the central carbon.

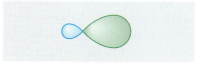

$$2s \pm 2p_x \longrightarrow \text{Two } sp \text{ hybrid orbitals}$$

One sp orbital in schematic form...

...and more correctly:

FIGURE 2.3 sp Hybridization.

the fat lobe is used. Contrast this improved overlap with that of a $2p$ orbital (Fig. 2.4). The better the overlap the greater the stabilization and the stronger the bond. In a sense, overlap with an unhybridized atomic $2p$ orbital "wastes" the back lobe of the $2p$ orbital. The process of hybridization improves overlap! The two new sp hybrids are also aimed 180° from each other. The electrons in the bonds are as remote from each other as possible, and destabilizing electron–electron repulsion is minimized.

FIGURE 2.4 A comparison between overlap of a hydrogen 1s orbital with an unhybridized 2p orbital (a) and with an sp hybrid orbital (b).

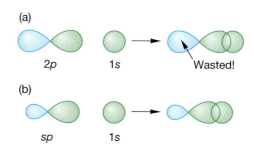

A number of compounds can be found in which the bonding can be nicely described by using *sp* hybridization. The classic example is H—Be—H, a linear hydride of beryllium. Beryllium ($_4$Be) has the electronic configuration $1s^2$, $2s^2$ and thus brings only two electrons to the bonding scheme (*Remember*: The $1s$ electrons are not used). We construct our picture of beryllium hydride (BeH$_2$) by making two *sp* hybrids from the Be $2s$ atomic orbital and one of beryllium's three equivalent $2p$ orbitals, arbitrarily taken as the $2p_x$ orbital. The overlap of these hybrids with a singly occupied hydrogen $1s$ orbital produces two Be—H bonds (*sp* + $1s$), directed at angles of 180° with respect to each other. All bonds of this type, with cylindrical symmetry, are called **sigma bonds** (σ bonds). The combination of Be (*sp*) and H ($1s$) orbitals must also yield an antibonding orbital (*sp* − $1s$), which in this case is empty, as shown in Figure 2.5. This antibonding orbital is also cylindrically symmetrical, and is therefore properly called a **σ** bond.

The antibonding orbital is designated **σ***, where the asterisk (*) means "antibonding." It is no accident that we use two atomic orbitals in our description of the bonding in BeH$_2$. *Two* hybrids are needed for bonding to *two* hydrogens, which means we must combine *two* unhybridized atomic orbitals ($2s$ and $2p_x$) to get them.

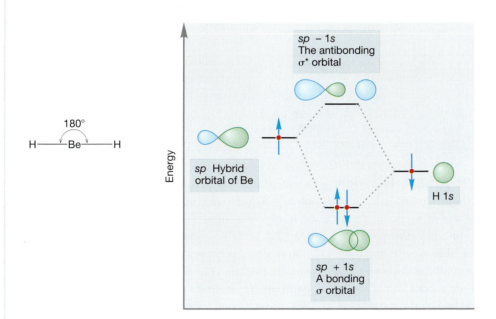

FIGURE 2.5 The formation of bonding (σ) and antibonding (σ*) orbitals from the overlap of an *sp* hybrid orbital with a hydrogen $1s$ orbital. Beryllium hydride is an *sp* hybridized compound. Note that the interaction diagram shows the $1s$ orbital at lower energy than the hybrid *sp* orbital.

PROBLEM 2.2 What happens to the $2p_y$ and $2p_z$ orbitals? Elaborate the picture in Figure 2.5 to show the unused, empty, $2p$ orbitals on Be.[†]

[†] This kind of thing generally drives physical chemists crazy. As an orbital is nothing more than the electron probability region, how can there be an "empty orbital"? The problem is at least partially avoided if the empty orbital is thought of as "the place the next electron would go."

WORKED PROBLEM 2.3 Use the hybridization model to build a picture of the linear molecule, H—C—H. Recall that a molecular orbital picture of this linear molecule was constructed in Chapter 1 (Problem 1.57, p. 46). If you didn't do that problem, you might go back, try it now, and contrast your "molecular orbital" answer with your "hybridization" answer.

ANSWER The hybridization part of this problem should be relatively easy. The carbon atom is attached to two other atoms, the two hydrogens. As for BeH_2 (p. 51), the central atom, here carbon, is hybridized sp. The $2s$ and $2p_x$ orbitals are combined to form two hybrid sp orbitals. Bonds to hydrogen are made through overlap of sp and 1s orbitals. The $2p_y$ and $2p_z$ orbitals remain unchanged. Carbon uses two of its four bonding electrons in the carbon–hydrogen bonds, leaving two for the remaining $2p$ orbitals. Each contains a single electron.

2.2c *sp²* Hybridization. Let's extend this way of thinking to analyze borane (BH_3), a slightly more complicated molecule, before we come to our attempt at a realistic picture of methane. In BH_3, we must make three bonds from boron to hydrogen. Boron's electronic configuration is $_5B = 1s^2 2s^2 2p_x$, and thus there are three electrons available to form bonds with the three singly occupied hydrogen 1s orbitals. This time we combine the boron $2s$, $2p_x$, and $2p_y$ orbitals to produce the three hybrids we need in our model. These are called, naturally, ***sp²* hybrids** (constructed from one s orbital and two p orbitals) and have 33% s character and 67% p character. They resemble the sp hybrids, shown in Figure 2.3, which gives both schematic and more accurate representations of an sp orbital. As with sp hybrids, the lopsided distribution allows for efficient overlap with a hydrogen 1s, or any other singly occupied orbital.

$$2s + 2p_x + 2p_y \longrightarrow \text{Three } sp^2 \text{ hybrid orbitals}$$

PROBLEM 2.4 Boron also forms a compound, BF_3. Create a bonding scheme and draw a Lewis structure for this molecule.

How are these new sp^2 hybrids arranged in space? Your intuition may well give you the answer. They are directed so that electrons in them will be as far as possible from each other. The angle between them is 120° and they are aimed toward the corners of a triangle (Fig. 2.6).

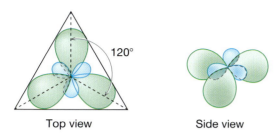

Top view   Side view

FIGURE 2.6 Top and side views of the arrangement in space of the three sp^2 hybrid orbitals.

PROBLEM 2.5 The $2p_z$ orbital on boron was not used in our construction of BH_3. Sketch it in, using the drawing of Figure 2.6.

2.2d sp^3 Hybridization: Methane.

Now we come again to methane. For CH_4 we need not two or three hybrids, but four—there are four hydrogens to be attached to carbon. Fortunately, we have just what we need in the $2s$, $2p_x$, $2p_y$, and $2p_z$ atomic orbitals of carbon. In our mathematical description we combine these orbitals to produce four new sp^3 **hybrids**, which also look similar to sp hybrids (Fig. 2.3).

$$2s + 2p_x + 2p_y + 2p_z \longrightarrow \text{Four } sp^3 \text{ hybrid orbitals}$$

The familiar lopsided shape appears again, and thus we can expect strong bonding to the singly occupied hydrogen $1s$ orbital. By now you will surely anticipate that these orbitals will be aimed in space so as to keep the orbitals, and any electrons in them, as far apart as possible. How are four objects arranged in space so as to maximize the distance between them? The answer is, at the corners of a tetrahedron (Fig. 2.2).

The carbon in methane is hybridized sp^3 (25% s character; 75% p character) and the orbitals are directed toward the corners of a tetrahedron. The four hydrogen atoms, each with an electron in a $1s$ orbital, overlap with the hybrids to form bonding and antibonding pairs (Fig. 2.7). The resulting H—C—H angles are 109.5°, and the length of a carbon–hydrogen bond in methane is 1.09 Å (Fig. 2.2).

The carbon–hydrogen bond in methane is very strong; its bond dissociation energy is 105 kcal/mol. This means that 105 kcal/mol must be applied in

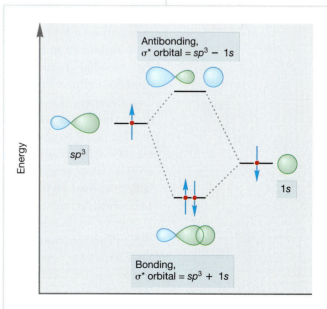

FIGURE 2.7 The formation of a carbon–hydrogen bond through the overlap of an sp^3 hybrid orbital with a hydrogen $1s$ orbital. Note the formation of both the bonding (σ) orbital and the antibonding (σ^*) orbital.

order to break one of the carbon–hydrogen bonds in methane homolytically. This reaction is endothermic by 105 kcal/mol (Fig. 2.8).

FIGURE 2.8 Homolytic cleavage of one carbon–hydrogen bond in methane.

$$\Delta H° = +105 \text{ kcal/mol}$$

2.2e Why Hybridization? Why do we need to develop this new structural model called hybridization? Why don't we just take what we already knew—the atomic structure of carbon and hydrogen—and let the $1s$ orbitals of four hydrogens overlap with the atomic orbitals of carbon? Well, we certainly could have done just that, but the results are not very satisfying—it just doesn't give us a decent approximation of the structure of methane. However, let's follow this flawed procedure in some detail because it yields very valuable insights. We will learn a lot from examining what is wrong with too simple models. Further, what we do here approximates what really happens in science in that we make a series of successive structural approximations in which we come increasingly close to the real methane. In Section 2.2d, we went right to a satisfactory model for methane, thus shortcutting the intellectual process.

A carbon atom has the electronic configuration $1s^2 2s^2 2p_x 2p_y$. The electrons in the $1s$ orbital are too low in energy to participate strongly in bonding, and the $2p_z$ orbital is empty. For now, we can imagine forming bonds between each of the four hydrogen $1s$ orbitals and the orbitals on carbon that contain electrons. In doing this, we mimic the construction of the hydrogen–hydrogen bond in H_2 from the overlap of a pair of singly occupied hydrogen $1s$ orbitals. Figure 2.9 shows the overlap of the available carbon $2s$, $2p_x$, and $2p_y$ orbitals with the four hydrogen $1s$ orbitals. Two predictions are clear: first, we should have two carbon–hydrogen bonds formed from the overlap of the hydrogen $1s$–carbon $2p_x$ and hydrogen $1s$–carbon $2p_y$ orbitals; second, these bonds must be at 90° to each other, because of the 90° angle between the carbon $2p$ orbitals. The third and fourth bonds are difficult to locate exactly, as they are formed from the overlap of a carbon $2s$ orbital with two hydrogen $1s$ orbitals. Both the $2s$ and $1s$ orbitals are spherically symmetric, and thus no directionality can be induced by the shape of the orbitals. We might guess that these two new carbon–hydrogen bonds would be aimed so as to keep the electrons in the bonds as far from each other as possible, thus minimizing repulsions.

FIGURE 2.9 A structure for methane, CH_4. Bonds are formed by the overlap of hydrogen 1s with carbon 2p and 2s orbitals. This requires some 90° bond angles in this model. Carbon's 1s electrons are shown in parentheses because they are not used in bonding.

PROBLEM 2.6 Sometimes people see a contradiction in this model. There appear to be too many electrons in the carbon 2s–hydrogen 1s system. We have spent some effort in Chapter 1 to explain the Pauli principle (p. 7), and to show why only two electrons may occupy any atomic or molecular orbital. You are used to thinking of two-electron bonds, and our new system is clearly more complicated than that. But there is no violation of the Pauli principle in the model above. Explain.

Hint: Remember that overlap of *n* orbitals produces *n* new orbitals. See Problem 1.50.

This model has a fatal flaw. Modern spectroscopic methods show that there is no 90° H—C—H angle in methane and that all carbon–hydrogen bonds are of equal length! The four carbon–hydrogen bonds are identical. So we're wrong, after all, even though the Pauli principle is not violated (see Problem 2.6). Knowing that we are wrong, we can begin to rationalize why, and try to construct a model that does a better job of describing Nature. Don't be offended by this—science works this way all the time. The goal is to use what we learn from our wrong guesses to come closer next time. We can already anticipate one problem in this model—the electrons in some of the bonds are only 90° apart. They could be farther apart, and thus electron–electron repulsion could be reduced.

PROBLEM 2.7 What about this repulsion? To be sure, the negatively charged electrons and positively charged nuclei will repel other like charges. But there is more to it than that. Consider the interaction through space of two bonds, each containing two electrons in a filled orbital, and show that the farther apart they are, the lower in energy the system must be.

There are other, more subtle problems. The H—C—H, 1s–2s–1s bonding system consists of three orbitals and four electrons. A system of three molecular orbitals emerges: one bonding, one nonbonding, and one antibonding. The two lowest energy molecular orbitals are occupied by the four available electrons (Fig. 2.10).

Although the lowest energy orbital is bonding, the middle orbital is essentially nonbonding. This use of electrons is not an efficient way to hold atoms together. Bonded systems are more stable than their separated constituent atoms because of

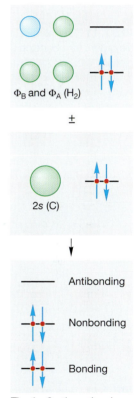

The 1s–2s–1s molecular orbital system

FIGURE 2.10 The orbital system made from the overlap of a carbon 2s atomic orbital with two hydrogen 1s orbitals consists of three molecular orbitals: one bonding, one nonbonding, one antibonding.

the stabilization originally shown in Figures 1.37 and 1.43 (pp. 33, 41), and repeated here in Figure 2.11. Electrons in nonbonding orbitals are by definition no lower in energy than those in separated atoms. If there is a better (lower energy) arrangement of atoms and orbitals, Nature will find it. It is our job to construct a model that reproduces Nature.

FIGURE 2.11 The interaction of two atomic orbitals produces a bonding and antibonding pair of molecular orbitals. An electron in the bonding molecular orbital will be stabilized (lower energy) relative to an electron in one of the separated atomic orbitals.

A third problem with this model might be labeled "inefficient overlap," or "wasted orbitals." Recall (p. 50) that we have already seen that unhybridized $2p$ orbitals are quite poorly designed for efficient end-on overlap, as overlap can take place only with one lobe. The back lobes of $2p$ orbitals are unused and thus wasted (Fig. 2.12). Our initial model for methane has two such overlapping pairs of orbitals.

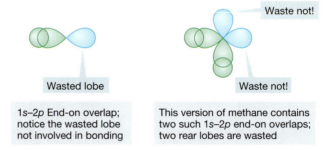

FIGURE 2.12 End-on overlap of a $2p$ atomic orbital with a $1s$ orbital.

Wasted lobe

Waste not!

Waste not!

$1s$–$2p$ End-on overlap; notice the wasted lobe not involved in bonding

This version of methane contains two such $1s$–$2p$ end-on overlaps; two rear lobes are wasted

PROBLEM 2.8 Why not allow a $2p$ orbital to overlap sideways with an s orbital? Why would this not solve the problem of waste of the back lobe (see below and Fig. 2.12)?

SUMMARY

There are three severe problems with this too simple model for methane. The hybridization model solves them all!

1. The electrons in the bonds were far from being as remote from each other as possible. The hybridization model yields four hybrid orbitals directed toward the corners of a tetrahedron, the best possible arrangement.

2. Two electrons occupied a nonbonding orbital and, therefore, did not contribute to holding the molecule together. In the hybridization model, all electrons are in bonding orbitals (Fig. 2.7).

3. Bonds were formed through overlap between carbon $2p$ and hydrogen $1s$ orbitals, thus wasting the rear lobes of the $2p$ orbitals. In the hybridization model, overlap is much improved in the bonds formed using the fat lobes of hybrid sp^3 orbitals and hydrogen $1s$ orbitals (contrast Figs. 2.12 and 2.7). Table 2.1 reviews the properties of sp, sp^2, and sp^3 hybrid orbitals.

TABLE 2.1 Properties of Hybrid Orbitals

Hybridization	Constructed from	Angle between Bonds (°)	Examples
sp	50% s, 50% p	180	BeH_2, linear HCH
sp^2	33.3% s, 66.7% p	120	BH_3, BF_3
sp^3	25% s, 75% p	109.5	CH_4, $^+NH_4$

So why didn't I just tell you that methane is tetrahedral and be done with it? We took several extra pages to rationalize and explore that experimental observation. First of all, learning science is not about gathering a collection of facts. Rather it is learning a way of thinking—a way of solving problems—of reaching an approximate idea of how Nature works. In science, we look at the known experimental data (methane has the formula CH_4, for example) and then postulate a model to explain those observations. In the beginning, that model is almost certain to be rather badly flawed, or at the very least severely incomplete. So we test it against new data—against what can be learned from new experiments, and modify our hypothesis. In this case, that process looks like this: We find out that methane has no 90° angles—and how can we change it to get rid of those errors and come to a better model of Nature? Ultimately, we come to the tetrahedral structure. We have learned much from looking at why our early, too simple models were inadequate, and from our process of investigation. You will forget most of the facts in this book, eventually. No matter, you can always look them up here or elsewhere. What I hope you will not forget is the investigative method used in organic chemistry and all other sciences—and many, many other disciplines as well.

But it is a utilitarian world these days, and I can hear cries of, "just gimme the facts, buster; I'm not in this course to learn about science, I just want to get on with my career goals." No! No! No! I would argue with you forever that if you feel this way you are making a profound, lifetime error, but even if I grant such a goal, there is a fatal flaw in just giving the facts. Put simply, there is too much material to be learned this year to memorize your way through it. You could survive at first, but somewhere along about the middle of the first semester you will run out of memory, and there is no way (yet) to install any more. If you try to get through this course by memorizing a set of facts, the bottom line in each chapter, you will almost certainly not succeed. Success in this course—and in much of life—depends on learning how to think, how to reason sensibly from new data. So that's what I will try to do throughout this book, and we have begun right here.

2.3   DERIVATIVES OF METHANE: METHYL (CH$_3$) AND METHYL COMPOUNDS (CH$_3$X)

We can now imagine moving to a less symmetrical situation. As soon as one of the four hydrogens surrounding the central carbon in methane is replaced with another atom, pure tetrahedral symmetry is lost. We might well anticipate that one of our newly constructed sp^3 hybrids could overlap in a stabilizing way with any atom X offering a singly occupied orbital. For maximum stabilization we want to fill the new, bonding orbital created from the overlap of an X orbital with one of the hybrids, which requires two and only two electrons (Fig. 2.13). All manner of derivatives of methane, called **methyl compounds** (CH$_3$—X), are possible.

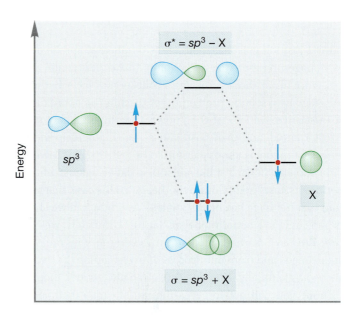

FIGURE 2.13 Formation of a carbon–X bond from overlap of an sp^3 hybrid with an s orbital of X.

Their names are formed by dropping the "ane" from the name of the parent, and appending "yl." The generic term for derivatives of alkanes is **alkyl**. Such compounds are often abbreviated R—X, where R stands for a general alkyl group.[*] Table 2.2 shows a number of methyl derivatives with their melting and boiling points.

CONVENTION ALERT

TABLE 2.2   Some Simple Derivatives of Methane, Otherwise Known as Methyl Compounds

H$_3$C—X	Common Name	mp (°C)	bp (°C)	Physical Properties
H$_3$C—H	Methane	−182.5	−164	Colorless gas
H$_3$C—OH	Methyl alcohol or methanol	− 93.9	65.0	Colorless liquid
H$_3$C—NH$_2$	Methylamine	− 93.5	− 6.3	Colorless gas
H$_3$C—Br	Methyl bromide	− 93.6	3.6	Colorless gas/liquid
H$_3$C—Cl	Methyl chloride	− 97.7	− 24.2	Colorless gas
H$_3$C—CN	Methyl cyanide or acetonitrile	− 45.7	81.6	Colorless liquid
H$_3$C—F	Methyl fluoride	−141.8	− 78.4	Colorless gas
H$_3$C—I	Methyl iodide	− 66.5	42.4	Colorless liquid
H$_3$C—SH	Methanethiol or methyl mercaptan	−123	6.2	Colorless gas/liquid

The bonding in methyl compounds closely resembles that in methane, and it is conventional to speak of the carbon atom in H$_3$C—X as being sp^3 hybridized, just as it is in the more symmetrical CH$_4$. Strictly speaking, this is wrong, although sp^3 is a good approximation. The four bonds to carbon in H$_3$C—X are not equivalent, as an X is not H. The bond from C to X is not the same length as that from C to H, and the H—C—H bond angle cannot be exactly the same as the X—C—H angle. As sp^3 hybridization yields four exactly equivalent bonds directed toward the corners of a tetrahedron, the bonds to H and X in H$_3$C—X cannot be *exactly* sp^3 hybrids, only approximately sp^3 ($sp^{2.8}$, say). This point is very often troubling to students, but it is really quite simple and, once one has seen it, even obvious. Consider converting an sp^2 hybridized carbon into an sp^3 hybridized carbon, as in Figure 2.14. We start with three sp^2 hybrids and a pure, unhybridized $2p$ orbital (H—C—H angle = 120°).

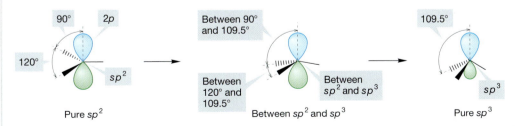

FIGURE 2.14 A thought experiment. The conversion of an sp^2 hybridized carbon into an sp^3 hybridized carbon. What is the hybridization at a point intermediate between the starting point (sp^2) and end point (sp^3)?

[*]Why R? In the old days R stood for *der Rest* (German for "residue" or "remains") or, in English, radical, a general name for groups that were commonly found in organic chemistry. These days "radical" is better reserved for the neutral species in which a single electron occupies a nonbonding orbital, R· (see Section 2.4 for a more detailed description), but the general use of R to mean "any alkyl group" remains.

As we bend the hybrid orbitals, we will come to a point at which the 120° H—C—H angle has contracted to 109.5°, or sp^3 hybridization. Our orbitals, originally ⅓ s and ⅔ p (sp^2: 33.3% s, 66.7% p character) have become ¼ s and ¾ p (sp^3: 25% s, 75% p character). *They have gained p character in the transformation from sp^2 to sp^3.* Now look at the $2p$ orbital. As we bend, it goes from pure p (100% p character) to ¼ s, ¾ p character (sp^3: 25% s, 75% p character). *It has gained s character in the transformation.* There is a smooth transformation as we pyramidalize the molecules from sp^2 to sp^3, and in between these two extremes, the hybridization of carbon is intermediate between sp^2 and sp^3, something like $sp^{2.8}$. The hybridizations sp^2 and sp^3 are *limiting* cases, applicable only in very symmetrical situations. The hybridization of a carbon (or other) atom is intimately related to the bond angles! If you know one, you know the other.

In general, methyl compounds (H_3C—X) will closely resemble methane in their approximately tetrahedral geometry; sp^3 is a good approximation, but they cannot be perfect tetrahedra.

So far, we have looked only at monosubstituted methanes. Of course, other substitution patterns are possible in which more than one hydrogen is replaced by another atom or group of atoms. Thus, even for one-carbon molecules, there are many possible substituted structures.

PROBLEM 2.9 Use the halogens (X = F, Cl, Br, or I) to build up all possible molecules CH_2X_2.

PROBLEM 2.10 Show all possible molecules of the formula CH_2X_2, CHX_3, and CX_4 (X = F or Cl).

2.4   THE METHYL CATION ($^+CH_3$), ANION ($^-:CH_3$), AND RADICAL ($\cdot CH_3$)

Table 2.2 could be augmented to include $^+CH_3$, $^-:CH_3$, and $\cdot CH_3$, three special cases of H_3C—X. The molecules in Table 2.2 are all stable species, and they can be purchased from any chemical supply company. The three new molecules are members of a class of compounds called **reactive intermediates**. This simply means that these molecules are too unstable to be isolable under normal conditions, and must usually be studied by indirect means, often by looking at what they did during their brief lifetimes, or sometimes by isolating them at very low temperature. These three species are especially important because they are prototypes—the simplest examples of whole classes of molecules to be encountered later when the study of chemical reactions takes over our attention.

In the cation $^+CH_3$, X is "nothing." To make this molecule, one could imagine removing **hydride** ($^-:H$) from methane to leave the positively charged **methyl cation** ($^+CH_3$), the simplest example of a **carbocation**, a molecule containing a positively charged carbon* [Fig. 2.15(a)].

---

*A topic of a long and too intense argument in the chemical community, a carbocation is sometimes called a carbonium ion (by traditionalists) or a carbenium ion by others. The compromise "carbocation" is both aptly descriptive and avoids the emotional reactions of the staunch defenders of the other terms.

In the **methyl anion** (⁻:CH₃), X is a pair of electrons. Removal of a proton (H⁺) from methane would leave behind a pair of nonbonding or lone-pair electrons in the negatively charged methyl anion, a simple example of a carbon-based anion, or **carbanion**. Both of these ways of producing ⁺CH₃ and ⁻:CH₃ involve the concept of breaking a carbon–hydrogen bond in unsymmetrical fashion, a process known as heterolytic bond cleavage (Chapter 1, p. 34; Fig. 2.15). Remember the arrow formalism—the arrows of Fig. 2.15 move the *pair* of electrons in the carbon–hydrogen bond to the hydrogen (a) or to the carbon (b).

(a) Carbocation formation   (b) Carbanion formation

The methyl   Hydride
cation

The methyl   A proton
anion

FIGURE 2.15 The formation of (a) the methyl cation (⁺CH₃) and (b) the methyl anion (⁻:CH₃) by two different heterolytic cleavages of a carbon–hydrogen bond in CH₄.

Recall that there is another way to think of breaking a two-electron bond, and that is to allow one electron to go with each atom at an end of the breaking bond. This homolytic bond cleavage (Chapter 1, p. 34) in methane would give a hydrogen atom (H•) and leave behind the neutral **methyl radical** (•CH₃). Here a single electron plays the part of X in CH₃—X, and the single-barbed "fishhook" arrow convention is used (Fig. 2.16).

The methyl   A hydrogen
radical   atom

FIGURE 2.16 The homolytic bond cleavage of a carbon–hydrogen bond in methane to give a hydrogen atom and the methyl radical.

All three of these species have been observed, although each is extremely reactive. They have structures, though, and we can make some predictions for at least two of them. In the methyl cation (⁺CH₃), carbon is attached to three other groups, suggesting the need for three hybrid orbitals (recall BH₃, p. 52) and that, in turn, suggests sp^2 hybridization (Fig. 2.17).

sp^2–1s Bond   Empty p orbital

FIGURE 2.17 The sp^2 hybridized methyl cation, ⁺CH₃.

1.10 Å

21.4°

107.5°

FIGURE 2.18 The structure of the methyl anion, $^-\!:CH_3$.

On the other hand, the carbon in the methyl anion is attached not only to three hydrogens but also to the pair of nonbonding electrons. In the methyl cation, with no lone-pair electrons, we have a pure p orbital (zero s character) and the species is as flat as a pancake (H—C—H angle = 120°). The methyl anion has two more electrons than the cation and we will have to consider them in arriving at a prediction of structure. Recall that s orbitals have density at the positively charged nucleus. Therefore, it would seem reasonable that a negatively charged electron would be more stable (lower in energy) in an orbital with a lot of s character. A pyramidal structure seems appropriate for the methyl anion, although of course the molecule cannot be a perfect tetrahedron. We can't predict exactly how pyramidal the species will be, and it's a difficult quantity to measure in any case, but recent calculations predict the structure in Figure 2.18.

It is harder to predict the structure of the neutral methyl radical ($\cdot CH_3$) in which there is only a single nonbonding electron. At present it is not possible to choose between a planar species and a rapidly inverting and very shallow pyramid, although it is clear that the methyl radical is close to planar (Fig. 2.19). Do not be disturbed by this! Our understanding of organic chemistry is still incomplete and many simple things are beyond us. There is still lots to do!

120°

H—C—H

or

sp^2

Two inverting shallow pyramids

FIGURE 2.19 The methyl radical ($\cdot CH_3$) is either flat or a rapidly inverting shallow pyramid.

PROBLEM 2.11 Draw a structure for the methyl radical at the halfway point for the inversion (Fig. 2.19). What is the hybridization of the carbon atom in the structure you drew?

SUMMARY

Methane, in principle, can be substituted in many ways through replacement of one or more hydrogens with another atom or groups of atoms. There may or may not be easy ways to carry out such substitutions in the real world—we will soon embark on an investigation of how to do such transformations.

In principle, removal of a hydrogen from methane can lead to the methyl anion ($^-\!:CH_3$), the methyl radical ($\cdot CH_3$), or the methyl cation ($^+CH_3$). In this chapter, we only discuss the structures of these intermediates—reactions are coming later.

2.5   ETHANE (C_2H_6), ETHYL COMPOUNDS (C_2H_5X), AND NEWMAN PROJECTIONS

There is a special substituent X that produces a most interesting H_3C-X. In this case, we let X = another methyl group. Now H_3C-X becomes H_3C-CH_3, methylmethane in the methyl-X terminology, but known always as ethane, C_2H_6. This compound is the second member of the class of alkanes (Fig. 2.20). You might try to anticipate the following discussion by building a model of ethane and examining its structure now.

FIGURE 2.20   The construction of ethane (H_3C-CH_3) by the replacement of X in H_3C-X with another methyl group.

We will soon come to a section (Section 2.6) devoted to the various drawing conventions (codes) but we must take a moment here to repeat one important drawing convention. The solid wedges of Figure 2.20 are taken as coming forward, out of the paper. The dashed wedges are read as retreating from you into the paper. It takes only a little practice before drawings such as the ones in Figure 2.20 come to life, and appear clearly three dimensional.

Even in a molecule as simple as ethane there are a number of interesting structural questions. From the point of view of one carbon, the attached methyl group takes up more room than the much smaller hydrogens. So we would expect the H_3C-C-H angle to be slightly larger than the $H-C-H$ angle. And so it is, $H-C-CH_3 = 111.0°$ and $H-C-H = 109.3°$ (Fig. 2.21).

CONVENTION ALERT

FIGURE 2.21   The detailed structure of ethane.

There are more complicated structural questions. How are the hydrogens of one end of the molecule arranged with respect to the hydrogens at the other end? There are two extreme, or limiting structures, called **conformations**, for ethane. Figure 2.21 shows ethane in only one conformation. Conformations are three-dimensional molecular structures interrelated by rotations about bonds. In this case, they are **eclipsed ethane** and **staggered ethane**, each shown in a side view in Figure 2.22. In the eclipsed conformation, each of the three hydrogens on one methyl stands directly in front of a hydrogen on the other methyl. There is a **dihedral angle** of 0° between the hydrogens. The dihedral angle is the torsional, or twist, angle between two carbon–hydrogen bonds on adjacent carbons; the angle between the H—C—C and C—C—H planes, and is shown in Figure 2.22 as θ. In the staggered conformation, the dihedral angle θ (pronounced Thā-ta) is 60° and the hydrogens are maximally far apart. Be sure you see how the two forms are interconverted by rotation about the carbon–carbon bond. Use a model!

FIGURE 2.22 Two conformations of ethane: the eclipsed and staggered forms.

The dihedral angle, θ, is the angle between the red C—H bonds

CONVENTION ALERT

A particularly effective way of viewing these structures is called a **Newman projection**, after its inventor, Melvin S. Newman (1908–1993) of Ohio State University. Like all devices, it contains arbitrary conventions, which can only be learned, not reasoned out. Its utility will repay you for the effort many times over, however, so do it! One first imagines looking down a particular bond, in this case the lone carbon–carbon bond in the molecule. The three carbon–hydrogen bonds attached to the "front" carbon are now drawn in as shown. Notice in Figure 2.23 that from this end-on view the angle between the carbon–hydrogen bonds on the front carbon is 120°. The rear carbon is next represented as a circle—a kind of exploded view—and the H atoms attached to it are put in. It is easy to construct the representation for the staggered form (Fig. 2.23), but not so easy for the eclipsed molecule. If we were strictly accurate in the drawing, we wouldn't be able to see the hydrogens attached to the rear carbon, because each is directly behind an H on the front carbon (the dihedral angle is 0°). So we cheat a little and offset these eclipsed hydrogens just enough so we can see them (Fig. 2.23). Newman projections are extraordinarily useful in making three-dimensional structures clear on a two-dimensional surface. Problem 2.12 offers a chance to practice visualizing and creating Newman projections. This problem is important.

Staggered ethane

Eclipsed ethane

FIGURE 2.23   A Newman projection for staggered ethane. The dihedral angle (θ) between two carbon–hydrogen bonds is 60°. A Newman projection for eclipsed ethane. The dihedral angle (θ) between two carbon–hydrogen bonds is 0°.

PROBLEM 2.12   Write Newman projections for the staggered conformations found by looking down the carbon–carbon bonds of ethyl chloride (CH_3-CH_2-Cl) and 1,2-dichloroethane ($Cl-CH_2-CH_2-Cl$). In the second case, there are two staggered forms of different energy. Can you estimate which will be more stable?

Which of these forms of ethane represents the more stable molecule? One's guess would surely be the staggered form, and that would be right. The electrons in the carbon–hydrogen bonds repel each other, and it is energetically best to have them as far from each other as possible, as in the staggered form. This repulsion can be effectively displayed in a plot of dihedral angle as a function of energy, as in Figure 2.24. Indeed, calculations show that the eclipsed form *is not even an energy minimum*. It is an energy maximum, the top of the barrier separating two staggered forms. Such a point is called a **transition state (TS)**.

FIGURE 2.24 A plot of the energy change in ethane as rotation around the carbon–carbon bond occurs.

PROBLEM 2.13 You have recently seen a transition state, although the term wasn't used. Where? *Hint*: Where have we described two forms of a molecule that interconvert by passing through a species **A**? This **A** turns out to be a transition state.

How big a difference is there between the two forms? This question asks how high the barrier is between two of the minimum energy staggered forms (see Fig. 2.24). The barrier turns out to be a small number, 2.9 kcal/mol, an amount of energy easily available at room temperature. As there are three eclipsed interactions in eclipsed ethane, we can see that each pair of eclipsed carbon–hydrogen bonds "costs" about 1 kcal/mol **torsional strain**. Under normal conditions, ethane is said to be freely rotating, as there is ample thermal energy to traverse the barrier separating any two staggered forms. In colder places, where there was much less thermal energy available, alien organic chemists would have to worry more about this rotational barrier. Thus, their lives would be even more complicated than ours.

PROBLEM 2.14 Draw a Newman projection for the structure depicted below by looking down the carbon–carbon bond.

$$H_3C - CH_2OH$$

Eye

PROBLEM 2.15 Imagine 1,2-dideuterioethane ($DCH_2 - CH_2D$), where D = deuterium, on some cold planet such as Pluto. If there is not enough energy available to overcome the barrier to rotation, how many 1,2-dideuterioethanes would there be?

The chemical and physical properties of ethane closely resemble those of methane. Both are low-boiling gases and both are quite unreactive under most chemical conditions. But not all! One need only light a match in a room containing ethane or methane and air to find that out. This reaction is actually very interesting. If we examine the debris after the explosion, we find two new molecules, water and carbon dioxide (Fig. 2.25).

A thermochemical analysis indicates that water and carbon dioxide are more stable than ethane and oxygen. Presumably, this is the reason ethane explodes when the match is lit. Energy is all too obviously given off as heat (Δ) and light ($h\nu$). But why do objects, stable in air, such as ethane or wood, explode or burn continuously when ignited? How are they protected until the match is lit?

$$2\ H_3C-CH_3 \quad \xrightarrow[\text{match}]{7\ O_2} \quad 4\ CO_2\ +\ 6\ H_2O\ +\ \underset{\text{Heat}}{\Delta}\ \text{and}\ \underset{\text{Light}}{h\nu}$$
Ethane

FIGURE 2.25 The combustion of ethane to give carbon dioxide, water, and energy.

We'll approach this question in Chapter 8, but it's worth some thought now in anticipation. This matter is serious because the molecules in our bodies are also less thermodynamically stable than their various oxidized forms, and we live in an atmosphere that contains about 20% oxygen. The question could be turned into, Why can humans live in such an atmosphere without spontaneously bursting into flame? This issue has been of some concern, not only for novelists.*

We can imagine making ethane by allowing two methyl radicals to come together. This process is by no means entirely imaginary; methyl radicals do react to form ethane. Figure 2.26 looks at the combination of two methyl radicals in some detail.

The two hybrid orbitals overlap to form a new carbon–carbon bond. The orbital diagram in Figure 2.26 shows this in graphic form. We produce a new bonding molecular orbital that is filled by the two electrons originally in the hybrid orbitals of the two methyl groups. This new bonding orbital is a σ orbital, as the required cylindrical symmetry is present. Of course, we simultaneously produce an antibonding molecular orbital (σ^*), which is empty in ethane. We might guess that the overlap of the two equal-energy hybrids would be very favorable, and it is. The carbon–carbon bond in ethane has a bond strength of

Two methyl radicals

Ethane

The antibonding molecular orbital, $\sigma*$

$H_3C\cdot$ $\cdot CH_3$

The bonding molecular orbital, σ

FIGURE 2.26 The formation of ethane through the combination of a pair of methyl radicals.

*Eliot . . . claimed to be deeply touched by the idea of an inhabited planet with an atmosphere that was eager to combine violently with almost everything the inhabitants held dear . . . "When you think of it, boys," he said brokenly, "that's what holds us together more than anything else, except maybe gravity. We few, we happy few, we band of brothers—joined in the serious business of keeping our food, shelter, clothing and loved ones from combining with oxygen."

—Kurt Vonnegut
God Bless You, Mr. Rosewater

about 90 kcal/mol. This value is the amount of energy by which ethane is stabilized relative to a pair of separated methyl radicals, and is therefore the amount of energy required to break this strong carbon–carbon bond. The formation of ethane from two methyl radicals is exothermic by 90 kcal/mol, and the dissociation of ethane into two methyl radicals is endothermic by 90 kcal/mol. In equation form,

$$H_3C\cdot + \cdot CH_3 \rightarrow H_3C-CH_3 \qquad \Delta H^\circ = -90 \text{ kcal/mol}$$

or, equivalently

$$H_3C-CH_3 \rightarrow H_3C\cdot + \cdot CH_3 \qquad \Delta H^\circ = +90 \text{ kcal/mol}$$

Note once again the sign convention for ΔH°. In the exothermic reaction the product H_3C-CH_3 is lower in energy than the starting material (two methyl radicals) by 90 kcal/mol; hence the minus sign. By contrast, in the endothermic reaction, 90 kcal/mol must be supplied to ethane to produce the higher energy pair of methyl radicals, and the sign becomes plus.

Just as we imagined replacing one of the hydrogens in methane with an X group to generate the series of methyl compounds, so we can produce **ethyl compounds** (CH_3CH_2-X) from ethane. Table 2.3 shows some common ethyl compounds and a few of their physical properties.

TABLE 2.3　Some Simple Derivatives of Ethane: Ethyl Compounds

CH_3CH_2-X	Common Name	mp (°C)	bp (°C)	Physical Properties
CH_3CH_2-H	Ethane	−183.3	−88.6	Colorless gas
CH_3CH_2-OH	Ethyl alcohol or ethanol	−117.3	78.5	Colorless liquid
$CH_3CH_2-NH_2$	Ethylamine	−81	16.6	Colorless gas/liquid
CH_3CH_2-Br	Ethyl bromide	−118.6	38.4	Colorless liquid
CH_3CH_2-Cl	Ethyl chloride	−136.4	12.3	Colorless gas/liquid
CH_3CH_2-CN	Ethyl cyanide or propionitrile	−92.9	97.4	Colorless liquid
CH_3CH_2-F	Ethyl fluoride	−143.2	−37.7	Colorless gas
CH_3CH_2-I	Ethyl iodide	−108	72.3	Colorless liquid
CH_3CH_2-SH	Ethanethiol or ethyl mercaptan	−144	35	Colorless liquid
$^+CH_2CH_3$	Ethyl cation (Reactive intermediate)			
$^-{:}CH_2CH_3$	Ethyl anion (Reactive intermediate)			
$\cdot CH_2CH_3$	Ethyl radical (Reactive intermediate)			

How many different CH_3CH_2-X compounds are there? From a two-dimensional, schematic picture of ethane there appear to be two ethyl compounds with the same formula but different structures. Compounds of this type are called **isomers** (from the Greek words meaning "the same parts"). If the two ethyl compounds of Figure 2.27 actually existed, they would be isomers. However, the corners apparent in the structures in Figure 2.27 do not in fact exist, but are artifacts of our attempt to represent a three-dimensional molecule in two dimensions. These corners show how important it is to know the three-dimensional structures of compounds. All six hydrogens in ethane are the same (equivalent) and therefore there is only one kind of ethyl compound.

Replace this H

Replace this H

Replace this H, or any other H

Are these two ethyl compounds the same? They don't look the same in this representation, but they are!

FIGURE 2.27 The two-dimensional representation of ethane suggests that there are two different kinds of ethyl compound, CH_3CH_2-X. We must look at ethane in three dimensions to see why there is only one CH_3CH_2-X.

SUMMARY

Even ethane, the two-carbon alkane, has a complex structure! For example, staggered and eclipsed forms exist, separated by a low barrier (~3 kcal/mol). Because this barrier to rotation is so low, there is rapid rotation around the carbon–carbon bond. All six hydrogens in ethane are the same (equivalent); there is only one ethane and only one kind of ethyl derivative.

2.6 STRUCTURE DRAWINGS

In the preceding sections, there have been many different representations of the very simple molecule, ethane. Figure 2.28 recapitulates them and adds some new ones. Methyl groups are sometimes written as Me and ethyl groups as Et, especially in colloquial use.

The difficulty of representing the real, dynamic, three-dimensional structure of ethane is more apparent now. In the real world, there is often not the time to

CONVENTION ALERT

draw out a good representation of even as simple a molecule as ethane carefully. The solid and dashed wedges of Figure 2.28 are the traditional attempts at adding a three-dimensional feel to the two-dimensional drawing. More complicated molecules can raise horrendous problems. Adequate codes are needed and you must learn to see past the coded structures to the real molecules both easily and quickly. It's worth considering here, at this very early point, some of the pitfalls of the various schemes.

FIGURE 2.28 Several different representations of ethane.

First of all, it is not simple to represent even ethane in a linear fashion. Clearly, neither C_2H_6 nor Et—H is as descriptive as H_3C—CH_3, as they don't even show the connectivity of the atoms in a general way. The representation H_3C—CH_3 is better, but even this picture lacks three dimensionality. We used this representation in this chapter, but you will sometimes see the variant CH_3—CH_3. The two formulations are equivalent, even though the second structural formula, CH_3—CH_3, seems to indicate that the bond to the right-hand carbon comes from one of the hydrogens on the left-hand carbon. Of course, it does not; it is the two carbons that are bonded, as H_3C—CH_3 shows. Yet it is easier to write CH_3—CH_3 and you will see it often.

PROBLEM 2.16 Draw three-dimensional structures for the following compounds: $(Me)_2CH_2$, $(CH_3)_4C$, $(CH_3)_3CH$, $EtCH_3$, $(Et)_2$, $CH_3CH_2CH_3$, EtMe, MeEt.

PROBLEM 2.17 Make three-dimensional drawings of the "two" propanes created by replacing the two "different" hydrogens in a flat structure of ethane. Convince yourself that there is but one propane. By all means use your models.

2.7   PROPANE (C_3H_8) AND PROPYL COMPOUNDS (C_3H_7X)

Propane, the third member of the alkane series, has many of the chemical and physical properties of ethane and methane. Figure 2.29 shows several representations for propane, including two new ones.

In one of these representations [Fig. 2.29(d)], only the carbon frame is retained, and all eight hydrogen atoms are implied. Not even the carbons are shown in Figure 2.29(e), and the reader is left to fill in mentally all the atoms. This most abstract representation is the scheme of choice for organic chemists and we will gradually slip into this way of drawing as we move along through the various structural types. Notice that many, but not all of the schematic representations for propane show that it is not a strictly linear species. All carbons are hybridized approximately sp^3, and the C—C—C angle will be close to 109° (in fact, it is 112°).

CONVENTION ALERT

(a)

CH_3—CH_2—CH_3

(b)

$CH_3CH_2CH_3$

(c)

(d)

(e)

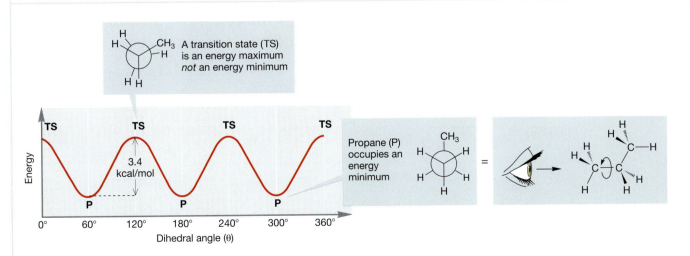

FIGURE 2.29 Several different representations of propane.

The Newman projection (p. 64) of Figure 2.30, constructed by looking down one of the two equivalent carbon–carbon bonds, shows the structure of propane particularly well. A plot of energy versus dihedral angle is similar to that for ethane, except that the rotational barrier is slightly higher, 3.4 kcal/mol (Fig. 2.30). In propane, it is a large methyl group that is eclipsed with a hydrogen at the top of the barrier (the transition state), whereas in ethane it is two small hydrogens that must pass each other.

A transition state (TS) is an energy maximum *not* an energy minimum

Propane (P) occupies an energy minimum

FIGURE 2.30   An energy versus dihedral angle (θ) plot for propane.

Propane One methylene group

Two equivalent (identical) methyl groups

FIGURE 2.31 Two types of hydrogen in propane.

Replacement of a hydrogen in propane with an X group yields propyl compounds, but the situation is more complicated than it was in methane or ethane. In both methane and ethane all hydrogens were equivalent, and there could be only one methyl or one ethyl derivative. In propane, there are two different kinds of hydrogen. There is a central CH_2 group, called a **methylene group**, as well as two equivalent methyl groups at the two ends of the molecule (Fig. 2.31).

Either of these two kinds of hydrogen can be replaced by an X to give a propyl compound (Fig. 2.32). The linear compound $CH_3-CH_2-CH_2-X$, in which X replaces an end methyl hydrogen is called **propyl** X (in the old days, *n*-propyl X). The branched compound $CH_3-CHX-CH_3$, in which X replaces a hydrogen on the central methylene carbon, is called **isopropyl** X (and sometimes, *i*-propyl X).

replace a methyl hydrogen

A series of propyl compounds

replace a methylene hydrogen

A series of isopropyl compounds

FIGURE 2.32 Propyl and isopropyl compounds.

PROBLEM 2.18 Make careful three-dimensional drawings of propyl alcohol ($CH_3-CH_2-CH_2-OH$) and the related isopropyl alcohol ($CH_3-CHOH-CH_3$).

2.8   BUTANES (C_4H_{10}), BUTYL COMPOUNDS (C_4H_9X), AND CONFORMATIONAL ANALYSIS

If we once again let X be methyl, we generate butane and isobutane from our two different propyl compounds (Fig. 2.33).

X = CH₃

Butane

X = CH₃

Isobutane

FIGURE 2.33 If X = CH₃, we produce butane and isobutane.

Butane is the next member of a series of straight-chain alkanes. Once we get beyond four carbons, a systematic naming protocol, the International Union of Pure and Applied Chemistry (IUPAC) system, is used. The straight-chain five-carbon alkane is pentane, the six-carbon version is hexane, seven carbon atoms is heptane, and so on. Some straight-chain alkanes are collected in Table 2.4.

TABLE 2.4 Some Straight-Chain Alkanes

Name	Formula	mp (°C)	bp (°C)
Methane	CH_4	−182.5	−164
Ethane	CH_3CH_3	−183.3	− 88.6
Propane	$CH_3CH_2CH_3$	−189.7	− 43.1
Butane	$CH_3(CH_2)_2CH_3$	−138.4	− 0.5
Pentane	$CH_3(CH_2)_3CH_3$	−129.7	36.1
Hexane	$CH_3(CH_2)_4CH_3$	− 95	69
Heptane	$CH_3(CH_2)_5CH_3$	− 90.6	98.4
Octane	$CH_3(CH_2)_6CH_3$	− 56.8	125.7
Nonane	$CH_3(CH_2)_7CH_3$	− 51	150.8
Decane	$CH_3(CH_2)_8CH_3$	− 29.7	174.1
Undecane	$CH_3(CH_2)_9CH_3$	− 25.6	195.9
Dodecane	$CH_3(CH_2)_{10}CH_3$	− 9.6	216.3
Eicosane	$CH_3(CH_2)_{18}CH_3$	36.8	343.0
Triacontane	$CH_3(CH_2)_{28}CH_3$	66	449.7
Pentacontane	$CH_3(CH_2)_{48}CH_3$	92	

Butane provides a nice example of **conformational analysis**, the study of the relative energies of conformational isomers. Figure 2.34 shows a variety of schematic representations for butane.

FIGURE 2.34
Different representations of butane.

Let's start our elaboration of these schematic pictures by constructing a Newman projection by looking down the C(2)—C(3) bond of butane (Fig. 2.35). The notation C(2), C(3), and so on will be used to refer to the carbon atom number. If we make the reasonable assumption that the relatively large methyl groups will best be kept as far from each other as possible, we come to conformation **A**, called the *anti* form. As in ethane (p. 63), a 120° clockwise rotation passes over a transition state (energy maximum) to a new minimum, **B**. Conformation **B** will be less stable than **A** because in **B**

CONVENTION ALERT

the two methyl groups are closer than they are in **A**. Conformation **B** is called *gauche* butane and the methyl–methyl interaction is a *gauche* interaction. *gauche* Butane (**B**) is about 0.6 kcal/mol higher in energy than the *anti* form (**A**).[*] The eclipsed transition state for the interconversion of **A** and **B** lies about 3.4 kcal/mol higher than **A**.

FIGURE 2.35 A dihedral angle (θ) versus energy plot for butane.

Another 120° clockwise rotation brings us to a second *gauche* butane, **B'**. The transition state for the conversion of **B** to **B'** involves an eclipsing of two C—CH_3 bonds and lies 3.8 kcal/mol above **B** and **B'**, a total of about 4.4 kcal/mol above **A**. A final 120° clockwise rotation returns us to **A** (Fig. 2.35).

[*]This number, 0.6 kcal/mol, is somewhat controversial. Most books give this value as 0.9 kcal/mol. That 0.9 figure is correct for the gas phase, but is not correct in solution. Indeed, the value is different in different solvents. It seemed best to give a number for a solution, in which most chemistry takes place. You will find slight differences in almost every version of the curve in Figure 2.35, but the big picture—the general shape of the curve—will not vary. The reasons for the difference between the solution and the vapor phase are complex, but if you would like to read more, see E. E. Eliel and S. H. Wilen, *Stereochemistry*, Wiley, 1994, pp. 600ff.

PROBLEM 2.19 Recall how much energy an eclipsed C—H/C—H interaction costs in ethane (p. 66). Given that value, and the information that the transition state for the interconversion of **A** and **B** is 3.4 kcal/mol higher in energy than **A**, calculate how much each eclipsed C—CH$_3$/C—H interaction in the transition state for the interconversion of **A** and **B** is worth. How much is the eclipsed C—CH$_3$/C—CH$_3$ interaction in the transition state for the conversion of **B** to **B'** worth (see Fig. 2.35)?

PROBLEM 2.20 Draw Newman projections constructed by looking down the indicated carbon–carbon bond in the following molecules. For the second molecule make an energy versus dihedral angle graph.

The branched compounds such as isobutane are more difficult to name systematically than their straight-chain isomers. The first members of this series have common names, but as the number of carbons increases, and complexity sets in, the system takes over. We start our analysis of branched compounds with the butyl compounds, C_4H_9—X. Butane contains two kinds of hydrogen atoms in its pairs of equivalent methylene (CH$_2$) and methyl (CH$_3$) groups, which should produce two butyl compounds, as shown in Figure 2.36. Similarly, isobutane contains a single **methine group** (CH) and three equivalent methyl groups and should also yield two butyl compounds.

FIGURE 2.36 Replacement of one of the hydrogens of butane with X yields two kinds of butyl derivatives, butyl—X and *sec*-butyl—X. Replacement of hydrogen with X in isobutane yields two more butyl compounds, isobutyl—X and *tert*-butyl—X.

It appears that there should be four isomers of C_4H_9—X. Indeed there are, and Figure 2.36 shows these butyl—X compounds. The four different types of butyl compound are called **butyl**, **isobutyl**, **sec-butyl** (short for *secondary*-butyl), and **tert-butyl** (short for *tertiary*-butyl). A new and important terminology has been introduced.

A **primary carbon** is attached to only one other carbon atom. Ethane contains two primary carbon atoms. **Secondary carbons** are attached to two other carbons and **tertiary carbons** to three. A **quaternary** (*not* quarternary) **carbon** atom is attached to four other carbons. Thus, the names for the butyl groups tell you something about the structure. These commonly encountered groups and names must be learned (Fig. 2.37).

PROBLEM 2.21 Draw two compounds containing quaternary carbons.

FIGURE 2.37 The four different kinds of butyl group.

2.9   PENTANES (C₅H₁₂) AND PENTYL COMPOUNDS (C₅H₁₁X)

One would now anticipate that four pentanes would emerge when we transform our four butyl-X compounds into five-carbon compounds by letting X = CH₃ (Fig. 2.38).

FIGURE 2.38 If we let X = CH₃, it appears as though four pentanes should exist. Yet there are only three pentanes.

These are the same compound, isopentane; rotate the top figure 180° as shown by the arrows

Yet, only three pentanes exist! Somehow the technique of generating larger alkanes by letting a general X become a methyl group has failed us. The problem lies again in the difficulty of representing three-dimensional structures on a two-dimensional surface. Two of the four structures in Figure 2.38 represent the same compound, even though in this two-dimensional code they look different. The three pentanes are called pentane, 2-methylbutane, and 2,2-dimethylpropane. These last two compounds are traditionally called "isopentane" and "neopentane." Section 2.10 will describe the naming system for alkanes.

One important lesson to be learned from Figure 2.38 is that the two-dimensional page will lie to you more often than not if you let it. As we have already seen, there is a tension between our need to talk quickly and efficiently to each other and the accurate representation of these complicated organic structures. This tension is at least partly resolved by using codes, but the price of this use is that you *always* must be able to see past the code to the real, three-dimensional structure if you expect to be able to think effectively about organic chemistry. These codes become quite abstract (see p. 69), and it is worth going through the process of developing the abstraction one more time. Consider the representations for isopentane in Figure 2.39. The first level of abstraction is to suggest only vaguely the real molecule, as in Figure 2.39(a). One then goes to the more schematic structure of Figure 2.39(b), and then to a structure in which the hydrogens are left out, Figure 2.39(c). Finally, one even removes the carbons, as in Figure 2.39(d), to give the ultimate abstraction for the structure of isopentane. Here nothing survives from the original picture except a representation of the backbone. You are left to supply the carbons, which must be put in at every vertex and terminus, as well as the attendant hydrogens. Most important, you must be able to translate these highly schematic structures into three dimensions, and to see the way these skeletal pictures transform into molecules.

PROBLEM 2.22 Draw schematic representations for the pentanes and hexanes.

Actually, there is even more to be done. These line drawings can be approximated with ball-and-stick models, in which balls represent atoms and sticks represent the bonds between atoms. Such representations do a relatively poor job of showing the **steric** (spatial) requirements of the real molecules. Chemists have therefore evolved space-filling models, which attempt to show the volumes of space carved out by the atoms. Figure 2.40 shows some representations of isopentane.

(a)

(b)

(c)

(d)

FIGURE 2.39 Four increasingly abstract representations for isopentane.

Isopentane

FIGURE 2.40 Views of isopentane.

So there are only three pentanes, but no fewer than eight pentyl derivatives (Fig. 2.41).

FIGURE 2.41 The eight pentyl
derivatives.

SUMMARY

In Sections 2.6–2.9, we explored the three-dimensional structures of substituted alkanes. We also encountered conformational analysis, the study of the relative energies of conformational isomers.

2.10   THE NAMING CONVENTIONS FOR ALKANES

Clearly, things are rapidly getting out of hand. Although it is not too difficult to remember the four butyl groups, the eight pentyls constitute a tougher task—and the problem rapidly gets worse as the number of carbons increases. A system is absolutely necessary. In practice, the old, non-systematic names are retained through the butanes, but once we reach five carbons the systematic naming protocol largely takes over, although a few old favorite names are still occasionally seen.

In general, the stem of the name tells you the number of carbon atoms (meth = 1, eth = 2, prop = 3, but = 4, and so on), and the suffix "ane" tells you that the molecule in question is an alkane and contains only single bonds.

Here are the most important rules of the naming convention:

1. Search for the longest straight chain of carbons. The structure is named as a derivative of this parent hydrocarbon. This rule can be a bit tricky at first.
   Remember: The corners apparent in many two-dimensional representations are
   not present in the real structures. In the examples in Figure 2.42 it is perfectly
   reasonable to count around the corner in order to find the longest chain. If this
   is not clear to you, by all means go back to Problem 2.17. Do not be reluctant to
   use your models. An atom or group attached to the longest straight chain is
   given a position number. The chain is numbered from the end that produces the
   lowest position number for the substituent.

3-Methylpentane   *not* **2-Ethylbutane**

3-Methylhexane   *not* **2-Ethylpentane**

not **4-Methylhexane**

FIGURE 2.42 In finding the longest
straight chain of carbons it is
permissible, indeed often necessary,
to count around corners.

2. In a substituted alkane, the substituent is given a number based on its position
   on the parent hydrocarbon. The longest chain is numbered so as to make the
   number of the substituent position as low as possible. Some examples are
   shown in Figure 2.43.

FIGURE 2.43 Number the chain so as to produce
the smallest number for a substituent.

2-Chlorobutane   *not* **3-Chlorobutane**

3-Fluorohexane   *not* **4-Fluorohexane**

3. In a multiply substituted alkane, all the substituents receive numbers. If the substituents are identical, the prefixes di-, tri-, tetra-, and so on are used. Figure 2.44 gives some examples:

2,3-Dimethylbutane   2,3,4-Trichloropentane   1,1,1,5,5,5-Hexafluoropentane

FIGURE 2.44   Polysubstituted alkanes.

If the substituents are not identical, they are ordered alphabetically, as shown in Figure 2.45. Prefixes such as *sec-* and *tert-* are ignored in determining alphabetical order. Thus "*tert*-butyl" would appear in the name earlier than "chloro."

2-Chloro-3-fluorobutane
not 3-chloro-2-fluorobutane
not 2-fluoro-3-chlorobutane

3-Bromo-4-methylhexane
not 4-bromo-3-methylhexane
not 3-methyl-4-bromohexane

4-*tert*-Butyl-5-chlorooctane
not 4-chloro-5-*tert*-butyloctane

FIGURE 2.45   Nonidentical substituents are incorporated into the name alphabetically.

4. When the rules above do not resolve the issue, the name that starts with the lower number is used, as demonstrated in Figure 2.46.

2,5,5-Trimethylheptane
not 3,3,6-trimethylheptane

FIGURE 2.46   In unresolvable cases, start with the lowest possible number.

PROBLEM 2.23 Write systematic names for the following compounds:

PROBLEM 2.24 Write structures for the following compounds: (a) 2-bromobutane, (b) decane, (c) *tert*-butyl chloride, (d) 1,4-difluoropentane, (e) 2,4,4-trimethylheptane.

Although the instruction to name substituents on alkanes in alphabetical order looks simple, the continued use of common names complicates matters a bit because some groups count in the alphabet and others do not. Table 2.5 contains a brief summary.

TABLE 2.5   Some Common Prefixes

Prefix	Example	Counts in Alphabetical Order?
di-	Any two identical groups	Yes
tri-	Any three identical groups	Yes
tetra-	Any four identical groups	Yes
iso	Isopropyl = $(CH_3)_2CH-$	Yes
neo	Neopentyl = $(CH_3)_3C-CH_2-$	Yes
sec-	*sec*-Butyl = $(CH_3)_2CH-CH_2-$	No
tert-	*tert*-Butyl = $(CH_3)_3C-$	No

PROBLEM 2.25 Use the information in Table 2.5 to name the following two compounds:

(a)

(b)

2.11   PROBLEM SOLVING: WRITING ISOMERS

A favorite question in courses on organic chemistry is, Write all the isomers of pentane (or hexane, or heptane, etc.). Such problems force one to cope with the translation of two-dimensional representations into three-dimensional reality. Figuring out the pentanes is easy; there are only 3 isomers. Even writing the hexanes (5 isomers) or heptanes (9 isomers) is not really difficult. But getting all the octanes (18 isomers) is a tougher proposition, and getting all the nonanes (35 isomers) with no repeats is a real challenge. Success depends on finding a systematic way to write isomers. Thrashing around writing structures without a system is doomed to failure. You will not get all the isomers and there is a high probability of generating repeat structures. Any system will work as long as it is truly systematic. One possibility is the following, shown generating the 9 heptanes in Figures 2.47–2.50:

1. Start with the longest straight chain possible. This system generates heptane itself (Fig. 2.47).

Heptane

Common repeats

Both still heptane!

FIGURE 2.47 Heptane.

Hexanes —
one C atom must
be added

2-Methylhexane

3-Methylhexane

Common repeats

Still 3-methylhexane

Still 3-methylhexane

FIGURE 2.48   Methylhexanes.

2. Shorten the chain by one carbon, and add the "extra" carbon as a methyl group at all possible positions, starting with the left-hand side carbon and moving to the right. This process generates the methylhexanes. You must check each isomer you make to be sure it is not a repeat. One way to be absolutely certain is to name each molecule as you generate it. If you repeat a name, you have repeated an isomer. Note that the name of this heptane isomer ends in "hexane." These molecules are named as substituted hexanes(Fig. 2.48).

3. Shorten the chain by one more carbon. The longest straight chain is now five carbons (pentanes). Two carbons are to be added as two methyl groups or a single ethyl group. Two methyl groups can be placed either on the same carbon or different carbons. Again, start at the leftmost carbon and move to the right, checking each isomer to be sure it is not a repeat. Be sure to see that when the two carbons are placed on the *same* carbon of the straight chain they can be added either as two methyl groups or as a single ethyl group. Take care to look carefully as you generate each structure to be sure you have not repeated an isomer. This step generates 3-ethylpentane and the four dimethylpentanes (Fig. 2.49).

4. Shorten the chain again, and try to fit the three "extra" carbons on a butane chain. There is only one way to do this (Fig. 2.50).

Pentanes —
two C atoms must be added

2,2-Dimethylpentane   **2,3-Dimethylpentane**   **2,4-Dimethylpentane**   **3,3-Dimethylpentane**   **3-Ethylpentane**

FIGURE 2.49   3-Ethylpentane and the dimethylpentanes.

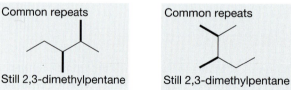

Common repeats

Still 2,3-dimethylpentane

Common repeats

Still 2,3-dimethylpentane

Butane—
The three C atoms
must be added

2,2,3-Trimethylbutane

Common repeats

Still 2,2,3-trimethylbutane

FIGURE 2.50   Trimethylbutane.

PROBLEM 2.26   Write and name all the isomeric octanes, C_8H_{18}.

2.12   RINGS

All the hydrocarbons we have met so far have the molecular formula C_nH_{2n+2}, and are called **saturated alkanes**. There is a class of closely related molecules that shares most chemical properties with the saturated alkanes, but not the general formula. These molecules have the composition C_nH_{2n} and are sometimes called **unsaturated**. In such compounds, we could find a way in principle (and often in practice as well) to attach two more hydrogen atoms to bring us up to C_nH_{2n+2}. Every pair of hydrogens that is "missing" constitutes one degree of unsaturation. What might the bonding in such molecules be? It probably will not deviate much from what we already know, because the chemical properties of these molecules resemble very closely those of the saturated alkanes. Recall how we first thought of making ethane. We allowed two roughly sp^3 hybridized carbons to approach and the overlap of the pair of singly occupied orbitals generated a bonding, filled σ carbon–carbon bond as well as its empty, antibonding counterpart, σ^* (Fig. 2.51).

$$H_3C\cdot + \cdot CH_3 \longrightarrow H_3C{-}CH_3$$

FIGURE 2.51 The formation of ethane through the overlap of two singly occupied hybrid orbitals.

Suppose we replace two of the hydrogens of ethane with carbons, here shown as methyl groups. This substitution corresponds to the construction of butane from two ethyl radicals. There is little difference between this process and the dimerization of two methyl groups. Now make one more small change. Instead of using methyl groups to replace the pair of hydrogens in Figure 2.51, use a chain of connected methylene groups to form a ring (Fig. 2.52).

FIGURE 2.52 The formation of butane and cyclopentane through the overlap of two singly occupied hybrid orbitals.

Methylcyclopentane

Chlorocyclohexane

1,1-Diethylcycloheptane

1-Fluoro-4-iodocyclooctane
not 4-iodo-1-fluorocyclooctane
or 4-fluoro-1-iodocyclooctane

FIGURE 2.53 Ring compounds are named and numbered in a similar fashion to acyclic alkanes.

There is no essential difference between this construction of a ring compound and the process used to make ethane and butane. Nor would we expect to see big differences in chemical properties. The bonding in pentane is not significantly different from that in cyclopentane (Fig. 2.52). These ring compounds are named by attaching the prefix "cyclo" to the name of the parent hydrocarbon (Table 2.6).

TABLE 2.6 Some Cyclic Alkanes (Cycloalkanes)

Name	Formula	mp (°C)	bp (°C, 760 mmHg)
Cyclopropane	$(CH_2)_3$	−127.6	− 32.7
Cyclobutane	$(CH_2)_4$	− 50	12
Cyclopentane	$(CH_2)_5$	− 93.9	49.2
Cyclohexane	$(CH_2)_6$	6.5	80.7
Cycloheptane	$(CH_2)_7$	− 12	118.5
Cyclooctane	$(CH_2)_8$	14.3	148–149 (749 mmHg)

Monosubstituted cycloalkanes require no numbers for naming. In multiply substituted compounds, substituents are assigned numbered positions, and the name is constructed so as to minimize the numbers. Multiple substituents are named alphabetically (Fig. 2.53).

There is one important new structural feature that appears in cycloalkanes. There is only a single methylcyclopropane, but there are *two* isomers of 1,2- dimethylcyclopropane. As shown in Figure 2.54, there is no "sidedness" to methylcyclopropane. The molecule with the methyl group "up" can be transformed into the molecule with the methyl group "down" by simply turning the molecule over.

FIGURE 2.54 There is only one isomer of methylcyclopropane. In the figure, a thicker line is used to show the side of the cyclopropane nearest to you.

Rotate 180°

However, no number of translational or rotational operations will suffice to change the 1,2-dimethylcyclopropane with both methyl groups on the same side of the ring, called **cis**, into the one with the methyl groups on opposite sides, called **trans**. The two are certainly different. The use of models is absolutely mandatory at this point. Make models of the two compounds and convince yourself that nothing short of breaking carbon–carbon bonds will allow you to turn *cis*-1,2-dimethylcyclopropane into the isomeric *trans*-1,2-dimethylcyclopropane (Fig. 2.55, and see Problem 2.27).

Methyl groups on the same side

Methyl groups on opposite sides

cis-1,2-Dimethylcyclopropane   **trans-1,2-Dimethylcyclopropane**

FIGURE 2.55 Two isomers of 1,2-dimethylcyclopropane.

PROBLEM 2.27 The situation is even more complicated, as you will see in Chapter 4. There are really two isomers of *trans*-1,2-dimethylcyclopropane! Use your models to construct the mirror image of the *trans*-1,2-dimethylcyclopropane you made and see if it is identical to your first molecule.

Ring compounds are extremely common. Both small and large varieties are found in Nature, and many kinds of exotic cyclic molecules not found so far outside the laboratory have been made by chemists. Moreover, rings can be combined in a number of ways to form polycyclic molecules. Here is an opportunity for you to think ahead. How might two rings be attached to each other? Some ways are obvious, but others require some thought. We will work through this topic as a series of problems. A number of different structural types can be created from two rings. These problems try to lead you through them.

WORKED PROBLEM 2.28 Find a compound of the formula $C_{10}H_{18}$ composed of two five-membered rings.

ANSWER Two five-membered rings can be joined in a very simple way to make bicyclopentyl. There is no real difference between this process and the formation of ethane from two methyl radicals (p. 67).

Cyclopentane
(C_5H_{10})

Cyclopentyl
(C_5H_9)

Bicyclopentyl
$(C_{10}H_{18})$

WORKED PROBLEM 2.29 Another structural type still containing two five-membered rings has the formula C_9H_{16}. Clearly, we are not dealing with a simple dimer here as we are short one carbon. The two rings must share one carbon somehow. Draw this compound.

ANSWER In this case, the two rings share a single carbon. The way to do it is to let one carbon be part of both five-membered rings.

(C_9H_{16})
The two rings share one carbon

PROBLEM 2.30 Two five-membered rings can share more than one carbon. Find a molecule of the formula C_8H_{14}. There are two similar examples of this kind of "fused" molecule. Try to find both. *Hint*: Focus on the two shared carbons and the attached hydrogens. (Make a model.) That's all the help you get from in-chapter answers. Use Problem 2.29 as a guide, and this one should be easy. The answer is, of course, in the Study Guide.

PROBLEM 2.31 Finally, and most difficult, there is a molecule, still constructed from five-membered rings, that has the formula C_7H_{12}. In this structural type, the two five-membered rings must share three carbons. Find this molecule.

Polycyclic compounds can be exceedingly complex. Indeed, much of the fascination that organic chemistry holds for some people is captured nicely by the beautifully architectural structures of these compounds. Figure 2.56 shows three molecules. Two molecules are "natural products" and one ([1.1.1]propellane) is not (yet) found outside the laboratory. Aflatoxin B is a highly toxic fungal metabolite. Progesterone is one of a class of molecules called steroids, and has an antiovulatory effect if taken during the middle days of the menstrual cycle.

Aflatoxin B₁
made in 1966 by G. Büchi
and his research group at MIT

Progesterone
made in 1967 by G. Stork and his
group at Columbia University

[1.1.1]Propellane
made in 1982 by K. B. Wiberg
and his group at Yale; this
molecule was also made
slightly later in a particularly
simple way by the group of
G. Szeimies at Munich

FIGURE 2.56 Some polycyclic molecules.

SUMMARY

Ring compounds are generated within the hybridization model in the same way that we built up larger alkanes from smaller molecules. Ring compounds have sides, which means that substituents can be on the same side (cis) or on opposite sides (trans). All manner of polycyclic molecules (two or more rings) exist.

2.13   PHYSICAL PROPERTIES OF ALKANES AND CYCLOALKANES

Simple saturated hydrocarbons and cycloalkanes are colorless gases, clear liquids, or white solids, depending on their molecular weight. To many people they smell bad, although I have long thought that these molecules have been the victims of a bad press, and I don't think they smell bad at all. The odor of cooking gas, mostly saturated hydrocarbons, comes from an additive, a mercaptan (R—SH) (*Remember*: R stands for a general alkyl group, p. 59), put in specifically so that escaping gas can be detected by smell. Tables 2.4 and 2.6 have collected some physical properties of straight-chain and cyclic alkanes. Why do the boiling points increase as the number of carbons in the molecule increases? The boiling point is a measure of the ease of breaking up intermolecular attractive forces. There is a factor that stabilizes the liquid phases of hydrocarbons called **van der Waals forces**. When two clouds of electrons approach each other dipoles are induced as the clouds polarize in such a fashion as to stabilize each other by opposing opposite charges (Fig. 2.57).

Of course, many alkanes have small dipoles to begin with, but they are very small and do not serve to hold the molecules together strongly. Other molecules are much more polar and this polarity makes a big difference in boiling point. Polar molecules can associate quite strongly by aligning opposite charges. This association has the effect of increasing the boiling point.

Separated molecules
(gas phase)

Aggregated molecules
(solution) held together
by opposite charges

δ−
δ+

δ−
δ+

Attraction between
induced opposite
charges in
liquid phase

δ−
δ+

FIGURE 2.57 The stabilization of molecules through van der Waals forces.

The more extended a molecule is, the more potent the induced dipole can be. More compact, more symmetric molecules have smaller induced dipoles, and therefore lower boiling points. A classic example is the difference between pentane and neopentane (Fig. 2.58). The more spherical neopentane boils about 25 °C lower than the straight-chain isomer. Isopentane is less extended than pentane but more extended than neopentane, and its boiling point is right between the two.

Symmetry is especially important in determining melting point, as highly symmetric molecules pack well into crystal lattices. It takes more energy to break up a well-packed lattice. Neopentane, for example, melts 113 °C higher than pentane does.

FIGURE 2.58 The more extended pentane boils at a higher temperature than the more compact neopentane does.

Highly symmetrical neopentane has a nearly spherical cloud of electrons (bp 9.5 °C; mp −16.5 °C)

The more extended molecule pentane has a much greater surface and has greater intermolecular interactions (bp 36.1 °C; mp −130 °C)

Minimal interaction between two spheres allows for relatively weak van der Waals forces

More extensive contact possible in the extended molecule allows more powerful van der Waals interactions

2.14   NUCLEAR MAGNETIC RESONANCE SPECTROSCOPY

Spectroscopy is the study of molecules through the investigation of their interaction with electromagnetic radiation. There are many kinds of spectroscopy (as we shall see mainly in Chapter 15). One particularly valuable version is called nuclear magnetic resonance (NMR). Although we won't go into much detail yet, this early introduction does allow us to address the critical question of "difference." When are two atoms the same and when are they different?

Like the electron, the nuclei of many atoms have a quantity called spin. The ^{13}C and ^1H nuclei are the most useful, and each has $\pm\frac{1}{2}$ values allowed for the spin, just like the electron. Although ^{13}C is present in only 1.1% abundance in ordinary carbon, which is mostly ^{12}C, that small amount can be detected. We shall devote much of Chapter 15 to a detailed discussion of NMR spectroscopy, but we'll take a moment here to say a bit about NMR, because it allows us to develop further the notions of identity and nonidentity of carbon atoms, and to hone our ability to see difference—an absolutely critical skill. You have heard of this form of spectroscopy before if you have ever read an article about "magnetic resonance imaging (MRI)." It is the same process, but the dreaded word "nuclear" must be hidden from public view.

Like the electron, the ^{13}C and ^1H nuclei can be thought of as spinning in one of two directions. In the presence of a strong magnetic field, those two states differ in energy but by only a tiny amount. Nonetheless, transitions between the two states can be detected by NMR spectrometers tuned to the proper frequency. We'll have more to say about those spectrometers and those transitions in Chapter 15, but there is really not much more to it than that. So we can see a signal whenever a transition between the lower and higher energy states is induced. So what? It would seem that we have simply built a (very expensive) machine to detect carbon or hydrogen in a molecule, and it would be hardly surprising to find such atoms in organic molecules!

The critical point is that every different carbon (or hydrogen)—every such atom in a different environment, no matter how slightly different—gives a different signal from that of the other carbons (or hydrogens) in the molecule. The NMR spectrometer can "count" the number of different carbons or hydrogens in a molecule by counting the number of signals. That ability can be enormously useful in structure determination, and NMR spectroscopy is very often used by "the pros" of structure determination in exactly that way.

Let's look at a few examples of ^{13}C NMR in Figure 2.59. How many signals will our ^{13}C NMR spectrometer "see" from methane? One, of course, as there is only one carbon.

FIGURE 2.59 Carbon-13 NMR signals for simple alkanes.

CH_4

One signal

H_3C—CH_3

One signal

Two signals, one for the CH_2 and another for the two identical CH_3 groups

How about ethane? There are two carbons here, but they are identically situated, and thus equivalent: one signal again. How about propane? Three carbons here, two methyl groups and a single methylene. Surely, we must see a different signal for that CH_2 than for the CH_3 groups. And we do—there are two signals, one for the CH_2 and another for the two equivalent CH_3 groups. Counting the number of signals is trivial if (and only if) we can discern whether the molecular environments of the carbons are the same or not. Here is a slightly harder example. The molecule furan shows two signals in its ^{13}C NMR spectrum (Fig. 2.60).

It is tempting to say that there are only methylene (CH_2) groups in this molecule so there should only be one signal. But those methylenes are not all the same. Two of them are adjacent to the oxygen, and two are not. So there are two signals, one for each set of two different methylenes, not one.

We will pick these points up again as we work through the various structural types in the next few chapters, but for the moment, try a few problems.

PROBLEM 2.32 How many ^{13}C signals will the NMR spectrometer "see" in the molecules of Figures 2.47–2.50?

One signal

Another signal

FIGURE 2.60 Furan has two different methylene groups.

PROBLEM 2.33 How many ^{13}C signals will the NMR spectrometer "see" in the molecules of Figure 2.53?

PROBLEM 2.34 How many ^{13}C signals will the NMR spectrometer "see" in the molecules of Figure 2.56?

Hydrogen NMR (^{1}H NMR) spectroscopy is basically similar to the ^{13}C NMR we have just discussed. The signals we observe in ^{1}H NMR spectroscopy are in the ratio of the numbers of different hydrogens; six hydrogens of one kind will give a signal three times as big as a signal for two hydrogens, and so on. There are other very useful complications introduced by the abundance of the NMR active isotope (^{1}H) in organic molecules, but we can leave them for now to Chapter 15. We will use some of the molecules introduced in this chapter to work through a bit of hydrogen NMR—we'll find a few more subtleties, but the overall picture will not be very different from what we have seen for ^{13}C.

PROBLEM 2.35 How many signals would be seen in the ^{1}H NMR spectra for the molecules in Figures 2.59 and 2.60?

PROBLEM 2.36 How many signals would be seen in the ^{1}H NMR spectrum for heptane (Fig. 2.47)? What would the relative sizes of these signals be?

PROBLEM 2.37 How many signals would be seen in the ^{1}H NMR spectrum for *cis*-1,2-dimethylcyclopropane (p. 85)? *Caution*: This question is a bit tricky—look carefully at a model, or use the website to see *cis*-1,2-dimethylcyclopropane in three dimensions.

2.15   ACIDS AND BASES REVISITED: MORE CHEMICAL REACTIONS

In Chapter 1 (p. 39), we introduced acids and bases. Now we know quite a bit more about structure, and can return to the important subject of acids and bases in greater depth. The Lewis definition of acids and bases is far more inclusive than the Brønsted* definition, which focuses solely on proton donation (**Brønsted acid**) and acceptance (**Brønsted base**). The archetypal Brønsted acid–base reaction is the reaction between KOH and HCl to transfer a proton from HCl to HO$^-$. This reaction is a competition between the hydroxide and the chloride for a proton. In this case, the stronger base hydroxide wins easily (Fig. 2.61).

FIGURE 2.61 The two Brønsted bases, HO$^-$ and Cl$^-$, compete for the proton, H$^+$.

*These acids and bases were named after Johannes Nicolaus Brønsted (1879–1947).

In the Lewis description, any reaction between a reactive pair of electrons in a filled orbital and an empty orbital is seen as an acid–base reaction. Another way of putting this is, the interaction of a filled orbital (**Lewis base**) with an empty orbital (**Lewis acid**) is stabilizing (Fig. 2.62).

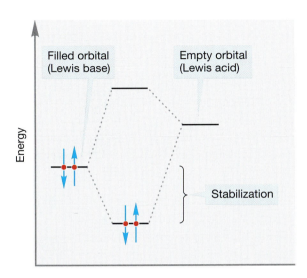

FIGURE 2.62   The interaction of a filled orbital (Lewis base) with an empty orbital (Lewis acid) is stabilizing.

Consider the methyl cation we drew on page 61. That empty $2p$ orbital makes $^{+}CH_3$ a powerful Lewis acid, very reactive toward all manner of electron pair bearing Lewis bases. Figure 2.63 gives three examples of such Lewis acid–Lewis base reactions, each of which fits the diagram of Figure 2.62.

FIGURE 2.63   Three Lewis bases reacting with the methyl cation, a Lewis acid.

The **arrow formalism** shows the electron flow as the new bond is formed in the following Lewis base–Lewis acid reaction (Fig. 2.64).

FIGURE 2.64   The arrow formalism for the formation of methyl chloride from the methyl cation and a chloride ion.

WORKED PROBLEM 2.38 Write arrow formalisms and products for the following reactions. Identify the participants as Lewis or Brønsted acids and bases. Sketch in the empty orbital for each Lewis acid, and fill in the complete Lewis dot structures for the Lewis bases. Be careful to remember that charge must be conserved in each of these reactions.

(a) $(CH_3)_3C^+$ Br^- $\longrightarrow$ (c) H^+ $^-NH_2$ $\longrightarrow$

(b) $(CH_3)_3C^+$ NH_3 $\longrightarrow$ (d) H_3B HO^- $\longrightarrow$

ANSWER (d) Hydroxide must be the Lewis base, but where is the empty orbital? Where is the Lewis acid? Recall the structure of the sp^2 hybridized borane (p. 52). Boron is neutral, but there is an empty $2p$ orbital on boron, and borane is very definitely a Lewis acid. Reaction with the Lewis base hydroxide looks just like the reaction of the methyl cation with hydroxide (Fig. 2.63) except that the product is still negatively charged.

These reactions, even the very simple ones, form the basis for understanding much of organic chemistry. Essentially all polar reactions can be looked at in exactly these terms—the ultimately stabilizing reaction of a filled orbital with an empty orbital. Figure 2.62 really does describe much of the next 1000 pages or so! But, of course, the details will change and there are surely devils in those details. Those all-too prolific details will be easier to manage if you keep Figure 2.62 in mind at all times—search for the Lewis base and the Lewis acid in every reaction, and you will be able to generalize.

2.16   SPECIAL TOPIC: ALKANES AS BIOMOLECULES

In truth, there really isn't very much biochemistry of alkanes. The carbon–carbon and carbon–hydrogen bonds of alkanes are just too strong to enter easily into chemical reactions. These molecules are not totally inert, as we shall see shortly, but saturated alkane chains serve more as frameworks in our bodies, holding less strongly bonded atoms in the proper place for reactions to occur, than as reactive species themselves. However, as we have seen already (p. 66), in the presence of oxygen and activating energy (a spark or a lighted match, for example), combustion occurs and heat is given off. We use that energy, that exothermicity of reaction, to warm our homes and propel our automobiles. Gasoline is largely a mixture of saturated hydrocarbons. And in the movie *Mad Max Beyond Thunderdome* the trading village Bartertown was powered by the methane produced by herds of pigs kept in the depths of Undertown.

How did those piggies produce that methane? Bacteria in the stomachs of both pigs and cattle are capable of producing methane from plant material, and the belching and flatulence of those animals is a major source of planetary methane.

Methane may also have been essential to the formation of life. The early atmosphere on Earth was relatively rich in methane and ammonia (as are the current atmospheres of most of the outer planets). Given an energy source such as ultraviolet (UV) radiation or lightning, methane and ammonia react to form hydrogen cyanide (HCN), a molecule that polymerizes to give adenine, a common component of ribonucleic acid (RNA) and other molecules of biological importance. In the presence of water and an energy source, methane and ammonia react to give amino acids, the building blocks of proteins.

2.17 SUMMARY

New Concepts

In this chapter, a bonding scheme for the alkanes is developed. We continue, as in Chapter 1, to form bonds through the overlap of atomic orbitals. Some atomic orbitals, $2p$ orbitals in particular, are not well suited to forming strong bonds through end-on overlap. For example, overlap with one lobe of a $2p$ orbital wastes the potential overlap with the other lobe. Moreover, $2p$ orbitals are directed at 90° angles, and this orientation does not keep the electrons in the bonds as far from each other as possible.

We describe a model in which the atomic orbitals of carbon are mathematically combined to form new, hybrid atomic orbitals. The results of a combination of three $2p$ orbitals and one $2s$ orbital are called sp^3 hybrids, reflecting the 75% p character and 25% s character in the hybrid. These hybrids solve the problems encountered in forming bonds between pure atomic orbitals. The sp^3 hybrids are asymmetric, and have a fat and a thin lobe. Overlap between a hydrogen $1s$ orbital and the fat lobe provides a stronger bond than that between $1s$ and carbon $2p$ orbitals. In addition, these hybrids are directed toward the corners of a tetrahedron, which keeps the electrons in the bonds as far apart as possible, thus minimizing destabilizing interactions. Other hybridization schemes are sp^2, in which the central atom is bonded to three other atoms, and sp, in which a central atom is bonded to two other atoms. Molecules hybridized sp^2 are planar, with the

three attached groups at the corners of an equilateral triangle, and sp hybridized molecules are linear.

Even simple molecules have complicated structures. Methane is a simple tetrahedron, but one need only substitute a methyl group for one hydrogen of methane for complexity to arise. In ethane, for example, we must consider the consequences of rotation about the carbon–carbon bond. The minimum energy conformation for ethane is the arrangement in which all carbon–hydrogen bonds are staggered. About 3 kcal/mol above this energy minimum form is the eclipsed form, the transition state, the high-energy point (not an energy minimum, but an energy maximum) separating two staggered forms of ethane. The 3 kcal/mol constitutes the rotational barrier between the two staggered forms. This barrier is small compared to the available thermal energy at room temperature, and rotation in ethane is fast.

This chapter introduces the concept of Lewis acids and bases. The familiar Brønsted bases compete for a proton, but Lewis bases are far more versatile. A Lewis base is defined as anything with a reactive pair of electrons, and a Lewis acid is anything that reacts with a Lewis base. We are paid back for these very general definitions with an ability to see as similar all manner of seemingly different reactions. These concepts will run through the entire book.

Key Terms

alkanes (p. 47)	hydride (p. 60)	quaternary carbon (p. 76)
alkyl (p. 59)	hydrocarbons (p. 47)	reactive intermediates (p. 60)
arrow formalism (p. 91)	isobutyl (p. 75)	saturated alkanes (p. 83)
Brønsted acid (p. 90)	isomers (p. 68)	secondary carbon (p. 76)
Brønsted base (p. 90)	isopropyl (p. 72)	sigma bonds (p. 51)
butyl (p. 75)	Lewis acid (p. 91)	spectroscopy (p. 88)
sec-butyl (p. 75)	Lewis base (p. 91)	sp hybrids (p. 49)
tert-butyl (p. 75)	methane (p. 47)	sp^2 hybrids (p. 52)
carbanion (p. 61)	methine group (p. 75)	sp^3 hybrids (p. 53)
carbocation (p. 60)	methyl anion (p. 61)	staggered ethane (p. 64)
cis (p. 84)	methyl cation (p. 60)	steric (p. 77)
conformational analysis (p. 73)	methyl compounds (p. 58)	tertiary carbon (p. 76)
conformations (p. 64)	methyl radical (p. 61)	torsional strain (p. 66)
dihedral angle (p. 64)	methylene group (p. 72)	trans (p. 85)
eclipsed ethane (p. 64)	Newman projection (p. 64)	transition state (p. 65)
ethyl compounds (p. 68)	primary carbon (p. 76)	unsaturated alkanes (p. 83)
hybridization (p. 49)	propyl (p. 72)	van der Waals forces (p. 87)

Reactions, Mechanisms, and Tools

In this chapter, new molecules are built up by first constructing a generic substituted molecule such as methyl—X (CH_3—X). The new hydrocarbon is then generated by letting X = CH_3. In principle, each different carbon–hydrogen bond in a molecule could yield a new hydrocarbon when X = CH_3. In practice, this procedure is not so simple. The problem lies in seeing which hydrogens are really different. The two-dimensional drawings are deceptive. One really must see the molecule in three dimensions before the different carbon–hydrogen bonds can be identified with certainty.

The coding or drawing procedures can get quite abstract. It is vital to be able to keep in mind the real, three-dimensional structures of molecules even as you write them in two-dimensional code. The Newman projection, an enormously useful device for representing molecules three dimensionally, is described in this chapter.

The naming convention for alkanes is introduced. There are several trivial names that are commonly used and which, therefore, must be learned.

Both ^{1}H and ^{13}C NMR spectroscopy are introduced in this chapter. At this point, the NMR spectrometer functions basically as a machine to determine the numbers of different hydrogens or carbons in a molecule. That function is important, however, both in these early stages of your study of chemistry and in the "real world" of structure determination.

Common Errors

It is easy to become confused by the abstractions used to represent molecules on paper or blackboards. Do not trust the flat surface! It is easy to be fooled by a "corner" that does not really exist, or to be bamboozled by the complexity introduced by ring structures. One good way to minimize such problems (none of us ever really becomes free of the necessity to think about structures presented on the flat surface) is to work lots of isomer problems, and play with models!

2.18 ADDITIONAL PROBLEMS

PROBLEM 2.39   Write Lewis structures for all the starred structural types in the list of functional groups on the inside front cover of this book. Show nonbonding electrons as dots and electrons in bonds as lines.

PROBLEM 2.40   Draw all the isomers of "bromopentane," $C_5H_{11}Br$. Give proper systematic names to all of them.

PROBLEM 2.41   How many signals would a ^{13}C NMR spectrometer "see" for each of the molecules in your answer to Problem 2.40?

PROBLEM 2.42   Draw all the isomers of "chlorohexane," $C_6H_{13}Cl$. Give proper systematic names to all of them. *Hint*: There are 17 isomers.

PROBLEM 2.43   What is the approximate hybridization of the indicated carbon in the following compounds?

(a)

H
 \
  C=O
 /
H

(b)

(c)

H—C≡C—H

(d)

        Cl
        |
Cl —— C —— Cl
        |
        Cl

PROBLEM 2.44   Use the boiling point data in Table 2.4 to estimate the boiling point of pentadecane, $C_{15}H_{32}$.

PROBLEM 2.45   Rationalize the differences in the melting points for the isomeric pentanes shown below.

Pentane
mp −129.7 °C

Isopentane
mp −159.9 °C

Neopentane
mp −16.5 °C

PROBLEM 2.46   Imagine replacing two hydrogens of a molecule with some group X. For example, methane (CH_4) would become CH_2X_2. (a) What compounds would be produced by replacing two hydrogens of ethane? (b) What compounds would be produced by replacing two hydrogens of propane? (c) What compounds would be produced by replacing two hydrogens of butane?

PROBLEM 2.47   Draw Newman projections of the eclipsed and staggered conformations of 1,2-dichloropropane by looking down the C(1)−C(2) bond.

PROBLEM 2.48   Draw Newman projections constructed by looking down the C(1)−C(2) bond of 2-methylpentane. Repeat this process looking down the C(2)−C(3) bond. In each case, indicate which conformations will be the most stable.

PROBLEM 2.49   Although a chlorine is about the same size as a methyl group, the conformation of 1,2-dichloroethane in which the two chlorines are eclipsed is higher in energy than the conformation of butane in which the two methyl groups are eclipsed. Explain. *Hint*: Size isn't everything!

PROBLEM 2.50   Write systematic names for all the pentanes and hexanes you drew in your answer to Problem 2.22.

PROBLEM 2.51   How many signals would a ^{13}C NMR spectrometer "see" for each of the molecules in your answer to Problem 2.50?

PROBLEM 2.52   Draw and write systematic names for all the isomers of nonane (C_9H_{20}). (*Hint*: There is a rumor that there are 35 isomers!) This problem is hard to get completely right.

 You may want to continue isomer drawing and naming on your own. But be careful, this is no trivial task. For example, there are 75 structural isomers of decane and over 6500 for tetradecane.

PROBLEM 2.53   Write systematic names for the following compounds:

(a)

(b)

(c)

(d)

PROBLEM 2.54   Draw structures for the following compounds:

(a) 4-Ethyl-4-fluoro-2-methylheptane
(b) 2,7-Dibromo-4-isopropyloctane
(c) 3,7-Diethyl-2,2-dimethyl-4-propylnonane
(d) 2-Chloro-7-iodo-5-isobutyldecane

PROBLEM 2.55   Although you should be able to draw structures for the following compounds, each is named incorrectly below. Give the correct systematic name for each of the following compounds:

(a) 2-Methyl-2-ethylpropane
(b) 2,6-Diethylheptane
(c) 1-Bromo-3-propylpentane
(d) 5-Fluoro-8-methyl-3-*tert*-butylnonane

PROBLEM 2.56   One of the octane isomers of Problem 2.26, 3-ethyl-2-methylpentane, could have been reasonably named otherwise. Draw this compound and find the other reasonable name. The answer will tell you why the name given is preferred.

PROBLEM 2.57   Name the following compound:

Alkenes and Alkynes

3

On the atomic and subatomic levels, weird electrical forces are crackling and flaring, and amorphous particles . . . are spinning simultaneously forward, backward, sideways, and forever at speeds so uncalculable that expressions such as "arrival," "departure," "duration," and "have a nice day" become meaningless. It is on such levels that magic occurs.

—Tom Robbins*
Skinny Legs and All

3.1   PREVIEW

Not all hydrocarbons have the formula $C_nH_{2n + 2}$. Indeed, we have already seen some that do not: the ring compounds, C_nH_{2n} (Fig. 3.1). We noted in Chapter 2 (p. 83) that the chemical properties of these ring compounds closely resemble those of

$$CH_3CH_2CH_2CH_3$$

Butane
(C_4H_{10})
a saturated alkane

$$H_2C - CH_2$$
$$|\qquad\quad|$$
$$H_2C - CH_2$$

Cyclobutane
(C_4H_8)
a cycloalkane

FIGURE 3.1 The chemical properties of saturated alkanes, $C_nH_{2n + 2}$, are very similar to those of the cycloalkanes, C_nH_{2n}.

the acyclic saturated hydrocarbons. There is another family of hydrocarbons that also has the formula C_nH_{2n}, many of whose chemical properties are sharply different from those of the ring compounds and saturated chains we have seen so far. These compounds are different from other hydrocarbons in that they contain carbon–carbon double bonds. They are called **alk<u>e</u>nes** to distinguish them from the saturated alk<u>a</u>nes.

*Thomas Eugene Robbins is an American author, born in 1936 in Blowing Rock, NC.

There are also hydrocarbons of the formula C_nH_{2n-2}, whose chemical properties resemble those of the alkenes but not those of the alkanes or cycloalkanes. These compounds are the **acetylenes**, or **alkynes**.

The structures and some of the properties of the families of **unsaturated** hydrocarbons, called alkenes and alkynes, are discussed in this chapter.

Our study of the structures of alkenes and alkynes allows us to begin to examine chemical reactivity in a serious way.

ESSENTIAL SKILLS

1. It is critical to be able to use the hybridization model to derive structures for the alkenes (double bonds) and the alkynes (triple bonds).

2. In this chapter, we encounter reactivity, as addition reactions to alkenes appear. It is most important to see these reactions not as the first of a near-endless series of different processes, but as the initial exemplars of classes of reactions. Be sure you can relate these additions to the discussion of acids and bases in Chapter 2.

IMPORTANT DETAILS

1. The Cahn–Ingold–Prelog priority system is essential for determining (*E*) and (*Z*), the arrangement of groups around a double bond.

2. It is important to see the relationship between the amount of *s* character of an orbital and the stability of an electron in that orbital.

3.2   ALKENES: STRUCTURE AND BONDING

The simplest alkene is properly called **ethene** but is almost universally known by its common, or "trivial" name, **ethylene**.*

Several spectroscopic measurements and chemical reactions show that ethylene has the formula C_2H_4 and is a symmetrical compound composed of a pair of methylene groups, H_2CCH_2. Earlier, when we encountered methane (CH_4), we constructed a bonding scheme designed to reproduce the way four hydrogens were attached to the central carbon. Here we have a slightly different problem. Each carbon atom in ethylene is not attached to four other atoms as in methane, but to three. It seems we will need a different bonding rationale with which to describe Nature now. Our strategy will be to develop a bonding scheme for the simplest trivalent compound of carbon, methyl (CH_3), and then to extend it to ethylene, in which each carbon is attached not to three hydrogens as in methyl, but to two hydrogens and the other methylene (CH_2) group (Fig. 3.2).

FIGURE 3.2 Replacement of one hydrogen in methyl (CH_3), with a methylene (CH_2) group leads to the framework of ethylene (H_2CCH_2). Note that each carbon so far has only three bonds. The full bonding scheme for ethylene is not yet in place.

*I like the old names and will use many of them. The newer, systematic names are fine when complexity develops—we obviously could not have a trivial name for every compound. Yet some of the flavor of organic chemistry is lost when the system is applied too universally. The trivial names connect to history, to the quite correct image of the bearded geezer slaving over the boiling retort, and I like that.

Three bonds are needed to make attachments to the three hydrogens in CH_3. Using the atomic orbitals of carbon to overlap with the three hydrogen $1s$ orbitals leads to problems, just as it did in our earlier construction of methane (see Section 2.2).

WORKED PROBLEM 3.1 Produce a model for neutral methyl ($\cdot CH_3$) using the unhybridized atomic orbitals of carbon and the hydrogen $1s$ orbitals. Critically discuss the shortcomings of this model. What's wrong with it?

ANSWER There is more than one way to construct such a model. Start by determining how many electrons are available. Carbon ($_6C$) supplies four (six less the two low-energy $1s$ electrons), and each hydrogen has one, for a total of seven. We might form three carbon–hydrogen bonds by overlap of the carbon $2p_x$, $2p_y$, and $2p_z$ orbitals with three hydrogen $1s$ orbitals, for example. This process uses six electrons (two each in the carbon–hydrogen bonds), and would leave the carbon $2s$ orbital to hold the remaining electron.

However, this model gives a "methyl" with carbon–hydrogen bonds 90° apart. It is possible for a central carbon to be surrounded by three hydrogens 120° apart, and so we might suspect that this unhybridized model with its 90° angles will be destabilized by electron–electron repulsion. Moreover, overlap of a $1s$ orbital with a $2p$ orbital is inefficient. The rear lobes of the $2p$ orbitals are wasted. Stronger bonds can be constructed from directed, hybrid orbitals.

Other possible answers include making two carbon–hydrogen bonds from $2p$–$1s$ overlap and the third from $2s$–$1s$ overlap. This model leaves the last electron in a $2p$ orbital. Consequently, this model has the same problems as noted previously.

As in our earlier discussion of sp^2 hybridization (Chapter 2, p. 52), we will create three new hybrid orbitals, which will do a better job of attaching the carbon atom to the three hydrogens than do the pure atomic orbitals. Three hybrids are needed, and our mathematical operations will therefore involve combining three wave functions (atomic orbitals) to produce the three new hybrid orbitals. Recall that in such quantum mechanical calculations the number of new orbitals always equals the number of orbitals originally combined in the calculation, here three. So, let's combine the carbon $2s$, $2p_x$, and $2p_y$ orbitals to produce three new, sp^2 **hybrids**. The $2p_z$ orbital unused in our calculation remains, unhybridized and waiting to be incorporated in our picture (Fig. 3.3). Remember that the choice of $2p$ orbitals to be combined is arbitrary. Any pair of p orbitals can be used, leaving the third left over.

FIGURE 3.3 The combination of three atomic orbitals of carbon.

$$C\ (1s^2)\ 2s^2\ 2p_x\ 2p_y\ 2p_z$$

Combine $2s$, $2p_x$, $2p_y$ $\longrightarrow$ three sp^2 hybrid orbitals; $2p_z$ is not used

The new sp^2 hybrids are generally shaped like the sp^3 hybrids constructed earlier. They are directed orbitals—there is a fat lobe and a thin lobe, so they lead to quite efficient overlap with the hydrogen $1s$ orbitals (Fig. 3.4). The directed sp^2 hybrid orbital overlaps with the spherical $1s$ orbital on hydrogen to produce a carbon–hydrogen σ bond. The antibonding orbital (σ*) is not shown.

FIGURE 3.4 The bonding C—H σ orbital.

An sp^2 hybrid H_{1s} C–H Bonding σ orbital

What do you guess is the angle between the hybrid orbitals? Remember that one of the problems with using unhybridized carbon atomic orbitals is that this process produces bonds that are too close together. Repulsions between filled orbitals are not minimized in the unhybridized structure. The hybridization scheme produces strong bonds (good overlap) that are directed so as to minimize repulsions. As Figure 3.5 shows, the best way to arrange three things in space (the hydrogens in this

FIGURE 3.5 In planar CH_3, the angle between the carbon–hydrogen bonds is 120°. Repulsive overlap between the filled carbon–hydrogen bonds is minimized in this arrangement.

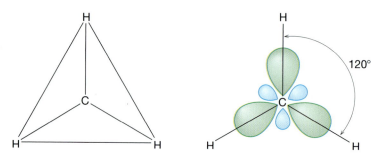

case) surrounding a central object (here, the carbon) so they are as far from each other as possible, is to put them at the corners of an equilateral triangle. And indeed, the angles between the three sp^2 hybrids are exactly 120°. Recall the earlier discussion of BH_3 in Section 2.2c.

Each of the three sp^2–$1s$ overlapping systems produces a bonding ($\sigma = sp^2 + 1s$) and an antibonding ($\sigma^* = sp^2 - 1s$) molecular orbital. Stabilization will be maximized if we form two-electron bonds, as the stabilized bonding molecular orbital will be filled and the antibonding, high-energy molecular orbital left empty (Fig. 3.6).

Each hydrogen supplies a single electron. Therefore, in order to form three, two-electron bonds we need a single electron from carbon for each of the three sp^2 hybrids (Fig. 3.7).

Remember that we did not use the carbon $2p_z$ orbital in creating the sp^2 hybrids. That's where the last electron of carbon goes. So our sp^2 scheme leads to the picture in Figure 3.8. There is a carbon atom, surrounded by a planar array of three hydrogen atoms, and the unhybridized $2p_z$ orbital extends

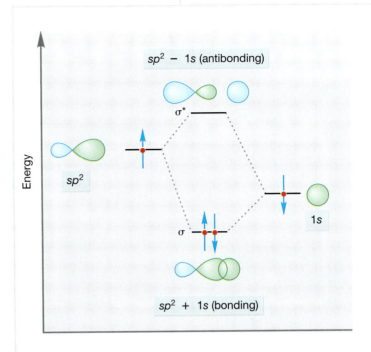

FIGURE 3.6 The overlap of sp^2 and $1s$ orbitals can occur in a bonding, stabilizing way ($sp^2 + 1s$), or an antibonding, destabilizing way ($sp^2 - 1s$).

FIGURE 3.7 Three two-electron carbon–hydrogen bonds can be formed using one electron from each hydrogen and three electrons from carbon (shown in red). This leaves one electron on carbon (shown in red) left over. In this figure, two of the sp^2 orbitals are shown schematically as wedges.

above and below the plane of the four atoms. The carbon–hydrogen bonds are familiar two-electron bonds (don't forget the empty antibonding orbitals, though) and there is a single electron in the $2p_z$ orbital. Don't be confused by the electron in one lobe of the $2p$ orbital. The two lobes are not separate, and the electron occupies the whole orbital, not just one lobe.

We have seen this molecule, CH_3, before! It is nothing more than a methyl group. In Problem 3.2, we develop a molecular orbital picture of methyl in which the delocalization of electrons throughout the molecule is emphasized. In the text, we have built up a picture that emphasizes localized bonds. There are advantages to both schemes. In one sense the molecular orbital picture is probably more "real," as electrons are not as limited to the regions of space as the hybridization scheme suggests. However, we do not make horrible energetic mistakes if we ignore the delocalization that the molecular orbital scheme shows so clearly, and the hybridization scheme is excellent for "bookkeeping" purposes. It helps us to keep track of electrons in the chemical reactions that follow in later chapters, for example.

FIGURE 3.8 A view of sp^2 hybridized CH_3.

PROBLEM 3.2 Construct the bonding molecular orbitals for planar methyl (CH_3). Use the molecular orbitals for cyclic H_3 (**A**, **B**, and **C**) given in the problem. Allow them to interact with the appropriate atomic orbitals of a carbon atom placed at the center of the triangle of hydrogens. Next, order the bonding molecular orbitals in energy. The dot in (**B**) shows the position of the third hydrogen atom.

It is important that we do not view the molecular orbital and hybridization schemes as being in conflict or as giving substantially different pictures of the molecule methyl, CH_3. Note, for example, that the geometry is exactly the same in the two schemes. We are humans, stuck with our inability to apprehend the properties of electrons easily, and needing approximations in order to represent Nature. Different approximate bonding schemes have been developed that emphasize different properties of the molecules. The molecular orbital picture does an excellent job of showing the distribution of electrons throughout the molecule. The hybridization picture sacrifices an ability to show this delocalization for the advantages of clarity, and ease of following the course of chemical reactions. We need to keep both representations in mind as we study chemical reactions.

We could easily imagine changing one of the three hydrogens of CH_3 for another atom or group. In constructing ethylene (H_2CCH_2), this other group would be another methylene (CH_2; Fig. 3.9).

FIGURE 3.9 Our first orbital picture of ethylene.

This transformation gives us our first orbital picture of ethylene ($H_2C=CH_2$), although the carbon–carbon double bond is not yet completely drawn. So our first picture of ethylene is derived from the joining of a pair of methylene groups. The two

sp^2 hybrids overlap to make an sp^2–sp^2 σ bond joining the two carbons of ethylene. The two $2p_z$ orbitals remain, and extend above and below the plane defined by the six atoms (Fig. 3.10).

FIGURE 3.10 Another orbital picture of ethylene.

This structure is quite analogous to that produced in the construction of ethane from a pair of sp^3 hybridized carbons (Section 2.5). In ethane, two sp^3 hybrids overlapped to form a bonding σ orbital and an antibonding σ* orbital (Fig. 2.26, p. 67). Construction of the carbon–carbon sigma bond system in ethylene begins with the similar overlap of a pair of sp^2 hybrids. Once more, a bonding σ orbital and antibonding σ* orbital are produced, this time by the constructive and destructive overlap of a pair of sp^2 hybrids (Fig. 3.11). As there are only two electrons, only the low-energy, bonding molecular orbital will be filled.

FIGURE 3.11 Overlap of the two sp^2 hybrids.

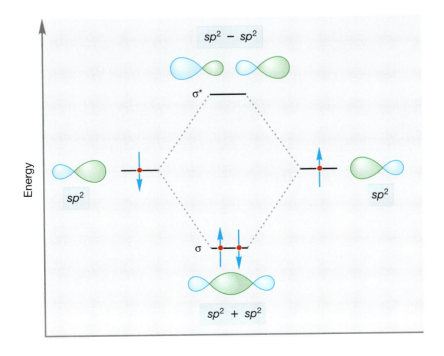

So far, we have ignored the $2p_z$ orbital on each carbon. It is time to take them into account. The two carbon $2p_z$ orbitals in question are only the distance of a single bond apart and will interact quite strongly. Moreover, they are of equal energy and that too will contribute to a strong interaction. Orbitals of equal energy interact most strongly (Chapter 1, p. 42). We know what happens when atomic orbitals overlap—bonding and antibonding molecular orbitals are created. The new orbitals formed from overlap of a pair of $2p$ orbitals are shown schematically in Figure 3.12. These new molecular orbitals are not σ orbitals as they do not have cylindrical symmetry. There is a plane of symmetry instead, and this makes them **π orbitals**, here called π (bonding) and π* (antibonding). There are only two electrons to be put into the new system of two molecular orbitals, one from each carbon $2p_z$ orbital. These are accommodated very nicely in the highly stabilized bonding π molecular orbital.

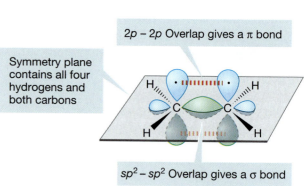

FIGURE 3.12 Overlap of a pair of $2p$ orbitals results in a lower energy, bonding π orbital and a higher energy, antibonding π* orbital. The stabilization amounts to about 66 kcal/mol when the π orbital is filled.

So the two carbons of ethylene are held together by two bonds. One is made up of sp^2–sp^2 σ overlap and the other of $2p$–$2p$ π overlap. There is a **double bond** between the two carbons. The convention is to simply draw two lines between the atoms, making no distinction between the two bonds, although, as you have just seen, this approximation is far from justified (Fig. 3.13). The overall bonding scheme for a carbon–carbon double bond includes both σ and π bonds (and their empty antibonding counterparts) and is also shown in Figure 3.13. Carbon–carbon double bonds are quite short, with a typical bond distance being 1.33 Å.

The coded representation for a double bond; note that there is no distinction made between the σ and π bonds; a typical C=C bond distance is 1.33 Å

FIGURE 3.13 The π and σ bonds of a carbon–carbon double bond are not differentiated in the coded representation of an alkene.

In ethane, there was nearly free rotation about the newly formed carbon–carbon bond (~3 kcal/mol, p. 66). Will the same be true for ethylene? Actually, the argument is a bit more subtle. There isn't really free rotation about the carbon–carbon bond in ethane. There is a 3-kcal/mol barrier, produced by the need to pass through a structure in which the carbon–hydrogen bonds are eclipsed (recall Fig. 2.24, p. 66). Our real question should be, How high is the barrier to rotation about the carbon–carbon double bond in ethylene? To begin the answer we construct Newman projections by sighting along the carbon–carbon bond (Fig. 3.14).

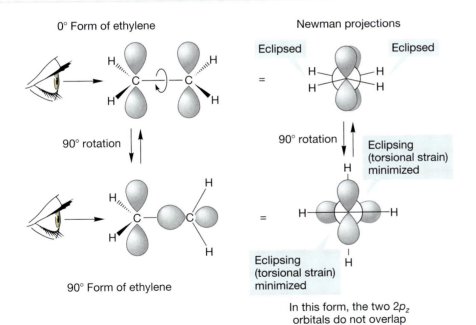

FIGURE 3.14 Although rotation about the carbon–carbon bond relieves torsional strain, it also destroys the overlap between the two $2p_z$ orbitals on the adjacent carbons. This decoupling leads to a high barrier to rotation. When one $2p$ orbital is rotated 90°, orbital overlap is lost.

Note that in the 0° conformation of ethylene there will be torsional strain induced by the eclipsing of the carbon–hydrogen bonds. This strain is removed in the 90° ro-tated arrangement in which the electrons are as far apart as possible. In our discussion of ethane, we found that each pair of eclipsed hydrogens induces approximately 1-kcal/mol destabilization. Therefore our initial guess might be that the 0° form of ethylene would be about 2 kcal/mol higher in energy than the 90° form. That estimate would be approximately right, *if torsional strain were the whole story*. But it is not, as we have ignored the pair of $2p_z$ orbitals, and their interaction completely overwhelms the small amount of torsional strain.

In the 0° form the two *p* orbitals overlap; in the 90° form they do not (Fig. 3.14). When rotation occurs, the $2p$–$2p$ overlap declines and with it, the stabilization derived from occupying the new bonding molecular orbital with two electrons. There is an enormous stabilizing effect in the 0° arrangement that overcomes the relatively minor effects of torsional strain (Fig. 3.14).

PROBLEM 3.3 Show that in the 90° rotated form there is zero overlap between the pair of carbon $2p_z$ orbitals.

Figure 3.15 plots the energy change as rotation occurs. The barrier to rotation here is no paltry 3 kcal/mol as it is in ethane, but 65.9 kcal/mol, an amount far too high to be available under normal conditions. Alkenes are "locked" in a planar conformation by this amount of energy. We will soon see some consequences of this locking.

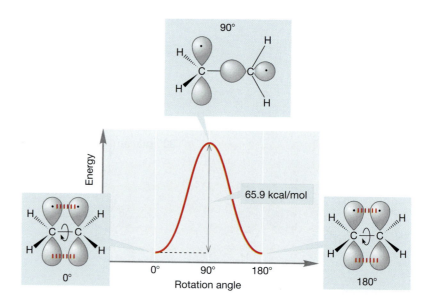

FIGURE 3.15 A plot of potential energy versus rotation angle in an alkene.

SUMMARY

Two double-bonded carbons are held together by a σ bond made up of overlapping sp^2 orbitals and a π bond made from $2p$ orbitals overlapping side to side. There is no free rotation about a carbon–carbon double bond, as such rotation diminishes overlap between the $2p$ orbitals making up the π bond and costs energy. A full 90° rotation requires about 66 kcal/mol of energy. Therefore, alkenes are planar.

In this hybridization model, we have constructed alkenes with both relatively strong σ bonds and weaker π bonds. The weaker, higher energy π bonds will be more reactive than the stronger, lower energy σ bonds. This point is critical, and is often a source of confusion. A bond with lower energy is more stable and less reactive than a higher energy bond. The weaker π bonds are likely to be the locus of reactivity of alkenes. In fact, all manner of reactions occur for alkenes and alkynes that are un-known for alkanes, with their exclusive set of strong, unreactive σ bonds. As soon as we work a bit more on nuts and bolts (nomenclature, isomers, and some thermochem-istry), we'll get to what is usually the most fun in organic chemistry—reactions—and they will be triggered by the relatively weak π bonds in alkenes.

3.3   DERIVATIVES AND ISOMERS OF ALKENES

After we developed a structure for ethane (p. 63), we examined substituted alkanes. Here we do the same thing for ethene, and discover a new kind of isomerism. There is but one ethylene (or ethene, if you insist). We can make derivatives of this mol-ecule by mentally replacing a hydrogen with some "X" group. Figure 3.16 shows a few such compounds. These are not called "ethenyl" or even "ethylenyl" com-pounds, but instead bear the delightfully trivial name, so far untampered with, of **"vinyl."**

X =	Br	Vinyl bromide	OH	Vinyl alcohol
	Cl	Vinyl chloride	CN	Vinyl cyanide
	F	Vinyl fluoride	SiH₃	Vinyl silane
	I	Vinyl iodide	NH₂	Vinylamine

FIGURE 3.16 Some substituted alkenes—vinyl compounds.

If we now let X = CH_3 (methyl), we get a compound, C_3H_6 (propene) or, com-monly, propylene (Fig. 3.17). Notice how the root prefix *prop* has been retained for this three-carbon species, and only the vowel changed to note the difference between saturated prop<u>a</u>ne and unsaturated prop<u>e</u>ne.

FIGURE 3.17 When X = CH_3 (methyl), we get propene (C_3H_6). Note the change of vowel from prop<u>a</u>ne to prop<u>e</u>ne.

We can now imagine replacing a hydrogen in propene with an "X" group, but the situation is not as simple as it was with ethylene. Although there is only one kind of hydrogen in ethylene, there are several different hydrogens in propene that might be replaced. A quick count is misleading and care must be taken from now on to draw out the three-dimensional structures of the alkenes on which we are doing our mental replacements. In practice, this is quite simple, because alkenes are planar, and most structures are easily drawn. If we just write a shorthand structure, we predict three new kinds of compounds when a hydrogen is replaced with an "X" group. As shown in Figure 3.18, these compounds are all known.

FIGURE 3.18 Substitution of three different hydrogens in propene appears to give three different propenyl–X compounds.

FIGURE 3.19 Some allyl compounds.

Note especially the use of the common name **allyl** for the $H_2C=CH-CH_2$ group, as this terminology is commonly encountered. Figure 3.19 shows some allyl compounds.

However, there is one severe problem with this analysis. In reality, there are *not* three kinds of substituted propenes, but four. The more detailed drawing in Figure 3.20 shows why. The coded structure of Figure 3.18 is inadequate to show that the two hydrogens on the end (called terminal) methylene group are *not* the same. One, H_a, is on the same side of the double bond as the methyl group, whereas the other, $H_{a'}$, is on the opposite side from the methyl. Replacement of hydrogens H_a and $H_{a'}$ with X gives two different compounds.

$H_2C{=}CH{-}CH_3$ This schematic drawing hides too much detail

replace H_a

replace $H_{a'}$

The two H_a atoms are really different! H_a is on the same side as the methyl group, but $H_{a'}$ is on the opposite side

These compounds are *not* the same!

FIGURE 3.20 A more detailed drawing shows the problem.

These molecules are interconvertible only by rotation about the double bond (Fig. 3.21). This rotation requires about 66 kcal/mol and cannot occur under normal conditions.

"*Zusammen*" (*Z*): Hydrogens on same side

"*Entgegen*" (*E*): Hydrogens on opposite sides

FIGURE 3.21 The (hypothetical) interconversion of two isomers of $CH_3CH{=}CH{-}X$ by rotation about the carbon–carbon double bond.

We know this rotation would require about 66 kcal/mol, too high an amount of energy for interconversion to be common. The molecule with the two remaining hydrogens on the same side of the double bond is designated as cis or (*Z*) (for *zusammen*, German for "together"). The compound with the hydrogens on the opposite side is called trans or (*E*) (for *entgegen*, German for "opposite"). Recall the use of cis and trans in our discussion of ring compounds (Fig. 2.55, p. 85).

If we let X = methyl we get the butenes (C_4H_8). The four butenes are 1-butene, *cis*-2-butene, *trans*-2-butene, and 2-methylpropene (or isobutene, but also sometimes called isobutylene) and are shown in Figure 3.22.

The numbers in the names 1-butene, *cis*-2-butene, and *trans*-2-butene are used to locate the position of the double bond. We will shortly discuss the naming convention for alkenes, but you might try to figure it out here from the names and structures of the butenes.

Not every alkene is capable of cis/trans (*Z/E*) isomerism! Ethylene and propene are not, and of the butenes only 2-butene has such isomers. Dealing with cis and trans isomers is an essential skill that can be "cemented" forever right now. Future difficulties can be avoided if you try Problem 3.4. It is simple, but worthwhile.

CONVENTION ALERT

CONVENTION ALERT

FIGURE 3.22 Replacement of the four different hydrogens in propene by a methyl group leads to the four butenes.

At this point, the systematic naming protocol takes over. Five-carbon compounds (C_5) are called pentenes, six-carbon compounds (C_6) are called hexenes, and so on. Be alert for isomerism of the cis/trans (Z/E) kind. It takes some practice to find it.

PROBLEM 3.4 Which of the following alkenes is capable of cis/trans (Z/E) isomerism?

PROBLEM 3.5 Write all the isomers of the pentenes (C_5H_{10}) and hexenes (C_6H_{12}). Be alert for isomerism of the cis/trans (Z/E) kind in these molecules.

3.4   NOMENCLATURE

Most of the rules for alkenes are similar to those used for alkanes, with a number attached to indicate the position of the double bond. Chains are numbered so as to give the double bond the smallest possible number. One important rule that is different in the alkenes is that the name is based on the longest chain containing a double bond, whether or not it is also the longest chain in the molecule. When molecules contain more than one double bond they are named as "dienes," "trienes," and so on. In these cases, the longest straight chain is defined as the one with the greatest number of double bonds. Cyclic molecules are named as "cycloalkenes." These rules are illustrated with the molecules in Figure 3.23.

1-Butene
not 3-butene

1,3-Butadiene

3-Methyl-1,3-heptadiene
not 2-vinyl-2-hexene

3-Propyl-1-nonene
not 4-vinyldecane

1-Methylcyclohexene

3-Methylcyclohexene
not 6-methylcyclohexene

2-Methyl-1,3-cyclohexadiene
not 3-methyl-1,3-cyclohexadiene

FIGURE 3.23 Examples of the naming protocols for alkene nomenclature

---

PROBLEM 3.6 Name the pentenes and hexenes of Problem 3.5.

PROBLEM 3.7 Recall Chapter 2 (p. 88) in which the ^{13}C NMR spectrometer was described. That machine will find a signal for each different carbon in a molecule. How many signals will the following molecules show in the ^{13}C NMR spectrum?

(a) ethylene   (b) propene   (c) 1-butene   (d) *cis*-2-butene   (e) *trans*-2-butene

(f) isobutene (2-methylpropene).

PROBLEM 3.8 How many signals will each of the molecules in Problem 3.4 show in its ^{13}C NMR spectrum?

Finally, there are examples in which the numbering of a substituent becomes important. Usually, it is the double bond numbering that must be considered first. The double bond is usually given the lowest possible number [Fig. 3.24 (a)]. An important exception is the hydroxyl (OH) group, which has priority over the double bond, and is given the lower number [Fig. 3.24 (b)]. Only if there are two possible names in which the double bond has the same number do you have to consider the position of the substituent. In such a case, give preference to the name with the lower number for the substituent [Fig. 3.24 (c)].

PROBLEM 3.9 Draw the following molecules: (*Z*)-2-pentene, 2-chloro-1-pentene, (*E*)-3-penten-2-ol, 4-bromocyclohexene, 1,3,6-cyclooctatriene.

(a)

trans-4-Chloro-2-pentene
not **trans** -2-chloro-3-pentene

(b)

trans-2-Buten-1-ol
not **trans**-2-buten-4-ol

(c)

cis-2-Chloro-3-hexene
not **cis**-5-chloro-3-hexene

FIGURE 3.24 The double bond takes precedence over most substituents, but when two names put the double bond at the same numbered position, we use the name in which the substituent has the lower number.

FIGURE 3.25 Sometimes the cis/trans convention is inadequate to distinguish two isomers. Which of these two isomers would you call cis and which trans?

Usually the cis/trans naming system is adequate to distinguish these pairs of isomers, but not always. Problem 3.10 and Figure 3.25 give examples in which cis/trans is inadequate. In the molecule 1-bromo-1-chloropropene there are clearly two isomers, but it is not obvious which is cis and which is trans. It was to solve such problems that the old cis/trans system was elaborated into the (Z/E) nomenclature by three European chemists, R. S. Cahn (1899–1981), C. K. Ingold (1893–1970), and V. Prelog (1906–1998). A priority system is set up, which ranks the groups attached to the double bonds. The compound with the higher priority groups on the same side of the double bond is (Z), and the other is (E). This priority system has other, very important uses to be encountered in Chapter 4, so we will spend some time here elaborating on it, even though we will see it again in another context.

WORKED PROBLEM 3.10 In Figure 3.23, the structure of 3-methyl-1,3-heptadiene is not completely specified by the name given. Even if the cis/trans nomenclature is applied, problems remain. What's the difficulty? Can you devise a system for resolving the ambiguity?

ANSWER In describing fully the structure of 3-methyl-1,3-heptadiene, we need to specify how groups are oriented in space (the stereochemistry), in order to distinguish the two possible forms. In one of the structures, two saturated alkyl groups (methyl and propyl) are cis while in the other they are trans.

trans CH_3 and $CH_2CH_2CH_3$

cis CH_3 and $CH_2CH_2CH_3$

Often, cis/trans will do the job, but here it fails. There is but one H, so you cannot find two hydrogens "on the same side" (cis), and two hydrogens "on opposite sides" (trans). That is the problem; it is up to you to devise a protocol for resolving it. For more about the device used by organic chemists, the (Z/E) system, read on in the text.

3.5 THE CAHN–INGOLD–PRELOG PRIORITY SYSTEM

CONVENTION ALERT

1. The first choice in setting priorities is made on the basis of atomic number. The atom of higher atomic number has the higher priority (Fig. 3.26).

   Thus, a methyl group, attached to the double bond through a carbon atom (atomic number = 6), has a higher priority than a hydrogen (atomic number = 1). Similarly, in the second compound, oxygen (atomic number = 8) has a higher priority than boron (atomic number = 5).

Higher priority
(atomic number = 6)

Higher priority
(atomic number = 8)

H₃C
      C=C
H

H₃CO
      C=C
H₂B

Lower priority
(atomic number = 1)

Lower priority
(atomic number = 5)

FIGURE 3.26 The substituent with the higher atomic number gets the higher priority.

How is this priority system used in practice? Consider the two isomers first encountered in Figure 3.25. Bromine has a higher atomic number than chlorine, so the isomer in which the higher priority methyl group and the bromine are on the same side is (*Z*), and the isomer in which the lower priority hydrogen is on the same side as bromine is (*E*) (Fig. 3.27).

(*Z*)-Isomer: The high priority groups are on the same side

(*E*)-Isomer: The high priority groups are on opposite sides

Higher priority (left C)   Higher priority (right C)

Higher priority (left C)   Lower priority (right C)

H₃C   Br
    C=C
H   Cl

H₃C   Cl
    C=C
H   Br

Lower priority (left C)   Lower priority (right C)

Lower priority (left C)   Higher priority (right C)

FIGURE 3.27 Use of the Cahn–Ingold–Prelog priority system.

2. For isotopes, atomic weight is used to break the tie in atomic number. Thus deuterium (atomic number = 1, atomic weight = 2) has a higher priority than hydrogen (atomic number = 1, atomic weight = 1) (Fig. 3.28).

Higher priority
(atomic number = 1,
atomic weight = 2)

Lower priority
(atomic number = 1,
atomic weight = 1)

D
    C=C
H

FIGURE 3.28 When the atomic numbers are the same, the heavier isotope gets the higher priority.

1-Chloro-2-methyl-1-butene

Same priority
(atomic number = 6)

H₃CCH₂   Cl

C=C

H₃C   H

Same priority
(atomic number = 6)

FIGURE 3.29 In this molecule, the two C atoms attached to the double bond of course have the same atomic number and, therefore, the same priority according to the Cahn–Ingold–Prelog Rule 1.

3. Nonisotopic ties are broken by looking at the groups attached to the tied atoms. For example, 1-chloro-2-methyl-1-butene has a double bond to which a methyl and an ethyl group are attached. Both these groups are connected to the double bond through carbons, so Rule 1 does not differentiate them (Fig. 3.29).

To break the tie, we look at the groups attached to these carbons. The carbon of the methyl group is attached to three hydrogens, whereas the "first" carbon of the ethyl group is attached to two hydrogens and a carbon. The ethyl group gets the higher priority (Fig. 3.30). If the groups are still tied after this procedure, one simply looks further out along the chain to break the tie, as in Problem 3.11.

H   H

H₃C—C   Cl

C=C

This C is attached to
two H atoms and
one C = **higher priority**   H₃C   H

This C is attached to three
H atoms = **lower priority**

FIGURE 3.30 The indicated methyl C is attached to three H atoms. The indicated ethyl C is attached to C, H, and H. The indicated ethyl C gets the higher priority.

PROBLEM 3.11 Determine which of the following molecules is (Z) and which is (E).

Cl   Cl

C=C   C=C

H   H

PROBLEM 3.12 Draw pictures of the following molecules:

(a) (E)-3-Fluoro-3-hexene

(b) (E)-4-Ethyl-3-heptene

(c) (Z)-1-Bromo-2-chloro-2-fluoro-1-iodoethylene

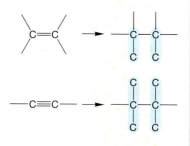

FIGURE 3.31 The priority of doubly bonded carbons is determined by adding two single carbon bonds as shown. A triple bond is treated similarly.

4. Multiple bonds are treated as multiplied single bonds. A double bond to carbon is considered to be two single bonds to the carbons in the double bond, as shown in Figure 3.31. As we will shortly see, it is also possible to connect two carbon atoms through a triple bond. The priority system treats triple bonds in a similar fashion (Fig. 3.31), which results in an isopropenyl group being of higher priority than a *tert*-butyl group, for example (Fig. 3.32).

FIGURE 3.32 An isopropenyl group has a higher priority than a *tert*-butyl group.

PROBLEM 3.13 Which of the following isomers is (Z), and which is (E)?

(a)

(b)

(c)

(d)

Higher priority (left C)   Higher priority (right C)

Lower priority (left C)   Lower priority (right C)

The two high priority groups are on the same side — this molecule is (Z)-2-methoxy-2-pentene

Higher priority (left C)   Lower priority (right C)

Lower priority (left C)   Higher priority (right C)

Here the situation is different — the two high priority groups are on opposite sides; this molecule is (E)-2-methoxy-2-pentene

FIGURE 3.33 Two examples of the Cahn–Ingold–Prelog priority system at work.

SUMMARY

The technique of assigning (Z) and (E) is to determine the priority (1 = high, 2 = low) at each carbon of the double bond. The isomer with the two high-priority groups on the same side of the double bond is (Z). The isomer with the two high-priority groups on opposite sides is (E). Two examples are given in Figure 3.33.

3.6   RELATIVE STABILITY OF ALKENES: HEATS OF FORMATION

The **heat of formation** (ΔH_f°) of a compound is the enthalpy of formation through the reaction of its constituent elements in their standard states. The standard state of an element is the most stable form of the element at 25 °C and 1-atm pressure. For an element in its standard state, ΔH_f° is taken as zero. For carbon, the standard state is graphite. For graphite, as well as simple gases (e.g., H_2, O_2, and N_2), the ΔH_f° is 0 kcal/mol.

The more negative—or less positive—a compound's ΔH_f° is, the more stable it is. A negative ΔH_f° for a compound means that its formation from its constituent elements would be *exothermic*—heat would be liberated. By contrast, a positive ΔH_f° means that the constituent elements are more stable than the compound and its formation would be *endothermic*—energy would have to be applied.

Remember: Bonding is an energy-releasing process. We would expect to find that the formation of a molecule from its constituent atoms is an exothermic reaction. For example, consider the simple formation of methane from carbon and hydrogen. Carbon in its standard state (graphite) and gaseous H_2 have $\Delta H_f^\circ = 0$ kcal/mol, whereas methane's $\Delta H_f^\circ = -17.8$ kcal/mol. The formation of methane from graphite and hydrogen releases 17.8 kcal/mol (is exothermic by 17.8 kcal/mol) (Fig. 3.34).

FIGURE 3.34 The formation of methane from graphite and gaseous hydrogen is exothermic.

C (Graphite)   + H_2 (Gas) $\longrightarrow$ CH_4 (Gas)

$\Delta H_f^\circ = 0$ kcal/mol $\Delta H_f^\circ = 0$ kcal/mol $\Delta H_f^\circ = -17.8$ kcal/mol

Table 3.1 gives the heats of formation of a number of hexenes.

TABLE 3.1   Heats of Formation for Some Hexenes

Isomer	ΔH_f° (kcal/mol)	
$CH_2{=}CHCH_2CH_2CH_2CH_3$	−10.0	Least stable
cis $CH_3CH_2CH{=}CHCH_2CH_3$	−11.2	
trans $CH_3CH_2CH{=}CHCH_2CH_3$	−12.1	
$(CH_3)_2C{=}CHCH_2CH_3$	−16.0	
$(CH_3)_2C{=}C(CH_3)_2$	−16.6	Most stable

First look at the pair of disubstituted isomers, *cis*- and *trans*-3-hexene. The trans isomer is the more stable compound (more negative ΔH_f°). Figure 3.35 shows the reason; in the cis isomer there is a cis ethyl–ethyl interaction, which is absent in the trans compound. In the trans isomer, both alkyls are eclipsed by hydrogen, and the destabilizing ethyl–ethyl repulsion is missing.

cis-**3-Hexene**

Newman projections

Here is the high-energy destabilizing interaction in the cis compound — the eclipsed ethyl–ethyl "bumping"

trans-**3-Hexene**

FIGURE 3.35 Newman projections show the unfavorable ethyl–ethyl eclipsing interaction in *cis*-3-hexene.

But both disubstituted isomers are more stable than the monosubstituted 1-hexene, and the trisubstituted isomer is more stable than either disubstituted molecule. Tetrasubstituted 2,3-dimethyl-2-butene has the lowest energy of all (most stable). From these observations, it would appear that the degree of substitution (the number of alkyl groups attached to the double bond) is important in stabilizing the molecule. In general, the more substituted the double bond the more stable it is (Table 3.1).

The reason for this has been the subject of some controversy. One good way to look at the question focuses on the different kinds of carbon–carbon bonds present in isomeric molecules of different substitution patterns. In all of these molecules, three kinds of carbon–carbon linkages are present, sp^2–sp^2, sp^2–sp^3, and sp^3–sp^3 bonds (Fig. 3.36).

Monosubstituted

σ Bond sp^2–sp^2

sp^2–sp^3 sp^3–sp^3

Disubstituted

σ Bond sp^2–sp^2

sp^3–sp^3 sp^2–sp^2 sp^3–sp^3

Trisubstituted

σ Bond sp^2–sp^2

sp^2–sp^3

sp^2–sp^3 sp^3–sp^3

Tetrasubstituted

σ Bond sp^2–sp^2

sp^2–sp^3 sp^2–sp^3

sp^2–sp^3

FIGURE 3.36 Three kinds of carbon–carbon bonds in the hexenes.

An electron in a 2*s* orbital is at lower energy than an electron in a 2*p* orbital. The more *s* character in an orbital, the more an electron in it is stabilized. In these isomeric molecules, the more substituted the carbon–carbon double bond is, the more relatively low-energy (strong) sp^2–sp^3 bonds are present. By contrast, less substituted alkenes have more relatively high-energy (weaker) sp^3–sp^3 bonds. The more strong bonds present, the more stable is the molecule. Figure 3.37 shows the number of sp^3–sp^3, sp^2–sp^3, and sp^2–sp^2 bonds in the isomeric hexenes of Table 3.1. 2,3-Dimethyl-2-butene, the molecule with the strongest bonds, is the most stable isomer of the set, and 1-hexene, with the weakest set of bonds, is the least stable.*

FIGURE 3.37 The isomeric hexenes have different numbers of these three types of carbon–carbon bonds.

PROBLEM 3.14 Carry out a similar analysis for the series of molecules: 1-pentene, 2-methyl-2-butene, (*Z*)-2-pentene.

3.7   DOUBLE BONDS IN RINGS

Even small rings can contain double bonds. Figure 3.38 shows a number of **cycloalkenes**.

FIGURE 3.38 Some cyclic alkenes.

*This analysis is somewhat primitive because it ignores the carbon–hydrogen bonds, which are also different in differently substituted alkenes. However, the changes in carbon–carbon bonds dominate. You might try to do a full comparison using detailed bond energies.

PROBLEM 3.15 Name all the compounds in Figure 3.38.

Once the ring becomes larger than cyclopropene, it becomes possible for it to contain more than one double bond. Several such molecules are shown in Figure 3.39.

Cyclobutadiene   Cyclopentadiene   **1,3-Cyclohexadiene**   **1,4-Cyclohexadiene**   **1,3,5-Cyclo-heptatriene**   **Cyclo-octatetraene**

FIGURE 3.39 Some cyclic polyenes.

PROBLEM 3.16 How many signals will appear in the ^{13}C NMR spectrum (Chapter 2, p. 88) of the compounds in Figure 3.39?

PROBLEM 3.17 Write the structures for 1,3,5-cyclohexatriene, 1,3,5,7-cyclooctatetraene, 3-methyl-1,4-cyclohexadiene, 2-fluoro-1,3-cyclohexadiene, and 2-bromo-1,4-cyclohepta-diene.

WORKED PROBLEM 3.18 Draw both (*E*)- and (*Z*)-1-methylcycloheptene. Which would you expect to be more stable? Explain.

ANSWER In (*Z*)-1-methylcycloheptene, the two higher priority groups will be on the same side of the double bond, whereas in (*E*)-1-methylcycloheptene they will be on opposite sides. On one side of the double bond, hydrogen is lower priority than CH_2, and on the other side, CH_3 is lower priority than CH_2—C. So, the "cis" compound, with the higher priority groups on the same side, is (*Z*), and the "trans" compound, with the higher priority groups on opposite sides, is (*E*).

Even the drawing gives a clue to the relative stabilities. It is trivial to draw the (*Z*) form, but the (*E*) form requires you to stretch bonds (arrow) in order to make the necessary connections. If you made a model, you saw that the (*E*) isomer contains a badly twisted double bond, with poor overlap between the 2*p* orbitals making up the π bond. There are no such problems in the (*Z*) isomer, which is much more stable.

(*Z*)-1-Methylcycloheptene   **(*E*)-1-Methylcycloheptene**

cis-Cyclopentene

The two hydrogens are on the same side of the double bond — the double bond in this molecule is cis, or (Z)

FIGURE 3.40 The double bond in cyclopentene is cis.

cis-Cyclooctene

trans-Cyclooctene

FIGURE 3.42
cis- and trans-Cyclooctene.

All the double bonds in Figures 3.38 and 3.39 are cis. For some reason this is often a sticky point for students. Make sure you understand why all the double bonds shown in the figures are appropriately called cis. To make the point absolutely clear we will use cyclopentene as an example. Note that the two hydrogens on the double bond are on the same side. The two alkyl groups (part of the ring) are on the other (Fig. 3.40). The molecule is properly called cis, or (Z).

It is extremely difficult to put a trans double bond in a small ring, and it's worth our time to see why. As we saw in Section 3.1, double bonds depend on overlap of p orbitals for their stability. There is no great problem in a ring containing a cis double bond (Fig. 3.40), but if the cycloalkene is trans, the double bond becomes severely twisted, and highly destabilized. Remember that π bonds derive their stability from overlap of 2p orbitals, and that a fully twisted (90°) π bond is about 66 kcal/mol higher in energy than the 0° form (p. 106). For a small or medium-sized ring, it is not possible to bridge opposite sides of the double bond without very severe strain. Let's use cyclopentene as an example again. A single CH_2 group cannot span the two carbons attached to opposite sides of a trans double bond (Fig. 3.41). This difficulty is easy to see in models. Use your models to try to make *trans*-cyclopentene. Hold the double bond planar, and then try to introduce the bridging methylene groups. They won't reach. Next, allow the π bond to relax as you connect the methylenes. The methylenes will reach this time, but the 2p orbitals making up the π bond will no longer overlap efficiently. Make sure you see this point.

To start a trans double bond, put the two hydrogens on opposite sides of the π bond

Now connect the carbons with a chain of methylene $(CH_2)_n$ groups; if the ring is large enough ($n \geq 6$), there is no great problem

To make *trans*-cyclopentene, the double bond carbons must be bridged by only three methylenes; this cannot be done without twisting the 2p orbitals out of overlap; this twisting is very costly in energy terms

FIGURE 3.41 It is difficult to incorporate a trans double bond in a small ring. Either the bridge of methylene groups is too small to span the trans positions or the 2p orbitals making up the π bond must be twisted out of overlap.

Of course the larger the ring, the less problem there is because more atoms are available to span the trans positions of the double bond. In practice, the smallest trans cycloalkene stable at room temperature is *trans*-cyclooctene. Even here, the trans isomer is 11.4 kcal/mol less stable than its cis counterpart (Fig. 3.42).

PROBLEM 3.19 Calculate the equilibrium distribution of *cis*- and *trans*-cyclooctene at 25 °C given an 11.4-kcal/mol energy difference between the two isomers (of course, this question assumes that equilibrium can be reached—something not generally true for alkenes). The relationship between the energy difference (ΔG) and the equilibrium constant (K) is $\Delta G = -RT \ln K$, or $\Delta G = -2.3RT \log K$, where R is the gas constant (1.986 cal/deg · mol) and T is the absolute temperature.

Many polycyclic compounds containing double bonds are known, including a great many natural products. Figure 3.43 shows some common structures incorporating cycloalkanes.

Limonene

Vitamin A

Cholesterol

Morphine

FIGURE 3.43 Some natural products incorporating cycloalkenes.

It was noticed in 1924 by the German chemist, Julius Bredt (1855–1937), that there was one structural type that was conspicuously absent among the myriad compounds isolated from sources in Nature. There seemed to be no double bonds attached to the **bridgehead position** in what are called "bridged bicyclic molecules" (Fig. 3.44). Apparently, great instability is caused by the incorporation of a double bond at the bridgehead position. You have met such structures before in Chapter 2 (p. 86), and they will reappear in some detail in Chapter 5, but you might make a model now of a simple bicyclic structure. Figure 3.44 shows two kinds of bicyclic molecules, "bridged" and "fused," and briefly introduces some terminology.

Fused bicyclic system   Bridged bicyclic system

FIGURE 3.44 Fused and bridged bicyclic molecules.

Bredt was unable to explain this absence of double bonds at the bridgehead position, but was alert enough to note the phenomenon, which has become deservedly known as **Bredt's rule**. What is the reason behind this rule? First of all, the rigid cage structure with its *pyramidal* bridgehead carbons requires that the π bond, the "double" part of the alkene, be formed not from $2p$–$2p$ overlap, but by $2p$–hybrid orbital overlap. Overlap is not as good as in a normal alkene π system (Fig. 3.45). But there is even more to this question, and a careful three-dimensional drawing of the molecule best reveals the story. As is so often the case, the Newman projection is the most informative way to look at the molecule (Fig. 3.46). Note that the orbitals making up the (hypothetical) double bond do not overlap at all well. Just as trans cycloalkenes are severely twisted, so are bridgehead alkenes. Moreover, the rigid structure of these compounds (make a model!) allows for no relief. This is a

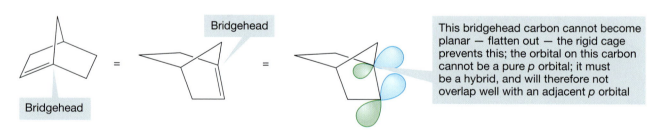

FIGURE 3.45 Three views of a bicyclic molecule containing a double bond at the bridgehead.

FIGURE 3.46 A Newman projection shows that in these bridgehead alkenes the orbitals making up the "π bond" cannot overlap well.

FIGURE 3.47 The smallest
bridgehead alkene stable
under normal conditions is
bicyclo[3.3.1]non-1-ene. This
molecule contains a *trans*-
cyclooctene, shown in color.

good example of how the two dimensions of the paper can fool you. There is no difficulty in drawing the lines making up the double bond on the paper, and unless you
can see the structure, you will almost certainly be fooled.

As the bridges get longer, flexibility returns and bridgehead "anti-Bredt" alkenes
become stable (Fig. 3.47). Clever syntheses have been devised so that even quite unstable compounds can be made and studied. In simple bridgehead alkenes, the limits of room temperature stability are reached with bicyclo[3.3.1]non-1-ene (do not
worry yet about the naming system; we will deal with it in Chapter 5), which contains a trans double bond in an eight-membered ring. In the simple trans cycloalkenes, it is also the eight-carbon compound that is the first molecule stable at
room temperature. It doesn't seem unreasonable that there should be a rough correspondence between the trans cycloalkenes and the bridgehead alkenes, which also
contain a trans double bond in a ring.

PROBLEM 3.20 Is the bicyclo[3.3.1]non-1-ene shown in Figure 3.47 (*Z*) or (*E*)?

PROBLEM 3.21 Draw the other form (*Z* or *E*). Why is it less stable than the one in Figure
3.47? *Hint*: Focus on the rings in which the double bonds are contained.

3.8 PHYSICAL PROPERTIES OF ALKENES

There is little difference in physical properties between the alkenes and their saturated relatives, the alkanes. Their odors are a bit more pungent and perhaps justify
being called "evil-smelling." In fact, the old trivial name for alkenes, **olefins**, readily evokes the sense of smell. Tables 3.2 and 3.3 collect some data for alkenes and
cycloalkenes.

TABLE 3.2 Some Simple Alkenes

Name	Formula	mp (°C)	bp (°C)
Ethene (ethylene)	$H_2C=CH_2$	−169	−103.7
Propene (propylene)	$H_2C=CH-CH_3$	−185.2	− 47.4
1-Butene	$H_2C=CH-CH_2-CH_3$	−185	− 6.3
cis-2-Butene	$CH_3-CH=CH-CH_3$	−138.9	3.7
trans-2-Butene	$CH_3-CH=CH-CH_3$	−105.5	0.9
Isobutene	$(CH_3)_2CH=CH_2$	− 40.7	− 6.6
1-Pentene	$H_2C=CH-(CH_2)_2-CH_3$	−138	30.1
1-Hexene	$H_2C=CH-(CH_2)_3-CH_3$	−139.8	63.3
1-Heptene	$H_2C=CH-(CH_2)_4-CH_3$	−119	93.6
1-Octene	$H_2C=CH-(CH_2)_5-CH_3$	−101.7	121.3
1-Nonene	$H_2C=CH-(CH_2)_6-CH_3$		146
1-Decene	$H_2C=CH-(CH_2)_7-CH_3$	− 66.3	170.5

TABLE 3.3 Some Simple Cycloalkenes

Name	mp (°C)	bp (°C)
Cyclobutene		2
Cyclopentene	−135	44.2
Cyclohexene	−103.5	83
Cycloheptene	−56	115
cis-Cyclooctene	−12	138
trans-Cyclooctene	−59	143
cis-Cyclononene		167–169
trans-Cyclononene		94–96 (at 30 mmHg)

3.9   ALKYNES: STRUCTURE AND BONDING

Like the simplest alkene, ethylene, the smallest alkyne is generally known by its trivial name, acetylene, not the systematic name, ethyne. Acetylene is a symmetrical compound of the formula HCCH in which the two carbon atoms are attached by a triple bond. Each carbon is attached to two other atoms: one hydrogen and the other carbon. Accordingly, a reasonable bonding scheme must yield only two hybrid bonds. As in the discussion of alkanes and alkenes (Chapter 2, p. 53; Chapter 3, p. 98), the atomic orbitals of carbon are combined to yield hybrid orbitals, which will do a better job of bonding than the unchanged atomic orbitals. Two bonds are needed, one for the bond to hydrogen, and another for the bond to carbon. Therefore, we need to combine only two of carbon's atomic orbitals to give us our hybrids. A combination of the $2s$ and $2p_x$ orbitals will yield a pair of **sp hybrids**. We can anticipate that these new sp hybrids will be directed so as to keep the bonds, and the electrons in them, as far apart as possible, which produces 180° angles. (Recall our discussion of BeH_2 in Section 2.2b, p. 49; Fig. 3.48.)

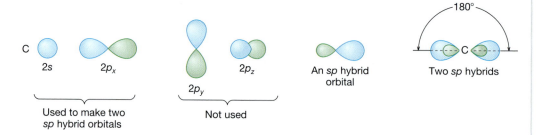

FIGURE 3.48 To make an sp hybrid we combine the wave functions for the carbon $2s$ and $2p_x$ orbitals. The $2p_y$ and $2p_z$ orbitals are not used in the hybridization scheme.

C   2s $2p_x$ | $2p_y$ $2p_z$ | An sp hybrid orbital | 180° Two sp hybrids

Used to make two sp hybrid orbitals | Not used

These new sp hybrid orbitals generally resemble the sp^3 and sp^2 hybrids we made before. As we used only two of carbon's four available atomic orbitals, the $2p_y$ and $2p_z$ orbitals are left over. Figure 3.49 shows an sp-hybridized carbon atom. (From now on, we will not show the small lobes.)

The bond to hydrogen is formed by overlap of one *sp* hybrid with the hydrogen 1*s* orbital, and the carbon–carbon bond is formed by the overlap of two *sp* hybrids. This process forms the σ system of the molecule (Fig. 3.50).

FIGURE 3.49 An *sp*-hybridized carbon atom. Note the leftover 2*p_y* and 2*p_z* orbitals.

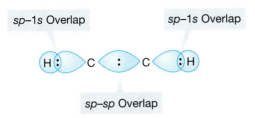

FIGURE 3.50 The σ bonding system of acetylene.

The figure shows the bonds formed by *sp–sp* and *sp–1s* overlap, but don't forget that these overlapping hybrid and atomic orbitals create both bonding and antibonding molecular orbitals. The empty antibonding orbitals are not in Figure 3.50, but are shown in the schematic construction of Figure 3.51.

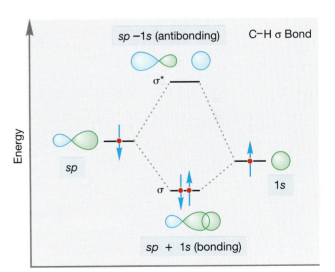

FIGURE 3.51 Orbital interaction diagrams showing the bonding and antibonding orbitals formed by *sp–1s* and *sp–sp* overlap.

As in the alkenes, the remaining unhybridized 2*p* orbitals also overlap to form π bonds. This time there are two *p* orbitals remaining on each carbon of acetylene, and we can form a pair of π bonds that are directed, as are the constituent 2*p* orbitals, at 90° to each other (Fig. 3.52).

FIGURE 3.52 Overlap of two $2p_y$ and two $2p_z$ orbitals forms a pair of π bonds in alkynes.

Both carbons have four valence electrons (*Remember*: We ignore the very low energy 1s electrons) and each can participate in four two-electron bonds. These are the σ bond to hydrogen and the carbon–carbon **triple bond** composed of one carbon–carbon σ bond and two carbon–carbon π bonds (Fig. 3.53).

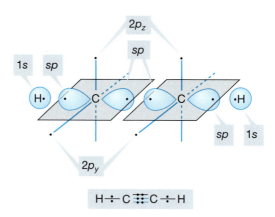

FIGURE 3.53 A highly schematic construction of the π and σ bonds in acetylene.

PROBLEM 3.22 Draw the orbital interaction diagrams for construction of the π bonds of acetylene from the 2p orbitals.

One of the structural consequences of triple bonding is an especially short carbon–carbon bond distance (~1.2 Å), considerably shorter than either carbon–carbon single or double bonds (Fig. 3.54).

$$H_3C—CH_3 \quad H_2C=CH_2 \quad HC≡CH$$

1.54 Å   1.33 Å   1.20 Å

FIGURE 3.54 Bond lengths in angstroms (Å) of simple two-carbon hydrocarbons.

SUMMARY

The σ bonding schemes we used to construct alkanes and cycloalkanes are not the only way that atoms can be held together to form molecules. In alkenes and alkynes, there is π bonding as well as σ bonding. In the construction of a π bond, $2p$ orbitals overlap side-to-side to form a bonding and antibonding pair of π molecular orbitals. Alkenes and cycloalkenes contain one π bond, whereas acetylenes (alkynes) have two π bonds.

3.10   RELATIVE STABILITY OF ALKYNES: HEATS OF FORMATION

Heats of formation show that alkynes are very much less stable than their constituent elements. Table 3.4 summarizes heats of formation for some alkynes, alkenes, and alkanes. The virtues of using hybrid orbitals for σ bonding rather than p orbitals for π bonding are apparent. The more π bonds in a molecule, the more positive the heat of formation, and the more endothermic the formation of the compound from its constituent elements. The very positive heats of formation for alkynes have their practical consequences. When a very high heat is desired, as for many welding applications, acetylene is the fuel of choice.

Disubstituted alkynes are more stable than their monosubstituted isomers. For example, we can see from Table 3.4 that the heat of formation of 1-butyne is about 5 kcal/mol more positive than that for 2-butyne. This phenomenon echoes the increasing stability of alkenes with increasing substitution (Section 3.6).

TABLE 3.4   Heats of Formation for Some Small Hydrocarbons

Alkanes	ΔH_f° (kcal/mol)	Alkenes	ΔH_f° (kcal/mol)	Alkynes	ΔH_f° (kcal/mol)
Ethane	−20.1	Ethylene	12.5	Acetylene	54.5
Propane	−25.0	Propylene	4.8	Propyne	44.6
Butane	−30.2	1-Butene	−0.1	1-Butyne	39.5
		2-Butene	−1.9 (cis)	2-Butyne	34.7
			−2.9 (trans)		

3.11   DERIVATIVES AND ISOMERS OF ALKYNES

Of course, we can imagine replacing one of the hydrogens of acetylene with another group X to give substituted alkynes. Figure 3.55 gives an example of an "ethynyl" compound. In practice, two naming systems are used here, and the figure shows them. Ethynyl chloride can also be called chloroacetylene, and both naming systems are commonly used.

H—C≡C—H ⟶ H—C≡C—X

H—C≡C—Cl

Ethynyl chloride
(or chloroacetylene)
X = Cl

FIGURE 3.55 An ethynyl compound, chloroacetylene.

We can start a construction of the family of alkynes by replacing X in Figure 3.55 with a methyl group, CH_3. This produces propyne (or methylacetylene), the lone three-carbon alkyne (Fig. 3.56).

FIGURE 3.56 Replacement of X with a methyl (CH_3) group gives propyne.

X = CH_3

H—C≡C—X ⟶ H—C≡C—CH$_3$

Propyne
(methylacetylene)

There are two different hydrogens in propyne, and therefore two different ways to produce a substituted compound (Fig. 3.57).

H_a

replace H_a ⟶ X—C≡C—CH$_3$
1-Propynyl compounds

H—C≡C—CH$_3$

replace H_b ⟶ H—C≡C—CH$_2$—X
3-Propynyl compounds
(propargyl compounds)

H_b

FIGURE 3.57 Examples of the two kinds of hydrogen in propyne, each replaced with X to give derivatives.

When the acetylenic hydrogen is replaced, the compounds can be named either as acetylenes or as "1-propynyl" compounds. 3-Propynyl compounds result from replacement of a methyl hydrogen with X. The common name **propargyl** is reserved for the HC≡C—CH_2 group and is often seen. Note the relationship between the allyl (Fig. 3.19, p. 108) and propargyl groups.

When X = CH_3 we produce the two four-carbon acetylenes, which must be butynes (Fig. 3.58).

FIGURE 3.58 If X = CH_3, we get the two butynes.

X = CH_3

H—C≡C—CH$_2$—X ⟶ H—C≡C—CH$_2$—CH$_3$

1-Butyne

X = CH_3

X—C≡C—CH$_3$ ⟶ H$_3$C—C≡C—CH$_3$

2-Butyne

The naming protocol is similar to that used for the alkenes. As with the alkenes, the triple bond is designated a position by assigning it a number, which is kept as low as possible. When both double and triple bonds are present, the compounds are named as "enynes," and the numbers designating the positions of the multiple bonds kept as low as possible. If the numbering scheme could produce two names in which the lower number could go to either the "-ene" or the "-yne," the "-ene" gets it (Fig. 3.59).

1-Butyne
not 3-butyne

trans-2-Hexen-4-yne
not trans-4-hexen-2-yne

3,3-Dimethyl-1-butyne
not 2,2-dimethyl-3-butyne

Ethynylcyclohexane
or cyclohexylacetylene

3-Propynylcyclopentane
or propargylcyclopentane

FIGURE 3.59 Some examples of the naming convention for alkynes.

It was at this point in the alkenes that we encountered cis/trans isomerism, and it's appropriate to search for it here as well. It takes only a short look to see that there can be no such isomerism in a linear compound (Fig. 3.60). The sp hybridization requires 180° angles, and therefore there can be no cis/trans isomerism in alkynes. The alkynes are simpler to analyze structurally than the alkenes.

FIGURE 3.60 There are 180° angles in 2-butyne.

PROBLEM 3.23 Draw all the pentynes, hexynes, and heptynes.

PROBLEM 3.24 Name all the isomers in Problem 3.23.

PROBLEM 3.25 How many signals will appear in the ^{13}C NMR spectra of the compounds in Figure 3.59?

3.12   TRIPLE BONDS IN RINGS

Like double bonds, triple bonds can occur in rings, although very small ring alkynes are not known. The difficulty comes from the angle strain induced by the preference for linear, 180° bond angles in an acetylene. When a triple bond is incorporated in a ring, it becomes difficult to accommodate these 180° angles. Deviation from 180° reduces the overlap between the p orbitals making up one of the π bonds of the acetylene and that raises the energy of the compound. The same problem exists for the alkenes, but it is less severe (Fig. 3.61); *Remember*: "R" is shorthand for a general group. In practice, the smallest ring in which an alkyne is stable under normal conditions is cyclooctyne. Thus, it and larger cycloalkynes (cyclononyne, etc.) are known, but the smaller cycloalkynes are either unknown or have only been observed as fleeting intermediates (Fig. 3.62).

Second π bond
in the acetylene

bend

Reduced
overlap

A cyclic acetylene

FIGURE 3.61 Incorporation of a triple bond in a ring is difficult because orbital overlap between the 2*p* orbitals of one of the π bonds is reduced by the bending required by the ring.

Cycloheptyne
(observable only at
low temperature)

Cyclooctyne
(stable at
room temperature
but reactive)

Cyclononyne
(a compound of
normal reactivity)

FIGURE 3.62 Some cycloalkynes.

3.13   PHYSICAL PROPERTIES OF ALKYNES

The physical properties of alkynes resemble those of alkenes and alkanes. Table 3.5 collects some data.

TABLE 3.5   Some Simple Alkynes

Name	Formula	mp (°C)	bp (°C)
Acetylene (ethyne)	$HC{\equiv}CH$	−80.8	−84
Propyne	$HC{\equiv}C-CH_3$	−101.5	−23.2
1-Butyne	$HC{\equiv}C-CH_2CH_3$	−125.7	8.1
2-Butyne	$CH_3-C{\equiv}C-CH_3$	−32.2	27
1-Pentyne	$HC{\equiv}C-(CH_2)_2CH_3$	−90	40.2
1-Hexyne	$HC{\equiv}C-(CH_2)_3CH_3$	−131.9	71.3
1-Heptyne	$HC{\equiv}C-(CH_2)_4CH_3$	−81	99.7
1-Octyne	$HC{\equiv}C-(CH_2)_5CH_3$	−79.3	125.2

3.14   ACIDITY OF ALKYNES

It is generally most difficult for a base ($B\!:\!^-$) to remove a proton from a hydrocarbon to give an anion. In other words, hydrocarbons are generally very weak acids. We have already had a brief discussion of the methyl anion in Chapter 2 (p. 60), but methane, the parent of this anion, is an extraordinarily weak acid (Fig. 3.63).

FIGURE 3.63 Strong bases ($B\!:\!^-$) can remove the terminal hydrogen of an acetylene. Terminal acetylenes are reasonably strong acids for hydrocarbons.

A remarkable property of terminal acetylenes is that they are reasonably strong Brønsted acids (proton donors). Compared to most hydrocarbons it is easy to remove the terminal acetylenic hydrogen as a proton, leaving behind an anion, called an **acetylide**. Acetylenes are stronger acids than alkanes by roughly a factor of 10^{30}! Be sure you are clear on this point! Acetylenes are not strong acids in the sense that traditional mineral acids such as hydrochloric acid (HCl) and sulfuric acid (H_2SO_4) are, but they are certainly strong in comparison to most other hydrocarbons.

Why should acetylenes be relatively strong hydrocarbon acids? Consider the orbital containing the residual pair of electrons in the two cases of Figure 3.63. For the methyl anion, this orbital is hybridized approximately sp^3, with roughly 25% s character. In the acetylide, the corresponding orbital is sp hybridized, and therefore has 50% s character. An electron in an s orbital is substantially lower in energy than one in a p orbital, and so the more s character the orbital has, the lower in energy an electron in it is. The nonbonding electrons in the acetylide ion ($HC\!\equiv\!C\!:\!^-$) are in a much lower energy orbital than those in the nonbonding orbital of the methyl anion ($H_3C\!:\!^-$), also called methide (Fig. 3.64). Therefore it is relatively easy to remove a proton to produce the acetylide ion.

If the hypothesis that the hybridization of the anion is critical is correct, then alkenes should be intermediate in acidity between alkanes and alkynes. This hypothesis is exactly right: alkenes are about 10^{10}–10^{12} *more* acidic than alkanes and 10^{18-20} *less* acidic than alkynes (Fig. 3.65).

C is hybridized approximately sp^3, therefore the electrons are in an orbital of about 25% s character

Now, C is hybridized approximately sp, and the electrons are in an orbital of approximately 50% s character

FIGURE 3.64 In an acetylide ion, the two nonbonding electrons are accommodated in an sp orbital.

FIGURE 3.65 Alkenes are intermediate in acidity between alkanes and alkynes.

3.15   MOLECULAR FORMULAS AND DEGREES OF UNSATURATION

How is one to determine the possible structures for a molecule of a given molecular formula? For example, what are the possibilities for C_6H_{10}? Cyclohexene, 1-hexyne, and 1,3-hexadiene are only a few of them. Before we begin to write down the possibilities, we need to know what are the general possible structural types. In this case, cycloalkenes, acyclic alkynes, and acyclic dienes are among the structures whose formula is C_6H_{10}. Given a formula for an unknown hydrocarbon, we are faced with the problem of determining the number of π bonds and rings. Saturated alkanes have the formula C_nH_{2n+2}. Both alkenes and cycloalkanes have the formula C_nH_{2n}. Such molecules are said to have "one degree of unsaturation," which simply means that they contain two fewer than the maximum number of hydrogens, as defined by the alkane formula, C_nH_{2n+2}. Alkynes and cycloalkenes have the formula C_nH_{2n-2}, and contain four fewer hydrogens than the C_nH_{2n+2} number. These compounds have two degrees of unsaturation. The **degree of unsaturation** (Ω) for a hydrocarbon is the total number of π bonds and rings, and can always be determined by calculating the number of hydrogens for the corresponding saturated alkane (H_{2n+2}), and then subtracting the number of hydrogens actually present and dividing by two. This procedure also works if hydrogens are replaced with other monovalent atoms such as the halogens. Such atoms are treated as if they were hydrogens. Later, when we see more complicated molecules containing nitrogen and oxygen, we will have to update this subject. Figure 3.66 works out two examples.

C_6H_{12} (C_nH_{2n})

For C_6H_{2n+2}, there would be 14 hydrogens.

Calculation:
14 − 12 = 2/2 = 1 degree of unsaturation, which means that there must be one π bond or one ring.

C_9H_{16} (C_nH_{2n-2})

For C_9H_{2n+2}, there would be 20 hydrogens.

Calculation:
20 − 16 = 4/2 = 2 degrees of unsaturation, which means that there must be a total of two π bonds and rings.

FIGURE 3.66 Some calculations of degrees of unsaturation.

PROBLEM 3.26 Calculate the degrees of unsaturation (Ω) for the following formulas. Give two possible structures for each one. (a) C_5H_6 (b) C_7H_8 (c) $C_{10}H_{10}$ (d) $C_5H_8Br_2$.

3.16   AN INTRODUCTION TO ADDITION REACTIONS OF ALKENES AND ALKYNES

Alkenes, and other molecules containing π systems such as the alkynes, undergo an addition reaction with Brønsted acids (proton donors) of the general formula HX. Figure 3.67 shows a general and specific reaction of this kind. In fact, there are many other related addition reactions, and we'll see several of them in Chapter 9.

FIGURE 3.67 Additions to alkenes.

We are going to look first at the addition of hydrogen chloride to 2,3-dimethyl-2-butene (Fig. 3.68). The π bond of the alkene and the σ bond of H—Cl have been converted into a pair of σ bonds attaching H and Cl to carbon atoms. We must lead into this reaction with a look at the structure and properties of molecules such as hydrogen chloride that contain bonds between two very different kinds of atoms.

2,3-Dimethyl-2-butene   **2-Chloro-2,3-dimethylbutane**

FIGURE 3.68 Addition of hydrogen chloride to 2,3-dimethyl-2-butene.

3.17   ELECTRONEGATIVITY

All bonds between different atoms are polar—the electrons in the bond cannot be shared equally in a covalent bond between different atoms. The limit of this phenomenon is an ionic bond in which two oppositely charged species are held together by the electrostatic attraction between them. Potassium chloride and sodium fluoride are examples (Fig. 3.69).

FIGURE 3.69 Two examples of charge-separated, ionic bonding.

Charges also exist in covalent bonds between different atoms. However, these are partial charges and are not fully developed. In such a bond, the electrons will be polarized toward the more electronegative atom, the atom with the greater attraction for the electrons in the shared orbital. As first mentioned in Chapter 1 (Table 1.8, p. 14), **electronegativity** is the ability of an atom to attract electrons.

Notice that the most electronegative elements are in the upper right of the periodic table. The ability to attract electrons is related to the effective nuclear charge of an atom, which in turn depends on the effectiveness of the electrons surrounding the nucleus in shielding an additional negatively charged electron from the nucleus and its positive charge. In any unsymmetrical covalent bond, there will be a dipole with the negative end on the more electronegative atom. A dipole in a polarized bond is indicated with an arrow pointing toward the more electronegative element and partial charges are shown as δ^+ and δ^- (Fig. 3.70).

Molecules such as hydrogen chloride and hydrogen bromide are very polar, even though they are covalently, not ionically, bonded molecules. The electrons in the bond joining the hydrogen and the halogen are not shared equally, and the dipole results from that unequal sharing. Hydrogen chloride has a dipole moment of 1.08 D. Although hydrogen chloride is not an ionic compound, it readily donates a proton, the positive end of its dipole. This familiar property is what makes hydrogen chloride a good Brønsted acid. Now let's apply these ideas to the reaction of alkenes with acids such as hydrochloric acid and hydrobromic acid.

FIGURE 3.70 The dipole in H—Cl.

3.18   MECHANISM OF THE ADDITION OF HYDROGEN HALIDES TO ALKENES

The addition process begins with a straightforward Brønsted acid–Brønsted base reaction, in which HX donates a proton to one of the two equivalent carbons of the π bond. The HX molecule acts as the Brønsted acid and the alkene acts as the Brønsted base in this reaction. The first protonation step in the reaction is followed by a second addition step in which the newly formed chloride ion captures the newly formed carbocation (Fig. 3.71).

FIGURE 3.71 The first step in this two-step reaction is the protonation of the alkene to give a carbocation. In the second step, a chloride ion adds to the cation to give the final product.

In looking at Figure 3.71, don't forget the curved **arrow formalism** (p. 91). Those double barbed arrows track the movements of pairs of electrons. The color-coding should help. Notice also the red and green equilibrium arrows. The different lengths reflect the exothermicity or endothermicity of each step. The first step is endothermic, and the second is exothermic.

Let's look at the intermediate carbocation in Figure 3.71. In Chapter 2 (p. 60), we described the methyl cation ($^+CH_3$). The cation formed by protonation of 2,3-dimethyl-2-butene is related to the methyl cation, but the three hydrogens of $^+CH_3$ have been replaced with three alkyl groups: two methyls and an isopropyl (Fig. 3.72).

The methyl cation is flat; the central carbon is hybridized sp^2

This cation will also be planar; the central carbon is approximately sp^2

FIGURE 3.72 Protonation of 2,3-dimethyl-2-butene gives a planar carbocation, closely related to the methyl cation. The central carbon of each species is hybridized sp^2.

No great structural differences appear, however. The trigonal (attached to three groups) carbon is hybridized sp^2 and the C—C—C angles will be roughly 120°. The empty carbon $2p_z$ orbital extends above and below the plane of the central carbon and the three carbons attached to it (Fig. 3.72). Let's even verify that the carbon is positively charged (Fig. 3.73).

In a neutral carbon atom, the six positive nuclear charges are balanced by six electrons. The central carbon in this cation has a pair of $1s$ electrons and a half-share in the electrons in the three covalent bonds to the alkyl groups, for a total of five. The six positive charges in the nucleus are balanced by only five electrons, and so the carbon atom is positively charged.

Polar reactions such as these are best understood in terms of **H**ighest **O**ccupied **M**olecular **O**rbital–**L**owest **U**noccupied **M**olecular **O**rbital (**HOMO–LUMO**) orbital interactions. As we have already seen (Section 1.7, p. 39), when a filled occupied orbital overlaps an empty orbital, the two electrons are stabilized in the new, lower energy molecular orbital. The words "Lewis bases react with Lewis acids" are essentially equivalent to saying, "The interaction of a filled and empty orbital is stabilizing." Indeed, this notion is one of the central unifying themes of organic reactivity, as essentially all polar reactions can be understood this way.

We will revisit this reaction in detail in Chapter 9, but see right now if you can identify the HOMOs and LUMOs in the two steps of this reaction, that is, do Problems 3.27 and 3.28.

$_6C\ (1s^2 2s^2 2p^2)$

Neutral (no charge)

This $_6C$ has only five electrons surrounding it: $1s^2$, and three shared in the bonds to the three R groups; therefore, it has a single positive charge

FIGURE 3.73 The determination of the charge on carbon in a carbocation.

WORKED PROBLEM 3.27 Identify the HOMO and LUMO for the second step of the reaction in Figure 3.68.

CONVENTION ALERT

WORKED PROBLEM 3.28 In this much harder problem, identify the HOMO and LUMO in the first step of the reaction shown in Figure 3.68.

ANSWER In the second step of the reaction, chloride ion acts as the Lewis base and reacts through one of the filled orbitals (the HOMO) containing the nonbonding pairs of electrons. The LUMO is easy to find; it must be the empty $2p$ orbital of the carbocation.

The first step of the reaction is much more difficult to deal with. Part of it is easy: The alkene is the base and supplies the HOMO. More precisely, the base is the filled π orbital of the alkene. What empty orbital acts as the acid, the LUMO? Focus on the requirement for an empty orbital. What can it be? There is only one, the empty σ^* orbital of HCl.

If you got this problem right, feel really good. If it was a struggle, that is alright as well. Not all problems are easy! If they were, Nature would be trivial to understand, and life would be boring indeed. In any case, try to learn from these in-chapter answered problems. There is a "take-home lesson" intended in each of them.

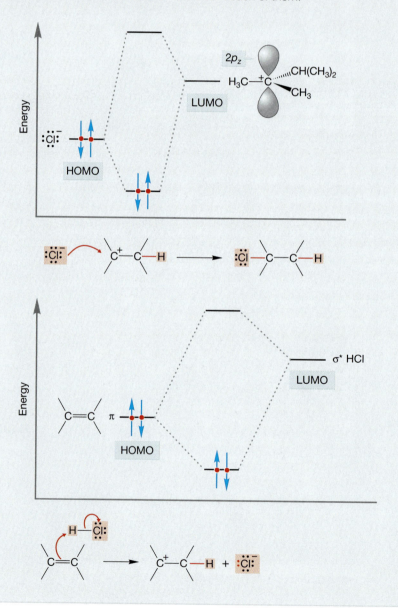

Now let's try to estimate qualitatively the overall change in energy as this reaction proceeds. We will plot energy on the vertical axis against something we might call "reaction progress" on the horizontal axis. Roughly speaking, we are asking how the energy changes as the reaction takes place. Notice that two bonds must be broken in the first step of the reaction. These are the carbon–carbon π bond (66 kcal/mol), and the bond in HCl (103.2 kcal/mol). In the final product, two bonds have been made, the new carbon–hydrogen bond (~96 kcal/mol), and the new carbon–chlorine bond (~85 kcal/mol). It requires almost 170 kcal/mol to break the bonds in this reaction, but that quantity is more than fully compensated by the roughly 181 kcal/mol released by the new bonds formed (Fig. 3.74).

Bond broken
(103.2 kcal/mol)

Bond made
(96 kcal/mol)

Bond made
(85 kcal/mol)

Overall: Bonds broken (103 + 66) = 169 kcal/mol
Bonds made (96 + 85) = 181 kcal/mol

Net: 181 − 169 = about 12 kcal/mol exothermic

Bond broken
(66 kcal/mol)

FIGURE 3.74 An estimation of the energetics of the addition of HCl to 2,3-dimethyl-2-butene.

In the chemistry trade, this kind of reaction, in which the products are more stable than the starting materials, is called an **exothermic reaction**. But this reaction actually takes place in two steps—we do not just suddenly arrive at a product from the starting material. There is an intermediate carbocation in the reaction, a transient species relatively high in energy. There are two factors that make this first step difficult. First, the bonds to be broken, the carbon–carbon π bond and the σ bond between hydrogen and chlorine, are collectively stronger than the bond made, the new carbon–hydrogen bond. There is another factor that makes the first step in this reaction difficult. Usually, charged species such as carbocations are high in energy, and not easily formed. Thus, the first step in this reaction is surely "uphill" in energy, which is called an **endothermic reaction**.

By contrast, the second step of this reaction, capture of the carbocation by chloride, is sure to be very energy releasing (an "exothermic reaction"). Charges are annihilated and a new carbon–chlorine bond is made, at no bond-breaking cost. If we plot these estimated energy changes, we get Figure 3.75, the prototype of many such diagrams to appear later in this book as we estimate the energetics of other reactions. The overall reaction is exothermic—the products are more stable than the starting materials, but an initial endothermic step, the formation of the intermediate carbocation, must occur before product can be formed. There is an energy hill—a barrier—to be passed.

The tops of the hills in Figure 3.75 are transition states—the high points in energy separating starting material and product. We encountered transition states earlier when we discussed the barrier to rotation in ethane and other alkanes (p. 66). Although these energy maxima cannot be isolated, we can describe their structures. In Figure 3.75, the making and breaking bonds are shown as dashed partial bonds. The developing partial positive and negative charges appear as δ^+ and δ^-.

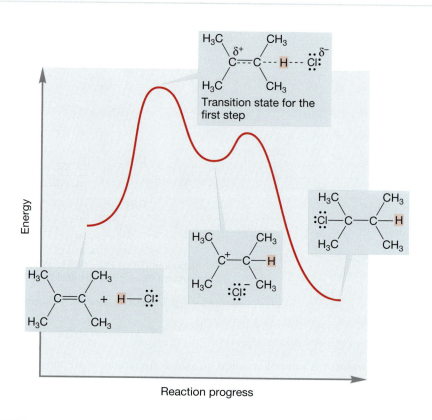

FIGURE 3.75 In this simple addition of hydrogen chloride to an alkene, the first step, the endothermic formation of a carbocation, is the slow step.

With just a little thought we have a very clear picture of the addition reaction. An unstable carbocationic intermediate is first formed and is then destroyed as the chloride ion adds to give the final product. Notice two important things about this reaction. First, it cannot proceed unless enough energy is supplied so that the first transition state can be reached. This amount of energy—that required to pass over the highest point, the highest transition state—is called the **activation energy** ($\Delta G^{\ddagger}$) for the reaction.

Second, we get a bonus from this analysis. If we understand the pathway from left to right, we also understand the reverse reaction—the one from right to left. It follows exactly the same energy curve, only in the opposite direction.

PROBLEM 3.29 Generalize from the discussion of HCl addition to predict the products of the following reactions:

3.19 THE REGIOCHEMISTRY OF THE ADDITION REACTION

In the example used in Figure 3.71, the addition of HCl to 2,3-dimethyl-2-butene, the two ends of the alkene are the same—there is no choice to be made in the protonation step. Addition to either end of the alkene gives exactly the same carbocation. Let's extend our discussion just a bit and examine an alkene in which the two ends are different. We'll do much more of this sort of thing in Chapter 9, when we take a long look at many kinds of addition reactions, but there is much to be learned from even a small extension of the symmetrical addition reaction.

So, let's allow HCl to react with isobutene, an alkene in which the two ends are certainly very different. Now, two different protonations are possible, and addition of chloride to the two carbocations will give two products (Fig. 3.76). This kind of distinction is called the **regiochemistry** of the reaction. In fact, only one of the two possibilities is formed. Let's see if we can guess why.

FIGURE 3.76 Two possible additions of HCl to isobutene.

Each potential reaction pathway involves a different carbocationic intermediate. Perhaps formation of a tertiary carbocation requires less energy than formation of a primary carbocation. If that idea were correct, the energy hill to climb to give the observed product would be lower than that leading to the product that is not observed (Fig. 3.77).

FIGURE 3.77 Relative stability of the *tert*-butyl cation (tertiary) and the isobutyl cation (primary).

This idea leads to a prediction. If our surmise that the stability of carbocations increases with substitution is correct, that is, a tertiary carbocation is more stable than a primary carbocation, presumably a tertiary carbocation will also be more stable than a secondary carbocation, and a secondary carbocation will be more stable than a primary carbocation. The complete order of cationic stability would be tertiary > secondary > primary > methyl (Fig. 3.78).

FIGURE 3.78 Predicted stability order for carbocations.

If the order of carbocation stability shown in Figure 3.78 is correct, we should always get the product resulting from formation of the more stable carbocation (Figure 3.79).* We can test this idea. Just allow HCl to add to 2-methyl-2-butene or propene, each an unsymmetrical alkene. If we are right in our ideas about carbocation

---

*I hope you are wondering why this order exists—we will get to that question in Chapter 9.

stability, the products will be 2-chloro-2-methylbutane and isopropyl chloride, respectively. When these experiments are run, those are exactly the products observed (Fig. 3.79).

FIGURE 3.79 Addition of HCl to two unsymmetrical alkenes.

The results are in accord with our mechanistic ideas; our proposed reaction mechanism is supported by the new data. Do we now know how this addition of HCl to alkenes works? Is our mechanism proved? Sadly the answer is "no" and in fact, we can never know for certain that we are absolutely right. A mechanistic hypothesis survives only pending the next clever test. Someone—you perhaps—may develop a new test of the ideas outlined in this section that reveals an error, or at least an area that needs further thinking and working out of details. It's sad but true: It is easy to disprove a scientific hypothesis—just find some contradictory data. It is impossible ever to prove a theory beyond a shadow of a doubt.

We all have to get used to this ambiguity—it is part of the human condition. We can be quite certain we are right in our mechanistic ideas, and the more time that goes by, and the more tests our ideas pass, the more confident we can be that we are basically right. However, there have been some big surprises, as conventional wisdom is reversed every once in a while, and there will be more. To be honest, I sort of like it when that happens.

PROBLEM 3.30 Predict the products of the following reactions:

$$H_3C \quad\quad H$$
$$C=C$$
$$H_3C \quad\quad H$$

(with HBr → ? and H_2SO_4 (HO—SO₂—OH) → ?)

3.20   A CATALYZED ADDITION TO ALKENES: HYDRATION

Not every HX is a strong enough acid to protonate an alkene. Water, HOH, is an example of such a weak acid. Yet, it would be very useful to be able to add the elements of water to an alkene, to "hydrate" an alkene as in Figure 3.80, because the products (R—OH, called **alcohols**) are useful in themselves and are also versatile starting materials for the synthesis of other molecules. The solution to this problem is to use a small amount of an acid that is strong enough to protonate the alkene. The added acid acts as a **catalyst** for the addition. A catalyst is a material that increases the rate of a reaction without being consumed. Let's work through the mechanism for the hydration of alkenes.

FIGURE 3.80 The highly desirable addition of water to an alkene.

The first step is protonation of the alkene by the catalyst, H_3O^+. As in the addition of HCl, this addition leads to a carbocation intermediate (Fig. 3.81). Now there is no halide ion (I^-, Br^-, Cl^-, F^-) to capture the carbocation, but there are lots of water molecules available. Capture by water leads not to the final product, the alcohol, but to an **oxonium ion**. In an oxonium ion, a trivalent oxygen bears a positive charge. Water deprotonates the oxonium ion to give the alcohol, and regenerate the catalyst, H_3O^+.

You can see from this last step why only a catalytic amount of acid is necessary. Although it is consumed in the first protonation step, it is regenerated in the last step and can recycle to start the hydration reaction again (Fig. 3.81).

FIGURE 3.81 The hydration of alkenes; the acid-catalyzed addition of water to an alkene.

WORKED PROBLEM 3.31   Only one alcohol is produced in the following reaction. What is it, and why is only one alcohol, not two, produced?

$$\underset{H_3C}{\overset{H_3C}{}}C{=}C\underset{H}{\overset{H}{}} \quad \xrightarrow{\ H_3O^+\ } \quad \text{A single alcohol product}$$

ANSWER   Formation of the more stable tertiary carbocation rather than the much less stable primary carbocation inevitably leads to the single product observed, *tert*-butyl alcohol.

3.21   SYNTHESIS: A BEGINNING

Organic chemistry is not only understanding how reactions occur. A very important part of chemistry is the use of those reactions to make molecules. The construction of target molecules from smaller pieces is called synthesis, and we are now able to do a surprising amount of this kind of thing. Let's just take stock of what kinds of transformations we can do—of what kinds of molecules we can make out of other kinds of molecules.

The addition of hydrogen halides to alkenes gives us a way to make alkyl bromides, chlorides, and iodides (fluorides don't work very well) as long as we have the appropriate starting alkene. We will have to be careful when using unsymmetrical alkenes, because there is a choice to be made here. This route will only lead to the more substituted alkyl halide, as the regiochemistry of the addition is controlled by the stability of the intermediate carbocation.

We can also make alcohols through the acid-catalyzed addition of water to alkenes.

In each chapter, the synthetic methods developed and discussed are collected on the Summary page. These reactions appear on page 147.

PROBLEM 3.32 Devise syntheses of the following three compounds, starting in each case with any alkene that contains four carbons or fewer. You do not have to write mechanisms, although at this point it may be very helpful for you to do so.

$$(CH_3)_3COH \qquad (CH_3)_2CHOH \qquad CH_3CH_2CHOHCH_3$$

3.22   SPECIAL TOPIC: ALKENES AND BIOLOGY

In sharp contrast to the alkanes, the unsaturated alkenes are highly active biologically. As we have seen in Section 3.2, alkenes contain π bonds that are substantially weaker than the σ bonds of alkanes. Accordingly, their reactivity is higher, and Nature "uses" this reactivity to accomplish things. One thing that can be accomplished is the formation of carbon–carbon bonds. It is extraordinarily fortunate for us that Nature is able to do that reaction easily, because our tissues are composed mainly of carbon–carbon bonds. I'll just sketch one example of alkene-mediated carbon–carbon bond formation here. We will return to more details of this and related transformations in later sections.

For its own purposes, the tuberculosis-causing organism *Mycobacterium tuberculosis* needs a molecule called 10-methylstearate. It makes this compound from a modified amino acid, *S*-adenosylmethionine, and oleate, a salt of oleic acid. The first interesting step in what turns out to be, ultimately, a reaction chock full of interesting steps, is the transfer of the methyl group to the alkene. The double bond acts as the trigger for the whole sequence. Note also that here the π bond is acting as a Lewis base, much as it does in its reaction with HCl (p. 134), where it acts as a Brønsted base (Fig. 3.82).

FIGURE 3.82 The reaction of an alkene (Lewis base) to produce a carbon–carbon bond. The new bond is shown in red.

It's amazing, but the simplest of all alkenes, ethylene, is an important plant hormone. Among other functions, ethylene acts to promote ripening of fruit. Moreover, production of ethylene is autocatalytic; that is, a little ethylene induces the formation of more from the amino acid methionine, and its effects are magnified. Tomatoes are now typically shipped green in well-ventilated containers so they will arrive unspoiled. Ripening can then be started by exposure to ethylene.

Ethylene also induces abscission, the falling of leaves or flowers, by promoting the formation of the enzyme cellulase that weakens cell walls by destroying the cellulose from which they are made. A weakened abscission layer forms at the base of the leaf or flower, and wind or rain then breaks the stem. Indeed, it was this phenomenon that indirectly led to the discovery of ethylene as a plant hormone. Trees surrounding gas lamps were often defoliated when gas leaks occurred. A principal component of the illuminating gas was ethylene, and a little investigation showed that this chemical was the cause of the defoliation.

3.23   SUMMARY

New Concepts

This chapter deals mainly with the structural consequences of sp^2 and sp hybridizations. The doubly bonded carbons in alkenes are hybridized sp^2 and the triply bound carbons in alkynes are hybridized sp. In these molecules, atoms are bound not only by the σ bonds we saw in Chapter 2 but by π bonds as well.

These π bonds, composed of overlapping $2p$ orbitals, have substantial impact upon the stereochemistry of the molecule. Alkenes can exist in cis (Z) or trans (E) forms—they have sides. Unlike the alkanes, which contain only σ bonds with very low barriers to rotation, there is a substantial barrier (~66 kcal/mol)

to rotation about a π bond. The linear alkynes do not have cis/trans isomers and the question of rotation does not arise.

Double and triple bonds can be contained in rings, if the ring size is large enough to accommodate the strain incurred by the required bond angles. The smallest trans cycloalkene stable at room temperature is *trans*-cyclooctene. The smallest stable cycloalkyne is cyclooctyne.

More substituted alkenes and alkynes are more stable than their less substituted relatives.

Key Terms

acetylenes (p. 98)
acetylide (p. 131)
activation energy ($\Delta G^{\ddagger}$) (p. 138)
alcohols (p. 142)
allyl (p. 108)
alkenes (p. 97)
alkynes (p. 98)
arrow formalism (p. 135)
Bredt's rule (p. 122)
bridgehead position (p. 121)
catalyst (p. 142)

cycloalkenes (p. 118)
degree of unsaturation (Ω) (p. 132)
double bond (p. 104)
electronegativity (p. 134)
endothermic reaction (p. 137)
ethene (p. 98)
ethylene (p. 98)
exothermic reaction (p. 137)
heat of formation (p. 115)
HOMO–LUMO (p. 135)
olefins (p. 123)

oxonium ion (p. 142)
π orbitals (p. 104)
propargyl (p. 128)
regiochemistry (p. 139)
sp hybrids (p. 124)
sp^2 hybrids (p. 100)
triple bond (p. 126)
unsaturated (p. 98)
vinyl (p. 107)

Reactions, Mechanisms, and Tools

The Cahn–Ingold–Prelog priority system is introduced. It is used in the (Z/E) naming system. We will encounter it again very soon in Chapter 4.

In this chapter, we continue the idea of constructing larger hydrocarbons out of smaller ones by replacing the different available hydrogens with an X group. When X is methyl (CH_3), a larger hydrocarbon is produced from a smaller one.

The reaction of alkenes with acids, H—X, to give the products of addition reactions is introduced. Many acids (H—Br, H—Cl, H—I, $HOSO_2OH$, generally, H—X) will add directly to alkenes. The first step is addition of a proton to give the more stable carbocation. Inductive and resonance effects influence the stability of the intermediate carbocations. In the second

step of the reaction, the negative end of the original H—X dipole usually adds to complete the addition process. The regiochemistry of the addition is determined by the formation of the more stable carbocation in the original addition.

Other H—X molecules are not strong enough acids to protonate alkenes. Water (HOH) is an excellent example. However, such molecules will add if the reaction is acid catalyzed. Enough acid catalyst is added to give the protonated alkene, which is attacked by water. The catalyst is regenerated in the last step and recycles to carry the reaction further.

The removal by a base of a hydrogen from the terminal position of acetylenes to give acetylides is mentioned. Terminal alkynes are moderately acidic molecules.

Syntheses

1. ACETYLIDES

$$CH_3-C\equiv C-H + B:^- \longrightarrow CH_3-C\equiv C:^- + B-H$$
Base

2. ALKYL HALIDES

X = Cl, Br, I, sometimes F

Addition gives the compound with the halide, X, on the more substituted carbon

3. ALCOHOLS

The addition reaction of water is catalyzed by acid

Common Errors

The concept of cis/trans (Z/E) isomerism is a continuing problem for students. Errors of omission (the failure to find a possible isomer, for example) and of commission (failure to recognize that a certain double bond, so easily drawn on paper, really cannot exist) are all too common. The idea begins with the notion that, unlike σ bonds, π bonds have substantial barriers to rotation (~66 kcal/mol). The barrier arises because of the shape of the p orbitals making up the π bond. Rotation decreases overlap and raises energy. Accordingly, double bonds have sides, which are by no means easily interchanged. Substituents cannot switch sides, and are locked in position by the high barrier to rotation. It is vital to see why there are two isomers of 2-butene, cis- and trans-2-butene, but only one isomer of 2-methyl-2-butene (Fig. 3.83).

This relatively easy idea has more complicated implications. For example, the need to maintain planarity (p–p overlap) in a π bond leads to the difficulty of accommodating a trans double bond in a small ring, or at the bridgehead position of bridged bicyclic molecules. In each case, the geometry of the molecule results in a twisted (poorly overlapping) pair of p orbitals.

FIGURE 3.83

3.24 ADDITIONAL PROBLEMS

In this chapter, we encounter the determination of molecular formulas. The following three review problems (3.33–3.35) deal with this subject.

PROBLEM 3.33 Alkanes combine with oxygen to produce carbon dioxide and water according to the following scheme:

$$C_nH_{2n+2} + (3n+1)/2\ O_2 \rightarrow (n+1)\ H_2O + n\ CO_2$$

This process is generally referred to as combustion. An important use of this reaction is the quantitative determination of

elemental composition (elemental analysis). Typically, a small sample of the compound is completely burned and the water and carbon dioxide produced are collected and weighed. From the weight of water the amount of hydrogen in the compound can be determined. Similarly, the amount of carbon dioxide formed allows us to determine the amount of carbon in the original compound. Oxygen, if present, is usually determined by difference. The determination of the relative molar proportions of carbon and hydrogen in a compound is the first step in deriving its molecular formula.

(Continued)

PROBLEM 3.33   *(Continued)*
If combustion of 5.00 mg of a hydrocarbon gives 16.90 mg of carbon dioxide and 3.46 mg of water, what are the weight percents of carbon and hydrogen in the sample?

PROBLEM 3.34   Calculate the weight percents for each element in the following compounds: (a) C_5H_{10} and (b) C_9H_6ClNO.

PROBLEM 3.35   A compound containing only carbon, hydrogen, and oxygen was found to contain 70.58% carbon and 5.92% hydrogen by weight. Calculate the empirical formula for this compound. If the compound has a molecular weight of approximately 135 g/mol, what is the molecular formula?

PROBLEM 3.36   How many degrees of unsaturation are there in compounds of the formula C_5H_8? Write at least eight isomers including ones with no π bonds and ones with no rings.

PROBLEM 3.37   Draw and write the systematic names for all of the acyclic isomers of 1-heptene (C_7H_{14}). There are 36 isomers, including 1-heptene.

PROBLEM 3.38   Draw and write the systematic names for all of the acyclic isomers of 1-octyne (C_8H_{14}) containing a triple bond. There are 32 isomers, including 1-octyne.

PROBLEM 3.39   Draw and write the systematic names for all of the isomers of dichlorobutene ($C_4H_6Cl_2$). There are 27 isomers, according to usually reliable sources.

PROBLEM 3.40   Write systematic names for the following compounds:

(a)

(b)

(c)

(d)

PROBLEM 3.41   Draw structures for the following compounds:
(a) (*Z*)-3-Fluoro-2-methyl-3-hexene
(b) 2,6-Diethyl-1,7-octadien-4-yne
(c) *trans*-3-Bromo-7-isopropyl-5-decene
(d) 5-Chloro-4-iodo-6-methyl-1-heptyne

PROBLEM 3.42   Although you should be able to draw structures for the following compounds, they are named incorrectly. Give the correct systematic name for each compound:
(a) 5-Chloro-2-methyl-6-vinyldecane
(b) 3-Isopropyl-1-pentyne
(c) 4-Methyl-6-hepten-1-yne
(d) (*E*)-3-Propyl-2,5-hexadiene

PROBLEM 3.43   Find all the compounds of the formula $C_4H_6Br_2$ in which the dipole moment is zero. It is easiest to divide this long problem into sections. First, work out the number of degrees of unsaturation so that you can see what kinds of molecules to look for. Then find the acyclic molecules. Next work out the compounds containing a ring.

PROBLEM 3.44   Find all the compounds of the formula C_4H_6 in which there are four different carbon atoms. There are only nine compounds possible, but two of them are quite strange at this point. *Hint*: One of these molecules contains a carbon that is part of two double bonds, and the other is a compound containing only three-membered rings.

PROBLEM 3.45   Find all the isomers of the formula C_4H_5Cl. As in Problem 3.43, it is easiest to divide this problem into sections. First, find the degrees of unsaturation so that you know what types of molecule to examine. There are 23 isomers to find. *Hint*: See the hint for Problem 3.44.

PROBLEM 3.46   Find the compound of the formula $C_{10}H_{16}$, composed of two six-membered rings, that has only three different carbons.

PROBLEM 3.47   Develop a bonding scheme for the compound $H_2C=O$ in which both carbon and oxygen are hybridized sp^2.

PROBLEM 3.48   (a) Develop a bonding scheme for a ring made up of six CH groups, $(CH)_6$. Each carbon is hybridized sp^2. (b) This part is harder. The molecule you built in (a) exists, but all the carbon–carbon bond lengths are equal. Is this consistent with the structure you wrote in (a), which almost certainly requires two different carbon–carbon bond lengths? How would you resolve the problem?

PROBLEM 3.49 In this chapter, we developed a picture of the methyl cation ($^+CH_3$), a planar species in which carbon is hybridized sp^2. In Chapter 2, we briefly saw compounds formed from several rings, bicyclic compounds. Explain why it is possible to form carbocation (a) in the bicyclic compound shown below, but very difficult to form carbocation (b). It may be helpful to use models to see the geometries of these molecules.

(a)　　　(b)

PROBLEM 3.50 Predict the products of the following reactions:

PROBLEM 3.51 Explain the formation of two products in the following reaction:

PROBLEM 3.52 Explain the formation of the two products in the reaction shown.

PROBLEM 3.53 In principle, there is a second "double addition" possible in Problem 3.52. Explain why only one compound is formed. *Hint*: Think "resonance."

This compound is formed　　This compound is not formed

Stereochemistry

4

I held them in every light. I turned them in every attitude. I surveyed their characteristics. I dwelt upon their peculiarities. I pondered upon their conformation.

—Edgar Allan Poe*

Berenice

4.1 PREVIEW

There are two isomers of *trans*-1,2-dimethylcyclopropane (p. 85). We will see here why two *trans*-1,2-dimethylcyclopropanes exist and continue with a detailed discussion of **stereochemistry**, the structural and chemical consequences of the arrangement of atoms in space. We have already seen several examples of stereoisomeric molecules. Substituents on both ring compounds and alkenes can be attached in cis [or (Z) = same side] and trans [or (E) = opposite side] arrangements (Fig. 4.1).

cis-1,2-Dimethylcyclopropane

Two representations of
trans-1,2-dimethylcyclopropane

cis-2-Butene
[or (Z)-2-butene]

trans-2-Butene
[or (E)-2-butene]

FIGURE 4.1 *cis*- and *trans*-1,2-Dimethylcyclopropane and *cis*- and *trans*-2-butene are pairs of stereoisomeric molecules.

*Edgar Allan Poe (1809–1849) was an American poet and writer probably most famous for his macabre stories and poems. As the quotation shows, he was also an early devotee of stereochemistry.

In this chapter, we will examine more subtle questions of stereoisomerism. Most molecules of Nature are **chiral**. The word "chiral" is derived from the Greek and means "handed." Chiral molecules are related to their mirror images in the same way that your left hand is related to your right. They are not identical! Indeed many molecules of Nature (for example, the amino acids) are handed in the same way: One might say that they are all left handed. The source of the ubiquity of left handedness is one of the great questions left in chemistry. Was it chance? Did an accidental left-handed start determine all that followed? When we encounter extraterrestrial civilizations will their amino acids (assuming they have any) also be left handed or will we find right-handed amino acids? As it now appears we will find no cohabitants within our own solar system, and as physical contact with our galactic neighbors is unlikely, you might consider how you would convey to an invisible resident of the planet Altair 4 our concepts of left and right.

This topic is a complicated but important one that is absolutely essential to organic and biological chemistry. When we return to our examination of reaction mechanisms—of the way molecules come together to make other compounds—we will often use chiral molecules to provide the level of detail necessary to work out how reactions occur. Moreover, the molecules of Nature are largely chiral, and this first look at chirality provides a base for virtually all of biochemistry and molecular biology.

You are very strongly urged to work through this chapter with models. It is a rare person who can see easily in three dimensions without a great deal of practice.

ESSENTIAL SKILLS

1. You have the broad outlines of structure under control now—alkanes, alkenes, and alkynes have appeared, as have both acyclic compounds and rings. Now we come to the details, to stereochemistry. "Sidedness"—cis/trans isomerism—is augmented by questions of chirality—"handedness." Learning to see one level deeper into three dimensionality is the next critical skill.

2. Although it also appears under the "Important Details" section, the question of difference and how it is determined arises in this chapter. The details may be nuts and bolts, and, indeed, any way that you work out to do the job will be just fine, but there is no avoiding the seriousness of the question. When are two atoms the same (in exactly the same environment) or different (not in the same environment)? This question gets to the heart of structure and is much tougher to answer than it seems. By all means, concentrate on this point. This chapter will help you out.

IMPORTANT DETAILS

1. There is no way out—the (R) and (S) priority system must be learned.

2. This chapter is filled with jargon: Be certain that you know the difference between enantiomers and diastereomers. Know also what "chiral," racemic," and "meso" mean.

3. The algorithm by which difference is assessed appears in this chapter. It is well worth knowing how it works.

4.2   CHIRALITY

The phenomenon of handedness, or **chirality**, actually surfaces long before a molecule as complicated as *trans*-1,2-dimethylcyclopropane. We went right past it when we wrote out the heptanes in Figure 2.48 (Chapter 2, p. 82). One of these heptanes, 3-methylhexane, exists in two forms. To see this, draw out the molecule in tetrahedral form using C(3) as the center of the tetrahedron. Next, draw the mirror image of this molecule. Contrast these structures with 3-methylpentane treated the same way, with C(3) still the center of the tetrahedron (Fig. 4.2).

3-Methylhexane and its mirror image

3-Methylpentane and its mirror image

FIGURE 4.2 3-Methylhexane and 3-methylpentane reflected to show their mirror images.

Now see if the reflections are the same as the originals. For 3-methylpentane, one need only rotate the molecule 180° and slide it over to superimpose the two mirror images (Fig. 4.3). The molecule and its mirror image are clearly identical, and hence, not chiral. However, Nature seems to lay a trap for us as there are many motions of the mirror image that will *not* directly generate a molecule superimposable on the original. In such cases, we may be led to think we have uncovered chirality. For example, try rotating the molecule as shown at the bottom of Figure 4.3. The original and the newly rotated molecule are not simply superimposable. But chiral? No, for as long as there is one motion (Fig. 4.3, top) that does generate the mirror image, the molecule is **achiral** (not chiral).*

*There is a language problem introduced by the choice of Greek for the word (chiral) to describe the phenomenon of handedness of molecules. In Greek, the word for "not chiral" is "achiral." This poses no problem in Greek, but, at least in spoken English it does. It is often difficult to distinguish between "Compound X is a chiral molecule" and "Compound X is achiral."

FIGURE 4.3 3-Methylpentane and its mirror image are identical. Rotate the mirror image at the top of the page 180° about the carbon–methyl bond to see this point. Although an unproductive motion tells us nothing about the chirality of the molecule, a motion that produces a superimposable mirror image tells us that the molecule is not chiral (achiral).

Now consider Figure 4.4. Here we apply to 3-methylhexane the same rotation that we applied before to 3-methylpentane. In this case, we cannot superimpose the original onto the newly rotated molecule no matter how many ways we try. The propyl group in one form winds up where the ethyl group is in the other. Indeed, no amount of twisting and turning can do the job: The two are irrevocably different.

FIGURE 4.4 3-Methylhexane and its mirror image are not identical.

Models are essential now, as you must be absolutely convinced of this point, and certain of the difference between 3-methylpentane and 3-methylhexane. The more symmetric 3-methylpentane is achiral, whereas 3-methylhexane and your hand are chiral. The two stereoisomers of 3-methylhexane are related in exactly the same way your two hands are. A chiral compound can simply be defined as a molecule whose mirror image is not superimposable on the original. The two versions of 3-methyl-hexane are called **enantiomers**.

Figure 4.5 shows schematically the difference between 3-methylhexane and 3-methylpentane. The difference in symmetry is obvious in these color-coded pictures. In 3-methylhexane, four different groups surround a central carbon atom, which is not the case for 3-methylpentane. Our next job is to see how these stereoisomers differ physically and chemically, and to explore the circumstances under which chirality will appear. What structural features will suffice to ensure chirality? Will, for example, the phenomenon of four different groups surrounding a carbon be a sufficient condition to ensure chirality? Will it be a necessary condition? This chapter discusses such questions.

FIGURE 4.5 3-Methylhexane, 3-methylpentane, and their mirror images

PROBLEM 4.1 Are the following molecules chiral? (a) Bromochlorofluoroiodomethane, (b) dibromochlorofluoromethane, (c) 3-methylheptane, (d) 4-methylheptane.

4.3   PROPERTIES OF ENANTIOMERS: PHYSICAL DIFFERENCES

Almost all physical properties of enantiomers are identical. Their melting points, boiling points, densities, and many other physical properties do not serve to distinguish the two. But they do differ physically in one somewhat obscure way: They rotate the plane of plane-polarized light to an equal degree but in *opposite* directions. Rotation of the plane of polarization is called **optical activity** and molecules causing such a rotation are optically active. The enantiomer that rotates the plane in a clockwise direction (as one faces a beam of light passing through the sample) is called **dextrorotatory**. If the rotation is counterclockwise, it is **levorotatory**. The direction of rotation is indicated by placing a (+) for dextrorotatory, or (−) for levorotatory, before the name (Fig. 4.6).

FIGURE 4.6 The (+)-enantiomer rotates the plane of plane-polarized light in a clockwise direction. The (−)-enantiomer—not shown here—will rotate the light an equal amount in the counterclockwise direction.

Plane-polarized light enters a tube containing one enantiometer of 3-methylhexane

The plane of light is rotated clockwise as viewed from the front

Your eye

Very often, equal amounts of the dextrorotatory and levorotatory stereoisomers are present. In such a case, no rotation can be observed as there are equal numbers of molecules rotating the plane in clockwise and counterclockwise directions, and the effects cancel. Such a mixture is called a **racemate**, or **racemic mixture**. The presence of a racemic mixture of enantiomers is often indicated by placing a (±) before the name as in Figure 4.7, which brings up a technical problem.

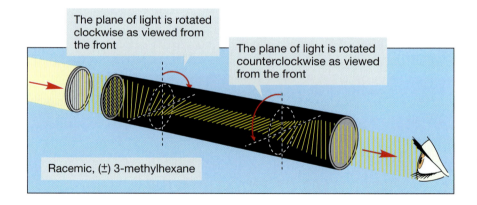

FIGURE 4.7 There is no net rotation by a racemic mixture of enantiomers. One isomer will rotate the plane of plane-polarized light to the right and the other an equal amount to the left.

The plane of light is rotated clockwise as viewed from the front

The plane of light is rotated counterclockwise as viewed from the front

Racemic, (±) 3-methylhexane

When we are talking about a chiral molecule, how do we know whether the subject of our discussion is a single enantiomer or the racemic mixture? Often the text will tell you by indicating (+), (−), or (±). If no indication is made, you are almost certainly talking about the racemic mixture. Drawings present other problems. If the subject is a single enantiomer, the drawing will certainly point this out. Sometimes an asterisk (*) is used to indicate optical activity.

There are times, however, when a three-dimensional representation is important, but the subject is a 50:50 racemic mixture of the two enantiomers rather than a single enantiomer. Unless both enantiomers are drawn, the figure appears to focus only on a single enantiomer. Our molecule, 3-methylhexane, illustrates this in Figure 4.8.

CONVENTION ALERT

If we want to show the three-dimensional structure of this molecule we must draw one of the enantiomers. To be strictly correct, both enantiomers should be drawn in any discussion of this molecule unless it really is a single enantiomer that is under discussion. In practice this is rarely done, and you must be alert for the problem. Unless optical activity is specifically indicated, it is usually the racemate that is meant.

CONVENTION ALERT

This is one enantiomer of 3-methylhexane; to emphasize chirality, a (+) [or (–)], or * is sometimes added

When it is the racemic mixture that is meant, sometimes (±) is added. However, this is not always done, and you must be alert for such times

Indicates the racemic mixture of the two enantiomers

FIGURE 4.8 A variety of conventions is used to indicate the presence of a single enantiomer.

Clearly, we desperately need a way to identify which enantiomer we mean when we talk to each other. In practice, we cannot take the time to describe in detail the orientation in space of the groups surrounding each atom in a molecule. We need a convention in order to specify the **absolute configuration**, and naturally one has been devised.

4.4 THE (R/S) CONVENTION

In order to specify the absolute configuration of a molecule, one first identifies the **stereogenic center**, very often a carbon atom.[†] A stereogenic atom can be simply defined as follows: An atom (usually carbon) of such nature and bearing groups of such nature that it can have two nonequivalent configurations.

In our example of 3-methylhexane, the stereogenic carbon, C(3), is the one upon which we have based the tetrahedral structures of Figures 4.2–4.5. Next, the four groups (atoms or groups of atoms) attached to the stereogenic carbon are identified and given priorities according to the Cahn–Ingold–Prelog scheme we first met in Chapter 3 (p. 112) when we discussed cis/trans or (Z/E) isomerism. The system is

CONVENTION ALERT

[†]The word "stereogenic," though officially suggested in 1953, came into general use only very recently when it was urged by Princeton professor Kurt M. Mislow (b. 1923) as a replacement for a plethora of unsatisfactory terms such as "chiral carbon," "asymmetric carbon (or atom)," and many others. These old terms are slowly disappearing, but will persist in the literature (and textbooks) for many more years.

FIGURE 4.9 The atom of lowest atomic number, often hydrogen, is given the lowest priority, **4**. The remaining priorities are assigned in order of increasing atomic number.

FIGURE 4.10 When the atomic numbers are equal, the atomic weights may be used to break the tie.

FIGURE 4.12 An example of a complicated tie-breaking procedure. In this case, we start by looking at the atoms attached to the three tied carbons. This procedure does not completely resolve the tie.

applied here so as to be able to identify any stereogenic atom as either (R) or (S) from a consideration of the priorities of the attached groups (1 = high, 4 = low). The application of the Cahn–Ingold–Prelog system is a bit more complicated than that for determining (Z) or (E) for alkenes, so the system will be summarized again here.

The atom of lowest atomic number is given the lowest priority number, **4**. In 3-methylhexane, this atom is hydrogen (Fig. 4.9). In some molecules, one can give priority numbers to all the atoms by simply ordering them by atomic number. Thus, bromochlorofluoromethane is easy: H is **4**, F is **3**, Cl is **2**, and Br is **1**. The priorities are assigned in order of increasing atomic number so that the highest priority, **1**, is given to the atom of highest atomic number, in this case bromine (Fig. 4.9).

A subrule is illustrated by another easy example, 1-deuterioethyl alcohol (Fig. 4.10). We break the tie between the isotopes H and D by assigning the lower priority number to the atom of lower mass, H. So H is **4**; D is **3**; C is **2**; and O, with the highest atomic number, is priority **1**.

There can be more difficult ties to break, however. The molecule *sec*-butyl alcohol (2-hydroxybutane) illustrates this point. The lowest priority, **4**, is H and the highest priority, **1**, is O. But how do we choose between the two carbons? The tie is broken by working outward from the tied atoms until a difference is found. In this example, it's easy. The methyl carbon is attached to three hydrogens and the methylene carbon to two hydrogens and a carbon. The methyl carbon gets the lower priority number **3** by virtue of the lower atomic numbers of the atoms attached to it (H,H,H = 1,1,1) and the ethyl carbon (attached to C,H,H = 6,1,1) gets the higher priority number, **2** (Fig. 4.11).

FIGURE 4.11 When both the atomic number and the atomic weights are equal, one looks at the atoms to which the tied atoms are attached and determines their atomic numbers.

C attached to (H,H,H) = **3**
C attached to (C,H,H) ⎫
C attached to (C,H,H) ⎬ Still tied
　　　　　　　　　　⎭

Sometimes there will be even more difficult ties to break, and another look at 3-methylhexane illustrates this problem well. It is easy to identify H as the atom with the lowest atomic number, but the other three atoms are all the same; they are carbons. How do we break this tie? We work outward from the identical atoms until a difference appears. In 3-methylhexane, the methyl group is attached to three hydrogens, but each methylene group is attached to two hydrogens and one carbon. As H is of lower atomic number than C, the methyl group gets priority number **3** (Fig. 4.12).

Now we are faced with yet another tie, and must find a way to distinguish the pair of methylene groups. Again we work outward. Look at the carbons attached to the two tied methylenes. One is a methyl carbon and is attached to three hydrogens. The other is a methylene and is attached to two hydrogens and a carbon. The methylene attached to the methyl gets priority **2** and the other methylene is left with the highest priority, **1** (Fig. 4.13). In practice, this detailed bookkeeping is often unnecessary.

In this example, it is really quite easy to see that the smaller ethyl group is of lower priority than the larger propyl group, and thus its carbon will get the lower priority number. More difficult examples can be devised, however!

C attached to (H,H,H) = **2**
C attached to (C,H,H) = **1**

To break this tie look at the carbons next further out along the chains

FIGURE 4.13 The final assignment of priority numbers in 3-methylhexane.

PROBLEM 4.2 Identify the priorities of the compounds in the following figure:

Note that in (d) there are two stereogenic carbon atoms

PROBLEM 4.3 Draw the mirror images of the chiral compounds (a–d) in Problem 4.2.

How do we deal with multiple bonds? Carbon is often doubly or triply bonded to another atom. Carbon–carbon double bonds, carbon–carbon triple bonds, and carbon–oxygen double bonds are treated as shown in Figure 4.14. In the example in Figure 4.14, hydrogen will obviously get the lowest priority, **4**, and the bromine the highest priority, **1**. But both the other substituents on the stereogenic carbon are carbons. The tie is resolved by treating the carbon atom in the carbon–oxygen double bond as being attached to two oxygens as shown in Figure 4.14. Therefore it gets the higher priority, **2**, and the carbon attached to only one oxygen gets the remaining priority, **3**.

FIGURE 4.14 Carbon–carbon double bonds and carbon–carbon triple bonds are elaborated by adding new carbon bonds as shown. The carbon–oxygen double bond is treated by adding a new bond from carbon to oxygen. Thus, the carbon is treated as being attached to one hydrogen and two oxygens and gets a higher priority number than the carbon that is attached to two hydrogens and one oxygen.

Once priority numbers have been established, we imagine looking down the bond from the stereogenic atom (usually carbon) toward the atom of lowest priority (**4**, often H). The other three substituents (**1**, **2**, **3**) will be facing you. Connect these three with an arrow running from highest to lowest priority number (**1**→**2**→**3**). If this arrow runs clockwise, the enantiomer is called (*R*) (Latin: *rectus*, "right"); if it runs counterclockwise it is called (*S*) (Latin: *sinister*, "left") (Fig. 4.15). The label (*R*) or (*S*) is added to the name as an italic prefix in parentheses. Figure 4.16 shows the (*R*) and (*S*) forms of some examples from the previous paragraphs.

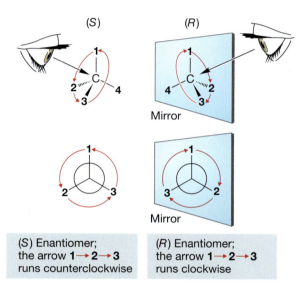

| (S) Enantiomer; the arrow **1**→**2**→**3** runs counterclockwise | (R) Enantiomer; the arrow **1**→**2**→**3** runs clockwise |

FIGURE 4.15 The (*R*) and (*S*) naming convention.

Note that your eye must look from the stereogenic carbon toward priority **4**; if you look in the other direction, or draw the arrow **3**→**2**→**1**, you will get the convention backward!

(R)-3-Methylhexane   **(R)-1-Deuteriopropyl alcohol**   **(S)-sec-Butyl alcohol**   **(R)-2-Bromo-3-hydroxy-propionaldehyde**

FIGURE 4.16 Some (*R*) and (*S*) molecules.

PROBLEM 4.4 Designate the stereogenic carbons of the compounds in Problem 4.2 as either (*R*) or (*S*).

PROBLEM 4.5 Draw out the enantiomers of the molecules in Figure 4.16.

Note also that there is no connection between (R) and (S) and the (+) and (−) used to denote the direction of rotation of the plane of plane-polarized light.

This convention looks a bit complicated. It is easier than it appears right now, but it just must be learned, and cannot be reasoned out. It's quite arbitrary at a number of points. For the price of learning this convention we get the return of being able to specify great detail most economically. Time and again we will use the subtle differences between enantiomers to test ideas about reaction mechanisms—of how chemical reactions occur. The study of the effects of stereochemistry on reaction mechanisms has been, and remains, one of the most powerful tools used by chemists to unravel how Nature works. Practice using the (R/S) convention and it will become easy. It's well worth it—indeed, it's essential.

4.5   THE PHYSICAL BASIS OF OPTICAL ACTIVITY

We have covered the way in which enantiomers differ physically. Enantiomers rotate the plane of plane-polarized light in opposite directions. Before we go on to talk about how enantiomers differ in their chemical reactivity, let's take a little time to say something about the phenomenon of optical activity. How does a collection of atoms (a molecule) interact with a wave phenomenon (light), and how does the property of chirality (handedness) induce the observed rotation?

Light consists of electromagnetic waves vibrating in all possible directions perpendicular to the direction of propagation (Fig. 4.17).

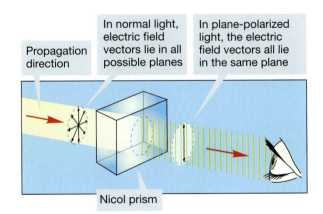

Propagation direction

In normal light, electric field vectors lie in all possible planes

In plane-polarized light, the electric field vectors all lie in the same plane

Nicol prism

FIGURE 4.17 A Nicol prism transmits only plane-polarized light.

Passing ordinary light through a Nicol prism (or a piece of Polaroid film) eliminates all waves but those whose electric vectors are vibrating in a single plane. The classic experiment to demonstrate this uses two pairs of Polaroid sunglasses. If we observe light through one pair, all is still visible, although the light has been plane polarized. The second pair of glasses only transmits light when oriented in the same way as the first pair. If it is turned 90° to the first pair none of the plane-polarized light can pass, and darkness results (Fig. 4.18).

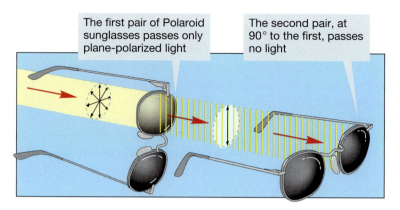

FIGURE 4.18 A classic experiment using Polaroid sunglasses: (a) The first pair of sunglasses acts as a Nicol prism and transmits only plane-polarized light. (b) If the second pair is held in the same orientation, light will still be transmitted. If it is held at 90° to the first pair, no light can be transmitted.

The electrons in a molecule, moving charged particles, generate their own electric fields, which interact with the field of the light. As we have already seen, many molecules are not highly symmetric, and the interactions of their unsymmetric electric fields with that of the plane-polarized light act to change the electric and magnetic fields. Thus, the plane of polarization is rotated slightly. Achiral molecules such as 3-methylpentane do not give a net rotation of plane-polarized light. Equivalent molecules exist in mirror-image positions, and thus the net rotation must be 0° (Fig. 4.19).

FIGURE 4.19 The net rotation of a racemic mixture of molecules must be 0°.

The result is no net rotation of the plane. However, when the interaction is with a chiral molecule, *there are no compensating mirror-image molecules* (Fig. 4.20). There must be a net change in the plane of polarization, and this is generally large enough to be observed by a device called a **polarimeter** (Fig. 4.21). The amount of rotation varies with the wavelength of light, so monochromatic light must be used to produce results easily reproduced in other laboratories. Generally,

Propagation of plane-polarized light

Interaction of the light with (R)-3-methylhexane rotates the plane

The solution contains only (R)-3-methylhexane — so the plane is further rotated!

FIGURE 4.20 The optically active molecules such as (R)-3-methylhexane will rotate the plane of plane-polarized light. There are no compensating mirror-image molecules for this kind of molecule.

the sodium D line is used, which is light of a wavelength of 589 nm (5890 Å). The light is passed through a Nicol prism to polarize it, and then through a tube containing a solution of the molecule in known concentration. The amount of rotation is proportional to the number of molecules in the path of the light, so the length of the tube and the concentration of the solution are important. A term called the specific rotation, $[\alpha]$, which is the rotation induced by a 1-g/mL solution 10 cm long, is reported. The specific rotation $[\alpha]$ is related to the observed rotation α in the following way:

$$[\alpha] = \alpha/cl, \text{ where } c = \text{concentration in grams per milliliter (g/mL)},$$
$$\text{and } l = \text{length of the tube in decimeters (dm)}$$

Sub- and superscripts are used to denote the wavelength and temperature of the measurement. Thus, $[\alpha]_D^{20}$ means the sodium D line was used and the measurement was made at 20 °C.

Light

Plane-polarized light

Tube containing sample of chiral molecule

Emerging light is rotated

Detector

Nicol prism

FIGURE 4.21 A highly schematic drawing of a polarimeter.

PROBLEM 4.6 Create a scheme like those in Figures 4.19 and 4.20 to show what happens when plane-polarized light is passed through a solution of racemic 3-methylhexane.

4.6   PROPERTIES OF ENANTIOMERS: CHEMICAL DIFFERENCES

Now let's look at how enantiomers differ chemically. We still can't discuss specific chemical reactions, but we can make some important general distinctions. Note that there is no difference in how the two enantiomers of 3-methylhexane interact with a "Spaldeen," an old-fashioned smooth pink rubber ball (Fig. 4.22).* Your left and right hands interact in exactly the same way with the featureless Spaldeen (Fig. 4.23). In both the right and left hand, the thumb and little finger are opposed by the same featureless pink rubber ball. Chemically, the Spaldeen represents an achiral molecule.

(R)-3-Methylhexane   Spaldeen   **(S)-3-Methylhexane**   Spaldeen

FIGURE 4.22 (R)- and (S)-3-methylhexane interact identically with the Spaldeen.

Spaldeen   Spaldeen

FIGURE 4.23 Right and left hands also interact identically with the Spaldeen.

*The inclusion of the term "Spaldeen" was the source of one of the few minor skirmishes between author and editor of this book. I maintained that a Spaldeen was part of our cultural heritage, and that all educated people would immediately know what one was. Perhaps I was overly optimistic. Here's what the editors of the *New Yorker* magazine replied when queried recently as to the meaning of the phrase "a kid bouncing a Spaldeen off a wall of the Boston Museum of Fine Arts":

"Spaldeens are the small rubber balls beloved by generations of New York stoopball and stickball players. In Boston, those balls are often known as "pinkies," but what would you have thought if we had written "bouncing a pinkie off a wall"?

WORKED PROBLEM 4.7 Actually, this is only true if the name has been worn off by too much stoopball playing. Show that the two enantiomers interact differently with a Spaldeen with the name "Spalding" still visible.

ANSWER In the approach shown, (R)-3-methylhexane interacts with the new Spaldeen to place the ethyl group near the "S" and the propyl group near the "G." These interactions are opposite in the (S) enantiomer; the propyl group approaches the "S" and the ethyl group approaches the "G."

(R)-3-Methylhexane

(S)-3-Methylhexane

You should be certain that this result is not caused by the particular approach chosen for the two objects. Try interacting 3-methylhexane by bringing it toward the Spaldeen "H and methyl first" or simply turn it over (H down, methyl up) in the figure above.

Figure 4.24 elaborates this picture by showing the two enantiomers of 3-methyl-hexane approaching an achiral molecule, propane.

FIGURE 4.24 The two enantiomers of 3-methylhexane interact in exactly the same way with the achiral molecule, propane.

In each instance, the 3-methyl group of 3-methylhexane approaches the hydrogen of propane and both the ethyl and propyl groups are opposite methyl groups of propane. One can *always* find exactly equivalent interactions for the two enantiomers and propane, no matter what approach you pick. Try it with models.

Now imagine your hands, or the two enantiomers of 3-methylhexane, interacting with a chiral object. The Spaldeen with the label will do, but let's use a more common object, such as the shell in Figure 4.25.

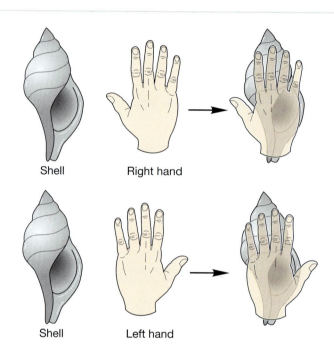

Shell   Right hand

Shell   Left hand

FIGURE 4.25 All sorts of common objects are chiral. A shell is one example. The interactions between a chiral object and two enantiomers—here a shell and two hands—are different. The enantiomeric right and left hand interact differently with the chiral shell.

PROBLEM 4.8 Convince yourself that the shell in Figure 4.25 is chiral.

As your right hand approaches the shell (Fig. 4.25), it is the little finger that opposes the opening, and the thumb that is directed away from it. For the left hand the situation is reversed: It is the thumb that is directed toward the opening and the little finger that is not. The interactions of the two hands with the shell are distinctly different.

PROBLEM 4.9 Convince yourself that turning the shell upside down doesn't change the situation.

For the molecular counterpart of this, look at the approach of the two enantiomers of 3-methylhexane to another chiral molecule, (*R*)-*sec*-butyl alcohol (Fig. 4.26). As the (*S*) enantiomer approaches, propyl is opposite methyl and ethyl is opposite H. For the other (*R*) enantiomer, the interactions are propyl with H and ethyl with methyl (see Problem 4.10). The two approaches are very different. Indeed, there is no approach to the chiral

FIGURE 4.26 The approach of (S)- and (R)-3-methylhexane to (R)-sec-butyl alcohol. The two approaches are very different and in sharp contrast to the interaction shown in Figure 4.24 with the achiral molecule propane.

alcohol that can give identical interactions for the two enantiomeric 3-methylhexanes. Try some. Contrast this situation with the approaches to the achiral propane shown before in Figure 4.24. The two stereoisomeric enantiomers interact *identically* with achiral propane but *differently* with the chiral molecule (R)-sec-butyl alcohol. This situation is general: Enantiomers have identical chemistries with achiral reagents but different chemistries with chiral ones.

WORKED PROBLEM 4.10 A reader of an early version of this chapter suggested that students would ask, quite correctly, Why can't I turn the picture of (R)-3-methylhexane upside down, and then the ethyl–hydrogen and propyl–methyl interactions with (R)-sec-butyl alcohol would be just like the interactions in the top half? Explain carefully why turning the molecule upside down doesn't produce superimposable mirror images.

ANSWER It is true that turning the molecule upside down will make two of the interactions the same. Propyl will be opposite CH_3 and C_2H_5 will be opposite H. However, *everything* else is wrong! No longer are CH_3 and OH being opposed, for example (red bonds). Simply turning (R)-3-methylhexane over does not produce the same interactions—or even similar interactions to those between (S)-3-methylhexane and (R)-sec-butyl alcohol.

A practical example of the consequences of chiral interactions is the different smells of enantiomeric compounds. One smells a molecule by binding it in a decidedly chiral receptor in the nose. There are many examples in which one enantiomer binds differently from the other or not at all. For example, (*R*)-carvone smells like spearmint and (*S*)-carvone like caraway (Fig. 4.27).

FIGURE 4.27
(*R*)- and (*S*)-Carvone.

(R)-(–)-Carvone
(spearmint)

(S)-(+)-Carvone
(caraway)

PROBLEM 4.11 Convince yourself that the assignment of (*R*) and (*S*) in Figure 4.27 is correct.

It is not at all easy to show (*R*)- and (*S*)-carvone fitting into different (and not yet exactly known) nasal receptors, but it is easy to show in principle how fit and non-fit works. Imagine a chiral binding site as the blue tetrahedron in Figure 4.28. The smaller red enantiomer fits perfectly inside, with the "proper" attachments, shown as 1–1, 2–2, 3–3, and 4–4. If we try the smaller green tetrahedron, the enantiomer of the red one, there is no possible way for the proper attachments to be achieved. The green tetrahedron cannot bind to the blue one with the proper 1–1, 2–2, 3–3, and 4–4 interactions.

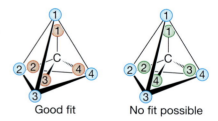

Good fit   No fit possible

FIGURE 4.28 The red tetrahedron fits inside the blue one with the proper fit but the green one does not.

4.7   INTERCONVERSION OF ENANTIOMERS BY MOBILE EQUILIBRIA: *GAUCHE* BUTANE

In Chapter 2 (Figs. 2.34 and 2.35, pp. 73–74), you analyzed the conformations of butane, finding an *anti* form lower in energy than a *gauche* form. Figure 4.29 shows the *anti* and *gauche* forms of butane in both framework and Newman projections. The

FIGURE 4.29

anti Form   *gauche* Form

anti- and *gauche*-Butane.

Newman projections clearly reveal the *gauche* methyl–methyl interaction. The *gauche* form of butane is chiral. It cannot be superimposed on its mirror image (Fig. 4.30).

FIGURE 4.30 The *gauche* form of butane is chiral! The mirror images are *not* superimposable. However, rapid rotation around carbon–carbon single bonds interconverts these "conformational enantiomers."

Why then is butane itself achiral? No matter how hard we try, no optical rotation can be observed for butane. The nearly free rotation about the carbon–carbon bond in butane serves to interconvert the two enantiomers of the *gauche* form. If we were able to work at the extraordinarily low temperatures required to "freeze out" the equilibrium between the two *gauche* forms we could separate butane into enantiomers (Fig. 4.30). The two *gauche* forms are **conformational enantiomers**. We will encounter this phenomenon again when we study the stereochemistry of cyclic compounds.

4.8 DIASTEREOMERS: MOLECULES CONTAINING MORE THAN ONE STEREOGENIC ATOM

We have just looked at the different interactions between a single enantiomer of a chiral molecule and an enantiomeric pair of molecules (Fig. 4.26). Let's take this interaction business to its limit and see what happens if we actually form bonds to produce new molecules. We can't worry yet about real chemical reactions—we are still in the tool-building stages—but that doesn't matter. We can imagine some process that will make an attachment, replacing the two X substituents with a chemical bond, shown in all succeeding figures with a blue screen (Fig. 4.31).*

If we follow this hypothetical process (which mimics some very real chemical reactions), we generate a new molecule containing two stereogenic carbons. First, let's see what happens when a single chiral molecule reacts with each isomer of a pair of enantiomers. It's worth taking a moment to see if we can figure out what's going to happen before writing out real molecules. Let's assume we are using the (*R*) form of our single enantiomer. It will react with the (*R*) form of the enantiomeric pair to generate a molecule we might call (*R*—*R*), and with the (*S*) form to generate (*R*—*S*) (Fig. 4.32).

FIGURE 4.31 A process that removes the two X substituents and joins the carbons. The new connection is shown in blue.

FIGURE 4.32 A schematic analysis of the reaction of one enantiomer with a racemic mixture of another molecule. The products of this reaction are stereoisomers but not enantiomers. They are called diastereomers.

The (**R**) form of a chiral molecule

A racemic mixture of another molecule 50% (*R*) and 50% (*S*)

The products are different molecules, diastereomers

(**R**) — (*R*) and (**R**)—(*S*) are stereoisomers, but not enantiomers!

PROBLEM 4.12 What is the mirror image of (*R*—*S*)? This question sounds insultingly simple, but most people get it wrong!

We can tell a lot even from the schematic coded picture of Figure 4.32. First of all, (*R*—*R*) and (*R*—*S*) are clearly neither identical nor enantiomers (nonsuperimposable mirror images), as the mirror image of (*R*—*R*) is (*S*—*S*), not (*R*—*S*).

*In this and many succeeding figures, the molecules are drawn in eclipsed forms for clarity. As you know, these arrangements are not stable and will be converted into staggered forms by rotation about the central carbon–carbon bond (Chapter 2, p. 64). None of the arguments presented is changed by this rotation, and in the staggered arrangements it is more difficult to visualize the relationships involved.

Yet these molecules are stereoisomers—they differ only in the arrangement of their constituent parts in space. There is a new word for such molecules—they are **diastereomers**, "stereoisomers that are not mirror images."

WORKED PROBLEM 4.13 If the two *gauche* forms shown in Figures 4.29 and 4.30 are enantiomers, what is the relationship between one *gauche* form and the *anti* form?

ANSWER The *gauche* forms are mirror images and, therefore, enantiomers. Each *gauche* form is a stereoisomer of the *anti* form, but not its mirror image. Stereoisomers that are not mirror images are diastereomers.

PROBLEM 4.14 We have seen diastereomers twice before. Show two different kinds of diastereomeric molecules. *Hint*: Start with a good definition of diastereomer.

Now let's make all this theory a little more real, and look at some molecules. Imagine a hypothetical reaction (X = Cl) in which the (*S*) enantiomer of 2-chlorobutane (*sec*-butyl chloride) reacts with the two enantiomers of 2-chloro-1,1,1-trifluoropropane (Fig. 4.33).

FIGURE 4.33 The hypothetical reaction of (*S*)-2-chlorobutane with two enantiomers of 2-chloro-1,1,1-trifluorochloropropane. Two stereoisomeric products, **A** and **B**, are formed. The products **A** and **B** are not mirror images (enantiomers) so they must be diastereomers.

Two new products, **A** and **B**, are produced from our hypothetical reaction. Look carefully at them. Are they identical? Certainly not. Are they enantiomers? No, again. They are stereoisomers, though, so they must be diastereomers.

PROBLEM 4.15 Verify the claims of the preceding paragraph. Are molecules **A** and **B** diastereomers?

The (R) form of 2-chlorobutane would also produce two new products, **C** and **D** (Fig. 4.34), which are neither identical nor enantiomeric, and thus must also be diastereomers.

FIGURE 4.34 The hypothetical reaction of (R)-2-chlorobutane with the same mixture of enantiomeric 2-chloro-1,1,1-trifluoropropanes to give another pair of diastereomers, **C** and **D**.

Now the question is, What are the relationships among **A**, **B**, **C**, and **D**? As Figure 4.35 shows, **A** and **D** are enantiomers and **B** and **C** are enantiomers. So all possible combinations of the two molecules have produced two pairs of enantiomers, or four stereoisomers in all. The (R) and (S) relationships are also shown in Figure 4.35.

FIGURE 4.35 Two pairs of enantiomers, **A** and **D** and **B** and **C**, are formed from this hypothetical reaction. Combinations **A** and **B**, **A** and **C**, **B** and **D**, and **C** and **D** are diastereomeric pairs. Each of these pairs is composed of two stereoisomers that are not mirror images.

As a simple theoretical analysis would predict (Fig. 4.36), we have formed all possible isomers in this combination [(S–R), **A**; (S–S), **B**; (R–S), **D**; and (R–R), **C**].

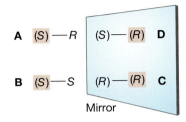

A (S)—R | (S)—(R) **D**

B (S)—S | (R)—(R) **C**

Mirror

FIGURE 4.36 (R)- and (S)-2-chlorobutane reacting with (R)- and (S)-2-chloro-1,1,1-trifluoropropane should give four isomers: two pairs of enantiomers.

PROBLEM 4.16 (S)-2-Chlorobutane and (S)-2-chloro-1,1,1-trifluoropropane (Fig. 4.33) are used to produce isomer **A**, which is properly designated as the (S,R) isomer (Fig. 4.35). Check that there is no mistake in this analysis, and then explain how the two (S) compounds can produce the (S,R) isomer, **A**.

We might be able to generalize even at this early stage. A molecule with one stereogenic center has two possible stereoisomers, a single pair of enantiomers, the (R) and (S) forms. A molecule containing two stereogenic centers gives two pairs of enantiomers, or four stereoisomers in all. It seems that for n stereogenic carbons, we get 2^n isomers. We can immediately predict that $n = 3$ stereogenic atoms will produce 2^3 or 8 isomers. This is easy to check—what isomers are possible? The easiest way to tell is not to examine some complicated molecule and thus incur the task of drawing a large number of stereoisomers, but to analyze the problem in a simpler fashion. A stereogenic carbon can only be (R) or (S). Figure 4.37 shows the possibilities for three such atoms.

(RRR)	All (R)
(RRS) (RSR) (SRR)	Two (R), one (S)
(SSR) (SRS) (RSS)	One (R), two (S)
(SSS)	All (S)

FIGURE 4.37 The eight possible stereoisomers for a molecule containing three stereogenic atoms.

The eight possible isomers appear easily in this coded form. Note that there is nothing in this analysis that requires the stereogenic atoms to be adjacent—they need only be in the same molecule.

This procedure reliably gives you the *maximum* number of possible stereoisomers. There are some interesting (and important) special cases, however, and this very general method won't spot them. Let's imagine using our hypothetical chemical coupling reaction again, but this time combining two molecules of *sec*-butylchloride. This procedure is exactly the same as the one we used before to generate the four stereoisomers of Figures 4.33 and 4.34, **A**, **B**, **C**, and **D**. We allow our (R) and (S) starting materials to combine in all possible ways (Fig. 4.38).

FIGURE 4.38 All possible combinations of (*R*)- and (*S*)-2-chlorobutane (*sec*-butyl chloride) produce four isomers, **E**, **F**, **G**, and **H**.

In the previous example, we found that we had made two pairs of enantiomers, (*S*–*R*) and (*R*–*S*), **A** and **D**, and (*S*–*S*) and (*R*–*R*), **B** and **C** (Figs. 4.35 and 4.36). This time one pair of enantiomers, **E** and **H**, does appear (Fig. 4.39).

FIGURE 4.39 Here **E** and **H** are enantiomers.

However, the second potentially enantiomeric pair of molecules, **F** and **G**, are in fact identical species—a **meso compound** (Fig. 4.40)! So, in this example we get only three stereoisomers, not the maximum number of $2^2 = 4$. Clearly, we are one stereoisomer short because of the identity of **F** and **G**. This molecule, **F** = **G**, contains stereogenic carbons but is not chiral. Such a molecule is called a meso compound.

FIGURE 4.40 The **F** and **G** molecules are identical. This compound is meso.

Twice before we have seen molecules that for different reasons give no observable rotation of plane-polarized light. Achiral molecules cannot induce optical rotation. A racemic mixture of chiral molecules contains equal numbers of right- and left-handed compounds, which rotate the light equally in clockwise and counterclockwise directions, respectively, for a net zero rotation. Although a meso compound does contain stereogenic atoms, it is achiral and induces no optical rotation.

We just answered the second part of our earlier (p. 155) question. No, "four different groups attached to a stereogenic carbon" is not a sufficient condition for optical activity. Our meso compound **F** = **G** certainly has carbons to which four different groups are bonded, but it is equally clearly not chiral. Even though searching for carbons attached to four different groups may be a good way to begin a hunt for chirality, one must be careful. Finding such a carbon does not guarantee we have found a chiral molecule, unless we are certain that the molecule contains one and only one such carbon. Alas, it is also the case that not finding such a carbon does not guarantee that there will be no chirality (for examples, see p. 184). The ultimate test remains the superimposability of mirror images. If the mirror images are superimposable, the molecule is not chiral; if the mirror images are different, nonsuperimposable, the molecule is chiral.

When will meso compounds appear? Meso compounds occur when there is a plane or point of symmetry in the molecule. In effect, this divides the molecule into halves, which contribute equally and oppositely to the rotation of plane-polarized light. Compound **F** = **G** provides an example of a molecule with a plane of symmetry in the eclipsed forms we have been using, and a point of symmetry if we draw the molecule in its energy minimum, staggered arrangement (Fig. 4.41).

FIGURE 4.41 Two views of the meso compound **F** = **G**.

PROBLEM 4.17 Draw all the stereoisomers of the following molecules in which there is free rotation about carbon–carbon single bonds. Some may be achiral:

(a) 2,2,3,3-tetrabromobutane

(b) 2,2-dibromo-3,3-dichlorobutane

(c) 2,3-dibromo-2,3-dichlorobutane

(d) 2,3-dibromo-2-chloro-3-fluorobutane

Hint: It is easier to analyze these molecules in their eclipsed forms.

PROBLEM 4.18 How does the situation change if there is no free rotation about carbon–carbon single bonds for 2,3-dibromo-2-chloro-3-fluorobutane (see Problem 4.17d)?

PROBLEM 4.19 Identify each stereogenic carbon in the compound of Problem 4.17c as either (R) or (S).

4.9   PHYSICAL PROPERTIES OF DIASTEREOMERS: OPTICAL RESOLUTION

The formation of diastereomers allows the separation of enantiomers. Separation of enantiomers, called **resolution**, is a serious experimental difficulty. So far we have ignored it. Enantiomers have identical physical properties (except for the ability to rotate the plane of plane-polarized light), and one might legitimately wonder how in the world we are ever going to get them apart. At several points we used a single enantiomer without giving any hint of how a pair of enantiomers might be separated. The key to this puzzle is that diastereomers, unlike enantiomers, have different physical properties—melting point, boiling point, and so on.

One general procedure for separating enantiomers is to allow them to react with a naturally occurring chiral molecule to form a pair of diastereomers. These can then be separated by taking advantage of one of their different physical properties. One typically can separate such a pair by crystallization, because the members of the diastereomeric pair will have different solubilities. Then, if the original chemical reaction can be reversed, we have the pair of enantiomers separated. Figure 4.42 outlines the general scheme and begins with a schematic recapitulation of Figure 4.33, which first described the reaction of a single enantiomer with a racemic mixture to give a pair of diastereomers. Be sure to compare the two figures.

One enantiomer

A racemic pair of molecules

A pair of diastereomeric products

separate physically

reverse

reverse

Now we have separated both enantiomers of the original racemic mixture

FIGURE 4.42 Resolution is a general method for separating the constituent enantiomers of a racemic mixture. A single enantiomer of a chiral molecule is used to form a pair of diastereomers, which can be separated physically. If the original chemical reaction can be reversed, the enantiomers can be isolated.

It is not even necessary to form covalent bonds. For example, in the traditional method for separating enantiomers of organic acids, optically active nitrogen-containing molecules, called **alkaloids** (see Section 6.11), are used to form a pair of diastereomeric salts, which can then be separated by crystallization.

The term alkaloid refers to any nitrogen-containing compound extracted from plants, although the word is used loosely and some compounds of nonplant origin are also commonly known as alkaloids. These biologically active compounds are basic like all amines, and it is this basicity that led to the name. Presumably, basicity was also important in the relative ease of extraction of alkaloids from the myriad compounds present in an organism as complex as a plant. Extraction of plant mass with acid yields any amines present as water-soluble ammonium ion salts from which they can easily be regenerated. Figure 4.43 shows this process for a simple alkaloid known as coniine. Coniine is the active ingredient in hemlock, and it is the physiological activity of coniine that led to the demise of Socrates.

Plant (hemlock) $\xrightarrow[\text{H}_3\text{O}^+]{\text{H}_2\text{O}}$ **Ammonium ion (water soluble)** — $\text{CH}_2\text{CH}_2\text{CH}_3$ **base**   **(S)-Coniine** — $\text{CH}_2\text{CH}_2\text{CH}_3$

FIGURE 4.43 A scheme for isolation of alkaloids from plant materials.

These days, this general procedure has been extended so that all manner of enantiomeric pairs can be separated by chromatography. In such a technique, covalent chemical bonds are not formed, as they are not in the salt formation shown in Figure 4.45. Rather, advantage is taken of the formation of partial bonds—complexes—as the pair of enantiomers passes over an optically active substrate. The optically active substrate (X*) forms a complex with the enantiomers A and A'. These complexes are diastereomers and have different physical properties, including bond strengths of A ⋯ X and A' ⋯ X. Both AX and A'X are in equilibrium with the free enantiomers, and these equilibria will be different for the two diastereomeric complexes. Therefore, A and A' will move through the column at different rates and emerge at different times (Fig. 4.44).

A racemic mixture of enantiomers A and A'

A column packed with X*, a chiral material

FIGURE 4.44 Separation of enantiomers through chromatography.

Other alkaloids have wondrously complex structures. Two further examples, brucine and the notorious strychnine, are shown in Figure 4.45 along with the general procedure for resolution.

FIGURE 4.45 Two alkaloids, brucine and strychnine, are commonly used to separate the enantiomers of chiral organic acids. Diastereomeric salts are first formed, then separated by crystallization, and the individual enantiomeric acids are regenerated.

WORKED PROBLEM 4.20 Identify with an asterisk (*) all the stereogenic carbons in brucine.

ANSWER

PROBLEM 4.21 Identify each stereogenic carbon in brucine as (R) or (S).

4.10   DETERMINATION OF ABSOLUTE CONFIGURATION (*R* OR *S*)

Now that we have achieved the separation of our racemic mixture of enantiomers into a pair of optically active stereoisomers, we face the difficult task of finding out which enantiomer is (*R*) and which is (*S*). This problem is not trivial! Indeed, in Chapter 23, when we deal with sugars, we'll find that until rather recently, there was *no* way to be certain, and one just had to guess (correctly in the case of the sugars, it turns out). One would like to peer directly at the structures, of course, and under some circumstances this is possible.

   X-ray crystallography can determine the relative positions of atoms in a crystal, and a special kind of X-ray diffraction called "anomalous dispersion" can tell the absolute configuration of the molecule. But this technique is not generally applicable—one needs a crystalline compound, for example. It does serve to give us some benchmarks, though. If we know the absolute configuration (*R* or *S*) of some compounds, we may be able to determine the absolute configurations of other molecules by relating them to the few compounds of known absolute configuration. We must be very careful, however. The chemical reactions that interconvert the molecules of known and unknown absolute configurations must not alter the stereochemical arrangement at the stereogenic atoms, or if they do, it must be in a known fashion. How do we know whether a given chemical reaction will or will not change the stereochemistry? We need to know the reaction mechanism—to know how the chemical changes occur—in order to answer this question. This reason is just one of many for the study of reaction mechanisms, and we'll take great pains to examine carefully the stereochemical consequences of the various chemical reactions we encounter. Figure 4.46 shows the general case, as well as examples of two chemical reactions that do not alter stereochemistry at the stereogenic carbon.

If we know the absolute configuration (*R* or *S*) of the starting material and the mechanism of the reaction, we can know the absolute configuration of the product

In both reactions, the absolute configuration (*R* or *S*) of the carbon marked with the asterisk is unchanged

FIGURE 4.46 Two reactions that do not change the configuration at an adjacent carbon (*).

SUMMARY

To know the absolute configuration of the product in the reaction in Figure 4.46 you must know two things: the absolute configuration of the starting material *and* the mechanism of the reaction with Y. Only if these two things are known will the absolute configuration of the product be known with certainty.

4.11   STEREOCHEMICAL ANALYSIS OF RING COMPOUNDS (A BEGINNING)

Most of the stereochemical principles in the previous sections apply quite directly to ring compounds. We'll start with a simple question. How many isomers of chlorocyclopropane exist? That looks like an easy question, and it is. There is only one (Fig. 4.47).

FIGURE 4.47 Different representations of the single isomer of chlorocyclopropane.

To be certain, however, we should examine this compound for chirality. Again this process is simple. As Figure 4.48 shows, the mirror image is easily superimposed on the original, once again through a 120° rotation. Chlorocyclopropane is an achiral molecule.

FIGURE 4.48 Chlorocyclopropane is superimposable on its mirror image and is, therefore, an achiral molecule.

How many isomers of dichlorocyclopropane exist? This question is much tougher than the last one. There are four isomers of dichlorocyclopropane. Most people don't get them all at this point, so it's worth going carefully through an analysis. Where can the second chlorine be placed in the single monochlorocyclopropane? As Figure 4.49 shows, there are only two places, the same carbon as the first chlorine, or one of the two equivalent adjacent carbons. This analysis yields two compounds called **structural isomers**.

FIGURE 4.49 Replacement of one hydrogen of chlorocyclopropane with another chlorine can occur in only two places, the position to which the first chlorine is attached or one of the other two equivalent carbons. This procedure generates the structural isomers 1,1-dichlorocyclopropane and 1,2-dichlorocyclopropane.

But that only brings us to two isomers; we're still two short of reality. As first mentioned in Chapter 2 (p. 84), rings have sides. We can find one more isomer if we notice that the second chlorine can be on the same side (the cis isomer) or the opposite side (the trans isomer) of the ring (Fig. 4.50). These compounds are stereoisomers, more precisely, diastereomers (stereoisomers that are not mirror images). Now we are just one isomer short.

Diastereomers

1,2-Dichloro-cyclopropane   ***cis*-1,2-Dichloro-cyclopropane**   ***trans*-1,2-Dichloro-cyclopropane**

FIGURE 4.50 1,2-Dichlorocyclopropane can exist in cis and trans forms. In the figure, all the ring hydrogens have been drawn in to help you see the difference between these two molecules. Be absolutely certain you see why these two molecules cannot be interconverted. Use models.

Once we are sure we have found all the different positions on which to put the two chlorines it's time to look for chirality. Figure 4.51 shows that neither 1,1-dichlorocyclopropane nor *cis*-1,2-dichlorocyclopropane is chiral. The mirror images are easily superimposable on the originals. Note that *cis*-1,2-dichlorocyclopropane is a meso compound. It contains stereogenic carbons but is superimposable on its mirror image, and therefore is achiral.

1,1-Dichloro-cyclopropane   Mirror

***cis*-1,2-Dichloro-cyclopropane**
(a meso compound)   Mirror

FIGURE 4.51 Both 1,1-dichlorocyclopropane and *cis*-1,2-dichlorocyclopropane are achiral.

trans-1,2-Dichlorocyclopropane is different. In this molecule, the mirror image is not superimposable on the original and this stereoisomer is chiral (Fig. 4.52).

FIGURE 4.52 *trans*-1,2-Dichlorocyclopropane is *not* superimposable on its mirror image. It is a chiral molecule and, therefore, two enantiomers exist. This stereochemical relationship is probably the most difficult you've encountered so far, so be sure you can see why no manipulation of one enantiomer will make it superimposable on the other.

***trans*-1,2-Dichloro-cyclopropane**

Mirror

So now we have the four isomers: 1,1-dichlorocyclopropane, *cis*-1,2-dichlorocyclo-propane, *trans*-(1*S*,2*S*)-dichlorocyclopropane, and *trans*-(1*R*,2*R*)-dichlorocyclopropane (Fig. 4.53).

FIGURE 4.53 The four isomers of dichlorocyclopropane.

1,1-Dichloro-cyclopropane   ***cis*-1,2-Dichloro-cyclopropane**   ***trans*-1,2-Dichlorocyclopropane**

PROBLEM 4.22 Verify the assignment of absolute configuration in the two enantiomers of *trans*-1,2-dichlorocyclopropane in Figure 4.53.

PROBLEM 4.23 Identify the stereochemical relationships between the four isomers of Figure 4.53. Find all the pairs of enantiomers and diastereomers.

This section illustrates one procedure for working out an isomer problem, but any rational technique will work. The important thing is to devise some systematic way of searching. I happen to like the one above in which one first finds all the structural isomers (in this case 1,1- and 1,2-dichlorocyclopropane), then searches for stereoisomerism of the cis/trans type, and finally examines each isomer for chirality. But it doesn't matter what procedure you use as long as you are systematic. What will not work is a nonrational scheme. So don't start a "find the isomers" problem by just writing out compounds without thinking first. No one can do it that way.

PROBLEM 4.24 There are six isomers of dichlorocyclobutane. Find them all and name them carefully.

We started with small rings for a reason—the inflexibility of small rings makes them flat, or very nearly so. Three points determine a plane, so cyclopropane must be planar. Cyclobutane need not be absolutely flat, and we can see that it is not if we take the time to make a model. At the same time, it cannot be far from planar. Larger rings are more complicated though, and we'll defer an examination of their stereochemistry until Chapter 5, which looks at a number of structural questions about rings.

4.12 SUMMARY OF ISOMERISM

The words isomer and stereoisomer are often tossed around quite loosely, and it's worth a bit of review. Structural isomers (often called *constitutional isomers*) are molecules of the same formula, but differing connectivity. Their constituent parts may well be different (but need not be). Butane (two methyl groups and two methylene groups) and isobutane (three methyl groups and one methine group) are typical examples of structural isomers (Fig. 4.54).

Stereoisomers have the same connectivity, but differ in the arrangement of their parts in space. Two kinds of stereoisomers exist: enantiomers and diastereomers. Enantiomeric molecules are nonsuperimposable mirror images of each other. Simple examples are (R)- and (S)-3-methylhexane. Slightly more complicated are (2S,3S)-dichlorobutane and its enantiomer, (2R,3R)-dichlorobutane. These pairs are not structural isomers, since they have the same connectivity (Fig. 4.55).

FIGURE 4.54 Two structural isomers.

FIGURE 4.55 Some examples of stereoisomers.

PROBLEM 4.25 Find a pair of structural isomers whose constituent parts are not different.

Diastereomers include all stereoisomers that are not mirror images; *cis*- and *trans*-2-butene are examples, as are *meso*-2,3-dichlorobutane and either the (R,R) or (S,S) isomer of 2,3-dichlorobutane in Figure 4.55 (Fig. 4.56).

FIGURE 4.56 Some examples of diastereomers.

Stereoisomers also include conformational isomers, in which different isomers are generated through rotations about bonds. Eclipsed and staggered ethane are typical examples. Note that a conformational isomer need not be an energy minimum—the eclipsed conformation of ethane is an energy maximum, for example. Conformational isomers can be either enantiomeric or diastereomeric. The two *gauche* forms of butane are conformational enantiomers, but the *gauche* and the *anti* forms of butane are conformational diastereomers (Fig. 4.57).

FIGURE 4.57 Some examples conformational isomers.

WORKED PROBLEM 4.26 Show that the two conformers labeled "enantiomers" in Figure 4.57 really are mirror images.

ANSWER The trick is to find a way to make an easy comparison. In this case, it takes only a simple 180° rotation to put the molecules into the proper orientation. Now it is easy to see that the two molecules from Figure 4.57 [labeled (a) and (b)] are indeed mirror images.

4.13 SPECIAL TOPIC: CHIRALITY WITHOUT "FOUR DIFFERENT GROUPS ATTACHED TO ONE CARBON"

Is the presence of a carbon atom attached to four different groups a necessary condition for chirality? No. We asked this question first on page 155. This section answers it by giving a number of examples of chiral molecules without this kind of atom.

trans-Cyclooctene is chiral, yet it contains no carbon atom attached to four other groups. Figure 4.58 gives one representation of the nonsuperimposable mirror images of this molecule, but you may need to use models to convince yourself that two enantiomers really do exist. Be careful! This molecule is racemized by rotation around carbon–carbon bonds, and your models are likely to be eager to do exactly that.

Two double bonds can be connected directly to each other. We will see these **allenes** again in detail in Chapter 12, but you are well equipped after this chapter to understand the bonding and geometry of an allene (Fig. 4.59). In summary, *trans*-cyclooctane and substituted allenes can be chiral, although these molecules have no "carbons attached to four different groups."

trans-**Cyclooctene**   Mirror

FIGURE 4.58 *trans*-Cyclooctene is chiral.

$$H_2C = C = CH_2$$
Allene

FIGURE 4.59

WORKED PROBLEM 4.27 What is the hybridization of each carbon of allene (see Fig. 4.59)?

ANSWER The end carbons are attached to three groups (two hydrogens and the other carbon), and are therefore hybridized sp^2. The central carbon is attached to two other groups (the two methylene groups), and therefore is hybridized sp.

WORKED PROBLEM 4.28 Make a careful, three-dimensional drawing of 1,3-dimethylallene. Pay special attention to the geometry at the ends of the molecule.

ANSWER The key to this problem is to notice that overlap of the $2p$ orbitals on the end sp^2 carbons with the $2p$ orbitals on the central, sp hybridized carbon requires that the groups on the end carbons be in perpendicular planes.

WORKED PROBLEM 4.29 Draw the mirror image of 1,3-dimethylallene and verify that the two enantiomers are not superimposable.

ANSWER The mirror images of 1,3-dimethylallene are not superimposable, although it may take some work with models before you are certain!

Top view

Side view

FIGURE 4.60 Two views of benzene.

There exists a flat molecule, C_6H_6, called benzene. Benzene consists of a six-membered ring of carbons to which hydrogens are attached, one at each vertex (Fig. 4.60). The key point is that benzene is flat as a pancake (for much more on benzenes see Chapters 13 and 14). Two benzenes can be attached to each other to make a molecule called "biphenyl" (Fig. 4.61).

Junction bond

Biphenyl

Mirror

FIGURE 4.61 Biphenyl and a substituted (chiral) biphenyl.

PROBLEM 4.30 Show that rotation about the junction bond will racemize the two enantiomers of Figure 4.61.

Benzene

Naphthalene

Anthracene

FIGURE 4.62 Some fused benzenes.

If we now substitute the four carbons directly adjacent to the junction between the two rings in the correct way, a chiral molecule can be achieved! It is necessary that the substituents be large enough so that rotation about the junction between the two rings is restricted (Fig. 4.61).

Finally, we can use benzene in another way to construct an odd chiral molecule. Two benzenes can be attached in what is called a "fused" fashion (Fig. 4.62).

We can imagine continuing this process in several ways. We might simply continue to add benzene rings in a linear fashion, as in Figure 4.62. Or, we might add the restriction that new rings be added so as to produce a curve (Fig. 4.63).

If we follow the latter course, the addition of the sixth ring produces a chiral molecule!

Benzene   **Naphthalene**   **Phenanthrene**   **3,4-Benzphen-anthrene**   **Dibenzo[c,g]phen-anthrene**   **Hexahelicene**
phenanthro[3,4-c]phenanthrene

FIGURE 4.63 More fused benzenes. Hexahelicene is chiral. Why?

PROBLEM 4.31 Explain why hexahelicene is chiral. You may need models to do this task.

4.14   SPECIAL TOPIC: STEREOCHEMISTRY IN THE REAL WORLD; THALIDOMIDE, THE CONSEQUENCES OF BEING WRONG-HANDED

You should be able to spot the stereogenic center in the formidably named (*N*-phthalidomido)glutarimide (thalidomide) shown in Figure 4.64. You should also be able to write structures for the (*R*) and (*S*) enantiomeric forms.

The stereogenic carbon

(*R*)-Thalidomide   (*S*)-Thalidomide

FIGURE 4.64 The two enantiomers of thalidomide.

Thalidomide was marketed as a sedative, and as a palliative for morning sickness in the 1950s, mostly in Europe. Let's see why this compound, ironically now used (carefully) in a number of other ways including cancer therapy and as an antileprosy agent, is so notorious. Much biological activity depends on fit—the accommodation of a molecule in a receptor into which it fits exactly. That perfect fit depends critically on the details of shape. As enantiomers have different shapes, they are often bound very differently by receptor sites. Sometimes the difference is not very important, as with carroway and spearmint, the enantiomers of carvone (p. 168). But sometimes the difference is critical. The defining example for our time is that of thalidomide. The (*R*) enantiomer is at least harmless, but the (*S*) enantiomer is quite another story. This enantiomer is a most potent teratogen, causing a very high incidence of fetal anomalies. At the same time it operates as an antiabortive agent, thus diminishing the body's tendency to purge abnormal fetuses. Why should this matter? Thalidomide was marketed as the racemic mixture, not the pure (*R*) enantiomer, and those taking the drug were automatically ingesting the bad along with the good.*

*There is much more to the story of thalidomide, and it is not a nice tale at all. If you would like to read more about the extraordinarily shabby science that went into the "testing" of this drug, and the even shabbier sociology, see Roald Hoffmann, *The Same and Not the Same*, Columbia University Press, 1995, Chapter 20. Here you will also encounter Dr. Frances Kelsey, the FDA physician who played a major role in preventing thalidomide from being licensed in the United States. She is a real hero of our time.

4.15   SUMMARY

New Concepts

This chapter deals almost exclusively with the concept and consequences of the "handedness" of molecules—chirality. Some molecules are related to each other as are your left and right hands. These are nonsuperimposable mirror images, or enantiomers.

An absolutely safe method of finding chiral molecules is to examine the mirror image to see if it is superimposable on the original. In particular, finding a carbon attached to four different groups is neither a necessary nor sufficient condition for chirality. A meso compound is an example of an achiral molecule containing carbons attached to four different groups.

Enantiomers have identical physical properties except that one stereoisomer rotates the plane of plane-polarized light by some amount to the right, whereas the other rotates the plane by the same amount but in the opposite direction. Enantiomers have identical chemistries with achiral molecules, but interact differently with other chiral molecules.

An examination of the consequences of the presence of more than one stereogenic carbon in a single molecule leads to the discovery of diastereomers; stereoisomers that are not mirror images.

Key Terms

absolute configuration (p. 157)
achiral (p. 153)
alkaloid (p. 177)
allenes (p. 185)
chiral (p. 152)
chirality (p. 153)
conformational enantiomers (p. 169)

dextrorotatory (p. 155)
diastereomers (p. 171)
enantiomers (p. 155)
levorotatory (p. 155)
meso compound (p. 174)
optical activity (p. 155)
polarimeter (p. 162)

racemate (p. 156)
racemic mixture (p. 156)
resolution (p. 176)
stereochemistry (p. 151)
stereogenic center (p. 157)
structural isomers (p. 180)

Reactions, Mechanisms, and Tools

The arbitrary, but important (*R/S*) convention is based upon the Cahn–Ingold–Prelog priority system and allows us to specify absolute configuration of molecules.

Resolution, or separation of enantiomers, is generally accomplished by allowing the racemic mixture of enantiomers to react with a single enantiomer of a chiral agent to form a pair

of diastereomers. Diastereomers, unlike enantiomers, have different physical properties and can be separated by crystallization or other techniques. If the original chemical reaction that formed the diastereomers can be reversed, the pure enantiomers can be regenerated.

Common Errors

This chapter continues the journey into three dimensions begun in Chapter 3. Most commonly encountered problems have to do with learning to see molecules in three dimensions and in particular, with the difficulty of translating from the two-dimensional page into the three-dimensional world. There can be no hiding the difficulty of this endeavor for many people, but it will yield to careful work and practice.

There are a few small errors to be careful to avoid. There are many arbitrary parts to the (*R/S*) convention, and there is no way to figure them out. The (*R/S*) convention is something that must be memorized.

When a single enantiomer is drawn, be sure to ask yourself whether this is done deliberately, in order to specify *that particular enantiomer*, or simply as a matter of convenience to avoid the work of drawing both enantiomers of the racemic pair. Usually some indication will be made if it is indeed a single enantiomer that is meant.

Remember: Finding a carbon with four different groups attached is a fine way to *start* looking for chirality, but it is neither a sufficient nor necessary condition. Problem makers love to find exotic molecules that prove this point!

4.16   ADDITIONAL PROBLEMS

PROBLEM 4.32   Find all the chiral isomers of the heptanes (Figs. 2.47–2.50, pp. 81–82). Designate the stereogenic atoms with an asterisk (*).

PROBLEM 4.33   Draw the (S) enantiomers of these compounds.

PROBLEM 4.34   Find all the chiral isomers of the octanes (Problem 2.26, p. 82).

PROBLEM 4.35   Draw the (R) enantiomers of all of these compounds whose name must end in "pentane."

PROBLEM 4.36   Now find all the chiral isomers of the nonanes (Problem 2.52, p. 95). Indicate those with more than one stereogenic carbon.

PROBLEM 4.37   Draw three-dimensional pictures of all the stereoisomers of 3,4-dimethylheptane and 3,5-dimethylheptane. It is probably easiest to draw them in the eclipsed arrangement, even though this is not a low-energy conformation. Determine the absolute configuration (R or S) of each stereogenic carbon. Designate the stereogenic carbons with an asterisk.

PROBLEM 4.38   You are given a bottle labeled 3,4-dimethylheptane. Presumably, all of the molecules you found in your answer to Problem 4.37 are in that bottle. How many signals will appear in the ^{13}C NMR spectrum of the contents of the bottle?

PROBLEM 4.39   You are given a bottle labeled 3,5-dimethylheptane. Presumably, all of the molecules you found in your answer to Problem 4.37 are in that bottle. How many signals will appear in the ^{13}C NMR spectrum of the contents of the bottle?

PROBLEM 4.40   Find all the chiral isomers of the hexenes (Problem 3.5, p. 110) and heptenes (Problem 3.37, p. 148).

PROBLEM 4.41   Find all the chiral isomers of the hexynes (Problem 3.23, p. 129), heptynes (Problem 3.23, p. 129), and octynes (Problem 3.38, p. 148).

PROBLEM 4.42   In Problem 2.42, you drew all the isomers of "chlorohexane" $C_6H_{13}Cl$. Which of these isomers contain stereogenic carbons? Be sure to be alert for molecules containing more than one stereogenic carbon. Designate the stereogenic carbons with an asterisk.

PROBLEM 4.43   Draw three-dimensional pictures of the stereoisomers of the molecules in your answer to Problem 4.42 that contain more than one stereogenic carbon.

PROBLEM 4.44   Which of the following molecules can exist as cis/trans (Z/E) isomers? Draw the (E) and (Z) forms. Which molecules possess stereogenic carbons? Designate the stereogenic carbons with an asterisk.

(a) $(CH_3)_2C\!=\!CHCH_2CH_2CH(CH_3)CH_2CH_2OH$

(b) $(CH_3)_2C\!=\!CHCH_2CH_2C(CH_3)\!=\!CHCH_2OH$

(c)   (d)

PROBLEM 4.45   Draw the (R) enantiomer for each molecule containing a stereogenic carbon in Problem 4.44.

PROBLEM 4.46   Draw all the stereoisomers of the following molecule. Identify all pairs of enantiomers and diastereomers.

PROBLEM 4.47   In Problem 3.45 (p. 148), you worked out all the isomers of the formula C_4H_5Cl. Find all the chiral isomers of the ring compounds of this formula. Indicate stereogenic carbons with an asterisk.

PROBLEM 4.48   Draw the four chiral isomers found in Problem 4.47, show the mirror images, and indicate the absolute configuration (R or S) of the stereogenic carbons.

PROBLEM 4.49   In Problem 3.43 (p. 148), you found 10 cyclopropanes of the formula $C_4H_6Br_2$. Find the chiral ones. Indicate the stereogenic carbons with an asterisk.

PROBLEM 4.50   One of the isomers in the answer to Problem 4.49 has a pair of bromines on one carbon. Draw the stereoisomers of this molecule, and use the (*R/S*) convention to show the absolute configuration of the stereogenic carbons.

PROBLEM 4.51   In Dorothy Sayers' and Robert Eustace's mystery novel, *The Documents in the Case* (Harper Paperbacks, 1995), mushroom expert George Harrison is murdered by Harwood Latham. Latham doses Harrison's stew of *Aminita rubescens* with muscarine, the toxin of the related, but deadly, *Aminita muscaria*. Lathom hopes to make it look like an accidental poisoning through a mistaken identification by Harrison. See if you can guess how Latham is caught. Here is a clue: Latham obtains the poison muscarine not from the mushroom itself, but from a laboratory in which it has been synthesized by an unwitting accomplice from simple starting materials. Thanks go to Ms. Dana Guyer, a former student of organic chemistry at Princeton, for finding this book.

Muscarine

Rings

5

Clearly the ring had an unwholesome power . . .

—J. R. R. Tolkien*
The Lord of the Rings

5.1   PREVIEW

In these early chapters, we have used our powers of imagination to picture the three dimensionality of organic molecules. But nowhere is thinking in three dimensions more important than in depicting the myriad structures formed from chains of atoms tied into rings. It's not always easy to see these structures clearly, so do not be reluctant to work with models, especially at the beginning. All organic chemists use them. Polycyclic compounds are especially complicated. No one can see all the subtlety of these compounds without models. Much of chemistry involves interactions of groups in proximity, and the two-dimensional page is often quite ineffective in showing which atoms are close to others. As you go on in this subject, your ability to use the two-dimensional surface to depict the three-dimensional molecular world will increase, but none of us will ever outgrow the need for models. Rings were first encountered in Section 2.12, and Figure 5.1 recalls some simple and complex ring compounds.

$H_2C \overset{\displaystyle CH_2}{\underset{\displaystyle CH_2}{}} CH_2$		
Two schematic views of cyclopentane	A complicated, polycyclic compound, bicyclo[2.2.1]heptane	A really complicated, polycyclic compound that nobody tries to name systematically — it's called "dodecahedrane"

FIGURE 5.1 Some simple and complex ring structures.

*John Ronald Reuel Tolkien (1892–1973) was Merton Professor of English language at Oxford University.

ESSENTIAL SKILLS

1. This chapter is entirely about the structures of rings. Drawing those rings accurately is an essential skill for any chemist. The reason is not only that ring compounds abound in Nature, but that the positions of groups attached to rings can be known quite well. If you know for certain the *shape* of the ring compound, you also know where the groups attached to it are located. The distance between groups and the orientation of groups can be controlled in this way . . . *if* you know the ring structure.

2. Many rings are highly mobile. One example of this phenomenon is the ring "flip" of cyclohexane. You must be able to "flip" cyclohexanes and draw the two interconverting chair forms well.

3. You have to be able to identify two sources of instability in rings. Eclipsing or torsional strain, and angle strain both act to make rings less stable and thus, more reactive.

IMPORTANT DETAILS

1. It is really more than a detail, but drawing the chair form of a cyclohexane perfectly (well, at least very accurately) is a necessary skill. Fortunately, this skill is one of those things that looks hard and is easy. Just follow the directions in Section 5.4.

2. The distinction between an intermediate and a transition state reappears in this chapter. An intermediate is a potentially isolable compound in an energy well. However, often it is high in energy and thus not easy to isolate. On the other hand, a transition state is the top of an energy hill—the top of the barrier separating two compounds in energy wells—and cannot be isolated.

5.2   RINGS AND STRAIN

Much can be learned from the simplest of all rings, cyclopropane, $(CH_2)_3$. Indeed, even at this early stage we can make some preliminary judgments about this compound. In doing so, we would be recapitulating the thoughts of Adolf von Baeyer (1835–1917) who in 1885 appreciated the importance of the deviation of the internal angles in most cycloalkanes from the ideal tetrahedral angle, 109.5°. He suggested that the instabilities of the ring compounds should parallel the deviations of their internal angles from the ideal; the further the angle from tetrahedral, the less stable the molecule. The idea that **angle strain** is an important factor in ring stability has continued to this day (Fig. 5.2). In fact, this smallest ring compound, cyclopropane, represents an

FIGURE 5.2 Ring compounds will have angle strain if angles are forced to be significantly smaller or larger than the ideal 109.5°.

A normal tetrahedron with a bond angle of 109.5°

A small ring will constrict this angle to less than 109.5°

A large ring will expand this angle to more than 109.5°

extreme example of the effects of angle strain. The strain induced by the reduction of the ideal tetrahedral angle of 109.5° to 60° is so great as to make other bonding arrangements possible. One way of looking at cyclopropane views the bonds between the carbons as "bent." The *internuclear* angle in an equilateral triangle is, of course, 60° and cannot be otherwise. However, the *interorbital* angle is apparently quite a bit larger (Fig. 5.3) and is estimated at 104°. Thus, the internuclear angle (60°) and the interorbital angle (~104°) are different for cyclopropane.

Cyclopropane
(internuclear angle is 60°)

Cyclopropane
(interorbital angle is 104°)

FIGURE 5.3 Bent bonds in cyclopropane.

Of course, the carbon skeleton of cyclopropane is flat—three points determine a plane—and planarity causes still other problems. Figure 5.4 shows a Newman projection looking down one of the three equivalent carbon–carbon bonds of cyclopropane. In cyclopropane, all carbon–hydrogen bonds are eclipsed. *Remember*: The solid wedges are coming toward you and the dashed wedges are retreating from you.

We have seen eclipsing before in the eclipsed form of ethane (and other alkanes). Ethane has a simple way to avoid this destabilizing opposing of atoms and electrons; it adopts the staggered conformation and minimizes the problem. There is no way for cyclopropane to do likewise. Its hydrogens are stuck in the eclipsed arrangement and there is no possible release. We can even make an estimate of how much damage this **torsional strain** (eclipsing strain) is causing. The rotational barrier in ethane is the result of three pairs of eclipsed hydrogens (Chapter 2, p. 66), and amounts to 3 kcal/mol (Fig. 5.5).

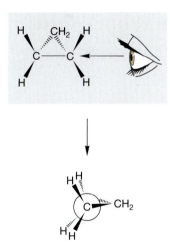

This Newman projection attempts to show the eclipsed C–H bonds; in cyclopropane all the C–H bonds are eclipsed

FIGURE 5.4 Eclipsing of C–H bonds in cyclopropane.

Torsional strain is 3 kcal/mol in the eclipsed form of ethane; this destabilization is about 1 kcal/mol per C–H bond pair

Eclipsed conformation

rotate 60°

Staggered conformation
(~3 kcal/mol lower in energy)

FIGURE 5.5 The destabilization introduced by a pair of eclipsed carbon– hydrogen bonds is about 1 kcal/mol.

The six C–H bonds are all eclipsed in cyclopropane; there should be about 6 kcal/mol of torsional strain

FIGURE 5.6 These are pairs of eclipsed C–H bonds in cyclopropanes.

As mentioned in Chapter 2 (p. 66), a reasonable guess would put the energy cost of each pair of eclipsed hydrogens at about 1 kcal/mol. The total strain of cyclopropane is composed of torsional strain (~6 × 1 = 6 kcal/mol) plus angle strain, which is much more difficult to estimate without the help of calculations or experiments (Fig. 5.6).

The effects of high strain in cyclopropane show up in a number of ways, including the unusual reactivity of the carbon–carbon bonds. Breaking the carbon–carbon bond in ethane requires 90 kcal/mol, but the carbon–carbon bond in cyclopropane needs only 65 kcal/mol (Fig. 5.7). This difference allows a first estimation of the strain energy of cyclopropane (see p. 199 for two more quantitative estimates). Strain destabilizes cyclopropane by about 25 (90 − 65) kcal/mol.

At first glance, the other cycloalkanes would seem to have similar difficulties. In cyclobutane, for example, angle strain is less of a problem as cyclobutane's 90° is a smaller deviation from the ideal 109.5° than is cyclopropane's 60°. Along with this relief, however, comes increased torsional strain, as there are four pairs of hydrogens eclipsed as opposed to cyclopropane's three pairs (Fig. 5.8). One might estimate the torsional strain in planar cyclobutane at 8 kcal/mol (1-kcal/mol strain for each pair of eclipsed carbon–hydrogen bonds).

Ethane
Bond dissociation energy about 90 kcal/mol

FIGURE 5.7 The bond dissociation energy of a carbon–carbon bond in cyclopropane is only 65 kcal/mol. Cyclopropane is strained by about 90 − 65 = 25 kcal/mol.

Cyclopropane
Bond dissociation energy about 65 kcal/mol

FIGURE 5.8 There are eight pairs of eclipsed carbon–hydrogen bonds in cyclobutane, and therefore about 8 kcal/mol of torsional strain.

Unlike cyclopropane, cyclobutane has a way to balance torsional and angle strain. The four-membered ring need not be planar, but can distort or "pucker" somewhat if this will result in an energy lowering. Let's look at the consequences of puckering the ring—of moving one methylene group out of the plane of the other three (Fig. 5.9).

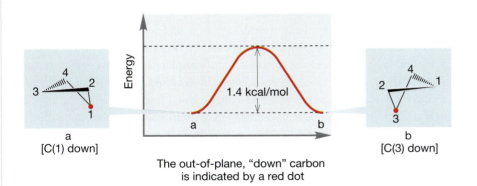

Puckering the ring involves keeping three carbons coplanar and moving the red carbon down

34°

In these Newman projections the red carbon is hidden

FIGURE 5.9 Puckering relieves torsional strain in cyclobutane.

As the ring puckers, torsional strain is reduced, but angle strain is increased. A balance between the two effects is struck in which the ring puckers about 34° and the C—C—C angle closes to about 88°. This form is not static, however; the ring is in motion, much as are the acyclic alkanes. The planar form of cyclobutane is like the eclipsed form of ethane and lies not in an energy well, but at the top of an energy barrier separating a pair of puckered cyclobutanes. This barrier is very small, about 1.4 kcal/mol (Fig. 5.10). Cyclobutane is a mobile, not static, molecule in which different nonplanar forms rapidly interconvert. In Figure 5.10, the bending is exaggerated—the flap is really only 34° out of plane.

Energy

1.4 kcal/mol

a   b

a
[C(1) down]

b
[C(3) down]

The out-of-plane, "down" carbon is indicated by a red dot

FIGURE 5.10 Mobile cyclobutane.

Let's now look at cyclopentane, $(CH_2)_5$, the next larger cycloalkane, and see how it distorts in a similar way. Were cyclopentane planar, it would suffer the torsional strain induced by five pairs of eclipsed hydrogens. As each pair of eclipsed hydrogens contributes about 1 kcal/mol, this amounts to the rather large figure of 10-kcal/mol destabilization. The internal angle in a planar pentagon is 108°, however, so angle strain would be quite small. As in cyclobutane, the ring distorts from planarity, relieving some eclipsing at the cost of increased angle strain. For nonplanar cyclo-pentane, there are two forms of comparable energy, the "envelope" and the "twist" molecules (Fig. 5.11).

Planar cyclopentane showing two pairs of eclipsed hydrogens

The two forms of nonplanar cyclopentane

FIGURE 5.11 Two distorted cyclopentanes.

PROBLEM 5.1 Draw a Newman projection looking down one of the carbon–carbon bonds in planar and envelope cyclopentane. Use models!

Pull up

Pull down

Planar cyclohexane:
120° C–C–C bond angles and about 12-kcal/mol torsional strain from C–H bond eclipsing; there will also be angle strain from the too wide 120° C–C–C internal angles

Chair cyclohexane:
111.5° C–C–C bond angles; about 0-kcal/mol torsional strain

A double Newman projection of the chair form of cyclohexane; note the perfect staggering of the C–H bonds

FIGURE 5.12 Planar and chair cyclohexane.

Neither the puckered envelope form nor the twist form of cyclopentane is static. In the envelope form the "flap" moves around the ring, generating the five possible puckered isomers. This motion requires only a series of rotations around carbon–carbon bonds and closely resembles the motions in cyclobutane shown in Figure 5.10. A model will help you to visualize this motion. Hold two adjacent carbons, sight down the bond that connects them, and convert one form of the envelope into another.

By now you must be expecting that a similar distortion from planarity will be present in cyclohexane, $(CH_2)_6$, but you are probably prepared for neither the magnitude of the effect nor its consequences for the structure and stability of cyclohexanes. In contrast to the small rings, distortion from planarity in cyclohexane relieves both the angle and torsional strain of the planar structure. The internal angle in a planar hexagon is 120°, larger, not smaller, than the ideal sp^3 angle of 109.5°. Deviation from planarity will *decrease* this angle and thus *decrease* angle strain. Torsional strain from the six pairs of eclipsed hydrogens in planar cyclohexane can be expected to contribute about 12 kcal/mol of strain, which is also decreased in a nonplanar form. So in cyclohexane, *both* angle and torsional strain will be relieved by relaxing from a planar structure. Remarkably, this relaxation produces a molecule in which essentially all of the torsional and angle strain is gone. The formation of the energy minimum cyclohexane, called the "chair" form, is shown in Figure 5.12. The internal angle in cyclohexane (111.5°) is close to the ideal C—C—C angle in a simple straight-chain alkane (112° in propane) and the carbon–hydrogen bonds are nicely staggered. Newman projections looking down the carbon–carbon bonds show this, as will (even better) a look at a model.

The details of cyclohexane stereochemistry are important enough to warrant a lengthy discussion (Section 5.4, p. 203), but we can do a few things here in preparation. First of all, it is necessary to learn to draw a decent cyclohexane. No person can truly be described as educated unless he or she can do this, and anyone can, regardless of artistic ability. So learn how to do it and the next time your roommate mentions some obscure European writer, impressing you with his or her erudition and calling into question your sophistication, confound your tormenter with a perfectly drawn cyclohexane!

There are a few tricks that make the construction of a perfect drawing easy. First of all, within the ring, opposite bonds are parallel (Fig. 5.13). With a little practice, keeping the proper bonds parallel should let you easily draw the carbon framework of cyclohexane.

It is a bit more difficult to get the hydrogens right. It helps to tip the ring a little and not to draw it with the "middle" bonds flat (Fig. 5.14). Now we see the molecule as it would rest on a flat surface. Six of the carbon–hydrogen bonds are easy to draw; three of them point straight up and three straight down. These are called the **axial hydrogens**.

FIGURE 5.13 In chair cyclo-hexane, there are three pairs of parallel carbon–carbon bonds.

tip

=

FIGURE 5.14 The set of six "straight up and down" or "axial" carbon–hydrogen bonds of chair cyclohexane are shown in red. All axial carbon–hydrogen bonds are parallel.

So far, so good, but it is the positioning of the last six hydrogens that gives people the most trouble. To get them right, take advantage of the parallel carbon–carbon bonds encountered before in drawing the cyclohexane ring. Each member of this second set of six **equatorial** carbon–hydrogen bonds is parallel to the two carbon–carbon bonds one bond away. Like the six axial carbon–hydrogen bonds, the equatorial bonds also alternate up and down but this set points only slightly up or slightly down (Fig. 5.15).

FIGURE 5.15 The equatorial carbon–hydrogen bonds are parallel to the ring carbon–carbon bonds one bond away. The drawing of a perfect chair cyclohexane is now complete.

The three pairs of in-plane or equatorial C–H bonds; note that they are parallel to the C–C bonds "one bond removed"

The six equatorial hydrogens of chair cyclohexane

All 12 C–H bonds of chair cyclohexane; axial hydrogens are shown in red, equatorial hydrogens in blue

Up hydrogens in green, down hydrogens in gray

Now you have a perfect cyclohexane ring in its energy-minimum chair form. This exercise in drawing lets us see one important thing immediately. Note again that there are two different kinds of hydrogen. There is the "straight up and down" set of six, which are called the axial hydrogens, and the other set of six, which lie roughly along the equator of the molecule and are called the equatorial hydrogens (Fig. 5.16).

The six axial hydrogens
of chair cyclohexane

The six equatorial hydrogens
of chair cyclohexane

FIGURE 5.16 Axial and equatorial hydrogens.

WORKED PROBLEM 5.2 What happens to a set of axial hydrogens when you convert one chair into another? Try it with models. This problem is important and simple.

Up axial becomes up equatorial

ring flip

Down axial becomes down equatorial

ANSWER The set of axial hydrogens becomes the set of equatorial hydrogens! An *up* hydrogen always remains *up*, however. Similarly, a *down* hydrogen is always *down*, whether axial or equatorial.

PROBLEM 5.3 How many carbons will the ^{13}C NMR spectrum of cyclohexane reveal?

PROBLEM 5.4 How many hydrogens will the ^{1}H NMR spectrum of cyclohexane reveal? **Caution:** This problem is much harder than Problem 5.3. Try to find two answers to this question, one correct at high temperature, the other correct at low temperature.

Larger rings will also be nonplanar, but no more spectacular surprises such as cyclohexane's energy minimum form remain. Planar heptagons and octagons have internal angles of 129° and 135°, respectively, and thus have substantial angle strain. Both seven- and eight-membered rings would have severe torsional strain were they to remain planar. There is relaxation to nonplanar forms but some amount of strain always remains in these medium-sized rings. It is a complicated business to analyze medium-sized rings completely, and we will not embark upon it. Rings with minimal angle strain will have severe interactions between the carbon–hydrogen bonds, and thus substantial torsional strain.

Relieving those eclipsings usually induces increased angle strain. As in cyclopentane, there is a compromise to be made between minimizing angle and torsional strain, and Nature does the best it can, finding the lowest energy structural compromise. Unlike cyclohexane, however, in the medium-sized rings there is no obvious low-energy solution. Often there are several somewhat different minima, close to each other in energy, and separated from each other by quite low barriers.

Fragments of chair cyclohexanes are sometimes seen in the structures of medium-sized ring compounds. Cyclodecane is an example, as Figure 5.17 shows. The structure of cyclodecane shows partial cyclohexane rings. Several destabilizing interactions between carbon–hydrogen bonds are apparent.

As rings get even bigger, the strain energy decreases. The limit is an infinitely large ring, which resembles an endless chain of methylene groups. In such a species, it is possible to stagger all carbon–hydrogen bonds, and, of course, in an infinite ring there is no angle strain. So after cyclohexane, the next strain-free species is reached when the ring size has grown large enough to approximate an infinitely large ring.

Cyclodecane

FIGURE 5.17 Cyclodecane.

5.3 QUANTITATIVE EVALUATION OF STRAIN ENERGY

Strain is an important factor in chemical reactivity. The more strained a compound, the higher its energy. The higher in energy a compound, the more likely it is to react. We have just seen how a combination of torsional and angle strains can act to change the energy of a ring compound, cyclopropane. Now we move on to some quantitative measures of the energies of ring compounds. We introduce some general techniques and look ahead to our detailed consideration of energy in Chapter 8.

5.3a Heats of Formation. To recapitulate material from Chapter 3 (p. 115), the heat of formation (ΔH_f°) of a compound is the enthalpy of its formation by the reaction of its constituent elements in their standard states. The standard state of an element is the most stable state at 25 °C and 1-atm pressure. For an element in its standard state, ΔH_f° is taken as zero. The more negative (or less positive) a compound's ΔH_f°, the more stable it is. A negative ΔH_f° for a compound means that its formation from its constituent elements would be *exothermic*—heat would be liberated. By contrast, a positive ΔH_f° means that the constituent elements are more stable than the compound and its formation would be *endothermic*—energy would have to be applied (Fig. 5.18).

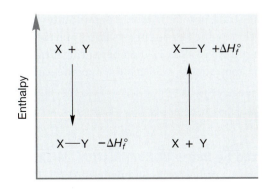

FIGURE 5.18 For a compound more stable than its constituent elements, ΔH_f° is negative. For a compound less stable than its constituent elements, ΔH_f° is positive.

The first two columns of Table 5.1 collect the heats of formation for some cycloalkanes and calculate the ΔH_f° *per methylene group* for these hydrocarbons.

TABLE 5.1   Strain Energies for Some Cycloalkanes

Molecule	1 ΔH_f° (kcal/mol)	2 ΔH_f° per CH_2 (kcal/mol)	3 Calcd ΔH_f° (kcal/mol)	4 Strain E (kcal/mol)	5 Strain E per CH_2 (kcal/mol)
Cyclopropane	+12.7	+4.2	−14.7	27.4	9.1
Cyclobutane	+ 6.8	+1.7	−19.6	26.4	6.6
Cyclopentane	−18.7	−3.7	−24.5	5.8	1.2
Cyclohexane	−29.5	−4.9	−29.4	0.1	0
Cycloheptane	−28.3	−4.0	−34.3	6.0	0.9
Cyclooctane	−29.7	−3.7	−39.2	9.5	1.2
Cyclodecane	−36.9	−3.7	−49.0	12.1	1.2
Cyclododecane	−55.0	−4.6	−58.8	3.8	0.3
$(CH_2)_\infty$		−4.9			

PROBLEM 5.5 How would one get the value for a CH_2 group in an infinite chain, $(CH_2)_\infty$? Use the following data to estimate the value of ΔH_f° of an individual methylene group in kilocalories per mole: ΔH_f° (hexane) = −39.9, (heptane) = −44.8, (octane) = −49.8, (nonane) = −54.5, (decane) = −59.6 kcal/mol.

Note that the ΔH_f° for a methylene (CH_2) group in a strain-free straight-chain alkane, $(CH_2)_\infty$, is −4.9 kcal/mol. This value is exactly the ΔH_f° we calculate for a methylene group in cyclohexane. Cyclohexane really is strain-free. We can now use these data to calculate some strain energies for the cycloalkanes. First, calculate (Table 5.1, column 3) the ΔH_f° for the ring constructed from strain-free methylene groups. The difference between the calculated, *strain-free* ΔH_f° and the real, *measured* ΔH_f° is the strain energy (strain E, column 4 in the table). For example, cyclopropane has a measured ΔH_f° of +12.7 kcal/mol. A strain-free cyclopropane ring would have a ΔH_f° of 3×-4.9 kcal/mol = −14.7 kcal/mol. The difference between these two values (27.4 kcal/mol) is the strain energy (column 4). The strain energy per CH_2 (column 5) is $(27.4)/3 = 9.1$ kcal/mol.

Now look at the other rings. Cyclopropane and cyclobutane are about equally strained, although the strain per methylene is significantly higher for cyclopropane. Cyclopentane and cycloheptane are slightly strained, and the strain gets worse in the medium-sized rings until we reach a 12-membered ring. Strain declines in the larger rings until (with some aberrations) we reach the hypothetical strain-free infinite ring.

PROBLEM 5.6 Use models to look for the sources of strain in medium-sized rings.

5.3b Strain Analyzed by Heats of Combustion.
Strain energies can also be determined from an analysis of heats of combustion (ΔH_c°). This method is attractive for analyzing strain because the measurements can be made and compared directly.

Combustion of a hydrocarbon is an exothermic reaction. The products, H_2O and CO_2, are more stable than the starting hydrocarbon and oxygen. This difference, the exothermicity of the reaction, appears as heat and light (Fig. 5.19), which is easily apprehended by looking at the light and feeling the heat evolved by a propane stove.

$$C_3H_8 + 5\,O_2 \rightarrow 3\,CO_2 + 4\,H_2O + Energy$$

$CH_3CH_2CH_3 \ + \ 5\,O_2$

530.6 kcal/mol = exothermicity
of this reaction

$3\,CO_2 \ + \ 4\,H_2O$

FIGURE 5.19 The combustion of propane to give CO_2, H_2O, and energy.

The conventions are confusing here. When writing energy as the product of a reaction it is conventional to show it as "+ energy." The enthalpy of this exothermic reaction, ΔH, is negative, however! So,

$$C_6H_{12} + 9\,O_2 \rightarrow 6\,CO_2 + 6\,H_2O \qquad +\,937\ \text{kcal/mol}$$

but,

$$C_6H_{12} + 9\,O_2 \rightarrow 6\,CO_2 + 6\,H_2O \qquad \Delta H = -937\ \text{kcal/mol}$$

CONVENTION ALERT

This method can often be used to establish energy differences between molecules. Compare, for example, the heats of combustion for octane and an isomer, 2,2,3,3-tetramethylbutane (Fig. 5.20).*

O_2

O_2

$CO_2 \ + \ H_2O$

(solid octane)

$C_8H_{18} + 12.5\,O_2 \longrightarrow 8\,CO_2 + 9\,H_2O$

FIGURE 5.20 Combustion of two isomeric octanes.

*A reviewer of this book, RMM, who sat with me in chemistry classes at Yale, many, many years ago, reminds me that we were once told that this molecule, the only isomer of octane with a melting point above room temperature (mp 104 ℃), was known as "solid octane." We were then told that this was likely to be the *only* fact we retained from our organic chemistry course! Now you are stuck with this knowledge.

The measured heats of combustion of octane and 2,2,3,3-tetramethylbutane are 1307.60 kcal/mol and 1303.04 kcal/mol, respectively. Figure 5.21 graphically shows the relationship between these values and the energies of the molecules. As these isomeric molecules produce exactly the same products on burning, the difference in their heats of combustion is the difference in energy between them. Of course, this analysis also tells you which isomer is more stable, and by how much.

FIGURE 5.21 A quantitative picture of the combustion of two octanes.

PROBLEM 5.7 The heats of combustion of heptane, 3-methylhexane, and 3,3-dimethylpentane, respectively, are 1149.9, 1148.9, and 1147.9 kcal/mol. Carefully draw a diagram showing the relative stabilities of these molecules.

We can also use heats of combustion to measure the relative strain energies of the cycloalkanes. The higher the ΔH_c°, the less stable the compound (Fig. 5.22).

FIGURE 5.22 The heats of combustion per CH_2 for a series of cycloalkanes and a strain-free sequence of methylene (CH_2) groups. The higher the heat of combustion, the less stable the compound.

Once more we need a baseline, and we use the ΔH_c° of a strain-free methylene group in an infinite chain of methylenes, 157.5 kcal/mol. This value is determined in much the same way as was used to get the ΔH_f° for a strain-free methylene (p. 200). Then we measure the ΔH_c° values for the series of cycloalkanes, finding the ΔH_c per methylene for each ring. Subtraction of 157.5 kcal/mol, the ΔH_c° of a strain-free methylene group, gives the strain energy per CH_2 and, therefore, the strain energy of the ring compound. As Table 5.2 shows, there is a good correspondence between the strain energies measured from ΔH_f° and ΔH_c°. Actually this correspondence is no surprise, as ΔH_f° values are often derived from ΔH_c°.

TABLE 5.2 Strain Energies for Some Cycloalkanes

Molecule	Experimental ΔH_c° per CH_2 (kcal/mol)	Strain E per CH_2 ($\Delta H_c^\circ - 157.5$) (kcal/mol)	Strain E from ΔH_c° (kcal/mol)	Strain E from ΔH_f° (kcal/mol)
Cyclopropane	166.3	8.8	26.4	27.4
Cyclobutane	163.9	6.4	25.6	26.4
Cyclopentane	158.7	1.2	6.0	5.8
Cyclohexane	157.4	−0.1	−0.6	0.1
Cycloheptane	158.3	0.8	5.6	6.0
Cyclooctane	158.6	1.1	8.8	9.5
Cyclodecane	158.6	1.1	11.0	12.1
Cyclododecane	157.8	0.3	3.6	3.8
$(CH_2)_\infty$	157.5			

5.4 STEREOCHEMISTRY OF CYCLOHEXANE: CONFORMATIONAL ANALYSIS

Earlier, when we were considering the ways in which ring compounds distort so as to minimize strain, we saw that cyclohexane adopted a strain-free chair conformation. Now it is time to look at cyclohexane in detail. Six-membered rings are most common in organic chemistry, and a great deal of effort has been made over the years at understanding their structure and reactivity. One chair form of cyclohexane can easily be converted into another. We are now going to look in detail at this transformation and estimate the energies of the various intermediates and transition states along the path from one chair to the other. This process is an example of **conformational analysis**.

A little manipulation of a model will show the overall conversion, but it's not easy to come to the actual pathway for the process. If we move carbons 1, 2, 3, and 4 into one plane, with carbon 6 above the plane and carbon 5 below the plane, we come to a "half-chair" structure (Fig. 5.23).

FIGURE 5.23 Conversion of the energy minimum chair cyclohexane into the half-chair and then the twist form.

Chair → Half-chair (transition state) → Twist

The half-chair conformation does not represent a stable molecule, but is instead a picture of the top of the energy mountain pass (a transition state) leading to a lower energy molecular valley called the twist conformation or, sometimes, twist-boat (Fig. 5.23). The half-chair contains many eclipsed carbon–hydrogen bonds, which become staggered somewhat in the twist conformation, and angle strain is partially relieved as well. The twist arrangement can pass through a second half-chair to give the other chair cyclohexane. Kinetic measurements allow an evaluation of the energies involved (Fig. 5.24). The half-chair and the twist lie 10.8 and 5.5 kcal/mol above the chair, respectively.

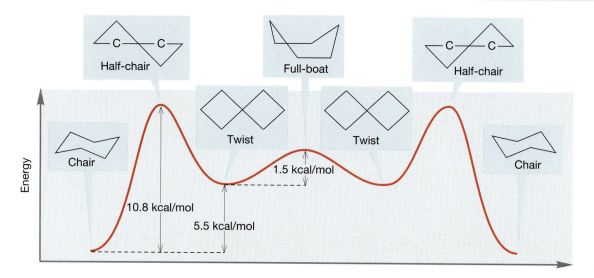

FIGURE 5.24 The interconversion of two chair cyclohexanes. The two chairs and the two twist forms are intermediates (energy minima), and the two half-chairs and the full-boat are transition states (energy maxima).

Do not confuse the twist with the full-boat (Fig. 5.24). The full-boat is not an energy minimum but is, like the half-chair, an energy maximum. It lies at the top of the barrier (like the half-chair, it, too, is a *transition state*) separating two twist forms. The boat is only 1.5 kcal/mol higher than the twist (7.0 kcal/mol higher than the chair), and can never be isolated because it is not an energy minimum.

Both the half-chair and the full-boat suffer from the kinds of strain we have seen before. In the half-chair, much of the ring is planar and there will be both angle and torsional strain. The full-boat also has torsional strain, but there is a further hydrogen–hydrogen interaction that is destabilizing. This interaction is be-tween the "prow" and "stern" carbons and between the two "inside" hydrogens at the prow and stern of the boat. This new kind of strain is induced when two atoms come too close to each other and is called **van der Waals strain** (Johannes Diderik van der Waals, 1837–1923) (Fig. 5.25).

Strain: van der Waals repulsion

"Boat" cyclohexane

Torsional strain

FIGURE 5.25 The full-boat form.

WORKED PROBLEM 5.8 Draw a Newman projection looking down the "side" H_2C–CH_2 bond of the full-boat cyclohexane.

ANSWER This task is relatively easy. Once again, set your eye slightly to one side so as to see the carbon–hydrogen bonds on the rear carbon.

One chair cyclohexane will equilibrate with another as long as the environment supplies the requisite 10.8 kcal/mol to traverse the barrier. Under normal condi-tions this is an easy process, but at very low temperature, one can "freeze out" one chair form. How much twist form is there at 25 °C? We will see many such calcula-tions in Chapter 8, but the chair–twist equilibrium can be treated like any equilib-rium process. A calculation shows that an energy difference of $\Delta G = 5.5$ kcal/mol results in an enormous preference for the more stable isomer of the pair, about 9999:1. The "take-home lesson" here is that small energy differences between two molecules in equilibrium result in a very large excess of the more stable isomer.

5.5 MONOSUBSTITUTED CYCLOHEXANES

Recall (Problems 5.2–5.4, p. 198) that "flipping" from one chair to the other interconverts the set of six axial hydrogens with the set of six equatorial hydrogens (Fig. 5.26).

flip

FIGURE 5.26 In cyclohexane, a ring flip interconverts axial and equatorial hydrogens.

The danger of relying too strongly on coded two-dimensional representations of molecules is shown nicely by methylcyclohexane. The coded structures give no hint of the richness of the complicated, three-dimensional structure of the cyclohexane; they even hide the presence of two conformational isomers of methylcyclohexane (Fig. 5.27). The methyl group, or any substituent, can be in either an axial or equatorial position, and these two molecules are diastereomers. (*Remember*: Diastereomers are stereoisomers that are not mirror images—see Chapter 4, p. 171.)

FIGURE 5.27 The uninformative skeletal structure hides the existence of two conformational isomers of methylcyclohexane. In one isomer the methyl group is equatorial, and in the other it is axial.

Equatorial methyl   Axial methyl

Given that two isomers of methylcyclohexane exist, our next job is to determine which is more stable. As it turns out, we can even make a reasonable guess at the magnitude of the difference in energy between the two diastereomers. First of all, notice that axial and equatorial methylcyclohexane are interconverted by a chair flip (Fig. 5.27).

What turns out to be the crucial factor in creating the energy difference between the two forms is shown by the Newman projections made by sighting down the bond attaching a ring methylene group to the carbon bearing the methyl group. In the axial form, there are two *gauche* interactions between the methyl group and the nearby ring CH_2 group (Fig. 5.28). Note the destabilizing "1,3-diaxial interaction" between the axial hydrogen and the axial methyl group.

Methylcyclohexane (methyl group axial)   *gauche*-Butane

FIGURE 5.28   Methylcyclohexane with the methyl group axial. The Newman projection shows the *gauche* interaction with one of the two equivalent ring methylene groups. Each of the two *gauche* interactions in this compound resembles a *gauche* interaction in butane.

Although it is easy to see one of the *gauche* interactions, the perspective of the drawing hides the other unless you are careful. No amount of words can help you here as much as working out this stereochemical problem yourself. It is very important to try Problem 5.9. The answer follows immediately.

WORKED PROBLEM 5.9 Draw the Newman projection of methylcyclohexane looking from the other adjacent methylene group toward the methyl-bearing carbon. Be sure that you see the second gauche methyl-ring interaction. Next, compare this gauche interaction to that in butane. Are the two exactly the same?

ANSWER It is difficult to place your "eye" correctly for this view, but if you look along the indicated carbon–carbon bond, the appropriate Newman projection can be constructed.

So, the two interactions are very similar, but not exactly the same. In butane two methyl groups bump, whereas in axial methylcyclohexane a methyl group bumps with a CH_2-C.

Methylcyclohexane (methyl group axial)

One of two *gauche* methyl-ring interactions; this involves bumping between CH_3 and the ring CH_2-CH_2

***gauche*-Butane**

Here, the *gauche* interaction is a methyl–methyl bumping

Recall our conformational analysis of butane in Chapter 2 (p. 72). The *gauche* form of butane was about 0.6 kcal/mol less stable than the *anti* form, in which the terminal methyl groups of butane were as far apart as possible.

A Newman projection of the isomer with the equatorial methyl group made from the same perspective reveals none of these *gauche* interactions—this arrangement resembles the *anti* form of butane (Fig. 5.29).

This ring carbon and the methyl group are as far apart as possible

Methylcyclohexane (methyl group equatorial)

The two methyl groups are as far apart as possible

***anti*-Butane**

FIGURE 5.29 Methylcyclohexane with an equatorial methyl group. There are no *gauche* methyl-ring interactions. This isomer is 1.74 kcal/mol more stable than the molecule with the methyl group axial. Note the resemblance to the *anti* form of butane.

Accordingly, we might make a guess that the isomer with the methyl group equatorial would be more stable than the axial isomer, and that the energy difference between the two would be approximately twice the difference between *gauche*- and *anti*-butane, $2 \times 0.6 = 1.2$ kcal/mol. We'd be close. The real difference is 1.74 kcal/mol.

PROBLEM 5.10 Why isn't the difference exactly 1.2 kcal/mol? Point out some of the differences between a *gauche* interaction in butane and that in axial methylcyclohexane.

Calculations of the percentage of isomers in the equatorial and axial forms of methylcyclohexane will become clearer in Chapter 8. For now it's fair to say that about 95% of the molecules will have the methyl group equatorial at any given instant. Once more, a small energy difference (1.74 kcal/mol) has resulted in a large excess of the more stable isomer. Don't make the mistake of thinking that methylcyclohexane consists of a mixture of 95% molecules locked forever in the conformation with the methyl group equatorial, and 5% of similarly frozen molecules with their methyl groups axial. These molecules are in rapid equilibrium. At any given moment 95% will have their methyl groups equatorial, but all molecules are equilibrating between equatorial and axial methyl isomers.

PROBLEM 5.11 The formula for doing these calculations is $\Delta G = -RT \ln K$, where R is about 2 cal/deg mol, and T is the absolute temperature (see Chapter 8 for more details). Calculate the amounts at equilibrium at 25 °C of two compounds differing in energy by $\Delta G = 2.8$ kcal/mol.

A group larger than methyl would result in an even greater dominance of the equatorial form over the less stable axial arrangement. The 1,3-diaxial "bumpings" responsible for the destabilizing *gauche* interactions will be magnified, and reflected in the equilibrium constant (Fig. 5.30; Table 5.3).

FIGURE 5.30 The larger R, the more favored will be the equatorial isomer.

If R is large, this *gauche* interaction will be more destabilizing than the interaction with a methyl group

In this form, in which the R is equatorial, there are no *gauche* interactions between R and the ring

Note the large axial–equatorial (ax/eq) energy difference for the *tert*-butyl group. The value is approximately 5–5.5 kcal/mol, which translates into an equilibrium constant of almost 12,000 at 25 °C. The situation is complicated for the *tert*-butyl group, because the conformation with an axial *tert*-butyl group is so greatly destabilized that a twist conformation, with an energy about 5 kcal/mol higher than the conformation with an equatorial *tert*-butyl group, may be lower in energy than the pure chair with an axial *tert*-butyl group (Fig. 5.31). Regardless, the conformation of *tert*-butylcyclohexane with the *tert*-butyl group equatorial is favored enormously.

TABLE 5.3 Axial–Equatorial Energy Differences for Some Alkylcyclohexanes at 25 °C

Compound	ΔG (ax/eq) (kcal/mol)	K
Methylcyclohexane	1.74	19.5
Ethylcyclohexane	1.79	21.2
Propylcyclohexane	2.21	43.4
Isopropylcyclohexane	2.61	86.0
tert-Butylcyclohexane	5.5	11,916

FIGURE 5.31 The large *tert*-butyl group distorts the equilibrium far toward the much more stable equatorial form. The geometrical relationships can be estimated with confidence.

WORKED PROBLEM 5.12 It is often said that the *tert*-butyl group locks the molecule into the form with the tert-butyl group equatorial. Is this a good way to put it? Why not?

ANSWER No! The molecule is not locked at all. The forms with the large *tert*-butyl group axial and equatorial are in rapid equilibrium, with the equatorial form greatly predominating. Very few molecules are in the axial form at any one time, but the equilibrium is mobile. This molecule is not locked into one form and immobilized forever.

PROBLEM 5.13 How many carbons will the ^{13}C NMR spectrum of *tert*-butylcyclohexane reveal?

PROBLEM 5.14 How many carbons will the ^{13}C NMR spectra of *trans*-1,4-di-*tert*-butylcyclohexane and all-*cis*-1,3,5-tri-*tert*-butylcyclohexane reveal?

In studies of reaction mechanisms it is often very helpful to know with some precision the spatial relationships between groups. Ring compounds are quite helpful in this regard, as, for example, the rigid frame of cyclopropane allows for quite accurate estimations of angles and distances to be made. But cyclopropanes are so strained as to be sometimes too reactive for mechanistic work. In fact, the ring is often opened in chemical reactions. For example, one wouldn't want one's rigid framework disappearing as the reaction occurred, but this is all too likely for cyclopropanes. Other rings are flexible and don't allow us the firm predictions of angles and distances we want. However, a large group such as *tert*-butyl enables us to prejudice the mobile equilibrium between cyclohexane rings so strongly in favor of the form with the *tert*-butyl group equatorial that we can determine with confidence the positions of other atoms in the molecule (Fig. 5.31). We will see later that this technique is frequently used.

SUMMARY

All cycloalkanes except cyclopropane distort from planarity so as to minimize strain. The amount of strain can be measured in a variety of ways including measurements of heats of combustion or heats of formation. Cyclohexane adopts an energy-minimum chair form in which there are six hydrogens in the axial position and six in the equatorial position. Cyclohexane is a mobile system, as two possible chair forms interconvert. This interconversion exchanges substituents in the axial and equatorial positions.

5.6   DISUBSTITUTED RING COMPOUNDS

We have already mentioned that cis and trans diastereomers exist for dichlorocyclopropanes (Chapter 4, pp. 181–182), and that the trans form is chiral (Fig. 5.32).

cis-1-Bromo-2-chlorocyclopropane is a chiral molecule—a pair of enantiomers exists

trans-1-Bromo-2-chlorocyclopropane is also chiral, and, of course, there is a pair of enantiomers

FIGURE 5.33 *cis*- and *trans*-1-Bromo-2-chlorocyclopropane.

cis-1,2-Dichlorocyclopropane
(an achiral molecule, a meso compound)

trans-1,2-Dichlorocyclopropane
(chiral—there is a pair of enantiomers)

FIGURE 5.32

When the two substituents on the ring are different, as in 1-bromo-2-chlorocyclopropane, two pairs of enantiomers are possible, and both the cis and trans forms of this molecule are chiral (Fig. 5.33).

This relatively simple situation becomes harder to see in cyclohexanes where the molecules are not planar, but it doesn't really change. We will first take a look at 1,1-dialkylcyclohexanes and then at the somewhat more complicated 1,2-dialkylcyclohexanes.

5.6a 1,1-Disubstituted Cyclohexanes.
1,1-Dimethylcyclohexane is a simple, achiral molecule in which the cyclohexane axial–equatorial equilibration induced by flipping one chair to the other interconverts equivalent molecules, each of which has one methyl group axial and one equatorial. The ring flip converts the axial methyl group into an equatorial methyl group and vice versa. As for cyclohexane itself, the equilibrium constant for the equilibration of the two equivalent forms of the 1,1-dimethyl isomer must be 1 (Fig. 5.34).

FIGURE 5.34 1,1-Dimethylcyclohexane.

1-Isopropyl-1-methylcyclohexane is only slightly more complicated. There are two isomers, nearly equal in energy, neither of which is chiral (Fig. 5.35).

FIGURE 5.35 Both isomers of 1-isopropyl-1-methylcyclohexane are achiral—the mirror images are superimposable.

PROBLEM 5.15 Use the data in Table 5.3 to calculate the energy difference between the possible isomers of 1-isopropyl-1-methylcyclohexane.

5.6b 1,2-Disubstituted Cyclohexanes.
Just like 1,2-dimethylcyclopropane, 1,2-dimethylcyclohexane can exist in cis and trans forms (as can any 1,2-disubstituted ring compound), but the chair form of cyclohexane makes the

compounds look somewhat different from the more familiar rigid isomers of cis and trans disubstituted cyclopropanes (Fig. 5.36).

FIGURE 5.36 Three-dimensional representations of *cis*- and *trans*-1,2-dimethylcyclohexane and *cis*- and *trans*-1,2-dimethylcyclopropane.

cis Methyl groups
cis hydrogens

trans Methyl groups
trans hydrogens

The words cis and trans refer to the "sidedness" of a molecule and do not depend strictly on the angles between the groups. In a cis disubstituted cyclopropane, the dihedral angle between the cis groups is 0°, whereas it is 60° in the disubstituted cyclohexane. The methyl groups are refered to as cis in either case. In *trans*-1,2-dimethylcyclopropane, the dihedral angle between the methyl groups is 120°, whereas in the six-membered ring it is 60° (Fig. 5.37).

Two views of *cis*-1,2-dimethylcyclopropane; the dihedral angle between the methyl groups is 0°

Two views of *trans*-1,2-dimethylcyclopropane; the dihedral angle between the methyl groups is 120°

Two views of *cis*-1,2-dimethylcyclohexane; the dihedral angle between the methyl groups is 60°

Two views of *trans*-1,2-dimethylcyclohexane; the dihedral angle between the methyl groups is also 60°

FIGURE 5.37 cis and trans Isomerism in the planar, inflexible cyclopropanes and the nonplanar, flexible cyclohexanes.

PROBLEM 5.16 Figure 5.37 shows only the equatorial, equatorial form of *trans*-1,2-dimethylcyclohexane. What is the dihedral angle between the methyl groups in the much less stable axial, axial dimethyl isomer?

In *cis*-1,2-dimethylcyclohexane, one methyl group is axial and the other is equatorial (Fig. 5.38). The conformational ring flip does not alter the structure—the axial methyl becomes equatorial, and the equatorial methyl becomes axial. The equilibrium constant between these two equivalent compounds must be 1.

FIGURE 5.38 *cis*-1,2-Dimethylcyclohexane.

Note that the two conformational isomers of *cis*-1,2-dimethylcyclohexane are enantiomers. The equilibration between the two conformations produces a racemic mixture of the two enantiomers. Have we seen a similar situation before? Indeed we have; just recall our discussion of the *gauche* forms of butane (Chapter 4, p. 168). The equilibration of the two *cis*-1,2-dimethylcyclohexanes is very similar to the equilibration of two *gauche* forms of butane (Fig. 5.39).

FIGURE 5.39 The ring flip in *cis*-1,2-dimethylcyclohexane converts one enantiomer into the other. A very similar process interconverts the two enantiomers of *gauche*-butane.

The trans form of this molecule presents a different picture. *trans*-1,2-Dimethylcyclohexane can either have both methyl groups axial or both equatorial (Fig. 5.40). The ring flip converts the diaxial form into the diequatorial form. Let's apply conformational analysis to predict which of these two diastereomers will be more stable.

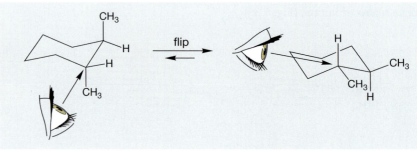

FIGURE 5.40 The ring flip in *trans*-1,2-dimethylcyclohexane interconverts a molecule with two axial methyl groups and one with two equatorial methyl groups.

As usual, Newman projections are vital, and in Figure 5.41 we sight along the carbon–carbon bond attaching the two methyl-substituted carbons. In the diequatorial isomer, there is one *gauche* methyl–methyl interaction costing 0.6 kcal/mol. In the diaxial isomer, the two methyl groups occupy *anti* positions and their interaction with one another is not destabilizing. However, for each axial methyl group we still have the two *gauche* interactions with the ring methylene groups (see p. 206), and this will cost the molecule 2 × 1.74 kcal/mol, or 3.48 kcal/mol. We can predict that the diequatorial isomer should be 3.48 − 0.6 = 2.88 kcal/mol more stable than the diaxial isomer. In turn, this calculation predicts that at 25 °C there will be more than 99% of the diequatorial form present.

FIGURE 5.41 Diequatorial *trans*-1,2-dimethylcyclohexane is more stable than the diaxial isomer.

PROBLEM 5.17 Show that the ring flip of the diequatorial form to the diaxial form will not racemize optically active *trans*-1,2-dimethylcyclohexane. What is the relation between the equatorial–equatorial and axial–axial forms of *trans*-1,2-dimethylcyclohexane?

In principle, both the diaxial and diequatorial forms should be separable into a pair of enantiomers, resolvable (p. 176)—they are both chiral molecules. In practice, we cannot isolate *trans*-diaxial-1,2-dimethylcyclohexane; it simply flips to the much more stable diequatorial form. The diequatorial form can be isolated and resolved.

SUMMARY

Neither *cis*-1,2-dimethylcyclopropane nor *cis*-1,2-dimethylcyclohexane can be re-solved, as the cyclopropane is a meso compound (Chapter 4, p. 174), and the cyclo-hexane flips into its mirror image. However, both *trans*-1,2-dimethylcyclopropane and *trans*-1,2-dimethylcyclohexane *can* be resolved; they are not superimposable on their mirror images. The point is that in a practical sense the planar cyclo-propanes and nonplanar cyclohexanes behave in the same way. This finding has important consequences. In deciding questions of stereochemistry, we can treat the decidedly nonplanar 1,2-dimethylcyclohexanes *as if* they were planar. Indeed, all cyclohexanes can be treated as planar for the purposes of stereochemical analy-sis, because the planar forms represent the average positions of ring atoms in the rapid chair–chair interconversions.

Now let's look at some slightly more complicated molecules in which the two substituents on the ring are different. In 1-isopropyl-2-methylcyclohexanes, cis and trans isomers exist, and once more we will have to worry about the effects of flipping from one chair form to the other. In the cis compound, there are two different con-formational isomers (Fig. 5.42). In one isomer, the methyl is axial and the isopropyl group is equatorial. When the ring flips, the axial methyl group becomes equatorial and the equatorial isopropyl group becomes axial.

FIGURE 5.42 *cis*-1-Isopropyl(ax)-2-methyl(eq)cyclohexane flips to *cis*-1-isopropyl(eq)-2-methyl(ax)cyclohexane.

Unlike the dimethyl case, these two compounds are not enantiomeric. They are conformational diastereomers—different conformational isomers with different physical and chemical properties. In principle, each could be isolated and resolved. So there is a total of four possible cis isomers, the two shown in Figure 5.42 and their mirror images (Fig. 5.43).

FIGURE 5.43 The four stereoisomers of *cis*-1-isopropyl-2-methylcyclohexane. Two pairs of enantiomers exist.

PROBLEM 5.18 Label the stereogenic carbons of Figures 5.43 and 5.44 as (*R*) or (*S*).

The rigid molecule *cis*-1-isopropyl-2-methylcyclopropane cannot undergo any ring flip and there is only a single pair of enantiomers. The lack of a possible ring flip reduces the total number of possible stereoisomers (Fig. 5.44).

FIGURE 5.44 The two enantiomers of *cis*-1-isopropyl-2-methylcyclopropane.

The situation is similar in the case of the trans form. Although the inflexible cyclopropane ring has but a single pair of enantiomers (Fig. 5.45), there are two pairs of enantiomers in the flexible cyclohexane (Fig. 5.46).

FIGURE 5.45 The single pair of enantiomers of *trans*-1-isopropyl-2-methylcyclopropane. The rigid cyclopropane ring prevents any ring flip.

FIGURE 5.46 The four stereoisomers of *trans*-1-isopropyl-2-methylcyclohexane. Two pairs of enantiomers exist.

PROBLEM 5.19 Point out the pairs of enantiomers and diastereomers in Figure 5.46. Assign all the stereogenic carbons as (R) or (S).

WORKED PROBLEM 5.20 Use the data of Table 5.3 to estimate the energy difference between *cis*- and *trans*-1-isopropyl-2-methylcyclohexanes.

ANSWER For the cis compound, ring flipping converts **A** with an axial isopropyl group and an equatorial methyl group into **B** in which the isopropyl group is equatorial and the methyl is axial. Table 5.3 tells you that an isopropyl group is more stable in the equatorial position by 2.61 kcal/mol. A methyl group is more stable in the equatorial position by only 1.74 kcal/mol. The more stable conformation will be **B** by 2.61 − 1.74 = 0.87 kcal/mol.

The trans compound has both groups axial (**C**) or both groups equatorial (**D**). Conformation **D** will be preferred by 2.61 + 1.74 = 4.35 kcal/mol. However, **D** suffers a methyl–isopropyl *gauche* interaction that will be destabilizing by somewhat more than 0.6 kcal/mol. So, our final answer would be approximately 4.35 − 0.6 = 3.75 kcal/mol.

PROBLEM 5.21 Estimate the energy difference between the two conformational isomers of *cis*-1,3-dimethylcyclohexane.

PROBLEM 5.22 Does ring flip racemize *trans*-1,3-dimethylcyclohexane, as it does for *cis*-1,2-dimethylcyclohexane (Fig. 5.39, p. 213)?

SUMMARY

The "take-home lesson" here is that it takes a careful analysis of the stereoisomers formed by ring flipping of substituted cyclohexanes to see all the possibilities. First, be sure you have made a *good* drawing (follow the procedure on p. 197). Next, determine whether or not the isomer is chiral by drawing the mirror image and seeing if it is superimposable on the original. Now do a ring flip of the original isomer. Is the ring-flipped isomer the same as the original? Is it the mirror image? Is it completely different? Differently substituted cyclohexanes give all three possibilities.

Now that we have dealt with both simple and complicated cycloalkanes (molecules containing a single ring), it is time to look briefly at molecules containing more than one ring.

5.7   BICYCLIC COMPOUNDS

Of course, one can imagine all sorts of molecules containing two separated rings. There is nothing special about such molecules. However, there are other molecules in which two rings share a carbon or carbons, and these are more interesting. There are three general ways in which two rings can be connected by sharing carbons: They can share either a single carbon or two or more carbons. These modes of attachment are called **spiro** (one carbon shared), **fused**, and **bridged** (more than one carbon shared). Fused is really a special case of bridged, with *n* = 0 (Fig. 5.47).

Spiro substitution is common in both "natural products" and in molecules so far encountered only in the laboratory. Note in Figure 5.48 how poorly a two-dimensional picture represents the real shape of such molecules. Even the simplest spirocyclic compound, spiropentane, is badly served by the flat page. The central carbon is approximately tetrahedral, and we really must use wedges to give a three-dimensional feel to this kind of molecule.

Spiro

Spiro substitution
(one carbon is shared)

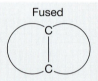

Fused

Fused substitution
(two carbons are shared)

Bicyclic

$(CH_2)_n$

Bridged substitution
(more than two carbons are shared; if *n* = 0, the molecule is fused)

FIGURE 5.47 Rings can share a single carbon (spiro), two carbons (fused), or more than two carbons (bridged).

FIGURE 5.48 A two-dimensional picture of spiropentane is misleading. The two rings lie in perpendicular planes.

The two-dimensional picture of spiropentane is highly misleading…

…as the central carbon must look like this… nearly tetrahedral

…this means that the two rings must lie in different, perpendicular planes

Figure 5.49 shows some "natural" and "unnatural" spiro compounds.

Agarospirol   Bakkenolide-A   A [5]triangulane   Tricyclo[4.1.0.0^{4,6}]heptane

FIGURE 5.49 Some spiro compounds. Don't worry about the exotic names.

Bicyclic molecules in which two rings share two or more carbons are even more important than the spirocyclic compounds. The simplest way in which two rings can share more than one carbon is for two adjacent carbons to be shared in a fused structure. The "fusion" positions, or "bridgeheads," are shown in red (Fig. 5.50).

FIGURE 5.50 Some schematic, two-dimensional structures for fused bicyclic compounds in which two rings share a pair of adjacent carbon atoms.

The hydrogens attached at the ring junction positions can be either on the same side (cis) or on opposite sides (trans). In practice, both stereochemistries are possible for larger rings, but for the small rings only the cis ring junction is stable (see Problem 5.23). Figure 5.51 shows two compounds that contain this kind of ring fusion.

PROBLEM 5.23 Use your models to convince yourself that the trans stereochemistry is not possible for the molecules at the left of Figure 5.50.

cis Ring fusion

Aflatoxin B$_1$

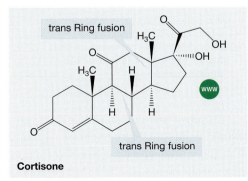

trans Ring fusion

trans Ring fusion

Cortisone

FIGURE 5.51
Two fused polycyclic molecules. Note the cis and trans ring fusions.

Particularly important are *cis*- and *trans*-decalin, the compounds formed by the fusion of two cyclohexanes. Although Figure 5.52 shows these compounds in schematic form, by now we can certainly guess that the real shape will be much more intricate.

FIGURE 5.52 In decalin, the two hydrogens at the ring fusion can be either on the same side (cis) or on opposite sides (trans) of the rings.

Decalin | Schematic picture of *cis*-decalin | Schematic picture of *trans*-decalin

It's easy enough to draw them out though, if one goes about it carefully. First of all, draw a single, perfect chair cyclohexane, and put in the axial and equatorial bonds at two adjacent carbons. As for any trans disubstituted cyclohexane (p. 214), the two rings of *trans*-decalin will be attached through either two equatorial or two axial bonds. As shown in Figure 5.53, attachment through two axial bonds is impossible—the distance to be bridged is too great (try it with models).

FIGURE 5.53 A futile attempt to construct *trans*-decalin by connecting two axial carbons with a pair of methylene groups. It is not possible to span the two axial positions with only four methylene groups; two methylene groups occupy axial positions leaving only two others (blue) to complete the ring.

So we are left with only the two equatorial bonds for trans attachment of a second ring, and this time the new chair fits in easily. *trans*-Decalin can be made by connecting two adjacent equatorial positions through a chain of four methylene groups (Fig. 5.54).

It is easy to connect two equatorial positions with a chain of four methylenes; the hydrogens at the fusion positions (bridgeheads) must be axial

trans-Decalin: two fused chair cyclohexanes

FIGURE 5.54 *trans*-Decalin.

Cis substitution must involve one axial and one equatorial bond in the attachments to the second ring (Fig. 5.55). If we pay attention to the rules for drawing perfect chairs, it's easy to produce a good drawing for this molecule. Remember that every ring bond in a cyclohexane is parallel to the bond directly across the ring (p. 197).

To make a cis junction we must connect one axial and one equatorial position with a four-carbon chain

=

cis-Decalin
The two bridgehead hydrogens occupy one axial and one equatorial position

FIGURE 5.55 cis-Decalin.

PROBLEM 5.24 Use models to convince yourself that cis-decalin is a mobile molecule, undergoing easy double chair–double chair interconversions, whereas trans-decalin is rigidly locked.

Now imagine two rings sharing three carbons. This sharing produces a different kind of structure, called a bridged bicyclic molecule (Chapter 2, p. 86). Two cyclopentanes, for example, can share two or three carbons. We have seen the first case before (Fig. 5.50) and the second is drawn out in Figure 5.56.

Bridgehead position

Bridgehead position

FIGURE 5.56 Two five-membered rings sharing three carbons.

WORKED PROBLEM 5.25 Design a molecule in which two rings share four carbons.

ANSWER One sure-fire way to do a problem such as this is to start by drawing the carbons to be shared (four, in this example). Then add the remainders of the other rings.

The four carbons to be shared

Here are those four carbons incorporated into one ring

Here they are incorporated into another ring

In each case, the carbons at the fusion points are called the **bridgehead positions**. The shared carbons are shown in red in Figure 5.56; the carbon–carbon bonds in blue complete one cyclopentane. A more complete structure for this bicyclic hydrocarbon is also shown. Such molecules are called "bridged" and are extremely common. Three compounds found in Nature are shown in Figure 5.57.

β-Pinene

Lycopodine

Calicheamicin

Several sugars

FIGURE 5.57 Three naturally occurring bicyclic molecules.

CONVENTION ALERT

Bicyclic compounds are named in the following way: One first counts the number of carbons in the ring system. The molecule in Figure 5.58 has eight ring carbons. Thus the base name is "octane." The molecule is numbered by counting from the bridgehead carbon [carbon number 1, C(1)] around the largest bridge first, proceeding to the other bridgehead. One then continues counting around the second longest bridge and finally numbers the shortest bridge. Any substituents can now be assigned a number. So the compound in Figure 5.58 is a 3,3-dimethylbicyclooctane.

FIGURE 5.58 Part of the naming protocol for a typical bicyclic compound.
1. Count from one bridgehead around the longest bridge to the other bridgehead (1, 2, 3, 4, 5).
2. Next, count around the second longest bridge (6, 7).
3. Last, number the shortest bridge (8). This compound is a 3,3-dimethylbicyclooctane.

The bridges are counted from the bridgeheads and assigned numbers equal to the number of atoms in the bridges, *not counting the bridgehead atoms themselves*. These numbers are enclosed in brackets, largest first, in between the designation "bicyclo" and the rest of the name (Fig. 5.59). So this compound is called 3,3-dimethylbicyclo[3.2.1]octane.

FIGURE 5.59 More of the naming protocol for bicyclic compounds. This molecule is 3,3-dimethylbicyclo-[3.2.1]octane. The red dots show the bridgehead atoms. These are not counted in sizing the bridges, but are counted in numbering the compound.

1 Atom bridge

3 Atom bridge

2 Atom bridge

3,3-Dimethylbicyclo[3.2.1]octane

PROBLEM 5.26 Make drawings of the following three compounds: (a) *cis*-bicyclo[3.3.0]-octane, (b) 1-fluorobicyclo[2.2.2]octane, (c) *trans*-9,9-dimethylbicyclo[6.1.0]nonane.

PROBLEM 5.27 Name the following compounds:

PROBLEM 5.28 How many signals would appear in the ^{13}C NMR spectrum for each of the molecules in Problem 5.27?

As with the 1,2-fused systems we must worry about the stereochemistry at the bridgehead positions. At first, this may seem trivial—where could these hydrogens be but where they are shown in the figures? The answer is, "inside the cage!" In the compounds we have already examined, it is not easy for hydrogens to occupy the inside position. As shown in Figure 5.60, all four valences (bonds) of the bridgehead carbons of bicyclo[3.2.1]octane would be pointing in the same direction if one or two bridgehead hydrogens were inside the cage. But now imagine increasing the size of the bridges. In principle, it should be possible to make the bridge chains long enough so that a normal, nearly tetrahedral, arrangement can be achieved with one or both bridgehead hydrogens inside the cage (Fig. 5.61). And so it is. The molecules shown in Figure 5.62 are both known, as are several other examples.

FIGURE 5.61 If the bridges are long enough, there is no angle strain in an "in, out" isomer.

FIGURE 5.62 Two known "in" or "in, out" compounds.

"Out, out" bicyclo[3.2.1]octane

"In, out" bicyclo[3.2.1]octane

Side view of "in, out" bicyclo[3.2.1]octane as seen by the eye

"In, in" bicyclo[3.2.1]octane

FIGURE 5.60 Three stereoisomers of bicyclo[3.2.1]octane. Only the first isomer is known.

5.8   SPECIAL TOPIC: POLYCYCLIC SYSTEMS

We needn't stop here. More rings can be fused or attached in fused or bridging fashion to produce wondrously complex structures. The naming protocols are complicated, if ultimately logical, and won't be covered here. Many of the compounds have common names that are meant to be evocative of their shapes. Prismane and cubane are examples. The molecular versions of the Platonic solids, tetrahedrane, cubane, and dodecahedrane are all known, although the parent tetrahedrane still evades synthesis (Fig. 5.63).

Prismane   **Cubane**

Cholesterol
a polycyclic compound
a steroid

Dodecahedrane

Tetra-*tert*-butyltetrahedrane

FIGURE 5.63 Polycyclic natural and "unnatural" products.

Tricyclo[1.1.1.0^{1,3}]pentane
([1.1.1]propellane)

Bicyclo[1.1.1]pentane

FIGURE 5.64 Two polycyclic small ring compounds.

Spectacular new molecules containing rings are always appearing. In 1984, Professor Kenneth B. Wiberg (b. 1927) of Yale University reported the construction of the polycyclic molecule, tricyclo[1.1.1.0^{1,3}]pentane (also known as [1.1.1]propellane, Fig. 5.64), a marvelously exciting molecule. You are now only five chapters into your study of organic chemistry, yet it is possible for you to appreciate why the chemical world was knocked out by this compound.

PROBLEM 5.29 Why is tricyclo[1.1.1.0^{1,3}]pentane unusual? Would you expect it to be especially stable or unstable with respect to its cousin bicyclo[1.1.1]pentane? Why?

5.9   SPECIAL TOPIC: ADAMANTANES IN MATERIALS AND BIOLOGY

Consider constructing a polycyclic molecule by expanding a chair cyclohexane. First, connect three of the axial bonds to a cap consisting of three methylene (CH_2) groups all connected to a single methine (CH) group. This process produces the beautiful molecule tricyclo[3.3.1.1^{3,7}]decane, better known as "adamantane," $C_{10}H_{16}$ (Fig. 5.65).

FIGURE 5.65 A schematic construction of the polycyclic molecule adamantane.

Chair cyclohexane → Add three methylene groups → Cap with a CH = **Adamantane** $(C_{10}H_{16})$

FIGURE 5.66 Adamantane is composed entirely of chair cyclohexanes. The first two (a and b) are easy to see as chairs, but (c) and (d) may require the use of models.

As Figure 5.66 shows, adamantane is composed entirely of perfect chair six-membered rings. As you would expect, this molecule is nearly strain-free, and constitutes the thermodynamic minimum for all the $C_{10}H_{16}$ isomers.

Adamantane was first found in the 1930s in trace amounts in petroleum residues by the Czech chemist Stanislav Landa (1898–1981), but can now be made easily in quantity by a simple process discovered by the American chemist Paul von R. Schleyer (b. 1930).

Consider what happens when we continue the capping process begun in our transformation of chair cyclohexane into adamantane in Figure 5.65. Adamantane itself is composed of only chair cyclohexanes, so we have a number of possible places to start the process (Fig. 5.67). Addition of one more four-carbon cap produces "diamantane," two caps, "triamantane," three, "tetramantane," and so on.

The ultimate result of the capping process is a molecule composed of a network of adamantanes, and this is the structure of diamond (Fig. 5.67).

FIGURE 5.67 The continuation of the capping process leads to polyadamantanes and, ultimately, to diamond.

Adamantane —cap→ **Diamantane** =

—cap→

Triamantane =

—cap→

One isomer of tetramantane =

Adamantane chemistry remains full of astonishing surprises. For years it has been known that synthetic diamonds could be made from another polymeric form of carbon, graphite, by treatment at high temperature (~2300 K) and very high pressure (7×10^4 kg/cm^2). However, in the 1980s in the Soviet Union and Japan, it was discovered that diamond films could be grown at low temperature and pressure by passing a stream of hydrogen containing a few percent methane through an electric discharge (Fig. 5.68). The mechanism of this reaction remains obscure. Any ideas?

$$H_2 \; + \; CH_4 \xrightarrow[\text{discharge}]{\text{electrical}} \text{Diamond}$$

FIGURE 5.68 A simple (and astonishing) synthesis of diamond.

PROBLEM 5.30 What is the empirical formula of diamond? For a large polymeric molecule such as diamond, the edges of the molecule are insignificant in figuring out the formula.

PROBLEM 5.31 Can you draw other isomers of tetramantane?

Adamantane remains a favorite of chemists because of its intrinsic symmetry and beauty, and probably because of the difficulty of working out the mechanism of its formation. It does not seem of much practical interest. However, some of its simple derivatives have remarkable properties and some are quite active biologically. For example, 1-aminoadamantane, a compound easily made from adamantane itself, is one of the very few antiviral agents known. This remarkable property was discovered during routine empirical screening at duPont in the 1960s, and 1-aminoadamantane has since been marketed, mostly as an agent against influenza A and C. It apparently works by migrating through the cell membrane to attack the virus within. It is the adamantane cage, acting as a molecular ball of grease, that helps in the membrane penetration. It has been speculated that other, large, symmetrical hydrocarbon "blobs" might act in the same way. Unfortunately, they are not so easy to make.

5.10   SUMMARY

New Concepts

This chapter deals exclusively with the structural properties of ring compounds. Two kinds of destabilizing effects on rings are discussed. Torsional strain, the destabilizing effect of eclipsed carbon–hydrogen bonds, has been mentioned before when the acyclic hydrocarbons were discussed. The planar forms of ring compounds are particularly subject to torsional strain. In addition, rings contain varying amounts of angle strain, depending on the ring size. Any deviation from the ideal tetrahedral angle of 109.5° will introduce angle strain. Rings adopt nonplanar forms in order to minimize the combination of torsional and angle strain.

One nonplanar ring deserves special mention. Cyclohexane avoids torsional and angle strain by adopting a chair conformation in which a nearly ideal tetrahedral angle is achieved and all carbon–hydrogen bonds are perfectly staggered. In cyclohexane, there is a set of six axial hydrogens that is converted into the set of six equatorial hydrogens through rotations about carbon–carbon bonds called a ring flip.

There are many possible quantitative evaluations of strain. This chapter uses related heats of formation (ΔH_f^o) and heats

of combustion (ΔH_c^o) to arrive at values for the strain energies of various rings. As rings increase in size, strain initially decreases, reaching a minimum at the strain-free cyclohexane. Strain then increases until large ring sizes are reached.

By combining rings in spiro, fused, and bridged fashion, very complex polycyclic molecules can be constructed.

Key Terms

angle strain (p. 192)
axial hydrogens (p. 197)
bridged (p. 218)
bridgehead positions (p. 222)

conformational analysis (p. 203)
equatorial (p. 197)
fused (p. 218)
spiro (p. 218)

torsional strain (p. 193)
van der Waals strain (p. 205)

Reactions, Mechanisms, and Tools

The few reactions encountered here are not really new. There is, however, a somewhat more detailed treatment of the formation of carbon dioxide and water from hydrocarbons and

oxygen (combustion), and a discussion of the effects of strain on the bond energy of the carbon–carbon bond.

Common Errors

You are presented in this chapter with the most difficult challenge so far in visualizing molecules. Moreover, the difficulty in translating from two dimensions to three is compounded by the mobile equilibria present in many ring compounds. In analyzing complicated phenomena, such as the equilibrations of disubstituted cyclohexanes, be sure to use models, at least at first.

Two small points seem to give students trouble. First, one hears constantly, I'm no artist, I can't draw a good chair cyclohexane! Nonsense. It does not take an artist, merely an artisan. Go slowly, do not scribble, and follow the simple procedure outlined in this chapter (p. 197). Drawing a perfect chair cyclohexane is one thing that everyone can do.

Determining which groups are cis and trans in ring compounds causes problems. *Remember*: cis means "on the same side," and trans, "on opposite sides." Bonds to cis groups do not have to be parallel, merely both up or both down. Figure 5.69 gives two common examples, one straightforward, the other not so obvious.

FIGURE 5.69 In both these compounds the two hydrogens shown are cis, as are the two X groups. This relationship is easy to see for the flat cyclopropane, but harder in the nonplanar cyclohexane.

5.11 ADDITIONAL PROBLEMS

PROBLEM 5.32   By now you have encountered most structural types and have some experience with rings. Write and name all the isomers of the formula C_5H_8. Stay alert for isomerism of the cis/trans (Z/E) and (R/S) types. Indicate chiral molecules with an asterisk. This problem is very hard at this point.[*] There are all sorts of odd structural types among these isomers. It should not be hard to get most of them, but getting them all is really tough. Think in three dimensions and *organize*. *Hint*: The total number of isomers, not including enantiomers, is 28.

[*]Indeed, this problem constituted the entire first hour examination in my first course in organic chemistry.

PROBLEM 5.33   Write out the pairs of enantiomers for the chiral isomers in Problem 5.32.

PROBLEM 5.34   Assign (R) and (S) configurations for the versions of the chiral cyclic molecules shown on the left in the answers to Problem 5.33 given in the Study Guide.

PROBLEM 5.35   Draw the two possible chair forms of *cis*- and *trans*-1,4-dimethylcyclohexane. Are the two forms identical, enantiomeric, or diastereomeric? In each case, indicate which chair form will be more stable and explain why. Is either of these molecules chiral?

PROBLEM 5.36   Draw the two possible chair forms of *cis*- and *trans*-1-isopropyl-4-methylcyclohexane. Are the two forms identical, enantiomeric, or diastereomeric? In each case, indicate which chair form will be more stable and explain why. Is either of these molecules chiral?

PROBLEM 5.37   Draw the two possible chair forms of *cis*- and *trans*-1,3-dimethylcyclohexane. Are the two forms identical, enantiomeric, or diastereomeric? In each case, indicate which chair form will be more stable and explain why. Is either of these molecules chiral?

PROBLEM 5.38   Draw the possible chair forms of *cis*- and *trans*-1-isopropyl-3-methylcyclohexane. Are the two forms identical, enantiomeric, or diastereomeric? In each case, indicate which chair form will be more stable and explain why. Is either of these molecules chiral?

PROBLEM 5.39   Use the data in Table 5.3 (p. 209) to calculate the energy difference between the possible isomers of *cis*- and *trans*-1,4-dimethylcyclohexane.

PROBLEM 5.40   Use the data in Table 5.3 (p. 209) to calculate the energy difference between the possible isomers of *cis*- and *trans*-1-isopropyl-4-methylcyclohexane.

PROBLEM 5.41   In Section 5.6b (p. 211), you saw that for purposes of stereochemical analysis you could treat the decidedly nonplanar *cis*- and *trans*-1,2-dimethylcyclohexanes as if they were planar. The planar forms represent the average positions of ring atoms in the rapid chair–chair interconversions. Use planar representations of the following cyclohexanes to determine which molecules are chiral.

(a) *cis*-1,3-Dimethylcyclohexane
(b) *trans*-1,3-Dimethylcyclohexane
(c) *cis*-1-Isopropyl-3-methylcyclohexane
(d) *trans*-1-Isopropyl-3-methylcyclohexane
(e) *cis*-1,4-Dimethylcyclohexane
(f) *trans*-1,4-Dimethylcyclohexane
(g) *cis*-1-Isopropyl-4-methylcyclohexane
(h) *trans*-1-Isopropyl-4-methylcyclohexane

PROBLEM 5.42   How many signals will appear in the ^{13}C NMR spectrum of 1,1-dimethylcyclohexane?
(a) At low temperature.
(b) At high temperature.
Hint: Start by figuring out what effect the increase in temperature will have.

PROBLEM 5.43   How many signals will appear for *cis*-1,3-dimethylcyclohexane in its ^{13}C NMR spectrum? How many signals will appear for *trans*-1,3-dimethylcyclohexane in its ^{13}C NMR spectrum?
(a) At low temperature.
(b) At high temperature.

PROBLEM 5.44   How many signals will appear in the ^{13}C NMR spectrum of adamantane (p. 225)?

PROBLEM 5.45 Draw both chair forms for the following cyclohexanes. Indicate the more stable chair form in each case. Not all examples will be obvious; some choices may be too close to call.

(a)

(b)

(c)

(d)

(e)

(f)

PROBLEM 5.46 Name the following compounds:

(a)

(b)

(c)

PROBLEM 5.47 Write structures for the following compounds:

(a) 2,2-Dibromo-3,3-dichloro-5,5-difluoro-6,6-diiodobicyclo[2.2.1]heptane
(b) Bicyclo[1.1.1]pentan-1,3-diol
(c) Hexamethylbicyclo[2.2.0]hexa-2,5-diene

PROBLEM 5.48 The sugar glucose is a six-membered ring containing an oxygen atom. There are groups attached at each of the five ring carbons as shown. The red "squiggly" bond merely means that the hydroxyl group at the 1-position may be either up or down. That is, there are two structures for glucose. Draw these two molecules in three dimensions.

PROBLEM 5.49 One of the isomers of methylbicyclo[2.2.1]heptane can exist in two diastereomeric forms (we are not counting mirror images, which increases the total number of stereoisomers to four). Write the possible isomers of methylbicyclo[2.2.1]heptane and explain how one of them can exist in two forms.

PROBLEM 5.50 (a) Can one of the isomers of methylbicyclo[2.2.2]octane also exist in two diastereomeric forms? That is, does it behave just like the methylbicyclo[2.2.1]heptane mentioned in Problem 5.49? (b) Can one of the isomers of methylbicyclo[2.2.2]octane exist in two enantiomeric forms?

Alkyl Halides, Alcohols, Amines, Ethers, and Their Sulfur-Containing Relatives

6

So, if Sunday you're free,
Why don't you come with me,
And we'll poison the pigeons in the park.

My pulse will be quickenin',
With each drop of strych-i-nin',
We feed to a pigeon,
(It just takes a smidgin),
To poison a pigeon, in the park,
Tra-La.

—Tom Lehrer*

6.1   PREVIEW

In this chapter, we will see some old friends, alkyl halides and alcohols, as well as some new molecules rather closely related to them, amines, ethers, thiols, and thioethers. Both alkyl halides and alcohols can be made through the addition reactions encountered in Chapter 3 (p. 133). Amines somewhat resemble alcohols, and understanding their chemistry is a necessary prelude to working through Chapter 23, which discusses the chemistry of the many biologically important molecules containing nitrogen. A nitrogen atom provides one of the crucial linking agents in the formation of polyamino acids, called **peptides** or **proteins** (Fig. 6.1). In this

*Thomas Lehrer (b. 1928) is an American mathematician and songwriter, most of whose work was done in the 1950s and 1960s. Despite the unfortunate fact that there has been little new material for many years, his songs, often sardonic and always very funny, retain their appeal.

chapter, we will examine the qualities that make these molecules similar and prepare for Chapter 7, in which we will take a long, detailed look at four prototypal new reactions in which alkyl halides and alcohols are especially important.

FIGURE 6.1 A nitrogen-based link is vital in the polymeric molecules known as proteins and peptides.

ESSENTIAL SKILLS

1. So much of chemistry, including organic chemistry, involves acids and bases. In this chapter, we review what we know and focus on a measure of acidity, pK_a. It is important to have a reasonable idea of the pK_a values of the general classes of compounds. There is a table of pK_a values on the inside of the back cover of this book.

2. We also encounter solvation in this chapter. Being able to choose an appropriate solvent for a reaction, and to understand how solvents operate, is an important skill both in the practical world and in this course.

IMPORTANT DETAILS

1. A low pK_a means that the molecule in question is a strong acid, a good proton donor. Conversely, a high pK_a means that the compound is a poor acid.

2. The synthetic sequence "alkene → halide → Grignard reagent" is one you will use many times in solving synthetic problems.

6.2   ALKYL HALIDES: NOMENCLATURE AND STRUCTURE

Alkyl iodides, bromides, chlorides, and fluorides are collected under the name of **alkyl halides**. Alkyl halides have the formula $C_nH_{2n+1}X$, where X = F, Cl, Br, or I. Quite appropriately, alkyl halides are named as fluorides, chlorides, bromides, and iodides. Both common and systematic (substitutive) nomenclature names are used, although as the complexity of the molecule increases, the system naturally takes over. The important rules to remember for naming saturated halides are (1) minimize the number given to the substituent, (2) name multiple substituents in alphabetical order, and (3) names are based on the longest straight chain that contains the halide. If a carbon–carbon double bond is present, the double bond takes precedence over the halogen when positional numbers are assigned. Table 6.1 gives some examples with both common and substitutive names given. There are some important common names that are used very often. For example, haloforms are CHX_3, vinyl halides are $H_2C=CHX$, and allyl halides are $H_2C=CH-CH_2X$.

TABLE 6.1   Some Alkyl Halides, Names, and Known Properties

Compound	Common Name	Substitutive Name	bp (°C)	mp (°C)
CH_3-F	Methyl fluoride	Fluoromethane	−78.4	−141.8
CH_3CH_2-Br	Ethyl bromide	Bromoethane	38.4	−118.6
$(CH_3)_2CH-I$	Isopropyl iodide	2-Iodopropane	89.4	−90.1
$(CH_3)_3C-Cl$	*tert*-Butyl chloride	2-Chloro-2-methylpropane	51.0	−25.4
CH_2Br_2	Methylene bromide	Dibromomethane	97	−52.7
$CHCl_3$	Chloroform	Trichloromethane	61.5	−63.5
$CH_2=CHBr$	Vinyl bromide	Bromoethene	15.8	−139.5
$CH_2=CH-CH_2Cl$	Allyl chloride	3-Chloropropene	45	−134.5
		2-Bromo-1-chlorobutane	147	
		3-Chloro-1-pentene	94	
		Bromocyclopentane	138	

Halides are classified as methyl, primary, secondary, tertiary, and vinyl (Fig. 6.2). Chapter 2 (p. 76) introduced this nomenclature.

$CH_3—I$
Methyl iodide (a methyl halide)

$CH_3CH_2—I$
Ethyl iodide (a primary halide)

$(CH_3)_2CHI$

Isopropyl iodide (a secondary halide)

$(CH_3)_3C—I$

tert-Butyl iodide (a tertiary halide)

Vinyl iodide

FIGURE 6.2 Some alkyl and vinyl halides at various levels of abstraction.

Remember: A primary carbon is attached to only one other carbon, a secondary carbon to two others, and a tertiary carbon to three others. In vinyl halides, there is always a double bond to which the halide is directly attached.

As pointed out in Chapter 2 (p. 87), substituted alkanes are polar molecules. The halogens are electronegative atoms and strongly attract the electrons in the carbon–halogen bond. Alkyl halides have dipole moments of approximately 2 D (Fig. 6.3; Table 6.2).

TABLE 6.2
Dipole Moments of Some Simple Alkyl Halides

Alkyl Halide	Dipole Moment (D)
Methyl fluoride	1.85
Ethyl fluoride	1.94
Methyl chloride	1.87
Ethyl chloride	2.05
Methyl bromide	1.81
Ethyl bromide	2.03
Methyl iodide	1.62
Ethyl iodide	1.91

FIGURE 6.3 A substantial dipole moment exists in the polar carbon–halogen bond, X = F, Cl, Br, and I.

Simple alkyl halides are nearly tetrahedral, and therefore the carbon to which the halogen is attached is approximately sp^3 hybridized. The bond to halogen involves an approximately sp^3 carbon orbital overlapping with an orbital on the halogen for which the principal quantum number varies from $n = 2$ for fluorine to $n = 5$ for iodine. The C—X bond lengths increase and the bond strengths decrease as we read down the periodic table. Some typical values for methyl halides are shown in Figure 6.4. The structures of alkyl halides at the bottom show both C—X bond lengths (in angstrom units) and C—X bond strengths (in kilocalories per mole). *Remember*: These values refer to homolytic bond breaking (formation of two neutral radicals).

FIGURE 6.4 The C—X bond of an alkyl halide is formed by the overlap of a singly occupied carbon sp^3 orbital with a singly occupied halogen orbital.

PROBLEM 6.1 Name the following compounds:

(a)

(b)

(c)

(d)

(e) CHI_3

(f)

(g)

PROBLEM 6.2 Draw structures for the following compounds:

(a) 3,3-Dichlorohexane

(b) 1-Bromocyclohexene

(c) 2-Bromopropene

(d) *meso*-3,4-Dichlorohexane (three dimensions!)

(e) 1 (*R*)-2(*R*)-Dichlorocyclopropane

6.3   ALKYL HALIDES AS SOURCES OF ORGANOMETALLIC REAGENTS: A SYNTHESIS OF HYDROCARBONS

Halides have all sorts of uses in the "real" practical world. They appear as anesthetics, insecticides, herbicides, the building blocks of polymers such as polyvinyl chloride (PVC), and myriad other products. I'll bet that as you read this paragraph you are thinking "Sure, but haven't there been serious problems with insecticides such as

DDT and herbicides like agent orange, and aren't chlorofluorocarbons eating up the ozone layer?" The answer is a resounding, Yes! There seems to be no "free lunch" as unintended consequences often appear when we try to solve problems with technology. The issues are complex, but it is clear that the chemists of today and the future need to supply the technical expertise to avoid these problems, while retaining the benefits.

Halides are important in synthetic chemistry because they allow for the formation of carbon–hydrogen and carbon–carbon bonds. Central to these uses of alkyl and other halides is their easy conversion into **organometallic reagents**, molecules containing carbon–metal bonds. We will look at only one limited use of these synthetically important compounds in this chapter, but many others will appear later.

When an alkyl halide (fluorides are generally exceptions) is added to a cold mixture of magnesium or lithium metal and an ether solvent [usually diethyl ether or tetrahydrofuran (THF)], the metal begins to disappear in an exothermic reaction. The end result is either a **Grignard reagent** or an **organolithium reagent** (Fig. 6.5).

Grignard reagent

Organolithium reagent

FIGURE 6.5 Formation of organometallic reagents.

R = alkyl or alkenyl
X = Cl, Br, or I

Grignard reagents were discovered by P. A. Barbier (1848–1922) and their chemistry was worked out extensively by his student, Victor Grignard (1871–1935), who received the 1912 Nobel prize for his work in this area. Neither the formation of the Grignard reagent nor its structure is an easy subject. The simplest formulation is "RMgX" and should be amplified by the addition of the ionic resonance form in Figure 6.5. The Grignard reagent is also in equilibrium with a mixture of the magnesium halide and dialkylmagnesium compound. Moreover, the ether solvent is essential for its formation. Grignard reagents incorporate two molecules of solvent, conveniently left out in the traditional formulation as RMgX (Fig. 6.6). Whatever the detailed structure, the result is a highly polar reagent that is a very strong Lewis base.

FIGURE 6.6 The Grignard reagent has a complex structure.

The mechanism of formation of the Grignard reagent involves radical-transfer reactions in which a transient alkyl radical is formed in the presence of a magnesium-centered radical (Fig. 6.7).

$$R\!-\!X \quad \overset{..}{Mg} \quad \longrightarrow \quad R \cdot \quad \cdot Mg\!-\!X \quad \longrightarrow \quad R\!-\!Mg\!-\!X$$

X = Cl, Br, or I

FIGURE 6.7 Formation of a Grignard reagent.

We met the methyl radical in Chapter 2 (p. 60). Radicals, sometimes called "**free radicals**," are single-electron neutral species that can recombine to form carbon–carbon bonds, or react with other single-electron species to form new bonds to other atoms. The arrow formalism for radical reactions uses single-barbed arrows.

CONVENTION ALERT

PROBLEM 6.3 Draw an interaction diagram to show the stabilization when two methyl radicals combine. How great is that stabilization? For an example of an interaction diagram see page 29.

Coupling generates RMgX, and further radical reactions can equilibrate this species with the dialkylmagnesium. Most halides (except fluorides) can be made into Grignard reagents.

Organolithium reagents are even more complex. Here the ether solvent is not essential, but the reagents themselves are known not to be monomeric and to form aggregates the size of which depends on the nature of the solvent and the structure of the R group. As in RMgX, the representation of organolithium reagents as RLi, even with the addition of the polar resonance form, is a convenient simplification (Fig. 6.8).

$$R\!-\!X \ + \ Li \ \longrightarrow \ (R\!-\!Li)_n$$

FIGURE 6.8 Organolithium reagents are not monomeric, but oligomeric. The exact value of n depends on solvent and the structure of R.

Both of these organometallic reagents are extraordinarily strong bases, and they must be carefully protected from moisture and oxygen. Chemists form them under a strictly dry and inert atmosphere. Although these reagents do not contain free carbanions, *they act as if they did*. They are sources of R:$^-$. The very polar carbon–metal bond attacks all manner of Lewis and Brønsted acids. Water is more than sufficiently strong enough to protonate a Grignard or organolithium reagent (Fig. 6.9).

$$R\!-\!Mg\!-\!X \ + \ H\!-\!\overset{..}{\underset{..}{O}}H \ \longrightarrow \ R\!-\!H \ + \ H\overset{..}{\underset{..}{O}}\!-\!Mg\!-\!X$$

$$R\!-\!Li \ + \ H\!-\!\overset{..}{\underset{..}{O}}H \ \longrightarrow \ R\!-\!H \ + \ Li\overset{..}{\underset{..}{O}}H$$

FIGURE 6.9 The powerfully basic organometallic reagents are easily protonated by water to give hydrocarbons and metal hydroxides.

The end product is a hydrocarbon. This sequence constitutes a synthesis of hydrocarbons from most halides (Fig. 6.10). This reaction can often be put to good use in the construction of isotopically labeled reagents, because D_2O can be used in place of water to produce a specifically deuteriolabeled hydrocarbon.

$$CH_3CH_2Br \ + \ Mg \xrightarrow{\text{ether}} CH_3CH_2MgBr \xrightarrow{D_2O} CH_3CH_2-D$$

FIGURE 6.10 Treatment of an organometallic reagent with D_2O gives deuteriolabeled molecules.

PROBLEM 6.4 Given inorganic reagents of your choice (including D_2O), devise syntheses of the following molecules from the indicated starting materials:

6.4   ALCOHOLS

6.4a Nomenclature. There are two ways to name alcohols: the systematic way, and the way chemists often do it, by using the splendidly vulgar common names. As usual, these common names are only used for the smaller members of the class, and the systematic system takes over as complexity increases.

Alcohols are named systematically by dropping the final "e" of the parent hydrocarbon and adding the suffix "ol." The functional group, OH, is given the lowest number possible, taking precedence over other groups in the molecule including the related thiol group, SH. The longest carbon chain is identified and the substituents numbered appropriately (Fig. 6.11).

The smaller alcohols, however, are commonly named by appending the word alcohol to the appropriate group name. Figure 6.11 gives some systematic and common names for the smaller alcohols. When we get past approximately five carbons, the systematic naming protocol takes over completely, and even the delightful name amyl for the five-carbon alcohols is disappearing.* However, some common names seem to be quite hardy. For example, the correct name for hydroxybenzene is benzenol, but it appears that the traditional phenol will survive for a long time (Fig. 6.12).

---

*Amyl derives from the Latin word for starch, "amylum." The first amyl alcohol was isolated from the fermentation of potatoes.

FIGURE 6.11 Some systematic and commonly used names for alcohols.

Phenol   **Amyl alcohol**   **Ethylene glycol**   **Glycerol (glycerin)**

FIGURE 6.12 Some widely used names for alcohols.

PROBLEM 6.5 Provide systematic names for the following alcohols:

6.4b Structure of Alcohols.

Alcohols are derivatives of water, and it will be no surprise to see that they rather closely resemble water structurally. The bond angle is expanded a bit from 104.5° in H—O—H to approximately 109° in simple alcohols R—O—H.

The oxygen–hydrogen bond length is little changed from that of water, but the carbon–oxygen bond is a bit shorter than the carbon–carbon bond in ethane (Fig. 6.13).

FIGURE 6.13 Bond lengths and angles for some hydroxy compounds.

PROBLEM 6.6 What is the approximate hybridization of the oxygen atom in methyl alcohol (H_3C—O—H angle = 108.9°). How do you know?

The electronegative oxygen atom results in a carbon–oxygen bond that is stronger and shorter than the carbon–carbon bond in ethane. An electron in an orbital near oxygen is more stable than it would be in an orbital near carbon. The orbital interaction diagram of Figure 6.14 shows the construction of the carbon–oxygen bond as well as typical bond energies.

FIGURE 6.14 A graphical description of the formation of the carbon–oxygen σ bond.

6.4c Physical Properties of Alcohols.

The presence of the electronegative oxygen atom ensures that bonds will be strongly polarized in alcohols, and that substantial dipole moments will exist. In solution, alcohols are strongly associated both because of dipole–dipole attractions and because of **hydrogen bonding**. The basic oxygen atoms form partial bonds to the acidic hydroxyl hydrogens. Such bonds can be quite strong, on the order of 5 kcal/mol, a value not high enough to make permanent dimeric or polymeric structures, but quite sufficient to raise the boiling points of alcohols far above those of the much less strongly associated alkanes. Figure 6.15 shows an ordered, hydrogen-bonded structure for water. In the absence of hydrogen bonding, water, with a molecular weight of only 18, would surely be a gas.

Water acts as Brønsted base (proton acceptor)

Water acts as Brønsted acid (proton donor)

etc.

FIGURE 6.15 An alcohol is both Brønsted acid and Brønsted base. This figure shows hydrogen bonding between molecules of water.

The polarity of alcohols makes them quite water soluble, and most of the smaller molecules are miscible with, or at least highly soluble in water. Table 6.3 gives some physical properties of alcohols. For comparisons with the parent alkanes see Table 2.4 (p. 73).

TABLE 6.3 Some Physical Properties of Alcohols

Compound	bp (°C)	mp (°C)	Density (g/cm³)	Dipole Moment (D)
CH_3OH Methyl alcohol	65.15	− 93.9	0.79	1.7
CH_3CH_2OH Ethyl alcohol	78.5	−117.3	0.79	1.69
$(CH_3)_2CH_2OH$ Isopropyl alcohol	82.4	− 89.5	0.80	1.68
$CH_3CH_2CH_2CH_2OH$ Butyl alcohol	117.2	− 89.5	0.81	1.66
$(CH_3)_2CHCH_2OH$ Isobutyl alcohol	108	−108	0.80	1.64
$CH_3CH_2CH(CH_3)OH$ sec-Butyl alcohol	99.5		0.81	
$(CH_3)_3COH$ tert-Butyl alcohol	82.3	25.5	0.79	
OH Phenol	181.7	43	1.06	1.45

Even though hydrogen bonds are relatively weak (~5 kcal/mol) they are critically important. For example, life as we know it is clearly not possible without water and there would be no liquid water without hydrogen bonds. Moreover, hydrogen bonds are used to maintain the proper structures in proteins and nucleic acids, polymeric structures essential to our existence that we shall meet later in this book.

6.4d Acid and Base Properties of Alcohols.

We have just seen water and alcohols acting as Brønsted acids and bases. Hydrogen bonding is a perfect example of such behavior. In Chapter 7, we are going to look closely at a number of reactions of alcohols. These processes depend on the ability of alcohols to act as both acids and bases. Accordingly, we will first make sure that we can see these roles clearly.

A Brønsted acid is any compound that can donate a proton. A Brønsted base is any compound that can accept a proton. Let's look first at a very simple process, the reaction of solid potassium hydroxide (KOH) with gaseous hydrochloric acid (HCl). This reaction is nothing but a competition of two Brønsted bases (Cl⁻ and HO⁻) for a proton, H^+. The stronger Brønsted base (HO⁻) wins the competition (Fig. 6.16).

FIGURE 6.16 Two Brønsted bases competing for a proton.

Hydrochloric acid (HCl) is called the **conjugate acid** of Cl⁻, and Cl⁻ is the **conjugate base** of hydrochloric acid. Conjugate bases and acids are related to each other through the gain and loss of a proton (Fig. 6.17).

H–B is the conjugate acid of B:⁻   B:⁻ is the conjugate base of H–B

Caution! This is a formalized picture and does not imply the reaction of bases with a bare proton, H^+. Outside of the gas phase, bare protons do not exist.

FIGURE 6.17 Conjugate acids and bases.

PROBLEM 6.7 What are the conjugate acids of the following molecules?
(a) H_2O (b) ⁻OH (c) NH_3 (d) CH_3OH (e) $H_2C{=}O$ (f) ⁻CH_3

PROBLEM 6.8 What are the conjugate bases of the following molecules?
(a) H_2O (b) ⁻OH (c) NH_3 (d) CH_3OH (e) CH_4 (f) $HOSO_2OH$ (g) ⁻OSO_2OH

A molecule may be *both* a Brønsted acid and a Brønsted base! Consider water, for example. Water can both donate a proton (act as a Brønsted acid) and accept a proton (act as a Brønsted base). Figure 6.18 shows two reactions, (a) the protonation of water by hydrogen chloride, in which water acts as a base, and (b) the protonation of hydroxide ion by water, in which water is acting as the acid.

Many reactions begin with a protonation step. As strong acids are better able to donate a proton than weaker acids, it will be important for us to know which molecules are strong acids and which are weak. The dissociation of an acid in water can be described by the equation

$$HA + H_2O \rightleftarrows H_3O^+ + A^-$$

This equation leads to an expression for the equilibrium constant, K.

$$K = [H_3O^+][A^-]/[HA][H_2O]$$

Because it is present as solvent, in vast excess, the concentration of water remains constant in the ionization reaction. This equation is usually rewritten to give the acidity constant, K_a,

$$K_a = [H_3O^+][A^-]/[HA]$$

where the square brackets indicate concentration.

By analogy with pH, we can define a quantity **pK_a**.

$$pK_a = -\log K_a$$

The stronger the acid, the lower the pK_a. Table 6.4 gives pK_a values for some representative compounds taken from the larger list on the inside back cover of this book.

(a)

(b)

FIGURE 6.18 Water is both a Brønsted acid and base.

TABLE 6.4 Some pK_a Values for Assorted Molecules^a

Compound	pK_a	Compound	pK_a
HI	−10	H$_4$N$^+$	9.2
H$_2$SO$_4$	−3	CH$_3$O$\underline{H}$	15.5
HBr	−9	H$_2$O	15.7
HCl	−8	RO$\underline{H}$	16–18
ROH$_2$$^+$	−2	HC≡CH	24
H$_3$O$^+$	−1.7	NH$_3$	38
HNO$_3$	−1.3	H$_2$C=CH$_2$	50
HF	3.2	(CH$_2$)$_3$ (cyclopropane)	46
RCOO$\underline{H}$	4–5	CH$_4$	50–60
H$_2$S	7.0	(CH$_3$)$_3$C$\underline{H}$	50–70

^{a}When there is a choice, it is the underlined $\underline{H}$ that is lost.

This short list spans about 80 powers of 10 (pK_a is a log function). In practice, any compound with a pK_a lower than about +5 is regarded as a reasonably strong acid; those with pK_a values below 0 are very strong. As we discuss new kinds of molecules more pK_a values will appear. Should you memorize this list? I think not, but you will need to have a reasonable idea of the approximate acidity of different kinds of molecules.

In alcohols and water, Brønsted basicity is centered on the lone-pair electrons of the oxygen atom, which form bonds with a variety of acids, including the O—H hydrogens of other water or alcohol molecules (hydrogen bonding, p. 240). Protonation of the oxygen gives an oxonium ion, as it converts an alcohol, ROH, into an oxonium ion (RO^+H_2). This conversion will have profound chemical consequences, which we can preview here. Breaking the R—OH bond into ions is very difficult, as both positive (R^+) and negative (HO^-) ions must be formed. Hydroxide, HO^-, is very difficult to form—one says that hydroxide is a poor leaving group—and alcohols do not ionize easily (Fig. 6.19). After protonation, however, the "leaving group" is no longer hydroxide, but water. Formation of the neutral molecule water is a much easier process (Fig. 6.19).

FIGURE 6.19 Protonation of an alcohol leads to an oxonium ion. In this process, the very poor leaving group OH is transformed into the good leaving group OH_2.

TABLE 6.5 Some pK_a Values for Simple Oxonium Ions

Compound	pK_a
$H_3\overset{+}{O}$	−1.74
$CH_3\overset{+}{O}H_2$	−2.2
$CH_3CH_2\overset{+}{O}H_2$	−1.9
$CH_3CH_2CH_2CH_2\overset{+}{O}H_2$	−2.3
$CH_3CH_2\overset{+}{C}HOHCH_3$	−2.2
$(CH_3)_3C\overset{+}{O}H_2$	−2.6
$R\overset{+}{O}HR'$	About −3.5

Protonated alcohols are very strong Brønsted acids. Therefore, in order to protonate an alcohol, it takes a very strong acid, indeed. The conjugate base ($B\!:^-$) of the protonating acid (HB) must be a weaker base than the alcohol. The base must be an ineffective competitor in the equilibrium of Figure 6.19.

Oxonium ions are much better proton donors than are the alcohols themselves. Table 6.5 gives the pK_a values for a few protonated alcohols. Oxonium ions formed from alcohols have pK_a values in the −2 range, and those from ethers are even more acidic.

Brønsted acidity, of course, is centered upon the hydroxyl hydrogen. Loss of this hydrogen to a Brønsted base acceptor (it is not just ionized as H^+) leads to the conjugate base of the alcohol, the **alkoxide ion** (RO^-) (Fig. 6.20).

FIGURE 6.20 Loss of a proton to a general Brønsted base leads to an alkoxide ion.

Alcohols are approximately as acidic as water. Table 6.6 gives the pK_a values for some simple alcohols and water.

Note that the acidity order in aqueous solution is $CH_3OH > CH_3CH_2OH > (CH_3)_2CHOH > (CH_3)_3COH$. Apparently, the more alkyl groups on the alcohol, the weaker an acid it is. For many years, this pK_a order was explained through an inductive effect. Alkyl groups were assumed to be intrinsically electron donating. If this were true, the product alkoxides would be destabilized by alkyl groups, and the acidity of the corresponding alcohol would be reduced (Fig. 6.21).

TABLE 6.6   Some pK_a Values for Simple Alcohols

Compound		pK_a (loss of underlined H)
Water	H_2O	15.7
Methyl alcohol	$CH_3O\underline{H}$	15.5
Ethyl alcohol	$CH_3CH_2O\underline{H}$	15.9
Isopropyl alcohol	$(CH_3)_2CHO\underline{H}$	16.5
tert-Butyl alcohol	$(CH_3)_3CO\underline{H}$	17

Destabilized by dipole in CH_3–C bond

FIGURE 6.21 If alkyl groups were electron releasing, formation of an alkoxide from an alcohol would be retarded by alkyl groups.

Inductive effects *are* important. For example, the pK_a of 2,2,2-trifluoroethanol ($CF_3CH_2O\underline{H}$) is 12.5, whereas the pK_a of ethyl alcohol is 15.9. The fluorinated alcohol is more than a thousand times as acidic as ethyl alcohol (remember the pK_a scale is logarithmic). Fluorinated alcohols are always stronger acids than their hydrogen-substituted counterparts.

WORKED PROBLEM 6.9 Explain in detail why 2,2,2-trifluoroethanol is a stronger acid than ethyl alcohol.

ANSWER As fluorine is very electronegative, the fluorines are strongly electron withdrawing and will help stabilize the alkoxide ion. They will also stabilize the transition state leading to the alkoxide because partial negative charge has begun to develop on the oxygen atom. So the fluorines will have an effect on both the kinetic and thermodynamic properties of the alcohol.

Yet this traditional, and simple, explanation that treats alkyl groups as intrinsically electron donating is not correct. In 1970, Professor John Brauman (b. 1937) and his co-workers at Stanford University showed that in the gas phase the *opposite* acidity order obtained. The intrinsic acidity of the four alcohols of Table 6.6 is exactly opposite to that found in solution. The acidity order measured in solution reflects a powerful effect of the solvent, not the natural acidities of the alcohols themselves. Organic ions are almost all unstable species, and the formation of the alkoxide anions depends critically on how easy it is to stabilize them through solvation. *tert*-Butyl alcohol is a weaker acid in solution than methyl alcohol because the large *tert*-butyl alkoxide ion is difficult to solvate. The more alkyl groups, the more difficult it is for the stabilizing solvent molecules to approach. Of course, in the gas phase where solvation is impossible, the natural acidity order is observed (Fig. 6.22).

FIGURE 6.22 The smaller the alkoxide ion, the easier it is for solvent molecules to approach and stabilize it.

In the gas phase, the alkyl groups are actually operating so as to *stabilize* the charged alkoxide ions, presumably by withdrawing electrons, exactly in the opposite sense to what had been thought (Fig. 6.23).

FIGURE 6.23 Alkyl groups can be electron withdrawing.

WORKED PROBLEM 6.10 Can you describe this phenomenon—the stabilization of an adjacent pair of electrons by an adjacent alkyl group—in orbital terms? No orbital construction or complicated argument is necessary. A simple statement is all that is needed.

ANSWER Alkyl groups have both filled and empty molecular orbitals (see the problems at the end of Chapter 1 for several examples). An adjacent pair of electrons can be stabilized through overlap with the LUMO of the alkyl group. (Similarly, an alkyl group stabilizes an adjacent empty orbital through overlap with the alkyl HOMO.)

What lessons are to be drawn from this discussion? First, solvation is important and not completely understood. We are certain to find other phenomena best explained in terms of solvation in the future. Second, the gas phase is the ideal medium for revealing the intrinsic properties of molecules. (Some would go further and say that calculation is the best way.) However, the practical world of solvated reactions is certainly real, and for us to understand reactivity we can no longer afford to ignore the solvent.

6.5 SOLVENTS IN ORGANIC CHEMISTRY

We have just seen an example of a solvent playing a crucial role in determining the properties of alcohols. The greater acidity of methyl alcohol over *tert*-butyl alcohol in polar solution is a result of the greater ease of solvating the conjugate base of methyl alcohol, the methoxide ion, over the *tert*-butoxide ion (Fig. 6.22). Let's take a quick look at the properties of solvents and the process of solvation.

6.5a Polar Solvents. We have just seen that stabilization by solvent—**solvation**— is a consequence of hydrogen bonding. For hydrogen bonding to be possible, there clearly must be both a proton donor (usually a hydroxy or amino group, OH or NH_2) and a proton acceptor available (a pair of electrons). Solvents that can donate a proton are called **protic** solvents. Protic solvents are usually also quite polar; that is, they have relatively high dielectric consonants, ε. Solvents without available protons, solvents that are not proton donors, are **aprotic** and can be either polar or nonpolar. It is distinctly possible for a solvent to be quite polar, to have a high ε, but not be a proton donor. Of course, many molecules are both nonpolar and aprotic; hydrocarbons are obvious examples you know. Table 6.7 collects a few common polar solvents and their descriptions.

TABLE 6.7 Some Common Polar Solvents

Compound	Name	Dielectric Constant (ε)	Description
H_2O	Water	78	Protic, polar
HCOOH	Formic acid	59	Protic, polar
CH_3OH	Methyl alcohol	33	Protic, polar
CH_3CH_2OH	Ethyl alcohol	25	Protic, polar
$(CH_3)_2SO$	Dimethyl sulfoxide (DMSO)	47	Aprotic, polar
CH_3CN	Acetonitrile	38	Aprotic, polar
$(CH_3)_2NCHO$	Dimethylformamide (DMF)	37	Aprotic, polar
CH_3NO_2	Nitromethane	36	Aprotic, polar

6.5b Solubility: "Like Dissolves Like." Oil spreads on water, but very little dissolves. Two solids, salt (the prototypal ionic solid sodium chloride) and the sugar sucrose ($C_{12}H_{22}O_{11}$, an organic material bristling with OH groups), both dissolve completely in water (Fig. 6.24). What is going on?

Sodium chloride, Na^+Cl^-, dissolves in the polar solvent water, H_2O, because both the positive sodium ion and the negative chloride ion are well solvated by the protic, polar solvent water (Fig. 6.25). Nonpolar solvents cannot solvate ions efficiently and do not dissolve sodium chloride.

FIGURE 6.24 Two very different solids that dissolve easily in water.

(Only two hydrogen bonds shown)

FIGURE 6.25 Solvation of the positive sodium cation, and hydrogen bonding to the negative chloride ion (only two hydrogen bonds shown).

Protic solvents are particularly good at dissolving molecules capable of hydrogen bonding. Formation of hydrogen bonds in solution is highly stabilizing. Thus sucrose, a molecule with many OH groups, dissolves easily in water. The polar solvent water stabilizes both through formation of many hydrogen bonds and stabilization of the polarized O—H groups of sucrose. Figure 6.26 shows a schematic picture of hydrogen bonding (only one of many shown) and dipolar stabilization of the polyhydroxy compound sucrose by water.

FIGURE 6.26 Two ways water can stabilize an O—H bond.

THF

FIGURE 6.27 Stabilization (solvation) of water by THF.

Polar aprotic solvents can help disperse the partial charges in other polar aprotic molecules, and thus, dissolve them well. Although they cannot act as proton donors, such solvents can align their negative ends toward the positive charges in a polar molecule, and their positive ends toward the negative charges (Fig. 6.27). For example, Figure 6.27 shows the polar solvent THF solvating a water molecule by aligning dipoles in a stabilizing way. Water and THF are completely miscible.

By contrast, hydrocarbons cannot form hydrogen bonds (they are neither proton donors nor acceptors) and are not very polar. Typical ε values for hydrocarbons are around 2. The more nonpolar and "greasy" the solvent, the more easily hydrocarbon molecules dissolve. Thus, hydrocarbons (oil is a mixture of many hydrocarbons) are almost completely insoluble in water, but very soluble in other hydrocarbons. Acetone, [dimethyl ketone, $(CH_3)_2CO$] with an ε value of 21, is completely miscible (soluble in all proportions) in water, but the alkene isobutene, with the same number of non-hydrogen atoms, is almost completely insoluble in water. Like does dissolve like.

SUMMARY

In the practical world of making chemical reactions work, we need to allow molecules to "find each other" and react. Dissolving molecules in fluid medium provides an opportunity for molecules to move around and do just that—find other reactive molecules. The choice of solvent is critical because we need to dissolve the molecules we want to react together. In making that choice, the rather simple notion that "like dissolves like" really works! As we work through chemical reactions in the next few chapters, we will have to take account of solvents and solvation.

6.6 DIOLS OR GLYCOLS

Although molecules containing two OH groups are logically enough called **diols**, there is an important common name as well, **glycol**.

Some 1,2-diols or 1,2-glycols are familiar and are often important molecules. 1,2-Ethanediol, or ethylene glycol, is the commonly used antifreeze, for example. Note that in naming these compounds, the final "e" of the alkane parent compound is not dropped, as it is in naming simple alcohols, and that the almost universally used common name can be very misleading. Ethylene glycol implies the presence of an alkene, and "propylene glycol" (1,2-propanediol) suggests a three-carbon unsaturated chain. In neither case is the implied unsaturation present (Fig. 6.28). Be aware that the real structures are based on *saturated* hydrocarbon chains.

Ethylene glycol or 1,2-ethanediol

Propylene glycol or 1,2-propanediol

FIGURE 6.28 1,2-Ethanediol and 1,2-propanediol are called ethylene glycol and propylene glycol, respectively.

6.7 AMINES

6.7a Nomenclature. Many systems for naming amines exist, and the chemical world seems to be resisting attempts to bring order out of this mini-chaos by keeping to the old common names. The parent of all amines, $:NH_3$, is called **ammonia**. Successive replacement of the hydrogens leads to **primary**, **secondary**, and **tertiary amines**. Quaternary nitrogen compounds are positively charged and are called **ammonium ions** (Fig. 6.29).

$:NH_3$	$:NH_2R$	$:NHR_2$	$:NR_3$	$\overset{+}{N}R_4$
Ammonia	Primary amine	Secondary amine	Tertiary amine	Ammonium ion

FIGURE 6.29 Substituted amines.

Primary amines are commonly named by using the name of the substituent, R, and appending the suffix, amine. Secondary and tertiary amines in which the R groups are all the same are simply named as di- and trialkylamines. Amines bearing different R groups are named by ordering the groups alphabetically. Ethylmethylamine is correct, methylethylamine is not, although both are surely intelligible (Fig. 6.30).

$CH_3\ddot{N}H_2$
Methylamine

$CH_3CH_2\ddot{N}H_2$
Ethylamine

$PhCH_2\ddot{N}H_2$
Benzylamine

$(CH_3)_2\ddot{N}H$
Dimethylamine

$(CH_3CH_2)_3N:$
Triethylamine

Ethylmethylamine,
(*not* methylethylamine)

Benzylethyl-methylamine

$(CH_3)_3C\ddot{N}H_2$
***tert*-Butylamine**

$CH_3CH_2CH_2\ddot{N}H_2$
Propylamine

Cyclohexylamine

Dicyclopentylamine

FIGURE 6.30 A naming scheme for primary, secondary, and tertiary amines.

$CH_3\ddot{N}H_2$
Methanamine

Cyclohexanamine

FIGURE 6.31 The official naming system for amines.

The naming system used by the official chemical world names amines analogously to alcohols. If CH_3OH is sometimes known as methan<u>ol</u>, then CH_3NH_2 is methan<u>amine</u>. Figure 6.31 gives two examples.

Still another method names amines as substituted alkanes. Now CH_3NH_2 becomes aminomethane. Figure 6.32 gives some examples of amines named this way.

$CH_3\ddot{N}H_2$
Aminomethane

$CH_3CH_2CH_2\ddot{N}H_2$
1-Aminopropane

2-Aminopropane

Aminocyclohexane

FIGURE 6.32 Still another method for naming amines.

The amino group is just behind the alcohol group in the naming protocol priority system. Figure 6.33 gives some correct and incorrect names of polyfunctional compounds containing amino groups.

FIGURE 6.33 Some polyfunctional amines and their names.

Isobutylamine
1-amino-2-methylpropane

2-Amino-3-chlorobutane
(*not* 3-amino-2-chlorobutane)

3-Amino-2-butanol
(or 3-Amino-2-hydroxybutane,
not 2-amino-3-butanol
or 2-amino-3-hydroxybutane)

3-Bromocyclopentylamine
(*neither* 3-aminobromocyclopentane
nor 4-bromoaminocyclopentane)

3-Aminocyclopentanol
(*not* 3-hydroxycyclopentylamine)

Simple, cyclic amino compounds are also widely seen in nature, and are almost invariably known by their common names. These compounds are often named as "aza" analogues of an all-carbon system. Thus **aziridine** becomes azacyclopropane. Methylamine could be called azaethane, although in point of fact this system is almost never used for acyclic compounds. Unless there is an oxygen atom in the ring, the numbering scheme places the nitrogen at 1 and proceeds so as to minimize the numbering of any substituents. Figure 6.34 gives a few important examples. Once again there is no way out other than to learn these names.

Aziridine
(azacyclopropane)

Azetidine
(azacyclobutane)

Pyrrolidine
(azacyclopentane)

Piperidine
(azacyclohexane)

Pyridine
(azabenzene)

Morpholine

3-Methylpiperidine
(*not* 5-methylpiperidine)

3-Bromopyrrolidine
(*not* 4-bromopyrrolidine)

4-Benzylmorpholine
(*N*-benzylmorpholine)

Pyrrole

FIGURE 6.34
Cyclic amines and their names.

Despite this seeming excess of nomenclature, another system must be mentioned. Substituted amines can also be named as *N*-substituted amines. In this system, the substituent on nitrogen is designated with the locating prefix *N*. Some examples are shown in Figure 6.35.

N-Benzylaziridine

N-Ethylpiperidine

N-Methylazacyclooctane
(heptamethylenemethylamine)

FIGURE 6.35 Still another naming system for amines.

Quaternary ammonium ions are named by alphabetically attaching the names of the substituents. Don't forget to append the name of the negatively charged counterion (Fig. 6.36).

FIGURE 6.36 Names for two quaternary ammonium ions.

$$Br^- \ ^+N(CH_2CH_3)_4$$

Tetraethylammonium bromide

$$CH_3CH_2 \overset{\overset{\displaystyle CH_3}{|}}{\underset{\underset{\displaystyle CH_2CH_2CH_2CH_3}{|}}{\overset{+}{N}}} CH_2CH_2CH_3 \ \ I^-$$

Butylethylmethylpropylammonium iodide

6.7b Structure and Physical Properties of Amines. In simple amines, the carbon–nitrogen bond distance is a little shorter than the corresponding carbon–carbon bond distance in alkanes. This phenomenon is similar to that found in the alcohols and ethers (Section 6.8) (Fig. 6.37).

FIGURE 6.37 The carbon–nitrogen bond in amines is shorter than a normal carbon–carbon bond.

$$\underset{H_3C-OH}{1.43 \text{ Å}} \quad \underset{H_3C-NH_2}{1.47 \text{ Å}} \quad \underset{H_3C-CH_3}{1.54 \text{ Å}}$$

PROBLEM 6.11 Why is the carbon–nitrogen bond distance in aniline slightly shorter than that in methylamine? (See below.)

Aniline NH_2 — 1.40 Å

NH_2 / CH_3 — 1.47 Å

Simple amines are hybridized approximately sp^3 at nitrogen and thus are pyramidal. The bond angles are close to the tetrahedral angle of 109.5°, but, of course, cannot be *exactly* the tetrahedral angle.

PROBLEM 6.12 Why not?

For amines in which all three R groups are identical, there will be only a single R—N—R angle, but for less symmetrical amines, the different R—N—R angles cannot be the same (Fig. 6.38).

FIGURE 6.38 Some bond angles in simple amines.

107.3°

108.4°

105.9°   112.9°

The most important structural feature of amines comes from the presence of the pair of nonbonding electrons acting as the fourth substituent on the pyramid. In alkanes, where there are four substituents, there is nothing more to consider once the pyramid is described. In amines there is, as the pyramid can undergo an "umbrella flip," **amine inversion**, forming a new, mirror-image pyramid (Fig. 6.39).

A carbon-based pyramid is static

The nitrogen-based pyramid can invert

Inversion produces the mirror-image pyramid

FIGURE 6.39 Amine inversion interconverts enantiomeric amines.

This inversion process is sometimes confusing when first encountered, so let's work through it, making comparisons to the alkanes as we go along. At the beginning of the process, the pyramid begins to flatten and the R—N—R bond angles increase. At the halfway point along the path to the inverted pyramid the transition state is reached. Here the molecule is flat, with 120° R—N—R bond angles. At this point, the hybridization of the central nitrogen is sp^2! The inversion process maintains a mirror plane of symmetry. In the alkanes, we could certainly imagine the beginning of the flattening process, but there is no way to force the fourth substituent through the molecule to the other side (Fig. 6.40).

FIGURE 6.40 Amine inversion has a planar transition state.

Planar transition state
sp^2

Inversion progress

This flattening will cost energy

This process can't go further

PROBLEM 6.13 In general terms, what is the hybridization of nitrogen at some point between the starting pyramidal amine and the transition state?

A most important concern is the rate of this inversion process. If it is slow at room temperature, then we should be able to isolate appropriately substituted optically active amines. If inversion is fast, however, racemization will also be fast as the inversion process interconverts enantiomers. In practice, this is what happens. Barriers to inversion in simple amines are very low, about 5–6 kcal/mol, and isolation of one enantiomer, even at very low temperature, is difficult. Inversion is simply too facile and the rapid inversion interconverts enantiomeric amines (Fig. 6.41).

FIGURE 6.41 Rapid inversion racemizes amines.

One enantiomer   The other enantiomer

PROBLEM 6.14 In contrast to simple amines, aziridines (three-membered rings containing a nitrogen) can often be separated into enantiomers. For example, the activation energy for the inversion of 1,2,2-trimethylaziridine is about 18.5 kcal/mol, much higher than that for simple amines (see below). Explain.

The barrier to this inversion is unusually high, $\Delta G^{\ddagger}$ = 18.5 kcal/mol

When the methyl on nitrogen is replaced
by phenyl, $\Delta G^{\ddagger}$ decreases to 11.2 kcal/mol

PROBLEM 6.15 If there is a phenyl group (Ph in the figure above) attached to the nitrogen atom of the aziridine, the barrier to inversion decreases (see figure in Problem 6.14). Explain.

Like alcohols, amines are much higher boiling than hydrocarbons of similar molecular weight, although the effect is not as great as with alcohols (Table 6.8). The reasons for the increased boiling points are the same as for alcohols. Amines are polar compounds and aggregate in solution. Like alcohols, small amines are miscible with water. Boiling requires overcoming the intermolecular attractive forces between the dipoles. Moreover, amines form hydrogen-bonded oligomers in solution, effectively increasing the molecular weight and further increasing the boiling point (Fig. 6.42).

When hydrogen bonding is impossible, as with tertiary amines, boiling points decrease (Table 6.8).

H_3C-NH_2

NH_2-CH_3

Association because of favorable dipole alignment

Oligomer formation through hydrogen bonding

FIGURE 6.42 The effective molecular weight of amines is increased through association and hydrogen bonding.

TABLE 6.8 Some Physical Properties of Amines and a Few Related Alkanes

Compound	bp (°C)	mp (°C)	Density (g/mL)
NH_3 Ammonia	−33.4	− 77.7	0.77
CH_3NH_2 Methylamine	− 6.3	− 93.5	0.70
CH_3CH_3 Ethane	−88.6	−183.3	0.57
$CH_3CH_2NH_2$ Ethylamine	16.6	− 81	0.68
$CH_3CH_2CH_3$ Propane	−43.1	−189.7	0.59
$(CH_3)_2NH$ Dimethylamine	7.4	− 93	0.68
$(CH_3)_3N$ Trimethylamine	2.9	−117.2	0.64
Aniline	184	− 6.3	1.02

It is impossible to discuss the properties of amines without mentioning smell! The smell of amines ranges from the merely fishy to the truly vile. Indeed, the odor of decomposing fish owes its characteristic unpleasant nature to amines.

Diamines are even worse. Some indication of how bad smelling they often are can be gained from their names. 1,4-Diaminobutane is called putrescine and 1,5-diaminopentane is called cadaverine.

PROBLEM 6.16 Many people use lemon when eating fish. This custom is a carryover from the days when it was difficult to preserve fish, and the lemon acted to diminish the unpleasant odor (if not the decomposition). Lemons contain 5–8% citric acid and this acidity contributes to their sour taste. Explain why lemon juice is effective at reducing the odor of fish.

6.7c Acid and Base Properties of Amines.

Like alcohols, amines are both Brønsted bases and nucleophiles. Primary and secondary amines are weak Brønsted acids as well. Brønsted basicity results in the formation of ammonium ions through proton transfer (Fig. 6.43).

FIGURE 6.43 Ammonia acting as a Brønsted base.

Ammonia acting as a Brønsted base in proton transfer

The better competitor the amine is in the proton-transfer reaction, the stronger is its Brønsted basicity. A way of turning this statement around is to focus not on the basicity of the amine, but on the Brønsted acidity of its conjugate acid, the ammonium ion. If the ammonium ion is a strong acid, the related amine must be a weak base. If it is easy to remove a proton from the ammonium ion to give the amine, the amine itself must be a poor competitor in the proton-transfer reaction, which is exactly what we mean by a weak Brønsted base. Strongly basic amines give ammonium ions from which it is difficult to remove a proton, ammonium ions with high pK_a values. So the pK_a of the ammonium ion has come to be a common measure of the basicity of the related amine. Ammonium ions with high pK_a values (weak acids) are related to strongly basic amines, and ammonium ions with low pK_a values (strong acids) are related to weakly basic amines (Fig. 6.44).

FIGURE 6.44 Ammonium ions with high pK_a values are related to strongly basic amines, and ammonium ions with low pK_a values are related to weakly basic amines.

A strongly basic amine is a good competitor for H$^+$ and produces…

… an ammonium ion that is a relatively weak Brønsted acid (high pK_a)

$$R-\ddot{N}H_2 + H-X \rightleftharpoons R-\overset{+}{N}H_3 + X\ddot{:}^-$$

A weakly basic amine is a less effective competitor for H$^+$ and produces an ammonium ion that is a relatively strong Brønsted acid (low pK_a).

Table 6.9 relates the basicity of ammonia and some very simple alkylamines by tabulating the pK_a values of the related ammonium ions.

TABLE 6.9 The pK_a Values for Some Simple Ammonium Ions

Amine	Formula	Ammonium Ion	pK_a (in aqueous solution)
Ammonia	NH_3	$^+NH_4$	9.24
Methylamine	CH_3NH_2	$CH_3\overset{+}{N}H_3$	10.63
Dimethylamine	$(CH_3)_2NH$	$(CH_3)_2\overset{+}{N}H_2$	10.78
Trimethylamine	$(CH_3)_3N$	$(CH_3)_3\overset{+}{N}H$	9.80

The general trend of the first three entries in Table 6.9 is understandable if we make the analogy between ammonium ions and carbocations. *Remember* (Chapter 3, p. 140): The more substituted a carbocation, the more stable it is. Similarly, the more substituted an ammonium ion, the more stable it is. The more stable the ammonium ion, the less readily it loses a proton, and the higher its pK_a. At least for the first three entries of Table 6.9, increasing substitution of the ammonium ion carries with it a decreasing ease of proton loss.

Actually, even the first three entries in Table 6.9 should make you suspicious. Why is there a much bigger change between ammonia and methylamine (1.4 pK_a units) than between methylamine and dimethylamine (only 0.15 pK_a units)? The structural change is the same in each case, the replacement of a hydrogen with a methyl group. When we get to the fourth entry, we find that things have truly gone awry; trimethylammonium ion is a stronger acid than the dimethylammonium ion and the methyl ammonium ion, and almost as strong as the ammonium ion itself, which implies that trimethylamine is a weaker base than dimethylamine or methylamine. There seems to be no continuous trend in these data.

Indeed there isn't, and it takes a look at matters in the gas phase to straighten things out. In the gas phase, the trend *is* regular. The basicity of amines increases with substitution, and the acidity of ammonium ions decreases with substitution (Fig. 6.45).

There must be some effect present in solution that disappears in the gas phase, and this effect must be responsible for the irregularities in Table 6.9. The culprit is the solvent itself. Ions in solution are strongly stabilized by solvation, by interaction of the solvent molecules with the ion. These interactions can take the form of electrostatic stabilization, or of partial covalent bond formation as in hydrogen bonding (Fig. 6.46).

$^+NH_4 < {}^+NRH_3 < {}^+NR_2H_2 < {}^+NR_3H$

Order of increasing gas-phase pK_a

Lowest pK_a Highest pK_a

FIGURE 6.45 The gas-phase acidity of ammonium ions.

Stabilization through dipole–dipole interaction with a polar solvent

Stabilization through hydrogen bonding

FIGURE 6.46 Stabilization of amines in solution.

In solution, an alkyl group has two effects on the stability of the ammonium ion. It stabilizes by helping it to disperse the charge, but it destabilizes the ion by interfering with solvation, by making intermolecular solvation (charge dispersal) more difficult. These two effects operate in different directions! There is little steric interference with solvation when we replace one hydrogen of the ammonium ion with a methyl group, and the pK_a change is more than a full pK_a unit. This change reflects the stabilizing effect of the methyl group on the ammonium ion. However, when a second hydrogen is changed for methyl, the stabilization and interference with solvation nearly balance. The result is practically no change in the overall stability of the ammonium ion. When a third hydrogen is replaced with methyl, the destabilizing effects outweigh the stabilizing forces, and the result is a less stable, more acidic ammonium ion. In the gas phase, where there can be no solvation, only the stabilizing effects remain, and each replacement of a hydrogen with a methyl is stabilizing.

In addition, as we progress along the series, $^+NH_4$, $^+NRH_3$, $^+NR_2H_2$, $^+NR_3H$, $^+NR_4$, there are fewer hydrogens available for hydrogen bonding, and therefore less stabilization of the ammonium ion in solution.

PROBLEM 6.17 Amines can be stabilized by other factors. Explain the pK_a data in the following figure.

| pK_a | 4.6 | 10.6 |

It must be admitted that this traditional discussion of base strength of amines in terms of the pK_a of the related ammonium ion is indirect. There is another, less common, but undeniably more direct method. That is to frame the discussion in terms of the pK_b ($-\log K_b$) of the amine (Fig. 6.47).

For　$RNH_2 + H_2O \underset{\longleftarrow}{\overset{K}{\longrightarrow}} \overset{+}{R}NH_3 + HO^-$

$$K_b = \frac{[\overset{+}{R}NH_3][HO^-]}{[RNH_2]}$$

FIGURE 6.47 The expression for pK_b.

The higher the pK_b, the weaker the base, just as the higher the pK_a, the weaker the acid. Typical amines have pK_b values of about 4.

Amines are much stronger bases than alcohols. The pK_a value for the ammonium ion is much higher than that for the oxonium ion. It is much harder to deprotonate $^+NH_4$ than $^+OH_3$ (Fig. 6.48).

Why is the oxonium ion much more acidic than the ammonium ion? Oxygen is a more electronegative atom than nitrogen and bears the positive charge less well. We can see the results of the increased stability of the ammonium ion in practical terms. Stable ammonium salts are common, but salts of oxonium ions, though known, are rare and usually unstable (Fig. 6.48).

+NH₄

Ammonium ion

$pK_a = 9.2$

+N(CH₃)₄ Cl⁻

Tetramethylammonium chloride, a stable solid, easily isolable

+OH₃

Oxonium ion

$pK_a = -1.7$

+O(CH₃)₃ BF₄⁻

Trimethyloxonium fluoroborate, not easily isolated, and very reactive

FIGURE 6.48 Ammonium ions are weaker acids than oxonium ions.

PROBLEM 6.18 The stability of oxonium ions depends on the nature of the negatively charged counterion. Fluoborate (⁻BF₄) is an especially favorable counterion. Why?

Hint: Lewis bases, B:⁻, can transfer methyl (and other) groups through a process called the S_N2 reaction. You will learn all about it in Chapter 7, but you might want to use it here in this problem.

B:⁻ H₃C—O⁺(CH₃)(CH₃) —S_N2→ B—CH₃ + :O:(CH₃)(CH₃)

Primary and secondary amines are Brønsted acids, but only very weak ones. Amines are much weaker acids than alcohols. Removal of a proton from an alcohol gives an alkoxide in which the negative charge is borne by oxygen. An amine forms an **amide*** in which the less electronegative nitrogen carries the minus charge (Fig. 6.49).

It takes a very strong base indeed to remove a proton from an amine to give the amide ion, which is itself a very strong base. Alkyllithium reagents (p. 236) are typically used (Fig. 6.50).

R—Ö—H + B⁻

Alcohol
$pK_a \sim 17$

R—Ö:⁻ + HB

Alkoxide ion

R—N̈H₂ + B⁻

Amine
$pK_a \sim 36$

R—N̈H⁻ + HB

Amide ion

FIGURE 6.49 Primary and secondary amines are weak Brønsted acids.

(CH₃CH₂)₂N̈H —BuLi→ (CH₃CH₂)₂N̈:⁻ Li⁺ + BuH

Amide ion

$pK_a \sim 36$ $pK_a > 50$

FIGURE 6.50 Very strong bases such as alkyllithium reagents can remove a proton from an amine to give an amide ion (Bu = CH₂CH₂CH₂CH₃).

*Be careful—there is another kind of amide that has the structure R—CO—NH₂. There is no written or verbal distinction made, so you need to know the context before you know which kind of amide is meant.

So, amines are good bases, but relatively poor acids. Amides, the conjugate bases of amines, are even stronger bases than the amines themselves (Fig. 6.51).

An exceptionally strong base

$$R-\ddot{N}H_2 \xrightarrow{\text{BuLi}} BuH + R-\overset{-}{\underset{..}{N}}H \quad Li^+$$

An amine (a good base and a poor acid)

An amide ion (an even stronger base than the related amine)

FIGURE 6.51 The amide ion is a far stronger base (and nucleophile) than the parent amine.

6.8   ETHERS

Alcohol (ROH) and amine (RNH$_2$) chemistry depends on both the OH or NH and the OR or N—R parts of the molecule. In this section, we look at molecules that no longer have any OH groups, but retain the R group. These compounds are called **ethers**.

6.8a Nomenclature. Naming ethers is simple. The two groups joined by oxygen are named in alphabetical order and the word "ether" appended. Figure 6.52 gives some simple examples, as well as some common names.

$$H_3C-\ddot{O}-CH_3$$
Dimethyl ether

$$H_3C-\ddot{O}-CH_2CH_3$$
Ethyl methyl ether

$$CH_3CH_2-\ddot{O}-CH_2CH_3$$
Diethyl ether (ether)

$$(CH_3)_3C-\ddot{O}-CH_3$$
***tert*-Butyl methyl ether**

$$CH_3CH_2-\ddot{O}-CH=CH_2$$
Ethyl vinyl ether

Furan

Tetrahydrofuran (THF)

Methyl phenyl ether (anisole)

FIGURE 6.52 Some ethers and their common names.

6.8b Physical Properties. Ethers are polar, but only for the smallest members of the class does this polarity strongly affect physical properties. Others are sufficiently hydrocarbon-like so as to behave as their all-carbon relatives. For example, diethyl ether has nearly the same boiling point as pentane, and is only modestly soluble in water (Table 6.10).

TABLE 6.10 Some Physical Properties of Ethers

Compound	bp (°C)	mp (°C)	Density (g/mL)
CH_3-O-CH_3 Dimethyl ether		−25	−138.5
$CH_3-O-CH_2CH_3$ Ethyl methyl ether	10.8		0.725
$CH_3CH_2-O-CH_2CH_3$ Diethyl ether	34.5	−116	0.714
$(CH_3)_3C-O-CH_3$ tert-Butyl methyl ether	55.2	−109	0.740
$CH_3CH_2-O-CH=CH_2$ Ethyl vinyl ether	35.5	−115.8	0.759
Furan	31.4	− 85.6	0.951
THF	67	−108	0.889
Anisole	155.0	− 37.5	0.996

6.8c Structure. The carbon–oxygen bond length in ethers is similar to that in alcohols, and the opening of the angle as we go from H—O—H (104.5°) to R—O—H (~109°), to R—O—R (~112°) continues. As the angle is approximately 112°, the oxygen is hybridized approximately sp^3.

6.8d Acidity and Basicity. Of course, ethers lack the OH group of their relatives water and the alcohols, and so are not Brønsted acids. The oxygen atom has two pairs of nonbonding electrons, however, and they make ethers weak Brønsted and Lewis bases. Ethers can be protonated, and the protonated form has a pK_a in the same range as those of the protonated alcohols, which means that ethers and alcohols are similarly strong bases (Fig. 6.53).

THE GENERAL CASE

A SPECIFIC EXAMPLE

$pK_a \sim -3.5$

$pK_a = -1.74$

$pK_a \sim -3.8$

FIGURE 6.53 Ethers are Brønsted bases and can be protonated. Protonated ethers (and protonated alcohols) are strong acids.

We have already seen one instance in which the ability of ethers to donate electrons was a critical property. Recall that an ether solvent is vital to success in the formation of the Grignard reagent (p. 236; Fig. 6.54). Ethers are stabilizing to Lewis acidic species. Boron trifluoride etherate is a commercially available complex of BF_3 and ether that can even be distilled. It is commonly used as a source of BF_3. For similar reasons, THF is used to stabilize the highly reactive molecule BH_3.

FIGURE 6.54 Ethers will react with Lewis acids as well as Brønsted acids. The BF_3–etherate complex is a stable source of diethyl ether and boron trifluoride.

6.9   SPECIAL TOPIC: THIOLS (MERCAPTANS) AND THIOETHERS (SULFIDES)

Thiols and **thioethers** are the sulfur-containing counterparts of alcohols and ethers, and their chemistries are generally similar to those of their oxygen-containing relatives. These compounds have a poor reputation, as some thiols are quite extraordinarily evil-smelling. The skunk uses a variety of four-carbon thiols in its defense mechanism, for example. But this bad press is not entirely deserved; not all thiols are malodorous. Garlic derives its odor from small sulfur-containing compounds. You may never need to ward off a vampire with it, but garlic contains many sulfur compounds that have quite remarkably positive qualities. Figure 6.55 shows some of the sulfur-containing molecules that can be found in garlic.

FIGURE 6.55 Some of the sulfur-containing molecules in garlic.

Ajoene is only one of the beneficial compounds present in garlic. Ajoene is lethal to certain tumor-prone cells, and promotes the antiaggretory action of some molecules on human blood platelets. It also seems effective in reducing repeat heart attacks among people who have already suffered an initial attack. Garlic contains many other compounds that also appear to have beneficial medicinal properties. And, of course, if a vampire is hot on your trail . . . .

6.9a Nomenclature. Thiols, or mercaptans, are named by adding the suffix *thiol* to the parent hydrocarbon name. Note that the final "e" is not dropped, as it is, for example, in alcohol nomenclature (Fig. 6.56).

Methanethiol
(methyl mercaptan)

Butanethiol
(butyl mercaptan)

Cyclohexanethiol

Phenyl mercaptan
(benzenethiol)
(thiophenol)

FIGURE 6.56 Some thiols, or mercaptans.

The sulfur counterparts of ethers are **sulfides**, which are named in the same way as ethers. The two groups attached to the sulfur atom are followed by the word sulfide. Disulfides are the counterparts of peroxides, R—O—O—R, but unlike the oxygen compounds, these molecules are not explosive. They are named in the same way as sulfides (Fig. 6.57).

Dimethyl sulfide

Ethyl methyl sulfide

***tert*-Butyl cyclopentyl sulfide**

Dicyclopropyl disulfide

FIGURE 6.57 Some thioethers (sulfides) and disulfides.

6.9b Acidity. Thiols are stronger acids than alcohols (pK_a = 9–12) and form **mercaptides**, the sulfur counterparts of alkoxides, when treated with base. The increased acidity of thiols makes formation of the conjugate base more favorable than from alcohols (Fig. 6.58).

FIGURE 6.58 Mercaptides are easier to form than alkoxides because thiols are much more acidic than alcohols.

$$H_3C—\overset{..}{\underset{..}{S}}—H \ + \ H_3C—\overset{..}{\underset{..}{O}}:^- \ Na^+ \ \rightleftharpoons \ H_3C—\overset{..}{\underset{..}{S}}:^- \ Na^+ \ + \ H_3C—\overset{..}{\underset{..}{O}}—H$$

pK_a = 10   Alkoxide   Mercaptide pK_a = 15.5
          (conjugate base   (conjugate base
          of CH₃OH)   of CH₃SH)

6.9c Reduction of Sulfur Compounds with Raney Nickel: A New Alkane Synthesis. Thiols and thioethers are reduced by a catalyst called **Raney nickel** (essentially just hydrogen absorbed on finely divided nickel) to give hydrocarbons (Fig. 6.59). This new synthetic method gives you a second way to make alkanes.

FIGURE 6.59 Desulfuration with Raney nickel.

PROBLEM 6.19 What reagents would you use to convert the indicated starting materials into cyclohexane?

6.10   SPECIAL TOPIC: CROWN ETHERS

Cyclic ethers are common solvents, and as we will see in Section 7.10 can often be prepared rather easily (Williamson ether synthesis, p. 331). It is possible to make rings in which there is more than one ether linkage. Figure 6.60 gives some examples of cyclic ethers and polyethers.

Ethylene oxide
(oxirane)

Trimethylene oxide
(oxacyclobutane)
(oxetane)

Furan

Tetrahydrofuran
(THF)

Tetrahydropyran
(pentamethylene oxide
(oxacyclohexane)

1,4-Dioxan
(1,4-dioxacyclohexane)

1,3-Dioxolane
(1,3-dioxacyclopentane)

1,3,5-Trioxane
(1,3,5-trioxacyclohexane)
(paraldehyde)

FIGURE 6.60 Some cyclic ethers and polyethers.

We have mentioned more than once the remarkable solvating powers of ethers. Ethers are nucleophiles and can donate electrons to Lewis acids, thus stabilizing them. One example of this is the requirement for an ether, usually diethyl ether or THF, in the formation of the Grignard reagent. Ethers are also polar compounds, and therefore ethers are able to stabilize other polar molecules through noncovalent dipole–dipole interactions.

This idea has been carried to extremes in work that ultimately led to the Nobel prize of 1987 for Charles J. Pedersen (1904–1989) of du Pont, Donald J. Cram (1919–2001) of UCLA, and Jean-Marie Lehn (b. 1939) of Université Louis Pasteur in Strasbourg for opening the field of "host–guest" chemistry. Pedersen discovered that certain cyclic polyethers had a remarkable affinity for metal cations. Molecules were constructed whose molecular shapes created different sized cavities into which different metal ions fit well. Because of their vaguely crown-shaped structures, these molecules came to be called **crown ethers**. Figure 6.61 shows two of them.

FIGURE 6.61 Two crown ethers.

12-Crown-4   **18-Crown-6**

The molecule 12-crown-4 is the proper size to capture lithium ions, and 18-crown-6 has a remarkable affinity for potassium ions. The first number shows the number of atoms in the ring, and the final number shows the number of heteroatoms in the ring. These crown ethers can stabilize positive ions of different sizes, depending on the size of the cavity (Fig. 6.62).

Three-dimensional versions of these essentially two-dimensional molecules have now also been made, using both oxygen and other heteroatoms such as sulfur and nitrogen as the Lewis bases. These molecules are called **cryptands**, and can incorporate positive ions into the roughly spherical cavity within the cage (Fig. 6.63).

FIGURE 6.62 Two crown ethers stabilizing cations.

A cryptand

FIGURE 6.63 Three-dimensional cryptands can incorporate metal ions into a central cavity.

There are chemical and larger implications of this work. Potassium permanganate is an effective oxidizing agent. Its great insolubility in organic reagents limits its use, however. For example, the deep purple $KMnO_4$ is completely insoluble in benzene. It simply forms a solid suspension when added to the benzene, which remains a colorless liquid. The addition of a little 18-crown-6 results in the dissolution of the permanganate, and the benzene takes on the deep purple color of the inorganic ion. The potassium ion has been effectively solvated by the crown ether and now dissolves in the organic solvent. The permanganate counterion, though not solvated by the crown compound, accompanies the positive ion into solution, as it must to preserve electrical neutrality. The purple benzene solution can be used as an effective oxidizing agent.

Other uses are more exciting, though they still lie in the future. One can imagine schemes where specially designed crown compounds are used to scavenge poisonous heavy metal ions from water supplies, for example. There have been less humanitarian, though admittedly clever schemes, for extracting the minute amounts of gold ions present in seawater, by passing vast amounts of water over some yet-to-be-discovered crown compound.

6.11   SPECIAL TOPIC: COMPLEX NITROGEN-CONTAINING BIOMOLECULES—ALKALOIDS

Alkaloids are naturally occurring amines that are powerfully active biologically. Alkaloids are among the most useful medicinal agents known. The painkiller morphine and the antimalarial agent quinine are typical examples. At the same time, some of the most destructively addictive substances known are also alkaloids. Heroin and cocaine are examples from our own time, and anyone at all familiar with popular literature knows of the notorious poison strychnine (Fig. 6.64).[*]

We have already seen brucine, a close relative of strychnine, in Figure 4.45, where it served as an reagent for the resolution of optically active carboxylic acids into separate (R) and (S) enantiomers (p. 178). As we work through the book, alkaloids will reappear now and again. The important thing to "take home" from this picture is not the detail, but a broad view of the great structural diversity—molecular architecture—of these bioactive molecules.

FIGURE 6.64 Some alkaloids.

Strychnine   Morphine   Heroin   Cocaine   Quinine

6.12   SUMMARY

New Concepts

A measure of a compound's acidity—its pK_a value—is introduced. The lower the pK_a value, the more acidic the compound in question.

Solvents appear as important players in questions of stability. For example, both the conjugate bases of alcohols (alkoxides) and the conjugate acids of amines (ammonium ions) depend greatly on solvation for stabilization. In assessing basicity or acidity, one must take account of the solvent.

A general discussion of solvents and solvation also appears. Protic solvents that can participate in hydrogen bonding dissolve other protic, polar molecules well. By contrast, polar solvents do a poor job of dissolving "greasy" nonpolar organic molecules such as hydrocarbons, which are dissolved by nonpolar, similarly "greasy" solvents. Like dissolves like—polar solvents dissolve polar molecules and nonpolar solvents dissolve nonpolar molecules.

Organic halogenated molecules can be used to form organometallic reagents. Many of the important synthetic reactions of organic chemistry use organometallic reagents. In this chapter, we see them only as very strong bases, and as sources of hydrocarbons through their reaction with water.

Amines, like alkanes with sp^3 hybridized carbons, are tetrahedral. Unlike alkanes, amines invert their "umbrellas" through an sp^2 hybridized transition state.

The hydrogen bond is introduced. This kind of bond is nothing more than a partially completed Brønsted acid–base reaction in which a partial bond is formed between a pair of electrons on one atom and a hydrogen on another (Fig. 6.65).

FIGURE 6.65 A hydrogen bond.

Key Terms

alkaloids (p. 266)
alkoxide ion (p. 244)
alkyl halides (p. 233)
amide (p. 259)
amine inversion (p. 253)
ammonia (p. 249)
ammonium ions (p. 249)
aprotic (p. 247)
aziridine (p. 251)
conjugate acid (p. 242)
conjugate base (p. 242)
crown ethers (p. 265)

cryptands (p. 266)
diols (p. 249)
ethers (p. 260)
free radical (p. 237)
glycol (p. 249)
Grignard reagent (p. 236)
hydrogen bonding (p. 240)
mercaptides (p. 264)
organolithium reagent (p. 236)
organometallic reagents (p. 236)
peptides (p. 231)

pK_a (p. 231)
primary amine (p. 249)
proteins (p. 231)
protic (p. 247)
Raney nickel (p. 264)
secondary amine (p. 249)
solvation (p. 247)
sulfides (p. 263)
tertiary amine (p. 249)
thioethers (p. 262)
thiols (p. 262)

Common Errors

People sometimes get the pK_a sign wrong. A strong acid has a low pK_a; a weak acid has a high pK_a.

It is easy to get confused when using the pK_a of an ammonium ion as a measure of the acidity of the related amine.

Review this section (p. 256) or use the more direct (but less common) pK_b values.

Syntheses

1. GRIGNARD REAGENTS

$$R—X \xrightarrow{\text{Mg}} R—MgX$$

R can be almost any organic
structure. X is I, Br, or Cl

3. ORGANOLITHIUM REAGENTS

$$R—X \xrightarrow{\text{Li}} R—Li$$

R can be almost any organic
structure. X is I, Br, or Cl

2. HYDROCARBONS

$$R—MgX \xrightarrow{\text{H}_2\text{O}} R—H$$

$$R—Li \xrightarrow{\text{H}_2\text{O}} R—H$$

If D_2O is used instead
of H_2O, deuterated
hydrocarbons can
be made

$$R—SH \xrightarrow[\text{nickel}]{\text{Raney}} R—H$$

Reduction of thiols by
Raney nickel

$$R—S—R \xrightarrow[\text{nickel}]{\text{Raney}} R—H$$

Reduction of sulfides
by Raney nickel

6.13 ADDITIONAL PROBLEMS

PROBLEM 6.20 Name the following compounds:

(a)

(b)

(c)

(h)

(i)

(d)

(e)

(j)

(k)

(l)

(f)

(g)

(m)

PROBLEM 6.21   Draw structures for the following compounds:
(a) 1,4-Butanediol
(b) 3-Amino-1-pentanol
(c) *cis*-4-*tert*-Butoxycyclohexanol
(d) 2,2,4,4-Tetramethyl-3-pentanol
(e) (*R*)-*sec*-Butyl alcohol
(f) *trans*-3-chlorocyclohexanol
(g) *tert*-Butyl cyclopentyl ether

PROBLEM 6.22   Draw structures for the following compounds:
(a) Neopentyl alcohol
(b) 18-Crown-6
(c) Tetrahydropyran
(d) THF
(e) Furan

PROBLEM 6.23   In the following molecules, there exists the possibility of forming more than one alkoxide or thiolate. In each case, explain which will be formed preferentially, and why.

(a)

OH   OH
(solution)

(b)

OH

(gas phase)

(c)

OH

SH
(solution)

PROBLEM 6.24   Predict the relative acidities of the following three alcohols:

Cl

Cl   OH

Cl

OH

Cl

OH

PROBLEM 6.25   Write all the isomers of the alcohols of the formula $C_6H_{14}O$.

PROBLEM 6.26   Name the following compounds (in more than one way, if possible). Identify each as a primary, secondary, or tertiary amine, or as a quaternary ammonium ion.

(a)

NH_2

(b)

NH
2

(c)

N
3

(d)

F

NH_2

(e)

N^+ I^-
4

(f)

NH_2

NH_2

PROBLEM 6.27   Name the following compounds (in more than one way, if possible). Identify each as a primary, secondary, or tertiary amine.

(a)

NH_2

(b)

Ph CH_3
N

CH_2CH_3

PROBLEM 6.28   Explain why compound **1** shows five signals in the ^{13}C NMR spectrum at low temperature but only three at higher temperature.

H_3C

CH_3
N
..

H_3C

1

PROBLEM 6.29   Write the expressions for K_a and K_b and show that $pK_a + pK_b = 14$.

PROBLEM 6.30   The pK_b values for a series of simple amines are given below. Explain the relationship between pK_b and base strength, and then rationalize the nonlinear order in the figure.

Amine	$:NH_3$	$:NH_2CH_3$	$:NH(CH_3)_2$	$:N(CH_3)_3$
pK_b	4.76	3.37	3.22	4.20

Substitution and Elimination Reactions: The S_N2, S_N1, E2, and E1 Reactions

7

Don't look back, something might be gaining on you.

—Leroy Robert "Satchel" Paige*

7.1 PREVIEW

We are about to embark on an extensive study of chemical change. The last six chapters mostly covered various aspects of structure, and still more analysis of structure will appear as we go along. Without this structural underpinning no serious exploration of chemical reactions is possible. The key to all studies of reactivity—reaction mechanisms—is an initial, careful analysis of the structures of the molecules involved. We have been learning parts of the grammar of organic chemistry. Now we are about to use that grammar to write some sentences and paragraphs. Let's begin with a reaction central to most of organic and biological chemistry, the replacement of one group by another (Fig. 7.1).

THE GENERAL CASE

$$R—L + N \rightleftharpoons R—N + L$$

$$R = \begin{array}{ll} H_3C & (CH_3)_2CH \\ \text{Methyl group} & \text{Isopropyl group} \\ CH_3CH_2 & (CH_3)_3C \\ \text{Ethyl group} & \textit{tert}\text{-Butyl group} \end{array}$$

FIGURE 7.1 This generic reaction presents an overall picture of the substitution process. A leaving group, L, is replaced by a new group, N, to give R–N. The box shows a standard set of alkyl groups that will be used throughout this chapter.

---

*Satchel Paige (1906–1982) was a pitcher for the Cleveland Indians and later the St. Louis Browns. He was brought to the major leagues by Bill Veeck after a long and spectacular career in the Negro Leagues. Some measure of his skill can be gained from realizing that Paige pitched in the major leagues after his 50th birthday.

Conversion of one alkyl halide into another is an example of the **substitution reaction** (Fig. 7.2). This reaction looks simple, and in some ways it is. The only change is the replacement of one halogen by another.

A SPECIFIC EXAMPLE

FIGURE 7.2 The replacement of one halogen by another is an example of a substitution reaction.

Yet, as we examine the details of this process we find ourselves looking deeply into chemical reactivity. This reaction is an excellent prototype that nicely illustrates general techniques used to determine reaction mechanism.

One reason that we spend so much time on the substitution reaction is that it *is* so general; what we learn about it and from it is widely applicable. Not only will we apply what we learn here to many other chemical reactions, but these processes have vast practical consequences as well. Substitution reactions, commonly (if sometimes inaccurately) called displacement reactions, are at the heart of many industrial processes and are central to the mechanisms of many biological reactions including some forms of carcinogenesis, the cancer-causing activity of many molecules. Biomolecules are bristling with substituting agents. It seems clear that when some groups on our DNA (deoxyribonucleic acid, see Chapter 23) are alkylated in a substitution reaction, tumor production is initiated or facilitated. Beware of alkylating agents.

This chapter is long, important, and possibly challenging. In it, we cover four of five building block reactions, two of them substitution reactions (S_N1 and S_N2) and two of them elimination reactions (E1 and E2). We have encountered another fundamental reaction, additions, already in Chapter 3, and it will reappear shortly, in Chapter 9. From now on, you really cannot memorize your way through the material; you must try to generalize. Understanding concepts and applying them in new situations is the route to success. This chapter contains lots of problems, and it is important to try to work through them as you go along. I know this warning is repetition, and I don't want to insult your intelligence, but it would be even worse not to alert you to the change in the course that takes place now, or not to suggest ways of getting through this new material successfully. You cannot simply read this material and hope to get it all. You must work with the text and the various people in your course, the Professor and the teaching assistants, in an interactive way. The problems try to help you do this. When you hit a snag, or can't do a problem, find someone who can and get the answer. The answer is more than a series of equations or structures and arrows; it also involves finding the right approach to the problem. There is much technique to problem solving in organic chemistry; it can be learned, and it gets much easier with practice. *Remember*: Organic chemistry must be read with a pencil in your hand.

ESSENTIAL SKILLS

1. We have seen it before, notably in Chapter 3, but in this chapter the curved arrow formalism becomes more important than ever. If you are at all uncertain of your ability to "push" arrows, now is definitely the time to solidify this skill. In the arrow formalism convention, arrows flow from electrons. But be careful about violating the rules of valence; arrows either displace another pair of electrons, or fill a "hole"—an empty orbital.

2. It is really important to be able to generalize the definition of Lewis bases and Lewis acids. All electron-pair donors are Lewis bases, and nearly any pair of electrons can act as a Lewis base under some conditions. Similarly, there are many very different appearing Lewis acids—acceptors of electrons.

3. The four fundamental reactions we study in this chapter, S_N2, S_N1, E2, and E1, are affected by the reaction conditions as well, of course, as by the structures of the molecules themselves. These four reactions often compete with each other, and it is important to be able to select conditions and structures that favor one reaction over the others. You will rarely be able to find conditions that give perfect specificity, but you will usually be able to choose conditions that give selectivity.

4. At the end of this chapter, we will sum up what we know about making molecules—synthesis. ALWAYS attack problems in synthesis by doing them backward. Do not attempt to see forward from starting material to product. In a multi-step synthesis, this approach is often fatal. The way to do it is to look at the ultimate product and ask yourself what molecule might lead to it in a single step. That technique will usually produce one or more candidates. Then apply the same question to those candidates, and before you know it, you will be at the indicated starting material. This technique works, and can make what is often a tough, vexing task easy.

IMPORTANT DETAILS

1. Alcohols are important starting materials for synthesis—they are cheap and easily available. Alcohols can be made even more useful by learning how to manipulate the OH group to make it more reactive—to make it a better "leaving group." Several ways to do that are outlined in Section 7.4f.

2. Keeping track of your growing catalog of useful synthetic reactions is not easy—it is a subject that grows rapidly. A set of file cards on which you collect ways of making various types of molecules is valuable. See page 337 for an outline of how to make this catalog.

7.2   REVIEW OF LEWIS ACIDS AND BASES

7.2a The Curved Arrow Formalism. Recall that there is a most important bookkeeping technique called the **curved arrow formalism** (p. 135), which helps to sketch the broad outlines of what happens during a chemical reaction. However, it does not constitute a full **reaction mechanism**, a description of how the reaction occurs. If a reaction mechanism is not an arrow formalism, what is it? A short answer would be a description, in terms of structure and energy, of the intermediates and transition states separating starting material and product. Implied in this description is a picture of how the participants in the reaction must approach each other. The arrow formalism is a powerful device of enormous utility in helping to keep track of bond makings and breakings. In this formalism, chemical bonds are shown as being formed from pairs of electrons flowing from a donor toward an acceptor, often, but not always, displacing other electron pairs. Figure 7.3 shows two examples. In the first, hydroxide ion, acting as a Brønsted base, displaces the pair of electrons in the hydrogen–chlorine bond. The reaction of ammonia ($:NH_3$) with hydrogen chloride is another example in which the electrons on neutral nitrogen are shown displacing the electrons in the hydrogen–chlorine bond.

FIGURE 7.3 Two demonstrations of the application of the curved arrow formalism. We will see many other examples.

WORKED PROBLEM 7.1 Draw an arrow formalism for the reversals of the two reactions in Figure 7.3.

ANSWER This problem may seem to be mindless, but it really is not. The key point here is to begin to see all reactions as proceeding in two directions. Nature doesn't worry about "forward" or "backward." Thermodynamics dictates the direction a reaction will run in a practical sense.

7.2b Lewis Acids and Bases. In Chapters 1 and 2 (p. 90), we began a discussion of acids and bases that will last throughout this book. Then we spent some considerable time in Chapter 3 reviewing Brønsted acids and bases, and developing the idea of pK_a as a measure of acidity. We extended the subject of acidity and basicity to **Lewis acids** and **Lewis bases**. These concepts will hold together much of the material in the remaining chapters.

Consider two conceptually related processes: the reaction of boron trifluoride with fluoride ion, F^-, and the reaction of the methyl cation with hydride, $H:^-$ (Fig. 7.4).

FIGURE 7.4 The fluoride and hydride ion are Lewis bases reacting with the Lewis acids BF_3 and $^+CH_3$.

In these two reactions there is no donating or accepting of a proton, and so by a strict definition of Brønsted acids and bases we are not dealing with acid–base chemistry. There is, however, a more general definition of acids and bases that deals with this situation. A Lewis base is defined as any species with a reactive pair of electrons (an electron donor). This definition may seem very general and indeed it is. It gathers under the "base" category all manner of odd compounds that don't really look like bases. The key word that confers all this generality is "reactive." Reactive under what circumstances? Almost everything is reactive under some set of conditions. And that's true; compounds that are extraordinarily weak Brønsted bases can still behave as Lewis bases.

If the definition of a Lewis base seems general and perhaps vague, the definition of a Lewis acid is even worse. A Lewis acid is anything that will react with a Lewis base (thus, an electron acceptor). Therein lies the power of these definitions. They allow us to generalize—to see as similar reactions that on the surface appear very different. To describe Lewis acid–Lewis base reactions in orbital terms, we need only point out that a stabilizing interaction of orbitals involves the overlap of a filled orbital (Lewis base) with an empty orbital (Lewis acid) (Fig. 7.5).

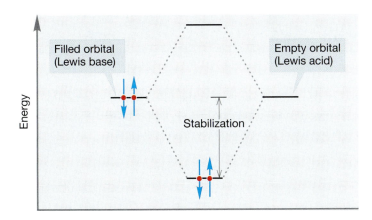

FIGURE 7.5 The interaction of a filled orbital with an empty orbital is stabilizing.

In the examples in Figure 7.4, the Lewis bases are the fluoride and hydride ions. In fluoride, the electrons occupy a $2p$ orbital and in hydride a $1s$ orbital. The Lewis acids are boron trifluoride and the methyl cation, each of which bears an empty $2p$ orbital. A schematic picture of these interactions is shown in Figure 7.6.

FIGURE 7.6 A schematic interaction diagram for the reactions of Figure 7.4.

7.2c HOMO–LUMO Interactions. It is possible to elaborate a bit on the generalization that "Lewis acids react with Lewis bases" by realizing that the strongest interactions (most stabilizing) between orbitals are between the orbitals closest in energy. For stabilizing filled-empty overlaps, this will almost always mean that we must look at the interaction of the highest occupied molecular orbital (**HOMO**) of one molecule (**A**) with the lowest unoccupied molecular orbital (**LUMO**) of the other (**B**) (Fig. 7.7).

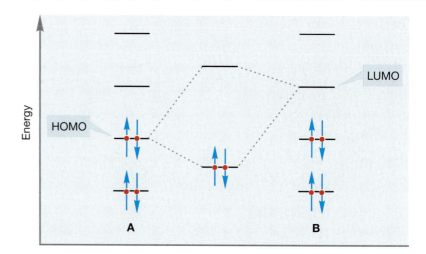

FIGURE 7.7 In two typical molecules, the strongest stabilizing interaction is likely to be between the HOMO of one molecule (**A**) and the LUMO of the other molecule (**B**), because the strongest interactions are between the orbitals closest in energy.

In the reaction of BF_3 with fluoride, it is the filled fluorine $2p$ orbital that is the HOMO and the empty $2p$ orbital on boron that is the LUMO (Fig. 7.8). *All* Lewis acid–Lewis base reactions can be described in similar terms.

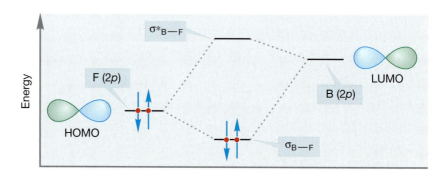

FIGURE 7.8 The interaction of the LUMO for BF_3 (an empty $2p$ orbital on boron) and the HOMO of F^-, a filled $2p$ orbital on fluorine, produces two new molecular orbitals: an occupied bonding σ orbital and an unoccupied antibonding σ* orbital.

PROBLEM 7.2 Identify the HOMO and LUMO in the reaction of the methyl cation with hydride in Figure 7.4.

WORKED PROBLEM 7.3 Point out the HOMO and LUMO in the following reactions:

(a) $HO:^- + Li^+ \longrightarrow HOLi$

(b) $H_2C=CH_2 + BH_3 \longrightarrow {}^+CH_2-CH_2-{}^-BH_3$

(c) $^-:CH_3 + H-Cl: \longrightarrow CH_4 + :Cl:^-$

(d) $H_3N: + H-Cl: \longrightarrow {}^+NH_4 + :Cl:^-$

(e) $HO:^- + H_3O:^+ \longrightarrow 2 H_2O:$

ANSWER

(a)
| HOMO: Filled nonbonding orbital on HO^-
LUMO: Empty $2s$ orbital on Li |

(b)
| HOMO: Filled π orbital of ethylene
LUMO: Empty $2p_z$ orbital on BH_3 |

(c)
| HOMO: Filled nonbonding orbital
of the methyl anion
LUMO: Empty σ^* orbital of H–Cl |

(d)
| HOMO: Filled nonbonding orbital
of ammonia, $:NH_3$
LUMO: Empty σ^* orbital of H–Cl |

(e)
| HOMO: Filled nonbonding orbital
of hydroxide
LUMO: Empty σ^* orbital of an O–H
bond of H_3O^+ |

7.3   REACTIONS OF ALKYL HALIDES: THE SUBSTITUTION REACTION

A great many varieties of substitution reactions are known, but all involve a competition between a pair of Lewis bases for a Lewis acid. Figure 7.9 shows five examples with the displacing group, the **nucleophile** (from the Greek, "nucleus loving"), and displaced group, the **leaving group**, identified. *Both* the nucleophile (usually Nu) and leaving group (usually L) are Lewis bases.

Nucleophile (Nu:)   Leaving group (:L)

(a) $HO:^-$ $K^+ + H_3C-I: \rightleftharpoons HO-CH_3 + K^+ :I:^-$

(b) $CH_3S:^-$ $Na^+ + H_3C-{}^+N(CH_3)_3 :I:^- \rightleftharpoons H_3CS-CH_3 + :N(CH_3)_3 + Na^+ :I:^-$

(c) $:NC:^-$ $K^+ + \begin{array}{c} H_3C \\ \\ H_3C \end{array}CH-Cl: \rightleftharpoons :NC-CH\begin{array}{c} CH_3 \\ \\ CH_3 \end{array} + K^+ :Cl:^-$

(d) $Li^+ :F:^- + CH_3CH_2CH_2-{}^+N_2 :Cl:^- \rightleftharpoons CH_3CH_2CH_2-F: + Li^+ :Cl:^- + N_2$

(e) $H_3N: + H_3C-I: \rightleftharpoons H_3N^+-CH_3 + :I:^-$

FIGURE 7.9 Five examples of the substitution reaction. The nucleophiles for the reactions reading left to right are (a) hydroxide (HO^-), (b) methyl mercaptide (CH_3S^-), (c) cyanide (NC^-), (d) fluoride (F^-), and (e) ammonia ($:NH_3$). The leaving groups are iodide (I^-), trimethylamine [$N(CH_3)_3$], chloride (Cl^-), nitrogen (N_2), and iodide (I^-) again.

Bimolecular nucleophilic substitution: S_N2

www

PROBLEM 7.4 Identify the HOMOs and LUMOs in the reactions of Figure 7.9.

Notice that these are all equilibrium reactions. We will consider equilibrium in detail in Chapter 8, but we need to see here that these competition reactions are all, in principle at least, reversible. A given reaction may not be reversible in a practical sense, but that simply means that one of the competitors has been overwhelmingly successful. There are many possible reasons why this might be so. One of the equilibrating molecules might be much more stable than the other. If one of the nucleophiles were much more powerful than the other, the reaction would be extremely exothermic *in one direction or the other*. The reactions (a–c) in Figure 7.9 are examples of this. We are used to thinking of reactions running from left to right, but this prejudice is arbitrary and even misleading. Reactions run *both ways*. There are other reasons why one of these reactions might be irreversible. Perhaps the equilibrium is disturbed by one of the products being a solid that crystallizes out of the solution, or a gas that bubbles away. If so, the reaction can be driven all the way to one side, even against an unfavorable equilibrium constant. Under such circumstances, an endothermic reaction can go to completion. In reaction (d) of Figure 7.9 both propyl fluoride (bp 2.5 °C) and nitrogen are gases, and their removal from the solution will drive the equilibrium to the right.

The second point to notice is that there is no obligatory charge type in these reactions. The displacing nucleophile is not always negatively charged [reaction (e)], and the group displaced does not always appear as an anion [reactions (b) and (d)].

If we examine a large number of substitution reactions, two categories emerge. The boundaries between them are not absolutely fixed, as we will see, but there are two fairly clear-cut limiting classes of reaction. One is a kinetically second-order process, common for primary and secondary substrates (Section 7.4). The other is a kinetically first-order reaction, common for tertiary substrates (Section 7.6). We will take up these two reactions in turn, examining each in detail.

7.4   SUBSTITUTION, NUCLEOPHILIC, BIMOLECULAR: THE S_N2 REACTION

7.4a Rate Law. For many substitution reactions, the rate v is proportional to the concentrations of both the nucleophile, [Nu:$^-$], and the substrate, [R—L]. This bimolecular reaction is first order in Nu:$^-$ and first order in R—L. The rate constant, k, is a fundamental constant of the reaction (Fig. 7.10). Here, R stands for a generic alkyl group. Don't confuse this R with the (R) used to specify one enantiomer of a chiral molecule. This bimolecular reaction is second order overall, and first order in both

CONVENTION ALERT

FIGURE 7.10 A bimolecular reaction.

$$:Nu \; + \; R—L \; \rightleftharpoons \; Nu—R \; + \; L:^-$$

$$\text{Rate } (v) = k \, [R—L] \, [:Nu]$$

R—L and Nu:⁻, which leads to the name **S**ubstitution, **N**ucleophilic, **bi**molecular, or **S$_N$2 reaction**.

7.4b Stereochemistry of the S$_N$2 Reaction. One of the most powerful of all tools for investigating the mechanism of a reaction is a stereochemical analysis. In this case, we need first to see what's possible, and then to design an experiment that will enable us to determine the stereochemical results. If we start with a single enantiomer (Chapter 4, p. 164) (*R*) or (*S*), there are several possible outcomes of a substitution reaction. We could get **retention of stereochemistry**, in which the stereochemistry of the starting molecule is preserved. The entering nucleophile occupies the same stereochemical position as did the departing leaving group and the product has the same handedness as the starting material (Fig. 7.11).

FIGURE 7.11 One possible stereochemical outcome for the S$_N$2 reaction. In this scenario, the product is formed with *retention of configuration*.

A second possibility is **inversion of stereochemistry**, in which the configuration of the starting material is reversed; (*R*) starting material going to (*S*) product and (*S*) starting material going to (*R*) product. In this kind of reaction, the pyramid of the starting material becomes inverted in the reaction like an umbrella in the wind (Fig. 7.12). In this and Figure 7.11, notice how the numbers enable us to keep track of inversions and retentions.

FIGURE 7.12 Another possible stereochemical result for the S$_N$2 reaction. The handedness of the starting material could be reversed in the product, in what is called *inversion of configuration*.

It is also possible that optical activity could be lost. In such a case, (*R*) or (*S*) starting material goes to a racemic mixture—a 50:50 mixture of (*R*) and (*S*) product (Fig. 7.13).

FIGURE 7.13 A third possible stereochemical result for the S$_N$2 reaction.

Finally, it is possible that some dreadful mixture of the (R) and (S) forms is produced. In such a case, it is very difficult to analyze the situation, as several reaction mechanisms may be operating simultaneously.

WORKED PROBLEM 7.5 Sometimes matters are even more complicated. It is possible, for example, that an (R) starting material could produce an (R) product by a process involving inversion. Here is a conceivable example. Explain.

ANSWER The designations (R) and (S) are determined for each molecule through the Cahn–Ingold–Prelog priority system (Chapter 3, p. 112; Chapter 4, p. 157). *If the groups attached to carbon change*, as in the example shown, there is no guarantee that inversion will change an (R) carbon to an (S) carbon. In the figure, the "umbrella" is inverted, but at the same time the substituents attached to the stereogenic carbon have also changed from (C, Cl, H, I) to (C, Cl, H, OH). Application of the priority system shows that both molecules are (R) despite the inversion.

For most secondary and primary substrates, substitution occurs with complete inversion of configuration. We now are faced with three questions: (1) How do we know the reactions go with inversion? (2) What does this result tell us about the reaction mechanism? (3) Why is inversion preferred to retention or racemization?

The answer to the first question is easy; we measure it. Suppose we start with the optically active iodide (R—I) in Figure 7.14 and monitor the reaction with radioactive iodide ion (¯I*).[†] We measure two things: the incorporation of radioactive iodide and the optical activity of R—I*. If radioactive iodide reacts with R—I with retention of configuration there will be no loss of rotation at all [I* and I are not significantly different in their contributions to the rotation of plane-polarized light (Fig. 7.14)].

FIGURE 7.14 A mechanism involving retention predicts that incorporation of radioactive labeled iodide will induce no change in optical rotation.

[†]In the following discussion, we have indicated radioactive iodine with an asterisk in both the text and the figures.

What happens if the reaction goes with inversion? The observed optical rotation of R—I is certainly going to decrease as, for example, (R)-iodide is converted into (S)-iodide. But it is usually not obvious how the rate of loss of optical activity will compare with the rate of incorporation of I*. If every replacement occurs with inversion, the rate of loss of optical activity must be twice the rate of incorporation of I*. Every conversion of an (R)-iodide (R—I) into an (S)-iodide (I*—R) generates a molecule, I*—R, whose rotation exactly cancels that of a molecule of starting R—I. In effect, the rotation of two molecules goes to zero for every inversion (Fig. 7.15), and this is exactly what is observed experimentally. Every incorporation of I* is accompanied by the inversion of one (R)-iodide to an (S)-iodide.

The boxed set of two molecules will contribute 0° to the rotation…

Net rotation of this set of molecules = 4(R)

…thus, the net rotation of this set of four molecules is 2(R)

FIGURE 7.15 A mechanism involving inversion. For the incorporation of one atom of radioactive iodide, the rotation of the set of molecules shown will be cut by a factor of 2.

Now, what does this result tell us about the mechanism? The observation of inversion tells us that the nucleophile, the incoming atom or molecule, must be approaching the substrate from the rear of the departing atom. Only this path can lead to the observed inversion. Figure 7.16 shows the reaction, along with an arrow formalism description of the bonds that are forming and breaking. The arrows show the path of approach of the entering nucleophile and the departing leaving group. The tetrahedral "umbrella" is inverted during the reaction.

FIGURE 7.16 The observation of inversion in the S_N2 reaction means that the entering nucleophile, radioactive (or labeled) iodide, must enter the molecule from the side opposite the departing leaving group, here unlabeled iodide.

Had the reaction occurred with retention, a frontside approach would have been demanded. The incoming nucleophile would have had to enter the molecule from the same side as the departing leaving group. The tetrahedral "umbrella" would have retained its configuration. This result is *not* observed (Fig. 7.17).

FIGURE 7.17 If retention had been the experimentally observed result, the nucleophile would have added from the same side that the leaving group departed.

All this analysis is useful; we now know the path of approach for the nucleophile in the S_N2 reaction, and so we know more about the mechanism of this process. But our new-found knowledge only serves to generate the inevitable next question. *Why* is inversion preferred? Why is frontside displacement—retention—never observed? Here comes part of the payoff for all the work we did in earlier chapters on molecular orbitals. One of the key lessons of those early chapters was that interactions of *filled* and *empty* orbitals will be stabilizing, as the two electrons originally in the filled orbital can be accommodated in the new bonding orbital, and no electrons need occupy the antibonding molecular orbital. Now let's look at this reaction in orbital terms. A filled nonbonding orbital "*n*" on the nucleophile interacts with an orbital on the substrate, R—L, involved in the bond from R to the leaving group, L. As the interaction of two filled orbitals is not stabilizing, the filled orbital *n* must overlap with the empty, antibonding σ^* orbital of R—L. This overlapping is nothing more than the interaction of the nucleophile's HOMO, *n*, with the sigma bond's LUMO, σ^*, as is shown schematically in Figure 7.18.

FIGURE 7.18 An orbital interaction scheme for the stabilizing interaction of a filled nonbonding orbital (*n*) on the nucleophile overlapping with the empty, antibonding (σ^*) orbital on the substrate.

Filled *n* (HOMO)

Empty σ^* (LUMO)

Energy

What does σ^* look like? Figure 7.19 shows us. Note that σ^* has one of its large lobes pointing to the *rear* of R—L. That is where overlap with the filled *n* orbital will be most efficient.

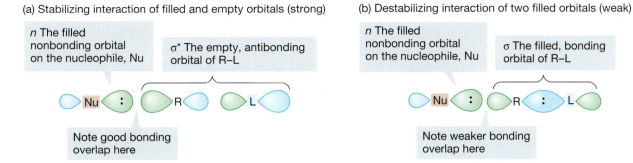

(a) Stabilizing interaction of filled and empty orbitals (strong)

n The filled nonbonding orbital on the nucleophile, Nu

σ* The empty, antibonding orbital of R–L

Note good bonding overlap here

(b) Destabilizing interaction of two filled orbitals (weak)

n The filled nonbonding orbital on the nucleophile, Nu

σ The filled, bonding orbital of R–L

Note weaker bonding overlap here

FIGURE 7.19 (a) The overlap of the filled nonbonding orbital (*n*) with the "outside" lobe of the empty σ* orbital of R–L takes advantage of good overlap with the lobe of σ*. (b) Contrast (a) with the destabilizing interaction of *n* and σ, the two filled orbitals. The interaction of two filled orbitals is destabilizing, but the overlap of the lobes involved, one fat, the other thin, is poor. This interaction is not important.

Now look what happens if the orbitals overlap in a way that would lead to retention through a frontside substitution. Not only is the magnitude of interaction reduced by the relatively poor overlap with the smaller lobe of σ*, but there is both a bonding and antibonding interaction, and these will tend to cancel each other (Fig. 7.20). Frontside displacement is a very unlikely process indeed.

We can now map out the progress of the bond making and breaking in the S$_N$2 reaction. The reaction starts as the nucleophile Nu:⁻ approaches the rear of the substrate. As the HOMO, the nonbonding orbital *n* on Nu:⁻, overlaps with the LUMO, the antibonding σ* orbital of C−L, the bond from C to L is weakened as the antibonding orbital begins to be occupied. The bond length of C−L increases as the new bond between Nu and C begins to form. As C−L lengthens, the other groups attached to C bend back, in umbrella fashion, ultimately inverting (just like the wind-blown umbrella) to form the product. For a symmetrical displacement such as the reaction of radioactive ⁻I* with R−I discussed before, the midpoint of the reaction will be the point at which the groups attached to C are exactly half-way inverted (Fig. 7.21).

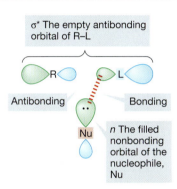

σ* The empty antibonding orbital of R–L

Antibonding   Bonding

Nu

n The filled nonbonding orbital of the nucleophile, Nu

FIGURE 7.20 Frontside substitution (retention) requires overlap with the inside part of σ*. There is poor overlap with the small lobes of σ* as well as offsetting bonding and antibonding interactions.

FIGURE 7.21 In a symmetrical inversion process, the midpoint, called the transition state, is the point at which the umbrella is exactly half-inverted.

I* + C−I ⟶ [I* ··· C ··· I]⁻ ⟶ I*−C + I⁻

sp^2

In the transition state for this symmetrical S$_N$2 reaction, the umbrella is half-way inverted and the carbon is sp^2 hybridized

What is the hybridization of the central carbon at this midpoint? It must be sp^2, as this carbon is surrounded by three coplanar groups. There are partial bonds to the incoming nucleophile, ⁻:I* and the departing leaving group, ⁻:I. No picture shows the competition between the two nucleophiles (⁻:I* and ⁻:I) as clearly as this one. What Lewis acid are they competing for? A carbon 2*p* orbital, as the figure shows. What is this structure? At least in the symmetrical displacement of ⁻:I by ⁻:I*, it is the halfway point between the equi-energetic starting material and product. This structure is the **transition state** for this reaction. Is this transition state something one could potentially catch and isolate? No! It occupies not an energy minimum, as stable compounds do, but it sits on an energy maximum along the path from starting material to product. We have seen such structures before: Recall, for example, the eclipsed form of ethane or the transition state for the interconversion of two pyramidal methyl radicals (Chapter 2, p. 62; Fig. 2.19; Problem 2.13; Fig. 7.22).

FIGURE 7.22 Two analogous diagrams. (a) Shows the rotation about the carbon–carbon bond in ethane; (b) shows the displacement of ⁻I by ⁻I*. In each case, two equi-energetic molecules are separated by a single energy barrier, the top of which is called the *transition state*, shown in brackets.

Now, let's consider the change in energy as the S_N2 reaction proceeds [Fig. 7.22 (b)]. The starting material and products are separated by the high-energy point of the reaction, the transition state. The energy difference between starting material and transition state is related to the rate of the reaction. The larger this difference, the higher the barrier, the slower the reaction. So, a quantity of vital importance to any study of a chemical reaction is the energy difference between starting material and the transition state, called the **activation energy**. This diagram is symmetrical only if the substitution reaction is symmetrical; if starting material and product are of equal energy.[†]

[†]Actually, even our picture of the displacement of ⁻:I by ⁻:I* is ever so slightly wrong. Normal iodine (I) is not *exactly* the same as radioactive iodine (I*), and even this reaction is not truly symmetrical. We can ignore the tiny differences in isotopes, however, as long as we keep the principle straight. If you want to be more correct, insert the words "except for a tiny isotope effect" wherever appropriate.

This specific picture is transferable to a general second-order substitution reaction, but some differences will appear. First of all, usually the starting material and product will not be at the same energy, and the transition state cannot be *exactly* half-way in between starting material and product. It will lie closer to one than the other. The transition state cannot be exactly *sp²*, only nearly *sp²* (notice its pyramidal shape). The reaction shown is exothermic as read from left to right, and endothermic as read from right to left (Fig. 7.23).

FIGURE 7.23 A general S$_N$2 reaction. If the nucleophile (Nu) and the leaving group (L) are different, the transition state is not exactly at the midpoint of the reaction, and the hybridization of carbon is only approximately *sp²*.

Figure 7.23 shows both an exothermic and endothermic S$_N$2 substitution reaction. If we read the figure conventionally from left to right, we see the exothermic reaction, but if we read it "backward," from right to left, we see the endothermic reaction. Notice that the activation energy for the forward process must be smaller than that for the reverse reaction. Thermodynamics will determine where the equilibrium settles out, and there may be practical consequences if one partner is substantially favored over the other, but one diagram will suffice to examine both directions. Many of us are prisoners of our language, which is read from left to right, and we tend to do the same for these energy diagrams. Nature is more versatile, or perhaps less prejudiced than we are.

The "take-home lesson" here is an important one: There is a strong preference for inversion in the S$_N$2 reaction. Indeed, there is *no* authenticated example of retention of configuration in this process, despite a great deal of searching by some very clever people. The principal actor in the early mechanistic work on the S$_N$2 reaction was Paul Walden, a chemist born in 1863 in what is now Latvia and who died in Tübingen, Germany, in 1957. Indeed, the inversion now known to occur universally in the S$_N$2 reaction is sometimes called "Walden inversion."

TABLE 7.1
Average Rates of S_N2
Substitution Reactions for
Different Groups

R	Average Rate
$CH_2=CHCH_2$	1.3
CH_3	1
CH_3CH_2	0.033
$CH_3CH_2CH_2$	0.013
$(CH_3)_2CH$	8.3×10^{-4}
$(CH_3)_3CCH_2$	2×10^{-7}
$(CH_3)_3C$	~0

No substitution

FIGURE 7.24 For tertiary substrates, approach from the rear is hindered by the alkyl groups, here all shown as methyls. This steric effect makes the S_N2 reaction impossible.

My professor, William Doering, told me a wonderful story about Paul Walden. Professor Doering was visiting Tübingen in 1956 to talk about his work and overheard a discussion by a now-famous German chemist who was lamenting the fact that Tübingen had not been bombed during the war. It seems that most of his colleagues in other "more fortunate" cities had newly built laboratories courtesy of Allied bombing. There was an old gentleman sitting in the corner who had been wheeled into the department from his nursing home to hear Doering's seminar. Paul Walden—for indeed it was he—was 92 at the time. He overheard the remark and replied, scathingly, "Es ist nicht der Käfig, sondern der kleine Vogel darin!" (It is not the cage that matters, but the little bird inside.)

Now let's examine the effects of structural change in the various participants in this reaction, the R group, the nucleophile, the leaving group, and the solvent.

7.4c Effects of Substrate Structure: The R Group.
The structure of the R group makes a huge difference in the rate of the S_N2 reaction. We anticipated this result when we mentioned earlier that the practically useful S_N2 reaction was restricted to methyl, primary, and secondary substrates. By implication, the rate of the S_N2 reaction with tertiary substrates is zero, or at least negligibly small. Table 7.1, which gives average relative rates of S_N2 reactions for a variety of R groups, shows that the preceding statement is correct. Why should this be? The simple answer seems to be that in a tertiary substrate the rear of the C—L bond is guarded by three alkyl groups, and the entering nucleophile can find no unhindered path along which to approach the fat lobe of σ^* (Fig. 7.24). So the S_N2 reaction is disfavored for tertiary substrates in which steric hindrance to the approaching nucleophile is especially severe. Another reaction, favorable for tertiary substrates, becomes possible. It is called the S_N1 reaction, and we will deal with its mechanism in Section 7.6.

If this steric argument is correct, secondary substrates should react more slowly than primary substrates, and primary substrates should be slower than methyl compounds. In general, this is the case (Table 7.1). In practice, the S_N2 reaction is usually useful as long as there is at least one hydrogen attached to the same carbon as the leaving group (methyl, primary, and secondary substrates). The small size of hydrogen opens a path at the rear for the entering nucleophile.

PROBLEM 7.6 Explain the following change in rate for the S_N2 reaction:

$$R-O^- + CH_3CH_2-I \longrightarrow R-O-CH_2CH_3 + I^-$$

Rate for R = CH_3 is much faster than for R = $(CH_3)_3C$

This picture, which emphasizes steric effects, allows us to make a prediction. In principle, there must be some primary group so gigantic that the S_N2 reaction would be unsuccessful.

In practice, it is rather easy to find such groups. Even the neopentyl group, $(CH_3)_3CCH_2$, is large enough to slow the bimolecular displacement reaction severely, because the *tert*-butyl group blocks the best pathway for rearside displacement of the leaving group (Fig. 7.25; Table 7.1).

FIGURE 7.25 Even neopentyl compounds, in which a *tert*-butyl group shields the rear of the C–L bond, are hindered enough so that the rate of the S$_N$2 reaction is extremely slow.

The S$_N$2 reaction also takes place when ring compounds are used as the substrates. Although the reaction proceeds normally with ring compounds, there are some interesting effects of ring size on the rate of the reaction (Table 7.2; Fig. 7.26).

TABLE 7.2
Relative Reactivities of Cycloalkyl Bromides in the S$_N$2 Reaction

Compound	Relative Rate
Cyclopropyl bromide	$< 10^{-4}$
Cyclobutyl bromide	8×10^{-3}
Cyclopentyl bromide	1.6
Cyclohexyl bromide	1×10^{-2}
Isopropyl bromide	1.0

FIGURE 7.26 The S$_N$2 reaction takes place normally with cyclic substrates.

PROBLEM 7.7 Use a ring compound to design a test of the stereochemistry of the S$_N$2 reaction.

Why should small rings be so slow? As with any question involving rates, we need to look at the structures of the transition states for the reaction to find an answer. In the transition state, the substrate is hybridized approximately sp^2, and that requires bond angles close to 120°. The smaller the ring, the smaller the C–C–C angles. For a three-membered ring, the optimal 120° must be squeezed to 60°, and for a cyclobutane, to 90°. This contraction introduces severe angle strain in the transition state, raises its energy, and slows the rate of reaction (Fig. 7.27).

FIGURE 7.27 A comparison of S$_N$2 transition states for isopropyl bromide and cyclopropyl bromide.

PROBLEM 7.8 Wait! The argument presented above ignores angle strain in the starting material. Cyclopropane itself is strained. Won't that strain raise the energy of the starting material and offset the energy raising of the transition state? Comment. *Hint*: Consider angle strain in both starting material and transition state—in which will it be more important?

In cyclohexane, the rate is apparently slowed by steric interactions with the axial hydrogens. There can be no great angle strain problem here (Fig. 7.28).

FIGURE 7.28 The slow S_N2 displacement of bromide by iodide in cyclohexyl bromide. Incoming iodide is apparently blocked somewhat by the axial carbon–hydrogen bonds.

WORKED PROBLEM 7.9 The axial cyclohexyl iodide shown below reacts more quickly in S_N2 displacements than the equatorial cyclohexyl iodide. Draw the transition states for displacement of iodide from axial cyclohexyl iodide by iodide ion and for the analogous reaction of equatorial cyclohexyl iodide. What is the relation between these transition states? Why does the axial iodo compound react faster than the equatorial compound? **Caution!** Hard, tricky question.

ANSWER A tricky problem indeed. The transition states are the same! There is no energy difference between the two transition states.

The transition state is the same for displacement of axial and equatorial cyclohexyl iodide from 4-iodo-*tert*-butylcyclohexane

Therefore, the reason that axial cyclohexyl iodide reacts more quickly than equatorial cyclohexyl iodide cannot lie in the transition state energies. However, the axial iodo group raises the energy of the *starting material*, thus lowering the activation energy.

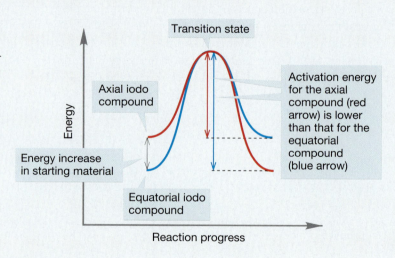

7.4d Effect of the Nucleophile. Some displacing agents are more effective than others. They are better players in the competition for the carbon $2p$ orbital, the Lewis acid. In this section, we examine what makes a good nucleophile; that is, what makes a powerful displacing agent.

Remember: Size can be important. We have already seen how the S_N2 reaction can be slowed, or even stopped altogether, by large R groups (Fig. 7.25). Presumably, large nucleophiles will also have a difficult time in getting close enough to the substrate to overlap effectively with σ^* (Problem 7.6, p. 286). Figure 7.29 shows two nucleophiles, carefully chosen to minimize all differences except size.

FIGURE 7.29 In both these molecules, the nucleophilic nitrogen atom is flanked by three two-carbon chains. In the cage compound, they are tied back by the CH group shown in red, and are not free to rotate. In triethylamine, they are free to rotate and effectively increase the bulk of the nucleophile.

The cage molecule, in which the three ethyl-like groups are tied back out of the way, is a more effective displacing agent than is triethylamine in which the three ethyl groups are freely rotating. Triethylamine is effectively larger than the cage compound in which the alkyl groups are tied back (Fig. 7.30).

FIGURE 7.30 The tied-back cage compound reacts faster with ethyl iodide than does triethylamine, which is effectively the larger compound. The larger nucleophile has a more difficult time in attacking the rear of the carbon–iodine bond in ethyl iodide.

One would expect there to be a general correlation of nucleophilicity with base strength. Brønsted basicity is a measure of how well a base competes for an empty hydrogen $1s$ orbital, and nucleophilicity is a measure of how well a nucleophile competes for an empty carbon $2p$ orbital. The two are not exactly the same thing, as $1s$ and $2p$ orbitals are different in energy and shape, but basicity and nucleophilicity are nevertheless related phenomena (Fig. 7.31).

FIGURE 7.31 Basicity and nucleophilicity are different, but related phenomena.

Actually, it is simple to refine our discussion of good nucleophiles. What do we mean by "a good competitor for a carbon $2p$ orbital"? We are talking about the overlap of a filled and empty orbital, and the resulting stabilization is a measure of how strong the interaction is. *Remember*: The strongest orbital interactions, and hence the greatest stabilizations, come from the overlap of orbitals close in energy (Fig. 7.32).

FIGURE 7.32 Stabilization is greater when filled and empty orbitals of equal or nearly equal energy interact than it is when orbitals that differ greatly in energy interact. The further apart two orbitals are in energy, the smaller the stabilization resulting from their overlap.

So we can anticipate that energy matching between the orbital on the nucleophile and an empty carbon orbital will be important in determining nucleophilicity.

Table 7.3 and Figure 7.33 show some nucleophiles segregated into general categories. It is not possible to do very much better than this, as nucleophilicity is a property that depends on the reaction partner. A good nucleophile with respect to carbon may, or may not, be a good nucleophile with respect to displacement on some other atom. Energy matching is important in the reaction. If we change the reaction partner, and thus the energy of the orbital involved, we will change the stabilization involved as well (Fig. 7.32).

$^-NH_2 > NH_3$ $\qquad$ $^-NH_2 > ^-OR > ^-OH$

$^-SH > SH_2$ $\qquad$ $H_2Se > H_2S > H_2O$

$^-OH > OH_2$ $\qquad$ $R_3P > R_3N$

$I^- > Br^- > Cl^- > F^-$

FIGURE 7.33 Some relative nucleophilicities. Be careful! Nucleophilicity (Lewis basicity) is a hard-to-categorize quantity. Relative nucleophilicity depends, for example, on the identity of the reaction partner, the Lewis acid, as well as the nature of the solvent.

Table 7.3 and Figure 7.33 reveal some spectacular exceptions to the "a good Brønsted base is a good nucleophile" rule. Look at the halide ions, for example. As we would expect from the pK_a values of the conjugate acids, HX (where X is a halide), the basicity order is $F^- > Cl^- > Br^- > I^-$. The nucleophilicity order is exactly opposite the basicity order (Fig. 7.34). Iodide is the *weakest* base but the *strongest* nucleophile. Fluoride is the *strongest* base but the *weakest* nucleophile. Why?

TABLE 7.3 Relative Nucleophilicities of Some Common Species

Species	Name	Relative Nucleophilicity
Excellent Nucleophiles		
NC^-	Cyanide	126,000
HS^-	Thiolate	126,000
I^-	Iodide	80,000
Good Nucleophiles		
HO^-	Hydroxide	16,000
Br^-	Bromide	10,000
N_3^-	Azide	8,000
NH_3	Ammonia	8,000
NO_2^-	Nitrite	5,000
Fair Nucleophiles		
Cl^-	Chloride	1,000
CH_3COO^-	Acetate	630
F^-	Fluoride	80
CH_3OH	Methyl alcohol	1
H_2O	Water	1

Increasing pK_a, decreasing acidity

HI	HBr	HCl	HF
−10	−9	−8	+3.2

Increasing basicity in solution

I^- $\quad$ Br^- $\quad$ Cl^- $\quad$ F^-

Increasing nucleophilicity in solution

F^- $\quad$ Cl^- $\quad$ Br^- $\quad$ I^-

FIGURE 7.34 For the halide ions in solution, the orders of basicity and nucleophilicity are opposite.

To answer this question, we need more data. It turns out that the effectiveness of these (and other) nucleophiles depends on the so-often neglected solvent. It is easy to forget that most chemical reactions are run in "oceans" of solvent, and it is perhaps not too surprising that these oceans of other molecules are not always without substantial effects on the reaction! Figure 7.35 examines rates for typical displacement reactions by halides in solvents of differing polarity and in the total absence of solvent (the gas phase). Note that iodide is a particularly effective nucleophile, relative to other halides, in water, a protic solvent. Recall our discussion in Chapter 6 (p. 247).

| | Relative Rates | | |
I	Br	Cl	F
160	14	1	---
1	5	11	---
---	<0.015	0.02	1

$$X^- + CH_3—I \xrightarrow[H_2O]{S_N2} X—CH_3 + I^-$$

$$X^- + CH_3—Br \xrightarrow[\substack{acetone \\ solvent}]{S_N2} X—CH_3 + Br^-$$

$$X^- + CH_3—Br \xrightarrow[\substack{gas\ phase \\ no\ solvent}]{S_N2} X—CH_3 + Br^-$$

$$X = Halogens$$

FIGURE 7.35 In the highly polar and protic solvent water, iodide is the best nucleophile (fastest reacting). In the polar, but aprotic solvent $(CH_3)_2C=O$ (acetone), the order of nucleophilicity is reversed. In the gas phase, where there is no solvent, fluoride emerges as the best nucleophile! Read the table across not down. The dashed line means "not measured."

Protic solvents are polar solvents containing protons (for example, water or alcohols). In polar solvents that do not contain hydroxylic protons, such as acetone, $(CH_3)_2C=O$, the reactivity order is reversed. In the absence of solvent, in the gas phase, the order is also reversed.

In a protic solvent, a halide can act as a *nucleophile* and react with the substrate through interaction with σ^* of the C—X bond. Alternatively, a halide can behave as a *base* by interacting with one of the protons of the solvent molecule. Chloride and fluoride are by far the better bases and thus hydrogen bond to protonic solvents much more strongly than iodide. The halides become highly solvated in protonic solvents and their relative sizes increase dramatically. They become more encumbered, and therefore less effective nucleophiles. Iodide, the poorer base, is less hydrogen bonded to protonic solvents and thus more free to do the S_N2 reaction (Fig. 7.36).

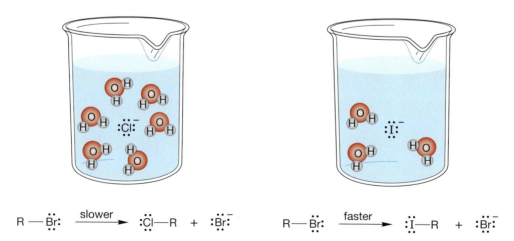

FIGURE 7.36 In a hydroxylated solvent, the highly solvated chloride ion reacts slower with R—Br than the less encumbered iodide ion.

As the polarity of the solvent decreases, fluoride becomes more competitive with iodide. What's the limit of this effect? A reaction with *no* solvent, of course. Pure nucleophilicity, with no competing effects of solvent, can only be examined in the absence of other bulk molecules. We can do that by running the reaction in the gas phase. It is no trivial matter to examine reactions of ions in the gas phase, as the stability of ions depends heavily on solvation. Yet, in the last two decades such experiments have become possible. In the gas phase, fluoride, the better base, is also the better nucleophile (Fig. 7.35).

7.4e Effect of the Leaving Group. Many leaving groups depart as anions.

$$Nu\!:^- + R-L \longrightarrow Nu-R + L\!:^-$$

It seems obvious that the more stable $L\!:^-$ is, the easier it will be to displace it. The stability of the anion $L\!:^-$ is related to the ease of dissociation of its conjugate acid, $H-L$, and thus to the pK_a of $H-L$. Acids ($H-L$) with low pK_a values (strong acids) are related to good leaving groups, $L\!:^-$ (weak bases). Acids with high pK_a values (weak acids) are related to poor leaving groups (strong bases). Table 7.4 shows some good and bad leaving groups and the pK_a values of their conjugate acids, $H-L$.

$$H-L + H_2O \rightleftarrows H_3O^+ + L\!:^-$$

TABLE 7.4 Some Good and Bad Leaving Groups and Their Conjugate Acids

Acid	pK_a	Leaving Group	Name
		Good Leaving Groups	
HI	−10	$^-$I	Iodide
HBr	− 9	$^-$Br	Bromide
HCl	− 8	$^-$Cl	Chloride
$HOSO_2R$	− 3	$^-OSO_2R$	Sulfonate
H_3O^+	− 1.7	OH_2	Water
		Bad Leaving Groups	
HF	+ 3.2	$^-$F	Fluoride
H_2S	+ 7.0	$^-$SH	Thiolate
HCN	+ 9.4	$^-$CN	Cyanide
H_2O	+15.7	$^-$OH	Hydroxide
$HOCH_2CH_3$	+15.9	$^-OCH_2CH_3$	Ethoxide
HOR	+16–18	$^-$OR	Alkoxide

7.4f How to Play "Change the Leaving Group." Many times it would be convenient to convert an alcohol ($R-OH$), which is generally cheap and available (in the chemical industry, those two terms are usually synonymous) into another compound, $R-Nu$. One naturally considers an S_N2 substitution reaction of $R-OH$ (Fig. 7.37) with some nucleophile $Nu\!:^-$.

FIGURE 7.37 A hypothetical S_N2 reaction converting an alcohol, $R-OH$, into a new compound, $Nu-R$.

The problem is that $^-$OH is an extraordinarily poor leaving group (Table 7.4). The pK_a of water is 15.7. Water is a rather weak acid, and its conjugate base, hydroxide ion, is a strong base. There are several ways to circumvent the problem, the simplest of which is to transfer a proton to the alcohol (protonate the alcohol) with a strong acid to give the conjugate acid of the alcohol, $R-\overset{+}{O}H_2$. Protonation converts the leaving group from the very poor leaving group, hydroxide, into the excellent leaving group, water (Fig. 7.38).

FIGURE 7.38 The first step in the reaction of an alcohol with the acid H—B is protonation of the Lewis base oxygen.

$$R-\overset{..}{\underset{..}{O}}H + H-B \rightleftharpoons R-\overset{+}{\underset{..}{O}}H_2 + B\overset{..}{\underset{..}{:}}^-$$

PROBLEM 7.10 What *will* happen when an alcohol is treated with a strong base? Think "simple."

Water is a weak base; its conjugate acid is H$_3$O$^+$, a strong acid with a pK_a of −1.7. Therefore, water is easily displaced by a good nucleophile. In the reaction of an alcohol with HBr, the strongest available nucleophile is Br$^-$, the conjugate base of the HBr used to protonate the alcohol. So, if we want to convert an alcohol into a bromide we need only treat it with HBr. The reaction and its mechanism are shown for the conversion of ethyl alcohol into ethyl bromide in Figure 7.39. Note especially the technique of changing the leaving group from the very poor hydroxide to the excellent water.

Bimolecular nucleophilic substitution: S$_N$2

Hydroxide
(a very poor leaving group)

Water
(an excellent leaving group)

$$CH_3CH_2-\overset{..}{\underset{..}{O}}H + H-\overset{..}{\underset{..}{Br}}: \rightleftharpoons CH_3CH_2-\overset{+}{\underset{..}{O}}H_2 + :\overset{..}{\underset{..}{Br}}:^-$$

Ethyl alcohol

$$:\overset{..}{\underset{..}{Br}}:^- \qquad \overset{H_3C}{\underset{H}{\overset{|}{\underset{H}{C}}}}-\overset{+}{\underset{..}{O}}H_2 \quad \xrightarrow[]{S_N2} \quad :\overset{..}{\underset{..}{Br}}-\overset{CH_3}{\underset{H}{\overset{|}{C}}}\cdots H + H_2\overset{..}{\underset{..}{O}}$$

Ethyl bromide

FIGURE 7.39 Playing "change the leaving group" with ethyl alcohol.

Another variant of the "change the leaving group" game is very closely related to the reactions of alcohols in halogenated acids to give halides. It is called ether cleavage, and it follows essentially the same mechanism. In strong halogenated acids such as HI or HBr, ethers can be cleaved to the corresponding alcohol and halide. In the presence of excess HBr, the alcohols formed in this reaction are often converted into the bromides (Fig. 7.40).

FIGURE 7.40 The mechanism of the cleavage of ethers by strong haloacids.

$$\text{(ether)} \quad \xrightarrow[\text{8 h}]{\text{excess 47\% HBr}} \quad \text{(protonated ether)} \xrightarrow{S_N2} \quad \text{(alcohol)} \overset{..}{\underset{..}{O}}H + :\overset{..}{\underset{..}{Br}}$$

In this reaction, the ether oxygen is first protonated, turning the potential leaving group in an S_N2 reaction from alkoxide, ^-OR, into the much more easily displaced alcohol, HOR. The displacement itself is accomplished by the halide ion formed when the ether is protonated. This reaction is quite analogous to the reaction of alcohols with haloacids. The halide must be a nucleophile strong enough to do the displacement, and the reaction is subject to the limitations of any S_N2 reaction (Fig. 7.40).

PROBLEM 7.11 Clearly, an unsymmetrical ether can cleave in two different ways. Yet often only one path is followed. Explain the specificity shown in reaction (7.1).

PROBLEM 7.12 Provide a mechanism for reaction (7.2).

Another technique for changing the leaving group uses a two-step process in which the alcohol is first converted into a sulfonate. Sulfonates are excellent leaving groups (Table 7.4) and can be displaced by all manner of nucleophiles to give desirable products (Fig. 7.41).

FIGURE 7.41 The conversion of ethyl alcohol into another ethyl derivative. In this sequence of reactions the poor leaving group, OH, is first converted into the good leaving group, sulfonate. The leaving group is then displaced through an S_N2 reaction with a nucleophile, $Nu\!:^-$.

A variation on the sulfonate reaction of Figure 7.42 involves the treatment of alcohols with **thionyl chloride (SOCl$_2$)**. The end result is the conversion of an alcohol into a chloride (Fig. 7.42). The initial product in this reaction is a chlorosulfite ester.

THE GENERAL CASE

A SPECIFIC EXAMPLE

$$CH_3S-CH_2CH_2-OH \xrightarrow[\substack{CHCl_3 \\ 100\ °C,\ 2\ h}]{SOCl_2} CH_3S-CH_2CH_2-Cl + SO_2 + HCl$$
(80%)

FIGURE 7.42 Treatment of an alcohol with thionyl chloride results in the formation of an alkyl chloride.

Mechanisms of breakdown of the chlorosulfite ester involve both S_N2 displacement by chloride ion to give the alkyl chloride, sulfur dioxide (SO$_2$), and a chloride ion, as well as direct decomposition to SO$_2$ and a pair of ions that recombine. In either case, the end product is the chloride (Fig. 7.43). Once again, the strategy in this reaction has been to convert a poor leaving group ($^-$OH) into a good one (SO$_2$ and Cl$^-$).

A chlorosulfite ester

FIGURE 7.43 Mechanisms of chloride formation from reaction of an alcohol with thionyl chloride.

PROBLEM 7.13 Treatment of an alkoxide with dimethyl sulfate leads to methyl ethers. Write a mechanism for this reaction.

(continued)

PROBLEM 7.13 *(continued)*

$$R—\ddot{\underset{..}{O}}{:}^{-} + CH_3—\ddot{\underset{..}{O}}—\overset{\displaystyle :\!O\!:}{\underset{\displaystyle :\!O\!:}{S}}—\ddot{\underset{..}{O}}—CH_3 \longrightarrow R—\ddot{\underset{..}{O}}—CH_3 + {:}\ddot{\underset{..}{O}}—SO_2OCH_3$$

Dimethyl sulfate

Several related reactions involve the use of phosphorus reagents. Alkyl halides can be made from the treatment of an alcohol with phosphorus halides, PX$_3$ or PX$_5$, as shown in Figure 7.44. Many of these reagents cause rearrangement, especially in secondary systems, and the stereochemical outcome, retention or inversion, depends on solvent and other reaction conditions. Accordingly, modifications have been worked out in recent years in order to avoid such problems.

FIGURE 7.44 Formations of alkyl halides with phosphorus reagents.

For example, alcohols can be converted into the related bromides and chlorides through treatment with triphenylphosphine and a carbon tetrahalide. The alkyl halide is generally formed with inversion, and a rough mechanism involving a series of S$_N$2 reactions is sketched out in Figure 7.45.

FIGURE 7.45 Halides are formed from the reaction of alcohols with the intermediate produced from triphenylphosphine and a carbon tetrahalide.

PROBLEM 7.14 Explain why sulfonates should give rather stable anions (weak nucleophiles) and are therefore good leaving groups.

7.4g Effect of Solvent. At first, the behavior of the S_N2 reaction as the polarity of solvent is changed is perplexing. As the solvent polarity is increased, some S_N2 reactions go faster, some slower. Two examples are shown in Figure 7.46. In the first reaction (a) a change to a more polar solvent results in a faster reaction, but in the second reaction (b), the rate decreases when the solvent is made more polar.

FIGURE 7.46 Opposite effects of a change in solvent polarity on the rates of two S_N2 reactions. Reaction (a) goes faster when the solvent is made more polar, whereas reaction (b) goes more slowly.

(a) $(CH_3CH_2)_3N\colon$ $CH_3CH_2 \!-\! \ddot{\underset{..}{I}}\colon$ ⟶ $(CH_3CH_2)_3\overset{+}{N} \!-\! CH_2CH_3$ + $\colon\!\ddot{\underset{..}{I}}\colon^{\!-}$
Rate increases as solvent polarity increases

(b) $H\ddot{\underset{..}{O}}\colon^{\!-}$ $CH_3 \!-\! \overset{+}{\underset{..}{S}}(CH_3)_2$ ⟶ $H\ddot{\underset{..}{O}} \!-\! CH_3$ + $\colon\!\ddot{\underset{..}{S}}(CH_3)_2$
Rate decreases as solvent polarity increases

There is a way to attack almost every problem that poses a question having to do with rates. The rate of a reaction is determined by the energy of the transition state—the high point in energy as the reaction proceeds from starting material to products. In order to solve a question involving rates, we must know the structure of the transition state of the reaction. Figure 7.47 shows the structures of the transition states for the two S_N2 reactions of Figure 7.46.

FIGURE 7.47 The transition state for reaction (a) is more polar than the starting material. The transition state for reaction (b) is less polar than the starting material.

In the first example (a), a more polar solvent will act to stabilize the charged products and the polar, charge-separated transition state. There will be very little effect on the uncharged starting materials. The energy diagram in Figure 7.48 shows what this differential solvation does to the rate of this reaction. The height of the transition state is decreased relative to the starting material and the reaction will go faster. It takes less energy for the molecules to traverse the barrier when the polarity of the solvent increases.

The second example (b) is different. Here the starting materials bear full charges. This charge is dispersed as the reaction goes on. A change to a more polar solvent will stabilize the fully charged starting materials more than the transition state or products. The activation energy increases and the reaction slows (Fig. 7.49).

FIGURE 7.48 In this reaction, the polar transition state will be stabilized by a change to a more polar solvent more than will the neutral starting materials. The result is a decreased activation energy for the reaction and a faster rate. Hydrocarbon solvent (nonpolar, blue line), alcohol solvent (much more polar, red line).

FIGURE 7.49 In this reaction, the transition state will be stabilized by a change to a more polar solvent less than will the charged starting materials. The result is an increased activation energy for the reaction and a slower rate. Alcohol solvent (blue line), alcohol and water solvent (more polar, red line).

SUMMARY

The S_N2 reaction—Substitution, Nucleophilic, bimolecular—is common for methyl, primary, and secondary substrates. It is disfavored by steric hindrance in either the substrate (the "R" group) or the entering nucleophile. It is favored by a strong Lewis base (nucleophile) and a good leaving group. The response of the S_N2 reaction to changes in solvent polarity varies with the charge type of the reaction. Each case must be examined individually. The S_N2 reaction always involves inversion of configuration at the center of substitution.

7.5   THE S_N2 REACTION IN BIOCHEMISTRY

As mentioned in Section 7.1, your vital biomolecules, DNA, for example, are covered with nucleophiles, each of them ready to do the S_N2 reaction if presented with a suitable alkylating agent. Once that alkylation takes place, two important changes are induced. First of all, the shape of the molecule is changed, and the critical fit, on which proper bioaction depends, is altered. Second, alkylation often removes the possibility for hydrogen bonding, and in turn that changes the ability of DNA to associate properly with other molecules. As we shall see in Chapter 23, DNA exists as pairs of associated "bases" and that association is achieved through hydrogen bonding (Fig. 7.50). Alkylation of one of the nucleophilic groups in DNA can disrupt base pairing and lead to destructive mutations.

FIGURE 7.50 Replacement of the N—H hydrogens by an alkyl group hinders—or even eliminates—hydrogen bonding. Thus, proper association of the bases is no longer possible.

Remember the methyl transfer from *S*-adenosylmethionine (p. 144)? Heinz Floss (b. 1934) and his co-workers have demonstrated that a similar transfer occurs through an S_N2 displacement. They did this experiment by showing that transfer took place with complete inversion of configuration—the hallmark of the S_N2 reaction. How can one show inversion in a simple methyl group, CH_3? That's not easy! Floss used two isotopes of hydrogen, deuterium (D) and tritium (T), so that his methyl was not CH_3 but CHDT. Use of a single enantiomer revealed the diagnostic inversion in the transfer shown in Figure 7.51. The work involved in synthesizing that single enantiomer and in analyzing the product was prodigious, but these prototypal techniques have been borrowed several times in similar studies of biochemical transformations.

FIGURE 7.51 Methyl transfer involves an S$_N$2 reaction with inversion.

The S$_N$2 reaction is by no means irrelevant to you—your life literally depends on it happening properly, as in the example studied by Floss, and not in the wrong place—as demonstrated by the sometimes lethal mutations induced by alkylating agents.

7.6 SUBSTITUTION, NUCLEOPHILIC, UNIMOLECULAR: THE S$_N$1 REACTION

7.6a Rate Law. *tert*-Butyl bromide does not undergo a substitution reaction when treated with the good nucleophiles listed in Table 7.3. The rate of the second order S$_N$2 displacement on *tert*-butyl bromide by hydroxide ion is effectively zero. Yet, *tert*-butyl bromide does react with the much less powerfully nucleophilic molecule water to form the product of substitution, *tert*-butyl alcohol (Fig. 7.52).

FIGURE 7.52 Despite the far greater nucleophilicity of hydroxide ion, the reaction of *tert*-butyl bromide with water is successful in producing *tert*-butyl alcohol, whereas reaction with hydroxide ion is not.

Unimolecular nucleophilic substitution: S_N1

PROBLEM 7.15 Although *tert*-butyl bromide does not give *tert*-butyl alcohol on reaction with hydroxide ion, an alkene product (C_4H_8) is produced. Provide a structure for the product and an arrow formalism for this bimolecular reaction.

The first clue to what's happening in Figure 7.52 comes from an examination of the rate law for the reaction. It turns out that the formation of *tert*-butyl alcohol in this reaction is a first-order process in which the concentration of added nucleophile is irrelevant to the rate. The rate law leads to the name of the **S**ubstitution, **N**ucleophilic, **uni**molecular, or the **S_N1 reaction**, where *k* is the rate constant, a fundamental constant of the reaction.

$$\text{Rate} = v = k\,[\textit{tert}\text{-butyl bromide}]$$

At this point, a reasonable mechanistic hypothesis might simply view the S_N1 reaction as beginning with the unimolecular ionization shown in Figure 7.53. The reaction continues as the carbocation is captured by the solvent water, in what is called a **solvolysis reaction**. Solvolysis simply means that the solvent, here water, plays the role of nucleophile in the reaction. In the first step of the S_N1 reaction, *tert*-butyl bromide slowly ionizes to a tertiary carbocation, which then is rapidly captured in a second step by the nucleophilic solvent water to give the protonated alcohol. Any base present, here likely to be bromide or water, can remove a proton (deprotonate the protonated alcohol) to give the alcohol itself (Fig. 7.53).

FIGURE 7.53 A schematic mechanism for the S_N1 reaction. A slow ionization is followed by a relatively fast capture of the carbocation by a nucleophile.

We can describe the reaction with an energy diagram as we did for the S_N2 reaction (Figs. 7.22 and 7.23, pp. 284–285). This diagram is more complicated, because of the formation of the intermediate carbocation. The presence of the intermediate means that there will be two transition states separating it from the starting material and the product. The transition state for the slow step, the initial endothermic ionization, is higher than the transition state for the second, faster step, the capture of the ion by the nucleophile (Nu:⁻) (Fig. 7.54).

This second step determines the structure of the product *but not the rate of the reaction*. If there are several nucleophiles present they will compete with each other in the capture of the carbocation to produce different products, but the rate of the

FIGURE 7.54 An energy diagram for the S_N1 reaction. Notice that the transition state for the slow step in the reaction— the ionization—is higher in energy than the transition state for the capture of the ion by a nucleophile.

reaction will not be affected (Fig. 7.55). The nucleophile determines the structures of the products, but not the rate of their formation. Note that the reverse of the ionization, recapture of the carbocation by bromide, is nothing more than another example of a reaction of this intermediate with a nucleophile.

FIGURE 7.55 This figure shows the capture of the carbocation by four nucleophiles ($^-$:Nu, $^-$:Nu', $^-$Br, and H_2O). However, the nucleophiles present have no effect on the rate of formation of the carbocation, the slow step in the reaction.

We will now see if this simple hypothesis allows a reasonable prediction of the stereochemical course of the reaction.

7.6b Stereochemistry of the S_N1 Reaction. If the S_N1 reaction were as straightforward as suggested by Figures 7.53–7.55, what would be expected if an optically active starting material were used? The crucial intermediate in the reaction is the planar carbocation, an achiral molecule! Once the free, planar carbocation is formed, optical activity is inevitably lost. Addition of the nucleophile must take place at both faces of the cation forming equal amounts of (R) and (S) product. This simple mechanism demands racemization, as Figure 7.56 shows.

FIGURE 7.56 Loss of optical activity occurs as soon as the free, planar carbocation is formed. Addition of any nucleophile will occur equally at the two available lobes of the 2p orbital to give a racemic mixture of products.

Sometimes racemization *is* the actual result. However, usually the stereochemical results are not so clean, and an excess of inversion is found. For example, the two optically active molecules of Figure 7.57 give 21 and 18% excess inversion in the S_N1 reaction with water.

FIGURE 7.57 Incomplete racemization is often found in S_N1 reactions. An excess of inversion is generally found.

What is the source of residual optical activity, and why is excess inversion (as opposed to retention) observed? These two questions are vitally important, as satisfactory answers must be found if we are not to be forced to abandon the proposed mechanism for the S$_N$1 reaction. The answers turn out to be simple. In the mechanism outlined in Figures 7.53–7.55, we assumed that the leaving group had diffused away from the cation, leaving behind a *symmetrical* ion. What if reaction with solvent occurs before the leaving group diffuses completely away? What if reaction occurs with the *pair* of ions shown in Figure 7.58? The leaving group guards the side of the cation from which it departed, and makes addition by solvent water more difficult at this "retention" side. The result is usually an excess of inversion.

Retention pathway is hindered by the bromide ion

<50% Retention

Inversion pathway is unhindered

>50% Inversion

Chiral, optically active

FIGURE 7.58 Unless the leaving group, here bromide ion, completely diffuses away, it shields the carbocation along the retention pathway and this results in an excess of inversion in most S$_N$1 reactions.

PROBLEM 7.16 The second reaction of Figure 7.57 shows an S$_N$1 reaction of a secondary halide. Can you see why the S$_N$1 reaction should be especially favorable for this secondary halide?

PROBLEM 7.17 Consider the role of solvent on the stereochemistry of the S$_N$1 reaction shown in Figure 7.58. A more polar solvent favors racemization. Why?

7.6c Effects of Substrate: The R Group. The rate of the S$_N$1 reaction is much greater for tertiary than for secondary substrates, and simple primary alkyl and methyl halides do not react at all by the S$_N$1 process. Our mechanistic hypothesis shows us why. If carbocation formation is the key step, only those compounds that can form relatively stable ions can react in S$_N$1 fashion.

Modern measurements show that the stability of carbocations increases with substitution (Chapter 3, p. 139). Table 7.5 shows the heats of formation of carbocations in the gas phase. *Remember*: The more positive a species' heat of formation the less stable it is (Chapter 3, p. 115). We can't make too much of the *quantitative* data, as these gas-phase measurements cannot take into account the effects of solvent, which are very large indeed. Nonetheless, they do give a good qualitative picture of carbocation stability: Entries 5, 6, and 7 in Table 7.5 show that tertiary carbocations (three alkyl groups) are more stable than secondary ones (two alkyl groups), and secondary carbocations are more stable than primary ones (one alkyl group). The methyl cation ($^+CH_3$), with no alkyl groups, appears to be the least stable of all. Do not confuse these relative stabilities with absolute stabilities. Carbocations are almost all very unstable species. A small atom such as carbon does not bear a positive charge with any grace, and a carbocation will react with available Lewis bases (for example, halide ion) very quickly. Do not say, "The *tert*-butyl cation is very stable," but rather, "*For a carbocation*, the *tert*-butyl cation is quite stable."

TABLE 7.5 Heats of Formation for Some Carbocations

Entry	Cation	Substitution	ΔH_f° (kcal/mol)
1	$^+CH_3$	Methyl	261.3
2	$^+CH_2CH_3$	Primary	215.6
3	$^+CH_2CH_2CH_3$	Primary	211
4	$(CH_3)_2\overset{+}{C}H$	Secondary	190.9
5	$^+CH_2CH_2CH_2CH_3$	Primary	203
6	$H_3C\overset{+}{C}HCH_2CH_3$	Secondary	183
7	$(CH_3)_3C^+$	Tertiary	165.8

There are many reactions in which tertiary and secondary carbocations are intermediates, but no simple reactions for which a primary or methyl cation is required. These intermediates are simply too unstable. In practice, this means that when you find yourself writing a mechanism using a primary carbocation, think again, it's probably wrong!

WORKED PROBLEM 7.18 We have again avoided (Chapter 3, p. 140) a detailed discussion of *why* more substituted carbocations are more stable than less substituted ones.* Nevertheless you can see the outlines of why this is so. Draw out a three-dimensional picture of the ethyl cation. (*Remember*: The positively charged carbon and the three surrounding atoms are coplanar—this part of the ethyl cation is flat.) Now consider the overlap of one of the carbon–hydrogen bonds of the methyl group with the empty $2p_z$ orbital. Will this interaction be stabilizing? Why?

(continued)

*If avoiding the question drives you crazy, and I think it should at least make you a little uncomfortable, be sure to do this problem—the answer falls right out. See also Chapter 9.

PROBLEM 7.18 *(continued)*

ANSWER Interactions between filled and empty orbitals are stabilizing. This fundamental tenet of chemistry can be applied in this case. If a filled carbon–hydrogen bond overlaps with the empty 2p orbital of the carbocation, the result is a stabilizing interaction.

This interaction can only occur if the appropriate carbon–hydrogen bonds exist. The more alkyl groups, the more stabilization there is. For example, in the methyl cation no such stabilization is possible without the carbon–hydrogen bonds of an alkyl group.

So, tertiary compounds, which form the most stable, *tertiary* carbocations, commonly react through the S_N1 mechanism, and methyl and primary compounds, which would have to form extremely unstable carbocations, do not. Naturally enough, secondary compounds react faster than primary ones in the S_N1 reaction, but more slowly than tertiary compounds (Fig. 7.59).

FIGURE 7.59 Relative reactivities in the S_N1 reaction.

PROBLEM 7.19 The heat of formation for cation **A** is 201 kcal/mol and that of cation **B** is 218.6 kcal/mol. Draw out the two cations carefully and explain why **A** is more stable than **B**.

7.6d Effects of the Nucleophile. We have already seen that the nucleophile does not affect the rate of the S$_N$1 reaction—the reaction is first order in substrate. Yet, the nucleophile does determine the product structure. Note the distinction between the **rate-determining step** (or rate-limiting step) and the **product-determining step**. For example, reaction of *tert*-butyl bromide in water containing a small amount of an added nucleophilic ion such as cyanide leads to substantial amounts of *tert*-butyl cyanide. Even though water is present as solvent, and thus has an immense advantage over cyanide ion because of concentration, the more powerfully nucleophilic cyanide still competes. Cyanide is many times more reactive toward the carbocation than the solvent water (Table 7.3, p. 291). In this reaction, the rate-determining step (the slow ionization) and the product-determining step (the faster capture of the intermediate ion by nucleophiles) are sharply separated (Fig. 7.60).

FIGURE 7.60 The rate- and product-determining steps are sharply distinguished in the S$_N$1 reaction.

7.6e Effect of the Leaving Group. The identity of the leaving group has a large influence on the rate of the S$_N$1 reaction because the structure of the departing group affects the ease of ionization. We have already discussed how the structure—and hence stability—of the carbocation affects the rate of ionization, but obviously the stability of the negatively charged counterion (L:$^-$) will be important as well. The more stable the anion, the more easily it will be lost (Fig. 7.61). Some good leaving groups were shown in Table 7.4 (p. 293).

FIGURE 7.61 The rate of ionization is affected by the stability of the leaving group, L:$^-$. The more stable the leaving group, the faster the ionization.

7.6f Effect of Solvent. For the S$_N$2 reaction, the effect of solvent was rather complicated and depended on the charge type of the reaction. Some S$_N$2 reactions were accelerated by a change to a more polar solvent; others were slowed down (p. 298). For most S$_N$1 reactions the situation is simpler. Ions are usually formed in the slow step of the reaction. In a polar solvent, represented by the dipole arrows in Figure 7.62, the charged species formed in the S$_N$1 reaction are stabilized. A nonpolar solvent cannot stabilize as well. One can expect that the S$_N$1 reaction will usually be favored by a polar solvent.

FIGURE 7.62 Stabilization of the ions formed in an S$_N$1 reaction.

7.7   SUMMARY AND OVERVIEW OF THE S$_N$2 AND S$_N$1 REACTIONS

These two substitution reactions are interconnected by more than the ultimate formation of a substitution product. They complement each other—when one is ineffective the other is likely to work well. For example, the S$_N$2 reaction works best for methyl and primary substrates, which is exactly the point at which the S$_N$1 reaction is least effective. When the S$_N$2 reaction fails, as with tertiary substrates, the S$_N$1 reaction is most effective (Fig. 7.63).

FIGURE 7.63 The S$_N$2 and S$_N$1 reactions are complementary.

So, at the extremes the situation is clear. The S_N2 dominates for methyl and primary substrates because there is little steric hindrance to the displacement of the leaving group and because the competing S_N1 reaction is disfavored by the necessity of forming the highly unstable methyl or primary carbocations (Fig. 7.64).

A difficult ionization to the hideously unstable primary carbocation

A relatively easy displacement at the unhindered primary carbon

FIGURE 7.64 A comparison of the S_N1 and S_N2 reactions for primary substrates.

The S_N1 dominates for tertiary substrates because ionization can produce the relatively stable tertiary carbocation and because the competing S_N2 reaction is strongly disfavored by the steric hindrance to the nucleophile approaching from the rear (Fig. 7.65).

An easy S_N1 reaction as the relatively stable tertiary carbocation is formed

The S_N2 is thwarted by the difficulty of displacement at the sterically hindered tertiary carbon

FIGURE 7.65 For tertiary systems, the S_N1 reaction dominates; the ionization produces the relatively stable tertiary carbocation and the S_N2 reaction is hindered by the three R groups that thwart approach of the nucleophile from the rear.

In the middle ground—secondary substrates—we might expect to be able to find both reactions. Small R groups, use of a powerful nucleophile, and a solvent of proper polarity for the charge type of the reaction can favor the S_N2 reaction. Large R groups, a weak nucleophile, and a polar solvent will favor the S_N1 reaction (Fig. 7.66).

FIGURE 7.66 For secondary substrates, both the S_N1 and S_N2 reactions may be possible.

We have given a picture of these reactions in which extremes are emphasized; in which an "either S$_N$2 or S$_N$1" view is developed. The situation is messier than this simple explanation would suggest, especially for secondary compounds. Imagine beginning to ionize a bond to a leaving group. As the bond stretches, positive charge begins to develop on carbon (δ^+) and rehybridization toward sp^2 begins as the groups attached to the carbon flatten out (Fig. 7.67).

FIGURE 7.67 The beginnings of an S$_N$1 reaction. The bond to the leaving group :L$^-$ has lengthened and the tetrahedron has begun to flatten out. The hybridization of the central carbon is changing from nearly sp^3 to the sp^2 of the planar carbocation.

If we simply continue this ionization process we have the pure S$_N$1 reaction. If we let a nucleophilic solvent molecule interact with the rear of the departing bond and ultimately form a bond to carbon, we have the S$_N$2 reaction with inversion (Fig. 7.68).

In the classic S$_N$1 reaction, this ionization continues to produce the planar carbocation. Notice, however, that the carbocation when first formed is not symmetrical. The leaving group lurks on one side, partially shielding the cation from solvent and other nucleophiles

However, solvation at the side away from the old leaving group can be converted into nucleophilic attack (S$_N$2) to give inverted product

FIGURE 7.68 Especially for secondary substrates, the difference between S$_N$2, displacement by solvent, and preferential solvation at the rear of the departing bond is a subtle one.

How strong an interaction with the rear lobe of the orbital is needed before we call it bonding? In polar solvents, ions are highly solvated and in an S$_N$1 reaction solvation will begin to be important as soon as a δ^+ charge begins to build up on carbon. Solvation will be especially easy at the rear of the departing group as the leaving group shields the frontside. We have already seen that the leaving group can hinder reaction at one side of the developing cation and produce some net inversion in the reaction (Fig. 7.58).

Now we begin to see the difficulty of telling the two mechanisms apart in some situations. When does "preferential solvation from the rear" become "displacement from the rear"? It becomes very hard to sort out in some cases. Organic chemistry is frustratingly, if delightfully, messy.

This messiness raises its head in other ways. Almost all substitution reactions are accompanied by some formation of alkene. In organic chemistry few reactions proceed in 100% yield, quantitatively producing one molecule from another. A large part of the business of organic synthesis is the search for specificity. Synthetic chemists seek reagents and reaction conditions that greatly favor one reaction pathway over all others. In particular, attempts to synthesize molecules through substitution reactions must take account of competitive reactions called eliminations. The elements of H—L are lost (eliminated) from the substrate to give an alkene (Fig. 7.69). Sections 7.8 and 7.9 will examine two general mechanisms for these elimination reactions.

FIGURE 7.69 In almost all S$_N$1 and S$_N$2 reactions, substitution is accompanied by some elimination of H—L to give an alkene.

PROBLEM 7.20 We have actually seen at least one S$_N$1 reaction before, although it was not labeled as such because we didn't yet know the name. Where was it? If you get this one, you have a great eye and memory for detail. *Hint*: It is in this chapter.

7.8   THE UNIMOLECULAR ELIMINATION REACTION: E1

7.8a Rate Law. If we measure the rate law for the formation of the alkene accompanying an S$_N$1 reaction, we find that this elimination reaction is also a first-order process. As in the S$_N$1 reaction, the slow step is the ionization of the substrate to give a carbocation. The carbocation intermediate can do two things: It can continue the S$_N$1 reaction by adding a nucleophile (Nu:$^-$) or, if a proton is available, an alkene can be formed in what is called the **E1 reaction**. Note that in the elimination phase the nucleophile is acting as a Brønsted base. The carbocation is an intermediate in *both* processes (Fig. 7.70).

Unimolecular elimination: E1

FIGURE 7.70
Competing E1 and S$_N$1 reactions.

THE GENERAL CASE

Favored by a highly *basic* nucleophile

Favored by a highly *nucleophilic* nucleophile

The mixture of E1 and S_N1 products depends primarily on the nucleophile's basicity. Strong Brønsted bases will be more effective at removing the proton than weak Brønsted bases. That's exactly what we mean by a strong Brønsted base—something that has a high affinity for a proton. Good nucleophiles will be less effective in the E1 reaction, but more effective in the S_N1 reaction (Fig. 7.71).

SOME SPECIFIC EXAMPLES

FIGURE 7.71 The E1 and S_N1 reactions compete. A strong base (ethoxide) favors the E1 reaction. For pure ethyl alcohol, the ratio of E1 product (isobutene) to S_N1 product (*tert*-butyl ethyl ether) is 0.25. In the presence of the much stronger base sodium ethoxide, this ratio increases to about 13.

7.8b Orientation of the E1 Reaction. In many E1 reactions, more than one hydrogen can be removed from the carbocationic intermediate to give an alkene. When there is a choice, the major product is the more substituted, more stable alkene (Chapter 3, p. 115). Of course, statistical effects will also influence the product mix, but the overwhelming factor is alkene stability. The more substituted alkene is favored strongly in the E1 reaction. For example, 2-bromo-2-methylbutane reacts in a solution of ethyl alcohol and water to give a mixture of 2-methyl-2-butene and 2-methyl-1-butene in which the more substitued alkene predominates by a factor of 4 (Fig. 7.72).

THE GENERAL CASE

A SPECIFIC EXAMPLE

FIGURE 7.72 The more substituted (more stable) alkene is favored in the E1 reaction, even when statistical factors favor the less substituted isomer. This reaction is known as Saytzeff elimination.

PROBLEM 7.21 Draw careful arrow formalisms for the reactions of Figure 7.72.

PROBLEM 7.22 The reaction in Figure 7.72 also produces two products of the S_N1 reaction: one $C_5H_{12}O$ and the other $C_7H_{16}O$. What are they and how are they formed?

To see why the more substituted alkene is favored, look at the transition state for proton removal. In the transition state the carbon–hydrogen bond is lengthened, and the double bond of the alkene will be partially formed (Fig. 7.73).

Transition states for proton removal by a base, :Nu to give the alkene and Nu–H

FIGURE 7.73 In the transition state, a double bond is partially formed. The increased stability of more substituted double bonds will be felt, and the transition state for formation of the more substituted alkene (loss of H_a) will be lower in energy than the transition state for formation of the less substituted alkene (loss of H_b).

The factors that operate to stabilize alkenes with fully formed double bonds will also be operative in the transition states in which there are partially formed double bonds. So, the more alkyl groups the better, and the E1 reaction preferentially produces the most substituted alkene possible. This phenomenon is called **Saytzeff elimination**, after Alexander M. Saytzeff (1841–1910), the Russian chemist who discovered it.*

*The name Saytzeff can be transliterated as Zaitzev, Saytzev, or Saytseff.

PROBLEM 7.23 Predict the products of the E1/S_N1 reactions of the following molecules in water:

(a)

$(CH_3CH_2)_3C$—Br

(b)

(c)

Br

Br

PROBLEM 7.24 It was noticed as early as 1862 that treatment of isobutyl halides, $(CH_3)_2CHCH_2$—X, with silver acetate, Ag^+ ^-OAc (^-OAc is "acetate," a weak nucleophile, and Ag^+ is a strong Lewis acid, especially toward chloride), produced *tert*-butyl acetate, $(CH_3)_3C$—OAc, and isobutene. Remember the stability order of carbocations (pp. 305–306) and provide an explanation for this behavior. *Hint*: Reason backward from the structure of the product.

7.9   THE BIMOLECULAR ELIMINATION REACTION: E2

As with the substitution reaction, there are two common mechanisms for the elimination reaction. We have just explored the first-order process, the E1 reaction. Now let's look at the bimolecular reaction, the **E2 reaction**.

7.9a Rate Law. The S_N2 reaction is also complicated by alkene formation. As for the S_N2 reaction, the typical rate law for the bimolecular E2 reaction shows that the alkene is formed in a second-order process, first order in both substrate and nucleophile.

$$v = k[\text{substrate}]\,[\text{nucleophile}]$$

The mechanism involves attack by the base at a proton attached to a carbon adjacent to the carbon bearing the leaving group. Figure 7.74 shows the arrow formalism pictures of the competing E2 and S_N2 reactions.

Bimolecular elimination: E2

FIGURE 7.74 The competing E2 and S_N2 reactions.

This picture gives us an immediate clue as to how to influence the product mixture from competing E2 and S_N2 reactions. As in the E1–S_N1 competition (p. 312), strong Brønsted bases (high affinity for hydrogen 1s orbital) should favor the E2 and strong nucleophiles (high affinity for carbon 2p orbital) the S_N2 (Fig. 7.75).

FIGURE 7.75 The E2 and S_N2 reactions compete. In the examples shown, the relatively good nucleophile, chloride, gives exclusively the S_N2 reaction, whereas the much more basic ethoxide ion gives mostly the product of the E2 reaction.

7.9b Effect of Substrate Branching on the E2–S_N2 Mix. The rate of the S_N2 reaction depends strongly on the structure of the substrate. The more highly branched the substrate, the more hindered is the obligatory approach from the rear of the leaving group and the slower the S_N2 reaction. This effect reaches a limit with tertiary substrates in which the S_N2 rate is effectively zero. The slower the S_N2 reaction, the more important is the competing E2. With a typical tertiary substrate such as *tert*-butyl bromide, the E2 reaction is the only effective reaction under E2–S_N2 conditions. The steric requirements for deprotonation are far less severe than for attack at the rear of the carbon–bromine bond of *tert*-butyl bromide (Fig. 7.76).

FIGURE 7.76 For tertiary substrates, the S_N2 reaction does not proceed, but the E2 reaction is relatively unhindered. Attack from the rear in S_N2 fashion is blocked by the three methyl groups. The E2 reaction is relatively easy.

With less hindered substrates the S_N2 reaction can compete much more effectively with the E2, and with primary substrates the S_N2 is dominant (Fig. 7.77).

For this secondary substrate, the S_N2 and E2 reactions are competitive

In this primary example, the S_N2 reaction is the only process

7.9c Stereochemistry of the E2 Reaction. When we explored the mechanism of the S_N1 and S_N2 reactions, an examination of the stereochemical outcome of the reactions was crucial to our development of their mechanisms. Stereochemical experiments are common in organic chemistry, and are one of the reasons we spent so much time and effort earlier on stereochemical analysis. In the E2 reaction, there are four possible extreme orientations. They can be classified by noting the dihedral angle between the two groups that are leaving, the hydrogen and the leaving group L (Fig. 7.78).

| 180° | 120° | 60° | 0° |

Dihedral angle between C–H and C–L

FIGURE 7.78 Four possible arrangements for the E2 reaction. The Newman projections emphasize the dihedral angle between the C–H and C–L bonds involved in the reaction.

As the C–H and C–L bonds lengthen in this reaction, the molecule flattens out as the $2p$ orbitals that will constitute the π system of the alkene are formed. *Remember* (Chapter 3, p. 105): In the product alkene the $2p$ orbitals are perfectly lined up, eclipsed, with a 0° dihedral angle (Fig. 7.79).

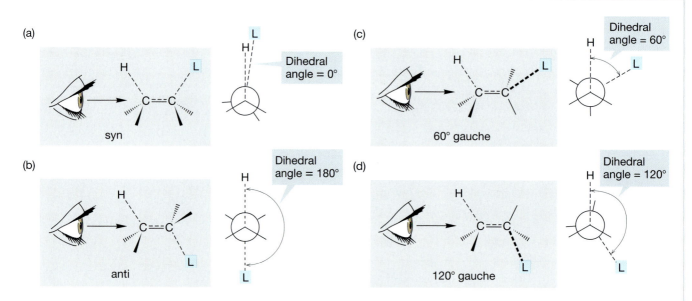

FIGURE 7.79 The generation of a π bond as the E2 reaction occurs. In the E2 reaction, the C—H and C—L bonds lengthen as the π bond between the carbons is created. Note especially the requirement for orbital line-up as the π bond is generated. The dihedral angle between the two *p* orbitals in the product alkene is 0°.

Transition state

Newman projection

Note that only the syn (0°) and anti (180°) arrangements shown in Figure 7.78 can lead smoothly to an alkene with perfectly overlapping $2p$ orbitals. The 120° and 60° forms cannot produce alkene without an additional rotation to line up the orbitals (Fig. 7.80).

(a) syn — Dihedral angle = 0°

(b) anti — Dihedral angle = 180°

(c) 60° gauche — Dihedral angle = 60°

(d) 120° gauche — Dihedral angle = 120°

FIGURE 7.80 Four views of the transition state for the E2 reaction. In examples (a) and (b), where the dihedral angle is 0° or 180°, the π bond can be generated without an extra rotation. In examples (c) and (d), where the dihedral angle is 60° or 120°, an extra rotation must take place for the π bond to be made.

Thus we can anticipate that the 60° and 120° *gauche* arrangements for elimination should be disfavored. It is much more difficult to decide which of the remaining two possibilities, syn (0°) or anti (180°) E2 will be preferred in the E2 reaction.

The E2 is a kind of remote S_N2 reaction. The Lewis base ($Nu{:}^-$) attacks the hydrogen and the pair of electrons in the carbon–hydrogen bond does the displacing of the leaving group. If we keep this analogy in mind we can see that the anti (180°) elimination involves displacement from the rear, as in the S_N2 reaction, but syn (0°) elimination requires frontside displacement. We might therefore expect the anti E2 to be favored (Fig. 7.81).

The favored backside displacement in the S_N2 reaction

The 180° E2 reaction: the electrons in the C–H bond displace the leaving group from the rear

The disfavored frontside S_N2 displacement

The 0° E2 mimics the frontside S_N2 reaction: the electrons in the C–H bond displace the leaving group from the front

FIGURE 7.81 The 0°, syn E2 reaction resembles a frontside S_N2 reaction, whereas the 180°, anti E2 reaction looks like the favored, backside S_N2 displacement reaction.

Now the problem is to find an experiment that will let us see this level of detail. As usual, we will use a stereochemical analysis, and this requires a certain complexity of substitution in the substrate molecule. For example, there is no use in examining the stereochemistry of the E2 reaction of a molecule as simple as 2-chloro-2,3-dimethylbutane because there is only one tetrasubstituted alkene possible (Fig. 7.82).

FIGURE 7.82 In this reaction there is only one possible tetrasubstituted alkene product.

PROBLEM 7.25 What other alkene could be formed in the reaction of Figure 7.82?

More highly substituted molecules can reveal the stereochemical preference, however. For example, consider the molecules of Figure 7.83. In diastereomer **A**, **syn (0°) elimination** will lead to a product in which the methyl groups are cis [this is, however, the (E) alkene], whereas **anti (180°) elimination** will give the compound with trans methyls [the (Z) alkene]. In **B**, these preferences are reversed: syn elimination produces the (Z) alkene, and anti elimination gives the (E) alkene. *Remember:* The syn form is an eclipsed energy maximum and the anti form is a staggered energy minimum. Many tests of this kind have been carried out, and it is certain that in acyclic molecules anti (180°) elimination is favored over the syn process (0°) by very large amounts, typically greater than 10^4 (Fig. 7.83).

FIGURE 7.83 The possibilities for 0° and 180° E2 reactions for the two diastereomers shown.

WORKED PROBLEM 7.26 In chemistry, it is customary to report one's research results by submitting papers to the chemical literature. As a kind of quality control, these papers are sent to experts in the field for comment and review before publication. Imagine that you are such a referee looking over the experiment described in Figure 7.83. Do the data absolutely require the conclusion that anti (180°) eliminations are preferred to syn (0°)? Obviously, I think not. What do you think?

(continued)

PROBLEM 7.26 *(continued)*

ANSWER There is a problem. The analysis of Figure 7.83 shows a 180° elimination proceeding from a staggered, energy minimum form. By contrast, the 0°, syn elimination of Figure 7.83 proceeds from a fully eclipsed arrangement, which is a transition state, not an energy minimum (Chapter 2, p. 66; Fig. 7.22, p. 284). This comparison is scarcely fair! In order to make a real comparison of the syn and anti E2 processes, we will have to be more clever.

Eclipsed, energy maximum
form—no lifetime!
syn = 0°

Staggered, energy
minimum form
anti = 180°

It is important to remember that orbital overlap is vital in this reaction, and a syn elimination in which the orbitals overlap well may compete effectively with an anti elimination in which the perfect 180° dihedral angle cannot be attained. For example, some polycyclic molecules give syn elimination. In the example shown in Figure 7.84, the 2-bicyclo[2.2.1]heptyl or "2-norbornyl" system, the *ideal* trans (180°) elimination (loss of HOTs) is impossible, as only a 120° arrangement of the departing groups can be achieved. By contrast, the syn elimination (loss of DOTs) has a perfect 0° dihedral angle between the two departing groups. The result is that elimination is very slow and that the perfect syn elimination is preferred to the 120° anti elimination (Fig. 7.84).

syn, Perfect 0° angle
between D and OTs

The dihedral angle between H and
OTs is 120°, far from a perfect 180°
anti elimination

FIGURE 7.84 Sometimes syn elimination can be preferred.

PROBLEM 7.27 Draw all the possible stereoisomers of 1,2,3,4,5,6-hexachlorocyclohexane. First, draw the molecules flat using some schematic device to indicate the stereochemistry of the chlorine atoms. Next, make three-dimensional drawings of these molecules.

PROBLEM 7.28 One of these isomers reacts much more slowly in the E2 reaction than all the others. Which one is it and why is it so slow to undergo an E2 reaction? *Hint*: Remember the steric requirements for the E2 reaction.

7.9d Orientation of the E2 Reaction.
Generally, Saytzeff elimination is the rule. The most substituted, and therefore most stable alkene is the major product in this regioselective reaction (Fig. 7.85). **Regiochemistry** simply refers to the outcome of a reaction in which more than one orientation of substituents is possible in the product.

FIGURE 7.85 In the E2 reaction, it is the more substituted alkene that is usually formed preferentially. Both cis and trans isomers are formed.

In the transition state for alkene formation, the π bond is partially developed. The factors that make more substituted alkenes more stable than less substituted alkenes will operate on the partially developed alkenes in the transition states as well. The more substituted the partially developed alkene is, the more stable it will be, and this effect will inevitably favor the more substituted alkene as the major product (Fig. 7.86).

In this transition state for the formation of the major product, there is a partially formed disubstituted double bond

FIGURE 7.86 In the E2 reaction, the more stable transition state leads to the major product, the more substituted alkene.

(continued)

FIGURE 7.86 *(continued)*

In this transition state for the formation of the minor product, there is a partially formed monosubstituted double bond

Minor product

There are many factors that influence the ratio of alkene products, however, and some are quite surprising at first. The size of the base makes a small difference. Larger bases favor the less stable alkene. Although there is only a very small change in the product distribution when the base is changed from methoxide to *tert*-butoxide in the reaction of 2-iodobutane with alkoxide (Fig. 7.87), when the base is made truly enormous these effects can be magnified. In order to form the more substituted alkene the base must penetrate further into the middle of the substrate. Formation of the less stable alkene requires attack by base further toward the periphery of the molecule and thus has lower steric requirements.

FIGURE 7.87 The effect of increasing base size is to increase the less stable "Hofmann" product somewhat.

THE GENERAL CASE

Size makes little difference in the approach to one of the hydrogens whose removal leads to the less substituted alkene

There are greater steric problems in approaching one of the hydrogens in the interior of the molecule; thus size of the base operates to favor the less substituted alkene (Hofmann elimination)

A SPECIFIC EXAMPLE

	Saytzeff elimination	Hofmann elimination
Na⁺ ⁻:ÖCH₃ / dimethyl sulfoxide (DMSO)	4.9 trans/cis = 3.1	1
K⁺ ⁻:ÖC(CH₃)₃ / DMSO	3.8 trans/cis = 3.0	1

Hofmann elimination

Formation of the less stable alkene is called **Hofmann elimination** after the man who discovered this preference, August W. von Hofmann (1818–1892). This kind of elimination contrasts with the more usual Saytzeff process in which the more stable alkene is the major product. Hofmann elimination is favored by certain leaving groups. Typical examples of Hofmann leaving groups are fluoride and the 'onium ions, such as ammonium and sulfonium (Fig. 7.88).

L	2-Alkene (%)	1-Alkene (%)
I	81	19 Saytzeff preferred
Br	72	28 Saytzeff preferred
Cl	67	33 Saytzeff preferred
F	30	70 Hofmann preferred

FIGURE 7.88 The regioselectivity of the E2 reaction depends on the identity of the leaving group. Leaving groups such as fluoride, ammonium (R_3N^+), and sulfonium (R_2S^+) lead to predominant Hofmann elimination (the less stable isomer is the major product).

Why is there a preference for the less-substituted alkene? And why does the leaving group make a difference? To answer these questions we first have to mention another kind of loss of H—L, the **E1cB reaction** (elimination, unimolecular, conjugate base). This reaction is in essence the opposite of the E1 reaction. It is a reaction in which the hydrogen is lost first to generate an anion. Subsequent ejection of the leaving group as an anion generates the alkene. In the E1 reaction, the slow step is ionization to give the carbocation. In the E1cB reaction, a proton is first removed to give an anion that subsequently eliminates the leaving group (Fig. 7.89).

FIGURE 7.89 The E1 and E1cB reactions contrasted.

This reaction is relatively uncommon, but some examples are known. The requirements for the E1cB reaction are a poor leaving group and a rather easily lost proton.

PROBLEM 7.29 Design a good E1cB reaction. *Hint*: What kind of leaving group do you need? What kind of carbon–hydrogen bond?

The E2 reaction of alkyl fluorides, ammonium ions, and sulfonium ions does not go by this mechanism but the transition state *is* "E1cB-like." In the transition state for any E2 reaction, the bonds to hydrogen and the leaving group are both breaking. In the E2 reaction, the transition state may or may not have the bonds to H and L breaking roughly simultaneously. If breaking of the C—H bond is further advanced in the transition state than is breaking of the C—L bond, the transition state will resemble the E1cB reaction. If the C—L bond is more broken in the transition state than is C—H, the transition state is E1-like (Fig. 7.90).

FIGURE 7.90 Three transition states for an E2 elimination.

In an E1cB-like transition state, the carbon adjacent to the carbon bearing the leaving group becomes partially negatively charged (Fig. 7.90). The highly polar carbon–fluorine bond will place a partial positive charge on the carbon to which the fluorine is attached. For an E1cB-like transition state, this situation is favorable. The same is true for the highly electronegative onium ions (Fig. 7.91).

The C–F bond is polarized; the very electronegative fluorine bears a partial negative charge, leaving the carbon partially positive

The partial positive charge on carbon favors a E1cB-like transition state for elimination, in which the C–H bond breaks ahead of the C–L bond because this kind of transition state places a δ^- near the existing δ^+

FIGURE 7.91 An E1cB-like transition state is favored by strongly electron-withdrawing groups such as fluorine.

In the example shown in Figure 7.92, the transition state for Hofmann elimination has a partial negative charge on a primary carbon, whereas the transition state for Saytzeff elimination has a partial negative charge on a secondary carbon (Fig. 7.92).

In this transition state, there is a partial negative charge on a primary carbon

Hofmann

This transition state, which leads to the more substituted alkene, has a partial negative charge on a secondary carbon

Saytzeff

cis/trans Isomers

FIGURE 7.92 In the transition state for Hofmann elimination, a partial primary carbanion is formed. In the transition state for Saytzeff elimination, it is a partial secondary carbanion.

In contrast to carbocations and radicals, less substituted carbanions are more stable than more substituted carbanions. Methyl and primary carbanions are more stable than secondary carbanions, and secondary are more stable than tertiary (Fig. 7.93).

Tertiary   Secondary   Primary   Methyl

Carbanion stability

FIGURE 7.93 The order of carbanion stability is the opposite of that for carbocations and radicals partly because of the greater ease of solvation for the smaller carbanions.

So, the more favorable transition state is the one with the developing negative charge on the primary carbon, the one leading to the less stable alkene (Fig. 7.92).

PROBLEM 7.30 How else can the nature of the leaving group determine the structure of the alkene? With the exception of F, the "Hofmann" leaving groups are all large. Consider the possible products of an E2 reaction from the compound in the figure at the right. First, show what the possible products are and identify the different hydrogens that must be lost to produce them. Then, draw Newman projections for the arrangements leading to the two products. Remember the preference for 180° anti elimination in the E2 process. Finally, use the Newman projections to explain why a large leaving group might favor Hofmann elimination.

PROBLEM 7.31 Predict the products of reaction of the following compounds with hydroxide ion. Be careful with the molecule in (c). The answer here is tricky.

(a)   Br

(b)   Br

(c)   Br

SUMMARY

We have already been introduced to addition reactions in Chapter 3 (p. 134). The opposite of addition is elimination, in which two groups or atoms are lost with the introduction of a π bond. There are two general mechanisms for elimination reactions. First, there is a bimolecular process in which starting material goes to product over a single transition state. This E2 process resembles (and competes with) the bimolecular S_N2 reaction, but is favored by a strong Brønsted base, rather than a strong Lewis base. The E2 elimination "prefers" an anti arrangement of the groups being eliminated. The regiochemistry of the product alkene depends on the nature of the leaving group. Most leaving groups favor the more substituted alkene (Saytzeff elimination) but a few large and/or highly electronegative leaving groups lead mainly to the less substituted alkene (Hofmann elimination).

Second, there is also a unimolecular elimination reaction (E1) that parallels (and competes with) the S_N1 substitution reaction. The two reactions share the same intermediate, the carbocation formed by ionization of the leaving group. Once this carbocation is formed it can be captured by nucleophiles (S_N1) or be deprotonated to give an alkene (E1). Saytzeff elimination is the rule.

7.10   WHAT CAN WE DO WITH THESE REACTIONS? HOW TO DO ORGANIC SYNTHESIS

Your ability to make molecules has just increased enormously. Starting from rather easily accessible compounds, the alkyl halides, and inorganic materials, myriad other structural types are suddenly available. Figure 7.94 summarizes the synthetic potential of the substitution reaction. Not all reactions will work for all R—L molecules and elimination reactions will surely also occur.

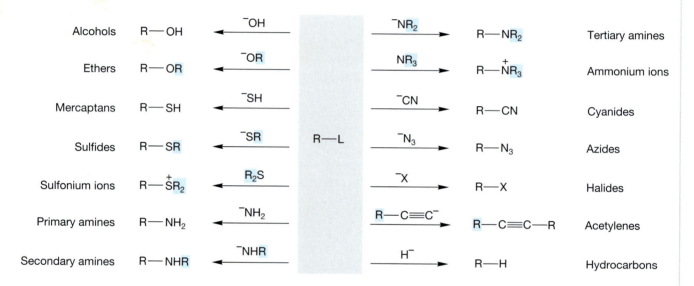

Alcohols	R—OH	$^-$OH		$^-$NR$_2$	R—NR$_2$	Tertiary amines
Ethers	R—OR	$^-$OR		NR$_3$	R—$\overset{+}{N}$R$_3$	Ammonium ions
Mercaptans	R—SH	$^-$SH		$^-$CN	R—CN	Cyanides
Sulfides	R—SR	$^-$SR	R—L	$^-$N$_3$	R—N$_3$	Azides
Sulfonium ions	R—$\overset{+}{S}$R$_2$	R$_2$S		$^-$X	R—X	Halides
Primary amines	R—NH$_2$	$^-$NH$_2$		R—C≡C$^-$	R—C≡C—R	Acetylenes
Secondary amines	R—NHR	$^-$NHR		H$^-$	R—H	Hydrocarbons

FIGURE 7.94 The synthetic potential of the substitution reaction.

Acetylide addition

The figure adds several reactions to the ones we have discussed specifically. For example, it sneaks in a brand-new synthesis of substituted acetylenes by using the acetylide ion as a nucleophile (usually only effective with primary R—L compounds). We mentioned the acidity of acetylenes in Chapter 3 (p. 131) and even warned that more was coming about this reaction. Monosubstituted (terminal) acetylenes are decent acids (pK_a ~ 25), which means that treatment of a terminal acetylene with a strong base can result in removal of a proton and formation of the acetylide. A favorite base for doing this is the amide ion, $^-$NH$_2$ (Fig. 7.95). There are some other new reactions in Figure 7.94. Go over each of them carefully and be sure you see how each works.

$$R—C≡C—H \xrightarrow[\text{:NH}_2\,/\,\text{:NH}_3]{\text{Na}^+} R—C≡C:^- \quad Na^+ + :NH_3$$

Acetylide ion

FIGURE 7.95 Formation of an acetylide from a terminal alkyne.

It is *relatively* easy (note that I didn't say it was easy!) to keep track of the available reactions, and therefore not too difficult to think up one-step transformations. Once synthetic chemistry goes beyond one-step reactions, it gets much harder. At the same time, the intellectual challenge of synthetic chemistry also increases sharply, and synthesis problems, whether in the real world or in this book, become not only more challenging, but much more fun. One can begin to see why the art of making molecules fascinates so many people. Let's look at a few examples.

7.10a Sulfur as Nucleophile. For example, suppose you are set the task of making propane-2-thiol from isopropyl bromide. We will use the nucleophilicity of sulfur to accomplish this transformation. The mercaptide ion itself ($^-$:SH), as well as alkyl mercaptides ($^-$:SR), is powerfully nucleophilic and can be used to great effect in the S_N2 reaction. This nucleophilicity leads to the most common synthesis of thiols and sulfides. The only serious restriction is that the alkyl halide must be active in the S_N2 reaction (Fig. 7.96).

THE GENERAL CASE

A SPECIFIC EXAMPLE

FIGURE 7.96 The S_N2 reactions of mercaptide and substituted mercaptides lead to thiols and sulfides.

Sulfides have two lone pairs of electrons and even these relatives of ethers (p. 202) are quite nucleophilic. Further S_N2 reactions give **sulfonium ions** (Fig. 7.97).

A sulfonium ion

FIGURE 7.97 Sulfides can also be alkylated to give sulfonium ions.

Sulfonium ions and the less stable oxonium ions are used as alkylating agents. Dimethyl sulfide and dimethyl ether are excellent leaving groups and can be displaced by nucleophiles in further S_N2 reactions (Fig. 7.98). As we have already seen (p. 114), Nature uses a variant of this reaction, displacement on the amino acid derivative *S*-adenosylmethionine, to do many methylation reactions.

FIGURE 7.98 Alkylation by a sulfonium ion.

So, Figure 7.99 gives one possible answer right away. Just treat isopropyl bromide with thiolate, HS$^-$. You might get a little elimination product as well, but the reaction should work quite well. That's an easy problem, at least if you know the reaction in Figure 7.99.

FIGURE 7.99 The conversion of isopropyl bromide into propane-2-thiol.

Isopropyl bromide   Propane-2-thiol

Now, however, suppose you had to make propane-2-thiol from propene? You have no direct way to do this transformation, and Figure 7.94 is no help. But you do have a way to convert propene into isopropyl bromide (Chapter 3, p. 134). So, a two-step synthesis is possible (Fig. 7.100), but it is *much* harder to find than the simple one-step method. By far, the best way to find it is to work backward from the desired product, asking yourself the question, From what immediate precursor molecule or molecules can I get the ultimate product? The answer to that question should lead you to isopropyl bromide, among others. Now ask again, From what immediate molecule or molecules can I get isopropyl bromide? The answer to that question should include propene plus HBr. Working backward in this way is the only successful route to solving problems in synthesis, and *all* successful chemists, both beginners and "pros," do it this way.

FIGURE 7.100 The two-step conversion of propene into propane-2-thiol.

Propene   Isopropyl bromide   Propane-2-thiol

Now let's make the sequence even harder, and ask how to make propane-2-thiol from propyl bromide. Now a three-step sequence is necessary, and one suggestion is shown in Figure 7.101. Be sure you go through the process of working backward for Figure 7.101.

FIGURE 7.101 A three-step sequence for converting propyl bromide into propane-2-thiol.

Propyl bromide   Propene   Isopropyl bromide   Propane-2-thiol

7.10b Oxygen as Nucleophile: The Williamson Ether Synthesis. Here's another synthetic challenge that uses propyl bromide. Make methyl propyl ether. Given our discussion of Figure 7.99, you have almost certainly come up with the idea of displacing the bromide with methoxide in a straightforward S_N2 sequence (Fig. 7.102). Good idea! In fact, you have just invented a good way to make ethers—well, not exactly, because A. W. Williamson (1824–1904) discovered this process, now called the **Williamson ether synthesis**, over 100 years ago (Fig. 7.103). There are both limitations and opportunities in this synthesis, and they rather nicely illustrate some of the kinds of thinking that synthetic chemists have to go through. So let's explore the Williamson synthesis a bit.

FIGURE 7.102 The Williamson ether synthesis.

FIGURE 7.103 Alkoxides can displace halides in an S_N2 reaction to make ethers. This reaction is the Williamson ether synthesis.

First of all, we need to be able to make the necessary alkoxides. As we saw in Chapter 6 (p. 242), when an alcohol is treated with a much stronger base, it can be converted entirely into the alkoxide. Note that the base used must be much stronger than the alkoxide or an equilibrium mixture of the base and the alkoxide will result. A favorite reagent is sodium hydride (Fig. 7.104). Sodium hydride has the advantage of producing hydrogen gas, which can easily be removed from the reaction mixture, thus driving the equilibrium toward the alkoxide.

FIGURE 7.104 Sodium hydride irreversibly removes a proton to give the alkoxide, the conjugate base of the alcohol.

Just as water reacts with metallic sodium or potassium to give the corresponding metal hydroxide, alkoxides can be made through a reaction in which sodium or potassium metal is dissolved directly in the alcohol. Hydrogen is liberated in what can be a most vigorous reaction indeed. Potassium is particularly active in all cases, and the reaction of sodium metal with smaller alcohols such as methyl and ethyl alcohol is also rapid and exothermic (Fig. 7.105).

$$2 \; CH_3 \!-\! \overset{..}{\underset{..}{O}} \!-\! H \; + \; 2 \, Na\cdot \longrightarrow H_2\!\uparrow \; + \; 2 \left[Na^+ \quad :\!\overset{..}{\underset{..}{O}}\!-\! CH_3 \right]$$

FIGURE 7.105 Alkoxides can be made through the reaction of metallic sodium or potassium with alcohols.

WORKED PROBLEM 7.32 Is an equivalent amount of sodium hydroxide an effective reagent for forming sodium ethoxide from ethyl alcohol? Explain your answer using Table 7.4.

$$CH_3CH_2 \!-\! \overset{..}{\underset{..}{O}} \!-\! H \; + \; Na^+ \; :\!\overset{..}{\underset{..}{O}}H \; \rightleftarrows \; CH_3CH_2 \!-\! \overset{..}{\underset{..}{O}}\!\!\!^- \; Na^+ \; + \; H_2\overset{..}{\underset{..}{O}}:$$

ANSWER The pK_a values for water (15.7) and ethyl alcohol (15.9) are very close, and the equilibrium concentration of sodium ethoxide will be slightly less than that of hydroxide ion. Hydroxide ion is a slightly weaker base than ethoxide in solution. Thus, hydroxide will not be effective in making large amounts of ethoxide. At equilibrium both species will be present.

$$\underset{\text{p}K_a = 15.9}{CH_3CH_2 \!-\! O \!-\! H} \; + \; ^-OH \; \rightleftarrows \; CH_3CH_2 \!-\! O^- \; + \; \underset{\text{p}K_a = 15.7}{H \!-\! O \!-\! H}$$

Alkoxides are named after the starting alcohol as metal alkoxides (Fig. 7.106).

$$H_3C \!-\! \overset{..}{\underset{..}{O}}\!\!\!^- \; Na^+ \qquad CH_3CH_2 \!-\! \overset{..}{\underset{..}{O}}\!\!\!^- \; K^+ \qquad \underset{H_3C}{\overset{H_3C}{>}}CH \!-\! \overset{..}{\underset{..}{O}}\!\!\!^- \; Cs^+ \qquad H_3C \!-\! \overset{CH_3}{\underset{CH_3}{\overset{|}{\underset{|}{C}}}} \!-\! \overset{..}{\underset{..}{O}}\!\!\!^- \; Rb^+$$

Sodium methoxide   **Potassium ethoxide**   **Cesium isopropoxide**   **Rubidium *tert*-butoxide**

FIGURE 7.106 Alkoxides are named as metal alkoxides.

Now let's use these alkoxides to make ethers—the Williamson ether synthesis. Several modifications of the original procedure have made this venerable method quite useful, although there are some restrictions. The reaction only works for alkyl halides that are active in the S_N2 reaction. Therefore, tertiary halides cannot be used. They are too hindered to undergo the crucial S_N2 reaction. Sometimes there is an easy way around this problem, but sometimes there isn't. For example, *tert*-butyl methyl ether cannot be made from *tert*-butyl chloride and sodium methoxide, but it can be made from *tert*-butoxide and methyl iodide. However, there is no way to use the Williamson ether synthesis to make di-*tert*-butyl ether (Fig. 7.107).

The Williamson ether synthesis cannot be used to make di-*tert*-butyl ether (CH₃)₃C–O–C(CH₃)₃

FIGURE 7.107 How to use the Williamson ether synthesis.

Even secondary halides are of little use in this reaction. Alkoxides are strong bases (high affinity for a hydrogen 1s orbital) and the E2 reaction is a prominent side reaction when a secondary halide is used (p. 315; Fig. 7.108).

FIGURE 7.108 Secondary halides undergo substantial E2 elimination with alkoxides.

Be alert for intramolecular, ring-forming versions of the Williamson ether synthesis. Cyclic ethers can be formed by intramolecular S_N2 reactions. There is no mystery in this reaction; it is simply the acyclic process applied in an intramolecular way (Fig. 7.109).

KOH
p-xylene
140 °C, 7 h

(33%)

FIGURE 7.109 An intramolecular Williamson ether synthesis.

Intramolecular S_N2

An important use of the intramolecular Williamson ether synthesis is in the making of **oxiranes**, ethers with the oxygen atom in a three-membered ring, also called **epoxides**. A 2-halo-1-hydroxy compound (a halohydrin) is treated with base to form the haloalkoxide, which then undergoes an intramolecular backside S_N2 displacement to give the three-membered ring (Fig. 7.110).

FIGURE 7.110 A common epoxide synthesis takes advantage of an intramolecular S_N2 reaction.

7.10c Nitrogen as Nucleophile: Alkylation of Amines. We have just examined a series of synthetically useful reactions that uses oxygen as the nucleophile. Nitrogen has a lone pair of electrons, and therefore is also nucleophilic. Let's look here at some synthetic reactions in which nitrogen is the nucleophile. Conceptually, the following reactions are rather closely related to the Williamson ether synthesis. Amines are good nucleophiles and can displace leaving groups in the S_N2 reaction. Of course, the limitations of the S_N2 reaction also apply—there can be no displacements at tertiary carbons. Successive displacement reactions yield primary, secondary, and tertiary amines. The initial products are alkylammonium ions that are deprotonated by the amine present. These S_N2 reactions produce amines that are also nucleophilic. Further displacement reactions are possible, which can lead to problems. Indeed, the product amine is often a better participant in the S_N2 reaction than the starting material. Therefore it can be difficult to stop the reaction at the desired point, even if an excess of the starting amine is used to minimize overalkylation (Fig. 7.111).

FIGURE 7.111 Alkylation reactions of amines.

Tertiary amines are also strong nucleophiles, and a final displacement reaction can occur to give a stable quaternary ammonium ion (Fig. 7.112). Thus, the alkylation process can be pushed to its logical conclusion in an effective synthesis of quaternary ammonium ions. In addition, Theodore Cohen (b. 1929) and his co-workers at the University of Pittsburgh have worked out a method that uses ammonia in excess under pressure to generate primary amines efficiently (Fig. 7.113).

$$(CH_3)_3N\colon \quad CH_3 - I \quad \xrightarrow{S_N2} \quad (CH_3)_3\overset{+}{N} - CH_3 \quad I^-$$

Tetramethylammonium iodide

FIGURE 7.112 The formation of an ammonium ion through alkylation of a tertiary amine.

$$H_3N\colon \xrightarrow{CH_3-I} H_2\ddot{N}-CH_3 \xrightarrow{CH_3-I} H\ddot{N}(CH_3)_2 \xrightarrow{CH_3-I} \colon N(CH_3)_3 \xrightarrow{CH_3-I} {}^+N(CH_3)_4$$
$$^-I$$

$$H_3N\colon + CH_3 - I \longrightarrow H_2\ddot{N} - CH_3$$
Excess

FIGURE 7.113 The synthesis of alkylamines through alkylation reactions (S_N2 reactions).

PROBLEM 7.33 What product do you expect from an attempt to make *tert*-butylamine from the reaction of the amide ion ($^-NH_2$) with *tert*-butyl bromide?

PROBLEM 7.34 Another nucleophilic synthesis of amines involves the reaction of simple amines with epoxides. Write a mechanism for the reaction below.

$$CH_3CH_2 - NH_2 + \triangle \xrightarrow{H_2O} CH_3CH_2NHCH_2CH_2OH \quad + \quad CH_3CH_2N(CH_2CH_2OH)_2$$

7.10d Summary and a Few "Real-World" Difficulties.

In the real world, the challenges of synthesis are great—one needs not only to find a reaction or more likely, a sequence of reactions that will make the desired molecule, but one must do so specifically and, in industry, economically. Specificity becomes increasingly important these days as we recognize how critical chirality is in biological action. And, of course, the more stereogenic centers (p. 157) a molecule has, the more difficult the synthesis. Imagine making a molecule with only a modest number of stereogenic atoms, say 4. There are 16 possible isomers and you almost certainly can use only 1. Every reaction you use must lead to the correct stereochemistry and no others.

Although it is easy to look down one's nose at questions of cost, one would be very wrong to do so. Chemists in industry must find sophisticated ways to make those stereochemically rich molecules without using super-expensive reagents that might be fine in very small scale work, but utterly useless in work designed to produce a molecule to be used by millions of people. Doing clever syntheses with cheap reagents is much more difficult, both intellectually and practically, than making the molecule regardless of cost.

PROBLEM 7.35 Predict the products of the following reactions:

(a) $HO^- + CH_3Br \longrightarrow$

(b) $HO^- + (CH_3)_3CBr \longrightarrow$

(c) $H_2O + (CH_3)_3CBr \longrightarrow$

PROBLEM 7.36 Devise syntheses for the following molecules. You may use any inorganic reagent, organic iodides containing up to four carbons, and the special reagents potassium cyanide and propyne.

(a)

(b)

(c)

(d) $(CH_3)_3C$—OCH_3

(e) $(CH_3)_3C$—N_3

PROBLEM 7.37 Indicate which member of the following pairs of compounds will react faster in (1) the S_N1 reaction and (2) the S_N2 reaction. Explain your reasoning.

(a) 1-Bromobutane and 2-bromobutane

(b) 1-Chloropentane and cyclopentyl chloride

(c) 1-Chloropropane and 1-iodopropane

(d) *tert*-Butyl iodide and isopropyl iodide

(e) *tert*-Butyl iodide and *tert*-butyl chloride

PROBLEM 7.38 Devise synthetic routes that will produce the following compounds as the major products from each of the indicated starting materials.

(a)

H_3C—$\overset{\overset{\textstyle CH_3}{|}}{\underset{\underset{\textstyle CH_3}{|}}{C}}$—$OH$ from H_3C—$\overset{\overset{\textstyle CH_3}{|}}{\underset{\underset{\textstyle CH_3}{|}}{C}}$—$Br$ $\overset{\textstyle H_3C}{\underset{\textstyle H_3C}{}}C{=}CH_2$ H_3C—$\overset{\overset{\textstyle CH_3}{|}}{\underset{\underset{\textstyle CH_2Br}{|}}{CH}}$

(continued)

PROBLEM 7.38 *(continued)*

(b)

(c)

Keeping track of synthetic methods can be difficult. You are right at the beginning of synthetic chemistry now, and here is a hint on how to keep track of it. Make a 4 × 6-in. file card for every structural type and then keep a short list of the available synthetic reactions on the card. To help in ordering things, and to keep as far as possible from the dreaded mindless memorization, you might make a mechanistic notation for each reaction as well. The reaction of acetylides with alkyl halides to make new acetylenes might get an "S_N2 with primary substrates" beside it, for example. You can start your catalog with the reactions in Figure 7.94. Be sure to keep up to date; synthetic ability is the kind of thing that sneaks up on you, like whatever it was that worried Mr. Paige.

PROBLEM 7.39 Use the following pK_a data to decide if sodium amide would be effective at producing vinyl anions: ammonia, 38; alkene, ~50.

Figure 7.94 also fails to show the potential complications we have talked about. *Remember*: There is always a competition between the substitution reactions, S_N1 and S_N2, and the E1 and E2 reactions. Under some circumstances the major products of the reactions summarized in Figure 7.94 will be alkenes. Your file card catalog should include the synthesis of alkenes through the E1 and E2 reactions. The identity of the leaving group, so conveniently avoided by the L in the figure, makes a difference, and the structure of R, also deliberately vague in the figure, is vital. In order to design a successful synthetic transformation, you have to think about all the components of the reaction. A sample file card for alkyl bromides is shown in Figure 7.114.

Alkyl Bromides		
Reaction	Mechanism	Notes
R — L + Br⁻	S_N2	Needs good leaving group
		Primary or secondary R
		E2 competes
R — L + Br⁻	S_N1	*tert* R group, polar solvent
		E1, E2 compete

FIGURE 7.114 A sample file card for alkyl bromides.

7.11 SUMMARY

New Concepts

The most important idea in this chapter is the continuing expansion of the concept of Brønsted acidity and basicity to embrace the species known as Lewis acids and Lewis bases. The relatively narrow notion that acids and bases compete for a proton (Brønsted acids and bases) is extended to define all species with reactive pairs of electrons as Lewis bases, and all molecules that will react with Lewis bases as Lewis acids. The key word here is "reactive." All bonds are reactive under some circumstances, and therefore all molecules can act as Lewis bases at times.

Another way to describe the reaction of Lewis bases with Lewis acids is to use the graphic device showing the stabilizing interaction of a filled orbital (Lewis base) with an empty orbital (Lewis acid) (Fig. 7.5). This idea can be generalized if you recall that the strength of orbital interaction depends on the energy difference between the orbitals involved. The closer the orbitals are in energy the stronger their interaction and the greater the resulting stabilization. Therefore the strongest interactions between orbitals are likely to be between the highest occupied molecular orbital (HOMO) of one molecule (**A**, a Lewis base) and the lowest unoccupied molecular orbital (LUMO) of the other (**B**, a Lewis acid) (Fig. 7.7). This notion is used in this chapter to explain the overwhelming preference for backside attack in the S_N2 reaction.

Stereochemical experiments are used here for the first time to probe the details of reaction mechanisms. The observation of inversion of configuration in the S_N2 reaction shows that the incoming nucleophile must approach the leaving group from the rear. Similarly, if the stereochemistry of the E2 reaction is monitored, it can be shown that there is a strong preference for the anti (180°) E2 reaction in acyclic molecules.

The curved arrow formalism is used extensively in this chapter. This bookkeeping device is extremely useful in keeping track of the pairs of electrons involved in the making and breaking of bonds in a reaction. It is important not to confuse this formalism with a reaction mechanism.

The terms "transition state" (the structure representing the high-energy point separating starting material and products) and "activation energy" (the height of the transition state above the starting material) appear several times in this chapter. Chapter 8 will continue a detailed discussion of these terms (Fig. 7.115).

FIGURE 7.115 A typical Energy versus Reaction progress diagram.

Key Terms

activation energy (p. 284)
anti elimination (p. 320)
curved arrow formalism (p. 273)
E1cB reaction (p. 324)
E1 reaction (p. 312)
E2 reaction (p. 315)
epoxides (p. 333)
Hofmann elimination (p. 324)
HOMO (p. 276)
inversion of stereochemistry (p. 279)
leaving group (p. 277)

Lewis acids (p. 274)
Lewis bases (p. 274)
LUMO (p. 276)
nucleophile (p. 277)
oxiranes (p. 333)
product-determining step (p. 308)
rate-determining step (p. 308)
reaction mechanism (p. 273)
regiochemistry (p. 322)
retention of stereochemistry (p. 279)
Saytzeff elimination (p. 314)

solvolysis reaction (p. 302)
S_N1 reaction (p. 302)
S_N2 reaction (p. 279)
substitution reaction (p. 272)
sulfonium ion (p. 329)
syn elimination (p. 320)
thionyl chloride (p. 296)
transition state (p. 284)
Williamson ether synthesis (p. 331)

Reactions, Mechanisms, and Tools

Four exceptionally important prototypal reactions are introduced in this chapter: the S_N1, S_N2, E1, and E2 reactions. The E1 and S_N1 reactions involve the same carbocationic intermediate. Addition of a nucleophile completes the S_N1 reaction; removal of a proton finishes the E1 (Fig. 7.70).

The S_N2 reaction and the E2 reaction are also competitive processes. In the S_N2 reaction, a nucleophile attacks the rear of a carbon-leaving group bond. In the E2 reaction, it is the adjacent hydrogen that is removed along with the leaving group (Fig. 7.74).

For primary substrates, it is the S_N2 and E2 reactions that compete. For tertiary compounds in polar solvents with weak bases, it is the S_N1 and E1 reactions that are important. If strong bases are employed in nonionizing solvents, it will be the E2 reactions that are most important. For secondary substrates all four reactions compete.

It is absolutely vital to have these four building block reaction mechanisms under complete control.

Syntheses

1. SUBSTITUTED ALKANES

R—L $\xrightarrow[\text{Nu:}]{\text{good nucleophile}}$ Nu—R

S_N2 reaction. The R must be methyl, primary, or secondary. Always goes with inversion. The E2 reaction is competitive.

R—L $\xrightarrow[\text{HOS}]{\text{polar solvent}}$ SO—R

S_N1 reaction. The R must be tertiary or secondary. Stereochemical result is largely racemization. The E1 reaction competes.

2. ALKENES

R—C(CH$_2$R)(R)—L $\xrightarrow{\text{polar solvent}}$ C=C (R,R / R,H)

E1 reaction. Must be a tertiary halide. Saytzeff elimination. Competes with the S_N1 reaction.

R—C(CH$_2$R)(R)—L $\xrightarrow[\text{nonpolar solvent}]{\text{good base}}$ C=C (R,R / R,H)

E2 reaction. Competes with S_N2, which dominates for primary substrates. anti Elimination favored, regiochemistry depends on leaving group.

3. ALKOXIDES

R—OH $\xrightarrow{\text{Na}^+\ ^-\text{H}}$ R—O$^-$ Na$^+$ + H$_2$

R—OH $\xrightarrow{\text{Na}}$ R—O$^-$ Na$^+$ + H$_2$

4. ALKYL HALIDES

R—OH $\xrightarrow{\text{HX}}$ R—X   This reaction can be S_N1 or S_N2, but the intermediate is always the protonated alcohol.

R—OH $\xrightarrow{\text{SOCl}_2}$ R—Cl   This reaction can be S_N1 or S_N2, but the intermediate is a chlorosulfite ester.

R—OH $\xrightarrow{\text{CCl}_4\ \text{(Ph)}_3\text{P}}$ R—Cl   Inversion is the usual result.

R—OH $\xrightarrow{\text{PCl}_5}$ R—Cl

R—OH $\xrightarrow{\text{PBr}_3}$ R—Br

5. ALKYL SULFONATES

R—OH $\xrightarrow{\text{Cl}-\text{SO}_2-\text{R}}$ R—O—SO$_2$—R

These sulfonates are good substrates for further conversions using S_N1 and S_N2 reactions.

6. AMINES

$$H_3N + CH_3-I \xrightarrow{S_N2} H_3\overset{+}{N}-CH_3 \; I^- \rightleftharpoons H_2NCH_3$$
Primary
amine

$$CH_3NH_2 + CH_3-I \xrightarrow{S_N2} CH_3\overset{+}{N}H_2-CH_3 \; I^- \rightleftharpoons HN(CH_3)_2$$
Secondary
amine

$$(CH_3)_2NH + CH_3-I \xrightarrow{S_N2} (CH_3)_2\overset{+}{N}H-CH_3 \; I^- \rightleftharpoons N(CH_3)_3$$
Tertiary
amine

Restrictions on S$_N$2 reaction apply;
overalkylation is a serious problem.

7. ETHERS

$$R-O^- + R-X \longrightarrow R-O-R$$

Williamson ether synthesis.
R must be able to undergo
the S$_N$2 reaction—it cannot
be tertiary.

The S$_N$1 and S$_N$2 reactions allow the conversion of alkyl halides into a host of new structures. These were summarized earlier in Figure 7.94. Remember that the S$_N$2 reaction works only for primary and secondary halides and the S$_N$1 reaction works only for tertiary and secondary molecules.

Alcohols can also be used as starting materials in these reactions, but the OH group is a very poor leaving group and must be transformed by protonation into water, or through other reactions into some other good leaving group.

Common Errors

Students of organic chemistry long for certainty. They have not yet learned to love the messiness of the subject, or at least to accommodate to it. That the answer to a question, What happens if we treat A with B? might be, All sorts of things, can be profoundly unsettling. In this chapter, we meet four basic, building blocks: the S$_N$1, S$_N$2, E1, and E2 reactions. It is a sad fact that in most instances they compete with each other. We can usually adjust reaction conditions to favor the reaction we desire, but it will be a rare occasion when we achieve perfect selectivity.

There is much detail in this chapter, but that is not usually a problem for most students. Good nucleophiles can be distinguished from good bases, good leaving groups can be identified, and so on. What is hard is to learn to adjust reaction conditions (reagents, temperature, and solvent) so as to achieve the desired selectivity; to favor the product of just one of these four reactions. That's hard in practice as well as in answering a question on an exam or problem set. To a certain extent, we all have to learn to live with this problem!

7.12 ADDITIONAL PROBLEMS

This chapter introduces organic synthesis. We still have very few reactions for making molecules at our disposal, basically just additions and the E1, E2, S$_N$1, and S$_N$2 reactions. Nonetheless, it is possible to devise simple routes to target molecules and even to craft a few multistep sequences. Organic synthesis is one of the most creative (and fun) parts of this science, and so it is important to start early. Here we group several synthesis-related problems before going to more mechanistic material. In just a few chapters, we will have many more reactions under control, and be able to do much fancier work. Stay tuned.

PROBLEM 7.40   What nucleophile would serve to effect the following conversions of 1-iodopropane?

PROBLEM 7.41 Explain with crystal clarity why treatment of propyl alcohol with bromide ion fails to give propyl bromide, but treatment with HBr succeeds.

PROBLEM 7.42 The contents of four flasks containing the compounds shown in (a), (b), (c), and (d) are treated with $(CH_3)_3C-O^-/(CH_3)_3C-OH$. In two cases, a product C_7H_{12} is formed that shows only five signals in its ^{13}C NMR spectrum. In two other cases, a product of the same formula, C_7H_{12}, is formed that shows seven signals in the ^{13}C NMR spectrum. What are the products, and which of the compounds, (a), (b), (c), and (d), leads to which product? Explain.

PROBLEM 7.43 Why are sulfonate esters (p. 295) such good participants in the S_N1 and S_N2 reactions?

PROBLEM 7.44 Write a mechanism for the formation of alkyl bromides from the reaction of alcohols and PBr_3.

PROBLEM 7.45 The reaction shown below was used to help clarify the S_N2 nature of the alkylation reactions of S-adenosylmethionine. Explain why the results are relevant to the mechanistic questions.

^3H = Tritium (T)
^2H = Deuterium (D)

PROBLEM 7.46 (a) Reaction of 1,2-dimethylpyrrolidine (**1**) with ethyl iodide leads to two products, $C_8H_{18}IN$. Explain.

1

(b) However, when 2-methylpyrrolidine (**2**) undergoes a similar reaction with ethyl iodide, followed by treatment with base, analysis of the product by ^1H NMR spectroscopy reveals only one set of signals for $C_7H_{15}N$. Explain.

2

PROBLEM 7.47 Treatment of 1,2-dibromoethane with the dithiolate shown below leads to two products, $C_4H_8S_2$ and $C_6H_{12}S_2Br_2$. Write structures for these products and explain how they are formed.

A dithiolate

$C_4H_8S_2$

+

$C_6H_{12}S_2Br_2$

PROBLEM 7.48 Devise a synthetic route that will convert (R)-sec-butyl alcohol into (S)-sec-butyl alcohol.

PROBLEM 7.49   Provide reagents that will effect the following changes:

(a)

(b)

(c)

PROBLEM 7.50   Devise S_N2 reactions that would lead to the following products:

(a)   (b)

(c)   (d)   (e)

PROBLEM 7.51   Predict the products of the following S_N2 reactions:

(a)

+ $CH_3CH_2—\ddot{I}:$ ⟶ ?

(b)

$CH_3CH_2—\ddot{O}:^-$ + ⟶ ?

(continued)

PROBLEM 7.51   *(continued)*

(c)

:$\ddot{I}:^-$ +   ⟶ ?

(d)

+ :CN: ⟶ ?

(e)

:CN: +   ⟶ ?

PROBLEM 7.52   Treatment of butyl ethyl ether with HI leads to both butyl iodide and ethyl iodide, along with the related alcohols. By contrast, when *tert*-butyl ethyl ether is treated the same way, only ethyl iodide and *tert*-butyl alcohol are initially formed. *tert*-Butyl iodide and ethyl alcohol are not produced. Explain.

$$CH_3CH_2CH_2CH_2—O—CH_2CH_3$$

⟍ HI

$$CH_3CH_2CH_2CH_2I \ + \ CH_3CH_2I$$

$$CH_3CH_2CH_2CH_2OH \ + \ CH_3CH_2OH$$

$$(CH_3)_3C—O—CH_2CH_3 \xrightarrow{\text{HI}} (CH_3)_3COH \ + \ CH_3CH_2I$$

PROBLEM 7.53   Write all the possible products resulting from treatment of alcohol **1** with HCl/H_2O.

$$\underset{\textbf{1}}{\text{(structure)}} \xrightarrow[\text{H}_2\text{O}]{\text{HCl}} ?$$

PROBLEM 7.54 Predict the products of the following S$_N$1 reactions:

(a)

(b)

CH$_3$CH$_2$ÖH ⟶ **?**

(c)

(d)

PROBLEM 7.55 You are given a supply of ethyl iodide, *tert*-butyl iodide, sodium ethoxide, and sodium *tert*-butoxide. Use the S$_N$2 reaction to make as many different ethers as you can. In principle, how many are possible? In practice, how many can you make?

PROBLEM 7.56 The following molecules can form conjugate bases by loss of a proton from more than one position (arrows). Start by drawing a good Lewis structure (electron dots are deliberately left out) and then choose which proton will be lost more easily. Explain your choice.

(a)

CH$_3$CH$_2$—CN

(b)

(c)

PROBLEM 7.57 The following molecules can accept a proton to form their conjugate acids by addition of a proton at more than one possible place (arrows). Choose the position at which protonation will be preferred and explain your choice. Once again, it makes sense to start by drawing a good Lewis structure for the molecule in question. *(continued)*

PROBLEM 7.57 *(continued)*

(a)

(b)

(c)

PROBLEM 7.58 Write arrow formalisms for the following reactions. Charges and electron dots are deliberately left out of the reagents and products, so you must start by putting them in.

(a)

(b)

(c)

PROBLEM 7.59 The S$_N$2 reactions at the allyl position are especially facile. For example, allyl compounds are even more reactive than methyl compounds (Table 7.1, p. 286). Explain this rate difference. Why are allyl compounds especially fast? *Hint*: This, like all "rate" questions, can be answered by examining the structures of the transition states involved.

PROBLEM 7.60 In the following pairs of compounds, rank the molecules in order of rate of reaction in the S_N2 process. Explain your reasoning.

(a)

(b)

(c)

(d)

PROBLEM 7.61 In the following pairs of compounds, rank the molecules in order of rate of reaction in the S_N1 process. Explain your reasoning.

(a)

(b)

(c)

PROBLEM 7.62 Each of the following molecules contains two halogens of different kinds. They are either in different positions in the molecule or are different halogens entirely.

(a)

(b)

(c)

In each case, determine which of the two halogens will be the more reactive in the S_N2 reaction. Explain.

PROBLEM 7.63 Each of the following molecules contains two halogens at different positions in the molecule. In each case, determine which of the two halogens will be the more reactive in the S_N1 reaction. Explain.

(a)

(b)

(c)

PROBLEM 7.64 You have two bottles containing the two possible diastereomers of the compound shown below ($C_8H_{11}IO_2$). When the salts of the carboxylic acids are made, only one of them is converted into a new compound, $C_8H_{10}O_2$. What is the new compound, and how does its formation allow you to determine the stereochemistry of the original two compounds? The squiggly bond indicates undefined stereochemistry.

Two isomers, each $C_8H_{11}IO_2$

Two isomers, each $C_8H_{10}IO_2Na$

$C_8H_{10}O_2$
+
One remaining $C_8H_{10}IO_2Na$

PROBLEM 7.65 You have two bottles containing the two possible diastereomers of the compound shown below (C_5H_9IO). When the molecules are treated with a strong base, such as sodium hydride, only one of them is converted into a new compound, C_5H_8O. What is the new compound, and how does its formation allow you to determine the stereochemistry of the original two compounds?

C_5H_8O + $C_5H_8IO^- Na^+$

C_5H_9IO

PROBLEM 7.66 Predict the products of the following E2 reactions. If more than one product is expected, indicate which will be the major compound formed.

(a)

(continued)

PROBLEM 7.66 (continued)

(b)

(c)

(d)

(e)

PROBLEM 7.67 Predict the products of the following E1 reactions. If more than one product is expected, indicate which will be the major compound formed.

(a)

(b)

(c)

(d)

PROBLEM 7.68 (a) Write the "obvious" arrow formalism for the following transformation. Of course it is going to be wrong; you are being set up here. But don't mind, it's all a learning experience.

(b) OK, what you wrote is almost certainly wrong. To find the correct mechanism, measure the kinetics of this process and discover that the reaction is bimolecular. Now write another arrow formalism.

(c) What's wrong with the mechanism suggested by the original arrow formalism? *Hint:* Think about the geometry of the transition state for the S_N2 reaction.

PROBLEM 7.69

(a) Menthyl chloride reacts with sodium ethoxide in ethyl alcohol to give a single product as shown. Analyze this reaction mechanistically to explain why this is the only alkene produced. *Hint:* Think in three dimensions.

Menthyl chloride

By contrast, neomenthyl chloride treated in the same fashion gives an additional, major, product. Moreover, menthyl chloride reacts much more slowly than neomenthyl chloride. Explain these observations mechanistically.

(continued)

PROBLEM 7.69 *(continued)*

Neomenthyl chloride

(b) When menthyl chloride is treated with 80% aqueous ethyl alcohol with no ethoxide present, once again, two products are formed in the ratio shown. Explain mechanistically.

Steric effects are important in solvolysis reactions. Problems 7.70 and 7.71 illustrate this point.

PROBLEM 7.70 Explain the following rates for S_N1 solvolysis. Construct an Energy versus Reaction progress diagram showing the two reactions.

R	Rate
CH_3	1
$(CH_3)_3C$	10^5

PROBLEM 7.71 Explain the relative rates of the following compounds in the S_N1 reaction:

R	Relative Rate
CH_3	540
CH_3CH_2	125
$(CH_3)_2CH$	27
$(CH_3)_3C$	1

PROBLEM 7.72 Rationalize the following curious experimental results. When the optically active chlorosulfite **1** is heated in toluene, the chloride with inverted configuration is obtained. However, when **1** is decomposed in 1,4-dioxane (**2**), retention of configuration in the resulting chloride is observed. *Hint*: See Figure 7.43, and also think about the differences between toluene and 1,4-dioxane. A good Lewis structure for 1,4-dioxane may also be helpful.

For Problems 7.73 and 7.74 you need to know that treatment of amines, $R-NH_2$ with nitrous acid, HONO, gives diazonium ions, $R-N_2^+$.

PROBLEM 7.73 Provide arrow formalism mechanisms for the following reactions:

(a)

(b)

PROBLEM 7.74 Rationalize the different products formed upon treatment of the stereoisomeric aminocyclohexanols **1–4** with nitrous acid.

PROBLEM 7.75   You are given a supply of 2-bromo-1-methoxybicyclo[2.2.2]octane. Your task is to prepare 1-methoxybicyclo[2.2.2]oct-2-ene (**A**) from this starting material. In your first attempt, you treat the bromide with sodium ethoxide in ethyl alcohol. Although a small amount of the desired alkene is formed, most of the product is 2-ethoxy-1-methoxybicyclo[2.2.2]octane (**B**).

Suggest a simple experimental modification that should result in an increase in the amount of the desired alkene, and a decrease in the amount of the unwanted substitution product, **B**.

PROBLEM 7.76   A common procedure to increase the rate of an S_N2 displacement is to add a catalytic (very small) amount of sodium iodide. Notice that even though the rate increases, the structure of the product is not the iodide. Explain the function of the iodide. How does it act to increase the rate, and why is only a catalytic amount necessary?

Equilibria

"I hope," the Golux said, "that this is true. I make things up, you know."

—James Thurber*
The 13 Clocks

8.1   PREVIEW

The study of how reactions occur depends on an analysis of the structures of the starting materials, the products, and the **transition states** separating them. We have spent much time on structure in the last several chapters. However, there are other, nonstructural factors that are important. To understand chemical reactivity, and to be able to use our knowledge in a predictive way, we also need to know something of the energetics involved in the reaction. Which are more stable, starting materials or products? Of course those designations—starting materials and products—are arbitrary. Nature doesn't care about such synthetic labels! How high is the transition state separating "starting materials and products"? If it is too high in energy, no reaction will occur. If there are several available low-energy pathways—several accessible transition states—there will be a mixture of products, and the synthetic chemists' longed for specificity will be impossible. So we need to know about energy.

In Chapter 7, we saw some examples of chemical equilibria. For instance, the S_N2 reaction involves an equilibrium in which two nucleophiles ($Nu{:}^-$ and $L{:}^-$) compete for the R group (Fig. 8.1).

FIGURE 8.1 The S_N2 reaction is an equilibrium in which two nucleophiles (Lewis bases) compete for a Lewis acid.

---

*James Grove Thurber (1894–1961) was an American writer and cartoonist, much of whose work first appeared in the *New Yorker* magazine. *The 13 Clocks* is a fairy tale of surpassing charm and wit written in Bermuda.

All chemical reactions can be regarded as equilibria. Multistep reactions are simply series of equilibria in which species occupying energy minima are separated by barriers, the tops of which are called transition states. The S_N1 solvolysis reaction of *tert*-butyl iodide in water makes a good example. Here we have three equilibria. The first is between *tert*-butyl iodide and the *tert*-butyl cation and iodide ion. This equilibrium is highly unfavorable, as the cation–anion pair is far less stable than the covalent iodide. The second equilibration represents the exothermic capture of the cation by water to give the protonated alcohol, an oxonium ion. Finally, the oxonium ion is deprotonated to give the ultimate product, *tert*-butyl alcohol (Fig. 8.2).

In practice, thermodynamics may dictate that one side of an equilibrium be overwhelmingly favored over the other. Nevertheless, we can still apply techniques for studying equilibria to the reaction, deriving information about the position of the equilibrium (thermodynamics), and the rate of reaction (kinetics). The concepts we will develop in this short chapter are extremely general and very important.

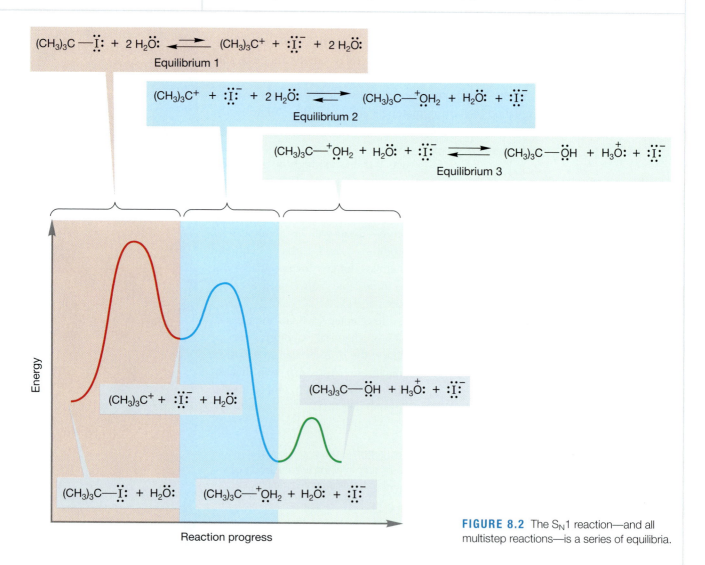

FIGURE 8.2 The S_N1 reaction—and all multistep reactions—is a series of equilibria.

ESSENTIAL SKILLS

1. Understanding the importance of the relative energies of starting materials and products, as well as the heights of the barriers separating these species, is critical to understanding reactivity. The relative energies of starting materials and products determine the ratio of products under conditions that allow all energy barriers to be passed.

2. The heights of the barriers separating starting materials and products determine the rates of formation of products. If there is insufficient energy available to return to starting materials from products, it is the heights of the accessible transition states that will determine the ratio of products.

IMPORTANT DETAILS

1. The outcome of a reaction at equilibrium is determined by the energy difference between the starting materials and products: $\Delta G = -RT \ln K$ is the operative equation.

2. However, the rate of a reaction has nothing to do directly with the energy difference between starting materials and products. Rates are determined solely by the energy difference between the starting molecules and the transition states separating those molecules from products.

8.2   EQUILIBRIUM

For a general chemical reaction at equilibrium,

$$aA + bB \rightleftarrows cC + dD \tag{8.1}$$

the concentrations of the species on the two sides of the arrows are related to each other by an **equilibrium constant**, K:

$$[C]^c [D]^d = K[A]^a [B]^b \tag{8.2}$$

or, more familiarly:

$$K = \frac{[C]^c [D]^d}{[A]^a [B]^b} \tag{8.3}$$

The small letter superscripts are the coefficients in the balanced Eq. (8.1), and the capital letters represent the molar concentrations of the products and reactants at equilibrium. The concepts of starting materials and products are arbitrary ones, arising from practical considerations—we generally are interested in making the reaction run one way or the other, and by convention the desired reaction is written left to right. But it need not be; it is just as valid to look at the reaction in the other direction.

The reaction will be a successful one if the compounds we seek, the products, are strongly favored at equilibrium. So we are very interested in what determines the magnitude of this equilibrium constant (K): the relative free energies ($\Delta G°$) of the compounds involved in the chemical reaction. The exact relationship between K and the relative energies of the starting materials and products is shown in Eq. (8.4):

$$\Delta G^\circ = -RT \ln K \qquad (8.4)$$

or, if we want to use base 10 rather than natural logarithms, in Eq. (8.5):

$$\Delta G^\circ = -2.3RT \log K \qquad (8.5)$$

R is the gas constant, 1.99×10^{-3} kcal/deg·mol, T is the absolute temperature, and ΔG°, the **Gibbs free energy change**, is the difference in energy between starting materials and products in their standard states.

Up to now, we have more or less equated energy with ΔH°, the **enthalpy**. Enthalpy, or the heat of reaction, is a measure of the bond energy changes in a reaction. The Gibbs free energy (ΔG°) is more complicated than this, as we will see in Section 8.3, and includes not only ΔH°, but **entropy** (ΔS°) as well. The entropy of a reaction is a measure of the change in amount of order in starting materials and products. The higher the degree of order, the greater the energy requirements for the reaction.

A number of times in Chapter 7 we drew diagrams that described the course of a reaction. We can now label those diagrams a bit more completely. The diagrams are plots of energy versus something called, most often, "reaction progress." Figures 8.3 and 8.4 show an exothermic reaction (products more stable than starting materials) and an endothermic reaction (starting materials more stable than products).

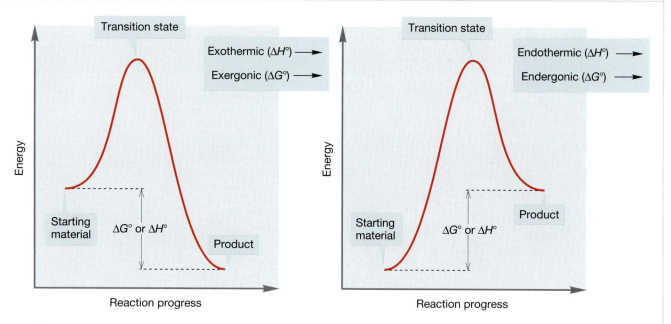

FIGURE 8.3 An exothermic (exergonic) reaction.

FIGURE 8.4 An endothermic (endergonic) reaction.

Actually, the term "exothermic" really refers to the enthalpy change, ΔH°, which may or may not be in the same direction as the free energy change, ΔG°, which involves entropy, ΔS°, as well as enthalpy, ΔH°. If we really mean free energy, ΔG, the proper terms are **exergonic** and **endergonic**.

For the exothermic reaction the difference in enthalpy between products and starting materials is negative, and the equilibrium constant (K) is greater than 1. When we look at an endothermic reaction (or read the first reaction backward to produce an endothermic process), the enthalpy difference is positive and K is less than 1.

More stable products are favored at equilibrium. By how much? Note that the relationship between the difference in free energy between starting material and products ($\Delta G°$) and the equilibrium constant K is logarithmic. We can rewrite Eqs. (8.4) and (8.5) to emphasize that point:

$$\Delta G° = -RT \ln K \qquad \text{or} \qquad K = e^{-\Delta G°/RT} \qquad (8.4)$$

or, in base 10:

$$\Delta G° = -2.3RT \log K \qquad \text{or} \qquad K = 10^{-\Delta G°/2.3RT} \qquad (8.5)$$

These equations show that a small difference in energy between starting materials and products results in a large excess of one over the other at equilibrium. Let's do a few calculations to show this. First, let's take a case in which products are more stable than starting materials by the seemingly tiny amount of 1 kcal/mol, $\Delta G° = -1$ kcal/mol,* so $K = 10^{1/(1.364)} = 10^{0.7331} = 5.4$.

An equilibrium constant of 5.4 translates into 84.4% product at equilibrium! One would very often be quite happy to find a reaction that would give about 85% product, and a mere 1 kcal/mol suffices to ensure this. Conversely, if we are "fighting" a 1-kcal/mol endothermic process it will be a futile fight indeed, because equilibrium will settle out at 84.4% starting material. *A little bit of energy goes a long way at equilibrium.* Table 8.1 summarizes a number of similar calculations.

PROBLEM 8.1 Calculate the energy difference between starting material and products at 25 °C given equilibrium constants of 7 and 14.

In practice, there often is something we can do to drive a reaction in the direction we want. **Le Chatelier's principle**† states that a system at equilibrium responds to stress in such a way as to relieve that stress. This statement means that if we increase the concentration of one of the starting materials, an equilibrium process will be driven toward products. In Eqs. (8.1–8.3), if we increase $[A]^a$, the quantity $[C]^c[D]^d/[B]^b$ must also increase to maintain K, and we will have more product. Similarly, if we can remove one product as it is formed, the equilibrium will be driven toward products. Perhaps either C or D is a solid that crystallizes out of solution or is a gas that can be driven off. As the concentration

TABLE 8.1 The Relationship between $\Delta G°$ and K at 25 °C

$\Delta G°$ (kcal/mol)	K	More Stable Isomer (%)
– 0.1	1.2	54.5
– 0.5	2.4	69.7
– 1	5.4	84.4
– 2	29.3	96.7
– 5	4631	99.98
–10	2.1×10^7	99.999996

*At 25 °C, $2.3RT = 1.364$ kcal/mol. This number is useful to remember because with it one can do calculations easily at odd moments without having to look up the value of R and multiply it by T.

†Henri Louis Le Chatelier (1850–1936). The mathematics of this principle had been worked out earlier by the American Josiah Willard Gibbs (1839–1903), who donated the G in Gibbs to ΔG. Le Chatelier knew of Gibbs's contributions, and was a great popularizer of Gibbs's work in France.

of this compound drops, the equilibrium will be disturbed and more C or D will be formed.

In a multistep reaction, it often does not matter that an early step involves an unfavorable equilibrium, as long as the overall reaction is exothermic. As long as all barriers can be traversed, the ratio of C and A will be governed by the $\Delta G°$ between C and A. So, A → B is endothermic, but, overall, A → C is exothermic ($\Delta G° < 0$). Figure 8.5 shows such a process for the conversion of A into C via an intermediate B.

FIGURE 8.5 As long as equilibrium is reached between A and C, it does not matter that formation of B from A is endothermic.

As long as some B is present in equilibrium with starting material A, the reaction will be successful. As the equilibrium with B is established, B will be converted into the much more stable final product C. As the small amount of B present in equilibrium with A is depleted, the A ⇌ B equilibrium will be reestablished and more B will be formed from A. In the long run, A will be largely converted into C, and as long as there is enough energy to reach the intermediate B, the final mixture of A and C will depend on the $\Delta G°$ between these two compounds.

8.3   GIBBS STANDARD FREE ENERGY CHANGE

The parameter $\Delta G°$ is composed of enthalpy ($\Delta H°$, the change in bond strengths in the reaction) and entropy ($\Delta S°$, the change in freedom of motion in the reaction). The exact relationship is,

$$\Delta G° = \Delta H° - T\Delta S° \tag{8.6}$$

where T is the absolute temperature. The enthalpy term ($\Delta H°$) is most directly related to the bond strengths in starting materials and products. We can make rather accurate estimates of this term by knowing the bond dissociation energies for a variety of bonds. We know a few of these values already from our discussion of alkanes. Table 8.2 gives several more.

TABLE 8.2   Some Homolytic Bond Dissociation Energies (BDE)

Bond	BDE (kcal/mol)	Bond	BDE (kcal/mol)
I—I	36.1	H—CH_2CH_3	101.1
Br—Br	46.1	H—$CH(CH_3)_2$	98.6
Cl—Cl	59.0	H—$C(CH_3)_3$	96.5
F—F	38.0	I—CH_3	57.6
HO—OH	51	Br—CH_3	72.1
$(CH_3)_3CO$—$OC(CH_3)_3$	38	Cl—CH_3	83.7
H—H	104.2	F—CH_3	115
H—I	71.3	HO—CH_3	92.1
H—Br	87.5	CH_3O—CH_3	83.2
H—Cl	103.2	H_2N—CH_3	85.2
H—F	136.3	H_3C—CH_3	90.1
H—OH	118.8	H_3C—CH_2CH_3	89.0
H—OCH_3	104.6	H_2C=CH_2 (π bond only)	66
H—NH_2	107.6	H_2C=CH_2 (total)	174.1
H—CH_3	105.0	HC≡CH (total)	230.7
		H_2C=O (total)	178.8

The entropy term ($\Delta S°$) is related to freedom of movement. The more restricted or "ordered" a molecule, the more negative the entropy change. Since it is the negative of $\Delta S°$ that is related to $\Delta G°$ [Eq. (8.6)], the more negative the entropy, the higher the free energy ($\Delta G°$) of the reaction. Note also that the entropy term is temperature-dependent. Entropy becomes more important at high temperature. Imagine the reaction in Figure 8.6, in which strong carbon–carbon σ bonds are broken and weaker carbon–carbon π bonds are made (Table 8.2). At the same time, one molecule is made into two other, smaller molecules.

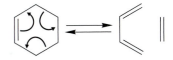

FIGURE 8.6   This reaction is a process in which one direction is favored by enthalpy and the other by entropy. Although ΔH favors cyclohexene, at very high temperature the formation of two compounds from one can drive this reaction to the right.

PROBLEM 8.2   Estimate the $\Delta H°$ for the following reactions:

(a) $CH_3OH + H_2C$=$CH_2 \longrightarrow H_3C$—$CH_2OCH_3$

(b) $H_2 + H_2C$=$CH_2 \longrightarrow H_3C$—$CH_3$

(c) $Br_2 + H_2C$=$CH_2 \longrightarrow BrCH_2$—$CH_2Br$

(d) $HCl + H_2C$=$CH_2 \longrightarrow ClCH_2$—$CH_3$

PROBLEM 8.3   The bond dissociation energies for the carbon–hydrogen bonds in Table 8.2 show a steady decline as the breaking carbon–hydrogen bond becomes more substituted. What can you infer from these data about the stability of the neutral species called "free radicals," R• ?

At low temperature, this reaction will certainly be endergonic because of the unfavorable $\Delta H°$ term. Yet at high temperature the $T\Delta S°$ term will become more important, and the reaction will become more favorable, largely because of the increase in the number of particles and the increased freedom of motion in the system. An unfavorable reaction can even become favorable at high temperature where the entropy term becomes more important in determining the value of $\Delta G°$.

WORKED PROBLEM 8.4 Use $\Delta H°$ values from Table 8.2, as well as the value for the BDE of $H_3C-CH_2CH=CH_2 \rightarrow H_3C\cdot \ \cdot CH_2CH=CH_2$ of 80 kcal/mol, to estimate $\Delta H°$ for the reaction in Figure 8.6.

ANSWER Compare the bonds that are breaking with those that are being made. Two carbon–carbon single bonds are being broken. As the problem says, each is worth 80 kcal/mol. In addition, a π bond (66 kcal/mol) is broken in the reaction. So, the bonds broken are worth a total of 226 kcal/mol. Three π bonds are made, each worth 66 kcal/mol, for a total of 198 kcal/mol. The left-to-right reaction is endothermic by about 28 kcal/mol, as the bonds made are weaker than those broken.

$$2 \times C-C \quad = +160 \text{ kcal/mol (bonds broken)}$$
$$C=C \ (\pi) \quad = + \ 66 \text{ kcal/mol (bond broken)}$$
$$3 \times C=C \ (\pi) = \underline{-198} \text{ kcal/mol (bonds made)}$$
$$\Delta H° = + \ 28 \text{ kcal/mol}$$

SUMMARY

Now we know something of the factors that affect equilibrium. Enthalpy, ΔH, is important. Basically, the side of the equilibrium with the stronger bonds will be favored by enthalpy. Also important is the temperature-dependent entropy, ΔS, term. The two terms may act in concert or oppose each other.

We have said nothing so far of the factors influencing the *rate* of establishing equilibrium.

8.4   RATES OF CHEMICAL REACTIONS

Just because a reaction is exothermic does not mean that it will be a rapid reaction, or even that it will proceed at all under normal conditions. First of all, an exothermic reaction can still be endergonic if entropy plays an especially important role. Entropy does not raise its ugly head in this way often, but there is another much more frequently encountered reason why an *exergonic* reaction might not be a *fast* reaction. As we have seen, there are barriers to chemical reactions. In principle, a very stable product can be separated from a much less stable starting material by a barrier high enough to stop all reaction (Fig. 8.7).

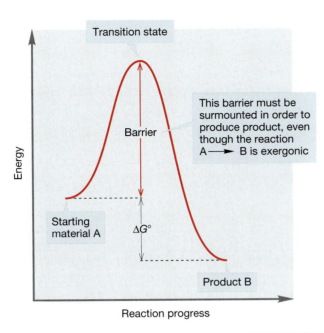

Transition state

Barrier

This barrier must be surmounted in order to produce product, even though the reaction A $\longrightarrow$ B is exergonic

Energy

Starting material A

$\Delta G°$

Product B

Reaction progress

FIGURE 8.7 Rates of reactions are not determined by the $\Delta G°$ between starting material and product, but by the height of the barrier separating starting material and product. Some very unstable materials are protected by high barriers, and thus can be isolated.

Collections of molecules exist at a given temperature in ranges of energies called **Boltzmann distributions**. The energies of most molecules cluster about an average value. As energy is provided to the molecules, usually through the application of heat, the average energy increases. Although there is always a small number of highly energetic molecules present (the high-energy tail of the Boltzmann distribution), in practice there may not be enough molecules of sufficient energy to make the reaction proceed at an observable rate if the barrier is high enough. The application of energy in the form of heat produces more molecules with sufficient energy to traverse the barrier (Fig. 8.8). A reaction rate approximately doubles for every increase in temperature of 10 °C.

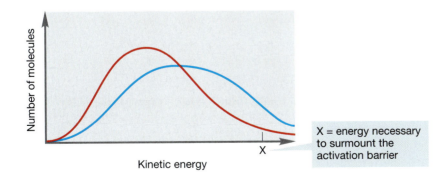

Number of molecules

Kinetic energy

X = energy necessary to surmount the activation barrier

FIGURE 8.8 Boltzmann distributions of molecules at two temperatures. In this figure, the blue line is the distribution at higher temperature and the red line is the distribution at lower temperature.

As we already saw in our discussion of the S_N2 and S_N1 reactions, rates of chemical reactions also depend on the concentrations of reactants. The rates of most organic reactions depend either only on the concentration of a single reactant (S_N1, for example) and thus are **first-order reactions**, or upon the concentrations of two species (S$_N$2) and are **second-order reactions** (Fig. 8.9). There are examples of higher order reactions, but they are relatively rare.

$$\text{Rate} = k\,[(CH_3)_3C-I]$$

$$\text{Rate} = k\,[CH_3CH_2CH_2CH_2-I][^-OH]$$

FIGURE 8.9 Rate laws for the S$_N$1 and S$_N$2 reactions.

Sometimes a reaction we expect to be second order appears not to be. The S$_N$2 solvolysis reaction of methyl iodide with solvent (therefore in great excess) ammonia would be an example. This reaction is first order in substrate, methyl iodide, because the concentration of one of the reactants, ammonia, does not change appreciably during the reaction. Such reactions are called **pseudo-first-order reactions** (Fig. 8.10).

Solvent (present in great excess)

$$\text{Rate} = k\,[CH_3-I]$$

FIGURE 8.10 A pseudo-first-order reaction.

8.5   RATE CONSTANT

In the rate laws mentioned so far, there has always been a proportionality constant, the **rate constant (k)**, which is a fundamental property of any given chemical reaction. A rate must have the units of concentration per unit time, or moles per liter per unit time. For a first-order process such as the S$_N$1 reaction, the rate constant k must have the units of reciprocal seconds (s^{-1}) (if we use the second as our time unit, Fig. 8.11).

FIGURE 8.11 For a first-order reaction, the rate constant (K) has the units of time^{-1} = s^{-1}.

$$\text{Rate} = k\,[A]$$

$$\text{or, } \underbrace{mol/L/s}_{} = \underbrace{k}_{} \underbrace{(mol/L)}_{}$$

concentration/time = (rate constant) (concentration)

$$mol/L/s \times L/mol = k$$

$$\text{and } 1/s = s^{-1} = k$$

For a second-order reaction, k must be in units of reciprocal moles per liter per second $(mol/L)^{-1}s^{-1}$ (Fig. 8.12).

Rate = k[A][B]

Rate = mol/L/s = k × mol/L × mol/L

Rate = concentration/time = (rate constant) (concentration) (concentration)

so, k = mol/L/s × L^2/mol^2 = L/mol/s = $(mol/L)^{-1}s^{-1}$

FIGURE 8.12 For a second-order reaction, the rate constant (k) has the units of $(m/L)^{-1}s^{-1}$.

We can determine the value of k by measuring how the reaction rate varies with the concentration(s) of the reactants.

It is easy to confuse the *rate* of a reaction, which is simply how fast the concentrations of the reactants are changing, with the *rate constant*, which tells us how the rate will change as a function of reactant concentration. The rate of a reaction will vary with concentration. This variation is easy to see by imagining the limiting case in which one reactant is used up. When its concentration goes to zero, the reaction rate must also be zero. On the other hand, the rate constant is an intrinsic property of the reaction. It will vary with temperature, pressure, and solvent, but does not depend on the concentrations of reactants.

Only when the concentrations of all reactants are 1 M are the rate and the rate constant the same. The rate constant is thus a measure of the rate under standard conditions (Fig. 8.13).

Rate = k [A][B][C][D] … [X]   if all [ ] = 1 M

Rate = k [1][1][1][1] … [1]   and therefore, rate = k

FIGURE 8.13 If the concentrations of the reactants are all 1M, the rate constant and the rates are the same.

8.6   ENERGY BARRIERS IN CHEMICAL REACTIONS: THE TRANSITION STATE AND ACTIVATION ENERGY

We have already mentioned that even exothermic reactions may be very slow, indeed immeasurably slow. And it's a good thing. Combustion is a very exothermic chemical reaction, and yet we and our surroundings exist more or less tranquilly in air without bursting into flame. Somehow we are protected by an energy barrier. Methane and oxygen do not react unless energy, called **activation energy**, is supplied. If we are considering free energy, the activation energy is $\Delta G^{\ddagger}$; if we are speaking of enthalpy, the activation energy is $\Delta H^{\ddagger}$. The double dagger (‡) is always used when we are referring to activation parameters involving the energy of a transition state. A match will suffice to provide the energy in this example. This highly exothermic process (see Chapter 2, p. 67) then provides its own energy as highly stable molecules are produced from less stable ones (Fig. 8.14).

CONVENTION ALERT

FIGURE 8.14 An exergonic reaction can provide the energy necessary to pass over an activation barrier. Once such a reaction is started, it continues until one reactant is used up.

In the figure: Energy (vertical axis), Reaction progress (horizontal axis)

$\Delta G^{\ddagger}$

This amount of energy must be supplied to start the reaction

$CH_4 + 2 O_2$

ΔG°

Very negative for this highly exergonic reaction; the energy given off can be used to supply $\Delta G^{\ddagger}$!

$CO_2 + 2 H_2O$

PROBLEM 8.5 Identify the activation energy for the reverse of the reaction in Figure 8.14, the formation of methane and oxygen from carbon dioxide and water.

We are protected from spontaneous combustion in the same way we are protected from hungry bears and tigers during a visit to the zoo. At the zoo the barrier is tangible and operates in a clear way to ensure our survival. Activation barriers are less visible but no less necessary for the continuance of our lives. In this important section, we will discuss these energy barriers to chemical reactions.

In a process such as the S_N1 reaction, a bond must be broken. Consider the ionization of *tert*-butyl iodide. The first step in this reaction is the ionization of the substrate, and surely this will involve an energy cost in breaking the carbon leaving group bond. The enthalpy change ΔH° (and ΔG°) is positive for this reaction (Fig. 8.15).

FIGURE 8.15 The enthalpy change ΔH° (and ΔG°) is positive for this endothermic reaction.

This cost is partially offset by an increase in entropy. We form two particles out of one and this will be favorable. But we have to be careful when we estimate entropy changes because there are invisible participants in most reactions, including this one. These are the solvent molecules. They will have to be disordered and reordered in any ionic reaction in solution and the net change in entropy may not be at all easy to estimate.

WORKED PROBLEM 8.6 Construct an Energy versus Reaction progress diagram (see Fig. 8.9) for the reaction of *tert*-butyl iodide and water, showing clearly the activation energies and transition states for all steps. A discussion of this reaction is coming, but try to anticipate it here.

WORKED PROBLEM 8.7 Show the activation energies for the reverse reactions involved in the formation of *tert*-butyl iodide from *tert*-butyl alcohol and hydrogen iodide (HI).

ANSWERS

PROBLEM 8.8 Show that at equilibrium, the equilibrium constant (K) is given by the ratio of the rate constants for the forward and reverse reactions. *Hint*: At equilibrium the rates of the forward and reverse reactions are equal.

The situation is a little different for the S_N2 reaction. Here we must also break the carbon leaving group bond, but this energy cost will be partially offset by the formation of the bond from carbon to the entering nucleophile. We might imagine that there would also be a substantial entropy cost in this reaction as we must order the reactants in a special way for the reaction to proceed. The S_N2 reaction can occur only if the nucleophile attacks from the rear of the departing leaving group and this requires a specific arrangement of the molecules and ions involved in the reaction (Fig. 8.16).

FIGURE 8.16 The transition state for the S_N2 reaction has severe ordering requirements. Attack of the nucleophile must be from the rear for the reaction to succeed.

We can follow the course of both these reactions in diagrams that plot Energy versus Reaction progress. One usually doesn't specify reaction progress in detail because there are many possible measures of this quantity, and for this kind of qualitative discussion it doesn't much matter what we pick. In the S_N2 reaction, for example, we might choose to plot energy versus the length of the carbon leaving group bond (which is increasing in the reaction), or the length of the carbon–nucleophile bond, which decreases through much of the process. The overall picture is the same.

The point of maximum energy in the reaction is called the transition state. If we have enough energy for molecules to cross the barrier in reasonable numbers, we will have a measurable rate of reaction and the final mixture of starting materials and products will depend on the $\Delta G°$ between starting materials and products. Figure 8.17 shows a diagram for the exothermic formation of methyl cyanide (CH_3CN) and potassium iodide (KI) from methyl iodide (CH_3I) and potassium cyanide (KCN). The transition state is the high point in energy for the reaction. The activation energy is the height of the transition state above the starting material. At equilibrium, the ratio of product and starting material will depend on $\Delta G°$, the difference in free energy between starting material and product. The reaction shown in Figure 8.17 has a negative $\Delta G°$, and will be successful in producing product.

FIGURE 8.17 An energy diagram for the S_N2 displacement of iodide by cyanide (and its reverse reaction, the much less favorable displacement of the strong nucleophile cyanide by iodide). The activation energy for the exergonic displacement of iodide is shown.

S_N2 with cyanide

In Figure 8.18, this same reaction is considered from the reverse point of view. This figure describes the endothermic formation of methyl iodide from methyl cyanide and iodide ion.

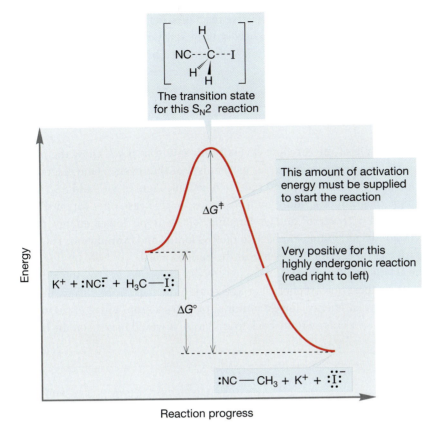

The transition state for this S_N2 reaction

This amount of activation energy must be supplied to start the reaction

Very positive for this highly endergonic reaction (read right to left)

$\Delta G^{\ddagger}$

$\Delta G°$

Energy

$K^+ + :NC^{\overline{\cdot}} + H_3C{-}\ddot{\underset{\cdot\cdot}{I}}:$

$:NC{-}CH_3 + K^+ + :\ddot{\underset{\cdot\cdot}{I}}:^{\overline{}}$

Reaction progress

FIGURE 8.18 This energy diagram for the same reaction emphasizes the reversal. Here the activation energy for displacement of cyanide by iodide is shown.

The activation energy is much higher than that for the forward reaction, and $\Delta G°$ is positive for the endergonic formation of methyl iodide. The transition state is exactly the same as that for the forward reaction. This reaction will not be very successful. It not only requires overcoming a high activation energy, but is "fighting" an unfavorable equilibrium as $\Delta G°$ is positive.

Of course, this separation of the two reactions is absolutely artificial—we are really describing a single equilibrium between methyl iodide and cyanide ion on one side and methyl cyanide and iodide ion on the other (Fig. 8.19).

$$K^+ + :NC^{\overline{\cdot}} + H_3C{-}\ddot{\underset{\cdot\cdot}{I}}: \;\rightleftharpoons\; :NC{-}CH_3 + K^+ + :\ddot{\underset{\cdot\cdot}{I}}:^{\overline{}}$$

FIGURE 8.19 The competition between cyanide and iodide for a Lewis acidic carbon.

The exergonic formation of methyl cyanide and the endergonic formation of methyl iodide are one and the same reaction. The reactions share exactly the same transition state; only the ease of reaching it is different ($\Delta G^{\ddagger}_{forward} < \Delta G^{\ddagger}_{reverse}$). The energy difference between the starting material and the product ($\Delta G°$) is the difference between the two activation energies.

The rates of the forward and reverse reactions, determined by the magnitudes of the activation energies for the two halves of the reaction, are intimately related to the equilibrium constant for the reaction K, determined by $\Delta G°$ [Eq. (8.7)].

$$\Delta G^{\ddagger}_{reverse} - \Delta G^{\ddagger}_{forward} = \Delta G° = -RT\ln K \qquad (8.7)$$

The principle of **microscopic reversibility** says that if we know the mechanism of a forward reaction, we know the mechanism of the reverse. One diagram suffices to tell all about both forward and reverse reactions. Figures 8.17 and 8.18 are in reality a single diagram describing both reactions, in this case one "forward" and exergonic, the other "reverse" and endergonic. There is no lower energy route from methyl cyanide and iodide ion back to methyl iodide and cyanide ion. The lowest energy path back is exactly the reverse of the path forward.

The classic analogy for this idea pictures the molecules as crossing a high mountain pass. The best (lowest energy, or lowest altitude) path from village A to village B is the best (lowest energy, lowest altitude) path back from village B to village A. If there were a lower energy route back from B to A, it would have been the better route from A to B as well (Fig. 8.20).

Best = lowest energy path from A to B. This is also the best path back from B to A.

Had there been a lower energy path (green) it would still be the best path in both directions.

FIGURE 8.20 Molecules, unlike explorers, take the best (lowest energy) path. You might take the red path from A to B and the orange path back, but a molecule wouldn't.

The Energy versus Reaction progress diagram for the S_N1 reaction is more complicated, because there are more steps. We dealt with the description of this reaction as a sequence of equilibria earlier (Fig. 8.2). Let's again take the reaction of *tert*-butyl iodide with solvent water as an example. The first step is the formation of an intermediate carbocation. We expect this first reaction to be quite endothermic as a carbon leaving group bond must be broken and ions must be formed. The high-energy point in this reaction, transition state 1, occurs as the carbon–iodine bond stretches (Fig. 8.21).

FIGURE 8.21 The first step in the S_N1 reaction of *tert*-butyl iodide in water. The formation of the cation is highly endothermic.

The second step involves the reaction of the nucleophilic solvent water with the carbocation, and is highly exothermic. A bond is made and none is broken. There is a second barrier, and therefore another transition state (2) for this second step of the reaction (Fig. 8.22).

FIGURE 8.22 By contrast, the second step in this S_N1 reaction, the capture of the cation by water, is highly exothermic. Transition state 2 is lower than transition state 1.

There is a third and final step in this particular S_N1 reaction, as a proton must be removed from the oxonium ion by a base to give the ultimate product, *tert*-butyl alcohol. So there is still another barrier and a third transition state (Fig. 8.23).

FIGURE 8.23 Transition states for the final step in this S_N1 reaction.

PROBLEM 8.9 Write an Energy versus Reaction progress profile for the S_N1 reaction of *tert*-butyl iodide in H_2O/KOH containing an additional nucleophile, Nu:⁻. Note that the major product is not necessarily *tert*-butyl alcohol (Chapter 7, p. 308).

There are six activation energies in the completed diagram for this S_N1 reaction, three for the forward reactions and three for the reverse reactions (Fig. 8.24).

FIGURE 8.24 In this figure, the black double-headed arrows represent activation energies for the forward reactions and the red double-headed arrows represent activation energies for the reverse reactions.

Now, concentrate on these barriers separating the various intermediates. The energy barriers are made up of the same quantities, enthalpy and entropy, that define the energy of any chemical species. Recall that $\Delta G° = \Delta H° - T\Delta S°$ [Eq. (8.6)]. Transition state theory assumes that we can treat the transition state as if it were a real molecule occupying a potential energy minimum. It is important to emphasize that it is *not,* however. It occupies an energy maximum and can never be bottled and examined as real molecules can. Nonetheless it has an energy, as indeed does every point in the energy diagrams on the preceding pages. The energy of the transition state, and thus the height of the energy barrier in the reaction, can be calculated using Eq. (8.8).

$$\Delta G^{\ddagger} = \Delta H^{\ddagger} - T\Delta S^{\ddagger} \tag{8.8}$$

Remember: The double daggers (‡) simply indicate that we are referring to a transition state, not to an energy minimum. The energy barrier, $\Delta G^{\ddagger}$, is the Gibbs free energy of activation, $\Delta H^{\ddagger}$ is the enthalpy of activation, and $\Delta S^{\ddagger}$ is the entropy of activation.

Clearly, the rate constant for any reaction will depend on the height of the barrier, and thus on $\Delta G^{\ddagger}$. The theoretical study of reaction rates gives us the exact equation relating the rate constant k and $\Delta G^{\ddagger}$ [Eq. (8.9)].

$$k = \nu^{\ddagger} e^{-\Delta G^{\ddagger}/RT} \quad \text{or} \quad k = \nu^{\ddagger} e^{-\Delta H^{\ddagger}/RT} e^{\Delta S^{\ddagger}/R} \tag{8.9}$$

It is worth dissecting this equation a bit further. The quantity $\nu^{\ddagger}$ is a frequency—its units are reciprocal seconds (s^{-1}) in this case. In qualitative terms, ν is the frequency with which molecules with enough energy to cross the barrier actually do so. So, the rate constant k is equal to $e^{-\Delta G^{\ddagger}/RT}$ (the fraction of molecules with enough energy to cross the barrier) times $\nu^{\ddagger}$ (the frequency with which such molecules actually do cross the barrier).

SUMMARY

Reaction rates are not determined by the relative energies of the starting material and product, but by the height of the transition state separating them. Even though it is an energy maximum, a transient species that cannot be isolated, the energy of the transition state is determined by enthalpy and entropy factors, just as is the energy of any stable compound.

8.7   REACTION MECHANISM

Throughout this book we will frequently be concerned with the determination of reaction mechanism. A mechanism is nothing less than the description of the structures and energies of the starting materials and products of a reaction, as well as of any reaction intermediates. In addition, all of the transition states (energy maxima) separating the stable molecules lying in energy minima must be described. It is relatively easy to deal with energy minima—we can often isolate the molecules themselves, take their spectra, and measure their properties. As we can by definition never isolate a transition state, we can never make a direct measurement on it. Yet, if we are to have a good picture of a reaction we must arrive at good descriptions of the transition states involved.

We can usually measure the rate law, which tells us the number and kinds of molecules involved in the transition state. The rate law does not tell us anything about the orientation of the molecules in the transition state. Usually, stereochemical experiments do that. Recall the S_N2 reaction, in which inversion of configuration is observed, thus specifying the direction of approach of the entering nucleophile in the transition state (Chapter 7, p. 279; Fig. 8.25).

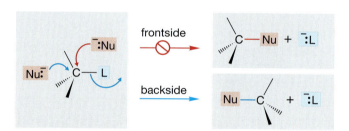

FIGURE 8.25 In the S_N2 reaction, kinetics cannot tell frontside from backside displacement. It takes a stereochemical experiment to do that.

This observation of inversion in the S_N2 reaction tells us that the incoming nucleophile must approach from the rear of the departing leaving group. The rate law, rate = $k[Nu\!:^-][R\!-\!X]$, does not tell us that, only that both the nucleophile and the substrate are involved in the transition state. As we work through the important reactions of organic chemistry we will see many similar experiments designed to determine the stereochemistry of the reaction. The energy of the transition state can be measured if good quantitative kinetic experiments can be performed. If the rate constant for the reaction is determined at a variety of temperatures, the height of the transition state can be calculated. But we can do this only for the step with the highest energy transition state. In a multistep reaction, we can only measure the rate for the slowest step in the sequence.

This slow step is called the **rate-determining step**, or, sometimes, the rate-limiting step. It is always the step in the reaction with the highest energy transition state. We cannot measure the rates of faster steps in the reaction. In the S_N1 reaction, for example, the slow step is the ionization of the substrate to a carbocation. This ion may then be captured by a host of nucleophiles in following faster steps. There will be a different product for every nucleophile (Chapter 7, p. 308; Fig. 8.26).

If we determine the rate of this process by measuring the disappearance of *tert*-butyl iodide over time, we can ultimately find the height of transition state 1 for the reaction. The rates of the faster captures of the carbocation *do not affect the rate of ionization of tert-butyl iodide*. The carbocation is gobbled up by the nucleophiles at a rate faster than that at which it is formed. In the limit, each molecule of the *tert*-butyl cation is captured before another is produced from *tert*-butyl iodide. The rate of formation of product depends on how fast the *tert*-butyl cation is produced in the slow, rate-determining step, but not on the rate of the fast capture by a nucleophile.

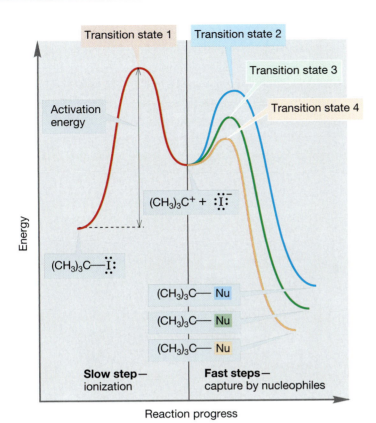

FIGURE 8.26 In this S_N1 reaction, the rate-determining step is the ionization of the starting iodide to give the *tert*-butyl cation. The faster, product-determining steps follow, but their rates cannot be measured kinetically (unless the intermediate carbocation can be isolated and investigated independently).

Imagine a cage of especially ravenous rats into which we are slowly dropping incredibly delicious food pellets. At first, the pellets are gobbled up by the hungry rats at a rate far faster than they are delivered to the cage. The slow step in this reaction, the step that determines the rate of disappearance of the food, is the rate at which the pellets are delivered to the cage. This process is analogous to a reaction in which the first step (food dropping into the cage) is slower than the second step (the rats eating the food). Later, however, a different situation obtains. Even these rats can become satiated and the food pellets accumulate in the cage as the no-longer starving rodents largely ignore them, only taking a nibble now and then. The second step is now slower than the first step. Now the step that determines the rate of disappearance of the food is the rate at which the rats eat the pellets. This second "reaction" has a different rate-determining step than the first.

8.8 THE HAMMOND POSTULATE: THERMODYNAMICS VERSUS KINETICS

We are used to saying, Product Y is more stable than product X, and therefore is formed faster. This statement has a comfortable sound to it, and intuitively makes sense. But it need not be true! The first half of the statement describes **thermodynamics**, the relationship of the energies of product X and product Y. The second half speaks of the *rates* of formation of X and Y, which is a statement about **kinetics**. Despite the good intuitive feel of the original statement, there need be no direct connection between thermodynamic stabilities and kinetics.

Figure 8.27 shows the relative energies of two products, X and Y, being formed from a starting material. One might expect that it would be the $\Delta G°$ between the two products X and Y that will determine the amount of the two compounds formed, but this is true only if X and Y are in equilibrium, via the starting material (Fig. 8.28).

FIGURE 8.27 The energy difference between X and Y does not necessarily affect their rates of formation.

Product X $\rightleftharpoons$ Starting material $\rightleftharpoons$ Product Y

FIGURE 8.28 Only if X and Y are in equilibrium does $\Delta G°$ determine their ratio.

It is conceivable that the products are *not* in equilibrium. There might not be enough energy for the products to go back to the starting material. Under these conditions, the amounts of X and Y will be determined not by their relative energies but only by the heights of the transition states leading to them (Fig. 8.29).

Figure 8.29 shows a conventional picture in which the transition state leading to the more stable product, Y, is lower in energy than the transition state leading to the higher energy product, X. Under such conditions, the major product will be the more stable compound, Y. We have seen many examples of reactions like this.

PROBLEM 8.10 Identify the activation energies for the reactions of X and Y to give starting material (Fig. 8.29).

FIGURE 8.29 The usual picture is this: The higher transition state leads to the less stable product. The lower transition state leads to the more stable product.

But it could be the case that the transition state leading to the more stable compound, Y, is actually higher in energy than the transition state leading to the less stable product, X (Fig. 8.30).

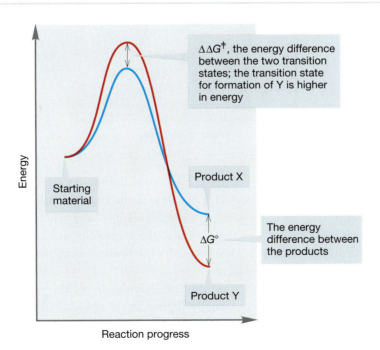

FIGURE 8.30 The situation in Figure 8.29 could be otherwise; the higher transition state could lead to the lower energy product, and the lower energy transition state to the higher energy product.

This inversion will not matter if X and Y equilibrate via the starting material. In such a case, the greater stability of Y will win out eventually, and the ratio of X and Y will depend on the $\Delta G°$ between them. Such a reaction is said to be under **thermodynamic control**. *But what if the two products do not equilibrate? What if there is enough energy so that the forward reactions are possible, but the backward reactions from X or Y to starting material are not?* In the exothermic processes shown, the forward reaction is easier—has a lower activation barrier—than the backward reaction (Fig. 8.31).

Under these circumstances, the relative amounts of X and Y depend only on the relative heights of the transition states leading to them. Such a reaction is said to be under **kinetic control**.

$\Delta\Delta G^{\ddagger}$, the energy difference between the two transition states; this quantity will determine the relative amounts of Y and X if these products cannot go back to starting material

$\Delta G^{\ddagger}$, activation energy for Y $\longrightarrow$ Starting material

$\Delta G^{\ddagger}$, activation energy for X $\longrightarrow$ Starting material

Product X

The energy difference between the products; this makes no difference *if the products cannot go back to starting material*

$\Delta G°$

Product Y

Energy

Starting material

Reaction progress

FIGURE 8.31 If there is enough energy so that all barriers can be crossed, the ratio of products will be determined by $\Delta G°$, and the reaction is said to be under thermodynamic control. If, however, there is not enough energy to cross the high barriers for the reverse reactions leading from X and Y to starting material, the ratio of products will be determined by the relative energies of the two transition states, and the reaction is said to be under kinetic control.

There is no *necessary* connection between thermodynamics and kinetics. In our original statement, product Y is more stable than product X, and *therefore* is formed faster; we are equating thermodynamics and kinetics. We tacitly assume that it will always be the case that the product of greater stability (lower energy) will be reached at a faster rate by the lower energy (more stable) transition state (Fig. 8.29).

In principle, it could be that the transition state for the reaction leading to the more stable product is actually higher in energy than the transition state leading to the less stable product. Such a situation can lead to the preferred formation of the less stable product, if the reaction is under kinetic control (Fig. 8.31).

Fortunately, there are factors that operate to preserve a parallelism between the energies of products and the transition states leading to them. There are examples in which the higher energy transition state leads to the lower energy product, and *vice versa,* but they are relatively rare. We will be at pains to point them out as they occur. The rest of this section will try to show you why the parallelism between product energy and transition state energy exists for many reactions. Consider, for a start, Problem 8.11.

WORKED PROBLEM 8.11 Alkenes can be protonated to give carbocations. Unsymmetrical alkenes, such as 2-methyl-2-butene, could be protonated at either end of the double bond to give two different carbocations. Predict the direction of protonation of 2-methyl-2-butene and explain your answer (Chapter 3, p. 139). See below.

ANSWER The choice is between the formation of a secondary and a tertiary carbocation. It is surely tempting to predict that the more stable tertiary carbocation will be formed faster than the less stable secondary carbocation.

This carbocation is tertiary

This carbocation is secondary

However, this comfortable reasoning equates thermodynamics (more stable) with kinetics (will be formed faster), and this equation is not necessarily valid. Read on.

In one sense, the answer to this problem is obvious; the more stable tertiary carbocation will be formed faster than the less stable secondary carbocation. But this reasoning again equates *thermodynamics* "the *more stable* tertiary carbocation . . . " with *kinetics* "will be formed *faster* than . . . ." Before we accept it, we had better think it through. Look first at the Energy versus Reaction progress diagram (Fig. 8.32).

FIGURE 8.32 The more substituted cation will be at lower energy than the less substituted cation. Note also that the transition state for formation of the tertiary cation is lower than the transition state for formation of the secondary cation. Why?

The tertiary cation is more substituted than the secondary cation and thus is more stable (Chapter 7, p. 305). Substitution stabilizes the positive charge on carbon in these species. In the transition states for formation of the carbocations, the new bond from carbon to hydrogen will be partially formed and the carbons will have developed partial positive charges (Fig. 8.33).

FIGURE 8.33 In the two transition states, partial positive charge has built up on carbon. A partial positive charge on a tertiary carbon is more stable than a partial positive charge on a secondary carbon.

The factors that operate to stabilize a full positive charge will also operate to stabilize a partial positive charge. One therefore expects the transition state for the formation of a tertiary carbocation to be more stable than the transition state for formation of a secondary carbocation, although the difference will not be as great as that for the fully developed cations (Fig. 8.34).

FIGURE 8.34 The energy difference between the tertiary and secondary cationic products will be reflected in the transition states for cation formation. There will be less difference in energy between the two transition states (bearing partial positive charges) than there is between the two products, in which there are full positive charges.

This specific situation has been generalized, in what is called the **Hammond postulate**, after its originator, G. S. Hammond (b. 1921): "For an endothermic reaction the transition state will resemble the product." Note immediately that this statement can be turned around. "For an exothermic reaction the transition state will resemble the starting material." Figure 8.35 shows an endothermic reaction, but we know that all we have to do to see an exothermic reaction is look at the endothermic reaction in the other direction! So these two versions of the Hammond postulate are really the same.

FIGURE 8.35 Equivalent energy diagrams for endothermic and exothermic reactions.

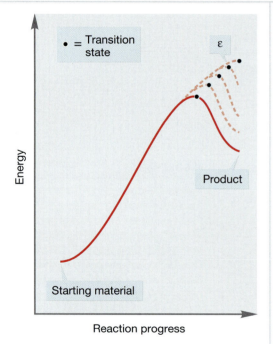

FIGURE 8.36 A thought experiment in which we raise the energy of the product without changing the energy of the starting material.

In the endothermic protonation of 2-methyl-2-butene, we rationalized the faster formation of the more stable tertiary cation by pointing out that in the transition state for its formation there was partial positive charge built up on a tertiary carbon. The transition state resembles the final product, just as the Hammond postulate would have it. We can also do a thought experiment to see that the more endothermic a reaction the more the transition state will look like the product. Imagine an endothermic reaction with the energy profile shown in Figure 8.35. Now apply a pair of molecule-grabbing chemical tweezers and raise the energy of the product without changing the energy of anything else. As we magically increase the energy of the product it becomes less and less stable. Eventually, it *becomes* the transition state, as it reaches a point where it is so unstable that it no longer lies in an energy minimum. Now think about the moment just before this happens. At this point (ε), the product and the transition state are so close to each other that the slightest increase in energy in the product makes them the same (Fig. 8.36).

By reflecting on this limiting situation, we can see how the transition state and the product approach each other as the energy of the product increases and the reaction becomes more endothermic.

It is a simple matter to turn this all around and examine the other side of the Hammond postulate, "The more exothermic a reaction the more the transition state resembles starting material." Figure 8.36 will do it for you if you simply read it backward, from right to left. See what happens to the transition state as the "right-to-left" reaction becomes less and less exothermic.

The take-home lesson of all this discussion is that although one cannot take the correspondence between thermodynamics and kinetics as a given—there will be counterexamples—in general, the statement "Y is more stable and thus formed faster" will be true. There is more to it than it appears at first. This statement is tricky indeed, and it is worth stopping and examining it every time we make it.

8.9   SPECIAL TOPIC: ENZYMES AND REACTION RATES

In this chapter, we discuss the interconversion of compounds by passing from starting material to product over a high-energy point called the transition state. The higher the transition state, the slower the reaction. In the laboratory, we affect the rate of reaction by adding acid or base catalysts, changing solvents, or sometimes simply by supplying lots of energy. Our bodies run on the same chemistry found in the laboratory. For example, the substitution and elimination reactions described in Chapter 7 are all reasonably common in the transformations of biomolecules carried out in us.

Nature can't overcome activation barriers to reactions important to us by adding catalysts such as sulfuric acid, and turning up the heat in our bodies to provide energy becomes uncomfortable very quickly indeed. So, Nature does it another way, and it is most successful. Nature uses enzymes, polyamino acid chains with catalytic activity, to lower the barriers to chemical transformations. Aconitase, an enzyme of molecular weight 89,000 (!) catalyzes the conversion of citrate into *cis*-aconitate, an elimination reaction. Aconitate is then hydrated to give isocitrate. So, this enzyme makes possible one of the key reactions of Chapter 7, elimination, and then reverses the process, but in the other direction, to give the isocitrate (Fig. 8.37).

Citrate   ***cis*-Aconitate**   **Isocitrate**

FIGURE 8.37 The interconversions catalyzed by the enzyme, aconitase.

The vast molecular architecture of this enormous molecule serves to bind the substrate precisely, holding it in perfect position for the side chains of the amino acids to act as catalysts, speeding the elimination and addition reactions. Spectacular rate accelerations can be achieved by enzymes, allowing all sorts of reactions to take place at 37 °C that could not occur otherwise.

8.10   SUMMARY

New Concepts

This chapter is devoted entirely to concepts. You won't find a new mechanism or synthetic route in it. Yet, it is extremely important because what we say here about equilibrium and rates is relevant to all the reactions that will appear in subsequent chapters.

All reactions are equilibria, or in the case of multistep reactions, sequences of equilibria between a starting material and a product. The equilibrium constant K is related to the difference in energy between starting materials and products in a logarithmic fashion ($\Delta G^\circ = -RT \ln K$ or $K = e^{-\Delta G^\circ/RT}$). Therefore, a small amount of energy difference (ΔG) has a great influence on the equilibrium constant K.

The ratios of products depend either on the energy difference between the products ΔG° (thermodynamic control), or on the relative heights of the transition states leading to products (kinetic control).

Differences in G, the Gibbs free energy, whether they are ΔG°, the difference in energy between starting material and product, or $\Delta G^\ddagger$, the difference in energy between

starting material and the transition state (activation energy), are made up of entropy (ΔS°) and enthalpy (ΔH°) terms ($\Delta G^\circ = \Delta H^\circ - T\Delta S^\circ$ or $\Delta G^\ddagger = \Delta H^\ddagger - T\Delta S^\ddagger$).

The rate of a reaction depends on temperature, the concentration of the reactant(s), and the rate constant for the reaction, k. The rate constant is a property of the reaction and depends on temperature, pressure, and solvent, but not upon the concentrations of the reactants.

The rate-determining step of a reaction is the step with the highest energy transition state. It may or may not be the same as the product-determining step. For example, in the S_N2 reaction it is the same, but in the S_N1 reaction it is not.

The principle of microscopic reversibility says that the lowest energy path in one direction of an equilibrium reaction is the lowest energy path for the reverse reaction.

The Hammond postulate says the transition state for an endothermic reaction resembles the product, or, equivalently, that the transition state for an exothermic reaction resembles the starting material.

Key Terms

activation energy ($\Delta G^\ddagger$) (p. 359)
Boltzmann distributions (p. 357)
endergonic (p. 352)
enthalpy (ΔH°) (p. 352)
entropy (ΔS°) (p. 352)
equilibrium constant (K) (p. 351)
exergonic (p. 352)

first-order reactions (p. 357)
Gibbs free energy change (ΔG°) (p. 352)
Hammond postulate (p. 375)
kinetic control (p. 372)
kinetics (p. 370)
Le Chatelier's principle (p. 353)
microscopic reversibility (p. 364)
pseudo-first-order reaction (p. 358)

rate constant (k) (p. 358)
rate-determining step (p. 368)
second-order reactions (p. 358)
thermodynamic control (p. 372)
thermodynamics (p. 369)
transition state (p. 349)

Common Errors

The most common problem experienced with the material in this chapter is the confusion between thermodynamics and kinetics. Rates of reactions (kinetics) are determined by the rela-

tive heights of the available transition states leading to products. If there is insufficient energy available for the products to re-form starting materials, it will be the relative heights of the transition states that determine the mixture of products, no matter what the energy differences among those products (Fig. 8.38).

FIGURE 8.38 Kinetic control of product distribution.

On the other hand, if equilibrium is established; if the products can revert to starting materials, it will be the relative energies of the product molecules that determine how much of each is formed (Fig. 8.39).

FIGURE 8.39 Thermodynamic control of product distribution.

Remember also to be careful to distinguish between a *rate* of a reaction and the *rate constant* for that reaction. The rate depends on the concentration(s) of the reacting molecules; the rate constant does not, and is a fundamental property of the reaction.

8.11   ADDITIONAL PROBLEMS

PROBLEM 8.12   The Finkelstein reaction, shown below, is often used to prepare alkyl iodides from alkyl chlorides or bromides through an S_N2 reaction:

$$R-X + NaI \rightleftarrows R-I + NaX$$

$$X = Cl \text{ or } Br$$

(a)   Suggest a simple way to drive this equilibrium toward the desired iodide.

(b)   In practice, the Finkelstein reaction is usually run in acetone solvent, as sodium iodide, but not sodium chloride or sodium bromide, is soluble in acetone.

Explain carefully how this difference in solubilities drives the equilibrium toward the desired iodide.

PROBLEM 8.13   Alcohol dehydration and alkene hydration is an equilibrium process.

(a)   Suggest experimental conditions that will favor cyclohexene.

(b)   Suggest experimental conditions that will favor cyclohexanol.

PROBLEM 8.14   Calculate the energy differences between the carbonyl (C=O) compounds and their hydrates at 25 °C.

(a)

$$K = 1.4 \times 10^{-3}$$

(b)

$$K = 2.8 \times 10^{4}$$

PROBLEM 8.15   Under certain highly acidic conditions, the three bicyclooctanes shown below can be isomerized. Given the following ratios at equilibrium, calculate the relative energies of the three isomers at 25 °C.

Bicyclo[2.2.2]octane
(3.66%)

Bicyclo[3.3.0]octane
(32.95%)

AlBr₃ | 25 °C

Bicyclo[3.2.1]octane
(63.35%)

PROBLEM 8.16   It is possible to equilibrate the following two ethers. Compound **B** is favored over **A** by 1.51 kcal/mol. What is the equilibrium constant for this interconversion at 25 °C?

A

Fe(CO)₅

B

PROBLEM 8.17   Use the BDE of Table 8.2 (p. 355) to estimate $\Delta H°$ for the following reactions. Reaction mechanisms are not necessary!

(a)

(b)

(c)

(d)

PROBLEM 8.18   Use the data in Table 8.2 (p. 355) to design two endothermic reactions. Don't worry about mechanism! This problem does not ask you to design a *reasonable* reaction—it merely asks you to find reactions that would be endothermic if they occurred.

PROBLEM 8.19 For the equilibrium A + B $\rightleftarrows$ C, calculate:

(a) $\Delta G°$ at 25 and 200 °C given $\Delta H° = -14.0$ kcal/mol and $\Delta S° = -15.8$ cal/deg·mol.

(b) $\Delta G°$ at 25 and 200 °C given $\Delta H° = -6.0$ kcal/mol and $\Delta S° = -15.8$ cal/deg·mol.

PROBLEM 8.20 Construct Energy versus Reaction progress diagrams for E1 and E2 reactions that are overall (a) exothermic, (b) endothermic. Use a tertiary iodide as a sample substrate molecule.

PROBLEM 8.21 For the following two reactions, described by the two Energy versus Reaction progress diagrams, (1) and (2):

(1)

(2)

(a) In each case, what compounds will be present at the end of the reaction?

(b) In each case, which compound will be present in the largest amount? Which will be present in the smallest amount?

(c) What is the rate-determining step for the left-to-right reaction in each case?

(d) Use the diagrams to point out in each case the activation energy for the reactions:

$$A \rightarrow B, \ A \rightarrow C, \ C \rightarrow B, \ \text{and} \ C \rightarrow A$$

(e) In each case, what is the rate-determining step for the reaction A → C?

PROBLEM 8.22 Explain in painstaking detail why one is justified in saying, "The reaction of 2-methyl-2-butene with HCl leads to the tertiary chloride because a tertiary carbocation is more stable than a secondary carbocation." You might start with the construction of Energy versus Reaction progress diagrams and good drawings for the transition states.

PROBLEM 8.23 Find the errors in the following Energy versus Reaction progress diagrams.

(a)

(b)

Additions to Alkenes 1

9

In the last third of his life, there came over Laszlo Jamf—so it seemed to those who from out in the wood lecture halls watched his eyes slowly granulate, spots and wrinkles grow across his image, disintegrating it towards old age—a hostility, a strangely *personal* hatred, for the covalent bond.

—Thomas Pynchon*
Gravity's Rainbow

9.1   PREVIEW

In this chapter, we revisit one of the building block reactions of organic chemistry—the addition to alkenes. We first saw this useful process in Chapter 3, where we used it to introduce reactivity. Here and in Chapter 10, we review very briefly the material from Chapter 3, and then expand on the addition process using the style of theme and variations. By all means, go back and reread the material in Chapter 3 (pp. 134–144) if this introduction is not completely obvious.

In Chapter 3, we restricted ourselves to a few examples of the addition of HX reagents to alkenes. Now we expand to a variety of old and new HX reagents, and in Chapter 10, we add X_2 and XY reagents as well (Fig. 9.1).

FIGURE 9.1 A variety of additions to alkenes. Many H—X, X_2, and X—Y reagents undergo the addition reaction.

*Thomas Pynchon is an American author born in 1937.

Our growing catalog of reactions will continue to add to your expertise in synthesis. The kinds of molecules we know how to make will increase sharply. Addition of hydrogen bromide and hydrogen chloride to alkenes provides access to alkyl halides. We already have the S_N1 and S_N2 reactions available for transforming these molecules into other compounds (Fig. 7.94, p. 328).

ESSENTIAL SKILLS

1. Many addition reactions, especially HX additions, proceed through formation of a carbocation that is captured in a second step by X⁻. You must be able to analyze the possibilities for cation formation and pick out the most stable one possible. In practice, this task is not especially difficult.

2. One of the important factors influencing cation stability is delocalization of charge through the phenomenon we call resonance. Writing resonance forms well is an essential skill, and this chapter provides a full review of how to do it properly.

IMPORTANT DETAILS

1. Two reactions—hydration and hydroboration—give you the ability to add water across an unsymmetrical double bond in either a Markovnikov or an anti-Markovnikov fashion. Although the mechanistic details of the hydroboration process are complex, the synthetic outcome is decidedly not. Be sure you know how to use these reactions to good effect in making alcohols.

2. Less stable carbocations rearrange to more stable ones through hydride (H:⁻) shifts. Stay alert—from now on, every time you see a carbocation, you have to consider whether it will rearrange into another, lower energy carbocation. Once again, you have to be able to judge and predict stability.

Alkene hydrohalogenation

9.2   MECHANISM OF THE ADDITION OF HYDROGEN HALIDES TO ALKENES

Consider once again the reaction of 2,3-dimethyl-2-butene with hydrogen chloride to give 2-chloro-2,3-dimethylbutane (Fig. 9.2).

2,3-Dimethyl-2-butene   **1** Protonation   **2** Addition of Cl⁻   **2-Chloro-2,3-dimethylbutane**

FIGURE 9.2 The first step in this two-step reaction is the protonation of the alkene to give a carbocation. In the second step, a chloride ion adds to the cation to give the final product.

The arrow formalism maps out the process for us and develops the picture of a two-step mechanism. In the first step, the alkene, with the filled π orbital acting as base, is protonated by hydrogen chloride to give a carbocation and a chloride ion. In the second step, chloride acts as nucleophile and adds to the strongly Lewis acidic cation to give the product (Fig. 9.2).

PROBLEM 9.1 Use the bond dissociation energies (BDE) of Table 8.2 (p. 355) to estimate the exothermicity or endothermicity of the reaction in Figure 9.2.

The first step in this sequence, the protonation of the alkene to give a carbocation, is the slow step in the reaction. The rate-determining step, the one with the highest energy transition state, is this addition of a proton. As this step is endothermic, the transition state will resemble the product carbocation, and it will have substantial positive charge developed on carbon (Fig. 9.3). That notion leads directly to the next topic.

FIGURE 9.3 In this simple addition of hydrogen chloride to an alkene, the first step, the endothermic formation of a carbocation, is the slow, rate-determining step. In the transition state for this step, the positive charge accumulates on one of the carbons of the starting alkene.

9.3   REGIOCHEMISTRY

The reaction of hydrogen chloride with a symmetrically substituted alkene such as 2,3-dimethyl-2-butene is relatively simple. Protonation is followed by chloride addition to give the alkyl chloride. There are no choices to be made about the position of protonation, and once the carbocation is made, chloride formation is easy to rationalize. What happens if the alkene is not so symmetrically substituted? What happens if the ends of the alkene are differently substituted? We discussed this subject briefly in Chapter 3 (p. 139), but Figure 9.4 poses the question again.

FIGURE 9.4 For an unsymmetrical alkene, there are two possible intermediate cations, and therefore two possible chloride products.

The answer is that it depends on *how* the ends of the alkene are different. The preferential formation of one isomer in those cases where a choice is possible is called regioselectivity, or if the choice is overwhelmingly in one direction, regiospecificity. The next few sections will discuss factors influencing **regiochemistry**.

WORKED PROBLEM 9.2 Chloride formation isn't the only process possible, however. In the early stages of this reaction, there is a small amount of an alkene isomeric with the 2,3-dimethyl-2-butene produced. What is this other alkene, and what is the mechanism of its formation?

ANSWER Remember the E1 reaction! Just because we are starting a new subject does not mean that old material disappears. Cations will undergo the E1 and S_N1 reactions as well as the addition reactions encountered in this chapter. The cation formed on protonation of 2,3-dimethyl-2-butene can lose a proton in E1 fashion in two ways. Path (a) simply reverses to give starting compounds back, whereas (b) leads to 2,3-dimethyl-1-butene.

(continued)

WORKED PROBLEM 9.2 *(continued)*

(a)

(b)

9.4   RESONANCE EFFECTS

When hydrogen chloride adds to vinyl chloride (chloroethylene), the major product is 1,1-dichloroethane, *not* 1,2-dichloroethane (Fig. 9.5).

Vinyl chloride (chloroethylene)   **1,1-Dichloroethane** *not* **1,2-Dichloroethane**

FIGURE 9.5 Addition of hydrogen chloride to vinyl chloride gives 1,1-dichloroethane, *not* 1,2-dichloroethane.

The best way to rationalize the regiospecificity of this reaction is to examine the two possible intermediate carbocations (Fig. 9.6) and see which of the two is more stable.

FIGURE 9.6 The two cations that could be formed by protonation of vinyl chloride.

Remember: The carbocation has an empty *p* orbital on carbon. In one of the two possible cations, this empty orbital overlaps a *p* orbital on chlorine containing two electrons. This overlap of filled and empty orbitals produces strong stabilization (Fig. 9.7). In this ion, the chlorine atom helps bear the positive charge. Only when the charge is adjacent to the chlorine is there stabilization through overlap of the empty 2*p* orbital on carbon with the filled 3*p* orbital on chlorine. This stabilization is not available to the other cation, which, therefore, is much harder to form.

FIGURE 9.7 Stabilization of a cation through filled orbital-empty orbital overlap.

Another way to represent this stabilization is through a resonance formulation. (For a brief earlier treatment of resonance see Section 1.4, p. 21.) There are two reasonable ways to draw a Lewis structure for the cation next to chlorine. In one the carbon is positively charged, and in the other it is the chlorine that bears the positive charge. The two representations differ only in their distribution of electrons. They are two different electronic representations of the same structure. By itself neither is exactly right. Neither by itself shows that two atoms share the charge. Taken together they do. These two representations are resonance forms. The real structure of this cation is not well represented by either form alone, but the two combined do an excellent job (Fig. 9.8).

FIGURE 9.8 Resonance stabilization of a carbocation.

In this resonance form, the carbon bears the positive charge

In this resonance form, the + resides on the chlorine

The molecular orbital picture we first developed does a good job of showing the stabilization, but it is not particularly well suited for bookkeeping purposes. In practice, the ease of bookkeeping—of easily drawing the next chemical reaction—makes the resonance picture simpler to use. The price is that you have to draw two structures (or more in other cases), or adopt some other code that indicates the presence of more than one representation for the molecule. We will see many examples in the next pages and chapters.

Here is a reaction that we will return to in some detail later, but can examine quickly now. Hydrogen chloride adds to 1,3-butadiene very easily to yield two products. One product you may be able to predict, but the other will probably not be immediately obvious (Fig. 9.9).

FIGURE 9.9 1,2- and 1,4-Addition of HCl to 1,3-butadiene.

Look at the possibilities. There are two possible cations that can arise from protonation of 1,3-butadiene by hydrogen chloride (Fig. 9.10). In one, the allyl cation, the positive charge is on the carbon adjacent to the double bond; in the other, it is isolated on a terminal primary carbon.

FIGURE 9.10 The two possible carbocations from protonation of 1,3-butadiene.

The choice between these two ions becomes clear if we draw out orbital pictures of the two carbocations. In the allyl cation, the *p* orbital on the positively charged carbon atom overlaps the π orbitals of the double bond (Fig. 9.11). We have seen the system of three parallel, overlapping *p* orbitals before (Problems 1.19c and 1.51).

FIGURE 9.11 In the allyl cation, the charge is shared by two carbons; in the primary cation it is localized on a single carbon.

Protonation at the end carbon gives a delocalized allyl cation

Protonation at an internal carbon gives a localized primary cation

Allyl cation

FIGURE 9.12 The molecular orbitals of the allyl cation.

Here we have an allyl cation again, and the molecular orbital description is just the same as before. The three 2*p* atomic orbitals can be combined to yield three molecular orbitals. There are only two electrons in an allyl cation and they will occupy the lowest, most stable molecular orbital (Fig. 9.12). The presence of the double bond is highly stabilizing for this cation, as the charge is delocalized over the allyl system.

PROBLEM 9.3 Derive the π orbitals of the allyl system shown in Figure 9.12. *Hint*: Do this by placing a carbon 2*p* orbital in between the two *p* orbitals of ethylene. Another *Hint*: Watch out for "net zero" interactions!

In the other possible cation in this reaction, the charge is localized at the end of the molecule and has no resonance stabilization. Protonation at the end carbon gives an ion (the allyl cation) that can be represented by more than one Lewis structure. The choice is easy once we have looked carefully at the structures. Formation of the lower energy delocalized allyl cation will be favored over formation of the higher energy localized primary cation (Fig. 9.13).

Protonation to give this delocalized allyl cation will be favored

Protonation to give this localized allyl cation will be disfavored

FIGURE 9.13 Preferential formation of the lower energy, delocalized cation.

The real structure is best represented as a combination of the two Lewis (resonance) forms in the figure. This way of looking at the allyl cation is especially helpful as it points out simply and clearly which carbons share the positive charge. When the chloride ion approaches the allyl cation it can add at either of the two carbons that share the positive charge to give the two products. Be careful not to think of this resonance-stabilized structure as spending part of its time with the charge on one carbon and part of the time with the charge on the other! The cation has a *single* structure in which the two end carbons share the positive charge.

This single, summary structure is often shown using dashed partial bonds to represent the bonds that are double in one resonance form and single in the other. The charge is placed at the midpoint of the dashes. Alternatively, one resonance form is drawn out in full and the other atoms sharing the charge are shown with a (+) or (−) (Fig. 9.14).

CONVENTION ALERT

Notice the partial (dashed) bonds used to indicate the delocalization

In an alternative formalism, the positions sharing the charge are shown by (+) placed at the appropriate sites

FIGURE 9.14 In the allyl cation, the positive charge is shared by two carbons. Note the summary structures in which dashed bonds or charges in parentheses are used to indicate delocalization without writing out all the resonance forms.

So, the resonance formulation is very useful and we are going to review it thoroughly in a moment. But first, let's ask if the molecular orbital description can predict both products. Of course it can, or it would hardly be very useful. What happens as the chloride ion, with its filled nonbonding orbitals, begins to interact with the allyl cation? Stabilizing interactions occur between filled orbitals and empty orbitals. Chloride bears the filled orbital; therefore we must look for the lowest *unoccupied* molecular orbital (LUMO) of allyl. Figure 9.15 shows it, Φ_2. There are two points at which chloride can add to Φ_2, and they lead to the two observed products. Note that the middle carbon, through which the node passes, cannot be attacked. So the molecular orbital description also explains the regiochemistry of the addition.

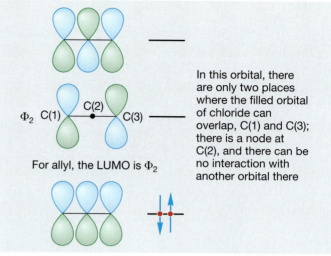

In this orbital, there are only two places where the filled orbital of chloride can overlap, C(1) and C(3); there is a node at C(2), and there can be no interaction with another orbital there

FIGURE 9.15 In this reaction, the HOMO–LUMO interaction is between a filled nonbonding orbital on chloride and the lowest energy unoccupied orbital of allyl, in this case, Φ_2.

Whether one uses resonance forms or molecular orbital descriptions is largely a matter of taste. Resonance forms are the traditional way, and they are extremely good at letting you see where the charge resides. We will use them extensively. It's worth taking time now to review the whole resonance system.

SUMMARY

Additions of many acids, HX, to alkenes really are straightforward. An initial protonation to give the most stable carbocation possible is followed by capture by X^-. The only difficulty is figuring out which carbocation is the lowest energy. Here, an analysis of the "usual suspects" is necessary. You have to consider the substitution pattern, resonance effects, and, as we shall soon see, inductive effects as well.

9.5 REVIEW OF RESONANCE: HOW TO WRITE RESONANCE FORMS

A molecule that is best described as a resonance hybrid, *does not*, repeat *not*, spend part of its time as one structure and part as another. That is chemical equilibrium not resonance (Fig. 9.16). The two are very different phenomena.

The real structure of a resonance-stabilized molecule is a hybrid, or combination, of the resonance forms contributing to the structure. These forms represent different electronic descriptions of the molecule. *Resonance forms differ only in the distributions of electrons and never in the positions of atoms.* If you have moved an atom you have written a chemical equilibrium (Fig. 9.17).

FIGURE 9.16 Arrows imply equilibrium, not resonance.

FIGURE 9.17 Resonance is always indicated by the special double-headed (↔) resonance arrow. Equilibrium is shown by two arrows headed in different directions (⇌).

This represents the combination of the two resonance forms at the right

These forms are different electronic representations of the allyl cation

equilibrium!

These figures do not represent resonance forms for the same allyl cation, but different, possibly equilibrating species

It is for this phenomenon of resonance that the special, double-headed arrow is strictly reserved. This special arrow is an important piece of the grammar of organic chemistry.

It is important to be able to estimate the relative importance of resonance forms in order to get an idea of the best way to represent a molecule. Some guidelines are listed below. Fortunately, this is an area in which common sense does rather well.

1. Identical resonance forms (**A** and **A′**) contribute equally. The allyl cation is a good example. Two equivalent forms are possible, and they contribute equally to the hybrid, real structure. In this case, the weighting factors, c_1 and c_2, are equal (Fig. 9.18).

CONVENTION ALERT

FIGURE 9.18 The allyl cation is an equal combination of **A** and **A′**.

c_1 c_2

A   **A′**

If we wrote an equation to represent the mathematics involved, it would be

$$\text{Allyl} = c_1 \text{ (form } \mathbf{A}) + c_2 \text{ (form } \mathbf{A}')$$

where $c_1 = c_2$. Here c represents the weighting factors, or coefficients. In this case, $c_1 = c_2$.

2. The more bonds in a resonance form the more stable it is. 1,3-Butadiene can be written as a resonance hybrid of four structures (Fig. 9.19).

FIGURE 9.19 Four resonance forms contributing to 1,3-butadiene. Form **A** is by far the best.

1,3-Butadiene is reasonably well described by the single form **A**. The structures **B**, **C**, and **C'** contain fewer bonds than structure **A**, and will contribute far less than **A** to the real 1,3-butadiene. Forms **C** and **C'** are equivalent and will contribute equally. This does not mean that **B**, **C**, and **C'** do not contribute at all! It simply means that in our equation

$$\text{Butadiene} = c_1 \text{ (}\mathbf{A}) + c_2 \text{ (}\mathbf{B}) + c_3 \text{ (}\mathbf{C}) + c_3 \text{ (}\mathbf{C}')$$

The weighting factor c_1 will be much larger than the other coefficients. 1,3-Butadiene looks much more like form **A** than forms **B**, **C**, or **C'**.

3. Resonance forms must have the same number of paired and unpaired electrons. As we saw in Chapter 1 (p. 7), two electrons occupying the same orbital must have opposite, or paired spin quantum numbers. Most organic molecules, including our sample molecule, 1,3-butadiene, have all paired electrons. All resonance forms for 1,3-butadiene must also have all paired electrons. Consider form **B** in the resonance hybrid shown in Figure 9.19. It is often the convention to emphasize that all electrons are paired in this molecule by appending little arrows to the single electrons in this form (Fig. 9.20). Opposed arrows ($\downarrow\uparrow$) indicate paired spins, identical arrows ($\uparrow\uparrow$, $\downarrow\downarrow$) indicate parallel (same direction) spins.

4. Separation of charge is bad. In our equation for butadiene in Figure 9.19, the neutral forms **A** and **B** will contribute more than **C** or **C'**, both of which require a destabilizing separation of charge. (Form **B** will still contribute less than form **A** because it has fewer bonds.)

5. In charged species (ions), delocalization is especially important. As we have mentioned a number of times, small charged atoms are usually most unstable.

Delocalization of electrons that allows more than one atom to share a charge is stabilizing in energy terms. The allyl cation is a good example—the two end carbons each bear one-half of the positive charge (Fig. 9.18).

The opposing arrows indicate that the spin quantum numbers of the two electrons are different; this is a resonance form of 1,3-butadiene

The identical arrows indicate that the spins of the two electrons are the same; this is *not* a resonance form of 1,3-butadiene

FIGURE 9.20 In 1,3-butadiene, all electrons are paired, and all resonance forms contributing to the structure must also have all-paired electrons.

The protonation of vinyl chloride is another good example. Here the charge is shared by the carbon and the chlorine, and the resonance forms show it (Fig. 9.21).

FIGURE 9.21 In this carbocation, the change is shared by carbon and chlorine.

PROBLEM 9.4 Use the molecular orbitals constructed in Problem 9.3 to show the electronic occupancy of the allyl cation, radical, and anion.

6. Electronegativity is important. In the allyl anion (Fig. 9.22), there are two equivalent resonance forms. As in the allyl cation, the charge resides equally on the two end carbons.

FIGURE 9.22 The allyl anion.

The related enolate anion, in which one CH_2 of allyl is replaced with an oxygen (Fig. 9.23), also has two forms, but they do not contribute equally because they are different. Each has the same number of bonds, so we cannot choose the better representation that way. However, one has the charge on the relatively electronegative oxygen while in the other form it is on carbon. Form **A** is the better representation for the enolate anion, although both contribute. Mathematically, we would say that the weighting factor for the first form, c_1, is larger than that for the second, c_2.

FIGURE 9.23 The enolate anion.

In this case, $c_1 > c_2$

7. Watch out! Geometry is important. Is the "enolate" anion in Figure 9.24 resonance stabilized? On paper it certainly looks like it is, and there is at least a typographical relationship to the enolate anion we saw before in Figure 9.23. But we have been fooled again by the two-dimensional surface. The three-dimensional structure of the cage molecule prevents substantial overlap of the orbitals. Even though the paper doesn't protest when you draw the second resonance form, in reality, it is not there. The charge is borne on only one carbon—the bridgehead—and is not shared by another. Build a model and you will see it easily.

FIGURE 9.24 In this "enolate" anion, the orbital containing the pair of electrons does not overlap well with the p orbitals of the carbonyl group. There is no resonance, no matter what the deceptive two-dimensional surface says.

WORKED PROBLEM 9.5 Add dots for the electron pairs and write resonance forms for the following structures:

(a)

$^-CH_2\!-\!C\!\equiv\!N$

(b)

(c)

(d)

(e) (f) (g)

ANSWER (f)

PROBLEM 9.6 How many signals will the following species show in their ^{13}C NMR spectra? In fact, not all of these species are stable enough to be detected by NMR, but in each case it is possible to predict what would be (will be?) seen.

(a)

(b)

(c)

(d) (e) (f)

PROBLEM 9.7 Write Lewis structures and resonance forms for the following compounds. If you have problems seeing the structures of some of these molecules, see inside the front cover of this book.

(a) $HONO_2$ (b) $^-OSO_2OH$ (c) CH_3COO^-

WORKED PROBLEM 9.8 Which of the following pairs of structures are not resonance forms? Why not? You may have to add dots to make good Lewis structures first.

(a)

(d)

(b)

(e)

(c)

(f)

ANSWER (b) These are not resonance forms. Atoms have been moved, not just electrons. These two molecules are in equilibrium. Molecules related through the change of position of a single hydrogen are called "tautomers."

PROBLEM 9.9 In the following pairs of resonance forms, assign weighting factors. In each case, tell us which form you think is more important, and therefore a better representation of the structure. You will have to add dots to make good Lewis structures first.

(a)

(c)

(b)

(d)

9.6 RESONANCE AND THE STABILITY OF CARBOCATIONS

We have already seen one example of how an atom capable of delocalizing electrons can help share a charge and influence the regiochemistry of an addition reaction. Addition of hydrogen chloride to vinyl chloride gives only 1,1-dichloroethane *not* 1,2-dichloroethane (Fig. 9.5). Protonation occurs so as to produce a carbocation adjacent to the chlorine, which is far more stable than the alternative in which the positive charge sits localized on a primary carbon (Fig. 9.25).

FIGURE 9.25 In the addition of hydrogen chloride to vinyl chloride, it is formation of the more stable, resonance-stabilized carbocation that determines the product.

In Section 9.2 (p. 384), we looked at the reaction of the symmetrically substituted 2,3-dimethyl-2-butene with hydrogen chloride. In the formation of the carbocation, there was no choice to be made—only one cation could be formed. When the less symmetrical alkene 2-methyl-1-butene reacts with hydrogen chloride there are two possible carbocations. The only chloride formed is the one produced from the more stable, tertiary carbocation. The product is the one in which the hydrogen has added to the less substituted end of the alkene and the chlorine to the more substituted end (Fig. 9.26).

Unsymmetrical alkenes such as 2-pentene, from which two secondary carbocations of roughly equal stability can be formed, do give two products in comparable amounts (Fig. 9.27).

These phenomena have been known a long time, and were first correlated by the Russian chemist Vladimir V. Markovnikov (1838–1904) in 1870. The observation that the more substituted halide is formed now bears his name in **Markovnikov's rule**. Rules are useless (or worse) unless we understand them, so let's see why Markovnikov's rule works.

There is only one cation possible when 2,3-dimethyl-2-butene is protonated by HCl

In the unsymmetrical system, it is the more stable carbocation that leads to product:

Tertiary carbocation (more stable)

Primary carbocation (less stable)

FIGURE 9.26 The regiospecific addition of HCl to 2-methyl-1-butene.

Secondary carbocations

(45–48%)

(52–55%)

FIGURE 9.27 Two products are formed from the reaction of HCl and 2-pentene.

The results of the previous reactions show that the mechanism of HX addition must involve formation of the more substituted cation (Fig. 9.26), as the less substituted cation would lead to a product that is not observed, or is only formed in minor amounts. In turn, a reasonable presumption is that the more substituted a carbocation is, the more stable it is. As first seen in Chapters 3 and 7 (pp. 140, 305), such is the case, and Figure 9.28 shows again the order of carbocation stability. *Remember*: These differences in stability are large. In practice, reactions in which tertiary and secondary carbocations are intermediates are common, but primary and methyl carbocations cannot be formed.

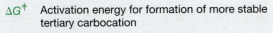

Methyl   Primary   Secondary   Tertiary

Cation stability (R = alkyl group)

FIGURE 9.28 Stability order for carbocations.

How does the difference in energy of these two carbocations, the tertiary and secondary cationic intermediates in the addition reaction, become translated into a preference for one product over the other? The relative rates of the two possible protonation steps will not depend directly on the relative energies of the two carbocations themselves, but on the energy difference between the two transition states leading to them ($\Delta\Delta G^{\ddagger}$ in Fig. 9.29). But both protonation steps are endothermic and it is likely that the transition states will bear a strong resemblance to the cations. (Remember the Hammond postulate, "For an endothermic reaction the transition state will resemble the product," Chapter 8, p. 369.) In these two transition states, positive charge is building up on carbon. There are partial positive charges in the transition states. *A tertiary partial positive charge will be more stable than a secondary partial positive charge, just as a full positive charge on a tertiary carbon is more stable than a full positive charge on a secondary carbon.*

$\Delta G^{\ddagger}$ Activation energy for formation of more stable tertiary carbocation

$\Delta G^{\ddagger}$ Activation energy for formation of less stable secondary carbocation

$\Delta\Delta G^{\ddagger}$ Difference in energies of transition states is the difference in activation energies

FIGURE 9.29 The transition states for formation of the high-energy carbocations will contain partial positive charges (δ^{+}). The same order of stability as holds for full positive charges will apply to partial positive charges. A δ^{+} on a tertiary carbon is more stable than a δ^{+} on a secondary carbon, and so on.

Reaction progress

Once the carbocation is formed, the structure of the product is determined. Capture of the tertiary carbocation leads to the observed product, in which X is attached to the more substituted carbon of the starting alkene (the site of the tertiary positive charge in the cation). Similarly, the secondary carbocation would lead inevitably to a product in which X is attached to the less substituted carbon (Fig. 9.29).

In Chapters 3 and 7, we avoided the issue, but it is now time to deal with the question of why alkyl substitution stabilizes carbocations. Our mechanistic hypothesis is consistent with the idea that the more substituted halide is produced because it is the result of the initial formation of the more substituted (and hence more stable) carbocation (Fig. 9.30).

FIGURE 9.30 For an unsymmetrical alkene, protonation to give the more stable carbocation will always lead to the product in which the nucleophile, here chloride, is attached to the more substituted carbon.

But this consistency doesn't answer anything; we have just pushed the question further back. *Why* are more substituted carbocations more stable than less substituted ones? To answer such questions, we start with structure. Carbocations are flat, and the central trigonal carbon is sp^2 hybridized. Figure 9.31 contrasts the methyl cation ($^+CH_3$) with the ethyl cation ($^+CH_2CH_3$).

Ethyl cation ($H_2\overset{+}{C}$—CH_3)

In the ethyl cation, a C—H bond can overlap with the empty $2p$ orbital; this filled–empty orbital interaction is stabilizing

Methyl cation ($\overset{+}{C}H_3$)

In the methyl cation, no such stabilization is possible!

2p

Empty

Energy

C—H

Filled σ bond

FIGURE 9.31 The ethyl cation is stabilized by a filled–empty orbital overlap.

Note that in the ethyl cation there is a filled carbon–hydrogen bond that can overlap with the empty *p* orbital. This phenomenon has the absolutely grotesque name of **hyperconjugation**, but there is nothing particularly "hyper" about it. It is a simple effect and can be rationalized in familiar molecular orbital or resonance terms. The empty–filled interaction is stabilizing (Fig. 9.31) and nothing like it is available to the methyl cation. The more interactions of this kind that are possible, the more stable the ion. Thus the *tert*-butyl cation is most stable, the isopropyl cation next, and so on.

Alternatively, we could write the following kind of resonance form, which shows the charge delocalization from carbon to hydrogen especially well (Fig. 9.32).

FIGURE 9.32 A resonance description of the ethyl cation ($^+CH_2CH_3$).

It's important to see that the second carbon–carbon bond in the resonance form of Figure 9.32 is *not* identical to the π bond of an alkene. It is not formed from $2p$–$2p$ overlap but rather from $2p$–sp^3 overlap (Fig. 9.33). Be sure you see the difference between these two fundamentally different, but similar-looking structures.

FIGURE 9.33 A detailed look at hyperconjugative stabilization of the ethyl cation.

The π bond in ethylene is made up of $2p$–$2p$ overlap

In the "hyperconjugative" resonance form for the ethyl cation, the "double" part of the double bond is not made up of $2p$–$2p$ overlap, but of $2p$–sp^3 overlap

PROBLEM 9.10 It is often claimed that the *tert*-butyl cation has 10 resonance forms, 1 in which the carbon is positive and 9 others in which each hydrogen bears the positive charge. Is this claim true?

PROBLEM 9.11 In Section 3.6 (p. 115), we described the stabilization of alkenes by attached alkyl groups in different terms, focusing on the σ bond system. We could construct a similar treatment here. Provide a justification of the relative stability of differently substituted carbocations using an analysis of the σ system.

PROBLEM 9.12 Do you think the π part of the double bond in the resonance form in Figure 9.33 is stronger or weaker than a normal π bond?

We have just seen that carbocations are stabilized by delocalization through the overlap of filled and empty orbitals. One example of this effect is fairly straightforward, as nonbonding, or lone-pair, electrons on adjacent atoms interact with an empty orbital. Less obvious is the last phenomenon, the overlap of filled σ orbitals of carbon–hydrogen bonds with empty orbitals, called hyperconjugation. Now we proceed to other factors influencing the stability of cations.

9.7   INDUCTIVE EFFECTS ON ADDITION REACTIONS

Polar bonds in a molecule can have a major effect on both the rate and the regiochemistry of an addition reaction. Let's look first at the series of ethylenes in Figure 9.34, which shows typical rates of addition for a series of substituted ethylenes. The reaction is, in fact, a hydration reaction in which HOH adds across the double bond.

FIGURE 9.34 Three addition reactions.

	Relative rate of addition at 25 °C
Ethylene	1
Propene	2×10^6
Methyl vinyl ether	5×10^{14}

Ethylene itself will protonate only very slowly. Propene will surely protonate so as to give the more stable secondary cation, and the methoxyl group of methyl vinyl ether will provide resonance stabilization to an adjacent positive charge. Resonance stabilization by the methyl and methoxyl groups of propene and methyl vinyl ether has a great accelerating effect on the rate of protonation, the rate-determining step in this reaction. Figure 9.34 shows the rates of protonation compared to that of unsubstituted ethylene itself.

PROBLEM 9.13 Why will the rate of protonation of ethylene be especially slow?

Substituents can slow reactions as well, especially if they are not in a position to provide resonance stabilization. Figure 9.35 shows typical rates for addition to some substituted alkenes. To see why bromine and chlorine so dramatically slow the rate of addition we must again look more closely at the bond from carbon to halogen (see p. 233 for an earlier look).

The two atoms in the bond, carbon and halogen, are different, so the bond between them must be polarized. The electrons in the bond cannot be equally shared. Bromine and chlorine are far more electronegative than carbon (see Table 1.8, p. 14) and they will attract electrons strongly. The carbon–halogen bond will be polarized as shown in Figure 9.36. The carbon will bear a partial positive charge (δ^+) and the more electronegative halogen a partial negative charge (δ^-).

For the molecules in Figure 9.35, protonation in the Markovnikov sense places a full positive charge adjacent to the already partially positive carbon. This interaction will surely be destabilizing, and accounts for the decrease in rate for addition reactions to the bromine- and chlorine-substituted molecules (Fig. 9.37). Effects such as these, which are transmitted through the σ bonds, are called **inductive effects.**

R =	H	Br	Cl
Rate	1	10^{-4}	10^{-4}

FIGURE 9.35 Rates of addition to three alkenes.

FIGURE 9.36 The polar C—X bond.

Notice the bad electrostatic interaction between δ^+ and the + charge

FIGURE 9.37 Protonation gives an intermediate destabilized by the electrostatic interaction between the newly formed full positive charge and the partial positive charge (δ^+).

But wouldn't the methoxyl group of methyl vinyl ether act in the same way? Oxygen is also more electronegative than carbon, and wouldn't this act to discourage any addition that puts more positive charge on the already partially positive carbon adjacent to oxygen? Why does addition to methyl vinyl ether occur so rapidly (Fig. 9.34) in the Markovnikov sense? Figure 9.38 frames the question, and answers it as well.

FIGURE 9.38 Stabilizing resonance effects and destabilizing inductive effects can "fight" each other. Generally, resonance stabilization is more important.

The C–O σ bond in methyl vinyl ether is polarized so that a partial positive charge is on carbon; this will make protonation on the adjacent carbon more difficult

This intermediate is strongly stabilized by resonance and the stabilizing resonance effect outweighs the inductive destabilizing effect

The answer is simply that the resonance effect of methoxy overwhelms the inductive effect. The two effects *are* in competition, and the stronger resonance effect wins out. Now let's summarize, then look at a number of related addition reactions.

SUMMARY

Protonation (or addition of any Lewis acid) to an unsymmetrical double bond leads preferentially to the more stable carbocation. The energy of carbocations is strongly influenced by stabilizing resonance effects. Substituents bearing lone-pair electrons are especially effective at stabilization, but even alkyl groups can donate electrons to the empty $2p$ orbital of a carbocation. In cases in which inductive and resonance effects oppose each other, resonance usually wins.

9.8   H—X ADDITION REACTIONS: HYDRATION

If H—Br and H—Cl add to double bonds, it should not be surprising that other acidic H—X compounds undergo similar addition reactions. Hydrogen iodide is a simple example, and H—I addition rather closely resembles the reactions we have already studied in this chapter. Markovnikov's rule is followed, for example (Fig. 9.39).

FIGURE 9.39 Addition of HI to 1-octene in Markovnikov fashion.

Water (H—O—H) is an H—X reagent, but, as we saw in Chapter 3 (p. 142) it is not normally a strong enough acid to undergo addition to simple alkenes at a useful rate. If one mixes water and a typical alkene, no reaction occurs. The difference is that H—Cl and H—Br are strong enough acids to protonate the alkene and start things going. Water is not. The pK_a values of H—Br, H—Cl, and H—O—H are −9, −8, and +15.7, respectively. Water is a less powerful acid than H—Br and H—Cl by factors of about 10^{25} and 10^{24}! Acid must be added to protonate the alkene and thus generate a carbocation that can continue the reaction. The acid **catalyst** in this reaction sets things going. The catalyst has no effect on the position of the final equilibrium, but it does lower the energy requirement for generating product. A catalyst provides a new mechanistic pathway for product formation. It is able to carve a new pathway along a lower slope of the mountain we first encountered in Section 8.6 (Fig. 8.20, p. 364; Fig. 9.40).

The lowest energy path from A to B

A catalyst provides a lower energy path, but it does not alter the energy of starting material and product, A and B; rather, it changes the energy of the transition state, or transition states, in the reaction

Village A

Village A

Village B

Village B

FIGURE 9.40 A catalyst does not change the energy of either starting material or product, but does provide a lower energy path between them.

As can be seen in Figure 9.41, the catalyst H_3O^+ is used up in the first step of the reaction, only to be regenerated in the last step. A number of acids can be used in the catalytic role; sulfuric (H_2SO_4 or HO—SO_2—OH) and nitric (HNO_3 or HO—NO_2) are commonly used to promote the addition reaction. In water containing a little sulfuric acid, 2,3-dimethyl-2-butene is protonated to generate the tertiary carbocation. This intermediate is then attacked by water (not hydroxide, $^-$OH) to give the **oxonium ion**, which can be deprotonated by water to regenerate the catalyst, H_3O^+, and produce a molecule of an **alcohol**, a compound of the general formula R—OH (Fig. 9.41).

Alkene hydration

FIGURE 9.41 The mechanism of hydration of 2,3-dimethyl-2-butene.

$H_3\overset{+}{\overset{..}{O}}$: Catalyst used up in this step

$H_3\overset{+}{\overset{..}{O}}$: Catalyst regenerated in this step

An oxonium ion

An alcohol

www

This mechanism would predict, indeed require, that the Markovnikov rule be followed, and that sets up a mechanistic test. We will examine the hydration of isobutene to probe the regiochemistry of the reaction. Before we even start, it is worth thinking a bit about the results. What will we know after we have done the experiment? There are only two general results possible: Either the Markovnikov rule is followed or it is not (Fig. 9.42).

FIGURE 9.42 Three possible regiochemical results from the hydration of isobutene.

If it is not, we can say with certainty that the mechanism we have suggested for this reaction is wrong. A crucial step in our mechanism involves the formation of a carbocation. When there is a choice, as there certainly is with isobutene, the more stable carbocation must be formed. Once formed, it must lead to the product in which the hydroxyl group is attached to the more substituted carbon (Fig. 9.43).

FIGURE 9.43 The mechanism for Markovnikov addition of water to isobutene. The product must be tert-butyl alcohol if this is indeed the mechanism.

If our postulated mechanism is correct, there is no other possible result. So if we get the "wrong" alcohol, or a mixture of alcohols [(b) and (c) in Fig. 9.42], we know the mechanism is wrong. It may need only a minor adjustment or it could require complete scrapping, but it cannot be correct as written.

More interesting, and less obvious, is what we know if our prediction turns out to be correct. Have we proved the mechanism? Absolutely not. Our mechanistic hypothesis remains just a hypothesis: a guess at how the reaction goes that is consistent with all experiments that have been done so far. There may be some other test experiment that would not work out as well. The sad fact is that no number of experiments we can do will suffice to prove our mechanism. We are doomed to be forever in doubt, and our mechanism to be forever suspect. There is no way out. Indeed all the mechanisms in this book are wrong, at least in the sense that they are broad-brush treatments that do not look closely enough at detail. Organic chemistry has come a long way since the days of the "rectangular hypothesis," which was little more than a bookkeeping device. In olden times it was common practice to write a reaction such as the S_N1 reaction of *tert*-butyl iodide with water in the way shown in Figure 9.44.

FIGURE 9.44 The "rectangular hypothesis" applied to the reaction of *tert*-butyl iodide with water.

To use the rectangular hypothesis, one simply surrounds the appropriate groups with a rectangle, which somehow removes the captured atoms and reconstitutes them, in this case as hydrogen iodide. This archaic device is scarcely a reasonable description of how the reaction occurs and has almost no predictive value. If you know one rectangle, you do not know them all! But let's not get too smug. We do know more than chemists of other generations. We know, for example, that carbocations are involved in the reactions we are currently studying, and molecular orbital theory gives us additional insight into the causes of these reactions. But we are still very ignorant, and I am certain that much of what we "know" is wrong, at least in detail. Chemists of the future will regard us as primitives, and they will be quite right. It doesn't matter, of course. Our job is to do the best we can; to add what we can to the fabric of knowledge and to provide some shoulders for those future chemists to stand on, as our earlier colleagues did for us.

In this case, things turn out well for our mechanistic hypothesis, and the Markovnikov rule is followed in hydration reactions.

There are variations on this reaction. Ethylene, for example, is hydrated only under more forcing conditions—concentrated sulfuric acid—and the initial product formed is not the alcohol, but ethyl hydrogen sulfate, which is converted in a second step into the alcohol (Fig. 9.45). It is not easy to be sure of the mechanism of this reaction. The concentrated sulfuric acid may be required to produce the very unstable primary carbocation, or a different, perhaps cyclic cation may be involved (see Problem 9.13).

FIGURE 9.45 Ethyl hydrogen sulfate is an intermediate in the hydration of ethylene.

SUMMARY

Water is not a strong enough acid to protonate a simple alkene. However, if a trace of acid catalyst is present, an initial protonation will be followed by addition of water to produce an oxonium ion. A final deprotonation gives the alcohol. As usual, you have to be careful to analyze what carbocation will be formed, and to consider structural, resonance, and inductive effects in making that determination.

9.9 DIMERIZATION AND POLYMERIZATION OF ALKENES

Let's look again at the hydration of isobutene (C_4H_8). What if we change the conditions a bit? What if both the temperature and concentration of sulfuric acid are increased? Figure 9.46 shows that two new products containing eight carbons are formed. These are dimeric products (C_8H_{16}), formed somehow by the combining of two molecules of isobutylene.

FIGURE 9.46 Dimeric products formed from isobutene.

Surely, increasing the acid concentration can only make protonation of the alkene more probable, so the first step of the reaction is unchanged. The initially formed carbocation finds itself in the presence of less water than it did in the reaction we looked at before. Accordingly, formation of the alcohol through reaction with water is less likely. We might ask what other Lewis bases are available for the strongly Lewis acidic cation? The most obvious one is isobutene itself. The *tert*-butyl cation can react with isobutene, just as did the proton. Both the proton (Brønsted) and the carbocation (Lewis) are acids and can add to the double bond to give a new cation (Fig. 9.47). There is no essential difference in the two reactions, and once more it will be the more stable, tertiary carbocation that will be formed on addition.

FIGURE 9.47 Many different acids add to isobutene (and other alkenes). Here we see H_3O^+ and the *tert*-butyl cation adding to isobutene to give new tertiary carbocations. In each case addition will be in the Markovnikov sense, as it will be the more stable carbocation that is preferentially formed.

Now we have a structure that contains eight carbons, as do the two new products. It looks as though we are on the right track. The only remaining trick is to see how to lose the "extra" proton and form the new alkenes. If a proton is lost from the two different positions (H_a and H_b) adjacent to the positive charge, we get the two products. We have seen similar deprotonations before in our discussion of the E1 reaction (Chapter 7, p. 312), and we have also seen the reverse reaction. Loss of these protons is just the reverse of protonation of an alkene (Fig. 9.48).

FIGURE 9.48 The two products are formed by deprotonation of the intermediate cation.

The proton can't just leave, however. It must be removed by a base. In concentrated sulfuric acid, strong bases do not abound. There are some bases present, however, and the carbocation is a high-energy species, and thus rather easily deprotonated. Under these conditions both water and bisulfate ion are capable of assisting the deprotonation (Fig. 9.48).

Note the exact correspondence of the steps involved in alkene protonation and deprotonation. The two reactions are just the two sides of an equilibrium (Fig. 9.49).

FIGURE 9.49 Protonation and deprotonation are just two sides of an equilibrium.

If the concentration of acid is even higher, the base concentration will be even lower and both alcohol formation and deprotonation will be slowed (both require water to act as Lewis base). There is no reason why the dimerization process cannot continue indefinitely under such conditions, leading first to a trimer (a molecule formed from three molecules of isobutylene) and, eventually, to a polymer (a molecule formed by combination of many starting alkene molecules). The cation-induced polymerization is called, logically enough, **cationic polymerization**, and ends only when the alkene concentration decreases sufficiently to slow the addition step (Fig. 9.50).

Alkene polymerization

FIGURE 9.50 The beginning stages of the cationic polymerization of isobutene.

PROBLEM 9.14 Treatment of 2-ethyl-1-butene with H_3O^+/H_2O leads to three products of the formula $C_{12}H_{24}$. Give structures and a mechanism for their formation.

9.10   REARRANGEMENTS DURING H−X ADDITION TO ALKENES

When H−Cl adds to isopropylethylene (3-methyl-1-butene), a quite surprising thing happens. We do get some of the product we expect, 2-chloro-3-methylbutane, but it's not even the major compound formed. The major product is 2-chloro-2-methylbutane (Fig. 9.51).

3-Methyl-1-butene   **2-Chloro-3-methylbutane**   **2-Chloro-2-methylbutane**
                          (40%)   (60%)

FIGURE 9.51 Reaction of HCl with 3-methyl-1-butene gives a new product in addition to the expected one.

The obvious question is, Why? The danger is that our mechanism for addition will have to be severely changed to accommodate these new data. In fact, we will be able to graft a small modification onto our general mechanism for addition reactions and avoid drastic changes. The way to attack a problem of this kind—the explanation of the formation of an unusual product—is to write out carefully the mechanism for the reaction as we have developed it so far, and see if we can see a way to rationalize the new product. In this case, protonation of 3-methyl-1-butene can give either a secondary or primary carbocation. There is no real choice here—

the secondary carbocation is far lower energy than the hideously unstable primary carbocation, and will surely be the one formed (Fig. 9.52).

FIGURE 9.52 No real choice here—the more stable secondary carbocation is preferred.

If the chloride ion simply adds to the positively charged carbon, we get the minor product of the reaction, and the one we surely expected. To get the major product, a **rearrangement** must take place. A hydrogen must move, with its pair of electrons, from the carbon adjacent to the cation to the positively charged carbon itself (Fig. 9.53). Such migrations of hydrogen with a pair of electrons are called **hydride shifts**.

FIGURE 9.53 The major product is formed through a mechanism involving the migration of hydrogen with its pair of electrons (hydride, H:⁻) to give the more stable tertiary carbocation from the less stable secondary ion.

Why should this rearrangement happen? Note that in such a process we have generated a lower energy tertiary carbocation from a higher energy secondary carbocation. The cation is certainly "better off" (lower in energy) thermodynamically if the hydride moves. In practice, such rearrangements are quite common. Hydride shifts to produce more stable cations from less stable ones are easy. Indeed, not only hydride shifts but alkyl shifts, called **Wagner–Meerwein rearrangements** (Georg Wagner, 1849–1903; Hans Lebrecht Meerwein, 1879–1965), are also common. For example, 3,3-dimethyl-1-butene adds hydrogen chloride to give not only 2-chloro-3,3-dimethylbutane but also 2-chloro-2,3-dimethylbutane. The new, rearranged compound results from an alkyl shift—in this case the migration of a methyl group with its pair of electrons (Fig. 9.54). Why is this reaction called an "alkyl shift," when by analogy to "hydride shift" it should be "alkide shift"? I don't know, but it is!

FIGURE 9.54 Formation of the major product involves a mechanism in which a shift of a methyl group with its pair of electrons generates a more stable tertiary carbocation from a less stable secondary carbocation.

The minor, unrearranged, product is formed from the secondary carbocation (path a)

2-Chloro-3,3-dimethylbutane
(25–40%)

(b) | :CH₃ shift

The major product is formed from the rearranged, tertiary carbocation (path b)

2-Chloro-2,3-dimethylbutane
(60–75%)

PROBLEM 9.15 Write a mechanism for the following reaction:

PROBLEM 9.16 A hydride shift to a positively charged carbon is just another example of a Lewis base–Lewis acid reaction. Although it looks very different from the more obvious Lewis acid–nucleophile reactions we have studied, it really is quite similar. In this case, there is a very strong and strategically located Lewis acid and a weak, but nonetheless reactive enough nucleophile. Identify the nucleophile and Lewis acid in this reaction.

Such rearrangements are so common that they have come to be diagnostic for reactions involving carbocations. Molecules prone to such rearrangements are used to test for carbocation involvement. Here is an example. Attempted S_N1 solvolysis of neopentyl iodide (1-iodo-2,2-dimethylpropane) in water leads not to neopentyl alcohol, as might have been anticipated, but instead to a rearranged alcohol, 2-methyl-2-butanol (Fig. 9.55).

Carbocation rearrangement E1

Neopentyl alcohol

Neopentyl iodide

2-Methyl-2-butanol
(97%)

FIGURE 9.55 The solvolysis of neopentyl iodide does not take a "normal" course.

These results imply the scenario outlined in Figure 9.56, in which an unstable primary carbocation rearranges to a much more stable tertiary carbocation through the shift of a methyl group. In fact, the presence of the primary ion is so unsettling—they are most unstable and not likely to be formed—that a variant of this mechanism has been proposed in which the methyl group migrates as the iodide departs (Fig. 9.56).

Neopentyl iodide   **Primary carbocation**   **Tertiary carbocation**   **2-Methyl-2-butanol**

Direct formation of the tertiary
carbocation; the methyl group
migrates as iodide leaves

FIGURE 9.56 Rearrangements are common in many reactions in which carbocations are involved.

Under certain conditions of very low temperature and highly polar but nonnucleophilic solvents, some carbocations can be observed spectroscopically. Under such conditions it is possible to determine that many carbocations are rapidly rearranging as we observe them (Fig. 9.57). For example, the two tertiary carbocations in the figure are rapidly interconverting at −60 °C.

FIGURE 9.57 Rearranging tertiary carbocations.

9.11 HYDROBORATION

Alkene hydroboration

Now we pass on to another addition reaction, **hydroboration**. The mechanism is not simple, and it will take some effort to become comfortable with all the details. The mechanistic discussion does make several worthwhile points, so it's worth your attention for that reason alone. But there is another, more compelling reason to master the hydroboration reaction—it is one of the most useful of synthetic reactions. Indeed H. C. Brown (b. 1912) won the Nobel prize of 1979 for the development of hydroboration. Consider the following seemingly simple synthetic task outlined in Problem 9.17.

WORKED PROBLEM 9.17 Try to provide a synthesis of 2-methyl-1-propanol. You may start from any alkene.

ANSWER Well, there isn't one (yet). As of now, you have no way to make this simple alcohol. If we start from the obvious alkene, isobutylene, the hydration reaction we learned in this chapter (p. 406) can only give the product of Markovnikov addition, *tert*-butyl alcohol. This nonanswer to such a simple question points out how limited our synthetic skills are so far.

The hydroboration reaction solves both Problem 9.17 and the general difficulty, so let's see how it works.

In the previous sections, we saw both the protonation of alkenes and the addition of a carbocation to an alkene to generate a new carbocation. The familiar theme of overlap between filled (alkene π) and empty (carbon $2p$) orbitals was recapitulated. "Lewis bases react with Lewis acids." Other reagents containing empty p orbitals, all good Lewis acids, might also be expected to add to alkenes, and so they do.

We know another kind of molecule that has an empty $2p$ orbital, the trigonal boranes, BF_3 and BH_3 (Chapter 2, p. 52; Fig. 9.58). There is no positive charge on these compounds, but they should still be strong Lewis acids.

Borane   **Boron trifluoride**

FIGURE 9.58 Both borane and boron trifluoride are Lewis acids and will react like cations.

PROBLEM 9.18 Verify that boron in BF_3 and BH_3 is neutral, and that each boron has an empty $2p$ orbital.

Although BF$_3$ is known and even commercially available, free BH$_3$ is unavailable as it spontaneously dimerizes to diborane (B$_2$H$_6$), an unpleasantly flammable and toxic gas. Diborane (H$_3$B—BH$_3$ = B$_2$H$_6$) does not have a structure similar to ethane, H$_3$C—CH$_3$. An ethane-like structure requires 14 electrons for 7 boron–hydrogen and boron–boron σ bonds. The 6 hydrogens contribute 6 electrons and the 2 borons another 6 to give a total of 12, so we are 2 electrons shy of the required number. The real structure is shown in Figure 9.59, along with a resonance formulation. The molecule B$_2$H$_6$ is a hybrid of the two resonance forms **A** and **A′**.

This structure cannot be B$_2$H$_6$ because there are only 12 electrons available for bonding in this molecule, and this formulation requires 14

A   **A′**

FIGURE 9.59 The real structure of diborane.

In this molecule, partial bonds (shown as dashed bonds) are used to connect the atoms. This kind of bonding is more common than once thought. It abounds in inorganic chemistry and in the organic chemistry of electron-deficient species such as carbocations as well.

When diborane dissolves in diethyl ether (CH$_3$CH$_2$—O—CH$_2$CH$_3$) a complex of ether and borane is formed. This complex of borane and an ether can be used as a source of BH$_3$. Often the cyclic ether THF is used (Fig. 9.60). The boron trifluoride–ether complex is so stable that it can even be distilled.

Furan   **Tetrahydrofuran (THF)**

FIGURE 9.60 Furan and THF.

WORKED PROBLEM 9.19 Propose a structure for the ether–borane complex, and suggest a mechanism for its formation.

ANSWER Borane is a Lewis acid, as the boron atom has an empty 2p orbital. An ether, here shown as THF, is a Lewis base. The reaction between the Lewis base and Lewis acid produces the borane–THF complex.

Lewis acid   Lewis base   **Borane–THF complex**

When the borane–ether complex is allowed to react with an alkene, there is a rapid addition of the borane across the double bond. The initial product is an alkylborane and the process is called hydroboration (Fig. 9.61).

FIGURE 9.61 A schematic hydroboration. The elements of BH_3 have been added across the double bond.

FIGURE 9.62 A possible first step in addition of BH_3 to an alkene. The π bond of the alkene acts as nucleophile and reacts with the strong Lewis acid, BH_3.

As in the addition of a carbocation to a double bond, the initial interaction is between the empty $2p$ orbital (in this case on neutral boron, not positive carbon) and the filled π orbital of the alkene (Fig. 9.62).

If the mechanism were directly parallel to that of cation addition, we would form the charge-separated molecule shown in Figure 9.62. A hydride (H:$^-$) could then be delivered from the negative boron to the positively charged carbon to complete the addition (Fig. 9.63).

FIGURE 9.63 A possible second step in the hydroboration reaction is transfer of a hydride ion (H:$^-$).

But the mechanism is *not* exactly parallel to what we have seen before. In this case, that hydride is transferred as the carbon–boron bond is made. This reaction occurs in a single step: There is no intermediate. In the transition state, partial bonds from carbon to boron and carbon to hydrogen are formed (Fig. 9.64). Let's see how we know this mechanism satisfies the experimental data, and, as we do this, why the more conventional mechanism is inadequate.

Transition state with partial bonds (dashed)

FIGURE 9.64 A one-step mechanism for hydroboration.

First of all, there are no rearrangements during hydroboration. The one-step mechanism outlined in Figure 9.64 cannot lead to rearrangements, but the open carbocation of Figures 9.62 and 9.63 would surely do so. For example, the hydroboration of 3,3-dimethyl-1-butene leads exclusively to borane **A**. Were a carbocation intermediate involved, rearrangement of the secondary carbocation to the more stable tertiary carbocation would seem inevitable, and product **B** should be observed (Fig. 9.65).

FIGURE 9.65 An open cation should lead to rearrangements in hydroboration, but none can be found.

But the reaction of Figure 9.65 shows a regiospecific hydroboration. Only one of two possible product boranes is formed. This phenomenon is general—in hydroboration the boron becomes attached to the less substituted carbon of the double bond. Our one-step mechanism must be able to explain this regiochemical preference.

There are at least two factors, one simple, the other more complicated, that influence the regiospecific addition in the hydroboration reaction. Consider the reaction of BH_3 with 2-methyl-1-butene. First of all, there are steric differences in the two possible concerted additions (Fig. 9.66). In the addition leading to the observed product, the larger end of the adding molecule, the BH_2 end, is nearer to the smaller hydrogens than to the larger alkyl groups. In the path leading to the product that is not observed, the larger end opposes the relatively large methyl and ethyl groups.

FIGURE 9.66 Steric factors favor addition of the larger BH_2 group to the less congested end of the alkene.

The sterically less congested arrangement leads to the observed product

Transition state

This sterically more congested arrangement leads to the product that is not observed

Transition state

Tertiary δ⁺
(more stable)

Observed product

Primary δ⁺
(less stable)

This product is not formed

FIGURE 9.67 In the two possible transition states for the concerted (one-step) addition, there will be partial positive charge developed on carbon. This δ⁺ will be more stable at the more substituted position. Polar factors contribute to the observed regiochemistry of the hydroboration reaction.

That's the simple part; the more subtle effect has to do with the way in which the two new bonds are made in the transition state for the hydroboration reaction. The two new bonds are formed in the same step, without a cationic intermediate, but there is no need for the new bonds to develop to the same extent as the reaction proceeds. Indeed, in a philosophical sense, they cannot, for they are different. In the transition state, one bond will be further along in its formation than the other. This differential progress can be symbolized by different length bonds as in Figure 9.67. In such a mechanism, a full positive charge is not developed on carbon, *but a partial one is.* All the factors that operate to stabilize a full positive charge will operate to stabilize a partial positive charge. A more substituted partial positive charge is more stable than a less substituted partial positive charge, and this favors the observed product. So, the initial product of hydroboration of an alkene is the monoalkylborane formed by syn addition in which boron becomes attached to the less substituted end of the alkene partner (Fig. 9.67). Is this anti-Markovnikov addition? No!

Here is an instance where rules can be confusing. When we say "Markovnikov addition," we are used to seeing the hydrogen attached to the less substituted end of the alkene and the X group to the other end (Fig. 9.68).

FIGURE 9.68 Typical Markovnikov addition in which the more stable cation is captured to give the product in which H is attached to the less substituted end of the original alkene, and X is attached to the more substituted end.

When $H-BH_2$ adds, it is the boron that is the positive end of the dipole, the Lewis acid in the addition. It is the boron, not the hydrogen, that becomes attached to the less substituted end of the alkene. As Figure 9.69 tries to show, it is the terminology that gets complicated, not the reaction mechanisms. There is danger in learning rules; it is much safer to understand the reaction mechanism.

FIGURE 9.69 This reaction is still Markovnikov addition. The positive end of the dipole, boron in this case, adds to the less substituted end of the alkene. Positive charge develops on the more substituted carbon.

Transition state

There is another reason we know that the one-step mechanism is correct, and it involves the stereochemistry of the addition reaction.

Hydroboration can be shown to proceed in syn fashion. The two pieces, H and BH_2, are delivered from the same side of the alkene. Figure 9.70 shows two examples, the hydroborations of 1-methylcyclopentene and *cis*-1,2-dideuterio-1-hexene. These reactions exclusively produce the products of syn addition shown in Figure 9.70.

cis-1,2-Dideuterio-1-hexene

1-Methylcyclopentene

FIGURE 9.70 These two stereochemical experiments show that addition of BH_3 proceeds in a syn fashion. The H and BH_2 groups are delivered to the same side of the alkene (R = cyclohexyl).

Of course, the one-step addition process must involve syn addition, as both new bonds, the one to boron and the one to hydrogen, are made at the same time. Figure 9.71 shows the observed syn addition to a deuterated 1-hexene. On the other hand, a mechanism involving an open cation has great problems accounting for syn addition.

Transition state

FIGURE 9.71 Syn addition is accommodated by a mechanism in which both new bonds are forming in the transition state. There are no intermediates, and therefore no chance for rotation and scrambling of stereochemistry.

If a planar cation were formed on addition to the deuterated 1-hexene, we would expect the rapid ($\sim 10^{11}$ s^{-1}) rotations about carbon–carbon single bonds to produce two products, the diastereomers shown in Figure 9.72.

FIGURE 9.72 Rotation about carbon–carbon bonds is very fast and the two-step mechanism predicts that two products should be formed from the hydroboration of *cis*-1,2-dideuterio-1-hexene. Both syn and anti addition products should be formed. However, only the product of syn addition is found (R = cyclohexyl).

SUMMARY

The current mechanistic hypothesis, a concerted, one-step syn addition to alkenes in which the two new bonds are partially formed to different extents in the transition state, nicely rationalizes all the experimental data. A mechanism involving formation of an open cation followed by hydride transfer fails to account for the observed syn stereochemistry of addition, which is demanded by the one-step mechanism.

WORKED PROBLEM 9.20 Draw Energy versus Reaction progress diagrams for the concerted and unmodified, two-step mechanisms for the addition of BH_3 to alkenes.

ANSWER The crucial difference is the presence of an intermediate in the two-step process. The one-step (concerted) reaction simply passes from starting material to product over a single transition state.

This example is typical of the use of stereochemistry in determining reaction mechanism. One reason we spent so much time on the stereochemical relationships of molecules of various shapes (diastereomers, enantiomers, and meso forms) was that we will use these relationships over and over again as we dig into questions of reaction mechanism. If it is not obvious what we are talking about in the last few paragraphs, and if Figures 9.70–9.72 are at all unclear, be sure to go back to Chapter 4 and work over the stereochemical arguments again. Otherwise, you run the risk of being lost in further discussion. The key points to see are the difference between the two hypothetical products shown in Figure 9.72 and their required formation from the rotating cation. A good test is to see if you can do Problems 9.22 and 9.23 easily. If you can, you are in good shape.

PROBLEM 9.21 Bromine adds to alkenes to give 1,2-dibromo compounds. From the following data on the reaction of bromine with the 2-butenes, determine if the addition is syn or anti. This problem does not pose a mechanism question—you do not have to draw arrow formalisms. You only need to determine from the structure of the product the direction from which the two bromines have added.

PROBLEM 9.22 Devise a way to determine the stereochemistry of the addition of hydrogen bromide to alkenes. Assume you can make any necessary starting material.

A single addition isn't the end of the hydroboration story. The initially produced monoalkylborane still has a trigonal boron; therefore, it still has an empty $2p$ orbital (Fig. 9.73). It's still a Lewis acid, and it can participate in the hydroboration reaction to give the product of two hydroborations of the alkene, a dialkylborane. Nor must the reaction stop here. The dialkylborane is also a Lewis acid, and it can do one more hydroboration to give a trialkylborane (Fig. 9.73). Now, however, all the original hydrogens are used up and there can be no further hydroborations.

Still a Lewis acid!

FIGURE 9.73 The initially formed alkylboranes are Lewis acids and can undergo further hydroborations.

In practice, the number of hydroborations depends on the size of the alkyl groups in the alkene. When these are rather large, further reaction is retarded by steric effects—the alkyl groups just get in the way. For example, 2-methyl-2-butene hydroborates only twice, and 2,3-dimethyl-2-butene only once (Fig. 9.74).

FIGURE 9.74 The number of hydroborations depends on steric effects.

WORKED PROBLEM 9.23 Provide a mechanism for the following reaction:

ANSWER The first hydroboration goes in normal fashion.

Now a second, intramolecular hydroboration takes place as the remaining double bond in the ring attacks the boron. It takes some unraveling to see the structure of the final product. It is important to begin to hone your skills at this kind of thing, and this problem provides practice.

9.12 HYDROBORATION IN SYNTHESIS: ALCOHOL FORMATION

It would be presumptuous to attempt to summarize fully the utility of the hydroboration reaction here. There are many variations of this reaction, each designed to accomplish a specific synthetic transformation. A book could be, and has been, written on the subject by Professor H. C. Brown. I'll only mention one especially important reaction here, which leads to a new synthesis of alcohols (recall Problem 9.17, p. 416). When one molecule of trialkylborane is treated with hydrogen peroxide (H_2O_2) and hydroxide ion (HO^-), three molecules of an alcohol are formed along with boric acid (Fig. 9.75).

$$R_3B \xrightarrow{H_2O_2 / \, \ddot{H O} \ddot{}^-} 3 \, R\!-\!\ddot{O}H \, + \, B(\ddot{O}H)_3$$

Alcohol   **Boric acid**

FIGURE 9.75 Alcohols are formed when alkylboranes are treated with basic hydrogen peroxide.

We are not yet up to a full discussion of the mechanism of this reaction, which is outlined in Figure 9.76, but a number of important points are within our powers.

FIGURE 9.76 The mechanism of alcohol formation from alkylboranes.

The first step, the addition of peroxidate ion (H—O—O⁻) to the borane, is just the reaction of a filled orbital—on oxygen—with the empty $2p$ orbital on boron. It's another nucleophile–Lewis acid reaction. A discussion of the later steps is deferred for now, but at least you can see that the boron–oxygen bonds in the boronic ester [B(OR)$_3$] should be especially strong ones and the reaction will be favored thermodynamically. Finally, in excess hydroxide the boronic ester equilibrates with boric acid [B(OH)$_3$] and the alcohol (ROH).

PROBLEM 9.24 Why are these boron–oxygen bonds especially strong?

PROBLEM 9.25 Write a mechanism for the formation of boric acid, B(OH)$_3$, from hydroxide ion, HO⁻, and a boronic ester, B(OR)$_3$.

For once, questions of mechanism are secondary to the utility of this reaction. Note one very important thing: In the overall reaction we have produced the *less* substituted alcohol from the alkene. The hydroxyl group enters at the position to which the boron was attached: the less substituted end of the alkene. This reaction is an overall anti-Markovnikov addition (Fig. 9.77).

FIGURE 9.77 Hydroboration–oxidation accomplishes the anti-Markovnikov addition of water to an alkene.

Our previous method for alcohol synthesis, hydration of alkenes, necessarily produced the *more* substituted alcohol. So now we have complementary synthetic methods for producing both possible alcohols from a given alkene. Direct hydration gives the more substituted product (Markovnikov addition) and the indirect hydroboration method gives the less substituted alcohol (anti-Markovnikov addition, Fig. 9.78). Be sure to add these reactions to your file-card collection of synthetically useful reactions.

FIGURE 9.78 Two addition reactions with different regiochemistries.

SUMMARY

Hydroboration/oxidation is a process that yields an alcohol that is the product of overall anti-Markovnikov addition. The mechanism of hydroboration is complex, but several lines of evidence have led to the picture we have of a one-step reaction with an unsymmetrical transition state in which one of the alkene's carbon atoms becomes partially positively charged (Figs. 9.64–9.69). The synthetic utility of this reaction is not complex at all. For unsymmetrical alkenes, hydroboration/oxidation leads to the less substituted alcohol.

PROBLEM 9.26 Devise syntheses for the following molecules from 2-methyl-2-butene. You may also use any inorganic reagent as well as organic compounds containing no more than one carbon.

9.13   SPECIAL TOPIC: REARRANGEMENTS IN BIOLOGICAL PROCESSES

The hydride shifts and alkyl migrations discussed in Section 9.10 are not arcane examples fit only for the nether regions of an organic text (and for especially vexing problems). They are going on all the time as part of many of the enzyme-driven reactions in your body. So, as you read this, hydrides are migrating somewhere deep inside you.

Here is an example of a biological reaction in which an intramolecular hydride shift must be occurring. Remember the section in Chapter 3 (p. 144) on methyl transfer to an alkene? We are a bit further along now and can flesh out the rather sketchy mechanistic description given earlier. In particular, Figure 9.79 shows the ultimate product of the reaction introduced in Figure 3.82 (p. 145).

FIGURE 9.79 The final product of the methylation of oleate is 10-methylstearate.

It looks as if intermediate **A** has somehow been reduced to give the final product. But a clever labeling experiment (Fig. 9.80) shows that the process is far more complex, and that a hydride shift is involved. It should be no surprise to you that the secondary carbocationic intermediate **A′** undergoes an intramolecular hydride shift to give the more stable tertiary carbocation **B**. A proton is then lost from the methyl group (E1 reaction, p. 312) to give an alkene. It is this alkene that is reduced to give the final product.

FIGURE 9.80 This deuterium labeling experiment shows that there is a "hydride"' shift at the center of this reaction.

9.14 SUMMARY

New Concepts

The central topic of this chapter is the addition of H—X molecules to alkenes. These reactions begin by the addition of the positive end of the dipole to the alkene in such a way as to form the more stable carbocation. The anion (X⁻) then adds to the cation to form the final addition product. The direction of addition—the regiochemistry of the reaction—is determined by the relative stabilities of the possible intermediate carbocations. Resonance and inductive effects are important factors that influence stability.

Alkyl groups stabilize carbocations by a resonance effect in which a filled σ orbital of the alkyl group overlaps with the empty $2p$ orbital on carbon. This effect is known by the curious name of "hyperconjugation" (Figs. 9.31 and 9.32).

In this chapter, we continue to use stereochemical experiments to determine reaction mechanisms. Generally, one-step (concerted) reactions preserve the stereochemical relationships of groups in the reacting molecules. Multistep processes introduce the possibility of rotation around carbon–carbon bonds and loss of the initial stereochemistry.

Remember: No experiment can prove a mechanism. We are always at the mercy of the next experiment, the results of which might contradict our current mechanistic ideas. Experiments can certainly disprove a mechanism, but they can never prove one.

Key Terms

alcohol (p. 407)
catalyst (p. 407)
cationic polymerization (p. 411)
hydride shifts (p. 413)
hydroboration (p. 416)

hyperconjugation (p. 403)
inductive effects (p. 405)
Markovnikov's rule (p. 399)
oxonium ion (p. 407)
rearrangement (p. 413)

regiochemistry (p. 386)
Wagner–Meerwein rearrangements (p. 414)

Reactions, Mechanisms, and Tools

Many acids (H—Br, H—Cl, H—I, $HOSO_2OH$, generally, H—X) will add directly to alkenes. The first step is addition of a proton to give the more stable carbocation. Inductive and resonance effects influence the stability of the intermediate carbocations. In the second step of the reaction, the negative end of the original H—X dipole adds to complete the addition process. The regiochemistry of the addition is determined by the formation of the more stable carbocation in the original addition (Fig. 9.26).

Other H—X molecules are not strong enough acids to protonate alkenes. Water (HOH) is an excellent example. However, such molecules will add if the reaction is acid catalyzed. Enough acid catalyst is added to give the protonated alkene, which is attacked by water. The catalyst is regenerated in the last step and recycles to carry the reaction further (Fig. 9.41).

If the concentration of alkene is high enough, it can compete with other Lewis bases in the system (for example, X⁻),

and dimerization or even polymerization of the alkene can take place. This reaction is not mysterious; the alkene is merely acting as Lewis base toward the carbocation as Lewis acid (Fig. 9.50).

In hydroboration, the boron atom of "BH_3" adds to an alkene to give an alkylborane. The mechanism involves a single step in which the new carbon–hydrogen and carbon–boron bonds are both made. Subsequent further hydroborations can give dialkyl- and trialkylboranes. Treatment of these boranes with basic hydrogen peroxide leads to alcohols (Figs. 9.75 and 9.76). Hydroboration is especially important because it gives access to the less substituted alcohol (anti-Markovnikov addition).

Additions that go through carbocationic intermediates can be complicated by rearrangements of hydride (H:⁻) or alkyl groups (R:⁻) to produce more stable carbocations from less stable ones (Figs. 9.53 and 9.54).

Synthesis

In this chapter, we introduce the synthesis of alkyl halides and alcohols from alkenes through addition reactions. Hydroboration allows us to do addition reactions in the anti-Markovnikov sense. The S_N2 and S_N1 reactions from earlier chapters allow us to do further transformations of the alcohols and halides. The new synthetic methods are summarized below.

1. ALCOHOLS

More highly substituted alcohol is formed
(Markovnikov addition)

Less highly substituted alcohol is formed
(anti-Markovnikov addition)

2. ALKYLBORANES

The number of hydroborations depends on the bulk of the alkene; boron becomes attached to the less substituted end of the alkene

3. ALKYL HALIDES

More highly substituted halide is formed
(Markovnikov addition)
(X = Br, Cl, or I)

4. ALLYL HALIDES

1,2- and 1,4-Addition compete

Common Errors

Keeping clear the difference between resonance and equilibrium is a constant difficulty for many students. Resonance forms are simply different electronic descriptions for a single molecule. The key word is "electronic." In resonance, only electrons are allowed to move to produce the various representations of the molecule. Atoms may not change their positions. If they do, we are no longer talking about resonance, but equilibrium. Be careful! This point is trickier than it sounds.

Remember also that the two-dimensional paper surface can fool us. Molecules are three dimensional. To have delocalization of electrons, orbitals must overlap. Sometimes it looks in two dimensions as though resonance forms exist when in fact they do not in the real world of three dimensions. Two excellent examples appear in Figure 9.24 (p. 396) and in Problem 9.8f (p. 398).

The grammar of chemistry—our use of arbitrary conventions—is important, because we need to be able to be precise in communicating with each other. The double-headed resonance arrow is reserved for resonance, and never used for anything else. There are several schemes for representing resonance forms, ranging from drawing them all out in full, to the summary structures of Figure 9.14 (p. 391).

Both mechanistic analysis and synthesis are gaining in complexity. There is much stereochemical detail to keep track of in many mechanistic analyses, for example. Perhaps the most complex mechanism considered so far is that for hydroboration. If you understand why it was necessary to modify the standard mechanism for the addition of Lewis acids to alkenes to accommodate the experimental observations, you are in fine shape so far.

There are no really complex synthetic procedures yet. However, even very simple steps taken in sequence can lead to difficulty. This area will rapidly proliferate and become more difficult, so do not be lulled by the deceptive simplicity of the reactions so far.

9.15 ADDITIONAL PROBLEMS

This chapter and Chapter 10 continue our cataloging of the basic reactions of organic chemistry. To the S_N1, S_N2, E1, and E2 reactions we now add a variety of addition reactions. Although there are several different mechanisms for additions, many take place through a three-step sequence of protonation, addition, and deprotonation. The following new problems allow you to practice the basics of addition reactions and to extend yourself to some more complex matters. Even simple additions become complicated when they occur in intramolecular fashion, for example. These problems also allow you to explore the influence of resonance and inductive effects, and to use the regiochemistry and stereochemistry of addition to help work out the probable mechanisms of reactions.

Your sophistication in synthesis is also growing, and the variety of addition reactions encountered in this chapter adds to the ways you have available to make differently substituted molecules. You are still not quite ready to undertake multistep syntheses, but you are now very close. In anticipation of these tougher problems, be sure to start working out synthetic questions backward. Always ask the question, What molecule is the immediate precursor of the target? Don't start, even with simple problems, thinking of how an ultimate starting material might be transformed into product. That approach will work in one- or two-step synthesis, but becomes almost impossible to do efficiently when we come to longer multistep syntheses.

PROBLEM 9.27 See if you can write the π molecular orbitals of allyl from memory. Show the electronic configuration (orbital occupancy) for the allyl cation, radical, and anion.

PROBLEM 9.28 The molecular orbitals for pentadienyl are shown in the next column.

Pentadienyl

Compare these molecular orbitals with those of allyl. What rules can you develop to predict the molecular orbitals of any odd-carbon, fully conjugated (2p orbital on each carbon) chain? It is probably easiest to work from the schematic top views at the right of the figure.

Views from the top of pentadienyl beneath every (+) is a (−) and beneath every (−) is a (+); nodes are shown as (0)

PROBLEM 9.29 Apply your rules to the molecular orbitals of heptatrienyl.

Heptatrienyl

PROBLEM 9.30 Draw a third resonance form for the allyl cation that involves a three-membered ring. Be careful not to move any atoms! What is the relative importance of this resonance form? Explain.

PROBLEM 9.31 Write resonance forms for the following species. You will first have to write good Lewis structures, as the electrons have been deliberately left out. You will also have to write a full Lewis structure for the nitro (NO₂) group.

(a)   (b)

(c)   (d)

PROBLEM 9.32 How many signals would appear in the ^{13}C NMR spectra of the compounds in Problem 9.31 (a) and (b)?

PROBLEM 9.33 How many signals would appear in the ^{13}C NMR spectra of the following compounds?

(a)   (b)

PROBLEM 9.34 Which of the following pairs represent resonance forms and which do not? Explain your choices.

(a)

(b)

$H_2\overset{+}{C}$—CH_2 ⟷ $H_2\overset{..}{C}$—$\overset{+}{C}H_2$

(c)

(d)

:C=Ö: ⟷ :C≡Ö:

PROBLEM 9.35 Analyze the relative importance of the following two resonance forms. What would you say to someone who claimed that form (b) must be unimportant because the positive charge is on the more electronegative atom? You will have to add electron dots to make good Lewis structures first.

(a)   (b)

PROBLEM 9.36 Predict the regiochemistry of the following addition reactions of a generic acid, H—X. Explain your predictions.

(a)   (b)   (c)

(d)   (e)

PROBLEM 9.37 What would be the structure of the polymer produced from the following two monomers through cationic polymerization?

PROBLEM 9.38 At −78 °C, hydrogen bromide adds to 1,3-butadiene to give the two products shown. Carefully write a mechanism for this reaction, and rationalize the formation of two products.

PROBLEM 9.39 Reaction of *cis*-2-butene with hydrogen bromide gives racemic 2-bromobutane.
(a) Show carefully the origin of the two enantiomers of 2-bromobutane.
(b) In addition, small amounts of two alkenes isomeric with *cis*-2-butene can be isolated. What are these two alkenes and how are they formed?

PROBLEM 9.40 Indene (**1**) can be hydrated to give a different product (**A** or **B**) under the following two conditions.

Reagents for the first route: H_3O^+/H_2O
Reagents for the second route: (1) BH_3, (2) $HOOH/HO^-$

One of these two routes leads to a compound, **A** ($C_9H_{10}O$), which has a ^{13}C NMR spectrum consisting of nine signals. The other route leads to a different compound, **B**, which is also $C_9H_{10}O$, and has a ^{13}C NMR spectrum consisting of only five signals. What are the structures of the two compounds **A** and **B**, and which route leads to which compound? Explain your reasoning in mechanistic terms for each compound (you do not have to sketch a mechanism for the oxidation step of route 2), and make very clear how you used the ^{13}C data in the structure determination.

1

In the following four problems, reactions appear that are quite similar to those discussed in this chapter, but are not exactly the same. Each raises one or more slightly different mechanistic points. In this book, we aim to develop an ability to understand new observations in the light of old knowledge. In a sense, you are being put in the place of the research worker being presented with new experimental results. In each case, explain the products mechanistically.

PROBLEM 9.41

(a)

(b)

PROBLEM 9.42

PROBLEM 9.43

PROBLEM 9.44

(a)

(b)

PROBLEM 9.45   Given the following experimental observation and the knowledge that the initial hydroboration step involves a syn addition, what can you say about the overall stereochemistry of the migration and oxidation steps of the organoborane (see Fig. 9.76, p. 426). Is there overall retention or inversion of configuration?

PROBLEM 9.46   Consider the syntheses of 2-chloro-3-methylbutane and 2-bromo-3-methylbutane suggested in the answer to Problem 9.26 (b) and (e), given in the Study Guide. Carefully consider the last step in the proposed syntheses, the reaction of 3-methyl-2-butanol with hydrogen bromide or hydrogen chloride. Do you see any potential problems with this step? What else might happen?

PROBLEM 9.47   Propose syntheses of the following molecules starting from methylenecyclohexane, alcohols, and iodides containing no more than four carbon atoms, and any inorganic materials (no carbons) you want.

In each of the following reactions, more than a simple addition takes place. Keep your wits about you, think in simple stages, and work out mechanisms.

PROBLEM 9.48   Provide a mechanism for the following change:

PROBLEM 9.49   Provide a mechanism for the following change:

PROBLEM 9.50   Provide a mechanism for the following change:

Hint for the hard part and for the next few problems: Think Wagner–Meerwein (p. 414).

PROBLEM 9.51   Provide a mechanism for the following transformations:

PROBLEM 9.52   What two products do you anticipate for the reaction shown below? Apply what you already know about the chemistry of alcohols in acid. There are two reasonable products, each with the formula $C_6H_{12}O$, but with very different structures. In fact, these compounds are formed in only minor amounts. The major product is pinacolone, shown in Problem 9.53.

PROBLEM 9.53   What is the mechanism of pinacolone formation?

Pinacolone

PROBLEM 9.54   Provide a mechanism for the following transformation of morphine into apomorphine. **Caution:** This problem is very hard.

Additions to Alkenes 2: Additions to Alkynes

10

A red dragonfly hovers above a backwater of the stream, its wings moving so fast that the eye sees not wings in movement but a probability distribution of where the wings might be, like electron orbitals: a quantum-mechanical effect that maybe explains why the insect can apparently teleport from one place to another, disappearing from one point and reappearing a couple of meters away, without seeming to pass through the space in between. There sure is a lot of bright stuff in the jungle. Randy figures that, in the natural world, anything that is colored so brightly must be some kind of serious evolutionary badass.

—Neal Stephenson
Cryptonomicon*

10.1   PREVIEW

We continue the discussion of additions to π systems in this chapter. Most reactions will follow the pattern seen in Chapter 9 in which a Lewis acid (usually a cation) is formed and subsequently is captured by a nucleophile, often an anion. Other reactions follow a different path. We begin with one of these, hydrogenation, the addition of H_2. We then move on to reactions of other X_2 reagents, which often proceed through the formation of an intermediate, but nonisolable, three-membered ring. Next, we take up additions in which the three-membered ring is stable, and finish with reactions forming different sized rings. We will also discuss the related additions of alkynes. Throughout this chapter, attention is paid to what all these reactions allow us to do in the way of synthesis, which is rather a lot.

*Neal Stephenson (b. 1959) is the author of the brilliant *Snow Crash*, as well as *The Diamond Age*, *Zodiac: The Eco-Thriller*, and, of course, *Cryptonomicon*.

ESSENTIAL SKILLS

1. Addition reactions have stereochemical consequences. In Chapters 3 and 9, where we first encountered additions, we focused mainly on regiochemistry, the direction of addition. Here, when we describe additions of X_2 and XY reagents, the stereochemistry, syn or anti, of the reaction can also be followed easily, and it is essential that you keep this factor in mind.

2. Addition reactions can take place through open or cyclic intermediates, and you will learn to assess the factors that control which way the reaction goes. If an open cation is well stabilized, usually by resonance, it is likely to be favored. If not, the cyclic intermediates generally prevail.

IMPORTANT DETAILS

1. Open cations permit addition of a nucleophile to each lobe of the empty orbital. Cyclic ions require opening in S_N2 fashion, leading to overall trans addition.

2. Epoxides open differently in acid and base. In base, the nucleophile adds to the less hindered position, whereas in acid the nucleophile becomes attached to the more substituted, more hindered side.

10.2   ADDITION OF X_2 REAGENTS

In this section, we see two very different reactions of X_2 reagents. In the first example, the one-step addition of hydrogen to alkenes, a catalyst is required to get things going. The second reaction, halogen addition, fits better into the pattern of Chapter 9, in which a two-step process is followed.

10.2a Hydrogenation. When an alkene is mixed with hydrogen, nothing happens (unless one is careless with oxygen and a match). If the two reagents are mixed in the presence of any of a large number of metallic catalysts, however, an efficient addition of hydrogen converts the alkene into an alkane. This reaction is called **hydrogenation**, and it can be achieved by two different catalytic processes. First, there is **heterogeneous catalysis**, in which an insoluble catalyst, often palladium (Pd) on charcoal, is used. Many other insoluble metallic catalysts work as well in heterogeneous catalysis. **Homogeneous catalysis**, in which a soluble catalyst, often Wilkinson's catalyst, after Geoffrey Wilkinson (b. 1921), can also be employed (Fig. 10.1).

PROBLEM 10.1 Use bond dissociation energies (BDE) (Table 8.2, p. 355) to calculate the exothermicity or endothermicity of the hydrogenation of ethylene to ethane.

FIGURE 10.1 Hydrogen and an alkene react very quickly in the presence of a metal catalyst such as palladium adsorbed on charcoal (Pd/C). Wilkinson's catalyst is typical of soluble molecules that also catalyze hydrogenation.

Other, nonmetallic, organic reducing agents can be used in hydrogenation as well. **Diimide** (HN=NH) is a good example. Diimide is not stable under normal conditions and is made only as it is needed. Diimide exists in two forms, syn and anti, but only the less stable syn form is active in hydrogenation (Fig. 10.2).

FIGURE 10.2 An effective nonmetallic reducing agent for alkenes is "diimide" (HNNH).

If hydrogenation is carried out with a ring compound, a cycloalkene, we can see that transfer of hydrogen takes place in predominately syn fashion. Both new hydrogens are delivered from the same side of the alkene (Fig. 10.3). In this example, D_2 is used so we can see the stereochemical outcome of the reaction.

FIGURE 10.3 Metal-catalyzed addition to cycloalkenes shows that deuterium is added in syn fashion.

The mechanistic details of this reaction remain vague. Presumably, the metal surface binds both hydrogen and alkene, weakening the π bond of the alkene and the σ bond of hydrogen. The alkane is formed by addition of a weakened hydrogen molecule to the coordinated alkene. As the transfer is completed, the product alkane detaches from the metal surface (Fig. 10.4).

FIGURE 10.4 Partial bonding to the metal surface is followed by hydrogen transfer to the alkene.

There are many different kinds of selectivity in the hydrogenation reaction. Not all carbon–carbon double bonds are hydrogenated at the same rate, and the differences in rate can often be used to hydrogenate one double bond in the presence of another. Generally, the more sterically hindered a double bond, the slower it is hydrogenated. This rate change is probably related to ease of adsorption on the catalyst surface. Large groups hinder adsorption, and therefore slow the rate of hydrogenation. One can often hydrogenate a less substituted double bond in the presence of more substituted double bonds, for example (Fig. 10.5).

Less substituted (less stable, faster to react)

+

More substituted (more stable, slower to react)

$$\xrightarrow[\text{Ru/C}]{\text{H}_2}$$

+

FIGURE 10.5 It is possible to hydrogenate less substituted double bonds in the presence of more hindered, more substituted double bonds.

We have to pay attention to the drawing conventions (codes) here. We do not always draw out all of the carbon and hydrogen atoms—we are beginning to adopt the convention most used by chemists—the "stick" drawings that show only the carbon framework of the molecule. This figure shows two levels of code, but we will often lapse into the most schematic representations of molecules.

PROBLEM 10.2 Hydrogenation of β-pinene gives only one of the possible diastereomeric products. Explain the direction of hydrogenation shown below.

Now let's consider the energetics of this reaction. Is hydrogenation likely to be exothermic or endothermic? As shown earlier in Problem 10.1, a simple and common technique for making such estimates uses bond energies. In the hydrogenation reaction, two bonds are breaking: the hydrogen–hydrogen σ bond and the π bond of the alkene. Two bonds are also being made in this reaction: the two new σ bonds in the alkane product. If we know bond energies (Table 8.2, p. 355), we can estimate $\Delta H°$ for this reaction. We made this estimate in Problem 10.1, and Figure 10.6 repeats the calculation for this reaction. The process should be exothermic by about 30 kcal/mol.

Table 10.1 collects some measured heats of hydrogenation. We can see that the crude bond energy calculation of Figure 10.6 has actually done quite well; the heats of hydrogenation of simple alkenes are quite close to our guess of 30 kcal/mol.

TABLE 10.1
Some Heats of Hydrogenation

Isomer	$\Delta H°_{H_2}$ (kcal/mol)
$H_2C=CH_2$	−32.7
$H_2C=CH-CH_3$	−30.0
$H_2C=CH-CH_2CH_3$	−30.3
cis $CH_3-CH=CH-CH_3$	−28.6
trans $CH_3-CH=CH-CH_3$	−27.6
$(CH_3)_2C=CH_2$	−28.4
$(CH_3)_2C=CHCH_3$	−26.9
$H_2C=CH-CH_2CH_2CH_2CH_3$	−29.6
cis $CH_3CH_2-CH=CH-CH_2CH_3$	−27.3
trans $CH_3CH_2-CH=CH-CH_2CH_3$	−26.4
$H_2C=C-CH_2CH_2CH_3$ (with CH₃ on the substituted carbon)	−28.0
$(CH_3)_2C=C(CH_3)_2$	−26.6

PROBLEM 10.3 Use the values in Table 10.1 to estimate the relative energies of the hexenes. Compare your estimates with those generated from heats of formation in Chapter 3 (p. 115).

PROBLEM 10.4 As a mixture of an alkene and hydrogen lies some 30 kcal/mol above the alkane product, why doesn't the reaction occur spontaneously? Use an Energy versus Reaction progress diagram to help explain *in general* the role of the metal catalyst (Chapter 8, p. 360).

Bonds broken

π bond = 66 kcal/mol
σ bond = 104 kcal/mol

Bonds made

Two C–H σ bonds =
~200 kcal/mol

200 − 170 = 30-kcal/mol exothermic
$\Delta H = -30$ kcal/mol

FIGURE 10.6 In a typical hydrogenation, it is the π bond of the alkene and the σ bond of hydrogen that are broken. Two new carbon–hydrogen bonds are made. The hydrogenation of an alkene should be exothermic by approximately 30 kcal/mol.

10.2b Additions of Halogens. Molecular bromine and chlorine (Br_2 and Cl_2) do not need a catalyst to add to alkenes to give **vicinal** dihalides (Fig. 10.7). "Vicinal" means "disubstituted on adjacent (1,2) positions."

FIGURE 10.7 Addition of Br_2 and Cl_2 to yield vicinal dihalides.

Addition is stereospecifically trans. For example, cyclopentene gives only *trans*-1,2-dibromocyclopentane (Fig. 10.8).

FIGURE 10.8 trans Addition of Br_2 to cyclopentene.

trans-**1,2-Dibromo-cyclopentane**   *cis*-**1,2-Dibromo-cyclopentane**

not

Alkene halogenation

If we take the addition of hydrogen bromide to an alkene as a model for this reaction, we can build a mechanism quite quickly. Just as the alkene reacts with hydrogen bromide to produce a carbocation and bromide (Br^-), so reaction with bromine–bromine might give a brominated cation and bromide (Fig. 10.9). The π bond of the alkene acts as a nucleophile, displacing Br^- from bromine. Readdition of bromide to the carbocation would give the dibromide product.

FIGURE 10.9 A first-guess mechanism for the reaction of an alkene and bromine. For analogy, it draws on the mechanism of hydrogen bromide addition to an alkene.

H—Br Addition

Possibly analogous Br_2 addition (?)

There are cases of bromination in which this mechanism *is* exactly correct, but for most alkenes it is not quite right. First, carbocation rearrangements (Chapter 9, p. 401) are not observed in most bromination and chlorination reactions. For example, addition of bromine to 3,3-dimethyl-1-butene gives only 1,2-dibromo-3,3-dimethylbutane, *not* 1,3-dibromo-2,3-dimethylbutane, as would be expected if an open cation were involved (Fig. 10.10).

FIGURE 10.10 Rearrangements of carbocations are not observed in most oxidation reactions.

We already know that if a cycloalkene is used, the product is exclusively the *trans*-1,2-dibromide. The cis compound cannot be detected (Fig. 10.8).

If the addition were to involve an open cation, we would expect the bromide ion to add from both sides to produce a mixture of products. The yields of cis and trans compounds would not be equal, but both compounds should be formed (Fig. 10.11). As only anti addition is observed, our initial mechanistic ideas cannot be completely correct. We need a new mechanism.

FIGURE 10.11 An open carbocation must lead to both syn and anti addition.

So our mechanism needs modification. How do we explain the simultaneous preference for anti addition, and absence of rearrangement? It seems that an open carbocation, which would lead to cationic rearrangements and permit both syn and anti addition, must be avoided in some way.

What if the initial displacement of Br$^-$ from Br$_2$ were to take place in a symmetrical fashion? Once again, the filled π orbital of the alkene acts as nucleophile in an S$_N$2 reaction in which Br$^-$ is displaced from Br$_2$ (Fig. 10.12). As the figure shows, a bromine-containing three-membered ring, a **bromonium ion** (brom = bromine; onium = positive), is formed. An open carbocation is avoided and rearrangements cannot occur.

FIGURE 10.12 A symmetrical attack of the alkene on bromine yields a cyclic bromonium ion along with bromide, Br⁻.

Symmetrical displacement   A bromonium ion

But what of the preference for anti addition? Look closely at the second step of the reaction, the open-ing of the bromonium ion by the nucleophile Br^-. This reaction is an S_N2 displacement, and therefore must take place with inversion, through addition from the rear as we saw so often in Chapter 7 (p. 279) (Fig. 10.13). Overall anti addition of the two bromine atoms is assured.

FIGURE 10.13 This anti addition of the two bromine atoms requires the formation of *trans*-1,2-dibromocyclopentane.

trans-**1,2-Dibromo-cyclopentane**

WORKED PROBLEM 10.5 In Figure 10.13, the reaction starts with achiral molecules. Yet the product, as drawn in the figure, is a single enantiomer. Clearly something is awry. What is it? *Hint:* Convention Alert! (Chapter 4, p. 156).

ANSWER You were warned about this! Of course there can be no optical activity produced from a reaction of two achiral reagents. In this reaction, opening must take place in two equivalent ways to produce a racemic pair of enantiomers. In situations such as this one, it is the custom not to draw both enantiomers.

CONVENTION ALERT

Enantiomers!

Mirror

So, a modification of our earlier addition mechanism suffices to explain the reaction. In contrast to the additions to alkenes we have seen earlier (H—X, carbocations, and boranes), in bromination and chlorination a three-membered ring is produced as an intermediate. In some, very specialized cases the bromonium ion can even be isolated, and it seems its intermediacy in most brominations is assured.

PROBLEM 10.6 One example in which the bromonium ion is stable to further addition of nucleophiles is the reaction of adamantylideneadamantane with bromine (see below). Can you see why this cyclic ion might be stable?

Adamantylideneadamantane

Br_3^-
A stable bromonium ion

PROBLEM 10.7 Addition of Br_2 to acenaphthylene gives large amounts of the product of cis addition along with the usual trans dibromide. Give a mechanism for this bromination and, more important, explain why the mechanism of the reaction shown below is different from the one we just outlined.

Acenaphthylene **A cis dibromide** **A trans dibromide**

Stabilized alkene halogenation

In the bromonium ion, there is the full complement of six electrons for the three σ bonds binding the two carbons and the bromine. The bromonium ion is a normal compound even though it may look odd at first. To see this clearly, imagine forming the bromonium ion from an open cation (Fig. 10.14).

Bromonium ion

FIGURE 10.14 The formation of a brominium ion from an open carbocation.

Normally, the nonnucleophilic solvent carbon tetrachloride (CCl_4) is used for the bromination or chlorination of alkenes. But the reaction will also work in protonic solvents such as water and simple alcohols. In this kind of reaction, new products appear that incorporate molecules of the solvent (Fig. 10.15).

FIGURE 10.15 In the presence of nucleophilic solvents, addition reactions of bromine and chlorine give products incorporating solvent molecules as well as dihalides.

How do the OH and OR groups get into the product molecule? To see the answer write out the mechanism and look for an opportunity to make the new products. The first step is just the same, in this case formation of the cyclic chloronium ion (Fig. 10.16). Normally, the chloronium ion is opened in S$_N$2 fashion by the nucleo-

THE GENERAL CASE

A SPECIFIC EXAMPLE

FIGURE 10.16 Once the chloronium ion is formed, it is vulnerable to attack by *any* nucleophile present. When a nucleophilic solvent is used in the reaction, it can attack the cyclic ion to give the solvent-containing product. In this case, addition of chloride accounts for 8% of the product, whereas the solvent methyl alcohol produces 91%. Notice that methyl alcohol can produce two products [(a) and (b)].

philic Cl⁻ to give the dichloride. This opening must be part of the mechanism of dichloride formation in these examples. But in these reactions there are other nucleophiles present: molecules of water or alcohol. Of course, they are weaker nucleophiles than the negatively charged Cl⁻, but they are present as solvent, and have an enormous numerical advantage over the more powerfully nucleophilic chloride, which is present only in relatively low concentration (Fig. 10.16).

If this mechanism is correct, we can predict that the required inversion in the opening of the cyclic ion must produce a trans stereochemistry in the products. So, we have the opportunity to test the mechanism, and it turns out that addition of bromine to cyclohexene in the solvent water exclusively produces the compound resulting from anti addition in which the bromine and hydroxyl groups have a trans relationship. The bromonium ion mechanism is correct (Fig. 10.17).

Halohydrin formation

FIGURE 10.17 This mechanism demands that the new products be formed by anti addition, as they are. This stereochemical experiment uses cyclohexene to show that the bromine and solvent molecule become attached from different sides of the ring through the anti addition enforced by the presence of the bromonium ion.

Addition of bromine in nucleophilic solvents presents us with another opportunity. We can measure the regioselectivity of the reaction. When two bromines add, as in the bromination of 2,3,3-trimethyl-1-butene in carbon tetrachloride, there is no way to determine the regiochemistry. Bromination in water allows us to see where the hydroxyl group goes as it opens the bromonium ion. There are two possibilities. The hydroxyl group can attach to the more [Fig. 10.18(a)] or less [Fig. 10.18(b)] substituted carbon of the original alkene.

2,3,3-Trimethyl-1-butene

FIGURE 10.18 A test of the regiochemistry of the addition of Br₂/H₂O to alkenes.

In practice, the hydroxyl group becomes attached to the more substituted carbon. It is as if the reaction were proceeding through a carbocation (Fig. 10.19). But we are quite certain that it does not! The problem now is to reconcile the observed regio-selectivity with the intermediacy of a bromonium ion (recall trans addition and the lack of carbocation rearrangements).

FIGURE 10.19 The bromine becomes attached to the less substituted carbon and the solvent to the more substituted carbon of the alkene. It is *as if* the reaction went through a carbocationic intermediate.

The bromonium ion is only *symmetrical* when formed from an alkene in which the two ends are *the same*. In all other cases, the two carbon–bromine bonds must be of different strength. In resonance terms, we would say that the weighting factor (coefficient) for the more stable form was greater than that for the less stable form ($c_1 > c_2$, Fig. 10.20). The two carbons share the positive charge, but not equally. The charge will be more stable on the more substituted carbon. Therefore, the more substituted carbon–bromine bond is longer and weaker than the less substituted bond in this positively charged intermediate.

FIGURE 10.20 A resonance description of the bromonium ion.

Accordingly, the transition state for breaking the more substituted bond will be lower in energy than that for breaking the less substituted bond. It is at the more substituted position that nucleophilic solvents such as methyl alcohol and water will add (Fig. 10.21).

FIGURE 10.21 Addition of nucleophiles breaks the longer, weaker C—Br bond.

SUMMARY

The critical difference between addition of halogen (X_2) and the many HX additions we saw in Chapter 9 is the presence of the intermediate cyclic halonium ion. This intermediate is demanded by the observed preference for anti addition, which is the mode generally observed. Whenever the related open carbocation is especially stabilized, generally by resonance, it can compete with the cyclic species. In such cases, there is no requirement for trans addition and both cis and trans modes of addition are observed.

10.3   HYDRATION THROUGH MERCURY COMPOUNDS: OXYMERCURATION

For small scale hydrations, it is convenient to use mercury salts such as mercuric acetate, $Hg(OAc)_2$, to carry out the reaction. In this two-step process, an alkene is first treated with mercuric acetate and then reduced with the reagent sodium borohydride (Na^+ $^-BH_4$) (Fig. 10.22).

FIGURE 10.22 Alcohol formation through oxymercuration.

The abbreviation AcO⁻ represents the acetate ion (CH₃COO⁻), the conjugate base of acetic acid, CH₃COOH (often abbreviated as AcOH). Lewis structures are given in Figure 10.23, which shows the relationship between the acetate ion and its conjugate acid (AcOH). Note the resonance stabilization of the acetate ion—the two oxygen atoms are equivalent.

Acetic acid, the
conjugate acid of...   ...the acetate ion, AcO⁻

FIGURE 10.23 Note the resonance stabilization of the acetate ion.

The first step in the reaction is the formation of a cyclic, mercurinium ion (Fig. 10.24). We can be quite certain of this because addition involves no carbocationic rearrangements. Open carbocations are unlikely intermediates, as they would surely undergo carbocationic rearrangements, and these are not seen even in systems normally prone to rearrangement.

PROBLEM 10.8 The pK_a of acetic acid is 4.8. The corresponding two-carbon alcohol, ethyl alcohol (CH₃CH₂OH), is a much weaker acid, as its pK_a of 15.9 demonstrates. Explain why acetic acid is much more acidic than ethyl alcohol.

WORKED PROBLEM 10.9 Design an experiment to test for carbocationic rearrangements in the **oxymercuration** reaction.

ANSWER It's simple. Just use any alkene in which migration of hydride (H:⁻) or R:⁻ would give a more thermodynamically stable cation. 3-Methyl-1-butene would work well. If addition were to give an open carbocation rather than the cyclic mercurinium ion, rearrangements would surely occur. That they are *not* observed provides strong evidence that open cations are not involved in oxymercuration.

With nonsymmetrical alkenes, the two bonds to mercury will not be equally strong. (Recall our discussion of asymmetrical bromonium ions just a few pages ago.) The bond between mercury and the more substituted carbon will be longer and weaker than the bond to the less substituted carbon. It is at this end that water opens the ring (Fig. 10.24). So, the first product of this reaction is the mercury-containing alcohol shown in Figure 10.24.

FIGURE 10.24 Oxymercuration begins by attack of the alkene on the Lewis acidic mercuric acetate to form a cyclic mercurinium ion. The nucleophilic solvent, water, then adds in S_N2 fashion to give the first product, a mercury-containing alcohol, where R = simple alkyl group.

Oxymercuration is synthetically useful because the reagent sodium borohydride ($Na^+ \ ^-BH_4$) efficiently replaces the mercury with hydrogen. Figure 10.25 shows the reaction with sodium borohydride and sodium borodeuteride ($Na^+ \ ^-BD_4$). The use of the deuterated molecule allows us to see with certainty the position of the entering deuterium.

So we now have a new hydration reaction; one that proceeds through a cyclic mercurinium ion but ultimately gives the product of Markovnikov addition, the more substituted alcohol.

FIGURE 10.25 Reduction of the mercury-containing products gives alcohols.

SUMMARY

Here are our three methods of hydration:

1. *Direct hydration* (Chapter 9, p. 409) proceeds through an intermediate carbocation that is captured by water to give the product of Markovnikov addition. The reaction is limited in utility because rearrangements of the initially formed cation to more stable species can lead to undesired, rearranged products.

2. *Hydroboration* (Chapter 9, p. 416) first generates an alkylborane that subsequently reacts with peroxide in base to give the product of anti-Markovnikov addition.

3. *Oxymercuration* proceeds through a cyclic mercury-containing ion and also gives the product of Markovnikov addition. No rearrangements take place, which is sometimes a great advantage.

PROBLEM 10.10 Work out the consequences of applying the three procedures outlined above to 3-methyl-1-butene.

10.4   OTHER ADDITION REACTIONS INVOLVING THREE-MEMBERED RINGS: OXIRANES AND CYCLOPROPANES

10.4a Oxiranes. We have seen that many addition reactions begin with formation of a three-membered ring. These intermediates have almost always been rapidly converted into the final products through addition of a nucleophile. This state of affairs will now change dramatically, as addition reactions are the best routes to certain stable three-membered rings: the cyclopropanes and **oxiranes** (**epoxides**).

When an alkene reacts with a perbenzoic acid or trifluoroperacetic acid, three-membered rings containing one oxygen atom are isolated. An organic acid has the formula R—COOH. A *per*acid has the formula R—COO$\underline{O}$H. Figure 10.26 shows the difference in more detail. The "per" indicates an "extra" oxygen is present. Thus hydrogen *per*oxide is HO$\underline{O}$H, whereas the normal oxide of hydrogen is water, HOH.

FIGURE 10.26 The prefix per- means extra. So a "peroxide" and a "peracid" each contain an extra oxygen atom, one more than the oxide or acid.

Hydrogen oxide (water)

A carboxylic acid

Trifluoroacetic acid

Hydrogen peroxide

A peracid

Trifluoroperacetic acid

The epoxidation reaction resembles the formation of a bromonium ion; a leaving group is displaced by the alkene acting as nucleophile in an S_N2 reaction (Fig. 10.27). At the same time, a hydrogen is transferred to the carbonyl (C=O) oxygen. The result is a stable three-membered ring, an epoxide.

Alkene epoxidation

FIGURE 10.27 Epoxidation is closely related to bromination. In epoxidation, there is an additional transfer of hydrogen to give the isolable, stable epoxide, or oxirane.

It is important to be clear why we are so sure that the ring is formed in a single step. Why not just form the carbocation and then close up to the oxirane (Fig. 10.28)?

This open cation must lead to rearrangements that are not observed

FIGURE 10.28 A stepwise mechanism for epoxidation.

If a free cation were involved, carbocation rearrangements would be expected, and they are not found. Moreover it is easy to monitor the stereochemistry of this reaction. Concerted (no intermediates) formation of the oxirane explains the observed syn addition very well. A reaction in which the two new bonds are formed at the same time cannot change the stereochemical relationships of the groups in the original alkene. On the other hand, a mechanism involving an open carbocation would very likely lead to a mixture of stereoisomeric oxiranes (Fig. 10.29). If only one bond were made, there would be a carbocationic intermediate in which rapid rotations about carbon–carbon single bonds would take place. These rotations would scramble the original stereochemical relationship present in the alkene, and this result is not observed. Oxirane formation retains the stereochemical relationships present in the starting alkene.

THE GENERAL CASE

FIGURE 10.29 A test of the mechanism of epoxidation. A stepwise mechanism predicts scrambling of stereochemistry, whereas a concerted, one-step, mechanism must retain the stereochemistry of the starting alkene. As cis alkenes give only cis epoxides, the reaction must be concerted.

10.4b Asymmetric (Sharpless) Epoxidation. In recent years, an enormously useful synthetic technique for producing enantiomeric oxiranes (epoxides) has appeared. Discovered by Barry Sharpless (b. 1941), now at the Scripps Research Institute, it uses a witches' brew of titanium isopropoxide, *tert*-butyl peroxide, and one enantiomer of a tartaric ester* to react with an allylic alcohol (R—CH=CH—CH₂OH). Part of the great utility of this procedure comes from the observation that the two enantiomers of tartaric ester lead to two different stereochemistries of product oxirane (Fig. 10.30). Professor Sharpless shared the Nobel prize in chemistry in 2001 for his discovery of chiral catalysts.

The mechanism of this complex reaction involves the titanium compound acting as a clamp, holding the alkene, peroxide, and tartaric ester together. Because the tartaric ester is asymmetric, the clamped combination of molecules is also asymmetric. In one cluster, the oxirane oxygen is delivered from one side, whereas in the enantiomeric cluster formed from the other tartaric ester, it comes from the other side. These epoxides can then be transformed into all manner of compounds, as we will now see.

*An ester (R—COOR′) is a derivative of a carboxylic acid (R—COOH).

FIGURE 10.30 Asymmetric (Sharpless) oxidation of alkenes (Et = CH₃CH₂).

PROBLEM 10.11 The other enantiomer of diethyl tartarate gives a diastereomer of the product shown in Figure 10.30. Draw this isomer.

10.4c Further Reactions of Oxiranes. Unlike the closely re-lated bromonium ions, oxiranes can be isolated under many reaction conditions. The bromonium ion is doomed to bear a positive charge, and is therefore a more powerful Lewis acid than the neutral oxirane.

Oxiranes will react, however, when treated in a second step with either acids or bases. For example, reaction with either H₃O⁺/H₂O or HO⁻/H₂O leads to opening of the three-membered ring and formation of a 1,2-diol, known as a **1,2-glycol** (Chapter 6, p. 249, Fig. 10.31).

The mechanisms of these ring openings are straightforward extensions of reac-tions you already know. In base, the strongly nucleophilic hydroxide ion attacks the oxirane and displaces the oxygen atom from one carbon in an S_N2 reaction (Fig. 10.32). In acid, there is no strong nucleophile, but protonation of the oxygen atom of the oxirane creates a good Lewis acid. Water is a strong enough nucleophile to open the protonated oxirane to give, after deprotonation, the 1,2-diol (Fig. 10.32).

FIGURE 10.31 Epoxides open in either acid or base.

Basic epoxide ring opening

The mechanism of epoxide opening in base

Acidic epoxide ring opening

The mechanism of epoxide opening in acid

FIGURE 10.32 Mechanisms for epoxide opening in base and acid.

There is a strong parallel here to reactions we have already seen. Hydroxide is a poor leaving group. Treatment with acid or formation of sulfonate esters converts hydroxide into water or another good leaving group, and many reactions require this transformation (see Chapter 7, p. 277 for examples). The opening of the oxirane in acid involves exactly the same kind of transformation; protonation facilitates the opening by making the leaving group ROH not RO⁻ (Fig. 10.33).

FIGURE 10.33 Water is not a strong enough nucleophile to open the unprotonated oxirane, as R—O⁻ is too poor a leaving group. Acid catalysis changes the leaving group to R—OH, which is much easier to displace.

Unsymmetrical oxiranes can open in two ways and the regiochemical result is generally different in acid and base (Fig. 10.34).

FIGURE 10.34 The regiochemistry of epoxide opening is different in acid and base.

In acid, the nucleophile generally becomes attached to the more substituted carbon

In base, the nucleophile generally becomes attached to the less substituted carbon

How does the difference in regiochemistry arise? Why should the two openings be different? In base, the ring opening is an S_N2 displacement by hydroxide on the unprotonated oxirane. As in any S_N2 reaction, steric matters will be important, and opening will occur at the sterically more accessible, less substituted carbon (Fig. 10.35).

PROBLEM 10.12 If hydroxide is such a poor leaving group, one would expect that RO⁻ would also be a bad leaving group, and it is. In fact, simple ethers (R—O—R) are not generally cleaved in strong base. Why, then, do oxiranes open so easily in base (Fig. 10.34)? What is special about the three-membered ring that facilitates the displacement?

FIGURE 10.35 In base, the nucleophile will add at the sterically less encumbered carbon.

THE GENERAL CASE

A SPECIFIC EXAMPLE

(41%)

In acid, the first step is the protonation of the oxygen of the three-membered ring. Recall the discussion of the regiochemistry of the opening of unsymmetrical bromonium ions (p. 444) and mercurinium ions (p. 450). The positive charge is borne not only by the oxygen but also by the two carbons of the oxirane. Most of the positive charge on carbon will reside at the more substituted position. The more substituted carbon–oxygen bond will be longer and weaker than the other, less substituted carbon–oxygen bond (Fig. 10.36). As in unsymmetrical bromonium ions, the transition state for addition at the more substituted position will be lower in energy than that for addition at the less substituted position.

THE GENERAL CASE

This weaker carbon–oxygen bond will be the major bond broken by an adding nucleophile

A SPECIFIC EXAMPLE

(55%)   (45%)

FIGURE 10.36 Both carbons help bear the positive charge in the protonated oxirane. Most of the positive charge on carbon will be on the more substituted position, and that is where the nucleophile will add. This reaction is much like the opening of an unsymmetrical bromonium ion by water or alcohol.

So, the apparently mysterious change in regiochemistry of the opening in acid and base can be reconciled easily. The trick is to learn to see new reactions in terms of what you already know. Books will always be organized in this way. Questions are set up by recently discussed phenomena. Nature is not always so cooperative and organized, however.

In Chapter 6, we met two organometallic reagents, the Grignard reagent, RMgX, and organolithium reagents, RLi. These compounds are very strong bases (p. 235), and are nucleophiles as well. Like other nucleophiles, organometallic reagents react with oxiranes to open the strained ring. The initial product is an alkoxide, which is protonated in a second step to give the alcohol. This reaction is a quite useful synthesis of alcohols in which the initial carbon atom frame of the organometallic reagent has been augmented by the two carbons originally in the oxirane (Fig. 10.37).

FIGURE 10.37 Opening of an epoxide by an organometallic reagent. The initial product is an alkoxide, which is protonated in a second step to give the alcohol.

The strongly basic organometallic reagents inevitably attack at the less substituted, less sterically hindered end of the oxirane, as did hydroxide ion (Fig. 10.38).

FIGURE 10.38 The S_N2 ring opening occurs at the less sterically demanding end of the epoxide.

The Grignard reagent and organolithium reagents are sources of "R⁻", they are carbanion equivalents. There are also reagents that are hydride (H:⁻) equivalents. We will see several later on in Chapter 16 and beyond, but it is useful to introduce lithium aluminum hydride (LiAlH$_4$, "LAH") now. Like the organometallic reagents, lithium aluminum hydride also opens oxiranes by nucleophilic addition, and this reaction provides you with another synthesis of alcohols. Addition of hydride is predominately from the less hindered side of the epoxide. Once again, an initially produced alkoxide is protonated by water in a second step (Fig. 10.39).

Organometallic reagents have drawbacks, however. Reduction of the oxirane (epoxide) by hydride transfer from the organometallic reagent sometimes occurs and rearrangements can be induced, especially by organolithium reagents. Moreover, Grignards and organolithium reagents will inevitably add to any carbonyl groups

FIGURE 10.39 Lithium aluminium hydride reduction of epoxides, followed by hydrolysis, is yet another alcohol synthesis.

present in the molecule (Chapter 16). By contrast, the organocuprates formed from RLi and CuCN open epoxides smoothly without rearrangement, and are not reactive enough to attack the carbon–oxygen double bond. The opening of the three-membered ring is from the rear, and occurs predominately at the less hindered position (Fig. 10.40).

FIGURE 10.40 Organocuprates also open epoxides.

SUMMARY

Epoxides are typically synthesized through peracid-mediated additions to alkenes. Opening in either acid or base is stereospecifically trans (here is the S_N2 reaction at work), but the regiochemistry of opening differs. In base, steric factors lead to addition of the nucleophile at the less hindered position. By contrast, in acid, the nucleophile adds predominantly to the protonated epoxide at the more sterically hindered position.

Openings of oxiranes by organometallic reagents or hydride reagents such as lithium aluminum hydride are valuable syntheses of alcohols.

10.4d Carbenes.

Additions to alkenes lead to cyclopropanes through intermediates called **carbenes**. Carbenes are neutral compounds containing divalent carbon, and are often formed from nitrogen-containing molecules called diazo compounds. Diazo compounds have the attractive feature (aside from being sources of carbenes) of often being quite handsomely colored. Diazomethane (CH_2N_2) is a bright yellow gas, diazocyclopentadiene an iridescent orange liquid, and 4,4-dimethyldiazocyclohexa-2,5-diene an almost blue liquid (Fig. 10.41). However, diazo compounds have a downside: They are certainly poisonous, demonstrably explosive, and probably carcinogenic. So one must be careful (and brave, and perhaps a little crazy) to do carbene chemistry!

Diazomethane (a yellow gas)

Diazocyclopentadiene
(an orange liquid)

Diazofluorene
(an orange solid)

4,4-Dimethyldiazocyclohexa-2,5-diene (a purple liquid)

Methyl diazoacetate
(a yellow liquid)

FIGURE 10.41 Some diazo compounds. Resonance forms are drawn only for diazomethane.

PROBLEM 10.13 Write resonance forms for 4,4-dimethyldiazocyclohexa-2,5-diene in Figure 10.41.

But this care is worth it because carbenes are fascinating species. Heat (Δ) or light ($h\nu$) liberates them from their parent diazo compounds through loss of nitrogen, and they react rapidly with alkenes, generally used as solvent for the reaction, to give cyclopropanes. Another common source of carbenes is the hydrolysis of halocarbons. Treatment of chloroform ($CHCl_3$) or bromoform ($CHBr_3$) with strong bases, such as hydroxide or potassium *tert*-butoxide gives dichlorocarbene or dibromocarbene (Fig. 10.42).

FIGURE 10.42 Loss of nitrogen from a diazo compound liberates a carbene that can add to alkenes to give a cyclopropane. Dihalocarbenes can be made by treatment of many halogenated alkanes with strong base.

PROBLEM 10.14 Propose a mechanism for the hydrolysis of chloroform ($CHCl_3$, $pK_a = 29$) to dichlorocarbene. *Hint*: When the reaction is carried out in D_2O/DO^-, recovered chloroform is partially deuterated ($CDCl_3$).

The additions of most carbenes occur in stereospecifically syn fashion, which means that cis alkenes react with carbenes to give only cis cyclopropanes and trans alkenes give the corresponding trans cyclopropanes (Fig. 10.43).

THE GENERAL CASE   SOME SPECIFIC EXAMPLES

(70–80%)

(70%)

FIGURE 10.43 Addition of carbenes to alkenes is a syn process with retention of stereochemistry.

The stereospecificity of the addition reactions of carbenes tells us that there can be no intermediates capable of rotation about carbon–carbon bonds in the ring-forming reaction. The mechanism involves a single step; in the addition both new ring bonds are formed at the same time. A two-step addition would lead to formation of both cis and trans cyclopropanes (Fig. 10.44).

Regardless of the mechanism, the reactions of carbenes with alkenes provide the most widely used synthetic route to cyclopropanes.

10.4e Triplet Carbenes. One of the reasons carbenes are so interesting is that unlike almost all other reactive intermediates, two reactive states exist, quite close in energy to each other. In Section 10.4d, we looked at reactions of a species in which two spin-paired electrons occupied the relatively low-energy hybrid orbital and the higher energy $2p$ orbital was empty. In fact, this species, called a **singlet carbene**, generally lies a few kilocalories per mole *higher* in energy than another form of divalent carbon called a **triplet carbene**. In the triplet species, one electron occupies the hybrid orbital and the other has been promoted to the previously empty $2p$ orbital. Of course, promotion incurs an energy cost—the approximately sp^2 hybrid orbital has more s character than the $2p$ orbital, and thus an electron in it is more stable than an electron in a pure $2p$ orbital. But there are energetic compensations. For example, the two nonbonding electrons are further apart in the triplet and thus electron–electron

One-step reaction

cis-1,1-Dibromo-2,3-
dimethylcyclopropane

Two-step reaction

cis-1,1-Dibromo-2,3-
dimethylcyclopropane

trans-1,1-Dibromo-2,3-
dimethylcyclopropane

FIGURE 10.44 The one-
and two-step mechanisms
contrasted.

repulsion is lowered. With some exceptions (R = halogen, for example), the triplets
are somewhat more stable than the corresponding singlets. For many carbenes the
two species are in equilibrium, and sorting out the chemistry can be complicated as
either spin state may undergo reactions (Fig. 10.45).

Singlet carbene

Triplet carbene

FIGURE 10.45 Both spin
states of a carbene may form
products.

PROBLEM 10.15 Singlet :CF$_2$ is much more stable than the corresponding triplet. Look
at the structure of this intermediate and explain why the singlet is especially favored in
this case.

Normally, even though the singlet is disfavored at equilibrium, its reactions are
observed. The singlet is usually *much* more reactive than the triplet. Even though there
is more triplet present at equilibrium, its reactions cannot compete with those of the
small amount of the much more reactive singlet. This point is important. A species does
not have to be present in great quantity for its chemical reactions to dominate those
of a more stable, but relatively unreactive equilibrium partner. The activation barriers

FIGURE 10.46 Even though the triplet is usually lower in energy, and thus favored at equilibrium, most of the chemistry comes from the singlet. The triplet is protected by quite high activation barriers. As long as some of the singlet is formed in equilibrium with the triplet, it will account for most or all of the chemistry.

surrounding the more stable partner may be higher than those surrounding the less stable partner. Figure 10.46 shows this situation in an energy diagram.

However, sometimes the barriers to triplet reactions are not effectively higher than those for singlets. Thus, triplet carbenes can sometimes be generated directly and their reactions observed. Like their singlet counterparts, they add to alkenes to give cyclopropanes (Fig. 10.47).

Now let's look at electron spin. For a singlet, the two nonbonding electrons on the carbene have paired (opposite) spins and can produce the two new cyclopropane carbon–carbon bonds without any problem by combining with the two paired electrons in the π bond. As we saw earlier (Fig. 10.43, p. 461), the reaction of a singlet with an alkene takes place in a single step, preserving the original stereochemical relationships of the groups on the alkene. Notice that the electrons involved have paired spins and can form two new cyclopropane bonds in a single step (Fig. 10.48).

What do we expect of the triplet? In particular, what will be the stereochemistry of the addition reaction? The first step in the addition of triplets to alkenes is shown in Figure 10.49.

In contrast to the singlet, the triplet cannot form two new bonds in a single step. A new species, called a **diradical**, is produced, with two nonbonding electrons on different carbons. Now what can happen? Why not simply close up to the three-membered ring and be done with it? Look closely at the spins of the electrons involved in the reaction. The second bond cannot close! The two electrons that would make up the new bond have the same spin and cannot occupy the same orbital (Fig. 10.50).

FIGURE 10.47 Triplet carbenes also add to alkenes to give cyclopropanes.

Singlet carbene   Spins in π bond are paired   All spins paired

FIGURE 10.48 In the addition of a singlet carbene to an alkene, two new bonds are formed in single step.

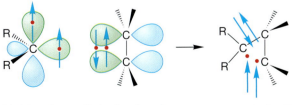

A triplet carbene   Spins in π bond are paired   A triplet 1,3-diradical (two electrons must have the same spin)

FIGURE 10.49 Only one bond can be formed in the reaction of the triplet with an alkene. The first step in triplet addition is reaction with the π bond of the alkene to give a 1,3-diradical.

FIGURE 10.50 The triplet diradical cannot close the second bond.

Triplet diradical

This bond cannot close in the triplet— both electrons have the same spin! In the singlet this bond can close because the two electrons have opposite spins

Singlet diradical

The triplet, with two nonbonding electrons of the same spin, must produce an intermediate that cannot close to a cyclopropane until a spin is somehow changed. Such spin "flips" are not impossible, but they are usually slow on the time scale of molecular processes. In particular, they are slow compared to the very rapid rotations about carbon–carbon single bonds. Therefore the stereochemistry of the original alkene need not be maintained in the eventual product of the reaction, the cyclopropane (Fig. 10.51). If rotation about the carbon–carbon bond is fast compared to changing of the spin quantum number of one electron (spin flip), it will not matter whether cis or trans alkene is used in the reaction with triplet methylene. The product will be a mixture of cis and trans 1,2-dialkylcyclopropanes in either case.

THE GENERAL CASE

cis Cyclopropane

rotation

trans Cyclopropane

FIGURE 10.51 Triplet carbene addition to alkenes. The formation of an intermediate 1,3-diradical induces scrambling of stereochemistry. Unlike singlets, which add in a single step to generate cyclopropanes stereospecifically, triplets give both cis and trans cyclopropanes.

(continued)

A SPECIFIC EXAMPLE

FIGURE 10.51 (continued)

It turns out that triplets do give mixtures of the cis and trans cyclopropanes, and the examination of the stereochemistry of the addition reaction of carbenes has become the method of choice for determining the spin state of the reacting carbene.

This discussion of triplet carbenes and diradicals leads directly to Chapter 11, which will concentrate on the chemistry of radicals: species with a single nonbonding electron.

SUMMARY

Carbenes come in two flavors—the spin-paired singlets and the spin-unpaired triplets. Often the triplet is the more stable form, but the singlet state is usually more reactive. Both singlet and triplet carbenes add to alkenes to give cyclopropanes. In the one-step reaction of the singlets, any stereochemistry present in the alkene is preserved. In the multistep additions of triplets, stereochemistry is lost.

10.5 DIPOLAR ADDITION REACTIONS: OZONOLYSIS AND THE SYNTHESIS OF CARBONYL ($R_2C=O$) COMPOUNDS

So far, we have seen a number of addition reactions leading to transient (bromonium ions, p. 442) or stable three-membered rings (oxiranes and cyclopropanes, p. 452). Now other additions leading to different sized rings appear. These reactions lead in new directions, and their products are not at all similar. Nonetheless they are tied together and to earlier material by a mechanistic thread—they involve an initial addition to an alkene to give a cyclic intermediate.

10.5a The Mechanism of Ozonolysis. The precursors of carbenes, diazo compounds, are members of a class of molecules called **1,3-dipoles**, or **1,3-dipolar reagents**. They are so-called because good neutral resonance forms cannot be written for them. Figure 10.52 shows a number of typical 1,3-dipoles (see Problem 10.16).

FIGURE 10.52 Some 1,3-dipolar reagents.

An important characteristic of 1,3-dipoles is that they add to alkenes and alkynes to give five-membered rings. Figure 10.53 shows the addition of a substituted diazomethane to an alkyne to give a pyrazole.

FIGURE 10.53 1,3-Dipolar addition to an alkyne.

A pyrazole

PROBLEM 10.16 Draw more resonance forms for the 1,3-dipoles in Figure 10.52.

PROBLEM 10.17 Use the arrow formalism to show the compounds formed from reaction of the azides, nitrile oxides, and carbonyl oxides shown in Figure 10.52 with a typical alkene and alkyne.

Ozone (O_3) is a particularly important 1,3-dipole (Fig. 10.54). The ozone molecule is often in the news these days. There is too little of it in the upper atmosphere, where it functions to screen the surface of the earth from much dangerous ultraviolet (UV) radiation. Yet, in all-too-many places near the surface, there is too much of it. Ozone is a principal component of photochemical smog.

FIGURE 10.54 A 1,3-dipolar addition of ozone to an alkene gives a primary ozonide, a five-membered ring containing three oxygen atoms in a row.

Ozone (O_3)

A primary ozonide

The first step in the process called **ozonolysis** involves the reaction of ozone with most alkenes to give the ozonide shown in Figure 10.54. The arrow formalism lets us map out the formation of the ozonide. The π bond of the alkene reacts with ozone to produce a five-membered ring containing three oxygens, the **primary ozonide** or **molozonide**. But these primary ozonides are extremely difficult to handle, and it took great experimental skill on the part of Rudolf Criegee (1902–1975) and his co-workers at the University of Karlsruhe in Germany to isolate the ozonide from addition to *trans*-di-*tert*-butylethylene. A concerted mechanism predicts that the stereochemical relationship of the alkyl groups in the original alkene will be preserved in the ozonide, and this is what happens (Fig. 10.55).

FIGURE 10.55 The stereospecific formation of a primary ozonide.

As implied previously, unless great care is taken these initially formed ozonides are not stable, but rearrange to other compounds, known simply as ozonides (Fig. 10.56).

FIGURE 10.56 The rearrangement of a primary ozonide into a new ozonide.

The rearrangement of primary ozonides into these other ozonides involves a pair of reverse and forward 1,3-dipolar additions. The first step in the rearrangement is the opening of the primary ozonide to give a molecule containing a carbon–oxygen double bond called a **carbonyl compound**, along with a molecule called a carbonyl oxide, a new 1,3-dipole (Fig. 10.52). These two species can recombine to give the ozonide (Fig. 10.57).

FIGURE 10.57 The mechanism of rearrangement involves two steps. In the first, a reverse 1,3-dipolar addition generates a carbonyl compound and a 1,3-dipole, a carbonyl oxide. The carbonyl oxide readds to the carbonyl in the other sense to give the new ozonide.

The mechanism of rearrangement from one ozonide to another may look a bit bizarre, and there can be no hiding the fact that it's not simple. It is not really far from things we already know, however. More frustrating is our distance from being able to predict the course of complicated reactions. At this point, you have little choice but to learn the course of the ozonolysis reaction, and that is not intellectually satisfying. It would be a far better situation if you were able to predict what must happen in this multistep, multi-intermediate process. Perhaps it is some comfort to know that you are not at all far behind the current frontier. Organic chemists have only recently emerged from the "gee whiz" phase to the point where advances in experimental skill (mostly in analytical techniques, if one is to be honest) and theoretical methods have begun to allow reliable predictions to be made in risky cases. All too often we are reduced to making explanations after the fact; to rationalizing, as we are just about to do with the rearrangement from one ozonide to the other.

We will take it one step at a time, working backward. First of all, the final addition of the carbonyl oxide, another 1,3-dipole, to a carbon–oxygen double bond is very little removed from the addition of ozone to an alkene. Like ozone, the carbonyl oxide can add to π bonds. In this case, such an addition leads to the observed ozonide (Fig. 10.58).

Ozone   A primary ozonide   A carbonyl oxide   The final ozonide

FIGURE 10.58 Like diazo compounds and ozone, a carbonyl oxide is a 1,3-dipole and adds to π systems to give five-membered ring compounds.

The first step in the conversion of one ozonide into the other is another 1,3-dipolar addition, but this time in reverse, and it is therefore a bit harder to see (Fig. 10.59). The reaction is completed when the carbonyl oxide turns over and readds to the carbonyl compound (Fig. 10.58).

FIGURE 10.59 The first step in the formation of the ozonide from the primary ozonide is a reverse 1,3-dipolar addition to generate a carbonyl compound and a carbonyl oxide.

reverse 1,3-dipolar addition

Carbonyl compound   +   Carbonyl oxide

The overall reaction is a series of three 1,3-dipolar additions: First, ozone adds to the alkene to give the primary ozonide. The primary ozonide undergoes a reverse 1,3-dipolar addition to give a carbonyl compound and a new 1,3-dipole, the carbonyl oxide. Finally, the carbonyl oxide readds to the carbonyl compound in the opposite sense to give the new ozonide (Fig. 10.60).

The reaction is a sequence of three reversible 1,3-dipolar additions driven by thermodynamics toward the relatively stable ozonide. We have now described the reaction; the tougher part is to explain why it runs the way it does. Why does thermodynamics favor the final, more stable ozonide over the primary ozonide (Fig. 10.61)?

The oxygen–oxygen bond is quite weak (~40 kcal/mol). Much more stable, though, is the carbon–oxygen bond (85–90 kcal/mol). The most important factor in

FIGURE 10.60 In the formation of the final ozonide from ozone and an alkene, Steps 1 and 3 are 1,3-dipolar additions. Step 2 is the reversal of a 1,3-dipolar addition.

making the final ozonide more stable than the primary ozonide is the difference in the numbers of weak oxygen–oxygen bonds and strong carbon–oxygen bonds. As long as the kinetic barriers to the forward and reverse 1,3-dipolar additions are not too high, thermodynamics will favor formation of the stronger carbon–oxygen bonds and the ultimate result will be the final, more stable, ozonide.

PROBLEM 10.18 Use the table of BDE (Table 8.2, p. 355) to do a full calculation of the bond energy difference between the primary and final ozonide.

PROBLEM 10.19 Why are oxygen–oxygen bonds so weak? *Hint*: It is not just electron–electron repulsion. Do a quick molecular orbital analysis.

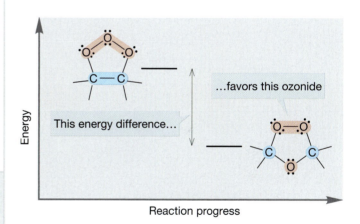

FIGURE 10.61 The ozonide contains fewer weak oxygen–oxygen bonds than the primary ozonide and is more stable.

PROBLEM 10.20 Complete the Energy versus Reaction progress diagram of Figure 10.61 to show the conversion of the primary ozonide into the final ozonide.

10.5b The Synthetic Potential of Ozonolysis: Carbonyl Compounds. Why do we spend so much time with this complicated reaction? We have already mentioned generality. A vast number of 1,3-dipolar agents—the charge-separated molecules known as **zwitterions**—have been made, and many will add to alkenes and alkynes to give five-membered rings.

Second, and more important, these ozonides are very useful compounds. Their further transformations give us entry into new classes of compounds. Depending on the structure of the starting alkene, reactions of the ozonide can yield **aldehydes**, **ketones**, or **carboxylic acids**. If the ozonide is reduced, the products are aldehydes or ketones, depending on the structure of the ozonide. Many reducing agents are known, among them H_2/Pd, Zn, and $(CH_3)_2S$ (dimethyl sulfide) (Fig. 10.62).

SPECIFIC EXAMPLES

FIGURE 10.62 Reductive decomposition of ozonides leads to aldehydes and/or ketones.

Oxidation of the ozonides, usually with hydrogen peroxide (H_2O_2), leads either to ketones or carboxylic acids. The structure of the product again depends on the structure of the ozonide, which itself is dependent on the structure of the original alkene (Fig. 10.63).

GENERAL CASES

SPECIFIC EXAMPLES

FIGURE 10.63 Oxidative decomposition of ozonides leads to ketones and/or carboxylic acids.

This reaction provides an important synthetic opportunity. We now have syntheses of aldehydes (RCH=O), ketones (R₂C=O), and carboxylic acids (RCOOH). In addition, the structures of the products of an ozonolysis reaction can be used to reason out the composition of an unknown alkene. Try Problem 10.21.

WORKED PROBLEM 10.21 Ozonolysis of alkenes can give the products shown below. Supply structures for the starting alkenes. Specify the nature of the workup step in the process (oxidation or reduction?).

(a)

(b)

(c)

[from two starting materials,
(1) C₆H₁₀ (2) C₁₂H₂₀]   *(continued)*

WORKED PROBLEM 10.21
(*continued*)

(d)

$$\text{HO} \overset{O}{\underset{}{\parallel}} \text{C} \cdots \text{CH}_2\text{CH}_2\text{CH}_2 \overset{O}{\underset{\parallel}{C}} \text{OH} \quad + \quad \text{H} \overset{O}{\underset{\parallel}{C}} \text{OH}$$

ANSWER (c) Ozonolysis converts carbon–carbon double bonds into two carbon–oxygen double bonds.

$$\begin{array}{c} \diagup \\ C=C \\ \diagup \quad \diagdown \end{array} \xrightarrow{\text{O}_3} \begin{array}{c} \diagup \\ C=O \end{array} + \begin{array}{c} O=C \\ \diagdown \end{array}$$

In this case, both new carbon–oxygen double bonds are within the same product molecule, which means that there must be a cyclic starting material. The product contains carboxylic acids, so we know that an oxidative workup is required. A reductive workup would give an aldehyde.

$$\begin{array}{c} H \quad\quad H \\ C=C \\ \end{array} \xrightarrow[\text{2. oxidative workup}]{\text{1. O}_3} \begin{array}{c} HO \quad\quad OH \\ C=O + O=C \\ \end{array}$$

There are several ways to construct an appropriate starting material. Two are shown below.

$$\begin{array}{c} \text{(C}_6\text{H}_{10}) \end{array} \xrightarrow[\text{2. HOOH}]{\text{1. O}_3} \begin{array}{c} \text{COOH} \\ \text{COOH} \end{array} \xleftarrow[\text{2. HOOH}]{\text{1. O}_3} \begin{array}{c} \text{HC}=\text{CH} \\ \text{HC}=\text{CH} \end{array} \quad \text{(C}_{12}\text{H}_{20})$$

Dihydroxylation of alkene

10.5c Oxidation with Permanganate or Osmium Tetroxide. Although they are not 1,3-dipolar reagents, both potassium permanganate (KMnO₄) and osmium tetroxide (OsO₄) add to alkenes to form five-membered rings in reactions that are related to the additions of ozone and other 1,3-dipoles (Fig. 10.64).

FIGURE 10.64 Addition of potassium permanganate or osmium tetraoxide to alkenes gives five-membered rings.

A cyclic osmate ester

The cyclic intermediate in the permanganate reaction cannot be isolated and is generally decomposed as it is formed to give vicinal diols (1,2-diols or 1,2-glycols) (Fig. 10.65). Good yields of cis-1,2-diols (glycols, Chapter 6, p. 249) can be isolated from the treatment of alkenes with basic permanganate. The cyclic osmate ester can be isolated, but it, too, is generally transformed directly into a diol product by treatment with aqueous sodium sulfite (Na_2SO_3). In both of these reactions it is the metal–oxygen bonds, not the carbon–oxygen bonds, that are broken. There are many variations of these reactions, which are among the best ways of synthesizing 1,2-diols.

GENERAL CASES

SPECIFIC EXAMPLES

FIGURE 10.65 These metal-containing five-membered rings can react further to generate 1,2-diols, which are also called 1,2-glycols.

PROBLEM 10.22 What mechanistic conclusions can be drawn from the observation that reaction of OsO_4 with cis-2-butene gives the single product shown below?

cis-2-Butene

meso-2,3-Dihydroxybutane
(shown in eclipsed
conformation for clarity)

SUMMARY

Ozonolysis of alkenes breaks both bonds in the carbon–carbon double bond and generates a pair of carbonyl groups. When there are two alkyl groups on one or both ends of the alkene, the product is a ketone ($R_2C=O$) regardless of workup conditions. When there is a hydrogen on one or both ends of the alkene, the product is an aldehyde ($RHC=O$) when the workup is done under reducing conditions and a carboxylic acid ($RCOOH$) when oxidizing conditions are used.

10.6   ADDITION REACTIONS OF ALKYNES: H—X ADDITION

Much of Chapters 9 and 10 has been devoted to the addition of various reagents to carbon–carbon double bonds. Alkynes contain two π bonds and it would be quite astonishing if they did not undergo similar addition reactions. If we keep in mind what we have learned about alkene additions, it is reasonably easy to work through alkyne additions. The presence of the second double bond will add mechanistic complications, however, and some synthetic opportunities as well.

Addition of H—Br or H—Cl to 3-hexyne gives the corresponding vinyl halides. It is tempting to begin with a protonation of the π system to give a carbocation and a halide ion. In this case, the positive ion would be a vinyl cation. Addition of the halide to the cation would give the vinyl halide. The stereochemistry of this addition is mixed, as would be expected from an open cation, although the trans diethyl compound is favored (Fig. 10.66).

FIGURE 10.66 Hydrogen halides add to alkynes as well as to alkenes.

A vinyl cation

(*Z*)-3-Chloro-3-hexene
(the ethyl groups in the product are mostly trans to each other)

sp Hybridized carbon

FIGURE 10.67 A viniyl cation contains an *sp* hybridized carbon.

However, there are problems with this simple extension to alkynes of the mechanism for additions to alkenes. Vinyl cations are unusual species and we must stop for a moment to consider their structures. The positive carbon in a vinyl cation is attached to two groups: the other carbon in the original triple bond and an R group. Accordingly, we would expect *sp* hybridization, and that leads to the structure in Figure 10.67.

Measurements in the gas phase show that vinyl cations are very unstable. Remember that these gas-phase values do not include the solvent, which will be very important to any cationic species. Remember also that direct comparisons can be made only between isomeric species. Nonetheless, one can get an approximate idea of the great instability of vinyl cations from this table. The absolute values of the energy differences are not important, but the relative stability order of the various carbocations is (Table 10.2).

TABLE 10.2 Heats of Formation for Some Carbocations

Cation	Substitution	ΔH_f° (kcal/mol)
$^+CH=CH_2$	Primary vinyl	285
$^+CH_3$	Methyl	261.3
$^+CH=CH-CH_3$	Primary vinyl	252 (calculation)
$CH_3-C\overset{+}{=}CH_2$	Secondary vinyl	231
$^+CH_2CH_3$	Primary	215.6
$^+CH_2CH_2CH_3$	Primary	211
$^+CH_2CH_2CH_2CH_3$	Primary	203
$H_3C\overset{+}{C}HCH_2CH_3$	Secondary	183
$(CH_3)_2\overset{+}{C}H$	Secondary	190.9
$CH_2=CH-\overset{+}{C}H_2$	Allyl	226
$(CH_3)_3C^+$	Tertiary	165.8

Table 10.2 shows that, just as with alkyl carbocations, secondary vinyl cations are more stable than the primary species. However, it also shows that even a secondary vinyl cation is only about as stable as a primary alkyl carbocation. Primary carbocations serve as a kind of mechanistic stop sign: They are generally considered too unstable to be viable intermediates in most reactions. Presumably, we should treat unstabilized vinyl cations the same way. What intermediate could we use to replace the highly unstable vinyl cations in additions to acetylenes? Perhaps cyclic intermediates are involved. Moreover, acetylenes are known to form complexes with hydrohalic acids (HX). A cyclic protonium ion or complex could accommodate the complicated kinetics, which show that more than one molecule of halide is involved, and also account for the generally observed predominance of trans addition (Fig. 10.68). We will write cyclic ions in the answers to problems, but you should know that the intermediacy of vinyl cations in these reactions is not a fully resolved issue. There is still lots to do in organic chemistry.

FIGURE 10.68 Possible cyclic intermediates in the reaction of an alkyne and hydrogen chloride.

This modified mechanism for addition to alkynes also predicts that Markovnikov addition should be observed: that the more substituted secondary vinyl halide should be formed whenever there is a choice. In the cyclic intermediate, there will be partial positive charge on the carbons of the original acetylene. The partial bond to hydrogen from the *more* substituted carbon of an unsymmetrical acetylene will be weaker than the bond from the *less* substituted carbon because the partial positive charge will be more stable at the more substituted position. Accordingly, addition of halide would be expected to take place more easily at the more substituted position leading to Markovnikov addition (Fig. 10.69).

FIGURE 10.69 Addition of the nucleophile to the cyclic ion will take place at the more substituted positions.

Let's look at the experimental data. There is no difference in the two ends of the molecule such as 3-hexyne, so let's examine the addition to a terminal acetylene—a 1-alkyne—to see if Markovnikov addition occurs. Our mechanism predicts that the more substituted halide should be the product, and it is. The products shown are the only addition products isolated (Fig. 10.70).

FIGURE 10.70 Markovnikov addition generally predominates in reactions of HX with alkynes.

In addition to the mechanistic complexity, additions of HX to acetylenes are usually not practical sources of vinyl halides. The vinyl halide products also contain π bonds and often compete favorably with the starting alkyne in the addition reaction. The vinyl halide products are not symmetrical and two products of further reaction are possible. It makes a nice problem to figure out which way the second addition of H—X should go.

As always in a mechanistic problem of this sort, the answer only appears through an analysis of the possible pathways for further reaction. Draw out both mechanisms, compare them at every point, and search each step for differences. We will use the reaction of propyne with hydrogen chloride as a prototypal example. The initial product is 2-chloropropene, as shown in Figure 10.70. There are two possible ions from the further reaction of the vinyl halide with hydrogen chloride. In (a), the positive charge is adjacent to a chlorine, but in (b) it is not (Fig. 10.71).

2-Chloropropene

(a)

2,2-Dichloropropane

(b)

1,2-Dichloropropane

Two possible carbocations

FIGURE 10.71 2-Chloropropene can protonate in two ways.

In (a), the cation is adjacent to chlorine, and there is a way for the chlorine to share the positive charge. The charge is shared by the two atoms through $2p$–$3p$ overlap. In molecular orbital terms, we see that a filled $3p$ orbital on chlorine overlaps with the empty $2p$ orbital on carbon, and this filled–empty overlap is stabilizing. The simple primary carbocation (b) is much less stable, and products from it are not observed (Fig. 10.72).

Resonance-stabilized cation

2,2-Dichloropropane

$3p$–$2p$ Overlap

FIGURE 10.72 The cation adjacent to chlorine is stabilized by resonance; the other is not.

So, addition of a second halogen will preferentially give the **geminal** dihalide (geminal means 1,1-disubstituted) not the vicinal (1,2-disubstituted) compound because the cation adjacent to chlorine is lower energy, and leads inevitably to the observed geminal product. The reactions of Figure 10.70 are complicated by the formation of compounds arising through double addition of hydrogen halide. The double addition always occurs so as to give the geminal dihalide (Fig. 10.73).

FIGURE 10.73 Addition of hydrogen halides to acetylenes gives mixtures of products of mono- and diaddition.

10.7   ADDITION OF X₂ REAGENTS TO ALKYNES

The pattern of alkene addition is also followed when X_2 reagents add to alkynes, although the mechanisms of these reactions are not well worked out. Both ionic and neutral intermediates (i.e., radicals, see Chapter 11) are involved. Alkynes add Br_2 or Cl_2 to give vicinal,1,2–dihalides. A second addition can follow to give the tetrahalides (Fig. 10.74).

FIGURE 10.74 Addition of Cl_2 (or Br_2) to alkynes gives both di- and tetrahalides.

10.8 HYDRATION OF ALKYNES

Like alkenes, acetylenes can be hydrated. The reaction is generally catalyzed by mercuric ions in an oxymercuration process (see Section 10.3, p. 449, for the related alkene reaction), although simple acid catalysis is also known (Fig. 10.75). In contrast to the oxymercuration of alkenes, no second, reduction step is required.

FIGURE 10.75 The oxymercuration of alkynes resembles the oxymercuration of alkenes. In the alkyne case, the product is an enol, which can react further.

By strict analogy to the oxymercuration of alkenes, the product should be a hydroxy mercury compound, but the second double bond exerts its influence and further reaction takes place. The double bond is protonated and mercury is lost to generate a species called an **enol**. This molecule is part alk*ene* and part alcoh*ol*, hence the name.

Enols are extraordinarily important compounds, and more than one chapter will be devoted to their chemistry. Here their versatility is exemplified by their conversion into ketones. Ketones are compounds in which the carbon of a carbon–oxygen double bond is attached to two other carbons. We have already seen a synthesis of ketones in this chapter through ozonolysis of alkenes (p. 465). Here, they are the final products of the hydration reactions of alkynes. The mechanism of their formation from enols is simple; the alkene part is protonated on carbon and deprotonated on oxygen to generate the compound containing the carbon–oxygen double bond. Note that this gives you another synthesis of carbonyl compounds (Fig. 10.76).

Nearly all ketones and aldehydes containing a hydrogen on the carbon adjacent to the carbon–oxygen double bond are in equilibrium with related enol forms. How much enol is present at equilibrium is a function of the detailed structure of the molecule, but there is almost always some enol present. In simple compounds, the ketone form is greatly favored (Chapter 17).

FIGURE 10.76 Protonation of carbon, followed by deprotonation at oxygen, generates the carbonyl compound. This sequence is a general reaction of enols.

PROBLEM 10.23 Draw a mechanism for the acid-catalyzed hydration of an acetylene.

WORKED PROBLEM 10.24 Why was I so sure the enol would protonate at the end to give the cation shown in Figure 10.76? There is another cation possible. Why is it not formed?

ANSWER Protonation in the sense shown leads to a resonance-stabilized carbocation. The adjacent oxygen shares the positive charge. Deprotonation gives the ketone. Protonation in the other sense leads to a carbocation that is not resonance stabilized, and which is therefore of much higher energy.

Resonance stabilized!

Not resonance stabilized!

PROBLEM 10.25 In general, hydration of unsymmetrical acetylenes such as 2-pentyne is not a practical source of ketones. Why not?

PROBLEM 10.26 Terminal acetylenes are certainly unsymmetrical, yet in these cases hydration is a useful process. In principle, either ketones or aldehydes could be formed, but in practice only the ketones are produced. Write mechanisms for both ketone and aldehyde formation and explain why the product is always the ketone. You may write open vinyl cations in your answer for simplicity, but be aware that the real intermediate(s) may be more complex cyclic species.

10.9   HYDROBORATION OF ALKYNES

Like alkenes, alkynes can be hydroborated by boranes. The mechanism is very similar to that for alkenes (Chapter 9, p. 416); the stereochemistry of addition of HBH_2 is cis, and it is the boron that becomes attached to the less substituted end of the starting alkyne (Fig. 10.77).

FIGURE 10.77 Hydroboration of alkynes gives vinylboranes that can react further (R = alkyl).

The initial products of hydroboration are prone to further hydroboration reactions. This complication can be avoided by using a sterically hindered borane so that further reaction is slowed. A favorite reagent is the borane formed by reaction of two equivalents of 2-methyl-2-butene with BH_3 (Fig. 10.78).

FIGURE 10.78 This sterically hindered dialkylborane reacts only once with alkynes to give an isolable vinylborane (R = alkyl).

Like alkylboranes, these alkenylboranes can be converted into alcohols by treatment with basic peroxide (Fig. 10.79). The products are enols, and they are in equilibrium with the corresponding carbonyl compounds (Fig. 10.79). Here's an important point: This reaction has the opposite regiochemistry from the hydration reaction of alkynes. This situation is exactly the same as the one that exists with the alkenes: Hydration of an alkene gives overall Markovnikov addition, whereas the hydroboration–oxidation sequence gives the anti-Markovnikov product. The ultimate formation of carbonyl compounds disguises the matter, but if we look back one step to the enols,

THE GENERAL CASE

Enol   Aldehyde

A SPECIFIC EXAMPLE

FIGURE 10.79 Oxidation of the vinylborane leads to an enol that can be hydrolyzed to the aldehyde.

(88%)

we can see that in alkyne reactions the same regioselectivity holds. Hydration gives Markovnikov addition and, ultimately, the ketone. The hydroboration–oxidation reaction gives anti-Markovnikov addition, and, ultimately, the aldehyde (Fig. 10.80).

Hydration

Ketone

Hydroboration

Aldehyde

FIGURE 10.80 It is possible to make two carbonyl compounds from a terminal alkyne. Mercuric ion-catalyzed hydration gives the ketone, and hydroboration–oxidation the aldehyde. Be sure you see the mechanistic differences that require the different products.

Be sure you see the similarities here as well as the mechanistic underpinnings of the different paths followed. There is too much to memorize. Be sure to annotate your file cards as you update your catalog of synthetic methods.

10.10   HYDROGENATION OF ALKYNES: AN ALKENE SYNTHESIS THROUGH SYN HYDROGENATION

Like alkenes, alkynes can be hydrogenated in the presence of many catalysts. Alkenes are the first products, but they are not stable under the reaction conditions and are usually hydrogenated themselves to give alkanes (Fig. 10.81). This reaction is your second synthesis of alkanes.

THE GENERAL CASE

FIGURE 10.81 Hydrogenation of an alkyne proceeds all the way to the alkane stage. The intermediate alkene cannot usually be isolated.

It would be extremely convenient to be able to stop at the alkene stage, and special, "poisoned" catalysts have been developed to do just this. A favorite is the Lindlar catalyst—Pd on calcium carbonate that has been treated with lead acetate in the presence of certain amines (H. H. M. Lindlar, b. 1909). For many years it was thought that the lead treatment poisoned the palladium catalyst through some sort of alloy formation, rendering it less active and the reaction more selective. It now appears that the treatment forms no alloys, but does modify the surface of the palladium. Regardless, the reaction can be stopped at the alkene stage, and the alkenes formed by hydrogenation with Lindlar catalyst have cis stereochemistry, as the usual syn addition of hydrogen has occurred (Fig. 10.82).

A SPECIFIC EXAMPLE

FIGURE 10.82 If the Lindlar catalyst is used, the cis alkene can be obtained.

10.11   REDUCTION BY SODIUM IN AMMONIA: AN ALKENE SYNTHESIS THROUGH ANTI HYDROGENATION

In a reaction that must at first seem mysterious, an alkyne dissolved in a solution of sodium in liquid ammonia forms the trans alkene (Fig. 10.83). The practical consequences of this are that you now have a way of synthesizing either a cis or trans alkene from a given alkyne. So, update your file cards.

$$CH_3(CH_2)_2-C{\equiv}C-(CH_2)_2CH_3 \xrightarrow[\substack{Pd/CaCO_3/Pb \\ \text{Lindlar catalyst}}]{H_2}$$

(structure: cis-alkene with H, H on top; $CH_3(CH_2)_2$ and $(CH_2)_2CH_3$ on bottom)

96% cis

FIGURE 10.83 An alkyne can give either the cis or trans alkene. Hydrogenation with Lindlar catalyst gives the cis alkene, whereas reduction with sodium in ammonia gives the trans.

$$CH_3CH_2-C{\equiv}C-(CH_2)_3CH_3 \xrightarrow{Na/NH_3}$$

(structure: trans-alkene with H and $(CH_2)_3CH_3$ on top; CH_3CH_2 and H on bottom)

98% trans

When sodium is dissolved in ammonia, it loses its odd electron to give a sodium ion and a **solvated electron** (Fig. 10.84).

$$Na{\cdot} + NH_3 \longrightarrow Na^+ + e^-[NH_3]_n$$

Solvated
electron

FIGURE 10.84 Formation of a solvated electron.

This complexed electron can add to an acetylene to give a species called a **radical anion** (Fig. 10.85).

FIGURE 10.85 A solvated electron can add to the alkyne to generate an acetylene radical anion.

$$H_3C-C{\equiv}C-CH_3 + e^-[NH_3]_n \longrightarrow NH_3 + \left[H_3C-C{\equiv}C-CH_3\right]^{\cdot-}$$

A radical anion

I vividly recall being quite uncertain about the structure of this species when I was first studying organic chemistry. I suspect I solved the problem of dealing with it by memorizing this reaction and hoping that no one would ever ask me to explain what was happening. But it's not so hard and you should not repeat my error. First of all, where is the "extra" electron in the radical anion? It is most certainly not in one of the bonding π orbitals of the acetylene, as they are each filled with two electrons already. If it is not there, it must be in one of the antibonding, π^*, orbitals (Fig. 10.86).

FIGURE 10.86 The alcohol and electron must go into an antibonding π^* orbital.

An added electron must go here

$$H_3C-C{\equiv}C-CH_3 = H_3C-C{=}C-CH_3$$

π^* _____

π _____

Why should the molecule put up with this electron in an anti-bonding orbital? The answer is twofold. First of all, the molecule is held together by many electrons occupying bonding orbitals, and a single electron in a high-energy orbital is not destabilizing enough to overcome all the bonding interactions. Figure 10.87 gives both molecular orbital and resonance pictures of the radical anion.

Second, the radical anion *is* quite unstable and not much of it will be formed at equilibrium. Its very instability contributes to its great base strength, however, and the small amount present at equilibrium is rapidly protonated by ammonia to give the amide ion ($^-NH_2$) and the vinyl radical (Fig. 10.88).

FIGURE 10.87 The radical anion.

FIGURE 10.88 The radical anion can abstract a proton from ammonia to generate a vinyl radical and the amide ion. The trans form, in which the alkyl groups are as far from each other as possible, will be favored.

There are two forms of the vinyl radical, and it will be the trans form, with the alkyl groups as far from each other as possible, that is the more stable. A solvated electron can be donated to this radical forming the vinyl anion, which is rapidly protonated by ammonia to give the final product, the trans alkene (Fig. 10.89).

FIGURE 10.89 Transfer of another electron and protonation completes the overall trans hydrogenation.

SUMMARY

Normal hydrogenation of an alkyne adds two equivalents of H_2 to give the alkane. Use of a "poisoned" Lindlar catalyst leads to absorption of one equivalent of H_2 and formation of the cis alkene. The trans alkene can be made by reducing the alkyne with Na in ammonia.

10.12 SPECIAL TOPIC: THREE-MEMBERED RINGS IN BIOCHEMISTRY

Despite the strain inherent in a cyclopropane, nature finds ways to make them. Bacteria, in particular, contain large amounts of cyclopropanated fatty acids. Even more remarkable is the further conversion of such molecules into cyclopropenes. The source of the single "extra" carbon is *S*-adenosylmethionine, the same agent involved in the methyl transfers discussed in Chapters 3 and 7 (pp. 145, 300; Fig. 10.90). The mechanism of this intriguing change isn't worked out yet.

FIGURE 10.90

Cyclopropanation of a fatty acid is followed by dehydrogenation.

PROBLEM 10.28 Devise syntheses for the following compounds, starting in each case from 2-butyne:

Cyclopropanes are occasionally found in Nature, and one class of naturally occurring cyclopropanes, the pyrethroids, are useful insecticides. These compounds are highly toxic to insects but not to mammals, and are rapidly biodegradable. These cyclopropanes do not persist in the environment. The naturally occurring pyrethroids are found in the members of the chrysanthemum family and are formally derived from chrysanthemic acid. Many modified pyrethrins not found in Nature have been made in the laboratory and several are widely used as pesticides. The molecule shown is pyrethrin I (Fig. 10.91), or, more systematically, 2,2-dimethyl-3-(2-methyl-1-propenyl)cyclopropanecarboxylic acid 2-methyl-4-oxo-3-(2,4-dipentadienyl)-2-cyclopenten-1-yl ester. See why the shorthand is used?

In Chapter 12, we will see other three-membered rings, epoxides, used as triggers in the construction of the biologically active molecules called steroids, and, amazingly enough, one of the most prominent molecules in interstellar space is not only a cyclopropene, but a carbene as well. This molecule, cyclopropenylidene, is shown in Figure 10.92.

Pyrethrin I

FIGURE 10.91 Structure of pyrethrin I.

Cyclopropenylidene

FIGURE 10.92
Cyclopropenylidene.

10.13 SUMMARY

New Concepts

This chapter continues our discussion of addition reactions, and many of the mechanistic ideas of Chapter 9 apply. For example, many polar addition reactions start with the formation of the more stable carbocation. The most important new concept in this chapter is the requirement for overall trans addition introduced by three-membered ring intermediates.

This concept appears most obviously in the trans bromination and chlorination of alkenes. The openings of the intermediate bromonium and chloronium ions by nucleophiles are S_N2 reactions in which inversion is always required. This process insures trans addition (Fig. 10.13).

Remember also the formation of oxiranes, stable three-membered rings containing an oxygen atom, and cyclopropane formation through carbene additions to alkenes (Fig. 10.93).

FIGURE 10.93 Stable three-membered rings can be formed in some cases. Two examples are epoxidation and carbene addition.

Alkenes and alkynes react in similar fashion in most addition reactions, although additional complications and opportunities are introduced by the second π bond in the alkynes.

Key Terms

aldehydes (p. 470)
bromonium ion (p. 444)
carbenes (p. 460)
carbonyl compound (p. 467)
carboxylic acids (p. 470)
diimide (HN=NH) (p. 439)
1,3-dipolar reagents (p. 465)
1,3-dipoles (p. 465)
diradical (p. 463)
enol (p. 479)

epoxides (p. 452)
geminal (p. 478)
1,2-glycol (p. 455)
heterogeneous catalysis (p. 438)
homogeneous catalysis (p. 438)
hydrogenation (p. 438)
ketones (p. 470)
molozonide (p. 466)
oxiranes (p. 452)
oxymercuration (p. 450)

ozonolysis (p. 466)
primary ozonide (p. 466)
radical anion (p. 484)
singlet carbene (p. 461)
solvated electron (p. 484)
triplet carbene (p. 461)
vicinal (p. 442)
zwitterions (p. 469)

Reactions, Mechanisms, and Tools

Both homogeneous and heterogeneous catalytic hydrogenation are syn additions, as both added hydrogens are transferred from the metal surface to the same side of the alkene.

In this chapter, a series of ring formations is described: the three-membered oxiranes, cyclopropanes, bromonium, chloronium, and mercurinium ions, as well as the five-membered rings formed through 1,3-dipolar additions, ozonolysis, and the additions of osmium tetroxide and potassium permanganate. Many of these compounds will react further to give the final products of the reaction (Fig. 10.94).

Cyclopropanes and oxiranes can be isolated, but the other rings react further under the reaction conditions to give the final products. Oxiranes can also be opened in either acid or base to give new products. Additions to alkynes resemble those of alkenes, but the second π bond adds complications, and further addition reactions or rearrangements occur. Alkynes can be reduced in either cis or trans fashion (Fig. 10.83).

FIGURE 10.94 Ring formation through addition reaction.

Synthesis

This chapter provides many new synthetic methods.

1. ACIDS

Initial product is the ozonide

2. ALCOHOLS

Markovnikov addition, no rearrangements

Organocopper reagents also work

3. ALDEHYDES

A vinyl borane and an enol are intermediates

4. ALKANES

$$R_2C=CR_2 \xrightarrow{H_2/Pt} R_2CH-CHR_2$$

$$R-C\equiv C-R \xrightarrow{H_2/Pt} RCH_2-CH_2R$$

syn Addition, many catalysts possible

5. ALKENES

$$R-C\equiv C-R \xrightarrow[\text{Lindlar catalyst}]{H_2} \text{(cis alkene, R and R on same side, H and H on same side)}$$

syn Addition

$$R-C\equiv C-R \xrightarrow{Na/NH_3} \text{(trans alkene, R and H arrangement)}$$

anti Addition

6. BROMO ALCOHOLS AND CHLORO ALCOHOLS

$$R_2C=CR_2 \xrightarrow{X_2/H_2O} R-\underset{HO}{\underset{|}{C}}(R)-\underset{R}{\underset{|}{C}}(X)-R$$

anti Addition, X = Br or Cl

7. BROMO ETHERS AND CHLORO ETHERS

$$R_2C=CR_2 \xrightarrow{X_2/ROH} R-\underset{RO}{\underset{|}{C}}(R)-\underset{R}{\underset{|}{C}}(X)-R$$

anti Addition, X = Br or Cl

8. CYCLOPROPANES

$$R_2C=CR_2 \xrightarrow[h\nu \text{ or } \Delta]{CH_2N_2} \text{(cyclopropane with } CH_2 \text{ bridge)}$$

cis Addition for singlet carbene, but both
cis and trans addition for triplet carbene

9. DIHALIDES

$$R_2C=CR_2 \xrightarrow{X_2} R-\underset{X}{\underset{|}{C}}(R)-\underset{R}{\underset{|}{C}}(X)-R$$

Vicinal dihalide, anti addition, X = Br or Cl

$$R-C\equiv C-R \xrightarrow{X_2} \text{(dihaloalkene)}$$

Vicinal dihalide, mostly anti addition,
X = Br or Cl, further reaction likely

$$R-C\equiv C-R \xrightarrow{H-X} R-\underset{H}{\underset{|}{C}}(H)-\underset{X}{\underset{|}{C}}(X)-R$$

Geminal dihalide, X = Br, Cl, or I,
double HX addition

10. 1,2-GLYCOLS

$$R_2C=CR_2 \xrightarrow[\substack{2.\ H_2O \\ Na_2SO_3}]{1.\ OsO_4} R-\underset{HO}{\underset{|}{C}}(R)-\underset{OH}{\underset{|}{C}}(R)-R$$

syn Addition

$$R_2C=CR_2 \xrightarrow[2.\ H_2O/H\bar{O}]{1.\ KMnO_4} R-\underset{HO}{\underset{|}{C}}(R)-\underset{OH}{\underset{|}{C}}(R)-R$$

syn Addition

$$R_2C\overset{O}{-}CR_2 \xrightarrow{H_3O^+/H_2O} R-\underset{HO}{\underset{|}{C}}(R)-\underset{R}{\underset{|}{C}}(OH)-R$$

anti Addition

(continued)

10. 1,2-GLYCOLS (continued)

anti Addition

11. HETEROCYCLIC FIVE-MEMBERED RINGS

This is only one example; there is a different product for every 1,3-dipole; alkynes work too

12. KETONES

An ozonide is an intermediate; other reagents will also decompose the ozonide

Forms methyl ketones, never the aldehyde

13. OXIRANES (EPOXIDES)

syn Addition

14. TETRAHALIDES

The second addition is slow; the intermediate vicinal dihalide can be isolated

15. VINYL HALIDES

Markovnikov addition and it's hard to stop here; a second addition of H–X occurs

Common Errors

What psychopathology exists in this chapter comes not so much from "something everyone always gets wrong" or from an overwhelmingly difficult concept, but from the mass of detail. There have been those who have succumbed to despair at sorting out all the stereochemical nuances, at dealing with the many new mechanisms and synthetic methods—at just making rational sense of all the material in this and Chapter 9. There certainly is beginning to be a lot of information, and you must be very careful to keep your new synthetic methods properly cataloged. There are unifying principles that may help you to keep from getting lost in the mechanism jungle. "Lewis acids react with Lewis bases." The Lewis base or nucleophile in most of the reactions in this chapter and Chapter 9 is the π bond of an alkene or an alkyne. The Lewis acids are

numerous, but most add to give intermediate, or even stable three-membered rings (not all: hydrogenation does not, and 1,3-dipolar addition leads to a host of five-membered rings). The three-membered rings are themselves often prone to further reaction with nucleophiles, and the final products of reaction may be quite far removed in structure from the starting material!

Moreover, not all the mechanistic details are known about all these processes. There are reactions about which we need much more information. The reaction of acetylenes with halogens and HX are examples, and even hydrogenation has a complicated mechanism, still somewhat obscure to organic chemists.

A good technique that helps one not to get too overwhelmed or lost is to anchor oneself in one or two basic reactions and then to generalize; to relate other reactions to the anchor reaction. For example, the polar addition of hydrogen chloride and hydrogen bromide to alkenes (Chapter 9, p. 384) is within anyone's ability to master. Extensions to hydration reactions of alkenes and alkynes, and to hydroboration become easier if the analogy with the anchor is always kept in mind. Similarly, use the reaction of alkenes with Br_2 as an anchor on which to hang the other ring-forming addition reactions. It is useful to start a set of mechanism cards to go along with your synthesis cards in order to keep track of the detail.

10.14 ADDITIONAL PROBLEMS

PROBLEM 10.29 Show the major organic product(s) expected when 1-methylcyclohexene reacts with the following reagents. Pay close attention to stereochemistry and regiochemistry where appropriate.
(a) $D_2/Pd/C$
(b) Br_2/CCl_4
(c) Br_2/CH_3OH
(d) $Hg(OAc)_2$, H_2O, then $NaBD_4$ in base (no stereochemical preference in this reaction)
(e) B_2H_6 (BH_3) in THF, then H_2O_2/HO^-
(f) CF_3COOOH
(g) CH_2N_2, $h\nu$
(h) O_3, then $(CH_3)_2S$
(i) OsO_4, then $NaHSO_3$, H_2O
(j) $HN=NH$

PROBLEM 10.30 Show in detail how both enantiomers of product are formed in Problem 10.29a.

PROBLEM 10.31 Show the major organic products expected when the following acetylenes react with the reagents shown. Pay attention to stereochemistry and regiochemistry where appropriate.

(a)

$C{\equiv}CH$ $\xrightarrow{\text{HBr}}$

(b)

$C{\equiv}CH$ $\xrightarrow{\text{Cl}_2}$

(c)

$C{\equiv}CH$ $\xrightarrow[\text{H}_2\text{O/H}_3\text{O}^+]{\text{Hg}^{2+}}$

(d)

$C{\equiv}CH$ $\xrightarrow[\text{2. H}_2\text{O}_2/\text{NaOH}]{\text{1.}}$

(e)

$C{\equiv}C$ $\xrightarrow[\text{Lindlar}\atop\text{catalyst}]{\text{H}_2}$

(f)

$C{\equiv}C$ $\xrightarrow[\text{NH}_3]{\text{Na}}$

PROBLEM 10.32 Devise syntheses for the following molecules starting from 3,3-dimethyl-1-butene (**1**). You may use any inorganic reagent (no carbons), and the following "special" organic reagents: diazomethane, carbon tetrachloride, *tert*-butyl alcohol, *tert*-butyl iodide, $Hg(OAc)_2$, carbon, $(CH_3)_2S$, trifluoroperacetic acid, strychnine, and Igor Likhotvorik's famous boiled goose-in-a-kettle. Your answers may be very short, and no mechanisms are required.

PROBLEM 10.33 What reagents would you use to convert the following starting materials into the products shown? There are no restrictions on reagents, but you must start from the molecule shown.

(continued)

PROBLEM 10.33 *(continued)*

PROBLEM 10.34 Devise syntheses for the following molecules starting from methylenecyclopentane (**1**). You may use any inorganic reagent (no carbons), and the following "special" organic reagents: diazomethane, carbon tetrachloride, *tert*-butyl alcohol, methyl alcohol, ethyl alcohol, *tert*-butyl iodide, Hg(OAc)$_2$, carbon, (CH$_3$)$_2$S, trifluoroperacetic acid, and the secret contents of Zhou Enlai's favorite veggie dumplings. Your answers may be very short, and no mechanisms are required.

PROBLEM 10.35 Provide syntheses for the following molecules starting from the indicated compound. You may use any appropriate reagents you need, including Mrs. Tao's incredible baby eels in white pepper sauce.

PROBLEM 10.36   In Section 10.2b (p. 442), we considered two possible mechanisms for the addition of Br_2 to alkenes. Ultimately, a stereochemical experiment that used a ring compound was used to decide the issue in favor of a mechanism in which addition proceeded through a bromonium ion rather than an open carbocation. Use a detailed stereochemical analysis to show how the experimental results shown below are accommodated by an intermediate bromonium ion but not by an open carbocation.

cis-2-Butene $\xrightarrow[\text{CCl}_4]{\text{Br}_2}$ Racemic dibromide

trans-2-Butene $\xrightarrow[\text{CCl}_4]{\text{Br}_2}$ Meso dibromide

PROBLEM 10.37   Explain the following regiochemical results mechanistically.

Major product

Major product

PROBLEM 10.38   In Section 10.4a (p. 452), you saw that epoxidation of an alkene with a peracid such as trifluoroperacetic acid results in addition of an oxygen atom to the alkene in such a way as to preserve the stereochemical relationships originally present in the alkene in the product epoxide. That is, cis alkene leads to cis epoxide, and trans alkene leads to trans epoxide. There is another route to epoxides that involves cyclization of halohydrins. Use cis-2-butene as a substrate to analyze carefully the stereochemical outcome of this process. What happens to the original alkene stereochemistry in the product?

A bromohydrin

PROBLEM 10.39   Treatment of cis-2-butene with Br_2/H_2O gives a product (C_4H_9OBr), that reacts with sodium hydride to give a meso compound of the formula C_4H_8O. trans-2-Butene reacts in the same sequence to give a different compound (a racemic mixture) of the same formula, C_4H_8O. Explain mechanistically, paying close attention to stereochemical relationships.

PROBLEM 10.40   The difference in nucleophilicity between differently substituted alkenes can lead to selectivity in the epoxidation reaction. Rationalize the position of faster epoxidation in the 1,4-cyclohexadiene below and explain your reasoning.

(73%)

PROBLEM 10.41   There is one alcohol that can be synthesized in one or two easy steps from each of the following precursors. What is the alcohol, and how would you make it from each starting material?

(a)   (b)   (c)

(d)

(cis or trans)

PROBLEM 10.42 Show the final products of the reactions below. Pay attention to stereochemistry.

(a)

$$CF_3COOOH \xrightarrow{} \xrightarrow[NaOH]{H_2O}$$

(b)

$$\xrightarrow[NaOH/H_2O]{KMnO_4}$$

PROBLEM 10.43 Glycols react with periodic acid to form a cyclic periodate intermediate. The periodate then decomposes to a pair of carbonyl compounds.

If the product from (b) in Problem 10.42 is cleaved with periodic acid, what would be the final product of the reaction sequence?

From a synthetic point of view, this two-step sequence—the treatment of an alkene with basic permanganate followed by periodate cleavage—is equivalent to what other direct method of cleaving carbon–carbon double bonds? What other way do you know to convert an alkene into a pair of aldehydes?

Periodate intermediate

PROBLEM 10.44 Why doesn't the product of Problem 10.42a react with periodate?

PROBLEM 10.45 Incomplete catalytic hydrogenation of a hydrocarbon, **1** (C_5H_8), gives a mixture of three hydro-

carbons, **2**, **3**, and **4**. Ozonolysis of **3**, followed by reductive workup, gives formaldehyde and 2-butanone. When treated in the same way, **4** gives formaldehyde and isobutyralde-hyde. Provide structures for compounds **1**–**4** and explain your reasoning.

PROBLEM 10.46 α-Terpinene (**1**) and γ-terpinene (**2**) are isomeric compounds ($C_{10}H_{16}$) that are constituents of many plants. Upon catalytic hydrogenation, they both afford 1-isopropyl-4-methylcyclohexane. However, on ozonolysis followed by oxidative workup, each compound yields different products. Provide structures for **1** and **2** and explain your reasoning.

PROBLEM 10.47 In practice, it is often very difficult to isolate ozonides from ozonolysis of tetrasubstituted double bonds. The product mixture depends strongly on reaction

(continued)

PROBLEM 10.47 *(continued)*
conditions, but the products shown below can all be isolated.
Construct reasonable arrow formalisms for all of them.

R = CH₃

O₃ | hexane

PROBLEM 10.48 Ozonolysis of **1** in solvent acetone leads
to **2** as the major product. Explain.

O₃
acetone

1   **2**

PROBLEM 10.49 *trans*-β-Methylstyrene reacts with phenyl
azide to give a single product, triazoline (**1**). What other
stereoisomeric products might have been produced?
Draw an arrow formalism for this reaction and explain what
mechanistic conclusions can be drawn from the formation
of a *single* isomer of **1**.

Ph—N₃

trans-β-Methylstyrene   **1**

Ph =

PROBLEM 10.50 Triazoline (**1**) (formed in Problem
10.49) decomposes upon photolysis or heating to give
aziridine (**2**). Write two mechanisms for the conversion
of **1** into **2**; one a concerted, one-step reaction, and the
other a nonconcerted, two-step process. How would you
use the stereochemically labeled triazoline (**1**) to tell which
mechanism is correct?

hν
or Δ

1   **2**
                        (no stereochemistry
                        implied in the drawing)

Ph =

PROBLEM 10.51 Explain the formation of aziridine (**2**) in
the following reaction of azide (**1**). *Hints*: Draw a full Lewis
structure for the azide, and see Section 10.4d (p. 460).

hν

1   **2**

PROBLEM 10.52 What does the formation of only cis aziridine (**2**) from irradiation of azide (**1**) in *cis*-2-butene tell you
about the nature of the reacting species in Problem 10.51?

hν

1   **2**

PROBLEM 10.53 A reaction of carbenes not mentioned in
the text is called "carbon–hydrogen insertion." In this startling
reaction, a reactive carbene ultimately places itself in between
the carbon and hydrogen of a carbon–hydrogen bond.

:CR₂   Old carbene

(a) Write two mechanisms for this reaction, one single step
and the other two steps.
(b) In 1959, in a classic experiment in carbene chemistry,
Doering and Prinzbach used ¹⁴C-labeled isobutene to

(continued)

PROBLEM 10.53 *(continued)* distinguish the two mechanisms. Explain what their results mean. The figure shows only one product of the many formed in the reaction. Focus on this compound only.

PROBLEM 10.54 Contrast the results of hydroboration–oxidation and mercury (Hg^{2+})-catalyzed hydration for 2-pentyne and 3-hexyne. Would any of these procedures be a practical preparative method? Explain.

PROBLEM 10.55 Recall that terminal alkynes are among the most acidic of the hydrocarbons (Chapter 3, p. 000), and that the acetylide ions can be used in S_N2 alkylation reactions with an appropriate alkyl halide. For example,

Provide syntheses for the following molecules, free of other isomers. You must use alkynes containing no more than four carbon atoms as starting materials. You may use inorganic reagents of your choice, and other organic reagents containing no more than two carbon atoms. Mechanisms are not required.

(a)

(b)

(c)

(d)

PROBLEM 10.56 Many additions of bromine to acetylenes give only trans dibromide intermediates. However, there are exceptions. Here is one. Explain why phenylacetylene gives both cis and trans dibromides on reaction with bromine.

Radical Reactions

11

Knowledge may have its purposes,
But guessing is always
More fun than knowing.
                    —W. H. Auden*

11.1   PREVIEW

Although there have been many variations in the details, with few exceptions all of the reactions we have studied so far have been polar ones. The S_N1 and S_N2 substitutions, the addition reactions of HX and X_2 reagents, and the E1 and E2 reactions all involve cationic and anionic intermediates, or at least have obviously polar transition states.

Now we come to a series of quite different processes involving the decidedly nonpolar, neutral intermediates called **radicals** or, sometimes, **free radicals**. We will also encounter **chain reactions**, in which a small number of radicals can set a cycling reaction in motion. In such a case, a few starter molecules can determine the course of an entire chemical process.

The question of selectivity is prominent in this chapter. What factors influence the ability of a reactive species to pick and choose among various reaction pathways? In particular, what is the connection between the energy of a reactive molecule and its reactivity?

*Wystan Hugh Auden (1907–1973) was one of the best known and most influential British poets of his generation.

CONVENTION ALERT

Don't forget—throughout our discussion of these radical reactions, the motion of single electrons is shown with single-barbed, "fishhook" arrows.

We begin with a section on the formation and simple reactions of radicals, and then move on to structure before considering the more complicated chain reactions.

WORKED PROBLEM 11.1 Find an exception to the generalization in the first paragraph. What reaction have we studied that involves no polar intermediates, and doesn't have a highly polar transition state? *Hint*: Think "recent."

ANSWER One example is the reaction recently studied in Section 10.4e (p. 461), the addition of a triplet carbene to an alkene. Only neutral intermediates are involved in this reaction. Another example is hydrogenation, the conversion of an alkene and hydrogen into an alkane (Chapter 10, p. 438).

PROBLEM 11.2 Find a reaction that involves no intermediates but certainly has a highly polar transition state.

ESSENTIAL SKILLS

1. You have to be able to sketch the steps of chain reactions. This chapter introduces this new process, which is quite common in radical reactions. In a chain reaction, a cycling process keeps the reaction going until starting materials are used up or termination occurs. It is important to be able to write initiation, propagation, and termination steps.

2. Another important skill is understanding the factors that influence selectivity. In synthesis, selectivity is everything, as the synthetic chemist seeks a single product over all others. In order to maximize the formation of that single product, we need to understand the factors that control selectivity. In this chapter, selectivity is discussed in the context of the photochemical halogenation of alkanes. Understanding selectivity is more important than the synthetic uses of this reaction.

IMPORTANT DETAILS

1. *N*-Bromosuccinimide (NBS) is a standard reagent for inducing bromine into the allylic position of alkenes.

11.2   FORMATION AND SIMPLE REACTIONS OF RADICALS

We have actually seen some radicals before (in Chapter 1) when we considered the formation of molecules such as hydrogen, methane, and ethane from their constituent parts: hydrogen atoms and methyl radicals (see Sections 1.5 and 2.3; Fig. 11.1). We also encountered them briefly in Section 2.4 in which we speculated a bit about the structure of radicals.

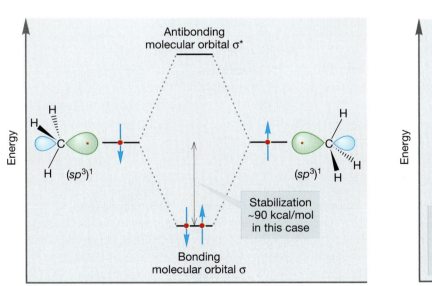

FIGURE 11.1 The formation of hydrogen, methane, and ethane through reactions of the radicals, H• and •CH$_3$.

Why can't we just reverse the bond-forming process to produce radicals from these molecules? Energy will surely have to be added in the form of heat, because bond forming is highly exothermic and bond breaking is highly endothermic. Figure 11.2 shows schematically the stabilization gained by bond formation from the overlap of two singly occupied sp^3 orbitals. In order to reverse the bond-making process and form two

FIGURE 11.2 A schematic picture of the formation of a covalent bond through overlap of two half-filled sp^3 orbitals. To reverse the process, energy must be supplied. For example, methyl radicals can be formed by supplying enough energy (90 kcal/mol) to break the carbon–carbon bond in ethane.

radicals, energy must be provided up to the amount of the stabilization achieved in the bond making. For example, we can apply heat to ethane in a process called **pyrolysis** or **thermolysis**, eventually providing 90 kcal/mol and breaking the carbon–carbon bond (Fig. 11.2). Carbon–carbon bond cleavage is the lowest energy process easily detectable in ethane.

The energy required to break the bond is called the **bond dissociation energy (BDE)**. A series of these values was collected in Table 8.2 (p. 355). Note that the bond breaks to give two neutral species, the two methyl radicals, and not to give two polar species, the methyl cation and the methyl anion.

WORKED PROBLEM 11.3 What unseen process is going on in ethane as we heat it?

ANSWER Don't forget rotation about the carbon–carbon bond! No new product is formed, but this invisible reaction is still occurring.

Bond breaking to give a pair of neutral radicals is called **homolytic bond breaking**, whereas bond breaking to make the ions is called **heterolytic bond breaking** (Fig. 11.3). The homolytic cleavage requires much less energy than the heterolytic formation of two ions, because no charges are generated. Recall the use of single-barbed fishhook arrows in writing arrow formalisms for reactions involving single electron species (radicals).

FIGURE 11.3 Two possible bond breakings in ethane.

Although this method is effective in producing methyl radicals from ethane, it is not generally useful. The carbon–carbon bond in ethane is strong, and therefore harsh conditions are required to break it, even in a homolytic fashion. Other, more delicate molecules will not survive such treatment. In practice, this means that if we want to generate radicals in the presence of other molecules we should not attempt to do so by heating simple hydrocarbons.

PROBLEM 11.4 Why does the carbon–carbon bond in ethane break rather than one of the carbon–hydrogen bonds?

The pyrolysis of alkanes can be effective in decreasing the overall chain length of the molecule and in introducing unsaturation. Consider the pyrolysis of butane. There are two possible homolytic cleaveages of carbon–carbon bonds in butane. The methyl, ethyl, and propyl radicals can be formed from breaking carbon–carbon bonds (Fig. 11.4).

$$CH_3CH_2 \frown CH_2CH_3 \longrightarrow CH_3\dot{C}H_2 \quad \dot{C}H_2CH_3$$

Two ethyl radicals

$$CH_3CH_2CH_2 \frown CH_3 \longrightarrow CH_3CH_2\dot{C}H_2 \quad \dot{C}H_3$$

Propyl   Methyl
radical   radical

FIGURE 11.4 There are two possible homolytic cleavages of carbon–carbon bonds in butane.

One reaction possible for these new radicals is abstraction of hydrogen from butane to make the smaller hydrocarbons methane, ethane, and propane, along with the butyl and *sec*-butyl radicals. There are two positions from which hydrogen abstraction can take place in butane, the primary C–H of the methyl group or the secondary C–H of the methylene group. We will consider the question of selectivity, of choosing between the two different positions, a little later in this chapter (Fig. 11.5).

Abstraction from the secondary position gives the *sec*-butyl radical

$$\longrightarrow CH_3 - CH_2 - \dot{C}H - CH_3$$

***sec*-Butyl radical**

$$CH_3 - CH_2 - CH_2 - CH_3$$

Abstraction from the primary position gives the butyl radical

$$\longrightarrow CH_3 - CH_2 - CH_2 - \dot{C}H_2$$

Butyl radical

$$\text{Abstracting radical} \begin{cases} \dot{C}H_3 \\ CH_3\dot{C}H_2 \\ CH_3CH_2\dot{C}H_2 \end{cases} \longrightarrow \begin{matrix} CH_4 \\ CH_3CH_3 \\ CH_3CH_2CH_3 \end{matrix} \left. \begin{matrix} \\ \\ \\ \end{matrix} \right\} \begin{matrix} \text{Products from} \\ \text{radical abstraction} \end{matrix}$$

FIGURE 11.5 Hydrogen abstraction from butane by radicals leads to the formation of lower molecular weight hydrocarbons and two different butyl radicals.

PROBLEM 11.5 Draw the transition states for the two possible abstractions of hydrogen from butane by a methyl radical.

It is also possible for one radical to abstract a hydrogen from the β-position of another radical in a process called **disproportionation**, which produces an alkane and an alkene. In the propyl radical, for example, a carbon–hydrogen bond could be broken by a methyl radical to give propene and methane. In the terminology of radical chemistry, it is the α–β bond that is broken. The point of reference is the radical itself; the adjacent atom, the carbon, is α, and the next atom, the hydrogen, is β. There are many radicals that can do the abstracting (Fig. 11.6).

FIGURE 11.6 The disproportionation reaction produces an alkane and an alkene from a pair of radicals. A hydrogen atom of one radical is abstracted by another radical.

PROBLEM 11.6 Draw the transition state for hydrogen abstraction in a disproportionation reaction.

Another reaction possible for many radicals is β-**cleavage** (Fig. 11.7). If an α–β carbon–carbon bond is present, it can break to give a new radical and an alkene. This cleavage is simply the reverse of an addition of a radical to an alkene (see Section 11.4, p. 511 for a fuller discussion of this process). In the *sec*-butyl radical, for example, cleavage between the methylene and methyl groups produces a methyl radical and propene. The β-cleavage reaction is favored at high temperature, and can be quite important in high-temperature reactions of alkanes.

FIGURE 11.7 β-Cleavage of the *sec*-butyl radical.

$$H_3C \cdots \overset{\beta}{\underset{}{}} CH_2 \overset{\alpha}{\underset{}{}} CH - CH_3 \longrightarrow H_3C\cdot + CH_2 = CH - CH_3$$

PROBLEM 11.7 Why will β-cleavage be favored at high temperature?

To make the issue even more complex, two radicals can occasionally react with each other (dimerize) in a very exothermic formation of new hydrocarbons (dimers) that can reenter the reaction sequence and provide new species. For example, two propyl radicals might combine to form hexane. Further pyrolysis of hexane would lead to *many* other products (Fig. 11.8).

$CH_3CH_2\overset{\bullet}{C}H_2$ $\overset{\bullet}{C}H_2CH_2CH_3$ $\xrightarrow{\text{dimerization}}$ $CH_3CH_2CH_2{-}CH_2CH_2CH_3$ $\xrightarrow{\Delta}$ New products

Two propyl radicals   **Hexane**

FIGURE 11.8 Two radicals can form a bond in a process called dimerization.

SUMMARY

The reactions available to an alkyl radical include dimerization, hydrogen abstraction to make a saturated compound and a new radical, as well as disproportionation (hydrogen abstraction from another radical to give an alkene and an alkane) and β-cleavage (the fragmentation of a radical into a new radical and an alkene) (Fig. 11.9).

$CH_3CH_2CH_2CH_2{-}CH_2CH_2CH_2CH_3 \longleftarrow$
Dimerization

$\longrightarrow 2\ CH_3CH_2CH_2CH_3$
Hydrogen abstraction

$CH_3CH_2CH_2\overset{\bullet}{C}H_2\ +\ \overset{\bullet}{C}H_2CH_2CH_2CH_3$

$CH_3\overset{\bullet}{C}H_2\ +\ CH_2{=}CH_2 \longleftarrow$
β-Cleavage

$\longrightarrow CH_3CH_2CH{=}CH_2\ +\ CH_3CH_2CH_2CH_3$
Disproportionation

FIGURE 11.9 Simple reactions of radicals.

Clearly, little specificity is possible in this pyrolytic process, and one might at first expect that there would be little utility in such a series of reactions. That's not quite right. The petroleum industry relies on this kind of **hydrocarbon cracking** to convert the high molecular weight hydrocarbons making up much of crude oil into the much more valuable lower molecular weight gasoline fractions. It is also possible to do the pyrolysis in the presence of hydrogen to help produce smaller alkanes from the radicals that are produced initially.

PROBLEM 11.8 What function does the hydrogen perform in such a reaction?

WORKED PROBLEM 11.9 What products do you expect from the thermal cracking of hexane?

ANSWER The point of this problem is to show you how incredibly complicated this "simple" process becomes even in a relatively small alkane such as hexane. First of all, carbon–carbon bond breakings give the methyl, ethyl, propyl, butyl, and pentyl radicals, which can abstract hydrogen to give methane, ethane, propane, butane, and pentane.

(continued)

WORKED PROBLEM 11.9 *(continued)*

Disproportionation gives the alkanes already mentioned, along with unsaturated compounds, ethylene, propene, 1-butene, and 1-pentene.

β-Cleavage of these radicals gives more ethylene and the methyl, ethyl, and propyl radicals.

Of course, the alkanes and alkenes formed in these reactions can react further. Propane, butane, and pentane as well as the alkenes formed in the disproportionation and β-cleavage steps can participate in similar radical reactions. Moreover, dimerization reactions have the potential of introducing larger hydrocarbons, only two of which are shown below. These dimers can go on to produce other molecules as they too begin to participate in the radical reactions. It's an incredible mess.

Now comes a big question, How can ethane be modified in order to make the carbon–carbon bond weaker, and thus more easily broken?

PROBLEM 11.10 Construct (on paper) a molecule with a carbon–carbon bond weaker than the one in ethane.

There are many possible answers to Problem 11.10. For example, we could induce some strain in the molecule, raising its energy and making bond breaking easier. Cyclopropane is a nice example. This small ring compound contains both severe angle strain and substantial torsional strain (Section 5.3). The strain energy of cyclopropane results in a low-bond dissociation energy of only 65 kcal/mol (Fig. 11.10).

FIGURE 11.10 Strain can raise the energy of a hydrocarbon making it easier to break a carbon–carbon bond.

Breaking a carbon–carbon bond in cyclopropane doesn't yield two new separated radicals, but a 1,3-**diradical**, in which two nonbonding electrons are present in one molecule. We haven't really done exactly what we wanted—produce two new radicals from a single molecule.

So let's try another way to modify ethane to make it easier to cleave. We just tried making the starting material *less stable* by introducing strain. How about taking the opposite approach by making the products of the reaction, the two radicals, *more stable*?

PROBLEM 11.11 Construct an Energy versus Reaction progress diagram for this second general approach of making the product of the reaction more stable.

The question now becomes, How can we make a radical more stable than the methyl radical? We know that delocalization of electrons is stabilizing, so one approach would be to construct a molecule that will yield resonance-stabilized radicals. That change should lower the bond strength of the crucial carbon–carbon bond. For example, introduction of a single vinyl group (as in 1-butene) lowers the bond dissociation energy by about 13 kcal/mol to a value of 76 kcal/mol. To calculate this number, the bond dissociation energy of propane and 1-butene are compared. Propane breaks a carbon–carbon bond to give a methyl radical and an ethyl radical. 1-Butene breaks a carbon–carbon bond to give a methyl radical and an allyl radical. The difference in the energy required for the two bond breakings is reasonably attributed mostly to the stabilization of the allyl radical, and, of course, of the transition state leading to it. Introduction of a second vinyl group, as in 1,5-hexadiene, reduces the BDE by approximately an additional 13 kcal/mol (Fig. 11.11). The key to lowering the bond dissociation energy is the formation of allyl radicals with their stabilizing delocalization of the nonbonding electrons.

	Bond Dissociation Energy (kcal/mol)

FIGURE 11.11 Stabilization of the products of bond breaking through delocalization can make the bond cleavage process easier.

PROBLEM 11.12 There is a lower energy process occurring in 1,5-hexadiene, although no *new* molecule appears! Can you draw an arrow formalism for the conversion of one molecule of 1,5-hexadiene into another?

Hint: 1,1,6,6-Tetradeuterio-1,5-hexadiene equilibrates on heating with 3,3,4,4-tetra-deuterio-1,5-hexadiene.

1,1,6,6-Tetradeuterio-1,5-hexadiene   **3,3,4,4-Tetradeuterio-1,5-hexadiene**

Another technique for producing radicals is to use molecules containing especially weak, and therefore easily broken bonds. Examples are acyl **peroxides** (Fig. 11.12). The oxygen–oxygen bond in simple peroxides is already quite weak (~38 kcal/mol; Table 8.2, p. 355). The addition of the carbon–oxygen double bond of the **acyl group** **(R—C=O)**, helps delocalize the nonbonding electron in the carboxyl radical, and further lowers the BDE to about 29 kcal/mol (Fig. 11.12).

	Bond Dissociation Energy (kcal/mol)
A peroxide	~38
An acyl peroxide → Two carboxyl radicals	~29

FIGURE 11.12 In alkyl peroxides, the weak oxygen–oxygen bond is easily cleaved. In acyl peroxides, the product radicals are stabilized and the oxygen–oxygen bond is even weaker.

Some molecules give stable gases (often nitrogen or carbon dioxide) on heating, and thus produce radicals irreversibly. **Azo compounds** and acyl peroxides are good examples (Fig. 11.13).

FIGURE 11.13 Some molecules give free radicals irreversibly through the formation of gases.

A particularly useful source of radicals is azobisisobutyronitrile (AIBN), a molecule that decomposes much more easily than simpler azo compounds (Fig. 11.14).

$$(CH_3)_3C-\underset{\cdot\cdot}{N}=\underset{\cdot\cdot}{N}-C(CH_3)_3 \xrightarrow{\Delta} 2\ (CH_3)_3C\cdot\ +\ N_2 \qquad \Delta H^{\ddagger} = 42.2\ \text{kcal/mol}$$

FIGURE 11.14
Azobisisobutyronitrile is an especially easily cleaved azo compound.

$$(CH_3)_2\underset{\underset{CN}{|}}{C}-\underset{\cdot\cdot}{N}=\underset{\cdot\cdot}{N}-\underset{\underset{CN}{|}}{C}(CH_3)_2 \xrightarrow{\Delta} 2\ (CH_3)_2\underset{\underset{CN}{|}}{C}\cdot\ +\ N_2 \qquad \Delta H^{\ddagger} = 31.9\ \text{kcal/mol}$$

AIBN

PROBLEM 11.13 Why does AIBN decompose much more easily than a simple azo compound?

11.3   STRUCTURE OF RADICALS

We have already encountered both carbocations and carbanions, and spent some time discussing their structures (Section 2.4, p. 60). The methyl cation, and carbocations in general, are flat, sp^2 hybridized molecules with an empty $2p$ orbital extending above and below the plane of the three substituents on carbon. By contrast, the methyl anion is pyramidal. One might well guess that the methyl radical, with a single nonbonding electron, would have a structure intermediate between that of the two ions (Fig. 11.15).

FIGURE 11.15 Radicals are intermediate in structure between the planar cations and the pyramidal carbanions.

Carbocation, planar   Radical—somewhere in between the cation and anion   Carbanion, pyramidal

120°   99.8°

Both theory and experiment agree that the structure of simple radicals is hard to determine! However, it is clear that these molecules are not very far from planar. If the molecules are not planar, the pyramid is *very* shallow and the two possible forms are rapidly inverting (Fig. 11.16).

FIGURE 11.16 Alkyl radicals are either flat or very shallow, rapidly interconverting pyramids.

Planar... or... very shallow pyramid

PROBLEM 11.14 Draw the transition state for the inversion process in Figure 11.16.

11.4   STABILITY OF RADICALS

Like carbocations, more substituted radicals are more stable than less substituted ones. Table 11.1 shows BDEs of four hydrocarbons that lead to a methyl, primary, secondary, and tertiary radical. The more substituted the radical, the easier it is to form, and the easier it is to break the bond that produces it. The data presented in the table are slightly suspect, as we are forming the radicals from different starting materials. Might not the different energies of the different starting materials influence the BDEs? Yes they might, and do. But this effect isn't large, and the data in the table can be used to show that more substituted radicals are more stable than less substituted radicals.

TABLE 11.1   Bond Dissociation Energies for Some Hydrocarbons

Hydrocarbon	Radical	BDE (kcal/mol)
		105.0
		101.1
		98.6
		96.5

Why should more substituted radicals be more stable than less substituted radicals? Let's start with an analogy. In explaining the order of stability for carbocations (tertiary > secondary > primary > methyl), we drew resonance forms in which a filled carbon–hydrogen bond overlapped with the empty $2p$ orbital (Fig. 11.17).

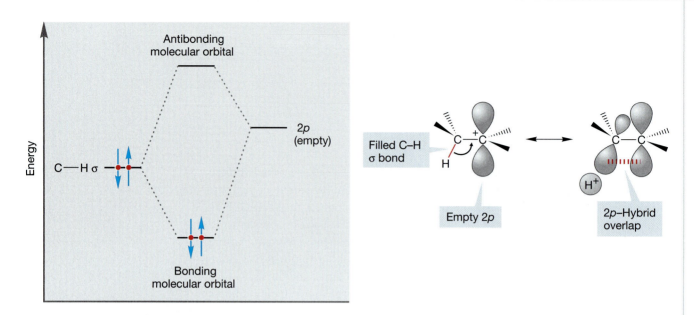

FIGURE 11.17 Resonance stabilization of a carbocation through "hyperconjugation," shown both by orbital overlap pictures and an interaction diagram.

We can draw similar resonance forms for radicals. As before, the more alkyl groups attached to the central carbon bearing the nonbonding electron, the more resonance forms are possible. But there is an important difference. Stabilization of the carbocation involves overlap of a filled carbon–hydrogen bond with an empty $2p$ orbital (Fig. 11.17), whereas in the radical, overlap is between a filled carbon–hydrogen bond and a *half-filled* $2p$ orbital (Fig. 11.18).

FIGURE 11.18 Resonance stabilization of a free radical.

Why should this interaction be stabilizing? The evident problem is that this is a *three*-electron system. One electron must occupy the antibonding orbital, and this is certainly destabilizing. Despite this, the *overall* system is still stabilized because two electrons are able to occupy the low-energy bonding orbital. Recall He_2^+ in which a similar orbital system was encountered. The He_2^+ molecule is held together (bound) by over 60 kcal/mol (Chapter 1, p. 38; Fig. 11.19).

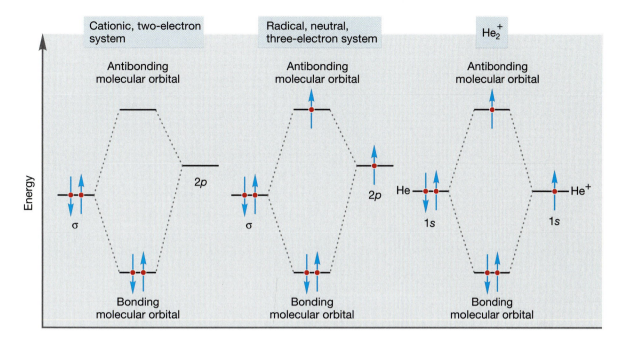

FIGURE 11.19 A comparison of the stabilizations of carbocations, radicals, and He_2^+. Notice the electron in the antibonding orbital of He_2^+ and the radical.

Consider the bond strengths in these radicals, as we did earlier in discussing alkenes and carbocations (Chapter 3, p. 115, Problem 9.11). In the 2-propyl radical, we have two $sp^2–sp^3$ carbon–carbon bonds, whereas in the less stable 1-propyl radical there is one $sp^2–sp^3$ bond and one weaker $sp^3–sp^3$ carbon–carbon bond (Fig. 11.20).

FIGURE 11.20 The 2-propyl radical contains two relatively strong $sp^2–sp^3$ bonds, whereas the 1-propyl radical has one strong $sp^2–sp^3$ bond and one relatively weak $sp^3–sp^3$ bond.

Of course, we should analyze the carbon–hydrogen bonds as well. If we do this, we find that the differences in the strengths of the C—H bonds should favor the 1-propyl radical. As in the example of carbocations, it is the C—C bonds that dominate and determine the issue. As a practical matter, it appears that we can "get away" with an analysis of the C—C bonds alone.

PROBLEM 11.15   Analyze the changes in bond strength of the C—H bonds in the 2-propyl and 1-propyl radical.

Radicals substituted with electron delocalizing groups such as vinyl are even more stable (p. 508). If they are immobilized, even quite simple radicals can be made and studied spectroscopically at low temperature. A few highly delocalized radicals are stable under normal, room temperature conditions. For example, the triphenylmethyl radical happily exists in solution as an equilibrium mixture with a dimer (Fig. 11.21).

FIGURE 11.21 The triphenylmethyl radical exists in equilibrium with a dimer.

Triphenylmethyl radical   Dimer

PROBLEM 11.16 As you can see from Figure 11.21, the dimer is not the "obvious" one, hexaphenylethane, but a much more complicated molecule. Can you see why hexaphenylethane is a most unstable molecule, and draw an arrow formalism mechanism for the formation of the real dimer?

SUMMARY

Radical stability parallels carbocation stability; tertiary more stable than secondary, secondary more stable than primary, and primary more stable than methyl. The determining factors are hyperconjugative stabilization and the presence of stronger carbon–carbon bonds in the more substituted, more stable radicals.

11.5   RADICAL ADDITION TO ALKENES

In Section 9.3 (p. 386), the addition of hydrogen bromide to alkenes was presented with the reaction always leading to Markovnikov addition. The historical reality was far from that simple. Chapter 9 stressed the initial formation of the more stable carbocation followed by addition of the residual bromide ion. The product of Markovnikov addition must be produced in this reaction (Fig. 11.22).

The regiochemistry of the reaction appeared strangely dependent on the source of the starting materials and even on the place in which the reaction was carried out. In some worker's hands, the reaction did follow the Markovnikov path; in other places, with other investigators, anti-Markovnikov addition was dominant. Confusion and argument was the rule of the day in the late 1800s. Oddly enough, the problem was

FIGURE 11.22 The addition of hydrogen bromide to an unsymmetrical alkene follows the Markovnikov rule because protonation of the alkene must give the more substituted (more stable) carbocation.

restricted to hydrogen bromide addition. Both hydrogen chloride and hydrogen iodide addition remained securely Markovnikov in everyone's hands (Fig. 11.23).

FIGURE 11.23 Both hydrogen chloride and hydrogen iodide addition always follow the Markovnikov rule, but hydrogen bromide addition is strangely capricious.

In the 1930s, largely through the efforts of M. S. Kharasch [Kharasch, 1895–1957, is reported by one of his students, F. R. Mayo (b. 1908), to have suggested that the direction of addition may have been affected by the phases of the moon], it was discovered that the purity of the compounds used was vital. Pure materials led to Markovnikov addition, and we can remain content with the mechanism for the polar reaction. Impure reagents led to predominately anti-Markovnikov addition. A special culprit was found to be peroxides, R—O—O—R (Fig. 11.24). Breaking the weak oxygen–oxygen bond in peroxides provides a relatively easy route to free radicals. The mechanism of addition in the presence of peroxides is different from the familiar polar one, and can be triggered by these small amounts of radicals.

FIGURE 11.24 Addition of hydrogen bromide to alkenes in the presence of peroxides is anti-Markovnikov.

11.5a The Radical Chain Mechanism of HBr Addition. Peroxides contain a very weak oxygen–oxygen bond and are easily cleaved by heat or light to give two R—O• radicals (Fig. 11.25).

FIGURE 11.25 The oxygen–oxygen bond is weak and can be easily cleaved to give a pair of RO• radicals.

$$R{-}\ddot{O}{-}\ddot{O}{-}R \xrightarrow[\text{or } h\nu]{\Delta} R{-}\ddot{O}{\cdot} + {\cdot}\ddot{O}{-}R$$

A free radical (for example, R—O•) can react with H—Br by hydrogen abstraction to form R—OH and a new radical, a bromine atom (Fig. 11.26, Step 2). Addition of the bromine atom to an alkene gives still another free radical (Step 3), which can play the role of R—O• and produce an alkyl bromide and another bromine atom (Step 4). The bromine atom can recycle to Step 3 and start the reaction over again. The overall result is the addition of hydrogen bromide to the alkene.

Step 1

Step 2

Step 3

Step 4

Recycle to Step 3 and start all over again

FIGURE 11.26 The radical chain mechanism for the addition of HBr to an alkene.

PROBLEM 11.17 Why is abstraction of H from H—Br preferred to abstraction of Br? *Hint*: The bond dissociation energy of HO—Br is 56 kcal/mol.

Of course, when two radicals meet they can react in a very exothermic fashion to form a new bond and end the reaction sequence. There are many possibilities for this kind of reaction using the radicals in Figure 11.26. A few ways are shown in Figure 11.27.

FIGURE 11.27 Whenever two radicals come together they can react to form a bond. This reaction removes two radicals and terminates this chain reaction.

We have just outlined the radical-induced addition of hydrogen bromide to an alkene. The steps of this chain reaction naturally divide into three kinds:

Step 1. **Initiation**. The reaction cannot start without the presence of a radical. Although there are now known examples of stable radicals, usually a radical must be formed to start the process, in a reaction known as an **initiation step**. Peroxides are good generators of radicals because of the weakness of their oxygen–oxygen bond. Once a radical is created in the presence of hydrogen bromide, a second initiation step, abstraction of hydrogen, takes place to form the chain-carrying bromine atom (Fig. 11.28).

FIGURE 11.28 Two initiation steps.

Step 2. **Propagation**. Once the reaction is started by the formation of a bromine atom in the initiation steps, the chain can begin to grow. The steps making up the chain are called **propagation steps**. In this example, the two chain-carrying propagation steps are the addition of the bromine atom to the alkene to give a new, carbon-centered alkyl radical and the subsequent abstraction of hydrogen from hydrogen bromide by this radical (Fig. 11.29).

Two propagation, or chain-carrying steps

An alkyl radical

Recycle to the first propagation step and start all over again

FIGURE 11.29 Two propagation steps.

The two radicals involved in the chain-carrying steps, the bromine atom and the alkyl radical, are called "chain-carrying radicals." The chain continues as the newly formed bromine atom adds to another alkene to start the process over again.

Step 3. **Termination**. Any time one of the chain-carrying radicals (in this case a bromine atom or one of the alkyl radicals) is annihilated by combination with another radical, the chain reaction is stopped. Without such steps a single radical could initiate a chain reaction that would continue until starting material was used up. In practice, this is not the case, and a competition is set up between the propagation steps carrying the chain reaction and the steps ending it. Such reactions are called **termination steps** and there are many possibilities (Fig. 11.30).

FIGURE 11.30
Two possible
termination steps.

Two termination steps

For a chain reaction to be effective in producing large amounts of products, the propagation steps must be much more successful than the termination steps. This is often the case, as for a termination step to occur two radicals must wander through the solution until they find each other. Typically, the chain-carrying propagation steps require reaction with a molecule present in relatively high concentration. So, even though termination steps often can take place very easily once the reactants find each other, propagation steps can compete successfully, and chain lengths of many thousands are common. Under such circumstances, the small amounts of product formed by the occasional termination step are inconsequential.

11.5b Inhibitors. Small amounts of substances that react with the chain-carrying radicals can stop chain reactions by removing the chain-carrying molecules from the process. These substances are called **inhibitors**. They act as artificial terminators of the chains (Fig. 11.31).

The first propagation step in a typical radical chain addition reaction

The chain can be stopped if the chain-carrying radical reacts with an inhibitor, X•

FIGURE 11.31 An inhibitor of radical chain reactions can act by intercepting one of the chain-carrying radicals.

Once again, a few molecules can have an impact out of proportion to their numbers, as stopping one chain can mean stopping the formation of thousands of product molecules. Oxygen (O_2), itself a diradical, can act in this way, and many radical reactions will not proceed until the last oxygen molecules are scavenged from the system, thus allowing the radical chains to carry on. Radical inhibitors are often added to foods to retard radical reactions causing spoilage. Figure 11.32 shows a few typical radical inhibitors. Both butylated hydroxy toluene (BHT) and butylated hydroxy anisole (BHA) are widely used in the food industry. It is not important that you memorize these structures. What is worth remembering is the general structural features of these materials. You should also think a bit about how they might function in intercepting radicals.

FIGURE 11.32 Some radical inhibitors.

Hydroquinone   BHT   BHA

11.5c Regiospecificity of Radical Additions: Anti-Markovnikov Addition.

When a radical (R•) adds to an unsymmetrical alkene, there are two possible outcomes: a more substituted or less substituted alkyl radical can be formed (Fig. 11.33).

FIGURE 11.33 When a radical adds to an unsymmetrical alkane, there are two possible radical products.

So, the regiospecificity of radical addition to isobutene is determined by the direction of radical attack on the alkene, and thus by the stability of the initially formed radical. The bromine atom adds to give the more substituted, more stable radical. The reaction continues when the newly formed radical abstracts a hydrogen atom from hydrogen bromide to give, ultimately, the compound in which the bromine is attached to the *less* substituted end of the original alkene. This reaction is in contrast to polar additions in which the first step is protonation of the alkene to give the more substituted carbocation, which is attacked by bromide ion to give the *more* substituted bromide (Fig. 11.34).

FIGURE 11.34 As in the cationic reaction, the first step in a radical addition to an alkene is the formation of the more substituted, and therefore more stable species. Notice, however, that in the cationic reaction it is the hydrogen atom that adds first, and in the radical reaction it is the bromine atom.

Radical alkene hydrohalogenation

The historic difficulty in determining the regiospecificity of hydrogen bromide addition had only to do with recognizing that radical initiators such as peroxides had to be kept out of the reaction. If even a little peroxide was present, the chain nature of the radical addition of hydrogen bromide could overwhelm the nonchain polar reaction, and produce large amounts of anti-Markovnikov addition.

This understanding of the causes of the regiochemical outcome of these reactions presents an opportunity for synthetic control. We now have ways to do Markovnikov addition of hydrogen chloride and hydrogen iodide, and either Markovnikov or anti-Markovnikov addition reactions for hydrogen bromide. Of course, the halides formed as products in these reactions can be used for further synthetic work.

PROBLEM 11.18 Show how to carry out the following conversions. You need not write mechanisms in such problems unless they help you to see the correct synthetic pathways.

Neither hydrogen chloride nor hydrogen iodide will add in anti-Markovnikov fashion, even if radical initiators are present. Now we have to see why this is so.

11.5d Thermochemical Analysis of HX Additions to Alkenes. Why does this chain reaction succeed with H—Br but not for other hydrogen halides? We can *write* identical radical chain mechanisms for H—Cl and H—I, but these must fail for some reason (Fig. 11.35).

Why does the following mechanism not operate for H–Cl or H–I?

Step 1, Step 2, Step 3, Step 4

Recycle to Step 3 and start all over again

FIGURE 11.35 Analogous mechanisms for radical-induced additions of HCl and HI to alkenes.

We can understand this effect if we examine the thermochemistry of the two propagation steps for the different reactions.

The energetics of the first propagation step are shown in Table 11.2.

TABLE 11.2   The Thermochemistry of the First Propagation Step;
The Addition of Halogen Radicals to Ethylene

$$X \cdot + H_2C=CH_2 \rightarrow X-CH_2CH_2 \cdot$$

X	Bond Strength (π bond)	Bond Strength (X–C in $X-CH_2CH_2 \cdot$)	ΔH (kcal/mol)
Cl	66	85	−19
Br	66	72	− 6
I	66	57	+ 9

These values are estimated by noting that in this first reaction a π bond is broken and a X–C bond is made. The overall exothermicity or endothermicity of the reaction will be a combination of these two bond strengths. A π bond is worth 66 kcal/mol, and we approximate the X–C bond strength in the product by taking the values for ethyl halides. We are making all sorts of assumptions, but the errors involved turn out to be relatively small. For both Br and Cl this first step is exothermic—the product is more stable than starting material and will be favored at equilibrium. For I, the reaction is substantially *endothermic*. The product will be strongly disfavored and the reaction will be very slow. So now we know why the addition of H–I fails; the first propagation step of the reaction is highly endothermic and does not compete with the termination steps. But both the bromine case, which we know succeeds, and the chlorine case, which we know fails, have exothermic first steps. Let's look at the second propagation step in Table 11.3.

TABLE 11.3   The Thermochemistry of the Second Propagation Step, the Abstraction of a Hydrogen Atom from H–X

$$H-X + \cdot CH_2-CH_2X \rightarrow \cdot X + H-CH_2-CH_2-X$$

X	Bond Strength (H–X)	Bond Strength (H–C in $H-CH_2-CH_2-X$)	ΔH (kcal/mol)
Cl	103	98	+ 5
Br	87	98	−11
I	71	98	−27

The second step is exothermic for Br and I, but *endothermic for Cl*. So, we can expect this reaction to succeed for Br and I, but fail for Cl. If we sum up the analysis for the two propagation steps we see that it is only for H–Br addition that *both* steps are exothermic. For H—Cl and H—I, one of the two propagation steps is endothermic, and the radical addition reaction cannot successfully compete with the termination steps. The polar addition of H—X wins out.

SUMMARY

The polar addition of HBr to alkenes leads to the more substituted bromide, whereas the radical-initiated addition of HBr gives the less substituted bromide. Thus, you have the ability to make either regioisomer when the alkene is unsymmetrically substituted. The addition of HCl and HI always gives the more substituted halide.

11.6   OTHER RADICAL ADDITION REACTIONS

Hydrogen bromide addition is only one example of a vast number of radical chain reactions. For example, in the presence of initiators carbon tetrahalides, such as CCl_4 and CBr_4, will add to alkenes. The steps in the peroxide-initiated reaction are shown in Figure 11.36.

1. $ROOR \longrightarrow RO\cdot + \cdot OR$ (initiation)

2. $RO\cdot + CCl_4 \longrightarrow ROCl + \cdot CCl_3$ (initiation)

3. $\cdot CCl_3 + CH_2{=}CH_2 \longrightarrow \cdot CH_2CH_2CCl_3$ (propagation)

4. $\cdot CH_2CH_2CCl_3 + CCl_4 \longrightarrow ClCH_2CH_2CCl_3 + \cdot CCl_3$ (propagation)
   (recycle to Step 3)

5. Many possible termination steps

FIGURE 11.36 The mechanism for the radical-induced addition of carbon tetrachloride to alkenes.

PROBLEM 11.19 Write a number of termination steps for the reaction of Figure 11.36.

Under some conditions, alkenes can be polymerized by free radicals. In the examples we have already discussed, radical addition to an alkene produced a new radical that abstracted hydrogen (or halogen in the case of CX_4) in a second propagation step to give the product of overall addition to the alkene. What if there were no easy second step? If the concentration of alkene were very high, for example, or if the addition step were especially favorable, the abstraction reaction might not compete effectively with addition. Under such conditions the newly formed alkyl radical can undergo repeated addition reactions until either a termination step intercedes or the supply of alkene is exhausted (Fig. 11.37).

FIGURE 11.37 In the radical-induced polymerization of alkenes, repeated additions grow a long chain. A high concentration of alkene favors this process (IN• = initiator radical).

PROBLEM 11.20 Styrene (vinylbenzene) is an example of a molecule in which the addition of a radical is especially favorable. Can you see why?

Styrene
(vinylbenzene)

Such polymerization reactions are common and have very long chain lengths. Very high molecular weight products can be formed. Table 11.4 shows a number of common polymers and the monomers from which they are built up.

TABLE 11.4 Some Monomers and Polymers

Monomer	Polymer Formula	Name	Monomer	Polymer Formula	Name
$H_2C{=}CH_2$ Ethylene	$(CH_2)_n$	Polyethylene, polymethylene	$F_2C{=}CF_2$ Tetrafluoroethylene	$(CF_2)_n$	Poly(tetrafluoro-ethylene), Teflon
Styrene	$-(CH_2-CH)_n-$	Polystyrene	$H_2C{=}CHCl$ Vinyl chloride	$-(CH-CH_2)_n-$ Cl	Poly(vinyl chloride) (PVC)
			$H_2C{=}CHCOOCH_3$ Methyl acrylate	$-(CH-CH_2)_n-$ COOCH$_3$	Polyacrylate

PROBLEM 11.21 Write a mechanism for the formation of polystyrene, the product of the polymerization of styrene (vinylbenzene).

11.7 RADICAL-INITIATED ADDITION OF HBr TO ALKYNES

Just as with alkenes, hydrogen bromide adds to alkynes in the anti-Markovnikov sense when free radicals initiate the reaction. The photoinitiated reaction of H–Br with propyne gives *cis-* and *trans-*1-bromopropene, *not* 2-bromopropene (Fig. 11.38).

FIGURE 11.38 The radical-induced addition of HBr to propyne.

The free radical chain mechanism for addition to alkynes is quite analogous to that operating in additions to alkenes. The crucial step is the formation of the more stable vinyl radical rather than the less stable vinyl radical (Fig. 11.39).

Initiation   Propagation

FIGURE 11.39 Addition of a bromine atom to 1-propyne gives the more stable vinyl radical, which then abstracts the hydrogen atom from hydrogen bromide to give the anti-Markovnikov product.

In principle, a second radical addition of hydrogen bromide could form either a **geminal** (1,1) or **vicinal** (1,2) dihalide. In contrast to ionic reactions, it is the vicinal dibromide that is preferred (Fig. 11.40).

FIGURE 11.40 A second, radical addition of hydrogen bromide to a vinyl bromide generates the 1,2-dibromo compound, *not* the 1,1-dibromo compound. Problem 11.22 asks you to explain why.

PROBLEM 11.22 Explain in resonance and molecular orbital terms why it is the vicinal dihalide that is formed. Be careful! This question is more subtle than it looks at first. Consider the stabilization of an adjacent radical (a half-filled orbital) by an adjacent bromine atom. Why is this a stabilizing interaction? *Hint*: Recall the molecule He$_2^+$ from Chapter 1 (p. 38) and see page 512 in this chapter.

PROBLEM 11.23 Now predict whether R• will abstract hydrogen from the methylene or methyl position of diethyl ether (CH$_3$CH$_2$OCH$_2$CH$_3$). Explain.

11.8 PHOTOHALOGENATION

So far, we have seen a variety of reactions of radicals with alkenes. Alkanes are much less reactive than alkenes, and we have not seen ways to induce these notoriously unreactive species into reaction in a specific way. We have seen pyrolysis, the thermal cracking of saturated hydrocarbons (p. 505), but this method was anything but useful in making specific molecules. Here we find a way to use alkanes in synthesis. There still will be little specificity here, and these reactions will not find a prominent place in your catalog of synthetically important processes. Yet they are occasionally useful, and provide a nice framework on which to base a discussion of selectivity.

11.8a Halogenation of Methane. A mixture of chlorine and methane is indefinitely stable at room temperature in the dark, but reacts vigorously to give a mixture of methyl chloride, methylene chloride, chloroform (trichloromethane), and carbon tetrachloride when the mixture is irradiated or heated (Fig. 11.41).

FIGURE 11.41 Photolysis of chlorine in methane gives a mixture of chlorinated methanes.

The product composition depends on the amount of chlorine available and the time the reaction is allowed to run. If we use a sufficient amount of chlorine and allow the reaction to run long enough, all the methane can be converted into carbon tetrachloride. By contrast, measuring the product composition as a function of time shows that the initial product of the reaction is methyl chloride. Methylene chloride and chloroform are formed on the way to carbon tetrachloride (Fig. 11.42).

FIGURE 11.42 The chlorinated products are formed by sequential chlorinations of methane.

Let's first examine the formation of the initial product, methyl chloride. The overall reaction is exothermic by about 23 kcal/mol (Fig. 11.43).

FIGURE 11.43 The formation of methyl chloride is exothermic.

The first step is the thermal or photochemical breaking of the chlorine–chlorine bond. This initiation reaction is followed by the first propagation step, the abstraction of a hydrogen atom from methane by a chlorine atom to produce a methyl radical and hydrogen chloride. In the second propagation step, the newly produced methyl radical abstracts a chlorine atom from a chlorine molecule to give methyl chloride and another chlorine atom that can carry the chain. There are many possible termination steps; Figure 11.44, which shows the overall mechanism, includes only one.

FIGURE 11.44 The mechanism of methyl chloride formation.

Now let's look at the thermochemistry of the propagation steps. Although the second step is **exothermic** by 25 kcal/mol and thus is very favorable, the first propagation step is **endothermic** by about 2 kcal/mol (Fig. 11.45).

First propagation step

$H_3C\!-\!H \quad \cdot\ddot{\underset{\cdot\cdot}{C}}l\!: \longrightarrow H_3C\cdot \ + \ H\!-\!\ddot{\underset{\cdot\cdot}{C}}l\!:$

105 kcal/mol   103 kcal/mol

This reaction is endothermic by 2 kcal/mol

Second propagation step

$H_3C\cdot \quad :\ddot{\underset{\cdot\cdot}{C}}l\!-\!\ddot{\underset{\cdot\cdot}{C}}l\!: \longrightarrow H_3C\!-\!\ddot{\underset{\cdot\cdot}{C}}l\!: \ + \ \cdot\ddot{\underset{\cdot\cdot}{C}}l\!:$

59 kcal/mol   84 kcal/mol

This reaction is exothermic
by 84 − 59 = 25 kcal/mol

FIGURE 11.45 The thermochemistry of the two propagation steps for methyl chloride formation.

Will this endothermic step be enough to stop the reaction? No. Although a 2-kcal/mol energy difference produces an unfavorable equilibrium mixture, the second highly exothermic step results in the overall reaction being favored. The small number of methyl radicals formed reacts rapidly to give methyl chloride, thus disrupting the slightly unfavorable equilibrium.

$$Cl\cdot + CH_4 \ \rightleftharpoons \ Cl\!-\!H + \cdot CH_3$$

Reestablishment of this equilibrium generates more methyl radicals, which go on to make more methyl chloride and chain-carrying chlorine atoms (Fig. 11.46).

FIGURE 11.46 The high exothermicity of the second step overcomes the slight endothermicity of the first step in the formation of methyl chloride from methane and a chlorine atom.

**Methylene
chloride**

FIGURE 11.47 The formation of methylene chloride (H$_2$CCl$_2$) from methyl chloride and a chlorine atom.

As methyl chloride builds up in the reaction, it, too, begins to react with chlorine atoms in another chain reaction (Fig. 11.47). Moreover, the carbon–hydrogen bond in methyl chloride is weaker than the one in methane, and thus easier to break. So methyl chloride competes favorably with methane and cannot build up in the reaction unless there is a vast excess of methane.

PROBLEM 11.24 Explain why the carbon–hydrogen bond in methyl chloride is easier to break than the carbon–hydrogen bond in methane itself.

Methylene chloride is more reactive than either methyl chloride or methane, and so it reacts to form chloroform as it builds up. Chloroform is more reactive still, and rapidly produces carbon tetrachloride through abstraction of its single hydrogen (Fig. 11.48).

Chloroform

**Carbon
tetrachloride**

FIGURE 11.48 The formation of chloroform from methylene chloride and of carbon tetrachloride from chloroform.

Methane can also be photohalogenated with bromine, but the overall reaction is much less exothermic than is chlorination. Moreover, the first propagation step is endothermic by the substantial amount of 17 kcal/mol (Fig. 11.49). As we will see, this strongly endothermic step will have important consequences for the selectivity of the reaction when a choice of carbon–hydrogen bonds exists.

FIGURE 11.49 The formation of methyl bromide from bromine and methane.

Br—Br: ⟶ 2 :Br·   Initiation

H₃C—H :Br: ⟶ H₃C· + H—Br:

105 kcal/mol   88 kcal/mol

This first propagation step is 105 − 88 = 17 kcal/mol endothermic!

H₃C· :Br—Br: ⟶ H₃C—Br: + ·Br:

⎫
⎬ Propagation
⎭

H₃C· + H₃C· ⟶ H₃C—CH₃   Termination

11.8b Halogenation of Other Alkanes. If we irradiate a mixture of chlorine and a hydrocarbon more complicated than methane or ethane, there is usually more than one possible initial product. For example, butane can give either butyl chloride or *sec*-butyl chloride. Figure 11.50 shows that the secondary carbon–hydrogen bonds of butane are more reactive than the primary bonds. Abstraction by a chlorine atom occurs more easily at the secondary position than at the primary position. The percentages of products would indicate that the preference is 72.2:27.8, or a factor of 2.6. However, this simple analysis ignores the fact that there are more primary carbon–hydrogen bonds than secondary. In order to correct this value for the statistical advantage of the primary bonds, we must multiply it by 6/4 to produce the true factor of 3.9. On a per hydrogen basis, secondary hydrogens are abstracted more easily than primary hydrogens by a factor of 3.9. Here is an example of selectivity (Fig. 11.50).

$$CH_3CH_2CH_2CH_3 \xrightarrow[h\nu,\ 35\ °C]{Cl_2} CH_3CH_2CHCH_3 + CH_3CH_2CH_2CH_2Cl$$

	Cl

sec-Butyl chloride (72.2%)   Butyl chloride (27.8%)

FIGURE 11.50 On a per hydrogen basis, abstraction of a secondary hydrogen is favored over abstraction of a primary hydrogen by $(72.2/27.8) \times 6/4$.

PROBLEM 11.25 Write a reaction mechanism for the photochlorination of ethane.

PROBLEM 11.26 Analyze the thermochemistry of the two propagation steps for this reaction. The bond dissociation energies of the chlorine–chlorine and carbon–chlorine bonds of ethyl chloride, and the carbon–hydrogen bond of ethane are 59, 84.8, and 101.1 kcal/mol, respectively.

Selectivity is increased when the possibilities include the formation of a tertiary radical. Photochlorination of isobutane gives a 64:36 mixture of isobutyl chloride and *tert*-butyl chloride. The primary/tertiary ratio is 64:36 = 1.78. However, there are nine primary hydrogens and only a single tertiary hydrogen in isobutane, so on a per hydrogen basis the primary/tertiary reactivity ratio is $1.78 \times \frac{1}{9} = 0.198$. Put another way, the tertiary hydrogen is favored by 1:0.198 = 5.1 (Fig. 11.51).

FIGURE 11.51 For a chlorine atom, abstraction of a tertiary hydrogen is favored over abstraction of a primary hydrogen by a factor of about 5 on a per hydrogen basis.

Alkane halogenation

Bromine is much more selective in the photohalogenation reaction than is chlorine. That is, bromine is able to choose among the various reaction possibilities with more discrimination; to pick out the most favorable abstraction pathway by a wider margin than chlorine. For example, a bromine atom prefers secondary to primary hydrogens by a factor of 82, and tertiary hydrogens to primary hydrogens by a factor of about 1600 (Fig. 11.52; Table 11.5).

FIGURE 11.52
Photobromination is far more selective than photochlorination.

TABLE 11.5 Selectivities of Halogen Atoms in Carbon-Hydrogen Bond Abstraction

Halogen	Primary	Secondary	Tertiary
F	1	1.2	1.4
Cl	1	3.9	5.1
Br	1	82	1600

PROBLEM 11.27 Use the bond dissociation energies of Table 8.2 (p. 355) to estimate the thermochemistry of the propagation steps for photochemical fluorination and iodination of methane.

Neither fluorine nor iodine undergoes a useful photohalogenation reaction. The reaction of fluorine with carbon–hydrogen bonds is overwhelmingly exothermic and can even occur explosively. Almost no selectivity is possible (Table 11.5). In direct contrast to the reaction of fluorine atoms, reaction of iodine atoms with hydrocarbons is highly endothermic. It is so slow as to be of no synthetic utility.

What is the reason for the high selectivity of bromine as compared to chlorine? Recall that the abstraction reaction is close to thermoneutral* for chlorine, but very endothermic for bromine (p. 529). The Hammond postulate (the transition state for an endothermic process resembles the product) tells us that the transition state for the abstraction of hydrogen by chlorine will be nearly halfway in between starting material and product, but very product-like for bromine. Because the reaction is more endothermic, the transition state for abstraction of hydrogen by bromine will be product-like. There is a strongly developed hydrogen–bromine bond and a large degree of radical character on carbon in the transition state (Fig. 11.53).

FIGURE 11.53 For the nearly thermoneutral abstraction of hydrogen by a chlorine atom, the transition state will be almost halfway between starting material and products. For the highly endothermic abstraction of hydrogen by bromine, the transition state will be quite product-like.

The transition states for abstraction of hydrogen from methane by chlorine and bromine are very different (Fig. 11.54).

*Watch out! One of the things chemistry professors laugh about is the number of people who call "thermoneutral" reactions "thermonuclear" reactions. Believe it or not, this happens a lot!

Transition state

This is approximately halfway between starting material and product; the C–H and H–Cl bonds are about equally formed

Transition state

In this transition state the bond between H and Br is largely made and the C–H bond is largely broken; in the transition state the radical on carbon is mostly developed; the stability of that radical (tertiary > secondary > primary) will be very important

FIGURE 11.54 In the transition state for abstraction by chlorine, the carbon–hydrogen and hydrogen–chlorine bonds will be approximately equally formed. For bromine, with a very product-like transition state, the hydrogen–bromine bond will be nearly completely formed and the radical on carbon will be well developed.

In the transition states for abstractions by bromine, the radical on carbon is far more developed than it is in the abstractions by chlorine, which means that differences in the stabilities of different radicals will be far more important in the bromine abstractions than in the chlorine abstractions (Fig. 11.55).

In these transition states the carbon–hydrogen bonds are only about half-broken

FIGURE 11.55 The "late" (product-like) transition state for bromine abstraction means that the radical on carbon will be well developed and the substitution pattern of the carbon atom (tertiary, secondary, primary, or methyl) will be relatively important.

(continued)

In these transition states, the carbon–hydrogen bonds are largely broken and the radicals are very well developed on the carbons; differences in radical stability will be important

FIGURE 11.55 (continued)

SUMMARY

When questions of selectivity arise, it is always important to consider what the transition state looks like. Selective radicals will have relatively product-like transition states for abstraction, and therefore the factors influencing the stability of those products will be important in the transition state as well.

11.9 ALLYLIC HALOGENATION: SYNTHETICALLY USEFUL REACTIONS

One of the goals of synthetic organic chemists is to increase selectivity—to be able to do chemical transformations at specific sites in a molecule without affecting the rest of the molecule. Selectivity is useful—even necessary—when one is dealing with a complicated molecule. We must have ways of inducing change at one place in a molecule without affecting others if we are to be able to manipulate molecules rationally. It's quite clear that the photochlorination reaction of alkanes is not a good way of doing this; it is far too unselective. Chlorination takes place all too easily at the various positions in the hydrocarbon.

Photobromination more closely approaches the goal of perfect selectivity because the bromine atom abstracts tertiary hydrogens much more readily than secondary or primary hydrogens. But even this reaction is not without its problems. It is a slow process and the side products of di- and polybromination are formed.

There are some instances where the chlorination or bromination reaction can be used to good effect, however. Alkenes containing *allylic* hydrogens can sometimes be halogenated specifically at the allylic position in a process called **allylic halogenation**, another free radical chain reaction. For example, when low concentrations of bromine are photolyzed in a complicated cyclohexene, the benzoate ester of cholesterol, the product is exclusively brominated in one allylic position (Fig. 11.56).

Cholesterol benzoate

Cyclohexene in red

Brominated only in this allylic position

(70%)

FIGURE 11.56 When low concentrations of bromine are photolyzed in the benzoate ester of cholesterol, the product is exclusively the allylic bromide.

The mechanism is a typical radical chain process in which a low concentration of bromine atoms is first produced by breaking of the weak bromine–bromine bond (initiation step). Abstraction of hydrogen is the first propagation step, but there are three possible kinds of hydrogen that could be removed from a cyclohexene, H_v (vinyl), H_a (allylic), and H_m (methylene) (Fig. 11.57).

FIGURE 11.57 In principle, a bromine atom could abstract a viniyl (H_v), allylic (H_a) or methylene (H_m) hydrogen.

Initiation step

abstract H_v

abstract H_a

abstract H_m

Three possible propagation steps

As shown in Figure 11.58, each of these radicals could lead to a bromocyclohexene through abstraction of a bromine atom in the second propagation step. The chain is then carried by the new bromine atom formed in this step. The formation of only the allylic bromide tells us that only the allylic radical is produced originally.

FIGURE 11.58 In a propagation step, any of the carbon radicals could abstract bromine to give bromides and a chain-carrying bromine atom.

As there is only one product formed, the allylic bromide, the allylic radical must be formed with high selectivity. Why? *It is only this radical that is resonance stabilized.* There are two equivalent resonance forms for the allyl radical and this delocalization makes it far more stable than the other two possibilities. Further reaction by bromine atom extraction leads to the product of allylic bromination and a new chain-carrying bromine atom (Fig. 11.59).

FIGURE 11.59 Only the resonance-stabilized allylic radical is formed.

One might hope that the photohalogenation reaction would be general; that photochlorination would be as useful as photobromination. Alas it is not, and Table 11.5 shows why. A chlorine atom is far less selective than a bromine atom. Utility in synthesis comes from selectivity—from doing one reaction, not several. Photobromination is useful because the bromine atom picks and chooses among the various possible carbon–hydrogen bonds with fairly high discrimination. A chlorine atom does not. For example, the allylic position in cyclohexene is only slightly more reactive toward a chlorine atom than the other methylene position (Fig. 11.60).

FIGURE 11.60 Chlorine is much less selective than bromine. The allylic chloride is not the only major product formed in this reaction.

In photohalogenation reactions, keeping the concentration of bromine low is essential to success. If there is too much halogen, polar addition of bromine or chlorine to give vicinal dihalides competes with formation of the allylic halide, and the desired selectivity is lost. Methods have been developed to ensure that this low concentration is maintained, and to lower the temperature necessary for the initiation step. A common and very effective method of allylic bromination involves the molecule *N*-**bromosuccinimide**, generally known as **NBS**, in the solvent carbon tetrachloride (CCl_4) (Fig. 11.61).

FIGURE 11.61 Allylic bromination with NBS.

The first step in this reaction is formation of HBr and bromine atoms (initiation step) (Fig. 11.62).

FIGURE 11.62 The first two steps in bromination of cyclohexene using NBS.

The hydrogen bromide then reacts with NBS to form bromine, which in turn reacts with the allyl radical to give the allyl bromide and a new, chain-carrying bromine atom. The NBS is only slightly soluble in carbon tetrachloride, thus ensuring the low concentration of bromine necessary for the success of this reaction (Fig. 11.63).

FIGURE 11.63 A molecule of bromine is formed, which then reacts with the cyclohexenyl radical to give the propduct and a new bromine atom.

PROBLEM 11.28 Suggest a mechanism for the formation of bromine from NBS and hydrogen bromide. *Hint*: The oxygen of a carbon–oxygen double bond is a base.

PROBLEM 11.29 Selectivity is not possible in a disubstituted alkene such as *trans*-2-pentene. Explain.

11.10   SPECIAL TOPIC: REARRANGEMENTS (AND NONREARRANGEMENTS) OF RADICALS

We have seen in Chapters 9 and 10 that polar additions to alkenes are bedeviled by shifts of hydride or alkyl groups when there is a possibility of forming a more stable cation from a less stable one. A typical example, the polar addition of hydrogen chloride to 3,3-dimethyl-1-butene, is shown again in Figure 11.64.

FIGURE 11.64
Rearrangements occur in the polar addition of hydrogen chloride to 3,3-dimethyl-1-butene.

Yet free radical additions are usually uncomplicated by similar rearrangements. For example, the peroxide-initiated addition of hydrogen bromide to 3,3-dimethyl-1-butene gives only the unrearranged product (Fig. 11.65).

FIGURE 11.65 The related radical addition gives no rearranged product.

We might have expected rearrangement through the shift of a methyl group to give the more stable tertiary radical from the less stable secondary one, yet this reaction doesn't happen. In fact, the situation is quite dramatic: 1,2-hydride shifts (migration of H:⁻) are among the fastest known organic reactions and the simple 1,2-shift of a hydrogen atom (or an alkyl group) is unknown (Fig. 11.66).

FIGURE 11.66 The simple 1,2-shift of a hydrogen atom (H•) is unknown.

This situation seems mysterious, but a look at the molecular orbitals involved in the transition state for the reaction shows the reason. There is more than one way to approach this problem, but I think an analysis of the transition state for the rearrangement is the easiest. In the transition state for any symmetrical 1,2-shift of hydrogen, the hydrogen atom is half-transferred from one carbon to the other (Fig. 11.67).

FIGURE 11.67 In the transition state for *any* symmetrical 1,2-shift, the hydrogen is halfway between the two carbons. Here the symbol * = + , • , or −.

What will the molecular orbitals of the transition state look like? To answer this question, we need only construct the molecular orbitals for the system of a hydrogen $1s$ orbital interacting with π and π^* orbitals of an alkene. The combination of $1s$, π, and π^* will yield three new molecular orbitals **A**, **B**, and **C**. They are constructed in Figure 11.68.

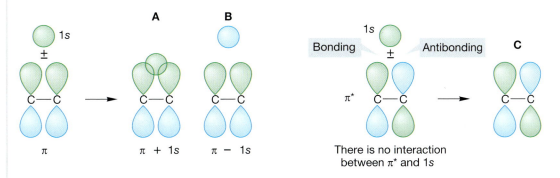

FIGURE 11.68 The three molecular orbitals for the transition state in this reaction can be easily constructed from H ($1s$), π, and π^*.

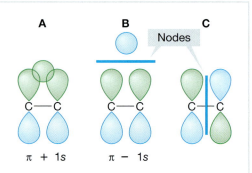

FIGURE 11.69 The new molecular orbitals (**A**–**C**) can be ordered in energy by counting nodes.

The $1s$ orbital interacts with π to form two new molecular orbitals (**A** and **B**), but there is no interaction between $1s$ and π^* in this geometry. Accordingly, π^* becomes **C**, the third molecular orbital for the transition state. To order the new molecular orbitals (**A**, **B**, **C**) in energy, we need only count nodes. Molecular orbital **A** has no new nodes, but **B** and **C** each has one additional node (Fig. 11.69).

The ordering of the new molecular orbitals is shown in Figure 11.70. Now, how many electrons do we need to put into the orbital system? For the cationic reaction, there are only two and they will go nicely into the new stabilized bonding molecular orbital, **A**. However, for the radical shift there is another electron, and it must be placed in one of the antibonding molecular orbitals, **B** or **C**. This high-energy electron is evidently destabilizing enough to completely shut off the reaction (Fig. 11.70).

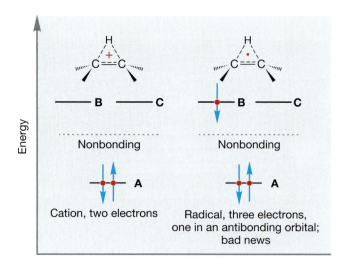

FIGURE 11.70 The transition state will have a single bonding molecular orbital (**A**) and two, equi-energetic antibonding molecular orbitals (**B** and **C**). For the cationic rearrangement, the two electrons involved nicely fill the bonding molecular orbital (**A**). In the radical case, there is another electron and it must occupy an antibonding molecular orbital (**B** or **C**) in the transition state. Occupancy of this antibonding orbital is unfavorable.

PROBLEM 11.30 Analyze the shift of hydrogen in the related anionic system.

WORKED PROBLEM 11.31 1,2-Shifts of vinyl groups are known in radicals. Can you see how this reaction might differ from the 1,2-shift of a hydrogen atom or a methyl radical, both of which are unknown?

(continued)

WORKED PROBLEM 11.31 *(continued)*

ANSWER A vinyl group, or any group with *p* orbitals, is not restricted to using the σ bond in the migration. In this case, the radical on C(2) could add to the vinyl group to give the symmetrical intermediate **A**.

Opening of intermediate **A** can occur in two directions (a and a′) to regenerate the starting material (a′) or complete the vinyl migration (a).

WORKED PROBLEM 11.32 We have already seen a reaction much like the opening of the cyclic intermediate shown in the answer to Problem 11.31. What is the name of this process?

ANSWER It is just a β-cleavage (p. 504).

11.11 SPECIAL TOPIC: RADICALS IN YOUR BODY; DO FREE RADICALS AGE US?

Free radicals are strongly implicated in the aging process. Cell membranes, which are made up of fatty acids, long-chain hydrocarbon carboxylic acids, are particularly vulnerable to oxidation through free radical attack. In time, the immune system comes to recognize damaged cells as "foreign" and attacks them. If radicals are involved in aging, then free radical trapping agents might be effective in arresting the aging process. Vitamin E (Fig. 11.71) is one such potential trapping agent.

Vitamin E is essential to rats for proper liver function as well as nerve and muscle condition. At least one observer has pointed out that as rats and humans are so alike, vitamin E might be vital for humans as well. Despite the obvious truth of the observation, there is no proven need for vitamin E in humans. Still, that lack is not sufficient to stop speculation, and it has been suggested that the ability of vitamin E to act as a free radical interceptor, an antioxidant, might make it effective in delaying aging. Vitamin E traps radicals through the red hydroxyl group on the six-membered ring. A radical (RO•) plucks a hydrogen atom from this hydroxyl to make RO—H, and destroy itself.

The long hydrocarbon chain of vitamin E also has its function. It aids in the delivery of vitamin E to the cell membrane by making the molecule soluble in fatty acids, which also contain long greasy hydrocarbon chains. *Remember:* "Like dissolves like" (Chapter 6, p. 247).

Vitamin E

FIGURE 11.71 Vitamin E.

11.12 SUMMARY

New Concepts

One of the most important concepts of this chapter is that of selectivity. It is not really new as you have seen many selective reactions before. A simple example is the protonation of isobutene to give the more stable tertiary carbocation rather than the less stable primary carbocation (Fig. 11.22).

In this chapter, you also see selectivity when a radical abstracts one of several possible hydrogens from an alkane. The ease of hydrogen abstraction is naturally determined by the energies of the transition states leading to the products. A tertiary radical is more likely to be formed than a primary radical. In the transition state for radical formation, the factors that make a tertiary radical more stable than a primary radical will operate to stabilize the partially formed tertiary radical over the partially formed primary radical (Fig. 11.72).

Transition state for formation of the primary radical (not favored)

Less stable primary radical

Transition state for formation of tertiary radical (favored)

More stable tertiary radical

FIGURE 11.72 The transition state with the partially formed tertiary radical will be more stable.

What is more subtle, and at least somewhat new, is the idea that the degree of selectivity in a reaction will be influenced by *how product-like* the transition state is. The more the transition state is like the product, the more important the relative stabilities of two possible products will be. Abstraction of a hydrogen by chlorine is endothermic by about 2 kcal/mol (p. 526). The same abstraction reaction by bromine is endothermic by about 17 kcal/mol. The transition state for the much more endothermic abstraction reaction by bromine will be more product-like than that for abstraction by chlorine. So we expect bromine to be more selective than chlorine, as the carbon–hydrogen bond will be more broken in the transition state. The stability of the product radical will be more important in the transition state for hydrogen abstraction by a bromine than it will in the transition state for abstraction by a chlorine (Fig. 11.55).

There are a few smaller concepts as well. Radical stability parallels carbocation stability. Tertiary radicals are more stable than secondary, secondary more stable than primary, and primary more stable than methyl. The differences aren't as great for radicals as for cations. Delocalization helps stabilize radicals by spreading the odd electron over two or more atoms.

Some reaction mechanisms are cyclic. Radical chain reactions produce a chain-carrying radical in their last, product-making step. The new radical can begin another series of propagation steps leading to another molecule of product. Chain reactions only stop when starting material is exhausted or a chain is terminated by bond formation by a pair of radicals.

Chains can be set going by small amounts of radicals released by radical initiators. In this way, a small number of chain-initiating radicals can have an impact far out of proportion to their numbers.

Key Terms

acyl group (R—C=O) (p. 509)
allylic halogenation (p. 534)
azo compounds (p. 509)
β-cleavage (p. 504)
bond dissociation energy (BDE) (p. 502)
chain reactions (p. 499)
diradical (p. 507)
disproportionation (p. 503)
endothermic (p. 526)

exothermic (p. 526)
free radicals (p. 499)
geminal (p. 525)
heterolytic bond breaking (p. 502)
homolytic bond breaking (p. 502)
hydrocarbon cracking (p. 505)
inhibitors (p. 518)
initiation step (p. 517)
N-Bromosuccinimide (NBS) (p. 536)

peroxides (p. 509)
propagation step (p. 517)
pyrolysis (p. 502)
radicals (p. 499)
termination step (p. 517)
thermolysis (p. 502)
vicinal (p. 525)

Reactions, Mechanisms, and Tools

The reactions in this chapter involve neutral free radicals, in contrast to the ionic species of earlier chapters. Chain reactions are characteristic of radical chemistry, and consist of several different kinds of steps. (1) Initiation: A small number of radicals serves to start the reaction. (2) Propagation: Radical reactions are carried out that generate a product molecule and a new, chain-carrying radical, which recycles to the beginning of the chain and starts a new series of product-forming steps.

(3) Termination: Radical recombinations destroy chain-carrying species and end chains. The success of a chain reaction depends on the relative success of the chain-carrying and termination steps. Figure 11.26 gives an example of a typical chain reaction.

Important reactions of radicals include abstraction, additions to alkenes and alkynes, β-cleavage, and disproportionation, but not simple 1,2-rearrangements (Fig. 11.73).

Abstraction

Addition

β-Cleavage

Disproportionation

1,2-Rearrangement

FIGURE 11.73 Some typical radical reactions.

Syntheses

The new synthetic reactions of this chapter are shown below.

1. ALKANES

Disproportionation gives both alkanes and alkenes;
this synthesis is not efficient

2. ALKENES

Disproportionation gives both alkanes and alkenes;
this synthesis is not efficient

3. ALKYL BROMIDES

$$R\!-\!H + Br_2 \xrightarrow{h\nu} R\!-\!Br + HBr$$

Also works for Cl_2; photobromination is more
selective than photochlorination

A chain reaction that works only for H—Br, not
other H—X molecules; note the anti-Markovnikov
regiochemistry

4. ALLYLIC HALIDES

Bromination is specifically at the allylic position

5. RADICALS

$$R\!-\!R \xrightarrow{\Delta} 2\,R\cdot$$

Pyrolysis of hydrocarbons is an unselective
method for making radicals; many different
radicals are typically produced

$$R\!-\!N\!=\!N\!-\!R \xrightarrow{\Delta} R\cdot + N_2 + R\cdot$$

Pyrolysis of azo compounds

6. VICINAL DIHALIDES

Vinyl halides (see 7) can add a second molecule of
HBr to give vicinal (not geminal) dihalides

7. VINYL BROMIDES

$$HC\!\equiv\!C\!-\!R \xrightarrow[\text{peroxides}]{HBr} BrHC\!=\!CHR$$

Radical-induced addition to alkynes gives
anti-Markovnikov addition; cis and trans
products are formed

Common Errors

A very common mistake is to confuse chain-carrying and termination steps in radical reactions. In the photobromination of an alkane such as methane, the product is formed in a radical chain process in which bromine and methyl radicals carry the chain (Fig. 11.75).

However, it is all too tempting to imagine that the product, methyl bromide, is formed by the combination of methyl and bromine radicals! In fact, although this step does produce a molecule of product, it destroys two chain-carrying radicals. This step terminates chains and does not contribute significantly to product formation (Fig. 11.76).

Product, along with a bromine atom that can recycle and start product formation again

FIGURE 11.75 The radical chain mechanism for the formation of methyl bromide from methane and bromine.

This step does form a molecule of CH₃Br, but it consumes two chain-carrying radicals; this is a termination step

FIGURE 11.76 Although the combination of a methyl radical and a bromine atom does form one molecule of methyl bromide, it terminates chains and does not contribute significantly to overall product formation.

11.13 ADDITIONAL PROBLEMS

PROBLEM 11.33 Let's build right away on Problem 11.30 and Section 11.10. Suppose you are walking down the street, and a stranger comes up to you and asks, Which is more stable, linear or triangular H_3^+? I maintain that you should be able to give him or her a quick answer to this seemingly bizarre question. *Obvious hint*: Look at Section 11.10. Here's an outline of how to do this problem:

First, build the molecular orbitals for linear and cyclic H_3. You did the linear molecule long ago in Chapter 1 (Problem 1.50, p. 45). Next, put the appropriate number of electrons in. Finally, look at the orbital containing those electrons in each species. Which is lower in energy?

PROBLEM 11.34 Give the major product(s) expected for each of the following reactions. Pay attention to regiochemistry and stereochemistry where appropriate. Mechanisms are not required, although they may be useful in finding the answers.

(a)

$$\xrightarrow[\text{ROOR}]{\text{HBr}}$$

(b)

$$\xrightarrow{\text{HBr}}$$

(c)

$$\xrightarrow[\text{ROOR}]{\text{HCl}}$$

(d)

$$\xrightarrow{h\nu}$$

(e)

$$\xrightarrow[h\nu]{\text{NBS, CCl}_4}$$

(continued)

PROBLEM 11.34 *(continued)*

(f)

(g)

PROBLEM 11.35 Predict the products of hydrogen bromide addition to 1-butyne. Consider both ionic and radical-initiated reactions. Rationalize your predictions with arrow formalism descriptions. Remember that a second molecule of hydrogen bromide can add to the initial product(s) in both cases.

(a)

$$CH_3CH_2—C≡CH \xrightarrow{HBr}$$

(b)

$$CH_3CH_2—C≡CH \xrightarrow[ROOR]{HBr}$$

PROBLEM 11.36 The relative rates of recombination of methyl and isopropyl radicals are quite different. Give two possible reasons for this difference.

			Relative Rate
$2\ H_3C•$	⟶	$H_3C—CH_3$	22
$2\ \begin{matrix}H_3C\\ \ \ \ CH•\\ H_3C\end{matrix}$	⟶	$\begin{matrix}H_3C\ \ \ \ \ \ \ \ \ \ CH_3\\ CH—CH\\ H_3C\ \ \ \ \ \ \ \ \ \ CH_3\end{matrix}$	1

PROBLEM 11.37 Decomposition of the following azo compounds at 300 °C occurs at quite different rates. Give two possible explanations for these differences.

$$\underset{R}{\overset{R}{R—C}}—N=N—\underset{R}{\overset{R}{C—R}} \xrightarrow{300\ °C} N_2\ +\ 2\ \underset{R}{\overset{R}{C•}}R$$

R	Relative Rate
CH₃, CH₃, CH₃	13,000
CH₃, CH₃, H	111
CH₃, H, H	11
H, H, H	1

PROBLEM 11.38 Predict the stereochemical results of the photochlorination of the optically active compound **1**. The (*) indicates a single enantiomer, here (*R*). It is the tertiary carbon–hydrogen bond that is chlorinated.

1

No stereochemistry implied in this drawing

PROBLEM 11.39 In Problem 11.25, we examine the photochlorination of ethane to give ethyl chloride. As was the case in the photochlorination of methane (p. 525), products containing more than one chlorine are also isolated in the photochlorination of ethane. What is the structure of the dichloroethane produced in this reaction? Rationalize your prediction with an analysis of the mechanism of the reaction.

PROBLEM 11.40 Write a detailed mechanism for the bromination of ethyl phenylacetate (**1**) with NBS.

PROBLEM 11.41 Write a detailed mechanism for the photochlorination of **1** to give **2**. Why is compound **1** reactive even though it contains no π bond?

PROBLEM 11.42 Here is a more complicated photochlorination of a compound containing a three-membered

(continued)

PROBLEM 14.42 *(continued)*

ring. Explain the formation of the following products from spiropentane (**1**).

1

Cl₂ | hv

PROBLEM 11.43   Photolysis of iodine in the presence of *cis*-2-butene leads to isomerization of the alkene to a mixture of *cis*- and *trans*-2-butene. Explain.

PROBLEM 11.44   Because of their high-potency, low-mammalian toxicity, and photostability, certain esters of halovinylcyclopropanecarboxylic acids (pyrethroids) are promising insecticides (see Chapter 10, p. 487). An important precursor to this class of insecticides is the carboxylic acid (**1**); prepared as follows:

(a) Write a mechanism for the formation of **2**. Be sure to account for the observed regiochemistry. *Hint*: The CuCl can abstract a chlorine atom.

(b) What transformations must be accomplished in the **2** → **1** conversion? Don't worry about mechanisms.

(c) Given the stereochemistry about the carbon–carbon double bond in **1**, how many stereoisomers of **1** are possible? The most potent insecticides are derived from the cis (1*R*, 3*S*) isomer of **1**. Draw this isomer.

PROBLEM 11.45   Write a mechanism for the following reaction of 1-octene:

PROBLEM 11.46   Provide mechanisms for the formation of the following products:

PROBLEM 11.47 Provide mechanisms for the formation of the following products:

$$CH_3(CH_2)_4 - C \equiv CH$$

$$CCl_4 \Big| \begin{array}{c} ROOR, \\ 77\ °C \end{array}$$

$$CH_3(CH_2)_4 \overset{Cl}{\underset{}{}}{=}CHCCl_3 \ +$$

PROBLEM 11.48 Propose a mechanism for the following reaction. Be sure to rationalize the preferred formation of diene **1**.

1
(major product)

PROBLEM 11.49 Write a mechanism for the formation of compound **1**.

ROOR, CCl₄
Δ

1   CH₂CCl₃

Dienes and the Allyl System: 2*p* Orbitals in Conjugation

12

Since chemists are not physicians, we shall scarcely benefit by their art, except by making the physician a chemist.

—George Rees*

12.1   PREVIEW

Sometimes the whole really is greater, or at least different, from the sum of the parts. In this chapter, we see just such a situation. One cannot always generalize from the properties of alkenes to those of polyalkenes. **Dienes**, compounds containing two double bonds (Chapter 3, p. 98), often have properties that are quite different from those of separated, individual alkenes. This difference depends on a phenomenon called conjugation. **Conjugated double bonds** are connected in 1,3-fashion. A molecule containing widely separated double bonds will generally react as would the individual alkenes. Double bonds in a 1,3-relationship react quite differently. A conjugated diene constitutes a separate functional group (Fig. 12.1).

Now we will generalize our discussion of the phenomenon of conjugation and its effect on chemical reactions, and we will develop further the chemistry of a conjugated molecule we already encountered, allyl. Figure 12.2 shows the neutral allyl radical, a system of three parallel 2*p* orbitals.

Finally, we will see the outlines of how large molecules are constructed in Nature from a rather simple building block, the diene isoprene (2-methyl-1,3-butadiene), shown in Figure 12.1.

*George Owen Rees was a nineteenth-century clinician who specialized in the analysis of blood and urine. A student of mine, Jordan M. Cummins, found this nice quote in the book by Stanley J. Reiser, *Medicine and the Reign of Technology*, Cambridge University Press, Cambridge, 1978.

FIGURE 12.1 Some conjugated and unconjugated double bonds.

FIGURE 12.2 Allyl has a 2p orbital on each carbon atom.

The allyl radical

A summary structure for the allyl radical

ESSENTIAL SKILLS

1. The overall skill that you will take away from this chapter is the ability to think about the structural and chemical consequences of conjugation—of placing two or more double bonds in a 1,3-relationship. Conjugation affects reactivity markedly, and an ability to see how and why is an essential skill to take into the following chapters.

2. The difference between kinetic and thermodynamic control must be understood. Kinetic control means that the products are determined by the relative heights of the transition states connecting starting materials and products, not by the energies of the products themselves. Under conditions that supply enough energy to go over *all* the transition states, thermodynamic control will operate. Under these conditions, the product mixture is determined by the relative energies of the products themselves.

3. The Diels–Alder reaction has been characterized as the most important synthetic reaction in organic chemistry. It is surely among the most important, as it makes the very common six-membered rings in a single step. Every cyclohexene is the potential product of a Diels–Alder reaction, and the potential source of a diene and a dienophile through a reverse Diels–Alder reaction. Be sure you can play the game of "find the cyclohexene"!

IMPORTANT DETAILS

1. The number of *sp* hybridized carbons in an allene or cumulene has stereochemical consequences: An even number of such carbons leads to cis/trans isomerism, whereas an odd number produces a pair of enantiomers.
2. Allylic resonance is important in reactions of cations, radicals, and anions. Be sure you can manipulate the allyl system easily.

12.2 ALLENES: 1,2-DIENES

The simplest diene possible is propadiene, almost always known as **allene*** (Fig. 12.3). Allene is the general term for molecules containing two double bonds that share a carbon atom. Allenes are commonly named as dienes or derivatives of allene (Fig. 12.3).

$H_2C=C=CH_2$

Allene
(propadiene)

$H_3C-HC=C=CH-CH_3$

1,3-Dimethylallene
(2,3-pentadiene)

$H_2C=C=C(CH_3)_2$

1,1-Dimethylallene
(3-methyl-1,2-butadiene)

1,2-Cyclononadiene

FIGURE 12.3 Some allenes.

The first problem is to work out a structure for allene. The key position in allene is certainly the central carbon atom shared by the two double bonds. We have not seen this kind of carbon before, and the bonding involved will have important structural consequences. The central carbon in any allene is attached to only two groups and, therefore, *sp* hybridization is appropriate. The central carbon uses two *sp* hybrid orbitals to form a pair of σ bonds with the flanking methylene groups, leaving the central carbon's remaining two *p* orbitals unused (Fig. 12.4).

Each end carbon, though, is attached to three groups and is therefore hybridized *sp²*, which leaves one unused 2*p* orbital on each end carbon. This easy approach leads immediately to a structure for allene (Fig. 12.5). Notice how each 2*p* orbital of the *sp²* hybridized end carbons overlaps perfectly with one of the perpendicular 2*p* orbitals of the central, *sp* hybridized carbon. The important thing is not to be panicked by the odd central carbon and to keep in mind how we dealt with new structures before. Assign an

2*p* Orbitals

$H_2C=C=CH_2 = H_2C-C-CH_2$

This carbon is attached to only two other groups (*sp* hybridization is appropriate)

FIGURE 12.4 The central carbon of an allene is *sp* hybridized.

*There is a chance for confusion here between the system of three parallel 2*p* orbitals called "allyl" and the allenes in which two double bonds are attached to one carbon. Be careful; these species are very different!

appropriate hybridization, be sure to pay attention to the geometrical requirements of the hybrid, and the structure will always work out.

FIGURE 12.5 An orbital picture of an allene. Notice the perpendicular π systems.

Notice that the two pairs of overlapping
2p orbitals are in *perpendicular* planes

But we do have to be careful and pay attention to detail. Finding a reasonable structure for allene is a classic example of a problem asked at this point in every course in organic chemistry (see Problems 12.1 and 12.2 and the following answers).

WORKED PROBLEM 12.1 It would appear that there could be cis and trans isomers of 1,3-dimethylallene, but there aren't. Explain.

cis ???   trans ???

1,3-Dimethylallene
(2,3-pentadiene)

ANSWER The two-dimensional paper will fool us all the time if we do not think in three dimensions. The end groups of an allene are not in the same plane, but lie in perpendicular planes. The words cis (same side) and trans (opposite sides) have absolutely no meaning for an allene.

The planes containing the two
H–C–CH₃ units are perpendicular

WORKED PROBLEM 12.2 Yet, two isomers of 1,3-dimethylallene do exist. What are they?

ANSWER There are two isomers of 1,3-dimethylallene, a chiral molecule. The mirror image of 1,3-dimethylallene is not superimposable on the original; therefore a pair of enantiomers exists.

Mirror

Allenes are high-energy molecules, just as are alkynes. The *sp* hybridization of the central carbon is not an efficient way of using orbitals to make bonds, as π bonds are higher energy than σ bonds. The π bond energy of about 66 kcal/mol is substantially lower than most σ bond energies. This inefficiency is reflected in relatively positive heats of formation (ΔH_f° values). Table 12.1 gives a few heats of formation for allenes and alkynes. Remember that heats of formation are strictly comparable only among isomers. As you can see, allenes are roughly of the same energy as alkynes. This similarity is not surprising; the two molecules contain the same number of π bonds.

TABLE 12.1 Heats of Formation for Some Small Hydrocarbons

Alkynes	ΔH_f° (kcal/mol)	Allenes	ΔH_f° (kcal/mol)
Propyne	44.6	Propadiene $H_2C{=}C{=}CH_2$	47.7
1-Butyne	39.5	1,2-Butadiene $H_2C{=}C{=}CH{-}CH_3$	38.8
2-Butyne	34.7	1,2-Pentadiene $H_2C{=}C{=}CH{-}CH_2CH_3$	33.6
		2,3-Pentadiene $H_3C{-}CH{=}C{=}CH{-}CH_3$	31.8

12.3 RELATED SYSTEMS: KETENES AND CUMULENES

In allenes, the double bonds share the central carbon atom. There are other molecules in which heteroatoms (non-carbon heavy atoms such as oxygen and nitrogen) take the place of one or more carbons of the allenes and/or in which there are more than two double bonds in a row. When one end carbon of allene is replaced with an oxygen, the compounds are called **ketenes**. If both end atoms are oxygens, the molecule is carbon dioxide. Hydrocarbons with three attached double bonds are called **cumulenes**, although they are more properly named as derivatives of 1,2,3-butatrienes. Cumulenes are very reactive and most difficult to handle. Figure 12.6 gives some examples of these and other molecules:

FIGURE 12.6 Some molecules containing cumulated double bonds.

$$O=C=O \qquad R_2C=C=O \qquad R_2C=C=C=CR_2$$

Carbon dioxide Ketenes Cumulenes (butatrienes)

PROBLEM 12.3 Draw Lewis dot structures for all the molecules in Figure 12.6 as well as for the related molecules, carbon disulfide (S=C=S), isocyanates (R—N=C=O), carbodiimides (RN=C=NR), and carbon suboxide (O=C=C=C=O).

PROBLEM 12.4 Unlike 1,3-dimethylallene, 2,3,4-hexatriene (1,4-dimethylcumulene) exists as a pair of cis and trans isomers. Explain.

PROBLEM 12.5 Demonstrate that 2,3,4-hexatriene is achiral.

12.4   ALLENES AS INTERMEDIATES IN THE ISOMERIZATION OF ACETYLENES

Now that we have the structures of allenes and cumulenes under control, it is time to look at reactivity. Our brief look at allenes permits us to understand one of the most astonishing reactions of alkynes: the base-catalyzed isomerization of internal triple bonds to the terminal position (Fig. 12.7).

FIGURE 12.7 The isomerization of 2-butyne to 1-butyne can be carried out in strong base.

$$H_3C-C\equiv C-CH_3 \xrightarrow[\text{2. } H_2O]{\text{1. } K^+ \ ^-:NH_2 / :NH_3} H_3C-CH_2-C\equiv C-H$$

2-Butyne   **1-Butyne**

Let's look at the isomerization of 2-butyne to 1-butyne in detail. We do not begin with allenes, rather we encounter them as intermediates in this reaction. When 2-butyne is treated with a very strong base such as an amide ion ($^-NH_2$), a proton can be removed from one of the methyl groups to give a resonance-stabilized carbanion (Fig. 12.8).

Ammonia and the carbon–hydrogen bonds of the methyl groups of 2-butyne both have pK_a values of about 38. Therefore, we can expect reasonable amounts of the carbanion to be produced on reaction with amide ion ($^-NH_2$). 2-Butyne is much more acidic than the parent hydrocarbon butane because the anion formed from 2-butyne is resonance stabilized, and that from butane is not (Fig. 12.8).

FIGURE 12.8 Deprotonation of 2-butyne to give a resonance-stabilized anion.

What can happen to this carbanion? It can be reprotonated by ammonia to re-generate 2-butyne, but this reaction surely gets us nowhere, even though it is reason-able, and must happen frequently. Suppose, however, that the anion reprotonates at the *other* position sharing the negative charge? The product is 1,2-butadiene, an allene (Fig. 12.9).

FIGURE 12.9 This anion can reprotonate in two ways. Reattachment of a proton at the end re-forms 2-butyne, but protonation at the other carbon sharing the negative charge produces an allene.

The pK_a of the allene at the terminal methylene position is also about 38, so the amide ion is strong enough to remove the terminal proton to generate a new, resonance-stabilized anion (Fig. 12.10).

FIGURE 12.10 Deprotonation of the allene at the terminal position leads to a new resonance-stabilized anion.

Reprotonation at one carbon sharing the negative charge regenerates the allene, but reprotonation at the other carbon yields a new, terminal acetylene, 1-butyne (Fig. 12.11).

FIGURE 12.11 Reprotonation either re-forms the allene or gives the isomerized alkyne, 1-butyne.

1-Butyne ($pK_a = 25$, Chapter 3, p. 131) is a *much* stronger acid than any other molecule present in the mixture and will be rapidly, and in a practical sense, irreversibly deprotonated to give the acetylide (Section 3.14). This deprotonation is the key step in this reaction. No matter where the equilibria in the earlier steps of this reaction settle out, the thermodynamic stability of the acetylide anion will drag the *overall* equilibrium far to the right. When water is added to neutralize the solution and end the reaction, the acetylide will be protonated and 1-butyne can be isolated (Fig. 12.12).

FIGURE 12.12 1-Butyne is by far the strongest acid in this equilibrating system. Removal of the acetylenic hydrogen gives an acetylide, the thermodynamic energy minimum in this reaction. When the reaction mixture is quenched by the addition of water, the acetylide protonates to give 1-butyne. It is critical to understand why it is not possible to re-enter the equilibrium at this point. After acidification there is no base present strong enough to remove the acetylenic hydrogen!

The overall reaction is summarized in Figure 12.13.

FIGURE 12.13 The full mechanism for the isomerization of 2-butyne to 1-butyne in strong base.

PROBLEM 12.6 Write a mechanism for the isomerization of 3-hexyne to 1-hexyne in the presence of sodium amide ($NaNH_2$).

WORKED PROBLEM 12.7 Di-*sec*-butylacetylene (3,6-dimethyl-4-octyne) fails to isomerize to another alkyne when treated with $NaNH_2/NH_3$. Explain.

ANSWER Isomerization is blocked by the alkyl groups. An allene can be formed, but the reaction can go no further as there is no allenic hydrogen that can be removed in base to produce a new resonance-stabilized anion.

There are no hydrogens on this carbon to remove and carry the isomerization further!

12.5   1,3-DIENES

The bonding in 1,3-dienes is not so obviously exotic as it is in 1,2-dienes, the allenes. Nevertheless, there are some subtle effects with far-reaching implications. The source can be seen from simple, schematic orbital drawings of 1,3-butadiene and 1,4-pentadiene. In a conjugated diene, the orbitals of the two double bonds overlap; in an unconjugated diene, they do not (Fig. 12.14).

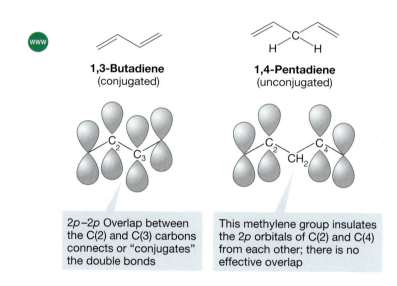

1,3-Butadiene
(conjugated)

1,4-Pentadiene
(unconjugated)

2p–2p Overlap between the C(2) and C(3) carbons connects or "conjugates" the double bonds

This methylene group insulates the 2p orbitals of C(2) and C(4) from each other; there is no effective overlap

FIGURE 12.14 A conjugated and an unconjugated diene.

In the 1,3-diene, the two double bonds are connected through 2p–2p overlap. These double bonds are called "conjugated." By contrast, in the 1,4-diene the two double bonds are insulated by a methylene group. These bonds are "not conjugated," or "unconjugated," and will behave much like two separate alkenes. The question we now have to examine is how the chemical behavior of a 1,3-diene will be influenced by the overlap between the 2p orbitals on C(2) and C(3). To start, let's take a closer look at the structure of 1,3-butadiene, giving special attention to the π system.

PROBLEM 12.8 Sketch out the σ system of 1,3-butadiene. What orbitals overlap to form the carbon–carbon and carbon–hydrogen bonds?

First, let's look at 1,3-butadiene from a resonance point of view. 1,3-Butadiene does have resonance forms that will contribute to the overall electronic structure of the molecule (Fig. 12.15).

Note the double-bond character between C(2) and C(3) as shown in resonance forms **b**, **c**, and **d**

FIGURE 12.15 Resonance forms contributing to the structure of 1,3-butadiene.

The best resonance form, and therefore the best single description of the molecule, is the diene structure we conventionally write. This structure has one more bond than the others and involves no charge separation. Thus, it wins by almost all of our criteria for estimating the relative energies of resonance forms (Chapter 9, p. 393). Nevertheless, the other resonance forms show that there is some "double-bond character" to the C(2)—C(3) bond (Fig. 12.15).

PROBLEM 12.9 For each of the other resonance forms in Figure 12.15 indicate what makes it less stable than the conventional structure, **a**.

Conjugation can be seen, perhaps even more clearly, if we examine the π molecular orbitals of 1,3-butadiene. In butadiene, we have a system of four $2p$ orbitals on four adjacent carbons. These four atomic $2p$ orbitals will overlap to produce four π molecular orbitals. We can get the symmetries of the four new molecular orbitals in a number of equivalent ways. It is always easiest to build up something new out of things we know already, so it's productive to construct the molecular orbitals of butadiene out of two sets of the π molecular orbitals of ethylene, π and π^*. One way to accomplish this is to allow the π molecular orbitals of ethylene to overlap end to end. In producing these π molecular orbitals in this way, we have made an important approximation. As often mentioned before, the closer in energy two orbitals are, the stronger the stabilization resulting from their overlap. In practice, all orbital overlap except for the strongest interactions can be ignored, at least at this level of theory. This observation means that we need to look only at the result of the $\pi \pm \pi$ and $\pi^* \pm \pi^*$ interactions and do not have to consider $\pi \pm \pi^*$.

The symmetries of the new butadiene orbitals can be easily derived from the symmetries of the starting ethylene orbitals (Fig. 12.16).

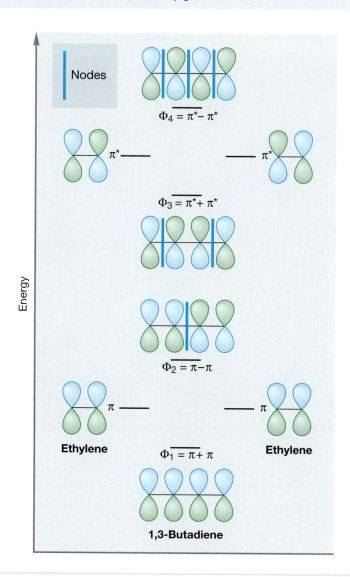

FIGURE 12.16 The schematic formation of the π molecular orbitals of 1,3-butadiene from the π molecular orbitals of ethylene.

To order the new molecular orbitals in energy, count nodes. Orbital Φ_1 has no new nodes, Φ_2 has one, Φ_3 has two, and Φ_4 has three. Not all methods of construction will give the correct molecular orbitals *in the correct order,* so you must always be careful to do an energy ordering.

PROBLEM 12.10 Construct the π molecular orbitals of 1,3-butadiene from those of ethylene by putting one ethylene "inside" the other.

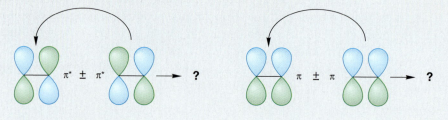

12.6   THE PHYSICAL CONSEQUENCES OF CONJUGATION

In 1,3-butadiene, or any conjugated diene, there are four electrons to accommodate in the π molecular orbital system. The lowest two bonding molecular orbitals are occupied and the two higher energy, antibonding molecular orbitals are empty. The HOMO for butadiene is Φ_2, and to a first approximation, Φ_2 will control the reactivity of the molecule. It is Φ_2 that looks most like the traditional representation of 1,3-butadiene. In the resonance formulation it is the traditional picture of 1,3-butadiene that is surely the lowest energy resonance form, and thus the best representation of the molecule. So we see the usual (and necessary) convergence between the resonance and molecular orbital way of looking at structures. Each method generates a picture of 1,3-butadiene that looks mostly like a molecule containing a simple pair of double bonds. Both methods warn us, however, that there are likely to be some consequences of conjugation in this molecule. This is shown by the C(1)—C(2)—C(3)—C(4) connection of Φ_1 in the molecular orbital picture, and by the existence of resonance forms in which there is double-bond character between C(2) and C(3) (Fig. 12.17).

Best resonance form

The best (most stable) resonance form is the best representation for 1,3-butadiene

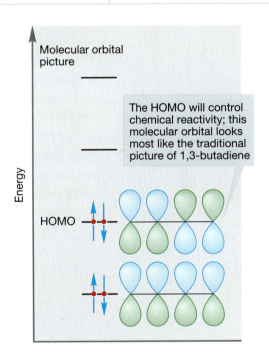

Molecular orbital picture

The HOMO will control chemical reactivity; this molecular orbital looks most like the traditional picture of 1,3-butadiene

Energy

HOMO

FIGURE 12.17 Both the resonance and molecular orbital views of 1,3-butadiene lead to the conclusion that the best view of 1,3-butadiene is of a molecule containing two traditional double bonds. However, both pictures make it clear that there will be some consequences of conjugation between the two double bonds.

What will be the effects of this C(2)—C(3) overlap on the structure of 1,3-butadiene? Carbon–carbon single bonds are relatively long; for example, in butane the carbon–carbon bond is 1.53 Å. The carbon–carbon double bond in 2-butene is only 1.32 Å, reflecting the presence of two bonds between the carbons, and the carbon–carbon triple bond in acetylene, 1.20 Å, is even shorter. To the extent that there is double-bond character between C(2) and C(3), the bond in 1,3-butadiene should be shortened. And so it is, the C(2)—C(3) bond distance is only 1.47 Å. It has been argued that the shortening is indicative of conjugation between the two double bonds (Fig. 12.18).

FIGURE 12.18 The central carbon–carbon bond length in 1,3-butadiene is in between that for an alkane and an alkene.

WORKED PROBLEM 12.11 Life is never simple. There is a persuasive counterargument that attributes the shortening of the C(2)—C(3) bond to other factors. Can you come up with another explanation? *Hint for this super-tough question:* We have ignored the σ system in this argument. What kind of a σ bond joins C(2) and C(3)? So what? See Problem 12.8 (p. 558).

ANSWER The two inner carbons of 1,3-butadiene are joined by an sp^2–sp^2 σ bond, in contrast to the sp^3–sp^3 σ bonds of butane. An sp^2 orbital has more s character than an sp^3 orbital, and bonds made from sp^2–sp^2 overlap will be stronger and shorter than those made from sp^3–sp^3 overlap. Accordingly, some of the shortening of the central σ bond in 1,3-butadiene may be attributed to the σ system and not to π orbital overlap.

Now the question becomes, How strong is the C(2)—C(3) overlap? What consequences will this overlap have for the structure of conjugated 1,3-dienes? Overlap between 2p orbitals on adjacent positions has profound consequences for the structure of alkenes, for example, and the same might be true for dienes. It is the 1,2-overlap between 2p orbitals that makes cis and trans isomers of alkenes possible (Fig. 12.19).

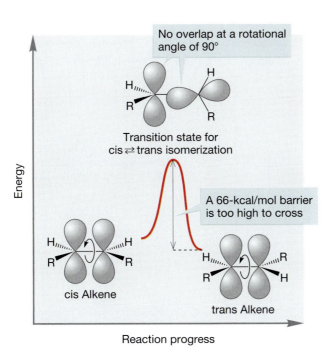

FIGURE 12.19 In alkenes, rotation about the carbon–carbon double bond requires overcoming a barrier of 66 kcal/mol. This barrier is too high to be surmounted under most conditions.

Might the same be true for dienes? Will overlap between the $2p$ orbitals at the C(2) and C(3) positions create a large enough barrier so that isomers, called the **s-cis** and **s-trans** conformations, could be isolated? The "s" nomenclature in dienes stands for "single" as these are two conformations involving stereochemistry about the single bond of the diene (Fig. 12.20).

 rotation   How easy is this interconversion?

FIGURE 12.20 Interconversion of the s-trans and s-cis conformations of 1,3-butadiene.

As we might guess, there is a barrier to rotation, but it's not very large. Overlap between C(2) and C(3) is not nearly as important as that between the orbitals at the two adjacent carbons in ethylene. At the transition state for the rotation of 1,3-butadiene about its central, C(2)—C(3) bond, overlap between the $2p$ orbitals on C(2) and C(3) vanishes, but C(1)—C(2) and C(3)—C(4) overlap is undisturbed (Fig. 12.21a).

The barrier to rotation in 1,3-butadiene turns out to be 4–5 kcal/mol, only a bit larger than the barrier to rotation about the carbon–carbon bond in an alkane. It is the trans form that is favored at equilibrium, by 2–3 kcal/mol. There is a simple steric effect favoring the s-trans conformation. In the s-cis conformation the two inside hydrogens are quite close to each other (make a model!) and in the s-trans conformation they are not. In order to make sense of these structures, we have to watch out not only for the locations of the double bonds, but for the positions of the hydrogens attached to them as well (Fig. 12.21b).

PROBLEM 12.12 Calculate the amount of the s-cis form present at equilibrium at 25 °C, assuming an energy difference between the s-cis and s-trans forms of 2.5 kcal/mol.

FIGURE 12.21 The barrier for the interconversion of the s-trans and s-cis conformations of 1,3-butadiene.

PROBLEM 12.13 There is another kind of rotation possible, that about the C(1)—C(2) or C(3)—C(4) bonds. We might guess that the barrier to this rotation would be little different from that for rotation in a simple alkene, but, as a former president of the United States once said, "that would be wrong." In 1,3-butadiene, it takes only 52 kcal/mol to do this rotation, some 14 kcal/mol less than the barrier of 66 kcal/mol for rotation in a simple alkene. Explain. *Hint*: Look at the transition state for C(1)—C(2) bond rotation in 1,3-butadiene.

PROBLEM 12.14 Design an experiment to determine whether rotation about the C(1)—C(2) bond is possible. Assume you have the ability to make any labeled molecules you need.

As both the resonance and molecular orbital pictures indicate, there is 2p–2p overlap between C(2) and C(3), but it does not dominate the structure of 1,3-butadienes the way 2p–2p overlap does for alkenes. We can get another idea of the magnitude of this interaction from measuring the heats of hydrogenation of a series of dienes (see Chapter 10, p. 438).

We begin here with one of the butene isomers, 1-butene. When 1-butene is hydrogenated to butane, the reaction is exothermic by 30.3 kcal/mol (Fig. 12.22). A reasonable assumption would be that a molecule containing two double bonds would be hydrogenated in a reaction twice as exothermic as the hydrogenation of 1-butene. And this idea is just right; the heat of hydrogenation is −60.8 kcal/mol for 1,4-pentadiene and −60.5 kcal/mol for 1,5-hexadiene (Fig. 12.22).

1-Butene

Find
$\Delta H = -30.3$ kcal/mol

1,4-Pentadiene

Predict
2×-30.3 kcal/mol $= -60.6$ kcal/mol
Find
$\Delta H = -60.8$ kcal/mol

1,5-Hexadiene

Find
$\Delta H = -60.5$ kcal/mol

Average $= \Delta H = -60.65$ kcal/mol

FIGURE 12.22 Hydrogenation of 1-butene is exothermic by 30.3 kcal/mol. Hydrogenation of unconjugated dienes is exothermic by almost exactly twice the heat of hydrogenation of 1-butene.

Both of our sample dienes were unconjugated. Now let's look at the heat of hydrogenation of the conjugated diene, 1,3-butadiene: ΔH is only −57.1 kcal/mol (Fig. 12.23).

$\Delta H = -57.1$ kcal/mol

Average ΔH for unconjugated dienes $= -60.7$ kcal/mol
1,3-Butadiene $= -57.1$ kcal/mol
Difference $=\quad 3.6$ kcal/mol

FIGURE 12.23 The heat of hydrogenation of 1,3-butadiene is 57.1 kcal/mol, smaller than expected by about 3.6 kcal/mol.

The heat of hydrogenation of the conjugated diene is about 3.6 kcal/mol lower than we would expect by analogy to the unconjugated dienes in Figure 12.22. This amount of energy can be attributed to the energy-lowering effects of conjugation. The situation is summarized in Figure 12.24.

So now we have two measures of the energetic value of conjugation in 1,3-butadiene, and they are remarkably close to each other. The first is the height of the barrier to rotation about the C(2)—C(3) bond (4–5 kcal/mol) and the other is the lowering of the heat of hydrogenation of a conjugated diene relative to that of an unconjugated diene (~3.6 kcal/mol).

Two conjugated double bonds are not utterly unconnected; the whole is somewhat different from the sum of the parts. Conjugation, the orbital overlap between C(2) and C(3) in a 1,3-diene, results in a shortening of the C(2)—C(3) single bond and in lower heats of hydrogenation for 1,3-dienes than for unconjugated systems. Now we move on to the consequences of further conjugation, and to an examination of the reactivity of these molecules. Here, too, we will see that conjugation influences the physical properties and chemical reactivity of these molecules.

FIGURE 12.24 A summary of the heat of hydrogenation data.

SUMMARY

Overlap between C(2) and C(3) of a 1,3-diene (conjugation) has structural consequences. The central sigma bond is slightly shorter than a normal carbon–carbon single bond. The magnitude of the conjugative stabilization is small, about 4 kcal/mol.

12.7   MOLECULAR ORBITALS AND ULTRAVIOLET SPECTROSCOPY

In the early days of organic chemistry, there were two extraordinarily difficult general problems. One was to separate mixtures into pure compounds, and the other was to determine the structures of the compounds, once pure samples were obtained. How simple these problems seem! Yet, much of the mechanistic and synthetic organic chemistry of the last decades would have been quite impossible without the progress in solving these two old difficulties. We will have little to do with separation techniques in this book (your laboratory will take another approach), but we must certainly spend some time here on structure determination.

About 50 years ago, structure determination was an entire branch of chemistry. People spent whole careers working out the structures of complicated molecules by running chemical reactions. The products of many reactions of the unknown compound were analyzed and inferences were made regarding the structure of the unknown. This task was time-consuming and extraordinarily demanding of experimental technique, intellectual rigor, and the development, over years of work, of an intuitive sense of what was right, of what a given experiment might mean. In a way, this work had all the satisfactions and charms of handwork, of a craft. Nowadays, the work of years can usually be done in hours, and structural problems can be attacked that would have been beyond any chemical determination. **Spectroscopy**, the study of the interactions between electromagnetic radiation and atoms and molecules, has made the difference. First came the development of **ultraviolet/visible (UV/vis) spectroscopy**. We will deal with that subject in this chapter. Then followed infrared (IR) spectroscopy and finally nuclear magnetic resonance (NMR). The power of NMR is so great that this technique has largely eclipsed all other forms of structure determination. We have already seen NMR spectroscopy in Section 2.14, and we will devote much of Chapter 15 to this subject.

The new spectrometers certainly brought progress, and no reasonable person would want to turn back the clock to the days of handcrafted structure determination. Yet, one can sympathize with the feeling that there has been some loss with the progress. We are somehow not as close to the molecules as we once were, even though we know much more about them. There is psychological distance involved here. One tends to believe what spectrometers say and to read too much into the data. In chemistry, nothing is more important than honing one's critical sense. It is somehow easier to ignore some small, but ultimately vital glitch on a piece of chart paper than to ignore the results of an experiment gone in an unexpected direction. The former somehow melts into the noise of the base line, while the latter teases the imagination.

One can also sympathize with those practitioners of the art of structure determination who saw the future when the first NMR machine was delivered, and realized that for them, "NMR" meant "No More Research." They were right.

Ultraviolet/visible spectroscopy is called **electronic spectroscopy** because it deals with the absorption of energy by electrons in molecules. The energy absorbed (ΔE) is used to promote an electron from one orbital to a higher energy one and the magnitude of ΔE is proportional to the frequency of the light, ν [Eq. (12.1); Fig. 12.25].

FIGURE 12.25 Absorption of UV light promotes an electron from the HOMO to the LUMO.

$$\Delta E = h\nu = hc/\lambda \qquad (12.1)$$

(h is Planck's constant, λ is the wavelength of the light, and c is the speed of light)

As electrons are restricted to orbitals of specific energy, ΔE is also quantized and may have only certain values. The distinction between ultraviolet and visible spectroscopy is an artificial one and relates only to the fact that the detectors built into our bodies, our eyes, are sensitive only to radiation of wavelength 400–800 nm (nm = nanometers; 1 nm = 10^{-9} m), which we call the visible spectrum. These machines are less limited in this respect than we are, and UV spectrometers easily detect radiation from 200–400 nm, as well as the entire visible range and beyond. At higher energy, below 200 nm, the atmosphere begins to absorb strongly, and spectroscopy in this range is more difficult, although by no means impossible.

A schematic spectrometer is shown in Figure 12.26. It consists of a source of radiation, a monochromator that is capable of scanning the range emitted by the source, a photomultiplier, and a detector.

Generally, the spectrometer determines A, the absorbance, and plots A versus wavelength (λ in nm). The absorbance is calculated according to the Beer–Lambert law, Equation (12.2).

$$A = \log I_0/I = \varepsilon lc \qquad (12.2)$$

FIGURE 12.26 A schematic picture of a UV/vis spectrometer.

The intensity of the light entering the sample is I_0, the intensity of the light exiting the sample is I, the concentration in moles per liter is c, the length of the solvent cell is l, and the proportionality constant ε is called the **extinction coefficient**. Every compound has a characteristic ε, which depends not only on A, l, and c [Eq. (12.2)], but also on wavelength, solvent, and temperature.

It is important to have a feel for the magnitudes of the energies involved in spectroscopy. Equation (12.1) repeats the familiar relationship between energy, Planck's constant h, and the frequency of light:

$$E = h\nu \tag{12.1}$$

As $\nu = c/\lambda$, and as we are interested in the energy per mole, Equation (12.3) is the important one, where N is Avogadro's number:

$$E = Nhc/\lambda \tag{12.3}$$

Because $h = 1.583 \times 10^{-37}$ kcal s/molecule, the quantity Nhc is equal to 28.6×10^3 (nm)(kcal/mol).

PROBLEM 12.15 Calculate the energies involved in the range 200–800 nm, the region of UV/vis spectroscopy.

For a given molecule there will be many possible UV (or vis) transitions, as there are many different kinds of electrons even in a simple molecule. For ethylene there are several possible absorptions, as σ or π electrons are promoted to antibonding σ^* or π^* orbitals. Figure 12.27 ignores the carbon–hydrogen bonds and shows only $\sigma \rightarrow \sigma^*$ and $\pi \rightarrow \pi^*$ transitions. There are other possible absorptions, even without including the carbon–hydrogen σ and σ^* orbitals: $\sigma \rightarrow \pi^*$ and $\pi \rightarrow \sigma^*$ are two examples. Such transitions between orbitals of different symmetry are usually relatively weak, however.

The starting point and final destination of the electron are used to denote the transition; hence $\sigma \rightarrow \sigma^*$, $\sigma \rightarrow \pi^*$, $\pi \rightarrow \pi^*$, and $\pi \rightarrow \sigma^*$ in this case. The energies, and therefore wavelengths of light involved, depend on the energy separating the

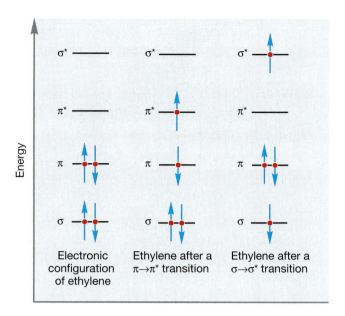

FIGURE 12.27 A $\pi \rightarrow \pi^*$ and $\sigma \rightarrow \sigma^*$ transition for ethylene.

orbitals. For ethylene, even the $\pi \rightarrow \pi^*$ separation is great enough so that detection is difficult with common spectrometers. The $\pi \rightarrow \pi^*$ absorption for ethylene occurs in the vacuum UV at about 165 nm, which corresponds to 173 kcal/mol, a high energy indeed. Little change is induced by alkyl substitution, and neither simple alkenes nor unconjugated dienes absorb strongly above 200 nm.

When a diene is conjugated, however, substantial changes occur in orbital energies, and therefore in the possible UV transitions. In general, the lowest energy (longest wavelength λ) transition possible for a polyene will be from the HOMO to the LUMO. In ethylene or butadiene, this is a $\pi \rightarrow \pi^*$ transition. Figure 12.28 shows the relative energies of the π molecular orbitals of ethylene and 1,3-butadiene.

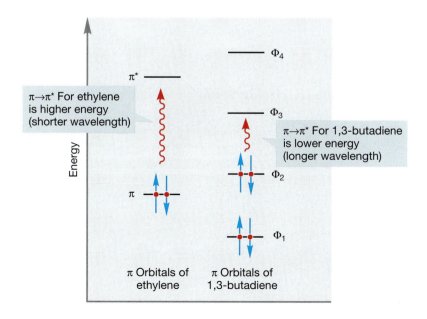

FIGURE 12.28 The HOMO–LUMO transition in ethylene is higher in energy (at shorter wavelength) than the HOMO–LUMO transition in 1,3-butadiene.

You can see that there is a much lower energy $\pi \to \pi^*$ (HOMO–LUMO) transition possible in the conjugated molecule 1,3-butadiene than there is in ethylene itself. Spectroscopy bears this out, and 1,3-butadiene absorbs at much longer wavelength (lower energy) than ethylene, 217 nm.

WORKED PROBLEM 12.16 How much energy is contained in light of wavelength 217 nm?

ANSWER $E = Nhc/\lambda$ and the quantity $Nhc = 28.6 \times 10^3$ (p. 568).
So, $E = (28.6 \times 10^3)/217 = 132$ kcal/mol at 217 nm.

Further conjugation does two things. First, it reduces the energy gap between the HOMO and LUMO, and this pushes the position of the lowest energy absorption to longer wavelength (lower energy). Eventually, the HOMO–LUMO gap becomes small enough so that absorption enters the visible range, and we perceive highly conjugated polyenes as colored. The first colored polyenes are yellow, as it is violet light (400–420 nm) that is first absorbed as the HOMO–LUMO gap shrinks. β-Carotene, an important precursor to vitamin A, absorbs at 453 and 483 nm, and is orange (Fig. 12.29).

FIGURE 12.29 As conjugation increases, the HOMO–LUMO gap decreases and the position of the $\pi \to \pi^*$ absorption shifts to longer wavelength (lower energy).

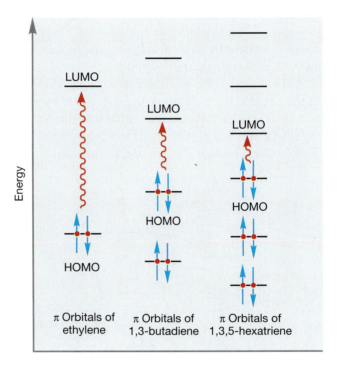

β-**Carotene**
$\lambda_{max} = 453$ and 483 nm

Second, as molecules become more complicated, and especially as conjugation is extended, the number of possible UV transitions increases and the spectra become complex. Figure 12.30 shows the spectrum of one moderately complicated molecule, which surely will resist a simple analysis, and which would be most difficult to predict in detail.

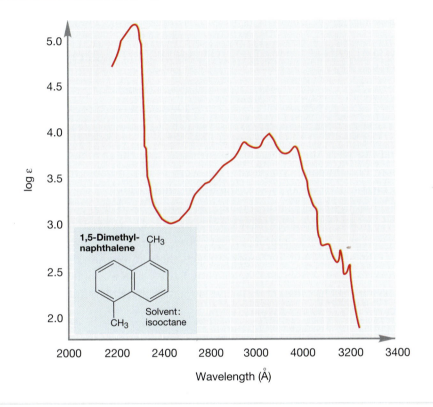

FIGURE 12.30 A complicated UV spectrum.

However, many important molecules contain simple diene or triene systems, either acyclic or contained in rings, and UV spectroscopy remains a useful tool for structure determination in such molecules.

The position of longest wavelength absorption (λ_{max}) depends in a predictable way on the substitution pattern of such molecules. A set of empirical correlations was collated by R. B. Woodward and was known then as "Woodward's rules" and now sometimes as "Woodward's first rules." These rules were extended by Louis Fieser (1899–1977) and Mary Fieser (1909–1997) to include cyclic dienes in the polycyclic compounds called steroids. Table 12.2 summarizes Woodward's and the Fiesers' rules. Woodward's very different second set of rules will appear in Chapter 20.

Robert Burns Woodward (1917–1979) was acknowledged as the greatest American organic chemist, and was certainly one of the greatest chemists of all time. There is remarkable agreement on the subjective evaluation of Woodward's preeminence among organic chemists, and it's worth taking a little time to try to see what set this man apart from his merely excellent colleagues. First of all, he was devoted to the subject and worked unbelievably hard at it. Although the main themes of his

TABLE 12.2   Woodward's and Fiesers' Rules for Ultraviolet Spectra of Dienes

Woodward's Rules for Acyclic Dienes	(nm)	Fiesers' Rules for Steroid Dienes	(nm)
Base diene value	217	Base diene with both double bonds in the same ring	253
Each added alkyl group	+5	Base diene with double bonds in different rings	214
Each double-bond exocyclic in a six-membered ring	+5	Each additional conjugated double bond	+30
		Each alkyl substituent	+ 5
		Each exocyclic double bond	+ 5

professional life were the elucidation of structure and the synthesis of molecules of ever more breathtaking complexity, his work is also notable for the attention he paid to mechanistic detail, and to the broad implications of the work. Working hard is not a matter of mere hours, but of maintaining focus as well. Woodward was able to *concentrate*, and had or developed a prodigious memory for minutiae. In another person that might amount to mere obsession, but Woodward was able to extract from the sea of detail the important facts that allowed him time and again to turn his work from the specific to the general. Early on he was able to propose correct structures for such disparate molecules as penicillin, tetracycline, and ferrocene (Fig. 12.31).

FIGURE 12.31 Some complicated molecules whose structures were proposed by Woodward and his co-workers.

Woodward was able to pull obscure facts out of remote chemical hats and apply them to the matter at hand in a way that often transformed the very nature of the problem. The molecules he considered worthy of synthesis grew in complexity throughout his career, but it was more than the difficulty of the target that made his work so universally impressive. Woodward's work led places—it had "legs." For example, Woodward, primarily a synthetic chemist, made the observations [with Roald Hoffmann (b. 1937)] that led to the development and exploitation of what is arguably the most important *theoretical* construct of recent times. Hoffmann and Kenichi Fukui (1918–1998) of Kyoto University received the Nobel prize for this work in 1981. Many believe Woodward would have shared this Nobel prize (it would have been his second; his first was awarded in 1965 for "his outstanding achievements in the art of organic synthesis") had he lived. Some of the molecules synthesized by Woodward and his co-workers are shown in Figure 12.32. Woodward said that the synthesis of quinine was conceived when he was 12 years old.

Should you memorize Woodward's rules? Frankly, I think not, although there are those who strongly disagree, and some of these people write examination questions. These rules were an important part of structure determination but are restricted in utility to relatively simple polyenes. You should know they exist, know what systems they apply to, and look them up when you need them. If you wind up in steroid research, you'll learn them.

Quinine, with W. von E. Doering, 1944

Patulin, 1950

Cholesterol, 1951

Strychnine, 1954

Reserpine, 1956

FIGURE 12.32 Some molecules synthesized by the Woodward group.

(continued)

Triquinacene, 1964

Cephalosporin C, 1966

Vitamin B$_{12}$ with A. Eschenmoser and a cast of thousands, 1972

FIGURE 12.32 *(continued)*

12.8   POLYENES AND VISION

When we hear or read the word "spectroscopy" we tend to imagine rooms full of expensive machines measuring away. It's quite an inhuman vision. Yet we are spectrometers ourselves! All of our sensory experiences, most obviously vision, are examples of spectroscopic measurements. Those machines in our imaginations are mere extensions of our biological, in-house equipment. When you see a red apple among the green ones, and decide that it is the ripe red one you will eat, you are acting on a spectroscopic measurement. Let's examine vision a little bit here, because it depends on polyene chemistry.

Many polyenes are important in biological processes, and as we might imagine, polyenes with visible spectra are important in human vision. Vitamin A is *trans*-retinol, for example (Fig. 12.33). We have two enzymes, one of which, retinol dehydrogenase, oxidizes retinol to retinal and another, retinal isomerase, which isomerizes one double bond of *trans*-retinal to produce *cis*-retinal (Fig. 12.33).

FIGURE 12.33 Vitamin A (*trans*-retinol) is oxidized to *trans*-retinal and then enzymatically isomerized to *cis*-retinal.

Vitamin A, trans-retinol   A trans double bond   CH₂OH

retinol dehydrogenase

Still a trans double bond

***trans*-Retinal**

retinal isomerase

A cis double bond

***cis*-Retinal**

The isomerized *cis*-retinal (but *not trans*-retinal) has the proper shape to react with a protein called opsin to form rhodopsin (Fig. 12.34). Rhodopsin has a strong absorption in the visible spectrum, which isomerizes the cis double bond back to trans. This absorption of visible light changes the shape of the molecule severely, destroying the crucial fit of molecule and protein. The attachment point is exposed, and the retinal is detached from the protein, opsin. A nerve impulse is generated and perceived by our brains as light (Fig. 12.34). The chemistry involved in the attachment and detachment of retinal and opsin is simple and will be covered in Chapter 16.

***trans*-Retinal**   retinal isomerase   ***cis*-Retinal**   **Opsin**   **Rhodopsin**   *hν*   ***trans*-Retinal**   **Opsin**

FIGURE 12.34 *cis*-Retinal has the correct shape to react with opsin to form rhodopsin.

1,2-Hydrohalogenation of dienes

1,4-Hydrohalogenation of dienes

12.9   THE CHEMICAL CONSEQUENCES OF CONJUGATION: ADDITION REACTIONS OF DIENES

Now we pass on to reactions (at last!). Conjugation affects not only the physical properties of molecules but their reactions as well. Although the basic chemistry of dienes does resemble that of alkenes, there are detailed, but important differences introduced by the presence of the *pair* of connected (conjugated) double bonds.

12.9a Addition of HBr and HCl. When hydrogen chloride adds to 1,3-butadiene at −78 °C, two major products, 3-chloro-1-butene (75.5%) and *trans*-1-chloro-2-butene (24%), are formed. There is a very small amount of *cis*-1-chloro-2-butene formed (0.5%). The main product is formed through 1,2-addition and the minor products by 1,4-addition (Fig. 12.35).

FIGURE 12.35 Addition of HCl to 1,3-butadiene produces the products of both 1,2- and 1,4-addition.

We have seen this reaction before briefly in the discussion of resonance in Chapter 9 (p. 389). The source of the two products becomes clear if we work through the mechanism step by step. The first reaction must be protonation of butadiene by hydrogen chloride. There is a choice of two cations, one of which is a resonance-stabilized allylic cation, and the other an unstabilized primary carbocation. The first of these will be much lower in energy and its formation will be greatly preferred (Fig. 12.36).

FIGURE 12.36 Formation of a resonance-stabilized allyl cation is preferred.

The resonance description of the allyl cation shows the source of the two products, as chloride can add to each of the two carbons sharing the positive charge to give the two observed alkyl chloride products (Fig. 12.37).

FIGURE 12.37 Addition of the nucleophile at the two positions (a and b) sharing the positive charge gives the products of 1,4- and 1,2-addition.

PROBLEM 12.17 When treated under S_N1 conditions, the two chlorides below give the same pair of alcohols in the same ratio. Provide structures for the two alcohols and explain mechanistically.

12.9b Addition of Br$_2$ and Cl$_2$. Other addition reactions of 1,3-dienes also produce two products. Like simple alkenes, dienes react with bromine or chlorine to give dihalides. As in hydrogen halide addition, products of both 1,2- and 1,4-addition are formed (Fig. 12.38).

FIGURE 12.38 Addition of bromine or chlorine to conjugated dienes also gives two products. The red numbers show the origins of the terms 1,2- and 1,4-addition.

WORKED PROBLEM 12.18 One potential source of the 1,4-addition product is the bromonium ion shown below. Look carefully at the products of the reaction of 1,3-butadiene and bromine (Fig. 12.38). Explain why the mechanism below is certainly wrong.

ANSWER Although there is nothing wrong structurally or electronically with the five-membered ring bromonium ion, it cannot be involved in this reaction. Opening of this ion must lead to an alkene containing a cis double bond, and the 1,4-product is, in fact, overwhelmingly trans (Fig. 12.38). The technique for solving this problem is to follow through using the five-membered bromonium ion as the intermediate, see what the structure of the product must be, and compare that to what is actually observed. If we pay careful attention to stereochemistry, we can see that the five-membered ring cannot be involved.

A cis double bond

12.10   THERMODYNAMIC AND KINETIC CONTROL OF ADDITION REACTIONS

The addition reactions in the previous sections allow us to look at general properties of chemical reactions and to address fundamental issues of reactivity. Just what are the factors that influence the direction a reaction takes? Why is one product formed more than another? Is it the energies of the products themselves, or something else?

There are some curious effects of temperature and time on the addition reactions of both H—X and X_2. If the reactions are run at low temperature, it is the product of 1,2-addition that is generally favored. The more severe the reaction conditions, the more product of 1,4-addition is formed. At high temperature (25 °C) the product of 1,4-addition is favored (Fig. 12.39).

FIGURE 12.39 The ratio of 1,2- to 1,4-addition is temperature dependent.

3-Bromo-1-butene   *trans*-1-Bromo-2-butene

Indeed, the products themselves will equilibrate if allowed to stand in solution. The same mixture of 1,2- and 1,4-addition products is formed from each compound, and the product of 1,4-addition is favored. Here is an example using the products of addition of Cl_2 to 1,3-butadiene (Fig. 12.40).

FIGURE 12.40 Equilibration of the products of 1,2- and 1,4-addition at 25 °C.

1,2-Addition product (100%)   1,2-Addition product (30%)   1,4-Addition product (70%)   1,4-Addition product (100%)

Several questions now confront us: (1) Why is the product of 1,4-addition favored? (2) How do the products of 1,2- and 1,4-additions equilibrate? (3) What are the reasons behind the curious temperature effect?

The first question is easy. The product of 1,4-addition contains a disubstituted double bond, whereas the 1,2-product is only monosubstituted, and therefore less stable. *Remember*: The more substituted an alkene, the more stable it is (see Chapter 3, p. 115; Fig. 12.41).

Monosubstituted double bond (less stable)

Disubstituted double bond (more stable)

1,2-Addition product   1,4-Addition product

FIGURE 12.41 The compound containing a disubstituted double bond is more stable than the compound with a monosubstituted double bond.

The second question isn't hard either. For example, S_N1 solvolysis of either of the two products in Figure 12.40 leads to the same, resonance-stabilized allylic cation. The chloride ion can add at the two positions sharing the positive charge and at equilibrium it will be the more stable product of 1,4-addition that dominates (Fig. 12.42).

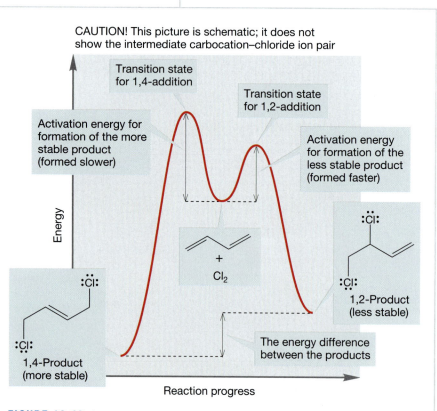

FIGURE 12.42 The 1,2- and 1,4-addition products equilibrate through the formation of a resonance-stabilized allyl cation.

The last question is tougher. Why should one product be favored under mild conditions and another at long time or under harsher conditions? At long time and relatively harsh reaction conditions, thermodynamics takes over and the more stable product prevails. At short time, or under mild conditions, it is not the *more stable* 1,4-compound that is favored, but the *less stable* product of 1,2-addition. But if the 1,2-product is formed *first*, the rate of its formation is by definition faster than that of the 1,4-product. Therefore, the transition state for formation of the less stable 1,2-product must be lower in energy than the transition state for formation of the more stable 1,4-product. The *lower energy* product is formed by passing over the *higher energy* transition state and the *higher energy* product by passing over the *lower energy* transition state. In Chapter 8, you were warned that such cases would occur; here is one. An energy diagram for the reaction of chlorine with 1,3-butadiene should make everything clear (Fig. 12.43).

The 1,2-addition product is formed under conditions of **kinetic control** (short time, mild conditions), and the 1,4-product is favored under conditions of **thermodynamic control** (long time, severe conditions). To form the 1,2-product only the lower transition state must be passed, but to produce the more stable 1,4-product it is the higher energy transition state that must be crossed (Fig. 12.43).

Now we have a fourth question, and it's a nice one. We know why thermodynamics favors the 1,4-product: a disubstituted alkene is more stable than a monosubstituted alkene. But why does kinetic control favor the product of 1,2-addition?

CAUTION! This picture is schematic; it does not show the intermediate carbocation–chloride ion pair

Transition state for 1,4-addition

Transition state for 1,2-addition

Activation energy for formation of the more stable product (formed slower)

Activation energy for formation of the less stable product (formed faster)

Energy

+

Cl₂

1,2-Product (less stable)

1,4-Product (more stable)

The energy difference between the products

Reaction progress

FIGURE 12.43 An energy diagram that summarizes 1,2- and 1,4-addition.

The simplest explanation focuses on distance. Where is the chloride ion after the initial protonation at C(1)? At the moment of its birth, the chloride ion is much closer to C(2) than to C(4). Perhaps addition of chloride is faster at the 2-position than at the 4-position by virtue of simple proximity (Fig. 12.44).

Chloride is closer to C(2) than to C(4) and perhaps this is the reason that 1,2-addition is favored

Addition to C(2)   Addition to C(4)

C(2) addition   C(4) addition

1,2-Product   1,4-Product

FIGURE 12.44 After protonation, the chloride ion is closer to C(2) than to C(4). Perhaps this proximity accounts for the preference for 1,2-addition.

A second possibility looks at the resonance forms for the allyl cation. The two resonance forms are not equivalent; one is a secondary carbocation, the other a primary carbocation. The first will be weighted more heavily than the second, and most of the positive charge in the ion will be at the secondary position. Perhaps it is this more positive position that is attacked by the anion under kinetic control to give the 1,2-addition product. It is this factor that is almost always cited as the reason 1,2-addition is preferred kinetically (Fig. 12.45).

Secondary carbocation (most of the positive charge is at this position); addition of Cl⁻ here gives the 1,2-product

+ HCl ⟶

Primary carbocation (relatively little positive charge here); addition of Cl⁻ gives the 1,4-product

Less stable, but formed first by addition of Cl⁻ to the position of more positive charge

More stable, but must be formed through addition to the position bearing less positive charge

FIGURE 12.45 The resonance forms for the allyl cation formed from the protonation of 1,3-butadiene. The secondary carbocation will contribute more to the hybrid than will the primary carbocation. Addition of chloride at the two positions bearing positive charge gives the two products. The site of more positive charge gives the major product.

There is a brilliant experiment that settles the issue, done by Erić Nordlander (1934–1984) at Case Western Reserve University. He studied the addition of D–Cl, not to 1,3-butadiene, but to 1,3-pentadiene. In this molecule, addition of D^+ yields a resonance-stabilized diene in which both resonance forms are secondary cations. Indeed these resonance forms are identical except for the remote isotopic label (Fig. 12.46).

FIGURE 12.46 Protonation (D^+ in this case) of 1,3-pentadiene gives a resonance-stabilized allylic carbocation in which both contributing resonance forms are secondary. Addition of chloride at the two positions must take place to give equal amounts of 1,2- and 1,4-adduct.

If the source of preference for 1,2-addition depends on the difference between a secondary and primary carbocation, that preference should disappear in the reaction with 1,3-pentadiene. If proximity is at the root of the effect, 1,2-addition should still be favored. For once simplicity wins out, and 1,2-addition is still preferred (Fig. 12.47).

FIGURE 12.47 1,2-Addition is still preferred in 1,3-pentadiene. The kinetic preference for 1,2-addition is a simple proximity effect.

If the proximity of chloride to C(2) is the reason for the kinetic preference for 1,2-addition, it should still be preferred in this electronically symmetrical cation, and it is!

1,2-Addition (shorter path for chloride) (75.5%)

1,4-Addition (longer path for chloride) (24%)

PROBLEM 12.19 In Figure 12.46, why is protonation preferred at the end carbon [C(1)] rather than at C(4)?

SUMMARY

Under thermodynamic (relatively high energy) conditions, the greater stability of the product of 1,4-addition to dienes suffices to make that molecule the major compound formed by addition to 1,3-dienes. Under kinetic conditions, the major product comes from 1,2-addition—it is the less stable product that dominates. Why? Proximity! In the formation of the allyl cation, the nucleophile is "born" closer to the 2-position than to the 4-position.

We will now continue with an examination of a few more reactions in which allylic stabilization is important.

12.11 THE ALLYL SYSTEM: THREE OVERLAPPING 2p ORBITALS

We have met the structure of the allyl system several times already (Problems 1.51, 9.3), but Figure 12.48 again summarizes both the molecular orbital and resonance treatment of this system of three contiguous 2p orbitals.

FIGURE 12.48 The resonance and molecular orbital treatment of the allyl system.

PROBLEM 12.20 Add electrons to both the resonance and molecular orbital descriptions in Figure 12.48 to form the allyl anion, radical, and cation.

12.11a The Allyl Cation: S$_N$1 Solvolysis of Allylic Halides. The resonance stabilization of the allyl cation makes allyl halides relatively reactive under S$_N$1 conditions (Fig. 12.49).

FIGURE 12.49 The S$_N$1 solvolysis of allyl iodide in water. This reaction proceeds much faster than the solvolysis of propyl iodide.

As mentioned earlier, the allyl cation can be captured at either carbon sharing the positive charge. Table 12.3 gives some average rates of S$_N$1 solvolysis for a few common structural types. Primary allyl halides react much faster than simple primary halides. Secondary and tertiary allyl halides also ionize faster than their non-allylic counterparts. Resonance stabilization does have an accelerating effect on the ionizations (Table 12.3).

PROBLEM 12.21 Explain why the last two entries in Table 12.3 on p. 585 react at nearly the same rate and are so much faster than the others.

Remember: We are doing a dangerous thing here. We are again equating thermodynamic stability (a secondary allyl cation is more stable than a simple secondary cation) with kinetics (therefore secondary allylic halides form cations faster than do simple secondary halides). Although justified here, we must be clear on the reason why thermodynamics and kinetics are closely related in this example. Work Problem 12.22.

PROBLEM 12.22 Explain carefully, with the use of an energy diagram, why the thermodynamic stability of the product cation is related to the ease of ionization of the corresponding halide. Remember the Hammond postulate (Chapter 8, p. 369), and consider the transition state for ionization.

TABLE 12.3   Relative Rates of S_N1 Solvolysis of RCl in 50% Ethyl Alcohol at 45 °C[a]

Molecule	Ion	Relative Rate	Molecule	Ion	Relative Rate
	Primary[a]	1.0[a]		Methallyl	1157
	Secondary	1.7		Tertiary	3×10^4
	Allyl	14.3		Dimethylallyl	1.9×10^6
	Allyl	21.4		Dimethylallyl	7.9×10^6
	Methallyl	1300			

[a]The value for the primary substrate is suspect, and may include some contribution from S_N2 reactions.

12.11b S_N2 Reactions of Allylic Halides.

Not only do allylic halides react faster than simple halides in the S_N1 reaction, but they are accelerated in S_N2 reactions as well. Table 12.4 gives some average rates of S_N2 reactions for a variety of halides.

TABLE 12.4   Relative Rates of S_N2 Reaction with Ethoxide in Ethyl Alcohol at 45 °C

Molecule	Substitution	Relative Rate	Molecule	Substitution	Relative Rate
	Primary	1.0		Methallyl	97
	Allyl	37.0		Methallyl	1.85
	Allyl	33.0		Dimethylallyl	556

When we are considering a question involving rates, the answer can always be found through an analysis of the transition states for the reactions. Rates are dependent on the activation energy for the reaction, the height of the transition state relative to the starting halide. Let's compare the transition state for the reaction of a 1-propyl halide with that for the reaction of an allyl halide (Fig. 12.50).

Transition state for S$_N$2 displacement
of a primary halide (higher energy)

Transition state for S$_N$2 displacement
of an allyl halide (lower energy)

FIGURE 12.50 A comparison of the transition states for S$_N$2 displacement of a primary and an allylic iodide.

In the transition state for the reaction of the allyl halide, there is overlap between the 2p orbital developed on the carbon at which displacement occurs and the 2p orbitals on the attached double bond. This transition state contains an allyl system. It will benefit energetically from the delocalization of electrons, and will lie at lower energy than the transition state for S$_N$2 displacement in the 1-propyl system, in which there is no delocalization.

12.11c The Allyl Radical. It is much easier to form a resonance-stabilized allylic radical than an undelocalized one. We have already seen the effects of delocalization in reactions involving allylic radicals in Chapter 11 (p. 533). For example, the regiospecific bromination of the allylic position of alkenes with NBS is possible because of the relative ease of forming the delocalized allyl radical (Fig. 12.51).

A resonance-stabilized radical

Recycles to continue
the chain reaction

FIGURE 12.51 Allylic bromination of cyclohexene using NBS. Only the allylic position is substituted in this reaction.

12.11d The Allyl Anion. Now let's look at the acidity of allylic hydrogens to see if the formation of a resonance-stabilized allyl anion has any effect. We would surely expect it to. If delocalization of electrons is important, then removal of a proton at the allylic position should be easier than at a position not leading to a delocalized anion. A comparison of the acidities of propane and propene should test this idea (Fig. 12.52).

In fact, the pK_a of propene is 43, some 15–20 pK_a units lower than that for propane ($pK_a \sim 50$–60). The uncertainty arises from the difficulty of measuring the pK_a of an acid as weak as a saturated hydrocarbon. In any event, propene is a much stronger acid than propane, and it is reasonable to suppose that the formation of the delocalized allyl anion is responsible for the increased acidity of the alkene.

FIGURE 12.52 Propene is far more acidic than propane.

12.12   THE DIELS–ALDER REACTION OF DIENES

In 1950, Otto Diels (1876–1954) and his student Kurt Alder (1902–1958) received the Nobel prize for their work on what has appropriately come to be known as the **Diels–Alder reaction**, which they discovered in 1928. (So much for timeliness; at least they lived long enough to outlast the Nobel committee's renowned conservatism.) We will have more to say about its mechanism in Chapter 20, but the Diels–Alder reaction is characteristic of 1,3-dienes and is of enormous synthetic utility. Therefore we must outline it here.

Diels–Alder reaction

When energy is supplied to a mixture of a 1,3-diene and a simple alkene (a **dienophile**), a ring-forming reaction takes place to produce a cyclohexene. Our discussion of mechanism begins with the arrow formalism, which points out which bonds are broken and shows where the new bonds are made in the forward reaction (Fig. 12.53).

FIGURE 12.53 The arrow formalism for the Diels–Alder reaction.

What orbitals are involved? The Diels–Alder reaction begins with the overlap of the π systems of the alkene (the dienophile) and the diene. The two reaction partners approach in parallel planes and as the bonds form, the end carbons of both the diene and alkene rehybridize from sp^2 to sp^3 (Fig. 12.54).

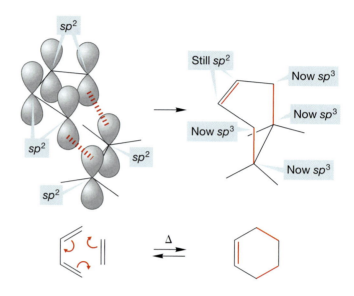

FIGURE 12.54 In the Diels–Alder reaction, the two participants approach each other *in parallel planes*. As the reaction occurs, the end atoms of the diene and the dienophile rehybridize from sp^2 to sp^3.

PROBLEM 12.23 Curiously, it matters just how energy is supplied to these molecules. If we heat the reactants up, the reaction occurs, but if we attempt to supply energy photochemically, it generally fails. We will examine such matters in detail in Chapter 20, but we might start to think about it now. When we think about reactions of atoms we do not consider every electron in the atom, just those most loosely held, the valence electrons. Similarly, in molecular reactions it is the most weakly held electrons, those in the HOMO, that control reactivity. The electrons in the HOMO might well be thought of as the valence electrons of molecules. Remember what we said about the interaction of light and molecules in Section 12.7 on UV/vis spectroscopy (Fig. 12.25, p. 567). First, consider the phase relationships in the possible HOMO–LUMO interactions in the transition state for a simple Diels–Alder reaction. Then, think about how the absorption of a *single* light photon might change those interactions.

PROBLEM 12.24 Write an arrow formalism for the reverse Diels–Alder reaction.

It is possible to estimate the thermochemistry in the Diels–Alder reaction by comparing the bonds made and broken in the reaction. Three π bonds in the starting material are converted into one π bond and two σ bonds in the product. Accordingly, the reaction is calculated to be exothermic by approximately 38 kcal/mol (Fig. 12.55).

3 π bonds =
3 × 66 = 198 kcal/mol

1 π bond =   66 kcal/mol
2 σ bonds = 170 kcal/mol
—————————————
                236 kcal/mol

The reaction is 236 −198 = 38 kcal/mol exothermic;
$\Delta H = -38$ kcal/mol

FIGURE 12.55 The simplest Diels–Alder reaction is exothermic by about 38 kcal/mol.

PROBLEM 12.25 Why is the bond strength of the new σ bonds in cyclohexene (85 kcal/mol) used in Figure 12.55 lower than that in ethane (90 kcal/mol)?

As shown in the previous figures, the Diels–Alder reaction requires the s-cis form of the diene component. However, it is the s-trans arrangement that is the more stable one, by approximately 3 kcal/mol (p. 563). If equilibrium favors the s-trans form strongly, why is the Diels–Alder reaction observed at all? Even though there is little s-cis form at equilibrium, reaction with the dienophile disturbs the equilibrium between the s-trans and s-cis form by depleting the s-cis partner. Reestablishment of equilibrium generates more s-cis molecules, which can react in turn. Eventually, all the diene can be converted into Diels–Alder product, even though at any moment there is very little of the active s-cis form present (Fig. 12.56).

FIGURE 12.56 Even though the unreactive s-trans form of 1,3-butadiene is favored at equilibrium, the small amount of the s-cis form present can lead to product. The vertical axis of the figure is not to scale. The squiggly lines indicate gaps.

no reaction
with dienophile

Transition state for the
Diels–Alder reaction

Energy

s-cis
Conformation

s-trans
Conformation

Reaction progress

When there is no possibility of any s-cis form there can be no Diels–Alder reaction. For example, although 1,2-dimethylenecyclohexane reacts normally with dienophiles, 3-methylenecyclohexene does not. The product of this reaction would contain a double bond far too strained to be formed (Fig. 12.57).

WORKED PROBLEM 12.26 Identify the source of strain in the putative product of Diels–Alder addition to 3-methylenecyclohexene in Figure 12.57. Use your models.

ANSWER Although the paper puts up with the double bond drawn in Figure 12.57, Nature won't. The orbitals making up this bridgehead double bond do not, in reality, overlap very well.

Another serious mechanistic question involves the timing of the formation of the two new σ bonds in the product cyclohexene. Are they made simultaneously in a one-step or "concerted" reaction, or are they formed in two separate steps? In principle the mechanism of the Diels–Alder reaction could either be concerted or involve two steps (Fig. 12.58).

A concerted mechanism; both new σ bonds are formed simultaneously

step one

step two

In this nonconcerted, two-step mechanism, the two σ bonds are formed in two separate steps

This question asks if there is an intermediate in the reaction or not. Figure 12.58 shows the two arrow formalisms and Figure 12.59 gives Energy versus Reaction progress diagrams for both the one- and two-step processes.

FIGURE 12.59 Energy versus Reaction progress diagrams for the concerted and two-step mechanisms.

We might immediately be suspicious of the two-step mechanism. It involves an intermediate allylic diradical (Figs. 12.58 and 12.59). Why don't we observe closure of the diradical at both possible positions? Should there not be a vinylcyclobutane produced as well as the cyclohexene (Fig. 12.60)?

FIGURE 12.60 The two-step reaction could lead to vinylcyclobutane. This product is not observed.

In fact, simple Diels–Alder reactions do not lead to vinylcyclobutanes, but give cyclohexenes exclusively. Although this would seem at first a powerful argument favoring the concerted, one-step mechanism, which can only give cyclohexenes, it could be countered that the transition state for closing the four-membered ring will contain much of the strain present in all cyclobutanes, and will therefore be high in energy.

PROBLEM 12.27 Write an Energy versus Reaction progress diagram for the two-step Diels–Alder reaction that includes formation of the vinylcyclobutane.

As is so often the case, the mechanistic question of the timing of new bond formation is finally answered with a stereochemical experiment. A one-step reaction must preserve stereochemical relationships present in the original alkene. For example, in a concerted reaction, a cis alkene must give a cis disubstituted cyclohexene. If the reaction is concerted, with the two new σ bonds formed simultaneously, there can be no change in the stereochemical relationship of groups on the alkene (Fig. 12.61).

FIGURE 12.61 A one-step, concerted mechanism must preserve the stereochemical relationships of the groups on the alkene.

If the bonds are formed in two steps, there should be time for rotation about the carbon–carbon σ bond in the original alkene. If the mechanism is stepwise, it should lead to stereochemical scrambling of the groups originally attached to the alkene (Fig. 12.62).

FIGURE 12.62 A two-step mechanism predicts a mixture of stereoisomers.

In fact, retention of stereochemistry is observed. The reaction is a concerted, one-step process (Fig. 12.63).

FIGURE 12.63 The preservation of stereochemistry in this reaction favors the concerted mechanism.

PROBLEM 12.28 The experiment described in Figures 12.61–12.63 is usually easy to understand, but there is another stereochemical experiment possible. It tests the same thing, but is harder to see without models. So, if this problem is difficult to follow, by all means work it through with models. Let's follow the stereochemical fate of groups attached

(continued)

PROBLEM 12.28 *(continued)*

to the ends of the diene, not the alkene. Analyze the reaction of a *trans,trans*-1,3-diene and ethylene. There is only a single cyclohexene product with the R groups cis (see below). Explain how this result shows that the mechanism of cycloaddition cannot be a two-step process.

trans,trans Diene   cis Cycloalkene

In the preceding figures, mechanistic points have been illustrated using very simple examples. In fact, the simplest Diels–Alder reaction, ethylene and 1,3-butadiene, has a rather high activation enthalpy of about 27 kcal/mol and requires high pressure and temperature for success. Even at 200 °C, a reaction at 4500 psi yielded only 18% cyclohexene after 17 h. Other Diels–Alder reactions proceed under milder conditions. Reactions are more facile when the dienophile bears one or more electron-withdrawing groups, and if the diene can easily achieve the required s-cis conformation. Two examples of these relatively fast Diels–Alder reactions are shown in Figure 12.64.

FIGURE 12.64 Two efficient Diels–Alder reactions.

Why is this reaction so important and so useful? Six-membered rings are very common, and the Diels–Alder reaction builds six-membered rings from relatively simple building blocks. Moreover, the reaction is very general. Most dienes react with most dienophiles to give Diels–Alder products. From now on, every cyclohexene you see must trigger the thought "can be made by a Diels–Alder reaction" in your mind. That's an important concept for both synthesis and problem solving.

Like most reactions in organic chemistry, the Diels–Alder reaction is not a one-way street. Reverse Diels–Alder reactions are common, especially at high temperature.

Remember: $\Delta G = \Delta H - T\Delta S$ (Eq. 8.6, p. 354), and the temperature-dependent ΔS term can overwhelm the temperature-independent ΔH term at high temperature. The reverse Diels–Alder reaction makes two molecules from one and will be favored by entropy, ΔS. Cyclohexenes are not only formed in a reaction of a diene and a dienophile, but can serve as sources of dienes through the reverse Diels–Alder reaction (Fig. 12.65).

THE GENERAL CASE   **A SPECIFIC EXAMPLE**

FIGURE 12.65 The reverse Diels–Alder reaction.

Now let's press on to look at other uses of this reaction, surely one of the most widely used in synthesis of all the reactions of organic chemistry. Chemists often use bicyclic molecules to investigate effects requiring a fixed geometrical arrangement of orbitals or groups. Consider the simple Diels–Alder reaction of ethylene or acetylene with a common cyclic diene, cyclopentadiene (Fig. 12.66).

FIGURE 12.66 The Diels–Alder reaction between cyclic dienes and alkenes (or alkynes) gives bicyclic molecules.

Presto! Bicyclic compounds! But this instant synthetic expertise is not without its price. There is a new stereochemical effect to be considered. When a substituted alkene reacts with a cyclic diene, there are two possible stereoisomeric products even though both new σ bonds are formed simultaneously. The substituents on the alkene must retain their stereochemical relationship during the reaction, but there are still two possible products. The substituents on the alkene can point toward the newly formed double bond or away from it (Fig. 12.67).

X pointing toward the double bond (called endo)

X pointing away from the double bond (called exo)

FIGURE 12.67 In principle, this kind of Diels–Alder reaction can give either endo or exo products.

The compound with the substituents aimed toward the double bond is called **endo** and that with the groups aimed away from the double bond, **exo**. The exo isomers are usually more stable, but the endo compounds are kinetically favored in the Diels–Alder reaction (Fig. 12.68).

FIGURE 12.68 It is the endo product that is favored in most Diels–Alder reactions. In each case, the percentage of endo product is greater than the percentage of exo product.

The exo compounds are more stable for steric reasons: the substituent points away from the more congested part of the molecule. Indeed, the transition state for exo product formation must also be less crowded than the transition state for endo product formation. But the transition state for endo product benefits from another factor and this overwhelms steric considerations. In the endo transition state, but not in the exo transition state, there can be "secondary orbital overlap." This fancy-sounding phrase simply means that there can be interaction between the "back" diene orbitals and orbitals on the substituent X only in the endo transition state. Figure 12.69 shows this effect for the first example of Figure 12.68.

FIGURE 12.69 Secondary orbital interactions stabilize the endo transition states. *(continued)*

FIGURE 12.69 *(continued)*

Even more complicated polycyclic structures can be made from the Diels–Alder reaction between two ring compounds. In the simplest case, the reaction of a cyclic 1,3-diene with a cycloalkene, endo addition is again preferred (Fig. 12.70).

endo Product
(98%)

exo Product
(2%)

FIGURE 12.70 The Diels–Alder reaction between a cyclic diene and a cyclic dienophile gives a polycyclic product. Once again, it is the endo form that is favored.

Two cyclic dienes can also react in Diels–Alder fashion. The self-condensation of cyclopentadiene is the classic example. We'll illustrate this reaction with a pair of problems.

WORKED PROBLEM 12.29 You attempt to prepare 1,3-cyclopentadiene from the base-catalyzed loss of hydrogen bromide (E2 reaction) from 3-bromocyclopentene (see below). Instead of isolating a compound of the formula C_5H_6, you get a high yield of a compound of the formula $C_{10}H_{12}$. Fame, fortune, and the Nobel prize await you if you can explain this result (at least if you were explaining it in 1928). Be careful—although this question is obviously set up in the text, it is not so easy to write a structure correct in every detail.

Br

base →

(C_5H_6)
The hoped-for product

base →   ($C_{10}H_{12}$)
The isolated product

(continued)

WORKED PROBLEM 12.29 *(continued)*

ANSWER The formula of the product C$_{10}$H$_{12}$ is a clue. Clearly, two C$_5$H$_6$ molecules have joined to make the C$_{10}$H$_{12}$ product. The reaction is, in fact, successful. The problem is that cyclopentadiene dimerizes in a Diels–Alder reaction very quickly. The molecule you isolate is the Diels–Alder dimer. Note formation (mainly) of the endo adduct. Anytime you see a 1,3-diene, especially a cyclic one, think Diels–Alder.

WORKED PROBLEM 12.30 In a fit of frustration or inspiration you decide to boil (reflux) your unanticipated C$_{10}$H$_{12}$, and you do so by attaching a long condenser to a receiving vessel kept very cold. Your C$_{10}$H$_{12}$ boils at about 165 °C, but you collect a product of the formula C$_5$H$_6$ boiling at 40 °C. Moreover, although this product is stable at low temperature for a long time, at room temperature it rapidly reverts to your starting C$_{10}$H$_{12}$. Explain.

ANSWER At 165 °C, the Diels–Alder reaction reverses, and the low-boiling cyclopentadiene can be collected and kept at low temperature. If it is allowed to warm up, it will again form the Diels–Alder dimer.

Cold

12.13   SPECIAL TOPIC: BIOSYNTHESIS OF TERPENES

When we discussed cationic addition reactions (Chapters 9 and 10), we included a section on the polymerization reactions of alkenes (Section 9.9, p. 410). Dienes form quite stable intermediates in addition reactions and are especially prone to polymerization.

PROBLEM 12.31 Write a mechanism for the radical-induced polymerization of 1,3-butadiene.

Nature uses related reactions (the cart is way before the horse here; we make approximations of Nature, not the other way round!). Natural rubber is formally polymerized **isoprene** (2-methyl-1,3-butadiene), for example (Fig. 12.71).

Polyisoprene
one form of natural rubber
with all double bonds (Z)

FIGURE 12.71 Natural rubber
is a polyisoprene.

PROBLEM 12.32 There is another, stereoisomeric form of natural rubber, also a polymer
of isoprene, called by the wonderful name gutta percha.* What is it?

Polyisoprene is a limiting case—a molecule composed of a vast number of isoprene units. Many important natural products called **terpenes** are constructed of
smaller numbers of isoprene units put together in various ways. Terpenes are widespread in nature and are still often extracted from plants. The smaller terpenes are
widely used as flavors and fragrances. Indeed, terpene chemistry is the very basis of
the fragrance industry.

Most of the reactions used to combine the diene units can be largely understood in
terms of reaction mechanisms we know well. 3-Methyl-3-buten-1-ol is converted first
into the corresponding 3-methyl-3-buten-1-ol **pyrophosphate** (isopentenyl pyrophosphate) by reaction with pyrophosphoric acid (Fig. 12.72).

3-Methyl-3-buten-1-ol   **Pyrophosphoric acid**   **3-Methyl-3-buten-1-ol pyrophosphate**
(isopentenyl pyrophosphate)

FIGURE 12.72 In the biosynthesis of terpenes, 3-methyl-3-buten-1-ol is first converted into
the corresponding pyrophosphate.

Recall the discussion of various ways to change OH, a very poor leaving group,
into much better leaving groups (Chapter 7, p. 293). Earlier, we saw examples of
protonation to make the leaving group water, or of formation of a sulfonate ester
(Fig. 12.73). Formation of the pyrophosphate ester serves exactly the same purpose
and is widely used in Nature to generate a good leaving group from a poor one.

PROBLEM 12.33 Explain why the pyrophosphate ester in Figure 12.73 is an excellent
leaving group.

*Gutta percha comes from the Malay, *getah percha* (gum of the percha tree).

FIGURE 12.73 Three ways of converting OH, a poor leaving group, into a good leaving group.

There is an enzyme that will equilibrate 3-methyl-3-buten-1-ol pyrophosphate with an isomeric pyrophosphate, 3-methyl-2-buten-1-ol pyrophosphate (Fig. 12.74).

3-Methyl-3-buten-1-ol pyrophosphate   **3-Methyl-2-buten-1-ol pyrophosphate**

FIGURE 12.74 An enzyme converts 3-methyl-3-buten-1-ol pyrophosphate into a new, allylic pyrophosphate.

This new pyrophosphate now has a good leaving group *in an allylic position*. The combination contributes to an easy ionization to produce an allyl cation that can be attacked by 3-methyl-3-buten-1-ol pyrophosphate to produce the C_{10} compound, geranyl pyrophosphate (Fig. 12.75).

FIGURE 12.75 Loss of the pyrophosphate gives an allylic cation that can be attacked by a molecule of 3-methyl-3-buten-1-ol to give a new cation. Deprotonation gives geranyl pyrophosphate.

Geranyl pyrophosphate itself has a good leaving group in an allylic position and this process can be repeated to produce the C_{15} compound, farnesyl pyrophosphate (Fig. 12.76).

FIGURE 12.76 This process can be repeated to give farnesyl pyrophosphate.

Hydrolysis of geranyl pyrophosphate and farnesyl pyrophosphate produces geraniol and farnesol (Fig. 12.77).

FIGURE 12.77 The alcohols geraniol and farnesol are formed by hydrolysis of the related pyrophosphates.

Notice that both geranyl and farnesyl pyrophosphate are formed by head-to-tail combinations of isoprene (Fig. 12.78).

FIGURE 12.78 Geranyl and farnesyl pyrophosphates are formed by head-to-tail attachment of isoprene units.

The limit of head-to-tail combining is polyisoprene, natural rubber. Figure 12.79 shows the schematic formation of the polymer of isoprene called gutta percha (p. 599).

Gutta percha

FIGURE 12.79 Gutta percha, one form of natural rubber, is a head-to-tail polymer of isoprene.

Molecules formed from isoprene units are called **isoprenoids**, or terpenes. Two units of isoprene combine to form terpenes, C_{10} compounds such as geraniol. Three isoprene units form sesquiterpenes, C_{15} compounds such as farnesol. Diterpenes are C_{20} compounds, triterpenes are C_{30} compounds, and so on. Myriad terpenoid compounds are known, and most can be traced to derivatives of isoprene as starting material. Exceptions are known, but most terpenes, including those in Figure 12.80, follow the **isoprene rule**, which dictates the head-to-tail formation described above.

*Limonene   Menthol   *(1R)-(+)-Camphor   (1S)-(−)-β-Pinene

Thujone   3-Carene   (1S)-(−)-Camphor   (1R)-(+)-α-Pinene

FIGURE 12.80 Some terpenes. (See Problem 12.34 for explanation of the use of the asterisk.)

WORKED PROBLEM 12.34 Dissect the molecules marked with an asterisk in Figure 12.80 into isoprene units.

ANSWER The isoprene units can usually be found by searching for the methyl groups first. One of these problems is easy, the other is more complicated.

PROBLEM 12.35 Devise a mechanism for converting geranyl pyrophosphate into racemic limonene (Fig. 12.80). Such processes are usually enzyme mediated in Nature. When you try this problem you should encounter a roadblock. It can't really be done without a cis/trans isomerization, very likely carried out in Nature by the enzyme.

12.14 SPECIAL TOPIC: STEROID BIOSYNTHESIS

There is an extraordinarily important class of molecules called **steroids**, which are formed from isoprene units. Two units of farnesyl pyrophosphate can be reductively coupled to give the symmetrical triterpene called squalene (Fig. 12.81).

FIGURE 12.81 Reductive coupling of two molecules of farnesyl pyrophosphate gives squalene, a triterpene.

Squalene can be epoxidized enzymatically to give squalene oxide, which is written in a suggestive way in Figure 12.82.

FIGURE 12.82 Enzymatic epoxidation of squalene gives squalene oxide.

Protonation of squalene epoxide leads to a relatively stable tertiary carbocation that is attacked by the proximate nucleophilic double bond to give a new cation. This cation is attacked in turn by another nearby alkene. The process continues until the molecule is sewn up in the four-ring steroid skeleton as shown in Figure 12.83.

Tertiary carbocation

The four-ring steroid skeleton is now constructed

FIGURE 12.83 Acid-catalyzed opening of the oxirane gives a tertiary cation that can undergo a series of ring closings to give a new, tertiary carbocation in which the four-ring steroid skeleton has been constructed.

The series of rearrangements summarized in Figure 12.84 ensues and a proton is finally lost to give the compound known as lanosterol.

FIGURE 12.84 A series of skeletal rearrangements, followed by removal of a proton, gives lanosterol.

Lanosterol

PROBLEM 12.36 Starting at the right-hand end of the molecule, write a stepwise mechanism for the conversion of the ion in Figure 12.84 to lanosterol. Take the several rearrangements one at a time and draw out each reaction intermediate.

Steroids are commonly found in Nature and are considered important primarily because of their powerful biological activities. They often function as regulators of human biology. They all share the four-ring skeleton of the molecule shown in Figure 12.84, lanosterol. The rings are designated A, B, C, and D, and numbered as shown in Figure 12.85. As in lanosterol, steroids have axial methyl groups attached to C(10) and C(13), and often carry an oxygen atom at C(3) and a carbon chain at C(17). The rings are usually, but not always, fused in trans fashion.

Steroids were, and still can be, isolated from natural sources. Such important biomolecules quickly became the subject of attempts at chemical modification and total synthesis in the hopes of uncovering routes to new molecules of greater and different bioactivities. Many naturally occurring steroids have been constructed in the laboratory and numerous "unnatural" steroids have been created as well. Figure 12.86 shows a few important steroids.

Whole industries have been derived from this work. Most famous are the steroids used in our attempts to control human fertility. "The pill" contains synthetic steroids that mimic the action of the body's natural steroids, the estrogens and progesterone, which are involved in regulation of egg production and ovulation. The

Lanosterol

FIGURE 12.85 The steroid ring system.

Cholesterol

Progesterone

Testosterone

Estrone

Cortisone

FIGURE 12.86 Some naturally occurring steroids.

body is essentially tricked into behaving as if it were pregnant, so that neither egg production nor development can take place. It should be no surprise that it has been difficult to use such powerfully biologically active molecules in a completely specific way. The goal of this work is to interfere specifically with fertility without doing *anything* else. Considering the activity of the molecules involved and the complexity of the mechanisms we are trying to regulate, it is remarkable that we have been so successful. This control is not totally without cost, however. In fertility, those costs appear as the side effects of the pill, which occur in a few women. Perhaps the most difficult of human problems is to find the balance between benefits and risks in cases such as this. Although such problems have been with us throughout our history, questions of biology are particularly vexing, it seems. It is easier to judge that the benefits of the automobile outweigh the societal costs than to make a decision regarding the steroids used in the pill. This problem is certain to become much worse as our ability to manipulate complicated molecules continues to grow.

12.15 SUMMARY

New Concepts

In this chapter, the chemical and physical consequences of the overlap of 2p orbitals (conjugation) are explored. In structural terms, conjugation appears as a short C(2)—C(3) bond and in a small barrier to rotation around this bond (Fig. 12.18).

Conjugation leads to the formation of allyl systems in addition reactions to dienes, and is a requirement for the Diels–Alder reaction (Fig. 12.87).

An important fundamental concept in this chapter involves the difference between thermodynamic and kinetic con-

trol of a reaction. When the product distribution depends on the relative energies of the products, the reaction is under thermodynamic control. Thermodynamic control occurs when there is sufficient energy available so that the activation barriers between starting materials and products can be crossed in both directions. By contrast, if products are not freely equilibrating with starting material, if there is not enough energy to carry out the reverse reactions, then it will be the energies of the transition states that control the distribution of products. This situation results in kinetic control.

In UV/vis spectroscopy, absorption of light results in the promotion of an electron from the HOMO to the LUMO.

An allyl cation

FIGURE 12.87 Two typical reactions of conjugated dienes: addition to give an allyl cation and the Diels–Alder reaction.

Key Terms

allene (p. 551)
conjugated double bonds (p. 549)
cumulenes (p. 554)
Diels–Alder reaction (p. 587)
dienes (p. 549)
dienophile (p. 587)
electronic spectroscopy (p. 567)
endo (p. 596)

exo (p. 596)
extinction coefficient (ε) (p. 568)
isoprene (p. 598)
isoprene rule (p. 603)
isoprenoids (p. 603)
ketenes (p. 553)
kinetic control (p. 580)
pyrophosphate (p. 599)

s-cis (p. 563)
spectroscopy (p. 566)
steroids (p. 604)
s-trans (p. 563)
terpene (p. 599)
thermodynamic control (p. 580)
ultraviolet (UV) spectroscopy (p. 567)
visible (vis) spectroscopy (p. 567)

Reactions, Mechanisms, and Tools

The base-catalyzed isomerization of disubstituted to terminal acetylenes takes place through allene intermediates (Fig. 12.13). Addition reactions to 1,3-dienes can give products of both 1,2- and 1,4-additions (Fig. 12.38). The Diels–Alder reaction forms cyclohexenes and cyclohexadienes from the ther-

mal reaction of 1,3-dienes and alkenes or alkynes. Stereo-chemical labeling experiments show that both new σ bonds are formed simultaneously (Figs. 12.63 and the figure in Problem 12.28). Many natural products (terpenes) are produced through the combination of various numbers of isoprene units.

Synthesis

The most important synthetic methods described here are a route to allylic halides through additions to conjugated dienes and the Diels–Alder synthesis of cyclohexenes and cyclohexadienes. Be especially careful about the Diels–Alder reaction: There are many structural possibilities for the products depending on the complexity of the diene and dienophile used as starting materials.

1. ALLYLIC HALIDES

The 1,2- and 1,4-addition; watch out, these allylic halides can be transformed further though S_N2 reactions

Both 1,2- and 1,4-addition takes place

2. CYCLOHEXADIENES

The dienophile is an alkyne; watch out for structural variation and the reverse reaction

3. CYCLOHEXENES

Diels–Alder reaction; watch out for structural variation; the reverse reaction is a synthesis of 1,3-dienes and alkenes from cyclohexenes

4. DIENES

The reverse Diels–Alder reaction; watch out for structural variation; this picture gives only the simplest version

Common Errors

Be sure that it is clear how the structures of allenes and cumulenes can be derived from an analysis of the hybridizations of the atoms involved. Although the structures all *look* flat on a two-dimensional page, they are not all planar by any means.

Although the Diels–Alder reaction can be quickly drawn using only three little arrows, and although it is not difficult to see that all cyclohexenes can be constructed on paper from the reaction of a diene and a dienophile, very complex structures can be built very quickly using Diels–Alder reactions. The simplicity can quickly disappear in the wealth of detail! Try to see to the essence of the problem; locate the cyclohexene, and then take it apart to find the building block diene and dienophile.

The reverse Diels–Alder reaction is even harder to find. Not only can cyclohexenes be constructed through Diels–Alder reac-

tions, but they can serve as sources of dienes and dienophiles as well. It is not simple to find reverse Diels–Alder reactions!

It is time to mention again another very common conceptual error that often traps students. Look, for example, at Figure 12.44. Protonation of 1,3-butadiene leads to an allyl cation in which two different carbons share the positive charge. Addition of chloride at these two positions leads to the two products. It is very easy to think of the resonance forms of the allyl cation as each having a separate existence; of chloride adding to one resonance form to give one product and to the other resonance form to give the other product. This notion is absolutely wrong! The allyl cation is a single species with two partially positive carbon atoms. Chloride adds to the allyl cation, not to one resonance form or the other. The figure tries to help out by showing a summary structure, but this is not always done, and you must be careful (Fig. 12.44).

12.16 ADDITIONAL PROBLEMS

PROBLEM 12.37 Use the rules given in Table 12.2 to calculate λ_{max} for the following steroids:

(a)

(b)

Ac = CH₃CO

PROBLEM 12.38 Write resonance forms for the following ions:

(a)　　　　　　　(b)

(c)　　　　　　　(d)

PROBLEM 12.39 Are any of the following compounds chiral? Explain with good drawings.

PROBLEM 12.40 Show the main products of the reactions of *trans,trans*-2,4-hexadiene under the following conditions. There may be more than one product in some cases.

(a) HCl　　(b) H₂/Pd　　(c) Cl₂/CCl₄

(d) Δ, H₃COOC—C≡C—COOCH₃

PROBLEM 12.41 Suggest a plausible arrow formalism mechanism to account for the products in the following reaction:

Br₂/CH₃OH −15 °C

1 (63%) 2 (22%) 3 (10%) 4 (4%)

PROBLEM 12.42 Predict the products of the reaction of hydrogen chloride with isoprene under kinetic and thermodynamic control.

Isoprene

HCl

PROBLEM 12.43 Devise syntheses for the following molecules. You may use 1-butyne and ethyl iodide as your only sources of carbon, as well as any inorganic reagents you need.

PROBLEM 12.44 Conversion of 2-butyne into 1-butyne is easy. The reverse reaction is more difficult. Suggest routes for doing both these conversions. Mechanisms are not required.

$$H_3C-C{\equiv}C-CH_3 \xrightarrow{?} CH_3CH_2-C{\equiv}C-H$$

$$CH_3CH_2-C{\equiv}C-H \xrightarrow{?} H_3C-C{\equiv}C-CH_3$$

PROBLEM 12.45 Predict the products of the following Diels–Alder reactions:

(a)

ether
0–25 °C

COOCH₃

(b)

toluene
150 °C

COOCH₃

COOCH₃

(c)

alcohol
0 °C

O

O

(d)

Ph
Ph
Ph
Ph

O

COOCH₃

COOCH₃

Ph =

dichlorobenzene Δ

$C_{10}H_6O_4Ph_4$

(e)

CH₃ C₈H₁₇

H₃C

toluene
Δ,10 h

O

O

O

(continued)

PROBLEM 12.46 *(continued)*

(f)

PROBLEM 12.47   In Problem 12.45(c), the initially formed adduct can be closed in a photochemical reaction to give a complex cage structure (shown below). Write an arrow formalism for this process and explain how the observation of this reaction was helpful in determining the stereochemistry (exo or endo) of the original Diels–Alder adduct.

(g)

PROBLEM 12.48   Use a carefully constructed Energy versus Reaction progress diagram to rationalize the following data. Be sure to explain why the product at low temperature is mainly endo and why the product at high temperature is mainly exo.

PROBLEM 12.46   Although simple alkyl and aryl nitriles are poor dienophiles, nitriles substituted with electron-withdrawing groups are more reactive. Provide mechanisms for the formations of the two products in the following reaction:

PROBLEM 12.49   In each of the following three reactions, two products are possible. Could ^{13}C NMR spectroscopy distinguish between the isomeric products formed? Explain. Can you remember or predict which products are actually favored in these reactions?

(a)

(continued)

PROBLEM 12.49 (continued)

(b)

(c)

endo or exo Product

PROBLEM 12.50
Here are two reactions of allyl systems related to the Diels–Alder reaction. Use a HOMO–LUMO molecular orbital analysis to show why reaction (a) succeeds whereas reaction (b) fails.

(a)

(b)

PROBLEM 12.51
In principle, one of the following reaction sequences could give two products, while the other must produce only one. Explain.

(a)

(b)

PROBLEM 12.52
In 1951, the great Kurt Alder himself examined the reactions of a mixture of dienes **1** and **2** with maleic anhydride. He discovered that diene **1** reacted at 35 °C to give a single adduct. Diene **2** reacted only at 150 °C to give a single, different, adduct. Write structures for the two products and explain the difference in the ease of reaction. Why is reaction of **1** so much easier than reaction with **2**? Alder never actually looked at the reaction with diene **3**, but if he had, I predict that no reaction would have occurred at 150 °C. Explain.

PROBLEM 12.53
Propose a synthesis of the powerful vesicant (blister inducer) cantharidin, the active ingredient in the putative aphrodisiac "Spanish fly." You may assume that organic starting materials containing no more than six carbon atoms are available, along with any inorganic materials you may need. *Hint*: The thermodynamically more stable exo Diels–Alder adducts of furan are generally isolable.

Cantharidin

PROBLEM 12.54 Write arrow formalism mechanisms for the following transformations:

(a)

(b)

(c)

(d)

PROBLEM 12.55 Allylic halides undergo the S_N2 reaction much faster than saturated halides (Section 12.11b). There is another substitution reaction pathway open to allylic halides called the S_N2' reaction.

As the figure shows, in this simple example it is not possible to distinguish the two reactions. Devise an experiment that would allow you to tell if the product were formed through the S_N2 or S_N2' reaction.

PROBLEM 12.56 Devise an experiment using isotopic labeling that would determine if the S_N2 or S_N2' reaction were occurring in the reaction of Problem 12.55.

PROBLEM 12.57 Now here's a much harder one. The stereochemistry of the S_N2 reaction has been well worked out and has been described in detail in Section 7.4b. The S_N2' reaction has also been investigated, and this problem asks you to put yourself in the place of R. M. Magid (b. 1938) and his co-workers. They used the following experiment to determine the stereochemistry of the S_N2' reaction. Use their data to determine whether the nucleophile approaches the

(*continued*)

double bond from the same side as the leaving group or from the opposite side.

(after deprotonation of the initially formed ammonium ions)

PROBLEM 12.58 Write an arrow formalism mechanism for the following bimolecular process. Be sure to account for the stereochemistry of the product.

Conjugation and Aromaticity

13

Curious, though, isn't it, um, Patwardhan, that the number, er, six should be, um, embodied in one of the most, er, er, beautiful, er, shapes in all nature: I refer, um, needless to say, to the, er, benzene ring with its single, and, er, double carbon bonds. But is it, er, truly symmetrical, Patwardhan, or, um, asymmetrical? Or asymmetrically symmetrical, perhaps. . . .

—Vikram Seth[*]
A Suitable Boy

13.1   PREVIEW

Chapter 12 was devoted to the consequences of conjugation—the overlap of $2p$ orbitals in acyclic systems. Now we will see what happens when conjugated systems are cyclized, turned into rings. Nothing much changes if the atoms of the rings do not maintain orbital overlap throughout the ring—if they are not fully conjugated. 1,3-Cyclohexadiene is not especially different from 1,3-pentadiene, for example. There are small differences in chemical and physical properties, but the two molecules clearly belong to the same chemical family. Hydrogenation is easy, addition reactions abound, the Diels–Alder reaction is common, and so on (Fig. 13.1).

FIGURE 13.1 Cyclic and acyclic dienes have similar chemical properties.

[*]Vikram Seth was born in Calcutta in 1952 and educated in India, Britain, and the United States. This brilliant novel gets the MJ 10-star recommendation.

When the conjugation is continued around the ring to give what we might call "1,3,5-cyclohexatriene," matters change. None of the reactions shown in Figure 13.1 is easy, and this molecule appears to be remarkably unreactive. It is clearly in a different class from 1,3-cyclohexadiene and the acyclic dienes (Fig. 13.2).

FIGURE 13.2 Benzene (1,3-cyclohexatriene) does not undergo most of the reactions of normal alkenes.

But not all ring compounds containing double bonds (cyclic polyenes) behave in this way. Some are normal cycloalkenes; cycloheptatriene and cyclooctatetraene are examples of this kind of molecule. Others are exceptionally unstable; cyclobutadiene is the archetype of this kind of cyclic polyene (Fig. 13.3).

1,3,5-Cycloheptatriene   **1,3,5,7-Cyclooctatetraene**   **Cyclobutadiene**

These are normal polyenes (for example, they undergo addition reactions and hydrogenate easily)

This molecule is extraordinarily unstable—it can only be isolated at very low temperature

FIGURE 13.3 Other polyenes are either normal in their chemical properties or exceptionally unstable.

In this chapter, we will uncover the sources of these differences in stability between cyclic polyenes and begin to see the chemical and physical consequences of this special kind of conjugative interaction of $2p$ orbitals. We will encounter the molecule **benzene** in which the overlap of six carbon $2p$ orbitals in a ring has great consequences for both structure and reactivity. We will also see a generalization of the properties that make benzene so stable, and will learn how to predict which cyclic polyenes should share benzene's stability and which should not.

ESSENTIAL SKILLS

1. The properties that determine whether or not a molecule is properly termed "aromatic" are simple to write down but are a bit more difficult to apply. It is very important that you be able to think about these qualities, to see why they apply to certain molecules and why they do not apply to others. Understanding the hard-to-define quality called aromaticity is critical.

Chemists have argued for years about the special stability called aromaticity that attends cyclic, planar polyenes with the proper numbers of π electrons. The wonderful quote that begins this chapter is not so far from an accurate representation of remarks heard at many a chemical conference! However, even though it is not easy to define this slippery concept precisely, especially in marginal cases, it is not difficult to understand the simple examples, to see why the number of π electrons is important, and to learn which molecules are likely to exhibit this property.

2. The generic aromatic substitution reaction is introduced. We will see this again many, many times in Chapter 14, but it would be a very good idea if you arrived at that chapter knowing the basics.

IMPORTANT DETAILS

1. The Frost circle allows you to determine quickly the relative energies of the π molecular orbitals for any planar, cyclic, fully conjugated polyene. *Remember*: The polygon must be inscribed vertex down.

2. Aromaticity is not magic! It is just another stabilizing effect that can be augmented or opposed by other effects.

13.2   THE STRUCTURE OF BENZENE

The molecule in Figure 13.2 that we called 1,3,5-cyclohexatriene was isolated from the thermal rendering of whale blubber in the nineteenth century. Although the formula was established as C_6H_6 [or better, $(CH)_6$], there remained a serious problem: What was its structure? There are many possibilities, among them the four shown in Figure 13.4. Two of these, Dewar benzene and Ladenburg benzene, are named for the people who suggested them, Sir James Dewar (1842–1923) and Albert Ladenburg (1842–1911). The third, benzvalene, is more cryptically named. The last compound, 3,3'-bicyclopropenyl, was first made only in 1989 by a group headed by Professor W. E. Billups (b. 1939) at Rice University.

The German chemist Friedrich August Kekulé von Stradonitz (1829–1896) suggested the ring structure. Legend has it that Kekulé was provoked by a dream of a snake biting its tail. Kekulé describes dozing and seeing long chains of carbon atoms twisting and turning until one "gripped its own tail and the picture whirled scornfully before my eyes." Kekulé's suggestion for the structure of benzene is shown in cyclohexatriene form in Figure 13.5.

"Dewar" benzene
(bicyclo[2.2.0]hexa-2,5-diene)

Benzvalene
(tricyclo[3.1.0.0²,⁶]hex-3-ene)

"Ladenburg" benzene
or prismane
(tetracyclo[2.2.0.0²,⁶.0³,⁵]hexane)

3,3'-Bicyclopropenyl

FIGURE 13.4 Some possible structures for benzene.

C_6H_6 $(CH)_6$

FIGURE 13.5 The structure of benzene, drawn as 1,3,5-cyclohexatriene.

PROBLEM 13.1 Suppose Dewar, Ladenburg, and Kekulé had had a ^{13}C NMR spectrometer. How many signals would the various candidates shown in Figure 13.4 show in their spectra? How many would cyclohexatriene show?

But there were problems with the Kekulé formulation[*] as well as with the others. Benzene didn't react with halogens or hydrogen halides like any self-respecting polyene did, for example (Fig. 13.2). Hydrogenation was much slower than for other alkenes and required severe conditions. Benzene was clearly a remarkably stable compound, and the cyclohexatriene structure couldn't account for this.

FIGURE 13.6 Josef Loschmidt's formulations of benzene structures (see footnote).

Traditional, or Kekulé, structures

The same structures written by J. J. Loschmidt in 1861

Moreover, it eventually became known that the structure of benzene was that of a *regular* hexagon. Cyclohexatriene is immediately ruled out as a structural possibility, because it must contain alternating long single bonds and short double bonds. We might expect normal double bonds of about 1.33 Å and single bonds of about the length of the C(2)—C(3) bond in 1,3-butadiene: 1.47 Å. Instead, benzene has a uniform carbon–carbon bond distance of 1.39 Å. Note that the real bond distance, 1.39 Å, is just about the average of the double-bond distance, 1.33 Å, and the single-bond distance, 1.47 Å (Fig. 13.7).

[*]Kekulé gets the lion's share of the credit, but in the mid-1800s there were others hot on the trail of structural theory in general and a decent representation of benzene in particular. One such chemist was Alexandr M. Butlerov (1828–1886) from Kazan, who may well have independently conceived of the ring structure for benzene, and who surely contributed greatly, if quite invisibly in the West, to the development of much of modern chemistry. Another, even more obscure contributor, was the Austrian chemist–physicist Johann Josef Loschmidt (1821–1895) who, four years before Kekulé's proposal in 1865, clearly wrote a ring structure for benzene. Figure 13.6 shows how forward-looking were the structures proposed by this too-little-known Austrian schoolteacher.

1.47 Å

1.33 Å

The structure expected for
1,3,5-cyclohexatriene

Real structure

www

Identical bond lengths (1.39 Å);
How can this be?

FIGURE 13.7
Any 1,3,5-cyclohexatriene
structure must contain alternating
long single bonds and short double
bonds. Such is not the case for
benzene, which is a regular
hexagon with a carbon–carbon
bond distance of 1.39 Å.

At this point, we can probably see the beginnings of the answer to these structural
problems. Kekulé's static cyclohexatriene formulation doesn't take account of resonance stabilization.

PROBLEM 13.2 One early problem with the Kekulé structure was that only one 1,2-disub-
stituted isomer of benzene could be found for any given substituent. That is, there is only
one 1,2-dimethylbenzene (*o*-xylene). Explain why this observation would have been hard
for Kekulé to explain.

13.3   A RESONANCE PICTURE OF BENZENE

Perhaps the remarkable stability of benzene can be explained simply by realizing
that delocalization of electrons is strongly stabilizing. There are really *two* **Kekulé
forms** for benzene. They are resonance forms, differing only in the distribution of
electrons but not in the positions of atoms. An orbital picture for "cyclohexatriene" should have influenced us in this direction. Each carbon is hybridized sp^2
and there is, therefore, a $2p$ orbital on every carbon. The structure in Figure
13.8(a) stresses overlap between the $2p$ orbitals on C(1) and C(2), C(3) and C(4),
and C(5) and C(6), but not between C(2) and C(3), C(4) and C(5), or C(6) and C(1).

FIGURE 13.8 These two
Kekulé forms together provide a
real structure for benzene.

(a)   (b)

This specificity is suspicious, as there is a $2p$ orbital on every carbon atom and the drawing of Figure 13.8(a) takes no account of the symmetry of the situation. If there is overlap between the orbitals on C(1) and C(2), then there must be equivalent overlap between the $2p$ orbitals on C(2) and C(3) as well, as in the structure in Figure 13.8(b). Drawing both Kekulé forms emphasizes this, and produces a regular hexagon as the combination of the two Kekulé resonance forms.

So, our picture of benzene has now been elaborated somewhat to show the orbitals. The $2p$ orbitals on each carbon, extending above and below the plane containing the six carbons and their attached hydrogens, overlap to form a circular cloud of electron density above and below the plane of the ring.

CONVENTION ALERT

A convention has been adopted that attempts to show the cyclic overlap of the six $2p$ orbitals. Instead of drawing out formal double bonds as in the Kekulé forms, a circle is inscribed inside the ring (Fig. 13.9). The cyclic overlap of $2p$ orbitals is nicely evoked by this formulation, but a price is paid. The Kekulé forms are easier for bookkeeping purposes: for keeping track of bonds making and breaking in chemical reactions. We will use both formulations in this book.

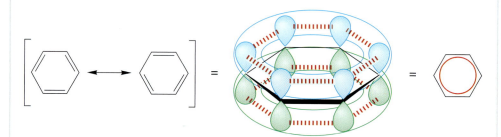

FIGURE 13.9 There is a ring of electron density above and below the six-membered benzene ring. The circle in the center of the benzene ring is evocative of the cyclic overlap of six $2p$ orbitals.

WORKED PROBLEM 13.3 There are other resonance forms for benzene, which can be included to get a slightly better electronic description of the molecule than is provided by the pair of Kekulé forms alone. In these other resonance forms, different sets of overlapping $2p$ orbitals from those shown in the Kekulé forms are emphasized. In these electronic structures, overlap between two p orbitals on "across the ring" or "para" carbons is taken into account. Draw these resonance forms, called **Dewar forms** after Sir James Dewar, including pictures of the orbital lobes involved in the "cross-ring" bond. Be careful! This problem is harder than it looks.

ANSWER Any resonance form represents one electronic description for a molecule and ignores all others. For example, in one resonance form for the allyl cation, overlap between the $2p$ orbitals on C(1) and C(2) is emphasized, and the third $2p$ orbital on C(3) is written as if it did not overlap. This error is taken into account in the other resonance form that shows overlap between C(3) and C(2), and ignores any overlap with the $2p$ orbital on C(1).

(continued)

WORKED PROBLEM 13.3 *(continued)*

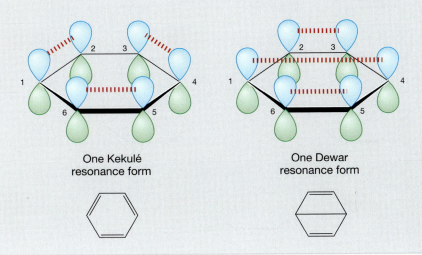

The two Kekulé forms of benzene make similar approximations and do a good job of giving the structure of benzene only when taken in combination (Fig. 13.8).

Other choices might have been made as to which orbital overlaps to emphasize. For example, in the Dewar forms, 1,4-overlap is explicitly emphasized. There are three Dewar forms that differ in which 1,4-overlap (1,4; 3,6; or 2,5) is counted.

But this business is tricky. By now, your eye is trained to see this central bond as a σ bond, *and it isn't*! As in the Kekulé forms, it is $2p$–$2p$, π overlap that is being considered here. For the central bond to be a σ bond, atoms would have to move, and that cannot be done in resonance forms. Only electrons can move. A drawing should make it clear.

One Kekulé
resonance form

One Dewar
resonance form

PROBLEM 13.4 Do you think these Dewar forms will be important, contributing strongly to the benzene structure? Why or why not?

PROBLEM 13.5 There is a real compound $(CH)_6$ called Dewar benzene containing a pair of fused (they share an edge) cyclobutene rings (Dewar benzene is correctly named bicyclo[2.2.0]hexa-2,5-diene). This molecule is not flat, and therefore *not* a resonance form of benzene, but a separate molecule (see figure below). Make a good three-dimensional drawing of Dewar benzene.

13.4   THE MOLECULAR ORBITAL PICTURE OF BENZENE

The raw material for a calculation of the molecular orbitals of benzene is six $2p$ orbitals in a ring. Linear combination of these atomic orbitals leads to a set of six molecular orbitals: three bonding and three antibonding. Their symmetries and ordering in energy are shown in Figure 13.10.

FIGURE 13.10 The π molecular orbitals of benzene.

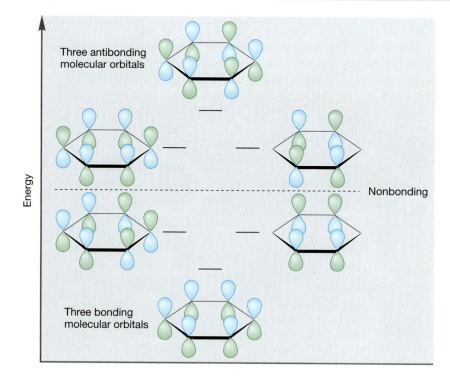

There are six electrons to go into the system, one from each carbon, and they naturally occupy the three lowest energy, bonding molecular orbitals. The antibonding molecular orbitals are empty. If we compare benzene to a system of three double bonds not tied back into a ring (1,3,5-hexatriene is a good model) we can see that there are substantial energetic savings involved in the ring structure. One can calculate the energy difference in terms of a quantity (β) whose value is approximately −18 kcal/mol (Fig. 13.11). If we sum all the energies of the electrons in Figure 13.11, we find an advantage for benzene over 1,3,5-hexatriene of 1β. The energetic advantage over three separated ethylenes is even greater: 2β (Fig. 13.11).

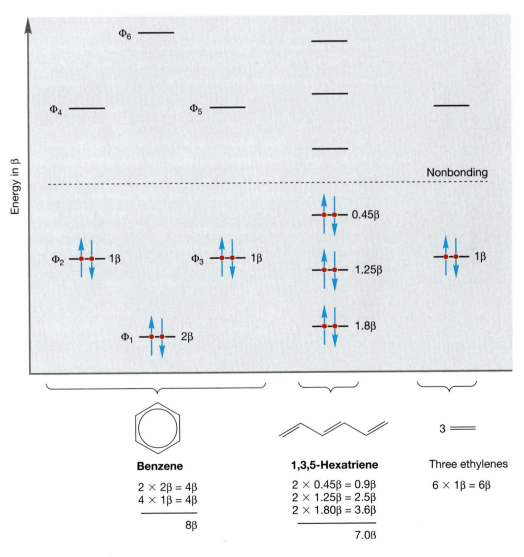

FIGURE 13.11 A comparison of the π molecular orbitals of benzene, 1,3,5-hexatriene, and ethylene.

Look carefully at how well the orbitals are used in benzene. There is maximum occupancy of bonding molecular orbitals and no electrons are placed in destructively antibonding molecular orbitals or wasted in nonbonding molecular orbitals. The molecular orbital picture gives a stronger hint of the special stability of benzene than does the resonance formulation. The stability comes not from a large number of resonance contributors, but from an especially good, especially low-energy, set of orbitals. We will see other examples. Now the question is, Just how good is this electronic arrangement? *How much* more stable is benzene than we might have expected from a cyclohexatriene model?

13.5   QUANTITATIVE EVALUATIONS OF RESONANCE STABILIZATION (DELOCALIZATION ENERGY) IN BENZENE

Although we can appreciate that delocalization is an energy-lowering phenomenon, it is not easy to see intuitively whether this effect is big or small. In this section, we see two experimental procedures for determining the magnitude of resonance stabilization.

13.5a Heats of Hydrogenation. The heat of hydrogenation of *cis*-2-butene is −28.6 kcal/mol (Table 10.1, p. 441). There is no change induced by incorporation of a double bond into a six-membered ring, as the heat of hydrogenation of cyclohexene is also −28.6 kcal/mol. A reasonable guess about the heat of hydrogenation of a compound containing two double bonds in a ring would be just twice the value for cyclohexene, or $2 \times -28.6 = -57.2$ kcal/mol. This guess is very close to correct; the heat of hydrogenation of 1,3-cyclohexadiene is −55.4 kcal/mol. There is no special effect apparent from confining two double bonds in a six-membered ring. The difference between the observed heat of hydrogenation for 1,3-cyclohexadiene (−55.4) and the predicted value (−57.2) is approximately what one would expect, as our prediction takes no account of the stabilizing effects of conjugation in 1,3-cyclohexadiene (Chapter 12, p. 561; Fig. 13.12).

FIGURE 13.12 Heats of hydrogenation for some alkenes and cycloalkenes.

Now let's guess (calculate) the heat of hydrogenation for 1,3,5-cyclohexatriene, the hypothetical molecule with three double bonds in a six-membered ring. We might make this estimate by assuming that the third double bond will increase the heat of hydrogenation of 1,3-cyclohexadiene by as much as the addition of a second double bond did for cyclohexene [−55.4 − (−28.6) = −26.8 kcal/mol]. This increment would give a predicted value of −55.4 + (−26.8) = −82.2 kcal/mol for the heat of hydrogenation of the hypothetical 1,3,5-cyclohexatriene. Now we can measure the heat of hydrogenation of the real molecule, benzene. Hydrogenation is extremely slow under normal circumstances but there are special catalysts that can speed the reaction up a bit, and eventually even the stable benzene will hydrogenate to give cyclohexane. The measured heat of hydrogenation for benzene, −49.3 kcal/mol, is much lower than the calculated value for 1,3,5-cyclohexatriene. Figure 13.13 combines all these numbers and shows graphically how this procedure yields the amount by which benzene is more stable than the hypothetical 1,3,5-cyclohexatriene, −32.9 kcal/mol (Fig. 13.13).

FIGURE 13.13 The value for the special stability of benzene (red arrow) as derived from measured and calculated heats of hydrogenation.

SUMMARY

We calculated the heat of hydrogenation of the unreal, unknown, hypothetical model molecule 1,3,5-cyclohexatriene by using values for the heats of hydrogenation of the real molecules cyclohexene and 1,3-cyclohexadiene. The heat of hydrogenation of 1,3-cyclohexadiene was calculated quite accurately, so we can be rather confident that our estimate for 1,3,5-cyclohexatriene won't be far off. That's important, because we can never make the cyclic triene to check the calculation. Any synthetic route will inevitably produce the much more stable real molecule, benzene. Next, measure the heat of hydrogenation for the real molecule, benzene, which turns out to be far lower than our estimate. This difference (Fig. 13.13) represents the amount of energy the special stabilization of benzene is worth. It is more than 30 kcal/mol and is called the **resonance energy** or **delocalization energy** of benzene.

13.5b Heats of Formation. The delocalization energy of benzene can be estimated in another way, using heats of formation (ΔH_f°). The chemistry of 1,3,5,7-cyclooctatetraene tells us there is no special stabilization in this molecule. Hydrogenation is fast and addition reactions occur with ease. The ΔH_f° for cyclooctatetraene is +71.23 kcal/mol, or +8.9 kcal/mol per CH unit. The ΔH_f° for benzene is +19.82 kcal/mol, or 3.3 kcal/mol per CH unit. Therefore, benzene is about 5.6 kcal/mol more stable per CH unit than a cyclic polyene with no special stability ($8.9 - 3.3 = 5.6$ kcal/mol). As there are six CH units in benzene we can estimate the special stability (resonance or delocalization energy) to be $6 \times 5.6 = 33.6$ kcal/mol. The two methods, which use the related heats of hydrogenation and heats of formation, are in rather good agreement (Fig. 13.14).

$\Delta H_f^\circ = +71.23$ kcal/mol

8.9 kcal/mol per CH unit

$\Delta H_f^\circ = +19.82$ kcal/mol

3.3 kcal/mol per CH unit

$6 \times (8.9 - 3.3) = 33.6$ kcal/mol delocalization energy — this value compares well with that calculated from heat of hydrogenation data (32.9 kcal/mol)

FIGURE 13.14 The delocalization energy of benzene as calculated from heats of formation.

This special stability has come to be known as benzene's **aromatic character** or **aromaticity**, a name derived in an obvious way from the pleasant (?) odor of many benzene derivatives. In the next few pages, we will expand on the theme of aromaticity, eventually ending up with a general method for finding aromatic compounds.

13.6   A GENERALIZATION OF AROMATICITY: HÜCKEL'S $4n + 2$ RULE

It would certainly be surprising if there were not other compounds sharing the special stability we have just called aromaticity. To begin our search for other aromatic compounds, and, ultimately, a general way of finding such compounds, let's review the structural features of the prototype, benzene, that lead to special stability.

1. The molecule is cyclic. Aromaticity is a property of ring compounds. Recall the discussion of the difference between the molecular orbitals for the cyclic molecule, benzene, and the acyclic molecule, 1,3,5-hexatriene (Fig. 13.11, p. 625).

2. The molecule is **fully conjugated**, which simply means that there is a $2p$ orbital on every atom in the ring, and that connectivity between adjacent $2p$ orbitals is

maintained throughout. 1,3,5-Cycloheptatriene is not aromatic, because the 2p orbitals at the end of the cycle do not overlap. Its properties resemble those of 1,3,5-hexatriene (Fig. 13.15).

1,3,5-Cycloheptatriene

2p Orbital connectivity at C(7) broken here by the CH$_2$ group (no 2p orbital)

FIGURE 13.15 In 1,3,5-cycloheptatriene, orbital connectivity is broken by the CH$_2$ group. This molecule is not aromatic, and behaves as a normal triene.

3. The molecule is planar. Why should this make a difference? It ensures optimal overlap of the 2p orbitals. If the ring is not planar, or at least quite close to planar, the 2p orbitals do not overlap well and some connectivity between orbitals is lost. Large deviations from planarity can nearly or completely uncouple the cycle of orbitals. An example is the tub-shaped molecule cyclooctatetraene (Fig. 13.16).

If the shape of the molecule uncouples the 2p orbitals, the molecule is effectively unconjugated

The tub-shaped cyclooctatetraene is an example of this decoupling phenomenon

=

Poor 2p connectivity here

FIGURE 13.16 The nonplanar cyclooctatetraene is not aromatic.

One might now be tempted to try a generalization. Might not all molecules that are cyclic, planar, and fully conjugated share benzene's aromatic character and stability? Apparently not. Recall that cyclobutadiene, a molecule that meets all of the three criteria mentioned above, is most *unstable* (Fig. 13.17).

There must be more to it, and in our search for further criteria, we must take into account the difference between benzene and cyclobutadiene. The crucial generalization was made by the German Erich Hückel (1896–1980) in the early 1930s. **Hückel's rule** can be added to our list of three criteria (also recognized by Hückel) developed earlier.

4. Aromatic systems will contain $4n + 2$ π electrons, where n is an integer, 0, 1, 2, 3, ... *Please note carefully that "n" has nothing to do with the number of atoms in the ring. We are considering only the number of π electrons here.*

A most unstable molecule even though it is planar, cyclic, and fully conjugated

FIGURE 13.17 Cyclobutadiene is cyclic, planar, and fully conjugated, but very unstable.

cis-**1,3,5-Hexatriene**

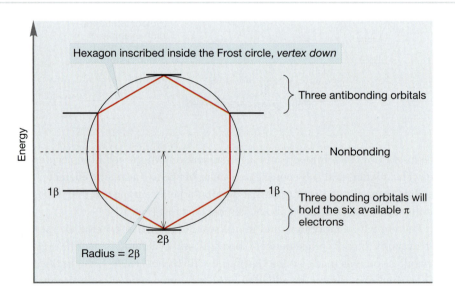

No strong overlap here; the molecule is not aromatic

FIGURE 13.18
cis-1,3-Hexatriene is fully conjugated and can be planar, but the lack of a ring structure means that overlap between the 2p orbitals on C(1) and C(6) is essentially zero.

We are now faced with the most important question of this section: Why should the number of π electrons make a difference? Why does Hückel's rule work? Briefly stated, it works because it is *only* those molecules containing $4n + 2$ π electrons (and satisfying the other three criteria for aromaticity) that will have molecular orbitals arranged like those of benzene. All bonding molecular orbitals must be fully occupied and there may be no electrons in destabilizing antibonding molecular orbitals or wasted in nonbonding molecular orbitals. It's best to look at some examples now.

cis-1,3,5-Hexatriene can serve as an example for all acyclic molecules. Although it is fully conjugated (there is a 2p orbital on every carbon), and could be planar without inducing great strain, it most certainly is not cyclic. The lack of a ring means that there is no chance of strong overlap between the 2p orbitals on C(1) and C(6) and, therefore, no chance of aromatic character (Fig. 13.18). 1,3,5-Cycloheptatriene is a cyclic molecule that is not fully conjugated. The connectivity of the 2p orbitals is broken by the methylene group at C(7) (Fig. 13.15). Cyclobutadiene is a planar, cyclic molecule that is fully conjugated. It fails only the fourth criterion as it has $4n$ ($n = 1$) π electrons, not a "Hückel number" of $4n + 2$.

Before we can go on, we need a reliable way of finding the molecular orbital patterns for these molecules. Consider a device called a **Frost circle**, or sometimes, a Frost device. It is named for the American Arthur A. Frost (b. 1909) who devised it in 1953. It is a method of finding, mercifully without the use of mathematics, the relative energies of the molecular orbitals for planar, cyclic, fully conjugated molecules. One merely inscribes the appropriate (same size as the ring) polygon in a circle (radius 2β), *vertex down,* and the intersections of the ring with the circle will mark the positions of the molecular orbitals.

For example, to locate the molecular orbitals of benzene, inscribe a hexagon in a circle, vertex down, with the nonbonding line dividing the hexagon exactly in half (Fig. 13.19).

FIGURE 13.19 The relative energies of the molecular orbitals of benzene derived from a Frost circle.

Hexagon inscribed inside the Frost circle, *vertex down*

Energy

Three antibonding orbitals

Nonbonding

1β 1β

Three bonding orbitals will hold the six available π electrons

2β

Radius = 2β

The familiar (p. 624) arrangement of molecular orbitals appears, and with some plane geometry we can even get the relative energies of the molecular orbitals exactly in terms of β (the radius of the Frost circle is 2β). For benzene, the available six π electrons completely fill the three available bonding molecular orbitals with no offending electrons remaining to be placed in the antibonding set. There are no nonbonding molecular orbitals.

If we follow this procedure for cyclobutadiene, a set of four molecular orbitals appears. One is bonding, one is antibonding, and two are equi-energetic (degenerate), nonbonding molecular orbitals (Fig. 13.20).

FIGURE 13.20 The relative energies of the molecular orbitals of cyclobutadiene derived from a Frost circle.

Four electrons are available and two of them will surely go into the bonding molecular orbital. There is no other bonding molecular orbital to accommodate the remaining two electrons. They will have to occupy the nonbonding degenerate pair, and will be most stable in this square arrangement if the spins are unpaired and a single electron is put into each orbital of the degenerate pair (Fig. 13.21).

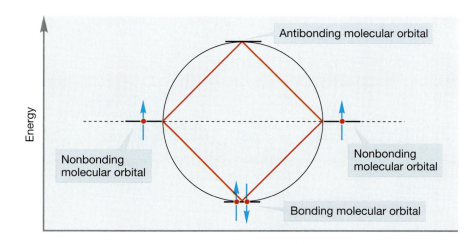

FIGURE 13.21 The electronic occupancy of the four molecular orbitals for square cyclobutadiene.

90° Bond angles = severe angle strain

FIGURE 13.22 Cyclobutadiene has severe angle strain.

This arrangement is surely not particularly stable, and we would not expect the properties of cyclobutadiene to resemble those of benzene, as they most emphatically do not. Moreover, cyclobutadiene must pay a severe price in strain as the internal angles must be 90° (Fig. 13.22).

Cyclobutadiene can be isolated, but only if it is kept very cold (its IR spectrum was measured at 8 K = −265 °C), and segregated from other reactive molecules, including itself. Theory and experiment agree that cyclobutadiene distorts from the square arrangement to form a rectangle with short carbon–carbon double bonds and longer carbon–carbon single bonds. This distortion minimizes the connection (conjugation) between the p orbitals (Fig. 13.23).

Square, delocalized cyclobutadiene   Rectangular cyclobutadiene

FIGURE 13.23 In rectangular cyclobutadiene, orbital connectivity is minimized.

1,3,5,7-Cyclooctatetraene is another example of a $4n$ molecule. Here we find neither the special stability of benzene, nor the special instability of cyclobutadiene. The molecule behaves like four separated, normal alkenes. Let's first examine planar cyclooctatetraene. The Frost circle device again determines the relative positions of the molecular orbitals (Fig. 13.24).

FIGURE 13.24 The molecular orbitals of planar cyclooctatetraene. Two electrons must occupy nonbonding orbitals.

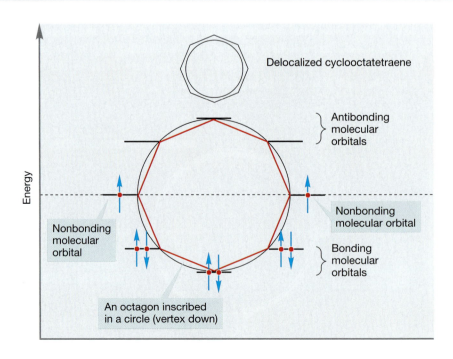

Delocalized cyclooctatetraene

Antibonding molecular orbitals

Nonbonding molecular orbital

Nonbonding molecular orbital

Bonding molecular orbitals

Energy

An octagon inscribed in a circle (vertex down)

Like cyclobutadiene, planar cyclooctatetraene must have two electrons in nonbonding molecular orbitals, and there will be no special stability. Like planar cyclobutadiene, planar cyclooctatetraene has both angle and torsional strain. The angle in a regular octagon is 135°, quite far from the optimum sp^2 value of 120°. Unlike cyclobutadiene, there is a way for cyclooctatetraene to get out of its misery. Cyclobutadiene must be planar, or very nearly so. By contrast, cyclooctatetraene can distort easily into a tublike form in which most of the angle and some of the torsional strain of the planar form is relieved (Fig. 13.25).

| Angle strain | Torsional strain | Angle strain relieved | Some (not all) torsional strain relieved |

FIGURE 13.25 In planar cyclooctatetraene, there is both angle and torsional strain. Both can be minimized by distorting into a tub.

Kinetic measurements reveal that two tubs interconvert rather easily. The energy difference between the tub and the planar, delocalized species can be extracted from the kinetic data. It turns out that the tub is 17 kcal/mol more stable than the planar, symmetrical, fully conjugated octagon (Fig. 13.26).

Energy

17 kcal/mol

FIGURE 13.26 The tub-shaped cyclotetraene is 17 kcal/mol more stable than the planar, delocalized structure.

PROBLEM 13.6 The process of interconversion is actually a bit more complicated. Two localized tubs (alternating single and double bonds) equilibrate without the double bonds changing position. The activation energy for this process is 14.6 kcal/mol. The conversion into the delocalized form (the symmetrical octagon of Fig. 13.26) requires an additional 2.4 kcal/mol. Neither the localized planar form nor the delocalized planar form is an energy minimum. Sketch out Energy versus Reaction progress diagrams for these processes.

PROBLEM 13.7 Which of the molecules below are properly termed "aromatic" and which are not?

(a)　　　(b)　　　(c)　　　(d)　　　(e)

CH₂

So we have now seen two examples of the lack of aromaticity in $4n$ molecules; it would seem to be time to look at some potentially aromatic $4n + 2$ molecules. It is in the stability of certain ions that the most spectacular examples have appeared. As we have stressed over and over, small carbocations are most unstable. But there are a few exceptions to this generality. One of them is the **cycloheptatrienylium ion**, or **tropylium ion** ($C_7H_7^+$), the ion derived from the loss of hydride ($H\!:^-$) from 1,3,5-cycloheptatriene, also called tropilidene (Fig. 13.27).

FIGURE 13.27 The tropylium ion ($C_7H_7^+$) can be made through transfer of hydride ($H\!:^-$) from 1,3,5-cycloheptatriene to the trityl cation.

| 1,3,5-Cycloheptatriene (tropilidine) | Trityl cation | Tropylium fluoborate | Triphenylmethane |

Hydride cannot be simply lost from cycloheptatriene, but it can be *transferred* to the **trityl cation**, which is itself a quite stable carbocation.

Tropylium fluoborate appears to be stable indefinitely under normal conditions. It sits uncomplaining on the desktop, for all the world resembling table salt. This stuff is no ordinary carbocation!

PROBLEM 13.8 Why is the trityl cation relatively stable?

Note first that the tropylium ion has one important difference from its parent, 1,3,5-cycloheptatriene. It is fully conjugated. The removal of a hydride from the cyclic triene generates a $2p$ orbital at the 7-position, which once insulated the ends of the π system from each other. In the cation they are connected; this molecule is fully conjugated (Fig. 13.28).

FIGURE 13.28 In the tropylium ion (the cycloheptatrienylium ion) orbital connectivity is continuous around the ring.

1,3,5-Cycloheptatriene is not fully conjugated

The cycloheptatrienylium (tropylium) ion is fully conjugated

The Frost circle device lets us see the relative energies of the molecular orbitals for this system (Fig. 13.29). How many electrons must go into the π system? In the cation, there is an empty 2p orbital on C(7) and the only π electrons are the six from the three original double bonds. These electrons can fit nicely into the three bonding molecular orbitals of the π system. Note how this system resembles the arrangement of benzene—all the bonding molecular orbitals are full, and no antibonding or nonbonding molecular orbitals are occupied (Fig. 13.29). This ion qualifies as aromatic, and the increased stability *relative to other carbocations* is dramatic indeed. Notice again the difference between the number of atoms involved in the ring—here seven—and the number of π electrons—here six $= (4n + 2)$, $n = 1$.

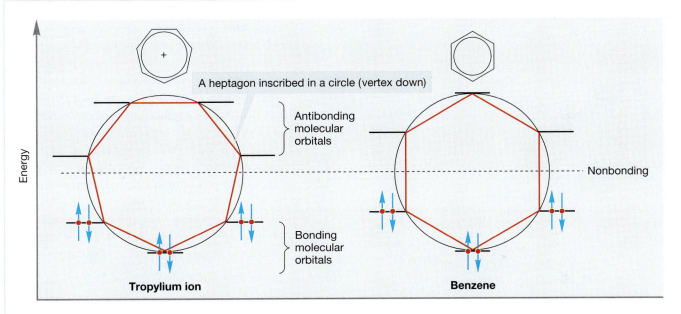

FIGURE 13.29 In the tropylium ion, the three bonding molecular orbitals are fully occupied and the antibonding molecular orbitals are unoccupied. The molecular orbital picture of the tropylium ion closely resembles that of benzene.

There is an anionic counterpart to this stable cation. The pK_a of propene is 43, and the pK_a of propane is greater than 60. The added delocalization in the allyl anion makes propene a stronger acid than propane by a factor of about 10^{20} (Fig. 13.30).

FIGURE 13.30 Resonance stabilization of its conjugate base makes propene much more acidic than propane.

But the pK_a of 1,3-cyclopentadiene is 15. Cyclopentadiene is a *much* stronger acid than propene. Look carefully at the structure of the **cyclopentadienyl anion**. Here too, we have a planar, cyclic, and fully conjugated system. The molecular orbitals can be derived from a Frost circle (Fig. 13.31).

There are six electrons to put into the molecular orbitals, and, as in the tropylium ion or benzene, they fully occupy the set of three bonding molecular orbitals without requiring antibonding or nonbonding molecular orbitals to be occupied (Fig. 13.31). The cyclopentadienyl anion can be described as aromatic, and *for an anion,* this species is remarkably stable. Do not fall into the trap of expecting this anion to be as stable as benzene. Benzene is a neutral molecule in which all of carbon's valences are satisfied. The anion is charged and contains unsatisfied valence. It is very stable, but only in comparison with the rest of the family of anions.

But wait. We have been at pains to equate stability with delocalization of electrons, which results in a distribution of charge to several atoms in the molecule (Fig. 13.30). Both the tropylium ion and the cyclopentadienyl anion can be described by a most impressive array of resonance forms (Fig. 13.32).

FIGURE 13.31 Cyclopentadiene is a quite strong acid. The cyclopentadienyl anion is an aromatic species. For an anion, it is extremely stable.

A pentagon inscribed in a circle (vertex down)

base, B:⁻

pK_a = 15

The cyclopentadienide anion is easily formed

Energy

Antibonding molecular orbitals

Nonbonding

Bonding molecular orbitals

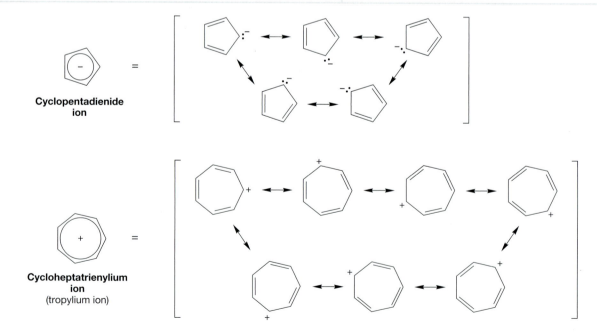

FIGURE 13.32 Resonance descriptions of the tropylium (cycloheptatrienylium) ion and the cyclopentadienyl anion.

Cyclopentadienide ion

Cycloheptatrienylium ion (tropylium ion)

Perhaps the aromaticity concept is unnecessary. Why attribute the stability of these species to some special effect when simple delocalization of electrons may do as well? That's a good question, and needs an answer. It is not necessary to invoke special effects if simpler concepts will suffice. The principle of "Ockham's razor"* says that when faced with a variety of equally attractive explanations, pick the simplest.

Wielding Ockham's razor, we might say that the cyclopentadienyl anion is very stable because of the five resonance forms so apparent in Figure 13.32. What then would we expect of the cycloheptatrienyl anion, whose seven resonance forms are shown in Figure 13.33?

Cycloheptatrienide
anion

FIGURE 13.33 The seven resonance forms for the cycloheptatrienyl anion.

If an ion with five resonance forms is good, surely one with seven forms will be better. Our explanation predicts a formidable acidity for 1,3,5-cycloheptatriene; in particular, it must be more acidic than cyclopentadiene. Its pK_a must be < 15. But the pK_a of cycloheptatriene is about 39. Cycloheptatriene is a *weaker* acid than cyclopentadiene by a factor of 10^{24} (Fig. 13.34)!

Five resonance forms

Seven resonance forms

FIGURE 13.34
Cycloheptatriene is a weaker acid than cyclopentadiene by a factor of 10^{24}.

*Named for the philosopher William of Ockham (1285–1349; he died in the plague), who put it better: "What can be done with fewer is done in vain with more."

So, the number of resonance forms is *not* always a predictor of stability. No longer are we faced with equally attractive explanations. The aromaticity concept is necessary to make sense of the data. A look at the molecular orbitals of cycloheptatriene shows the source of the instability. The cycloheptatrienyl anion must contain electrons in antibonding orbitals (Fig. 13.35)!

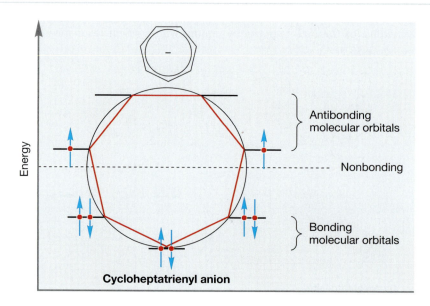

FIGURE 13.35 In the cycloheptatrienyl anion, two electrons must occupy antibonding orbitals.

PROBLEM 13.9 How many signals do the ^{13}C NMR spectra of the cyclopentadienyl anion and the tropylium ion show?

Now let's search for other neutral compounds exhibiting aromaticity. Of course, all manner of substituted benzenes exist, but these hardly constitute exciting new examples. It is possible to combine benzene rings so that they share an edge in what is called a fused relationship. These molecules are also aromatic. Atoms other than carbon can be introduced into the ring while maintaining the sextet of electrons to produce **heterobenzenes**, fully conjugated, six-membered rings containing one or more non-carbon atoms (N, O, S, etc.). Some smaller rings also contain a sextet of π electrons; pyrrole is an example. Molecules of these types are shown in Figure 13.36, and will be considered further in later sections.

FIGURE 13.36 Some simple extensions of aromaticity: substituted, fused, and heterobenzene compounds, as well as a heteroaromatic compound, pyrrole.

Toluene
(a simple substituted benzene)

Naphthalene
(a fused, two-ring aromatic compound)

Pyridine
(a six-membered heterobenzene)

Pyrrole
(a five-membered aromatic compound)

PROBLEM 13.10 It is startlingly easy to form the cyclopropenyl cation. Explain, and classify the cyclopropenyl cation in terms of n.

$(C_6H_5)_3C^+ \ ^-BF_4$

$+ \ (C_6H_5)_3CH$

C_6H_5 = benzene ring

The cyclopropenium
ion

PROBLEM 13.11 Do you expect the corresponding anion to be stable? Explain.

What we are looking for here are not examples of relatively straightforward extensions, but new $4n + 2$ molecules. For benzene, and even the stable ions we have been considering, n has always been equal to 1, and $4n + 2 = 6$. If the concept of aromaticity is valid, there should be examples where $n = 0$, 2, 3, and so on.

Let's test this notion. Consider the cyclodecapentaenes, cyclic molecules with 10 π electrons ($4n + 2$, $n = 2$) (Fig. 13.37). Are they as stable as the six-electron molecules we have been examining?

***cis,cis,cis,cis,cis-*
Cyclodecapentaene**

***trans,cis,cis,cis,cis-*
Cyclodecapentaene**

***trans,cis,trans,cis,cis-*
Cyclodecapentaene**

FIGURE 13.37 Some cyclodecapentaenes. These molecules have 10π electrons: $4n + 2 = 10$, $n = 2$.

PROBLEM 13.12 The several cyclodecapentaenes of Figure 13.37 differ in the stereochemistries of the double bonds. Why isn't there more than one isomer of benzene: a six-membered ring containing three double bonds?

PROBLEM 13.13 How many signals would be seen in the ^{13}C NMR spectra of the three cyclodecapentaenes in Figure 13.37?

A remarkable amount of effort was expended on the synthesis of the cyclodecapentaenes before they were successfully made. The magnitude of this effort should make us suspicious; benzenes are everywhere and extraordinarily easy to make. Moreover, the known cyclodecapentaenes are not only not especially stable, they are instead strikingly *un*stable. So, we have some explaining to do. We have been relying on the theory of special stability conferred by the aromatic $4n + 2$ number

These two hydrogens occupy the same space!

trans,cis,trans,cis,cis-Cyclodecapentaene

FIGURE 13.38 In *trans,cis,trans,cis,cis-* cyclodecapentaene, two inside hydrogens occupy the same space.

of π electrons, and the cyclodecapentaenes are members of the $4n + 2$ club. If we cannot explain why these molecules are unstable, our theory must be abandoned. It turns out not to be too difficult to resolve the problem. Aromatic stability is not some magical force. It is just one kind of stabilizing effect, which in the face of stronger destabilizing effects will not prevail. Put simply, the planar cyclodecapentaenes are destabilized by an amount greater than aromaticity can overcome. What might those destabilizing effects be? Look first at the molecule in Figure 13.38, the trans,cis,trans,cis,cis isomer. The two hydrogens inside the ring occupy the same space. This crowding induces extraordinary strain that is greater than any resonance stabilization. Aromaticity cannot compensate for this strong destabilizing effect (Fig. 13.38).

The trans,cis,cis,cis,cis isomer has some of the same problems, although here only one hydrogen is inside the ring. In this molecule, however, there is severe angle strain, and the planar form, required for aromaticity, is badly destabilized. These compounds are polyenes, not aromatic molecules. The all-cis isomer has even more angle strain, and it too can be planar, and therefore aromatic, only at a prohibitively high energy cost. The benefits of delocalization are overwhelmed by the angle strain (Fig. 13.39).

FIGURE 13.39 *trans,cis,cis,cis,cis-* Cyclodecapentaene and the all-cis isomer have severe angle strain and cannot be planar.

trans,cis,cis,cis,cis-Cyclodecapentaene
(angle and other strain caused by the inside hydrogen)

cis,cis,cis,cis,cis-Cyclodecapentaene
(angle strain)

A good set of models will show this strain easily. Construct the planar, all-cis cyclodecapentaene and the strain will be apparent. If you are lucky, the plastic will mimic the molecule and the model will snap into a pretzel shape, which approximates the energy minimum form of the molecule. These molecules are best described as strained polyenes. They fail the test of planarity for aromatic molecules because the planar structures are much more strained than the nonplanar forms.

PROBLEM 13.14 The cyclodecapentaenes actually escape their misery by forming a new, cross-ring bond (see below). As they warm up, a bicyclo[4.4.0]decatetraene is formed. Write an arrow formalism for this reaction.

cis,cis,cis,cis,cis -Cyclodecapentaene

Δ

Bicyclo[4.4.0]deca-2,4,7,9-tetraene

So we haven't really judged the limits of the theory of aromaticity by using the cyclodecapentaenes. We might be able to if the offending strain could somehow be removed. On paper, this transfomation is simple. For example, we might just erase the two inside hydrogens in one of the molecules (Fig. 13.40). Unfortunately, this doesn't give a real molecule—there are two trivalent carbons at the "dotted" positions.

A particularly clever German chemist, Emanuel Vogel (b. 1927), found a way to do this erasure chemically. He used a remarkable synthesis to replace the offending inside hydrogens with a bridging methylene group. The two inside hydrogens are gone, replaced with the red methylene group. The removal of the offending hydrogens eliminates the strain that comes from their "bumping" (Fig. 13.41).

***trans,cis,trans,cis,cis*-Cyclodecapentaene** (strain caused by the inside red hydrogens)

magic hydrogen eraser

(much less strain! But... what is going on at the red dotted positions?)

FIGURE 13.40 The strain caused by two inside hydrogens in one cyclodecapentaene can be removed by removing the offending hydrogens.

FIGURE 13.41 Vogel's bridged cyclodecapentaene.

The resulting molecule, a bridged 10 π electron system, isn't a perfect cyclodecapentaene. The overlap between the $2p$ orbitals at the bridgehead and adjacent positions isn't optimal, for example (Fig. 13.42). Still, the molecule maintains enough orbital overlap to remain fully conjugated and satisfy the criteria for aromaticity. Both its chemical and physical properties allow it to be classified as aromatic.

FIGURE 13.42 Vogel's molecule is fully conjugated, but orbital overlap is not optimal, as it is, for example, in benzene.

SUMMARY

Planar, cyclic, fully conjugated molecules containing $4n + 2$ π electrons are especially stable and are called "aromatic." The "$4n + 2$ rule" works because such molecules have a set of fully filled strongly bonding molecular orbitals, and no electrons in non-bonding or antibonding orbitals.

13.7   ANNULENES

The generic term for monocyclic, fully conjugated molecules is **annulene**. Benzene could be called [6]annulene, square cyclobutadiene is [4]annulene, and planar cyclooctatetraene is [8]annulene (Fig. 13.43).

[6]Annulene   [4]Annulene   [8]Annulene   etc.

FIGURE 13.43 Some simple annulenes.

[18]Annulene

FIGURE 13.44 [18] Annulene.

Other annulenes appear to be aromatic. These annulenes are large enough so that the destabilizing steric effects present in the cyclodecapentaenes ([10]annulenes) are diminished, and severe angle strain is avoided. A nice example is [18]annulene, a planar molecule in which all carbon–carbon bond lengths are very similar. It contains $4n + 2$ π electrons, where $n = 4$ (Fig. 13.44).

The description of this molecule as aromatic derives largely from an examination of its nuclear magnetic resonance (NMR) spectrum, and that must wait for Chapter 15, but [18]annulene does appear to be aromatic.

SUMMARY

We have concentrated on the phenomenon of aromaticity as applied to benzene itself and larger and smaller polyenes. We saw two experimental methods of evaluating the stabilization of aromaticity quantitatively, and with the aid of Hückel's insight, we found a generalization of the phenomenon. Now we move on to ways of varying the structure of our prototypal example, benzene, and then take a quick look at reactivity.

13.8   SUBSTITUTED BENZENES

13.8a Monosubstitution. Monosubstituted benzenes are usually named as derivatives of benzene, and the names are written as single words. There are numerous common names that have survived a number of attempts to systemize them out of existence. You really do have to know a number of these names in order to be literate in organic chemistry. I don't think the systematic "benzenol" will replace "phenol" in your lifetime, for example. At least I hope it doesn't. Figure 13.45 shows some simple benzenes with both the systematic and common names.

(flashcards!)

FIGURE 13.45 Some simple monosubstituted benzenes. Some common names are widely used.

Two of the most often used common names are **phenyl** (sometimes even **Ph**), which stands for a benzene ring, and **benzyl**, which stands for the Ph—CH$_2$ group. Now is a good time to summarize the various abbreviations for benzene rings, and Figure 13.46 does just that.

CONVENTION ALERT

FIGURE 13.46 Abbreviations for benzene and benzyl groups.

A phenyl group
Ph = C$_6$H$_5$ = φ
A benzyl group
PhCH$_2$ = C$_6$H$_5$CH$_2$ = Bz

There are three possible isomers of disubstituted benzenes. 1,2-Substitution is known as **ortho** (= o-), 1,3-substitution as **meta** (= m-), and 1,4-substitution as **para** (= p-).

PROBLEM 13.15 Indeed, the isolation of three, and only three isomers of disubstituted benzenes was a crucial factor in the assignment of the Kekulé form as the correct structure of benzene. Use the fact that there are three and only three achiral isomers of a disubstituted benzene ($C_6H_4R_2$) to criticize the suggestions that benzene might have the following structures:

In practice, either the numbers or the "ortho, meta, para" designations are used. If there is a widely used common name, it is used as a base for the full name. For example, two of the molecules of Figure 13.47 are quite properly named as o-bromotoluene and p-nitrophenol, respectively.

ortho- (o-) 1,2-Disubstituted	meta- (m-) 1,3-Disubstituted	para- (p-) 1,4-Disubstituted

| o-**Bromotoluene** or
1-bromo-2-methylbenzene | m-**Dichlorobenzene** or
1,3-dichlorobenzene | p-**Nitrophenol** or
4-nitrophenol | p-**Dimethylbenzene** or
1,4-dimethylbenzene or
p-xylene |

FIGURE 13.47 The three possible substitution patterns for disubstituted benzenes and some disubstituted benzenes.

13.8b Polysubstitution. There are polysubstituted benzenes that are also known by common names. Figure 13.48 shows a sampling of these molecules.

FIGURE 13.48 Some common names for disubstituted benzenes.

PROBLEM 13.16 How many signals will appear in the ^{13}C NMR spectra of the three dihydroxybenzenes at the bottom of Figure 13.48?

The substituents on polysubstituted benzenes are numbered and named alphabetically. Figure 13.49 gives a few examples.

FIGURE 13.49 Some named trisubstituted benzenes.

13.9   PHYSICAL PROPERTIES OF SUBSTITUTED BENZENES

The physical properties of substituted benzenes resemble those of alkanes and alkenes of similar shape and molecular weight. Table 13.1 collects some physical properties for a number of common substituted benzenes. Notice the effects of symmetry. *p*-Xylene melts at a much higher temperature than the ortho or meta isomer, for example. Many para isomers have high melting points, and crystallization can sometimes be used as a means of separating the para regioisomer from the others.

Aromatic compounds are called **arenes** and the general abbreviation "Ar = aryl" is often used, in analogy to the R used for a general alkyl species.

TABLE 13.1   Physical Properties of Some Substituted Benzenes

Name	bp (°C)	mp (°C)	Density (g/mL)
Cyclohexane	80.7	6.5	0.78
Benzene	80.1	5.5	0.88
Methylcyclohexane	100.9	−126.6	0.77
Toluene	110.6	− 95	0.87
o-Xylene	144.4	− 25.2	0.88
m-Xylene	139.1	− 47.9	0.86
p-Xylene	138.3	13.3	0.86
Aniline	184.7	− 6.3	1.02
Phenol	181.7	43	1.06
Anisole	155	− 37.5	1.0
Bromobenzene	156.4	− 30.8	1.5
Styrene	145.2	− 30.6	0.91
Benzoic acid	249.1[a]	122.1	1.1
Benzaldehyde	178.6	− 26	1.0
Nitrobenzene	210.8	5.7	1.2
Biphenyl	255.9	71	0.87

[a] At 10-mm pressure.

Pyridine

Benzene

FIGURE 13.50 Pyridine, a heterobenzene.

13.10   HETEROBENZENES AND OTHER HETEROCYCLIC AROMATIC COMPOUNDS

13.10a Contrasting Structures of Pyridine and the Five-Membered Rings: Pyrrole, Furan, and Thiophene.
If a CH unit of benzene is replaced with a nitrogen, the result is **pyridine**, another aromatic six-membered ring (Fig. 13.50). From the provincial view of organic chemistry all heavy atoms except carbon are called "heteroatoms," and pyridine is a heterobenzene, a member of the general class of compounds called **heteroaromatic compounds**, aromatic molecules containing at least one non-carbon heavy atom.

The electron counting for pyridine can be confusing at first because of the extra pair of electrons on nitrogen. Doesn't pyridine have eight π electrons and thus violate the $4n + 2$ rule? Put a better way, we have to worry whether antibonding orbitals are occupied in pyridine. A good orbital drawing shows that the "extra" two electrons are not in the π system, and pyridine is not in violation of the Hückel rule (Fig. 13.51).

sp^2 Orbital perpendicular to the π system

The bonds made from sp^2–sp^2 overlap are not in the π system

FIGURE 13.51 Pyridine has only 6 π electrons. The lone-pair electrons on nitrogen are in an sp^2 orbital perpendicular to the π system, not in the π system itself.

Other six-membered rings containing one heteroatom are known, but none is very stable. Oxabenzene must be cationic and the charge will contibute to making it a reactive molecule. Aromaticity must overcome the destabilizing influences of the charged atom and doesn't succeed in this case (Fig. 13.52).

FIGURE 13.52 Oxabenzene must be charged and is unlikely to be very stable.

The neutral, five-membered heterocyclic ring compounds, **pyrrole** and **furan**, also show aromatic character. Even **thiophene** is apparently aromatic (Fig. 13.53).

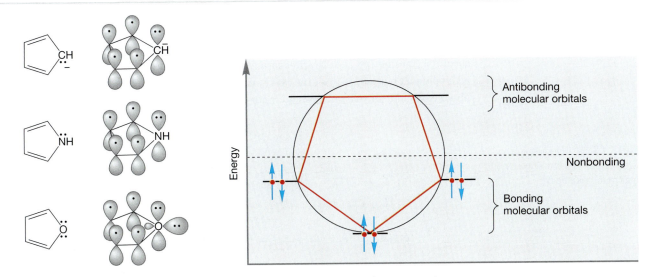

Cyclopentadienyl anion **Pyrrole** **Furan** **Thiophene**

FIGURE 13.53 Four five-membered aromatic rings. There are still $4n + 2$ π electrons in the three neutral molecules, which are isoelectronic (same number of electrons) with the cyclopentadienyl anion.

Furan and pyrrole can complete an aromatic sextet of π electrons by using a pair of nonbonding electrons in a $2p$ orbital. These molecules have the same number of π electrons as the cyclopentadienyl anion, but none of the problems induced by the charge on the all-carbon molecule. In thiophene and furan, there is a second pair of electrons remaining in an sp^2 orbital perpendicular to the π orbital system (Fig. 13.54).

FIGURE 13.54 Orbital pictures of the isoelectronic five-membered aromatic compounds.

Pyridine and pyrrole have contrasting resonance descriptions (Fig. 13.55), and this difference results in some very different physical properties. For example, the direction of the dipole moment is different in the two aromatic molecules.

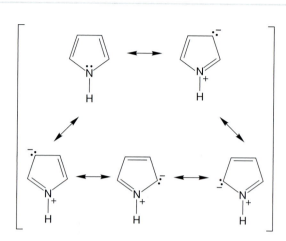

FIGURE 13.55 Resonance descriptions of pyrrole and pyridine.

1.8 D   2.2 D

Pyrrole   **Pyridine**

FIGURE 13.56 The dipole moments of pyrrole and pyridine are in different directions.

In pyridine, the electronegative nitrogen is the negative end of the dipole, but in pyrrole nitrogen is the positive end (Fig. 13.56).

SUMMARY

As we have just seen, both pyridine and pyrrole qualify as aromatic. However, we have also seen sharp differences. For example, the lone-pair electrons on nitrogen are in the σ system in pyridine, whereas in pyrrole they are in the π system, completing the aromatic sextet.

13.10b Acid and Base Properties. Pyridine can be protonated in acid, but pyridine is a weaker base than a simple cyclic secondary amine such as piperidine. The pK_a values of the conjugate acid ammonium ions show this clearly, as the pyridinium ion is a much stronger acid ($pK_a = 5.2$) than the ammonium ion formed from a simple secondary amine ($pK_a \sim 11$). Alternatively, we might make this argument in terms of pK_b values. The pK_b of pyridine is 8.8, whereas that for a simple secondary amine is about 3. So, the simple amine is a far stronger base than pyridine (Fig. 13.57).

FIGURE 13.57 Pyridine is a weaker base than the similar, but nonaromatic piperidine.

Pyridine
$pK_b = 8.8$

$pK_a = 5.2$

Piperidine
$pK_b \sim 3$

$pK_a \sim 11$

The reason for the decreased basicity of pyridine is that the lone-pair electrons of the sp^2 hybridized nitrogen of pyridine are held more tightly than those of the approximately sp^3 nitrogen of a simple secondary amine like piperidine, and therefore are less basic.

Pyrrole is even less basic than pyridine. One can estimate the pK_a of the conjugate acid at about −4! When pyrrole is protonated, aromaticity is lost as the aromatic sextet is disrupted (*not* true for pyridine). The lone pair of electrons is tied up in the new bond to hydrogen and no longer contributes to the completion of the aromatic sextet (Fig. 13.58)!

FIGURE 13.58 Pyrrole is a very weak base because protonation on nitrogen destroys aromaticity.

PROBLEM 13.17 Notice in Figure 13.58 that pyrrole is protonated on carbon, not nitrogen. Explain why.

For an organic compound, pyrrole is a rather strong Brønsted acid, $pK_a \sim 23$; but it is much weaker than cyclopentadiene, $pK_a = 15$. This difference may seem curious at first, because removal of a proton from pyrrole leaves a negative charge on a relatively electronegative nitrogen, whereas cyclopentadiene gives a carbon anion, generally a much less happy event (Fig. 13.59).

FIGURE 13.59 Pyrrole is a relatively weak acid compared to cyclopentadiene.

However, cyclopentadiene *becomes* aromatic when a proton is removed, and pyrrole is *already* aromatic. Removal of a proton from cyclopentadiene is made easier by the gain in aromaticity, which does not happen in pyrrole.

13.10c Pyridine as Nucleophile.
Like other amines, pyridine can be alkylated in an S_N2 reaction in which the lone-pair electrons on nitrogen displace a leaving group (Chapter 7, p. 334). Treatment of pyridine with a primary or secondary alkyl iodide leads to alkyl **pyridinium ions**. Pyridine gives **pyridine *N*-oxide** when treated with hydrogen peroxide (Fig. 13.60).

FIGURE 13.60 Pyridine can act as a nucleophile in the formation of pyridinium salts and pyridine *N*-oxides.

Methyl pyridinium iodide

Pyridine *N*-oxide (90%)

13.11 POLYNUCLEAR AROMATIC COMPOUNDS

13.11a All-Carbon Polynuclear Aromatic Compounds.
At this point, we can see benzene as a building block from which more complicated, polycyclic molecules can be constructed. For example, benzene and other aromatic rings can share edges to form fused, **polynuclear aromatic compounds**. The simplest of these is naphthalene, in which two benzene rings share an edge. In quinoline and isoquinoline, it is a benzene ring and a pyridine that share an edge (Fig. 13.61).

FIGURE 13.61 Three two-ring fused aromatic compounds.

Naphthalene   **Quinoline**   **Isoquinoline**

PROBLEM 13.18 Draw schematic orbital pictures of naphthalene and isoquinoline. Put in the π electrons as dots. It is not necessary to draw the carbon–hydrogen σ bonds.

PROBLEM 13.19 Can two benzene rings share a single carbon to form a spiro substituted compound?

Clearly, we can continue this building process. Three fused benzene rings produces either anthracene or phenanthrene (Fig. 13.62).

FIGURE 13.62 Aromatic hydrocarbons containing three fused benzene rings.

Anthracene   **Phenanthrene**

WORKED PROBLEM 13.20 Why is the following molecule not stable? *Hint*: Kekulé forms are often useful.

ANSWER The circles are misleading. Drawing the molecule in Kekulé form allows you to see that there is no way for each carbon to be hybridized *sp²* if each carbon has four valences.

These two rings are fine: Each carbon is hybridized *sp²*

But what about this ring, and this carbon?

PROBLEM 13.21 Draw all the fused polynuclear aromatic compounds made from four benzene rings. Be careful. This question is much harder than it seems.

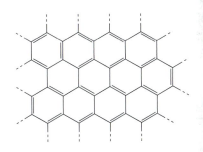

FIGURE 13.63 Graphite, an infinitely extended planar array of fused benzene rings.

The limit of the ring fusing business is reached in graphite, a polymer composed of fused benzene rings extended in what amounts to an infinite plane. Of course, the plane is not really infinite, but it is large enough so that the edges do not affect the physical or chemical properties of graphite. Graphite's lubricating properties are related to the ease of sliding the planar layers of graphite over each other (Fig. 13.63).

In the mid-1980s, R. F. Curl (b. 1933), H. W. Kroto (b. 1939), R. E. Smalley (b. 1943), and their co-workers were attempting to explain the striking abundance of certain fragments observed when a carbon rod was vaporized. In the face of much skepticism in the chemical community, they proposed that the major fragment, C₆₀, had the structure of a soccer ball of pure carbon. They turned out to be absolutely right, and named this molecule "buckminsterfullerene" after R. Buckminster Fuller (1895–1983), the inventor of the geodesic dome. This compound, which can now be made easily in multigram lots through the vaporization of carbon rods, is a new allotrope of carbon, and a testimony to the proposition that organic chemistry is hardly played out as a provider of exciting new molecules (Fig. 13.64). Curl, Kroto, and Smalley shared the 1996 Nobel prize in chemistry for this discovery.

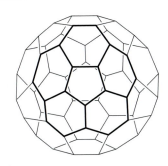

FIGURE 13.64
Buckminsterfullerene; C₆₀, a soccer ball shaped form of carbon. The double bonds (or circles) are left out for clarity.

PROBLEM 13.22 The ring-fusing process produces some interesting molecules along the way to graphite (see above). "Hexahelicene" and "twistoflex" are chiral, for example. Why?

Hexahelicene

Twistoflex

Benz[a]pyrene

Cholanthrene

Dibenz[a,h]anthracene

FIGURE 13.65 Some carcinogenic aromatic hydrocarbons.

Expanding this building block process to generate ever larger and varied polycyclic aromatic compounds may be fun, but there is a downside to the process. Many of the polycyclic aromatic compounds (and some of the simple ones as well) are carcinogenic. Some of the worst are shown in Figure 13.65. A prudent person avoids exposure to the known strong carcinogens, and minimizes contact with related aromatic compounds. It is less easy to avoid such compounds than it sounds. It is clearly simple to avoid bathing in benzene, and usually (but not always) possible to substitute other molecules when benzene is called for in a chemical recipe. However, one of the bad carcinogens, **benz[a]pyrene**, is a product of the internal combustion engine as well as combustion (cigarettes, charcoal-broiled meat) in general. The control of such molecules is not only physically difficult, but intersects with politics and economics as well. This situation creates serious problems of social policy because, beyond the questions of public health, there are fortunes to be made and elections to be won. Such matters seem often to complicate the purely scientific questions!

13.11b Related Heterocycles: Indole, Benzofuran, and Benzothiopene.

The benzo-fused counterparts of the simple heterocycles pyrrole, furan, and thiopene are **indole**, **benzofuran**, and **benzothiopene**. These compounds are related to the two-ring, all-carbon aromatic compound naphthalene (Fig. 13.66).

FIGURE 13.66 Naphthalene and three related heteroaromatic molecules.

Naphthalene   **Indole**   **Benzofuran**   **Benzothiophene**

They are important chiefly because of the large variety of natural products containing these ring systems. Some examples of natural products containing simple and complex five-membered heterocyclic rings are given in Figure 13.67.

Biotin
(a growth factor present in every cell)

Reserpine
(antihypertensive)

FIGURE 13.67 Two natural products containing five-membered heterocyclic rings.

13.12 INTRODUCTION TO THE CHEMISTRY OF BENZENE

We will take a quick look here at reactions. We will first introduce a general reaction that preserves the aromatic character of benzene. It is risky even to suggest that molecules have desires, but the imperative in benzene chemistry definitely is, Preserve the Aromatic Sextet! A less flamboyant way of saying this is to point out that thermodynamics will surely favor very stable molecules, and therefore benzene chemistry is likely to lead to more very low energy aromatic compounds. However, occasionally there are useful reactions in which the aromatic stabilization is lost, and we will see one of them soon in Section 13.12b.

13.12a Aromatic Substitution.
We have already mentioned that benzene does not react with Br_2 or HBr, as do almost all alkenes (p. 618) to give addition products. This apparent lack of reactivity results from the stabilization of benzene by aromaticity, which produces activation barriers that are too high for the addition reactions (Fig. 13.68).

FIGURE 13.68 Benzene does not react with Br_2 or HBr. Simple addition reactions do not occur.

If we use a deuterated acid we can see that there is a reaction taking place, at least in strong acid, but the product is merely the exchanged, deuterated benzene, not an addition product (Fig. 13.69).

FIGURE 13.69 The hydrogens of benzene can be exchanged for deuterium in deuterated acid.

What is happening? Why is the reaction diverted from ordinary addition? The answers to these questions will serve to summarize much of the chemistry of aromatic compounds, which will be encountered in detail in Chapter 14. In deuterio acid, addition of a deuteron gives a resonance-stabilized carbocation, but aromaticity is lost (Fig. 13.70).

FIGURE 13.70 Addition of a deuteron to benzene gives a resonance-stabilized, but nonaromatic, carbocation.

A pentadienyl cation (well stabilized but not aromatic)

WORKED PROBLEM 13.23 Be sure you see that the ion in Figure 13.70 is not aromatic. Why is aromaticity lost?

ANSWER

~sp^3 Hybridized carbon: No $2p$ orbital

This ion is cyclic, could be planar, but is definitely not aromatic, as the CHD group interrupts the connectivity of $2p$ orbitals around the ring; this molecule is not fully conjugated

In the normal addition reaction of alkenes or dienes, a nucleophile such as bromide would attack this carbocation to give the final product. In this case, we would anticipate the two products of 1,2- and 1,4-addition (Fig. 13.71).

FIGURE 13.71 Addition of Br⁻ to this ion would give two products.

(b) 1,4-addition

(a) 1,2-addition

But aromaticity, the more than 30 kcal/mol of delocalization energy, has been lost in each of these products. Their formation from benzene is generally endothermic. A far better reaction that regenerates the benzene ring is possible. Loss of a proton (or deuteron) regenerates the aromatic sextet of π electrons (Fig. 13.72).

**Cyclohexadienyl
cation**

FIGURE 13.72 Loss of a proton or deuteron from the carbocation regains aromaticity.

WORKED PROBLEM 13.24 Construct an Energy versus Reaction progress diagram for the conversion of benzene into deuteriobenzene.

ANSWER The first step is protonation (using a deuteron, not a proton) of the benzene ring to give the resonance-stabilized cyclohexadienyl cation. This process is surely very endothermic, as aromaticity is lost. The second half of this symmetrical reaction (except for the difference between the isotopes H and D) involves the deprotonation of the intermediate to produce monodeuterated benzene. If there is a sufficient source of deuterium, the reaction can continue until all H atoms are replaced with D atoms.

The second step is the reverse of the first, and simply is the exothermic deprotonation of the intermediate cation to regenerate the aromatic ring.

If a deuteron is lost, we see no reaction, as benzene is simply regenerated. If a proton is lost, we see the product of an exchange reaction in which deuteriobenzene is formed (Fig. 13.72). If a large excess of deuterio acid is used, all of the hydrogens can be exchanged to form hexadeuteriobenzene. We will elaborate on this reaction in Chapter 14.

13.12b Birch Reduction of Aromatic Compounds.

Now we come to a reaction in which aromaticity *is* lost, the Birch reduction. Alkynes can be treated with sodium in liquid ammonia to give, ultimately, trans alkenes (see Chapter 10, p. 483). Although catalytic hydrogenation of benzene is difficult (p. 626), treatment of benzene with sodium in liquid ammonia and a little alcohol does lead to a reduced product, in this case, 1,4-cyclohexadiene (Fig. 13.73). The reaction is named after the Australian chemist Arthur J. Birch (1915–1995).

Immediately note the important difference between this method of reducing the benzene ring and a typical catalytic reduction in which all carbon–carbon π bonds are reduced. In this case, the ring is not totally reduced; the reaction stops at the diene stage.

The mechanism of this **Birch reduction** is similar to that for dissolving metal reduction of alkynes (Chapter 10, p. 483), and starts in the same way, with a transfer of an electron from the metal to one of the antibonding orbitals of benzene. The product is a resonance-stabilized **radical anion** (Fig. 13.74).

**1,4-Cyclo-
hexadiene**

FIGURE 13.73 Birch reduction of aromatic compounds leads to 1,4-cyclohexadienes.

FIGURE 13.74 Transfer of an electron from sodium to benzene leads to a resonance-stabilized radical anion.

A resonance-stabilized radical anion

Protonation by alcohol gives a radical that can undergo a second electron transfer–protonation sequence to complete the reduction (Fig. 13.75).

FIGURE 13.75 The radical anion is protonated to give a resonance-stabilized cyclohexadienyl radical that goes on to produce the 1,4-cyclohexadiene through another reduction–protonation sequence (Et = CH₂CH₃).

A resonance-stabilized radical

A resonance-stabilized anion

Substituted benzenes can also be reduced. Electron-releasing groups such as alkyl or alkoxy always wind up on one of the double bonds, but electron-withdrawing groups such as acid (COOH) invariably appear on the methylene positions of the product dienes (Fig. 13.76).

CH$_3$O⟶⟶   1. Na/NH$_3$, EtOH   2. H$_2$O/H$_3$O$^+$ ⟶   CH$_3$O⟶

COOH   1. Na/NH$_3$, EtOH   2. H$_2$O/H$_3$O$^+$ ⟶   COOH

FIGURE 13.76 In the Birch reduction of substituted benzenes, electron-releasing groups appear on one of the double bonds, but electron-withdrawing groups are located at the methylene position (Et = CH$_2$CH$_3$).

13.13   THE BENZYL GROUP AND ITS REACTIVITY: ACTIVATION OF THE ADJACENT (BENZYL) POSITION BY BENZENE RINGS

Thus far we have been looking at processes directly involving the benzene ring itself. Here we sketch out some reactions in which benzene plays a major role in the progress of the reaction, but in which the action takes place more on the outskirts of the ring. The carbon adjacent to a benzene ring is called the benzylic carbon, or the benzylic position. The **benzyl group** is Ph—CH$_2$. In many ways, the reactivity of the benzylic position resembles that of the allylic position. Indeed, an orbital at the benzylic position is stabilized by overlap with adjacent orbitals in much the same way as is an orbital at the allyl position (Fig. 13.77).

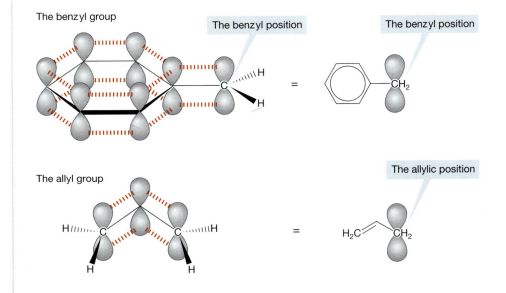

The benzyl group

The benzyl position

The benzyl position

The allyl group

The allylic position

FIGURE 13.77 The position adjacent to a benzene ring is quite analogous to the allylic position.

13.13a Substitution Reactions of Benzyl Compounds.
Benzyl compounds are unusually active in both the S_N1 and S_N2 reactions. Clearly, the benzene ring helps enormously to stabilize the carbocation formed in the rate-determining ionization of an S_N1 reaction. For example, benzyl chloride reacts 145 times faster than isopropyl chloride in a typical S_N1 process (Fig. 13.78).

FIGURE 13.78 The resonance stabilization of the benzyl cation makes its formation relatively easy. Accordingly, S_N1 reactions of benzyl compounds containing good leaving groups are quite fast.

The reason is simple: The benzene ring helps to stabilize the positive charge on the intermediate carbocation and, of course, in the transition state leading to it. In the benzyl cation itself, the positive charge is borne by 4 carbons, in the benzhydryl cation (Ph_2C^+H)—the **benzhydryl group** is Ph_2CH—by 7 carbons, and in the trityl cation (Ph_3C^+) by 10 carbons (Fig. 13.79).

PROBLEM 13.25 Why is triptycenyl chloride not nearly as reactive in the S_N1 reaction as trityl chloride?

Trityl chloride   **Triptycenyl chloride**

The benzyl cation – 4 resonance forms

A composite picture in which the positions sharing the positive charge are shown as (+)

In this composite representation of the benzyhydryl cation, the positions sharing the positive charge are shown as (+)

The benzyhydryl cation—7 resonance forms

In this composite representation of the trityl cation, the positions sharing the positive charge are shown as (+)

The trityl cation—10 resonance forms

FIGURE 13.79 Resonance stabilization of the benzyhydryl and trityl cations.

Benzyl halides are also especially reactive in the S_N2 reaction. The reason is the same as that for the enhanced reactivity of allyl chloride (Chapter 12, p. 585). The transition state for S_N2 displacement is delocalized, and therefore especially stable. If the transition state for a reaction is at relatively low energy, the activation energy for the reaction will also be relatively low and the reaction itself will be fast (Fig. 13.80).

FIGURE 13.80 The transition state for S_N2 displacement at the benzyl position is stabilized through orbital overlap with the ring.

Transition state for S_N2 displacement of an allyl halide

Transition state for S_N2 displacement of a benzyl halide

TABLE 13.2
Some Relative Rates of S_N2 Displacements in the Reaction of R—I with Iodide Ion at 50 °C in Ethyl Alcohol Solvent

R group	Relative Rate
Ethyl	1.0
Propyl	0.6
Butyl	0.4
Isopropyl	30.04
Allyl	33
Benzyl	78

Table 13.2 compares the rates of reaction of benzyl iodide with other alkyl iodides in an S_N2 reaction.

13.13b Radical Reactions at the Benzyl Position. A benzene ring will also stabilize an adjacent half-filled orbital (a free radical) through resonance (Fig. 13.81).

FIGURE 13.81 The resonance forms for the benzyl radical.

In Chapter 11 (p. 533), we discussed radical reactivity at the allylic position, and very little need be added to get a description of benzylic reactivity. Toluene, for example, can be brominated by *N*-bromosuccinimide (NBS) to give benzyl bromide (Fig. 13.82).

PROBLEM 13.26 Write a mechanism for this reaction. *Hint*: See Chapter 11 (p. 536) for a description of the reaction of NBS with allyl compounds.

NBS = (structure shown)

Toluene → Benzyl bromide

CCl₄, hν

FIGURE 13.82 The radical bromination of toluene.

FIGURE 13.83 The thermal or photochemical decomposition of bromine in toluene leads to benzyl bromide through radical bromination at the benzyl position.

PROBLEM 13.27 Ethylbenzene yields only the product of bromination at the α-position (the position adjacent to the ring) when treated with NBS in CCl₄. Explain.

As your mechanism for Problem 13.27 would indicate, abstraction of hydrogen is especially easy at the α-position of ethylbenzene. This abstraction occurs easily because of the resonance stabilization of the benzyl radical (Fig. 13.81). Because bromine will not attack the benzene ring without a catalyst, benzyl compounds can often be brominated by simply heating or photolyzing bromine in their presence (Fig. 13.83).

13.13c Oxidation at the Benzyl Position. Both free radicals and the special activity of the benzyl position play a role in the reaction of alkylbenzenes with KMnO₄ or H₂Cr₂O₇ to give **benzoic acids**. Side chains are "chewed down" to carboxylic acids no matter what their length. The details are vague, but the key observation is that tertiary side chains are untouched by oxidizing agents. There must be at least one hydrogen in the benzylic position for the reaction to succeed (Fig. 13.84).

Benzylic oxidation

FIGURE 13.84 Oxidizing agents convert most alkyl side chains on benzene rings into the acid group, COOH. However, for the oxidation to succeed there must be at least one benzyl hydrogen.

(Reaction scheme showing various alkylbenzenes converting to **Benzoic acid** via KMnO₄, with C(CH₃)₃ blocked)

SUMMARY

The chemistry of benzenes is summed up by the imperative: "Preserve the circle!" In plain English, that statement means that there is a strong thermodynamic driving force to preserve the delocalization energy—more than 30 kcal/mol—that the circle represents. In this chapter, the classic aromatic substitution reaction in which an electrophile replaces (substitutes for) one of the hydrogens on the ring is introduced. In Chapter 14, this prototypal reaction will be expanded in many ways.

Other aromatic chemistry includes a few reactions in which aromaticity is lost (the Birch reduction is one example), and the special influence of the aromatic ring upon the adjacent, "benzyl" position.

13.14   SPECIAL TOPIC: THE BIO-DOWNSIDE, THE MECHANISM OF CARCINOGENESIS BY POLYCYCLIC AROMATIC COMPOUNDS

We mentioned earlier that some polycyclic aromatic hydrocarbons were strongly carcinogenic. Although all the details of this carcinogenesis are not worked out, the broad outlines are known for some molecules. Deoxyribonucleic acid (DNA) and ribonucleic acid (RNA) are huge biomolecules that contain and help transcribe genetic information (Chapter 23). Both DNA and RNA are phosphate-linked sequences of nucleotides, which are sugar molecules attached to heterocyclic bases. These molecules bear large numbers of nucleophilic groups that can be alkylated by other molecules in what seem to be simple S_N2 reactions (Fig. 13.85).

The structures of the four bases attached to sugars

Adenine   Guanine   Cytosine   Thymine

FIGURE 13.85 The DNA polymer is composed of phosphate-linked sugar-base units. Two strands of DNA are held together by hydrogen bonding between pairs of bases. Alkylation of the bases changes the size and shape of one of the units. This change can interfere with hydrogen bonding and base pairing. In turn, this can lead to mutations and cancer.

S_N2 alkylation

Alkylation changes the size and shape of the DNA molecule and makes mismatched base pairing through hydrogen bonding more likely (see Chapter 23 for more details of base pairing). This mismatch can lead to errors in replication and to mutations. Most mutations are harmless to the whole organism, as they lead to the death of only the individual cell. On rare occasions, however, a mutation can lead to uncontrolled cell division and cancer.

The hydrocarbons themselves are only indirectly responsible for the chemical processes that lead to cancer. Aromatic hydrocarbons are not very reactive and there is no obvious way for them to interact chemically with the nucleophilic bases in DNA or RNA (Fig. 13.86).

Guanine   **Benz[a]pyrene**

FIGURE 13.86 Aromatic hydrocarbons do not react with nucleophiles, and there is no way for a molecule like benz[a]pyrene to alkylate one of the bases directly.

However, there are enzymes that can modify benzene rings. Nonpolar molecules such as aromatic hydrocarbons accumulate in fat cells and methods have been evolved to purge our bodies of such molecules. These methods generally involve making the hydrocarbons more water soluble by adding polar groups. For example, benz[a]pyrene is epoxidized by the enzyme P-450-mono-oxygenase and the product epoxide is then opened enzymatically to give a trans diol (Fig. 13.87).

Benz[a]pyrene

FIGURE 13.87 Benz[a]pyrene is enzymatically epoxidized and ring opened to give a polar trans 1,2-diol.

However, a second enzymatic epoxidation, again by P-450-mono-oxygenase, gives a molecule that reacts with guanine, one of the DNA bases, to give a molecule of DNA that is severely encumbered by the large, alkylated guanine (Fig. 13.88). This modified base has difficulty in maintaining normal hydrogen bonding with another complementary base, and this mismatch can lead to mutations.

This alkylation is not a new reaction. You already know of the opening of epoxides in both base and acid (Chapter 10, p. 455), and this process is nothing more than a complicated version of that simple reaction.

FIGURE 13.88 A second epoxide is formed. This epoxide is opened by guanine to give a highly encumbered molecule, **A**, that has problems maintaining proper hydrogen bonding.

13.15 SUMMARY

New Concepts

The material in this chapter is composed almost entirely of concepts. There are few new reactions or synthetic procedures. We concentrate here on the special stability of some planar, cyclic, and fully conjugated polyenes. The special stability called *aromaticity* is encountered when the polyene has a molecular orbital system in which all bonding molecular orbitals are completely filled and there are no electrons in antibonding or nonbonding molecular orbitals.

These especially stable molecular orbital systems will be found in planar, cyclic, fully conjugated polyenes that contain $4n + 2\,\pi$ electrons (Hückel's rule).

Heats of hydrogenation or heats of formation can be used to calculate the magnitude of the stabilization. For benzene, the delocalization energy or resonance energy amounts to more than 30 kcal/mol.

It's vital to keep clear the difference between resonance forms and molecules in equilibrium. In this chapter, that difference is exemplified by Dewar benzene, bicyclo[2.2.0]hexa-2,5-diene, and the Dewar resonance forms contributing slightly to the structure of benzene. These are shown in Problems 13.3–13.5. As always, resonance forms are related only by the movement of electrons, whereas equilibrating molecules have different geometries—different arrangements of atoms in space. Benzene and Dewar benzene are different molecules; the Kekulé and Dewar resonance forms contribute to the real structure of benzene (Fig. 13.89).

FIGURE 13.89 Equilibrium versus resonance.

Key Terms

annulene (p. 642)
arenes (p. 646)
aromatic character (p. 628)
aromaticity (p. 628)
benz[*a*]pyrene (p. 652)
benzene (p. 618)
benzhydryl group (p. 658)
benzofuran (p. 652)
benzoic acids (p. 661)
benzothiophene (p. 652)
benzyl (p. 643)
benzyl group (p. 657)
Birch reduction (p. 656)

cycloheptatrienylium ion (p. 634)
cyclopentadienyl anion (p. 636)
delocalization energy (p. 627)
Dewar forms (p. 622)
Frost circle (p. 630)
fully conjugated (p. 628)
furan (p. 647)
heteroaromatic compounds (p. 646)
heterobenzenes (p. 638)
Hückel's rule (p. 629)
indole (p. 652)
Kekulé forms (p. 621)
meta (p. 644)

ortho (p. 644)
para (p. 644)
phenyl (p. 643)
polynuclear aromatic compounds (p. 650)
pyridine (p. 646)
pyridine *N*-oxide (p. 650)
pyridinium ions (p. 650)
pyrrole (p. 647)
radical anion (p. 656)
resonance energy (p. 627)
thiophene (p. 647)
trityl cation (p. 634)
tropylium ion (p. 634)

Reactions, Mechanisms, and Tools

In this chapter, the first reaction mechanism encountered is the important and general electrophilic substitution of benzene. A host of substitution reactions will be studied in Chapter 14 and are exemplified here by deuterium exchange. The aromatic ring is destroyed by an endothermic addition of D^+, but reconstituted by an exothermic loss of H^+ (Fig. 13.72).

Synthesis

There is rather little in the way of new synthetic procedures in this chapter. You might remember the formation of the tropylium ion by hydride abstraction and the Birch reduction of benzenes to 1,4-cyclohexadienes (Fig. 13.90). You also have a method of synthesizing deuteriobenzene through the acid-catalyzed exchange reaction of benzene. This reaction will serve as the prototype of many similar substitution reactions to be found in Chapter 14.

1. Benzyl compounds

S_N1 and S_N2 reactions

Radical bromination at the benzyl position

Oxidation of benzyl positions bearing at least one hydrogen

2. Cyclohexadienes

1,4-Cyclohexadiene

Birch reduction of benzene

3. Deuteriobenzenes

Acid-catalyzed exchange of hydrogens on a benzene ring; all hydrogens can eventually be exchanged

4. Tropylium ions

Tropylium fluoborate

Hydride transfer from cycloheptatriene to the trityl cation ($^+CPh_3$)

Common Errors

Certainly the most common error is to overgeneralize Hückel's rule. We often forget to check each potentially aromatic molecule to be certain it satisfies all the criteria. The molecule must be cyclic or there cannot be the fully connected set of p orbitals necessary (Fig. 13.18). The molecule must be planar so that the p orbitals can overlap effectively (Fig. 13.16). There may be no positions in the ring at which p–p overlap is interrupted (Fig. 13.15). Finally, there must be the proper number of $(4n + 2)$ π electrons. Be certain you are counting π electrons, and not including in your count electrons

not in the π system. The classic example is pyridine, which contains two paired electrons in an sp^2 orbital that are often mistakenly counted as π electrons (Fig. 13.51). If, and only if, *all* of these criteria are satisfied do we find aromaticity.

Do not misapply the Frost circle device. The most common error is to forget to inscribe the circle vertex down. This requirement is hard to remember because at this level of discussion this point is arbitrary—there is no easy way to figure it out if you don't remember the proper convention.

Do not confuse absolute and relative stability. For example, the cyclopentadienide anion is aromatic, and therefore exceptionally stable, *but only within the family of anions*. The negatively charged cyclopentadienide anion is not as stable as benzene.

As always, you must be clear as to the difference between resonance and equilibrium. Be certain that you know the difference between the two Kekulé resonance forms for benzene and the hypothetical molecule 1,3,5-cyclohexatriene.

13.16 ADDITIONAL PROBLEMS

PROBLEM 13.28   Give names for the following compounds:

(a)

(b)

(c)

(d)

(e)

(f)

PROBLEM 13.29   Write structures for the following compounds:
(a) 2,4,6-Trinitrotoluene (TNT)
(b) 4-Bromotoluene
(c) *p*-Bromotoluene
(d) *m*-Dihydroxybenzene
(e) 3-Hydroxyphenol
(f) *m*-Hydroxyphenol
(g) 1,2,4,5-Tetramethylbenzene (durene)
(h) *o*-Deuteriobenzenesulfonic acid (Note the "i" in "deuterio". It is not "deutero".)
(i) *p*-Aminoaniline
(j) *m*-Chlorobenzoic acid

PROBLEM 13.30   Write arrow formalisms for the conversions of Dewar benzene, Ladenburg benzene (prismane), benzvalene, and 3,3′-bicyclopropenyl (Fig. 13.4) into benzene. Some of these may be quite challenging tasks.

PROBLEM 13.31   Which of the following heterocyclic compounds might be aromatic? Explain. It will help if you start by drawing good Lewis structures for all the compounds.

PROBLEM 13.32   Which of the following hydrocarbons might be aromatic? Explain.

PROBLEM 13.33   Which of the following ions might be aromatic? Explain.

PROBLEM 13.34   Design an aromatic heterocyclic compound that
(a) Contains one boron atom
(b) Contains one boron and one oxygen atom in a six-membered ring
(c) Contains one oxygen and one nitrogen atom in a five-membered ring
(d) Contains two nitrogen atoms in a five-membered ring
(e) Contains three boron and three nitrogen atoms in a six-membered ring

PROBLEM 13.35   Try to construct an aromatic heterocycle in which there is an oxygen and a nitrogen in a neutral six-membered ring.

PROBLEM 13.36   Carefully explain why azirines of structure **1** have never been isolated, whereas the isomeric azirines of structure **2** are well known.

1   **2**

PROBLEM 13.37   The dipole moment of THF is 1.7 D, but that of furan is only 0.7 D. Explain.

PROBLEM 13.38   The heat of hydrogenation of 1,3,5,7-cyclooctatetraene (COT) is about 101 kcal/mol. The heat of hydrogenation of cyclooctene is about 23 kcal/mol. Is COT aromatic? Explain.

PROBLEM 13.39   As early as 1891 the German chemist G. Merling showed that the bromination of 1,3,5-cycloheptatriene leads to a liquid dibromide. When the dibromide is heated, HBr is evolved and a yellow solid of the formula C_7H_7Br (mp 203 °C) can be isolated. This compound is soluble in water, but cannot be reisolated from water solution. Instead, ditropyl ether is obtained. Merling could not explain what was happening, but perhaps you can.

Ditropyl ether

PROBLEM 13.40   Most diazo compounds are exceedingly, and sometimes disastrously unstable. They are notorious explosives. By contrast, the brilliant, iridescent orange compound diazocyclopentadiene is quite stable. Explain.

Diazocyclopentadiene

PROBLEM 13.41   Construct the π molecular orbitals for benzene (p. 624) from the π molecular orbitals of allyl. *Remember*: Only combinations of orbitals of comparable energy need be considered.

PROBLEM 13.42   Vogel and Roth's synthesis of the bridged cyclodecapentaene (**1**) is outlined below. Supply the missing reagents (a, b, c, and d) and an arrow formalism for the last step.

1

PROBLEM 13.43   Pyridine and imidazole are modest Brønsted bases at nitrogen, whereas pyrrole is not.

Pyridine   Pyrrole   Imidazole

In fact, pyrrole is protonated only in strong acid, and protonation occurs at C(2), not on nitrogen. First, explain why pyridine and imidazole are basic and pyrrole is not. Then, explain why eventual protonation of pyrrole is at carbon, not nitrogen. **Caution!** You will have to write full Lewis structures to answer this question.

PROBLEM 13.44   Explain why the 1,2-bond of anthracene is shorter than the 2,3-bond.

Anthracene

PROBLEM 13.45   The nitronium ion ($^+NO_2$) is known and can react with simple benzenes to give nitrobenzenes. Write an arrow formalism mechanism for this reaction.

PROBLEM 13.46   The molecule shown below undergoes reaction with cupric nitrate in acetic anhydride (a source of $^+NO_2$) to give a high yield of the substituted product shown. Write a mechanism for this reaction.

PROBLEM 13.47   The compound shown below was synthesized by Professor I. M. Dimm who was astonished to find that it was exceedingly stable. Prof. Dimm had expected world renown for making what he thought would be a most unstable compound. Dimm reasoned that with six double bonds and a triple bond in the molecule, there would be 16 π electrons and thus, no aromatic character ($4n$, $n = 4$, *not* $4n + 2$). Where was his reasoning wrong?

PROBLEM 13.48   Provide a mechanism for the following reaction sequence:

PROBLEM 13.49 Even though naphthalene and anthracene are polynuclear aromatic compounds, they undergo the Diels–Alder reaction. For example, anthracene readily reacts with maleic anhydride. However, the reaction of naphthalene with maleic anhydride proceeds efficiently only under high pressure.

$$\text{(anthracene)} + \text{(maleic anhydride)} \xrightarrow[\text{140 °C, 10 min}]{\text{xylene}} \;?$$

$$\text{(naphthalene)} + \text{(maleic anhydride)} \xrightarrow[\text{9.6 kbar, 17 h}]{\text{100 °C}} \;?$$

(a) Draw the structures of the Diels–Alder adducts formed in these reactions.

(b) Explain why anthracene is so much more reactive than naphthalene in the Diels–Alder reaction. *Hint*: The calculated resonance energies of naphthalene and anthracene are 61 and 84 kcal/mol, respectively.

PROBLEM 13.50 A certain alkylbenzene was known to have one of the isomeric structures shown below.

(a) CH_3 — ring — $C(CH_3)_3$

(b) CH_2CH_3 — ring — $CH(CH_3)_2$

(c) $H_3C\!-\!\overset{CH_3}{\underset{}{C}}\!-\!CH_2CH_3$ with phenyl

Explain how the structure of the unknown compound could be determined by oxidation with $KMnO_4$.

Substitution Reactions of Aromatic Compounds

14

Even electrons, supposedly the paragons of unpredictability, are tame and obsequious little creatures that rush around at the speed of light, going precisely where they are supposed to go. They make faint whistling sounds that when apprehended in varying combinations are as pleasant as the wind flying through a forest, and they do exactly as they are told. Of this, one can be certain.

—Mark Helprin*
Winter's Tale

14.1 PREVIEW

In Chapters 12 and 13, we examined the structural consequences of conjugation, the overlap of $2p$ orbitals. This story culminates in the enormously stable aromatic molecules, with the parent compound, benzene, being the prototype for all these "Hückel" compounds. In this chapter, we will focus mostly on substitution reactions of aromatic compounds, an idea already introduced in Section 13.12a (p. 653). The chemistry of aromatic compounds is dominated by a tendency to retain the stable aromatic sextet of electrons. It takes a great deal of energy to destroy this arrangement, and it is very easily regained.

Because aromatic compounds are so stable, so low in energy, they are often surrounded by especially high energy barriers, and the activation energies for their reactions are usually formidably large. The transformation of an aromatic compound into a nonaromatic compound is most often highly endothermic (Fig. 14.1), so the transition state will lie close to the relatively high energy nonaromatic product (recall the Hammond postulate, Chapter 8, p. 387). By contrast, the conversion of a nonaromatic com-

*Mark Helprin (b. 1947) is an American author. His wonderfully imaginative novels are *Winter's Tale, A Soldier of the Great War,* and *Memoir from Antproof Case.*

FIGURE 14.1 A typical Energy versus Reaction progress diagram for the conversion of an aromatic compound into a nonaromatic compound. Aromaticity is lost in this reaction, and there is a high activation energy.

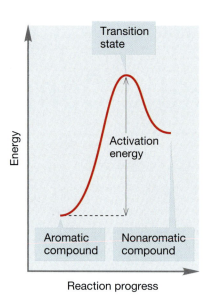

pound into an aromatic molecule is likely to be very exothermic (Fig. 14.2). In such a case, the transition state will resemble the starting material. So, if we again recall the Hammond postulate (and why it works), we will expect that only small activation barriers will need to be surmounted in such reactions. A complete energy diagram for a typical conversion of one aromatic compound into another is shown in Figure 14.3.

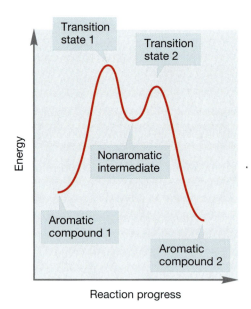

FIGURE 14.2 A typical Energy versus Reaction progress diagram for the conversion of a nonaromatic compound into an aromatic compound. Aromaticity is regained in this reaction, and there is a correspondingly low activation energy.

FIGURE 14.3 A typical Energy versus Reaction progress diagram for the conversion of one aromatic compound into another.

In this chapter, we will see many examples of a general reaction of aromatic compounds in which E^+, an **electrophile**, a "lover of electrons" (in other words, a Lewis acid), substitutes for one of the hydrogens on a benzene ring. Note that the aromatic sextet, represented by the circle in the hexagon, is retained in the overall reaction (Fig. 14.4).

An electrophile = Lewis acid

FIGURE 14.4 A generic aromatic substitution reaction.

Later in this chapter, we take up the question of how an initial substitution on the ring affects further reaction. Some groups make a second substitution reaction easier, some make it harder. We begin with reactions that are rather unusual for benzene rings, simple additions in which aromaticity is lost.

ESSENTIAL SKILLS

1. The great thermodynamic and kinetic stability of benzenes results in reactions that ultimately preserve the aromatic sextet of π electrons. The classic aromatic substitution reaction, introduced in Chapter 13, is greatly elaborated here. Knowing the mechanism of electrophilic aromatic substitution and manipulating the details in order to synthesize polysubstituted compounds is the central theme of this chapter.

2. Some substituents increase the rate of a subsequent substitution and generally direct it to the ortho and para positions. Other substituents slow the rate and direct further substitution to the meta position. Understanding the dependence of position and rate on the structure of the initial substituent is essential.

IMPORTANT DETAILS

1. The resonance-stabilized cyclohexadienyl cation is an intermediate in the central reaction of this chapter. Be sure you understand why three—and only three—of the carbons in the ring share the positive charge.

2. The nitrobenzene–aminobenzene (aniline) pair is one important key to doing aromatic syntheses. Be sure you exit this chapter being able to interconvert these two compounds and to use both nitrobenzene and aniline in synthesis.

3. Making bonds between carbon and the aromatic ring generally requires either Friedel–Crafts alkylation or Friedel–Crafts acylation. There are subtle differences between the two reactions, and you'll need to know them to take advantage of Friedel–Crafts reactions in synthesis.

14.2   ADDITION REACTIONS OF BENZENES TO GIVE NONAROMATIC COMPOUNDS: HYDROGENATION

This section will be short simply because there are so few of these reactions! If one were to try to predict the reactions of benzene from first principles, one would probably start by examining the reactions of other π systems and applying them to aromatic compounds. For example, we know that hydrogenation of cyclohexene is exothermic by 28.6 kcal/mol (Section 13.5a, p. 626). It is no great surprise to find that absorption of 3 mol of hydrogen by benzene to give cyclohexane is exothermic as well; indeed we examined this reaction in some detail in Chapter 13. Although the reaction is still very exothermic, it is far less exothermic than a guess based on the value of −28.6 kcal/mol for a simple alkene would suggest. The difference between the estimated value and the measured value was taken as a measure of the delocalization energy of benzene, approximately 33 kcal/mol (Fig. 14.5).

FIGURE 14.5 The determination of the delocalization, or resonance energy, of benzene.

Often, conditions under which alkenes are swiftly transformed into alkanes are quite without effect on benzene and other simple aromatic compounds. Indeed, benzene can be used as solvent in many hydrogenation reactions. We can eventually make the reaction occur: high temperature and pressure will help, as does the use of especially active catalysts, such as rhodium or ruthenium (Fig. 14.6).

FIGURE 14.6 Benzene does not hydrogenate under the same conditions as alkenes.

The reaction is highly exothermic, by 49.3 kcal/mol in the case of the parent benzene. The combination of benzene and three hydrogen molecules does lie above cyclohexane, the product of hydrogenation, by a substantial amount, but benzene is protected by high activation barriers. The heat, pressure, and special catalysts are needed to help overcome these barriers. Benzene rests in a valley, separated from other valleys—even lower energy valleys—by especially high mountains.

In the language of chemistry, we say that benzene is both thermodynamically and kinetically very stable. The valley containing benzene is low (thermodynamics), and the mountains surrounding it (the activation barriers, kinetics) are very high. Benzene is even protected by these barriers from undergoing reactions that are, ultimately, exothermic. The example of benzene plus hydrogen nicely illustrates this point, as the combination of these molecules lies higher in energy than the product of hydrogenation, cyclohexane (Fig. 14.7).

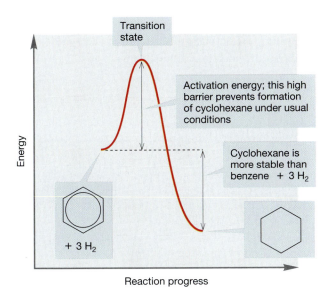

FIGURE 14.7 The activation barrier for cyclohexane formation is high, and makes this reaction difficult under normal conditions.

PROBLEM 14.1 Explain why the only product of hydrogenation of benzene is cyclohexane. Cyclohexadienes and cyclohexene must be formed first. Why can they never be isolated, even if the reaction is interrupted before completion?

14.3   DIELS–ALDER REACTIONS

Benzene and other aromatic compounds contain a conjugated π system. Such molecules are known to undergo the Diels–Alder reaction (Section 12.12, p. 587). Yet, it is extremely difficult to induce the Diels–Alder reaction for simple benzenes, and yields are often poor. Harsh conditions (temperature as high as 200 °C, for example) and very reactive dienophiles such as hexafluoro-2-butyne are necessary (Fig. 14.8).

FIGURE 14.8 Benzenes will undergo the Diels–Alder reaction, but only with reactive dienophiles under forcing conditions.

The reason for the lack of reactivity is simple: Benzene is too stable (lies in a deep energy well) and the activation energies for this reaction are very high.

To see if all this makes sense, let's try a thought experiment. Suppose we were set the task of finding a way to make the Diels–Alder reaction possible for benzenes. We might first try to use as reactive (high energy) a dienophile as possible, as we just saw in Figure 14.8. But there is another way. We might try instead to destabilize the benzene so that surmounting the activation barriers for Diels–Alder reaction becomes easier. Figure 14.9 shows this idea in graphic form.

FIGURE 14.9 If benzene is destabilized, the activation barrier for reaction should be reduced.

There are several nice practical examples of this idea. Imagine a benzene ring bridged by a chain of methylene groups between the 1- and 4-positions (Fig. 14.10). Such a molecule is called an **[*n*]paracyclophane**, where *n* gives the number of methylene groups in the bridge. As long as there are enough methylene groups (at least 7) there is no problem, and the molecule behaves like a simple benzene. But as [*n*] grows small (molecules with *n* = 5 and 6 are known and the *n* = 4 isomer has been detected), the ring becomes warped, and must assume a boat shape (Fig. 14.10), which severely destabilizes the molecule. In this case, the models tell all. Make a model of a paracyclophane and reduce the number of bridging methylenes one by one. The benzene ring will increasingly deform as this happens. The source of the destabilization should also become apparent. Try Problem 14.2.

[*n*]Paracyclophanes

For large *n*, the ring can be flat

A small *n* forces the ring into a boat form

PROBLEM 14.2 Why are these boat-shaped benzenes less stable than flat ones?

FIGURE 14.10 1,4-Bridged benzenes: paracyclophanes.

As anticipated, these destabilized benzenes are much more reactive in the Diels–Alder reaction than their undistorted relatives (Fig. 14.11).

[7]Paracyclophane

160 °C
F₃C—C≡C—CF₃

(CH₂)₇
(>52%)

FIGURE 14.11 [7]Paracyclophane undergoes the Diels–Alder reaction.

PROBLEM 14.3 Use an arrow formalism to show how [7]paracyclophane forms the product shown in Figure 14.11 through a Diels–Alder reaction.

We really needn't go to the trouble of making those cyclophanes (unless one is interested in what happens to benzene's π system, as it deforms, as I am), because there are naturally occurring molecules that are less aromatic than benzene and thus should be more reactive. Recall how we reached a quantitative determination of the resonance energy for benzene through measurements of the heats of hydrogenation (or combustion) for the model compounds cyclohexene and cyclohexadiene. We then used these data to estimate the heat of hydrogenation of the *hypothetical* molecule cyclohexatriene. Finally, we compared our estimate with the experimental value determined by hydrogenating the *real* molecule benzene (Section 13.5a, p. 626). Table 14.1 gives similarly determined values for a variety of heteroaromatic compounds.

Furan will undergo addition reactions in which aromaticity is not regained more easily than benzene. There is less delocalization energy in furan than benzene (Table 14.1) and preservation of aromaticity is less important. Diels–Alder reactions are common for furans, which make excellent trapping agents for reactive π systems (Fig. 14.12).

TABLE 14.1
Resonance Energies of Some Heteroaromatic Compounds and Benzene

Compound	Resonance Energy (kcal/mol)
Benzene	33
Thiophene	29
Pyridine	23
Pyrrole	21
Furan	16

FIGURE 14.12 Two Diels–Alder reactions of furans.

Pyrroles are also 1,3-dienes and one might suspect that the Diels–Alder reaction would occur as with furan. This suspicion is correct, but the aromaticity of pyrrole introduces complications. Such reactions are often slow, and aromaticity is sometimes regained after loss through a retro Diels–Alder reaction. With those hints, you should be able to tackle Problem 14.4.

WORKED PROBLEM 14.4 Write a mechanism for the following reaction:

ANSWER All those hints in the text surely point directly at a Diels–Alder reaction. In this case, that leads to intermediate **A**.

(continued)

Now, with luck, you can see the symmetry of the azanorbornadiene, **A**. As Diels–Alder reactions are reversible, there are two possible routes for **A** to take. In one, the reaction simply reverses to the starting materials. In the other, a reverse Diels–Alder reaction proceeds in the other direction, and acetylene is lost to give the target molecule. One imagines that it is the irreversible loss of the gas acetylene that drives the reaction toward the product shown. The arrows show the product-forming pathway.

SUMMARY

It is not easy to disrupt the aromatic sextet, and few reactions of benzene do so permanently. It takes a very reactive reagent to overcome the high kinetic barriers surrounding the benzene ring and usually there is a path available that regains the lost aromaticity. In this section, you saw two reactions that do "dearomatize" benzene; in the next sections you will see many that don't.

14.4 SUBSTITUTION REACTIONS OF AROMATIC COMPOUNDS

14.4a Why Benzene Doesn't Undergo Simple Electrophilic Addition Reactions.
We are first going to do some thought experiments. We are going to ask some "What if" questions. This process will lead us to an understanding of why benzene reacts so differently from simple alkenes. In particular, we are going to see why benzene does not undergo simple addition of HX as alkenes do. Then we will go on to see why benzene reacts as it does. What might we anticipate if benzene reacted as the simple alkenes do to give addition products with X_2 and HX reagents? The answer to this hypothetical question is more complicated than it appears at first. Let's work through a prediction for the addition of HBr to benzene, a reaction that in fact does not occur. By analogy to alkene reactions, the first step should be the addition of a proton to the π system of benzene to give an intermediate cyclohexadienyl carbocation. Note that aromaticity is lost (Fig. 14.13).

A nonaromatic
cyclohexadienyl cation

FIGURE 14.13 Protonation of benzene to give a cyclohexadienyl cation.

Simple addition of the bromide ion to the cation, again by analogy to the addition of HBr to an alkene, would generate a cyclohexadienyl bromide (Fig. 14.14).

FIGURE 14.14 Addition of bromide would give a cyclohexadienyl bromide.

But we have forgotten to look closely at the structure of the carbocation formed by the addition of a proton. Just as is the case with simple dienes, the positive charge is not localized on one atom, but instead is shared by two other carbons. The resonance picture of Figure 14.15 shows a better representation of the cyclohexadienyl cation.

Summary structures

FIGURE 14.15 The cyclohexadienyl cation is resonance stabilized. The positive charge is shared by three carbons (dots).

CONVENTION ALERT

Look carefully again at the shorthand used to indicate the delocalized species. The dashed loop of Figure 14.15 indicates delocalization quite clearly, but it does not do a good job of showing which atoms share the positive charge. To get this information, you must draw out the resonance forms, or use a single resonance form with the other positive positions labeled with (+). Keep in mind that the dashed loop does not mean that *all* the atoms surrounding the loop are partially positive. In Figure 14.15, the positions sharing the positive charge are shown with dots.

So, addition of bromide would take place at the three carbons sharing the positive charge to produce two different products (Fig. 14.16).

FIGURE 14.16 Capture of the cyclohexadienyl cation at the two different positions bearing positive charge must give two cyclohexadienyl bromides. These additions are shown using both of the summary structures of Figure 14.15.

Yet, neither of these compounds is the product of the reaction! In fact, there is no apparent reaction at all, and benzene is recovered unchanged from the attempted addition reaction. One searches in vain for the predicted cyclohexadienyl bromides.

SUMMARY

Benzene fails to react with most HX reagents under conditions that easily induce addition to alkenes and dienes. The X_2 reagents, such as bromine and chlorine, are similarly unreactive. Benzene lies in too deep a well, and is protected by activation barriers that are too high for these reactions to occur.

14.4b Electrophilic Deuteration of Benzene. In this section, we see how benzene *does* react. What we will see is the result of an energetic imperative to preserve the aromatic sextet. Benzene reacts not by addition, which would destroy aromatic stabilization, but by substitution reactions that preserve the aromatic sextet and stabilization. The prototypal example of the substitution process is H/D exchange. If sufficiently acidic conditions are used, we can see that the apparent nonreaction of HX with benzene is an illusion. When benzene is treated with D_2SO_4, the recovered benzene has incorporated deuterium, with the amount of incorporation depending on the strength of acid used and the time of the reaction. Ultimately, excess deuteriosulfuric acid will produce fully exchanged benzene (Fig. 14.17).

FIGURE 14.17
Hydrogen–deuterium exchange in benzene.

As deuterium must become incorporated onto the ring, perhaps the reaction shown in Figure 14.13 is not wrong after all. If benzene is protonated to produce the cyclohexadienyl cation shown in Figure 14.18, there are two possible further reactions.

FIGURE 14.18 Protonation of benzene by sulfuric acid, followed by addition of the bisulfate ion, would produce a pair of addition products.

Bisulfate ion

As shown in Figure 14.18, the anion produced in the reaction, bisulfate, could add to give two cyclohexadienyl compounds. We imagined a related reaction in Section 14.4a, where the Lewis base was bromide, Br⁻. In this reaction, aromaticity is lost, and we would not expect this process to be favorable energetically. Alternatively, a proton might be lost to reverse the initial addition reaction. The aromatic sextet is regenerated. In the absence of an isotope such as deuterium, this reaction is invisible as it simply regenerates starting material (Fig. 14.19).

FIGURE 14.19 The protonation reaction is reversible. Bisulfate could remove a proton to regenerate benzene.

If it is a deuteron that is added to produce an isotopically labeled cyclohexadienyl cation, either a deuteron or a proton could be lost to regenerate benzene. If the proton is lost, deuteriobenzene will appear as product, as one hydrogen will have been exchanged for a deuterium. Repetition of the reaction can serve to exchange all the hydrogens for deuterons (Fig. 14.20).

FIGURE 14.20 If deuteriosulfuric acid is used, the initial addition must generate a deuterated cyclohexadienyl cation. Reversal of the reaction will regenerate benzene, but loss of a proton will give the exchanged benzene. In an excess of D_2SO_4, all the protons of benzene can be exchanged for deuterons.

If the cyclohexadienyl cation is formed, why aren't normal addition reactions observed? Loss of a proton (or deuteron) regenerates an aromatic system in a highly exothermic reaction. Were addition to occur, the destruction of the relatively high-energy carbocation would still be exothermic, but not nearly as much as the substitution reaction (red); no aromatic molecule would be generated. The addition reaction (green) is much less exothermic than the rearomatization reaction and cannot compete with it (Fig. 14.21).

FIGURE 14.21 Both proton removal and addition are exothermic reactions for the cyclohexadienyl cation, but the regaining of aromaticity is much more so.

SUMMARY

We have just seen one example of substitution of a benzene ring in which aromaticity is initially lost only to be regained later. Now we generalize, discovering many variations on this theme. The main points to remember are, first, that aromaticity is not easily disrupted, and second, that once the aromatic stabilization is lost, it is easily regained.

14.4c Other Electrophilic Substitution Reactions of Benzene: The General Reaction.
The exchange reaction of benzene in deuterated acid serves as a prototype for a host of similar reactions in which an electrophile, E^+, is substituted for a hydrogen on the ring (Fig. 14.22).

FIGURE 14.22 A generic aromatic substitution reaction.

The reaction is always referred to as **electrophilic aromatic substitution**, since the emphasis is upon the agent doing the substituting, the electrophile. Fair enough, but in every electrophilic reaction there must be a nucleophile as well. In this case, the π system of the benzene ring acts as the nucleophile. In orbital terms, it is one of the degenerate (equi-energetic) filled molecular orbitals of benzene that overlaps with the empty orbital on the electrophile (Fig. 14.23).

FIGURE 14.23 One of benzene's equi-energetic (degenerate) HOMOs overlaps with the empty orbital on E$^+$ in a Lewis base–Lewis acid reaction.

As in the exchange reaction of D for H, the general reaction is completed by the removal of a proton from the intermediate cyclohexadienyl cation by a base, regenerating the aromatic benzene ring (Fig. 14.24).

FIGURE 14.24 The general aromatic substitution mechanism involves formation of a cyclohexadienyl cation, followed by proton removal, regenerating the aromatic ring.

14.4d Sulfonation. When benzene is treated with very concentrated sulfuric acid, or a solution of sulfur trioxide in sulfuric acid, called "oleum," benzenesulfonic acid (Fig. 14.25) is the product. In a mixture of SO_3 and H_2SO_4, the sulfur trioxide is protonated to produce another E$^+$ reagent, HO—$\overset{+}{S}O_2$.

FIGURE 14.25 Sulfonation of benzene.

Benzene-sulfonic acid
(100%)

PROBLEM 14.5 Write a mechanism for the formation of benzenesulfonic acid when benzene is treated with $SO_3/HOSO_2OH$ (H_2SO_4). *Hint*: Sulfur trioxide (SO_3) can be written as a sulfur atom doubly bonded to each of the three oxygens.

When SO_3 is used in an aprotic medium, it is a strong enough electrophile to react with benzene without protonation (Fig. 14.26). Under these circumstances the reaction looks a little different. Addition leads to a cyclohexadienyl cation, but there is also a negative charge in the molecule (Fig. 14.26). As in the general mechanism for aromatic substitution, a base now removes a proton to give the product. Lewis bases abound: The SO_3 itself is one, as is sulfuric acid. Proton loss might even be intramolecular as shown in the second part of Figure 14.27. The cyclohexadienyl cation can recover aromaticity by donating a proton to one of the bases present. The structure of the product of this reaction, benzenesulfonic acid, gives students problems because the coded structures, SO_3H or SO_2OH hide the detail. The full structure is drawn out in Figure 14.27.

The resonance-stabilized cyclohexadienyl cation intermediate

FIGURE 14.26 The SO_3 molecule acts as an electrophile and adds to benzene.

Three representations of benzenesulfonic acid

FIGURE 14.27 Removal of a proton from the cyclohexadienyl cation intermediate regenerates the aromatic ring.

Sulfonation of benzene is reversible, especially at high temperature. Treatment of benzenesulfonic acid in acidic steam regenerates the parent benzene (Fig. 14.28).

FIGURE 14.28 Sulfonation is reversible.

WORKED PROBLEM 14.6 Write a detailed mechanism for the reaction in Figure 14.28.

ANSWER The mechanism for the transformation of benzenesulfonic acid into benzene is the reverse of the formation of benzenesulfonic acid from benzene. We have already written something very much like this mechanism in the earlier figures when we drew out the mechanism for the forward version of this reaction. Protonation gives a cyclo-hexadienyl cation that can either lose a proton to revert to benzenesulfonic acid, or lose the sulfonic acid group to give benzene.

14.4e Nitration.

In concentrated nitric acid ($HONO_2$), often in the presence of sulfuric acid (H_2SO_4), benzene is nitrated, converted into nitrobenzene (Fig. 14.29).

FIGURE 14.29 Nitration of benzene.

In strong acid, nitric acid is protonated to give $H_2\overset{+}{O}NO_2$. Loss of water can generate a powerful Lewis acid, the nitronium ion, $^+NO_2$, that can act as the electrophile in an aromatic substitution reaction. Indeed, stable nitronium salts are known and can also effect nitration (Fig. 14.30).

FIGURE 14.30 Two nitration reactions. In each case the electrophile is the nitronium ion, $^+NO_2$.

As usual, the reaction is completed by deprotonation of the cyclohexadienyl cation by one of the Lewis bases present in the mixture.

14.4f Halogenation.

Unlike simple alkenes, benzene fails to react with Br_2 or Cl_2. It takes a catalyst, usually $FeBr_3$ or $FeCl_3$, to initiate the reaction. Chlorine is not a strong enough electrophile (Lewis acid) to react with the weakly nucleophilic benzene by itself. The function of the catalyst is to convert chlorine into a stronger Lewis acid that can react with the aromatic compound. Ferric chloride ($FeCl_3$) does this by forming a complex with chlorine (Fig. 14.31).

FIGURE 14.31 Complex formation from Cl_2 and $FeCl_3$.

Benzene is nucleophilic enough to displace the good leaving group $FeCl_4^-$ from the Cl_2-FeCl_3 complex. Displacement generates a cyclohexadienyl cation that can be deprotonated to give chlorobenzene. This reaction may look strange, but it has a most familiar analogy in the protonation of an alcohol to convert OH into the good leaving group $^+OH_2$ (Fig. 14.32). The $FeCl_3$ plays the role of the proton in the alcohol reaction. The mechanism of bromination using Br_2 and $FeBr_3$ is similar.

FIGURE 14.32 Chlorination of benzene.

PROBLEM 14.7 Write a mechanism for the bromination of benzene using $Br_2/FeBr_3$.

These reactions (sulfonation, nitration, chlorination, and bromination) serve to underscore the general principles of electrophilic aromatic substitution. In each, an electrophile powerful enough to react with the weakly nucleophilic benzene ring must be generated. The addition reaction first generates a resonance-stabilized (but not aromatic) cyclohexadienyl cation that can be deprotonated to regenerate an aromatic system. The general mechanism is shown again in Figure 14.33.

FIGURE 14.33 Once again, here is the general mechanism for aromatic substitution, where $E^+ = D^+$, $^+SO_2OH$, $^+NO_2$, ^+Cl, or ^+Br.

We are now able to make a few substituted aromatic compounds, and this capability will shortly prove quite useful. But much of the interest in synthetic chemistry involves the construction of carbon–carbon bonds. We have made no progress at all in finding ways to make bonds from benzene to carbon. In Sections 14.5 and 14.6, we do just that by making use of another electrophilic aromatic substitution reaction quite closely related to the bromination and chlorination reactions we have just seen.

14.5 CARBON–CARBON BOND FORMATION: FRIEDEL–CRAFTS ALKYLATION

The **Friedel–Crafts alkylation** reaction is named for Charles Friedel (1832–1899) and James M. Crafts (1839–1917), who discovered it by accident when they tried to synthesize pentyl chloride from pentyl iodide through reaction with aluminum chloride in an aromatic solvent. Instead of the hoped-for chloride, substituted aromatic hydrocarbons appeared. The reaction is closely related to bromination and chlorination of benzene. If we treat benzene with isopropyl bromide in the hope that a nucleophilic attack of benzene on isopropyl bromide will produce isopropylbenzene (cumene), we are sure to be disappointed. Benzene is by no means a strong nucleophile and isopropyl bromide can scarcely be described as a powerful electrophile. The proposed reaction is utterly hopeless, and fails to give product (Fig. 14.34).

FIGURE 14.34 Benzene fails to react with isopropyl bromide.

So why even bring the subject up? It does look a bit like another reaction that fails to give product—the attempt to form chlorobenzene or bromobenzene from chlorine or bromine and benzene. Those halogenations are transformed into a success by the addition of a catalyst, $FeCl_3$ or $FeBr_3$, which converts the halogen, a weak electrophile, into a much stronger Lewis acid (Figs. 14.31 and 14.32). Perhaps a related technique will work here. Indeed it does, as the addition of a small amount of the catalyst $AlBr_3$ or $AlCl_3$ to a mixture of benzene and isopropyl bromide or chloride results in the formation of isopropylbenzene. The mechanism should be easy to write now. A complex is first formed between isopropyl bromide and $AlBr_3$. This complex can be attacked by benzene to give a resonance-stabilized cyclohexadienyl cation. Deprotonation of the intermediate cyclohexadienyl cation gives isopropylbenzene (Fig. 14.35).

~100% in 0.0005 s

FIGURE 14.35 The Friedel–Crafts alkylation of benzene.

Whether or not the complex in Figure 14.35 can actually form a free carbocation depends on the alkyl halide used in the reaction. Friedel–Crafts alkylation succeeds with a variety of alkyl halides, including methyl-, ethyl-, isopropyl-, and *tert*-butyl chlorides and bromides (Fig. 14.36).

We would not expect the methyl or ethyl cation to be formed in these reactions, and surely these alkylations take place by attack of benzene on the complexed halide. *tert*-Butyl chloride, on the other hand, can produce the relatively stable *tert*-butyl cation, not an unlikely intermediate at all. So, the detailed mechanism of the Friedel–Crafts alkylation depends on the nature of the R group.

FIGURE 14.36 Friedel–Crafts alkylation succeeds for a wide variety of alkyl halides.

PROBLEM 14.8 Why is attack of benzene on complexed *tert*-butyl bromide to give a cyclohexadienyl cation an unlikely mechanistic step?

PROBLEM 14.9 The synthesis of toluene by the aluminum chloride–catalyzed Friedel–Crafts alkylation of benzene with methyl chloride is badly complicated by the formation of di-, tri-, and polymethylated benzenes. It appears that the initial product of the reaction, methylbenzene (toluene), is more reactive in the Friedel–Crafts reaction than is benzene. Analyze the mechanism of electrophilic aromatic substitution to see why toluene is more reactive than benzene. *Hint*: Look carefully at substitution in the position directly across the ring from the methyl group (the para position) for both benzene and toluene, and look for differences.

PROBLEM 14.10 How would you minimize the overalkylation mentioned in Problem 14.9 and achieve a reasonable synthesis of toluene from benzene?

Not only do the mechanistic details of the reaction depend on the alkyl group (R), but even the gross outcome of the reaction depends on the structure of R. Although the alkyl halides we have mentioned so far all succeed in making the related alkylated benzenes, there are many that do not. Suppose we set out to make propylbenzene by the Friedel–Crafts alkylation of benzene using propyl chloride, benzene, and aluminum chloride. If we run this reaction, all looks well at first. The products formed are of the correct molecular weight, but a closer look uncovers a problem. Some propylbenzene is formed, but the major product is isopropylbenzene (Fig. 14.37).

FIGURE 14.37 An attempt to make propylbenzene by Friedel–Crafts alkylation gives isopropylbenzene as the major product.

Isopropylbenzene (55%)   Propylbenzene (45%)
(Total yield ~ 50%)

A close look at the mechanism of the reaction shows the problem. In the complex between the alkyl halide (propyl chloride) and aluminum chloride (AlCl$_3$), the carbon–chlorine bond is weakened and the positive charge is shared by the chlorine and the adjacent carbon. The resonance form shown in Figure 14.38 shows this clearly.

FIGURE 14.38 In the complex formed between propyl chloride and aluminum chloride, the positive charge is partially on the primary carbon adjacent to the chlorine.

The complex formed from propyl chloride and aluminum chloride contains a partial positive charge on a primary carbon, a most unhappy state of affairs. When this happens in reactions of methyl chloride or ethyl chloride, there is nothing that can be done about it, but propyl chloride has a way out. A rearrangement of hydride generates the much more stable complex, or even a free cation in which a secondary carbon bears the positive charge. Straightforward alkylation now gives the observed major product, isopropylbenzene (Fig. 14.39).

In this complex no rearrangement is possible, but…

…in this one, hydride migration can convert a partial positive charge on a primary carbon into a partial positive charge on a secondary carbon — a much more energetically favorable situation

FIGURE 14.39 A hydride shift can convert a less stable complex with a partial positive charge on a primary carbon into a more stable complex in which the partial positive charge is on a secondary carbon. The substitution reaction follows.

These rearrangements are very common, and are encountered whenever a hydride or alkyl shift will generate a more stable cation, or complex (Fig. 14.39). This phenomenon is not new. Recall, for example, the rearrangements of hydride and alkyl groups encountered in reactions involving carbocations (Fig. 14.40; also Section 9.10).

FIGURE 14.40 Hydride shifts are commonly encountered in addition reactions whenever a more stable carbocation can be generated from a less stable one.

SUMMARY

Friedel–Crafts alkylation is limited to those alkyl halides for which rearrangement is not a problem. In all other cases, rearrangements can be expected. You just cannot make pure propylbenzene by a Friedel–Crafts alkylation. But there is a way out; you can use another reaction discovered by Friedel and Crafts.

PROBLEM 14.11 *tert*-Butylbenzene is converted into benzene on treatment with AlCl$_3$ and a trace of HCl. Write a mechanism for this reaction.

14.6   FRIEDEL–CRAFTS ACYLATION

A variation of the Friedel–Crafts alkylation (**Friedel–Crafts acylation**) introduces a new group, called an **acyl group** (RC=O) (Fig. 14.41). An acyl chloride (RCOCl) is also known as an **acid chloride**. We will take up the chemistry of all manner of acyl derivitives in Chapter 19, but we need to use them here. In practice, they are rather easy to make from the relatively available carboxylic acids, RCOOH, by treatment with thionyl chloride (SOCl$_2$) (Fig. 14.42).

The acyl group   An acyl chloride or an acid chloride

FIGURE 14.41 The acyl group.

FIGURE 14.42 Acid chlorides can be made by the reaction of carboxylic acids with thionyl chloride.

A carboxylic acid   An acid chloride $+ \; SO_2 \; + \; HCl$

Acyl chlorides react with benzene under the influence of an equivalent (not a catalytic amount) of AlCl$_3$. A molecule of AlCl$_3$ is needed for each molecule of product. By now it should be easy to write a general mechanism. A complex is first formed with AlCl$_3$. But which complex? Unlike alkyl chlorides, acyl chlorides contain two nucleophilic sites, the oxygen and chlorine atoms. The evidence is that both complexes are formed, and that the two are in equilibrium. The resonance-stabilized **acylium ion** can be formed by dissociation of either complex (Fig. 14.43).

FIGURE 14.43 The acid chloride–AlCl$_3$ complex can dissociate to give the acylium ion.

The acylium ion

The acylium ion can then add to the benzene ring to generate an intermediate cyclohexadienyl cation. Deprotonation by a Lewis base completes the reaction and produces acyl benzenes (Fig. 14.44). As mentioned above, unlike Friedel–Crafts alkylation, Friedel–Crafts acylation is not a catalytic reaction. The product, an acyl benzene, is also reactive toward $AlCl_3$ and forms its own 1:1 complex as it is produced. So, for every molecule of acyl benzene formed in the reaction, one molecule of $AlCl_3$ is used up in complex formation. A stoichiometric amount of $AlCl_3$ is required in this reaction. At the end of the reaction, addition of water destroys the complex and generates the free, acylated benzene (Fig. 14.44).

Friedel–Crafts acylation

FIGURE 14.44 The mechanism of Friedel–Crafts acylation. Aromatic ketones can form complexes with $AlCl_3$, which is the reason why a full equivalent of $AlCl_3$ is necessary for Friedel–Crafts acylation.

A final hydrolysis step liberates the ketone, an acyl benzene

Despite the mechanistic complexity, in practice the Friedel–Crafts acylation reaction works very well, and a variety of phenyl-substituted ketones can be produced in good yield. Be certain to add this reaction to your collection of "synthesis" file cards (Fig. 14.45).

(97%)

(77%)

FIGURE 14.45 Two typical Friedel–Crafts acylations.

Figure 14.46 shows another successful acylation, but this reaction contains a surprise. The reaction of butyryl chloride with benzene yields phenyl propyl ketone (butyrophenone). This result implies that acyl complexes are not prone to rearrangement (Fig. 14.46).

FIGURE 14.46 Friedel–Crafts acylation is not complicated by hydride rearrangements.

Phenyl propyl ketone
(butyrophenone)
(77%)

Butylbenzene

The four-carbon chain is intact in the product, and if we had a way to transform the carbon–oxygen double bond into a methylene group, we would have a way to make butylbenzene without rearrangement (Fig. 14.46).

WORKED PROBLEM 14.12 Explain why these acylium ions (or acyl complexes) are not likely to rearrange. Two reasons should be apparent. First, what differentiates the acylium ion from simple alkyl carbocations? Next, look closely at the product of rearrangement and consider the polarity of the carbon–oxygen double bond.

ANSWER The first reason has to do with the stability of the acylium ion itself. The acylium ion is well stabilized by resonance, and is therefore less likely to rearrange than simple alkyl carbocations.

An acylium ion

Second, consider the structure of the putative rearranged cation.

The carbonyl group is strongly polarized in the sense shown in the drawing. The carbon is δ^+ and the oxygen is δ^-. Rearrangement opposes a full and partial positive charge. This situation is very bad energetically and will act to retard migration of hydrogen or an alkyl group.

In fact, there are many such ways, two of which are shown in Figure 14.47. The Clemmensen reduction, named for its discoverer E. C. Clemmensen (1876–1941), takes place under strongly acidic conditions, whereas the Wolff–Kishner reduction (Ludwig Wolff, 1857–1919; N. Kishner, 1867–1935) requires strong base.

Clemmensen reduction

Zn/Hg, HCl

EtOH, 79 °C, 24 h

CH$_3$

CH$_3$
(93%)

Wolff–Kishner reduction

HO$\qquad$O$\qquad$OH
(a high-boiling-point
alcohol used to provide
a high temperature)

H_2N—NH_2, KOH, 200 °C

H$_3$C $\quad$ S $\quad$ CH$_2$CH$_3$
(~90%)

FIGURE 14.47 Two reductions of ketones to hydrocarbons.

PROBLEM 14.13 Other reagents besides acid chlorides can lead to acylation reactions. Write a mechanism for the following reaction:

H$_3$C $\qquad$ O $\qquad$ CH$_3$ $\qquad$ +

1. AlBr$_3$
2. H$_2$O

C—CH$_3$

Acetic anhydride

Acetophenone
(methyl phenyl ketone)

SUMMARY

Strong electrophiles, E^+ reagents, will add to benzene in what is called electrophilic aromatic substitution. Aromaticity is regained when the intermediate cyclohexadienyl cation is deprotonated. That's all there is to it—all the rest is details. Identifying the E^+ electrophile is the major task. Remember to watch out for rearrangements if a carbocation is the electrophile.

14.7 SYNTHETIC REACTIONS WE CAN DO SO FAR

We have uncovered six synthetically important reactions so far. We can brominate or chlorinate benzene. We can nitrate and reversibly sulfonate benzene. The two Friedel–Crafts reactions allow the construction of carbon–carbon bonds to give some alkyl benzenes and give you another synthetic route to compounds containing carbon–oxygen double bonds. Figure 14.48 summarizes the situation.

FIGURE 14.48 Aromatic substitution reactions encountered so far.

FIGURE 14.49 The reduction of nitrobenzene to aniline.

Although there are relatively few groups we can attach to a benzene ring directly, we can often transform one group into another through chemical reactions. It is nitrobenzene that is the key to many of these transformations. Nitrobenzene itself is relatively unreactive, but the nitro group can be reduced in many ways to the amino group. This transformation opens the way for the synthesis of a host of new molecules (Fig. 14.49).

Aminobenzene, generally called by its common name **aniline**, can be treated with nitrous acid (HONO) or its sodium salt ($Na^+ \ ^-ONO$) and HCl, to produce benzenediazonium chloride, which is explosive when dry, but a relatively safe material when wet (Fig. 14.50).

FIGURE 14.50 Reaction of aniline with nitrous acid to give benzenediazonium chloride.

PROBLEM 14.14 Draw resonance forms for benzenediazonium chloride.

The mechanism for this reaction contains a lot of steps, but is really within our range now. A critical intermediate is dinitrogen trioxide, formed in equilibrium with nitrous acid by loss of water. When nitrous acid is protonated, a good leaving group, water, is created and can be displaced by nitrite ion (Fig. 14.51).*

FIGURE 14.51 The formation of dinitrogen trioxide, N_2O_3.

Displacement of the good leaving group nitrite (^-ONO) from N_2O_3 by the Lewis basic nitrogen of aniline leads, after removal of a proton, to *N*-nitrosoaniline (Fig. 14.52).

FIGURE 14.52 Formation of a nitrosamine by displacement of nitrite from N_2O_3.

The nitrosamine equilibrates with a molecule called a diazotic acid, which loses water in acid to generate the **diazonium ion** (Fig. 14.53).

FIGURE 14.53 Loss of water from the protonated diazotic acid gives the diazonium ion, benzenediazonium chloride.

*This reaction is actually a bit more complicated than is shown in the figure. It involves addition of nitrite to the nitrogen–oxygen double bond followed by loss of water. We will explore this kind of mechanism in detail beginning in Chapter 16.

These transformations are probably bewilderingly complex right now, even though they are made up of very simple reactions in sequence. It certainly takes some practice to get them right. However, many reaction mechanisms involve similar sequences of protonations, deprotonations, and additions and losses of water molecules. Now is the time to start practicing them. We will encounter many more when we discuss carbonyl chemistry in Chapters 16–19.

The diazonium ion is important because it is the critical intermediate in a number of transformations lumped under the heading of the **Sandmeyer reaction**, named for Traugett Sandmeyer (1854–1922). Cuprous salts convert the diazonium ion into the related halide or cyanide. In recent times, numerous variations on the basic theme of the Sandmeyer reaction have improved yields and reduced side products. Figure 14.54 shows three typical Sandmeyer reactions.

FIGURE 14.54 Three Sandmeyer reactions. Decomposition of the diazonium ion with cuprous salts leads to substitution.

Reagents other than cuprous salts are also effective. Treatment with potassium iodide (KI) or fluoboric acid (HBF$_4$) gives the iodide or fluoride (the Schiemann reaction; G. Schiemann 1899–1969), respectively, and copper-catalyzed hydrolysis gives **phenols**. Heating with hypophosphorous acid (H$_3$PO$_2$) removes the diazonium group and replaces it with hydrogen (Fig. 14.55).

FIGURE 14.55 Other substitution reactions of diazonium ions.

Of course, you are now clamoring for the mechanisms for all these changes. Sadly, only the outlines of the mechanisms are known, and much remains to be filled in. The details of many reactions are imperfectly understood, and there remains a delightful (to some) atmosphere of magic about parts of organic chemistry. What are we to make of the admonition in one procedure for the Sandmeyer reaction to "stir the mixture with a lead rod." What happens if we use a glass rod? Although one would think that the reaction might still succeed, there are examples to the contrary, and perhaps microscopic pieces of lead do affect the success of the reaction. I'd use lead.*

PROBLEM 14.15 We reject out of hand the notion that any of these net displacement reactions could go by something as simple as an S_N2 reaction. Why?

In the reactions involving copper, electron-transfer-mediated radical reactions are certainly involved. In HCl, cuprous chloride (CuCl) is in equilibrium with the dichlorocuprate ion $^-CuCl_2$ (Fig. 14.56). The dichlorocuprate ion can transfer an electron to the diazonium ion, which can then lose molecular nitrogen to give a phenyl radical. In turn, the phenyl radical can pluck a chlorine atom from a copper chloride molecule to give the chlorobenzene and a molecule of cuprous chloride. The mechanism for the formation of chlorobenzene in the Sandmeyer reaction is shown in Figure 14.57.

$$CuCl + HCl \rightleftharpoons H^+\ ^-CuCl_2$$

The dichlorocuprate ion

FIGURE 14.56 Formation of the dichlorocuprate ion from CuCl and HCl.

FIGURE 14.57 The mechanism of chlorobenzene formation.

Other Sandmeyer reactions apparently involve a phenyl cation formed by the irreversible S_N1 ionization of benzenediazonium chloride (Fig. 14.58).

It doesn't matter that some mechanistic details are sketchy; the formation of the diazonium ion from aniline (and therefore ultimately from the easily made nitrobenzene) and its further transformations gives you the ability to make a great number of substituted benzenes from very simple precursors.

FIGURE 14.58 Other conversions of benzenediazonium chloride involve a phenyl cation.

*Speaking of lead and magic, two of my chemical relatives, Milton Farber and Adnan Abdul-rida Sayigh (~1922–1980) once discovered a useful reaction of lead dioxide, and the work was duly published as an effective way to make alkenes from 1,2-diacids. However, the reaction only worked when one particular bottle of lead oxide was used, and once that bottle was used up no one could reproduce the yields achieved earlier. All other sources of lead dioxide failed! Apparently, the particle size of the lead dioxide was critical. (What's a "chemical relation"? That's someone who did his or her graduate work with the same professor as you. Farber and Sayigh both worked with the same person I did and hence are my chemical brothers.)

PROBLEM 14.16 Draw resonance forms for the phenyl radical Ph• (be careful).

PROBLEM 14.17 In the formation of diazonium ions from aromatic amines, side products are occasionally encountered, as shown below. Give a mechanism for the formation of these compounds.

An azo compound

PROBLEM 14.18 Suggest synthetic procedures leading to the following compounds. You may start with benzene, any inorganic reagent, and organic compounds containing no more than four carbon atoms.

(a) SO_2OH

(b) $C(CH_3)_3$

(c) Cl

(d) $CCH_2CH_2CH_3$ (with $=O$)

(e) $CH_2CH_2CH_2CH_3$

(f) NH_2

(g) F

(h) I

(i) OH

PROBLEM 14.19 The phenols that can be made by the Sandmeyer reaction can be used further in synthesis. First, explain why phenol ($pK_a = 10$) is a much stronger acid than cyclohexanol ($pK_a = 16$).

PROBLEM 14.20 Aryl ethers can be made by the following reaction. Write an arrow formalism mechanism.

PROBLEM 14.21 Varying amounts of a side product are often formed in the reaction in Problem 14.20. Explain the formation of the compound shown below in the reaction of phenoxide with allyl bromide. You will need the additional information that ketones are transformed in base (or acid) into "enols."

14.8   ELECTROPHILIC AROMATIC SUBSTITUTION OF HETEROAROMATIC COMPOUNDS

Although heteroaromatic compounds generally have lower resonance energies than benzene (Table 14.1, p. 677), they do undergo electrophilic aromatic substitution.

14.8a Reactions of Pyridine and Pyrrole. The heteroatom introduces a complication in aromatic substitution not present in benzene. Benzene has six equivalent positions, but substitution could occur at two positions in pyrrole and three in pyridine (Fig. 14.59).

FIGURE 14.59 Possible positions for electrophilic aromatic substitution in pyridine and pyrrole.

As for benzene, electrophilic attack on pyridine gives an intermediate in which three atoms share the positive charge. Substitution ortho or para to the nitrogen in pyridine places a partial positive charge on the relatively electronegative nitrogen atom. Substitution at the meta, or 3-position does not, and is accordingly favored (Fig. 14.60).

GENERAL CASES

SPECIFIC EXAMPLES

FIGURE 14.60 In pyridine, substitution at the 3-position is favored, although the reaction is slow compared to that of benzene.

This effect is only accentuated if the nitrogen is protonated, as it very often is under the strongly acid conditions necessary to carry out substitution on pyridine. Therefore it is not surprising that substitution of pyridine takes place more slowly than analogous reactions of benzene. Pyridine itself substitutes about as rapidly as nitrobenzene, 10^6 more slowly than benzene itself. The pyridinium ion substitutes an impressive 10^{18} times more slowly than benzene.

A similar analysis serves to locate the more favorable position for substitution of pyrrole. Addition at either carbon 2 or 3 allows the nitrogen to help bear the positive charge. However, in the intermediate produced by substitution at carbon 2, two ring carbons share the charge, whereas substitution at carbon 3 leads to an intermediate in which only one carbon shares the charge. Addition to the 2-position is favored, although the formation of mixtures is commonly observed (Fig. 14.61).

GENERAL CASES

A SPECIFIC EXAMPLE

FIGURE 14.61 In pyrrole, substitution at the 2-position is favored.

Pyrrole is more reactive than benzene or pyridine, as pyrrole is less well stabilized by resonance than the six-membered ring molecules (Table 14.1).

14.8b Reactions of Furan and Thiophene. These compounds are very reactive toward electrophilic reagents, E+. As in pyrrole, aromatic substitution is directed first to the 2-position, although mixtures are sometimes obtained in which some 3-substitution appears (Fig. 14.62).

FIGURE 14.62 Electrophilic substitution reactions of furan and thiophene.

PROBLEM 14.22 Explain why 2-substitution is preferred to 3-substitution in furan.

PROBLEM 14.23 There is another reasonable mechanism for aromatic substitution of furan. Suggest one. *Hint*: Think simple! What are the possible reactions between electrophiles and double bonds (Chapters 9 and 10)? Use the chlorination of furan as an example.

PROBLEM 14.24 There is a difficulty with the answer to Problem 14.23. Find what is wrong (or unlikely) and suggest an alternative mechanism. *Hint*: Recall what you know of the stereoelectronic requirements of the E2 reaction (Section 7.9, p. 315).

14.9   DISUBSTITUTED BENZENES: ORTHO, META, AND PARA SUBSTITUTION

We have already mentioned (Problem 14.9) that the synthesis of methylbenzene (toluene) from benzene is made difficult because once a significant amount of toluene is formed, it competes successfully with the unsubstituted benzene in further methylation reactions. Unless one is careful to use a vast excess of benzene, a horrible mixture of benzene, toluene, dimethylbenzenes (xylenes), and even more substituted molecules results.

WORKED PROBLEM 14.25 If one attempts to methylate benzene completely with $CH_3Cl/AlCl_3$, the product contains a greenish solid of the formula $C_{13}H_{21}AlCl_4$. Propose a structure for the compound and a mechanism for its formation.

ANSWER This problem involves a sequence of Friedel–Crafts alkylations. Toluene is made first, then the xylenes, then trimethylbenzenes, and so on, until hexamethylbenzene is formed.

(continued)

What happens if another Friedel–Crafts reaction occurs? The intermediate is the heptamethylcyclohexadienyl cation tetrachloroaluminate salt, and this material is stable under the reaction conditions. Why? The resonance forms contributing to the structure are all tertiary carbocations, and there is no proton that can be lost to give a new aromatic system. All positions are occupied by methyl groups.

The general question now arises of how the substituted benzenes we have learned to make will react with electrophilic reagents. What are the possibilities? There are three kinds of disubstituted benzene, called **ortho** (1,2-disubstituted), **meta** (1,3-disubstituted), and **para** (1,4-disubstituted), as shown in Figure 14.63.

ortho	meta	para
(1,2-substitution)	(1,3-substitution)	(1,4-substitution)

FIGURE 14.63 The three possible disubstituted benzenes.

Substituents influence the distribution of the three isomeric disubstituted products dramatically. Further substitution of monosubstituted benzenes by no means proceeds to give a statistical distribution of two parts ortho product, two parts meta product, and one part para product. In fact, if one looks at the reactions of a number of monosubstituted benzenes, a pattern emerges. Some groups produce predominantly ortho and para disubstituted compounds while for others meta is dominant. There is also a correlation of the position of further substitution with the rate of the reaction. Substituted benzenes that give mainly ortho and para products usually react faster than benzene itself. Substituted benzenes that give mainly meta product react more slowly than benzene itself (Table 14.2).

ortho + para   meta

TABLE 14.2	Patterns of Further Reactions of Substituted Benzenes	

R	Position[a]	Relative Rate
H (benzene)		1
NH$_2$, NHR, NR$_2$	o/p	Very fast
OH	o/p	Very fast
OR	o/p	Very fast
Alkyl	o/p	Fast
NH—C(=O)R'	o/p	Fast
I, Cl, Br, F	o/p	Slow
—C(=O)R' (R' = OH, OR, NH$_2$, Cl, alkyl, aryl)	m	Slow
SO$_2$OH	m	Slow
NO$_2$	m	Slow
CN	m	Slow
$^+$NR$_3$	m	Slow

[a]Ortho = o, para = p, meta = m.

A look at the general mechanism of substitution shows how different groups influence the position of subsequent reaction. We can also see how these orientation (regiochemical) effects are connected to the rates of the reactions. Let's look first at methoxybenzene (**anisole**). Methoxy is one of the groups most effective at directing further substitution to the ortho and para positions and at increasing the rate of reaction. We will work through the mechanism for both meta and para substitution and look for a point of difference. Figure 14.64 shows the mechanism for meta substitution of anisole by a generic E$^+$ reagent, and compares the reaction to that of benzene itself. As usual, a resonance-stabilized cyclohexadienyl cation is formed first, and then deprotonated to give the substituted product.

Substitution of benzene

Meta substitution of anisole

FIGURE 14.64 The intermediate cyclohexadienyl cations involved in the substitution of benzene and meta substitution of anisole.

At first glance, substitution at the para position seems similar (Fig. 14.65).

FIGURE 14.65 The intermediate cyclohexadienyl cation involved in para substitution of anisole.

But look carefully at the intermediate in para substitution, the resonance-stabilized cyclohexadienyl cation. Remember our discussion of the stabilization of adjacent cations by oxygen (Chapter 9, p. 416). In this intermediate, there is a fourth resonance form in which oxygen shares the positive charge. This fourth form stabilizes the cation relative to that formed in the meta substitution or in substitution of benzene itself (Fig. 14.66).

FIGURE 14.66 In the intermediate for para substitution, there is a fourth resonance form in which oxygen bears the positive charge.

WORKED PROBLEM 14.26 Work out the structure of the intermediate formed by substitution of anisole at the ortho position.

ANSWER The figure shows the four resonance forms of the cyclohexadienyl cation intermediate. As in the para case, the methoxyl group is in a position to help share the positive charge. The resulting delocalization contributes to the stability of the intermediate and makes it easy to form, relative to the intermediate formed in meta substitution, or from benzene itself (Fig. 14.64).

The intermediate formed in para substitution of anisole is more stable than that formed in meta substitution of anisole or in substitution of benzene itself. Accordingly, the transition state leading to the para intermediate will be more stable than the transition states for meta substitution or substitution of benzene itself. We can expect para substitution to occur more rapidly than the other two reactions (Fig. 14.67).

FIGURE 14.67 The intermediate for para substitution, with four resonance forms, is lower in energy than the intermediate for meta substitution (three forms) or that for substitution of benzene (also three forms). The transition state for para substitution of anisole will also be lower in energy than the transition state for meta substitution.

In fact, the chlorination of anisole gives only ortho and para substituted product (Fig. 14.68).

FIGURE 14.68 The chlorination of anisole.

The relatively low energy of the intermediate formed by addition at the ortho or para substitution (four resonance forms) means that this reaction will prevail over meta attack (the intermediate has only three resonance forms). Moreover, if we compare the rate of formation of this intermediate to that of addition to benzene (also only three resonance forms), we would expect the lower energy intermediate from anisole to be formed faster, as is indeed observed.

PROBLEM 14.27 Why is ortho substitution often less favorable than para substitution, even though ortho substitution is favored statistically by a 2:1 factor?

PROBLEM 14.28 Hold it! Explain *exactly why* we "expect" that attack on anisole by an electrophilic reagent to form a relatively low energy, resonance-stabilized intermediate will be faster than attack on benzene? Are we not confusing thermodynamics with kinetics?

Let's now look at the substitution of toluene. Figure 14.69 shows the intermediates formed by meta and para substitution using a generic electrophile, E⁺.

The meta substitution of toluene

All resonance forms secondary

The para substitution of toluene

This resonance form is a tertiary carbocation

There *is* another resonance form for this intermediate! Remember hyperconjugation

FIGURE 14.69 The intermediates for para and meta substitution of toluene. In the intermediate for para substitution, two resonance forms are secondary but one resonance form is a tertiary carbocation. In the intermediate for meta substitution, all three resonance forms are secondary.

This time there is no obvious fourth resonance form, as there was in the para (and ortho) substitution of anisole. Why is ortho/para substitution still preferred, and why does toluene react faster than unsubstituted benzene (Table 14.1, p. 677)? Look at the positions sharing the positive charge. In the intermediate formed from meta substitution, the three positions sharing the positive charge are all secondary. In the intermediate formed by para substitution, one of the three positions is a tertiary carbon and will be especially effective in stabilizing a positive charge. So there really is a fourth resonance form! Recall the description of the stabilization of adjacent positive charge by alkyl groups (Chapter 9, p. 411). An alkyl group can stabilize an adjacent positive charge in a hyperconjugative way (Fig. 14.69).

The intermediate for para substitution, *and the transition state leading to it*, will be lower in energy than those involved in meta substitution (Fig. 14.70).

FIGURE 14.70 The intermediate and transition states for para substitution of toluene are lower in energy than those for meta substitution.

This theoretical analysis is borne out in practice. Figure 14.71 shows some typical substitution reactions of toluene.

FIGURE 14.71 Some substitution reactions of toluene.

So, for many groups, meta substitution does not present a problem. However, separation of ortho and para products *is* a problem, especially as the two are often formed in comparable amounts. Happily, physical methods such as crystallization and distillation can generally be used successfully to separate the products. Electrophilic aromatic substitution is a practical source of many ortho and para disubstituted benzenes.

PROBLEM 14.29 Write a mechanism for the ortho substitution of toluene. Compare this reaction to meta substitution.

PROBLEM 14.30 Explain why bromobenzene and aniline direct subsequent substitution reactions ortho/para.

What about those groups that direct further substitution to the meta position and slow the rate? Why are the effects we have seen so far not general? Once more, the way to attack this problem is to write out the various possibilities and look for points of difference. Look first at the intermediates formed by reaction of the trimethylanilinium ion in Figure 14.72.

Meta substitution of the trimethylanilinium ion

Para substitution of the trimethylanilinium ion

In this resonance form, two positive charges are very close; this form is very high energy and will contribute little

FIGURE 14.72 Meta and para substitution of the trimethylanilinium ion. Both intermediates have three resonance forms, but the adjacent positive charges in the intermediate for para substitution are especially destabilizing.

Both meta and para substitution give an intermediate with three resonance forms, so there is no gross difference here. Note that we are introducing a second positive charge into the molecule when we do *any* electrophilic aromatic substitution reaction of this molecule. This second charge is bound to slow the reaction relative to substitution of benzene itself. Now we know why this substitution reaction is especially slow, but we can't yet explain why the meta position is preferred. The difficulty in para substitution is that the reaction places the two positive charges especially close together. In para substitution, the new positive charge resides partly on the carbon bearing the original substituent. When this substituent can stabilize the charge through resonance, as in anisole or toluene, para substitution will be favored. When it brings two positive charges together, as in the trimethylanilinium ion, para substitution is decidedly disfavored for electrostatic reasons.

PROBLEM 14.31 Work out the structure of the intermediate formed by ortho substitution of the trimethylanilinium ion.

PROBLEM 14.32 Use an Energy versus Reaction progress diagram to compare ortho and meta substitution of the trimethylanilinium ion.

So the situation is clear for the trimethylanilinium ion. Meta substitution, which does not place two like charges on adjacent carbons, is favored over para (or ortho) substitution, which does. Figure 14.73 shows one example: bromination of the trimethylanilinium ion.

FIGURE 14.73 Bromination of the trimethylanilinium ion.

What about the other meta directing groups? They do not bear obvious positive charges. Nitrobenzene is a good example. Nitrobenzene is so deactivated that it can be used as a solvent for Friedel–Crafts reactions of other aromatic compounds and not suffer attack itself. Further substitution on nitrobenzene is invariably largely in the meta position (Table 14.2; Fig. 14.74).

FIGURE 14.74 Nitrobenzene nitrates very slowly at 0 °C, and almost entirely in the meta position.

Yet, there is no obvious positive charge as there was in the trimethylanilinium ion we analyzed before. A close look at the structure of the nitro group reveals the problem, which has been hidden by the too-schematic rendering of nitro as NO_2. If you draw out a nitro group carefully, you will see that there is no good uncharged structure, and that the nitrogen always bears a positive charge (Fig. 14.75). So there really is a positive charge adjacent to the ring in nitrobenzene and the same factors that lead to meta substitution and a slow rate in the trimethylanilinium ion will operate for nitrobenzene as well.

In nitrobenzene, there is a positive charge adjacent to the ring

So,

Meta substitution of nitrobenzene

is favored over

Para substitution of nitrobenzene

In this resonance form, two positive charges are very close; this form is very high energy and will contribute little to the stability of the intermediate

FIGURE 14.75 The resonance description for the nitro group reveals that nitrogen bears a full positive charge. Further substitution will be directed to the meta position.

But other meta directors do not have hidden positive charges. Ethyl benzoate is a good example. It too substitutes slowly, and predominantly in the meta position (Fig. 14.76).

Ethyl benzoate   (68.4%)   (3.3%)   (28.3%)

FIGURE 14.76 The nitration of ethyl benzoate gives mainly meta product.

The answer again comes from a close look at the structure of the group substituting the benzene ring, in this case a carbon–oxygen double bond, a carbonyl group. Although there are no hidden full positive charges as in nitro, the carbon–oxygen double bond is highly polar. Because oxygen is much more electronegative than carbon, the dipole in the bond will place a partial positive charge on the carbon next to the ring (Fig. 14.77).

FIGURE 14.77 The carbonyl group is polar, with the carbon bearing a partial positive charge.

So, once again, ortho or para substitution will oppose like charges. This time it is not two full positive charges that must be adjacent, but a full and partial positive charge. No matter—meta substitution, which keeps the charges further apart, will still be favored, and the rate of the reaction will still be slow relative to that for benzene (Fig. 14.78).

The meta substitution of ethyl benzoate

The para substitution of ethyl benzoate

In this resonance form, a positive charge and a partial positive charge are very close; this form is very high energy and will contribute little

FIGURE 14.78 Meta substitution will be preferred to para substitution. In the intermediate for para substitution, a full and partial positive charge are opposed on adjacent carbons.

PROBLEM 14.33 Explain why cyanobenzene (benzonitrile) substitutes preferentially in the meta position.

SUMMARY

Substituents on the benzene ring determine the position of the next substitution reaction. Generally, substituents that direct ortho/para also increase the rate of substitution. If further substitution is in the ortho or para position (but not the meta position), they provide additional stabilization to the cyclohexadienyl cation intermediate, and thus increase the rate. Meta-directing substituents make the best of an energetically bad set of choices. Meta substitution is slow and wins by default—it is the least-bad alternative for some substituents.

14.10 INDUCTIVE EFFECTS IN AROMATIC SUBSTITUTION

So far, we have mainly considered the way in which groups attached to a benzene ring can affect further reaction through resonance stabilization or electrostatic destabilization of an intermediate cyclohexadienyl cation. We should also consider the inductive effects of groups, which act by withdrawing or donating electrons through the σ bonds. For example, both the trimethylammonium group and the nitro group will be strongly electron withdrawing (Fig. 14.79).

These σ bonds are strongly polarized so as to make introduction of a positive charge into the ring difficult

↕ Shows a dipole in the direction indicated by the δ^+ and δ^-

FIGURE 14.79 In the trimethylanilinium ion and nitrobenzene, the σ bond from the ring to the substituent will be polarized by the electron-withdrawing group. Further substitution is difficult.

These subsituents will reduce the electron density in the ring making the aromatic compound a weaker nucleophile, and thus a less active participant in any reaction with an electrophilic reagent. More important will be the destabilizing effect of electron withdrawal on the charged intermediate and the partially charged transition state leading to the intermediate. So, it is no surprise to find that both trimethyl-anilinium ion and nitrobenzene are very slow to react in electrophilic aromatic substitution. Similar rate-retarding effects might be expected to operate whenever an atom more electronegative than carbon is attached to the ring.

If that is the case, what is one to make of the rate *accelerating* effect of the methoxyl group? Anisole (methoxybenzene) reacts much faster than benzene itself, and yet the methoxyl group is certainly withdrawing electrons from the ring through induction. The dipole in the carbon–oxygen σ bond surely points toward the electronegative oxygen atom (Fig. 14.80).

FIGURE 14.80 The carbon–oxygen bond in anisole is also electron withdrawing, yet substitution is directed ortho/para and the reactions are fast. The resonance effect outweighs this inductive effect.

This σ bond is also strongly polarized so as to make introduction of a positive charge into the ring difficult; yet, the reactions of anisole are very fast

The answer to this conundrum is simply that the resonance effect must substantially outweigh the inductive effect. The fourth resonance form in Figure 14.66 helps more in energy terms than the inductive withdrawal of σ electrons by methoxy hurts.

Thus, often we have two effects operating in opposition. In anisole, the resonance effect stabilizes the intermediate for ortho or para substitution and the transition states leading to them, accelerating the reaction and directing ortho/para. The inductive effect is decelerating for any substitution reaction. For anisole and most groups, the resonance effect wins out, but we can surely find examples where the two effects are largely in balance; where substitution is still ortho/para because of a resonance effect, but where a strong inductive effect slows the reaction relative to unsubstituted benzene. A look at Table 14.2 shows that there are such cases. For example, halobenzenes direct substitution to the ortho/para positions but substitution proceeds at a slower rate than that for reaction of benzene itself. In halobenzenes, *all* electrophilic aromatic substitution is retarded by the strong electron-withdrawing effect of the electronegative halogens (Fig. 14.81).

FIGURE 14.81 Chlorine and the other halogens are strongly electron withdrawing and will slow the rate of further substitution.

This bond is strongly polarized and the δ+ on carbon acts to retard any substitution reaction that would place more positive charge on the ring carbons

However, ortho/para substitution is still favored over meta because the halogens can stabilize an adjacent positive charge by resonance (Fig. 14.82).

The meta substitution of chlorobenzene — three resonance forms

The para substitution of chlorobenzene — there are four forms because the chlorine can help share the positive charge

There is another resonance form for this intermediate! Here it is.

FIGURE 14.82 Chlorine and other halogens can help stabilize the intermediate for para substitution through a fourth resonance form. The halogens strongly withdraw electrons from the ring through an inductive effect, however, and substitution is slow relative to benzene.

The net result is an overall slowing of the rate. The rate of meta substitution, in which the stabilizing resonance effect cannot operate, is slowest of all, and so meta product is greatly disfavored (Fig. 14.83).

(0.9%)   (69.5%)   (29.6%)

FIGURE 14.83 The nitration of chlorobenzene gives almost entirely ortho and para products.

14.11 POLYSUBSTITUTION OF AROMATIC COMPOUNDS AND SYNTHESIS OF MULTIPLY SUBSTITUTED BENZENES

Sometimes it is easy to predict the position of attack on a multiply substituted benzene ring. For example, if we have a ring substituted 1,3 with meta directing groups, further substitution will surely occur at the position meta to both groups, although the rate will be severely retarded by the presence of a pair of deactivating groups (Fig. 14.84).

FIGURE 14.84 On a benzene ring substituted 1,3 with a pair of meta directing groups, both substituents direct further substitution to the same position.

For example, nitration of 1,3-dinitrobenzene gives only meta substitution (Fig. 14.85).

FIGURE 14.85 Nitration of 1,3-dinitrobenzene gives a single product, 1,3,5-trinitrobenzene.

Further substitution of a molecule containing an ortho/para director and a meta director at the 1- and 4-positions is similarly easy to predict. Both groups direct the next substitution to the position ortho to the ortho/para directing group. An example of this kind of reaction is the nitration of 4-nitrotoluene (Fig. 14.86).

FIGURE 14.86 Substitution is also directed to a single position if a benzene ring is substituted 1,4 with one ortho/para directing group and one meta directing group.

In the example in Figure 14.86, we really didn't have to worry much about the directing effect of the nitro group, a rate-retarding substituent, which will be overwhelmed by the activating and ortho/para directing methyl group. However, there are less obvious situations. If the benzene ring bears different ortho/para directing groups at the 1- and 4-positions, the position of further substitution is not immediately clear. Sometimes steric effects determine the outcome. For example, Friedel–Crafts acylation of 4-isopropyltoluene gives reaction only at the less encumbered position adjacent to the methyl group (Fig. 14.87). In other cases, electronic factors determine the outcome, and further reaction will be at the position activated by the more strongly activating group. For example, the Friedel–Crafts alkylation of 4-methylphenol (*p*-cresol) gives almost entirely substitution ortho to the strongly ortho/para directing hydroxyl group, and only traces of substitution ortho to the less activating methyl group (Fig. 14.87).

FIGURE 14.87 Two examples of reactions of benzenes substituted 1,4 with two different ortho/para directing groups. Steric effects decide the first case, electronic effects the second.

Other situations are more difficult to predict. Generally, the result will be determined by the sum of the steric and electronic effects of groups already present on the ring.

Some substituents are so strongly activating that no catalyst is needed, and it is often difficult to stop substitution before more than one new substituent has been added. For example, many anilines are brominated with Br_2 alone to give 2,4,6-tribromoanilines. Phenols are also highly activated toward further substitution. Mild conditions are needed to restrict the reaction to monosubstitution (Fig. 14.88).

FIGURE 14.88 The bromination of anilines and phenols often leads to multiple additions of several bromines. The strongly activating groups, amino and hydroxyl, direct substitution to the ortho and para positions.

Arene halogenation

The hydroxyl group and the amino group are so strongly activating that neither monobromo nor dibromo compounds can be isolated unless great care is taken. Note that when the first bromine is added in the para position we have the general situation of Figure 14.87. Here, the OH and the Br direct further substitution to different positions. The more powerfully activating OH and NH_2 groups win out (bromine is actually _de_activating), and the second and third bromines add ortho to the amino or hydroxyl group.

It is possible to deactivate the amino group by acylating it through reaction with acetyl chloride to give acetamidobenzene, also known as acetanilide (Fig. 14.89). _N_-Acylated aromatic amines still direct ortho/para, but substitution is slower than for the unsubstituted amine, and the reaction can be stopped after a single new substitution. The protecting acyl group can be removed later by treatment with acid or base. We will see the mechanisms for these reactions in Chapter 19 (Fig. 14.89).

FIGURE 14.89 Acylated aniline is a less activated molecule than aniline. Monosubstitution can be achieved.

PROBLEM 14.34 Through a mechanistic analysis of the substitution of acetamidobenzene (acetanilide), explain why this compound substitutes mainly para, but at a rate slower than aniline itself.

Effective use can sometimes be made of removable blocking groups on the ring. Suppose you are set the task of making pure _o_-bromophenol. Simple bromination of phenol will fail because of the dominant formation of the para isomer and polybrominated compounds (Fig. 14.90).

FIGURE 14.90 Direct bromination of phenol cannot give *o*-bromophenol.

The solution is first to sulfonate to give the 2,4-disulfonic acid of phenol. The deactivating effect of the sulfonic acid groups prevents further sulfonation. This compound can now be brominated at the position ortho to the activating hydroxyl group. The sulfonic acids have acted to block substitution at the other ortho and para positions. Removal of the blocking sulfonate groups with acidic steam completes the reaction (Fig. 14.91).

FIGURE 14.91 The para and one ortho position can be reversibly blocked by use of the sulfonic acid group. Once the sulfonic acid is attached, only one ortho position remains for further substitution. After bromination, the sulfonic acid groups can easily be removed.

The ability to interconvert nitro and amino groups is also useful. The synthesis of 4-bromonitrobenzene from nitrobenzene is an example. Nitrobenzene cannot be brominated in the para position because the meta-directing nitro group will force formation of the meta isomer (Fig. 14.92).

FIGURE 14.92 Nitrobenzene cannot be brominated in the ortho or para positions. Substitution is entirely meta.

The solution is to convert nitrobenzene into aniline by a reduction step, acetylate, and then brominate. Bromination now goes 100% in the para position. Now the amino group can be regenerated by removal of the acyl group and then oxidized to a nitro group (Fig. 14.93). The initial reduction of nitro to amino can be carried out catalytically, or by a variety of other reducing agents. The oxidation step is usually accomplished with trifluoroperacetic acid or *m*-chloroperbenzoic acid.

FIGURE 14.93 Conversion of the nitro group into the ortho/para-directing amino group followed by acylation allows the introduction of a single para bromine. Regeneration of the amine, followed by oxidation back to nitro, completes the reaction sequence.

The synthesis of a disubstituted benzene in which two meta-directing groups are substituted ortho or para to each other constitutes a general synthetic problem. To solve such a problem, the ability to convert a meta-directing group into an ortho/para director *and back again* is crucial. A similar difficulty is encountered in the construction of disubstituted benzenes bearing ortho/para-directing groups in meta positions.

WORKED PROBLEM 14.35 The nitration of aniline takes place in acid. Under acidic conditions the amino group must be largely protonated to give an anilinium ion, a meta-directing group. In fact, if the amino group is not acetylated, the product does contain substantial amounts of *m*-nitroaniline. Why is the product not entirely *m*-nitroaniline? *Hint*: Think about the relative rates of the reactions involved.

(continued)

WORKED PROBLEM 14.35 *(continued)*

Anilinium ion

ANSWER Aniline will protonate in acid, and the anilinium ion will substitute meta. However, as long as there is some unprotonated aniline present at equilibrium, it will continue to substitute ortho and para. It is estimated that some anilines react at 10^{19} times the rate of the deactivated anilinium ions! So, even if there is only a little free amine present there can still be a great deal of product formed from it.

This arrow is the key; as long as some free amine is present, the high rate of its substitution reactions will lead to the formation of ortho and para product

PROBLEM 14.36 Devise syntheses of the following substituted benzenes. You may start from benzene, organic reagents containing no more than two carbons, and any inorganic reagents you may need (no carbons).

(a)

(b)

(c)

(d)

SUMMARY

When the reacting benzene ring is polysubstituted, the original substituents may reinforce each other or compete. One has to determine the relative influences, both resonance and inductive, of each substituent in assessing where the next substitution will occur. Resonance effects usually dominate inductive effects.

14.12 NUCLEOPHILIC AROMATIC SUBSTITUTION

14.12a Nucleophilic Additions to Benzenes. Normally, alkenes are unaffected by treatment with nucleophiles, and it is no surprise to find that benzene also survives treatment with strong nucleophiles. In particular, no reaction occurs when simple substituted benzenes are treated with base (Fig. 14.94).

FIGURE 14.94 Benzenes do not normally react with nucleophiles to give substituted benzenes.

A few specially substituted benzenes do undergo reaction in base. For example, 2,4-dinitrochlorobenzene is converted into 2,4-dinitrophenol by treatment with sodium hydroxide in water (Fig. 14.95).

FIGURE 14.95 Some substituted benzenes are reactive toward nucleophiles. Electron-withdrawing groups are particularly effective, and 2,4-dinitrochlorobenzene can be swiftly transformed into the corresponding phenol by hydroxide ion in water.

Nucleophilic Aromatic Substitution

The nitro group is essential, and it must be substituted in the right place relative to the chlorine. Chlorobenzene and *m*-nitrochlorobenzene do not react under these conditions, but *p*-nitrochlorobenzene does, although it substitutes much more slowly than the 2,4-dinitro compound (Fig. 14.96).

Relative Rate		

115,000

NO₂ / NO₂ / Cl → Na⁺ ⁻OCH₃, HOCH₃, 50 °C → NO₂ / NO₂ / OCH₃

3.4

NO₂ / Cl → Na⁺ ⁻OCH₃, HOCH₃, 50 °C → NO₂ / OCH₃

1.0

Cl / NO₂ → Na⁺ ⁻OCH₃, HOCH₃, 50 °C → OCH₃ / NO₂

0

NO₂/Cl or Cl → Na⁺ ⁻OCH₃, HOCH₃, 50 °C → No reaction

FIGURE 14.96 For the reaction to succeed, at least one nitro group must be present. 2,4-Dinitro compounds (represented by the first molecule in this figure) react much faster than mononitro compounds.

Neither an S_N2 nor an S_N1 reaction seems likely. Attack from the rear is impossible, as the ring blocks the path of any entering nucleophile, and ionization would produce a most unstable carbocation (Fig. 14.97).

Nu:⁻

The path of addition in the S_N2 reaction is blocked

S_N1

An unstable phenyl cation — the S_N1 reaction will not be easy

FIGURE 14.97 Neither an S_N2 nor S_N1 reaction is likely in this system.

The mechanism of this reaction involves addition to the benzene ring by methoxide to generate a resonance-stabilized anionic intermediate, called a **Meisenheimer complex**, after Jakob Meisenheimer (1876–1934) (Fig. 14.98).

FIGURE 14.98 Addition to the ring by methoxide leads to a highly resonance-stabilized anion. Both nitro groups help bear the negative charge.

The nitro groups stabilize the negative charge through resonance, and therefore will exert their effect *only* when they are substituted at a carbon that helps to bear the negative charge. This stabilization by nitro is the reason that *o*- or *p*-nitrochlorobenzene undergoes the reaction, but the meta isomer does not (Fig. 14.99).

FIGURE 14.99 The stabilizing nitro groups must be in the proper position. Nitro groups meta to the point of attack cannot stabilize the negative charge by resonance.

The aromatic system can be regenerated by reversal of the attack of methoxide or by loss of chloride to give 2,4-dinitroanisole (Fig. 14.100).

FIGURE 14.100 After addition, the reaction is completed by loss of chloride and regeneration of the aromatic ring. Loss of methoxide only leads back to starting material.

14.12b Nucleophilic Additions to Heteroaromatics. Pyridine is more active in nucleophilic aromatic substitution than benzene. Not only is pyridine less well resonance stabilized than benzene (Table 14.1), but the electronegative nitrogen atom provides an attractive place for an incoming negative charge (Fig. 14.101).

The classic example of nucleophilic aromatic substitution of pyridine is the **Chichibabin reaction**, named for Alexei E. Chichibabin (1871–1945), in which pyridine is converted into 2-aminopyridine through treatment with potassium amide in liquid ammonia (Fig. 14.102).

The first step of the mechanism is conventional, as the strongly basic amide ion adds to pyridine. Addition at the 2- or 4-position makes it possible for the negative charge to be shared by nitrogen and so is preferred to attack at the 3-position (Fig. 14.103).

FIGURE 14.101 Nucleophilic attack on pyridine.

(78%)

FIGURE 14.102 The Chichibabin reaction.

FIGURE 14.103 Addition to the 3-position does not allow the nitrogen to help stabilize the negative charge. Addition to the 2- or 4-position does.

This conventional beginning is followed by a rather unconventional conclusion, as aromaticity is regained through loss of hydride. Aminopyridine and potassium hydride are the initially formed products. These react rapidly and irreversibly to form hydrogen gas and a nicely resonance-stabilized amide ion. A final hydrolysis step regenerates aminopyridine. It is presumably the irreversible nature of hydrogen gas formation that helps drive this reaction to completion (Fig. 14.104).

FIGURE 14.104 Loss of hydride is followed by formation of H_2. Hydrolysis regenerates 2-aminopyridine.

Other strong nucleophiles such as organolithium reagents also form substituted pyridines. In these reactions, hydrogen gas cannot be formed until the final hydrolysis step (Fig. 14.105).

FIGURE 14.105 Addition of alkyllithium reagents also leads to substituted pyridines.

Nucleophilic substitution provides a means of converting pyridines bearing a leaving group at the 2- or 4-position into differently substituted molecules. Addition to the position bearing the leaving group (called **ipso attack**) leads to an intermediate that can either revert to starting material or eliminate the leaving group to give a new compound (Fig. 14.106).

FIGURE 14.106 Substitution of a pyridine through ipso attack. A leaving group is displaced through an addition–elimination reaction.

WORKED PROBLEM 14.37 Explain why this procedure works only for pyridines sub-stituted at the 2- or 4-position and fails for 3-substituted pyridines.

ANSWER The key to the relatively easy nucleophilic additions to pyridines is the place-ment of the negative charge on nitrogen. If the leaving group is attached to the 3-position, ipso addition of a nucleophile does not place the negative charge on the electronegative nitrogen atom. Only if the leaving group is at the 2- or 4-position does this happen.

The *N*-oxides mentioned in Section 13.10c are especially active in nucleophilic aromatic sustitution because the incoming nucleophile annihilates the positive charge on nitrogen as it adds. Addition destroys aromaticity, but at least for the *N*-oxides, loss of a leaving group regenerates the aromatic sextet. This reaction is gen-erally carried out in the presence of an acid chloride (here, SO_2Cl_2), in order to pro-duce the necessary good leaving group (Fig. 14.107).

FIGURE 14.107 Pyridine *N*-oxides are relatively easily attacked by nucleophiles. Loss of a leaving group leads to substituted pyridines.

Nature uses a similar reaction, the transfer of hydride to the "para" position of the pyridinium ion, NAD^+, the oxidized form of nicotinamide adenine dinucleotide, in biological oxidation. We will see more of this reaction in Chapter 16 (Fig. 14.108).

FIGURE 14.108 Reduction of NAD^+ through hydride transfer.

SUMMARY

There are similarities between nucleophilic aromatic substitution and its more usual electrophilic counterpart. Each involves the formation of a resonance-stabilized intermediate, and each involves a temporary loss of aromaticity that is regained in the final step of the reaction. But the similarities are only so deep. One reaction is cationic; the other involves anions. Use the differing effects of a nitro group, strongly decelerating in the electrophilic substitution and strongly accelerating in the nucleophilic reaction, to keep the two mechanisms distinct in your mind.

14.13 SPECIAL TOPIC: STABLE CARBOCATIONS IN "SUPERACID"

The Friedel–Crafts reactions succeed because relatively stable carbocations can be formed by the reaction of aluminum chloride (an exceptionally strong Lewis acid with respect to halide as Lewis base) in solvents that are nonnucleophilic. The strongest nucleophile available is benzene, which attacks the cation to start the Friedel–Crafts reaction. If stable carbocations cannot be formed, as with methyl chloride or ethyl chloride, the $AlCl_3$ complexes can also serve to alkylate benzenes.

How would we modify the reaction conditions so as to maximize the possibility of generating the carbocation and minimize the chances for its further reaction? Antimony pentafluoride (SbF_5) is an even stronger Lewis acid than $AlCl_3$ toward halogen, particularly fluorine. When SbF_5 removes a fluoride ion from an alkyl compound the anion antimony hexafluoride (SbF_6^-) is produced (Fig. 14.109). This anion is nonnucleophilic at antimony, and a poor nucleophile at fluorine as well.

FIGURE 14.109 Cation formation through reaction of an alkyl fluoride with SbF_5.

PROBLEM 14.38 Explain why SbF_6^- is nonnucleophilic at antimony and a very poor nucleophile at fluorine.

Treatment of alkyl fluorides with SbF_5 at very low temperatures in highly polar, nonnucleophilic solvents such as SO_2, SO_2ClF, or FSO_3H (so-called **superacid** conditions) can generate many stable carbocations. This process is Friedel–Crafts chemistry carried to extremes. Methyl and primary carbocations cannot be generated this way, but tertiary and secondary cations not prone to rearrangement to more stable cations can be (Fig. 14.110).

FIGURE 14.110 Stable secondary and tertiary carbocations can be formed in superacid.

Many carbocations, once thought impossible species even as reaction intermediates, can be generated and studied under superacid conditions. This work, pioneered by George A. Olah (b. 1927) now at the University of Southern California, has been one of the most exciting research areas of recent times and earned Olah the Nobel prize in 1994. We will see a number of examples of the utility of looking at cations in superacid in future chapters.

PROBLEM 14.39 When butyl fluoride is dissolved in superacid at low temperature, it is possible to observe a stable species. However, the ^{13}C NMR spectrum of that stable species shows only two signals.

(a) Why doesn't the "expected" spectrum appear? What is the species that is being observed?

(b) Although part (a) is quite easy, this reaction actually poses a vexing problem. Try to write a mechanism for formation of the stable species. What problems appear when you do?

14.14   SPECIAL TOPIC: BENZYNE

In Section 14.12, we mentioned the lack of reactivity of halobenzenes with bases. It took the activation of a nitro group for substitution to occur, for example (Figs. 14.94–14.96). In truth, if the base is strong enough, even chlorobenzene will react. In the very strongly basic medium potassium amide (K^+ $^-NH_2$) in liquid ammonia (NH_3), chlorobenzene is converted into aniline (Fig. 14.111).

FIGURE 14.111 Aniline can be formed from chlorobenzene by reaction with very strong base, potassium amide in ammonia.

$$\text{(chlorobenzene, Cl)} \xrightarrow[\text{NH}_3,\ -33\,°C]{K^+\ {}^-NH_2} \text{(aniline, NH}_2) \quad + \quad KCl$$

(50%)

The mechanism is surely not an S_N2 displacement, and there is no reason to think nucleophilic addition would be possible to an unstabilized benzene ring. This difficult mechanistic problem begins to unravel with the observation by John D. Roberts (b. 1918) and his co-workers that labeled chlorobenzene produces aniline *labeled equally in two positions* (Fig. 14.112).

FIGURE 14.112 A clue to the mechanism is provided by Roberts' observation of isotopic scrambling when labeled chlorobenzene is used in the reaction. The red dot = ^{14}C, an isotopic label.

$$\text{(chlorobenzene, Cl, •)} \xrightarrow[\text{NH}_3]{Na^+\ {}^-NH_2} \text{(aniline, NH}_2, •) \quad + \quad \text{(aniline, •, NH}_2)$$

Benzyne formation

What reactions of halides do you know besides displacement? The E1 and E2 reactions compete with displacement reactions (Chapter 7, p. 312). If HCl were lost from chlorobenzene, a cyclic alkyne, called **dehydrobenzene** or **benzyne**, would be formed. This symmetrical bent acetylene might be reactive enough to undergo an addition reaction with the amide ion. If this were the case, the labeled material must produce two differently labeled products of addition (Fig. 14.113).

$$\text{(chlorobenzene, Cl, •, H)} \xrightarrow{Na^+\ {}^-NH_2} \text{NH}_3\ +\ NaCl\ +\ \text{(Benzyne, •)}\ \overset{(a)}{\underset{(b)}{\rightleftharpoons}}\ :\!NH_2$$

www

Benzyne

$$(a)\downarrow \qquad (b)\downarrow$$

FIGURE 14.113 If the symmetrical intermediate benzyne were formed, readdition of the amide ion must produce the observed labeling results.

$$\text{(ÑH}_2,\ •,\ H) \quad + \quad \text{(H, •, ÑH}_2) \xleftarrow{NH_3} \text{(ÑH}_2,\ •,\ {}^-) \quad + \quad \text{(•, }{}^-,\ ÑH_2)$$

$$+\ {}^-\!:\!NH_2$$

Benzyne is surely an unusual species and deserves a close look. Although this intermediate retains the aromatic sextet, the triple bond is badly bent, and there must be very severe angle strain indeed. Remember, the optimal angle in triple bonds is 180° (Chapter 3, p. 124). Moreover, the "third" bond is not composed of $2p$–$2p$ overlap like a normal alkyne, but of overlap of two hybrid orbitals (Fig. 14.114).

These π bonds are the remainder of the π system

These orbitals are part of the π system

These orbitals are not in the π system — they make up the "extra" bond in benzyne

Benzyne

FIGURE 14.114 The structure of benzyne.

One would therefore not be too surprised to find that this molecule is able to add strong bases such as the amide ion. In fact, one would expect very high reactivity for this strained acetylene. Problem 14.40 gives you a chance to see some of the reactions of benzyne.

PROBLEM 14.40 Provide arrow formalisms for the formation of the following products:

14.15   SPECIAL TOPIC: BIOLOGICAL SYNTHESIS OF AROMATIC RINGS; PHENYLALANINE

Some of the amino acids from which our tissues are constructed contain benzene rings, and so nature must have a way of making them. You might actually set your-self that problem right now; you have at your disposal the means to generate ben-zene from simple hydrocarbon precursors.

PROBLEM 14.41 Design a synthesis of benzene starting from inorganic compounds, and organic compounds containing only hydrogen and no more than four carbon atoms. *Hint*: Your primary goal is to construct that six-membered ring. You have only one easy way to do that (so far).

Nature can't use the method you designed in Problem 14.41, because too much energy is needed to do the crucial reaction that constructs the six-membered ring. Nature starts from simple sugars, which contain the six carbons necessary for the ring, along with many hydroxyl groups. (For much more on carbohydrates, wait for Chapter 22.) The overall strategy, if one may speak of Nature having a strategy, is to build the ring, then use those OH groups in various ways to do elimination reactions and put in the necessary double bonds. Here is how it works for phenylalanine, one of your amino acids.

The six-carbon sugar is transformed enzymatically into dehydroquinate (**A**), which undergoes an elimination reaction, also enzyme mediated, to give dehydroshikimate (**B**) (Fig. 14.115).

FIGURE 14.115 Enzymes transform a sugar into dehydroquinate (**A**) and then dehydroshikimate (**B**).

A sugar   Dehydroquinate (**A**)   Dehydroshikimate (**B**)

The carbon–oxygen double bond in **B** is reduced and phosphorylated to give shikimate-3-P (**C**). Another enzyme adds a three-carbon fragment to one of the hydroxyl groups (you'll see why this is critical in a moment) to give **D**, which undergoes a second elimination to place the second double bond in the ring in **E**, chorismate (Fig. 14.116).

FIGURE 14.116 Dehydroshikimate (**B**) is transformed into chorismate (**E**).

Dehydroshikimate (**B**)   Shikimate-3P (**C**)

Chorismate (**E**)   D

So far the strategy we guessed at is working—Nature has effectively done two elimination reactions and seems poised to do a third. But now that side chain that was added in **D** comes into play, and chorismate mutase induces a transformation of **E** to prephenate, **F** (Fig. 14.117). You will see this process in some detail much later if you do Chapter 20, but see if you can write an arrow formalism for this transformation.

Chorismate
(E)

enzyme

Prephenate
(F)

FIGURE 14.117 Chorismate (**E**) is made into prephenate (**F**).

PROBLEM 14.42 Write an arrow formalism for the transformation of **E** into **F**. Don't worry about the role of the enzyme.

The final elimination occurs in prephenate, with the carboxylate (COO⁻) acting as trigger. Elimination leads to phenylpyruvate (**G**). Now the second role of the three-carbon chain is revealed, as it is elaborated into the side chain of phenylalanine (Fig. 14.118).

Prephenate
(F)

Phenylpyruvate
(G)

Phenylalanine

FIGURE 14.118 Finally, prephenate (**F**) is transformed into phenylalanine.

This sequence is far from simple, but even though it is laden with bells and whistles, some of which we surely could not anticipate, the general adherence to our early strategy—build the six-membered ring and then use the hydroxyls as leaving groups—is apparent.

PROBLEM 14.43 There is another amino acid, tyrosine, that is formed from prephenate. At first this transformation seems surprising, perhaps even impossible. Analyze what must be done and explain why it will not be easy.

Tyrosine

14.16 SUMMARY

New Concepts

This chapter involves the chemistry of aromatic compounds and is dominated by two basic interrelated notions. First, it is simply difficult to disturb the aromatic sextet, and second, once aromaticity is disrupted, it is easily regained. In chemical terms, this means that reactions of benzene and other simple aromatic compounds generally involve a highly endothermic first step with formation of a high-energy intermediate. In electrophilic aromatic substitution, the most common reaction of aromatic compounds, this intermediate is the cyclohexadienyl cation. As this intermediate lies far above benzene in energy, the transition state leading to it is also high in energy, and the reaction is difficult. Reactions of the cyclohexadienyl cation that regenerate aromatic systems are highly exothermic, and therefore have low activation barriers.

Further substitution reactions of an already substituted benzene are controlled by the properties of the first substituent. Both resonance and inductive effects are involved, but the resonance effect usually dominates. Many substituents can stabilize an intermediate cyclohexadienyl cation through resonance if substitution is in the ortho or para position. The extra stabilization afforded by the substituent leads to relatively high rates of reaction. Positively charged substituents generally substitute meta, as ortho or para substitution places two positive charges on adjacent positions. The introduction of a second positive charge means that these substitutions are relatively slow.

At low temperature, in sufficiently polar, but nonnucleophilic solvents, some carbocations are stable. Even under these superacid conditions, rearrangements through hydride or alkyl shifts to the most stable possible carbocation are common.

Key Terms

acid chloride (p. 692)
acyl group (p. 692)
acylium ion (p. 692)
aniline (p. 696)
anisole (p. 706)
benzyne (p. 734)
Chichibabin reaction (p. 729)

dehydrobenzene (p. 734)
diazonium ion (p. 697)
electrophile (p. 673)
electrophilic aromatic substitution (p. 683)
Friedel–Crafts acylation (p. 692)
Friedel–Crafts alkylation (p. 688)
ipso attack (p. 730)

Reactions, Mechanisms, and Tools

By far the most important reaction in this chapter is electrophilic aromatic substitution by a variety of E^+ reagents. This reaction involves the initial formation of a resonance-stabilized, but not aromatic, cyclohexadienyl cation. Deprotonation regenerates an aromatic system (Fig. 14.119).

Nucleophilic aromatic substitution occurs only when the ring carries substituents that can stabilize the negative charge introduced through addition of a nucleophile. Departure of a leaving group regenerates the aromatic system (Fig. 14.100).

FIGURE 14.119 The general mechanism for electrophilic aromatic substitution.

Syntheses

This chapter provides many new syntheses of aromatic compounds. Although we can now make all sorts of differently substituted benzenes, there is still a unifying theme: each product ultimately derives from a reaction of benzene with an E^+ reagent.

1. ALKANES

Wolff–Kishner reduction

Clemmensen reduction

2. ALKYLBENZENES

Friedel–Crafts alkylation; carbocationic rearrangements occur to give the most stable ion; the possible structures of R are therefore limited

Wolff–Kishner or Clemmensen

See item 1

3. ANILINES

Reduction of nitro groups

4. BENZENES

SO₂OH

Reversal of sulfonation reaction

₊N₂

The diazonium ion can be
made from nitrobenzene

5. BENZENESULFONIC ACIDS

Simple sulfonation

6. CYANOBENZENES

₊N₂ Sandmeyer reaction CN

7. CYCLOHEXANES

High-pressure catalytic hydrogenation

8. DEUTERATED BENZENES

See also methods for making benzenes in item 4

9. DIAZONIUM IONS

10. HALOBENZENES

Aromatic halogenation

Aromatic halogenation

Sandmeyer reaction

Sandmeyer reaction

Schiemann reaction

11. NITROBENZENES

Simple nitration

Oxidation of aniline

12. PHENOLS

13. PHENYL KETONES

Friedel–Crafts acylation; no rearrangements occur

Common Errors

The most common conceptual error is to attempt to analyze the course of aromatic substitution reactions through looking at ground-state molecules (the starting materials and products). The various resonance and inductive effects discussed in this chapter will be far more important in the charged intermediates and in the transition states leading to those intermediates. When analyzing one of these reactions, always look first at the cyclohexadienyl intermediate (a cation in electrophilic aromatic substitution and an anion in the less common nucleophilic reactions), and then consider the transition state leading to it.

Another potential error is failing to keep up with the proliferating synthetic detail. Be sure to look at each new reaction in this and subsequent chapters with an eye to using it in synthesis. Keep up your file cards!

14.17 ADDITIONAL PROBLEMS

PROBLEM 14.44 Explain why the trifluoromethyl group is meta directing in electrophilic aromatic substitution. Would you expect CF_3 to be activating or deactivating? Why?

PROBLEM 14.45 Show where the following compounds would undergo further substitution in electrophilic aromatic substitution. There may be more than one position for further substitution in some cases. Explain your reasoning.

(continued)

PROBLEM 14.45   *(continued)*

(d)

(e)

(f)

(g)

(h)

Ac =

PROBLEM 14.46   Predict the position of further substitution in biphenyl. Explain your reasoning.

Biphenyl

PROBLEM 14.47   Provide syntheses for the following compounds, as free of other isomers as possible. You do not need to write mechanisms, and you may start from benzene, inorganic reagents of your choice, organic reagents containing no more than four carbons, and pyridine.

(a)

(b)

(c)

(free of para)

(d)

PROBLEM 14.48   Provide syntheses for the following compounds, free of other isomers. You do not need to write mechanisms, and you may start from benzene, inorganic reagents of your choice, organic reagents containing no more than two carbons, and pyridine.

(a)

(b)

(c)

PROBLEM 14.49   How many signals would appear in the ^{13}C NMR spectra of the following four compounds? Try to use an analysis of symmetry in your answer.

(a)

(b)

(c)

(d)

PROBLEM 14.50   Provide syntheses for the following compounds, as free of other isomers as possible. You do not need to write mechanisms, and you may start from benzene, inorganic reagents of your choice, organic reagents containing no more than three carbons, and pyridine.

(a)

(b)

(c)

PROBLEM 14.51 Provide structures for compounds **A–E**. Mechanisms are not required, but a mechanistic analysis of the various reactions will probably prove helpful in deducing the structures. *Hint*: You might look at the figure in Problem 14.13 (p. 695).

PROBLEM 14.52 One substituent you do not know how to attach to a benzene ring is the aldehyde group, HC=O. Here is a way. The price of this knowledge is that you must write a mechanism for the following aromatic substitution reaction.

PROBLEM 14.53 Lies! lies! lies! I can think of at least one way you *can* put an aldehyde group on a benzene ring with your current knowledge. It is a roundabout process, and so makes a good "roadmap" problem. Try this one. Provide structures for compounds **A–D**.

PROBLEM 14.54 Explain why it is possible to alkylate a phenol selectively in the presence of an alcohol.

PROBLEM 14.55 Write structures for compounds **A, B**, and **C**.

PROBLEM 14.56 How would you convert aniline into each of the following molecules? Mechanisms are not necessary.

PROBLEM 14.57   Write mechanisms for the following two related reactions:

(43%)

75–80%
H₂SO₄

(75%)

PROBLEM 14.58   Rationalize the following data. An Energy versus Reaction progress diagram will be useful. Be sure to explain why the low-temperature product is mainly naphthalene-1-sulfonic acid, and why the high-temperature product is mainly naphthalene-2-sulfonic acid. Note that naphthalene-2-sulfonic acid is more stable than naphthalene-1-sulfonic acid. You do not have to explain the relative energies of the two naphthalene sulfonic acids, although you might guess, and the answer will tell you.

concd H₂SO₄ | 40 °C

concd H₂SO₄ | 160 °C

SO₂OH

(major)
+

SO₂OH

(minor)

SO₂OH

(minor)
+

SO₂OH

(major)

PROBLEM 14.59   Write a mechanism for the following chloromethylation reaction:

H₂C=O
HCl

CH₂Cl

(75%)

PROBLEM 14.60   The following wonderful question appeared on a graduate cumulative examination at Princeton in January, 1993. With a little care you can do it easily. Write a mechanism for this seemingly bizarre change. *Hint*: It is not bizarre at all. Think simple! First consider what HCl might do if it reacts with the starting material.

AlCl₃/H₂O/HCl

benzene,
80 °C

• = ¹³C

PROBLEM 14.61   At low temperature, only a single product (**A**) is formed from the treatment of *p*-xylene with isopropyl chloride and AlCl₃, which is surely no surprise, as there is only one position open on the ring. However, at higher temperature some sort of rearrangement occurs, as two products are formed in roughly equal amounts. First, write a mechanism for formation of both products. *Hint*: Note that HCl is a product of the first reaction to give **A**. Second, explain why product **A** rearranges to product **B**. What is the thermodynamic driving force for this reaction?

CH₃

(CH₃)₂CHCl
AlCl₃, 70 °C,
2 h

CH₃

p-Xylene

CH₃
CH(CH₃)₂

A
(48%)

CH₃

H₃C
CH(CH₃)₂

B
(52%)

CH₃

PROBLEM 14.62 Provide mechanisms for the following reactions:

(a)

(b)

+ CO₂

PROBLEM 14.63 Construct an Energy versus Reaction progress diagram for the reaction of 1-chloro-2,4-dinitrobenzene with sodium methoxide.

PROBLEM 14.64 When 1,2-dichloro-4-nitrobenzene is treated with an excess of sodium methoxide, three different substitution products (two monosubstituted, one disubstituted) could, in principle, be formed. In fact, only a single product is obtained.
(a) Draw the three possible products.
(b) Carefully analyze the mechanism for this nucleophilic aromatic substitution reaction. What single product will be formed and why?

PROBLEM 14.65 A useful benzyne precursor is benzenediazonium-2-carboxylate (**1**), a molecule readily formed from anthranilic acid. Upon heating, **1** affords benzyne, which can be trapped by anthracene to give the architectural marvel, triptycene (**2**).
(a) Provide an arrow formalism for the formation of benzyne on decomposition of benzenediazonium-2-carboxylate.
(b) Similarly, provide an arrow formalism for the formation of triptycene from benzyne and anthracene.

(continued)

PROBLEM 14.65 *(continued)*

PROBLEM 14.66 Years before J. D. Roberts' classic labeling experiment (see Section 14.14, p. 733), it was known that nucleophilic aromatic substitution occasionally resulted in what was termed *cine* (from the Greek *kinema* meaning "movement") substitution. In this type of substitution reaction, the nucleophile becomes attached to the carbon adjacent to the carbon bearing the leaving group. Here are two examples of this kind of substitution:

Only isomer formed

Rationalize these *cine* substitutions through use of a substituted benzyne as an intermediate. Explain why only a single isomer is formed in the second experiment.

Analytical Chemistry: Spectroscopy

15

After a great expenditure of life and treasure a Daring Explorer had succeeded in reaching the North Pole, when he was approached by a Native Galeut who lived there.

"Good morning," said the Native Galeut. "I'm very glad to see you, but why did you come here?"

"Glory," said the Daring Explorer, curtly.

"Yes, yes, I know," the other persisted; "but of what benefit to man is your discovery? To what truths does it give access which were inaccessible before?—facts, I mean, having scientific value?"

"I'll be Tom scatted if I know," the great man replied frankly; "you will have to ask the Scientist of the Expedition."

But the Scientist of the Expedition explained that he had been so engrossed with the care of his instruments and the study of his tables that he had found no time to think of it.

—Ambrose Bierce*

"At the Pole," from *Fantastic Fables*

15.1   PREVIEW

Years ago, all universities gave one or several courses in the subject of analytical chemistry. Sadly, in recent times this material has sometimes been folded into courses in physical and organic chemistry, and the subject "lost." Sometimes also lost, I fear, is the sense of how important the subject is. Nothing has influenced the development of organic chemistry more strongly over the last 30 years than progress in analytical chemistry. After a chemical reaction is run, the experimentalist is faced

*Ambrose Bierce (1842–1914?) was an American journalist and writer of bitter, sardonic tales who vanished during the Mexican civil war at the start of this century.

with two general problems before he or she can begin to think about what the results mean. The products of the reaction must first be separated, then identified. Separation was a severe problem in the early days of organic chemistry. For many compounds, the only available methods were distillation or, for solids, fractional crystallization. Fractional crystallization works well (if done properly, which it rarely is), but this method requires reasonably large amounts of material and, of course, works only for compounds that are solids at or near room temperature. By now, your laboratory experience has likely revealed that distillation is not often the final answer to securing large amounts of very pure materials. Distillation works well if there is a large difference between the boiling points of all the desired components of a mixture, and if substantial amounts of material are available. It seems that reactions in the research laboratory are all too prone to give several isomeric products, and distillation and fractional crystallization are often impractical methods. The vexing problem of separation was relieved somewhat by the advent of column chromatography, then more or less solved, at least in the research lab, by the discovery of gas and liquid chromatography. Other chromatographic techniques include thin-layer, thick-layer, and flash chromatography. Nowadays, it is the chemist's realistic expectation that any mixture can be separated. Even enantiomers are now often separable by chromatography.

PROBLEM 15.1 How would a separation of enantiomers be possible? Are not all physical properties (except, of course, for the direction of rotation of the plane of plane-polarized light) identical for enantiomers? Can you guess the principle upon which the chromatographic separation of enantiomers must rest? *Hint*: What is true for enantiomers is not necessarily true for diastereomers (Chapter 4, p. 176).

ESSENTIAL SKILLS

1. "All" there is to understanding nuclear magnetic resonance (NMR) is to know how the chemical shift δ arises, and how the coupling constant *J* works. That's only two things, but both require a lot of experience before using them becomes second nature. Understanding the chemical shift δ requires no more than realizing that "different" hydrogens will resonate at different frequencies. That's not a difficult concept, but understanding what makes two hydrogens "different" is less simple! The coupling constant *J* means that hydrogens on adjacent carbons "talk" to each other. The state of one influences the properties of the other. There is a simple method for translating the electron "talk" but, again, it requires an understanding of difference.

2. Problem solving. Good spectroscopy problem solvers are good at solving spectroscopy problems because they have solved lots of spectroscopy problems. There is no way out! To be a decent spectroscopist you have to do many problems; the skill is highly experience-intensive and there is only one way to gain that experience. Section 15.11 will give you one way to approach spectroscopy problems, but it is certainly appropriate to develop your own.

Even given a successful separation, the chemist is now faced with the daunting task of working out the structures of the various components. Just how do we find out what's in that bottle? In the old days, structure determination was a respected profession, and many careers were built through the chemical determination of complex molecular structures. Diagnostic reactions were run, and the presence or absence of functional groups inferred from the results. Gradually, the results from many chemical experiments allowed the formulation of a reasonable structural hypothesis that could be tested through further chemical reactions. The intellectual content of such work was substantial, and the experimental skill required was prodigious. These days, the correlation of the results of chemical reactions with chemical structure has been largely replaced with the spectroscopic identification of compounds, and *every* chemist is an analytical chemist. The profession of analytical chemist has by no means disappeared, but rather has expanded to include us all. Ultraviolet/visible (UV/vis) spectroscopy (see Chapter 12, p. 566) was an early arrival, followed by mass spectrometry (MS), which is capable of determining molecular weights, and infrared (IR) spectroscopy, in which the vibrations of atoms in bonds are detected. At the end of the 1950s, nuclear magnetic resonance (NMR) spectroscopy appeared,* and the real revolution in the determination of the structures of organic molecules began. In this chapter, we will briefly discuss chromatography, MS, and IR spectroscopy, and then proceed to a more detailed investigation of NMR.

15.2   CHROMATOGRAPHY

In Chapter 4 (p. 177), we briefly encountered column chromatography. All chromatography works in much the same way. The components of a mixture are forced to equilibrate between an immobile material, the stationary phase, and a moving phase. The more time one component spends adsorbed on the stationary phase, the more slowly it moves. If it is possible to monitor the moving phase as it emerges from the chromatograph, each component of the original mixture can be detected and collected as it appears. The efficiency of column chromatography can be greatly increased by a technique called **high-performance liquid chromatography (HPLC)**. In this variation, high-pressure pumps force the liquid phase through a column of densely packed uni-

*Professor Martin Saunders (b. 1931) told me of the arrival of the long and eagerly awaited first NMR spectrometer at Yale in the late 1950s. The magnet arrived in a plywood carton labeled in bold letters, "2.5 tons." Unfortunately, the workman who was to bring the machine into the laboratory didn't believe that such a heavy object could be in such a flimsy box and attempted to use his forklift, rated at 1 ton, to bring it in. The forklift did its best and labored mightily to lift the object into the air. It succeeded at this task, but then, apparently exhausted, gave out and dropped the magnet to the stone floor, where it burst its packing and lay blocking the entrance. The workman then attached a chain to the magnet and dragged it about 30 feet across the cobblestones to the side. Naturally Professor Saunders, whose professional fate depended to a fair extent on the performance of this instrument, was less than happy about this state of affairs. However, the story has a happy ending. The machine functioned brilliantly after minor repairs, and Professor Saunders has had a long and outstanding career at Yale. He claims that he advocated to Varian Associates, the manufacturer of the instrument, that they add a 5-foot drop to a cobblestone floor to the manufacturing process for their magnets, in order to bring all their magnets up to the standard of this one. Apparently, they ignored his advice.

form microspheres, which provide an immense surface for adsorption. As the liquid elutes, a detector, often a small UV/vis spectrometer, monitors the composition. Figure 15.1 shows a schematic instrument as well as a nice separation of a series of organic compounds.

FIGURE 15.1 (a) A schematic view of a liquid chromatograph. (b) A chromatogram showing the separation of a mixture of organic compounds into its components.

For separating many organic molecules, **gas chromatography (GC)** is the technique of choice. Gas chromatographs are remarkably simple and it is still possible to build one on your own that will function well in a research setting. The stationary phase consists of a high-boiling material adsorbed on a solid support. High molecular weight silicone oil polymers are often used, as well as many other organic chemicals. The moving phase is a gas, usually helium or hydrogen. Typically, the solid support is packed in a coiled column through which the carrier gas flows. The mixture to be separated is injected at one end. As the carrier gas moves the mixture through the column, the various components are differentially adsorbed by the stationary phase and exist in an equilibrium between the gas phase and solution. The more strongly a component is adsorbed, the more slowly it moves through the column. At the end of the device, there is a detector that plots the amount of material emerging against time. A schematic instrument and a typical chromatogram are shown in Figure 15.2.

FIGURE 15.2 (a) A schematic view of a gas chromatograph. (b) A chromatogram showing the separation of a mixture of complex alkaloids into its components.

There are many kinds of detectors, but two common ones use thermal conductivity or flame ionization. In a thermal conductivity detector, the carrier gas passes over two sets of hot filaments whose temperature depends on the composition of the gas flowing over them. The detector can be set to zero for pure carrier gas passing through the instrument over both filaments. As the components of the mixture elute over one set of filaments, their temperature will change and the current needed to bring them back to the zero point can be measured and plotted. A flame ionization detector measures the current induced by ions created as the effluent is burned. Again, a zero point for pure carrier gas is determined and deviations from zero detected and plotted as the components of the mixture elute. A typical gas chromatogram is shown in Figure 15.2(b).

All sorts of fancy bells and whistles are possible, and the art of creating efficient columns has progressed immensely from the days when one packed large-bore columns by hanging them down a stairwell, and pouring in the stationary phase. The resulting columns were then tested for efficiency in the sometimes vain hope that one would do a good enough job of separating the mixture of products. One especially effective modification is the construction of capillary columns. Such columns can be several hundred feet long and contain only a tiny amount of stationary phase coated on the walls. They often have immense separating powers. Yet, for the practicing chemist, the old-style simple packed columns are still very important, as they allow for the collection of the components of the mixture as they pass out of the column.

Not only can the mixtures be separated, but the pure components can be isolated for identification by the simple means of attaching a cooled receiver to the end of the chromatograph (Fig. 15.2).

A particularly useful modification of a gas chromatograph is to attach it to a mass spectrometer (**GC/MS**) or infrared spectrometer (**GC/IR**) so that the various components can be analyzed directly as they emerge from the column. In the following sections, we will see what kinds of information MS and IR spectroscopy can give about molecules.

15.3   MASS SPECTROMETRY

First of all, never make the mistake of calling it "mass spectroscopy." Spectroscopy involves the absorption of electromagnetic radiation and as we will see later, **mass spectrometry** (**MS**) is different. The mass spectrometrists sometimes get upset if you confuse the issue. In MS, a compound is bombarded with very high energy electrons. Typically, electrons of 70 eV are used.

WORKED PROBLEM 15.2 How many kilocalories per mole does 70 eV represent? See Chapter 1, page 3.

ANSWER One electron volt (eV) is 23.06 kcal/mol, so 70 eV is 1614 kcal/mol. The point of this problem is to show that the incoming electron is carrying a *lot* of energy!

One thing these very high energy electrons can do is eject an electron from a molecule, M, to give the corresponding **radical cation**, $M^{\cdot+}$ (Fig. 15.3). Now what can happen to this radical cation? In a typical mass spectrometer, it is first accelerated toward a negatively charged plate (Fig. 15.4). There is a slit in the plate, and some of the accelerated ions do not hit the plate but pass through the slit to form a beam. The beam of ions then follows a curved path between the poles of a magnet. In a magnetic field, a moving charged particle is deflected by an amount proportional to the strength of the magnetic field. In this case, too little deflection (too weak a magnetic field) results in the ion hitting the far wall of the tunnel and in destruction of the ion. If there is too much deflection (too strong a magnetic field), the ion impinges on the near wall and is also lost. If, however, the strength of the magnetic field is just right, the ion will follow a curved path to the detector and be "seen." The magnetic field can be tuned so that ions of different mass are curved just the right amount to be detected. For ions of low mass, a relatively low magnetic field will be necessary; for ions of higher mass, the field will have to be stronger to do the required bending of the path.

Mass spectra are reported as graphs of the mass-to-charge, m/z, ratio versus intensity. As long as we are dealing with singly charged ions, the positions of the signals will correspond to the molecular weights of the ions detected.

So far, we have a device that can detect the ions, $M^{\cdot+}$, formed by bombardment of a molecule M. In our detector, we should see an ion p, the **parent ion**

The in-coming electron and the ejected electron

$$M + e^- \longrightarrow 2\,e^- + M^{\cdot+}$$

A high-energy electron

The residual radical cation

FIGURE 15.3 When bombarded with high-energy electrons, molecules can lose an electron to give a radical cation.

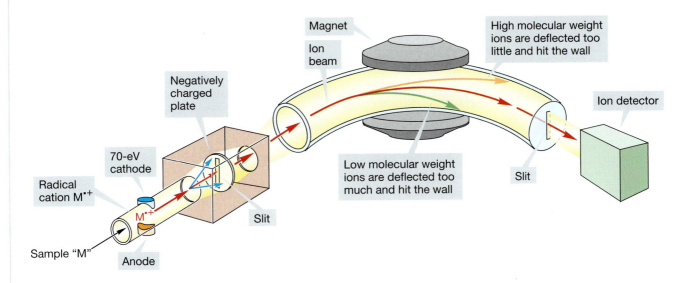

FIGURE 15.4 A schematic view of a mass spectrometer.

of mass m of M. For cyclohexane, we should see a parent ion of 84 mass units, as the molecular weight of cyclohexane (C_6H_{12}) is 84. In the mass spectrum of cyclohexane, there is a small peak above the parent ion, at m/z = 85. It is only about 7% of the parent ion, and therefore one must look closely in order to see it. This **p + 1 peak** is not the result of an impurity, however, as it is always there no matter how carefully the cyclohexane is purified. The reason is that carbon and hydrogen have naturally occurring isotopes, and the mass spectrometer can detect molecules containing these heavy isotopes. Table 15.1 gives the relative abundance of some common isotopes.

TABLE 15.1 Some Common Isotopes and Their Abundances

Isotope	Abundance	Isotope	Abundance
^1H	99.985	^{17}O	0.038
^2H (D)	0.015	^{18}O	0.200
^{12}C	98.90	^{35}Cl	75.77
^{13}C	1.10	^{37}Cl	24.23
^{14}N	99.63	^{79}Br	50.69
^{15}N	0.37	^{81}Br	49.31
^{16}O	99.762	^{127}I	100

For cyclohexane, there is a 1.1% probability that *each* carbon will be a ^{13}C, and therefore about 7% probability that the mass of cyclohexane will be one unit high. Deuterium (^2H or D) has a natural abundance of only 0.015%, and therefore contributes relatively little to the p + 1 peak. Sometimes the presence of isotopes is useful in the diagnosis of structure. For example, the observation of a roughly 1:1 doublet of peaks indicates the presence of bromine in the molecule, as the pool of natural bromine consists of an almost 50:50 mixture of ^{79}Br and ^{81}Br.

But how do we know that the mass spectrum with $m/z = 84$ is really cyclohexane, C_6H_{12}? If the starting material is of unknown structure, we probably do not know the formula of the molecule. The detection of a parent ion of $m/z = 84$ only tells us that the molecular weight of the unknown is 84, nothing more. Any other molecule of $m = 84$ could give the parent ion. In this case, we might be looking at 3,4-dihydro-2*H*-pyran, for example ($C_5H_8O = 84$, Fig. 15.5).

FIGURE 15.5 Many different ions can have a molecular weight of 84. Two possibilities are cyclohexane and 3,4-dihydro-2*H*-pyran.

PROBLEM 15.3 Find other possible structures that might fit the observed *m/z* of 84.

These days, mass spectrometers are so accurate that we can actually tell C_6H_{12} from C_5H_8O. The two species do not have the same molecular weight *if we measure molecular weights accurately enough.* Table 15.2 gives the accurate atomic weights for a variety of isotopes as well as the average values.

When a molecular formula is calculated from combustion data, it is the average values in Table 15.2 that must be used, because unless there has been a deliberate introduction of an isotope during the synthesis of the material, it will be the natural

TABLE 15.2   Some Atomic Weights

Element	Atomic Weight	Element	Atomic Weight
^1H	1.008	^{19}F	19.998
^2H	2.010	^{28}Si	27.977
H (av)	1.008	^{29}Si	28.975
^{10}B	10.013	^{30}Si	29.974
^{11}B	11.009	Si (av)	28.086
B (av)	10.810	^{31}P	30.974
^{12}C	12.000	^{32}S	31.972
^{13}C	13.003	^{33}S	32.972
C (av)	12.011	^{34}S	33.968
^{14}N	14.003	^{36}S	35.967
^{15}N	15.000	S (av)	32.066
N (av)	14.007	^{35}Cl	34.969
^{16}O	15.995	^{37}Cl	36.966
^{17}O	16.999	Cl (av)	35.453
^{18}O	17.999	^{79}Br	78.918
O (av)	15.999	^{81}Br	80.916
		Br (av)	79.904
		^{127}I	126.905

isotopic mixture that appears in the compound. The mass spectrometer, on the other hand, is able to separate ions containing different isotopes. Each peak represents a particular set of isotopes, for example, $^{12}C_6{}^1H_{12}$ or $^{12}C_5{}^1H_8{}^{16}O$. So, a good mass spectrometer can give us the molecular formula of an unknown, as long as we can detect the parent ion (Fig. 15.6).

C_6H_{12}

6 ^{12}C = 6 × 12.000 = 72.000	
12 ^1H = 12 × 1.008 = 12.096	
84.096	

m of C_6H_{12}

C_5H_8O

5 ^{12}C = 5 × 12.000 = 60.000	
8 ^1H = 8 × 1.008 = 8.064	
1 ^{16}O = 15.995 = 15.995	
84.059	

m of C_5H_8O

FIGURE 15.6 If molecular weights are measured carefully enough, mass spectrometry can differentiate between different molecules of the nominal molecular weight of 84.

As the full mass spectra of the two $m = 84$ molecules show (Fig. 15.7), mass spectra are not simple composites of parent and p + 1 peaks. There are many lower molecular weight ions formed besides the parent and p + 1 peaks, to give a characteristic array of ions known as the **fragmentation pattern** for each molecule.

FIGURE 15.7 The mass spectra of cyclohexane and 3,4-dihydro-2H-pyran.

The original radical cation, M$^{\cdot+}$, can undergo many fragmentation reactions before it is accelerated into the magnetic field and passed on to the detector. The parent ion can react to give many **daughter ions**, and these appear as peaks in the final mass spectrum—the fragmentation pattern. Often these peaks are more intense than the parent ion. In all cases, the intensities of mass spectral peaks are given as percentages of the largest peak in the spectrum, which may be the parent ion, but usually is not. The largest peak is called the **base peak** of the spectrum.

Is there any way to tell how a given molecular ion will fragment? Yes and no. There is no way to predict the detailed pattern of daughter ions. Yet we can make general predictions by applying what we know about reaction mechanisms and the stabilities of ions. Fragmentations usually take place to give the most stable possible cations, and we are now able to make some reasonable guesses as to what these might be. First of all, note that there are two ways for a typical parent radical cation to produce two fragments (Fig. 15.8). When the radical cation $A-B^{•+}$ produces A and B by fragmentation, the charge may go with either A or B. The detector can see only positively charged ions, and so any neutral species formed are invisible. In the fragmentation process, the charge will generally wind up where it is more stable, and we are often able to predict where it will go. For example, *tert*-butyl chloride gives large amounts of the *tert*-butyl cation, $m/z = 57$, in its mass spectrum, not Cl^+, $m/z = 35$ (Fig. 15.8). The charge is more easily borne on the tertiary cation than on the relatively electronegative chlorine. Nevertheless, it must be admitted that predicting is not always a straightforward process. High energies are involved, and there is plenty of energy available to do a lot of bond breaking. Rearrangements are also common, especially if a particularly stable ion can be reached.

FIGURE 15.8 Two possible fragmentations for a general molecule A—B and the mass spectrum of *tert*-butyl chloride.

Some other examples will probably be useful. Figure 15.9 shows the mass spectrum of toluene.

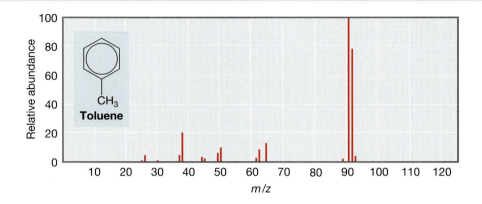

FIGURE 15.9 The mass spectrum of toluene.

The parent ion is visible as a strong peak at $m/z = 92$, but it is not the base peak. The largest peak in the spectrum is the p − 1 peak at $m/z = 91$. What can this be? A little thought brings us quickly to a reasonable answer. If a hydrogen atom (not a proton, which would be positively charged and leave a neutral fragment behind) is lost from the parent ion, it leaves behind a benzyl cation that is well stabilized by resonance (Fig. 15.10). Accordingly, loss of the hydrogen atom from the first-formed radical cation is relatively easy, and the largest peak in the spectrum belongs to the ion $C_7H_7^+$, $m/z = 91$, the benzyl cation.

FIGURE 15.10 The molecular ion of toluene loses a hydrogen atom to give the benzyl cation.

PROBLEM 15.4 There is some evidence that the $m/z = 91$ peak may not be the benzyl cation but another species. Can you (a) think of a $C_7H_7^+$ species that might be even more stable than the benzyl cation, and (b; much harder) write an arrow formalism for the conversion of the benzyl cation into this new species? *Hint*: See Chapter 13 (p. 634).

Similarly, the mass spectra of compounds containing carbon–oxygen double bonds (carbonyl compounds) often reveal large peaks for the resonance-stabilized acylium ions (Chapter 14, p. 692). For example, 2,2,4,4-tetramethylpentan-3-one (di-*tert*-butyl ketone) ionizes to give large amounts of the acylium ion of $m/z = 85$.

Acylium ions can break down further to give carbon monoxide and carbocations (Fig. 15.11).

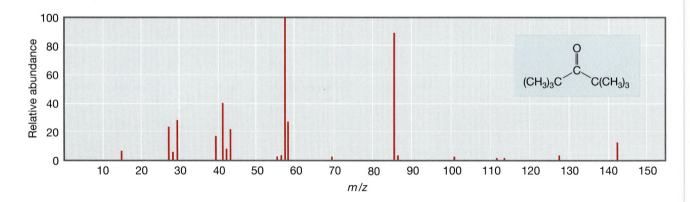

FIGURE 15.11 Carbonyl compounds usually give large amounts of a resonance-stabilized acylium ion.

Alkanes cleave so as to give the most stable possible carbocation (Fig. 15.12). For example, 2,2-dimethylpropane (neopentane) fragments to give an ion of $m/z = 57$, the *tert*-butyl cation, much more easily than it gives the far less stable methyl cation, $m/z = 15$.

There are other fragmentation reactions that resemble solution chemistry. For example, an important peak in the mass spectrum of 1-butanol is at $m/z = 56$. High-resolution analysis of this peak shows that its formula is C_4H_8. Apparently, the parent

FIGURE 15.12 As illustrated by the mass spectrum of 2,2-dimethylpropane, alkanes fragment to give the most stable carbocation possible.

ion eliminates water to give a butene radical cation (Fig. 15.13). This behavior is typical for alcohols and has clear roots in the solution chemistry described in our sections on elimination reactions (Chapter 7, p. 312).

FIGURE 15.13 In the mass spectrometer, alcohols such as 1-butanol can eliminate water to give alkene radical cations.

PROBLEM 15.5 Draw a schematic molecular orbital diagram for the π orbitals of the butene radical cation, $C_4H_8^{\bullet+}$.

WORKED PROBLEM 15.6 What do you suggest as a structure for the base peak ($m/z = 31$) in the mass spectrum of 1-butanol?

ANSWER The base peak is $m/z = 31$. The mass of 1-butanol ($C_4H_{10}O$) is 74, so some quite serious fragmentation must have taken place. A mass of 31 can only have two heavy (non-hydrogen) atoms, so it must be either a C_2 ($m = 24$) or CO ($m = 28$) compound. If the ion contained two carbons, it would have to be C_2H_7, an impossible formula, so it can only be CH_3O. The problem now is to find a reasonably stable ion of this formula that can be made from a fragmentation of 1-butanol. A simple carbon–carbon bond breaking leads to a resonance-stabilized cation that certainly is a good candidate.

As mentioned earlier, rearrangements are common. Perhaps the most famous is named for Professor F. W. McLafferty (b. 1923) of Cornell University. He noticed that when a hydrogen was available at the γ-position, carbonyl compounds underwent what has come to be known as the **McLafferty rearrangement**. An alkene is produced as the neutral partner and the positive fragment is the radical cation of an alkenol called an enol. Note that when there is a hydrogen in the γ-position, it can be transferred to the Lewis basic oxygen atom through a relatively strain-free six-membered transition state. For example, the base (most prominent) peak in the mass spectrum of butyraldehyde is an ion of $m/z = 44$ produced by the McLafferty rearrangement (Fig. 15.14).

FIGURE 15.14 If there is a carbon–hydrogen bond in the γ-position, carbonyl compounds can fragment through the McLafferty rearrangement.

A SPECIFIC EXAMPLE OF THE McLAFFERTY REARRANGEMENT

There are still plenty of mysteries in mass spectrometry, however. For example, my research group first made the molecule bicyclo[4.2.2]deca-2,4,7,9-tetraene ($m = 130$; Fig. 15.15) over 20 years ago, and were startled then to find that the base peak in the mass spectrum was at $m/z = 115$. It seemed likely that this ion (p − 15) represented the loss of a methyl group from the parent ion. We have been thinking about this process (not constantly!) over the ensuing two decades and still have no reasonable explanation for the loss of a methyl group. Any ideas?

Bicyclo[4.2.2]deca-2,4,7,9-tetraene

$m/z = 130$

70 eV

$m/z = 115$???

FIGURE 15.15
Bicyclo[4.2.2]deca-2,4,7,9-tetraene apparently loses a methyl group in the mass spectrometer. How?

Despite occasional mysteries, mass spectrometry is still highly useful, even in the bare-bones form described here. It gives us a way to determine molecular composition through detection of a parent ion, and can give hints as to structure from the fragment ions produced. Even though it is difficult to predict the fragmentation patterns in more than a general way, it is possible to compare a fragmentation pattern with that of a known compound in order to prove identity.

SUMMARY

Mass spectrometry usually provides the molecular weight of an unknown compound. If run at sufficiently high resolution, MS can even give you the molecular formula of a compound. Some information can be gathered about structure from the ions formed on fragmentation.

15.4 INFRARED SPECTROSCOPY

Infrared (IR) spectroscopy is true spectroscopy—we are studying what happens when electromagnetic radiation is absorbed by a molecule. In Chapter 12, we discussed UV/vis spectroscopy, and IR spectroscopy is qualitatively similar. A schematic picture of the electromagnetic spectrum (Fig. 15.16) identifies the regions used for UV/vis, IR, and the spectroscopy we will study in detail in the second half of this chapter, nuclear magnetic resonance (NMR).

The wavelength, λ (cm), is the length of a wave as measured from peak to peak or trough to trough (Fig. 15.16), and the frequency (ν) is the number of cycles passing a given point per unit time, usually seconds. So, the units of frequency (ν) are cycles per second (usually written as reciprocal seconds, or s^{-1}), or more simply, hertz (Hz). Wavelength and frequency are related by Eq. (15.1).

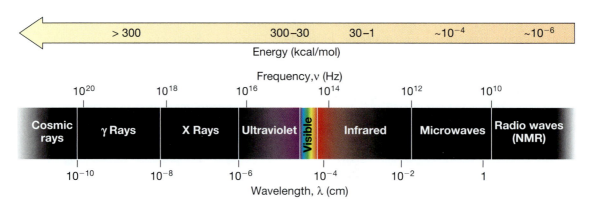

FIGURE 15.16 The electromagnetic spectrum.

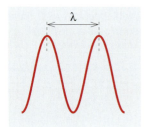

$$v = \frac{c}{\lambda} \quad c = \text{speed of light in centimeters per second (cm/s)} \quad (15.1)$$

The quantity $1/\lambda = \tilde{v}$ is called the **wavenumber** and its units are expressed in terms of reciprocal centimeters (cm^{-1}). Physically, $\tilde{v}$ is the number of wavelengths contained in a centimeter (cm^{-1}) and is related to the more familiar frequency, v, by the relationship in Eq. (15.2).

$$v = \frac{c}{\lambda} \quad \text{or} \quad v = c\tilde{v} \quad (15.2)$$

Infrared spectra are quoted these days in wavenumbers (cm^{-1}), but in the older literature one also sees values in microns (μ). One micron (a micrometer, μm) is 10^{-4} cm.

Absorption of IR light involves energies of roughly 1–10 kcal/mol, and is associated with molecular vibrations. A chemical bond can be approximated by a picture of two weights connected by a spring. Hooke's law [Eq. (15.3)] covers weights connected by real springs.

$$v = k\sqrt{f(m_1 + m_2)/m_1 m_2} \quad (15.3)$$

The frequency v is related by a constant k to the masses of the weights (m_1 and m_2) and another constant, f, which is the **force constant** of the spring. Chemical bonds do behave rather like classical springs. So, like a real spring, the frequency of vibration depends on the masses of the atoms and the strength of the bond between them. Bond strength turns out to be the dominant feature, although mass does make a difference. We would expect strong bonds, with higher force constants, to absorb at higher wavenumbers. The family of carbon–carbon single bonds (~90 kcal/mol, ~1450 cm^{-1}), double bonds (~168 kcal/mol, ~1650 cm^{-1}), and triple bonds (~230 kcal/mol, ~2180 cm^{-1}) illustrates this point well.

However, not all vibrations can be detected. For a vibration to be infrared active, it must produce a net change in the dipole moment of the molecule, which means that symmetrical vibrations are weak or invisible in the IR. For example, ethylene itself shows no signal for the carbon–carbon double-bond stretch in the IR. However,

Raman spectroscopy, which is related to IR spectroscopy, does detect symmetrical vibrations, and shows a band at 1623 cm^{-1} for ethylene.

15.4a Measuring an Infrared Spectrum.

In practice, an IR spectrum is obtained by passing IR radiation through a sample. If no light is absorbed at a given wavelength, transmission is 100%. If light is absorbed at a certain wavelength, transmission is less than 100% and the sample must be absorbing some of the energy. A plot of absorption versus wavenumber (cm^{-1}) is thus generated and constitutes the IR spectrum of the sample. Spectra of gases are measured in cells with nonabsorbing windows. Spectra can be obtained from liquids either as pure samples or in solution. To measure the spectrum of a solid, the sample is usually compressed into a disk with KBr and placed in the beam of IR radiation.

Spectrometers are routinely attached to computers that can search for matches between the spectrum of an unknown and a library of known spectra. As with mass spectrometry, gas chromatographs can be attached to IR spectrometers and spectra can be determined as the individual components of a mixture elute from a column. This technique is called gas chromatography/infrared spectroscopy, or GC/IR.

15.4b Characteristic Infrared Absorptions.

From the foregoing discussion, one might expect that IR spectroscopy would be a simple matter of noting what characteristic absorptions appear in the spectrum. An IR spectrum might be a very simple thing, composed only of signals for each kind of bond present in the sample molecule. There certainly are characteristic regions of absorption for certain bonds, but the practical situation is not quite so simple. Connected vibrating springs interact. Two springs attached to the same weight (or atom) do not vibrate independently—try it. If you set one spring in motion, the other moves as well. In fact, this is true even if the two springs (bonds) are attached to different atoms within the same molecule. The closer in energy the two vibrations are, the more strongly they affect each other. A molecule is like a collection of interacting springs, and it is often not possible to make a simple assignment of all the bands in an IR spectrum, although computational methods are becoming more proficient at this task. Most IR spectra are complicated series of bands from which we can extract information, but which we cannot completely rationalize in more than a general way. Figure 15.17 shows the IR spectrum of a typical organic molecule and includes a few band assignments.

FIGURE 15.17 A typical IR spectrum.

Although this complexity may seem confusing at first, it is usually possible to gain a lot of information about the groups present in a molecule, even if we cannot assign all the bands, or draw a complete structure of the sample molecule. It is a great help in structure assignment to know what types of bonds are present in a molecule. There is also an important side benefit to these complicated IR spectra. The very complexity of the spectrum means that IR spectra of quite similar molecules are different. Each spectrum serves as a fingerprint of the molecule. If two IR spectra are identical (not similar—my old boss insisted on there being no difference greater than the width of the pen line drawn by the recorder) the compounds must be the same.

Table 15.3 gives the general positions of absorptions of a variety of functional groups, and once again we come to the question of memorization. Should one learn this chart by heart? Personally, I think not. It is important to know that this kind of general correlation exists, and you should have a rough idea of where some important functional groups absorb. If you come to use IR often you will automatically learn the relevant details of the chart, as you work out what the signals in your IR spectra tell you.

TABLE 15.3   Typical Infrared Absorptions of Functional Groups[a]

Functional Group	Position (cm^{-1})	Intensity[b]	
Alkanes			
C–H	2980–2850	m–s	(stretch)
C–C	1480–1420	m	(bend)
Alkenes			
=C–H	3150–3000	m	(stretch)
C=C	1680–1620	m–w	(stretch)
(conj) C=C	1630–1600	m–w	(stretch)
R$_2$C=CH$_2$ (R, H)	995–985 / 915–905	s	(out-of-plane bend)
(R, H)C=C(H, R)	980–960	s	(out-of-plane bend)
(R, H)C=C(R, H)	730–665	s	(out-of-plane bend) (br, variable)
R$_2$C=CH$_2$	895–885	s	(out-of-plane bend)
(R, R)C=C(R, H)	840–790	m	(out-of-plane bend)

(continued)

TABLE 15.3 Typical Infrared Absorptions of Functional Groups[a] (*Continued*)

Functional Group	Position (cm^{-1})	Intensity[b]	
Alkynes			
$\equiv$C—H	3350–3300	s	(stretch)
C$\equiv$C	2260–2100	m–w	(stretch)
Alcohols			
O—H			
free	3650–3580	m	(stretch)
hydrogen bonded	3550–3300	br, s	(stretch)
C—O	1260–1000	s	(stretch)
	1150–1050		
Amines			
N—H	3500–3100 (two bands for primary amines, one band for secondary amines)	br, m	(stretch)
C—N	~1200	m	(stretch)
Aromatic compounds			
=C—H	3080–3020	m–w	(stretch)
C=C	1600–1580	m–w	(stretch)
C—H			
mono	770–730	s	(out of plane bend)
	710–690		
ortho	770–735	s	(out-of-plane bend)
meta	900–860	m	(out of plane bend)
	810–750	s	(out-of-plane bend)
	725–680	m	(out-of-plane bend)
para	860–800	s	(out-of-plane bend)
Carbonyl compounds aldehydes, ketones			
C=O	1730–1700 (higher in strained cyclic molecules)	s	(stretch)

(continued)

TABLE 15.3   Typical Infrared Absorptions of Functional Groups[a] (*Continued*)

Functional Group	Position (cm^{-1})	Intensity[b]	
Carbonyl compounds aldehydes, ketones			
	1680–1660	s	(stretch)
C—H (aldehydes)	2900–2700 (two bands)	w	(stretch)
Esters			
	1750–1735 1300–1000	s (C=O) s (C—O)	(stretch) (stretch)
Acids			
	1730–1700 3200–2800	s (C=O) s, br (O—H)	(stretch) (stretch)
Acid chlorides			
	1820–1770	s (C=O)	(stretch)
Anhydrides			
	1820–1750 (two bands) 1150–1000	s (C=O) s (C—O)	(stretch) (stretch)
Imines			
C=N	1680–1640	m	(stretch)
Cyanides (nitriles)			
C≡N	~2250	s	(stretch)

[a]**CAUTION!** There certainly is some subjectivity in this table, and the values represent average positions for "normal" compounds. Conjugation generally lowers double-bond stretching vibrations by about 20 cm^{-1}.

[b]Medium = m, strong = s, weak = w, broad = br.

WORKED PROBLEM 15.7 Use the data in Table 15.3 to identify the functional groups in the IR spectra below. Perhaps, given the molecular formulas, you can make some guesses as to structure.

(continued)

ANSWER (a) In the C—H region almost all the bands are below 3000 cm^{-1}, indicating that there are only alkane-like carbon–hydrogen bonds. No hydrogens are attached to double bonds, as they would absorb above 3000 cm^{-1}. There is a strong absorption at about 1740 cm^{-1}, too high to be an aldehyde, acyclic ketone, or acid. The compound must be an ester. Finally, there is a strong band at 1200 cm^{-1}, appropriate for the C—O stretch for an ester. Given the formula of $C_4H_8O_2$, one can make two guesses at the structure.

Methyl propionate or **Ethyl acetate**

This problem tries to show that without further information there is no reliable way to choose between these two possibilities. Given the information so far, either answer is possible. We could either compare the IR spectrum with those of known samples, or obtain further spectra. An NMR spectrum would distinguish the two possibilities easily, as you will soon see. In fact, this compound is methyl propionate.

SUMMARY

The IR spectrum of a compound provides information on the kinds of functional groups in the molecule. In addition, if standard spectra of known compounds are available, comparison of the highly detailed IR spectra can be used as a proof of identity—two identical IR spectra can only come from identical compounds.

15.5 NUCLEAR MAGNETIC RESONANCE SPECTROSCOPY

In contrast to IR spectroscopy, which gives an overall view of the structure of a molecule, **nuclear magnetic resonance (NMR)** can give a wealth of specific detail. It is the general feeling among organic chemists that given good NMR spectra of a molecule, the structure must follow. Today, the frontier of structure determination by NMR lies in immense biomolecules, and even these structures are yielding to an increasing variety of sophisticated NMR techniques.

Nuclear magnetic resonance was introduced very early on in Section 2.14 (p. 88). There we just developed a bare-bones outline—just enough to let you work out the numbers of different carbon or hydrogen atoms in a molecule, and to get used to working with symmetry. Here, we go on to many of the gory details of this incredibly powerful spectroscopic technique. We will start with ^{1}H NMR (hydrogen nuclear magnetic resonance), and then pick up ^{13}C NMR (carbon nuclear magnetic resonance) again.

As you can see from Figure 15.16, NMR spectroscopy involves energies much smaller than those in IR, UV, or vis spectroscopy. What happens when radiation of the wavelength 1–100 m ($\sim 10^{-6}$ kcal/mol) is absorbed? What change can such a tiny amount of energy induce? Even molecular vibrations require much higher energies (1–10 kcal/mol). Molecular rotations demand much less energy ($\sim 10^{-4}$ kcal/mol), but even these motions require about 100 times more energy than that of the radio waves used for NMR. To see what happens when radio waves are absorbed by molecules, we must first go back to the beginning of our discussion of atomic structure, and learn a bit more about the structure of the atomic nucleus.

Like the electron (Chapter 1, p. 6), the nucleus has spin. For some nuclei (^{1}H, ^{13}C, ^{15}N, ^{19}F, ^{29}Si), the value of the nuclear spin (I) is ½. For others, the spin can take different values, such as 0 (^{12}C, ^{16}O), 1 (^{2}H, ^{14}N), ³⁄₂ (^{11}B, ^{35}Cl), ⁵⁄₂ (^{17}O), … . A nonzero spin is a requirement for the NMR phenomenon. It is convenient that some very common isotopes have zero spin (^{12}C, ^{16}O) and will be NMR inactive. However, the most common nucleus in organic chemistry, hydrogen, is one of the nuclei with a spin of ½ and will be NMR active. The NMR effect arises as follows: Any spinning charged particle generates a magnetic field, and therefore we can think of the proton as a bar magnet. In the absence of a magnetic field, these bar magnets will be oriented randomly (Fig. 15.18).

In the absence of an applied
magnetic field

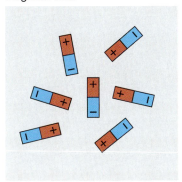

In the presence of an applied
magnetic field, B_0

Aligned against
the field: higher
energy β hydrogens

Aligned with the
field: lower energy
α hydrogens

+Pole

B_0 Applied magnetic field

−Pole

Energy

FIGURE 15.18 In the absence of an applied magnetic field, nuclear spins will be randomly oriented. When an external magnetic field is applied, alignment with the field will be slightly favored energetically over alignment against the field; there will be a small excess of molecules aligned with the field.

However, when we apply a magnetic field (B_0), the proton can align either with the field or against it (Fig. 15.18). Alignment with the field ($I = +$ ½, α hydrogen) will be slightly more favorable energetically than alignment against the field ($I = -$½, β hydrogen), and there will be an excess of nuclei in the lower energy, more favorable orientation. As the energy difference between the two orientations is very small, there will only be a slight excess, but it will exist, and there is the possibility of inducing transitions between the two orientations if the proper amount of energy is supplied. The function of the radio waves is to supply the energy necessary to change the orientation of the nuclear spin (often called "flipping" the spin). The disturbed equilibrium is reestablished when this absorbed energy is returned to the environment as heat (Fig. 15.19).

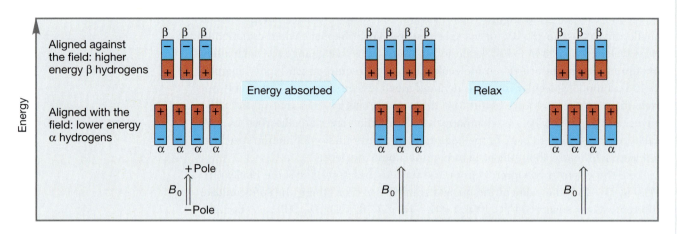

FIGURE 15.19 Absorption of energy can "flip" a nuclear spin, converting the low energy orientation into the higher one.

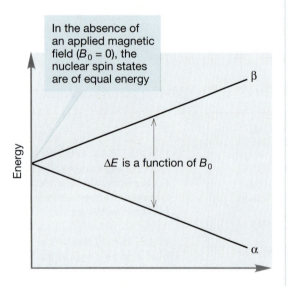

FIGURE 15.20 The energy difference between the two spin states depends on the strength of the applied magnetic field.

The magnitude of the separation of the two states—alignment with and against the applied field—depends on the strength of the applied magnetic field, B_0, as shown in Eq. (15.4) and Figure 15.20. The higher the field strength, the greater the energy difference. For values of B_0 in the range commonly used, the frequencies ν for common isotopes are in the 60–750-MHz range. The gyromagnetic ratio, γ, is characteristic of the nucleus; there is a different γ for every NMR active nucleus.

$$\nu = \gamma B_0 / 2\pi \qquad (15.4)$$

If we can build an instrument capable of delivering the required radio waves and of detecting the absorption of tiny amounts of energy, we should be able to find a signal from any hydrogen-containing material. But won't there be problems from the other nuclei? Many common nuclei are NMR active, and there might be too many overlapping signals for us to make sense of the spectrum. We are saved from this fate by the gyromagnetic ratios, γ [Eq. (15.4)]. Gyromagnetic ratios are sufficiently different so that for the usual field strengths, there is a wide separation of signals from different nuclei. For example, signals from ^{13}C or ^{19}F nuclei appear far from those for ^{1}H.

WORKED PROBLEM 15.8 Calculate the energy involved in an NMR transition at a field strength of 4.7 T. For H, γ is 2.7×10^8 rad T^{-1} s^{-1}, and Planck's constant is 9.5×10^{-14} kcal/mol·s.

(continued)

ANSWER Energy equals $h\nu$, Planck's constant (h) × frequency (ν): ($E = h\nu$). The frequency is given by Equation (15.4), $\nu = \gamma B_0/2\pi$, so the useful relationship for this problem is

$$E = h\gamma B_0/2\pi$$
$$E = (9.5 \times 10^{-14})(2.7 \times 10^8)(4.7)/2(3.14) = 1.9 \times 10^{-5}$$

1.9×10^{-5} whats? Figure out the units using the same equation.

$$E = (kcal/mol)(s)(rad\ T^{-1}\ s^{-1})(T)/rad = kcal/mol$$

The answer is 1.9×10^{-5} kcal/mol or 0.019 cal/mol. Notice that this value is *calories* per mole! These NMR transitions involve *tiny* energies.

15.5a The Nuclear Magnetic Resonance Spectrometer.

In the early NMR spectrometers, the magnetic field was varied, while the radio frequency (rf) was held constant. The field was swept over the appropriate range for the nucleus in question, most often hydrogen or carbon. As the chemically different nuclei absorbed energy (came into resonance) they would be detected one by one. It took several minutes to acquire the spectrum and rather large samples—several milligrams—were necessary. Today a different and more sensitive technique is used. Superconducting magnets are used that supply far higher magnetic fields than did the old permanent magnets, and the magnetic field is not varied—swept over the range. Instead, the sample is irradiated with a very short pulse of rf energy over the entire range for the nucleus in question. All hydrogen atoms or all carbon atoms absorb energy at once. They return to equilibrium over time, and a computer is used to transform the intensity versus time data into intensity versus frequency data through what is called a Fourier transform (FT). An FT spectrum can be obtained in a few seconds on very small samples, and the results from many pulses are averaged to increase the signal-to-noise ratio.

Resonance positions are always referenced to a standard signal, and given in parts per million (ppm) of applied magnetic field. The standard chosen is **tetramethylsilane (TMS)**, which shows a sharp signal in a position somewhat removed from most other resonance positions of hydrogens in organic molecules. The position of TMS is defined as the zero point of the **ppm scale**. Why is the somewhat odd ppm scale used? Let's assume that we are using a 60-MHz spectrometer and that the resonance frequency of the sample molecule appears at 30 Hz from TMS. If we measured the NMR spectrum of the same sample on one of the higher field instruments common today, say, 300 MHz, the resonance position would be different because the splitting between the two nuclear spin states depends on the strength of the applied magnetic field [Eq. (15.4); Fig. 15.20]. At 300 MHz, our sample hydrogen signal would appear at 150 Hz from TMS. Although the measured resonance position is different for every applied magnetic field, *the position is always the same in ppm*. In the example above, it appears at 0.5 ppm each time: $30/(60 \times 10^6)$, or $150/(300 \times 10^6)$ (Fig. 15.21). The ppm scale was devised so that chemists could report the *same* values for resonance positions regardless of the magnetic field strength of their spectrometer.

Spectrum taken at 60 MHz

30 Hz from TMS or
0.5 ppm from TMS

Signal

Spectrum taken at 300 MHz

150 Hz from TMS but
still 0.5 ppm from TMS

Signal

FIGURE 15.21 The ppm scale ensures that peak positions will be the same regardless of field strength.

15.5b The Chemical Shift. Up to this point, NMR spectroscopy might seem to be merely an expensive way of seeing if a compound contained hydrogen. We pay some hundreds of thousands of dollars for a 300-MHz machine, put a sample of an organic compound in it, and if we see absorption of radio waves at an appropriate frequency, we know the sample contains hydrogen. That's usually not very surprising if we are dealing with organic molecules. So what's the big deal?

It turns out that we don't see just one big signal from a hydrogen-containing molecule. Instead, there is a signal at a different position for every different hydrogen in the molecule. The resonance frequency for each hydrogen in the molecule is strongly dependent on its local environment.

WORKED PROBLEM 15.9 The higher the applied field, the greater the dispersion between the different signals, and therefore the greater the resolving power of the instrument. This resolution is one of the reasons that chemists are willing to spend extra thousands of dollars to get extra megahertz. For example, two signals appearing at 30 and 36 Hz in a spectrum taken on a 60-MHz machine will be much better separated if the spectrum is obtained at 300 MHz. Show that this last statement is true.

ANSWER At 60 MHz, the signals at 30 and 36 Hz are at 0.5 and 0.6 ppm, respectively, [$30/(60 \times 10^6)$ and $36/(60 \times 10^6)$]. At 300 MHz, 0.5 and 0.6 ppm are 150 and 180 Hz, respectively. So, the two signals are separated by 6 Hz at 60 MHz, but by 30 Hz at 300 MHz. That's quite an improvement in separation.

We now have two questions to answer: First, we need to know how the dependence of the **chemical shift** (δ) on the local environment arises. In a sense, this is a nonquestion. If two hydrogens are different, they *must* give different NMR signals. It is in fact just a question of whether our ability to detect the difference is great enough, because the difference must be there. Electrons in molecules occupy regions of space defined by the molecular orbitals of the molecule. When an external magnetic field is applied (B_0), the electrons will circulate within those regions of space so as to create an induced magnetic field (B_i) that opposes the applied field B_0 (Fig. 15.22).

The net field ($B_0 \pm B_i$) will be slightly different at every different hydrogen in the molecule, and therefore the resonance frequency will also be different. Relative to the hydrogens in TMS, for most hydrogens it is necessary to decrease the external field strength because of the field induced by the electrons in the molecule. Such hydrogens are said to be "deshielded" by the induced field and will appear at a lower field than TMS. In a few instances, we will find that hydrogens are "shielded" and appear upfield of TMS. In these cases, the applied field, B_0, is opposed by the induced field, B_i, and it takes more applied field than usual to flip the nuclear spin.

So we can expect to see a signal for each different hydrogen in a molecule, and this leads to the second, and sometimes thorny question, What constitutes different hydrogens? There is a simple scheme for finding identical (chemically and magnetically equivalent) hydrogens. First, carry out mental substitutions of the hydrogens in question with a phantom group X. If the resulting molecules are identical, the hydrogens are also identical, or **homotopic**. Figure 15.23 gives an example using methylene chloride. Replacement of each hydrogen of methylene chloride with X (red and green in the figure) gives the same $CHCl_2X$. The two hydrogens of methylene chloride are homotopic and equivalent in all chemical and spectroscopic operations.

Hydrogens in more complicated molecules are often not homotopic. Chlorofluoromethane is a good example. Replacement of the two hydrogens by the phantom X does not yield identical molecules, but enantiomers (Fig. 15.24).

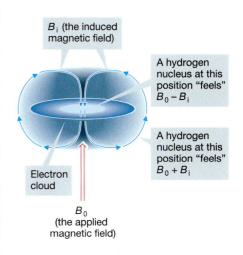

FIGURE 15.22 In an applied magnetic field (B_0), electrons will circulate so as to generate an induced magnetic field (B_i) that will oppose the applied field.

FIGURE 15.23 The two hydrogens in methylene chloride are homotopic.

These molecules are enantiomers; the two hydrogens in chlorofluoromethane are enantiotopic

FIGURE 15.24 These two hydrogens are enantiotopic.

These methylene hydrogens are **enantiotopic**. Enantiotopic hydrogens are still equivalent in chemical reactions and in the NMR experiment.

PROBLEM 15.10 Explain clearly why the enantiotopic hydrogens of chlorofluoromethane will show different signals if the NMR spectrum is obtained in a solvent consisting of a single enantiomer.

Finally, consider 1,2-dichlorofluoroethane. Now replacement of the two methylene hydrogens gives neither identical molecules nor enantiomers, but diastereomers (stereoisomers, but not mirror images, Chapter 4, p. 170; Fig. 15.25). These hydrogens are **diastereotopic**, and will react differently in all circumstances and will give different signals in the NMR spectrum of this molecule.

FIGURE 15.25 These two hydrogens are diastereotopic.

These molecules are diastereomers; the two methylene hydrogens are diastereotopic

PROBLEM 15.11 Identify the indicated hydrogens in the following molecules as homotopic, enantiotopic, or diastereotopic.

There is a general correlation between the resonance frequency of hydrogens in similar positions in different molecules. All methyl groups attached to double bonds appear at roughly the same position, all aromatic hydrogens at another, and so on.*

CONVENTION ALERT

*To say "aromatic hydrogen" is certainly wrong; the hydrogen isn't aromatic, it is the ring to which it is attached that is. Nonetheless, one must admit that it is slightly awkward to say "hydrogens attached to aromatic systems," and this kind of loose talk is common. One sees vinyl or olefinic hydrogen, cyclopropyl hydrogen, and so on.

Table 15.4 gives a correlation chart for the peak positions of many different kinds of hydrogen.

TABLE 15.4	Chemical Shifts of Various Hydrogens[a,b]	
Hydrogen		δ **(ppm)**
CH_3		0.8–1.0
CH_2		1.2–1.5
CH		1.4–1.7
C=C—CH (allylic hydrogens)		1.8–2.3
O=C—CH		2.0–2.5
Ph—CH (benzylic hydrogens)		2.3–2.8
≡C—H		2.5
R_2N—CH		2.0–3.0
I—CH		2.8–3.3
Br—CH		2.8–3.5
Cl—CH		3.1–3.8
F—CH		4.1–4.7
O—CH		3.1–3.8
=CH_2 (terminal alkene)		5.0
C=CH (internal alkene)		4.5–5.5
Ph—H (aromatic hydrogens)		7.0–7.5
O=CH (aldehyde hydrogens)		9.0–10.0
RCOOH		10–13

[a]These values are approximate. There will surely be examples that lie outside the ranges indicated. Use them as guidelines, not "etched in stone" inviolable numbers.

[b]Watch out for loose talk. For example, "aromatic hydrogen" means a hydrogen attached to a benzene ring.

It is also possible to estimate the number of hydrogens giving rise to a particular peak by measuring the relative intensities of the signals in a spectrum. This process is called **integration** of a signal. As an example, look at the ^{1}H NMR spectrum of 4,4-dimethyl-2-pentanone in Figure 15.26. Three signals appear, and the integral shows that they are in the ratio 9:3:2. These days, integrals are calculated electronically and printed out as numbers on the chart paper. In older spectra, one sees a line traced by a pen, as in Figure 15.26. The integral is determined by measuring the vertical (not horizontal or diagonal) distance the pen moves as the signal appears. In this spectrum, the assignment is easy: The nine hydrogens of the *tert*-butyl group give rise to the highest field signal,

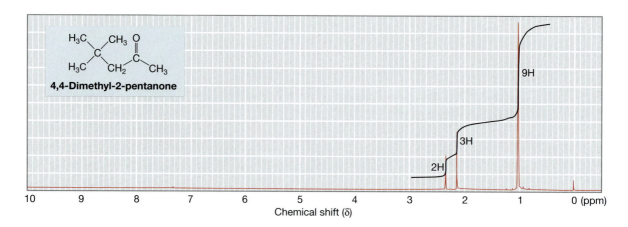

FIGURE 15.26 The ^{1}H NMR spectrum of 4,4-dimethyl-2-pentanone.

WORKED PROBLEM 15.12 Which of the molecules below give rise to which spectra?

ANSWER (a) A signal integrating for nine hydrogens at about δ 1 ppm is a sure sign of the presence of a *tert*-butyl group. In the first spectrum, there is only one other signal, at δ 3.2 ppm. This signal is in an appropriate position for a methyl group next to an oxygen, and the compound is *tert*-butyl methyl ether. In the other compound, 2,2-dimethyl-1-propanol, we see two other signals: two hydrogens at δ 3.3 ppm (the methylene next to the oxygen) and one hydrogen at δ 1.7 ppm, the O<u>H</u>.

(b) The key to this problem is the position of the methyl signal. In the first spectrum, it is far downfield of the position in the second spectrum. The two *tert*-butyl groups are in about the same place. Hydrogens adjacent to oxygen appear well downfield of hydrogens next to a carbon–oxygen double bond (see Table 15.4), so it is easy to make the assignment of methyl trimethylacetate, $(CH_3)_3CCOOCH_3$, to the first spectrum and *tert*-butyl acetate, $CH_3COOC(CH_3)_3$, to the second.

(c) Here symmetry allows a quick assignment. 1,4-Dicyanobenzene has only one kind of hydrogen and must give rise to the second spectrum. 1,3-Dicyanobenzene has three different hydrogens and has the far less symmetrical first spectrum.

the methyl group produces the signal for three hydrogens, and the methylene group gives the peak for two hydrogens. From now on, in the figures integrations will often be indicated by a number followed by H, as in 1H, 2H, etc. There is no difficulty in making assignments for this simple example, but you should check the positions of the peaks against those predicted by Table 15.4 to be certain there is no error.

The examples in Problem 15.12 are simple, and are designed to ease you into analyzing NMR spectra. In no area of organic chemistry is practice so necessary in the development of skill. You simply cannot become proficient at this kind of structure determination by reading about how it is done and studying Table 15.4. Working problems is an absolute must. Do the problems at the end of this chapter and seek out others. Other textbooks have good ones; don't be reluctant to dig them out.

15.6 SURVEY OF NMR SPECTRA OF ORGANIC MOLECULES

Table 15.4 gives a summary of the chemical shifts for hydrogens in the vicinity of a variety of common organic functional groups (p. 775). Use this table when working problems and you will gradually become familiar with the chemical shifts of commonly encountered hydrogens.

15.6a Saturated Alkanes. Signals from saturated alkanes (and the simplest cycloalkanes) are at relatively high field. Note that the signals for cyclopropanes are at unusually high field, sometimes even appearing above TMS (Fig. 15.27).

15.6b Substitution by Electronegative Groups. Electron-withdrawing groups will reduce the electron density at neighboring hydrogens, which decreases the ability of nearby electrons to shield the nucleus, causing the chemical shifts to appear at relatively low field. Notice, for example, that as fluorine is the most electronegative halogen, alkyl fluorides appear at the lowest field of all the halogen-substituted molecules. Of course, the more electronegative groups in the neighborhood of a hydrogen, the greater the electron-withdrawing effect becomes and the resonance position shifts to

FIGURE 15.27 The ^1H NMR chemical shifts of alkanes.

lower field. The chemical shifts of a variety of halogenated and oxygenated compounds are shown in Figure 15.28.

	X=F	X=Cl	X=Br	X=I
CH_3X	4.26	3.05	2.68	2.16
CH_2X_2		5.31	4.96	3.88
CHX_3		7.28	6.86	5.36

FIGURE 15.28 The ^1H NMR chemical shifts of alkanes substituted with electron-withdrawing groups.

FIGURE 15.29 The ^1H NMR chemical shifts of hydrogens adjacent to double and triple bonds.

15.6c Allylic Hydrogens. A carbon–carbon double or triple bond is electron withdrawing, and therefore will exert a deshielding effect, producing relatively low-field chemical shifts for hydrogens in the allylic position. The carbon–oxygen double bond exerts a similar, and greater effect (Fig. 15.29).

15.6d Alkenes. Hydrogens attached directly to a carbon–carbon double bond, "olefinic hydrogens," appear at fairly low field. Two effects are operating to deshield such hydrogens. In an applied magnetic field, B_0, the electrons in the π bond will circulate so as to create a magnetic field B_i in opposition to B_0. As shown in Figure 15.30, this opposition results in an *augmented* applied field in the region of the hydrogens attached to the double bond. In addition, the sp^2 carbon of the double bond has high s character and attracts electrons, thereby removing electrons from the vicinity of the hydrogen and deshielding it (Fig. 15.30).

Induced magnetic field, B_i

At this point, B_0 is augmented by B_i. The hydrogen will "feel" a net $B_{net} = B_0 + B_i$, and require a reduced B_0 to come into resonance. It appears at relatively low field

B_0

The sp^2 carbon has high s character and withdraws electrons, deshielding the hydrogen

FIGURE 15.30 A carbon–carbon double bond acts in two ways to deshield attached hydrogens. Such deshielded hydrogens appear at relatively low field.

Accordingly, olefinic hydrogens are deshielded and appear at relatively low field (Fig. 15.31). There are important differences between the various kinds of hydrogens attached to double bonds. For example, terminal methylene hydrogens resonate at unusually high field for olefinic hydrogens and in a region of few other absorptions. Hydrogens attached to a double bond, and β to a carbonyl group, absorb at exceptionally low field, and the hydrogens of aldehydes are even lower (Fig. 15.31).

FIGURE 15.31 The ^{1}H NMR chemical shifts of alkenes and alkynes.

PROBLEM 15.13 Can you explain why vinyl hydrogens β to a carbonyl group are at very low fields?

WORKED PROBLEM 15.14 Hydrogens attached directly to the carbon of a carbonyl group (aldehyde hydrogens) appear at especially low field, and must be strongly deshielded. Explain why.

ANSWER In the absence of other effects, we would expect to see the hydrogen of an aldehyde group at about the positon of hydrogens attached to carbon–carbon double bonds. However, the carbonyl group is very polar. The oxygen atom is much more electronegative than the carbon and will attract electrons strongly, becoming partially

(continued)

(continued)

negative (δ^-) and leaving the carbon partially positive (δ^+). The same effect can be seen from a resonance formulation.

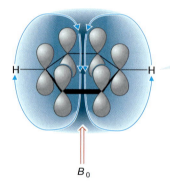

The electron density around carbon is depleted and the attached hydrogen less effectively shielded than usual. Accordingly, a less strong applied magnetic field is necessary to reach the resonance frequency, and the hydrogen appears at very low field.

15.6e Aromatic Compounds.

Hydrogens on aromatic rings are found at even lower field than hydrogens attached to double bonds. In fact, they can quite reliably be diagnosed by the existence of signals in the extended range δ 6.5–8.0. The reason for the low-field resonance is similar to that for hydrogens attached to double bonds. The applied magnetic field induces a circulation of electrons in the ring (a "ring current") that creates its own magnetic field B_i opposing B_0. At the edge of the ring where the hydrogens are, the induced magnetic field augments B_0. Therefore, such hydrogens require a lower B_0 to bring them into resonance (Figs. 15.31 and 15.32).

At this point the induced magnetic field, B_i augments the applied field B_0; the hydrogen will "feel" a net magnetic field, $B_{net} = B_0 + B_i$, and require a reduced applied field to come into resonance

FIGURE 15.32 The field induced (B_i) by the ring current of an aromatic ring in an applied magnetic field (B_0).

PROBLEM 15.15 Explain the unusually high chemical shifts (above TMS) of the hydrogens indicated in the following molecules:

δ –3 δ –0.5 δ –0.01

(these negative chemical shifts mean
that the signals are above that for TMS)

15.6f Alkynes. Although the position of hydrogens α to a carbon–carbon triple bond (the propargyl position) is similar to that of allylic hydrogens, those hydrogens directly attached to the triply bonded carbon appear at substantially higher field than hydrogens attached to double bonds, δ 2.2–2.8 (Fig. 15.31). The upfield shift results from a strong shielding effect induced by circulation of electrons in the triple bond. Compare Figures 15.33 and 15.30 to be certain you see the different shielding effects of π bonds at work.

15.6g Alcohols and Amines. Hydrogens attached to a carbon bearing the electronegative oxygen or nitrogen atom of an alcohol or a primary or secondary amine absorb, as we might expect, in the same region as the related hydrogens in ethers and tertiary amines (Fig. 15.28). But what about the OH and NH hydrogens themselves? Here we do not see what a simple analysis would lead us to expect. First of all, the chemical shifts of OH and NH vary greatly from sample to sample. The problem is that these molecules are extensively hydrogen bonded, and the extent of hydrogen bonding depends on the concentration of the alcohol or amine. The chemical shift depends strongly on the chemical environment and this effect changes depending on the extent of hydrogen bonding. Generally, the less hydrogen bonding, the further upfield the resonance appears. In the gas phase, where hydrogen bonding does not occur, the chemical shifts of these hydrogens are quite high, about 1 ppm.

15.7 SPIN–SPIN COUPLING: THE COUPLING CONSTANT, *J*

The utility of NMR spectroscopy goes far beyond that of a simple hydrogen-detecting machine. We can determine the number of different kinds of hydrogen in a molecule, and gain a general idea of what chemical environments those hydrogens might occupy. But there is much more information available. Most NMR spectra are not as simple as those of Figure 15.26 or Problem 15.12. The signals in typical NMR spectra are not all single peaks, but are often multiplets containing much fine structure. Ethyl iodide makes a nice example. Our initial expectation might be that the spectrum would contain only two signals, one for the three equivalent hydrogens of the methyl (CH_3) group and another for the two hydrogens of the methylene (CH_2) group. A look at Table 15.4 would allow us to estimate the general position for each peak. What we see is somewhat different. There are signals at the positions we estimated, but they are composed of several lines, and are not singlets as were most of the signals of the molecules in Problem 15.12. The signal for the methyl hydrogens shows three lines (a triplet) in a 1:2:1 ratio centered at δ 1.85, and the methylene appears as four lines (a quartet) in the ratio 1:3:3:1, centered at δ 3.2. The spacings between the lines are all the same, 7.6 Hz (Fig. 15.34).

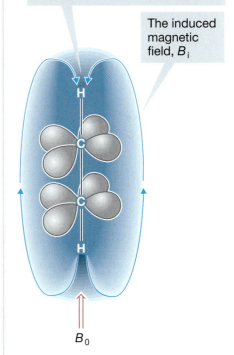

At this point the induced magnetic field B_i opposes the applied field B_0; a hydrogen here "feels" a net magnetic field, $B_{net} = B_0 - B_i$. A relatively high B_0 will have to be applied to bring this shielded hydrogen into resonance

The induced magnetic field, B_i

B_0

FIGURE 15.33 Acetylenic hydrogens are strongly shielded by the induced magnetic field.

FIGURE 15.34 The ^{1}H NMR spectrum of ethyl iodide.

Multiple line signals are called "doublets" (d; two lines), "triplets" (t; three lines), "quartets" (q; four lines), and so on. Our job is now to see how these multiple signals arise, and then to see how we can use that information.

15.7a Spin–Spin Coupling: The (n + 1) Rule. Let's examine the three-line signal (triplet) for the methyl hydrogens first. To do this, we look first at the neighboring methylene hydrogens. Each of the two equivalent methylene hydrogens has a spin of either $+\frac{1}{2}$ or $-\frac{1}{2}$. Therefore, there are four different combinations of these spins, $(+\frac{1}{2} +\frac{1}{2})$, $(+\frac{1}{2} -\frac{1}{2})$, $(-\frac{1}{2} +\frac{1}{2})$, and $(-\frac{1}{2} -\frac{1}{2})$ (Fig. 15.35).

To explain the three-line signal for the methyl hydrogens of ethyl iodide, look first at the neighboring methylene hydrogens

The possible combinations for the methylene groups are

Equivalent

Means spin $= -\frac{1}{2}$

Means spin $= +\frac{1}{2}$

FIGURE 15.35 For two equivalent hydrogens there are four different combinations of the nuclear spins of the two hydrogens. Two of these combinations are equivalent.

The net magnetic field experienced by a methyl hydrogen will be modified by each of the different combinations. Each methyl hydrogen will experience an applied magnetic field (B_0) modified by each of the different spin combinations of the neighboring methylene hydrogens. So we now expect to see four lines, one for each possible spin combination. But two of these combinations, $(+\frac{1}{2} -\frac{1}{2})$ and $(-\frac{1}{2} + \frac{1}{2})$, are equivalent and will give rise to the same net magnetic field, which is the reason we see only a three-line signal in a 1:2:1 ratio for the methyl hydrogens (Fig. 15.36).

The methyl hydrogens are said to be coupled to the adjacent methylene hydrogens. The splitting between the lines is the **coupling constant** (J), and is measured in hertz. In ethyl iodide, this splitting is 7.6 Hz (Fig. 15.36).

Now let's rationalize the four-line signal for the methylene hydrogens of ethyl iodide. For the neighboring methyl group there are eight possible combinations of the nuclear spins of the three equivalent hydrogens (Fig. 15.37). Note that there are now two sets of three equivalent spin combinations (two up and one down or two down

FIGURE 15.36 The methyl group adjacent to the two methylene hydrogens will be split into three lines in the ratio 1:2:1. The splitting between the lines is called the coupling constant, *J*.

and one up). So the hydrogens of the methylene group feel an applied magnetic field modified by the four different combinations of spins of the adjacent methyl hydrogens. Accordingly, the spectrum consists of four lines in a 1:3:3:1 ratio, a quartet. The methylene hydrogens are coupled to the methyl hydrogens by the same coupling constant, $J = 7.6$ Hz, by which the methyl hydrogens are coupled to the methylene hydrogens (Fig. 15.38).

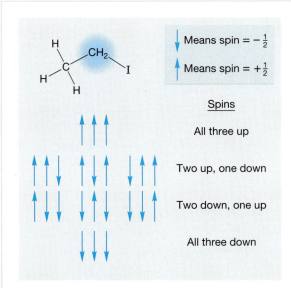

FIGURE 15.37 For the three methyl hydrogens there are eight possible spin combinations, but only four different energy combinations.

FIGURE 15.38 The methylene group of ethyl iodide will be split into four lines in the ratio 1:3:3:1.

In the general situation, for a given hydrogen, H_a, the NMR spectrum will consist of $n + 1$ lines where n is the number of equivalent adjacent hydrogens, H_b. The lines are separated by J Hz. The intensity of the lines can be determined by analyzing the various possible spins of the coupled hydrogens as we have been doing, or by using a device called "Pascal's triangle." In this device, the value (intensity) of a number is the sum of the two numbers above it. Figure 15.39 explains it better than these words. Pascal's triangle allows the rapid determination of the intensities of lines in NMR signals. Note that we are speaking here of the relative intensities of the lines *within a given signal,* not of the intensities of signals for different hydrogens in the molecule.

THE GENERAL CASE

Pascal's triangle: Each number is the sum of the numbers above (see examples shown with dashed lines); the intensities of NMR signals follow this pattern

FIGURE 15.39 In the general case, there will be $n + 1$ lines for a hydrogen adjacent to n equivalent hydrogens.

15.7b Multiple Couplings. We have just developed the "$n + 1$" rule. The number of lines for a given hydrogen is equal to the number of adjacent equivalent hydrogens plus 1. But what happens when a hydrogen is flanked by more than one kind of hydrogen, a very common situation in complex organic molecules? The answer is quite simple—just apply the "$n + 1$" rule for each set of equivalent adjacent hydrogens. Consider the situation in Figure 15.40 in which H_a is adjacent to two different hydrogens, H_b and H_c. The hydrogen H_a will be split into two lines by H_b and each of these lines will be further split into two by coupling to H_c. Let's say that J_{ab} is 2 squares and J_{ac} is 1 square. The result is shown in Figure 15.40(a): four lines, a doublet of doublets. But does it matter which way to do it? In Figure 15.40(a) we started with J_{ab}. In Figure 15.40(b), we do it the other way, starting with J_{ac}. As you can see, the order makes no difference, we get the same four lines either way.

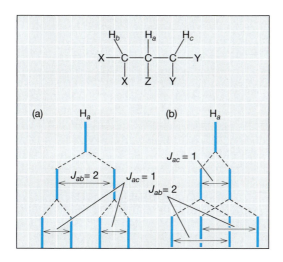

FIGURE 15.40 The pattern for a single hydrogen, H_a, is determined by coupling to two different adjacent single hydrogens, H_b and H_c.

Similarly, Figure 15.41 shows what happens if H_a has one adjacent H_b and two adjacent H_c's, with $J_{ab} = 3$ squares and $J_{ac} = 1$ square. A doublet of triplets results.

PROBLEM 15.16 What happens in general if H_a is flanked by one H_b and three H_c's?

PROBLEM 15.17 What happens to the spectrum of the molecule in Figure 15.41 if $J_{ab} = 2$ squares and $J_{ac} = 1$ square?

SUMMARY

The number of lines resulting from multiple couplings can be determined by constructing "tree" diagrams such as those in Figures 15.40 and 15.41 in which the different couplings are worked out in sequence. The order in which one applies the different couplings never matters.

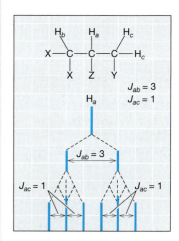

FIGURE 15.41 The splitting pattern for a single hydrogen, H_a, is determined by coupling to a single adjacent hydrogen, H_b, and two equivalent adjacent hydrogens, H_c.

15.7c The Magnitude of J: The Karplus Curve.

There is no coupling between equivalent nuclei. But why is coupling observed only to hydrogens on the same or adjacent atoms and not to all other hydrogens in a molecule? Usually, only "two-bond" or "three-bond coupling" is large enough to be detectable (Fig. 15.42). In some cases, **long-range coupling** can be observed, especially if the hydrogens are connected in allylic fashion. The coupling constant between hydrogens on adjacent carbons is most sensitive to the dihedral angle between those hydrogens, and this can often be used in assigning structure. Figure 15.43 shows the **Karplus curve**, named for Professor Martin Karplus (b. 1930) of Harvard University, which shows the dependence of J on the dihedral angle for two hydrogens on adjacent sp^3 carbons. Figure 15.43 also shows the dependence of the coupling constant on the angle between hydrogens attached to sp^2 hydridized carbons.

FIGURE 15.42 Coupling can be measured over a three-bond distance, but is not usually easily detectable over longer distances.

	Magnitude of J (Hz)
Identical hydrogens do not couple	0
H_a is two bonds away from H_b; coupling over this distance is usually observable	2–30
H_a is three bonds away from H_c; coupling over this distance is usually observable	6–8
H_a is four bonds away from H_d; coupling is not usually observable	0–1
H_a is in the allylic position relative to H_e; coupling is usually observable	2–3

FIGURE 15.43 Coupling between two adjacent hydrogens is sensitive to the dihedral angle between them. The Karplus curve shows the calculated relationship between J and the dihedral angle.

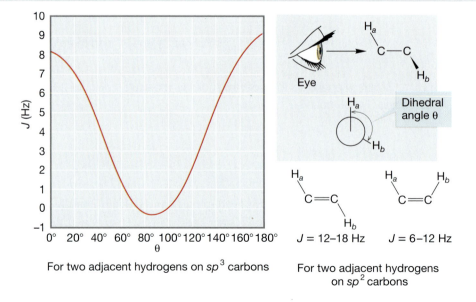

For two adjacent hydrogens on sp^3 carbons

$J = 12–18$ Hz $J = 6–12$ Hz

For two adjacent hydrogens on sp^2 carbons

PROBLEM 15.18 How many lines do you expect to see for the central hydrogen in 1,3-dichloro-2-methoxypropane? *Hint*: Be careful! See page 774.

1,3-Dichloro-2-methoxypropane

PROBLEM 15.19 The two molecules in the figure below have coupling constants for the hydrogens attached to the three-membered ring of 8.4 and 5.3 Hz. Which compound has which coupling constant?

The cis coupling constant, J_{cis}, is smaller than the trans coupling constant, J_{trans}. In alkenes, J_{cis} is in the range 6–12 Hz and J_{trans} is in the range 12–18 Hz. Notice that the ranges slightly overlap. One cannot always assign stereochemistry from just one stereoisomer. What would you make of a coupling constant of 12 Hz, for example? But if both isomers are available, the cis stereochemistry can be reliably assigned to the one with the smaller *J* value. Figure 15.44 gives some typical coupling constants in alkenes and aromatic compounds.

FIGURE 15.44 Some coupling constants for hydrogens attached to double bonds.

15.7d "Decoupling" through Chemical Exchange. The splitting patterns of alcohols and amines are often not what we would expect. Consider the common alcohol, ethyl alcohol, whose ^1H NMR spectrum is shown twice in Figure 15.45. Notice also that these two spectra, taken from different catalogs in the chemical literature, show the concentration dependence of the O<u>H</u> signal. The methyl and methylene hydrogens are in the same place in the two spectra, but the two different samples reveal different chemical shifts for the O<u>H</u>. The spectrum of ethyl alcohol does show the expected triplet for the methyl group, but the methylene and hydroxyl hydrogens do not appear as a first-order analysis would predict. The most accurate description of them might be "blobs." What is going on?

FIGURE 15.45 Two ^{1}H NMR spectra of ethyl alcohol.

Here the problem is one of chemical exchange. In the presence of either an acidic or a basic catalyst, alcohols and amines exchange the OH or NH hydrogens. Figure 15.46 gives the mechanism for base-catalyzed exchange of an alcohol and acid-catalyzed exchange of an amine, using deuterated materials in order to be able to detect the exchange reaction. This exchange reaction can be used to locate the signals for

FIGURE 15.46 The base-catalyzed exchange reaction of alcohols, and the acid-catalyzed exchange reaction of amines.

OH and NH groups. The spectrum is taken, then a drop of D_2O is added to the sample tube. The tube is shaken and the spectrum taken again. All OH and NH peaks will disappear, as their protons are exchanged for deuterons.

PROBLEM 15.20 Write a mechanism for acid-catalyzed exchange of ethyl alcohol and the base-catalyzed exchange of diethylamine.

In IR spectroscopy, the absorption of energy is essentially instantaneous. The IR spectrometer takes a stop-action picture of the molecule. Nuclear magnetic resonance spectroscopy is different. Acquisition of the signal is much slower, about 10^{-3} s, and the exchange reaction is fast compared to the time it takes to make the NMR measurement. During this time the oxygen of a given molecule will be bonded to many different hydrogens. The two methylene hydrogens of ethyl alcohol "see" only an average of all spins of the hydroxyl hydrogen, not the two different $+\frac{1}{2}$ and $-\frac{1}{2}$ states. The same phenomenon serves to remove coupling between alkyl hydrogens and the NH hydrogens of primary and secondary amines.

If this explanation is correct, the coupling should be detectable in scrupulously purified ethyl alcohol because the exchange reaction will be stopped or, at least, greatly slowed. It is very hard to remove the last vestiges of acidic or basic catalysts from these polar molecules, but if the effort is made, the coupling returns, as it should. Absolutely pure ethyl alcohol does give the anticipated first-order triplet for the OH hydrogen (Fig. 15.47).

FIGURE 15.47 The ^{1}H NMR spectrum of absolutely pure ethyl alcohol. See Problem 15.51.

This section gives the first hint that NMR spectra are dependent on the rates of reactions. If the rate of the exchange reaction had not been fast, if it were slow on the NMR time scale, we would not see the averaged spectra that we do for alcohols and amines. We will return to this notion in Section 15.13.

SUMMARY

In the best of all worlds (which means in this case that your budget includes enough money to buy the highest field spectrometer available), we will see a different signal for every different hydrogen in our molecule, and we will be able to determine the number of equivalent adjacent hydrogens from the $n + 1$ rule. In practice, things may not be quite so simple. We can see some possible complications right away. Perhaps the signals will not be sufficiently separated, and the spectrum may consist of complicated overlapping multiplets. Here is an example that shows why chemists are so anxious to spend taxpayers' money on ever higher field spectrometers. The higher the field, the better the dispersion of the signals (recall Problem 15.9, p. 772). Remember that chemical shifts are dependent on the strength of the applied field and the higher the field, the more separated the signals. But there are more complex problems as well, and Section 15.8 touches upon one of them. Once again, we'll find that the solution to the difficulty involves using the highest possible magnetic field.

15.8   MORE COMPLICATED SPECTRA

Even at high field, NMR spectra are often not the simple "first-order" collections of lines we might derive from an analysis of the molecule using the $n + 1$ rule. A classic example of this phenomenon occurs in *AB* signals. For the two-spin system in Figure 15.48, we would expect to see a pair of doublets. Each hydrogen is different, and coupled to only one hydrogen on the adjacent carbon (Fig. 15.48).

FIGURE 15.48 Two different hydrogens adjacent to each other will give rise to a pair of doublets as long as the hydrogens are very different.

In fact, we see this simple spectrum only when the two hydrogens, H_a and H_b, are *very* different: when they have very different chemical shifts. The more alike the two hydrogens, the more the spectrum deviates from first order, the spectrum predicted by application of the $n + 1$ rule. As the chemical environments of the two hydrogens become more similar, the chemical shifts for the corresponding NMR signals of course also become more alike. The more alike they are, the closer the signals are and the more pronounced the leaning of the peaks toward the center (Fig. 15.49).

In the limit, H_a and H_b become the same, and have the same chemical shift. As there is no coupling between equivalent hydrogens, the signal for them collapses to a sharp line. Figure 15.49 shows the gradual transition between the case in which the two hydrogens are very different and that in which they are identical. In the language of NMR, the very different system is called *AX*, the identical system A_2, and an intermediate case *AB* or *AM*.

As we have seen (Section 15.7a, p. 782) it is often simple to derive the expected multiplicity of the NMR peak from the $n + 1$ rule. For example, in a molecule such as 2-furoic acid, the two different hydrogens on C(3) and C(5) (H_A and H_X) are each coupled to the single hydrogen on C(4) (H_M) to give a pair of doublets. This system of three very different hydrogens is called *AMX* (Fig. 15.50).

An *AMX* spin system

2-Furoic acid

Predicted spectrum for H_A and H_X

FIGURE 15.50 In 2-furoic acid, H_A and H_X will appear as doublets. Each is strongly coupled to a single adjacent hydrogen, H_M. Each line will be further split by allylic coupling (J_{AX}) into narrow doublets.

FIGURE 15.49 As the two hydrogens become more similar, the inner lines grow at the expense of the outer lines. The signals increasingly lean toward each other. In the limit, the two hydrogens become identical, and there is only a single line.

Hydrogen H_M itself is coupled differently to H_A and H_X by the coupling constants J_{AM} and J_{MX}. How can we figure out the pattern for H_M that should be observed in the spectrum? The first-order pattern can be determined by working out the number of lines from each coupling one at a time (p. 785). We first determine the number of lines produced by coupling to H_A and then apply the coupling to H_X to each of the lines in a "tree diagram." For example, coupling of H_M to H_A will produce a doublet, separated by J_{AM}. Each of these two lines will be further split into two by coupling to H_X. The result is four lines, a doublet of doublets. Figure 15.51 shows the theoretical prediction and the real spectrum of 2-furoic acid.

2-Furoic acid

Predicted spectrum for H_M, four lines

Here is the real spectrum of 2-furoic acid; the acid hydrogen is off-scale

FIGURE 15.51 The first-order splitting pattern can be determined by calculating the couplings one at a time. In this case, H_M is split into a doublet by H_A and each line of the doublet is split again into doublets by coupling to H_X. Note the allylic coupling, J_{AX}.

First-order spectra are quite easy to predict. For example, according to the $n + 1$ rule, the two different methylene groups of a generic molecule $X-CH_2-CH_2-Y$ should appear as a pair of 1:2:1 triplets. 2-Chloropropionic acid fits the pattern, and, as Figure 15.52 shows, satisfactorily produces the expected pair of methylene triplets in its ^1H NMR spectrum at δ 2.89 and 3.78.

FIGURE 15.52 In 2-chloropropionic acid, each methylene group appears as a simple triplet.

However, when the molecules become more complex we can anticipate complicated spectra, even for first-order systems. Methyl butyrate is a simple molecule, and we can predict its first-order spectrum quite easily. The methyl group attached to oxygen has no neighboring hydrogens and appears as a singlet at δ 3.68. The other methyl group is adjacent to a pair of methylene hydrogens and appears as a triplet at δ 0.94. Similarly, the methylene group adjacent to the carbon–oxygen double bond is flanked by the same methylene group. It, too, is a triplet, this time at δ 2.30. But what about the other methylene group? It is split by three methyl hydrogens, and then split again by two methylene hydrogens. The result should be a quartet of triplets, or, equivalently, a triplet of quartets. Either way you work it out, the result is a 12-line pattern, a dodectuplet! It would be quite amazing if these lines did not overlap to give a pattern difficult to analyze. So they do, as Figure 15.53 shows. A higher field spectrometer would probably reveal all 12 lines.

Triplet at δ 0.94

Triplet at δ 2.30

Singlet at δ 3.68

Methyl butyrate

These hydrogens are split into a triplet by the two adjacent methylene hydrogens, and into a quartet by the three adjacent methyl hydrogens; the result is a 12-line signal at δ 1.66

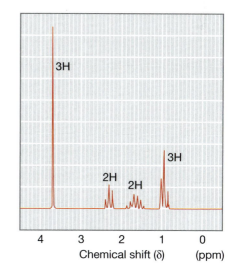

FIGURE 15.53 The multiplet at δ1.66 results from the splitting of one methylene group by the adjacent methyl and methylene groups. The result is 12 lines that overlap in a complex pattern.

PROBLEM 15.21 In Figure 15.53, the expected dodectuplet has simplified to six lines. The methylene group giving rise to that signal is coupled to both the adjacent methyl and methylene groups. Use a "tree" diagram such as the one in Figure 15.51 to show that, if the coupling constants J to the methyl group and the methylene are fortuitously the same, the observed six lines should appear. Graph paper is very useful!

PROBLEM 15.22 Explain why the methylene group adjacent to the carbonyl group is downfield of the methylene group adjacent to the methyl group.

Clearly, even if the spectrum is first order we are likely to be faced with a seriously overlapping set of lines in a case such as this one, and interpretation may not be simple. Moreover, as we have already seen, NMR spectra are often not first order. Even in the simple system of two adjacent different hydrogens, the observed spectrum is often more complicated than what a first-order analysis would predict. First-order spectra are only observed when all the hydrogens in a molecule are very different. As they become less different, more complicated spectra appear. Finally, if we imagine carrying the process of making the hydrogens more and more alike to its logical conclusion, we get a single signal for equivalent hydrogens. But when are hydrogens "very different" and when are they not? In a practical sense, for hydrogens to give rise to first-order spectra, the difference in hertz in their chemical shifts ($\Delta\delta$) must be at least 10 times greater than their coupling constant, J_{ab} [Eq. (15.5)].

$$\Delta\delta > 10 \, J_{ab} \qquad (15.5)$$

That requirement is another reason why high-field NMR spectrometers are so much more useful than lower field instruments. Although the coupling constant J does not depend on the field strength, the value (in hertz) of the chemical shifts does (p. 770). The greater the field strength, the greater the chemical shift difference ($\Delta\delta$), and therefore the greater the likelihood that the condition of Eq. (15.5) will be satisfied. Figure 15.54 shows the ^{1}H NMR spectrum of butyl bromide at 60, 90, and 300 MHz. Even the modest change from 60 to 90 MHz results in substantial improvement in the spectrum. You get what you pay for.

FIGURE 15.54   The ^{1}H NMR spectra of butyl bromide taken at 60, 90, and 300 MHz.

15.9 DECOUPLED SPECTRA

When we discussed the NMR spectra of alcohols and amines, we encountered the phenomenon of chemical exchange of the OH and NH hydrogens. Rapid exchange of these hydrogens resulted in an averaging of the coupling to adjacent hydrogens. Exchange effectively **decoupled** the OH and NH hydrogens from hydrogens at adjacent positions. This same trick can be accomplished electronically, and it can be very useful, or even essential in the interpretation of NMR spectra. Suppose we are set the task of determining the stereochemistry of ethyl crotonate from its ^{1}H NMR spectrum (Fig. 15.55)?

FIGURE 15.55 The ^{1}H NMR spectrum of ethyl crotonate. The coupling between H_a and H_b is obscured by the coupling to the adjacent methyl groups.

We might hope to make this distinction by measuring the coupling constant between the two vinyl hydrogens. If the compound is trans, we would expect J to be 12–18 Hz; if it is cis, J would be smaller, 6–12 Hz (Fig. 15.43). However, the rather complex spectrum makes the determination of J difficult. The problem is that both vinyl hydrogens are coupled to the methyl group, and this coupling leads to the complex signals we see. If we could somehow get rid of the coupling to the adjacent methyl group, the spectrum would surely simplify, and it might become possible to pick out the couplings between H_a and H_b. It is possible to accomplish this simplification through electronic decoupling. The spectrometer is modified to allow a second application of radiofrequency waves during the NMR experiment. While the spectrum is being measured, radio frequency at the resonance frequency of the methyl hydrogens is applied. The result is to induce rapid transitions between the two possible spin states for each methyl hydrogen. As in chemical exchange of ethyl alcohol, the spin of the methyl hydrogens is averaged to zero and there is no observable coupling to the adjacent hydrogen H_a. Figure 15.56 shows a blow-up of the signals for the two vinyl hydrogens with and without irradiation at the methyl group. It is now easy to pick out the coupling between H_a and H_b and the stereochemistry of the molecule can be determined with ease from the observed J of 17 Hz. The compound is trans.

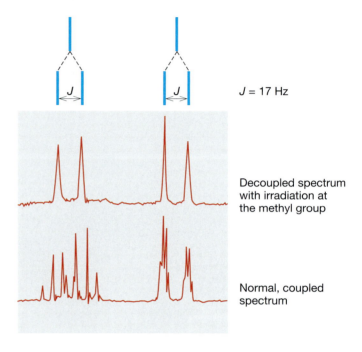

$J = 17$ Hz

Decoupled spectrum
with irradiation at
the methyl group

Normal, coupled
spectrum

FIGURE 15.56 Irradiation at the resonance frequency of the allylic methyl group averages the spins of the methyl hydrogens and decouples them. The coupling constant J between H_a and H_b is now easier to determine.

There are many similar applications, and modern NMR spectrometers are all equipped to do decoupling experiments routinely.

15.10   NUCLEAR MAGNETIC RESONANCE SPECTRA OF ^{13}C AND SOME OTHER NUCLEI

Several other nuclei are of interest to the organic chemist, mainly ^{19}F, ^{15}N, ^{2}H (deuterium), and ^{13}C, an isotope of carbon, which is present in 1.1% natural abundance. Even though all organic compounds contain carbon, the natural abundance of ^{13}C is low enough so that coupling to ^{13}C is not observed in hydrogen spectra. These other nuclei can be used either in natural abundance or in enriched samples to provide their own NMR spectra. Recall that the constant γ in Eq. (15.4) will be different for ^{19}F, ^{2}H, ^{15}N, or ^{13}C, and their NMR signals will not overlap with the hydrogen spectra.

One other nucleus, ^{13}C, is comparable in importance to ^{1}H. Like hydrogen, ^{13}C has a spin of ½, and coupling between ^{13}C and ^{1}H will complicate ^{13}C spectra. Normally, ^{13}C spectra are run using a broad-band decoupling so that all hydrogens in the molecule are decoupled. All ^{13}C$-^{1}$H coupling is removed in this way. What about coupling between two ^{13}C atoms? Won't this coupling complicate the spectrum? The chance of finding a ^{13}C at any given position is only 1.1% or 0.011. The chance of finding two ^{13}C atoms on adjacent positions is therefore $0.011 \times 0.011 = 0.00012$! At least for natural abundance spectra we can ignore ^{13}C$-^{13}$C coupling. Accordingly, ^{13}C spectra are collections of sharp lines, singlets. If the broad-band decoupler is turned off, a spectrum can be obtained through a technique called **off-resonance decoupling**, which allows the ^{13}C signals to be split according to the $n + 1$ rule, only by the *directly attached* (not adjacent) hydrogens. As there are no attached hydrogens, quaternary carbons will still appear as singlets, but methine carbons (one attached hydrogen) will be doublets, methylene carbons (two attached hydrogens) triplets, and methyl carbons (three attached hydrogens) will appear as quartets. Figure 15.57 summarizes this notion and shows the decoupled and off-resonance decoupled spectra of ethyl maleate.

FIGURE 15.57 Decoupled and off-resonance decoupled ^{13}C NMR spectra of ethyl maleate.

Several other techniques now allow for the easy determination of the number of hydrogens attached to a carbon. One of the most convenient techniques goes by the acronym **DEPT** (distortionless enhancement with polarization transfer). In this technique, several separate spectra are determined for a molecule under conditions that allow for the appearance of only methine (CH), methylene (CH_2), or methyl (CH_3) carbons. Quaternary carbons can be determined by difference: Any signal in the full spectrum that does not appear in the separate spectra for CH, CH_2, and CH_3 carbons must belong to a quaternary carbon. Figure 15.58 shows the DEPT technique applied by Kathryn Williams and Roy W. King of the University of Florida to ethyl benzoate. The full ^{13}C spectrum is shown on the bottom line, (e). On the line (d) above, which shows the signals for all carbons attached to any number of hydrogens, note the absence of the signals for the quaternary carbons at 130 and 166 ppm. The top three lines (c, b, a) show the methine, methylene, and methyl carbons.

FIGURE 15.58 The DEPT technique applied to ethyl benzoate.

Table 15.5 gives a series of chemical shift correlations for ^{13}C NMR. Notice the greatly expanded range of chemical shifts extending over 200 ppm, as opposed to the mere 10 ppm for ^1H. Dispersion is much greater for ^{13}C NMR than for ^1H NMR. Once again, the question of memorization arises. Use Table 15.5 to work problems, and you will automatically memorize the information you really need (if you do enough problems!).

TABLE 15.5 Some ^{13}C Chemical Shifts

Type of Carbon	Chemical Shift $(\delta)^a$	Type of Carbon	Chemical Shift $(\delta)^a$
Alkanes		Alcohols, ethers	
Methyl	0–30	C—O	50–90
Methylene	15–55	Amines	
Methine	25–55	C—N	40–60
Quaternary	30–40	Halogens	
Alkenes		C—F	70–80
C=C	80–145	C—Cl	25–50
Alkynes		C—Br	10–40
C≡C	70–90	C—I	−20–10
Aromatics	110–170	Carbonyls, C=O	
Benzene	128.7	R_2C=O	190–220
		RXC=O (X = O or N)	150–180

^{a}The chemical shift δ is in parts per million (ppm) from TMS.

15.11 PROBLEM SOLVING: HOW TO USE SPECTROSCOPY TO DETERMINE STRUCTURE

The task I set myself here is an almost impossible one—to show in a practical way how to use the various kinds of spectral data to come up with a structure. Every compound is different, of course, and so there can be no absolutely general method. Even worse, there are many different approaches; what works for you may not be the optimal way for others. Yet, there are some general techniques that are useful, and I'll try to set them out here. Remember, though, as the old lady in New York answered, when asked for directions on how to get to Carnegie Hall, "practice, practice, practice." For you, that old joke means, "do lots and lots of problems." Nowhere else is that advice so essential.

First of all, when confronted by a set of spectral charts or data derived from them, try to determine the molecular formula. Sometimes it will be given, sometimes you can work it out. Other times, it may be obvious or at least likely from the chemical reactions involved. Sometimes you may only have a molecular weight. Either way, having a formula or molecular weight lets you see what's left once you have determined part of the structure.

Next, I use any available IR data, along with NMR chemical shift data, to get an idea of what functional groups are present. There are many items in the Additional Problems section that let you practice this technique.

Surprisingly, symmetry is often enough to solve a structural problem—a simple counting of signals may be sufficient to make a choice between several alternatives, so I always look hard at symmetry before I start on any detailed analysis. Here is an example:

WORKED PROBLEM 15.23 You have bottles containing three isomeric compounds, **A**, **B**, and **C**, each of the formula $C_{10}H_{14}$. The only significant bands in the IR spectra of all three compounds are at about 3050, 2950, and 1610 cm^{-1}. The ^1H NMR spectrum of **A** shows signals at δ 2.13 (s, 6H) and 6.88 (s, 1H); **B** shows δ 2.18 (s, 3H), 2.25 (s, 3H), and 6.89 (s, 1H); and **C** shows δ 2.11 (s, 3H), 2.19 (s, 3H), 2.22 (s, 6H), and 6.80 (s, 2H). The ^{13}C NMR spectrum of **A** shows lines at δ 131.02, 133.52, and 19.04 ppm. Compound **B** shows signals at δ 126.88, 133.76, 133.79, 20.65, and 15.78 ppm. Compound **C** shows signals at δ 128.29, 131.69, 134.43, 136.15, 20.74, 20.20, and 14.86 ppm. What are the structures?

ANSWER First of all, these compounds seem to be aromatic hydrocarbons. There are four degrees of unsaturation, and the IR shows both aromatic and aliphatic CH stretches, as well as a band at 1610 cm^{-1} appropriate for an aromatic double-bond stretch. If six of the available carbons are tied up in an aromatic ring, the other four must be attached in some way. A look at the ^1H NMR spectrum shows only singlets, and so the four remaining carbons are certainly methyl groups. Thus, putting the data together leads to the idea that **A**, **B**, and **C** are the three possible tetramethylbenzenes. But which is which?

A   B   C

To answer this question, there is no need to do anything more elaborate than look at the number of signals. There is more than one way to do this, but let's use the methyl signals in the ^1H NMR spectrum. Compound **A** has only one methyl signal and must be the left-hand molecule. Compound **B** has two methyl signals, and must be in the middle. Of course, we now know that the remaining isomer must be **C**, but it is gratifying that it has the appropriate three methyl signals. There was no need to consider any details, only count.

PROBLEM 15.24 Could we have used the aromatic CH signals to do the same determination?

PROBLEM 15.25 Could we have used the ^{13}C NMR spectra to do the same determination?

Chemical shift data are often decisive. After considering symmetry, that is what I look at next. Your correlation chart, Table 15.4, will let you make many decisions. See Problem 15.26 for an example.

WORKED PROBLEM 15.26   You have four bottles, each containing one of the four compounds shown below, **E**, **F**, **G**, and **H**. From the spectral data given, determine which bottle holds which compound.

All compounds have signals for 4H in the aromatic region of the ^{1}H NMR spectrum at δ 7–8 ppm, so that information is of no use.

Bottle 1   IR 1720 cm^{-1} ^{1}H NMR δ 1.38 (t, 3H), 2.40 (s, 3H), 4.36 (q, 2H)

Bottle 2   IR 1688 cm^{-1} ^{1}H NMR δ 1.12 (t, 3H), 2.81 (q, 2H), 3.76 (s, 3H)

Bottle 3   IR 1725 cm^{-1} ^{1}H NMR δ 1.21 (t, 3H), 2.65 (q, 2H), 3.85 (s, 3H)

Bottle 4   IR 1680 cm^{-1} ^{1}H NMR δ 1.40 (t, 3H), 2.50 (s, 3H), 4.07 (q, 2H)

ANSWER   First, use Table 15.3 to determine that the two esters are in bottles 1 and 3 (IR conjugated C=O stretch at about 1720 cm^{-1}) and the two ketones are in bottles 2 and 4 (IR conjugated C=O stretch much lower, at about 1680 cm^{-1}).

Now the chemical shift data allow a quick determination of the rest of the problem. Again, there are several ways to do this last part, but the easiest is to look at the chemical shift of the singlet methyl group (you could also use the methylene quartet). A methyl group next to an oxygen has a chemical shift δ of about 4.5, and a "benzylic" methyl group (directly attached to a benzene ring) has a δ of about 2.3, so Compound **E** is in bottle 1, and compound **F** is in bottle 2. Similarly, compound **G** is in bottle 3, and compound **H** is in bottle 4.

Sometimes, however, even the chemical shift is insufficient to make the determination. In such cases, the coupling constant J often does the trick. Sometimes it is the magnitude of J that is determining. The molecules in Problem 15.19 are a perfect example of this kind of use of J. More often, it is the splitting pattern that works. An examination of J reveals who is adjacent to whom in the hydrogen world, and thus provides deep insight into structure. Here's another problem, this time one that requires you to think about J.

WORKED PROBLEM 15.27   Again you have bottles 1 and 2, this time containing **I** and **J**. Explain how you would use ^{1}H NMR spectroscopy to tell which molecule is in which bottle.   *(continued)*

(continued)

ANSWER In Compound **J**, the hydrogens adjacent to the benzene rings and the chlorines will couple. Each will appear as a doublet, with a small coupling constant of a few hertz (Hz). But those same hydrogens in Compound **I** are too far from each other to couple; they will appear as singlets as identical hydrogens do not couple. Be sure you see that there is no other easy way to do this determination.

Now we pass on to some "bells and whistles" of NMR.

15.12   SPECIAL TOPIC: DETERMINATION OF THE STRUCTURE OF A COMPLEX BIOMOLECULE, MAITOTOXIN

These days sophisticated NMR techniques seem capable of any structure determination. This section is almost pure "gee-whiz" in that I am trying to dazzle you with the complexity of the molecular structures that can be solved by the application of spectroscopy (and a lot of hard work). The example I chose is Maitotoxin, shown in Figure 15.59.

FIGURE 15.59 Maitotoxin.

Maitotoxin, a molecule for which I of course maintain a certain affection, is extraordinarily lethal [for half the mice treated the lethal dose is only 50 ng/kg (LD_{50})] and is one of the largest natural products known. In 1993, a group of Japanese chemists reported a variety of complex NMR studies that established the general structure, and in 1996, chemists at Harvard used much spectroscopy, along with organic synthesis, to establish the complete relative stereochemistry. That same year the absolute stereochemistry was determined in Japan by Professor K. Tachibana (b. 1949) and his research group at the University of Tokyo. How many stereogenic carbons are there in Maitotoxin? How many possible stereoisomers are there? To get this structure is a remarkable accomplishment.

Why show this molecule? No one is ever going to ask you to reproduce the structure, I hope, and there are not likely to be any immediate practical consequences. I think that the answer is simply that the sheer difficulty of the enterprise is overwhelming. We humans can do some quite amazing things. There are times when that gee-whiz factor must be confronted directly, when we are allowed to sit back and marvel at some accomplishment, and this is one of those moments.

15.13 SPECIAL TOPIC: DYNAMIC NMR

We have already seen that rates of hydrogen exchange are important in NMR spectroscopy. Recall that fast exchange of hydroxyl and amine hydrogens generally results in a decoupling of the OH and NH hydrogens (p. 787). An adjacent hydrogen "feels" only an averaged perturbation of the applied magnetic field if exchange is fast enough. If exchange is slowed by careful removal of all acidic or basic catalysts, the coupling can be detected. The *rate* of exchange is important to the spectrum. If it is fast, the NMR experiment detects an averaged spectrum; if it is slow, it sees a static picture.

The classic example of this phenomenon involves the spectrum of cyclohexane. As you know, there are two different kinds of hydrogen in cyclohexane, the axial and equatorial hydrogens, and these are interconverted by a ring flip, with an activation energy of about 11 kcal/mol (Chapter 5, p. 203). Despite the existence of two kinds of hydrogen, the room-temperature spectrum of cyclohexane shows only a single line at δ 1.4 ppm (Fig. 15.60). This simple spectrum appears because the rate

FIGURE 15.60 The ^{1}H NMR spectrum of cyclohexane at 25 °C.

of interconversion of axial and equatorial hydrogens is so fast, with respect to the time necessary for the NMR measurement, that an averaged spectrum appears. Suppose we decrease the energy available to the system by lowering the temperature. If we can decrease the rate of the ring flip sufficiently, the spectrometer will no longer detect an average of the two possible positions but will be able to see both axial and equatorial hydrogens. Figure 15.61 shows the ^{1}H NMR spectrum of undecadeuteriocyclohexane (cyclohexane-d_{11}) at -90 °C. Two equal signals appear, centered upon the position of the room temperature averaged signal.

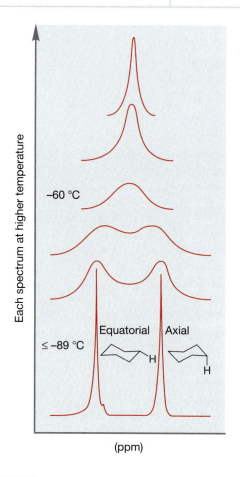

FIGURE 15.61 At low temperature, ring flipping is slowed and the axial and equatorial hydrogens of cyclohexane can be resolved. As the temperature is raised, the -90 °C spectrum approaches the 25 °C spectrum.

At -90 °C, the interconversion between the two forms of cyclohexane is slow enough so that the NMR machine no longer detects an averaged signal. In practice, for a signal to be detectable for a given species, it must have a lifetime of about 1 s. As the temperature is raised to room temperature, the lifetime of individual cyclohexanes decreases as the rate of ring flipping increases, and an averaged spectrum appears.

Now, what happens at an intermediate temperature? Figure 15.61 shows you. A common sense answer would be correct here. If two signals are found at −90 °C, and a single signal at 25 °C, at intermediate temperatures the two kinds of signal must approach each other. The two sharp peaks of the −90 °C spectrum will slowly merge into the single peak of the 25 °C spectrum. In fact, it is possible to derive kinetic information from the temperature dependance of NMR spectra, and this is still another use of NMR spectroscopy.

More molecules than you might at first expect give this kind of averaged spectrum. It takes a detailed look at structure to see why. Consider ethyl alcohol. If we ignore the OH signal, which we know to be concentration and catalyst dependent, we expect to see only one peak for the three equivalent methyl hydrogens and another for the two equivalent methylene hydrogens. But wait—look at the Newman projection in Figure 15.62, which shows the most stable staggered conformation of ethyl alcohol.

FIGURE 15.62 In static ethyl alcohol, there are three different kinds of hydrogen attached to carbon. However, rapid rotation around the carbon–carbon bond interconverts the methyl hydrogens and makes them equivalent. The two methylene hydrogens are enantiotopic.

There are three kinds of hydrogen (H_a, H_b, and H_c), not two. The three methyl hydrogens are *not* the same. Yet, we do not see separate signals for them. By now you should be able to see the answer to this conundrum. As shown in Figure 15.62, rotation about the carbon–carbon bond of ethyl alcohol interconverts the three methyl hydrogens, rendering them equivalent. Moreover, we know that the low barrier for rotation, about 3 kcal/mol, ensures a rapid rate for this simple process. In Figure 15.62, H_a and the two H_c hydrogens are individualized with colored squares. Note how each hydrogen occupies one of the three possible positions as rotation occurs. The NMR spectrometer detects only the averaged signal.

What about the two methylene hydrogens, H_b? The Newman projection of Figure 15.62 shows that even in the absence of rapid rotation they are equivalent, and therefore must *always* give rise to one signal. These hydrogens are enantiotopic. As we saw earlier, the origin of the name lies in the mirror-image relationship of the two to each other.

So, as long as rotation is fast, neither the individual methyl hydrogens nor the individual methylene hydrogens of ethyl alcohol are distinguishable by NMR spectroscopy.

PROBLEM 15.28 In a solvent consisting of a single enantiomer, the two enantiotopic hydrogens of ethyl alcohol (H_b in Fig. 15.62) appear as separate signals. Explain.

WORKED PROBLEM 15.29 Use the device of replacing the methylene hydrogens (H_b in Fig. 15.62) with an X group to show that they are enantiotopic (p. 773).

ANSWER

FIGURE 15.63 In propyl alcohol, the methyl hydrogens are homotopic and there are two pairs of enantiotopic methylene hydrogens.

Now let's change some things about ethyl alcohol. Let's first convert it into propyl alcohol by replacing one methyl hydrogen with a methyl group. There are now three sets of hydrogens, the methyl hydrogens and two sets of different methylene hydrogens. Rapid rotation ensures that the three methyl hydrogens will be equivalent, and each pair of methylene hydrogens consists of two magnetically equivalent, enantiotopic hydrogens (Fig. 15.63).

Now let's replace a hydrogen of propyl alcohol with a chlorine to make 2-chloro-1-propanol. Look at the Newman projection of this molecule in Figure 15.64. Are the two methylene hydrogens equivalent? No. The Newman projections of Figure 15.64 show the three rotational isomers and in *all* of them the two hydrogens are different. *Never* does H_x become equivalent to H_y. They are forever different no matter how fast rotation about the carbon–carbon bond is, and so must give rise to separate signals. These hydrogens are diastereotopic, and will give rise to distinct signals. This point is particularly difficult, so let's anticipate the Common Errors section a little bit and deal with the problem right here. Is not H_y in drawing **2** of Figure 15.64 in the same position as H_x in drawing **1**? Are the two not equivalent? Both are 120° from the hydroxyl group, and both lie in between a methyl group and a hydrogen (Fig. 15.64).

FIGURE 15.64 In 2-chloro-1-propanol, the two methylene hydrogens, H_x and H_y, are neither equivalent (homotopic) nor enantiotopic.

No, H_x in drawing **1** of Figure 15.64 is different from H_y in drawing **2**. A short demonstration shows that the two OH groups in drawings **1** and **2** are not identical. One is between hydrogen and chlorine, the other is between methyl and chlorine. Both H_x and H_y are 120° from *different* OH groups in the two forms. A more elaborate demonstration takes advantage of the device mentioned earlier (p. 773). Replace H_x with a group X and determine the relationship between this compound and the one produced by replacing H_y with X. These two compounds are different, neither identical nor enantiotopic (Fig. 15.65). There is no way out, H_x and H_y are diastereotopic.

These two structures are neither identical nor enantiomers; they are diastereomers, and H_x and H_y are diastereotopic hydrogens

FIGURE 15.65 The two hydrogens are diastereotopic and will give different signals in the NMR spectrum.

PROBLEM 15.30 The methyl group of *trans*-1-*tert*-butyl-4-methylcyclohexane appears as a single doublet, but in *cis*-1-isopropyl-3-phenylcyclohexane there are two methyl doublets. Explain.

15.14   SUMMARY

New Concepts

This chapter recapitulates the ideas first introduced in Chapter 4 in the discussion of column chromatography, and introduces the much more versatile separation methods of gas chromatography and high-performance liquid chromatography. All chromatography works in much the same way: The components of a mixture are forced to partition themselves between a moving phase and a stationary phase. The more a given component exists in the moving phase, the faster it moves through the column; the more strongly it is absorbed, the longer it remains on the column.

In mass spectrometry, a molecule is first ionized by bombardment with high-energy electrons. The parent ion so formed can undergo numerous fragmentation reactions before it passes, along with the daughter ions formed through fragmentation, into a region between the poles of a magnet. Ions are separated by mass, and focused on a detector. The mass spectrometer gives the molecular weight of a molecule, provided that the parent ion can be detected, and high-resolution mass spectrometry can determine the molecular formula of an ion. Structural information can often be gained through an analysis of the important daughter ions in the fragmentation pattern.

Infrared spectroscopy detects vibrational excitation of chemical bonds. Functional groups have characteristic vibrational frequencies in the IR that depend on the masses of the atoms in the bond and, importantly, on the force constant of the bond. Analysis of an IR spectrum gives information on the kinds of functional groups present.

Nuclear magnetic resonance detects the energy absorbed in the transition from a low-energy state in which the nuclear spin is aligned with an external magnetic field to a slightly higher energy state in which it is aligned against the applied field. Hydrogens, ^{13}C, and some other nuclei have characteristic absorptions (the chemical shift, δ), which depend critically on the surrounding chemical environment. Hydrogen nuclei are coupled to adjacent hydrogens through a coupling constant, J. The number of lines observed for a particular hydrogen depends on the number of other nuclei to which it is coupled. For first-order spectra, the number of lines equals $n + 1$, where n is the number of equivalent coupled nuclei. The NMR spectra can be integrated to reveal the relative number of hydrogens in each peak.

Key Terms

base peak (p. 756)
chemical shift (p. 773)
coupling constant (J) (p. 782)
daughter ions (p. 755)
decoupled (p. 795)
DEPT (p. 798)
diastereotopic (p. 774)
enantiotopic (p. 774)
force constant (p. 762)
fragmentation pattern (p. 755)
gas chromatography (GC) (p. 750)
GC/IR (p. 752)
GC/MS (p. 752)
high-performance liquid chromatography (HPLC) (p. 749)
homotopic (p. 773)

infrared (IR) spectroscopy (p. 761)
integration (p. 775)
Karplus curve (p. 786)
long-range coupling (p. 786)
mass spectrometry (MS) (p. 752)
McLafferty rearrangement (p. 760)
nuclear magnetic resonance (NMR) (p. 768)
off-resonance decoupling (p. 797)
p + 1 peak (p. 753)
parent ion (p. 752)
ppm scale (p. 771)
radical cation (p. 752)
tetramethylsilane (TMS) (p. 771)
wavenumber (p. 762)

Common Errors

Practice, practice, practice! The successful interpretation of spectra requires it. The most common error is not to realize this fact. The memorization of tables of data will not make you into a spectroscopist, but working lots and lots of problems will. Also, see Section 15.11 for some hints.

A central question in this chapter, and in all of chemistry, is, What's different? When are two hydrogens equivalent and

when are they not? This chapter offers a device that is useful in making such distinctions.

It is often difficult to keep track of the interplay of the applied magnetic field, B_0, the chemical shift, δ, and the coupling constant, J. Even at very high fields, first-order spectra are not always seen unless the coupled hydrogens have very different chemical shifts.

15.15   ADDITIONAL PROBLEMS

General advice: Although this section is longer than most problem sections, it cannot do more than illustrate some of the difficulties involved in going from a set of lines on chart paper, or a table of numbers, to a structure. Moreover, in the real world of chemistry there is close interaction between the reactions involved and the interpretation of spectra. The chemist has some idea (usually!) of what is likely to be present. Some of the later problems must be solved in an interactive way. The chemistry involved will suggest structures, and you will have to see if they are consistent with the data available. All of the following problems involve manipulating spectral data. You should start well equipped with a table of IR frequencies and ^1H NMR and ^{13}C NMR correlation charts for chemical shifts. A table of coupling constants will also be essential, and a copy of the Karplus curve is often useful. Do not be shy about looking values up. Most chemists analyze their "real world" spectra that way, and *everyone* starts that way.

Specific advice: In Chapter 3 we first encountered Ω, the number of "degrees of unsaturation." In this chapter, we see nitrogen-containing compounds and will have to modify our formula a bit. Here is the protocol for nitrogen and some other heteroatoms. Be sure to review Ω in Chapter 3 (p. 132).

1. Oxygen and divalent sulfur can be ignored.
2. Halogens (F, Cl, Br, and I) are counted as if they were hydrogens.
3. Treat nitrogen as if it were a carbon. However, as nitrogen is trivalent, not tetravalent, subtract one hydrogen per nitrogen from the total number of hydrogens computed for the saturated hydrocarbon.

We begin with some relatively straightforward problems largely involving infrared spectroscopy and mass spectrometry, and then move on.

PROBLEM 15.31 Here is a practical problem of no intellectual content whatsoever. In doing problems involving IR, it is often much easier to read the values on the chart paper in microns (μ), rather than wavenumber ($\bar{\nu}$). Unfortunately, most of us have learned to think in wavenumbers. There is a simple conversion. Convert 3.00, 5.00, 7.00, 9.00, and 11.00 μ into wavenumbers. *Hint*: See p. 762.

PROBLEM 15.32 Calculate the degrees of unsaturation in the following molecules:

PROBLEM 15.33 The IR spectrum of a molecule known to have structure **A**, **B**, **C**, or **D** is shown below. Which structure is most consistent with the spectrum? Explain your reasoning.

PROBLEM 15.34 The IR spectra of compounds **1–5** are shown on the next page. Each compound is either an aldehyde, an anhydride, a carboxylic acid, an ester, or a ketone. Identify to which class each compound belongs based on its IR spectrum. Explain your reasoning by assigning characteristic and/or distinguishing absorptions. Each class is represented by one compound.

(continued)

PROBLEM 15.34 (CONTINUED)

1

2

3

4

5

1 Alcohol
  Carboxylic acid
  Phenol

2 Aldehyde
  Ester
  Ketone

3 1-Alkyne
  Symmetrical disubstituted alkyne
  Cyanide

4 Primary amide
  Primary amine
  Nitro compound

5 Anhydride
  Carboxylic acid
  Ester

PROBLEM 15.35   For each of the following IR spectra (**1–5**) there is a choice of three possible types of compounds. For each spectrum, choose the most appropriate class of compound. Explain your reasoning by noting the presence or absence of characteristic bands in the spectrum.

PROBLEM 15.36 Assign the IR spectra below to the correct octene isomer. Explain your reasoning.

1-Octene

***trans*-2-Octene**

***cis*-2-Octene**

PROBLEM 15.38 Assign the IR spectra below to the correct chlorotoluene isomer.

PROBLEM 15.37 The IR spectra below belong to 5-phenyl-1-pentyne and 6-phenyl-2-hexyne. Which is which? Explain your reasoning.

$PhCH_2CH_2CH_2C{\equiv}CH$ $PhCH_2CH_2CH_2C{\equiv}CCH_3$

PROBLEM 15.39 Compound **A** has the formula $C_8H_7BrO_2$. The two most intense peaks in the mass spectrum are at $m/z = 216$ and 214, and they appear in almost equal amounts. Why? Compound **A** is an aldehyde. Explain why the second-most intense peaks in the mass spectrum come at m/z values of 215 and 213.

PROBLEM 15.40   Compound **A**, containing only C, H, and O, gave 80.00% C and 6.70% H upon elemental analysis. Summaries of the mass spectrum and the IR spectrum of compound **A** are shown below. Deduce the structure of A and explain your reasoning.

m/z	Relative Abundance
105	100
77	88
51	40
120	29
50	21
43	17
78	10
39	9

PROBLEM 15.41   Compound **A**, containing only C, H, and O, gave 70.60% C and 5.90% H upon elemental analysis. Summaries of the mass spectrum and the IR spectrum of compound **A** are shown below. Deduce the structure of **A** and explain your reasoning.

m/z	Relative Abundance
135	100
136	71
77	28
92	15
107	15
39	8
65	8
63	8

PROBLEM 15.42   Compounds **A**, **B**, and **C** all gave 71.17% C, 5.12% H, and 23.71% N upon elemental analysis. Summaries of the mass spectra and the IR spectra of these compounds are shown below. Deduce the structures of **A–C** and explain your reasoning.

A

m/z	Relative Abundance
118	100
96	20
119	9
64	8
90	7
63	6
39	5
117	5

B

m/z	Relative Abundance
118	100
96	40
119	22
64	22
90	19
63	14
39	14
117	12

m/z	Relative Abundance
118	100
96	35
119	12
64	7
90	7
63	7
39	6
117	4

PROBLEM 15.43   Extraction of black Hemiptera bugs affords a colorless oil **A** of the formula $C_7H_{12}O$ possessing a sickening odor. The IR spectrum of **A** displays characteristic peaks at 2833 (w), 2725 (w), 1725 (s), 1380 (m), and 975 (m) cm^{-1}. Upon catalytic hydrogenation, the peak at 975 cm^{-1} disappears. Ozonolysis of compound **A**, followed by an oxidative workup, gives succinic acid ($HOOC-CH_2CH_2-COOH$) and propionic acid ($HOOC-CH_2CH_3$). Deduce a structure for **A** and explain your reasoning.

PROBLEM 15.44   The O—H portion of the IR spectrum changes with concentration. As the solution becomes more dilute, the broad band centered at roughly 3300 cm^{-1} sharpens. Explain.

PROBLEM 15.45   The O—H stretching IR band for *tert*-butyl alcohol is much sharper than that for methyl alcohol. Explain.

PROBLEM 15.46   The IR stretching frequency for the O—H group in the syn unsaturated molecule in the next column is sharp at all concentrations and shows no change on dilution, whereas that for the related saturated molecule or the anti unsaturated molecule is broad and sharpens with dilution. Explain. *Hint*: Oxygen is not the only base that can participate in hydrogen bonding.

syn-Unsaturated   anti-Unsaturated   Saturated

Now we move on to NMR spectroscopy. Again, we start with simple exercises:

PROBLEM 15.47   Explain carefully why the following two spectra of 1-bromobutane look so different:

(a) Spectrum taken at 60 MHz.

(b) Spectrum taken at 300 MHz.

PROBLEM 15.48   Are the underlined hydrogens in the following molecules homotopic, enantiotopic, or diastereotopic? Justify your answers.

PROBLEM 15.49 Assign the following chemical shifts to the hydrogens in compounds in (a–d). Use the $(n + 1)$ rule to predict the multiplicities of the peaks.

(a) $Cl_2CH—CHCl—CH_3$

δ
5.88
4.37
1.70

(b) $ClCH_2—CH(OCH_3)_2$

δ
4.53
3.53
3.43

(c) $(CH_3)_2CH—CH_2I$

δ
3.16
1.72
1.03

(d) $BrCH_2CH_2CH_2Br$

δ
3.59
2.36

PROBLEM 15.50 Predict the first-order spectra of the indicated hydrogens in the molecules in (a–c).

PROBLEM 15.51 Figure 15.47 shows an enlarged view of the multiplet centered at roughly δ 3.6 in the ^1H NMR spectrum of pure ethyl alcohol. Use a tree diagram to analyze this complex multiplet.

PROBLEM 15.52 The ^1H NMR spectra of the two scrupulously dry alcohols shown below have signals (among others) at δ 2.60 (d, $J = 3.8$ Hz, 1H) (**A**) and δ 2.26 (t, $J = 6.1$ Hz, 1H) (**B**). Assign the spectra, and explain your reasoning.

PROBLEM 15.53 When a drop of aqueous acid is added to the samples described in Problem 15.52, the signals described collapse to broad singlets. Explain.

PROBLEM 15.54 In Sections 15.5 and 15.10, it was claimed that the NMR signals for ^2H (deuterium) and ^{13}C will not overlap ^1H spectra. Verify that this statement is true. Show that the frequency (v) will be different for ^1H, ^2H, and ^{13}C at a field strength (B_0) of 4.7 T. The gyromagnetic ratios for ^1H, ^2H, and ^{13}C are 2.7×10^8, 0.41×10^8, and 0.67×10^8 rad T^{-1} s^{-1}, respectively.

PROBLEM 15.55 Match the following ^1H NMR absorptions to the compounds shown below, each of which has a one-signal spectrum: δ 2.25, 0.04, 1.42, 0.90, 7.37 ppm.

PROBLEM 15.56 Match the following ^{13}C NMR absorptions to the following compounds, each of which has a one-signal spectrum: δ 26.9, −2.9, 59.7, 128.5.

CH_3OCH_3

PROBLEM 15.57 Explain why the hydrogens α to an amino group ($R—C\underline{H}_2—NH_2$) appear at higher field in the ^1H NMR spectrum than those α to a hydroxyl group ($R—C\underline{H}_2—OH$).

PROBLEM 15.58 How many signals for the methylene hydrogens (underlined) of $CH_3C\underline{H}_2NHPh$ (**1**) will appear in the ^1H NMR spectrum? Analyze **1** exactly as drawn in the figure.

PROBLEM 15.59 The carbon-bound hydrogens of pyrrole do not appear at especially low field in the ^1H NMR spectrum (δ 6.05 and 6.62 ppm). Does this observation mean that pyrrole is not aromatic?

PROBLEM 15.60   At room temperature the pyridinophane shown below has a signal in its ^1H NMR spectrum at about δ −0.2 ppm. What relevance does this observation have to the question of the aromaticity of pyridine? *Hint*: See p. 780.

PROBLEM 15.61   Predict a structure for the product and outline the steps in this change.

IR: (CCl$_4$) 1729 cm^{-1}
^1H NMR, δ 9.5 (1H)

$(C_7H_{12}O)$

PROBLEM 15.62   The following spectrum appears in an old catalog of ^1H NMR spectra of organic compounds. Is the assigned structure correct? What's wrong?

PROBLEM 15.63   In the synthesis of the compound described in Problem 15.62, HCl is produced. Use this information to postulate a reasonable structure for the compound whose ^1H NMR spectrum is shown above.

PROBLEM 15.64   Isobutene can be hydrated in acid, or under hydroboration–oxidation conditions to give two different alcohols. Give structures for these alcohols and assign the peaks in the following ^1H NMR spectra (**1** and **2**). Which peaks would vanish if a drop of D$_2$O were added?

1

2

PROBLEM 15.65   2-Methyl-2-butene can be hydrated in acid, or under hydroboration–oxidation conditions to give two different alcohols. Give structures for these alcohols and assign the following ^1H NMR spectra (**1** and **2**) to these isomers. Explain your answers.

1

(continued)

PROBLEM 15.65 (CONTINUED)

2

PROBLEM 15.66
On careful examination, it can be seen that the two upfield signals in spectrum 2 of Problem 15.65 (at δ 0.90 and 0.93) are both doublets. Explain.

PROBLEM 15.67
The ¹H NMR spectra for the following eight-membered ring compounds are summarized below. Assign structures, and explain your reasoning.

A: δ 1.54 (s), **B:** δ 5.74 (s), **C:** δ 2.39 (m, 8H), 5.60 (m, 4H), **D:** δ 1.20–1.70 (m, 4H), 1.85–2.50 (m, 4H), 5.35–5.94 (m, 4H).

PROBLEM 15.68
You are given three bottles containing *o*-dichlorobenzene, *m*-dichlorobenzene, and *p*-difluorobenzene, along with broad-band decoupled ¹³C NMR spectra of the three isomers. Assign the three spectra to the three compounds and explain your reasoning.

A: δ 127.0, 128.9, 130.6, 135.1 ppm. **B:** δ 127.7, 130.5, 132.6 ppm. **C:** δ 116.5, 159.1 ppm

PROBLEM 15.69
Could this same technique (looking at symmetry) allow you to assign the spectra of 4-methylcyclohexanone, 2-methylcyclohexanone, and 2,3-dimethylcyclopentanone, **A**, **B**, and **C**? If not, suggest an experiment, involving only ¹³C NMR spectroscopy, that would let you finish the job. (*Hint*: See p. 799.)

PROBLEM 15.70
The three isomers of pentane, C_5H_{12}, have the following ¹³C NMR spectra. Assign the spectra of these three compounds.

A: δ 13.9, 22.8, 34.7, **B:** δ 22.2, 31.1, 32.0, 11.7, **C:** δ 31.7, 28.1

PROBLEM 15.71
Assign the following ¹H NMR spectra to the isomers of trichlorophenol shown below and explain your reasoning.

(a)   (b)   (c)   (d)

1

2

3

4

PROBLEM 15.72   Isomeric compounds **A** and **B** have the composition $C_{11}H_{12}O_4$. Spectral data are summarized below. Deduce structures for **A** and **B** and explain your reasoning.

Compound **A**

IR (KBr): 1720 (s), 1240 (s) cm^{-1}

^{1}H NMR (CDCl$_3$): δ 2.46 (s, 3H)
3.94 (s, 6H)
8.05 (d, $J = 2$ Hz, 2H)
8.49 (t, $J = 2$ Hz, 1H)

Compound **B**

IR (Nujol®): 1720 (s), 1245 (s) cm^{-1}

^{1}H NMR (CDCl$_3$): δ 2.63 (s, 3H)
3.91 (s, 6H)
7.28 (d, $J = 8$ Hz, 1H)
8.00   (dd, $J = 8$ Hz, 2Hz, 1H),
dd = doublet of doublets
8.52 (d, $J = 2$ Hz, 1H)

PROBLEM 15.73   Compounds **A** and **B**, containing only C, H, and O, gave 63.15% C and 5.30% H upon elemental analysis. Their mass spectra do not show a parent ion, but freezing point depression data indicate a molecular weight of 150 g/mol. Catalytic hydrogenation of both **A** and **B** leads to **C**. Spectral data for compounds **A**, **B**, and **C** are summarized below. Assign structures for compounds **A**, **B**, and **C** and explain your reasoning.

Compound **A**

IR (Nujol): 1845 (m), 1770 (s) cm^{-1}

^{1}H NMR (CDCl$_3$): δ 2.03–2.90 (m, 2H)
3.15–3.70 (m, 1H)
5.65–6.25 (m, 1H)

Compound **B**

IR (Nujol): 1825 (m), 1755 (s) cm^{-1}

^{1}H NMR (CDCl$_3$/DMSO-d_6): δ 1.50–2.10 (m, 1H)
2.15–2.70 (m, 1H)

Compound **C**

IR (neat): 1852 (m), 1786 (s) cm^{-1}

^{1}H NMR (CDCl$_3$/DMSOd_6): δ 1.10–2.30 (m, 4H)
3.05–3.55 (m, 1H)

PROBLEM 15.74   Spectral data for isomeric compounds **A** and **B** are summarized below. Assign structures for compounds **A** and **B**, and explain your reasoning.

Compound **A**

Mass spectrum: m/z = 148 (p, 7%), 106 (8%), 105 (100%), 77 (29%), 51 (8%)

IR (neat): 1675 (s), 1220 (s), 980 (s), and 702 (s) cm^{-1}

^{1}H NMR (CDCl$_3$): δ 1.20 (d, $J = 7$ Hz, 6H)
3.53 (septet, $J = 7$ Hz, 1H)
7.20–7.60 (m, 3H)
7.80–8.08 (m, 2H)

Compound **B**

IR (neat): 1705 (s), 738 (m), and 698 (s) cm^{-1}

^{1}H NMR (CDCl$_3$): δ 0.95 (t, $J = 7$ Hz, 3H)
2.35 (q, $J = 7$ Hz, 2H)
3.60 (s, 2H)
7.20 (s, 5H)

PROBLEM 15.75   Upon direct photolysis or heating at 220 °C, the dimer of 2,5-dimethyl-3,4-diphenyl-2,4-cyclopentadien-1-one (**A**) yields compound **B**, mp 166–168 °C.

A
(dimer)

Spectral data for compound **B** are summarized below. Deduce the structure of compound **B** and explain your reasoning. (You may have to make an educated guess about the actual stereochemistry of **B**.)

Compound **B**

Mass spectrum: $m/z = 492$ (p)

IR (CCl_4): 1704 cm^{-1}

^{1}H NMR ($CDCl_3$): δ 0.73 (s, 3H)
0.92 (s, 3H)
1.51 (s, 3H)
1.88 (s, 3H)
6.6–7.5 (m, 20H)

PROBLEM 15.76   An appreciation of the concepts of electronegativity (inductive effects) and of electron delocalization (resonance effects), combined with an understanding of ring current effects, permits both rationalization and prediction of many chemical shifts.

(a) The carbons of ethylene have a ^{13}C NMR chemical shift of δ 123.3 ppm. Rationalize the ^{13}C NMR chemical shifts of the β carbons of 2-cyclopenten-1-one (**A**) and methyl vinyl ether (**B**). (*Hint*: Think resonance!)

(b) The ^{1}H NMR chemical shifts of hydrogens ortho, meta, and para to a substituent on a benzene ring can also be correlated in this way. For reference sake, note that the

hydrogens in benzene appear at about δ 7.37 ppm. Aniline (**C**) exhibits three shielded aromatic hydrogens (i.e., δ < 7.0), whereas the other two aromatic hydrogens appear in the normal range. On the other hand, two of the aromatic hydrogens in acetophenone (**D**) are deshielded (δ 7.9–8.1 ppm), whereas the other three aromatic hydrogens have a normal chemical shift. Finally, all five aromatic hydrogens in chlorobenzene (**E**) appear at δ 7.3. Assign the ^{1}H NMR chemical shifts of the aromatic hydrogens for **C**, **D**, and **E** by considering inductive, resonance, and ring current effects. Additionally, the ^{13}C NMR chemical shifts for the aromatic carbons of **C** and **E** are shown in color below. (The ^{13}C NMR chemical shift for benzene is δ 128.5 ppm.) Are these ^{13}C NMR chemical shifts consistent with the ^{1}H NMR chemical shifts for **C** and **E**?

C   **D**   **E**

PROBLEM 15.77   Compound **A** was isolated from the bark of the sweet birch (*Betula lenta*). Compound **A** is soluble in 5% aqueous NaOH solution but not in 5% aqueous NaHCO$_3$ solution. The spectral data for compound **A** are summarized below. Deduce the structure of compound **A**. (Be sure to assign the aromatic resonances in the ^{1}H NMR spectrum. Note that J_{meta} and J_{para} were not observed.) *Hint*: The pK$_a$ of phenol is about 10, and that of benzoic acid is about 4.2.

Compound **A**

Mass spectrum: $m/z = 152$ (p, 49%), 121 (29%), 120 (100%), 92 (54%)

IR (neat): 3205 (br), 1675 (s), 1307 (s), 1253 (s), 1220 (s), and 757 (s) cm^{-1}

^{1}H NMR ($CDCl_3$, 300 MHz): δ 3.92 (s, 3H)
6.85 (t, $J = 8$ Hz, 1H)
7.00 (d, $J = 8$ Hz, 1H)
7.44 (t, $J = 8$ Hz, 1H)
7.83 (d, $J = 8$ Hz, 1H)
10.8 (s, 1H)

^{13}C NMR ($CDCl_3$): δ 52.1 (q), 112.7 (s), 117.7 (d), 119.2 (d),
(hydrogen coupled)   130.1 (d), 135.7 (d), 162.0 (s), 170.7 (s)

PROBLEM 15.78   Dehydration of α-terpineol (**A**) affords a mixture of at least two isomers, one of which, **B**, reacts with maleic anhydride to give **C**.

Isomer **B**, which exhibits the pleasant odor of lemons and can be isolated from the essential oils of cardamom, marjoram, and coriander, reacts with maleic anhydride to yield compound **C**, mp 64–65 °C. Spectral data for compound **C** are summarized below. Deduce the structures of compounds **B** and **C**. (You may have to make an educated guess about the stereochemistry of compound **C**.)

Compound **C**

Mass spectrum: $m/z = 234$ (p)

IR (Nujol): 1870 (sh), 1940 (m), and 1780 (s) cm^{-1}

^{1}H NMR (CDCl$_3$): δ 0.98 (d, $J = 6.8$ Hz, 3H)
1.08 (d, $J = 6.8$ Hz, 3H)
1.36 (m, 4H)
1.46 (s, 3H)
2.60 septet ($J = 6.8$ Hz, 1H)
2.80 (d, $J = 9$ Hz, 1H)
3.20 (d, $J = 9$ Hz, 1H)
6.01 (d, $J = 8.5$ Hz, 1H)
6.10 (d, $J = 8.5$ Hz, 1H)

PROBLEM 15.79   Isoquinoline (**1**) undergoes the Chichibabin amination reaction when treated with potassium amide in liquid ammonia to yield 1-aminoisoquinoline (**2**) (Chapter 14,

p. 729). When the reaction mixture containing **1**, potassium amide, and liquid ammonia was examined at –10 °C by ^{1}H NMR spectroscopy, the following signals were observed (solvent hydrogens and NH$_2$ hydrogens not reported):

δ 4.87 (d, $J = 5.5$ Hz, 1H), 5.34 (t, $J = 7.0$ Hz, 1H; signal collapses to a singlet when amide ion is present in excess), 6.35–7.3 (br m, 5H)

Rationalize this low-temperature ^{1}H NMR spectrum. *Hint*: For reference sake, the ^{1}H NMR spectrum of isoquinoline itself (**1**) is summarized below.

δ 7.3–8.2 (m, H$_4$, H$_5$, H$_6$, H$_7$, H$_8$), 8.61 (d, $J = 6$ Hz, H$_3$), 9.33 (s, H$_1$).

PROBLEM 15.80   When reactions are run in THF, it is often possible to isolate traces of an impurity, compound **1**. Compound **1** gives 81.76% C and 10.98% H upon elemental analysis. Spectral data for **1** are summarized below. Deduce the structure of compound **1** and explain your reasoning.

Compound **1**

Mass spectrum: $m/z = 220$ (p, 23), 205 (100), 57 (27)

IR (Nujol): 3660 (m, sharp) cm^{-1}

^{1}H NMR (CDCl$_3$): δ 1.43 (s, 18H), 2.27 (s, 3H), 5.00 (s, 1H), 6.98 (s, 2H)

^{13}C NMR (CDCl$_3$): δ 21.2 (q), 30.4 (q), 34.2 (s), 125.5 (d), 128.2 (s), 135.8 (s), 151.5 (s)

Carbonyl Chemistry 1: Addition Reactions

16

"You have two choices," Jamf cried, his last lecture of the year: outside were the flowery strokings of wind, girls in pale colored dresses, oceans of beer, male choruses intensely, movingly lifted as they sang *Semper sit in flores!* Semper sit in flo-ho-res . . . "stay behind with carbon and hydrogen, take your lunch-bucket in to the works every morning with the faceless droves who can't wait to get in out of the sunlight—or move *beyond*. Silicon, boron, phosphorus—these can replace carbon, and can bond to nitrogen instead of hydrogen.—" a few snickers here, not unanticipated by the playful old pedagogue, be he always in flower: his involvement in getting Weimar to subsidize IG's Stickstoff Syndikat was well known—"move beyond life, toward the inorganic. Here is no frailty, no mortality—here is Strength, and the Timeless." Then his well-known finale, as he wiped away the scrawled C–H on his chalkboard and wrote, in enormous letters, Si–N.

—Thomas Pynchon*
Gravity's Rainbow

16.1 PREVIEW

In Chapters 3 and 9, we examined the structure of alkenes and the addition reactions of their carbon–carbon double bonds. Indeed, we have been exploring various aspects of carbon–carbon double-bond chemistry since Chapter 9. Now it is time to extend what we know to the chemistry of carbon–oxygen double bonds, the **carbonyl group**. Although it bears many similarities to carbon–carbon double bonds, the carbon–oxygen double bond is not "just another double bond," as there are also many important differences. The chemistry of the carbonyl group is the cornerstone of synthetic chem-

---

*Our first Pynchon quote was in Chapter 9. Same man, same book.

istry. Much of our ability to make molecules depends on manipulation of carbonyl groups because they allow us to make carbon–carbon bonds. There are several centers of reactivity in a molecule containing a carbon–oxygen double bond, and this versatility contributes to the widespread use of carbonyl compounds in synthesis. There is the π bond itself (don't forget the antibonding π* orbital), and the lone-pair electrons on oxygen. In addition, there is special reactivity associated with the carbon–hydrogen bonds in the position adjacent (α) to the carbon–oxygen double bond (Fig. 16.1). The centerpiece reaction of this chapter is the acid- or base-induced addition to the carbon–oxygen double bond. The chemistry of the α-position will wait until Chapter 17.

FIGURE 16.1 A generic carbonyl group has three loci of reactivity: the nonbonding, or lone-pair electrons; the π bond; and the α carbon–hydrogen bonds.

ESSENTIAL SKILLS

1. Almost this entire chapter involves additions of nucleophiles to carbon–oxygen double bonds. Many reactions are reversible, others are not. It is the reversible reactions that trouble students most, as there seems to be a bewildering array of additions, proton transfers, and eliminations. So there is, but keep your wits about you, and you'll learn to push those protons and waters around properly.

2. In this chapter, you will learn a lot about synthesis, especially alcohol synthesis. It is not hard to do these syntheses, but you have to learn the basic reaction—the irreversible addition of organometallic reagents and metal hydrides to carbonyl compounds—very well.

IMPORTANT DETAILS

1. A compound does not have to be favored at equilibrium to be the reactive ingredient in a chemical reaction. Even if it is disfavored, it can still have low barriers to further reaction. In such a case, it is used up, then regenerated at equilibrium. In such a way, all of the product can come from the minor partner in an equilibrium.

2. It is very useful to be comfortable with the redox interconversion of aldehydes and ketones with the related alcohols. There is a lot of detail here, and a none-too-simple mechanism as well.

3. Protecting groups are introduced in this chapter. A reactive functional group can be converted into a nonreactive group and stored in this fashion. Later, when the reactivity of the functional group is required, it can be regenerated as needed. There are many protecting groups, and it is very helpful to be able to use them in synthesis.

16.2   STRUCTURE OF THE CARBON–OXYGEN DOUBLE BOND

By analogy to the carbon–carbon double bond, we might expect sp^2 hybridization for the carbon atom. The $2p$ orbitals on carbon and oxygen form the π bond, and two lone pairs of electrons remain on oxygen. Figure 16.2 gives a schematic orbital picture of both the σ and π systems for the simplest carbonyl compound, formaldehyde.

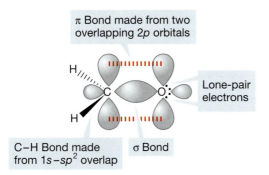

FIGURE 16.2 One orbital picture of the simplest carbonyl compound, formaldehyde.

PROBLEM 16.1 Make schematic, graphical orbital interaction diagrams for each different kind of bond in the molecule formaldehyde, shown in Figure 16.2.

The structures of some known carbonyl compounds show that this picture is a reasonable one. Bond angles are close to the 120° required by pure sp^2 hybridization, and of course we would not expect a molecule of less than perfect trigonal symmetry to have exactly sp^2 hybridization. Bond lengths are similar to those in simple alkenes, although the carbon–oxygen double bond is a bit shorter and stronger than its carbon–carbon counterpart (Fig. 16.3).

FIGURE 16.3 A comparison of the structures of formaldehyde, acetaldehyde, and ethylene.

The π bond is constructed in the same way as the related carbon–carbon π bond. The 2*p* orbitals on the adjacent carbon and oxygen atoms are allowed to interact in a constructive way to form a bonding π molecular orbital, and in a destructive, antibonding way to form π*. But there is a difference. In the alkenes, the constituent 2*p* orbitals are of equal energy, and therefore contribute equally to the formation of π and π*. In a carbonyl group, the two 2*p* orbitals are not identical. The 2*p* orbital on the more electronegative oxygen atom is lower in energy than the 2*p* orbital on a carbon atom, so the oxygen atom contributes more to π than the more energetically remote carbon 2*p* orbital. The bonding π molecular orbital will be constructed from more than 50% of the oxygen 2*p* atomic orbital and less than 50% of the carbon 2*p* atomic orbital. In a similar way, we can see that π* will be made up of more than 50% of the energetically close carbon 2*p* orbital and less than 50% of the energetically more remote 2*p* orbital on oxygen (Fig. 16.4).

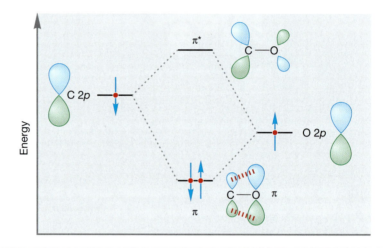

FIGURE 16.4 A graphical construction of π and π* for a carbonyl compound. The π orbital is constructed from more than 50% of the oxygen 2*p* orbital, and π* from more than 50% of the carbon 2*p* orbital.

So π and π* bear a general resemblance to π and π* of the alkenes, but are less symmetrical. The two electrons naturally occupy the lower energy bonding molecular orbital. In the π molecular orbital, there is a greater probability of finding an electron in the neighborhood of the relatively electronegative oxygen atom than near the more electropositive carbon, and this unequal sharing of electrons contributes to the dipole moment in molecules containing carbonyl groups. As can be seen from Figure 16.5, carbonyl compounds do have substantial dipole moments and are quite polar molecules.

FIGURE 16.5 Carbonyl compounds are polar molecules with substantial dipole moments [measured in debye (D) units]. Notice the small bond dipole arrow that points from the positive end toward the negative end of the dipole.

PROBLEM 16.2 It is not just the π bonds that are unsymmetrical. Use the hybrid orbitals of oxygen and carbon to construct the σ and σ* orbitals.

PROBLEM 16.3 One way to construct a carbon–oxygen double bond within the hybridization formalism is to use *sp* hybridized oxygen. Make a drawing of this model in the style of Figure 16.2.

In resonance terms, we can describe the carbonyl group as a combination of the two major contributing forms shown in Figure 16.6. This picture emphasizes the Lewis acid character of the carbon of the carbon–oxygen double bond. Note the polar resonance form.

FIGURE 16.6 A resonance formulation of a carbonyl group.

WORKED PROBLEM 16.4 There is another possible polar form. What is it, and why is it not considered a major contributor to the carbonyl group?

ANSWER The other polar resonance form has the positive charge on oxygen and the negative charge on carbon. Oxygen is more electronegative than carbon and the negative charge will be more stable on oxygen than on carbon. Similarly, the relatively electropositive carbon should have the positive charge, not the negative. This form is of high energy and cannot be a major contributor to the structure.

WORKED PROBLEM 16.5 Why is the dipole moment in acetone larger than that in acetaldehyde (Fig. 16.5)?

ANSWER As the negative end of the dipole is the same for each species, there can be no answer there. However, the two methyl groups in acetone stabilize the positive charge much better than the methyl and hydrogen of acetaldehyde. The polar form is more stable in acetone than it is in acetaldehyde, and the compound has a larger dipole.

16.3   NOMENCLATURE OF CARBONYL COMPOUNDS

In a generic carbonyl compound $R_2C{=}O$, there is a vast array of possible substitution patterns, and there is a separate name for each one of them. Simple compounds in which both R groups are hydrocarbons such as alkyl, alkenyl, or aryl are called

ketones. If one R group is hydrogen, the compound is an **aldehyde**, and the molecule in which both R groups are hydrogen is called **formaldehyde** (Fig. 16.7).

FIGURE 16.7 Different substitution patterns for simple carbonyl compounds.

Formaldehyde   **An aldehyde**   **A ketone**

Common names are normally used for the aldehydes containing four carbons or fewer, but thereafter the system takes over, and the naming process becomes rational. The "e" of the parent alkane's name is dropped and the suffix "al" is added. In the priority system the aldehyde group is always number 1, and higher numbers are given to any substituents. When aldehydes are attached to rings, the term "carbaldehyde" or "carboxaldehyde" is used, as in "cyclopropanecarboxaldehyde." Some aldehydes and their common names are given in Figure 16.8.

Acetaldehyde   **Propionaldehyde**   **Butyraldehyde**   **Isobutyraldehyde**   **Pentanal**
(ethanal)   (propanal)   (butanal)   (2-methylpropanal)

Hexanal   **2-Methylpentanal**   **3-Chloropropanal**   **Cyclohexanecarboxaldehyde**

FIGURE 16.8 Some simple aldehydes and their names.

Compounds containing two aldehydes are **dials**. In this case, the final "e" of the parent name is not dropped before adding "dial" (Fig. 16.9).

Propanedial   **Butanedial**

FIGURE 16.9 Dialdehydes are named as "dials."

Ketones, too, have common names, but most have been discarded. An exception is the name for the simplest member of the class, dimethyl ketone, which is still called **acetone**. Ketones are named in several ways. For example, acetone is also 2-oxopropane, propanone, or dimethyl ketone (Fig. 16.10). The rules are fairly simple: the "oxo" system treats the oxygen atom as a substituent that is given a number like any other attached group. In using the "one" system, the final "e" of the parent alkane is dropped, and the suffix "one" is appended. The number of the carbon atom bearing the carbon–oxygen double bond begins the name.

Acetone
dimethyl ketone
propanone
2-oxopropane

Ethyl methyl ketone
butanone
2-oxobutane

3-Methyl-2-pentanone
sec-butyl methyl ketone
3-methyl-2-oxopentane

FIGURE 16.10 Simple ketones can be named in many ways.

PROBLEM 16.6 Why is acetone properly named propanone and *not* 2-propanone?

Compounds containing two ketones are **diones**. As for dials, the final "e" of the parent framework is retained, and the name begins with the position numbers of the two carbonyl groups (Fig. 16.11). Ketoaldehydes are named as aldehydes, not ketones, because the aldehyde group has a higher priority than a keto group (Fig. 16.11).

2,4-Pentanedione   **1,4-Cyclohexanedione**   **3-Oxohexanal**   **2-Oxopropanal**
(pyruvic aldehyde)

FIGURE 16.11 Diketones are named as "diones." As for dials, the final "e" of the hydrocarbon backbone is retained. Ketoaldehydes are named as aldehydes, not ketones.

A great many common names are retained for aromatic aldehydes and ketones. For example, the simplest aromatic aldehyde is not generally called "benzenecarboxaldehyde" but **benzaldehyde**. That one's quite easy to figure out, but most are not. A few commonly used names are given in Figure 16.12.

Benzaldehyde   **4-Methoxybenzaldehyde**   ***trans*-Cinnamaldehyde**   **4-Methylbenzaldehyde**
                           (anisaldehyde)   (*p*-tolualdehyde)

FIGURE 16.12 Some common names for aromatic aldehydes.

Aromatic ketones are often named by treating the benzene ring as a **phenyl group**, as in isopropyl phenyl ketone (Fig. 16.13). A few can be named through a system in which the framework Ph—C=O is indicated by the term ophenone and the remainder of the molecule designated by one of the alternative names in Figure 16.13. Thus, diphenyl ketone is also benzophenone, methyl phenyl ketone is acetophenone, ethyl phenyl ketone is propiophenone, and phenyl propyl ketone is butyrophenone (Fig. 16.13).

Cyclopropyl phenyl ketone   **Isopropyl phenyl ketone**   **Diphenyl ketone**   **Methyl phenyl ketone**
                                                                         (benzophenone)   (acetophenone)

FIGURE 16.13 Ketones containing simple benzene rings are "phenyl" ketones. This figure also shows the "ophenone" method of naming these molecules.

Figure 16.14 gives the names of some compounds containing carbon–oxygen double bonds in which one of the R groups is neither hydrogen nor a hydrocarbon group. We will meet these compounds in later chapters and deal with nomenclature more extensively there.

FIGURE 16.14 Some names for other carbonyl compounds in which one R group is neither hydrogen nor a hydrocarbon group.

Carboxylic acid   Ester   Amide   Acid chloride   Anhydride

16.4   PHYSICAL PROPERTIES OF CARBONYL COMPOUNDS

Carbonyl compounds are all somewhat polar, and the smaller, less hydrocarbon-like molecules are all at least somewhat water soluble. Acetone and acetaldehyde are miscible with water, for example. Their polarity means that they can profit energetically by associating in solution, and this means that their boiling points are considerably higher than those of hydrocarbons containing the same number of heavy atoms (carbons and oxygens). For example, propane and propene are gases with boiling points of −42.1 and −47.4 °C, respectively. By contrast, acetaldehyde boils at +20.8 °C, about 63 °C higher. Acetone boils more than 60 °C higher than isobutene.

Table 16.1 gives some physical properties of a number of common carbonyl-containing compounds (and some related hydrocarbons in parentheses).

TABLE 16.1   Some Physical Properties of Carbonyl Compounds

Compound	bp (°C)	mp (°C)	Density (g/mL)	Dipole Moment (D)
Formaldehyde	− 21	− 92	0.815	2.33
Acetaldehyde	20.8	−121	0.783	2.69
Propionaldehyde	48.8	− 81	0.806	2.52
(Propane)[a]	− 42.1			
(Propene)[a]	− 47.4			
Benzaldehyde	178.6	− 26	1.04	
Acetone	56.2	− 95.4	0.789	2.88
(Isobutene)[a]	− 6.9			
Ethyl methyl ketone	79.6	− 86.3	0.805	
Cyclohexanone	155.6	− 16.4	0.948	
Methyl phenyl ketone	202.6	20.5	1.03	
Diphenyl ketone	305.9	48.1	1.15	2.98

[a] Hydrocarbons for comparison purposes.

16.5   SPECTROSCOPY OF CARBONYL COMPOUNDS

16.5a Infrared Spectroscopy. Carbonyl compounds are ubiquitous in nature. The C=O bond is one of the most widely found of the "functional groups." Spectroscopy, especially IR spectroscopy, played an important role in identifying many medicinally and otherwise important compounds. The carbon–oxygen double bond has a diagnostically useful, and quite strong stretching frequency in the IR, generally near 1700 cm^{-1}. This absorption is at substantially higher frequency than that of carbon–carbon double bonds. This shift should be no surprise, as we already know that carbon–oxygen double bonds are stronger than carbon–carbon double bonds, and therefore require more energy for stretching (Chapter 15, p. 761). A useful diagnostic for aldehydes is the pair of bands observed in the C—H stretching region at about 2850 and 2750 cm^{-1}. Table 16.2 gives some IR stretching frequencies and some other data.

TABLE 16.2   Some Spectral Properties of Carbonyl Compounds

Compound	IR (CCl$_4$, cm^{-1})	^{13}C NMR (δ C=O)	UV [nm (ε)]
Formaldehyde	1744 (gas phase)		270
Acetaldehyde	1733	199.6	293.4 (11.8)
Propionaldehyde	1738	201.8	289.5 (18.2)
Acetone	1719	205.1	279 (14.8)
Cyclohexanone	1718	207.9	285 (14)
Cyclopentanone	1751	213.9	299 (20)
Cyclobutanone	1788	208.2	280 (18)
Ethyl methyl ketone	1723	206.3	277 (19.4)
Methyl vinyl ketone	1684	197.2	219 (3.6)
2-Cyclohexenone	1691	197.1	224.5 (10,300) ($\pi{\rightarrow}\pi^*$)
Benzaldehyde	1710	191.0	244 (15,000) ($\pi{\rightarrow}\pi^*$)
			328 (20) ($n{\rightarrow}\pi^*$)
Diphenyl ketone	1666	195.2	252 (20,000) ($\pi{\rightarrow}\pi^*$)
			325 (180) ($n{\rightarrow}\pi^*$)
Methyl phenyl ketone	1692	196.8	240 (13,000) ($\pi{\rightarrow}\pi^*$)
			319 (50) ($n{\rightarrow}\pi^*$)

Note that conjugated carbonyl compounds are always found at lower frequency than their unconjugated relatives. For example, cyclohexanone appears at 1718 cm^{-1} and 2-cyclohexenone at 1691 cm^{-1}. Thus there is reduced double-bond character in the conjugated molecules. The resonance formulation of 2-cyclohexenone clearly shows the partial single-bond character of the carbon–oxygen bond in these compounds (Fig. 16.15).

FIGURE 16.15 Conjugation reduces the double-bond character of a carbonyl group. The result is a shift of the IR stretching band to lower frequency.

IR C=O stretching frequency   1718 cm^{-1} 1691 cm^{-1}

16.5b Nuclear Magnetic Resonance Spectroscopy.
The carbonyl group inductively withdraws electrons from the nearby positions, and this inductive effect results in a deshielding of adjacent hydrogens. The further away from the carbonyl group the less effective the deshielding. This effect is well illustrated by methyl propyl ketone (Fig. 16.16).

In aldehydes, hydrogens directly attached to the carbonyl group resonate at very low field, $\delta \sim 8.5$–10 ppm. Not only does the carbon bear a partial positive charge, but the applied magnetic field B_0 creates an induced field B_i in the carbonyl group that augments the applied field in the region of a directly attached hydrogen (Fig. 16.17). Accordingly, a relatively small applied B_0 is required to bring these hydrogens into resonance, and they appear at unusually low field (Chapter 15, p. 775).

FIGURE 16.16 The closer a hydrogen is to the electron-withdrawing carbonyl group, the lower field is its signal in the ^{1}H NMR spectrum (δ in ppm).

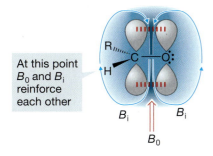

At this point B_0 and B_i reinforce each other

FIGURE 16.17 Aldehyde hydrogens are in a strongly deshielded region in which the induced magnetic field B_i augments the applied field B_0. These hydrogens resonate at exceptionally low field in ^{1}H NMR spectra.

The carbon atoms of carbonyl groups carry partial positive charges and are greatly deshielded by their relative lack of surrounding electrons. They, too, resonate at especially low field in the ^{13}C NMR. Table 16.2 shows typical resonance positions.

16.5c Ultraviolet Spectroscopy.
Simple aldehydes and ketones, like simple alkenes, do not have a π–π^* absorption in the region of the UV spectrum accessible to most spectrometers. However, unlike alkenes, there is another possible absorption which, though weak, is detectable. This absorption involves promotion of a nonbonding, or n, electron, to the antibonding π^* orbital, the $n \rightarrow \pi^*$ absorption. Table 16.2 gives a few absorptions.

As for alkenes (Chapter 12, p. 569), conjugation greatly reduces the energy necessary to promote an electron from the highest occupied π molecular orbital to the LUMO. So, the strong $\pi \rightarrow \pi^*$ transition in unsaturated carbonyl compounds can be detected because of the conjugation-induced shift to the longer wavelengths (lower energy) observable by normal UV/vis spectrometers (Table 16.2; Fig. 16.18).

This lower energy absorption can be observed

FIGURE 16.18 Conjugation leads to lower energy (longer wavelength) $\pi \rightarrow \pi^*$ absorptions that are observable by UV/vis spectrometers.

16.5d Mass Spectrometry. There is a diagnostic cleavage reaction that occurs for many simple carbonyl-containing compounds. It involves breaking the bond attached directly to the carbonyl carbon, with the charge remaining on the carbonyl group. Cleavage of this bond generates a resonance-stabilized acylium ion. It is this resonance stabilization that accounts for the prominence of this particular fragmentation. Figure 16.19 shows the mass spectrum for acetone, the archetypal methyl ketone.

FIGURE 16.19 In the mass spectrum of acetone, the major peak results from formation of the resonance-stabilized acylium ion.

16.6 REACTIONS OF CARBONYL COMPOUNDS: SIMPLE REVERSIBLE ADDITIONS

In simple addition reactions, carbonyl compounds behave as both Lewis acids and Lewis bases depending on the other reagents present. As we might expect, there is some correspondence between addition reactions of carbon–oxygen double bonds and those of the carbon–carbon double bonds we saw in Chapters 9 and 10. As we might also expect, what complications exist are introduced by the presence of two *different* atoms in the carbon–oxygen double bond. Figure 16.20 compares the reaction between water and a generic carbonyl compound (the formation of a **hydrate**) and the hydration of an alkene.

FIGURE 16.20 Addition of water to an alkene to give an alcohol and addition to a ketone to give a hydrate are analogous reactions.

Alkene

Alcohol

Carbonyl

Hydrate

Water acts as a nucleophile and adds to the Lewis acid carbonyl group. However, there is a choice to be made. In principle, water could add to either the carbon or oxygen end of the carbonyl group (Fig. 16.21).

FIGURE 16.21 In principle, addition of water to a carbonyl group could occur in two ways.

Three factors must be considered in predicting which way the addition will go. First of all, addition to the carbon of the carbon–oxygen double bond places the negative charge on the highly electronegative oxygen atom, whereas addition at oxygen generates a much less stable carbanion. Second, addition to oxygen forms a weak oxygen–oxygen bond, whereas addition to carbon makes a stronger oxygen–carbon bond. Formation of the alkoxide must be vastly preferred (Fig. 16.22).

FIGURE 16.22
Addition to the carbon end, which puts the negative charge on the relatively electronegative oxygen atom, will be greatly preferred.

WORKED PROBLEM 16.7 Explain how this thermodynamic preference for placing a negative charge on oxygen, not carbon, can influence the rates of addition to the two different ends of the carbon–oxygen double bond.

ANSWER The rates of reaction depend, of course, on the heights of the transition states, not on the energies of the products in which a full negative charge resides on either oxygen or carbon. In the transition states for addition, there will be a partial negative charge on either oxygen or carbon. The factors that favor the presence of a full negative charge on oxygen rather

Transition state

than carbon (electronegativity) will also favor the presence of a partial negative charge (δ^-) on oxygen. In the transition state, a partial negative charge will be more stable on oxygen than on carbon.

Transition state

But there is more to consider. If we look at the molecular orbitals involved, we can see how they influence the direction in which the reaction occurs. As in all polar reactions, we look for the combination of a nucleophile (here, water) and a Lewis acid. The Lewis acid must be the *empty π* orbital* of the carbonyl group. Figure 16.23 shows the orbital interactions graphically, and gives pictures of the orbital lobes involved.

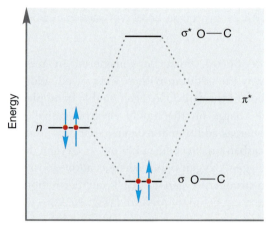

FIGURE 16.23 In the addition of water to a carbonyl, a filled *n* orbital of water overlaps with the empty π* orbital of the carbonyl group. This interaction between a filled and an empty orbital is stabilizing.

As we saw in our earlier section on structure (p. 823), neither π nor π* of the carbonyl group is symmetric with respect to the midpoint of the carbon–oxygen bond. In particular, π* is much larger in the region of the carbon atom than it is near the oxygen. As the magnitude of stabilization depends on the amount of orbital overlap, it is no surprise to see that addition at the carbon, where π* is largest, dominates the reaction. So, not only is the direction of the addition affected by the electronegativities of the two different atoms in the double bond, but it's affected by the shapes of the orbitals involved as well.

After the addition of water, protons are lost and added to give the hydrate. It is not necessary that the proton lost from one molecule be readded to the same molecule. The reaction is very likely to be intermolecular, as shown in Figure 16.24.

FIGURE 16.24 The hydration reaction is completed by a series of proton transfers. The order of the two steps is not certain.

Now consider how we might make the addition reaction easier. Two ways seem possible—we could make either the Lewis acid or nucleophile stronger. Let's start with the acid-catalyzed reaction in which the carbonyl group is converted into a stronger Lewis acid. If we take as our model for the mechanism the acid-catalyzed additions of alkenes, a reasonable first step would seem to be protonation of the carbonyl group. But here enters another complication. Which end is protonated? Again, there are two possiblities (Fig. 16.25).

Carbonyl hydration

Protonation of an alkene (one possibility)

Protonation of a carbonyl (two possibilities)

Resonance stabilized

FIGURE 16.25 Protonation of the carbonyl group can occur in two ways. The placement of a proton on oxygen to give a resonance-stabilized cation will be greatly preferred.

The choice is easy this time. First, the oxygen atom is a strong Lewis base, as it has two lone pairs of nonbonding electrons. Moreover, it is surely better to protonate on oxygen, and thus place the positive charge on the relatively electropositive carbon rather than to protonate on carbon, which would place the positive charge upon the highly electronegative oxygen atom. Finally, protonation at oxygen gives a resonance-stabilized cation, whereas protonation at carbon does not.

So there is not really much choice in the direction of the protonation step, which creates a strongly Lewis acidic species, the protonated carbonyl compound. In the second step of the reaction, water acts as a nucleophile and adds to the Lewis acid (the protonated carbonyl) to generate a protonated 1,1-diol (a geminate diol). In the final step of the reaction, this species is deprotonated by a water molecule to generate the 1,1-diol itself (the hydrate) and regenerate the acid catalyst, H_3O^+. Note how closely the acid-catalyzed additions of water to an alkene and a carbonyl group are related. Each reaction is a sequence of three steps: protonation, addition of water, and deprotonation (Fig. 16.26).

The case for an alkene...

...and for a carbonyl

FIGURE 16.26 A comparison of the acid-catalyzed addition reactions of an alkene and a carbonyl compound.

What about making the nucleophile more reactive? There is another mechanism for hydration that operates under basic conditions. Here the active agent is the hydroxide ion, ⁻OH. The first step in the base-catalyzed hydration reaction is the addition of the strong nucleophile hydroxide to the carbonyl group. Unlike protonation, the first step in the acid-catalyzed reaction, this reaction has no counterpart in the chemistry of simple alkenes. After addition, a proton is transferred from water to the newly formed alkoxide, generating the hydrate and regenerating the hydroxide ion (Fig. 16.27).

The case for an alkene...

FIGURE 16.27 Although strong bases such as hydroxide ion do not add to alkenes, they do attack carbonyl groups. The final step in the base-catalyzed hydration reaction is protonation of the newly formed alkoxide ion by water. Hydroxide is regenerated.

...and for a carbonyl

Hydrate

Hydroxide ion regenerated

PROBLEM 16.8 Explain why bases such as hydroxide ion do not add to the oxygen of the carbon–oxygen double bond, and do not add at all to alkenes.

SUMMARY

The carbonyl group is hydrated under acidic, basic, and neutral conditions. In acid, the mechanism is accelerated by protonation of the carbon–oxygen double bond to generate a strong Lewis acid. The relatively weak nucleophile, water, then adds. In base, the strong nucleophile hydroxide adds to the relatively weak Lewis acid carbonyl group.

16.7   EQUILIBRIUM IN ADDITION REACTIONS

These hydration reactions consist of several consecutive equilibria. In basic, acidic, or neutral water, simple carbonyl compounds are in equilibrium with their hydrates, as Figure 16.28 shows. Even for molecules as similar as acetone and acetaldehyde, these equilibria settle out at widely different points. If we include formaldehyde, the difference is even more striking. For these three molecules, which differ only in the relative number of methyl groups and hydrogens, there is a difference in K of about a factor of 10^6 (Fig. 16.28).

FIGURE 16.28 The equilibrium constant for hydration of a carbonyl-containing molecule can favor either the hydrate or the carbonyl compound.

Why should there be such great differences among these rather similar compounds? First of all, remember that a little energy goes a long way at equilibrium

(Chapter 8, p. 365). Recall also that alkyl groups stabilize double bonds (Chapter 3, p. 115). The more highly substituted an alkene, the more stable it is. The same phenomenon holds for carbon–oxygen double bonds. So, the stability of the carbonyl compound increases along the series: formaldehyde, acetaldehyde, acetone as the substitution of the carbon–oxygen double bond increases (Fig. 16.29, left).

FIGURE 16.29 Like alkenes, more substituted carbonyl groups are more stable than their less substituted counterparts. In the hydrates, increasing substitution results in increasing destabilization through steric interactions.

If we now look at the molecule at the other side of the equilibrium, the hydrate, we can see that increased methyl substitution will introduce steric destabilizations. The more substituted the hydrate, the less stable it is (Fig. 16.29, right).

We can immediately see the sources of the difference in K of 10^6. They are the increased stabilization of the starting carbonyl compounds and destabilization of the hydrates, both of which are induced by increased substitution by methyl groups (Fig. 16.30).

Acetone is more stable and the hydrate is less stable
$K < 1$

The aldehyde and hydrate are of comparable energy
$K \sim 1$

Formaldehyde is less stable and the hydrate is more stable
$K > 1$

FIGURE 16.30 The magnitude of the equilibrium constant (K) depends on the relative energies of the carbonyl compounds and their hydrates. We know that for acetaldehyde the energies of hydrate and aldehyde are comparable, as $K \sim 1$. Acetone is more stable than acetaldehyde and formaldehyde is less stable. Steric factors make the hydrate of acetone less stable than the acetaldehyde hydrate. By contrast, the hydrate of formaldehyde is more stable than the acetaldehyde hydrate. The equilibrium constants reflect these changes.

We can see other striking effects on this equilibrium process in Table 16.3. For example, benzene rings, which can interact with the carbon–oxygen double bond by resonance, decrease the amount of hydrate at equilibrium. However, electron-withdrawing groups in the position adjacent to the C=O bond, the α-position, strongly favor the hydrate. There also appears to be a steric effect. Large alkyl groups favor the carbonyl compound, as can be seen from the sequence in Table 16.3: propionaldehyde, butyraldehyde, and *tert*-butyl-carboxaldehyde. Each of these phenomena can be understood by considering the relative stabilities of the starting materials and products.

Resonance stabilization of the carbonyl group is generally not accompanied by a resonance stabilization of the product hydrate. For example, the aromatic aldehyde, *m*-chlorobenzaldehyde (Table 16.3), is more strongly stabilized by resonance than acetaldehyde. There is no corresponding difference in the hydrates, and therefore the aromatic aldehyde is less hydrated at equilibrium than acetaldehyde (Fig. 16.31).

TABLE 16.3 Equilibrium Constants for Carbonyl Hydration

Compound		
Structure	Name	K
⬡–CHO (Cl)	*m*-Chlorobenzaldehyde	0.022
CH_3CH_2CHO	Propionaldehyde	0.85
$CH_3CH_2CH_2CHO$	Butyraldehyde	0.62
$(CH_3)_3CCHO$	*tert*-Butylcarboxaldehyde	0.23
CF_3CHO	Trifluoroacetaldehyde	2.9×10^4
CCl_3CHO	Trichloroacetaldehyde	2.8×10^4
$ClCH_2CHO$	Chloroacetaldehyde	37

Resonance stabilization of the aromatic aldehyde
[note use of (+) to show resonance forms]

Hydrate (*not* stabilized by resonance)

Model acetaldehyde (carbonyl and hydrate are of comparable stability K ~ 1)

Aldehyde below hydrate

FIGURE 16.31 Resonance stabilization of the aromatic carbonyl compound is not accompanied by resonance stabilization of the product hydrate. Accordingly, the carbonyl compound is relatively favored at equilibrium.

Electron-withdrawing groups such as chlorine or fluorine sharply destabilize carbonyl groups when attached at the α-position. The energy of an α-halo carbonyl compound is raised by opposed dipoles. The carbon atom is already electron deficient, and further inductive removal of electrons is costly in energy terms. There is relatively little effect in the product hydrate, although there will be some destabilization in this molecule as well. So, as the diagram in Figure 16.32 shows, the strong destabilization of the carbonyl group favors the hydrate at equilibrium (Fig. 16.32).

FIGURE 16.32 Addition of a chlorine in the α-position of acetaldehyde raises the energy of the unhydrated carbonyl compound.

Model acetaldehyde (carbonyl and hydrate are of comparable stability)

The inductive effect of the electronegative chlorine atom destabilizes the aldehyde relative to the hydrate

The steric effect is slightly more subtle. In the carbonyl group, the carbon is hybridized sp^2 and the bond angles are therefore close to 120°. In the product hydrate, the carbon is nearly sp^3 hybridized, and the bond angles must be close to 109°. So, in forming the hydrate, groups are moved closer to each other, and for a large group that will be relatively destabilizing (Fig. 16.33).

FIGURE 16.33 In the hydrate, groups are closer together than they are in the carbonyl compound. This proximity will destabilize the hydrate relative to the carbonyl compound.

It is less costly in energy terms to move a methyl group closer to a hydrogen than to move a *tert*-butyl group closer to a hydrogen. Accordingly, acetaldehyde is more strongly hydrated than is *tert*-butylcarboxaldehyde (pivaldehyde) (Fig. 16.34).

FIGURE 16.34 The larger the R group, the more important the steric effect. For example, *tert*-butylcarboxaldehyde is less hydrated at equilibrium than is acetaldehyde.

These equilibrium effects are important to keep in mind. Equally important is the notion that as long as there is *some* of the higher energy partner present at equilibrium, it can be the active ingredient in a further reaction. For example, even though it is a minor component of the equilibrium mixture, the hydrate of acetone could go on to other products (X), with more hydrate being formed as the small amount present at equilibrium is used up. Similarly, even though it is extensively hydrated in aqueous solution, reactions of formaldehyde to give Y can still be detected. Figure 16.35 illustrates the point with Energy versus Reaction progress diagrams.

FIGURE 16.35 Molecules not favored at equilibrium may nonetheless lead to products. The two examples show a pair of carbonyl compound–hydrate equilibria in which it is the higher energy, *unfavored* partner that leads to product. As the small amount of the higher energy molecule is used up, equilibrium will be reestablished and more reactive compound formed.

As usual, mechanisms stick in the mind most securely if we are able to write them backward. "Backward" is an arbitrary term anyway, as by definition an equilibrium runs both ways. Carbonyl addition reactions are about to grow more complex, and you run the risk of being overwhelmed by a vast number of proton additions and losses in the somewhat more complicated reactions that follow. Be sure you are completely comfortable with the acid-catalyzed and base-catalyzed hydration reactions *in both directions* before you go on. That advice is important.

16.8 OTHER ADDITION REACTIONS: ADDITIONS OF CYANIDE AND BISULFITE

Many other addition reactions follow the pattern of hydration. Two typical examples are **cyanohydrin** formation and bisulfite addition. Both of these reactions are generally base catalyzed. Because of an understandable and widespread aversion to working with the volatile and notorious HCN, cyanide additions are generally carried out with cyanide ion. Potassium cyanide (K^+ ^-CN) is certainly still poisonous, but it's not volatile, and one knows where it is at all times, at least as long as basic conditions are maintained. Cyanide ion is a good nucleophile and attacks the Lewis acidic carbonyl compound to produce an alkoxide. This alkoxide is protonated by solvent, often water, to give the final product, a cyanohydrin (Fig. 16.36).

FIGURE 16.36 Cyanohydrin formation.

PROBLEM 16.9 Why does acetaldehyde in an aqueous solution form an excellent yield of cyanohydrin when we know that acetaldehyde is substantially hydrated in water ($K \sim 1$)? Might it not be argued that, as about one-half of the aldehyde is present in hydrated form at equilibrium, no more than about 50% cyanohydrin can be formed. Why is this argument wrong? *Hint*: See Figure 16.35.

Bisulfite addition is similar, although the nucleophile in this case is sulfur. Sodium bisulfite adds to many aldehydes and ketones to give an addition product, often a nicely isolable solid (Fig. 16.37).

THE GENERAL CASE

A SPECIFIC EXAMPLE

(98%)

FIGURE 16.37 Bisulfite addition.

The reactions we have seen so far have all been simple additions of a generic HX to the carbon–oxygen bond. Now we move on to related reactions in which there is one more dimension. In these new reactions, addition of HX is accompanied by the loss of some other group.

PROBLEM 16.10 Bisulfite addition is generally far less successful with ketones than it is with aldehydes. Explain why.

16.9   ADDITION REACTIONS FOLLOWED BY WATER LOSS: ACETAL AND KETAL* FORMATION

If aldehydes and ketones form hydrates in water (HOH), shouldn't there be a similar reaction in the presence of alcohols (ROH)? Indeed there should be, and there is. The reactions are slightly more complicated than simple hydration, at least under acidic conditions. In base, the two reactions are entirely analogous, so let's look at that reaction first.

Figure 16.38 contrasts the base-catalyzed reaction in water to give a hydrate with the related formation of a **hemiacetal** (from aldehydes) or hemiketal* (from ketones) in an alcohol and its corresponding alkoxide.

FIGURE 16.38 The reaction of a carbonyl group with hydroxide parallels the reaction with an alkoxide ion. In water, a hydrate is formed; in alcohol, the analogous product is called a hemiacetal.

As hydration and hemiacetal formation are comparable, it is no surprise that Section 16.7, on equilibrium effects, is quite directly transferable to this reaction. Aldehydes, for example, are largely in their hemiacetal forms in basic alcoholic solutions (in acid the reaction goes further, as we will see), whereas ketones are not (Fig. 16.39).

FIGURE 16.39 Hemiacetals formed from aldehydes are generally favored at equilibrium, but hemiacetals formed from ketones are not.

Although most simple hemiacetals are not isolable, often the cyclic versions of these compounds are. For example, consider combining the carbonyl and hydroxyl components of the reaction within the same molecule. Now hemiacetal formation is an intramolecular process. If the ring size is neither too large nor too small, the hemiacetals are often more stable than the open-chain molecules (Fig. 16.40).

*Strictly speaking, neither ketal nor hemiketal is correct, as the system has dropped the names in favor of using acetal and hemiacetal for the species formed from both aldehydes and ketones. We will not use ketal or hemiketal, but be alert, because one still sees these sensible terms occasionally.

THE GENERAL CASE

SPECIFIC EXAMPLES

Open-chain
hydroxy–aldehyde

Cyclic
hemiacetal

(11.4%)

(88.6%)

(6.1%)

(93.9%)

FIGURE 16.40 Cyclic hemiacetals are often more stable than their open-chain counterparts.

PROBLEM 16.11 Write a mechanism for the intramolecular formation of one of the hemi-acetals of Figure 16.40.

The most spectacular examples of this are the sugars, polyhydroxylated compounds that exist largely, though not quite exclusively, in their cyclic hemiacetal forms (Chapter 22; Fig. 16.41).

The sugar, D-glucose

Open form

Cyclic hemiacetal form

Watch out, there is a convention at work here; there is no "corner" in the oxygen bridge; sugar chemists have learned to ignore the apparent right angle in the bottom part of the structure

FIGURE 16.41 Sugars exist mainly in their cyclic hemiacetal forms.

PROBLEM 16.12 Construct a three-dimensional picture of D-glucose in its cyclic hemiacetal form. Note that the position of the OH on the carbon marked in red in Figure 16.41 can be either axial or equatorial.

Now, what about the acid-catalyzed version of this reaction? The process begins with hemiacetal formation, and the mechanism parallels that of acid-catalyzed hydration. A three-step sequence of protonation, addition, and deprotonation appears in each case (Fig. 16.42).

Hydration reaction in acid

Hemiacetal formation in acid

FIGURE 16.42 The acid-catalyzed formation of a hemiacetal parallels the acid-catalyzed hydration reaction.

Now, what can happen to the hemiacetals in acidic solution? One thing you know happens is the set of reactions indicated by the reverse arrows in the equilibria of Figure 16.42. In this case, the oxygen atom of the OR is protonated and alcohol is lost to give the resonance-stabilized protonated carbonyl compound. This intermediate is deprotonated to give the starting ketone or aldehyde. There is no need for us to write out the mechanism because it already appears exactly in the reverse steps of Figure 16.42. It is simply the forward reaction run directly backward.

But suppose it is not the OR that is protonated, but instead the new OH? Surely, if one of these reactions is possible, so is the other. If we keep strictly parallel to the reverse process, we can now lose water to give a resonance-stabilized species (Fig. 16.43), which is quite analogous to the resonance-stabilized protonated carbonyl compound of Figure 16.42.

FIGURE 16.43 Protonation of the OH in the hemiacetal, followed by water loss, leads to a resonance-stabilized species that is analogous to the protonated carbonyl.

But now deprotonation is not possible—*there is no proton on oxygen to lose,* as its place is taken by an R group. Another reaction occurs instead, the addition of a second molecule of alcohol. Deprotonation gives a full **acetal** (Fig. 16.44).

FIGURE 16.44 A full acetal is formed by addition of a second molecule of alcohol to the resonance-stabilized intermediate.

Acetal formation

Notice that the second addition of alcohol in acetal formation is just like the first; the only difference is the presence of an H or R group in the species to be attacked (Fig. 16.45).

FIGURE 16.45 The two alcohol additions are very similar.

PROBLEM 16.13 Write the mechanism for the formation of the acetal from treatment of cyclopentanone in ethylene glycol (1,2-dihydroxyethane) with an acid catalyst.

PROBLEM 16.14 Write a mechanism for the reverse reaction, the regeneration of the starting ketone on treatment of the acetal with aqueous acid.

WORKED PROBLEM 16.15 Explain clearly why hemiacetals, but not full acetals, are formed under basic conditions.

ANSWER If we work through the mechanism we can see why. The first step is addition of an alkoxide to the carbonyl group. Protonation then gives the hemiacetal and regenerates the base.

Now what? In order to get a full acetal, RO⁻ must displace hydroxide! This reaction is most unfavorable. The more favorable reaction between RO⁻ and the hemiacetal will be removal of a proton, *the reverse of the last step in the figure at the bottom of the previous page.* In base, the reaction cannot go beyond hemiacetal formation.

PROBLEM 16.16 When acetone is treated with $H_2O^{18}/H_3\overset{+}{O}^{18}$, the acetone oxygen atom is found to be labeled with O^{18}. Explain.

$$O = {}^{18}O$$

SUMMARY

By now, you may be getting the impression that almost any nucleophile will add to carbonyl compounds. If so, you are quite right. So far, we have seen water, cyanide, bisulfite, and alcohols act as nucleophiles toward the Lewis acidic carbon of the carbon–oxygen double bond. Addition can take place under neutral, acidic, or basic conditions. All of the addition reactions are similar, with only the details being different. Those details are important though, as well as highly effective at camouflaging the essential similarity of the reactions. We'll now see some examples.

16.10   PROTECTING GROUPS IN SYNTHESIS

16.10a Acetals as Protecting Groups for Aldehydes and Ketones.
By all means work out the mechanism of acetal formation by setting yourself the problem of writing the mechanism for the *reverse* reaction (Go back and do Problem 16.14!), the formation of a ketone or aldehyde from the reaction of an acetal in acidic water. Acetal formation is an equilibrium process, and conditions can drive the overall equilibrium in the forward or backward direction. In an excess of alcohol, an acid catalyst will initiate the formation of the acetal, but in an excess of water, that same acid catalyst will set the reverse reaction in motion, and we will end up with the carbonyl compound. This point is important and practical. Carbonyl compounds can be hidden, stored or "protected," by converting them into their ketal or acetal forms, and

then regenerating them as needed. The acetal constitutes your first **protecting group**. For example, it might be desirable to convert group X into group Y in a base-catalyzed reaction. But we know that carbonyl compounds are sensitive to base, and the starting material may be destroyed in the attempted conversion of X into Y. The solution is first to convert the carbonyl group into an acetal, then carry out the conversion of X to Y, and finally regenerate the carbonyl group (Fig. 16.46).

Desired conversion

FIGURE 16.46 A protecting group can be used to mask a sensitive functional group while reactions are carried out on other parts of the molecule. When these are completed, the original functionality can be regenerated.

An acetal (stable in base); the carbonyl group is protected, or masked

Here the C=O is unmasked (regenerated)

16.10b Protecting Groups for Alcohols. It is often desirable to protect a hydroxyl group so that further reactions in base can take place without complication. One widely used method is to convert the hydroxyl group into a **tetrahydropyranyl (THP) ether** through reaction with dihydropyran (Fig. 16.47). The mechanism in-

THE GENERAL CASE

Dihydropyran

Alcohol to be protected

The regenerated alcohol

A THP ether — an acetal and a protected alcohol

FIGURE 16.47 The use of THP ethers as protecting groups for alcohols.

A SPECIFIC EXAMPLE

acid catalyst
15.5 h, 25 °C

(97%)

FIGURE 16.47

volves protonation of dihydropyran to give the resonance-stabilized cation, which is captured by the alcohol. The result is an acetal, a molecule stable to base, but not to acid (p. 843). Regeneration of the alcohol can be accomplished through acid-catalyzed hydrolysis.

The reaction of an alcohol with a trialkylsilyl halide leads to a trialkylsilyl ether in near-quantitative yield. The reaction depends for its success on the enormous strength of the silicon–oxygen bond (~110 kcal/mol). The silyl group can be removed through reaction with fluoride ion or hydrogen fluoride. The silicon–fluorine bond is even stronger than the silicon–oxygen bond (~140 kcal/mol), and the formation of the trialkylsilyl fluoride provides the thermodynamic driving force for the reaction (Fig. 16.48).

FIGURE 16.48   The use of trialkylsilyl ethers as protecting groups for alcohols.

So, alcohols can be stored or protected by converting them into THP or silyl ethers, and regenerating them when needed.

16.11   ADDITION REACTIONS OF NITROGEN BASES: IMINE AND ENAMINE FORMATION

The nitrogen atom of an amine (RNH_2) is a nucleophile, and, like the related oxygen bases, it can add to carbonyl compounds. Indeed, the acid-catalyzed reaction between a carbonyl compound and an amine to give a **carbinolamine** is quite analogous to the reactions of water and alcohols we have just seen (Fig. 16.49).

FIGURE 16.49 Reactions of carbonyl compounds to give hemiacetals and hydrates are completely analogous to carbinolamine formation. The usual sequence of protonation, addition, and deprotonation steps is followed.

PROBLEM 16.17 Write the base-catalyzed versions of the reactions in Figure 16.49.

Further steps in the reaction are all similar as well. An OH is protonated, and then lost as water (note that ⁻OH is *not* the leaving group) to give a resonance-stabilized cation (Fig. 16.50).

In the first example in Figures 16.49 and 16.50, the adding nucleophile is an alcohol, ROH, and R^+ cannot usually be lost. Instead, the cation goes on to full acetal formation by a second addition of alcohol. For the second and third examples of Figures 16.49 and 16.50, however, there remains an available hydrogen that can be lost as a proton. In the third reaction (hydration), this loss simply completes the re-formation

Resonance-stabilized
cations

FIGURE 16.50 Loss of water leads to a resonance-stabilized intermediate in all three reactions.

of the carbonyl group of the starting material, but in the second example it leads to the formation of a carbon–nitrogen double bond, called, in simple cases, a **Schiff base** or an **imine** (Fig. 16.51).

There is a vast variety of imines formed from specialized primary amines (RNH_2, monosubstituted derivatives of ammonia, NH_3), and each has its own common name. Many of them were once useful in analytical chemistry because they are gen-

Here there is *no proton* that
can be lost from oxygen

Imine formation

FIGURE 16.51 In the second
two reactions, loss of a proton
leads to an imine or carbonyl
compound. In the first reaction,
there is no proton to be lost, and
a second molecule of alcohol
adds to give an acetal.

In these cases, there is a proton
that can be lost from N or O

erally solids, and could be characterized by their melting points. For example, formation of a 2,4-dinitrophenylhydrazone or a semicarbazone from an unknown was not
only diagnostic for the presence of a carbonyl group, but yielded a specific derivative
that could be compared to a similar sample made from a compound of known structure. Figure 16.52 gives a few of these substituted imines. Be sure to note that all of
these reactions are the same—the only difference is the identity of the R group. If
you know one, you know them all—but you have to *know* that one.

THE GENERAL CASE

SPECIFIC EXAMPLES

H—NH₂ → Imine

Ph—NH₂ → Phenyl imine

H₂N—NH₂ → Hydrazone

PhHN—NH₂ → Phenylhydrazone

2,4-Dinitrophenylhydrazone

Semicarbazone

HO—NH₂ → Oxime

FIGURE 16.52 Some substituted imines.

PROBLEM 16.18 Write a mechanism for the general reaction of Figure 16.52, the acid-catalyzed reaction of a carbonyl compound with RNH_2.

Note also that the success of this reaction relies on the presence of at least two hydrogens in the starting amine. One is lost as a proton in the first step of the reaction, the formation of the carbinolamine. The second proton is lost in a deprotonation step as the final imine is produced (Fig. 16.53).

FIGURE 16.53 In order to form an imine, there must be *two* (red) hydrogens on the starting amine. That is, the amine must be primary (RNH_2) or ammonia ($:NH_3$).

What happens if one or both of these hydrogens is missing? Tertiary amines (R_3N) carry no hydrogens and undergo no visible reaction with carbonyl groups. But this masks reality. In fact, addition to the carbon–oxygen double bond does take place, but the unstable intermediate has no available reaction except reversal to starting material (Fig. 16.54).

FIGURE 16.54 Tertiary amines ($R_3N:$) can add to carbonyl groups, but the only available further reaction is reversion to starting materials.

Secondary amines (R_2NH) have one hydrogen, and so carbinolamine formation is possible. This reaction can be followed by protonation and water loss to give a resonance-stabilized cation, called an **immonium ion**, or, sometimes, an imminium ion. But now imine formation is *not* possible as there remains no second proton on nitrogen to be lost. Figure 16.55 shows the reaction for secondary amines.

An immonium ion (no proton can be lost from nitrogen to give an imine because there is no H left attached to N)

FIGURE 16.55 For secondary amines (R_2NH) carbinolamine formation is possible, but there is no hydrogen remaining on nitrogen that can be lost to give an imine. Instead, water is lost to give a resonance-stabilized immonium ion.

Think now about what you know of the possibilities for proton loss by carbocations. The E1 reaction (Chapter 7, p. 312) is a perfect example. In the E1 reaction, a leading group departs to give a carbocation, which reacts further by deprotonating at all possible adjacent positions. If an adjacent carbon-attached hydrogen is available, the immonium ion can do this same reaction. Figure 16.56 makes the analogy.

FIGURE 16.56 Deprotonation of an immonium ion to give an enamine is exactly like deprotonation of a carbocation to give an alkene.

This reaction gives a molecule called an **enamine**, which is the final product of the reaction of secondary amines and simple aldehydes and ketones. Now comes an important question: Why doesn't the immonium ion formed from primary amines or ammonia also lose a proton from carbon to give an enamine? Why is imine formation preferred when there is a choice (Fig. 16.57)?

No choice here—base can only remove H⁺ from C; there are no H atoms on N

Enamine

Here there is a choice; base can remove H⁺ from either C (red) or N (green)

Enamine (less stable)

Imine (more stable)

Immonium ion from a secondary amine

Immonium ion from a primary amine

FIGURE 16.57 Why don't immonium ions formed from primary amines give enamines rather than imines?

The answer is extraordinarily simple: The imine is the more stable molecule and so is preferred when possible. It is only in the cases where imine formation is not possible, as when secondary amines are used, that enamines are formed.

PROBLEM 16.19 Many enamines and imines are in equilibrium. Write a mechanism for the acid- and base-catalyzed reactions (both directions) below.

Imine   Enamine

SUMMARY

All manner of oxygen and nitrogen bases will add reversibly to carbonyl compounds. The ultimate fate of the reaction and the ultimate structure of the product depend on concentration, and, for nitrogen bases, on the number of hydrogens available on nitrogen.

16.12   ORGANOMETALLIC REAGENTS: A GENERAL SYNTHESIS OF HYDROCARBONS

Not all additions to carbonyl compounds are reversible. Some of the irreversible reactions are of great importance in synthetic chemistry, as they involve the creation of carbon–carbon bonds. In order to discuss these, we must first talk a bit about the formation of **organometallic reagents**.

We first met organometallic reagents in Section 6.3, page 235. It would be a very good idea to go back and re-read those few pages carefully. There, we saw that most alkyl halides (usually not fluorides) react with some metals to give compounds containing a metal–carbon bond. The most common examples are the **Grignard reagent**, "RMgX," and the **organolithium reagent**, "RLi." Remember that the polar resonance forms are very important for these species, especially for the Grignard reagent (Fig. 16.58).

Grignard reagent

Organolithium reagent

R = alkyl, alkenyl, or aryl
X = Cl, Br, or I

FIGURE 16.58 Treatment of alkyl chlorides, bromides, and iodides with Mg or Li leads to the formation of an organometallic reagent.

Remember also that the Grignard reagent is in equilibrium with a mixture of magnesium halide MgX_2 and the dialkylmagnesium reagent, R_2Mg, and that the ether solvent is critical for formation of this organometallic reagent (Fig. 16.59).

FIGURE 16.59 The Grignard reagent has a complex structure in which the ether solvent is important. The Grignard reagent (RMgX) is also in equilibrium with a dialkylmagnesium reagent (R_2Mg) and magnesium halide (MgX_2).

The mechanism of formation of the Grignard reagent involves radical-transfer reactions in which a transient alkyl radical is formed in the presence of a magnesium-centered radical (Fig. 16.60). Coupling generates RMgX, and further radical reactions can equilibrate this species with the dialkyl magnesium. Most alkyl, alkenyl, and aryl halides (except fluorides) can be made into Grignard reagents.

FIGURE 16.60 The mechanism of formation of the Grignard reagent involves a radical coupling reaction.

Organolithium reagents are not monomeric; the formula RLi is a simplification; they are more correctly represented as $(RLi)_n$ (Fig. 16.61).

FIGURE 16.61 Organolithium reagents are not monomeric, but oligomeric. The exact value of n depends on solvent and the structure of R.

Although the Grignard reagent and the organolithium reagent do not contain free $R:^-$, they act as if they did and react vigorously with Brønsted acids such as water. Thus, they must constantly be protected from moisture (Fig. 16.62).

FIGURE 16.62 The powerfully basic organometallic reagents are easily protonated by water to give hydrocarbons and metal hydroxides.

The reaction with water is a synthetic opportunity, however, as hydrocarbons, R—H, are formed. If the organometallic reagents are allowed to react with D_2O, the result is a deuterated hydrocarbon (Fig. 16.63).

$$CH_3CH_2Br + Mg \xrightarrow{\text{ether}} CH_3CH_2MgBr \xrightarrow{D_2O} CH_3CH_2\!-\!D$$

FIGURE 16.63 Reaction with D_2O gives deuterated hydrocarbons.

PROBLEM 16.20 Given inorganic reagents of your choice (including D_2O), devise syntheses of the following molecules from the indicated starting materials:

Neither Grignard reagents nor organolithium reagents are good nucleophiles toward organic halides. If they were, the syntheses we use to generate the organometallic reagents in the presence of the starting organic halide would be quite ineffective, as hydrocarbons would be the major products. Small amounts of coupling are sometimes observed, especially with highly reactive halides such as allyl bromides (Fig. 16.64).

$$R\!-\!X + Li \longrightarrow R\!-\!Li \quad R\!-\!X \xrightarrow{\;\oslash\;} R\!-\!R + LiX$$

FIGURE 16.64 Organometallic reagents are strong bases, but not strong nucleophiles.

There are other organometallic reagents that can be used to displace halides to good effect. Examples are the **organocuprates** formed from organocopper reagents and organolithium compounds. The organocopper reagents themselves can be made from organolithium compounds and cuprous halides (Fig. 16.65).

$$RLi + CuI \longrightarrow RCu + LiI \xrightarrow{RLi} R_2Cu^- \, ^+Li$$

An organocuprate

FIGURE 16.65 The synthesis of lithium organocuprates from alkyllithiums and cuprous halides.

The organocuprates are quite versatile reagents, and undergo a variety of synthetically useful reactions. We will meet them again a number of times, but for now, note that they are able to displace chloride, bromide, and iodide ions, as well as other good leaving groups, from primary, and often secondary alkyl halides. This capability gives you another quite effective synthesis of hydrocarbons (Fig. 16.66).

THE GENERAL CASE

$$R_2CuLi + R{-}X \longrightarrow R{-}R + R{-}Cu + LiX$$

X = I, Cl, Br, or other good leaving groups
R = Primary or secondary alkyl group

SPECIFIC EXAMPLES

$$\xrightarrow[\text{0 °C, ether}]{(CH_3)_2CuLi}$$

(77%)

$$\xrightarrow[\text{−75 °C, ether}]{(CH_3CH_2CH_2CH_2)_2CuLi}$$

(98%)

OTs = tosylate, an excellent leaving group

FIGURE 16.66 Lithium organocuprates will react with primary and secondary alkyl halides to give hydrocarbons.

PROBLEM 16.21 Why is tosylate such a good leaving group? If you can't recall the structure of tosylate—and you need that first—see page 295.

PROBLEM 16.22 One possible mechanism for hydrocarbon formation by organocuprates is an S_N2 displacement by one of the R groups on the alkyl halide. Can you devise an experiment to test this possibility?

Many metal hydrides such as lithium aluminum hydride undergo related reactions. Such reagents react violently with moisture to liberate hydrogen. Lithium aluminum hydride and some other metal hydrides are strong enough nucleophiles to displace halides from most organic halides. Lithium triethylborohydride, $LiHB(Et)_3$, is an especially effective displacing agent (Fig. 16.67).

FIGURE 16.67 Metal hydrides react with water to generate hydrogen, and are often strong enough nucleophiles to displace halides and generate hydrocarbons.

Like other Lewis bases, Grignard reagents, organolithium reagents, and metal hydrides are also able to undergo addition reactions to carbonyl-containing compounds. These additions lead directly to the next subject.

16.13 IRREVERSIBLE ADDITION REACTIONS: A GENERAL SYNTHESIS OF ALCOHOLS

Now that we know something of the structures of these organometallic reagents, it is time to look at their chemistry. This subject is important, because these reagents are able to form carbon–carbon bonds, a central problem in synthetic chemistry. Like oxygen- and nitrogen-centered nucleophiles, organometallic reagents (and metal hydrides) add to carbonyl compounds. Although there is no free $R:^-$ (or $H:^-$) in these reactions, the organometallic reagents are reactive enough to deliver an $R:^-$ group, acting as a nucleophile, to the Lewis acidic carbon of the carbonyl group (Fig. 16.68).

FIGURE 16.68 Organometallic reagents are strong enough nucleophiles to add to carbonyl compounds (red). When water is added in a second step (green), alcohols are produced.

R—Li $\xrightarrow{\text{H}_2\ddot{\text{O}}:}$ R—H + LiOH

R—MgX $\xrightarrow[\text{H}_2\ddot{\text{O}}:]{}$ R—H + HOMgX

FIGURE 16.69 If water had been present at the beginning of the reaction, the organometallic reagent would have been destroyed through hydrocarbon formation.

The initial product is a metal alkoxide, which is protonated when the solution is neutralized by addition of aqueous acid in a second step. Water cannot be present at the start of the reaction because it would destroy the organometallic reagent (Fig. 16.69).

It is easy to make the mistake of writing this reaction sequence as in Figure 16.70 [reaction (16.1)], which means, "Treat the ketone with a mixture of alkyllithium reagent and water." This procedure is impossible, as water would destroy the alkyllithium immediately (Figs. 16.62 and 16.69). The correct way to write the sequence is shown in Figure 16.70 [reaction (16.2)] and means, "First allow the ketone to react with the alkyllithium reagent, and then, in a second step, add water."

$$\diagdown_\diagup C=\ddot{\text{O}}: \xrightarrow[\text{H}_2\ddot{\text{O}}:]{\text{RLi}} \text{RH} + \text{Li}\ddot{\text{O}}\text{H} + \diagdown_\diagup C=\ddot{\text{O}}: \quad (16.1)$$

$$\diagdown_\diagup C=\ddot{\text{O}}: \xrightarrow[\text{2. H}_2\ddot{\text{O}}:]{\text{1. RLi}} \text{R}-\overset{|}{\underset{|}{\text{C}}}-\ddot{\text{O}}\text{H} \quad (16.2)$$

FIGURE 16.70 Reaction (16.1) means, Add a mixture of R—Li and water to the carbonyl compound. Under such conditions the organometallic reagent will be destroyed. Reaction (16.2) means, First add the organometallic reagent to the carbonyl compound and then in a second step, add water.

This procedure gives us a new and most versatile alcohol synthesis. New primary alcohols can be made through the reaction of an organometallic reagent with formaldehyde (Fig. 16.71). Figure 16.71 also shows the shorthand form of the reaction, which emphasizes the synthetic utility. Be careful to insert the missing details.

Similarly, reactions of organometallic reagents (RLi or RMgX) with aldehydes give secondary al-

Grignard reaction

FIGURE 16.71 Reaction of organometallic reagents with formaldehyde, followed by hydrolysis, leads to primary alcohols. The reaction of aldehydes with organometallic reagents, followed by hydrolysis, gives secondary alcohols.

cohols. Two kinds are possible, depending on the relationship between the two R groups. The R of the organometallic reagent can either be the same as that of the aldehyde or different (Fig. 16.71).

Ketones can react to give tertiary alcohols. Three different substitution patterns are possible depending on the number of different R groups in the reaction (Fig. 16.72).

THE GENERAL CASE

$$\text{RLi (or RMgX)} + \underset{R}{\overset{R}{\diagdown}}C{=}\overset{..}{\underset{..}{O}} \longrightarrow R{-}\underset{R}{\overset{R}{\underset{|}{\overset{|}{C}}}}{-}\overset{..}{\underset{..}{O}}{:}^{-}\ Li^{+} \xrightarrow{H_2\overset{..}{O}{:}} R{-}\underset{R}{\overset{R}{\underset{|}{\overset{|}{C}}}}{-}\overset{..}{O}H \quad (16.3)$$

$$\text{RLi} + \underset{R}{\overset{R}{\diagdown}}C{=}\overset{..}{\underset{..}{O}} \longrightarrow R{-}\underset{R}{\overset{R}{\underset{|}{\overset{|}{C}}}}{-}\overset{..}{\underset{..}{O}}{:}^{-}\ Li^{+} \xrightarrow{H_2\overset{..}{O}{:}} R{-}\underset{R}{\overset{R}{\underset{|}{\overset{|}{C}}}}{-}\overset{..}{O}H \quad (16.4)$$

$$\text{RLi} + \underset{R}{\overset{R}{\diagdown}}C{=}\overset{..}{\underset{..}{O}} \longrightarrow R{-}\underset{R}{\overset{R}{\underset{|}{\overset{|}{C}}}}{-}\overset{..}{\underset{..}{O}}{:}^{-}\ Li^{+} \xrightarrow{H_2\overset{..}{O}{:}} R{-}\underset{R}{\overset{R}{\underset{|}{\overset{|}{C}}}}{-}\overset{..}{O}H \quad (16.5)$$

$$\text{RLi} + \underset{R}{\overset{R}{\diagdown}}C{=}\overset{..}{\underset{..}{O}} \longrightarrow R{-}\underset{R}{\overset{R}{\underset{|}{\overset{|}{C}}}}{-}\overset{..}{\underset{..}{O}}{:}^{-}\ Li^{+} \xrightarrow{H_2\overset{..}{O}{:}} R{-}\underset{R}{\overset{R}{\underset{|}{\overset{|}{C}}}}{-}\overset{..}{O}H \quad (16.6)$$

A SPECIFIC EXAMPLE OF REACTION (16.5)

$$\xrightarrow[\substack{\text{ether} \\ 2.\ 0\ °C,\ H_2\overset{..}{O}{:}}]{1.\ CH_3MgBr}$$

(95%)

FIGURE 16.72 The reaction of ketones with organometallic reagents, followed by hydrolysis, yields tertiary alcohols.

Metal hydrides (LiAlH$_4$, NaBH$_4$, and many others) react in a similar fashion to deliver hydride (H:$^-$) as the nucleophilic agent. Formaldehyde is reduced to methyl alcohol, aldehydes to primary alcohols, and ketones to secondary alcohols. The alcohols are formed when water is added in a second, quenching step. There is no way to make a tertiary alcohol through this kind of reduction (Fig. 16.73).

Carbonyl reduction

THE GENERAL CASE

Formaldehyde → **Methyl alcohol**

LiAlH$_4$ (or NaBH$_4$)

LiAlH$_4$ → A primary alcohol

LiAlH$_4$ → A secondary alcohol

A SPECIFIC EXAMPLE

1. NaBH$_4$
   CH$_3$OH, 20–30 °C
2. H$_2$Ö:

(97%)

FIGURE 16.73 Metal hydride reduction, followed by hydrolysis, gives methyl alcohol from formaldehyde, primary alcohols from aldehydes, and secondary alcohols from ketones. Tertiary alcohols cannot be made this way.

We have seen reduction of carbonyl compounds before. In Chapter 14 (p. 694), we encountered the Clemmensen and Wolff–Kishner reductions in which a carbon–oxygen double bond was reduced all the way to a methylene group (Fig. 16.74).

Clemmensen reduction

$$\text{C}=\text{Ö:} \xrightarrow[\text{HCl}]{\text{Zn/Hg}}$$

Wolff–Kishner reduction

$$\text{C}=\text{Ö:} \xrightarrow[\substack{\text{KOH, }\Delta \\ \text{HOCH}_2\text{CH}_2\text{OH}}]{\text{NH}_2-\text{NH}_2}$$

FIGURE 16.74 The Clemmensen and Wolff–Kishner reductions of the carbonyl group.

16.14 RETROSYNTHETIC ALCOHOL SYNTHESIS

It's worth taking a moment to say a word again about strategy in doing synthesis problems, and to reinforce an old idea. It is almost never a good idea to try to see all the way back from the target molecule to the initial starting materials. Chemists have learned to be much less ambitious and to work back one step at a time. This process has been dignified with the title "retrosynthetic analysis" (encountered first in Chapter 10), which makes the idea seem far more complicated than it is. The idea behind this strategy is simply that we can almost always see the synthetic step immediately leading to the product. Indeed, the problem is more likely to be one of sorting among various possibilities than of finding a possible reaction. We have then created a new problem in synthesis for ourselves, and this one is treated in exactly the same way. We just ask ourselves for a set of reactions leading to the new target. Eventually, this technique will lead back to reagents that are readily available. Don't be too proud to work back step by step, even in simple problems. For example, the tertiary alcohol of Figure 16.75 could be made in three ways, starting from each of the three ketones shown in the figure. Now the problem is to devise a synthesis of the required ketone. Unfortunately, you don't yet have versatile ways to make ketones, and so it is time to remedy that deficiency.

Remember also that retrosynthetic analysis has appropriated a special arrow for its own use. The double-lined arrow always points *from* the target molecule *to* the immediate precursor (Fig. 16.75).

CONVENTION ALERT

Now the problem is, How to make the ketones, that is,

You know how to make the alkyllithium reagents

RLi $\Rightarrow$ R—X + Li

X = Cl, Br, or I

FIGURE 16.75 The start of a retrosynthetic analysis of ways of making a tertiary alcohol.

SUMMARY

The irreversible addition of organometallic reagents and metal hydrides provides a simple synthesis of primary, secondary, and tertiary alcohols. But you have to be careful to choose your reagents properly, so that the desired substitution pattern is achieved. Be sure to work these syntheses backward. It is hard to make a mistake if you do.

PROBLEM 16.23 Starting from inorganic reagents, tosyl chloride, formaldehyde, acetaldehyde, acetone, and butyl alcohol, devise syntheses of the following molecules:

(a)

OH

(b)

OH

(c)

OH

(d)

OTs

(e)

Cl

(f)

16.15   OXIDATION OF ALCOHOLS TO CARBONYL COMPOUNDS

16.15a Oxidation of Simple Monoalcohols. So far, we have generated two new and very general alcohol syntheses, the reaction of aldehydes and ketones with organometallic reagents, and the reduction of carbonyl compounds by metal hydrides. There is a complement to the reduction reaction in the oxidation of alcohols to aldehydes and ketones (Fig. 16.76). These oxidations extend the utility of the alcohol syntheses we have just learned, and complicate your life by increasing the complexity of the molecules you are able to make.

FIGURE 16.76 Alcohols can be oxidized to carbonyl compounds. This reaction is the reverse of what you have just learned—the reduction of carbonyl compounds to alcohols through reactions with metal hydride reagents.

Now your collection of syntheses can be extended by using the product alcohols as starting materials in the oxidation reaction. Tertiary alcohols cannot be simply oxidized further, but primary and secondary ones can be. As shown in Figure 16.77, secondary alcohols can only give ketones, but primary alcohols can give either aldehydes or carboxylic acids (RCOOH) depending on the oxidizing agent used.

R—C—OH with R groups →⊘→ No simple oxidation products

2-hexanol (OH) → $K_2Cr_2O_7$ / H_2SO_4 → **3-Hexanone** (80%)

$HOCH_2$—benzene ring—isopropyl → CrO_3 / pyridine → **4-Isopropylbenzaldehyde** (83%)

1-heptanol OH → H_2CrO_4 → **Heptanoic acid** (70%)

FIGURE 16.77 Tertiary alcohols cannot be easily oxidized, but secondary or primary alcohols can be. It is logical to infer that for the oxidation reaction to succeed there must be a carbon–hydrogen bond available.

How do these oxidation reactions work? A typical oxidizing agent for primary alcohols is chromium trioxide (CrO_3) in pyridine. As long as the reagent is kept dry, good yields of aldehydes can often be obtained (Fig. 16.77). Although a metal–oxygen bond is more complicated than a carbon–oxygen bond, the double bonds in CrO_3 and R_2C=O react in similar ways. For example, the alcohol oxygen is a nucleophile and can attack the chromium–oxygen bond to give, ultimately, a chromate **ester**. This reaction of a Cr=O with an alcohol is analogous to the formation of a hemiacetal from the reaction of an alcohol and a C=O (Fig. 16.78).

Alcohol oxidation

R—CH_2—ÖH + C=O ⇌ R—CH_2—Ö⁺—C—Ö⁻ (with H) —proton transfers→ RCH_2—Ö—C—ÖH

Hemiacetal

R—CH_2—ÖH + Cr=O (with :Ö: and :O:) ⇌ R—CH_2—Ö⁺—Cr—Ö⁻ (with H and :O:) —proton transfers→ RCH_2—Ö—Cr—ÖH (with :O: groups)

A chromate ester

FIGURE 16.78 Nucleophilic alcohols can add to Cr=O bonds just as they do to C=O bonds. This reaction is the first step in the oxidation process, and in this case gives a chromate ester.

Pyridine (or any other base) now initiates an E2 reaction to give an aldehyde (an oxidized alcohol) and HOCrO$_2^-$ (a reduced chromium species) (Fig. 16.79).

FIGURE 16.79 The second step in the oxidation reaction is a simple E2 reaction to generate the new C=O double bond.

It should now be clear why tertiary alcohols resist this reagent. The initial addition reaction can proceed, but there is no hydrogen to be removed in the elimination process (Fig. 16.80).

No H available
for an E2 reaction!

FIGURE 16.80 A tertiary alcohol can add to the Cr=O double bond, but there is no hydrogen available to be lost in an E2 reaction.

Notice also how this seemingly strange reaction is really made up of two processes we have already encountered. The first step is addition of an oxygen base to a chromium–oxygen double bond. The second step is nothing more than a garden-variety elimination. The lesson here is not to be bothered by the identity of the atoms involved. When you see a reaction that looks strange, rely on what you already know. Make analogies, and more often than not you will come out alright. That is exactly what all chemists do when we see a new reaction. We think first about related reactions and try to find a way to generalize.

Chromium trioxide (CrO$_3$) will also oxidize secondary alcohols to ketones, but a more usual reagent is sodium chromate (Na$_2$CrO$_4$) or sodium dichromate (Na$_2$Cr$_2$O$_7$) in aqueous strong acid (Fig. 16.81).

FIGURE 16.81 A typical reagent for oxidation of secondary alcohols to ketones is sodium dichromate in acid. Other oxidizing agents also work, however.

Under these conditions, or any aqueous conditions, it is difficult to stop oxidation of primary alcohols at the aldehyde stage. Overoxidation to the carboxylic acid is common. The difficulty is that aldehydes are hydrated in water (p. 832). These hydrates are themselves alcohols and can be oxidized further (Fig. 16.82).

FIGURE 16.82 Oxidation of a primary alcohol leads to an aldehyde. Aldehydes are hydrated in the presence of water, and the hydrates are themselves alcohols. Further oxidation of the hydrate leads to the carboxylic acid.

PROBLEM 16.24 Write a mechanism for the oxidation of a secondary alcohol to a ketone by chromic oxide.

There are many other oxidizing agents of varying strength and selectivity. Manganese dioxide (MnO_2) is effective at making aldehydes from allylic or benzylic alcohols, nitric acid (HNO_3) will oxidize primary alcohols to acids, and potassium permanganate ($KMnO_4$) is an excellent oxidizer of secondary alcohols to ketones and of primary alcohols to acids (Fig. 16.83).

FIGURE 16.83 Some other oxidizing agents.

SUMMARY

Now we can reduce carbonyl compounds to the alcohol level with hydride reagents, and oxidize them back again using the various reagents of this section. Although one probably wouldn't want to spend one's life cycling a compound back and forth between the alcohol and carbonyl stage, these reactions are enormously useful in synthesis. Many alcohols are now the equivalent of aldehydes or ketones, because they can be oxidized, and carbonyl compounds are the equivalent of alcohols because they can be reduced with hydride or treated with an organometallic reagent.

PROBLEM 16.25 Starting from alcohols containing no more than three carbon atoms and inorganic reagents of your choice, devise syntheses of the following molecules:

16.15b Oxidative Cleavage of 1,2-Diols. There is another oxidation that works only for 1,2-diols (1,2-glycols). When a 1,2-diol is oxidized by periodic acid

(HIO$_4$) it is cleaved to a pair of carbonyl compounds. This reaction has long been used as a test for 1,2-diols (Fig. 16.84).

Cleavage of vicinal diols

THE GENERAL CASE

A SPECIFIC EXAMPLE

FIGURE 16.84 The cleavage of 1,2-diols by periodic acid.

The mechanism involves the formation of a cyclic periodate ester in a reaction that looks much like the formation of a cyclic acetal from a 1,2-diol and a ketone (Fig. 16.85). The intermediate breaks down to form the two carbonyl compounds.

Ester of periodic acid

FIGURE 16.85 The cleavage reaction involves a cyclic ester of periodic acid that breaks down to give a pair of carbonyl compounds.

PROBLEM 16.26 *trans*-1,2-Cyclohexanediol is much more reactive toward periodic acid than the cis isomer. Explain why.

16.16 OXIDATION OF THIOLS AND OTHER SULFUR COMPOUNDS

Like alcohols, thiols can be oxidized, but oxidation takes place not at carbon, as with alcohols, but on sulfur to give **sulfonic acids** (Fig. 16.86).

Butanethiol → **Butanesulfonic acid**
(85%)

FIGURE 16.86 Oxidation of mercaptans with nitric acid gives sulfonic acids.

Milder oxidation, using halogens (I_2 or Br_2) in base, is a general source of disulfides. This reaction probably involves a sequence of displacement reactions (Fig. 16.87).

FIGURE 16.87 Milder oxidation with bromine or iodine in base gives disulfides.

Unlike ethers, sulfides can easily be oxidized to **sulfoxides**, usually with hydrogen peroxide, H_2O_2. Further or more vigorous oxidation gives **sulfones** (Fig. 16.88).

THE GENERAL CASE

Sulfide → Sulfoxide → Sulfone

FIGURE 16.88 Sulfides can be oxidized to sulfoxides with hydrogen peroxide. Further oxidation gives sulfones.

FIGURE 16.88 (*continued*)

SPECIFIC EXAMPLES

(90%)

(76%)

Sulfoxides and sulfones are more complicated structurally than they appear at first. First, they are not planar, but approximately tetrahedral. Second, the uncharged resonance forms are not the major contributors to the structures. Better representations involve charge-separated resonance forms in which the sulfur octet is not expanded (Fig. 16.89).

Sulfoxides

Sulfones

FIGURE 16.89 A resonance description of sulfoxides and sulfones.

SUMMARY

Thiol and sulfide chemistry resembles alcohol and ether chemistry. There are differences, but many of the differences are quantitative, not fundamental. For example, mercaptides are better nucleophiles than alkoxides, but both species undergo the S_N2 reaction. There are areas in which substantial differences do appear; for example, sulfur is far more prone to oxidation than oxygen.

16.17   THE WITTIG REACTION

As we have seen, all manner of nucleophiles add to carbon–oxygen double bonds, often leading to useful compounds such as alcohols. Here is an example of a synthetically useful carbonyl addition reaction that produces certain kinds of alkenes. This reaction was discovered in the laboratories of Georg Wittig (1897–1987) and is called the **Wittig reaction**. Wittig shared the 1979 Nobel prize largely for his work on this reaction. The process begins with the reaction between phosphines and alkyl halides to give phosphonium halides. Although we have not encountered reactions of phosphorus yet, it sits right below nitrogen in the periodic table and, like nitrogen, is a good nucleophile. Formation of the phosphonium halide takes place through an S_N2 displacement (Fig. 16.90).

FIGURE 16.90 Displacement of iodide by the nucleophilic phosphorus atom of triphenylphosphine leads to phosphonium ions.

The phosphonium halides contain acidic hydrogens that can be removed by strong bases such as alkyllithium reagents. The product is an **ylide**, a compound containing opposite charges on adjacent atoms (Fig. 16.91).

FIGURE 16.91 Protons adjacent to the positively charged phosphorus atom can be removed in strong base to give ylides.

The carbon of the ylide is a nucleophile, and like other nucleophiles adds to carbon–oxygen double bonds. Intramolecular closure of the intermediate leads to a four-membered ring, an **oxaphosphetane**. Oxaphosphetanes are strained and unstable relative to their quite stable constituent parts, phosphine oxides and alkenes. Therefore, it is easy for these compounds to fragment (Fig. 16.92).

FIGURE 16.92 Ylides are nucleophiles and will add to carbonyl compounds to give intermediates that can close to oxaphosphetanes. These four-membered ring compounds can open to give triphenylphosphine oxide and the product alkenes.

The Wittig synthesis does contain some unfamiliar intermediates. You have not seen either an ylide or an oxaphosphetane before, for example. Yet the key reactions, the S_N2 displacement to form the phosphonium halide (Fig. 16.90) and the addition to the carbonyl group (Fig. 16.92), are simple variations on basic processes. In a practical sense, it is worth working through the new parts of the Wittig reaction because it is so useful synthetically. Note that the Wittig reaction is, in a formal sense, the reverse of ozonolysis (Chapter 10.5a, p. 465). Figure 16.93 shows some specific examples.

Wittig reaction

SPECIFIC EXAMPLES

$(Ph)_3P=CH_2$
12 h, 65 °C
ether

(67%)

$(Ph)_3P=CHCOOC_2H_5$
10 h, 170 °C

(43%)

GENERAL INTERCONVERSIONS

$$\underset{R}{\overset{R}{>}}C=O \quad \underset{\text{ozonolysis}}{\overset{\text{Wittig}}{\rightleftarrows}} \quad \underset{R}{\overset{R}{>}}C=CH_2$$

FIGURE 16.93 Some typical synthetic uses of the Wittig reaction.

The Wittig reaction is very useful. How else could we convert cyclohexanone into pure methylenecyclohexane? We can use the reactions of this chapter to devise a different synthesis. Some thought might lead to a sequence in which cyclohexanone reacts with methyllithium to give, after hydrolysis, a tertiary alcohol. An acid-catalyzed elimination reaction would give some of the desired product, but there is no easy way to avoid the predominant formation of the undesired isomer, 1-methylcyclohexene (Fig. 16.94). The Wittig reaction solves this synthetic problem and is an important constituent of any synthetic chemist's bag of tricks.

1. CH_3Li
2. H_2O

H_3O^+
H_2O

Major   +   Minor

FIGURE 16.94 This approach fails to give very much of the desired product. The Wittig reaction is the usual alternative.

PROBLEM 16.27 Explain why the acid-catalyzed dehydration reaction of Figure 16.94 will give mostly the undesired 1-methylcyclohexene. .

PROBLEM 16.28 Starting with cycloheptanone, cyclopentanone, triphenylphosphine, propyl iodide, butyllithium, and inorganic reagents of your choice, devise synthesis of the following molecules:

(a)

(b)

16.18 SPECIAL TOPIC: BIOLOGICAL OXIDATION

We have seen several examples of redox reactions in the last few sections. Related processes are of vital importance in biological systems. We humans, for example, derive much of our energy from the oxidation of ingested sugars and fats. However, we do not use chromium reagents or nitric acid to accomplish these reactions, for these reagents are far too unselective and harsh. They would surely destroy most of our constituent molecules; a quite uncomfortable process one imagines. Instead, a series of highly selective biomolecules has evolved, each dedicated to a particular purpose. One of these, nicotinamide adenine dinucleotide (**NAD⁺**)* is the subject of this section.

Nicotinamide adenine dinucleotide is a pyridinium ion, and its full structure is shown in Figure 16.95.

FIGURE 16.95 Nicotinamide adenine dinucleotide, NAD⁺.

*A glance at the structure of this molecule and some brief thoughts on how complicated its systematic name must be quickly reveal the obvious reasons why biochemists are addicted to acronyms.

The business end of the molecule, the nicotinamide, is connected through a sugar (ribose) and a pyrophosphate linkage to another ribose which is, in turn, attached to the base adenine. The molecule is a **coenzyme,** meaning that it can only operate in connection with an enzyme, in this case, alcohol dehydrogenase. The enzyme binds both ethyl alcohol, the molecule to be oxidized, and NAD^+, which will be reduced.

The enzyme serves to bring the acid, NAD^+, and the base, ethyl alcohol, together, but this bringing together is of immense importance to the reaction, as it allows the redox process to take place without the requirements that molecules in solution have for finding each other and orienting properly for reaction.

The NAD^+ is a strong acid by virtue of the quaternary nitrogen with its positive charge. It can be reduced by transfer of a hydrogen, *with its pair of electrons* (a **hydride** reduction) from ethyl alcohol (Fig. 16.96). The reaction is completed by deprotonation to give the product acetaldehyde and removal of the products from the enzyme. This removal is easy because the enzyme has evolved to bind ethyl alcohol and NAD^+, not acetaldehyde and the reduced form of NAD^+, NADH.

FIGURE 16.96 The NAD^+ is reduced by transfer of hydride ($H\mathbf{:}^-$) from ethyl alcohol.

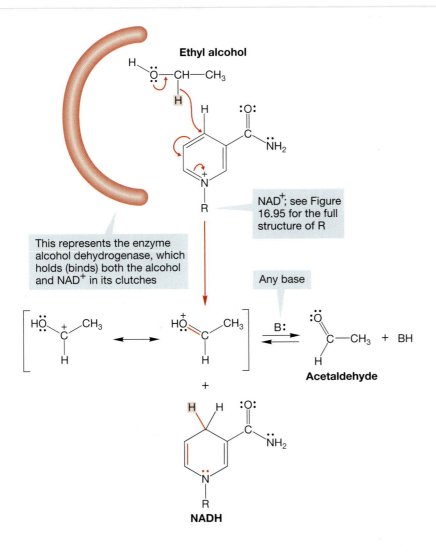

This hydride reduction may seem strange, but it is not. There is a close relative in the intramolecular hydride shifts in carbocation chemistry. These, too, are redox reactions. The originally positive carbon is reduced and the carbon from which the hydride moves is oxidized (Fig. 16.97).

FIGURE 16.97 Hydride shifts in carbocations are also redox reactions.

The end products of the biological reaction are NADH, the reduced NAD^+, and the oxidized ethyl alcohol, acetaldehyde (Fig. 16.96).

WORKED PROBLEM 16.29 Each of the two methylene hydrogens of ethyl alcohol can be replaced by deuterium to give a pair of chiral deuterioethanols. Show this process and indicate which of the new compounds is (R) and which is (S).

ANSWER Replacement of one methylene H with D gives the (S) enantiomer; replacement of the other gives the (R) enantiomer.

This structure shows the priorities for the (R) compound;
the 1 ... 2 ... 3 arrow is clockwise

PROBLEM 16.30 Oxidation of (R)-1-deuterioethanol produces deuterio-NADH (NADD) and acetaldehyde. Predict the products of the oxidation of (S)-1-deuterioethanol with NAD⁺. Explain carefully.

16.19 SUMMARY

New Concepts

The centerpiece of this chapter, and its unifying theme, is the reaction in which nucleophiles of all kinds add to carbonyl groups. This process can be either acid or base catalyzed, and starts with the overlap of a filled orbital on the nucleophilic Lewis base with the empty π^* orbital of the carbonyl to give an alkoxide in which the highly electronegative oxygen atom bears the negative charge (Fig. 16.23).

Many addition reactions are equilibria, and in these cases the position of the equilibrium will depend on the relative stabilities of the starting materials and products. These stabilities depend, as always, on a variety of factors including electronic structure, substitution pattern, and steric effects.

The π system of the carbonyl group is constructed through taking combinations of carbon and oxygen $2p$ orbitals. This process leads to the π orbitals of Figure 16.4.

The concept of retrosynthetic analysis reappears in this chapter. Retrosynthetic analysis simply means that it is best to approach problems of synthesis by searching not for the *ultimate* starting material, but rather to undertake the much easier task of finding the *immediate* precursor of the product. This process can be repeated until an appropriate level of simplicity is reached (Fig. 16.98).

Target molecule $\Longrightarrow$ Next precursor molecule $\Longrightarrow$ Immediate precursor molecule $\Longrightarrow$ Simple starting material

FIGURE 16.98 Retrosynthetic analysis.

Key Terms

acetal (p. 847)
acetone (p. 827)
aldehyde (p. 826)
benzaldehyde (p. 828)
carbinolamine (p. 852)
carbonyl group (p. 821)
coenzyme (p. 879)
cyanohydrin (p. 842)
dials (p. 826)
diones (p. 827)
enamine (p. 858)
ester (p. 869)
formaldehyde (p. 826)
Grignard reagent (p. 859)
hemiacetal (p. 844)
hydrate (p. 832)
hydride (p. 879)

imine (p. 853)
immonium ion (p. 857)
ketones (p. 826)
NAD^+ (p. 878)
organocuprates (p. 861)
organolithium reagent (p. 859)
organometallic reagents (p. 859)
oxaphosphetane (p. 876)
phenyl group (p. 828)
protecting group (p. 850)
Schiff base (p. 853)
sulfones (p. 874)
sulfonic acids (p. 874)
sulfoxides (p. 874)
tetrahydropyranyl (THP) ether (p. 850)
Wittig reaction (p. 876)
ylide (p. 876)

Reactions, Mechanisms, and Tools

As mentioned in the New Concepts section, there is a single, central reaction in this chapter, the addition of a nucleophile to the Lewis acid carbonyl group. The details of the mechanism vary, depending on whether the reaction is acid or base catalyzed, and on the substitution patterns of the nucleophile and the carbonyl-containing compound. Here are some general examples.

A simple case is the acid- or base-catalyzed reversible addition reaction (Fig. 16.36). Examples are hydration or cyanohydrin formation.

A slightly more complicated reaction involves an addition followed by loss of water. An example is the reaction of primary amines with carbonyl groups to give substituted imines (Fig. 16.53).

If there is no proton that can be removed after loss of water, a second addition reaction can occur, as in acetal formation (Fig. 16.45).

There are also irreversible additions, as in the reactions of organometallic reagents or metal hydrides with carbonyl compounds. Protonation of the initial products, alkoxides, gives alcohols (Fig. 16.68).

This chapter also introduces the oxidation reactions of alcohols. These reactions involve addition reactions in their first steps. For example, in the oxidation of primary alcohols by CrO_3, the first step is addition of the alcohol to the chromium–oxygen double bond (Figs. 16.78 and 16.79). The oxidation is completed by an elimination in which the new carbon–oxygen double bond is constructed.

Syntheses

This chapter contains lots of new syntheses.

1. ACETALS

Acetals can be used as protecting groups for aldehydes and ketones; treatment with H_2O/H_3O^+ regenerates the aldehyde. The hemiacetal is an intermediate

2. ACIDS

Other oxidizing agents such as $KMnO_4$ work also. The aldehyde is an intermediate

3. ALCOHOLS

Primary alcohols

The alkoxide is an intermediate; $LiAlH_4$ can also be used

The alkoxide is an intermediate

Secondary alcohols

The alkoxide is an intermediate; the R groups can be the same or different

The alkoxide is an intermediate; the R groups can be the same or different, depending on the structure of the ketone; $LiAlH_4$ can also be used

Tertiary alcohols

The product can be R_3C-OH, R_2RC-OH, or $RRRC-OH$, depending on the structures of the starting ketone and organometallic reagent

4. ALDEHYDES

Hydrolysis of an acetal

Oxidation of a primary alcohol; water must be absent

Periodate cleavage of a 1,2-glycol; a cyclic ester is an intermediate

5. ALKENES

The Wittig reaction

6. ALKYLITHIUM REAGENTS

$$R—X \xrightarrow{Li} RLi$$

X = Cl, Br, or I

RLi is a simplistic picture of the reagent, which is *not* monomeric

7. BISULFITE ADDITION PRODUCTS

Also works with some ketones

8. CYANOHYDRINS

Cyanohydrin formation is more favorable for aldehydes than for ketones

9. DISULFIDES

$$RSH \xrightarrow[base]{I_2} RSSR$$

10. ENAMINES

At least one α-hydrogen must be available; the amine must be secondary

11. GRIGNARD REAGENTS

$$R—X \xrightarrow[ether]{Mg} RMgX$$

The structure of the organometallic reagent is complicated, as RMgX is in equilibrium with other molecules; the ether solvent is also critical to the success of the reaction

12. HEMIACETALS

Rarely isolable; exceptions are cyclic hemiacetals, especially sugars; the reaction usually goes further to give the full acetal

13. HYDRATES

Unstable; for ketones the starting material
is usually favored

More likely to be favored at equilibrium than
the hydrated ketones; not usually isolable

14. HYDROCARBONS

$$RMgX \xrightarrow{H_2O} RH$$

$$RLi \xrightarrow{H_2O} RH$$

$$R_2Cu^- Li^+ \xrightarrow{R-X} R-R$$

R—X must be primary or secondary

15. IMINES

This reaction works for ketones as well;
many varieties of imine are known,
depending on the structure of R in the
amine, which must be primary

16. KETONES

The hydrolysis of an acetal

Many other oxidizing agents will oxidize secondary
alcohols to ketones

Periodate cleavage of a 1,2–diol: A cyclic ester
is an intermediate

17. LITHIUM ORGANOCUPRATES

$$RLi + CuX \longrightarrow R_2Cu^- Li^+ + LiX$$

$$X = I, Br, or Cl$$

18. SULFONES

$$RSR + H_2O_2 \longrightarrow$$

The sulfoxide is an intermediate

19. SULFONIC ACIDS

$$RSH + HNO_3 \longrightarrow$$

20. SULFOXIDES

$$RSH + H_2O_2 \longrightarrow$$

Further oxidation gives the sulfone

Common Errors

By far the hardest thing about this chapter is the all-too-plentiful detail. To get this material under control it is absolutely necessary to be able to generalize. The easiest mistake to make is to memorize the details and lose the broad picture. Although there is one general principle in this chapter (nucleophiles add to carbon–oxygen double bonds), it is easy to get lost in the myriad protonation, addition, and deprotonation steps. The reversible addition reactions of this chapter are the hardest to keep straight. The following analysis is a model for the process of sorting necessary to focus on the fundamental similarities while being mindful of the small, but important differences introduced by structural changes in the molecules involved in these reactions.

In base, hydroxide, alkoxides, and amide ions all add to the carbonyl group to give alkoxides that can be protonated to produce hydrates, hemiacetals, and carbinolamines. The three reactions are very similar (Fig. 16.99).

FIGURE 16.99 Base-catalyzed additions to carbonyl groups.

In acid, similar reactions occur, and the three processes begin with closely related sequences of protonation, addition, and deprotonation (Fig. 16.100).

FIGURE 16.100 Acid-catalyzed additions to carbonyl groups.

In the acid-catalyzed reactions, water can be lost from the hydrate, hemiacetal, and carbinolamine to give new, resonance-stabilized intermediates. In the hydrate, this results only in re-formation of the original carbonyl group, but new compounds can be produced in the other two cases. The carbinolamines formed from primary amines follow a similar path that leads ultimately to imines. Hemiacetals can lose water but do not have a second proton that can be lost. Instead, they add a second molecule of alcohol to give full acetals (Fig. 16.101).

FIGURE 16.101 Further reactions in acid of hydrates, carbinolamines, and hemiacetals.

The carbinolamines formed from secondary amines and carbonyl compounds also have no second proton to lose and instead are deprotonated at carbon to give enamines (Fig. 16.102).

FIGURE 16.102 Further reactions of carbinolamines formed from secondary amines.

16.20 ADDITIONAL PROBLEMS

PROBLEM 16.31   Name the following molecules:

(a)

(b)

(c)

(d)

(e)

(f)

PROBLEM 16.32   Draw structures for the following compounds:
(a) *p*-Nitrobenzaldehyde
(b) 4-Methylhexanal
(c) 2-Methylpropiophenone (isobutyrophenone)
(d) *cis*-2-Bromocyclopropanecarboxaldehyde

PROBLEM 16.33   Name the following compounds:

(a)

(b)

(c)

PROBLEM 16.34   Draw structures for the following compounds:
(a) Methyl propyl ketone
(b) 2-Pentanone
(c) 2,3-Hexanedione
(d) Butyrophenone
(e) 3-Bromo-4-chloro-5-iodo-2-octanone

PROBLEM 16.35   Draw structures for the following compounds:
(a) 3,5-Di-*tert*-butyl-4-hydroxybenzaldehyde
(b) 2-Acetylcyclopentanone
(c) 2,4′-Dichloro-4-nitrobenzophenone
(d) 6-Methyl-2-pyridinecarboxaldehyde
(e) *trans*-4-Phenyl-3-buten-2-one
(f) Aminoacetaldehyde dimethyl acetal

PROBLEM 16.36   Use the indicated spectroscopy to distinguish the following pairs of compounds:

(a)

(IR)

(b)

(IR)

(c)

(^{13}C NMR)

(d)

(IR, ^{1}H NMR)

PROBLEM 16.37   Give the major organic products expected in each of the following reactions:

(a)

$$HO\!\!\diagdown\!\!\diagup\!\!\diagdown\!\!OH$$
benzene, H_3O^+

(b)

$PhNH_2$

(c)

(continued)

PROBLEM 16.37 *(continued)*

(d)

1. PhMgBr
2. H$_2$O/H$_3$O$^+$

(e)

1. CH$_3$Li
2. H$_2$O/H$_3$O$^+$

(f)

NaBH$_4$

CH$_3$OH

(g)

1. LiAlH$_4$
2. H$_2$O/H$_3$O$^+$

(h)

Na$_2$Cr$_2$O$_7$

H$_2$SO$_4$

(i) PhCH$_2$OH

H$_2$CrO$_4$

(j)

CrO$_3$

pyridine

(k)

1. Ph$_3$P
2. BuLi
3. Δ

Hint: Part (k) involves an intramolecular reaction.

PROBLEM 16.38 Provide an argument not involving molecular orbitals to explain why a carbon–oxygen double bond protonates on O, not C.

PROBLEM 16.39 Write products and arrow formalism mechanisms for the following simple changes:

(a)

H$_2$O/H$_3$O$^+$

(b)

H$_2$O

(c)

CH$_3$OH

CH$_3$OH$_2^+$

(d)

KOH/H$_2$O

(e)

KOH/H$_2$O

(f)

$^+$NH$_3$OH

NH$_2$OH

PROBLEM 16.40 The reduction of 4-*tert*-butylcyclohexanone with lithium aluminum hydride affords two isomeric products. What are the structures of these two products? Which isomer do you expect to predominate?

PROBLEM 16.41 As we saw on page 863, benzyl halides can be reduced with lithium aluminum hydride.

Why can compound **A** not be reduced successfully with lithium aluminum hydride? What reaction will complicate matters?

PROBLEM 16.42 Devise a scheme for avoiding the problem you discovered in your answer to Problem 16.41. *Hint*: See page 850.

PROBLEM 16.43 Write an arrow formalism for the following reaction:

PROBLEM 16.44 Starting from isopropyl alcohol as your primary source of carbon, incidental organic reagents such as bases and solvents, as well as inorganic reagents of your choice (including labeled materials as needed), devise syntheses of the following labeled compounds. It is not necessary to write mechanisms.

PROBLEM 16.45 Starting from propyl alcohol, organic reagents containing no more than one carbon, and incidental organic reagents such as bases and solvents, as well as inorganic materials, devise syntheses of the following compounds. Some of the necessary transformations will require conversions covered in the first semester. See, for example, figures on page 295.

PROBLEM 16.46 Starting from alcohols containing no more than four carbon atoms, inorganic reagents of your choice, and "special" organic reagents (such as ether and pyridine) provide syntheses of the following molecules:

(continued)

PROBLEM 16.46 *(continued)*

(c)

(d)

PROBLEM 16.47 Propose an arrow formalism mechanism for the following reaction:

PROBLEM 16.48 Propose an arrow formalism mechanism for the following reaction. What is the driving force for this reaction? Why does it go from the seven-membered ring to the five-membered ring?

PROBLEM 16.49 Propose an arrow formalism mechanism for the following reaction:

PROBLEM 16.50 Rationalize the following somewhat surprising result. *Hint:* This problem is not as difficult as it looks. First, consider what the reaction of the alcohol starting material with the chromium reagent will give. Second, what will happen when this compound reacts with more alcohol? Finally, use the chromium reagent again.

PROBLEM 16.51 The reaction of acetone with methylamine reaches equilibrium with a carbinolamine and, eventually, an imine. There is another participant in the equilibrium called an aminal. Its formula is $C_5H_{14}N_2$. Provide a structure and a mechanism.

PROBLEM 16.52 Treatment of aromatic compound **A** as shown below leads to two products of the same formula, C_9H_{10}, **B** and **C**. The ^1H NMR spectra of **B** and **C** are very similar except that the coupling constant between two hydrogens at δ 6.3 is 8 Hz in **B** and 14 Hz in **C**. Treatment of both **B** and **C** with ozone followed by oxidative workup leads to benzoic acid and acetic acid. Provide structures for **B** and **C** and explain your reasoning.

PROBLEM 16.53 Les Gometz, a professor from New York who is very often disappointed in September, was interested in investigating the Diels–Alder reaction of methyl coumalate (**A**) and the enamine **B**. The professor attempted to prepare **B** by allowing propionaldehyde and morpholine (**C**) to react.

After an appropriate time, Professor Gometz assayed the reaction mixture by ^{1}H NMR spectroscopy and was disappointed to find that only a small portion (<5–10%) of the desired enamine **B** had formed. Ever the optimist, he ran the Diels–Alder reaction with diene **A** anyway, using the reaction mixture containing what he knew to be only a small amount of **B**. He was delighted and astonished to obtain an 80% yield of cycloadducts.

Your task is to explain how Professor Gometz could get such a good yield of products when only a small amount of enamine **B** was present in the reaction mixture. *Hint*: See Problem 16.9.

PROBLEM 16.54   Hydrogens adjacent to carbonyl groups, C=O, are relatively acidic (pK_a ~ 20). That is, they are relatively easily removed by a base to leave an anion. Explain briefly, then do Problem 16.55.

PROBLEM 16.55   Remember that essentially all nucleophiles add to carbonyl groups. Given that notion, and your answer to Problem 16.54, provide a mechanism for the following reaction:

PROBLEM 16.56   Provide structures for compounds **A–D**. Spectral data for compound **D** are shown below.
Compound **D**
Mass spectrum: m/z = 332 (p, 82%), 255 (85%), 213 (100%), 147 (37%), 135 (43%), 106 (48%), 77 (25%), 43 (25%).
IR (KBr): 3455 (s) and 1655 (s) cm^{-1}.
^{1}H NMR (CDCl$_3$): δ 2.58 (s, 3H), 2.85 (s, 1H, vanishes when a drop of D$_2$O is added), 3.74 (s, 3H), 6.77–7.98 (m, 13H).
^{13}C NMR (acetone-d_6): δ 27.0, 55.8, 82.0, 128.1–159.9 (12 lines), 197.8.

PROBLEM 16.57   Provide structures for compounds **A–D**. Spectral data for compound **D** are shown below. Although mechanisms are not required, a mechanistic analysis may prove helpful in deducing the structures.
Compound **D**
IR (neat): 2817 (w), 2717 (w), 1730 (s) cm^{-1}.
^{1}H NMR (CDCl$_3$): δ 0.92 (t, J = 7 Hz, 6H), 1.2–2.3 (m, 5H), 9.51 (d, J = 2.5 Hz, 1H).
^{13}C NMR (CDCl$_3$): δ 11.4, 21.5, 55.0, 205.0.

PROBLEM 16.58   Write a mechanism for the acid-catalyzed transformation of glycerol into acrolein.

PROBLEM 16.59   This problem introduces another alcohol protecting group, methoxymethyl (MOM), particularly popular in the month of May. The MOM protecting group is stable to base, but can be cleaved upon treatment with mild acid. Provide an arrow formalism for the introduction of the MOM protecting group. Also explain why this protecting group is readily cleaved upon treatment with aqueous acid.

$$R—OH \xrightarrow[\text{2. CH}_3\text{OCH}_2\text{Cl}]{\text{1. NaH}} R—O—CH_2OCH_3$$

$$\downarrow \begin{array}{c} H_3O^+ \\ H_2O \end{array}$$

$$R—OH$$

PROBLEM 16.60   Metaproterenol (orciprenaline, **1**) is a β-adrenoreceptor stimulant used therapeutically as a bronchodilator. In the synthesis of the hydrobromide salt of racemic **1**, outlined below, supply the appropriate reagents.

1
(as HBr salt)

PROBLEM 16.61   Treatment of 2-thiabicyclo[2.2.1]-heptane (**1**) with 1 equiv of hydrogen peroxide affords a mixture of diastereomers **2** and **3** of molecular formula $C_6H_{10}OS$. Both diastereomers **2** and **3** yield the same product **4** ($C_6H_{10}O_2S$) upon treatment with peracetic acid. Compound **4** is also produced when **1** reacts with 2 equiv of peracetic acid. Propose structures for compounds **2**, **3**, and **4**. *Hint*: Start by drawing a good Lewis structure for the starting material **1** (see Fig. 16.89).

PROBLEM 16.62   Provide structures for compounds **A–C**.

PROBLEM 16.63   Provide structures for compounds **A–C**.

Carbonyl Chemistry 2: Reactions at the α-Position

17

I want to beg you, as much as I can, dear sir, to be patient towards all that is unsolved in your heart and try to love the *questions themselves*.

—Rainer Maria Rilke*
Letters to a Young Poet

17.1 PREVIEW

In this chapter, we begin an exploration of the chemistry of the α-position in carbonyl compounds that will continue for the better part of Chapters 17–19. Our ability to understand complex reactions and to construct complicated molecules will greatly increase. By this time, there really is little fundamentally new chemistry—there is some, but much of what we find from now on will be applications of reactions we already know, placed in more complicated settings. For example, in Chapter 16 we saw a vast array of additions of nucleophiles to carbon–oxygen double bonds. Here that same reaction appears, when a new nucleophile, an enolate, adds to carbonyl groups to yield products of remarkable complexity. The reaction is the same old addition, but the context makes it look complicated (Fig. 17.1).

The chemistry in the next few chapters can be either daunting or exhilarating. Much depends on how well you have the basic reactions under control. If you do, you will be able to apply them in the new situations we are about to explore. If not, the next chapters will seem crammed with new and complicated reactions, each of which requires an effort to understand. Like anything else, learning chemistry can have a strong psychological component. When you are able to think, "That's not so bad, it's just an application of . . . ," you gain confidence and what you see on the page is likely to stick in your mind.

*Rilke (1875–1926) was a German lyric poet much influenced by the French sculptor, Rodin.

FIGURE 17.1 Additions of nucleophiles to carbonyl compounds.

It is now possible to ask questions that stretch your knowledge and talent. It's hard to teach this kind of problem solving, as success has a lot to do with having worked lots of related problems in the past. Nevertheless, there are some techniques that I can try to pass on. Don't worry if you don't get every problem! My colleagues can write problems that I can't get. It's no disgrace to have trouble with a problem. There are several favorite problems that I've been working on (not continuously) for many years. You'll see some of them.

Success in the chemistry business has relatively little to do with how *fast* you can solve problems, at least not problems that require thought and the bringing together of information from different areas. People think in different ways and answers to problems appear at different rates. There are those who are blindingly fast at this kind of thing and others who reach answers more slowly. There is nothing wrong with chewing over a problem several times, approaching it in several ways. Letting a particularly vexing problem rest for a while is often a good idea. Of course, this is the bane of exam writers (and even more so of exam answerers). How do you write an exam that allows for thought and doesn't simply select those who can answer quickly? It's hard, especially if you don't have the luxury of asking a small number of questions and giving lots of time.

What does make for success in the real world of organic chemistry? I think nothing correlates better than love of the subject, and by "subject" I do not necessarily mean what you are doing now, but rather really *doing* organic chemistry. Doing organic chemistry means not only liking the manipulative aspects of the subject, although that is a great help, but also, as Rilke said, of "loving the *questions themselves*." At any rate, it is not the number of A's on your transcript that correlates best with your later success in research!

We will start with an examination of the behavior of aldehydes and ketones as Brønsted acids, and then go on to see the many reactions that are possible once a proton has been lost.

ESSENTIAL SKILLS

1. Two reactions critical to understanding the material in this chapter are enol formation in acid and enolate formation in base. Be sure these two basic reactions are under absolute, 100% complete control—they lead into a consideration of the next point.

2. The central reaction of this chapter is the aldol condensation. Both acid- and base-catalyzed aldol condensations exist. Although each is just an elaborate example of reactions studied in Chapter 16, the increased complexity of the reagents involved means that you have to pay close attention here. Understanding the aldol condensation really is an essential skill—without that skill, one cannot deal with the next few chapters.

IMPORTANT DETAILS

1. Although both the acid- and base-catalyzed aldol condensations lead to the same initial product, a β-hydroxy carbonyl compound, the reaction usually goes further in acid (sometimes in base) to give an α,β-unsaturated carbonyl compound. *All* α,β-unsaturated carbonyl compounds can, in principle, be made through aldol or closely related condensations, which is an important problem-solving tool. Always look for the α,β-unsaturated carbonyl compound.

2. Watch out for the addition of nucleophiles to α,β-unsaturated carbonyl compounds to give enolates. This reaction, called the Michael reaction, is very common, both in Nature and in organic chemistry problems. Magid's Second Rule, which says, "Always try the Michael first!" is more than a joke—it's really good advice.

17.2   ALDEHYDES AND KETONES ARE WEAK BRØNSTED ACIDS

One of the remarkable things about carbonyl compounds is that they are not only Lewis acids (see the addition reactions throughout Chapter 16), but are proton donors as well. The pK_a values of simple carbonyl compounds are generally in the high teens (Table 17.1).

It is the hydrogen at the α-position, the position adjacent to the carbon–oxygen double bond, that is lost to a base as a proton (Fig. 17.2). Our first task is to see why these compounds are weak acids at their α-positions, and then to see what new chemistry results from this acidity.

TABLE 17.1
Some pK_a Values for Simple Ketones and Aldehydes

Compound[a]	pK_a
$CH_3CH_2COC\underline{H}_2CH_3$	19.9
$CH_3COC\underline{H}_3$	19.3
$PhCOC\underline{H}_3$	18.3
$PhCH_2COC\underline{H}_3$	18.3
$PhC\underline{H}_2COCH_3$	15.9
Cyclohexanone (α-hydrogen)	18.1
$C\underline{H}_3CHO$	16.7
$(CH_3)_2C\underline{H}CHO$	15.5

[a]The proton to be lost is underlined when there is a choice.

FIGURE 17.2 Carbonyl compounds bearing hydrogen at the α-position are weak acids, with pK_a values in the high teens.

$pK_a = 15\text{–}20$

We'll start with butyraldehyde (pK_a = 16.7) and look at the three possible anions we might form from it by breaking an sp^3–$1s$ carbon–hydrogen bond (Fig. 17.3).

FIGURE 17.3 There are three possible anions that can be formed from butyraldehyde by breaking an sp^3 – $1s$ carbon–hydrogen bond.

First of all, the dipole in the carbon–oxygen bond will stabilize an adjacent anion more than a more remote anion. Loss of the α-hydrogen will lead to a more stable anion than loss of either the β- or γ-hydrogen (Fig. 17.4).

FIGURE 17.4 The dipole in the carbon–oxygen bond will stabilize an adjacent anion more than a more distant anion.

Moreover, one of these anions, but only one of them, is stabilized by resonance, as the oxygen atom can share the negative charge. It is this stabilization that makes the α-position, but only the α-position, a proton source. The intermediate is called an **enolate** anion, as it is part alkene (*ene*) and part alcoholate (*olate*). Please note again that the enolate anion is *not* part of the time an alkoxide and part of the time a carbanion, but all of the time a single species, the resonance-stabilized enolate (Fig. 17.5).

FIGURE 17.5 Loss of the α-hydrogen leads to a resonance-stabilized enolate anion.

Resonance-stabilized enolate

The enolate anion can reprotonate in two ways. If it reacts with a proton source such as water at carbon, the original carbonyl compound is regenerated. If it protonates at oxygen, the neutral **enol** is formed. In base, carbonyl compounds are in equilibrium with their enol forms (Fig. 17.6).

FIGURE 17.6 Reprotonation of the enolate can lead either to the original carbonyl compound or to the related enol.

WORKED PROBLEM 17.1 Write a mechanism for the base-catalyzed equilibration of the carbonyl and enol forms of acetone.

ANSWER This problem is not hard, but it needs to be done right now. To be able to do the more complicated material that will appear later, these simple interconversions must become second nature. This problem is drill, but it is nevertheless very important.

Loss of either the α-hydrogen from acetone or the hydroxyl hydrogen of the enol form of acetone leads to the same enolate. Protonation at carbon gives the ketone; protonation at oxygen gives the enol. It is formation of the enolate that achieves the equilibration.

(continued)

ANSWER *(continued)*

An orbital picture clearly shows the source of the stabilization of the enolate. The three 2p orbitals overlap to form an allyl-like system (Chapter 12, p. 583). Two of the four π electrons are accommodated in the lowest bonding molecular orbital, Φ_1, and two in the nonbonding orbital, Φ_2. The presence of the oxygen atom makes the enolate less symmetrical than an allyl system formed from three carbon 2p orbitals (Fig. 17.7).

FIGURE 17.7 A comparison of the enolate and allyl anions.

How important is the electronegative oxygen atom to the stabilization of the enolate? The pK_a of propene is 43 and that of acetaldehyde is 16.7. There is a 10^{26} difference in acidity in the two molecules, which can be attributed to the influence of the oxygen (Fig. 17.8).

$pK_a = 43$

$pK_a = 16.7$

FIGURE 17.8 The oxygen atom of the enolate plays a crucial role in promoting the acidity at the α-position. Acetaldehyde is much more acidic than propene.

WORKED PROBLEM 17.2 Propionaldehyde (propanal) can form two enols. What are they?

ANSWER There are stereochemical differences between the two enols. Like any unsymmetrical alkene, this enol can exist in (Z) or (E) forms.

(Z)-Enol   (E)-Enol

Enolate formation reveals itself most commonly through the exchange of α-hydrogens in deuterated base. For example, treatment of acetaldehyde with D_2O/DO^- results in exchange of all three α-hydrogens for deuterium (Fig. 17.9).

α-Carbon   α-Carbon

FIGURE 17.9 In D_2O/DO^-, the three α-hydrogens of acetaldehyde are exchanged for deuterium.

In hydroxide (here deuteroxide), not all of the acetaldehyde will be transformed into the enolate. The pK_a of acetaldehyde is 16.7 and that of water is 15.7. So, hydroxide is a *weaker* base than the enolate anion and at equilibrium only a small amount of the enolate anion will be present. Enolate formation is endothermic in this case (Fig. 17.10).

FIGURE 17.10 Enolate formation is an equilibrium reaction and is endothermic in the case of acetaldehyde.

$pK_a = 16.7$
Weaker acid

Weaker base

Enolate is stronger base

$pK_a \sim 15.7$
Stronger acid

Moreover, these exchange reactions are typically run using catalytic amounts of base, DO^-, in a large excess of D_2O. Under such conditions only a small amount of enolate can be formed. However, any enolate present will react with the available water, here D_2O, to remove a deuteron and produce a molecule of exchanged acetaldehyde. A new molecule of DO^- that can carry the reaction further is also formed (Fig. 17.11).

Eventually all the available α-hydrogens will be exchanged for deuterons.

FIGURE 17.11 The exchange reaction is a catalytic process, with deuteroxide ion ($^-$OD) acting as the catalyst.

Catalyst regenerated

$D_2O/ :\ddot{O}-D$
(repeat) two times

Fully exchanged acetaldehyde

Enolate

PROBLEM 17.3 Explain why the aldehydic hydrogen in acetaldehyde does not exchange in D_2O/DO^-.

PROBLEM 17.4 Explain why the bicyclic ketone in the following reaction exchanges only the two α-hydrogens shown (H$_\alpha$) and not the bridgehead hydrogen, which is also "α." *Hint:* Use models and examine the relationship between the bridgehead carbon–hydrogen bond and the π orbitals of the carbonyl group.

Bridgehead
hydrogen

The α-hydrogens of acetaldehyde can also be exchanged under acidic conditions (Fig. 17.12).

FIGURE 17.12 Exchange can also be carried out in deuterated acid, D_3O^+/D_2O.

This result illustrates a general principle: In carbonyl chemistry, there will usually (not quite always) be an acid-catalyzed version for every base-catalyzed reaction. The mechanisms of the two reactions will be different, however, even though the end results are similar or even identical. For example, in this acid-catalyzed reaction there is no base strong enough to remove an α-hydrogen from the starting acetaldehyde. The only reasonable reaction is protonation of the Brønsted basic oxygen by D_3O^+ to give a resonance-stabilized cation (Fig. 17.13).

Resonance-stabilized intermediate

FIGURE 17.13 The first step in the acid-catalyzed exchange is addition of a D^+ to the carbonyl oxygen. A resonance-stabilized cation results.

Now we have created a powerfully acidic species, the protonated carbonyl compound. Removal of a deuterium from the oxygen simply reverses the reaction to regenerate acetaldehyde and D_3O^+, but deprotonation at carbon generates the enol. Note that the product is not an anionic *enolate*, but the neutral *enol*. Note also that in acid, as in base, carbonyl compounds that have α-hydrogens are in equilibrium with their enol forms (Fig. 17.14).

Acetaldehyde

Product from removal of D⁺ from oxygen

Protonated (D⁺) acetaldehyde

Enol

Product from removal of H⁺ from carbon

FIGURE 17.14 Removal of a proton from carbon (green) generates the neutral enol form. Removal of a deuteron from oxygen (red) regenerates starting material.

We will soon see that although enols are in equilibrium with the related keto forms, it is usually the keto forms that are favored. If this enol regenerates the carbonyl compound in D_3O^+, exchanged acetaldehyde will result. Notice that the steps in the enol → acetaldehyde reaction are simply the reverse of the acetaldehyde → enol reaction (Fig. 17.15).

Resonance-stabilized intermediate

protonate (D⁺) on carbon

deprotonate (D⁺) on oxygen

repeat exchange two times

Once-exchanged acetaldehyde

FIGURE 17.15 Reaction of the enol with D_3O^+ will generate exchanged acetaldehyde.

WORKED PROBLEM 17.5 Write a mechanism for the equilibration of acetaldehyde and its enol form in acid. This problem may seem mindless, and it certainly has little intellectual content at this point, but it is an important drill problem. Doing this problem is part of "reading organic chemistry with a pencil."

(continued)

ANSWER PROBLEM 17.5 *(continued)* This process must become part of your chemical vocabulary from now on. The carbonyl oxygen is protonated and the resulting cation deprotonated at carbon to give the enol.

Recently, it has become possible to generate the simplest enol, vinyl alcohol (hydroxyethylene), in such a fashion that the acid or base catalysts of the previous reactions are absent. As long as it is protected from acids or bases, the enol is stable and can be isolated and studied (Fig. 17.16).

Vinyl alcohol
(the simplest enol)

FIGURE 17.16 The parent of all enols, vinyl alcohol, is stable as long as it is protected from acid or base catalysts.

PROBLEM 17.6 It is impossible to protect vinyl alcohol from *all* acidic and basic materials, and so it eventually reverts to the carbonyl form, acetaldehyde. Explain what the unavoidable acid and base catalysts are.

The detailed structure of vinyl alcohol provides no surprises. Bond lengths and angles are not unusual (Fig. 17.17).

FIGURE 17.17 The structure of vinyl alcohol.

So far we have seen that α-hydrogens can be removed, and that carbonyl compounds are often in equilibrium with enol forms. The next question to address is the position

FIGURE 17.18 Where does the equilibrium between carbonyl compound and enol lie?

of the equilibrium between the carbonyl compound and its enol form. Where does the equilibrium lie, and how does it depend on structure (Fig. 17.18)?

Simple aldehydes and ketones are predominantly in their carbonyl forms. Figure 17.19 shows the equilibrium constants, K, for acetone and acetaldehyde. These numbers are typical for other simple aldehydes and ketones.

Acetaldehyde $K \sim 6 \times 10^{-7}$

Acetone $K \sim 5 \times 10^{-9}$

FIGURE 17.19 For simple aldehydes and ketones, it is the carbonyl form that is favored at equilibrium.

Neither simple ketones nor simple aldehydes are very highly enolized, although aldehydes generally exist somewhat more in their enol forms than ketones. The second methyl group stabilizes the carbon–oxygen double bond more than the carbon–carbon double bond (Fig. 17.20).

FIGURE 17.20 Acetone is less enolized than acetaldehyde because of the greater stabilization provided to the keto form by the second methyl group.

More complex carbonyl compounds can be much more strongly enolized. β-Dicarbonyl compounds (1,3-dicarbonyl compounds) exist largely in their enol forms, for example (Fig. 17.21).

FIGURE 17.21 β-Dicarbonyl compounds are more enolized than are monoketones.

PROBLEM 17.7 From a bond strength analysis (ΔH), decide whether the keto or enol form of acetone is more stable. Use the data from Table 8.2 (p. 355). These molecules both contain sp^2–sp^3 C–C bonds. Use a value of 101.4 kcal/mol for these bonds.

In the enol forms of these compounds, it is possible to form an intramolecular hydrogen bond, thus forcing the equilibrium position away from the diketo forms. There is also conjugation between the carbon–carbon double bond and the remaining carbonyl group in these enols (Fig. 17.22).

FIGURE 17.22 The formation of an intramolecular hydrogen bond, as well as conjugation between the carbon–carbon double bond and the carbonyl group, contributes to the increased stability of enols in β-dicarbonyl compounds.

The ultimate enol is phenol. It is estimated that the equilibrium constant for formation of the aromatic enol form from the nonaromatic ketone is greater than 10^{13} (Fig. 17.23).

1,3-Cyclohexadienone   **Phenol**

FIGURE 17.23 The equilibrium constant for enolization of cyclohexadienone is estimated to be greater than 10^{13}.

SUMMARY

Ketones and aldehydes bearing α hydrogens are in equilibrium with their enol forms, although for simple ketones and aldehydes the carbonyl forms are greatly favored. Equilibration with the enol form can be either acid- or base-catalyzed. The enol form can be favored in special cases.

17.3   REACTIONS OF ENOLS AND ENOLATES

Now we know something of the origins of enols and enolates, and we have seen that enols are in equilibrium with carbonyl compounds. Now it is time to move on to a look at reactions of these species.

17.3a Exchange Reactions. We saw in the previous sections that as long as an enol or enolate can be formed, hydrogens in the α-position can be exchanged for deuterium through treatment with deuterated acid or base (Fig. 17.24).

FIGURE 17.24 Exchange reactions of carbonyl compounds bearing α-hydrogens can be either acid or base catalyzed.

17.3b Racemization. Treatment of a simple optically active aldehyde or ketone with acid or base results in loss of optical activity, racemization, *as long as the stereogenic carbon is in the α-position to the carbon–oxygen double bond, and as long as there is an α-hydrogen* (Fig. 17.25).

FIGURE 17.25 Optically active carbonyl compounds are racemized in acid or base, as long as an α-hydrogen is present on the stereogenic carbon.

Once more, the mechanism depends on enol formation in acid, or enolate formation in base. In acid, an achiral, planar enol is formed. Reformation of the keto form
occurs by equal protonation from either side of the enol (Fig. 17.26).

FIGURE 17.26 Protonation of the planar enol must
result in formation of equal amounts of two enantiomers.
An optically active carbonyl compound is racemized on
enol formation.

In base, removal of the α-hydrogen generates an enolate anion that is itself achiral.
For overlap between the 2p orbitals on carbon and oxygen to be maximized, the enolate must be planar, and the planar structure is not chiral (Fig. 17.27).

FIGURE 17.27 Resonance-stabilization of the enolate depends on overlap of the 2p orbitals
on carbon and oxygen. Maximum overlap requires planarity, and the planar enolate is necessarily
achiral. Racemization has been accomplished at the enolate stage.

A racemic mixture is formed when the enolate reprotonates. Addition of a proton *must* take place with equal probability at the two equivalent faces ("top" = green; "bottom" = red) of the planar molecule (Fig. 17.28).

FIGURE 17.28 Reprotonation of the enolate from the two equivalent faces must give a pair of enantiomers.

17.3c Halogenation in the α-Position. Halogenation at the α-position is one of the relatively rare examples of a reaction that takes a different course in its acid- and base-catalyzed versions. As you will see, though, the two reactions are closely related.

Treatment of a ketone containing an α-hydrogen with halogen (X_2, X = I, Cl, or Br) in acid results in the replacement of one of the methyl hydrogens with a halogen atom (Fig. 17.29).

FIGURE 17.29 Treatment of a ketone containing an α-hydrogen with iodine, bromine, or chlorine in acid leads to α-halogenation.

The enol is formed first, and like most alkenes, reacts with iodine to displace iodide and form a resonance-stabilized cation (Fig. 17.30). The iodide ion deprotonates the intermediate cation generating the α-iodo carbonyl compound, and ending the reaction.

Resonance-stabilized intermediates

FIGURE 17.30 Under acidic conditions the enol is formed, and then reacts with iodine to give the resonance-stabilized cation. Deprotonation leads to the α-iodide.

Enol formation

Iodination

Earlier, we saw how deuterium replaces *all* α-hydrogens. Why doesn't the α-iodo compound react further to add more iodines? The way to answer a question such as this is to work through the mechanism of the hypothetical reaction and see if you can find a point at which it is disfavored. In this case, that point appears right away. It is enol formation itself that is slowed by the introduction of the first halogen. The introduced iodine inductively withdraws electrons and this disfavors the first step in enol formation, protonation of the carbonyl group (Fig. 17.31).

It is not easy to *predict* the extent of this sequence of enol formations and halogenations. It is not reasonable to expect you to anticipate that further enol formation would be sufficiently slowed so as to make the monoiodo compound the end product. On the other hand, it is not so difficult to *understand* this effect. Rationalization of results is much easier than predicting new reactions (or in this case, nonreactions).

Enol halogenation

The dipole in the C—I bond destabilizes the protonated carbonyl and makes it difficult to form

FIGURE 17.31 Protonation of the α-iodoketone is disfavored by the electron-withdrawing character of the halogen.

Under basic conditions matters are quite different. In fact, it is very difficult to stop the reaction at the monoiodo stage. The general result is that all α-hydrogens are replaced with halogen. For molecules with three α-hydrogens, methyl compounds, the reaction goes even further in what is called the **haloform reaction**.

In base, it is not the enol that is formed, but the enolate. This anion reacts in S$_N$2 fashion with iodine to generate the α-iodo carbonyl compound. Now, however, the electron-withdrawing inductive effect of iodine makes enolate formation easier, not harder, as for enols. The negative charge of the enolate is *stabilized* by the electron-withdrawing effect (Fig. 17.32).

FIGURE 17.32 The initially formed α-iodo carbonyl compound is a stronger acid than the carbonyl compound itself. The introduced iodine makes enolate formation easier.

The C–I dipole stabilizes the iodonated enolate

A second and third displacement reaction on iodine generates the α,α,α-triiodo carbonyl compound. Similar reactions are known for bromine and chlorine (Fig. 17.33).

FIGURE 17.33 Sequential enolate formations and iodinations lead to the α,α,α-triiodo carbonyl compound.

In NaOH/H_2O, these trihalocarbonyl compounds react further to give, after a final acidification, a molecule of a carboxylic acid and a molecule of a **haloform** (trihalomethane) (Fig. 17.34).

FIGURE 17.34 Trihalocarbonyl compounds react in base to give a molecule of a carboxylate anion and a haloform (trihalomethane).

Haloform formation is a new reaction whose mechanism is well within your powers to figure out. First, ask yourself what bonds must be broken and made in the reaction. Then ask what reactions you know that can make this happen. Almost always there will be a connection between the answers to these two questions. First of all, the formation of a carboxylic acid means that a new oxygen–carbon bond must be made. Similarly, the formation of a haloform, a one-carbon fragment, means that a carbon–carbon bond must have been broken. Those are the bond-making and bond-breaking requirements, and any mechanistic hypothesis must allow for them to take place in a reasonable way. It also seems obvious that the "CI_3" portion of iodoform (HCI_3) will come from the CI_3 present in the starting trihaloketone (Fig. 17.35).

FIGURE 17.35 The bond-making and bond-breaking requirements for this reaction.

Now what's possible? The easiest possible connection would be to make the oxygen–carbon bond at the same time that the carbon–carbon bond is broken. But this process requires an S_N2 displacement at an sp^2 hybridized carbon, and that is not a reaction for which there is any precedent (Fig. 17.36).

FIGURE 17.36 There is no precedent for this hypothetical S_N2 displacement of an sp^2 carbon.

PROBLEM 17.8 Explain why S_N2 displacement at an sp^2 hybridized atom is difficult.

What other reaction can happen between a carbonyl group and a base? Essentially all nucleophilic reagents add to carbonyl groups. Moreover, the addition reaction is strongly favored by a halogen substitution (Chapter 16, p. 840)! So, the addition of hydroxide to the carbonyl looks like a reaction that is almost certain to occur. Once the addition reaction has taken place, there is an opportunity to generate the acid and the haloform if the triiodomethyl anion can be lost as the carbonyl group re-forms. A proton transfer completes the reaction (Fig. 17.37).

FIGURE 17.37 Addition of hydroxide to the carbonyl group leads to a tetrahedral intermediate that can lose triiodomethide anion to generate the carboxylic acid. Transfer of a proton completes the reaction.

Is this mechanism reasonable? Is the triiodomethyl anion a good enough leaving group to make this step a sensible one? The iodines are strongly electron withdrawing, and that property will stabilize the anion. As a check we might look up the pK_a of iodoform to see how easily the molecule is deprotonated to form the anion. The value is about 14, and so iodoform is a relatively strong acid, at least compared to water ($pK_a = 15.7$). So the loss of $^-CI_3$ in Figure 17.37 does look reasonable (Fig. 17.38).

FIGURE 17.38 The pK_a of iodoform is about 14. Iodoform is a relatively strong acid, and the loss of $^-CI_3$ as a leaving group is a reasonable step.

$$I_3C-H \quad :\ddot{O}H \quad \rightleftharpoons \quad I_3C:^- \quad + \quad H-\ddot{O}H$$

$$pK_a \sim 14 \qquad\qquad\qquad pK_a = 15.7$$

Deprotonation of the carboxylic acid, $pK_a \sim 4.5$, gives iodoform and the carboxylate anion. A final acidification step gives us back the carboxylic acid. In fact, this reaction serves as a test for methyl ketones. Formation of iodoform, a yellow solid, is a positive test for a molecule containing a methyl group attached to a carbonyl carbon.

PROBLEM 17.9 If we measure the rates of three reactions of the ketone shown below, exchange of the α-hydrogen for deuterium in D_2O/DO^-, racemization in H_2O/HO^-, and α-bromination using $Br_2/H_2O/HO^-$, we find that they are identical. How can the rates of three such different reactions be the same? Explain, using an Energy versus Reaction progress diagram.

D₂O/NaOD
dioxan
35 °C

A
Deuterium exchange

Br₂
NaOH/H₂O

B
α-Bromination

H₂O/NaOH
35 °C

C
Racemization

* Means optically active

SUMMARY

We have now developed three reactions taking place at the α-position of carbonyl compounds. We will use these as prototypes on which to base further, more complicated processes. The unifying theme is formation of an enolate ion (in base) or an enol (in acid) that can act as a nucleophile. What other reactions might such a nucleophile undergo?

17.3d Alkylation Reactions. In the previous sections, we have seen several examples of enolate anions acting as bases and nucleophiles. A logical extension would be to attempt to use enolates as nucleophiles in S_N2 displacements. If we could do this reaction, we would have a way to alkylate the α-position of carbonyl compounds, and a new and most useful carbon–carbon bond-forming reaction would result (Fig. 17.39).

Enolate   The methylated
                           enolate, the
                           product ketone

FIGURE 17.39 If the enolate could act as a nucleophile in the S_N2 reaction, we might have a way of alkylating at the α-position.

We can see some potential difficulties and limitations right away. Alkylation of an enolate is an S_N2 reaction, and that means we could not use a tertiary halide, a species too hindered for participation in the S_N2 reaction.

The resonance formulation of enolate anions clearly shows that the negative charge is shared between an oxygen and a carbon atom. At which atom will alkylation be faster? If there is little or no selectivity in the alkylation reaction, it will surely be of limited use (Fig. 17.40).

C-Alkylation

O-Alkylation

FIGURE 17.40 In principle, alkylation of the enolate could take place at either carbon or oxygen.

It turns out that most enolates are better nucleophiles at carbon. So, alkylation takes place faster at carbon than at oxygen (Fig. 17.41).

C-Alkylation   + :I:⁻   *not*   O-Alkylation

FIGURE 17.41 In practice, alkylation generally takes place at carbon.

Even though alkylation at oxygen is not common, there are still other problems, and this alkylation reaction of simple ketones is not very useful in practice. Consider 2-methylcyclohexanone as an example. Both α-positions are active, and two enolates will be produced (Fig. 17.42).

Both α-positions are active

Two enolates

FIGURE 17.42 For many ketones there are at least two possible enolates, and therefore mixtures are obtained in the alkylation reaction.

Moreover, the once-alkylated compounds can undergo further alkylation. Reactions of the initial products compete with the desired single alkylation reaction. Mixtures of products are generally formed, and that is not a very useful situation (Fig. 17.43).

FIGURE 17.43 For ketones, there can often be more than one alkylation. Mixtures of products can easily result.

Aldehydes, which can only contain a single kind of α-hydrogen, are also problematic in this reaction. Not only do we have to deal with this problem of overalkylation, but several side reactions (see Section 17.4) tend to dominate the reaction.

Only if we restrict ourselves to carbonyl compounds containing a single α-hydrogen is some measure of specificity found, and we are now very far indeed from the general alkylation reaction we had hoped for. Strongly basic, but poorly nucleophilic species are used to form the enolate. Sodium hydride is effective, as is **lithium diisopropylamide (LDA)**. Both of these strong bases are only weakly nucleophilic, and addition to the carbonyl group is minimized in favor of deprotonation (Fig. 17.44).

FIGURE 17.44 Strong bases such as LDA or NaH are effective at forming enolates.

Note again the use in Figure 17.44 of the shorthand notation for resonance forms. Instead of always drawing out each important resonance form, one summary form is drawn. The positions sharing the charge or electron are indicated with parentheses, (+), (−), or (•).

CONVENTION ALERT

PROBLEM 17.10 Explain why LDA is a poor nucleophile.

Perhaps not surprisingly, a way around some of these problems has been found, and it uses **enamines** (Chapter 16, p. 852) as substitutes for carbonyl compounds. Enamines can be formed by the addition of secondary amines (not primary or tertiary amines) to carbonyl compounds. Two favorite secondary amines are pyrrolidine and morpholine, shown in Figure 17.45.

FIGURE 17.45 Enamine formation with the secondary amines pyrrolidine and morpholine.

PROBLEM 17.11 Why can't primary or tertiary amines be used to make enamines (Chapter 16, p. 852)?

Enamines are electron-rich and nucleophilic and will attack S_N2-active halides (no tertiary halides!) to form immonium (also called imminium) ions. The reaction is ended by adding water to hydrolyze the immonium ions to the corresponding ketones (Fig. 17.46).

FIGURE 17.46 Alkylation of an enamine initially gives an immonium ion. The immonium ion can be hydrolyzed to give the alkylated carbonyl compound.

PROBLEM 17.12 Write the mechanism for the hydrolysis of the immonium ion to the ketone. *Hint*: You have seen this reaction before, in reverse, in the mechanism for acid-catalyzed enamine formation (Chapter 16, pp. 852–3).

SUMMARY

Figure 17.47 shows the overall procedure: The ketone is first converted into the enamine, the enamine is alkylated, and then the ketone is regenerated by hydrolysis. There are still some problems, however. It is not easy to use alkylations of enamines when more than one possible enamine can be formed, and there are occasionally difficulties with overalkylation or substitution on nitrogen.

FIGURE 17.47 An overview of the three-step alkylation process using an enamine.

17.4   CONDENSATION REACTIONS OF CARBONYL COMPOUNDS: THE ALDOL CONDENSATION

So far, we have seen the reactions of enolates and enols with a variety of Brønsted and Lewis acids, for example: D_2O, H_2O, Br_2, Cl_2, I_2, and alkyl halides. What happens if these acids are absent, or, as is the case for water, simply regenerate starting material and do not result in new products? Suppose, for example, we treat acetaldehyde with KOH/H_2O. First of all, the answer to the question of what happens when a base and a carbonyl compound are combined is almost never, "nothing." Nucleophiles add to carbonyl groups, and hydroxide will surely initiate hydrate formation. The hydrate is usually not isolable, but it is formed in equilibrium with the aldehyde. Moreover, enolate formation is also possible and the enolate, as well as the hydrate, will be in equilibrium with the starting aldehyde. Be very careful about "nothing happens" answers. They are almost always wrong (Fig. 17.48).

FIGURE 17.48 Two reactions of acetaldehyde with hydroxide ion: addition (hydrate formation) and enolate formation.

Although hydrate formation does not lead directly to an isolable product, formation of the enolate does. Over the last few pages we have seen several examples of the reactions of nucleophilic enolates with Lewis and Brønsted acids. What acids are present in this solution? Water is surely one, but reaction with water simply generates the enol in equilibrium with acetaldehyde (Fig. 17.49).

FIGURE 17.49 The enolate can be reprotonated at either carbon or oxygen. Reaction at oxygen usually dominates, but the resulting enol equilibrates with the more stable carbonyl form.

There is another acid present, the carbonyl group of acetaldehyde. Nucleophiles add to carbonyl groups, and the enolate anion is surely a nucleophile. There is no essential difference between the addition of hydroxide to the carbonyl group of acetaldehyde and the addition of the enolate anion (Fig. 17.50).

Aldol condensation

FIGURE 17.50 Addition of hydroxide and the enolate anion to the carbonyl group are simply two examples of the addition reaction of nucleophiles to carbonyl groups.

Nor is there much structural difference between the hydrate formed by protonation of the hydroxide addition product, and the molecule known as "aldol" formed by protonation of the enolate addition product. Note that hydroxide ion, the catalyst, is regenerated in the final step of hydrate formation or of aldol formation (Fig. 17.51).

FIGURE 17.51 Protonation of these intermediates gives the hydrate, or, in the enolate case, a compound known as "aldol", a β-hydroxy aldehyde.

Aldol
a β-hydroxy aldehyde

But the product itself is surely more complex than the relatively simple hydrate! The conversion of an aldehyde or ketone into a β-hydroxy carbonyl compound is called the **aldol condensation*** after the product of the reaction of acetaldehyde, and is quite general for carbonyl compounds with α-hydrogens.

As usual, this base-catalyzed reaction has its acid-catalyzed counterpart. For the acid-catalyzed reaction the catalyst is not hydroxide, but H_3O^+, and the active ingredient is not the enolate anion, but the enol itself. The first step in the reaction is acid-catalyzed enol formation (Fig. 17.52).

FIGURE 17.52 The acid-catalyzed aldol condensation begins with enol formation.

Enol

The enol, though nucleophilic, is not nearly as strong a nucleophile as hydroxide. However, the Lewis acid present is the protonated carbonyl, and it is a far stronger Lewis acid than the carbonyl group itself. The reaction between the enol and the protonated carbonyl group is analogous to the reaction of the enol with

---

*Credit for discovery of the aldol condensation generally goes to Charles Adolphe Wurtz (1817–1884) who coined the word "aldol" in 1872. However, equal billing at least should also go to Alexandr Borodin, the Russian composer–physician–chemist (1833–1887). Borodin was the illegitimate son of Prince Gedianov, and actually spent his formative years as his father's serf. Happily, he was given his freedom and became an eminent chemist who is far better known for the somewhat schmaltzy products of his avocation, music, than for his excellent chemistry.

H_3O^+. Once again, the final step of the reaction regenerates the catalyst, this time H_3O^+ (Fig. 17.53).

Protonation/deprotonation of the enol to regenerate acetaldehyde

Reaction of the enol with the protonated carbonyl to give aldol

Aldol

FIGURE 17.53 Two reactions of the weakly nucleophilic enol with Lewis acids. In the first case, it is protonated to regenerate acetaldehyde; in the second the enol adds to the strongly Lewis acidic protonated carbonyl group to give aldol.

The acid- and base-catalyzed aldol condensations of acetaldehyde give the same product, the β-hydroxy carbonyl compound called aldol. The aldol condensation is a general synthesis of β-hydroxy ketones and aldehydes.

WORKED PROBLEM 17.13 Write the products of the aldol condensations of the following compounds. Write both acid- and base-catalyzed mechanisms for reactions of (b).

ANSWER (b) As usual, in base the enolate is first formed and then adds to the carbonyl group of another molecule. A final protonation generates the condensation product and regenerates the catalyst, here hydroxide.

(continued)

ANSWER *(continued)*

Watch out! In this figure, arrows emanate from one of a pair of resonance forms, which is done for simplicity's sake only. Write the arrow formalism using the other resonance form.

In acid, the protonated carbonyl group reacts with the enol form of the ketone to give a resonance-stabilized intermediate. Deprotonation gives the product and regenerates the acid catalyst, H_3O^+.

In acid, a second reaction, the dehydration of the β-hydroxy carbonyl compound to the α,β-unsaturated aldehyde or ketone is very common, and it is generally the dehydrated products that are isolated (Fig. 17.54).

Aldol, a β-hydroxy aldehyde

An α,β-unsaturated aldehyde

FIGURE 17.54 The acid-catalyzed dehydration of aldol.

Hydroxide is a poor leaving group. Therefore, dehydration is more difficult in base and the β-hydroxy compounds can often be isolated. Even under basic conditions dehydration can be achieved at relatively high temperature. The mechanism is not a simple E2 reaction, however. Removal of the α-hydrogen is relatively easy, as this process gives a resonance-stabilized enolate anion. In a second step, hydroxide is lost and the α,β-unsaturated compound is formed. This kind of elimination mechanism, called the E1cB reaction, was the subject of discussion in Chapter 7 (p. 324) (Fig. 17.55).

Resonance-stabilized enolate

FIGURE 17.55 The base-catalyzed elimination of water from aldol. The reaction mechanism is E1cB.

PROBLEM 17.14 What are the products of dehydration of the condensation products of Problem 17.13?

PROBLEM 17.15 Figure 17.54 shows protonation of the hydroxyl oxygen, not the carbonyl. That picture is correct as the pK_a of a protonated alcohol is about -2, and the pK_a of a protonated aldehyde is about -10. First explain the relevance of the pK_a data, and then explain why the hydroxyl oxygen is protonated in preference to the aldehyde. *Hint*: Think hybridization.

Like many reactions, the aldol condensation is a series of equilibria. It is not always clear what the slow step in the aldol condensation, the rate-determining step, will be. For example, consider the reaction of acetaldehyde in D_2O/DO^-. Deuterium incorporation at carbon in the aldol formed is a function of acetaldehyde concentration. At high acetaldehyde concentration the aldol product initially formed is not deuterated at carbon, but at low acetaldehyde concentration it is. The reason for the concentration dependence is that the rates of the first two steps of the reaction are similar. Once the enolate is formed it can either revert to acetaldehyde, which will result in exchange in D_2O, or go on to aldol product in a bimolecular reaction with a molecule of acetaldehyde (Fig. 17.56).

FIGURE 17.56 A high concentration of acetaldehyde results in a relatively rapid aldol condensation and little or no exchange at carbon. If, however, the concentration of acetaldehyde is low, the bimolecular condensation reaction will be slowed and the enolate can revert to acetaldehyde. In D_2O, this reaction leads to exchange of hydrogens attached to carbon. The deuterium can appear in either or both of the two positions shown as (D).

If the concentration of acetaldehyde is high, the rate of the bimolecular reaction (reaction 1) is faster ($v = k$[enolate][acetaldehyde]), and acetaldehyde is not reformed from the enolate. Under these conditions, it is the formation of the enolate that is the slow step in the reaction, the rate-determining step (Chapter 8, p. 368). If the concentration of acetaldehyde is low, the enolate is reconverted to acetaldehyde (reaction 2), and in D_2O results in exchange. When exchanged acetaldehyde goes on to form aldol, the product also contains deuterium attached to carbon. At low acetaldehyde concentrations it is the condensation step itself that is the slower one, and therefore the rate-determining step.

Ketones can undergo the aldol condensation too, and the mechanism is similar to that for the aldol condensation of aldehydes. We will use acetone as an example. In base, the enolate is formed first, and adds to the Lewis acidic carbonyl compound. Protonation by water yields a molecule once known as diacetone alcohol, 4-hydroxy-4-methyl-2-pentanone (Fig. 17.57).

Diacetone alcohol
(4-hydroxy-4-methyl-2-pentanone)

FIGURE 17.57 The base-catalyzed aldol condensation of acetone.

In acid, the enol is the intermediate, and it adds to the very strong Lewis acid, the protonated carbonyl group, to give initially the same product as the base-catalyzed reaction, diacetone alcohol. Dehydration generally follows to give the α,β-unsaturated ketone, 4-methyl-3-penten-2-one, which is still often called mesityl oxide (Fig. 17.58).

4-Methyl-3-penten-2-one
(mesityl oxide)

FIGURE 17.58 The acid-catalyzed aldol condensation of acetone. The first product, diacetone alcohol, is generally dehydrated in acid to give mesityl oxide, 4-methyl-3-penten-2-one.

We have been at pains to point out the similarities between these aldol condensations and other addition reactions of carbonyl compounds. Like simple additions, such as hydration, the aldol condensations are series of equilibria. They are affected by structural factors in the same way as the simple reactions. Product formation is unfavorable in base-catalyzed aldol reactions of ketones, just as it is in hydration (Fig. 17.59). In acid, dehydration generally occurs and formation of the α,β-unsaturated ketone pulls the equilibrium to the right.

FIGURE 17.59 In the hydration of ketones, the hydrate is usually not favored at equilibrium. Similarly, in the aldol condensation of ketones, the product molecule is not usually favored.

If so, how can the aldol condensation of ketones be useful? If equilibrium favors starting material, how can the product be isolated? Ingenious ways have been found to thwart the dictates of thermodynamics. In one, a spectacularly clever piece of apparatus called a Soxhlet extractor is used. The ketone is boiled in a flask separated from the catalyst, usually barium hydroxide, $Ba(OH)_2$, which is insoluble in organic materials. The ketone distills to the condenser, off of which it drips to come in contact with the catalyst (stage 1, Fig. 17.60). Now the aldol condensation takes place to give an equilibrium mixture of starting ketone (favored) and the product of aldol condensation (unfavored) (stage 2). As the starting ketone continues to boil, and the level of the liquid in the trap rises, the whole solution eventually is siphoned from the chamber into the flask that previously contained only ketone (stage 3). There is still no catalyst in the flask, and so the aldol product *cannot revert to starting material.* The mixture continues to boil, the remaining ketone, which has a much lower boiling point than the product, continues to come in contact with the catalyst and slowly the concentration of product in the flask increases to a useful amount (stage 4) (Fig. 17.60).

FIGURE 17.60 The operation of a Soxhlet extractor.

All β-hydroxy carbonyl compounds are potential products of an aldol condensation. Whenever you see one, your thoughts about synthesis must turn first to the aldol condensation. The same is true for the dehydration products, the α,β-unsaturated carbonyl compounds. It is important to be able to go quickly to the new bond, to find the one formed during the condensation, and to be able to dissect the molecule into its two halves. In doing this operation, you are merely carrying out a retrosynthetic analysis and following the mechanism backward (Fig. 17.61).

FIGURE 17.61 A retrosynthetic analysis for the product of an aldol condensation followed by dehydration.

PROBLEM 17.16 Write mechanisms for the acid- and base-catalyzed reverse aldol condensation of diacetone alcohol.

Diacetone alcohol

WORKED PROBLEM 17.17 Perform retrosynthetic analyses on the following three molecules:

(a)

(b)

(c)

ANSWER (b) In any aldol, or aldol-like condensation, the carbon–carbon π bond is formed through an elimination reaction (Fig. 17.58, p. 928). So, the first part of our retrosynthetic analysis recognizes that the precursor to the final product is β-hydroxy ketone **A**.

A

(continued)

ANSWER *(continued)*

Now, where does **A** come from? All β-hydroxy carbonyl compounds are aldol products, so the σ bond joining the two rings must come from the following aldol:

A

SUMMARY

Base- and acid-catalyzed condensation reactions of aldehydes generally lead to useful amounts of aldol product, the β-keto alcohol, or the product of dehydration, the α,β-unsaturated carbonyl compound. Aldol condensations of ketones are less useful because equilibrium does not usually favor product. Special techniques can be used (Soxhlet extractor) to thwart thermodynamics. If dehydration to the α,β-unsaturated carbonyl compound occurs, as it often does in acid-catalyzed aldols, the reaction may be practically successful (Fig. 17.62).

FIGURE 17.62 Although equilibrium generally favors the condensation product in the aldol reaction of aldehydes, the reaction of ketones generally favors starting material.

17.5 REACTIONS RELATED TO THE ALDOL CONDENSATION

In the next few pages, we will use what we know of the simple aldol condensation to understand some related reactions. We will start with reactions that are very little different from simple aldols, and slowly increase the complexity. The connection to the simple, prototypal acid- and base-catalyzed aldols will always be maintained.

17.5a Intramolecular Aldol Condensations. Like most intermolecular reactions, the aldol condensation has an intramolecular version. If a molecule contains both an enolizable hydrogen and a receptor carbonyl group, an intramolecular aldol condensation will be theoretically possible. Particularly favorable are intramolecular condensations that form the relatively strain-free five- and six-membered rings. The products are still β-hydroxy ketones (or the dehydration products, α,β-unsaturated ketones), and the mechanism of the reaction remains the same (Fig. 17.63).

THE GENERAL BASE-CATALYZED CASE

SPECIFIC BASE- AND ACID-CATALYZED EXAMPLES

FIGURE 17.63 The mechanism for intramolecular aldol condensations.

PROBLEM 17.18 Write the mechanisms for the acid- and base-catalyzed aldol condensations of the molecule at the bottom of Figure 17.63.

PROBLEM 17.19 Write the mechanisms for the acid- and base-catalyzed reverse aldol condensations for the following molecule:

17.5b Crossed Aldol Condensations. In practice, not all β-hydroxy ketones can be formed efficiently by aldol condensations. Suppose we were set the task of synthesizing the molecule in Figure 17.64. We recognize it as a potential aldol product through the dissection shown in the figure, which suggests that an aldol condensation of diethyl ketone and acetone should be a good route to the molecule. An aldol condensation between two different carbonyl compounds is called a **crossed aldol condensation**.

FIGURE 17.64 A retrosynthetic analysis suggests that a condensation between diethyl ketone and acetone should give the β-hydroxy ketone shown.

However, a little thought about the mechanism reveals potential problems. Two enolates can be formed and each enolate has two carbonyl groups to attack. Thus, four products, **A**, **B**, **C**, and **D**, are possible (more if any of the β-hydroxy ketones are dehydrated to α,β-unsaturated ketones), and all are likely to be formed in comparable yield. The enolate of diethyl ketone can add to the carbonyl of acetone or diethyl ketone to give **A** and **B**. Similarly, the enolate of acetone gives **C** and **D** by reaction with the two carbonyl compounds (Fig. 17.65).

FIGURE 17.65 When this synthetic route is attempted, four β-hydroxy ketones, **A**, **B**, **C**, and **D**, are likely to be produced.

There are some simple ways to limit the possibilities in a crossed aldol reaction. If one partner has no α-hydrogens, it can function only as an acceptor, and not as the nucleophilic enolate partner in the reaction. For example, we might hope for better luck in a reaction pairing *tert*-butyl methyl ketone and benzaldehyde. In principle, however, there are still two possibilities, as the enolate can add to either the ketone or aldehyde to generate two different β-hydroxy ketones (Fig. 17.66).

FIGURE 17.66 The crossed aldol reaction of *tert*-butyl methyl ketone and benzaldehyde can give only two products. Benzaldehyde has no α-hydrogens and cannot form an enolate.

When this reaction is run, however, only one product is formed in substantial yield (Fig. 17.67).

FIGURE 17.67 In fact, there is only one major product in the aldol condensation of *tert*-butyl methyl ketone and benzaldehyde.

PROBLEM 17.20 There is a reaction between benzaldehyde and base. What is it, and why does it not interfere with the aldol condensation?

There are two reasons that this crossed aldol is successful. First, the rate of addition of the enolate to the carbonyl group of benzaldehyde is much greater than that of addition to *tert*-butyl methyl ketone. The carbonyl groups of aldehydes are more reactive than those of ketones. Second, the equilibrium constant for the addition to an aldehyde carbonyl is more favorable than that for addition to a ketone carbonyl (Chapter 16, p. 837; Fig. 17.68).

This reaction is important enough to have been given its own name, the **Claisen–Schmidt condensation** (Ludwig Claisen, 1851–1930).

Mixed aldol condensation

FIGURE 17.68 The carbonyl group of benzaldehyde is more reactive than that of *tert*-butyl methyl ketone, and equilibrium favors the product in the reaction with benzaldehyde, but not for the reaction with *tert*-butyl methyl ketone.

Earlier in this chapter we saw LDA, a base that was effective in producing enolates without the complications of addition reactions to the carbonyl group. These enolates can be used to do crossed aldol reactions too. Lithium diisopropylamide is first used to create a large amount of enolate to which a carbonyl-containing compound is added. Addition of the enolate followed by protonation gives the aldol product. When there is a choice, LDA forms the more accessible, less substituted enolate. Lithium diisopropylamide is a large, sterically encumbered base and removes a proton from the sterically less hindered position (Fig. 17.69).

FIGURE 17.69 A crossed aldol condensation with LDA as base.

(65%)

17.5c Knoevenagel Condensations and Related Reactions. Condensation

reactions are not limited to simple aldehydes and ketones. Any compound that can yield an anion can attack a carbonyl group to give a condensed product. Diketones and similar compounds are quite strong acids (pK_a = 5–13, Table 17.2) and readily form anions when treated with bases.

TABLE 17.2   Some pK_a Values for Some Dicarbonyl Compounds

Compound	Conjugate Base (enolate)	pK_a
		13.3
		11
		10.7
		8.9
		8.5
		5

PROBLEM 17.21 Draw resonance forms for the anions of the first two entries of Table 17.2.

Bases, often amines, can generate significant amounts of enolates from the compounds in Table 17.2, and similar molecules. In the presence of an acceptor aldehyde, condensation occurs to give, ultimately, the α,β-unsaturated diester shown in Figure 17.70. This reaction is called the **Knoevenagel condensation** after Emil Knoevenagel (1865–1921).

THE GENERAL CASE

(R, R are anion-stabilizing groups)

(dehydration may or may not occur)

SPECIFIC EXAMPLES

FIGURE 17.70 The Knoevenagel condensation.

Many other reactions, some with names, many without, are conceptually similar, in that they involve the addition of a complex anion to a carbonyl group. But, in these other reactions the sources of the nucleophilic anionic partner are not always aldehydes or ketones. For example, cyclopentadiene ($pK_a = 15$) is a strong enough acid to give an anion when treated with a base, and the cyclopentadienide ion (Chapter 13, p. 636) can add to aldehydes or ketones. The end result is a synthesis of the beautifully colored compounds known as **fulvenes** (Fig. 17.71).

6,6-Dimethylfulvene www
(75%)

FIGURE 17.71 A condensation reaction leading to fulvenes.

So now your analytical task is more complicated. It is quite easy to learn to dissect β-hydroxy ketones or α,β-unsaturated ketones into their components, but how is one to deal with the less obvious compounds like fulvenes? There is no easy answer to this question. A double bond is always the potential result of a condensation reaction followed by a dehydration. You must learn to look on them that way, and to evaluate the pieces thus created as parts of a possible synthesis (Fig. 17.72).

R must be an
anion-stabilizing group

FIGURE 17.72 Any double bond is the formal result of a condensation reaction if followed by dehydration.

The following problem gives some practice at devising syntheses of molecules formed by condensation reactions.

PROBLEM 17.22 Propose syntheses for the following molecules:

(a)

Ph   CN

Ph   COOCH$_2$CH$_3$

(b)

(CH$_3$)$_3$C   COOCH$_2$CH$_3$

H   COOCH$_2$CH$_3$

(c)

Ph

H

Michael reaction

17.5d More Related Condensations: The Michael Reaction. We now have

a synthetic route to α,β-unsaturated carbonyl compounds, the aldol (and related) condensation reactions. There is some especially important chemistry associated with these compounds, called the **Michael reaction**, after its discoverer, the American chemist Arthur Michael (1853–1942).

Ordinary carbon–carbon double bonds are unreactive toward nucleophiles. In particular, addition is not observed, because it would require formation of an unstabilized anion, the conjugate base of an alkane. Alkanes are extraordinarily weak acids. Their pK_a values are difficult to measure, but numbers as high as 70 have been suggested (Fig. 17.73). In answering problems, an unstabilized carbanion is the functional equivalent of a methyl or primary carbocation. It is a stop sign: A signal that you are almost certainly wrong!

FIGURE 17.73 Addition of a base such as alkoxide to a simple alkene would yield an unstabilized anion. A measure of the difficulty of this reaction can be gained by examining the acidity of the related hydrocarbon. Such species are extraordinarily weak acids and their pK_a values are very difficult to determine.

By contrast, many nucleophiles add easily to the double bonds of α,β-unsaturated carbonyl compounds. Notice the resonance stabilization of the resulting enolate anion (Fig. 17.74).

FIGURE 17.74
Additions to
α,β-unsaturated
carbonyl compounds
are common.

An α,β-unsaturated
carbonyl compound

An enolate
(resonance stabilized)

The reason is the initial formation, not of a bare, unstabilized carbanion as was the case in Figure 17.73, but of a nicely resonance-stabilized species. The carbonyl group stabilizes any α-anion, no matter how it is formed. The Michael reaction is another way of generating an enolate.

When the species adding to the α,β-unsaturated compound is itself an enolate, the reaction is properly called the Michael reaction, but more simple related additions are often also lumped under this heading (Fig. 17.75).

FIGURE 17.75 Two Michael additions.

Enolate

A real Michael reaction

A nucleophile

Still usually called a Michael reaction

There is also an acid-catalyzed version of the Michael reaction, in which 1,4-addition first occurs to the α,β-unsaturated system. The carbonyl group is initially protonated, and then the carbon–carbon double bond is attacked by a nucleophile to give the enol (Fig. 17.76).

The enol equilibrates with the more stable keto compound to produce the final product of addition.

THE GENERAL CASE

FIGURE 17.76 An acid-catalyzed Michael reaction.

Enol

Ketone

A SPECIFIC EXAMPLE

PROBLEM 17.23 Write a mechanism for the following reaction:

These α,β-unsaturated compounds present a general problem. There are too many reactions possible. How, for example, can we decide whether a nucleophile will add to the carbonyl group or the double bond? This question captures in microcosm the difficulties presented by condensation problems in particular, and of the chemistry of polyfunctional compounds in general. How is it possible to decide which of a variety of reactions is the best one (either in energy terms or in the more practical sense of finding the route to the end of an exam problem)?

At least the specific question about the Michael addition has a rational answer. First, notice that the product of Michael addition is likely to be more stable than the product of addition to the carbonyl group. The product of Michael addition retains the strong carbon–oxygen double bond, and the simple addition product does not (Fig. 17.77).

FIGURE 17.77 The two possible reactions of a nucleophile (Nu:⁻) with an α,β-unsaturated carbonyl compound. Michael addition preserves the carbonyl group and is usually favored thermodynamically.

But this is a thermodynamic argument, and as we have often seen, the site of re-activity is many times determined by kinetics, not the thermodynamics of the overall reaction. However, most nucleophiles add *reversibly* to the carbonyl group, and this will be as true for α,β-unsaturated carbonyls as for others. So even if addition to the carbonyl group occurs first, for most nucleophiles the thermodynamic advantage of Michael addition should ultimately win out. This notion suggests that strong nucleo-philes such as hydride (H:⁻) and alkyllithium compounds, which surely add *irre-versibly*, might be found to give the products of attack at the carbonyl group, and such is the case'(Fig. 17.78).

FIGURE 17.78 Two irreversible addition reactions to the carbonyl group of an α,β-unsaturated carbonyl compound.

There are exceptions to the general rule that weak bases add to the double bond and strong bases add to the carbonyl group. Cuprates, for example, give exclusively Michael addition, and Grignard reagents generally give mixtures of 1,2- and 1,4-addition unless they are copper catalyzed, in which case Michael addition is again the rule (Fig. 17.79).

Cuprate additions

FIGURE 17.79 Cuprates (and often Grignard reagents) add in Michael fashion to α,β-unsaturated carbonyl compounds.
Bu = $CH_3CH_2CH_2CH_2$

The Michael reaction is critical to the operation of a rather promising anticancer drug, calicheamicin (Fig. 17.80). In the first step in its operation, the trisulfide bond is broken in **A**. The nucleophilic sulfide then adds in Michael fashion to give the enolate (**B**). This addition changes the shape of the molecule, bringing the ends of the two acetylenes closer together. A cyclization occurs to give a diradical (**C**), and this diradical abstracts hydrogen from the cancer cell's DNA, killing it. Calicheamicin depends for its action on the change of shape. Before the Michael reaction, the ends of the two acetylenes are too far apart to cyclize. They are freed to do so only when the sulfur adds to the α,β-unsaturated carbonyl.

FIGURE 17.80 The operation of the antitumor agent calicheamicin.

SUMMARY

In this section, we have seen several variations on the aldol theme (intramolecular aldol, crossed aldol, Knoevenagel condensation, and the Michael reaction). In all of these reactions, a single nucleophile adds to a single Lewis acid. The reactions are fundamentally the same, the variations come mainly in the structural differences in the reacting partners. Now we move to a more exciting subject: condensation reactions in sequence.

17.6   PROBLEM SOLVING: CONDENSATION REACTIONS IN COMBINATION; MAGID'S SECOND RULE

Alas, neither Nature nor a typical chemistry teacher shows much mercy, and one is more likely than not to find condensation reactions in combination. None of these reactions is particularly difficult to understand when encountered by itself. Similarly, the synthetic consequences of the reactions are relatively easy to grasp when they are encountered one at a time, but are by no means simple when combinations of these reactions are involved. Condensation problems, occasionally called "Fun in Base" problems (a term that seems to encompass reactions run in acid as well as base; there are no "Fun in Acid" problems) can be very hard. The best way to become proficient is to work lots of them, but there are also some practical hints that may help.

One common sequence is a combination of the Michael and aldol condensations. Indeed, there is a synthetic procedure for constructing six-membered rings, invented by Sir Robert Robinson (1886–1975) and called the **Robinson annulation**, that uses exactly that combination. In it, a ketone is treated with methyl vinyl ketone in a base-catalyzed reaction. A Michael addition is followed by an intramolecular aldol condensation to give the new six-membered ring (Fig. 17.81).

FIGURE 17.81 The Robinson annulation.

(continued)

A SPECIFIC EXAMPLE OF THE ROBINSON ANNULATION

(52%)

(85%)

FIGURE 17.81 *(continued)*

PROBLEM 17.24 Although this is the way the Robinson annulation is always described mechanistically, there is another (harder) way to write the reaction. Write a mechanism for the general reaction in Figure 17.81 that involves doing the aldol condensation first.

Problem 17.24 brings up a serious practical question. You may well be able to analyze a problem by figuring out the different possible reactions, but choosing among them is more difficult. How do you find the easiest route (there are very often several) to the product and the solution to the problem? Clearly, experience is invaluable and you are by definition short on that commodity. In such a situation, it is wise to be receptive to good advice, and here is a practical hint, from one of the best problem solvers I know, Ronald M. Magid (b. 1938) of the University of Tennessee, in the form of **Magid's Second Rule,*** which says: "Always try the Michael first."

Magid's Second Rule is more than a wisecrack. In fact, there are sound reasons for doing the Michael reaction first. Aldol condensation products of ketones are not generally favored at equilibrium, for example, and doing the Michael reaction makes the subsequent aldol condensation intramolecular, and thus more favorable.

There is no substitute for practice, though, and Problem 17.25 offers some good opportunities to practice solving Fun in Base problems.

PROBLEM 17.25 Provide mechanisms for the reactions shown below.

(a)

(continued)

---

*Magid's First Rule is archaic, and we will shortly encounter Magid's Third Rule.

PROBLEM 17.25 *(continued)*

(b)

(c)

(d)

(e)

(f) *Hint*: There is a 10-membered ring involved.

17.7   SPECIAL TOPIC: ALKYLATION OF DITHIANES

Here is a clever way to alkylate carbonyl compounds at the carbonyl carbon atom. Thioacetals, also called **dithianes**, can be made from carbonyl compounds and the sulfur counterparts of glycols (Fig. 17.82).

FIGURE 17.82 Formation af a dithiane (a thioacetal).

Dithianes are rather strong Brønsted acids and can be easily deprotonated with alkyl-lithium reagents (Fig. 17.83).

1,3-Dithiane
(1,3-dithiacyclohexane)
$pK_a = 31.1$

$Bu = CH_2CH_2CH_2CH_3$

FIGURE 17.83 Deprotonation of 1,3-dithiacyclohexane with butyllithium.

PROBLEM 17.26 Explain how the sulfur atoms operate to increase the acidity of the adjacent hydrogens in 1,3-dithiane (1,3-dithiacyclohexane).

The resulting anion is a powerful nucleophile and can undergo both displacement reactions (subject, of course, to the usual restrictions of the S_N2 reaction) and additions to carbonyl compounds. The products of these reactions are thioacetals and can be hydrolyzed, usually by mercury salts, to aldehydes or ketones. The dithiacyclohexane functions as a masked carbonyl group! Here is another synthesis of aldehydes and ketones (Fig. 17.84).

FIGURE 17.84 A synthesis of carbonyl compounds using alkylation of 1,3-dithiane (1,3-dithiacyclohexane).

THE GENERAL CASE

1. BuLi
2. RX

(continued)

SPECIFIC EXAMPLES

(84%)

(70%)

FIGURE 17.84 *(continued)*

WORKED PROBLEM 17.27 Suggest a synthetic route to alkanes starting from 1,3-dithiane.

ANSWER Easy (and effective in practice). First alkylate the dithiane, then desulfurize with Raney nickel (p. 264).

17.8   SPECIAL TOPIC: AMINES IN CONDENSATION REACTIONS; THE MANNICH REACTION

In the acid-catalyzed version of this condensation reaction, named for Carl Mannich (1877–1947), an aldehyde or ketone is heated with an acid catalyst in the presence of formaldehyde and an amine. The initially formed ammonium ion is then treated with base to liberate the final condensation product, the free amine (Fig. 17.85).

THE GENERAL CASE

FIGURE 17.85 The Mannich reaction.

(continued)

A SPECIFIC EXAMPLE

FIGURE 17.85 *(continued)*

This reaction presents a nice mechanistic problem because two routes to product seem possible, and it is the simpler (and incorrect) process that is the easier to find. The difficulty is that we are now experienced in working out aldol condensations and a reasonable person would probably start that way. An acid-catalyzed crossed aldol condensation (p. 933) leads to a β-hydroxy ketone. Displacement of the hydroxyl group (after protonation, of course) would lead to the initial product, the ammonium ion. Treatment with base would surely liberate the free amine (Fig. 17.86).

FIGURE 17.86 An attractive, but incorrect, mechanism for the Mannich reaction.

However, it is possible to make the β-hydroxy ketones in other ways and show that they are not converted into Mannich products under the reaction conditions. Sadly, this simple mechanism cannot be correct. What else can happen? Amines are nucleophiles and react rapidly with carbonyl compounds in acid to give immonium ions (Chapter 16, p. 852). It is this immonium ion that is the real participant in the Mannich condensation reaction. Formaldehyde is more reactive than the ketone and the immonium ion will be preferentially formed from this carbonyl compound. In an aldol-like process, the enol of the other carbonyl compound reacts with the immonium ion to give the penultimate product, the ammonium ion. In the second step, the free amine is formed by deprotonation in base (Fig. 17.87).

FIGURE 17.87 The correct mechanism of the Mannich reaction.

17.9   SPECIAL TOPIC: CARBONYL COMPOUNDS WITHOUT α-HYDROGENS; MAGID'S THIRD RULE

Throughout this chapter we have concentrated on molecules capable of forming enolates or enols, and the subsequent chemistry has been dependent on these intermediates. It is a legitimate question to ask if there is a base-induced chemistry of carbonyl compounds that do *not* have α-hydrogens, and that cannot form enolate anions.

17.9a The Cannizzaro Reaction. Hydroxide ion will give products from nonenolizable carbonyl compounds. If we examine a sample of benzaldehyde that has been allowed to stand in the presence of strong base, we find that a redox reaction has occurred. Some of the aldehyde has been oxidized to benzoate and some reduced to benzyl alcohol (Fig. 17.88).

Benzaldehyde   **Potassium benzoate**   **Benzyl alcohol**
                              (oxidized)   (reduced)

FIGURE 17.88 Benzaldehyde gives benzoate and benzyl alcohol on treatment with hydroxide ion.

This process is known as the **Cannizzaro reaction**, for Stanislao Cannizzaro (1826–1910). If the reaction is carried out in deuterated solvent, no carbon deuterium bonds appear in the alcohol (Fig. 17.89).

FIGURE 17.89 A labeling experiment shows that no solvent deuterium becomes attached to carbon. The hydrogen that reduces the aldehyde must come from another aldehyde, not the solvent.

Benzaldehyde is especially prone to hydrate formation, and the first step of the reaction is addition of hydroxide to benzaldehyde. Much of the time this reaction simply reverses to re-form hydroxide and benzaldehyde, but some of the time the intermediate finds a molecule of benzaldehyde in a position to accept a transferred hydride ion (Fig. 17.90).

FIGURE 17.90 A hydride shift is at the heart of the mechanism of the Cannizzaro reaction.

Do not confuse this step with the expulsion of hydride as a leaving group. Hydride is not a decent leaving group and *cannot* be lost in this way. However, if there is a Lewis acid positioned in exactly the right place to accept the hydride, it can be transferred in a step that simultaneously reduces the receptor aldehyde and oxidizes the donor aldehyde. It is very important to be aware of the requirement for the Lewis acid acceptor of hydride in the Cannizzaro reaction. The hydride must be *transferred,* and in the transition state for the reaction there is a partial bond between the hydride and the accepting carbonyl group. A deprotonation of the acid and protonation of the alkoxide formed from benzaldehyde completes the reaction. Notice how critical the observation about the nonparticipation of the deuterated solvent is. It identifies the source of the reducing hydrogen atom. As it cannot come from solvent (all the "hydrogens" in the solvent are deuteriums) it must come from the aldehyde.

PROBLEM 17.28 Draw a transition state for the hydride shift reaction.

If there were a way to hold the hydride acceptor in the right position, the Cannizzaro reaction would be much easier, as the requirements for ordering the reactants in just the correct way (entropy) would be automatically satisifed. The two partners in the reaction would not have to find each other. This notion leads to the next reaction in this section, which is largely devoted to hydride shift reactions with wonderful names.

17.9b The Meerwein–Ponndorf–Verley–Oppenauer Equilibration.
In the Meerwein–Ponndorf–Verley–Oppenauer (MPVO) reaction, an aluminum atom is used to clamp together the two halves (oxidation and reduction) of the Cannizzaro reaction. An aluminum alkoxide, typically aluminum isopropoxide, is used as a hydride source to reduce a second carbonyl compound (Fig. 17.91).

FIGURE 17.91 The reduction of benzaldehyde using the Meerwein–Ponndorf–Verley–Oppenauer reaction.

The first step in this reaction is the formation of a bond between the carbonyl oxygen and aluminum. Aluminum is a very strong Lewis acid with respect to oxygen as Lewis base, and this bond forms easily. An analogous reaction is the protonation of the carbonyl oxygen (Fig. 17.92).

FIGURE 17.92 The formation of a bond between the nucleophilic carbonyl oxygen and the Lewis acid aluminum alkoxide to give complex **A**.

This new molecule (**A**) is set up for an intramolecular transfer of hydride, which simultaneously reduces a molecule of the carbonyl compound and oxidizes an isopropoxide group to acetone (Fig. 17.93).

FIGURE 17.93 The crucial intramolecular hydride transfer is facilitated by the aluminum atom chemical clamp.

PROBLEM 17.29 Aluminum alkoxides may look very strange to you. Compare the structure of Al(OR)$_3$ to that of a trivalent boron compound such as B(OH)$_3$.

Repetition leads to a new aluminum oxide that is hydrolyzed at the end of the reaction to give the corresponding alcohol (Fig. 17.94).

FIGURE 17.94 Repetition and hydrolysis lead to more acetone and the new alcohol.

PROBLEM 17.30 Write a mechanism for the second stage in the reaction, the formation of (R$_2$CH—O)$_2$Al—O—CH(CH$_3$)$_2$.

PROBLEM 17.31 Write a mechanism for the final hydrolysis reaction of Al(OR)$_3$ to Al(OH)$_3$.

The effect of the clamp on the rate of the reaction is profound. This reaction, here shown as a reduction, takes place under very mild conditions and with few side reactions.

The reaction can also be used to oxidize alcohols to carbonyl compounds. In this case, a simple ketone such as acetone is used as the acceptor, and the alcohol to be oxidized is used to form the aluminium oxide (Fig. 17.95).

FIGURE 17.95 The Meerwein–Ponndorf–Verley–Oppenauer equilibration can be used as an oxidation reaction.

17.9c The Bartlett–Condon–Schneider Reaction.*

Strictly speaking, this reaction does not belong here, as it does not involve a carbonyl compound. However, it surely fits logically in any section devoted to hydride shifts. This reaction shows that hydride shifts can also occur under acidic conditions. We are already quite familiar with an intramolecular version of this reaction. For example, in the addition of hydrogen chloride to 3-methyl-1-butene, the major product results from addition of chloride to a rearranged carbocation formed by an intramolecular shift of hydride (Fig. 17.96).

FIGURE 17.96 The major product from the addition of hydrogen chloride to 3-methyl-1-butene is formed by an intramolecular hydride shift.

There are intermolecular versions of this reaction and one of them makes an exquisite problem. Isobutane exchanges all the primary hydrogens in D_2SO_4, but the tertiary hydrogen remains H, not D (Fig. 17.97).

FIGURE 17.97 A strange-looking exchange reaction.

The beauty of this problem lies in its ability to misdirect. When one sees acidic conditions and a tertiary hydrogen, one is naturally led to think of the tertiary cation. In turn, that idea leads one to think of the tertiary position as the center of reactivity. The exchange of all hydrogens *except* for the tertiary one is seemingly contradictory!

*Paul D. Bartlett (1907–1998), Francis E. Condon (b. 1919), Abraham Schneider (b. 1919).

The exchange actually requires a small amount of alkene impurity to initiate the reaction. We'll use a molecule of isobutene, but any alkene really will do. Transfer of D^+ takes place to give the more stable tertiary cation in which one deuterium has been incorporated into a methyl group (Fig. 17.98).

FIGURE 17.98 The reaction starts with the addition of D^+ to isobutene to give the relatively stable tertiary carbocation.

Loss of a proton in either of two ways leads back to an isobutene in which one deuterium is retained (Fig. 17.99).

FIGURE 17.99 Loss of a proton (H_a or H_b) leads to deuterated isobutene.

PROBLEM 17.32 What happens if D^+ is removed from the intermediate in Figure 17.99?

Continued reaction with D_2SO_4 leads to further exchange until all nine of the methyl hydrogens have been washed out of the methyl groups by deuterium (Fig. 17.100).

C_4H_7D
(one D incorporated)

$C_4H_6D_2$
(two D's incorporated)

C_4D_8
(eight D's incorporated)

$C_4D_9^+$

FIGURE 17.100 Repetition of this process leads to the fully deuterated *tert*-butyl cation.

Now nothing further can happen unless a hydride transfer occurs from isobu-tane. The hydride can *only* come from the tertiary position because only this transfer will generate a tertiary cation. This reaction opens another molecule for exchange (Fig. 17.101).

FIGURE 17.101 Hydride transfer can only come from the *tertiary* position of isobutane, because only this transfer leads to a relatively stable tertiary carbocation.

PROBLEM 17.33 Write a mechanism for the second stage in which a monodeuterio-isobutene forms a dideuterioisobutene.

Actually, there is no need to wait for full exchange before the hydride transfer. It can occur at any point in the reaction, and doubtless there is a soup of partially exchanged isobutane molecules and *tert*-butyl cations throughout the reaction. No matter, however; the exchanges will continue and as long as there is excess D_2SO_4, all primary hydrogens, but *never* the tertiary hydrogen, will be exchanged. It's a beautiful problem.

Problems involving hydride transfers tend to be hard, to thwart even excellent problem solvers. That leads to **Magid's Third Rule**, which states, "When all else fails and desperation is setting in, look for the hydride shift." Problem 17.34 provides some chances to practice hydride shifts.

PROBLEM 17.34 Write mechanisms for the following reactions:

(a)

(b)

(continued)

PROBLEM 17.34 *(continued)*

(c)

Hint: Draw a good three-dimensional picture of this molecule first!
Hint: The other stereoisomer exchanges only five hydrogens for deuteriums.

17.10   SPECIAL TOPIC: THE ALDOL CONDENSATION IN THE REAL WORLD; AN INTRODUCTION TO MODERN SYNTHESIS

In this section, we are going to introduce some of the difficulties of real world chemistry. The detail isn't so important here, although it will be if you go on to become a synthetic chemist! What is important is to get an idea of the magnitude of the problems, and of the ways in which chemists try to solve them. The general principles and relatively simple reactions we have studied (for example, the straightforward aldol condensation) set the stage for a glance at the difficulties encountered when chemists try to do something with these reactions.

In the real world of practical organic synthesis, one rarely needs to do a simple aldol condensation between two identical aldehydes or two identical ketones. Far more commonly, it is necessary to do a crossed aldol between two different aldehydes, two ketones, or an aldehyde and a ketone. As noted earlier (Section 17.5b, p. 933), there are difficulties in doing crossed aldol reactions. Suppose, for example, that we want to condense 2-pentanone with benzaldehyde. Benzaldehyde has no α-hydrogen, so no enolate can be formed from it. Some version of the Claisen–Schmidt reaction p. 936) seems feasible. But 2-pentanone can form two enolates, and the first problem to solve is the specific formation of one or the other enolate (Fig. 17.102).

The two enolates are quite different in stability; one contains a disubstituted double bond and will be more stable, whereas the other has only a monosubstituted double bond and will be less stable (Fig. 17.103).

Effective methods have evolved to form the less stable enolate by taking advantage of the relative ease of access to the less hindered α-hydrogen. Lithium diisopropylamide (p. 918) is one base especially effective at forming these less stable **kinetic enolates**. The key point is that this strong but large base has difficulty in gaining access to the more hindered parts of the carbonyl compound.

FIGURE 17.102 In principle, an unsymmetrical ketone leads to two enolates, and therefore two products.

Less stable (monosubstituted) enolate

More stable (disubstituted) enolate

FIGURE 17.103 The two enolates will differ in stability, depending on the number of alkyl groups attached to the double bond.

Removal of the less hindered proton leads to the less stable, or "kinetic" enolate (Fig. 17.104).

Other methods allow formation of the more stable, **thermodynamic enols**. In one of these, enol borinates are formed. Variation of the R group allows formation of either the kinetic (less stable, but more easily formed) or thermodynamic (more stable) enol (Fig. 17.105).

FIGURE 17.104 Formation of the "thermodynamic" enolate involves making the more stable enolate. The "kinetic" enolate is the one formed more easily (faster). It need not be the more stable enolate.

FIGURE 17.105 Methods have been found that favor either the thermodynamic or kinetic enolates and enols.

We have described only one of the difficulties encountered in just one important synthetic reaction, the aldol condensation. One goal of all synthetic chemists is specificity. How do we do only one reaction? For example, how do we form one, and only one, stereoisomer of the many often possible? Molecules found in nature are sometimes quite spectacularly complicated. Yet we are now able to modify the simple reactions we have been describing so as to be able to make breathtakingly complicated molecules. Here is just one example, from the laboratory of Professor Yoshito Kishi (b. 1937) of Harvard University. Professor Kishi and his co-workers set out to make a molecule called palytoxin, a compound of the formula $C_{129}H_{223}N_3O_{54}$, found in a coral resident in a small tidal pool on the island of Maui in Hawaii. A derivative of this extraordinarily toxic molecule, palytoxin carboxylic acid, is shown in Figure 17.106. Palytoxin is difficult to make, not so much because of its size, but because of its stereochemical complexity and delicacy. It contains no fewer than 61 stereogenic atoms, each of which must be generated specifically and perfectly in order to produce the real palytoxin.

FIGURE 17.106 The incredibly complicated molecule, palytoxin carboxylic acid. The box shows a link forged through an aldol-like reaction.

Several aldol-related reactions were necessary in Kishi's spectacular synthesis. For example, one of the last critical reactions in making palytoxin was a variation of the aldol condensation, a so-called Horner–Emmons reaction, which sewed the molecule together with an α,β-unsaturated carbonyl group that was later reduced (stereospecifically) to give palytoxin. The box shows the site of condensation and subsequent reduction (Fig. 17.107).

FIGURE 17.107 A critical aldol-like condensation finishes the sewing together of palytoxin.

17.11 SUMMARY

New Concepts

It becomes harder and harder to introduce a truly new concept. This chapter deals almost exclusively with the consequences of the acidity of a hydrogen α to a carbonyl group. This concept is really not new, however. You have encountered the idea of resonance stabilization of an enolate before (for example, Problem 9.5), and could surely have dealt with the question of why an α-hydrogen is removable in base, whereas other carbon–hydrogen bonds are not active.

In base, most carbonyl compounds containing α-hydrogens equilibrate with an enolate anion. In acid, it is the enol that is formed in equilibrium with the carbonyl compound (Fig. 17.108).

FIGURE 17.108 Enolate formation in base and enol formation in acid are typical reactions of carbonyl compounds bearing α-hydrogens.

The consequences of enol and enolate formation are vitally important for understanding the mechanisms of reactions of aldehydes, ketones, and other carbonyl-containing compounds, as well as for doing synthesis. Enol or enolate formation leads to exchange, halogenation, and alkylation at the α-position, as well as the more complicated condensation reactions. All these reactions involve fundamentally simple Lewis base–Lewis acid chemistry (Fig. 17.109).

FIGURE 17.109 Reactions of enolates and enols with various Lewis acids.

This chapter also introduces the notion of forming an enolate indirectly, through Michael addition to an α,β-unsaturated carbonyl group (Fig. 17.74).

The idea of transferring a hydride ion (H$\colon^-$) is elaborated in Section 17.9. *Remember*: Hydride is not a good leaving group; it cannot be displaced by a nucleophile, but it can sometimes be transferred, providing only that a suitable acceptor Lewis acid is available.

Key Terms

aldol condensation (p. 922)
Cannizzaro reaction (p. 951)
Claisen–Schmidt condensation (p. 936)
crossed aldol condensation (p. 933)
dithianes (p. 947)
enamines (p. 918)
enol (p. 899)

enolate (p. 898)
fulvenes (p. 939)
haloform reaction (p. 912)
haloform (p. 913)
kinetic enolates (p. 958)
Knoevenagel condensation (p. 938)
lithium diisopropylamide (LDA) (p. 918)

Magid's Second Rule (p. 946)
Magid's Third Rule (p. 957)
Michael reaction (p. 940)
Robinson annulation (p. 945)
thermodynamic enols (p. 959)

Reactions, Mechanisms, and Tools

The two central reactions in this chapter, from which almost everything else is derived, are enolate formation in base and enol formation in acid (Fig. 17.108).

The enols and enolates are capable of undergoing many reactions, among them α-exchange, racemization, α-halogenation, and α-alkylation. Reaction of the enolate, a strong nucleophile, with the Lewis acidic carbonyl compound, or of the less strongly nucleophilic enol with the powerful Lewis acid, the protonated carbonyl, leads to the aldol condensation. These reactions have been previously summarized in Figure 17.109.

Intramolecular aldols, crossed aldols, and the related Knoevenagel condensations involve similar mechanisms.

The Michael reaction, in which a nucleophile adds to the carbon–carbon double bond of an α,β-unsaturated carbonyl group, is another way of forming an enolate anion (Fig. 17.75).

A variety of reactions involving hydride shifts is described. The Cannizzaro reaction is the most famous of these and involves, as do all hydride shift reactions, a simultaneous oxidation of the hydride donor and reduction of the hydride acceptor (Fig. 17.88).

Enamines, formed through reaction of carbonyl groups with secondary amines, can be used to alkylate α-positions. The carbonyl group can be regenerated through hydrolysis of the immonium ion product (Fig. 17.47).

Finally, in a little reaction that leads us to Chapter 18, the idea of addition–elimination reactions of carbonyl groups is introduced. The example in this chapter involves trihalomethyl groups, which can be lost as anions in the elimination phase of the process known as the haloform reaction (Fig. 17.37). In Chapters 18 and 19, we will see just how general and important this process is.

Syntheses

The new synthetic procedures of this chapter are summarized below.

1. ACIDS

The Cannizzaro reaction; R may not have an α-hydrogen

The haloform reaction; the haloform (CHX$_3$) is also formed; works for methyl compounds, X = Cl, Br, and I

2. ALCOHOLS

The MPVO reaction; the mechanism involves complex formation followed by hydride shift

See also the Cannizzaro reaction under "acids," and "β-hydroxy aldehydes and ketones"

3. ALKYLATED ALDEHYDES AND KETONES

1. R$_2$NH
2. R—X
3. H$_2$O

The enamine is an intermediate and R cannot be tertiary

4. β-AMINOKETONES (AND ALDEHYDES)

1. HCl/EtOH/ (CH$_3$)$_2$NH CH$_2$=O
2. H$_2$O NAOH

Mannich reaction; there is an intermediate immonium ion; of course this is also an amine synthesis

5. DEUTERATED ALDEHYDES AND KETONES

KOD/D$_2$O or D$_3$O$^+$

All α-hydrogens will be exchanged in either acid or base

6. α-HALO ALDEHYDES AND KETONES

H$_3$O$^+$ / X$_2$

In acid, the reaction stops after one halogenation; X = Cl, Br, and I

base / X$_2$

In base, up to three α-hydrogens can be replaced by X; the trihalo carbonyl compounds react further in the haloform reaction

7. HALOFORMS

1. base, X$_2$
2. H$_3$O$^+$

+ HCX$_3$

Starting carbonyl compound must have a methyl group; X = Cl, Br, or I

8. β-HYDROXY ALDEHYDES AND KETONES

acid or base

The aldol condensation; dehydration often occurs in acid, and on occasion in base

9. KETONES

1. AlCl$_3$
2.

The MPVO reaction; the first step is formation of a new aluminum alkoxide; the mechanism involves complex formation and a hydride shift

10. α, β-UNSATURATED ALDEHYDES AND KETONES

The dehydration of the products of
aldol condensations is usually acid-catalyzed

Common Errors

Condensation reaction problems can be daunting. At the same time, many chemists agree that they provide a great deal of fun, and almost everyone has his or her favorites. A common error is to set out without a plan. Don't start on a complicated problem without an analysis of what needs to be accomplished. Is the reaction done in acid or base? What connections (bonds) must be made? Are rings opened or closed in the reaction? These are the kinds of questions that should be asked before starting on a problem.

Two conventions, used widely in this chapter, have the potential for creating misunderstanding. In the less dangerous of the two, charges or dots in parentheses are used to indicate the atoms sharing the charge, electron, or electrons in a molecule best described by more than one electronic description (resonance form).

The second involves drawings in which arrows of an arrow formalism emanate from only one of several resonance forms. Although extraordinarily convenient, this practice is dangerous, as it carries the inevitable implication that the resonance form used has some individual reality. It does not. What can be done with one resonance form can always be done from all resonance forms. For simplicity, and bookkeeping purposes, we often draw arrow formalisms using only one of several resonance forms. This practice can also be dangerous unless we are very clear about the shortcuts we are taking.

17.12 ADDITIONAL PROBLEMS

PROBLEM 17.35   Write mechanisms for the following acid- and base-catalyzed equilibrations. There is repetition in this drill problem, but being able to do these prototypal reactions quickly and easily is an essential skill. Be certain you can solve this problem easily before going on to more challenging ones.

(a)

(b)

(c)

PROBLEM 17.36   The following compound contains two different types of α-hydrogens. Removal of which α-hydrogen will yield the *more stable* enolate? Which α-hydrogen is likely to be removed *faster*? Explain why these questions need not have the same answer. That is, why might the less stable enolate be formed faster than the more stable enolate and *vice versa*?

PROBLEM 17.37 Which positions in the following molecules will exchange H for D in D_2O/DO^-?

(a)

(b)

(c)

(d)

(e)

PROBLEM 17.38 As we saw in Problem 17.37c, methyl vinyl ketone exchanges the three methyl hydrogens for deuterium in D_2O/DO^-. Why doesn't the other formally "α" hydrogen exchange as well?

This H does not exchange

PROBLEM 17.39 Which positions in the following molecules will exchange H for D in D_2O/DO^-?

(a)

(b)

(c)

(d)

PROBLEM 17.40 The proton-decoupled ^{13}C NMR spectrum of 2,4-pentanedione consists of six lines (δ 24.3, 30.2, 58.2, 100.3, 191.4, and 201.9 ppm), not the "expected" three lines. Explain.

PROBLEM 17.41 Provide simple synthetic routes from cyclohexanone to the following compounds. *Hint*: These questions are all review.

(a)

(b)

(c)

(d)

(e)

PROBLEM 17.42 Provide simple synthetic routes from cyclohexanone to the following compounds:

(a)

(b)

(two ways)

(c)

(d)

PROBLEM 17.43 Write mechanisms for the following isomerizations:

(a)

(b)

PROBLEM 17.44 Perform retrosynthetic analyses on the following molecules, each of which, in principle, can be synthesized by an aldol condensation–dehydration sequence. Which syntheses will be practical? Explain why some of your suggested routes may not work in practice. What are some of the potential problems in each part?

(a)

(b)

(c)

(d)

PROBLEM 17.45 Treatment of compound **1** with an aqueous solution of bromine and sodium hydroxide affords, after acidification, pivalic acid (**2**) and bromoform. Deduce the structure of **1** and write an arrow formalism mechanism for its conversion into **2** and bromoform. Use your mechanism to predict the number of equivalents of bromine and sodium hydroxide necessary for this reaction.

$$\textbf{1} \xrightarrow[\text{2. H}_2\text{O/H}_3\text{O}^+]{\text{1. NaOH/Br}_2\text{/H}_2\text{O}} (CH_3)_3C-COOH + CHBr_3$$

$$\textbf{2}$$

PROBLEM 17.46 Treatment of ketone **1** with LDA in dimethylformamide (DMF) solvent, followed by addition of methyl iodide leads to two methylated ketones. Write arrow formalisms for their formations and explain why one product is greatly favored over the other.

PROBLEM 17.47 As we have seen, alkylation of enolates generally occurs at carbon (see Problem 17.46 for an example). However, this is not always the case. First, use the following estimated bond strengths to explain why treatment of an enolate with trimethylsilyl chloride leads to substitution at oxygen (Si—C ~69 kcal/mol; Si—O ~109 kcal/mol).

Second, explain the different regiochemical results under the two reaction conditions shown. *Hint:* Remember that enolate formation with LDA is irreversible.

PROBLEM 17.48 When bromo ketone **1** is treated with potassium *tert*-butoxide in *tert*-butyl alcohol at room temperature, it gives exclusively the 5,5-fused bicyclic ketone **2**. In contrast, when **1** is treated with LDA in tetrahydrofuran (THF) at –72 °C, followed by heating, the product is predominately the 5,7-fused ketone **3**. Write arrow formalism mechanisms for these cycloalkylation reactions and explain why the different reaction conditions favor different products.

PROBLEM 17.49   Condensation of 1,3-diphenylacetone (**1**) and benzil (**2**) in the presence of alcoholic potassium hydroxide affords tetraphenylcyclopentadienone (**3**), a dark purple solid. Write an arrow formalism mechanism for the formation of **3**.

PROBLEM 17.50   Aldehydes, but not ketones, readily react with 5,5-dimethyl-1,3-cyclohexanedione (dimedone, **1**) in the presence of a base (such as piperidine) to give crystalline dimedone derivatives, **2**. These compounds can be converted into octahydroxanthenes, **3**, upon heating in ethyl alcohol with a trace of acid. Provide arrow formalism mechanisms for the formation of **2** from **1** and an aldehyde (RCHO), as well as for the conversion of **2** into **3**.

PROBLEM 17.51   Write an arrow formalism mechanism for the following reaction (the Darzens condensation):

PROBLEM 17.52   Propose a mechanism for the following reaction:

$$PhHC=\!\!\!=N-Ph \ + \ ClCH_2COOCH_2CH_3$$

PROBLEM 17.53   Account for the formation of both stereoisomers in the Darzens condensation shown below:

PROBLEM 17.54   Provide structures for intermediates **A–D**. Mechanisms are not necessary.

PROBLEM 17.55   Provide a mechanism for the following reaction:

PROBLEM 17.56   Here is a pair of relatively simple Knoevenagel condensations. Write arrow formalism mechanisms for the following conversions:

(a)

(81%)

(b)

(79%)

PROBLEM 17.57   Now here is a pair of slightly more complicated Knoevenagel reactions: Provide arrow formalism mechanisms.

(a)

(>50%)

(b)

PROBLEM 17.58   Write an arrow formalism mechanism for the conversion of diketone **1** into the important perfumery ingredient *cis*-jasmone (**2**). There is another possible cyclopentenone that could be formed from **1**. What is its structure? Write an arrow formalism mechanism for its formation.

PROBLEM 17.59 When benzaldehyde and an excess of acetone react in the presence of aqueous sodium hydroxide, compound **1** is obtained as the major product as well as a small amount of compound **2**. However, when acetone and an excess (at least 2 equiv) of benzaldehyde react under the same conditions, the predominant product is compound **2**. Spectral data for compounds **1** and **2** are summarized below. Deduce structures for compounds **1** and **2** and rationalize their formation.

Compound **1**
Mass spectrum: m/z = 146 (p, 75%), 145 (50%), 131 (100%), 103 (80%)
IR (Nujol); 1667 (s), 973 (s), 747 (s), 689 (s) cm^{-1}
^{1}H NMR (CDCl$_3$): δ 2.38 (s 3H), 6.71 (d, J = 16 Hz, 1H), 7.30–7.66 (m, 5H), 7.54 (d, J = 16 Hz, 1H)

Compound **2**
Mass spectrum: m/z = 234 (p, 49%), 233 (44%), 131 (42%), 128 (26%), 103 (90%), 91 (32%), 77 (100%), 51 (49%)
IR (melt): 1651 (s), 984 (s), 762 (s), 670 (s) cm^{-1}
^{1}H NMR (CDCl$_3$): δ 7.10 (d, J = 16 Hz, 1H), 7.30–7.70 (m, 5H), 7.78 (d, J = 16 Hz, 1H)

PROBLEM 17.60 Write an arrow formalism mechanism for the following transformation:

PROBLEM 17.61 Reaction of 1-morpholinocyclohexene (**1**) and β-nitrostyrene (**2**), followed by hydrolysis, yields the nitro ketone **3**. Write an arrow formalism mechanism for this reaction sequence and be sure to explain the observed regiochemistry.

PROBLEM 17.62 Diisobutylaluminum hydride (DIBAL-H), like lithium aluminum hydride (LiAlH$_4$) and sodium borohydride (NaBH$_4$), can reduce aldehydes and ketones to alcohols.

Write a mechanism for this reaction. Interestingly, 1 mol of DIBAL-H can reduce up to 3 mol of aldehyde despite possession of only one apparent hydride equivalent. Propose a mechanism that accounts for the second and third equivalents of hydride in these aldehyde reductions. *Hint*: Remember the Meerwein–Ponndorf–Verley–Oppenauer equilibration.

PROBLEM 17.63 Provide structures for intermediates **A–D**. A mechanistic analysis is not required for this problem, but will almost certainly be of use in developing an answer. *Hint*: You will discover that there are certain structural ambiguities in this problem. The following experimental observation should prove helpful. Mild hydrolysis of compound **B** affords compound **E** (C$_{12}$H$_{18}$O$_3$), which is characterized by the following spectra:
IR (CHCl$_3$): 1715 cm^{-1}
^{1}H NMR (CDCl$_3$): δ 1.55–1.84 (m, 9H), 2.60 (s, 4H), 3.77 (s, 3H), 5.52 (br d, 1H), 6.08 (d, 1H).

(continued)

PROBLEM 17.63 *(continued)*

Thiazolium ion

Deprotonated thiazolium ion

PROBLEM 17.64 When treated with KCN in methyl alcohol, benzaldehyde is converted into a molecule called benzoin. Write a mechanism for this change.

1. KCN/CH₃OH
   30 h, 65 °C

2. H₂O/HCl

Benzoin
(80%)

CAUTION! This problem is hard. Do not succumb to the temptation to remove that aldehydic proton—it is *not* acidic. Something else must happen. *Hints*: What will the nucleophile cyanide do to benzaldehyde? How does that reaction alter the nature of the formerly "aldehydic" hydrogen?

PROBLEM 17.65 Show clearly why hydroxide or ethoxide is ineffective in promoting the benzoin condensation. These anions do add to the carbonyl group of benzaldehyde to form compounds much like a cyanohydrin.

PROBLEM 17.66 A few other catalysts will promote benzoin-like condensations. Deprotonated thiazolium ions are examples. Show how the compound shown in the next column can function like cyanide and catalyze the benzoin condensation of pivaldehyde (*tert*-butylcarboxaldehyde).

PROBLEM 17.67 Acetolactate (**2**) is a precursor of the amino acid valine in bacterial cells. Acetolactate is synthesized by the condensation of two molecules of pyruvate (**1**).

The enzyme that catalyzes this reaction (acetolactate synthetase) requires thiamine pyrophosphate (TPP) as a coenzyme.

TPP

Suggest a plausible mechanism for the synthesis of acetolactate (**2**). [*Hints*: Review Problem 17.66. Note that this condensation is accompanied by a decarboxylation (loss of CO₂) that actually precedes coupling. Finally, you may use the enzyme to do acid- or base-catalyzed reactions as required.]

PROBLEM 17.68 Provide mechanisms for the following changes. *Hint*: Definitely do (a) before (b).

(a)

(b)

PROBLEM 17.69 Write a mechanism for the following transformation:

PROBLEM 17.70 Explain the striking observation that methylation of the alcohol **1** leads to a methoxy compound in which the stereochemistry of the oxygen has changed. *Hint*: Note that **1** is a β-keto alcohol.

PROBLEM 17.71 Write a mechanism for the following transformation:

PROBLEM 17.72 A synthesis of pyridines involves the condensation of an α,β-unsaturated ketone (**1**) with a methyl ketone (**2**). This reaction gives a pyrilium salt (**3**), which then reacts with ammonia to produce a 2,4,6-trisubstituted pyridine.

Propose mechanisms for the formation of pyrilium salt **3** from the condensation of **1** and **2**, and for the formation of the pyridine **4** from **3** and ammonia. *Hints*: The formation

(continued)

PROBLEM 17.72 *(continued)*

of **3** occurs best when 1.5–2.0 equivalents of **1** are used. The ketone **5** is a byproduct of the condensation; how can it be formed? Think "redox."

5

PROBLEM 17.73   In 1917, Robinson reported a synthesis (called "von bewundernswerter Eleganz" by the German chemist Willstätter) of tropinone (**3**) involving the condensation of succindialdehyde (**1**), methylamine, and 1,3-acetone dicarboxylic acid (**2**). Schöpf subsequently improved the yield of this reaction to about 90%. Propose a plausible mechanism for this reaction. *Hints*: This reaction involves a double condensation of some kind. Acetone can be used in place of **2**, although the yield is reduced.

1

2

3

Carboxylic Acids

If I mix CH_2 with NH_4 and boil the atoms in osmotic fog, I should get speckled nitrogen!

—Donald Duck*

He's talking chemical talk!

—Huey

But he knows nothing

—Dewey

About chemicals!

—Louie

18.1   PREVIEW

In this chapter, we explore the properties and chemistry of carboxylic acids (general formula: R—COOH). As with other functional groups, we will begin with a naming system, and then move on to structure. Carboxylic acids have a rich chemistry; they are both acids (hence the name) and bases, and we have seen in Chapters 16 and 17 how the carbonyl group is a locus of reactivity. Although many new reactions will appear, most important will be the emergence of a general reaction mechanism, addition–elimination, that we have only seen in one or two examples. Chapter 19 will extend the exploration of this process.

*This quotation comes from a 1944 issue of *Walt Disney's Comics and Stories,* "Donald Duck the Mad Chemist." It was discovered by Peter P. Gaspar, then a postdoctoral fellow at Caltech, now a professor at Washington University. Those of us who work with CH_2 have still not reproduced Professor Duck's experiment with osmotic fog, but a few of us are still trying.

WORKED PROBLEM 18.1 How many answers can you find to the question, What happens when acetone is treated with hydroxide ion in water?

ANSWER This open-ended problem is hard to answer because so much is possible. Don't worry if you find an extra answer (be sure that it is right, though). In base, the chemistry of carbonyl groups is dominated by addition of nucleophiles to the carbon–oxygen double bond (hydrate formation in this case) and removal of α-hydrogens (enolate formation). Both these reactions are possible for acetone, and these will be the most important processes going on.

Once the enolate is formed, further reaction can lead to aldol products. Of course, one is at the mercy of thermodynamics, and for ketones such as acetone the aldol product is not favored. Still, it is a reaction that happens and is a legitimate answer to the problem.

Resonance-stabilized enolate

Enol

Diacetone alcohol

One could imagine many other answers. The alkoxide **A**, formed as an intermediate in the hydration reaction, could add to a second molecule of acetone, for example.

ESSENTIAL SKILLS

1. This chapter's centerpiece is the Fischer esterification—acid hydrolysis equilibrium. If you know this pair of reactions well, you will have much of this chapter under control. The two mechanisms are exactly the reverse of each other—one is the "forward" reaction, the other is the "backward" reaction.

IMPORTANT DETAILS

1. The reaction of a carboxylic acid with two equivalents of an alkyllithium reagent, followed by hydrolysis, yields a ketone. This reaction is a very effective way to make these carbonyl compounds. Be sure you understand why two equivalents of alkyllithium are necessary.

18.2 NOMENCLATURE AND PROPERTIES OF CARBOXYLIC ACIDS

In the systematic nomenclature, the final "e" of the parent alkane is dropped and the suffix "oic" is added. Diacids are named similarly, except that the suffix is "dioic," and the final "e" is not dropped. Cyclic acids are named as "cycloalkanecarboxylic acids." Table 18.1 shows some common mono- and diacids along with some of their physical properties and common names, many of which are still used.

TABLE 18.1 Carboxylic Acids and Their Properties

Structure	Systematic Name	Common Name	bp (°C)	mp (°C)	pK_a
HCOOH	Methanoic acid	Formic acid	100.7	8.4	3.77
CH_3COOH	Ethanoic acid	Acetic acid	117.9	16.6	4.76
CH_3CH_2COOH	Propanoic acid	Propionic acid	141	−20.8	4.87
$CH_3CH_2CH_2COOH$	Butanoic acid	Butyric acid	165.5	− 4.5	4.81
$CH_3CH_2CH_2CH_2COOH$	Pentanoic acid	Valeric acid	186	−33.8	4.82
$CH_3CH_2CH_2CH_2CH_2COOH$	Hexanoic acid	Caproic acid	205	− 2	4.83
$CH_2{=}CHCOOH$	Propenoic acid	Acrylic acid	141.6	13	4.25
	Cyclopentanecarboxylic acid		216	− 7	4.91
	Cyclohexanecarboxylic acid		232	31	4.88

(continued)

TABLE 18.1   Carboxylic Acids and Their Properties (continued)

Structure	Systematic Name	Common Name	bp (°C)	mp (°C)	pK_a
(benzene ring)—COOH	Benzenecarboxylic acid	Benzoic acid	249	122	4.19
HOOC—COOH	Ethanedioic acid	Oxalic acid		190	1.23
					4.19[a]
HOOC—CH$_2$—COOH	Propanedioic acid	Malonic acid		136	2.83
					5.69[a]
HOOC—(CH$_2$)$_2$—COOH	Butanedioic acid	Succinic acid		188	4.16
					5.61[a]
HOOC—(CH$_2$)$_3$—COOH	Pentanedioic acid	Glutaric acid	~300	99	4.31
					5.41[a]
HOOC—(CH$_2$)$_4$—COOH	Hexanedioic acid	Adipic acid	>300	156	4.43
					5.41[a]
HOOC—CH=CH—COOH cis Isomer	cis-Butenedioic acid	Maleic acid		140	1.92
					6.23[a]
HOOC—CH=CH—COOH trans Isomer	trans-Butenedioic acid	Fumaric acid		~300	3.02
					4.38[a]

[a]These are the pK_a values for loss of the second proton.

For substituted acids, the numbering of the longest chain begins with the acid itself, which is given the number "1." In the cyclic compounds as well, it is the "COOH" that gets the first priority in naming. Polyfunctional acids are generally named as acids (Fig. 18.1).

3-Bromobutanoic acid
(3-bromobutyric acid)

3,5-Dichlorohexanoic acid

***m*-Ethylbenzoic acid**
(3-ethylbenzoic acid)

***cis*-4-Aminocyclohexane-
carboxylic acid**

FIGURE 18.1 Some carboxylic acids and their names.

PROBLEM 18.2 Name the following compounds:

18.3   STRUCTURE OF CARBOXYLIC ACIDS

Figure 18.2 shows the structure of formic acid and compares it to formaldehyde. The structures of acids are essentially what one would expect by analogy to other carbonyl compounds. The carbonyl carbons are sp^2 hybridized, and the molecules are flat. The strong carbon–oxygen double bonds are appropriately short, about 1.23 Å.

FIGURE 18.2 The structure of formic acid compared to that of formaldehyde.

There are two complications with the structural picture. First, in solution the simplest carboxylic acids are largely dimerized. It is possible to form two rather strong hydrogen bonds (~7 kcal/mol, each) in the dimeric forms, and this energy gain accounts for the ease of dimer formation. In turn, the ease of dimer formation helps explain the high boiling points (Table 18.1) of carboxylic acids (Fig. 18.3).

Second, there are two energy minima for simple carboxylic acids. The hydroxylic hydrogen can point away (anti) from the carbonyl group, or toward it as in the syn structure of Figure 18.4. These are real energy minima separated in the gas phase by a barrier of about 13 kcal/mol. The syn form for formic acid is more stable by about 6 kcal/mol.

Hydrogen bond

Hydrogen bond

FIGURE 18.3 A dimeric carboxylic acid.

FIGURE 18.4 The syn and anti forms of a simple carboxylic acid.

syn Form   anti Form

PROBLEM 18.3 Explain why the syn form of the carboxylic acid is more stable than the anti form.

PROBLEM 18.4 Calculate the equilibrium constant for the equilibrium between syn and anti formic acid at 25 °C.

WORKED PROBLEM 18.5 Interconversion of the anti and syn forms can occur very easily in water solution. Devise a mechanism not involving simple rotation about a bond.

ANSWER A series of protonations and deprotonations will do the job.

syn   anti

18.4   INFRARED AND NUCLEAR MAGNETIC RESONANCE SPECTRA OF CARBOXYLIC ACIDS

The most prominent features of the IR spectra of carboxylic acids are the strong C=O stretch for the dimeric form at about 1710 cm^{-1} and the strong and very broad band of O—H stretching frequencies centered at about 3100 cm^{-1}. As usual, conjugation shifts the C=O stretch to lower frequency by about 10–20 cm^{-1}. The O—H stretching band is broad for the same reasons that the O—H stretching bands of alcohols are broad (Chapter 15, p. 764). Many hydrogen-bonded dimers and oligomers with different O—H bond strengths are present. Figure 18.5 shows the IR spectrum of acetic acid.

FIGURE 18.5 The IR spectrum of acetic acid.

Hydrogens in the α-position to the carbonyl group of carboxylic acids are deshielded by both the electron-withdrawing nature of the carbonyl group and by local magnetic fields set up by the circulation of electrons in the π bond. They appear in the same region as other hydrogens α to carbonyl groups, δ 2–2.5 ppm.

The carbon of the carboxylic acid group is also strongly deshielded and appears at low field, as do other carbonyl carbons. The position is about 180 ppm, slightly higher field than the chemical shifts of the corresponding carbons of aldehydes and ketones.

18.5   ACIDITY AND BASICITY OF CARBOXYLIC ACIDS

Because carboxylic acids are both acids and bases, we might well expect that a rather diverse chemistry would be found, as with the aldehydes and ketones we saw in Chapters 16 and 17. Carboxylic acids wouldn't be called acids if they were not quite strong Brønsted acids. Table 18.1 gave the pK_a values for some mono- and dicarboxylic acids.

Organic acids ($pK_a = 3–5$) are much stronger acids than alcohols ($pK_a = 15–17$) or the related aldehydes or ketones ($pK_a = 15–20$). Reasons for this increased acidity are not hard to find, but assessing their relative importance is more difficult. One explanation focuses on the formation of a resonance-stabilized **carboxylate anion** after proton loss (Fig. 18.6).

Resonance-stabilized
carboxylate anion

FIGURE 18.6 The formation of a resonance-stabilized carboxylate anion through removal of a proton from the hydroxyl group of an acid.

Deprotonation of an alcohol gives an oxyanion, an alkoxide, but it is not stabilized by resonance. Loss of a proton from the α-position of other carbonyl compounds does give a resonance-stabilized species, the enolate, but the negative charge is shared between oxygen and carbon, not a pair of highly electronegative oxygen atoms as in the carboxylate anions (Fig. 18.7). It is reasonable to accept the notion that the formation of the highly stabilized carboxylate is responsible for the high acidity of carboxylic acids

FIGURE 18.7 A comparison of carboxylate, enolate, and alkoxide anions.

FIGURE 18.8 The highly polar carbonyl group contributes to the stability of the carboxylate anion.

Recently, however, this traditional view has been sharply challenged by at least three research groups in both theoretical and experimental arguments. In various ways, the research groups of Andrew Streitwieser (b. 1927) at Berkeley, Darrah Thomas (b. 1932) at Oregon State, and Kenneth Wiberg (b. 1927) at Yale pointed out the importance of the inductive effect of the highly polar carbonyl group. In their view, the increased acidity of acetic acid over ethyl alcohol derives not from resonance stabilization of the product carboxylate anion, but from the electrostatic stabilization afforded the developing negative charge by the adjacent polar carbonyl group in which the carbon bears a partial positive charge. They reckoned that resonance could account for no more than about 15% of the total stabilization (Fig. 18.8).

Now we should ask *why* resonance might not be so important. The carboxylate anion is surely delocalized, whereas the alkoxide ion is not. The key point is that the carbonyl group of the acid is *already* so polar that little further delocalization can occur as the anion is formed. The negative charge on the *carbonyl* oxygen isn't developed as the oxygen–hydrogen bond breaks; it is already largely there (Fig. 18.9)!

Both factors surely contribute to the acidity of carboxylic acids; the argument is only over their relative importance.

WORKED PROBLEM 18.6 Explain why fluoroacetic acid has a lower pK_a than acetic acid (2.66 vs. 4.76).

ANSWER The structure of the anion formed by removal of a proton tells the story. The dipole in the carbon–fluorine bond stabilizes the carboxylate anion **A** and makes removal of the proton relatively easy.

The δ$^+$ stabilizes the carboxylate anion

The stabilization by electron-withdrawing groups pointed out in Problem 18.6 is a general phenomenon. As Table 18.2 shows, acids bearing electron-withdrawing groups are stronger acids than their parent compounds. The further the electron-withdrawing group from the acid, the smaller the effect (Fig. 18.10).

TABLE 18.2
Acidities of Some Substituted Carboxylic Acids

Acid	pK_a
Acetic	4.76
α-Chloroacetic	2.86
α,α-Dichloroacetic	1.29
α,α,α-Trichloroacetic	0.65
α-Fluoroacetic	2.66
α,α-Difluoroacetic	1.24
α,α,α-Trifluoroacetic	−0.25
α-Bromoacetic	2.86
α-Iodoacetic	3.12
α-Nitroacetic	1.68

| pK_a = 4.81 **Butyric acid** | pK_a = 4.52 **4-Chlorobutyric acid** | pK_a = 4.06 **3-Chlorobutyric acid** | pK_a = 2.84 **2-Chlorobutyric acid** |

FIGURE 18.10 The greater the distance (number of bonds) between an electron-withdrawing group, here chlorine, and the point of ionization, the less effect an electron-withdrawing group has.

Carboxylic acids are Lewis acids as well as Brønsted acids. The presence of the carbonyl group ensures that. Remember all the addition reactions of carbonyl groups encountered in Chapters 16 and 17. However, expression of this Lewis acidity is often thwarted because the easiest reaction with base is not addition to the carbonyl group, but removal of the acid's hydroxyl hydrogen to give the carboxylate anion. The anion is far more resistant to addition than an ordinary carbonyl because addition would introduce a *second* negative charge (Fig. 18.11).

FIGURE 18.11 The fastest reaction in base is usually removal of the acidic hydroxyl hydrogen to give the carboxylate anion. This anion is resistant to addition reactions that would introduce a second negative charge.

Nevertheless, some especially strong nucleophiles are able to do this second addition. An example is the organolithium reagent, RLi, and shortly we will see the synthetic consequences of this reaction. The formation of the "dianion" of Figure 18.12 is not as bad as it looks because the O—Li bond is substantially covalent. It is not completely ionic (Fig. 18.12).

FIGURE 18.12 Some strong nucleophiles can add to the carboxylate anion. The organolithium reagent is one example.

WORKED PROBLEM 18.7 Anticipate a little. What synthetic use can you see for the reaction in Figure 18.12? *Hint*: What will happen when water is added at the end of the reaction?

ANSWER When water is added, a hydrate will be formed. You know from Chapter 16 that simple hydrates are generally unstable relative to a carbonyl compound and water.

Hydrate

So, this reaction should be a good synthesis of ketones. It is, as we will see later on page 1004.

Finally, organic acids are Lewis bases, and can react with Lewis and Brønsted acids. The simplest reaction is the protonation of a carboxylic acid. There are two possible sites for protonation, the carbonyl or hydroxyl oxygen (Fig. 18.13). Which will it be?

FIGURE 18.13 The two possible sites for protonation of a carboxylic acid.

Protonation of the carbonyl oxygen

Protonation of the hydroxyl oxygen

The compound in which protonation has taken place at the carbonyl oxygen is resonance stabilized, whereas that in which the hydroxyl oxygen is protonated is not. Moreover, protonation of the hydroxyl oxygen is destabilized by the dipole in the carbon–oxygen double bond that places a partial positive charge on carbon. It will not be energetically favorable to introduce a positive charge adjacent to this already partially positive carbon. The more stable cation, in which the carbonyl oxygen is protonated, will be preferred (Fig. 18.14).

Resonance stabilization of the product of protonation of the carbonyl group

Unstabilized by resonance, destabilized by the C=O dipole

FIGURE 18.14 Delocalization stabilizes the product of protonation of the carbonyl oxygen. The carbon–oxygen dipole destabilizes the product of protonation of the hydroxyl group.

Accordingly, carboxylic acids are more strongly basic at the carbonyl oxygen than at the hydroxyl oxygen. Does this mean that the hydroxyl group is never protonated? Certainly not, but reaction at the more basic site is favored and will be faster.

PROBLEM 18.8 Explain why the thermodynamic stability of the product (the protonated carbonyl in this case) should influence the *rate* (a kinetic parameter) of protonation.

SUMMARY

With so many possible sites for reaction we might expect a rich chemistry of carboxylic acids. That would be exactly right (Fig. 18.15).

FIGURE 18.15 The various sites of reactivity for a carboxylic acid.

Fischer esterification

18.6 REACTIONS OF CARBOXYLIC ACIDS

18.6a Formation of Esters: Fischer Esterification. An **ester** is a derivative of a carboxylic acid in which the hydrogen of the hydroxyl group has been replaced with an R group. We will examine the nomenclature, properties, and chemistry of esters in detail very soon in Chapter 19, but one of the most important syntheses of these species involves the acid-catalyzed reaction of carboxylic acids with excess alcohol, and must be given here. The reaction is called **Fischer esterification** after the great German chemist Emil Fischer (1852–1919) (Fig. 18.16).

FIGURE 18.16 The acid-catalyzed reaction of a carboxylic acid with an alcohol—Fischer esterification.

We can write the mechanism for this reaction rather easily, as its important steps are quite analogous to reactions of other carbonyl-containing compounds. The first step is surely protonation of the carbonyl group, as it is in the reaction of an aldehyde or ketone in acid. A resonance-stabilized intermediate is formed (Fig. 18.17).

Protonation of the carbonyl part of the carboxylic acid

Protonation of a carbonyl (from Chapter 16)

FIGURE 18.17 The first step in the mechanism is protonation of the carbonyl oxygen to give a resonance-stabilized intermediate.

PROBLEM 18.9 Explain why treatment of acetic acid with $^{18}\overset{+}{O}H_3/^{18}OH_2$ leads to exchange of both oxygen atoms.

Labeled O appears
in both positions

The second step in Fischer esterification is also completely analogous to aldehyde or ketone chemistry. A molecule of alcohol adds to the protonated carbonyl group. For a ketone, this addition merely gives the hemiacetal; for the acid, a somewhat more complicated intermediate with one more hydroxyl group is formed. In both reactions, however, a planar, sp^2 hybridized species has been converted into a tetrahedral intermediate (Fig. 18.18).

FIGURE 18.18 The second step in Fischer esterification involves addition of the alcohol to the protonated acid to give a tetrahedral intermediate.

Now what can these tetrahedral intermediates do in acid? The hemiacetal has only two options. It can reprotonate the OR group and then revert to the starting ketone, or it can protonate the OH group and proceed to the full acetal (Fig. 18.19).

FIGURE 18.19 For a ketone, there are two further possibilities: protonation of the OH leads to water loss and formation of the full acetal, whereas protonation of the OR leads back to the original starting ketone.

The intermediate formed from the acid is more versatile, as it has two OH groups. Protonation of the OR simply takes us back along the path to the starting acid. Protonation of one of the OH groups leads to an intermediate that can lose water, deprotonate, and give the ester. In excess alcohol, the reaction is driven in this direction (Fig. 18.20).

FIGURE 18.20 For an acid, there are also two further possibilities: protonation of the OR leads back to the starting acid, whereas protonation of the OH leads to water loss and formation of the ester.

The mechanism for Fischer esterification is repeated below in Figure 18.21. Note the symmetry of the process—it is symmetrical about the tetrahedral intermediate **A**. Note also the similarity of **B** and **C**, the intermediates on either side of **A**.

Centerpoint of this mechanism

FIGURE 18.21 The full mechanism for Fischer esterification as well as for the reverse reaction, hydrolysis of an ester to a carboxylic acid.

In one of these intermediates, **B**, the OR is protonated. This pathway leads back to the starting acid. In the other, **C**, it is an OH that is protonated. This path leads to the ester. Flanking **B** and **C** are **D** and **E**, the two resonance-stabilized intermediates resulting from protonation of the acid or ester carbonyl.

PROBLEM 18.10 Write the mechanism for the acid-catalyzed formation of an organic acid from the reaction of an ester with excess water

PROBLEM 18.11 Ester formation is not new to you. Give products and mechanisms for the following reactions, which appear in earlier chapters.

We have spent a lot of time and space on Fischer esterification and its reverse, acid hydrolysis. We did this because this reaction is prototypal, an anchor for further discussion. As for other anchor reactions (for example, the S_N2 reaction and addition of Br_2 to alkenes), knowing this reaction, quite literally backward and forward, allows you to generalize and to organize the many following reactions that show a similar pattern, although the details may be different. Fischer esterification is definitely a reaction to know well.

Note here also what makes the chemistry of aldehydes and ketones different from that of acids (or esters). In the aldehydes and ketones there is no leaving group present, whereas in an acid (or ester) there is. The chemistries of aldehydes and ketones on the one hand, and acids and esters on the other, diverge only after the addition to the carbonyl. In the acids and esters there is a group that can be lost, providing a route to other molecules; in the aldehydes and ketones there isn't, which means that the acids and esters can undergo an **addition–elimination reaction** that the aldehydes and ketones can't (Fig. 18.22).

FIGURE 18.22 Unlike aldehydes and ketones, acids and esters bear OH and OR groups that are potential leaving groups.

When Fischer esterification takes place in intramolecular fashion, the result is a cyclic ester, known as a **lactone**. As usual, the relatively unstrained five- and six-membered rings are formed more easily than other sized rings. Figure 18.23 gives an example of lactone formation.*

*Lactones figure in the first of Magid's Rules, which says, "All neutral products are lactones." This statement refers to a kind of problem now out of fashion. The word "neutral" was a clue that a carboxylic *acid* was no longer present. Problem 18.13 gives an example.

FIGURE 18.23 Lactone formation involves an intramolecular Fischer esterification.

A lactone

PROBLEM 18.12 Write a detailed mechanism for the reaction in Figure 18.23.

PROBLEM 18.13 Treatment of the compound in the figure below with I_2/KI and Na_2CO_3 in water leads to a neutral molecule of the formula $C_8H_{11}IO_2$. Give a structure for the product and a mechanism for its formation.

$$\xrightarrow[\text{Na}_2\text{CO}_3/\text{H}_2\text{O}]{\text{I}_2/\text{KI}} C_8H_{11}IO_2$$

Lactones are commonly named in reference to the acid containing the same number of carbon atoms. Thus a butanolide has four carbons, a pentanolide has five, and so on. A Greek letter is used to designate the size of the ring. An α-lactone has one carbon joining the oxygen and carbonyl carbon, a β-lactone has two carbons, a γ-lactone three carbons, and so on. In the systematic naming protocol, lactones are called "oxacycloalkanones" (Fig. 18.24).

Acetic acid
(ethanoic acid)

Propionic acid
(propanoic acid)

Butyric acid
(butanoic acid)

Pentanoic acid

An ethanolide
(an α-lactone)
(2-oxacyclopropanone)

A propanolide
(a β-lactone)
(2-oxacyclobutanone)

A butanolide
(a γ-lactone)
(2-oxacyclopentanone)

A pentanolide
(a δ-lactone)
(2-oxacyclohexanone)

FIGURE 18.24 A naming protocol for lactones.

But we should wait a moment. Other mechanisms for Fischer esterification can be imagined, and it is not fair to dismiss them without evidence. For example, why not use the protonated alcohol as the Lewis acid rather than the protonated carboxylic acid (Fig. 18.21)? Displacement of water using the acid as nucleophile could give the ester (Fig. 18.25).

FIGURE 18.25 A different potential mechanism for Fischer esterification. Here the acid is the nucleophile and the protonated alcohol is the Lewis acid. Watch out for conventions! The figure shows proton removal from one resonance form, which is purely for convenience sake—the resonance form has no individual existence.

The key difference between the mechanism for Fischer esterification we started with in Figure 18.21 (which is correct) and this hypothetical one in Figure 18.25 is the site of carbon–oxygen bond breaking. In the first (correct) mechanism, it is a carbon–oxygen bond in the *acid* that is broken; in the new (incorrect) mechanism, it is the carbon–oxygen bond in the *alcohol* that breaks (Fig. 18.26).

FIGURE 18.26 The two possible mechanisms for Fischer esterification compared (A here and B on the next page). Notice the difference in the carbon–oxygen bonds that are being broken in the two mechanisms.

(continued)

(B)

FIGURE 18.26 *(continued)*

WORKED PROBLEM 18.14 Design an experiment to tell these two mechanisms (Fig. 18.26) apart. Assume access to any labeled compounds you may need.

ANSWER Using labeled alcohol will do the trick. If the mechanism involved carbon–oxygen bond breaking in the alcohol (it does not), the use of ^{18}O-labeled alcohol would not lead to ^{18}O incorporation in the product ester.

However, if the mechanism involves breaking of a carbon–oxygen bond in the original acid (it does) ^{18}O will be incorporated.

Why don't esters react further under these conditions? Aldehydes and ketones form acetals when treated with an acid catalyst and excess alcohol. Why don't esters go on to form **ortho esters** (Fig. 18.27)?

FIGURE 18.27 The conversion of an ester into an ortho ester is analogous to the formation of acetals from aldehydes and ketones.

Figure 18.28 outlines a reasonable mechanism for the conversion of an ester into an ortho ester. Ortho esters are known compounds, and there is nothing fundamentally wrong with the steps outlined in Figure 18.28. The steps are quite analogous to those for acetal formation.

FIGURE 18.28 A reasonable mechanism for the conversion of an ester into an ortho ester.

However, stable ortho esters cannot be made this way. The problem is one of thermodynamics. The sequence of Figure 18.28 is a series of equilibria, and at equilibrium it is the ester that is greatly favored, not the ortho ester. Note that the starting material in this reaction, the ester, is stabilized by resonance and the product ortho ester is not. The ortho ester is analogous to the thermodynamically unstable intermediate **A** in Figure 18.21. Aldehydes and ketones lack this ester resonance, and are therefore less stable, and more prone to further addition reactions. Figure 18.29 makes this point with a partial Energy versus Reaction progress diagram. The acetal and the ortho ester are roughly equivalent in energy, but the ester lies well below the ketone. The result is that acetals are thermodynamically accessible from ketones, but ortho esters are disfavored relative to esters.

FIGURE 18.29 A truncated Energy versus Reaction progress diagram comparing acetal formation from a ketone with ortho ester formation from an ester.

There are other mechanisms for ester formation, and some of them do not involve breaking a carbon–oxygen bond in the starting acid. Sometimes the carboxylate anion is nucleophilic enough to act as the displacing agent in an S_N2 reaction. The partner in the displacement reaction must be especially reactive. In practice, this means that a primary halide or even more reactive species must be used (Fig. 18.30).

FIGURE 18.30 Carboxylate anions can be used to displace very reactive halides in an ester-forming S_N2 reaction.

THE GENERAL CASE

A SPECIFIC EXAMPLE

(91%)

Another reaction in which the carboxylate acts as a nucleophile is the ester-forming reaction of acids with diazomethane, CH_2N_2. Diazomethane, though easily made, is quite toxic and a powerful explosive. So, this method is generally used only when small amounts of the methyl esters are needed (Fig. 18.31).

FIGURE 18.31 Diazomethane can be used to make methyl esters from acids in yields as high as 100%.

The first step in this reaction is deprotonation of the acid by the basic carbon of the diazo compound. The result is a carboxylate anion and a diazonium ion, an extraordinarily reactive alkylating agent (Fig. 18.32).

FIGURE 18.32 In the first step, diazomethane acts as a Brønsted base and deprotonates the acid.

FIGURE 18.33 In the second step, the carboxylate anion displaces nitrogen to give the methyl ester.

Now the alkylation of Figure 18.33 takes place, and nitrogen, a superb leaving group, is displaced to give the methyl ester (recall Figure 18.30). In practice, this method of esterification is restricted to the synthesis of methyl esters. The reason is the accessibility of diazomethane and the relative difficulty of making other simple diazo compounds.

18.6b Formation of Amides. Although carboxylic acids do react with amines to form **amides,*** in general, this reaction is not a very useful process. The dominant reaction between the basic amine and the carboxylic acid must be proton transfer to give the ammonium salt of the carboxylic acid. Heating of the salts has been used as a source of amides (Fig. 18.34).

*There is potential for confusion here. There are two kinds of amides, the carbonyl derivative shown here (R—CO—NHR) and the negatively charged ions, ⁻NR₂.

FIGURE 18.34 The formation of amides through the heating of ammonium salts of carboxylic acids.

Note that the reaction of Figure 18.34 is a dehydration. Water is lost in the reaction. Several dehydrating agents have been developed that greatly facilitate amide formation. In one, **dicyclohexylcarbodiimide (DCC)** is used (Fig. 18.35).

FIGURE 18.35 The use of DCC, a dehydrating agent, to produce amides.

The strategy is to convert the poor leaving group, OH, into a better one. We used this technique extensively in Chapter 7 (p. 293), when we discussed the transformation of alcohols. Here, the acid adds to one carbon–nitrogen double bond of DCC, accomplishing the transformation of the leaving group (Fig. 18.36).

FIGURE 18.36 The first step in the mechanism is addition of the acid to one carbon–nitrogen double bond. The overall transformation involves a change of leaving group from OH to OR.

Two mechanistic pathways are now possible. In the simpler one, the amine adds to the carbon–oxygen double bond to give a tetrahedral intermediate that expels a relatively stable ion to generate the amide (Fig. 18.37).

FIGURE 18.37 Addition of the amine to the newly formed intermediate leads to a tetrahedral intermediate that can decompose to give the amide. Can you see why the leaving group is so good in this reaction?

Good leaving group

Amide

PROBLEM 18.15 There is another more complicated mechanism for this reaction. The initially formed intermediate reacts not with the amine, as in Figure 18.37, but with another molecule of acid. The result is an anhydride. The anhydride is the actual reagent that reacts with the amine to give the product amide (see below). Sketch mechanisms for anhydride and amide formation.

An anhydride

Amide

protonate

SUMMARY

Although this amide synthesis may seem complicated, it is in fact merely a collection of steps closely resembling those in Fischer esterification or acid hydrolysis. We are beginning to see the generality of the addition–elimination process. The addition–elimination reaction will continue to be prominent throughout this and other chapters.

18.6c Polyamides (Nylon) and Polyesters. The reaction shown at the start of Section 18.6b, the heating of the ammonium salt of a carboxylic acid to give an amide (Figure 18.34, p. 997), is at the heart of the synthesis of a set of polyamides collectively known as Nylons. This reaction is certainly not trivial, as recently 4 million tons of Nylon were produced worldwide. In one example, the double-headed amine, hexamethylenediamine, reacts with a double-headed acid, adipic acid, to form the salt (Fig. 18.38). Pyrolysis of the salt produces the long-chain polymer known as Nylon 6,6.

FIGURE 18.38 Formation of Nylon 6,6.

Polyesters are similarly made through the reactions of diacids with diols—simple Fischer esterification reactions that produce a polymeric structure (Fig. 18.39).

FIGURE 18.39 Polyester synthesis.

18.6d Formation of Acid Chlorides. Treatment of a carboxylic acid with thionyl chloride or phosphorus pentachloride gives a good yield of the corresponding **acid chloride**. This reaction closely resembles the formation of chlorides by the reaction of alcohols with thionyl chloride (Chapter 7, p. 296). Once again, the key concept in each case is the transformation of a very poor leaving group, OH, into a better one (Fig. 18.40).

Acid choride formation

FIGURE 18.40 The reactions of alcohols and acids with thionyl chloride (SOCl₂) or phosphorus pentachloride (PCl₅) to give chlorides and acid chlorides are similar.

$$\xrightarrow[100\ °C]{SOCl_2}$$

(>90%)

$$\xrightarrow[100\ °C]{PCl_5}$$

(93%)

(Recall from Chapter 6: $R\text{–}\ddot{O}H \xrightarrow{SOCl_2} R\text{–}\ddot{C}l\!:$)

In the first step an ester is formed, in this case an ester of chlorosulfinic acid (Fig. 18.41).

[chemical reaction scheme showing carboxylic acid + thionyl chloride forming a chlorosulfite ester + HCl]

A chlorosulfite ester

FIGURE 18.41 A chlorosulfite ester is formed in the first step of the reaction.

PROBLEM 18.16 Write a mechanism for the formation of the product in Figure 18.41. Watch out for the details! Almost everyone makes a small error the first time on this problem.

The hydrogen chloride generated in the formation of the product next protonates the carbonyl group and the chloride ion adds to the strong Lewis acid. The intermediate so-formed then breaks down into chloride, sulfur dioxide, and the acid chloride. Notice the change in leaving group induced by the overall transformation of OH into OSOCl (Fig. 18.42).

[mechanism scheme: Chlorosulfite ester → protonation → addition → Good leaving group → elimination → deprotonation → acid chloride + SO₂ + Cl⁻]

Chlorosulfite ester

protonation

addition

Good leaving group

elimination

deprotonation

$+ SO_2 + $:Cl:⁻

FIGURE 18.42 Hydrogen chloride then protonates the carbonyl group, and chloride attacks this strong Lewis acid. In the crucial step, sulfur dioxide and chloride ion are lost as the acid chloride is formed.

Acid chlorides contain a good potential leaving group, the chloride, and the addition–elimination process can lead to all manner of acyl derivatives, R—CO—X, as we will see soon in Chapter 19 (Fig. 18.43).

[scheme showing acid chloride reacting to form esters (ROH), amides (NH₃), carboxylic acids (H₂O), and alcohols (1. LiAlH₄ 2. H₂O)]

Esters

Amides

H_2O → Carboxylic acids

1. LiAlH₄
2. H_2O → R—CH₂OH Alcohols

Acid chloride aminolysis

FIGURE 18.43 Some compounds formed from addition–elimination reactions of acid chlorides. See Problem 18.17.

PROBLEM 18.17 Write a mechanism for the last transformation of acid chlorides shown in Figure 18.43.

PROBLEM 18.18 Another excellent synthesis of acid chlorides from carboxylic acids uses phosgene (Cl—CO—Cl) or oxalyl chloride (Cl—CO—CO—Cl) as the reactive agent. Suggest a mechanism for the second transformation in the following figure:

PROBLEM 18.19 Phosgene (see Problem 18.18) is a most effective poison. Can you guess its mode of action? What happens when phosgene is absorbed by moist lung tissue?

18.6e Anhydride Formation. **Anhydrides** are, as the name suggests, related to acids by a formal loss of water. An anhydride is "two acids less a molecule of water" (Fig. 18.44).

FIGURE 18.44 Anhydrides.

Carboxylic acids and their conjugate bases, the carboxylate anions, react with acid halides to give anhydrides (Fig. 18.45).

FIGURE 18.45 Anhydride formation through the reaction of a carboxylate salt with an acid chloride.

The mechanisms for these reactions involve addition of the carboxylate anion to the carbonyl group of the acid chloride to give a tetrahedral intermediate that can lose the good leaving group, a chloride ion (Fig. 18.46).

FIGURE 18.46 The mechanism of anhydride formation involves the generation of a tetrahedral intermediate and loss of chloride ion.

Other **dehydrating agents** such as DCC or P_2O_5 can be used to form anhydrides. Cyclic anhydrides can be formed through reaction with another anhydride, or even by heating of the diacid (Fig. 18.47).

FIGURE 18.47 Two formations of phthalic anhydride, a cyclic anhydride.

18.6f Reactions with Organolithium Reagents and Metal Hydrides. In a useful synthetic reaction, organic acids react with 2 equivalents of an organolithium reagent to give the corresponding ketone (Fig. 18.48).

THE GENERAL CASE   **A SPECIFIC EXAMPLE**

FIGURE 18.48 A general ketone synthesis from carboxylic acids and organolithium reagents. Note again the convention that differentiates sequential reactions (1. RLi, 2. H₂O) from "dump 'em all together" reactions (RLi, H₂O).

The first step in the reaction is the formation of the lithium salt of the carboxylic acid. Although this species already bears a negative charge, the organolithium reagent is a strong enough nucleophile to add to it to give the "dianion" (Fig. 18.49).

FIGURE 18.49 The first step in the reaction is formation of the carboxylate anion. The organolithium reagent is strong enough to add to the carboxylate anion to give a lithium-stabilized dianion.

Now what? What can happen to the product dianion in base? This question is one of the rare instances when the answer really is, "nothing." There is no possible leaving group, as neither R:⁻ nor O²⁻ can be lost. So, the dianion remains in solution until the reaction mixture is hydrolyzed, at which point a hydrate is formed. Hydrates are unstable relative to their ketone precursors (Chapter 16, p. 837), and so the end result is the ketone (Fig. 18.50).

FIGURE 18.50 When the dianion is protonated in the second step, a hydrate is formed. Hydrates are generally unstable compared to the related ketones.

The key to the formation of the ketone is the inability of the dianionic intermediate to expel a leaving group. Compare this nonreaction with a general process in which a leaving group is present. We have already seen many reactions of this type, and in Chapter 19 we will see many more (Fig. 18.51).

This reaction is fine, as long as L is a good leaving group

FIGURE 18.51 The dianion is foiled in any attempt to expel a negative charge by the lack of a possible leaving group.

Carboxylic acids are reduced by lithium aluminum hydride, followed by hydrolysis, to give the corresponding primary alcohols (Fig. 18.52).

THE GENERAL CASE   **A SPECIFIC EXAMPLE**

FIGURE 18.52 The reduction of carboxylic acids to alcohols.

There is no unanimity on the details of the reaction mechanism, but the first step must be formation of the carboxylate anion and hydrogen. Anion formation is followed by addition of hydride to the carbonyl group. The result is another dianion. Now, however, a molecule of metal oxide can be lost to generate the aldehyde (Fig. 18.53).

FIGURE 18.53 Deprotonation to give the carboxylate anion is followed by hydride addition to give a dianion. Loss of $LiOAlH_2$ leads to an intermediate aldehyde.

The aldehyde is formed in the presence of several reducing agents, and cannot survive for long (Chapter 16, p. 866). Another addition of hydride gives the alkoxide, and subsequent hydrolysis forms the alcohol (Fig. 18.54).

FIGURE 18.54 Reduction of the aldehyde leads to an alkoxide. Protonation by water gives the alcohol.

18.6g Decarboxylation. Some carboxylic acids easily lose carbon dioxide. The best examples are β-keto acids, and 1,3-diacids such as malonic acid (propanedioic acid) in which there is also a carbonyl group β to the acid. The reactions often take place at room temperature or on gentle heating. Note that these reactions provide new syntheses of ketones and acids (Fig. 18.55).

GENERAL REACTIONS

SPECIFIC EXAMPLES

FIGURE 18.55 Three kinds of carboxylic acid that decarboxylate (lose CO$_2$) easily.

We will defer discussion of the **decarboxylation** of β-keto acids until we learn how to make such compounds in Chapter 19, but carbonates and malonic acids will be mentioned here.

Carbonic acid (H_2CO_3 or HOCOOH) is itself unstable, although its salts are isolable and familiar as **sodium bicarbonate** and **sodium carbonate** (Fig. 18.56).

Carbonic acid
(unstable)

Sodium bicarbonate
(stable)

Sodium carbonate
(stable)

FIGURE 18.56 Carbonic acid is unstable, but sodium bicarbonate and sodium carbonate are stable.

Monoesters of carbonic acid lose carbon dioxide easily. The loss of carbon dioxide is irreversible, as the gaseous carbon dioxide escapes. The diesters of carbonic acid (carbonates) are stable (Fig. 18.57).

Similarly, **carbamic acid**, the amide derivative of carbonic acid, is also unstable, and loses carbon dioxide to give amines. The diamide, **urea**, and **carbamates**, which cannot lose carbon dioxide, are stable (Fig. 18.58).

$$CO_2\uparrow + R\ddot{O}H$$

Stable carbonate
(no H to be removed)

FIGURE 18.57 Monoesters of carbonic acid decarboxylate easily, but carbonates do not.

$$\longrightarrow CO_2\uparrow + :NH_3$$

Carbamic acid

Urea
(stable)

Carbamate
(stable)

FIGURE 18.58 Carbamic acids decarboxylate, but urea and carbamates are stable.

Acids can be electrochemically decarboxylated to give, ultimately, a hydrocarbon composed of two acid R groups. This reaction, called the **Kolbe electrolysis** after Hermann Kolbe (1818–1884), involves an electrochemical oxidation of the carboxylate anion to give the carboxyl radical. Loss of carbon dioxide gives an alkyl radical that can dimerize to give the hydrocarbon (Fig. 18.59).

FIGURE 18.59 The Kolbe electrolysis is a source of molecules made up of coupled R groups of R—COOH.

THE GENERAL CASE

electrochemical
oxidation

$$\longrightarrow CO_2 + R\cdot \longrightarrow R\!-\!R$$

A SPECIFIC EXAMPLE

$$+ \quad 2\,CO_2$$

(75%)

18.6h Formation of Alkyl Bromides: The Hunsdiecker Reaction. The carboxyl radical just encountered is also the active ingredient in the **Hunsdiecker reaction** in which alkyl bromides are formed from acids (Heinz Hunsdiecker, 1904–1981; Cläre Dieckmann Hunsdiecker, 1903–1995). Although this reaction is named for the Hunsdieckers, here is another case in which the perspicacious Borodin (see p. 922) was far ahead. This reaction was first done by Borodin in 1861, 81 years before the Hunsdieckers! Typically, the silver salt of the acid is used and allowed to react with bromine. The first step is the formation of the hypobromite. On heating, the weak oxygen–bromine bond breaks homolytically to give the carboxyl radical and a bromine atom. As in the Kolbe reaction, the carboxyl radical loses carbon dioxide to give an alkyl radical. In this case, however, the alkyl radical can carry the chain reaction further by recycling to react with the hypobromite to produce an alkyl bromide and another molecule of carboxyl radical (Fig. 18.60). For a review of radical chain reactions, see Chapter 11, pp. 514–524.

FIGURE 18.60 The decarboxylation of the carboxylate radical is also involved in the Hunsdiecker reaction, a synthesis of alkyl bromides from acids. The recycling, chain-carrying carboxyl radical is boxed.

18.7   REACTIVITY OF CARBOXYLIC ACIDS AT THE α-POSITION

As we saw in Chapter 17, in base, aldehydes and ketones can be brominated in the α-position. The enolate anions are the active ingredients (Fig. 18.61).

Enolate

FIGURE 18.61 An enolate is the intermediate in the base-catalyzed α-bromination of aldehydes and ketones.

A similar reaction is possible with carboxylic acids, but two equivalents of strong base are necessary, as this reaction proceeds through a dianion. The first equivalent of base removes the carboxyl hydrogen to give the carboxylate salt. If the base used is a strong *base* and a weak *nucleophile,* a second hydrogen can be removed, this time from the α-position. The typical base used is lithium diisopropylamide (LDA), which is a strongly basic but poorly nucleophilic species because of its steric demands. The dianion can now be brominated or alkylated at the α-position, as long as the alkylating agent is reactive. Primary halides work well in alkylation reactions because the S_N2 reaction is especially facile, but more substituted halides lead mostly to the products of an E2 reaction (Fig. 18.62).

THE GENERAL CASE

A SPECIFIC EXAMPLE

(63%)

FIGURE 18.62 Dianions of carboxylic acids can function as enolates. They can be brominated or alkylated in the α-position with primary or other reactive halides.

PROBLEM 18.20 Draw the resonance forms for the dianion. Which is the best representation?

PROBLEM 18.21 Provide mechanisms for the following reactions that show two of the reactions possible at the α-position of carboxylic acids:

18.8   SYNTHESES OF CARBOXYLIC ACIDS

We already know a number of routes to carboxylic acids, and this section will add an important new one, the reaction of organometallic reagents with carbon dioxide.

18.8a Oxidative Routes. As we know (Chapter 16, p. 868), alcohols and aldehydes can be oxidized to acids. Alkenes can also be oxidatively cleaved to acids by basic potassium permanganate. We also already know that alkenes form 1,2-diols when treated with basic permanganate (Chapter 10, p. 472). The reactions of alkenes with $KMnO_4$ can go further to form two acids. The trick is to use 18-crown-6 (Chapter 6, p. 265) to make $KMnO_4$ soluble in benzene. The crown ether has a great ability to bind potassium ions, and this allows the negatively charged permanganate ion (MnO_4^-) to follow the bound cation, K^+(crown), into solution (Fig. 18.63).

FIGURE 18.63 Some oxidative routes to carboxylic acids.

Many alkenes can be ozonized using an oxidative workup to give acids (Chapter 10, p. 465), and the side chains of alkyl aromatic compounds can be oxidized to acid groups (Chapter 13, p. 661; Fig. 18.64).

FIGURE 18.64 More oxidative routes to acids.

18.8b Haloform Reaction.

The reaction in base of methyl ketones with iodine, bromine, or chlorine gives the acid and the haloform (Chapter 17, p. 912; Fig. 18.65).

FIGURE 18.65 The haloform reaction is another source of acids.

18.8c Reaction of Organometallic Reagents with Carbon Dioxide.

Like any other compound containing a carbonyl group, carbon dioxide reacts with organometallic reagents through an addition reaction. The initial product is a carboxylate salt that is converted into the acid when the reaction mixture is acidified. This reaction is very general and a source of many acids (Fig. 18.66).

FIGURE 18.66 A general route to carboxylic acids uses the carboxylation of Grignard reagents with carbon dioxide.

THE GENERAL CASE

A SPECIFIC EXAMPLE

PROBLEM 18.22 Starting from inorganic reagents, butyl alcohols, CO_2, tosyl chloride and ethyl alcohol as your only other source of carbon, devise syntheses of the following molecules:

18.9   SPECIAL TOPIC: ANOTHER BROMINATION AT THE α-POSITION, THE HELL–VOLHARD–ZELINSKY REACTION

Treatment of carboxylic acids with Br_2 and PBr_3, or the equivalent, a mixture of phosphorus and bromine, leads ultimately to formation of the α-bromo acid (Fig. 18.67).

FIGURE 18.67
The Hell–Volhard–Zelinsky reaction.

The process is known as the **Hell–Volhard–Zelinsky (HVZ) reaction** after Carl M. Hell (1849–1926), Jacob Volhard (1834–1910), and Nicolai D. Zelinsky 1861–1953). The first step is formation of the acid bromide through reaction with PBr_3. The acid bromide is in equilibrium with its enol form. Bromination of the enol form with Br_2 gives an isolable α-bromo acid bromide. Recall the bromination of carbonyl compounds (Chapter 17, p. 910), a reaction that also proceeds through the enol form (Fig. 18.68).

Acid bromide Enol form α-Bromo acid bromide

FIGURE 18.68 An intermediate in the Hell–Volhard–Zelinsky reaction is the α-bromo acid bromide. This compound can be isolated if a full equivalent of PBr_3 is used in the reaction.

An α-bromo acid halide reacts like any acid halide, and can be used to generate the α-bromo acid, ester, and amide, for example (Fig. 18.69).

THE GENERAL REACTIONS

A SPECIFIC EXAMPLE

(>90%)

FIGURE 18.69 Compounds available from reactions of the α-bromo acid bromide.

PROBLEM 18.23 Write a general mechanism for the reactions of Figure 18.69.

If a catalytic amount of PBr$_3$ is used, and not the full equivalent of the reaction in Figure 18.68, the product of the reaction is the α-bromo acid, not the α-bromo acid bromide. The α-bromo acid bromide is still an intermediate, but only a small amount of it can be made at any one time, as it depends on PBr$_3$ for its formation. The small amount of α-bromo acid bromide produced is attacked by a molecule of the acid in an anhydride-forming reaction (Section 18.6e, p. 1002). The anhydride is opened by bromide to give the carboxylate anion and a new molecule of α-bromo acid bromide that recycles. A final hydrolysis gives the α-bromo acid itself. This reaction surely involves a complicated mechanism (Fig. 18.70).

FIGURE 18.70 The complex mechanism of the HVZ reaction using a catalytic amount of PBr$_3$.

The HVZ reaction is limited to the formation of α-bromo compounds and is, in truth, sometimes awkward to carry out. The reagents, bromine and phosphorus, are noxious and reaction times are often long and reaction conditions are harsh.

Methods have evolved to extend and replace the classic HVZ process. For example, David Harpp (b. 1937) and his colleagues at McGill University use the reaction of acid chlorides and *N*-bromosuccinimide (NBS, p. 536), in the presence of a catalytic amount of acid, to form α-bromo acid chlorides conveniently and in excellent yields (Fig. 18.71).

FIGURE 18.71 The Harpp modification of the HVZ reaction.

N-Chlorosuccinimide (NCS) can be used in place of NBS to form the α-chloro compounds, and I_2 gives the α-iodo compounds (Fig. 18.72).

FIGURE 18.72 Formation of α-chloro and α-iodo compounds.

The mechanism of Harpp's modification is an ionic one, and uses many of the steps we have seen already. The acid catalyst functions to protonate the carbonyl group of the acid halide. Either NBS or NCS deprotonates the α-position to form an enol. The enol now functions as base to remove a Br (or Cl) from NBS or NCS. Deprotonation then gives the final product and a new molecule of acid catalyst (Fig. 18.73).

FIGURE 18.73 The mechanism of the Harpp modification of the HVZ reaction.

Like the α-halo aldehydes and ketones, the α-bromo acids are very reactive in displacement reactions. They serve as sources of many α-substituted acids. A particularly important example is the formation of α-amino acids through reaction with ammonia (Fig. 18.74).

THE GENERAL CASE

SPECIFIC EXAMPLES

FIGURE 18.74 Some reactions of α-bromo acids.

That's a lot of material. Many new kinds of compounds have appeared and there may seem to be an overwhelming number of new reactions. Don't be dismayed! Almost all of these last few sections involve versions of the addition–elimination reaction. Keep this generalization in mind and you will be able to figure out what is happening in these reactions relatively easily. It is not necessary, or productive, to try to memorize all this material. What is far more effective is to learn the basic process well, and then practice applying it in different circumstances.

18.10   SPECIAL TOPIC: FATTY ACIDS

We have spent many pages looking at the reactions of carboxylic acids. Now let's see a reaction of molecules containing long hydrocarbon chains and a single acid. The properties of such molecules are strongly influenced by the presence of both the long chain and the acid group. Even-numbered long-chain acids are called **fatty acids** because they are formed by the base-induced **ester hydrolysis (saponification)** of fats (Fig. 18.75). Fatty acids are synthesized in nature from acetic acid, a two-carbon acid, and this accounts for their containing even numbers of carbons.

$n = 10$	Lauric acid
$n = 12$	Myristic acid
$n = 14$	Palmitic acid
$n = 16$	Stearic acid
$n = 18$	Arachidic acid

FIGURE 18.75 Some fatty acids that can be made from the hydrolysis of fats.

PROBLEM 18.24 Give a mechanism for the hydroxide ion induced hydrolysis (saponification) of a fat to a fatty acid.

Fatty acids are most important molecules. For example, part of our energy supply comes from the hydrolysis of fats to fatty acids, which are further broken down. The fatty acids are converted into a molecule called acetyl-CoA,* and fed into the Krebs cycle, a series of reactions in which acetyl-CoA is oxidized to CO_2 and H_2O. Palmitic acid, for example, ultimately leads to eight molecules of acetyl-CoA (Fig. 18.76).

FIGURE 18.76 The compound CoASH converts fatty acids such as palmitate into acetyl-CoA.

*CoA stands for coenzyme A, a molecule necessary to promote an enzymatic reaction.

The long-chain fatty acid is decomposed in a sequence of two-carbon cleavages (Fig. 18.77).

Same carbon

Acetyl-CoA

CoASH

Palmitate

FIGURE 18.77 Fatty acids are deconstructed two carbons at a time.

Thus it seems that the methyl group of acetyl-CoA must be the α-carbon of the original fatty acid. We will come back to this reaction in Chapter 19, but you should be able to appreciate both the extent and difficulty of the task that Nature faces here—How to break the α–β bond in palmitate? That's no simple task, and we shall need the material in Chapter 19 to solve it.

Soaps are the salts of long-chain fatty acids, and work in the following way. These compounds contain both a highly polar end, the acid salt, and a most nonpolar end, the long-chain hydrocarbon end. The polar ends will be soluble in polar solvents such as water, but the nonpolar hydrocarbon ends will not be. In aqueous solution, the **hydrophobic** (water-hating) hydrocarbon ends cluster together at the inside of a sphere, protected from the aqueous environment by the **hydrophilic** (water-loving) polar acid salt groups forming the interface with the polar water molecules. Such aggregates are called **micelles** (Fig. 18.78).

Similar micellar species are formed from **detergents** (synthetic soaps), which are long-chain sulfonic acid salts (Fig. 18.79).

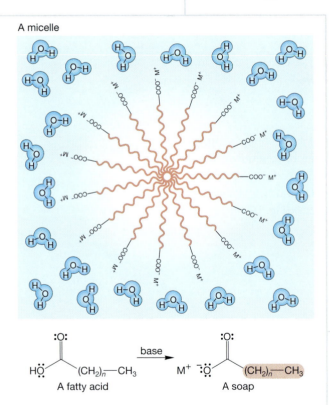

A micelle

FIGURE 18.78 The formation of a soap from a fatty acid. In polar media, soaps form micelles, with the hydrophobic hydrocarbon chains directed toward the center of a sphere, away from the polar solvent. The surface of the sphere contains the polar carboxylate anions and their positive counterions.

A long-chain hydrocarbon

FIGURE 18.79 A synthetic soap, a detergent.

Organic dirt (grease) is not water soluble and is not easily removed without the aid of a soap or a detergent. However, when a detergent is used, the grease is dissolved by the hydrocarbon-like internal portion of the micelle and made soluble in water by the polar acid salts outside (Fig. 18.80).

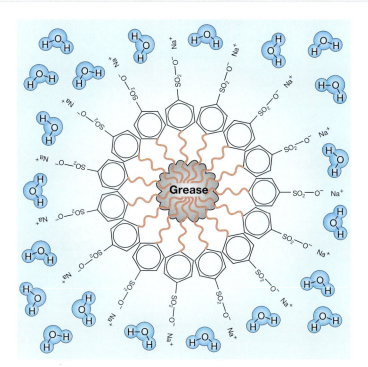

FIGURE 18.80 A micelle of a detergent solubilizing a grease molecule.

18.11 SUMMARY

New Concepts

Carboxylic acids exhibit several different kinds of reactivity. So far, we have seen many kinds of molecules (aldehydes and ketones, for example) that can be both acids and bases. But carboxylic acids are even more polyreactive. They are Brønsted acids through proton loss from the hydroxyl group, or sometimes from loss of an α-hydrogen. They are also Lewis acids through reaction at the carbonyl group. Carboxylic acids act as

Brønsted bases and Lewis bases through reaction at the more basic carbonyl oxygen or, occasionally, at the less basic hydroxyl oxygen.

The Brønsted acidity of acids, long thought to rest largely upon the delocalization of the carboxylate anion produced by ionization, is also dependent on the inductive effect of the polarized carbonyl group.

Key Terms

acid chloride (p. 1000)
addition–elimination reaction (p. 990)
amides (p. 996)
anhydrides (p. 1002)
carbamates (p. 1007)
carbamic acid (p. 1007)
carbonic acid (p. 1007)
carboxylate anion (p. 981)
decarboxylation (p. 1006)
dehydrating agent (p. 1003)
detergents (p. 1018)
dicyclohexylcarbodiimide (DCC) (p. 997)
ester (p. 986)
ester hydrolysis (saponification) (p. 1017)

fatty acids (p. 1017)
Fischer esterification (p. 986)
Hell–Volhand–Zelinsky (HVZ) reaction (p. 1012)
Hunsdiecker reaction (p. 1008)
hydrophilic (p. 1018)
hydrophobic (p. 1018)
Kolbe electrolysis (p. 1007)
lactone (p. 990)
micelles (p. 1018)
ortho esters (p. 993)
soaps (p. 1018)
sodium bicarbonate (p. 1007)
sodium carbonate (p. 1007)
urea (p. 1007)

Reactions, Mechanisms, and Tools

Most of the new reactions in this chapter are examples of addition–elimination processes. This mechanism has appeared occasionally before but it is emphasized in this chapter for the first time. Carbonyl compounds bearing leaving groups can be attacked by nucleophiles to give tetrahedral intermediates (addition) that can then expel the leaving group (elimination) (Fig. 18.81).

FIGURE 18.81 The addition–elimination reaction.

For acids, the success of an addition–elimination process depends on the transformation of the carboxyl hydroxyl group into a better leaving group. This conversion is often done through protonation in acid-catalyzed reactions. The best

example is Fischer esterification, and its cyclic variation, lactone formation (Fig. 18.23).

The carboxyl OH can also be made into a better leaving group through other chemical conversions. An excellent example is acid chloride formation in which a chlorosulfite ester is first produced. Addition of chloride to the carbonyl group of the ester leads to an intermediate that can eliminate sulfur dioxide and chloride, thus producing the acid chloride (Figs. 18.41 and 18.42).

The acid chlorides contain an excellent leaving group, and addition–elimination reactions are common (Fig. 18.43).

Occasionally, the carboxylate anion can be used as a displacing agent in an S_N2 reaction. In order for this reaction to be successful, the Lewis acid partner in the reaction must be very reactive (Fig. 18.30).

The electrochemical reaction of carboxylate anions leads to hydrocarbons (Kolbe electrolysis). Other reactions involving decarboxylations are briefly mentioned (Fig. 18.55).

Syntheses

1. ACID HALIDES

A chlorosulfite ester is an intermediate

Hell–Volhard–Zelinsky reaction; a full equivalent of PBr_3 is used to make the α-bromo acid bromide

2. ACIDS

Other oxidizing agents also work; the aldehyde is an intermediate

Other oxidizing agents work

Note the oxidative workup; the aldehyde can be obtained by using dimethyl sulfide or H_2/Pd

The crown ether is essential, as it brings the $KMnO_4$ into solution

Acid-catalyzed ester hydrolysis

3. ALCOHOLS

cis Diol is formed

An aluminum alkoxide is a leaving group in this reaction

Decarboxylation of a bicarbonate ester
See also other "hydroxy" compounds in this section

4. ALKYL HALIDES

The Hunsdiecker reaction is used almost exclusively for bromides

5. AMIDES

An addition–elimination process
See also "imides" and "lactams"

6. AMINES

Decarboxylation of a carbamic acid

7. α-AMINO ACIDS

An S_N2 displacement reaction

8. ANHYDRIDES

This addition–elimination process works even
better with the carboxylate anion as nucleophile

Other dehydrating agents also work; intramolecular
reactions can often be accomplished by simple heating

9. α-BROMO CARBOXYL COMPOUNDS

Hell–Volhard–Zelinsky reaction; a full equivalent
of PBr_3 is used to make the α-bromo acid bromide

X = OH, OR, or NR_2

This addition–elimination reaction
makes α-bromo acids, esters, and amides

10. α-CYANO ACIDS

This reaction is an S_N2 displacement

11. ESTERS (FOR CYCLIC ESTERS, SEE "LACTONES")

Fischer esterification

For this S_N2 reaction to succeed,
R–I must be very reactive

Only common for methyl esters

12. α-HALO ACIDS

To get the α-bromo acid, a catalytic amount of PBr_3
must be used in the Hell–Volhard–Zelinsky reaction

The intermediate is the dianion, and two equivalents
of the hindered base LDA must be used

13. HYDROCARBONS

The Kolbe electrolysis, a radical dimerization process

14. α-HYDROXY ACIDS

This reaction is an S_N2 displacement

15. IMIDES

A double amide formation

16. KETONES

The hydrate is an intermediate,
2 mol of RLi are required

Decarboxylation of β-keto acids

17. LACTAMS

This formation of cyclic amides works
best for unstrained rings

18. LACTONES

Intramolecular Fischer esterification

Common Errors

Addition–elimination reactions are basically simple: The nucleophile adds to the carbonyl, then the carbonyl is reconstituted as a leaving group is displaced. But there is such variety! That's where the difficulty lies—in seeing through all the "spinach" hanging on the nucleophile and hiding the carbonyl to find the simple mechanism at the heart of matters.

Also, there are vast numbers of small steps—proton transfers mostly—in these mechanisms, and it is hard to get them all right.

Finally, there is a small error that all of us seem to make on occasion. This mistake is to add an electrophile to the wrong oxygen of a carboxylic acid. It is the carbonyl oxygen that is the more basic and nucleophilic site, not the hydroxyl oxygen.

18.12 ADDITIONAL PROBLEMS

PROBLEM 18.25   Write names for the following structures:

(a)

(b)

(c)

(d)

(e)

PROBLEM 18.26   Draw structures for the following molecules:
(a) 2-Aminopropanoic acid (alanine)
(b) 2-Hydroxypropanoic acid (lactic acid)
(c) (Z)-2-Methyl-2-butenoic acid
(d) 2-Hydroxy-1,2,3-propanetricarboxylic acid (citric acid)
(e) 2-Hydroxybenzoic acid (o-hydroxybenzoic acid or salicylic acid)
(f) 2-Oxopropanoic acid (pyruvic acid)

PROBLEM 18.27   Give reagents for converting butyric acid (butanoic acid) into the following compounds:

(a)

(b)

(c)

(d)

(e)

(f)

(g)

(h)

(i)

(j)

PROBLEM 18.28   Give the major organic products expected in each of the following reactions:

(a)

$$\text{Br}\,\diagup\!\!\diagdown\!\!\diagup\text{C(=O)OH} \xrightarrow[\text{CH}_3\text{OH, }\Delta]{\text{CH}_3\overset{+}{\text{OH}}_2}$$

(b)

$$\text{RCH}_2\text{C(=O)OH} \xrightarrow[\text{CH}_3\text{CN}]{\text{DCC}}$$

(c)

$$\xrightarrow[\substack{2.\ \text{H}_3\overset{+}{\text{O}}/\text{H}_2\text{O}}]{\substack{1.\ \text{NaOH}/\\ \text{H}_2\text{O, }\Delta}}$$

(d)

$$\text{RCH}_2\text{C(=O)OH} \;+\; \text{RRCHNH}_2 \xrightarrow[\text{CH}_3\text{CN}]{\text{DCC}}$$

(e)

$$\xrightarrow[\substack{2.\ \text{H}_3\overset{+}{\text{O}}/\text{H}_2\text{O}}]{\substack{1.\ \text{LiH}}}$$

(f)

$$\xrightarrow[\substack{2.\ \text{CH}_3\text{Li}\\ 3.\ \text{H}_2\text{O}}]{\substack{1.\ \text{LiH}}}$$

(g)

$$\xrightarrow[\text{PBr}_3\ (\text{cat.})]{\text{Br}_2}$$

(h)

$$\xrightarrow[\substack{2.\ \text{CO}_2\\ 3.\ \text{H}_2\text{O}/\text{H}_3\overset{+}{\text{O}}}]{\substack{1.\ \text{Mg, ether}}}$$

PROBLEM 18.29 Provide structures for compounds **A–F**. Mechanisms are not required, but may be helpful in your analysis.

$(CH_3)_3C-C(=O)-OH$

↓ CH_2N_2

A
$(C_6H_{12}O_2)$

↓ KOH | H_2O

B
$(C_5H_9O_2^-\ ^+K)$ → CH₃I (DMSO) → **C** $(C_6H_{12}O_2)$

C $(C_6H_{12}O_2)$

↑ H_2O | H_3O^+

D
$(C_5H_{10}O_2)$

↑ $SOCl_2$

E
(C_5H_9OCl)

↑ CH_3OH

F
$(C_6H_{12}O_2)$

PROBLEM 18.30 We begin this problem with cyclohexanol. Your job is to determine the structures of the compounds in between.

OH (cyclohexanol)

↓ HBr

A

↓ 1. Mg 2. CO_2 3. H_3O^+/H_2O

B
$(C_7H_{12}O_2)$

→ 1. LiAlH₄ 2. H_2O/H_3O^+ → **C** $(C_7H_{14}O)$

C $(C_7H_{14}O)$

↑ H_3O^+ **B**

D
$(C_{14}H_{24}O_2)$

↑ H_3O^+/H_2O

B + C

PROBLEM 18.31 Provide structures for compounds **A–D**.

(benzyl)CH₂COOH → SOCl₂ → **A** (C_8H_7ClO)

↓ NCS | HCl

B $(C_8H_6Cl_2O)$

→ H_2O → **C** $(C_8H_7ClO_2)$

→ NH₃ → **D** $(C_8H_9NO_2)$

PROBLEM 18.32 Provide a mechanism for this unusual esterification process.

$H_3C-C(=O)-OH$ + $CH_2=C(CH_3)_2$ → H_3O^+ → $H_3C-C(=O)-OC(CH_3)_3$

PROBLEM 18.33 Provide brief explanations for the pKₐ differences noted below.

(a) 3.77   4.76
$H-C(=O)-OH$ $H_3C-C(=O)-OH$

(b) 4.76   1.68
$H_3C-C(=O)-OH$ $NO_2CH_2-C(=O)-OH$

(c) 4.19   4.88
(phenyl)—COOH   (cyclohexyl)—COOH

(d) 1.92   3.02
HOOC—CH=CH—COOH (cis)   HOOC—CH=CH—COOH (trans)

(e) 6.23   4.38
HOOC—CH=CH—COO⁻ (cis)   HOOC—CH=CH—COO⁻ (trans)

PROBLEM 18.34 In Section 18.6b, we saw how dicyclohexylcarbodiimide (DCC) could be used as a dehydrating agent for the formation of amides from carboxylic acids and amines. Another reagent that can be used to good effect is the phosgene derivative N,N′-carbonyldiimidazole (CDI). Carboxylic acids and CDI react under mild conditions to form imidazolides, **1**, which then easily react with amines to give amides. Propose mechanisms for the formation of **1** from carboxylic acids and CDI, and for the formation of amides from **1** and amines.

(continued)

PROBLEM 18.34 *(continued)*

PROBLEM 18.35 In Section 18.6e and Figure 18.47, we saw that one method of forming an anhydride is to allow a dicarboxylic acid to react with another anhydride such as acetic anhydride, Ac$_2$O. Propose a mechanism for this process.

PROBLEM 18.36 In Problem 18.12, you worked out the mechanism for lactone formation from an hydroxy acid. Here is a related reaction, the formation of a cyclic amide, a lactam. Provide a mechanism.

A lactam
(97%)

PROBLEM 18.37 There is a compound called the Vilsmeier reagent (we will ask you about its formation in Chapter 19) that reacts with carboxylic acids to give acid chlorides and the molecule dimethylformamide (DMF). Provide a mechanism for this process.

Vilsmeier reagent   DMF

PROBLEM 18.38 Propose syntheses of the following molecules from the readily available fatty acid, lauric acid (**1**):

$$CH_3(CH_2)_{10}COOH \longrightarrow$$
1

(a) $CH_3(CH_2)_{11}Br$

(b) $CH_3(CH_2)_{10}Br$

(c) $CH_3(CH_2)_{11}COOH$

PROBLEM 18.39 Propose an arrow formalism mechanism for the following reaction. *Hint*: See Section 18.6g and Figure 18.59.

PROBLEM 18.40 Explain the following observations. Note that the only difference in the starting materials is the stereochemistry of the carbon–carbon double bond.

PROBLEM 18.41 Here are some somewhat more complicated syntheses. Devise ways to make the compounds shown below from cyclopentanecarboxylic acid.

PROBLEM 18.42 Estragole (**1**) is a major constituent of tarragon oil. Treatment of **1** with hydrogen bromide in the presence of peroxides affords **2**. Compound **2** reacts with Mg in ether, followed by addition of carbon dioxide and acidification, to give **3**. Spectral data for compound **3** are summarized below. Deduce the structure of compound **3** and propose structures for estragole and compound **2**.

Compound **3**

Mass spectrum: $m/z = 194$ (p)
IR (Nujol): 3330–2500 (br, s), 1696 (s) cm^{-1}
^{1}H NMR (CDCl$_3$): δ 1.95 (quintet, $J = 7$ Hz, 2H),
2.34 (t, $J = 7$ Hz, 2H), 2.59 (t, $J = 7$ Hz, 2H), 3.74 (s, 3H),
6.75 (d, $J = 8$ Hz, 2H), 7.05 (d, $J = 8$ Hz, 2H), 11.6 (s, 1H)

PROBLEM 18.43 Reaction of cyclopentadiene and maleic anhydride affords compound **1**. Hydrolysis of **1** leads to **2**, which gives an isomeric compound **3** upon treatment with concentrated sulfuric acid. Spectral data for compounds **1–3** are shown below. Propose structures for compounds **1–3** and provide mechanisms for their formations.

Compound **1**

Mass spectrum: $m/z = 164$ (p, 3%), 91 (23%), 66 (100%), 65 (22%)
IR (Nujol): 1854 (m) and 1774 (s) cm^{-1}
^{1}H NMR (CDCl$_3$): δ 1.60–1.70 (m, 1H), 3.40–3.50 (m, 1H), 3.70–3.80 (m, 1H), 6.20–6.30 (m, 1H)

Compound **2**

IR (Nujol): 3450–2500 (br, s), 1710 (s) cm^{-1}
^{1}H NMR (CDCl$_3$): δ 0.75–0.95 (m, 1H), 2.50–2.60 (m, 1H), 2.70–2.80 (m, 1H), 5.60–5.70 (m, 1H), 11.35 (s, 1H)

Compound **3**

IR (KBr): 3450–2500 (br, s), 1770 (s), 1690 (s) cm^{-1}
^{1}H NMR (CDCl$_3$): δ 1.50–1.75 (m, 3H), 1.90–2.05 (m, 1H), 2.45–2.55 (m, 1H), 2.65–2.75 (m, 1H), 2.90–3.05 (m, 1H), 3.20–3.30 (m, 1H), 4.75–4.85 (m, 1H), 12.4 (s, 1H)

PROBLEM 18.44 When N-nitroso-N-methylurea (NMU) is treated with a two-phase solvent system, H$_2$O/KOH and ether, the yellow color of diazomethane (p. 996) rapidly appears in the ether layer. Explain mechanistically.

N-Nitroso-N-methylurea
(NMU)

Diazomethane
(as a yellow solution in ether)

Derivatives of Carboxylic Acids: Acyl Compounds

19

Strange shadows from the flames will grow,
Till things we've never seen seem familiar . . .

—Jerry Garcia and Robert Hunter*
Terrapin Station

19.1 PREVIEW

In this chapter, we focus upon things that should seem familiar even though we have never seen them in detail. Many of the molecules we saw in Chapter 18 will reappear, for example. Acid halides, anhydrides, esters, and amides are all **acyl compounds** of the general structure R—CO—L. These compounds are shown in Figure 19.1 along with the related nitriles (cyanides) and ketenes.

FIGURE 19.1 Some acyl and related derivatives of carboxylic acids.

*Jerry Garcia (1942–1995) played lead guitar with the Grateful Dead, and Robert Hunter (b. 1941) was a frequent songwriting collaborator.

Not only are the structural types of these molecules related, but their chemistries are also closely intertwined—a reaction of one acyl compound is generally a synthesis of another, for example. Luckily, there are unifying properties of these seemingly disparate compounds. The structures of the acyl compounds are determined largely by the interaction between the L group and the carbonyl, and their chemistries are dominated by addition–elimination mechanisms in which the L group is replaced with some other substituent (Chapter 18). Note the tetrahedral intermediate at the center of this generic mechanism. This addition–elimination process is an equilibrium, and can proceed in both the "forward" and "backward" directions. The old leaving group $L:^-$ is also a nucleophile (Fig. 19.2).

FIGURE 19.2 The most important mechanism for reactions of acyl compounds is the addition–elimination process.

Esters, in particular, have a rich chemistry of the enolate anions produced by removal of an α-hydrogen. We will see some new condensation reactions that will greatly expand your ability to do "Fun in Base" problems.

We will first go over some common ground—nomenclature, structure, and spectra—and then go on to examine reactions and syntheses for the different derivatives.

ESSENTIAL SKILLS

1. The Claisen condensation bears the same relation to the addition–elimination mechanisms of Chapter 18 as the aldol condensation of Chapter 17 does to the reversible additions to aldehydes and ketones described in Chapter 16. Be sure you see the relationship.

2. As always, we need to be able to generalize. This chapter describes the chemistry of a set of related acyl compounds. It will be very hard to memorize the vast array of addition–elimination processes, but it should be relatively easy to see them all as the same process repeated over and over. Only the detailed structure changes—the overall process does not.

IMPORTANT DETAILS

1. One such detail is a clear understanding of why the Claisen condensation requires a full equivalent of base, not just a catalytic amount. If you understand this point, you understand the Claisen condensation.

2. The synthetic procedures known as the acetoacetic ester synthesis and the malonic ester synthesis are very useful. The former is a synthesis of ketones; the latter is a synthesis of acids.

3. In addition reactions, carbon–nitrogen triple bonds (nitriles or cyanides) behave much as do carbonyl compounds.

19.2 NOMENCLATURE

All the names of these compounds are based on the parent carboxylic acid. For molecules containing up to five carbons, common names prevail, but after that the system takes over. For small **esters** (R—COOR′), the R group is named first and the "ic" of the parent acid is replaced with the suffix "ate." So we have alkyl (methyl, ethyl, etc.) formates, acetates, propionates, and butyrates before we get to the systematic pentanoates. In the systematic naming protocol used for larger esters, the R′ group of the OR′ is again first named and then the suffix "oate" replaces the final "e" of the parent alkane. As usual, the carbonyl group of the acid or ester gets top priority in the naming scheme and substituents are numbered with respect to the carbonyl as position-1 (Fig. 19.3).

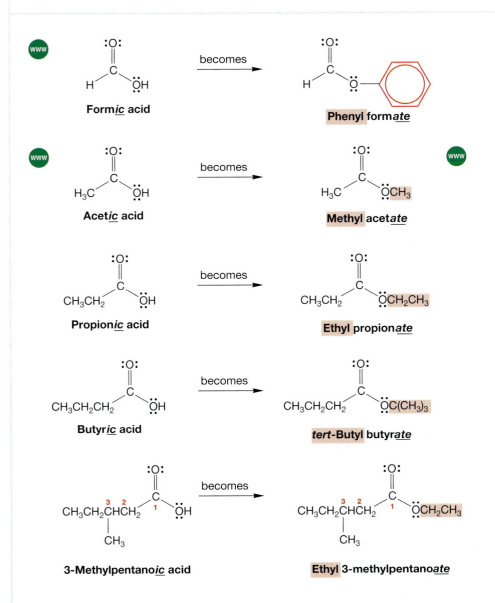

FIGURE 19.3 Some examples of the naming protocol for esters.

Cyclic esters are called **lactones** (Chapter 18, p. 990). The system names these compounds as "oxa" cycloalkanones. The ring size of a lactone can be described by starting at the carbonyl carbon and designating the other carbons in the ring with Greek letters until the oxygen atom is reached. A three-membered ring is an α-lactone, a four-membered ring is a β-lactone, and so on (Fig. 19.4).

2-Oxacyclopentanone **2-Oxacycloheptanone** **α-Lactone** **β-Lactone** **γ-Lactone** **δ-Lactone**

FIGURE 19.4 Cyclic esters are lactones or oxacycloalkanones.

Acid halides are named systematically by replacing the final "e" of the parent alkane with "oyl halide." In practice, the names for the smaller acid halides are based upon the common names for the related acids. Thus, "acetyl chloride," "propionyl bromide," and "butyryl fluoride" would be understood everywhere. Formyl chloride is most unstable and efforts to make it are rewarded only with HCl and CO. The diacid chloride of carbonic acid (Cl—CO—Cl) is known, however, and this deadly poison is called phosgene (Fig. 19.5).

Methanoic acid
(formic acid) *becomes* **Methanoyl chloride**
(formyl chloride) *becomes* **Phosgene**

Ethanoic acid
(acetic acid) *becomes* **Ethanoyl chloride**
(acetyl chloride) **Propanoic acid**
(propionic acid) *becomes* **Propanoyl bromide**
(propionyl bromide)

Butanoic acid
(butyric acid) *becomes* **Butanoyl fluoride**
(butyryl fluoride) **Pentanoic acid** *becomes* **Pentanoyl chloride**

FIGURE 19.5 Examples of the systematic and common naming protocols for acid halides.

Anhydrides are named by reference to the acid from which they are formally derived by loss of water. Both systematic and common names are used in this process (Fig. 19.6).

Ethanoic acid
(acetic acid)

becomes

Ethanoic anhydride
(acetic anhydride)

Benzoic acid

becomes

Benzoic anhydride

FIGURE 19.6 Two examples of the naming protocol for symmetrical anhydrides.

Mixed anhydrides present more problems. Here, both constituent acids must be named alphabetically. Two examples appear in Figure 19.7.

Ethanoic propanoic anhydride
(acetic propionic anhydride)

Benzoic butanoic anhydride
(benzoic butyric anhydride)

FIGURE 19.7 Named unsymmetrical anhydrides.

Cyclic anhydrides are named as derivatives of the related carboxylic acids (Fig. 19.8).

Phthalic acid

becomes

Phthalic anhydride

Succinic acid

becomes

Succinic anhydride

Maleic acid

becomes

Maleic anhydride

FIGURE 19.8 Some cyclic anhydrides.

Amides, the nitrogen counterparts of esters, are systematically named by dropping the final "e" of the parent alkane and adding the suffix "amide." As usual, the smaller members of the class have retained their common names, which are based on the common names for the smaller carboxylic acids. Substitution on nitrogen is indicated with a prefix *N*, as in *N*-methyl, *N*,*N*-diethyl, and so on (Fig. 19.9).

Ethanoic acid
(acetic acid)

becomes

Ethanamide
(acetamide)

3-Methylbutanoic acid

becomes

3-Methylbutanamide

Benzoic acid

becomes

Benzamide

CH_3CH_2 $N(CH_2CH_3)_2$
***N*,*N* - Diethylpropanamide**

FIGURE 19.9 Examples of the naming protocol for amides.

Cyclic amides are known as **lactams**, and the naming protocol closely follows that for lactones, the cyclic esters (p. 1032). Lactams are named systematically as "azacyclo-alkanones." **Imides** are the nitrogen counterparts of anhydrides (Fig. 19.10).

2-Azacyclohexanone
(δ-valerolactam)

2-Azacyclobutanone
(β-propiolactam)

Phthalic acid

becomes

Phthalimide

FIGURE 19.10 Cyclic amides are called lactams, and imides are the nitrogen counterparts of anhydrides.

Compounds containing carbon–nitrogen triple bonds, **nitriles**, are also known as **cyanides**. Systematically, these compounds are alkanenitriles. Notice that the final "e" of the alkane is not dropped. Small nitriles are commonly named as derivatives of the parent carboxylic acid or as cyanides (Fig. 19.11).

FIGURE 19.11 Some examples of the naming protocol for nitriles, or cyanides.

Ketenes (not to be confused with ket*o*nes), are named simply as derivatives of the parent unsubstituted ketene. Recall the allenes, compounds with two directly attached carbon–carbon double bonds and carbon dioxide (Fig. 12.3, p. 551; Fig. 19.12).

FIGURE 19.12 Ketenes are named as derivatives of the parent unsubstituted ketene.

19.3   PHYSICAL PROPERTIES AND STRUCTURES OF ACYL COMPOUNDS

Acyl compounds are quite polar and have boiling points substantially higher than those of the alkanes from which they are formally derived. Amides, like carboxylic acids, form hydrogen-bonded dimers and oligomers, and are exceptionally high boiling. Table 19.1 gives some physical properties of the acyl compounds related to acetic acid.

TABLE 19.1   Some Properties of Acyl Compounds Related to Acetic Acid

Formula	Name	bp (°C)	mp (°C)
CH$_3$COOH	Acetic acid	117.9	16.6
CH$_3$COOCH$_3$	Methyl acetate	57	− 98.1
CH$_3$COOCH$_2$CH$_3$	Ethyl acetate	77	− 83.6
CH$_3$COCl	Acetyl chloride	50.9	−112
CH$_3$COBr	Acetyl bromide	76	− 98
CH$_3$CO—O—COCH$_3$	Acetic anhydride	139.6	− 73.1
CH$_3$CONH$_2$	Acetamide	221.2	82.3
CH$_3$CN	Acetonitrile	81.6	− 45.7
CH$_2$=C=O	Ketene	−56	−151

Acyl compounds lack the hydroxyl group of carboxylic acids, and so are not strong Brønsted acids. However, hydrogens in the α-position are acidic for the same reasons that α-hydrogens are acidic in aldehydes and ketones. The conjugate bases are resonance stabilized by the adjacent carbonyl or nitrile group (Fig. 19.13).

FIGURE 19.13 A hydrogen at the α-position of an acyl compound or nitrile is acidic and can be removed by base to give an enolate, or, in the case of nitriles, a resonance-stabilized species much like an enolate.

The ease of loss of a proton at the α-position varies, however, depending on the ability of the adjacent carbonyl group to provide stabilization (Table 19.2).

Esters and amides are more stabilized by resonance than are aldehydes and ketones. Therefore their carbonyl groups have less double-bond character, and they are weaker acids at the α-position. Notice the connection between resonance stabilization and acidity. The more resonance stabilized the starting material, the more difficult it will be to remove a proton from it.

Amides are the most resonance stabilized of acyl compounds, as in the polar resonance form it is the relatively electropositive nitrogen atom that bears the positive charge. For this reason these compounds are the weakest acids (Table 19.2; Fig. 19.14).

TABLE 19.2 The Acidity of Some Acyl Compounds[a]

Compound	pKa
Acetaldehyde	16.7
Acetone	19.3
Ethyl acetate	24
Acetamide	25
Acetonitrile	24

[a]The arrow shows the proton lost.

FIGURE 19.14 Esters and amides are more stabilized by resonance than aldehydes and ketones—it is harder to remove an α-hydrogen from the R group of these more stable species. They are weaker acids as shown by their higher pKa values.

PROBLEM 19.1 Draw structures for the following compounds: cyclopropyl 2-methylbutyrate, 3-chlorobutanoyl chloride, benzoic propionic anhydride, *N,N*-diethyl-4-phenylpentanamide, and ethylpropylketene.

Acyl compounds are also Lewis acids. The carbonyl group provides the locus of Lewis acidity for most of them, but nitriles, even though they have no carbon–oxygen double bond, are still Lewis acids. The electronegative nitrogen strongly polarizes the carbon–nitrogen triple bond, which can react like a carbonyl group in reactions with nucleophiles (Fig. 19.15).

Acyl compounds are Brønsted bases and Lewis bases as well, with the carbonyl oxygen or the nitrile nitrogen the center of basicity (Fig. 19.16).

FIGURE 19.15 The carbonyl group of acyl compounds is the center of Lewis acidity and can be attacked by nucleophiles (Nu:⁻) just as are aldehydes and ketones. An analogous reaction occurs with nitriles.

FIGURE 19.16 As these protonation reactions show, acyl compounds and nitriles can act as Brønsted bases.

The carbonyl-containing molecules vary in their base strengths in the following order: amides > esters > acid halides. As we have seen, all these compounds are stabilized by dipolar resonance forms in which the L group bears a positive charge and the carbonyl oxygen a negative charge (Fig. 19.14). Acid chlorides are the least stabilized by this resonance, and therefore have the least basic carbonyl oxygen atoms.

Esters, anhydrides, and acids are similarly stabilized by resonance, but in amides it is a nitrogen atom, not an oxygen atom, that bears the positive charge. Nitrogen is less electronegative than oxygen and therefore bears this positive charge with more grace than oxygen. The result is a more polar carbonyl group and a more basic carbonyl oxygen atom (Fig. 19.17).

FIGURE 19.17 The carbonyl oxygens of amides are stronger Lewis bases (and nucleophiles) than are those of esters. Acid halide carbonyls are the weakest nucleophiles of all.

Stronger Lewis base
More charge on oxygen
More resonance stabilization

Poor (high-energy) resonance forms

Resonance stabilization influences the structures of these compounds as well. Although the carbon–chlorine bond distance is little changed in acid chlorides from that in alkyl chlorides, the carbon–oxygen and carbon–nitrogen bond distances in esters and amides are shortened, reflecting some double-bond character in these compounds. In the much less well resonance stabilized acyl halides there is little double-bond character in the carbon–halide bond and little shortening (Fig. 19.18).

1.20 Å	1.22 Å	1.19 Å

$$:O:$$ $$:O:$$ $$:O:$$

H—C—O—CH₃   H—C—NH₂   H₃C—C—Cl:

1.33 Å	1.35 Å	1.78 Å

H₃C—Ö—H   H₃C—NH₂   H₃C—Cl:

1.43 Å	1.47 Å	1.78 Å

FIGURE 19.18 The structures of some simple acyl compounds. Notice how the carbon–oxygen and carbon–nitrogen single bonds are shortened by their double bond character.

Nitriles (cyanides) are linear molecules with short carbon–nitrogen bond distances. The *sp* hybridization of the carbon and nitrogen atoms ensures this. Figure 19.19 gives orbital pictures of both the σ and π systems, which closely resemble those of the alkynes (Chapter 3, p. 124).

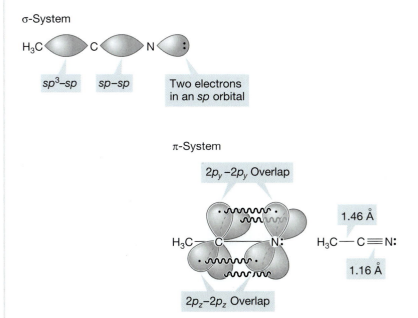

σ-System

H₃C   C   N:

sp³–sp   *sp–sp*   Two electrons in an *sp* orbital

π-System

2*p_y*–2*p_y* Overlap

H₃C—C   N:

2*p_z*–2*p_z* Overlap

1.46 Å

H₃C—C≡N:

1.16 Å

FIGURE 19.19 The σ and π orbital systems of acetonitrile.

19.4   SPECTRAL CHARACTERISTICS

19.4a Infrared Spectra. Characteristic strong carbonyl stretching frequencies are observed for all these compounds (except, of course, for the cyanides, which contain no carbonyl group). The contribution of the dipolar resonance forms can be seen in the position of the C=O stretch. The acid chloride carbonyl is the strongest, with the most double-bond character, so its C=O stretch appears at the highest frequency. Esters, acids, and anhydrides are next. Amides, in which the contribution from the dipolar resonance form is strongest, and the *single*-bond character of the carbonyl group is greatest, have the lowest carbonyl stretching frequencies (Table 19.3).

TABLE 19.3   Infrared Stretching Frequencies of Some Carbonyl Compounds

Compound	Frequency of C=O in CCl_4 (cm^{-1})	Compound	Frequency of C=O in CCl_4 (cm^{-1})
$H_2C=C=O$ **Ketene**	2151	**Acetaldehyde**	1733
Acetic anhydride	1833, 1767	**Acetone**	1719
Acetyl chloride	1799	**Acetic acid**	1717
Methyl acetate	1750	**N-Methyl acetamide**	1688

The green and red π bonds are not as alike as it seems at first

FIGURE 19.20 The bonding scheme in ketene.

Ketenes have very strong carbonyl groups, as the inner carbon is hybridized *sp*, not the usual *sp²* (Fig. 19.20). The more *s* character in a bond, the stronger it is. Therefore, ketenes absorb in the IR at very high frequency (~ 2150 cm^{-1}).

In the anhydrides, which contain two carbonyl groups, there are usually two carbonyl stretching frequencies. This doubling is not the result of independent stretching of the separate carbonyls, because symmetrical anhydrides as well as unsymmetrical anhydrides show the two IR bands.

Instead, the two bands are caused by the coupled symmetrical and unsymmetrical stretching modes of these compounds (Fig. 19.21). Remember, vibrations of bonds in molecules are no more independent of each other than are the vibrations of attached real springs (Chapter 15, p. 761).

Nitriles show a strong carbon–nitrogen triple-bond stretch at characteristically high frequency (2200–2300 cm^{-1}). Notice that this is quite near the frequency of the related carbon–carbon triple bonds (2100–2300 cm^{-1}).

19.4b Nuclear Magnetic Resonance Spectra. Hydrogens in the α-position are deshielded by the carbonyl or nitrile group, and therefore appear at relatively low field. The carbonyl carbons of acyl compounds, and the nitrile carbons of cyanides bear partial positive charges and appear at relatively low field in the ^{13}C NMR spectra (Table 19.4).

Coupled symmetrical stretch

Coupled unsymmetrical stretch

FIGURE 19.21 The two carbonyl stretching frequencies of anhydrides.

TABLE 19.4 Some NMR Properties of Carbonyl Compounds and CH$_3$CN (ppm, CDCl$_3$)

Compound	δ^{13}C (C=O, C≡N)	δ (H$_\alpha$)	Compound	δ^{13}C (C=O, C≡N)	δ (H$_\alpha$)
Acetaldehyde	199	2.25	Ethyl acetate	169	2.05
Acetone	205	2.18	N,N-Dimethyl acetamide	170	2.09
Acetyl chloride	169	2.66	Acetonitrile	117	2.05

PROBLEM 19.2 In the room temperature ^{1}H NMR spectrum of dimethylformamide (DMF), *two* methyl resonances appear. As the temperature is raised they merge into a single signal. Explain.

Dimethylformamide (DMF)

19.5   REACTIONS OF ACID CHLORIDES: SYNTHESIS OF ACYL COMPOUNDS

As mentioned before, all acyl compounds participate in the addition–elimination process. Acid chlorides are especially reactive toward nucleophiles. The carbonyl group, being the least stabilized by resonance, has the highest energy and is the most reactive. So, an initial addition reaction with a nucleophile is relatively easy. The chloride atom of acid chlorides is an excellent leaving group, and sits poised, ready to depart once the tetrahedral addition product has been formed. The result is an exceedingly facile and common example of the addition–elimination mechanism (recall Chapter 18, p. 990; Fig. 19.22).

THE GENERIC ADDITION–ELIMINATION REACTION

Tetrahedral intermediate

GENERAL REACTIONS

Acid chloride aminolysis

FIGURE 19.22 Addition–elimination reactions of acid chlorides. Note the synthetic potential.

(continued)

SOME SPECIFIC EXAMPLES

Amide (80%)

Ester (100%)

Acid (82%)

FIGURE 19.22 *(continued)*

PROBLEM 19.3 Acid chlorides often smell like hydrochloric acid. Explain.

Many nucleophiles are effective in this reaction, and a great many acyl compounds can be made using acid halides, usually acid chlorides, as starting materials.

Of course, strong nucleophiles such as organolithium compounds, Grignard reagents, and the metal hydrides also react rapidly with acid halides. The problem here is one of control of secondary reactions. For example, reduction of an acid chloride with lithium aluminum hydride initially gives an aldehyde. But the newly born aldehyde finds itself in the presence of a reducing agent easily strong enough to reduce it further to the alkoxide, and hence to the primary alcohol after water is added (Fig. 19.23).

THE GENERAL CASE

The aldehyde is born in the presence of LiAlH₄ and *must* react further

Final product primary alcohol

FIGURE 19.23 The reaction of an acid chloride with lithium aluminum hydride leads to a primary alcohol.

(continued)

A SPECIFIC EXAMPLE

FIGURE 19.23 *(continued)*

Similarly, reaction of an acid halide with an organometallic reagent initially gives a ketone that usually reacts further to give the tertiary alcohol. Note that this reaction must give a tertiary alcohol in which at least two R groups are the same (Fig. 19.24).

THE GENERAL CASE

This ketone is born in the presence of an organolithium reagent and *must* react further

A SPECIFIC EXAMPLE

FIGURE 19.24 The reaction of an acid chloride with an organometallic reagent proceeds all the way to the tertiary alcohol.

All is fine if we are trying to make the alcohols, but often we are not. How to stop the process at the aldehyde or ketone stage is the problem, and several solutions have been found over the years. Less reactive organometallic reagents or metal hydrides allow the isolation of the intermediate aldehydes or ketones. The reactivity of lithium aluminum hydride can be attenuated by replacing some of the hydrogens with other groups. For example, use of lithium aluminum tri-*tert*-butoxyhydride allows isolation of the aldehyde. Presumably, the bulky metal hydride is unable to reduce the aldehyde, which is less reactive than the original acid halide (Fig. 19.25).

THE GENERAL CASE

Lithium aluminum tri-*tert*-butoxyhydride

A SPECIFIC EXAMPLE

1. LiAl[OC(CH$_3$)$_3$]$_3$H
   −78 °C
2. H$_2$O

(85%)

FIGURE 19.25 Lithium aluminum tri-*tert*-butoxyhydride is not reactive enough to add to the initially formed aldehyde.

Catalytic reduction using a deactivated, or "poisoned" catalyst is called the **Rosenmund reduction** after Karl W. Rosenmund (1884–1964) (Fig. 19.26). In this reaction, only the more reactive acid chloride is attacked. The more stable, less reactive aldehyde can be isolated.

THE GENERAL CASE

Quinoline
a catalyst poison

A SPECIFIC EXAMPLE

H$_2$/BaSO$_4$/Pd
quinoline
145 °C

(77%)

FIGURE 19.26 The Rosenmund reduction of acid chlorides to aldehydes.

Organocuprates react with acid chlorides but are not reactive enough to add to the product ketones (Fig. 19.27).

FIGURE 19.27 Cuprates are reactive enough to add R⁻ to the carbonyl group of the acid chloride, but not reactive enough to attack the product ketones.

Finally, don't forget that acid chlorides are used in the synthesis of aromatic ketones through Friedel–Crafts acylation (Chapter 14, p. 692; Fig. 19.28).

FIGURE 19.28 Friedel–Crafts acylation of benzene.

PROBLEM 19.4 Write a mechanism for the formation of the product of Figure 19.28.

PROBLEM 19.5 Starting with acetyl chloride, devise syntheses of the following molecules. You may also use dimethylcuprate.

SUMMARY

The presence of the good leaving group chloride attached directly to the carbon–oxygen double bond makes all manner of addition–elimination reactions possible for acid chlorides.

19.6   REACTIONS OF ANHYDRIDES

The reactivity of anhydrides is qualitatively similar to that of acid chlorides. A carboxylate salt is the leaving group in a variety of syntheses of acyl derivatives, all of which are examples of the addition–elimination process. One reaction of this kind is the basic hydrolysis of phthalic anhydride to phthalic acid. Here the leaving group is an internal carboxylate anion (Fig. 19.29).

THE GENERAL CASE

FIGURE 19.29 Addition–elimination reactions of an anhydride.

(continued)

A SPECIFIC EXAMPLE

Phthalic anhydride

Phthalic acid

FIGURE 19.29 *(continued)*

PROBLEM 19.6 Provide a mechanism for the formation of *N*-phenylmaleimide from maleic anhydride and aniline. Be sure you give a structure for the intermediate **A**, and account for the role of the acetic anhydride in the second step.

Maleic anhydride   **Aniline**   **A**   acetic anhydride   **N-Phenylmaleimide**

Like acid chlorides, anhydrides can be used in the Friedel–Crafts reaction. Once again, a strong Lewis acid catalyst such as aluminum chloride is used to activate the anhydride. The mechanism follows the pattern of the Friedel–Crafts reactions we saw earlier (Chapter 14, p. 692; Fig. 19.30).

THE GENERAL CASE   A SPECIFIC EXAMPLE

(96%)

FIGURE 19.30 A Friedel–Crafts reaction using an anhydride as the source of the acyl group.

PROBLEM 19.7 Write a mechanism for the reaction of Figure 19.30.

19.7 REACTIONS OF ESTERS: ADDITION–ELIMINATION

Esters are less reactive than acid chlorides in addition reactions, but more reactive than amides. Esters can be hydrolyzed to their parent acids under either basic or acid conditions in a process called, logically enough, **ester hydrolysis**. In base, the mechanism is the familiar addition–elimination one: Hydroxide ion attacks the carbonyl group to form a tetrahedral intermediate. Loss of alkoxide then gives the acid, which is rapidly deprotonated to the carboxylate anion in basic solution. Notice that this reaction, **saponification**, is not catalytic. The hydroxide ion used up in the reaction is not regenerated at the end. To get the acid itself, a final acidification step is necessary (Fig. 19.31).

Ester hydrolysis

FIGURE 19.31 The base-induced ester hydrolysis reaction, saponification.

Esters can also be converted into carboxylic acids in acid. The carbonyl oxygen is first protonated to give a resonance-stabilized cation to which water adds. Water is by no means as nucleophilic as hydroxide, but the protonated carbonyl is a very strong Lewis acid and the overall reaction is favorable (Fig. 19.32).

FIGURE 19.32 The acid-catalyzed ester hydrolysis reaction.

FIGURE 19.33 Acids and esters are usually interconvertable.

The reaction concludes with proton transfers and loss of alcohol. Have you seen this reaction before? You certainly have. It is the exact reverse of the mechanism for Fischer esterification (Chapter 18, p. 986), so there is nothing new here at all. What determines where the overall equilibrium settles out? The structure of the ester is important, but much more important are the reaction conditions. Excess water favors the acid; excess alcohol favors the ester. LeChatelier's principle is at work here, and the chemist has the ability to manipulate this equilibrium to favor either side. In practice, most acids and esters are easily interconvertible (Fig. 19.33).

PROBLEM 19.8 Where appropriate, draw resonance forms for the intermediate species in Figure 19.33. You will have to write mechanisms for the reactions first.

A very closely related reaction is acid-catalyzed **transesterification**. In this reaction, an ester is treated with an excess of an alcohol and an acid catalyst. The result is replacement of the ester OR group with the alcohol OR group (Fig. 19.34). Like ester hydrolysis, transesterification can be carried out under either acid or basic conditions. The mechanisms are uncomplicated extensions of those you have already seen in the hydrolysis reaction or Fischer esterification.

FIGURE 19.34 Base- and acid-catalyzed transesterification.

PROBLEM 19.9 Write mechanisms for acid- and base-induced transesterification.

WORKED PROBLEM 19.10 There is an important difference between the reaction of an ester with hydroxide ($^-$OH) and alkoxide ($^-$OR). The reaction with hydroxide is neither catalytic nor reversible, whereas the reaction with alkoxide is both catalytic and reversible. Analyze the mechanisms of these reactions and explain these observations.

ANSWER Look at the final products of these two reactions. Reaction with hydroxide leads to a carboxylic acid ($pK_a \sim 4.5$) and an alkoxide ion. These two species *must* react very rapidly to make the more stable carboxylate anion (resonance stabilized) and the alcohol ($pK_a \sim 17$). The catalyst, hydroxide, is destroyed in this reaction, and the overall process is so thermodynamically favorable that it is irreversible in a practical sense.

(continued)

(continued)

With an alkoxide, things are different. Now the product is not a carboxylic acid but another ester. There is no proton that can be removed, and there is a new molecule of alkoxide generated. The reaction is approximately thermoneutral.

WORKED PROBLEM 19.11 Given that there are acid- and base-catalyzed transesterification reactions, should there not be both acid- and base-catalyzed Fischer esterifications as well? Explain carefully why there is no base-catalyzed version of Fischer esterification (RCOOH + RO⁻ → RCOOR + HO⁻).

ANSWER The reaction of a carboxylic acid with an alkoxide can't proceed by addition–elimination to give an ester because there is another much easier reaction available; that reaction is simple removal of the carboxylic acid hydroxyl proton to give the resonance-stabilized carboxylate anion. They don't call these compounds "acids" for nothing! The lesson in this problem is that you have to "think simple." Look first for "trivial" reactions (loss of the proton) before proceeding on to more complicated processes (addition–elimination).

$pK_a \sim 4.5$ $pK_a \sim 17$

PROBLEM 19.12 Esters are also able to react with amines to give amides. Write a mechanism for the reaction of methyl acetate (CH_3COOCH_3) and ammonia to form acetamide (CH_3CONH_2).

SUMMARY

There are both acid- and base-catalyzed versions of the reaction of esters with water or alcohols (hydrolysis or transesterification). In the last chapter (Chapter 18, p. 986), we saw the acid-catalyzed version of the conversion of an acid into an ester, called Fischer esterification.

Organometallic reagents and metal hydrides both react with esters in still another example of the addition–elimination process. Usually, the organometallic reagents react further with the ketones that are the initial products of the reactions. The ketones are not as well stabilized by resonance as are the esters and so are more reactive in the addition reaction (Fig. 19.35).

FIGURE 19.35 The reaction of esters with organolithium reagents to form complex alcohols.

This reaction is another general synthesis of complex tertiary alcohols. Here are the possibilities. First, tertiary alcohols in which all three R groups are the same can be made if the R group of the organometallic reagent is the same as that of the ester alkyl group. Second, the R of the organometallic reagent can be different from that of the ester. In this case, the tertiary alcohol will contain two different R groups (Fig. 19.36).

GENERAL CASES

FIGURE 19.36 Tertiary alcohols in which all three R groups are the same, or molecules in which only two R's are the same can be made, but the structural type in which all three R groups are different cannot be made this way.

(continued)

SPECIFIC EXAMPLES

(82%)

(83%)

FIGURE 19.36 (continued)

Because two R groups must come from the organometallic reagent, there is no way to use this ester-based synthesis to make a tertiary alcohol with three different R groups.

PROBLEM 19.13 Dialkyl carbonates can be made from the reaction of phosgene with alcohols. These carbonates are converted into tertiary alcohols with three identical R groups on reaction with Grignard reagents or organolithium reagents. Provide mechanisms for the formations of carbonates and tertiary alcohols in these reactions.

Phosgene   **Carbonates**   **Alcohols**

Metal hydride reduction of esters is difficult to stop at the intermediate aldehyde stage, and the usual result is further reduction to the primary alcohol. Either LiAlH$_4$ or LiBH$_4$ (but not NaBH$_4$, which will not reduce esters) is the reagent of choice (Fig. 19.37).

THE GENERAL CASE

SPECIFIC EXAMPLES

(90%)

(90%)

FIGURE 19.37 The reduction of esters with powerful hydride donors such as lithium aluminum hydride and lithium borohydride leads to primary alkoxides and then, after acidification, to primary alcohols.

As with acid chlorides, other metal hydrides that permit isolation of the intermediate aldehydes have been developed. For example, diisobutylaluminum hydride (DIBAL-H) reduces ketones to alcohols, but can be used to isolate the intermediate aldehydes from esters if care is taken (Fig. 19.38).*

FIGURE 19.38 If DIBAL-H is used as the hydride donor, the reduction of esters can often be stopped at the aldehyde stage.

Esters also have a substantial chemistry involving the related enolate anions. We will develop this subject in detail later in this chapter.

PROBLEM 19.14 Suggest one reason why DIBAL-H might be slower than $LiAlH_4$ to react with the aldehyde of Figure 19.38.

19.8   REACTIONS OF AMIDES

These least reactive of the acyl compounds can nevertheless be hydrolyzed in either acid or base. The mechanisms of these reactions are directly related to those of acid- or base-induced ester hydrolysis. In base, the carbonyl is attacked to give tetrahedral intermediate **A**. Loss of amide ion ($^-NH_2$) gives the carboxylic acid, and deprotonation of the acid gives the carboxylate salt. Loss of amide ion is surely difficult, as $^-NH_2$ is a much stronger base than hydroxide. However, when amide is lost, a carboxylic acid is formed, and is quickly converted into the carboxylate. So, loss of amide, though slow, is essentially irreversible. A final acidification regenerates the carboxylic acid in isolable form. Overall, it's addition–elimination once again (Fig. 19.39).

FIGURE 19.39 The base-induced hydrolysis of amides gives carboxylate anions. Acidification gives the acid.

*This reaction is actually complicated and tricky. See Problem 19.71 in the Additional Problems section.

PROBLEM 19.15 In strong base, the hydrolysis of an amide is second order in hydroxide. Two molecules of HO⁻ are required to do the hydrolysis. Write a modified mechanism to account for the need for two molecules of hydroxide.

In acid, the carbonyl oxygen is protonated creating a Lewis acid strong enough to react with the relatively weak nucleophile water. The leaving group in the acid-induced reaction is the amine (Fig. 19.40).

FIGURE 19.40 The acid-induced hydrolysis of amides gives the carboxylic acid directly.

WORKED PROBLEM 19.16 Write resonance forms for the structures in Figure 19.40 where appropriate.

ANSWER This problem is very similar to Problem 19.8. In this case, it is the protonated amide and protonated acid that are most stabilized by resonance.

The starting amide and acid are also resonance stabilized, but this kind of stabilization is more important for the charged intermediates. For example, the resonance forms for the amide and acid are charge-separated species.

Similar acid- or base-catalyzed reactions of amides with alcohols lead to esters (Fig. 19.41).

FIGURE 19.41 Amides can be converted into esters in acid or base.

Ester   Amide   Ester

PROBLEM 19.17 Write detailed mechanisms for the reactions of Figure 19.41.

Metal hydrides reduce amides to amines. Clearly, something new is happening here, as analogy to the reactions of acids, esters, and acid chlorides would lead us to expect alcohols as products (Fig. 19.42).

FIGURE 19.42 By analogy to what we know of other acyl compounds, we might well expect the conversion of amides to alcohols to be the result of treatment of an amide with a metal hydride. In fact, this reaction does *not* take place.

The mechanism of this reaction is difficult, and may vary with the number of alkyl groups on nitrogen. However, the first step in this reaction is the same as in the other reductions by metal hydrides; the hydride adds to the carbonyl to give a complex alkoxide. Now, however, the metal oxide is lost with formation of an **immonium ion**. There are several possible ways to describe the formation of the immonium ion, but once it is produced, reduction by the metal hydride to give the amine is sure to be rapid (Fig. 19.43).

THE GENERAL CASE

Immonium ion   Amine

A SPECIFIC EXAMPLE

(88%)

FIGURE 19.43 Amines are the real products of the reduction of amides with metal hydrides.

There is no substantial chemistry of amide enolates. The problem is that the α-hydrogens on the amide nitrogen are more acidic than the α-hydrogens on carbon, and therefore are more easily removed by base to give anions (Fig. 19.44).

FIGURE 19.44 There are two kinds of α-hydrogen in most amides. It is much easier to remove the proton attached to nitrogen than the proton attached to carbon.

PROBLEM 19.18 Explain why α-hydrogens on the nitrogen of amides are more acidic than α-hydrogens on carbon (Fig. 19.44).

The resulting anions can be alkylated by reactive alkyl halides (Fig. 19.45).

THE GENERAL CASE

A SPECIFIC EXAMPLE

FIGURE 19.45 Alkylation of amidate anions gives substituted amides. This reaction is just an S$_N$2 substitution in which the amidate acts as nucleophile.

If there are no hydrogens on the nitrogen, it becomes possible to form enolates. Strong, hindered bases such as lithium diisopropylamide (LDA; Chapter 18, p. 1009) are generally used (Fig. 19.46).

FIGURE 19.46 Amides in which there are no hydrogens attached to nitrogen can form enolates.

No α-hydrogens on nitrogen

Enolate anion

19.9   REACTIONS OF NITRILES

Of course nitriles, or cyanides, are not carbonyl compounds. But they are closely related, and an examination of their chemistry naturally falls into this section. Like carbonyl compounds, nitriles are both Lewis acids and Lewis bases. As in carbonyl compounds, a polar resonance form contributes to the structure of nitriles. Consequently, the triple bond is polarized and can act as a Lewis acid at carbon. One might anticipate that nucleophiles would add to nitriles, and that idea would be correct (Fig. 19.47).

FIGURE 19.47 Nitriles are Lewis acids and nucleophiles add to them to give imine anions. There is a close analogy to carbonyl chemistry here.

An imine anion

The nitrogen atom bears a pair of nonbonding electrons and is therefore the center of Lewis basicity and nucleophilicity. The nitrogen is *sp* hybridized, however, and is a relatively weak Lewis base.

PROBLEM 19.19 Explain why an *sp* hybridized nitrogen should be a weaker base than an sp^2 hybridized nitrogen. What about sp^3 hybridized nitrogens?

Nitriles are also Brønsted acids, as removal of an α-hydrogen gives a resonance-stabilized anion (Fig. 19.48). As a result, the pK_a values of nitriles are similar to those of esters (~24). Be sure to compare the resonance-stabilized species in Figure 19.48 to the more conventional enolates formed from aldehydes, ketones, and esters.

FIGURE 19.48 In another reaction reminiscent of carbonyl chemistry, an α-hydrogen of a nitrile can be removed in base to give a resonance-stabilized species closely related to an enolate.

Resonance stabilized
Note the similarity to an enolate

These similarities mean that many of the reactions of carbonyl groups are possible for nitriles, and the mechanisms are generally closely related. Students sometimes have trouble with these reactions, however, so it is important not to dismiss them with only a cursory look. For some reason, the analogy between the carbon–oxygen double bond and the carbon–nitrogen triple bond is not always easy to keep in mind.

Both acid- and base-induced hydrolysis of a nitrile gives the amide, but rather severe conditions are required for the reactions, and further hydrolysis of the intermediate amides gives the carboxylic acids (p. 1055). In acid, the first step is protonation of the nitrile nitrogen to give a stong Lewis acid (**A**), which is attacked by water. A series of proton shifts gives an amide that is hydrolyzed to the acid (Fig. 19.49).

Nitrile hydrolysis

FIGURE 19.49 The acid-induced hydrolysis of a nitrile to a carboxylic acid.

In base, the nucleophilic hydroxide ion adds to the carbon–nitrogen triple bond. Protonation on nitrogen and deprotonation on oxygen leads to the amidate anion, which is protonated to give an intermediate amide. Basic hydrolysis of the intermediate amide leads to the carboxylate salt (Fig. 19.50).

FIGURE 19.50 The base-induced hydrolysis of a nitrile to give a carboxylate anion.

Stronger bases also add to nitriles. Organometallic reagents give ketones, although the product before the hydrolysis step is the imine. Addition to the nitrile first takes place in normal fashion to give the imine salt. When the reaction is worked up with water, protonation gives the imine, which in acid undergoes further hydrolysis to the ketone. Note the difference between this reaction and those of alkyllithium or Grignard reagents with acid chlorides or esters, which go all the way to alcohols. With nitriles the reaction can be stopped at the carbonyl stage. With acid chlorides and esters the intermediate carbonyl compounds are born in the presence of the organometallic reagent and must react further. In the reaction with nitriles, there is no leaving group present and the imine salt is stable until hydrolysis (Fig. 19.51).

THE GENERAL CASE

A SPECIFIC EXAMPLE

(55%)

FIGURE 19.51 Reaction of a nitrile with an organometallic reagent gives an imine salt. Acidification first generates the imine and then the ketone.

PROBLEM 19.20 Write a mechanism for the hydrolysis of imines under neutral conditions.

Metal hydrides reduce nitriles to give primary amines. Two equivalents of hydride are delivered, and a final addition of water yields the amine (Fig. 19.52).

THE GENERAL CASE

A SPECIFIC EXAMPLE

1. LiAlH₄
   THF, 0 °C, 1 h
2. H₂O

(72%)

FIGURE 19.52 The metal hydride reduction of nitriles followed by hydrolysis yields primary amines.

Catalytic hydrogenation of nitriles also gives primary amines. Notice that only primary amines can be formed this way (Fig. 19.53).

THE GENERAL CASE   **A SPECIFIC EXAMPLE**

FIGURE 19.53 Catalytic hydrogenation of nitriles also gives primary amines.

19.10   REACTIONS OF KETENES

Ketenes are strained compounds and are very reactive even at low temperature. Simple ketenes are so reactive that they must be carefully protected from atmospheric moisture. Most nucleophiles add to the carbonyl group to give acyl derivatives. This addition is another reaction that gives students trouble, but shouldn't. The ketene reacts with nucleophiles like any other carbonyl compound to give an addition product. In this case, the addition product is an enolate, and protonation at oxygen gives an enol, which equilibrates with the more stable carbonyl compound. Protonation at carbon (probably the slower process) gives the carbonyl compound directly. It's all routine addition-to-carbonyl chemistry. Figure 19.54 gives some examples.

THE GENERAL CASE

An enolate   Acyl compounds

FIGURE 19.54 Ketenes react with nucleophiles to give acyl compounds.

(continued)

SPECIFIC EXAMPLES

FIGURE 19.54 *(continued)*

19.11   ENOLATE CHEMISTRY OF ESTERS: THE CLAISEN AND RELATED CONDENSATION REACTIONS

Esters are not as acidic as aldehydes and ketones; their pK_a values are about 24 (Table 19.2). Esters are therefore some five pK_a units less acidic than their less well resonance-stabilized relatives. Nevertheless, it is possible for some enolate to be formed in the presence of a base such as an alkoxide ion, even though the reaction must be substantially endothermic (Fig. 19.55).

FIGURE 19.55 Formation of an enolate anion from an ester. As *tert*-butyl alcohol has a pK_a of about 17, and the ester a pK_a of about 24, this reaction must be highly endothermic.

Why doesn't the alkoxide add to the carbonyl group and start a base-catalyzed transesterification reaction (Section 19.7)? It does! If one is not careful to use the same OR group in the alkoxide and the ester, then mixtures of products are found. As long as the OR groups match, however, the base-catalyzed transesterification doesn't change anything—the product is the same as the starting material (Fig. 19.56).

FIGURE 19.56   A base-catalyzed transesterification reaction will generate a new ester unless the OR group of the alkoxide is identical to the OR group of the ester.

What can the ester enolate do? It can react with the acidic alcohol to reprotonate, regenerating the starting ester. However, it can also react *with any other Lewis acid in the system.* The carbonyl group of the ester is just such a Lewis acid, and it can react with the enolate in an addition reaction. Although the product is more complex, the addition reaction is no different from any other addition of a nucleophile to an ester carbonyl (Fig. 19.57).

FIGURE 19.57   In the first step of the Claisen condensation, the nucleophilic ester enolate adds to the Lewis acid ester carbonyl.

The next reaction is also analogous to reactions of esters with simpler nucleophiles. The addition product can do two things. It can expel the enolate in a simple reverse of the original addition reaction, or it can lose alkoxide to give a molecule of **acetoacetic ester**, a β-keto ester (Fig. 19.58). This reaction is called the **Claisen condensation** after Ludwig Claisen (1851–1930), who has many reactions named for him. Recall the Claisen–Schmidt condensation, for example (Chapter 17, p. 936). More are to come.

FIGURE 19.58 In the second step of the Claisen condensation, an alkoxide ion is lost and acetoacetic ester is formed. This process closely resembles the loss of alkoxide to give acyl compounds.

The Claisen condensation bears the same relationship to the aldol condensation that the addition of a nucleophile to an ester does to the addition to an aldehyde or ketone. Whereas the alkoxides formed in the aldol condensation can only revert to starting material or protonate, both ester reactions have the added option of loss of the ester alkoxide. What is added in ester chemistry is the possibilty of the elimination phase of the addition–elimination process (Fig. 19.59).

FIGURE 19.59 The aldol condensation is just an example of the addition of a nucleophile to a carbonyl group followed by protonation. The Claisen condensation follows the addition–elimination sequence available to esters because of the presence of a leaving group, OR, within the molecule.

As in aldol condensations of ketones, in the Claisen condensation it is the traditional starting materials that are favored thermodynamically. Esters are stabilized by resonance and there are two ester groups in the starting materials but only one in the product. The result is a thermodynamic favoring of the two separated esters (Fig. 19.60).

Two esters stabilized by ester resonance

Only one ester stabilized by ester resonance

FIGURE 19.60 Thermodynamics favors the pair of esters (starting material) in the Claisen condensation, not the β-keto ester product. The reaction to form the β-keto ester is endothermic.

However, there is an easy way out of this thermodynamic bind. In the last step of the Claisen condensation a molecule of alkoxide catalyst is regenerated. The catalyst is regenerated in the aldol condensation as well, and in the aldol condensation the catalyst goes on to initiate another condensation. In the Claisen condensation, however, there is another, faster reaction possible. The β-keto ester product is by far the strongest acid in the system as it contains hydrogens that are α to *two* carbonyl groups. The anion formed by deprotonation of the β-keto ester is resonance stabilized by *both* carbonyl groups (Fig. 19.61).

Claisen condensation

$pK_a \sim 10$
"Doubly α" hydrogens

$pK_a \sim 17$ HOR

FIGURE 19.61 The removal of a doubly α proton by alkoxide is an exothermic reaction.

The pK_a values of β-keto esters are about 10, and therefore these molecules are about 10^{14} more acidic than the starting esters! By far, the best reaction for the alkoxide regenerated at the end of the condensation step is removal of one of the doubly α hydrogens to give the highly resonance-stabilized salt of the β-keto ester (Fig. 19.62).

exothermic

Catalyst (RO⁻) is destroyed!

endothermic

$pK_a \sim 10$
Doubly α hydrogen

+ HOR

$pK_a \sim 17$

Note resonance stabilization

FIGURE 19.62 Alkoxide is destroyed in the last step of the Claisen condensation as a doubly α proton is removed to give a stable anion.

PROBLEM 19.21 There is another reaction, encountered earlier in this chapter, that also relies for its success on formation of a resonance-stabilized anion. What is it?

If the doubly α proton is removed, all is well thermodynamically; now the product is more stable than the starting material, *but the catalyst for the reaction has been destroyed!* A Claisen condensation using a catalytic amount of base must fail, as it can only generate an amount of product equal to the original amount of base. The simple remedy is to use a full equivalent of alkoxide, not a catalytic amount. A thermodynamically favorable Claisen condensation is not a catalytic reaction, but requires a full equivalent of base. Be certain you are clear as to the reason why the aldol condensation requires only a catalytic amount of base but the Claisen needs a full equivalent (Fig. 19.63).

FIGURE 19.63 The aldol condensation can be carried out using a catalytic amount of base because the final protonation step regenerates a molecule of hydroxide. In the last step of the Claisen condensation, the catalyst alkoxide ion is destroyed by removal of the doubly α proton.

Acidification of the reaction mixture leads to protonation of the anion to form the β-keto ester. What prevents this product, thermodynamically unstable with respect to the starting ester molecules, from re-forming the esters? There is no basic catalyst present! Without the catalyst, there can be no reverse Claisen condensation, and the product ester is trapped in its relatively high-energy thermodynamic valley (Fig. 19.64).

FIGURE 19.64 The final neutralization step produces a β-keto ester *in the absence of base.*

Figure 19.65 shows Energy versus Reaction progress diagrams for the aldol and Claisen condensations.

Reaction progress

The aldol condensation for acetaldehyde is approximately thermoneutral

Reaction progress

FIGURE 19.65 Energy versus Reaction progress diagram for the aldol and Claisen condensations.

SUMMARY

A full equivalent of base makes the Claisen condensation a practical source of aceto-acetic esters, β-keto esters. The five steps of the process are summarized in Figure 19.66:

1. Enolate formation.
2. The condensation itself (addition).
3. Loss of alkoxide from the tetrahedral intermediate (elimination).
4. Removal of the doubly α-hydrogen.
5. Acidification to produce the β-keto ester.

The Claisen condensation

FIGURE 19.66 The Claisen condensation.

WORKED PROBLEM 19.22 Even when a full equivalent of sodium ethoxide is used, ethyl 2-methylpropanoate fails to give useful amounts of a condensation product. Explain with a careful mechanistic analysis.

$$CH_3CH_2\ddot{O}-\overset{\overset{\displaystyle :O:}{\|}}{C}-CH(CH_3)_2 \xrightarrow[\substack{H\ddot{O}CH_2CH_3}]{\substack{Na^+ \ :\ddot{O}CH_2CH_3 \\ \text{(full equiv)}}} \text{No useful amount of Claisen condensation product}$$

Ethyl 2-methylpropanoate

(continued)

(continued)

ANSWER In this classic problem, Claisen condensation leads to product **A** through the usual addition–elimination scheme:

No hydrogens here

elimination

addition

A

However, the product **A** contains no doubly α hydrogens! There can be no formation of the stable, highly resonance-stabilized anion that accounts for the success of the Claisen condensation. Thermodynamics dictates that the starting materials are more stable than the product, and the reaction settles out at a point at which there is little product. The mechanism of the reverse reaction is written for you: Just read the scheme backward.

19.12 VARIATIONS ON THE CLAISEN CONDENSATION

So, that's the Claisen condensation: two esters combine with an equivalent of base to make an acetoacetic ester. Now it's time to look at how changes in structure of the starting materials affect the structure of the product and the mechanism of its formation.

19.12a Cyclic Claisen Condensations: The Dieckmann Condensation. The

Dieckmann condensation, the cyclic version of the Claisen condensation, is named after Walter Dieckmann (1869–1925). Five- and six-membered rings are favored, and there is little beside the intramolecular nature of the reaction to distinguish the Dieckmann from the standard Claisen condensation. Like the Claisen, the Dieckmann condensation is thermodynamically unfavorable if only a catalytic amount of base is used. Unless there is sufficient base available to remove the doubly α hydrogen and drive the reaction to a thermodynamically favorable conclusion, the reaction is not synthetically useful. A final acidification liberates the β-keto ester in the absence of base. Figure 19.67 gives an example of the Dieckmann condensation.

The Dieckmann condensation—a cyclic Claisen condensation

THE STANDARD DIECKMANN

enolate formation

Removable doubly α hydrogen

intramolecular condensation (addition phase)

Final product— the β-keto ester

acidification

Resonance-stabilized anion

deprotonation

β-Keto ester

elimination phase

A MORE COMPLEX EXAMPLE

1. NaOCH₃ benzene

2. HCl/H₂O:

(97%)

FIGURE 19.67 The Dieckmann condensation—an intramolecular Claisen condensation.

19.12b Crossed Claisen Condensations. A naively designed crossed Claisen condensation is doomed to the same kind of failure as is the crossed aldol condensation. There are four possible products, and therefore no easy source of the selectivity we need for a practical synthetic reaction (Fig. 19.68).

FIGURE 19.68 There are four possible—and likely—products from a crossed Claisen condensation.

PROBLEM 19.23 Show in a general way how these products are produced and write a mechanism for the formation of one of the four products shown in Figure 19.68.

As in the crossed aldol, however, there are some simple steps we can take to improve matters. If one partner has no α-hydrogen, there can be no enolate formation from it, and it can act only as a Lewis acid, never as a nucleophile. It is helpful to add the ester with an α-hydrogen to a mixture of the partner without an α-hydrogen and the base. In this case, the enolate will be very likely to react as it is formed with the large amount of the first ester, and a successful crossed Claisen can be achieved. Benzoates (PhCOOR), formates (HCOOR), and carbonates (ROCOOR) are examples of esters without α-hydrogens and are often used (Fig. 19.69).

THE "CROSSED CLAISEN"—THE GENERAL MECHANISM

X = Ph Benzoates
 = H   Formates
 = RO Carbonates

A SPECIFIC EXAMPLE

NaH/benzene
78 °C

(56%)

A RELATED EXAMPLE

(78%)

FIGURE 19.69 Successful crossed Claisen condensations.

19.12c Condensations of Esters with Ketones. A combination aldol-Claisen condensation can occur between an ester and a ketone. It is unlikely that formation of the product of aldol condensation of the ketone will be a problem. Aldol condensations of ketones are reversible, and usually endothermic. So even though the product of the aldol condensation *will* be formed, starting ketone will be favored. If ketone is drained away through reaction with the ester, the reversibly formed aldol product merely serves as a storage point for the ketone. The normal Claisen condensation of two esters is usually not a problem because the ketone is a much stronger acid than the ester (Table 19.2, p. 1037), and formation of the ketone enolate will be much preferred to formation of the ester enolate. The ketone enolate adds to the ester carbonyl, alkoxide is lost, and a β-dicarbonyl compound is formed. If a full equivalent of base is used, the doubly α hydrogen is removed to give a nicely resonance-stabilized anion. The reaction is completed by an acidification step that regenerates the β-dicarbonyl compound in the absence of base. Note that the reaction contains the same five steps as the Claisen condensation itself (p. 1068, Fig. 19.70).

FIGURE 19.70 The mechanism of a crossed condensation reaction between a ketone and an ester.

PROBLEM 19.24 Why are esters less reactive in addition reactions than aldehydes and ketones?

Frequently, the strong base sodamide ($NaNH_2$), or a dialkylamide such as LDA, is used in this reaction. It is generally the kinetically favored, more accessible hydrogen that is removed when there is a choice of possible enolates (Fig. 19.71).

FIGURE 19.71 Amide bases are often used in this reaction and lead to the formation of the "kinetic" enolate through removal of the most accessible hydrogen.

19.12d Reverse Claisen Condensations.

As we have mentioned, formation of the initial condensation product in the Claisen condensation is usually endothermic. If no doubly α hydrogens exist for salt formation, β-keto esters will revert to their component ester starting materials when treated with base. Ethoxide ion is not a strong enough base to remove the lone singly α hydrogen in the product (recall Problem 19.22). However, if a strong enough base is used, enolates *can* be formed through removal of the singly α hydrogen. The classic example is the Claisen condensation induced by the very strong base, triphenylmethyl sodium (Fig. 19.72).

Reverse Claisen

Normal Claisen condensation

Singly α hydrogen removed by *very* strong base

FIGURE 19.72 If no doubly α hydrogens are available, a β-keto ester reacts with base to regenerate a pair of esters. Very strong bases such as triphenylmethyl sodium can remove the singly α hydrogen.

(35%)

(Ph)$_3$C—H

19.13   REACTIONS OF β-KETO ESTERS

19.13a Alkylation Reactions. As we have seen, β-keto esters are quite strong acids. The pK_a of acetoacetic ester itself is about 10 and other β-dicarbonyl compounds, β-cyano carbonyl compounds, and dicyano compounds are similarly acidic (Table 19.5).

TABLE 19.5   The Acidity of Some β-Dicarbonyl and Related Compounds

Compound	pK_a	Compound	pK_a
Malonaldehyde	5	Ethyl malonate	13
2,4-Pentandione	9	Ethyl cyanoacetate	9
Methyl acetoacetate	10	Malononitrile	11

The anions formed from these compounds can be alkylated in S_N2 reactions. The once-alkylated compounds still contain one doubly α hydrogen, and therefore remain acidic. A second alkylation reaction is often possible, as long as the requirements of the S_N2 reaction are kept in mind. **Malonic esters** are sufficiently acidic ($pK_a \sim 13$) so that alkylation through the enolate is easy. Both mono- and dialkylation reactions of β-keto esters and related 1,1-diesters are common (Fig. 19.73).

FIGURE 19.73 Alkylations of some β-dicarbonyl compounds.

PROBLEM 19.25 Write a mechanism for the base-induced alkylation of 1,1-dicyanomethane (malononitrile) with ethyl iodide.

19.13b Hydrolysis and Decarboxylation of β-Dicarbonyl Compounds.
When a β-keto ester such as acetoacetic ester is hydrolyzed in acid, the resulting β-keto acid is most unstable and decarboxylates easily (Section 18.6g; Fig. 19.74).

THE GENERAL CASE

An acetoacetic ester → β-Keto acid → Acetone + CO_2

A SPECIFIC EXAMPLE

FIGURE 19.74 The acid-catalyzed hydrolysis and decarboxylation of acetoacetic esters.

The end result is one molecule of acetone and one of carbon dioxide. This reaction is a roundabout method of making this simple ketone. However, the combination of the alkylation reactions of Figure 19.73 followed by acid- or base-hydrolysis and **decarboxylation** is a useful source of complicated ketones (Fig. 19.75).

THE GENERAL PROCESS

FIGURE 19.75 The decarboxylation route to substituted ketones.

(continued)

A SPECIFIC EXAMPLE

$$H_3C-\overset{:O:}{\overset{\|}{C}}-\underset{\underset{CH_2CH_2CH_2CH_3}{|}}{CH}-\overset{:O:}{\overset{\|}{C}}-\overset{..}{O}Et \xrightarrow[\substack{+ \\ 2.\ H_3O:/H_2O:}]{1.\ NaOH/H_2O:\ 25\ °C} \left[H_3C-\overset{:O:}{\overset{\|}{C}}-\underset{\underset{CH_2CH_2CH_2CH_3}{|}}{CH}-\overset{:O:}{\overset{\|}{C}}-\overset{..}{O}H \right] \longrightarrow H_3C-\overset{:O:}{\overset{\|}{C}}-\underset{\underset{CH_2CH_2CH_2CH_3}{|}}{CH_2} + CO_2$$

Not isolated   (57%)

FIGURE 19.75 *(continued)*

In the decarboxylation reaction, the carbonyl oxygen acts as base and removes the hydroxyl hydrogen. Carbon dioxide is lost and the enol of the ketone is formed. Normal equilibration with the more stable ketone gives the final product (Fig. 19.76).

Decarboxylation

$$\underset{\beta\text{-Keto acids}}{R-\overset{:O:}{\overset{\|}{C}}-\underset{CH_2}{}-\overset{\overset{H}{\diagdown}{\overset{..}{O}:}}{\overset{\|}{C}}-\overset{..}{O}:} \xrightarrow{-CO_2} \underset{Enol}{R-\overset{\overset{..}{O}\diagdown{H}}{\overset{|}{C}}=CH_2} \rightleftharpoons \underset{Ketone}{R-\overset{:O:}{\overset{\|}{C}}-CH_3}$$

FIGURE 19.76 The mechanism for the decarboxylation of β-keto acids.

Be sure to update your file cards for ketone syntheses with this reaction. Moreover, this is not such a simple reaction to see, as it requires a number of steps. In practice, you must dissect a potential target molecule into its parts to see if it is available through this route. You must identify the R groups to be attached through alkylation, then piece together the original Claisen condensation. It's not always easy.

Some examples are probably useful here, and as usual they will appear as problems. For instance, see Problem 19.26.

WORKED PROBLEM 19.26 Show how the following ketones can be made through acetoacetic ester syntheses. Mechanisms are not required.

(a)

$$H_3C-\overset{O}{\overset{\|}{C}}-\underset{\underset{CH_3}{|}}{CH}CH_2CH_3$$

(b)

$$CH_3CH_2-\overset{O}{\overset{\|}{C}}-\underset{\underset{CH_3}{|}}{CH}CH_2CH_3$$

(c)

$$\text{(benzene ring)}-CH_2-\overset{O}{\overset{\|}{C}}-CH_2-\text{(benzene ring)}$$

(continued)

(continued)

ANSWER (b) A Claisen condensation using a propionic ester gives a β-keto ester that can be alkylated with ethyl iodide. Hydrolysis and decarboxylation gives the desired product. For clarity, the following answer shows the isolation of the product of the Claisen condensation, but this step is not necessary. The anion formed at the end of the Claisen condensation could be alkylated directly.

Diesters such as malonic ester can also be used in this kind of synthetic sequence. The diester is first hydrolyzed in acid or base to the diacid, and then heated to induce loss of carbon dioxide. This process is called the **Malonic ester synthesis**. The end product is acetic acid, not the ketones formed in the related reactions of β-keto esters (Fig. 19.77).

FIGURE 19.77 The formation of acetic acid from malonic acid.

Malonic ester   Malonic acid   Acetic acid

Note an intuitive point here. The acetoacetic esters, which contain a ketone, are the sources of ketones, and the malonic esters, which contain only ester carbonyls, are the source of acids (Fig. 19.78).

FIGURE 19.78 Hydrolysis and decarboxylation of acetoacetic esters gives substituted *ketones*, whereas the same sequence applied to malonic esters gives substituted acetic *acids*.

Ketone   Ketone

Ester   Acid

Here is how the reaction works. The hydrolysis converts the diester into a diacid. These compounds are similar to β-keto acids and will decarboxylate on heating to give the enols of acetic acids. The enols are in equilibrium with the more stable acids, which are the isolated products (Fig. 19.79).

FIGURE 19.79 The mechanism of decarboxylation of 1,1-diacids such as malonic acid.

When this reaction is carried out on the parent malonic ester, acetic acid results, and the process is of no practical synthetic utility. As with the synthesis of ketones starting from acetoacetic esters, this process is far more useful when combined with the alkylation reaction. One or two alkylations of malonic ester give substituted diesters that can be hydrolyzed and decarboxylated to give rather complicated acids (Fig. 19.80).

Substituted acetic acids are potential sources of esters, acid halides, and any other compound that can be made from a carboxylic acid.

Malonate alkylation

THE GENERAL SEQUENCE

Et = CH₂CH₃

A SPECIFIC EXAMPLE

FIGURE 19.80 The alkylation–decarboxylation sequence for the synthesis of substituted acetic acids from malonic acids.

WORKED PROBLEM 19.27 How can the following molecules be made through a malonic ester synthesis?

(a)

HOOC

(b)

HOOC

(c)

CH$_3$CH$_2$OOC

ANSWER (a) Malonic ester is alkylated, hydrolyzed, and decarboxylated.

$$\xrightarrow[\text{2. PhCH}_2-\text{I}]{\substack{\text{1. NaOR/HOR} \\ \text{(full equiv)}}}$$

$$\xrightarrow[\text{2. }\Delta]{\text{1. H}_2\text{O/H}_3\text{O}^+}$$

+ CO$_2$

19.13c Direct Alkylation of Esters.
It is possible to alkylate esters directly and thereby avoid the extra steps of the malonic ester synthesis. Very strong bases such as dialkyl amides (LDA, for example) can be used to remove the α-hydrogens of esters. When simple esters are added to these bases at low temperature, an α-hydrogen is removed to give the enolate anion. This anion can then be alkylated by any alkyl halides able to participate in the S$_N$2 reaction. The end result is an alkylated ester (Fig. 19.81).

THE GENERAL CASE

A SPECIFIC EXAMPLE

$$\xrightarrow[\substack{\text{2. CH}_3\text{CH}_2\text{Br} \\ -78\ ^\circ\text{C}}]{\substack{\text{1. LDA, }-78\ ^\circ\text{C} \\ \text{THF, 30 min}}}$$

(92%)

FIGURE 19.81 The low-temperature alkylation of an ester using LDA as base.

SUMMARY

This section gives you new ways of making complex ketones and acetic acids through the acetoacetic ester and malonic ester syntheses. Although these routes are somewhat roundabout—acetoacetic ester or malonic ester must be made, alkylated, and finally decarboxylated—in a practical setting they are efficient and solve some of the synthetic problems posed earlier. You might want to revisit our earlier discussion of the difficulties involved in straightforward alkylation reactions of aldehydes, ketones, and acids (Chapter 17, p. 916).

19.14   COMPLICATED "FUN IN BASE" PROBLEMS

We now have some new reactions to use in problems. Put another way, your lives are now even more complicated than before, as even more sadistically difficult problems are within your abilities. There really aren't too many new things to think about. You only have to add the Claisen-like processes that involve the loss of a leaving group from a carbonyl group that has been attacked by an enolate. In other words, you always have to keep the addition–elimination reaction in mind.

But be careful—reactions are driven by thermodynamics, not our concepts of forward and backward. The reverse Claisen is much harder to detect in problems than the forward version, but no less reasonable mechanistically. As always in matters such as these, experience counts a lot, and there is no way to get it except to work a lot of problems.

PROBLEM 19.28 Provide mechanisms for the following reactions:

PROBLEM 19.29 Provide a mechanism for the following transformation. Be careful. This problem is **much** harder than it looks at first.

Mixture of stereoisomers
(35%)

19.15 SPECIAL TOPIC: OTHER SYNTHETIC ROUTES TO ACYL COMPOUNDS

As mentioned in the introduction, reactions and syntheses are even more inter-twined than usual with these compounds. A reaction of one is a synthesis of another. The various interconversions of acyl compounds discussed in this chapter (and in Chapter 18) are summarized in Section 19.19. Here we first discuss two rearrangement reactions that are useful as syntheses of esters and amides, respectively, then move on to a brief discussion of routes to nitriles (cyanides) and ketenes.

19.15a The Baeyer–Villiger Reaction. Named by Kurt Mislow (p. 157) for the great Adolf von Baeyer (p. 192) and his student Victor Villiger (1868–1934), the Baeyer–Villiger reaction is an oxidation of aldehydes and ketones, usually with a peracid, that results in the insertion of an oxygen atom. Seen in its bare form, the reaction appears quite remarkable—how can that "extra" oxygen atom insert itself into a carbon–carbon bond (Fig. 19.82)?

FIGURE 19.82 Baeyer–Villiger oxidation of aldehydes and ketones.

Fleshed out in a real mechanism, however, the reaction becomes intelligible. The first stage is a straightforward addition of the peracid to the carbonyl group. Now look at that weak oxygen–oxygen bond, and at the good leaving group formed when it breaks. Migration of an R group as the oxygen–oxygen bond breaks gives the product directly (Fig. 19.83).

Baeyer–Villiger reaction

overall transformation
—insertion of one oxygen atom

Weak bond

R moves as the O—O bond breaks

proton transfer

Good leaving group

FIGURE 19.83 The mechanism of the Baeyer–Villiger reaction.

Of course when there are two different R groups in the ketone, two products can result, as each R group can migrate. However, there is a well-established order of migration for different groups, H > tertiary alkyl > secondary alkyl > primary alkyl > methyl, and there is quite substantial selectivity in product formation. That is why aldehydes give the acid, as shown in Figure 19.82. The hydrogen migrates in preference to the alkyl or aryl group (Fig. 19.84).

PhCOOOH

proton transfer

(90% of product)

FIGURE 19.84 Hydrogen migrates especially easily in the Baeyer–Villiger reaction.

When the ketone is cyclic, the Baeyer–Villiger reaction becomes a synthesis of cyclic esters, called lactones (p. 1032). Watch out for this reaction. The change from cyclic to acyclic adds apparent complexity, and this reaction is a favorite of problem writers (Fig. 19.85).

FIGURE 19.85 When a cyclic ketone is used, the Baeyer–Villiger reaction becomes a synthesis of lactones.

(71%)

PROBLEM 19.30 Write a mechanism for the reaction in Figure 19.85.

PROBLEM 19.31 In the paper that named the Baeyer–Villiger reaction, Mislow and his student Joseph Brenner showed that in the reaction below, the product was formed with *retention* of optical activity. What does that observation tell you about the details of the mechanism?

19.15b The Beckmann Rearrangement.

Here is a related reaction, named for Ernst Otto Beckmann (1853–1923), in which the end product is not an ester, but an amide. It starts with an oxime (p. 855), the product of the addition of hydroxylamine to a ketone. Treatment of the oxime with any of a variety of strong acids leads to the amide (Fig. 19.86).

THE GENERAL CASE

An oxime   An amide

A SPECIFIC EXAMPLE

FIGURE 19.86 The Beckmann rearrangement.

(100%)

The mechanism of the Beckmann rearrangement involves a migration much like the one in the Baeyer–Villiger reaction, but the reaction is not triggered by the breaking of a weak oxygen–oxygen bond. It is initiated by the displacement of a good leaving group, water. That good leaving group is formed in the first step of the reaction as the OH of the oxime is protonated by acid. Intramolecular rearrangement leads to a nicely resonance-stabilized intermediate that is captured by water. Proton shifts lead to the final product, the amide (Fig. 19.87).

FIGURE 19.87 The mechanism of the Beckmann rearrangement.

In its cyclic version, the Beckmann rearrangement leads to lactams, cyclic amides (p. 1034, Fig. 19.88).

A lactam

(90%)

FIGURE 19.88 When cyclic oximes are used, the Beckmann rearrangement becomes a synthesis of lactams.

PROBLEM 19.32 Provide a mechanism for the reaction shown in Figure 19.88.

PROBLEM 19.33 Oximes come in syn/anti pairs, and the two stereoisomers react differently. Explain the results in the following figure.

Nitriles are most often made by displacement reactions of alkyl halides by cyanide ion or by dehydration of amides, generally with P_2O_5 (Fig. 19.89).

THE GENERAL CASE

$$:N\equiv C:^- \quad R\!-\!X \quad \xrightarrow{S_N2} \quad :N\equiv C\!-\!R \ + \ X:^-$$

A SPECIFIC EXAMPLE

$$CH_3CH_2CH_2CH_2CH_2\!-\!\ddot{C}\ddot{l}: \quad \xrightarrow[\text{20 min, <160 °C}]{\substack{\text{NaCN:}\\ \text{dimethyl sulfoxide}}} \quad CH_3CH_2CH_2CH_2CH_2\!-\!CN:$$

(93%)

THE GENERAL CASE

A SPECIFIC EXAMPLE

(~75%)

FIGURE 19.89 Nitriles are usually made through S_N2 reactions using cyanide as the nucleophile, or by dehydrations of amides.

Ketenes are available through elimination reactions of acyl chlorides using tertiary amines as bases (Fig. 19.90).

THE GENERAL CASE

A SPECIFIC EXAMPLE

FIGURE 19.90 One common synthesis of ketenes involves an elimination of acyl halides.

19.16   SPECIAL TOPIC: THERMAL ELIMINATION REACTIONS OF ESTERS

Esters that contain a hydrogen in the β-position of the OR group give alkenes when heated to high temperatures (> 400 °C) (Fig. 19.91).

For this reaction to succeed, there must be a hydrogen here, in the "β"-position

FIGURE 19.91 The thermal elimination reaction of esters containing a hydrogen in the β-position.

In this reaction, the carbonyl group acts as a base, attacking the β-hydrogen intra-molecularly. At the same time the ester acts as a leaving group, and acetic acid is lost in a single, concerted step (Fig. 19.92).

FIGURE 19.92 An arrow formalism description of the thermal elimination reaction of esters.

In contrast to most E2 reactions, this pyrolytic, intramolecular version proceeds through loss of a syn hydrogen. The reason is simple: Only the syn hydrogen is within reach of the carbonyl group. There is no possibility of doing an anti elimination in this reaction. The elimination reactions of the deuterium-labeled compounds shown in Figure 19.93 provide an example of the stereospecificity of this elimination reaction. In diastereomer **A**, it is only the syn hydrogen that is lost as *trans*-stilbene is formed. By contrast, in **B** it is the deuterium atom that must occupy the syn position in the formation of *trans*-stilbene, and this time it is D, not H, that is lost.

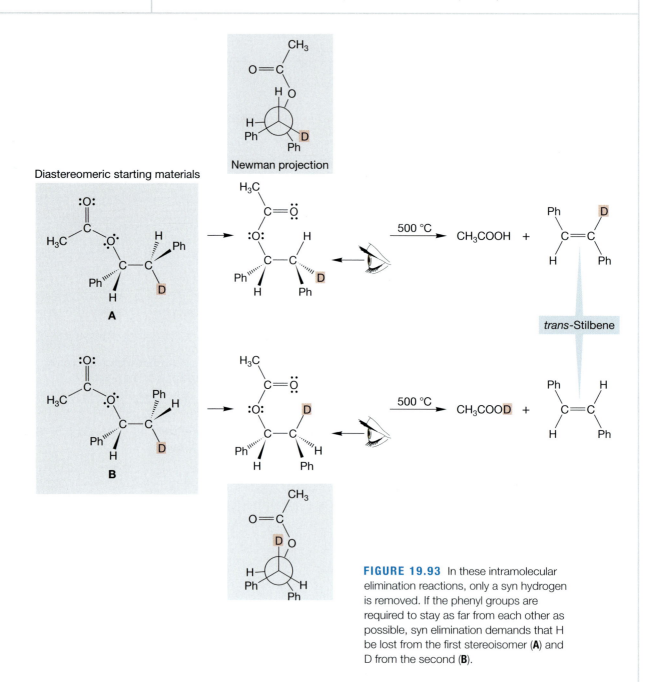

FIGURE 19.93 In these intramolecular elimination reactions, only a syn hydrogen is removed. If the phenyl groups are required to stay as far from each other as possible, syn elimination demands that H be lost from the first stereoisomer (**A**) and D from the second (**B**).

There are other versions of this new elimination process, some of which proceed under much milder conditions than those necessary for ester pyrolysis. For example, the sulfur-containing **xanthate esters** undergo elimination especially easily (Fig. 19.94). Xanthate esters can be prepared as shown in the figure by addition of alkoxide to CS_2 (recall the addition of nucleophiles to carbon dioxide, Chapter 18, p. 1011). These intramolecular thermal elimination reactions provide a new route to alkenes, so update your file cards.

THE GENERAL CASE

A SPECIFIC EXAMPLE

FIGURE 19.94 The formation and thermal fragmentation of a xanthate ester.

PROBLEM 19.34 Write arrow formalisms for the reactions of Figure 19.94.

PROBLEM 19.35 Provide mechanisms for the following reactions. The first reaction below is called the Cope elimination, after Arthur C. Cope (1909–1966).

(a) Cope elimination

(b) Ester pyrolysis

One of the themes of this and Chapter 18 has been the use of enolates formed from ketones and esters on reaction with strong bases such as LDA. For example, condensation reactions can be directed (p. 1073), and alkylations of simple esters and ketones achieved (p. 1080; Chapter 17, p. 918). Here we combine the alkylation of enolates with a Cope-like elimination reaction to do most useful chemistry. The four-step combination of enolate formation, thioether or selenoether formation, oxidation, and heating can be used to convert saturated ketones and esters into α,β-unsaturated compounds in very high yield (Fig. 19.95).

FIGURE 19.95 The potential transformation of a saturated ketone into an α, β-unsaturated ketone.

The base LDA or a close relative, lithium *N*-cyclohexyl-*N*-isopropylamide, can be used to form an enolate. The enolate reacts in S_N2 fashion with a disulfide (R—S—S—R), or if you can stand the extremely bad smell, a diselenide (R—Se—Se—R), or Ph—Se—Br to give a thioether or a selenoether. These ethers can be oxidized to **sulfoxides** or **selenoxides** with sodium metaperiodate $NaIO_4$ (Fig. 19.96).

FIGURE 19.96 The formation of sulfoxides and selenoxides from thioethers and selenoethers.

The sulfoxides undergo a Cope-like elimination at about 100 °C, and the selenoxides decompose at about room temperature (Fig. 19.97).

Recall the Cope elimination from Problem 19.35

FIGURE 19.97 Thermal decomposition of sulfoxides and selenoxides resembles that of *N*-oxides.

The result is a simple, high-yield introduction of a double bond at the α,β-position of a ketone or ester (Fig. 19.98).

(98%)   (92%)   (87%)

Not isolated

FIGURE 19.98 Formation of alkenes by heating sulfoxides and selenoxides.

19.17   SPECIAL TOPIC: FORWARD AND REVERSE CLAISEN CONDENSATIONS IN BIOLOGY

Both forward and reverse Claisen condensations feature prominently in the biochemical synthesis and degradation of fatty acids (p. 1017). A critical step in the construction of these long-chain carboxylic acids involves enzyme-mediated Claisen condensations in which activated two-carbon fragments are sewn together. But, of course, Nature has a thermodynamic problem here—how is the endothermicity of the Claisen condensation to be overcome? The trick is to use an activated malonate in the condensation. Loss of carbon dioxide is used to drive the equilibrium toward the product (Fig. 19.99).

FIGURE 19.99 Loss of CO_2 drives this reaction toward product.

Nature's work is not yet done; first, a reduction of the ketonic carbonyl group occurs. The reducing agent is NADPH, a molecule closely related to NADH, which we saw in Chapter 16 (p. 878). Next, there is an enzyme-mediated elimination reaction. Finally, NADPH participates again, reducing the newly formed double bond to give the final product. Repetition of the steps in Figures 19.99 and 19.100 eventually leads to the fatty acid (Fig. 19.100).

FIGURE 19.100 A series of enzyme-mediated reactions leads to an overall reduction of one carbonyl group.

In Chapter 18 (pp. 1017–1018), we discussed briefly the metabolism of fatty acids, and raised a problem; how was a fatty acid such as palmitate cleaved at the α—β bond to give myristate (Fig. 19.101)?

FIGURE 19.101 The formation of myristate from palmitate—loss of two carbon atoms.

The problem is that the α–β bond needs to be activated in some way. If the β-carbon were oxidized, we would have a β-keto ester, and we would be all set up for a reverse Claisen condensation. That is exactly the pathway Nature uses. Here, palmitate is first derivatized as the S—CoA derivative. Then three enzymes act in sequence to dehydrogenate, hydrate, and finally dehydrogenate again to give the β-keto ester (Fig. 19.102).

A β-keto ester

FIGURE 19.102 Oxidation of palmitate to a β-keto ester.

Now the reverse Claisen does the first two-carbon cleavage reaction, and the process of metabolism is started (Fig. 19.103).

FIGURE 19.103 A reverse Claisen condensation "clips" two carbons.

SUMMARY

In fact, both construction and deconstruction of fatty acids are Claisen condensations; one operates in the "forward" direction, the other in "reverse."

19.18   SPECIAL TOPIC: A FAMILY OF CONCERTED REARRANGEMENTS OF ACYL COMPOUNDS

In this section, we take up a few rearrangement reactions of some new kinds of molecules related to acyl compounds. The mechanisms are not always fully worked out, but the reactions are of interest and are often synthetically useful.

19.18a Diazo Ketones: The Wolff Rearrangement.
When an acid chloride is allowed to react with a diazo compound, the result is formation of a **diazo ketone** (Fig. 19.104).

FIGURE 19.104 Diazo ketone formation.

The mechanism of this reaction is only slightly more complicated than the usual re-action of nucleophiles with acid chlorides (p. 1042). Diazo compounds are nucleophiles at carbon and add to the carbonyl group of the reactive acid chloride. Chloride ion is ex-pelled in the elimination phase of this normal addition–elimination process. In the only new step of the reaction, chloride removes the newly acidic hydrogen to give the diazo ketone (Fig. 19.105).

FIGURE 19.105 The mechanism of diazo ketone formation.

Diazo compounds are generally rather dangerous, as they are not only poisonous but are explosives as well. The resonance-stabilized diazo ketones are far less explosive than simple diazo compounds, but they must still be treated with the greatest respect.

PROBLEM 19.36 Draw resonance forms for a diazo ketone.

Like other diazo compounds, diazo ketones are sensitive to both heat and light (Chapter 10, p. 460). Nitrogen is lost and a carbene intermediate, in this case, a ketocarbene, can be formed (Fig. 19.106).

A ketocarbene

Addition to alkenes

Insertion into C–H bonds

FIGURE 19.106 Diazo ketones decompose to give ketocarbenes on heating or photolysis. The reactions of ketocarbenes include additions to π systems and carbon–hydrogen insertion.

Ketocarbenes are capable of the usual carbene reactions, addition to multiple bonds and insertion into carbon–hydrogen bonds chief among them (Fig. 19.106). In addition, there is a possible intramolecular carbon–carbon insertion reaction, often silver catalyzed, and generally described as migration of the R group. The product of this reaction is a ketene. This reaction is known as the **Wolff rearrangement**, after Ludwig Wolff (1857–1919; remember the Wolff–Kishner reduction, Chapter 14, p. 695? Same Wolff; Fig. 19.107).

FIGURE 19.107 Intramolecular ketene formation from a ketocarbene.

The mechanistic complication comes from the notion that it may not be the carbene, but the diazo compound itself that gives the ketene in the Wolff rearrangement. The ketocarbene is not necessarily involved (Fig. 19.108). The question isn't completely sorted out yet, but the best evidence currently favors at least some direct formation of the ketene from most diazo ketones. This evidence doesn't exclude the reaction of the carbene to give the ketene as well.

FIGURE 19.108 In the Wolff rearrangement, the C—R bond can break as the nitrogen departs.

WORKED PROBLEM 19.37 Much experimentation indicates that only one of the two possible stereoisomers of the diazo ketone can give ketene directly. What are the two stereoisomers of the diazo ketone?

ANSWER There are syn and anti isomers of diazo ketones.

syn   anti

WORKED PROBLEM 19.38 Explain why only one of these isomers can rearrange easily to a ketene.

ANSWER The syn isomer is set up for migration of R with displacement from the rear in what is really an intramolecular S_N2 reaction. By contrast, the anti isomer would have to do a frontside S_N2 for R to migrate as the nitrogen leaves.

If the mechanistic questions still are not completely resolved, the synthetic applications certainly are. The Wolff rearrangement generates ketenes in the absence of nucleophiles, and therefore they can be isolated in this reaction. If the product ketene is deliberately hydrolyzed to the acid, the overall sequence is called the **Arndt–Eistert reaction** after Fritz Arndt and Bernd Eistert. It constitutes a most useful chain-lengthening reaction of the original acid used to make the acid chloride (Fig. 19.109).

FIGURE 19.109 The Arndt–Eistert synthesis adds a methylene group to a carboxylic acid.

PROBLEM 19.39 Devise a synthetic scheme for converting each of the starting materials shown below into 3-methylbutyric acid.

(a)

(c)

(b)

(d)

3-Methylbutyric acid

19.18b Acyl Azides: The Curtius Rearrangement. Acyl azides are related to diazo ketones, and can be made by reaction of acid chlorides with the azide ion, $^-N_3$ (Fig. 19.110).

FIGURE 19.110 Acid chlorides react with azide ion ($^-N_3$) to give acyl azides.

An acyl azide

Like diazoketones, acyl azides lose nitrogen when heated or irradiated. The end result is an **isocyanate**, formed either by intramolecular rearrangement of a **nitrene** (the nitrogen analogue of a carbene) or directly from the acyl azide in a reaction called the **Curtius rearrangement** after Theodore Curtius (1857–1928) (Fig. 19.111).

migration of R as N_2 is lost

An isocyanate

loss of N_2

A nitrene

FIGURE 19.111 The Curtius rearrangement leads to isocyanates. The mechanism probably involves a concerted rearrangement, not a nitrene.

PROBLEM 19.40 Can you guess what nitrene chemistry will be like? What products do you anticipate from the reaction of a nitrene with the carbon–carbon double bond? What product would you get from reaction with a carbon–hydrogen bond? Remember that a nitrene is the nitrogen analogue of a carbene (Chapter 10, p. 460).

For simple acyl azides, it is known that the isocyanate is formed by direct re-arrangement of the acyl compound and that the nitrene, though also formed from the azide, is not involved in isocyanate production.

Isocyanates, like ketenes, are very sensitive to nucleophiles. For example, alcohols add to isocyanates to give carbamate esters (Chapter 18, p. 1007; Fig. 19.112).

FIGURE 19.112 Isocyanates react with alcohols to give carbamate esters.

PROBLEM 19.41 Write a mechanism for the addition of an alcohol to an isocyanate to give a carbamate ester (Fig. 19.112).

PROBLEM 19.42 Explain the formation of the amine in the following reaction:

19.18c Hofmann Rearrangement of Amides. Amides are able to take part in a reaction similar to the Wolff and Curtius rearrangements. This process occurs in two steps. The amide is first brominated to give the *N*-bromoamide (Fig. 19.113). Base then removes the very acidic α-hydrogen and rearrangement to the isocyanate ensues. The question of whether a nitrene is involved in this **Hofmann rearrangement*** (August W. von Hofmann, 1818–1892)† is not completely settled, but I would bet on the direct rearrangement shown in Figure 19.113.

---

*In earlier, more colorful times, this reaction was known as the "Hofmann degradation."
†Do you remember the Hofmann elimination (Chapter 7, p. 324)? It's the same Hofmann.

migration of R

An isocyanate
(but *not* isolable under these conditions)

FIGURE 19.113 The first stages in the Hofmann rearrangement. An isocyanate is an intermediate, but cannot be isolated.

The isocyanate is not isolable, as it is in the Curtius rearrangement. In this reaction, it is born *in the presence of base* and is hydrolyzed to the carbamic acid, which is unstable and decarboxylates (recall Problem 19.42). The end product is the amine (Fig. 19.114).

addition

H₂O
protonation

A carbamic
acid

decarboxylation

An amine—
the end product
of the Hofmann
rearrangement

H₂O
protonation

FIGURE 19.114 Under the reaction conditions the isocyanate reacts with hydroxide or water and decarboxylates to give the amine.

SUMMARY

There is an essential difference between the Curtius rearrangment and the Hofmann rearrangement. Both involve the formation of isocyanates, but only in the Curtius rearrangement is this intermediate isolable. In the Hofmann rearrangement, base is present and the isocyanate cannot survive. This synthesis of amines is not easy to remember because it involves many steps (Fig. 19.115), thus making it a great favorite of problem writers.

Curtius

Isolable! No nucleophiles present

But Hofmann

Not isolable — born in the presence of ⁻OH

FIGURE 19.115 The Curtius and Hofmann rearrangements contrasted.

This section has a set of reactions united by formation of an intermediate containing two cumulated (adjacent) double bonds. Their common ancestor is allene (Chapter 12, p. 551). Are these molecules mere curiosities, chemical oddities of no real importance outside a textbook? Not at all. In 1992, broccoli was shown to contain a sulfur-containing isocyanate, an iso<u>thio</u>cyanate, that induces formation of something called a "phase II detoxication enzyme," which is involved in the metabolism of carcinogens (Fig. 19.116). Eat your broccoli and study your orgo.

1-Isothiocyanato-(4*R*)- (methylsulfinyl)butane (an anticancer agent found in broccoli)

FIGURE 19.116 Eat your broccoli!

19.19 SUMMARY

New Concepts

This chapter is filled with detail, but there are no new basic concepts. Most important, although not really new, is the continuation of the exploration of the addition–elimination reaction of carbonyl compounds begun in Chapter 18. When nucleophiles add to carbonyl compounds that do not bear a leaving group (for example, aldehydes or ketones) reversal of the addition or protonation are the only two possible reactions. When there is a leaving group attached to the carbonyl, as there is with most acyl compounds, another option exists, loss of the leaving group (Fig. 19.117).

FIGURE 19.117 Carbonyl groups bearing good leaving groups can undergo the addition–elimination reaction; others cannot.

The more stable a carbonyl compound, the less easy will be the endothermic loss of an α-hydrogen to give an enolate.

Well resonance-stabilized carbonyl compounds such as esters and amides are weaker acids than aldehydes and ketones.

Key Terms

acetoacetic ester (p. 1063)
acid halides (p. 1032)
acyl compounds (p. 1029)
amides (p. 1034)
anhydrides (p. 1033)
Arndt–Eistert reaction (p. 1097)
Claisen condensation (p. 1063)
Curtius rearrangement (p. 1098)
cyanides (p. 1035)
decarboxylation (p. 1076)
diazo ketone (p. 1094)
Dieckmann condensation (p. 1069)
esters (p. 1031)
ester hydrolysis (p. 1049)
Hofmann rearrangement (p. 1099)
imides (p. 1034)

immonium ion (p. 1056)
isocyanate (p. 1098)
ketenes (p. 1034)
lactams (p. 1034)
lactones (p. 1032)
malonic esters (p. 1075)
malonic ester synthesis (p. 1078)
nitrene (p. 1098)
nitriles (p. 1035)
Rosenmund reduction (p. 1045)
saponification (p. 1049)
selenoxides (p. 1090)
sulfoxides (p. 1090)
transesterification (p. 1050)
Wolff rearrangement (p. 1096)
xanthate esters (p. 1089)

Reactions, Mechanisms, and Tools

The addition–elimination reaction, introduced in Chapter 18, dominates the reaction mechanisms discussed in this chapter.

The chemistry of nitriles (cyanides) is similar to that of carbonyl compounds. The carbon–nitrogen triple bond can act as an acceptor of nucleophiles in addition reactions, and will provide resonance stabilization to an α-anion, much as will a carbonyl group (Figs. 19.47 and 19.48).

Dicarbonyl compounds, such as 1,1-diesters and β-keto esters (both species are β-dicarbonyl compounds), are quite

strong acids and can be alkylated through formation of the anion produced by loss of the doubly α proton (Fig. 19.73).

There is a class of intramolecular thermal elimination reactions that provides a new route to alkenes. Esters, xanthates, amine oxides, sulfoxides, and selenoxides are commonly used in this reaction. The reactions are concerted (one-step), and steric requirements dictate that a syn hydrogen must be lost in the reaction, as the carbonyl group cannot reach a hydrogen in an anti position (Fig. 19.93).

Syntheses

Here are the many important synthetic reactions found in this chapter. A few have been touched upon already in Chapter 18, but most are new, at least in a formal sense. Many of the following reactions share a common (addition–elimination) mechanism.

1. ACIDS

Acid- or base-induced hydrolysis of acid halides, anhydrides, esters, and amides; X = Cl, Br, I, O–CO–R, OR, NH$_2$

Acid- or base-induced hydrolysis of nitriles; the amide is an intermediate

Hydration of ketenes

If the 1,3-diester is alkylated first, substituted acetic acids can be made; the diacid is an intermediate

2. ACYL AZIDES

Addition–elimination reaction

3. ALCOHOLS

The ketone is an intermediate but cannot be isolated; all three R groups may be the same, but all three R's cannot be different

Other metal hydride donors also work; the aldehyde is an intermediate, but cannot be isolated

(continued)

3. ALCOHOLS (CONTINUED)

$$\text{R}-\overset{\overset{\displaystyle O}{\|}}{\text{C}}-\text{OR} \quad \xrightarrow[\text{2. H}_2\text{O}]{\text{1. LiAlH}_4} \quad \text{R}-\overset{\overset{\displaystyle \text{OH}}{|}}{\underset{\displaystyle \text{H}}{\text{C}}}-\text{H}$$

Other metal hydride donors also work; the aldehyde is an intermediate, but cannot be isolated

$$\text{RO}-\overset{\overset{\displaystyle O}{\|}}{\text{C}}-\text{OR} \quad \xrightarrow[\text{2. H}_2\text{O}]{\text{1. RLi}} \quad \text{R}-\overset{\overset{\displaystyle \text{OH}}{|}}{\underset{\displaystyle \text{R}}{\text{C}}}-\text{R}$$

The ester and ketone are intermediates, but cannot be isolated; all three R groups must be the same

4. ALDEHYDES

$$\text{R}-\overset{\overset{\displaystyle O}{\|}}{\text{C}}-\text{Cl} \quad \xrightarrow[\text{2. H}_2\text{O}]{\text{1. LiAlH[OC(CH}_3)_3]_3} \quad \text{R}-\overset{\overset{\displaystyle O}{\|}}{\text{C}}-\text{H}$$

Other encumbered metal hydrides also work

$$\text{R}-\overset{\overset{\displaystyle O}{\|}}{\text{C}}-\text{OR} \quad \xrightarrow[\text{2. H}_2\text{O}]{\text{1. DIBAL-H}} \quad \text{R}-\overset{\overset{\displaystyle O}{\|}}{\text{C}}-\text{H}$$

Diisobutylaluminum hydride (DIBAL-H), will not reduce the aldehyde if the reaction is carefully run

$$\text{R}-\overset{\overset{\displaystyle O}{\|}}{\text{C}}-\text{Cl} \quad \xrightarrow[\text{poisoned catalyst}]{\text{H}_2} \quad \text{R}-\overset{\overset{\displaystyle O}{\|}}{\text{C}}-\text{H}$$

Rosenmund reduction

5. ALKENES

Note that only a syn hydrogen can be removed

Xanthate ester pyrolysis requires lower temperature than ester pyrolysis

A Cope elimination

Sulfoxide decomposition

Selenoxide decomposition

6. AMIDES

$$\text{R}-\overset{\overset{\displaystyle O}{\|}}{\text{C}}-\text{Cl} \quad \xrightarrow{\text{NH}_3} \quad \text{R}-\overset{\overset{\displaystyle O}{\|}}{\text{C}}-\text{NH}_2$$

Substituted amines give substituted amides

$$\text{R}-\overset{\overset{\displaystyle O}{\|}}{\text{C}}-\text{OR} \quad \xrightarrow[\text{Na}^+ \,^-\text{NH}_2]{\text{NH}_3} \quad \text{R}-\overset{\overset{\displaystyle O}{\|}}{\text{C}}-\text{NH}_2$$

Acid catalysis also works

$$\text{R}-\overset{\overset{\displaystyle O}{\|}}{\text{C}}-\text{NH}_2 \quad \xrightarrow[\text{2. R}-\text{X}]{\text{1. base}} \quad \text{R}-\overset{\overset{\displaystyle O}{\|}}{\text{C}}-\text{NHR}$$

The amidate is an intermediate; two alkylations can be carried out if S_N2-active alkyl halides are used

6. AMIDES (CONTINUED)

Substituted amines give substituted amides

The Beckmann rearrangement

7. AMINES

Catalytic hydrogenation

Loss of a metal oxide from the initially formed intermediate is the key to this reaction

Loss of a metal oxide from the initially formed intermediate is the key to this reaction, too

The Hofmann rearrangement; an isocyanate is an intermediate

8. ANHYDRIDES

9. CARBAMATES

10. CARBONATES

The mixed carbonate cannot be made this way

11. DIAZO KETONES

Addition–elimination mechanism followed by removal of an acidic hydrogen

12. ESTERS

Fischer esterification

Acid-catalyzed transesterification; this transformation also works with base catalysis

Direct alkylation at low temperature using the hindered strong base, LDA

(continued)

12. ESTERS (CONTINUED)

Reverse Claisen condensation

The Baeyer–Villiger reaction

13. ISOCYANATES

Curtius rearrangement

14. KETENES

Elimination using a tertiary amine as base

Wolff rearrangement

15. β-KETO ESTERS

Claisen condensation

16. KETONES

The cuprate will react with the acid chloride but not with the product ketone

Friedel–Crafts acylation; this reaction works with the anhydride as well

The imine is an intermediate

Hydrolysis gives the β-keto acid which decarboxylates; more complicated ketones can be made by alkylating the β-keto ester first

17. LACTAMS

Amide-forming reactions applied in an intramolecular way will work

Lactams are formed by the cyclic version of the Beckmann rearrangement

18. LACTONES

Ester-forming reactions applied in an
intramolecular way will work

Lactones are formed by the cyclic version
of the Baeyer–Villiger reaction

19. NITRILES

$$NC^- + R-X \longrightarrow NC-R$$

S_N2 displacement

Dehydration of amides

20. SELENOXIDES

21. SULFOXIDES

22. XANTHATE ESTERS

Common Errors

Try to analyze problems before starting. Do not be too proud to do the obvious. Use the molecular formula to see what molecules must be combined. In acid (H_3O^+, HCl, etc.), carbocations are the likely intermediates. In base ($^-$OH, $^-$OR, etc.), carbanions are probably involved. Try to identify what must be accomplished in a problem. Although sometimes it will be hard to plan, and you will have to try all possible intermediates [Problem 19.28(c) is a good example], usually a problem will tell you much of what you must do. Ask yourself questions. What rings must be opened? What rings must be closed? What atoms become incorporated in the product? Pay attention to "stop signs." Primary carbocations are stop signs. Unstabilized carbanions are stop signs. If you are forced to invoke such a species in an answer, it is almost certainly wrong, and you must backtrack. Hard problems become easier if you keep such questions in mind. All of this advice sounds obvious, but it's not. Thinking a bit about what must be done in a problem saves much time in the long run and avoids the generally hopeless random arrow pushing that can trap you into a wrong answer.

19.20 ADDITIONAL PROBLEMS

Let's warm up with some simple problems, just to be certain that the basic (and acidic) mechanisms of this chapter are under control. Then we can go on to other things. Please be sure you can do the "easy" problems before you try the more difficult ones at the end of this section.

PROBLEM 19.43 Write mechanisms for the following conversions in base:

(a)

(b)

PROBLEM 19.44 Write mechanisms for the following conversions in acid:

(a)

(b)

PROBLEM 19.45 Now try a slightly more complicated acid-catalyzed hydrolysis.

PROBLEM 19.46 Write a mechanism for the following Dieckmann condensation:

PROBLEM 19.47 Here is a reverse Dieckmann. Write a mechanism.

PROBLEM 19.48 Write good Lewis structures for the following compounds:
(a) Propionic acid (propanoic acid)
(b) Propionyl chloride (propanoyl chloride)
(c) N,N-Diethyl-2-phenylpropanamide
(d) Methylketene
(e) Propyl cyclopropanecarboxylate
(f) Phenyl benzoate

PROBLEM 19.49 Give reasonable names for the following molecules:

(a)

(b)

(continued)

(c)

CH₃CH₂—C(=O)—OCH₃

(d)

PROBLEM 19.51 Provide structures for compounds **A–E**. Mechanisms are not required.

(e)

—C(=O)—OH

(f)

—C(=O)—Cl

(g)

—C(=O)—NH₂

(h)

O=C—NH₂ (with Br-substituted benzene ring)

PROBLEM 19.50 Provide structures for compounds **A–D**. Mechanisms are not required.

PROBLEM 19.52 Provide structures for compounds **A–F** in this series of reactions.

PROBLEM 19.53 Analysis of the IR stretching frequencies of acyl compounds involves an assessment of resonance and inductive effects. This assessment is a somewhat risky business, and it must be admitted that near-circular reasoning is sometimes encountered. Nonetheless, see if you can make sense of the following observations taken from Table 19.3. In each case, it will be profitable to think of all the important resonance forms of the acyl compound.

(a) Acetone absorbs at 1719 cm⁻¹, whereas acetaldehyde absorbs at 1733 cm⁻¹.

(continued)

PROBLEM 19.53 *(continued)*

(b) Explain why methyl acetate absorbs at higher frequency (1750 cm^{-1}) than acetone (1719 cm^{-1}).

(c) *N*-Methyl acetamide absorbs much lower (1688 cm^{-1}) than methyl acetate (1750 cm^{-1}).

PROBLEM 19.54 Is your explanation in Problem 19.53(b) consistent with the observation that the carbonyl carbon of methyl acetate appears at δ 169 ppm in the ^{13}C NMR spectrum, whereas the carbonyl carbon of acetone is far downfield at δ 205 ppm? Explain.

PROBLEM 19.55 Give the major organic products expected in each of the following reactions or reaction sequences. Mechanisms are not required.

(a)

1. SOCl$_2$
2. HN(CH$_3$)$_2$
3. LiAlH$_4$
4. H$_2$O

(b)

CH$_3$OH
Δ

(c)

CH$_3$ONa
CH$_3$OH, Δ

(d)

HCl
H$_2$O, Δ

(e)

1. NaH
2. CS$_2$
3. CH$_3$I
4. 210–235 °C

(f)

1. LICA, THF, −78 °C
2. I$_2$
3. (CH$_3$CH$_2$)$_3$N

LICA =

PROBLEM 19.56 As you know, most β-keto acids decarboxylate very easily (p. 1006). Compound **1** is an exception, as it survives heating to very high temperature without losing carbon dioxide. Explain.

1

PROBLEM 19.57 In Section 19.15, we saw that primary amides could be dehydrated to nitriles (cyanides) with P$_2$O$_5$. Nitriles can also be prepared by treating aldoximes (**1**) with dehydrating agents such as acetic anhydride. Propose a mechanism for the formation of benzonitrile from the reaction of **1** with acetic anhydride. (How are oximes prepared?)

PROBLEM 19.58 Reaction of benzonitrile (**1**) and *tert*-butyl alcohol in the presence of concentrated sulfuric acid, followed by treatment with water, gives *N-tert*-butylbenzamide (**2**). Provide an arrow formalism mechanism for this reaction.

Ph—C≡N + (CH$_3$)$_3$COH

1

1. concd H$_2$SO$_4$ 2. H$_2$O

2

PROBLEM 19.59 Devise syntheses for the following molecules. You may start with benzene, methyl alcohol, sodium methoxide, ethyl alcohol, sodium ethoxide, butyl alcohol (BuOH), phosgene, and propionic acid as your sources of carbon. You may also use any inorganic reagent and solvents as needed. Mechanisms are not required.

(a)

Bu_3C—OH

(b)

CH_3CH_2—C(=O)—Bu

(c)

CH_3CH_2—C(OH)(Bu)—Bu

(d)

benzene—C(=O)—CH_2CH_3

(e)

CH_3CH_2CN

(f)

CH_3CH_2—C(=O)—CH(CH_3)—C(=O)—OCH_2CH_3

(g)

CH_3CH_2—C(=O)—CH(CH_3)—Bu

Bu = $CH_3CH_2CH_2CH_2$

PROBLEM 19.60 Propose syntheses for the following target molecules starting from the indicated material. You may use any other reagents that you need. Mechanisms are not required. Use a retrosynthetic analysis in each case.

(a)

$\text{allyl—CH}_2\text{—C(=O)—CH}_3$ with isopropyl substituent **from** H_3C—C(=O)—OCH_2CH_3

(b)

$\text{long chain—CH(CH}_3)\text{—COOH}$ **from** $CH_2(COOCH_3)_2$

(c)

cyclopentanone—CH_2—Ph **from** cyclohexene

(d)

cyclobutyl—COOH from $CH_2(COOCH_3)_2$ and $BrCH_2CH_2CH_2Cl$

PROBLEM 19.61 Propose syntheses of the following molecules starting from cyclopentanone:

(a)

(b)

PROBLEM 19.62 Provide an arrow formalism mechanism for the following reaction:

benzaldehyde (Ph—CHO) + H_3C—C(=O)—O—C(=O)—CH_3

1. NaOAc, Δ
2. H_2O/H_3O^+

→ **Cinnamic acid** (Ph—CH=CH—COOH)

This problem is harder than it looks. Your answer must take into account the fact that compound **B** is an intermediate.

Ph—CH(—O—C(=O)—CH_3)—CH_2—COO^-

B

PROBLEM 19.63   Here is the problem concerning the formation of the Vilsmeier reagent we promised you in Chapter 18. As we saw in Section 18.6d, acid chlorides can be prepared from carboxylic acids and reagents such as thionyl chloride, phosphorus pentachloride, phosgene, and oxalyl chloride. The conditions for these reactions are sometimes quite severe. It is possible to prepare acid chlorides under appreciably milder conditions by using the Vilsmeier reagent (Problem 18.37). The Vilsmeier reagent is prepared from DMF and oxalyl chloride. Write an arrow formalism mechanism. *Hint*: DMF acts as a nucleophile, but not via the nitrogen atom.

DMF   **Oxalyl chloride**

Vilsmeier reagent

PROBLEM 19.64   Provide an arrow formalism for the following reaction:

1. NaOCH$_3$/CH$_3$OH, Δ | 2. H$_3$COOC $\diagdown$ Ph
3. CH$_3$COOH

PROBLEM 19.65   Rationalize the following transformation. *Hint*: See Problem 19.28(a).

1. NaOCH$_3$/CH$_3$OH, Δ | 3. Br $\diagdown$ COOCH$_2$CH$_3$, Δ
2. remove CH$_3$OH

PROBLEM 19.66   Provide an arrow formalism mechanism for the following reaction:

+   CH$_2$(COOCH$_2$CH$_3$)$_2$

CH$_3$CH$_2$OH, Δ

The following two problems should be done together.

PROBLEM 19.67   Provide an arrow formalism for the following reaction:

HOCH$_3$
NaOCH$_3$

PROBLEM 19.68 Here is an answer to Problem 19.67, which appears in the original paper describing the reaction, as well as at least one compilation of problems. Perhaps it is the answer you came up with. First, criticize this proposal—what makes it unlikely? Next, you can either feel good about finding another mechanism already, or, if this was your answer, set about finding a better mechanism. *Hint*: Find a way to remove strain in the starting material.

PROBLEM 19.69 Reaction of salicylaldehyde (**1**) and acetic anhydride in the presence of sodium acetate gives compound **2**, a fragrant compound occurring in sweet clover.

Spectral data for compound **2** are collected below. Deduce the structure of **2**. Naturally, it is an interesting question as to how **2** is formed. This problem turns out to be harder than it looks at first. Nonetheless, you might want to try to formulate a mechanism. However, note that *O*-acetylsalicylaldehyde (**3**) gives only traces of **2** in the absence of acetic anhydride.

Compound **2**
Mass spectrum: $m/z = 146$ (p, 100%), 118 (80%), 90 (31%), 89 (26%)
IR (Nujol): 1704 cm^{-1} (s)
^{1}H NMR (CDCl$_3$): δ 6.42 (d, $J = 9$ Hz, 1H), 7.23–7.35 (m, 2H), 7.46–7.56 (m, 2H), 7.72 (d, $J = 9$ Hz, 1H)
^{13}C NMR (CDCl$_3$): δ 116.6 (d), 116.8 (d), 118.8 (s), 124.4 (d), 127.9 (d), 131.7 (d), 143.4 (d), 154.0 (s), 160.6 (s)

PROBLEM 19.70 Provide structures for compounds **A–F**. Spectral data for compounds **E** and **F** are summarized below.

Compound **E**
Mass spectrum: $m/z = 162$ (p, 100%), 106 (77%), 104 (51%), 43 (61%)

(continued)

IR (melt): 1778 cm^{-1} (s)

^{1}H NMR (CDCl$_3$): δ 2.1–2.15 (m, 1H), 2.5–2.7 (m, 3H), 5.45–5.55 (m, 1H), 7.2–7.4 (m, 5H)

^{13}C NMR (CDCl$_3$): δ 28.9 (t), 30.9 (t), 81.2 (d),125.2 (d), 128.4 (d), 128.7 (d), 139.3 (s), 176.9 (s)

Compound **F**

Mass spectrum: m/z = 162 (p, 28%), 104 (100%)

IR (film): 1780 cm^{-1} (s)

^{1}H NMR (CDCl$_3$): δ 2.67 (dd, J = 17.5 Hz, 6.1 Hz, 1H), 2.93 (dd, J = 17.5 Hz, 8.1 Hz, 1H), 3.80 (m, 1H), 4.28 (dd, J = 9.0 Hz, 8.1 Hz, 1H), 4.67 (dd, J = 9.0 Hz, 7.9 Hz, 1H), 7.22–7.41 (m, 5H)

^{13}C NMR (CDCl$_3$): δ 35.4 (t), 40.8 (d), 73.8 (t), 126.5 (d), 127.4 (d), 128.9 (d), 139.2 (s), 176.3 (s)

PROBLEM 19.71 In Section 19.7, we saw that aldehydes are available from the reduction of esters at low temperature (–70 °C) by diisobutylaluminum hydride (DIBAL-H), (see Fig. 19.38). If the reaction mixture is allowed to warm before hydrolysis, the yield of aldehyde drops appreciably. For example, reduction of ethyl butyrate with DIBAL-H at –70 °C followed by hydrolysis at –70 °C, affords a 90% yield of butyraldehyde. However, if the reaction mixture is allowed to warm to 0 °C before hydrolysis, the yield of aldehyde is less than 20%. Among the products under these conditions are ethyl butyrate (recovered starting material) and butyl alcohol (CH$_3$CH$_2$CH$_2$CH$_2$OH), formed from hydrolysis of CH$_3$CH$_2$CH$_2$CH$_2$O—AlR$_2$ (R = isobutyl). Rationalize these observations with arrow formalism mechanisms. *Hint*: Recall the MPVO equilibration and the Cannizzaro reaction, Chapter 17.

PROBLEM 19.72 γ-Aminobutyrate transaminase (GABA-T) is the key enzyme controlling the metabolism of γ-aminobutyric acid (GABA), a mammalian inhibitory neurotransmitter. In the following synthesis of the hydrochloride salt of the putative GABA-T inhibitor, 3-amino-1,4-cyclohexadiene-1-carboxylic acid (**G**), propose structures for intermediates **A–F**. Mechanisms are not necessary, but may be helpful in some cases.

PROBLEM 19.73 Treatment of phthalimide (**1**) with bromine in aqueous sodium hydroxide, followed by acidification with acetic acid, affords compound **2**. Spectral data for **2** are summarized on the next page. Deduce the structure of **2**, and then propose a mechanism for its formation.

1

Compound 2

Mass spectrum: $m/z = 137$ (p, 59%), 119 (100%), 93 (79%), 92 (59%)

IR (KBr): 3490 (m), 3380 (m), 3300–2400 (br), 1665 (s), 1245 (s), and 765 (m) cm^{-1}

^{1}H NMR (DMSO-d_6):* δ 6.52 (t, $J = 8$ Hz, 1H), 6.77 (d, $J = 8$ Hz, 1H), 7.23 (t, $J = 8$ Hz, 1H), 7.72 (d, $J = 8$ Hz, 1H), 8.60 (br s, 3H, washes out with D$_2$O)

^{13}C NMR (DMSO-d_6): δ 109.5 (s), 114.5 (d), 116.2 (d), 131.1 (d), 133.6 (d), 151.4 (s), 169.5 (s)

*Note that J_{meta} and J_{para} were not observed under these conditions.

PROBLEM 19.74   Sketch the transition state for the Cope elimination (p. 1089) shown below.

PROBLEM 19.75   Design an experiment to test the stereochemistry of the Cope elimination. How would one determine that it is a cis hydrogen that is lost? Assume you can make any labeled compounds you need.

PROBLEM 19.76   Write a mechanism for the Schmidt reaction shown below. *Hint*: The first step is probably the formation of an acylium ion, R–C=O. Note also that the reagent is HN$_3$, not NH$_3$.

PROBLEM 19.77   Outline mechanisms for the following transformations leading to quinine*:

(a)

1. NaOEt/HOEt, Δ
2. neutralize

(continued)

*Named for the Countess del Chinchon, wife of a seventeenth century Peruvian viceroy, supposedly cured of the vapors by use of the bark.

PROBLEM 19.77 *(continued)*

(b)

The structure of this complex molecule presents an extraordinary synthetic challenge, and was accomplished in the early 1940s by R. B. Woodward and William von E. Doering. It's worth a look at this synthesis. It was not only an achieve-

ment of brilliance, but used many of the reactions we have studied in a mechanistic way.

PROBLEM 19.78 The reactions shown below were important in the synthesis of LSD. Provide structures for the indicated compounds.

(a)

(b)

(c)

F
(C$_{15}$H$_{16}$N$_2$O) →[Ac$_2$O] **G**
(C$_{17}$H$_{18}$N$_2$O$_2$) →[1 NaBH$_4$ / 2 H$_2$O] **H**
(C$_{17}$H$_{20}$N$_2$O$_2$) →[SOCl$_2$] **I**
(C$_{17}$H$_{19}$N$_2$OCl)

(d)

I →[NaCN] **J**
(C$_{18}$H$_{19}$N$_3$O) →[CH$_3$OH / H$_2$O/H$_2$SO$_4$] **K**
(C$_{17}$H$_{20}$N$_2$O$_2$) →[1. NaOH / 2. H$_3$O$^+$/H$_2$O] **L**
(C$_{16}$H$_{18}$N$_2$O$_2$) →[several steps] **LSD**

PROBLEM 19.79   Provide structures for compounds **A–E**.
Mechanisms are not necessary.

COOH
→[SOCl$_2$ / 150 °C] **A**
(C$_7$H$_{11}$ClO) →[(CH$_3$)$_2$NH / benzene] **B**
(C$_9$H$_{17}$NO) →[1. LiAlH$_4$ / ether, 35 °C / 2. H$_2$O] **C**
(C$_9$H$_{19}$N) →[30% H$_2$O$_2$ / 25 °C] **D**
(C$_9$H$_{19}$NO) →[100 °C / 2 h] **E**
(C$_7$H$_{12}$)

PROBLEM 19.80   In the following synthesis of racemic
nicotine (**1**), propose structures for intermediates **A–C**. Although mechanisms are not required, a mechanistic analysis
will probably prove helpful, as there are several nonisolable
intermediates involved.

COOH
→[CH$_3$CHN$_2$] **A**
(C$_8$H$_9$NO$_2$) →[1. (pyrrolidinone/NaOCH$_2$CH$_3$, benzene, Δ) / 2. CH$_3$COOH, 0 °C] **B**
(C$_{11}$H$_{12}$N$_2$O$_2$) →[HCl/H$_2$O / 130–135 °C] **C**
(C$_{10}$H$_{14}$N$_2$O·2HCl)

→[1. KOH/CH$_3$OH / 2. H$_2$/PtO$_2$] **1**

PROBLEM 19.81   Reaction of 1,3-diphenylisobenzofuran
(**1**) and vinylene carbonate (**2**) gives compound **3**. Acid hydrolysis of **3** affords **4**. Propose a structure for **3** and write a
mechanism for its formation and hydrolysis to **4**.

PROBLEM 19.82   Write an arrow formalism mechanism for
the following reaction:

Special Topic: Aromatic Transition States; Orbital Symmetry

20

The fascination of what's difficult
Has dried the sap of my veins, and rent
Spontaneous joy and natural content
Out of my heart.

—William Butler Yeats*
"The Fascination of What's Difficult"

20.1   PREVIEW

In Chapters 12–14, we explored the consequences of conjugation—the sideways overlap of $2p$ orbitals. This study led us to the aromatic compounds, whose great stability can be traced to an especially favorable arrangement of electrons in low-lying bonding molecular orbitals. It can surely be no surprise to find that transition states can also benefit energetically through delocalization, and that most of the effects, including aromaticity, that influence the energies of ground states are important to transition states as well.

In this chapter, we will encounter reactions that were frustratingly difficult for chemists to understand for many, many years. By now, you are used to seeing acid- and base-catalyzed reactions, in which an intermediate is first formed and then produces product with the regeneration of the catalytic agent. Acid-catalyzed additions to alkenes are classic examples, but there are already many other such reactions in your notes. There is a class of reactions that is extraordinarily insensitive to catalysis. Bases, acids, and radicals are largely without effect. Even the presence of solvent seems of little relevance, as the reactions proceed as well in the gas phase as in solution. What is one to make of such reactions? How does one describe a mechanism?

*William Butler Yeats (1865–1939) was an Irish poet who received the Nobel prize for literature in 1923.

1,5-Hexadiene　　Still
1,5-Hexadiene

FIGURE 20.1 A single-barrier, one-transition-state, "no-mechanism" reaction.

We are used to speculating on the structure of an intermediate and then using the postulated picture to predict the structures of the transition states surrounding it. In these uncatalyzed reactions in which starting material and product are separated by a single transition state (called **concerted reactions**), there is very little with which to work in developing a mechanism. Indeed, one may legitimately ask what "mechanism" means in this context. Such processes became known as "no-mechanism" reactions. Some, in which starting material simply rearranges into itself (a **degenerate reaction**) can even be called "no-mechanism, no-reaction, reactions." Figure 20.1 shows an arrow formalism for a typical no-mechanism, no-reaction, reaction.

In 1965, R. B. Woodward (1917–1979) and Roald Hoffmann (b. 1937), both then at Harvard, began to publish a series of papers that ventured a mechanistic description of no-mechanism reactions, and gathered a number of these seemingly different processes under the heading of **pericyclic reactions**. Their crucial insight that bonding overlap must be maintained between orbitals during the course of a concerted, pericyclic reaction now seems so simple that you may find it difficult to see why it eluded chemists for so many years. All I can tell you is that simple things are sometimes very hard to see, even for very smart people. **Woodward–Hoffmann theory** had been approached very closely before without the crucial "aha!"; without the lightbulb over the head turning on. Perhaps what was lacking was the ability to combine a knowledge of theory with an awareness of the chemical problem, which is precisely what the brilliant experimentalist Woodward and the young theorist Hoffmann brought to the problem. Ultimately, these papers and their progeny resulted in the Nobel prize of 1981 for Hoffmann [together with Kenichi Fukui (1918–1998) of Kyoto], and would probably have given Woodward his second Nobel prize, had he lived long enough.

The reaction of the chemical community to these papers ranged from enthusiastic admiration to "We knew all this trivial stuff already." I would submit that admiration was by far the more appropriate reaction, and suspect that behind much of the carping was a secret smiting of many foreheads. Woodward–Hoffmann theory is brilliant. It not only answered long-standing and difficult questions, but it had remarkable "legs." Its implications run far into organic and the rest of chemistry, and it makes what are called risky predictions. It is important to separate theory that merely rationalizes—that which explains known phenomena—from that which demands new experiments. The success of Woodward–Hoffmann theory can be partly judged from the flood of experiments it generated. Of course, not all the experiments were incisive, and not all the interpretations were appropriate—the area entered a rococo phase quite early on. Still, the best of the work generated by the early Woodward–Hoffmann papers stands today as an example of the combination of theory and experiment that characterizes the best of our science. In this chapter, we will see electrocyclic reactions in which rings open and close, cycloaddition reactions in which two partners come together to make a new cyclic compound, and sigmatropic shifts in which one part of a molecule flies about, coming to rest at one specific position, and no other. Specificity is the hallmark of all these reactions; atoms move as if in lock step in one direction or another, or move from one place to one specific new place, but no other. The chemical world struggled mightily to understand these mys-

teriously stereospecific reactions, and finally figured it out in a marvelously simple
way. This chapter will show you some wonderful chemistry.

ESSENTIAL SKILLS

1. To be able to understand the course of "no-mechanism" reactions, you need only
   remember that in any concerted (one barrier, one transition state) reaction, a bonding
   overlap must be maintained between orbitals. The rest is all detail.

IMPORTANT DETAILS

1. The opening and closing reactions of rings—electrocyclic reactions in the jargon of this
   chapter—are best understood through an analysis of the highest occupied molecular
   orbital (HOMO) of the open-chain partner. Ring closing will involve only bonding interactions
   for allowed reactions.

2. Cycloaddition reactions require an analysis of the interactions of the HOMO and lowest
   unoccupied molecular orbital (LUMO) of the two reacting species. Allowed reactions will
   have two bonding interactions between the reacting partners; "forbidden" reactions will not.

3. In sigmatropic shift reactions, one portion of a molecule detaches from one atom and
   reattaches at another position of the same molecule. The transition state is best described
   by an atom (or group of atoms) moving across a polyenyl radical. Once again, an analysis
   of the symmetry of the HOMO of the polyenyl radical is the key to understanding. A
   bonding interaction must be achieved at the arrival point for the reaction to be allowed.

20.2   CONCERTED REACTIONS

Woodward and Hoffmann provided insights into concerted reactions. A reaction is
concerted if starting material goes directly to product over a single transition state.
Such reactions are "single-barrier" processes. The S_N2 and Diels–Alder reactions (Chapters 7 and 12) are nice examples of concerted reactions you know well (Fig. 20.2).

FIGURE 20.2 The S_N2 and Diels–Alder reactions are
concerted, single-barrier processes.

Of course many reactions involve intermediates, and in these reactions more than one transition state is traversed on the way from starting material to product. These are called nonconcerted or stepwise reactions. Typical examples are the S_N1 reaction or the polar addition of HBr to an alkene, two reactions that go through the same intermediate, the *tert*-butyl cation (Chapters 7 and 3; Fig. 20.3).

FIGURE 20.3 Two reactions in which starting material and product are separated by a pair of transition states surrounding an intermediate.

It is important to recognize that nonconcerted reactions are made up of series of single-barrier, concerted reactions, each of which could be analyzed independently of whatever other reactions followed or preceded (Fig. 20.4).

Woodward and Hoffmann offered explanations of why some reactions were concerted and others not; they found a way to analyze reactions by determining the pathways required to maintain bonding interactions between orbital lobes as the reaction progressed. The following sections take up some examples in which Woodward–Hoffmann theory is particularly useful.

20.3   ELECTROCYCLIC REACTIONS

In 1957, a young German chemist named Emanuel Vogel (b. 1927) was finding his way through the labyrinth of the German academic system, working on the experiments necessary for the entrance ticket for a German university: a super Ph.D. degree called the "Habilitation." His topic included the thermal rearrangements of cyclobutenes. Although cyclobutene itself had been studied by R. Willstätter (1872–1942, Nobel prize in 1915) as early as 1905, and the rearrangement to 1,3-butadiene noted by the American chemist John D. Roberts (b. 1918)

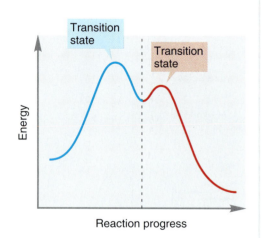

FIGURE 20.4 Any reaction can be separated into a series of single-barrier, concerted processes.

in 1949, only Vogel was alert enough to notice the startling stereospecificity of the thermal reaction. For example, the cis 3,4-disubstituted cyclobutene shown in Figure 20.5 rearranges *only* to the cis,trans 1,3-butadiene and no other stereoisomer.

The arrow formalism of Figure 20.5 is easy to write, but the stereochemical outcome of the reaction is anything but obviously predictable. It's worthwhile to see exactly why. There are three possible stereoisomers of the product, and a naive observer would certainly be forgiven for assuming that thermodynamic considerations would dictate the formation of the most energetically favorable isomer, the trans,trans diene, in which the large ester groups are as far apart as possible. Notice that the opening of the cyclobutenes must initially give the dienes in their s-cis conformations rather than in the lower energy s-trans arrangements. Of course, they will rapidly rotate to the lower energy conformations (Fig. 20.6).

A cis 3,4-disubstituted cyclobutene

A cis,trans 1,3-butadiene

FIGURE 20.5 The thermal opening of a cyclobutene to a 1,3-butadiene. The cis disubstituted cyclobutene rearranges to the cis,trans butadiene. The Δ means heat.

E = COOCH₃   *E* = Entgegen

FIGURE 20.6 The three possible stereoisomers of the diene product. But, only one, the cis,trans (*Z,E*) diene of intermediate thermodynamic stability, is formed in the reaction.

As Vogel first found, and as others showed later for many other systems, this is not the actual result at all; nor is the reaction stereorandom, as there is only a single stereoisomer found. Vogel well understood that this reaction was rather desperately trying to tell us something. In the intervening years, many similar reactions were found and many a seminar hour was spent in a fruitless search for the explanation. What the reaction had to say to us waited for the papers of Woodward and Hoffmann.

Remarkably, it was discovered that the stereochemical outcome of the reaction was dependent on the energy source. Heat gave one stereochemical result and light another. Figure 20.7 sums up the results of the thermal and photochemical interconversions of cyclobutenes and butadienes. In the thermal reaction, a cis cyclobutene is related to the cis,trans diene. By contrast, in the photochemical process the cis cyclobutene is related to a pair of molecules, the cis,cis diene and the trans,trans diene (Fig. 20.7).

FIGURE 20.7 The stereochemical outcomes of the thermal and photochemical reactions of cyclobutenes and butadienes are different.

Notice that the two reactions of Figure 20.7 run in different directions. The thermal reaction favors the more stable 1,3-butadiene isomer, but the photochemical process uses the energy of the photon to convert the 1,3-butadiene into the less stable, more energetic cyclobutene.

Let's start our analysis by thinking about the thermal opening of cyclobutene to 1,3-butadiene. Here is a general practical hint. It is usually easier to analyze these processes, called **electrocyclic reactions**, by examining *the open-chain partner*, regardless of which way the reaction actually runs. Figure 20.8 shows the molecular orbitals involved in the cyclobutene to butadiene conversion.

In this reaction, the four π molecular orbitals of 1,3-butadiene are converted into the σ, σ*, π, and π* orbitals of cyclobutene. Now comes a most important point. *It is assumed that it will be the HOMO of the system that controls the course of this reaction.* This assumption is not ultimately necessary, but it makes working out the results of this and similar reactions very easy. It can be defended in the following way. It is the electrons of highest energy, the valence electrons located in the highest energy orbital, that control the course of atomic reactions, and it should be no surprise that the same is true for molecular reactions. To do a complete analysis of any reaction, atomic or molecular, we really should follow the changes in energy of all electrons. That is difficult or impossible in all but a very few simple reactions. Accordingly, theorists have sought out simplifying assumptions, one of which says that it will be the highest energy electrons, those most loosely held, that will be most important in the reaction. These are the electrons in the Highest Occupied Molecular Orbital, the HOMO. The HOMO of butadiene is Φ_2, and we will assume that it is the controlling molecular orbital. Now ask how Φ_2 must move as 1,3-butadiene becomes cyclobutene. Don't worry that we are analyzing the reaction in the counterthermodynamic sense. If we know about the path in one direction, we know about the path in the other direction as well. In order to create the bond between the end carbons, the *p* orbitals on the end carbons must rotate, and this rotation must be in a fashion that creates a *bond* (σ) between the atoms, not an *antibond* (σ*) (Fig. 20.9).

FIGURE 20.8 The molecular orbitals involved in the interconversion of cyclobutene and butadiene.

FIGURE 20.9 The molecular orbitals involved in the interconversion of the cyclobutene and butadiene.

Now look at the symmetry of the end orbitals of Φ_2, the controlling HOMO of butadiene. There are two ways to make a bond: in one the plus (blue) lobes overlap, in the other it is the minus (green) lobes that overlap. In each case we have created a bonding interaction. In this example, the plus/plus overlap requires that the two ends of the molecule each rotate in a clockwise fashion. The minus/minus overlap requires that each end rotate counterclockwise (Fig. 20.10).

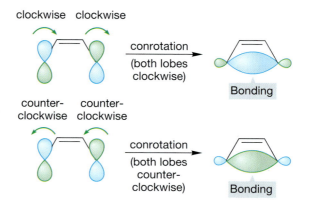

FIGURE 20.10 Formation of a bonding interaction requires what is called conrotation of the ends of the diene. There are two, equivalent conrotatory modes.

These two modes of rotation, in which the two ends rotate in the *same* direction, are called conrotatory motions, and the whole process is known as **conrotation**.

There is another rotatory mode possible, called **disrotation**, in which the two ends of the molecule rotate in a different sense. Figure 20.11 shows the two possible disrotatory motions. In both, an antibond, not a bond, is created between the two end carbons. Disrotation cannot be a favorable process in this case.

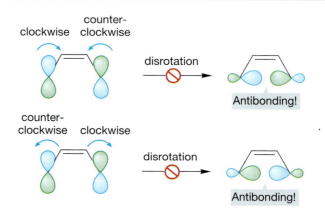

FIGURE 20.11 Disrotation creates an antibonding interaction between the lobes at the ends of the diene. Neither of the two possible disrotatory modes can be a favorable process.

This reaction is as easy to analyze as that. Identify the HOMO, and then see what sort of motion is demanded of the end carbons by the lobes of the molecular orbital. All electrocyclic reactions can be understood in this same simple way. The theory tells us that the thermal interconversion of cyclobutene and 1,3-butadiene must take place in a conrotatory way. For the cyclobutene studied by Vogel, conrotation requires the stereochemical relationship that he observed. The cis 3,4-disubstituted cyclobutene can only open in conrotatory fashion, and conrotation *forces* the formation of the cis,trans diene. Note that there are always two possible conrotatory modes (Fig. 20.12).

FIGURE 20.12 For the cis disubstituted cyclobutene studied by Vogel, both conrotatory modes demand the formation of the diene with one cis and one trans double bond.

WORKED PROBLEM 20.1 Show through an equivalent analysis that a mixture of the cis,cis and trans,trans dienes must be formed in the thermal opening of trans 3,4-disubstituted cyclobutenes.

ANSWER As the opening is taking place thermally, it must be conrotatory. There are two conrotatory modes for the trans compound. One leads to the cis,cis (Z,Z) isomer and the other to the trans,trans (E,E) isomer.

Now what about the photochemical reaction? Look again at the molecular orbitals of the open partner, 1,3-butadiene. Recall that absorption of a photon promotes an electron from the HOMO to the LUMO (Chapter 12, p. 566), in this case from Φ_2 to Φ_3, creating a new, "photochemical HOMO" (Fig. 20.13).

FIGURE 20.13 The HOMO involved in the photochemical reaction of butadiene is Φ_3.

No longer will conrotation in the HOMO, now Φ_3, create a bond between the end carbons. For an orbital of this symmetry, *conrotation creates an antibond*. For the photochemical case, it is disrotation that creates the bonding interaction (Fig. 20.14).

FIGURE 20.14 The new, "photochemical HOMO" (Φ_3) demands disrotation. In this molecular orbital, conrotation produces an antibond.

Φ_3 of butadiene, the "photochemical HOMO"

disrotation → Bonding

conrotation → Antibonding

PROBLEM 20.2 Figure 20.14 shows only one disrotatory and one conrotatory mode. Verify that the other possible disrotation also produces a bond between the two carbons, and that the other possible conrotation produces an antibond.

Two disrotatory modes interconvert the cis 3,4-disubstituted cyclobutene and the cis,cis and trans,trans isomers of the butadiene (Fig. 20.15).

FIGURE 20.15 Disrotation interrelates the cis disubstituted cyclobutene with the cis,cis and trans,trans dienes.

CH$_3$OOC   COOCH$_3$
cis

disrotation
$h\nu$

H   H
E   E
cis,cis (Z,Z)

CH$_3$OOC   COOCH$_3$
cis

disrotation
$h\nu$

E   E
H   H
trans,trans (E,E)

E = COOCH$_3$ E = Entgegen

PROBLEM 20.3 Show that disrotation in the photochemical reaction of the trans 3,4-disubstituted cyclobutene leads to the cis,trans isomer of the butadiene.

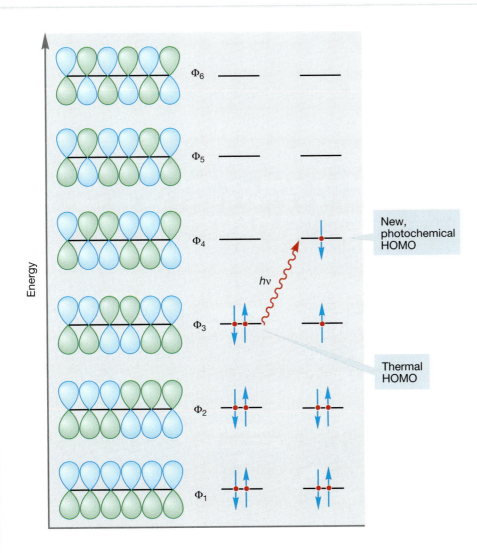

cis,trans (Z,E)

Let's now use this technique for analyzing electrocyclic reactions to make some predictions about the 1,3,5-hexatriene–1,3-cyclohexadiene system (Fig. 20.16).

As it is always easier to look first at the open polyene system, Figure 20.17 gives the π molecular orbitals for hexatriene. The HOMO for the thermal reaction will be Φ_3, and that for the photochemical reaction, Φ_4.

1,3,5-Hexatriene

1,3-Cyclohexadiene

FIGURE 20.16 Another electrocyclic reaction interconverts 1,3,5-hexatriene and 1,3-cyclohexadiene.

FIGURE 20.17 The π molecular orbitals for hexatriene. The thermal HOMO is Φ_3 and the photochemical HOMO is Φ_4.

New, photochemical HOMO

Thermal HOMO

As the open triene closes to the cyclohexadiene, the only way to create a bonding interaction between the end carbons is to close in a disrotatory fashion from Φ_3, and in a conrotatory fashion from Φ_4. So, we predict that the thermal interconversion will occur in a disrotatory way, and that the photochemical reaction must involve conrotation (Fig. 20.18).

Thermal reactions

Φ_3 of hexatriene
(HOMO)

Photochemical reactions

Φ_4 of hexatriene
(photochemical HOMO)

FIGURE 20.18 For this system, in order to produce a bonding interaction, the thermal reactions must take place in a disrotatory fashion and the photochemical reactions in a conrotatory way.

To test these predictions we need some labeled molecules. The *trans,cis,trans*-2,4,6-octatriene shown in Figure 20.19 will do nicely. A disrotatory thermal closure to *cis*-5,6-dimethylcyclohexa-1,3-diene and a conrotatory photochemical closure to the trans isomer are both known.

FIGURE 20.19 The experimental results bear out the predictions. The thermal reaction is disrotatory and the photochemical reaction is conrotatory.

Figure 20.19 shows the apparent formation of a single enantiomer of *trans*-5,6-dimethylcyclohexa-1,3-diene from an achiral precursor, which cannot be possible. We have not taken the trouble to draw both enantiomers of the racemic mixture of the *trans*-5,6-dimethylcyclohexa-1,3-diene formed in the reaction. This omission leads directly to Problem 20.4.

PROBLEM 20.4 Figure 20.19 shows only one conrotatory and one disrotatory mode. Show that the other disrotation and conrotation give the same stereochemical results.

PROBLEM 20.5 Why are cyclohexadienes formed only from the trans,cis,trans isomer shown? Why does the trans,trans,trans molecule not give a similar reaction?

PROBLEM 20.6 Work out the predicted products from the cis,cis,trans and cis,cis,cis isomers of 2,4,6-octatriene for both photochemical and thermal reactions.

SUMMARY

Table 20.1 gives the results so far. It also points out that the cyclobutene–1,3-butadiene interconversion is a four-electron process, whereas the hexatriene–1,3-cyclohexadiene reaction involves six electrons. In the cyclobutene–butadiene reaction, the four electrons in σ and π of the cyclobutene come to occupy Φ_1 and Φ_2 of 1,3-butadiene. Similarly, the six electrons in σ, Φ_1, and Φ_2 of 1,3-cyclohexadiene are interconverted with the six electrons in Φ_1, Φ_2, and Φ_3 of the 1,3,5-hexatriene molecule.

TABLE 20.1   Rotatory Motions in Two Electrocyclic Reactions

Reaction	Number of Electrons	Thermal	Photochemical
Cyclobutene–butadiene	4	Conrotation	Disrotation
Hexatriene–cyclohexadiene	6	Disrotation	Conrotation

It turns out that all $4n$ systems will behave like the four-electron case (thermal reactions conrotatory and photochemical reactions disrotatory) and all $4n + 2$ systems behave like the six-electron case (thermal reactions disrotatory and photochemical reactions conrotatory). So, Table 20.1 can be generalized to Table 20.2, which gives the rules for all electrocyclic reactions. Now, if you can count the number of electrons correctly you can easily predict the stereochemistry of any electrocyclic reaction without even working out the molecular orbitals.

TABLE 20.2   Rotatory Motions in All Electrocyclic Reactions

Reaction	Number of Electrons	Thermal	Photochemical
Cyclobutene–butadiene	4 (4n)	Conrotation	Disrotation
Hexatriene–cyclohexadiene	6 (4n + 2)	Disrotation	Conrotation
All 4n electrocyclic reactions	4n	Conrotation	Disrotation
All 4n + 2 electrocyclic reactions	4n + 2	Disrotation	Conrotation

PROBLEM 20.7 Provide mechanisms for the following reactions:

(a)

"Dot" means "H up"

trans

(b) Dewar benzene lies perched some 60 kcal/mol above its isomer benzene. It would seem that a simple stretching of the central bond would inevitably and easily give benzene. Yet Dewar benzene certainly does exist and in fact rearranges to benzene only quite slowly. Why?

(c)

X

trans

(d)

cis,cis

trans,trans

20.4   CYCLOADDITION REACTIONS

Cycloaddition reactions are another kind of orbital symmetry controlled (pericyclic) reaction. We have seen one already, the Diels–Alder reaction, and will use it as our prototype. We found the Diels–Alder cycloaddition to be a thermal process and to take place in a concerted (one-step) fashion. Several stereochemical labeling experiments were described in Chapter 12 (p. 587), all of which showed that the reaction involved neither diradical nor polar intermediates. This stereospecificity is important because orbital symmetry considerations apply only to single-step reactions. Of course, all reactions can be subdivided into series of single-step, single-barrier processes, and each of these steps could be analyzed separately by orbital symmetry. However, it makes no sense to speak of the overall orbital symmetry control of a multistep process.

When the diene and dienophile approach each other in the Diels–Alder reaction, the important orbital interactions are between the HOMOs and LUMOs. As Figure 20.20 shows, there are two possible HOMO–LUMO interactions, $HOMO_{(diene)}$–$LUMO_{(dienophile)}$ and $HOMO_{(dienophile)}$–$LUMO_{(diene)}$.

FIGURE 20.20 The two possible HOMO–LUMO interactions in the prototypal Diels–Alder reaction between butadiene and ethylene.

The stronger of these two will control the reaction. In this case, the stronger interaction is between the HOMO and LUMO pair closer in energy. *Remember*: The strength of orbital overlap, and the magnitude of the resulting stabilization, depends on the relative energies of the two orbitals. The closer the two are in energy, the stronger the interaction.

In the simplest, symmetric Diels–Alder reaction between ethylene and 1,3-butadiene, the two HOMO–LUMO interactions are of equal energy and we must look at the orbital symmetry of both. Figure 20.21 shows the symmetry relationships for this reaction. Don't forget that in the Diels–Alder (and related) reactions the two participants approach each other in parallel planes (Chapter 12, p. 588).

FIGURE 20.21 The orbital overlaps in the two HOMO–LUMO interactions of the Diels–Alder reaction. These HOMO–LUMO interactions produce two new bonds.

Both interactions involve bonding overlaps at the points of formation of the two new σ bonds. Accordingly, it's no surprise to find that the thermal Diels–Alder reaction occurs. One says that it is "allowed by orbital symmetry."

What about the photochemical Diels–Alder reaction? This reaction is most uncommon, and this observation might lead to the immediate suspicion that there is something wrong with it. Usually, the absorption of a photon will promote an electron from the HOMO to the LUMO. In this case, the lower energy HOMO–LUMO gap is that in the diene partner. Absorption of light creates a new photochemical HOMO for the diene, Φ_3, and now the HOMO–LUMO interaction with the dienophile partner involves one antibonding overlap. Both new bonds cannot be formed at the same time (Fig. 20.22). So this photochemical Diels–Alder reaction is said to be "forbidden by orbital symmetry."

FIGURE 20.22 Absorption of a photon creates a new, photochemical HOMO Φ_3. Now the HOMO–LUMO interaction involves one antibond. The reaction is forbidden by orbital symmetry.

PROBLEM 20.8 What would happen if the light were used to promote an electron from the HOMO of ethylene, π, to the LUMO of ethylene, π^*? Would the Diels–Alder reaction be allowed or forbidden?

Let's now analyze another potential cycloaddition, the simple joining together of a pair of ethylenes to give a cyclobutane (Fig. 20.23).

Two ethylenes   A cyclobutane

FIGURE 20.23 The hypothetical combination of two alkenes to give a cyclobutane.

PROBLEM 20.9 Will the 2 + 2 reaction of a pair of ethylenes to give a cyclobutane be exothermic or endothermic? Explain.

The HOMO and LUMO are easy to identify, and it is quickly obvious that here, too, the apparently easy bond formation shown by the arrow formalism is thwarted by an antibonding interaction as the two molecules come together (Fig. 20.24).

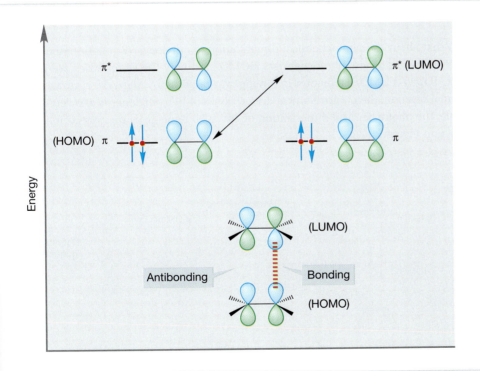

FIGURE 20.24 An analysis of the HOMO and LUMO for the reaction of two alkenes to give a cyclobutane shows that it is forbidden by orbital symmetry. There is an antibonding interaction.

Therefore we might guess that the thermal dimerization of ethylenes would be a rare process and we would be exactly correct. Moreover, it is clear in most of the reactions of this kind that do occur that the two new bonds are formed sequentially, and *not* in a concerted manner (Fig. 20.25).

FIGURE 20.25 The known dimerizations of substituted ethylenes can be shown to involve diradical intermediates.

PROBLEM 20.10 Predict the stereochemistry of the 2 + 2 cycloaddition of a pair of substituted ethylenes, X—CH=CH—X (see Fig. 20.25).

What about the photochemical dimerization of ethylenes? Promotion of a single electron creates a new photochemical HOMO, and now the symmetries are perfect for a HOMO–LUMO interaction involving two bonding overlaps. In contrast to the thermal reaction, the photochemical dimerization of two ethylenes is allowed. For example, the photochemical dimerization of *trans*-2-butene gives the two cyclobutanes shown in Figure 20.26.

FIGURE 20.26 The photochemical dimerization of two butenes.

Again, it's time for generalization. Table 20.3 shows what we have so far.

TABLE 20.3 Rules for Two Cycloaddition Reactions

Reaction	Number of Electrons	Thermal	Photochemical
Ethylene + Ethylene ⇄ Cyclobutane	4 ($4n$)	No	Yes
Ethylene + Butadiene ⇄ Cyclohexene (Diels–Alder)	6 ($4n + 2$)	Yes	No

It is possible to expand Table 20.3 to encompass all $4n$ and $4n + 2$ reactions. Other $4n$ processes will follow the rules for the 2 + 2 dimerization of a pair of ethylenes, and $4n + 2$ processes will resemble the 4 + 2 cycloaddition we know as the Diels–Alder reaction (Table 20.4).

TABLE 20.4 Rules for All Cycloaddition Reactions

Reaction	Number of Electrons	Thermal	Photochemical
Ethylene + Ethylene ⇄ Cyclobutane	4 ($4n$)	No	Yes
Ethylene + Butadiene ⇄ Cyclohexene (Diels–Alder)	6 ($4n + 2$)	Yes	No
All $4n$ reactions	$4n$	No	Yes
All $4n + 2$ reactions	$4n + 2$	Yes	No

Now compare Table 20.2 with Table 20.4. This comparison should be a bit disquieting to you. Notice how much more information is available for electrocyclic reactions than for cycloadditions. For the electrocyclic reactions, we have a clear stereochemical prediction of how things must happen. For $4n$ reactions, thermal closures will be conrotatory and photochemical reactions disrotatory. On the other hand, for cycloadditions, we get a crude "thermal no," "photochemical yes" message; a much less detailed picture. Moreover, electrocyclic reactions are not really different from cycloadditions. Figure 20.27 compares the equilibration of 1,3-butadiene and cyclobutene with the 2 + 2 dimerization of a pair of ethylenes. The only difference is the extra σ bond in butadiene, and this bond is surely not one of the important ones in the reaction—it seems to be just going along for the ride. Why should its presence or absence change the level of detail available to us through an orbital symmetry analysis? It shouldn't, and in fact, it doesn't.

Look again at Figure 20.24, from which we derived the information that the thermal 2 + 2 dimerization of a pair of ethylenes was forbidden. It's only forbidden if we insist on smushing the two ethylenes together head to head exactly as

σ Bond   σ Bond

FIGURE 20.27 The equilibration of cyclobutene and 1,3-butadiene and the 2 + 2 dimerization of a pair of ethylenes are very closely related reactions.

shown in the figure. If we are somewhat more flexible, and allow one ethylene to rotate about its carbon–carbon σ bond as the reaction takes place, the offending antibonding overlap vanishes. What orbital symmetry really says is that for the thermal dimerization of two ethylenes to take place, there must be this rotation (Fig. 20.28).

FIGURE 20.28 The thermal 2 + 2 dimerization is allowed if one end of one ethylene can rotate as the reaction occurs.

In practice, this rotation is a very difficult thing to achieve. It is this difficulty that accounts for the absence of the thermal 2 + 2 reaction, not the absolute "no" of Table 20.4. Similarly, the photochemical 4 + 2, Diels–Alder reaction also becomes allowed if we permit one of the partners to undergo a rotation during the reaction (Fig. 20.29).

FIGURE 20.29 The photochemical 4 + 2 reaction (the Diels–Alder reaction) is also allowed if a rotation occurs during the cycloaddition.

So we should amend our summary Table 20.4 to produce a new Table 20.5.

TABLE 20.5   Rules for All Cycloaddition Reactions

Reaction	Number of Electrons	Thermal	Photochemical
Ethylene + Ethylene ⇄ Cyclobutane	4 ($4n$)	One rotation	No rotation
Ethylene + Butadiene ⇄ Cyclohexene (Diels–Alder)	6 ($4n + 2$)	No rotation	One rotation
All $4n$ reactions	$4n$	One rotation	No rotation
All $4n + 2$ reactions	$4n + 2$	No rotation	One rotation

Don't forget; these rules do not include any consideration of thermodynamics. The processes allowed with a rotation may well not be very favorable thermodynamically, even though they are permitted electronically.

SUMMARY

Cycloaddition reactions involve the coming together of two (or, rarely, even more) molecules to make a ring. The easiest analysis involves examining the interactions of the HOMOs and LUMOs of the two fragments forming the ring compound.

20.5 SIGMATROPIC SHIFT REACTIONS

When one heats 1,3-pentadiene nothing obvious happens, as the starting material is recovered, apparently unchanged. Finally, at very high temperatures, radical reactions begin as carbon–hydrogen bonds are broken. However, a deuterium-labeling experiment reveals that much is happening long before the energies necessary for the formation of radicals have been reached (Fig. 20.30).

FIGURE 20.30 A labeling experiment reveals the degenerate thermal rearrangement of 1,3-pentadiene.

This kind of process, in which starting material rearranges into itself, is called a degenerate reaction. Isotopically labeled molecules reveal the degenerate reaction and allow a determination of the kinetic activation parameters. The reaction requires an activation energy of 36–38 kcal/mol. Other labels are possible, of course, and substituted molecules will reveal the reaction as well as the more sophisticated isotopic labels. In most cases, the reaction is no longer degenerate, and the equilibrium constant K for the reaction can no longer equal 1 (Fig. 20.31).

FIGURE 20.31 Substituted molecules also reveal the reaction, but the rearrangement is no longer degenerate, as starting material and product are different.

PROBLEM 20.11 Which of the molecules of the drawings of Figure 20.31 will be favored at equilibrium? Explain your choice.

WORKED PROBLEM 20.12 Estimate the bond dissociation energy (BDE) for the migrating carbon–hydrogen bond. Why is the observed activation energy of 36–38 kcal/mol for the [1,5] shift of hydrogen so much lower than your answer?

ANSWER The carbon–hydrogen bond in methane has a BDE of 105 kcal/mol. However, this is a poor model for the breaking carbon–hydrogen bond in Figure 20.31, because the radical formed from the larger molecule is tertiary, not methyl. The BDE for the tertiary carbon–hydrogen bond in isobutane is about 96 kcal/mol.

But the cyclohexadienyl radical is resonance stabilized and that will further decrease the BDE. If allylic resonance is worth about 12 kcal/mol, we might expect the BDE to be about 96 − 12 = 84 kcal/mol.

Yet the real BDE is much, much lower than this, about 37 kcal/mol. The reason is that as the carbon–hydrogen bond begins to break, the new carbon–hydrogen bond at C(5) is forming. It is *not* necessary to break the carbon–hydrogen bond fully.

As the C(1)−H bond breaks, the new carbon–hydrogen bond at C(5) begins to form; this bond formation decreases the energy required to break the C(1)−H bond.

The deuterium labeling experiment reveals that much is happening before the onset of radical reactions. A hydrogen (deuterium, in this case) atom is moving from one position in the molecule to another. This kind of reaction has come to be known as a **sigmatropic shift.*** The starting point of the shift is called atom 1. The rest of the atoms are labeled by counting toward the terminus of the migrating group, and the two numbers denoting the starting and arriving points of the migrating species are enclosed in brackets. In this case, we have a [1,5] shift (Fig. 20.32). Be sure to note that the labeling need not follow the numbering convention used in the naming of the compound.

FIGURE 20.32 In a [1,5] sigmatropic shift of hydrogen, the migrating hydrogen atom moves from carbon 1 to carbon 5.

PROBLEM 20.13 Classify the following reactions as [x,y] sigmatropic shifts.

(a)

(b)

(c)

(d)

Be careful: parts (c) and (d) are *not* [1,x] shifts!

(e)

This reaction responds to attempted mechanistic experiments in true no-mechanism fashion. Neither acids nor bases strongly affect the reaction, and the polarity, or even the presence of solvent, is unimportant. The reaction proceeds quite nicely in the gas phase. It is simple to construct an arrow formalism picture of the reaction, but this

*It must be admitted that the Woodward–Hoffmann theory is filled with jargon. There is nothing to do but learn it.

FIGURE 20.33 An arrow formalism description of the [1,5] shift of hydrogen in 1,3-pentadiene.

FIGURE 20.34 An arrow formalism for the [1,3] shift of hydrogen that does not occur.

device does little more than point out the overall change produced by the migration. In these figures, the arrow formalism is even more of a formalism than usual. For example, the arrows of Figure 20.33 could run in either direction, clockwise or counterclockwise. That is not true for a polar reaction in which the convention is to run the arrows from pairs of electrons (the Lewis base) toward the Lewis acid.

It is already possible to identify one curious facet of this reaction, and a little experimentation reveals others. The product of a [1,3] shift is absent. Why, if the hydrogen is willing to travel to the 5-position, does it never stop off at the 3-position? An arrow formalism can easily be written, and it might be reasonably argued that the [1,3] shift, requiring a shorter path than the [1,5] shift, should be easier (Fig. 20.34). Why are [1,3] shifts not observed?

A second strange aspect of this reaction comes from photochemical experiments. As you already know, it is possible to deliver energy to a molecule in more than one way. One way is to heat up the starting material. However, conjugated molecules absorb light (Chapter 12, p. 566), light quanta contain energy, and this gives us another way to transmit energy to the molecule. Curiously, when 1,3-pentadienes are irradiated, the products of the reaction include the molecules formed through [1,3] shifts, but *not* those of [1,5] shifts (Fig. 20.35). The reactions are sometimes complicated, as many other photochemical reactions take place, but if we just concentrate on the products of shifts, this strange dependence on the method of energy delivery appears.

THE GENERAL CASE

Product of
a [1,3] shift

not

Product of
a [1,5] shift

A SPECIFIC EXAMPLE

hv

[1,3] H
shift

FIGURE 20.35 Photolysis of 1,3-dienes induces [1,3] shifts, but not [1,5] shifts.

WORKED PROBLEM 20.14 Perhaps it could be argued that the loss of conjugation in the product of the [1,3] shift results in an energetic favoring of the [1,5] process. Design a substituted 1,3-pentadiene that would test this (incorrect) surmise.

ANSWER This problem is hard because you are not used to thinking in the correct way. The hypothesis is that the [1,5] shift is preferred to the [1,3] shift because it preserves the conjugation present in a 1,3-diene. In order to test this surmise, we need to design a molecule in which both the [1,5] and [1,3] shifts preserve the same amount of conjugation. The deuterium label is needed to reveal the [1,5] or [1,3] nature of the reaction.

These are the same molecule except for the isotopic label

SUMMARY

Thermal energy induces [1,5] shifts and photochemistry allows [1,3] shifts (Fig. 20.35).

So, any mechanism must include an explanation of why thermal shifts are [1,5], whereas photochemically induced shifts are [1,3].

Let's start our analysis by examining the early stages of the reaction. As the reaction begins, a carbon–hydrogen bond of the methyl group begins to stretch. If this stretching were continued to its limit, it would produce two radicals, a hydrogen atom and the pentadienyl radical (Fig. 20.36). We know this cleavage does not happen because the products of recombination of these radicals are not found.

FIGURE 20.36 In the earliest stages of the reaction, a carbon–hydrogen bond begins to stretch. Ultimately a pair of radicals would result. We know this bond breaking does not take place because the radical dimer is not found.

We should stop right here to question the assumption that the carbon–hydrogen bond is breaking in homolytic (radical) fashion. Why not write a heterolytic cleavage to give a pair of ions rather than radicals (Fig. 20.37)?

FIGURE 20.37 There are two possible ways to break the carbon–hydrogen bond heterolytically to give a pair of ions. Either way would be strongly dependent on solvent polarity. As there is essentially no dependence of the rate of the reaction on solvent polarity, this mechanistic possibility is excluded.

The stability of ions depends greatly on solvation. It is inconceivable that a heterolytic bond breaking would not be highly influenced by solvent polarity. As we have just seen, a change in solvent polarity affects the rate of this reaction only very slightly. Only homolytic bond breaking is consistent with the lack of a solvent effect.

As the methyl carbon–hydrogen bond begins to stretch, something must happen that thwarts the formation of the pair of radicals. Perhaps the developing hydrogen atom is intercepted by reattachment at atom 5. In orbital terms, we would say that as the C(1)—H bond breaks, the developing 1s orbital on hydrogen begins to overlap with the p orbital on the carbon at position 5. Now we have a first crude model for the transition state for this sigmatropic shift (Fig. 20.38).

FIGURE 20.38 Radical formation can be thwarted by partial bonding of the migrating hydrogen to the orbitals on carbon atoms 1 and 5 in the transition state. As the C(1)–H bond stretches, the hydrogen 1s orbital begins to overlap with an orbital on C(5).

This model does describe an intramolecular reaction, but it leaves unanswered the serious questions of mechanism raised before: Why doesn't the migrating hydrogen reattach at the 3-position in thermal reactions? This process does happen in photochemical reactions. We have answered neither the questions of regiochemistry ([1,5] or [1,3]) nor of dependence on the mode of energy input (thermal or photochemical).

Let's look more closely at our transition state model. In essence, it describes a hydrogen atom moving across the π system of a pentadienyl radical (Fig. 20.39).

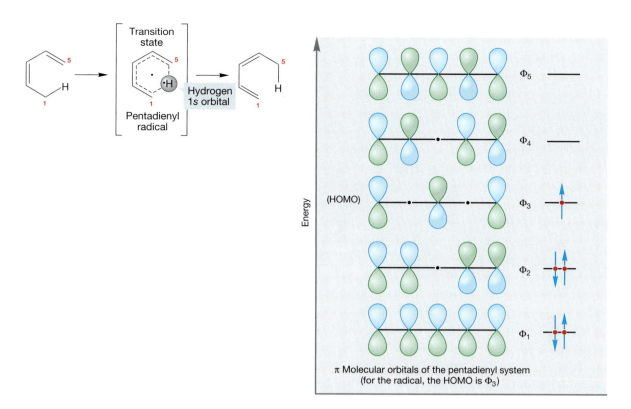

FIGURE 20.39 A model for the transition state shows a hydrogen 1s orbital migrating across a pentadienyl radical, for which the π molecular orbitals are shown.

The hydrogen atom is simple to describe—it merely consists of a proton and a single electron in a 1s atomic orbital. The pentadienyl molecular orbital system isn't difficult either. There will be five π molecular orbitals and these can be constructed by taking combinations of five carbon 2p orbitals. We assume again that it will be the highest energy electrons, those most loosely held, that will be most important in the reaction. These are the electrons in the HOMO. If we accept this assumption, we can elaborate our picture of the transition state for the reaction to give Figure 20.40.

FIGURE 20.40 An elaboration of our model for the transition state, showing a crude outline of the interaction of the hydrogen 1s orbital with the lobes of the HOMO for pentadienyl, Φ_3.

Notice a very important thing about the transition state outlined in Figure 20.40. In the starting pentadiene, the bond made between the $1s$ orbital of hydrogen and the hybrid orbital on carbon atom 1 must of course be made between lobes of the same sign. As this bond stretches, the signs of the lobes do not change: The $1s$ orbital is always bonded or partially bonded to a lobe of the same symmetry (Fig. 20.41).

FIGURE 20.41 As the carbon–hydrogen bond lengthens, the phase relationships do not change.

In this C–H bond, the overlapping lobes are of the same sign

Sign?

Partial bond— same sign

In the transition state, the C and H are still partially bonded, so the lobes still have the same phase

In this way, we know the relative symmetries of some of the lobes in the transition state for the reaction. We also know the symmetry (from Fig. 20.39, or by figuring it out) of Φ_3, the HOMO of pentadiene. So, we can elaborate the picture of the transition state to include the symmetries of all the orbital lobes. The symmetry of the lobes allows the [1,5] shift to take place as shown (Fig. 20.42).

Φ_3 of Pentadienyl (HOMO)

FIGURE 20.42 Now we can elaborate our transition state model by filling in the lobes of the HOMO of pentadienyl, Φ_3.

Positive overlap is possible at C(5) — the orbital phases are correct!

$1s$

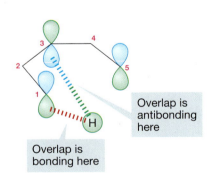

Overlap is antibonding here

Overlap is bonding here

FIGURE 20.43 By contrast, the [1,3] shift involves the formation of an antibond as the hydrogen reattaches to carbon 3.

Notice that the process through which the hydrogen detaches from carbon 1 and reattaches to carbon 5 from the same side allows the maintenance of bonding (same sign) lobal interactions at all times. This reaction is "orbital symmetry allowed." Now look at the [1,3] shift. To do the analogous [1,3] shift, rebonding requires the overlap of lobes of different symmetry—the formation of an antibond. This reaction would be "orbital symmetry forbidden" (Fig. 20.43)!

More important jargon enters here: The motions we have been describing in which the migrating atom leaves and reattaches from the *same* side of the π system are called **suprafacial**. If the migrating atom or group of atoms leaves from one side of the π system and arrives at the other, the process is called **antarafacial** motion (Fig. 20.44). In the [1,5] shift, suprafacial motion is allowed by the symmetries of the orbitals; in [1,3] shift, it is not (Figs. 20.42 and 20.43).

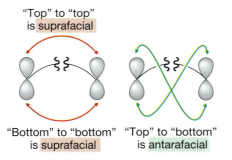

"Top" to "top"
is suprafacial

"Bottom" to "bottom"   "Top" to "bottom"
is suprafacial   is antarafacial

FIGURE 20.44 Suprafacial motion involves bond breaking and bond making on the *same* side of the π system. Antarafacial motion involves bond breaking on one side and bond making on the other.

Why not migrate in a different fashion? Why can't the hydrogen depart from one side and reattach from the other? This antarafacial migration would result in an orbital symmetry allowed [1,3] migration (Fig. 20.45). The problem is that the $1s$ orbital is small and cannot effectively span the distance required for an antarafacial migration. There is nothing electronically disallowed or forbidden about the antarafacial [1,3] migration, but there is an insuperable steric problem.

Bonding

Φ_3 of Pentadienyl (HOMO)

FIGURE 20.45 The antarafacial [1,3] shift is allowed by orbital symmetry, but the small size of the $1s$ orbital makes the reaction impossible.

SUMMARY

The [1,5] shift is intramolecular, and the lack of any dependence on solvent polarity makes intermediate ions or even polar transition states unlikely. In our model of the transition state, we focus on the maintenance of bonding interactions, which explains why [1,5] shifts are possible and [1,3] shifts are exceedingly rare. We have not yet been able to deal with the dependence of the course of this reaction on the mode of energy input—thermal or photochemical.

PROBLEM 20.15 Do you expect the [1,3] shift shown below to occur when propene is heated? Explain carefully, using a molecular orbital argument.

$$H_2C \underset{3}{\overset{2}{\diagup}} \overset{1}{\diagdown} \overset{H}{\underset{H}{\diagup}} \quad \underset{[1,3]}{\overset{\Delta}{\rightleftharpoons}} \quad \overset{H}{\underset{H}{\diagup}} \overset{H}{\diagdown} CH_2$$

Now comes a most important point. Our model not only rationalizes known experimental data, but includes within itself the roots of a critical prediction. The Woodward–Hoffmann explanation predicts—requires—that the [1,5] shift occur in a suprafacial fashion. That is the only way in which bonding interactions can be maintained throughout the migration of a hydrogen atom from carbon 1 to carbon 5. There is *nothing* so far in the available experimental data that allows us to tell whether the observed shifts are suprafacial, antarafacial, or a mixture of both modes.

Let's set about finding an experiment to test the requirement of the theory for suprafacial [1,5] motion. What will we know at the end of the experiment? If we find that the [1,5] shift is indeed strictly suprafacial, the theory will be supported (not proved!), and we will certainly feel better about the mechanistic hypothesis. If we find antarafacial motion, or both suprafacial and antarafacial motions, the theory will be proved (yes, proved) wrong. There is no way our hypothesis can accommodate antarafacial motion; there is an absolute demand for suprafaciality, which nicely illustrates the precarious life of a theory. It can always be disproved by the next experiment, and it can never become free of this state of affairs.

The crucial test was designed by a German chemist, W. R Roth (1930–1997).* It is the first of a series of beautiful experiments you will encounter in this chapter. Roth and his co-workers spent some years in developing a synthesis of the labeled molecule shown in Figure 20.46. Notice that the molecule has the (S) configuration at carbon 1, and the (E) stereochemistry at the C(4)—C(5) double bond.

Now let's work out the possible products from suprafacial and antarafacial [1,5] shifts in this molecule. The situation is a little complicated because there are two forms of this molecule that can undergo the reaction. These are the two conformations (rotational isomers) shown in Figure 20.47. In one, the methyl group lies inside, aimed at the C(4)—C(5) double bond. In the other, it is the ethyl group that is directed toward the C(4)—C(5) double bond. In suprafacial migration of hydrogen in one conformation, the product has the (R) configuration at C(5) and the (E) stereochemistry at the new double bond located between C(1) and C(2). Another possible suprafacial migration starts from the second conformation. Here it is the opposite stereochemistries that result. The configuration of C(5) is (S) and the new double bond is (Z) (Fig. 20.47).

FIGURE 20.46 The 1,3-diene used by Roth and his co-workers to see whether the [1,5] shift proceeded in a strictly suprafacial manner.

FIGURE 20.47 The products formed by suprafacial migration in the two rotational isomers of Roth's diene.

*No typo here. There is no period after the R. Roth claimed that he was advised by his postdoctoral advisor at Yale, William Doering, to take a middle initial in order not to be lost among the myriad "W. Roth's" in German chemical indexes. Others with common last names, and who were at Yale at the time, rejected similar advice.

In antarafacial migrations, the hydrogen leaves from one side and is delivered from the other. Figure 20.48 shows the results of the two possible antarafacial migrations. Note that the products are different from each other and also different from those formed from suprafacial migration.

FIGURE 20.48 The possible products of antarafacial migration in Roth's diene.

Roth and his co-workers had the experimental skill necessary to sort out the products and found that only the products of suprafacial migration were formed. The theory is beautifully confirmed by this experiment.

At last we come to the remaining question. Why should the source of energy make a difference in the mode of migration? Remember, the common, thermally induced shifts are [1,5], whereas the relatively rare photochemical shifts are [1,3]. Consider again what happens when a molecule absorbs a photon of light. If you recall Chapter 12, you know that the packet of energy is used to promote an electron from the HOMO to the LUMO. After the absorption of light, there is a new HOMO, what we might call the photochemical HOMO (Fig. 20.13, p. 1128).

The new photochemical HOMO will have different symmetry relationships of the lobes than the old thermal HOMO. Figure 20.49 shows the situation for the simplest [1,3] shift in propene. Migration using Φ_3, the photochemical HOMO, can occur in a suprafacial fashion, and the steric barrier to thermal, necessarily antarafacial [1,3] shifts is gone. By using Φ_3, the hydrogen can migrate in an easy suprafacial manner.

FIGURE 20.49 Suprafacial motion of a hydrogen is possible in Φ_3, the photochemical HOMO for this reaction.

SUMMARY

Sigmatropic shifts of hydrogen are quite simple to understand. As long as you keep in mind the necessity to form bonds and not antibonds, analysis of the HOMO gives you the positions available for reattachment of a migrating hydrogen. One can determine whether the reaction is allowed in a suprafacial or antarafacial fashion. Add to this analysis the idea that hydrogen has only a small 1s orbital, and therefore any short antarafacial shifts are generally impossible. That is all there is to it.

PROBLEM 20.16 Provide mechanisms for the following reactions:

(a)

(b)

20.6 THE COPE REARRANGEMENT

Not all shift reactions are [1,x] shifts. One very common sigmatropic shift reaction is called the **Cope rearrangement**, and is named for Arthur C. Cope (1909–1966, the same Cope of the Cope elimination, Chapter 19, p. 1089) who spent most of his professional life at MIT.* The Cope rearrangement is a [3,3] shift of carbon. This terminology is almost universally confusing, but it is not really so bad. The two parts of the molecule are attached at their respective 1-positions (remember, sigmatropic shift terminology has no relationship to the protocol for naming the molecule). In the reaction, this 1—1 bond breaks and the two parts (allyl groups) are reattached through their respective 3-positions. In its simplest form, the Cope rearrangement is degenerate and must be revealed through a labeling experiment (Fig. 20.50).

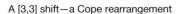

A [3,3] shift—a Cope rearrangement

This degenerate, "invisible" reaction…   …is revealed by this labeling experiment

FIGURE 20.50 The degenerate Cope rearrangement, a [3,3] shift, can be revealed by a labeling experiment.

Like the [1,5] shifts we looked at first, the Cope rearrangement is intramolecular, uncatalyzed, and shows no strong dependence on solvent polarity. The activation energy for the Cope rearrangement is about 34 kcal/mol, a value far below what one would estimate for the simple bond breaking into two separated allyl radicals. If the mechanism of the Cope rearrangement did involve two independent allyl radicals, the labeled 1,5-hexadiene would have to give two products, which it does not (Fig. 20.51).

recombine 1,1

recombine 3,3

recombine 1,3

Two independent allyl radicals

Not found

FIGURE 20.51 A labeling experiment shows that the Cope rearrangement does not involve formation of a pair of free allyl radicals.

*Cope didn't actually discover this reaction; C. Hurd of Northwestern University (1897–1998) apparently did. It was Cope's research group that developed the reaction, however.

The Cope rearrangement has a nonpolar transition state, but this time it is not a simple hydrogen atom that migrates but a whole allyl radical. An analysis of the symmetry of the orbitals involved shows why this reaction is a relatively facile thermal process but is not commonly observed on photochemical activation. As we break the C(1)—C(1) bond, the phases of the overlapping lobes must be the same. The HOMO of the allyl radical is Φ_2, and that information allows us to fill in the symmetries of the two allyl radicals making up the transition state (Fig. 20.52).

Reattachment at the two C(3) positions is allowed because the interaction of the two lobes on the two C(3) carbons is bonding. This picture is consistent with the near ubiquity of the Cope rearrangement when 1,5-dienes of all kinds are heated (Fig. 20.53).

Transition state
(migration of an
allyl radical)

FIGURE 20.52 The transition state for the Cope rearrangement involves the migration of one allyl radical across another.

FIGURE 20.53 An orbital symmetry analysis shows that the thermal Cope rearrangement is allowed.

Transition state (two partially bonded allyl radicals — Φ_2 is the HOMO for each)

So we now have a general picture of the transition state for the Cope rearrangement. But much more detail is possible if we work a bit at it. How might the two migrating allyl radicals of the transition state arrange themselves in space during the reaction? Two things are certain. First, the two C(1) atoms must start out close together, as they are bonded. Second, the two C(3) positions must also be within bonding distance, as in the transition state there is a partial bond between them (Fig. 20.53).

WORKED PROBLEM 20.17 Use orbital symmetry to analyze the photochemical [3,3] shift in 1,5-hexadiene. Be careful! Photons are absorbed *one at a time*. It is most unlikely that a singly excited molecule will live long enough to absorb a second photon.

ANSWER The transition state for the [3,3] shift involves two interacting allyl radicals.

Transition state
for the [3,3] shift

If one electron is promoted from the HOMO to the LUMO, the orbital picture will be as follows:

(continued)

(continued)

Now an orbital analysis shows that there is no easy path for the photochemical [3,3] shift. One antibonding interaction is always present.

If a rotation were energetically feasible; if one end of one of the partners could turn over, a bonding interaction could be achieved where the figure notes the antibonding interaction. However, for the Cope rearrangement this rotation involves a large energy cost.

The question now comes down to determining the position of the two C(2) atoms in the transition state. Two possibilities are shown in Figure 20.54. The two C(2) atoms might be close together as in the boat-like transition state (originally called the **six-center transition state**), or rather far apart, as in the chair, or **four-center transition state**. The figure shows a number of views of these two possibilities, but this is probably a time to get out the models.

Boat-like, or "six-center" arrangement

Chair-like, or four-center arrangement

• = C(2)

FIGURE 20.54 Several views of the chair-like (four-center) and boat-like (six-center) arrangements for the transition state of the Cope rearrangement.

In 1962, W. von E. Doering and W. R Roth carried out a brilliant experiment that determined which of these two possible transition states was favored. They used a labeled molecule, 3,4-dimethyl-1,5-hexadiene. Notice first that there are three forms of this molecule, a meso, achiral form, and a racemic pair of enantiomers. It all depends on how the methyl labels are attached (Fig. 20.55).

FIGURE 20.55 The meso and chiral forms of 3,4-dimethyl-1,5-hexadiene.

Meso form   Racemic mixture

CONVENTION ALERT

In the following figures, we have drawn only one enantiomer of the racemic pair. Be sure to keep in mind that the actual experiment used a 50:50 mixture of the two enantiomers, the racemate (Chapter 4, p. 156).

Both the meso and racemic forms of 3,4-dimethyl-1,5-hexadiene can adopt either the boat (six-center) or chair (four-center) arrangement. Figure 20.56 shows the results for the meso compound. The key to Doering and Roth's experiment is that the chair and boat arrangements give different products for both the meso and chiral forms (Fig. 20.56).

Meso compound in the boat-like arrangement   Meso compound in the chair-like arrangement

FIGURE 20.56 The products formed from the boat-like (six-center) and chair-like (four-center) arrangements of *meso*-3,4-dimethyl-1,5-hexadiene.

The results showed that it is the chair-like four-center form that is favored, and that it is preferred to the boat-like six-center arrangement by at least 6 kcal/mol. Presumably, some of the factors that make chair cyclohexane more stable than the boat form (Chapter 5, p. 203) operate to stabilize the related chair-like transition state.

There are many things that are important about this work. First of all, it is a prototypal example of the way labeling experiments can be used to dig out the details of a reaction mechanism. The second reason is more subtle, but perhaps more important. As we will see in Section 20.8, the study of the Cope rearrangement leads directly to one of the most intellectually creative ideas of the last few years.

Cope rearrangement

20.7 A BIOLOGICAL COPE REARRANGEMENT

The Cope rearrangement is by no means restricted to the laboratory. Nature uses Cope-like processes as well. For example, in bacteria and plants a critical step in the production of the essential amino acids tyrosine and phenylalanine is the rearrangement of chorismate to prephenate, catalyzed by the enzyme, chorismate mutase. At the heart of this rearrangement is a Cope-like process called the Claisen* rearrangement. The research groups of Jeremy Knowles (b. 1935) at Harvard and Glenn Berchtold (b. 1932) at MIT collaborated on a tritium (^{3}H = T, a radioactive isotope of hydrogen) labeling experiment, related to that used by Doering and Roth, to determine that the enzymatic reaction also proceeded through a chair-like transition state, not a boat-like one. Notice that the chair-like transition state must take the compound with a (Z) double bond to a product in which the stereogenic carbon is (S), not (R) (Fig. 20.57).

FIGURE 20.57 The conversion of chorismate into prephenate is a Cope-like rearrangement.

*The same Claisen of the Claisen condensation (Chapter 19, p. 1063) and the Claisen–Schmidt reaction (Chapter 17, p. 936).

20.8   A MOLECULE WITH A FLUXIONAL STRUCTURE

We spent some time in Section 20.6 working out the details of the structure of the transition state for the Cope rearrangement, which is a most general reaction of 1,5-hexadienes, and appears in almost every molecule incorporating the 1,5-diene substructure. This section will examine the consequences of the Cope rearrangement and will work toward a molecule whose structure differs in a fundamental way from anything you have seen so far in this book. At room temperature it has no fixed structure; instead it has a **fluxional structure** in which nearest neighbors are constantly changing. Along the way to this molecule there is some extraordinary chemistry and some beautiful insights. This section shows not only clever ideas and beautiful chemistry, but also how much fun this subject can be.

The simple 1,5-hexadiene at the heart of the Cope rearrangement can be elaborated in many ways, and in fact we've already seen one. If we add a central π bond we generate *cis*-1,3,5-hexatriene, a molecule we saw when we considered electrocyclic reactions (p. 1122). The arrow formalism of the Cope rearrangement leads to 1,3-cyclohexadiene (Fig. 20.58).

The degenerate Cope rearrangement of 1,5-hexadiene

cis-**1,3,5-Hexatriene**   **1,3-Cyclohexadiene**

A modified Cope rearrangement

FIGURE 20.58 The electrocyclic ring closure of *cis*-1,3,5-hexatriene to 1,3-cyclohexadiene is a modified Cope rearrangement.

trans-**1,3,5-Hexatriene**

FIGURE 20.59 *trans*-1,3,5-Hexatriene cannot undergo this rearrangement because the ends of the molecule (dots), where bonding must occur, are too far apart.

But there is a stereochemical problem. This reaction is only possible for the stereoisomer with a cis central double bond. In the trans isomer, the ends of the triene system are too far apart for bond formation (Fig. 20.59).

That's an easy concept, and it appears again in a slightly different modified 1,5-hexadiene, 1,2-divinylcyclopropane. The Cope rearrangement is possible in this molecule and here, too, it is only the cis molecule that is capable of rearrangement. In the trans compound, the ends cannot reach (Fig. 20.60).

FIGURE 20.60 *cis*-1,2-Divinylcyclopropane can undergo the Cope rearrangement, but the trans isomer cannot.

cis-**1,2-Divinyl-cyclopropane**   **1,4-Cycloheptadiene**

No Cope

trans-**1,2-Divinyl-cyclopropane**

PROBLEM 20.18 The Cope rearrangement of Figure 20.60 is very easy. It is rapid at room temperature. Why should it be so much easier than the other versions we have seen?

There is another stereochemical problem in this reaction that is a bit more subtle. The most favorable arrangement for *cis*-1,2-divinylcyclopropane is shown in Figure 20.61. The two vinyl groups are directed away from the ring, and in particular, away from the syn ring hydrogen. Yet this extended form cannot undergo the Cope rearrangement at all! To do the Cope, a less favored coiled form must be adopted. Now the two vinyl groups point back toward the ring and are closer to the offending syn hydrogen.

Less stable, "coiled" form   More stable, "extended" form

Both cis   Both trans

FIGURE 20.61 Only the less stable coiled arrangement of *cis*-1,2-divinylcyclopropane can undergo the Cope rearrangement. The more stable, extended form must produce a hideously unstable *trans,trans*-cyclohepta-1,4-diene.

The problem is that the extended form must produce a 1,4-cycloheptadiene product with two trans double bonds. Although the paper puts up with these trans double bonds without protest, the molecule will not. Remember, trans double bonds in small rings are most unstable. *trans*-Cycloheptene itself is a barely detectable reactive intermediate that can be intercepted only at low temperature. The constraints of the ring lead to severe twisting about the π bond and destabilization. A seven-membered ring containing two trans double bonds is unthinkably unstable.

By contrast, the higher energy coiled form yields a seven-membered ring containing two cis double bonds, and that presents no problem. Be certain you see why the two forms give different stereochemistries. Figure 20.62 shows the Energy versus Reaction progress diagram for the Cope rearrangement of *cis*-1,2-divinyl-cyclopropane.

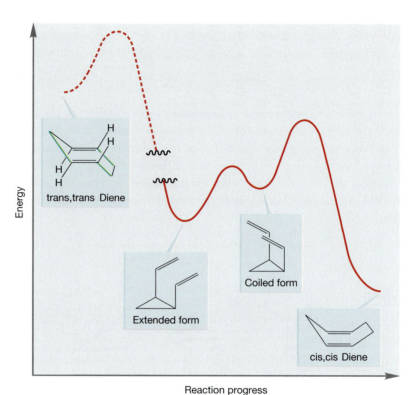

FIGURE 20.62 The Cope rearrangement of *cis*-1,2-divinylcyclopropane.

Now let's make another addition to the system. Let's add a single methylene group joining the two double bonds to give the molecule named "homotropilidene." We can still draw the arrow formalism, and the Cope rearrangement still proceeds rapidly. As Figure 20.63 shows, we are back to a degenerate Cope rearrangement, as the starting material and product are the same.

FIGURE 20.63 The degenerate Cope rearrangement of homotropilidene.

Homotropilidene   Still **homotropilidene**

PROBLEM 20.19 This rapid Cope led to problems for W. R Roth who first made the molecule. Can you explain in general terms the perplexing ¹H NMR spectra for this molecule (see next page)? Note that at both low and high temperature the spectrum is sharp, whereas at an intermediate temperature, 20 °C, it consists of two "blobs." You do not have to analyze the spectra in detail—just give the general picture of what must be happening.

(continued)

(continued)

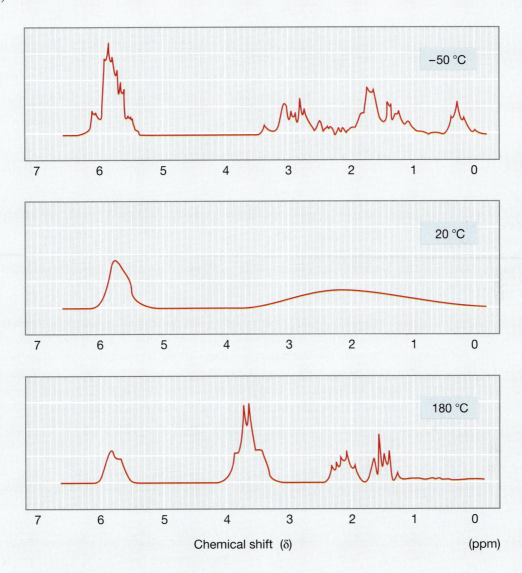

There are also stereochemical problems for this molecule. Like 1,2-divinyl-cyclopropane, the most stable extended form of homotropilidene is unable to undergo the Cope rearrangement, and it must be the higher energy coiled arrangement that is actually active in the reaction. Here the destabilization of the coiled arrangement is especially obvious, as the two methylene hydrogens badly oppose each other. As before, the extended form must give an unstable molecule with a pair of trans double bonds incorporated into a seven-membered ring. The less stable coiled arrangement has no such problem, as it can give a product containing two cis double bonds (Fig. 20.64).

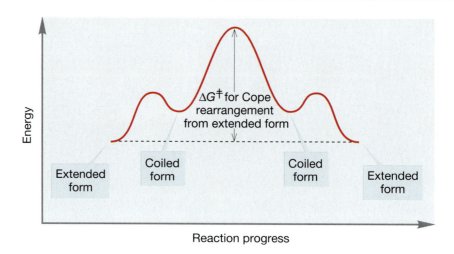

FIGURE 20.64 Only the higher energy coiled form of homotropilidene can give a stable product in the Cope rearrangement. The lower energy extended form must produce a product containing two trans double bonds in a seven-membered ring.

Although *cis*-1,2-divinylcyclopropane and homotropilidene undergo the Cope rearrangement quite rapidly even at room temperature and below, one might consider how to make the reaction even faster. For the reaction to occur, an unfavorable equilibrium between the more stable, but unproductive extended form, and the higher energy, but productive coiled form must be overcome. The activation energy for the reaction includes this unfavorable equilibrium (Fig. 20.65).

FIGURE 20.65 An Energy versus Reaction progress diagram for the Cope rearrangement of homo-tropilidene. The activation energy ($\Delta G^{\ddagger}$) for the reaction includes the energy difference between the extended and coiled forms.

If this unfavorable equilibrium could be avoided in some way, and the molecule locked into a coiled arrangement, the activation energy would be lower, and the reaction faster (Fig. 20.66).

FIGURE 20.66 An Energy versus Reaction progress diagram for the Cope rearrangement of a homo-tropilidene locked into the coiled arrangement. The activation energy is lower, as it is no longer necessary to fight an unfavor-able equilibrium from the extended form, shown here as a phantom (dashes) just to remind you of the previous figure.

Our next variation of the Cope rearrangement includes the clever notion of bridg-ing the two methylene groups of homotropilidene to lock the molecule in a coiled arrangement (Fig. 20.67).

FIGURE 20.67
A homotropilidene locked into the coiled arrangement can be achieved if the two methylene groups are connected with a bridge.

This modification greatly increases the rate of the Cope rearrangement as the activation energy is decreased (Fig. 20.66). No longer is it necessary to fight an unfavorable equilibrium in order for the Cope rearrangement to occur; the coiled form is built into the structure of the molecule itself. Many bridged homotropilidenes are now known, and some are shown in Figure 20.68.

Now that's a clever idea, and it has led to some fascinating molecules and some very nice chemistry. But it's only a good idea. What comes next is an extraordinary idea; one of those insights that anyone would envy, and which define through their example what is meant by creativity. Suppose we bridge the two methylene groups with a double bond to give the mol-ecule known as **bullvalene**.* At first, this molecule seems nothing special. As Figure 20.69 shows, the Cope rearrangement is still possible in this highly modified 1,5-hexadiene.

Dihydro-bullvalene **Barbaralone**

Barbaralane **Semibullvalene**

FIGURE 20.68 Four bridged homo-tropilidenes. All undergo very rapid Cope rearrangements.

*What an odd name! If you are curious about its origin, see Alex Nickon and Ernest F. Silversmith's nice book, *The Name Game* (Pergamon, New York, 1987) on chemical naming. You won't find the sys-tematic naming scheme in here, but you will find hundreds of stories about the origins of the names chemists give to their favorite molecules, including bullvalene.

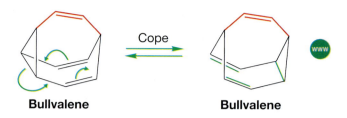

FIGURE 20.69 If the new bridge is a carbon–carbon double bond, the resulting molecule is known as bullvalene.

Bullvalene   **Bullvalene**

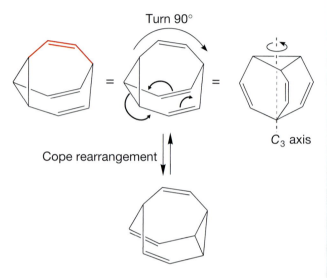

FIGURE 20.70 Bullvalene has a threefold rotational axis: All three double bonds are equivalent. The Cope rearrangement can occur in each of the three faces of the molecule and is degenerate in every case.

But now perhaps you notice the threefold rotational axis in bullvalene. Remove the color and the three double bonds become indistinguishable. In fact, there is a Cope rearrangement possible in each of the three faces of the molecule, and every Cope rearrangement is degenerate; each re-forms the same molecule (Fig. 20.70).

That's quite interesting, but now begin to look at the structural consequences for bullvalene. As Cope rearrangements occur, *the carbons begin to wander apart*! Figure 20.71 only begins the process, but it can be demonstrated that the Cope rearrangement is capable of completely scrambling the 10 CH units of the molecule.* As long as the Cope rearrangement is fast, bullvalene *has no fixed structure*; no carbon has fixed nearest neighbors, but is instead bonded on time average to each of the 9 other methine carbons. It has a fluxional structure (Fig. 20.71).

FIGURE 20.71 A sequence of Cope rearrangements serves to move neighboring carbon atoms apart.

*In an early paper, Doering noted that a sequence of 47 Copes sufficed to demonstrate that two adjacent carbons could be exchanged without disturbing the rest of the 10-carbon sequence, and that therefore all 1,209,600 possible isomers could be reached. No claim was made that this represented the minimum path. A later computer-aided effort by two Princeton undergraduates (Allan Fisher and Karl Bennett) located a sequence of 17 Copes, which could do the same thing.

WORKED PROBLEM 20.20 At somewhat above room temperature the ^{1}H NMR spectrum of bullvalene consists of a single broad line at δ 4.2 ppm. Explain.

ANSWER If all the carbon–hydrogen bonds are the same on time average, the NMR spectrum must consist of a single line. We see the averaged spectrum.

WORKED PROBLEM 20.21 Explain why the ketone below exchanges *all* its hydrogens for deuteriums when treated with deuterated base.

ANSWER Exchange at the α-position is routine (Chapter 17, p. 908).

However, simple ketones are in equilibrium with their enol forms, and the enol of this molecule is a substituted bullvalene. The Cope rearrangement will make all the carbons equivalent, and thus, on time average, all carbon–hydrogens will come to occupy the exchangeable α-position.

Now this hydrogen can exchange

Repetition leads to exchange of all the hydrogens for deuterium.

Naturally, efforts at synthesis began as soon as the prediction of the properties of bullvalene appeared in the chemical literature. Long before the efforts at a rational synthesis were sucessful, the molecule was discovered by Gerhard Schröder (b. 1929), a German chemist then working in Belgium. Schröder was working out the structures of the several dimers of cyclooctatetraene, and in the course of this effort irradiated one of the dimers. It conveniently fell apart to benzene and bullvalene (Fig. 20.72).

FIGURE 20.72 Schröder's accidental synthesis of bullvalene.

Schröder was alert enough to understand what had happened, and went on to a highly successful career, which included an examination of the properties of bullvalene and substituted versions of this molecule. Doering's predictions were completely confirmed by Schröder's synthesis.

PROBLEM 20.22 Why is the second step of Figure 20.72, the photochemical fragmentation, so easy?

Other neutral, completely fluxional organic molecules have not appeared, although the phenomenon of fluxionality appears to be rather more common in organic cations and organometallic compounds.

PROBLEM 20.23 Provide mechanisms for the following reactions:

SUMMARY

The Cope reaction is far more than just a chemical curiosity. It is a general reaction of almost all 1,5-hexadienes, and appears almost no matter how the basic hexadiene form is perturbed and disguised. All sorts of chemical "spinach" can be added, often making it quite difficult to find the bare bones of the hexadiene system, but the reaction will persist. Moreover, as we have just seen, digging deep into the Cope reaction, following the intellectual implications of substitution, leads to the marvelous molecule bullvalene, without precedent in organic chemistry.

20.9   HOW TO WORK ORBITAL SYMMETRY PROBLEMS

All people and all problems are, of course, different; nonetheless, there are some general principles that apply, and it is worth summarizing ways to approach—and hopefully solve—orbital symmetry problems, which are separable into three varieties.

20.9a Electrocyclic Reactions

1. Look for rings opening or closing within a single molecule. Sound like simple advice? Yes, it is, but in all these problems the major difficulty is one of recognition, and it is well worth your time to be a little "too straightforward," and to "think simple."

2. Once you are sure that the question involves an electrocyclic reaction, draw an arrow formalism.

3. Next, work out the molecular orbitals *of the open-chain partner*.

4. Count the number of electrons involved in the reaction. Be careful to take account of any charges.

5. Knowing the number of electrons allows you to find the HOMO of the open-chain partner.

6. Finally, work out the motion, conrotatory or disrotatory, that allows the ring to be closed through a bonding—not antibonding—interaction.

20.9b Cycloaddition Reactions

1. Look for a ring forming from two molecules or breaking open to give two new molecules. When you find this situation, you know you have a cycloaddition reaction.

2. Carefully draw an arrow formalism.

3. Work out the π molecular orbitals for the two partners.

4. Identify the HOMOs and LUMOs.

5. Examine either possible HOMO–LUMO interaction. It is generally easier to look at ring closing, rather than ring opening. Your final answer will be good for either direction (of course!). If the cycloaddition can occur with two bonding overlapping interactions, it is allowed; if not, it is forbidden, unless an unlikely rotation occurs in one partner.

20.9c Sigmatropic Shift Reactions. These reactions are by far the hardest for students to "get."

1. The major difficulty is one of recognition. Is some piece of the molecule flying around, departing from one place and reattaching somewhere else? If so, that's probably a sigmatropic shift reaction. Once you are sure it is a shift, identify—mark—the shifting atom or group.

2. Where does it depart from? Where does it arrive? Draw an arrow formalism.

3. Now let's work out a picture of the transition state. Break the bond attaching the migrating group to the position from which it departs in a *homolytic* fashion. What species is left behind? What species is migrating?

4. Work out the molecular (or atomic) orbitals for both partners.

5. Remember that the migrating group was attached through a bonding interaction. That fact allows you to sketch in the symmetry of the HOMO of the frame along which the atom or group is migrating.

6. Can the migrating group reattach reasonably (can it reach?) in a bonding fashion? If so, the reaction is allowed. If it must reattach through an antibonding interaction, the reaction is forbidden.

Now practice! Use the schemes above to work the following three problems.

PROBLEM 20.24 Is the following reaction allowed or disallowed? Give a detailed analysis.

PROBLEM 20.25 Is the following thermal reaction allowed or disallowed? Give a detailed analysis.

PROBLEM 20.26 Is the following thermal reaction allowed suprafacially or antarafacially? Give a detailed analysis.

20.10 SUMMARY

New Concepts

The central insight of Woodward–Hoffmann theory is that one must keep track of the phase relationships of orbital interactions as a concerted reaction proceeds. If bonding overlap can be maintained throughout, the reaction is allowed; if not, the reaction is forbidden by orbital symmetry. This notion leads to a variety of mechanistic explanations for reactions that can otherwise be baffling. Some of these so-called pericyclic reactions are discussed in the section on Reactions, Mechanisms, and Tools.

The notion of a fluxional structure is introduced through the $(CH)_{10}$ molecule bullvalene. In this molecule, a given carbon atom does not have fixed nearest neighbors, but instead is bonded, over time, to each of the other nine carbons in the molecule. In bullvalene, the fluxionality is the result of a rapid degenerate Cope rearrangement.

Many stereochemical labeling experiments are described in this chapter. A careful labeling experiment is able to probe deeply into the details of a reaction mechanism. Examples include W. R Roth's determination of the suprafacial nature of the [1,5] shift of hydrogen and Doering and Roth's experiments on the structure of the transition state for the Cope rearrangement.

Key Terms

antarafacial (p. 1146)
bullvalene (p. 1161)
concerted reaction (p. 1120)
conrotation (p. 1126)
Cope rearrangement (p. 1151)
cycloaddition reactions (p. 1133)
degenerate reaction (p. 1120)
disrotation (p. 1126)

electrocyclic reactions (p. 1124)
fluxional structure (p. 1156)
four-center transition state (p. 1153)
pericyclic reaction (p. 1120)
sigmatropic shift (p. 1141)
six-center transition state (p. 1153)
suprafacial (p. 1146)
Woodward–Hoffmann theory (p. 1120)

Reactions, Mechanisms, and Tools

Certain polyenes and cyclic compounds can be interconverted through a pericyclic process known as an electrocyclic reaction. Examples include the 1,3-butadiene–cyclobutene and 1,3-cyclohexadiene–1,3,5-hexatriene interconversions (Figs. 20.5 and 20.16).

Orbital symmetry considerations dictate that in $4n$-electron reactions the thermal process must use a conrotatory motion, whereas the photochemical reaction must be disrotatory. Just the opposite rules apply for reactions involving $4n + 2$ electrons. The key to analyzing electrocyclic reactions is to look at the way the p orbitals at the end of the open-chain π system must move in order to generate a bonding interaction in the developing σ bond.

Cycloaddition reactions are closely related to electrocyclic reactions. The phases of the lobes in the HOMO and LUMOs must match so that bonding interactions are preserved in the transition state for the reaction. Sometimes a straightforward head-to-head motion is possible in which the two π systems approach each other in parallel planes to produce two bonding interactions. Such reactions are typically easy and are quite common. An example is the Diels–Alder reaction (Fig. 20.21).

In other reactions, the simple approach of HOMO and LUMO involves an antibonding overlap. In such reactions, a rotation is required in order to bring the lobes of the same sign together (Fig. 20.28). Although in principle this rotation maintains bonding interactions, in practice it requires substantial amounts of energy, and such reactions are rare.

Sigmatropic shifts involve the migration of hydrogen or another atom along a π system. Allowed migrations can be either suprafacial (leave and reattach from the same side of the π system) or antarafacial (leave from one side and reattach from the other) depending on the phase of the orbitals involved in the migration. The important point is that in an allowed shift, bonding overlap must be maintained at all times both to the lobe from which the migrating group departs and to that to which it reattaches. Steric considerations can also be important, especially for hydrogen, which must migrate using a small $1s$ orbital (Fig. 20.73).

FIGURE 20.73 Suprafacial and antarafacial migratory paths for sigmatropic shifts.

Syntheses

The photochemical 2 + 2 dimerization of alkenes.

$$\| + \| \xrightarrow{h\nu} \square$$

Common Errors

For many students the big problem in this material is to determine the kind of reaction involved. The way to start a problem in this area is to spend some time and thought identifying what kind of reaction is taking place. Cycloadditions are generally easy to find, although even here confusion exists between these reactions and electrocyclic processes. Sigmatropic shifts are even harder.

Are two molecules coming together to form a new ring? If so, it's a cycloaddition. Is a ring opening or closing in an in-

tramolecular reaction? That's an electrocyclic process. In sigmatropic shifts, something moves from one point in the molecule to another.

Often it is not easy to identify exactly what has happened in a sigmatropic reaction. When one atom or a group of atoms goes flying from one part of the molecule to another, there is often a rather substantial structural change. The product sometimes doesn't look much like the starting material.

20.11 ADDITIONAL PROBLEMS

PROBLEM 20.27   Analyze the photochemical and thermal [1,7] shifts of hydrogen in 1,3,5-heptatriene. Be sure to look at all possible stereoisomers of the starting material. What kind of [1,7] shift will be allowed thermally? What kind photochemically? How will the starting stereochemistry of the triene affect matters? Top views of the π molecular orbitals of heptatrienyl are given.

$$H_2C{=}CH{-}CH{=}CH{-}CH{=}CH{-}CH_3$$

1,3,5-Heptatriene

Energy ↑

```
+ - + - + - +
+ - + 0 - + -
+ - - + - - +
+ 0 - 0 + 0 -
+ + - - - + +
+ + + 0 - - -
+ + + + + + +
```

Top views of the π molecular orbitals of heptatrienyl

PROBLEM 20.28   Vitamin D₃ (**3**) is produced in the skin as a result of UV irradiation. It was once believed that 7-dehydrocholesterol (**1**) was converted directly into **3** upon photolysis. It is now recognized that there is an intermediate, previtamin D₃ (**2**), involved in the reaction. This metabolic process formally incorporates two pericyclic reactions, **1** → **2** and **2** → **3**. Identify and analyze the two reactions.

PROBLEM 20.29   The thermally induced interconversion of cyclooctatetraene (**1**) and bicyclo[4.2.0]octa-2,4,7-triene (**2**) can be described with two different sets of arrows. Although these two arrow formalisms may seem to give equivalent results, one of them is, in fact, not a reasonable depiction of the reaction. Explain.

PROBLEM 20.30   Analyze the thermal ring opening to allyl ions of the two cyclopropyl ions shown below. What stereochemistry do you expect in the products? Will the reactions involve conrotation or disrotation? Explain. Do not be put off by the charges! They only serve to allow you to count electrons properly.

PROBLEM 20.31   Orbital symmetry permits one of the two following photochemical cycloaddition reactions to take place in a concerted fashion. What would be the products, which reaction is the concerted one, and why?

PROBLEM 20.32   The following two cyclobutenes can be opened to butadienes thermally. However, they react at very different rates. Which is the fast one, which is the slow one, and why?

PROBLEM 20.33   These same two cyclobutenes (Problem 20.32) react photochemically to give two monocyclic (one ring) compounds isomeric with starting material. Predict the products, and explain the stereospecificity of the reaction.

PROBLEM 20.34   Benzocyclobutene (**1**) reacts with maleic anhydride when heated at 200 °C to give the single product shown. Write a mechanism for this reaction.

200 °C

PROBLEM 20.35   Given your answer to Problem 20.34, explain the stereochemistry observed in the following reaction:

PROBLEM 20.36   Write a mechanism for the following reaction. Note that no similar reaction appears for *cis*-1,3,5-hexatriene. It may be useful to work backward from the product. What molecules must lead to it?

Δ

PROBLEM 20.37   Now, here's an interesting variation on the reaction of Problem 20.36. Diels–Alder additions typically take place as shown in Problem 20.36, with the ring-closed partner (**A** in the answer to Problem 20.36) actually doing the addition reaction. However, the molecule shown below is an exception, as Diels–Alder reaction takes place in the normal way to give the product shown. Draw an arrow formalism for the reaction and then explain why **1** reacts differently from other cycloheptatrienes.

NC━━━CN

Δ

PROBLEM 20.38 Write an arrow formalism mechanism for the following thermal reaction and explain the specific trans stereochemistry. Careful here—note the acid catalyst for this reaction.

PROBLEM 20.39 Write an arrow formalism mechanism for the following reaction and explain the observed trans stereochemistry:

PROBLEM 20.40 Explain the following differences in product stereochemistry:

PROBLEM 20.41 Provide mechanisms for the following transformations:

(a)

(b)

PROBLEM 20.42 Provide mechanisms for the following transformations. Part (a) is harder than it looks (*Hint*: There are two intermediates in which the three-membered ring is gone) and (b) is easier than it looks (*Hint*: What simple photochemical pericyclic reactions do you know?).

(a)

(b)

PROBLEM 20.43 In Problem 20.23(b), we saw an example of a [3,3] sigmatropic shift known as the Claisen rearrangement. This reaction involves the thermally induced intramolecular rearrangement of an allyl phenyl ether to an o-allylphenol.

(a) It might be argued that this rearrangement is not a concerted process, but instead involves a pair of radical intermediates. Write such a mechanism and design a labeling experiment that would test this hypothesis.

(b) If both the ortho positions of the allyl phenyl ether are blocked with methyl groups, the product of this rearrangement is the p-allylphenol. This version of the Claisen reaction is often called the para-Claisen rearrangement.

Explain why the rearrangement takes a different course in this case. In addition, would the label you suggested in (a) be useful in this case for distinguishing concerted from nonconcerted (radical) mechanisms?

PROBLEM 20.44 Propose an arrow formalism mechanism for the following reaction. Be sure that your mechanism accommodates the specific labeling results shown by the dot ($\bullet = {}^{14}C$). Also, be sure that any sigmatropic shifts you propose do not involve impossibly long distances.

PROBLEM 20.45 We saw on page 1155 that chorismate undergoes a [3,3] sigmatropic shift (Claisen rearrangement) to prephenate catalyzed by the enzyme chorismate mutase. This enzyme-catalyzed reaction occurs more than 1 million times faster than the corresponding nonenzymatic reaction. How does the enzyme effect rate enhancement? Both enzymatic and nonenzymatic reactions proceed through a chair-like transition state. However, the predominant conformer of chorismate in solution, as determined by ^{1}H NMR spectroscopy, is the pseudodiequatorial conformer **2**, that cannot undergo the reaction.

Perhaps the role of the enzyme is to bind the predominant pseudodiequatorial conformer **2**. The enzyme–substrate complex then undergoes a rate-determining conformational change to convert the substrate into the pseudodiaxial form **1**, that rapidly undergoes rearrangement.

(a) Explain why **1** can rearrange to prephenate but **2** cannot.

(b) Suggest a reason why conformer **2** is preferred over conformer **1**. *Hint*: Focus on the OH. What can it do in **2** that it cannot do in **1**?

(c) Draw an appropriate Energy versus Reaction progress diagram that illustrates the situation outlined above. That is, a reaction in which the rate-determining step is the conversion of **2** into **1**. What would the Energy versus Reaction progress diagram look like if the rate-determining step were the conversion of **1** into prephenate?

PROBLEM 20.46 In 1907, Staudinger reported the reaction of diphenylketene (**1**) and benzylideneaniline (**2**) to give β-lactam **3**. Although this reaction is formally a 2 + 2 cycloaddition reaction, there is ample experimental evidence to suggest that the reaction is not concerted, but is instead a two-step process involving a dipolar intermediate.

(a) Propose an arrow formalism mechanism for this cyclo-addition reaction. Don't forget to include the dipolar intermediate.

(b) Although cycloaddition reactions of this type are not concerted, they are often stereospecific, as in the following example:

Only isomer formed

Attempt to explain the stereochemistry observed in the reaction above. Start by minimizing the steric interactions in your proposed dipolar intermediate. Then, see if you can find a concerted process that converts your proposed intermediate into the observed β-lactam. *Hint*: A *tert*-butyl group is much bigger than a cyano group.

PROBLEM 20.47 Heating α-phenylallyl acetate (1) affords isomeric compound 2. Spectral data for compound 2 are summarized below. Deduce the structure of 2 and provide an arrow formalism mechanism for this isomerization.

Compound 2
IR (film): 1740 (s), 1225 (s), 1025 (m), 960 (m), 745 (m), 690 (m) cm^{-1}
^{1}H NMR (CDCl$_3$): δ 2.08 (s, 3H), 4.71 (d, J = 5.5 Hz, 2H), 6.27 (dt, J = 16, 5.5 Hz, 1H), 6.63 (d, J = 16 Hz, 1H), 7.2–7.4 (m, 5H)
^{13}C NMR (CDCl$_3$): δ 20.9 (q), 65.0 (t), 123.1 (d), 126.5 (d), 128.0 (d), 128.5 (d), 134.1 (d), 136.1 (s), 170.6 (s)

PROBLEM 20.48 Reaction of bicyclopropenyl (1) and *s*-tetrazine (2) gives the semibullvalene (3). Provide an arrow formalism mechanism for this transformation. *Hint*: There is at least one nonisolable intermediate involved. Also, explain the relatively simple ^{1}H and ^{13}C NMR spectra of compound 3, summarized below.

Compound 3
^{1}H NMR (CDCl$_3$): δ 1.13 (s, 3H), 3.73 (s, 3H), 4.79 (s, 2H)
^{13}C NMR (CDCl$_3$): δ 14.9 (q), 51.4 (q), 60.6 (s), 93.7 (d), 127.2 (s), 164.7 (s)

PROBLEM 20.49 Figure 20.56 shows the products of the Cope rearrangement of *meso*-3,4-dimethyl-1,5-hexadiene from both the chair-like (four-center) and boat-like (six-center) transition states. Work out the products expected from the other possible form of the diene, racemic 3,4-dimethyl-1,5-hexadiene. Predict what products will actually be formed.

Special Topic: Intramolecular Reactions and Neighboring Group Participation

21

Our race would not have gotten far,
Had we not learned to bluff it out
And look more certain than we are
Of what our motion is about.

—W. H. Auden*

21.1   PREVIEW

Here we introduce the chemistry of polyfunctional molecules. This chapter by no means contains an exhaustive treatment. The chemistry of di- and polyfunctional compounds could easily be the subject of an advanced course in organic chemistry, and we will only catch some highlights here.

In principle, two functional groups in a single molecule might not interact at all, and the chemistry of the molecule would be the simple combination of the chemistries of the two separate functional groups. This occurrence is rare, as groups in molecules often do interact, even if they are quite remote from each other. Through inductive and resonance effects, one group can affect the chemistry of another group at a distant point in a molecule. Sometimes, groups interact even more strongly, and the chemistry is best described not as a sum of the two individual chemistries, but as something quite new. These are the most interesting examples.

A **neighboring group effect** can change the rates of formation of products as well as the very structures of the molecules produced. Don't be put off by the fancy name. All of intramolecular chemistry involves interactions between neighboring groups. You have already seen lots of intramolecular chemistry. Figure 21.1 gives some familiar, or at least old, examples of such reactions, and we will see more as we go along.

*Wystan Hugh Auden (1907–1975) was one of the most influential British poets of his generation.

Intramolecular S$_N$2

(a)

KOH / H$_2$O → + KBr + H$_2$O

(see Chapter 7)

(b)

+ H—Cl

(see Chapter 18)

(c)

Δ

(see Chapter 19)

FIGURE 21.1 Neighboring group interactions.

WORKED PROBLEM 21.1 Write mechanisms for the reactions of Figure 21.1.

ANSWER (a) The first step is certainly loss of the hydroxyl proton to give an alkoxide. Intramolecular S$_N$2 displacement gives the product.

Alkoxide

(b) Addition–elimination gives the lactone (a cyclic ester).

(c) This reaction is an intramolecular elimination with the carbonyl oxygen playing the part of the nucleophile (neighboring group).

The exploitation of the intuitively obvious notion that groups might interact in a synergistic fashion constituted one of the most fascinating areas of chemical exploration during the middle of the twentieth century, and it continues even today. As we examine this area, we will encounter some sociology as well as chemistry. One of the most heated arguments in the history of organic chemistry centered upon this subject, and it is worth our time to see what we can learn about science from a look at this intensely emotional issue. There will be some opinion here, but I'll try to keep it labeled as such.

ESSENTIAL SKILLS

1. There are three clues to the operation of a neighboring group effect. Learning to recognize them is critical to understanding this chemistry. Fortunately, one or more of those clues is always present, and they are usually quite easy to find. This material is of the "looks hard—is easy" variety.

   The clues are
   a. An unusual stereochemical result, usually retention where inversion is expected.
   b. A strange rearrangement, often of a label.
   c. An unexpectedly fast rate.

IMPORTANT DETAILS

1. Be certain you are completely clear on the difference between an intermediate described by a set of equilibrating structures and an intermediate described by a set of resonance forms. Figure 21.43 gives an example.

21.2   HETEROATOMS AS NEIGHBORING GROUPS

In its simplest form, the study of neighboring group effects notes that a heteroatom with a pair of nonbonding electrons can act as a nucleophile in an intramolecular displacement reaction, just as external nucleophiles can participate in intermolecular S_N2 reactions. In an example we have seen, oxiranes can be formed from halohydrins in base (Fig. 21.2).

First, base removes the hydroxyl hydrogen to make the alkoxide. Intramolecular S_N2 displacement of the bromide generates the oxirane. Here the alkoxide is the neighboring group, and the whole process couldn't be simpler.

Intermolecular S_N2

Intramolecular S_N2 by "neighboring" alkoxide

FIGURE 21.2 Formation of an oxirane by treatment of a bromohydrin with hydroxide involves an alkoxide acting as a neighboring group.

Another example is intramolecular hemiacetal formation. If a relatively unstrained ring can be formed, compounds containing both a carbonyl and a hydroxyl group exist in equilibrium with the cyclic hemiacetal (Chapter 16, p. 844; Fig. 21.3).

(6.1%)   A hemiacetal
                  (93.9%)

FIGURE 21.3 Cyclic hemiacetal formation involves a hydroxyl group acting as a neighboring group.

PROBLEM 21.2 Write a mechanism for the reaction of Figure 21.3.

Other reactions are more subtle than these straightforward intramolecular processes, but there are always clues to the operation of a neighboring group effect that you can find. For example, consider the reaction of the optically active α-bromocarboxylate in methyl alcohol (Fig. 21.4).

FIGURE 21.4 Treatment of α-bromopropionate in methyl alcohol. Notice that the reaction involves retention of configuration.

The * indicates a stereogenic carbon

Still (R) retention

The product is formed with the stereochemistry opposite to what we might naively expect. An S_N2 displacement of bromide by methyl alcohol or methoxide would give an optically active product, but the stereochemistry would be exactly the reverse of what is observed (Fig. 21.5)!

FIGURE 21.5 A straightforward mechanism in which the bromide is displaced by methyl alcohol gives the wrong stereochemical result—inversion.

The observation of a strange stereochemical result is the first kind of clue to the operation of a neighboring group effect. The simple mechanism of Figure 21.5 must be wrong, as it cannot accommodate the observed retention of configuration in the product. In this reaction, the carboxylate salt acts as an internal nucleophile, displacing bromide to give an intermediate called an **α-lactone** (Chapter 19, p. 1032; Fig. 21.6).

FIGURE 21.6 If the carboxylate anion acts as a neighboring group and displaces the bromide in an intramolecular fashion, methyl alcohol can open the resulting α-lactone. The overall result of this double inversion mechanism is retention.

These smallest cyclic esters cannot be isolated, but they are known as reactive intermediates. In this case, methoxide or methyl alcohol adds to the α-lactone in an intermolecular S_N2 displacement to give the real product of the reaction. Now examine the stereochemistry of the reaction with care. The first displacement is an *intra*molecular S_N2 reaction with the carboxylate anion acting as the nucleophile and the bromide as the leaving group. As with all S_N2 reactions, inversion must be the stereochemical result. A second S_N2 displacement, this time *inter*molecular, occurs as the α-lactone is opened by methyl alcohol. Once again inversion is required. The end result of two inversions (or any even number of inversions) is retention of configuration (Fig. 21.7).

FIGURE 21.7 Net retention is the result of two (or any even number of) inversions.

SUMMARY

Whenever you see a backward stereochemical result (usually retention where inversion would be expected), look for an intramolecular displacement reaction—a neighboring group effect. You will almost always find it.

Oxygen need not be negatively charged to participate in an intramolecular displacement reaction. The reaction of the "brosylate" (*p*-bromobenzenesulfonate) of 4-methoxy-1-butanol in acetic acid (HOAc = CH$_3$COOH) gives the corresponding acetate (Fig. 21.8).

FIGURE 21.8 Acetolysis of the brosylate of 4-methoxy-1-butanol gives the corresponding acetate.

CONVENTION ALERT

Remember that the acetate group (CH$_3$COO) is often abbreviated OAc. The acetyl group (Ac) is CH$_3$CO.

Sulfonates are excellent leaving groups and a reasonable mechanism for the reaction involves intermolecular displacement by acetic acid or acetate (Fig. 21.9).

FIGURE 21.9 A simple S$_N$2 displacement of the brosylate by acetic acid might explain the reaction.

WORKED PROBLEM 21.3 There are two possible nucleophiles for the S_N2 displacement in Figure 21.9, the acetate ion or acetic acid itself. In the displacement using acetic acid, why was the carbonyl oxygen, not the hydroxyl oxygen, used as the displacing agent?

ANSWER The carbonyl oxygen is the better nucleophile. Recall that protonation of acetic acid goes faster on the carbonyl oxygen (Chapter 18, p. 985). A resonance-stabilized intermediate is formed in this case, but not from protonation of the hydroxyl oxygen. Reaction of a Lewis acid with acetic acid will be similar. In the figure, only the "new" resonance form is shown; in each case there is another form in which the carbon atom bears the positive charge—ordinary carbonyl resonance.

Resonance stabilized!

Not resonance stabilized

When this reaction is run with a similar, but methyl-substituted brosylate, we can see that something more complicated than the simple S_N2 reaction of Figure 21.9 is going on. Our first mechanism can accommodate one of the products but not the other (Fig. 21.10), which is the second clue to neighboring group participation—the observation of an unusual rearrangement reaction.

This product could be formed by a simple S_N2 displacement...

...but *not* this one

FIGURE 21.10 Reaction of the methyl-labeled compound shows that the simple mechanism of Figure 21.9 is inadequate.

In this case, it is the ether oxygen that is acting as a neighboring group. The sulfonate is an excellent leaving group and a five-membered ring is relatively unstrained, so the intramolecular displacement of *p*-bromobenzenesulfonate ($^-$OBs) to give a cyclic oxonium ion takes place (Fig. 21.11).

FIGURE 21.11 Here it is the methoxyl oxygen that acts as neighboring group. A cyclic ion is formed that can open in two ways. "Rearrangement" is the result of one mode of opening of the intermediate oxonium ion.

PROBLEM 21.4 There is actually a third product in this reaction, 2-methyltetrahydrofuran (see below). Can you explain the formation of a small amount of this molecule? *Hint*: The intermediate ring is obviously retained in the formation of this product. Think about how this reaction can happen.

2-Methyltetrahydrofuran

The cyclic oxonium ion can be opened by acetic acid in two ways (Fig. 21.11) to give the two observed products. The neutral ether oxygen has acted as a neighboring group to provide an intermediate that can react further in more than one way. The result is an apparent rearrangement.

Sulfur atoms are often powerful nucleophiles, and the participation of sulfur as a neighboring group is common. Here is just one example. Both hexyl chloride and 2-chloroethyl ethyl sulfide react in water to give the corresponding alcohols (Fig. 21.12). The rate of reaction of the sulfur-containing compound is about 10^3 greater than that of the alkyl chloride, however.

	Relative rate
	1
	~700

FIGURE 21.12 If a sulfur atom is substituted for the C(3) of hexyl chloride, the formation of the related alcohol is greatly accelerated.

The mechanism of alcohol formation from the primary halide must be a simple S_N2 displacement with water acting as nucleophile (Fig. 21.13). It is difficult to see why the sulfide should be so much faster if the reaction mechanism is analogous.

FIGURE 21.13 The reaction of hexyl chloride must be a simple S_N2 displacement of chloride by water, followed by deprotonation.

The third kind of clue to the operation of a neighboring group effect is the observation of an unexpectedly rapid rate.

In the reaction of the sulfide, it is the sulfur atom that displaces the leaving chloride. This intramolecular reaction is much faster than the intermolecular displacement of chloride by water either in the alkyl chloride or the chlorosulfide. The product is an **episulfonium ion**. This intermediate is opened by water in a second, intermolecular S_N2 displacement step to give the product (Fig. 21.14).

FIGURE 21.14 The mechanism of the sulfur reaction must be different. Here, sulfur can act as a neighboring group, displacing chloride in an intramolecular S_N2 reaction. An episulfonium ion is formed and can be opened in a second, intermolecular S_N2 reaction to give the product.

PROBLEM 21.5 There are many possible labeling experiments that would support (not prove) the mechanism suggested for the reaction of the chlorosulfide. Design two, one using a methyl group as a marker, and the other using an isotopic label.

It is important to be certain you understand the source of the rate increase in the sulfide. Intramolecular reactions often have a large rate advantage over intermolecular reactions as there is no need for the nucleophile and substrate to find each other in solution. Intramolecular assistance in the rate-determining step of a reaction is called **anchimeric assistance**. Once again, it is important not to be put off by the elaborate name. The concept is quite simple: Intramolecular nucleophiles are often more effective displacing agents than intermolecular nucleophiles. What is somewhat more difficult to anticipate is that an intramolecular displacement by a very weak, but perfectly situated nucleophile can often still compete effectively with intermolecular reactions.

If we make a seemingly trivial change in our sulfur-containing molecule, to add another chlorine (Fig. 21.15), we come to bis(2-chloroethyl)thioether. This seemingly routine molecule exposes the dark side of our science; indeed, a review of the history of this compound causes one to reflect on the depths to which human behavior can sink. For this molecule is mustard gas, developed in the early part of this century as a replacement for other less efficient toxic agents for which defenses had been worked out. Mustard gas penetrated the rubber in use in gas masks at the time and thus was more effective. This mustard is lethal in a few minutes at 0.02–0.05% concentration. Less severe exposure nonetheless leads to irreversable cell damage and only delays the inevitable. Introduced by the Germans at Ypres in the first world war in 1917, it was also employed in the China campaign by the Japanese in the second world war. Stories of experiments on prisoners of war are easy to find, if not easy to read. Nor is there any reason for national pride in not being the first to use this agent, for we in America were the developers of napalm, essentially jellied hydrocarbons, and surely easily defined as a chemical agent.

I leave it to you to consider whether it is morally superior to shoot people, or drop an atomic bomb on them, rather than to dose them with mustard gas or burn them with napalm. Even more complicated for us is the question of the participation by chemists in the development of such agents. Of course, often it is not easy to see the consequences of one's research. But what if one can? Where are the limits?

Bis(2-chloroethyl)thioether

FIGURE 21.15
Bis(2-chloroethylthioether).

SUMMARY

There are three clues that point to the operation of a neighboring group effect. The first is an unusual, often "backward" stereochemical result (clue 1). Generally, this involves retention of configuration when inversion might have been anticipated. Second, the operation of a neighboring group effect can often present the opportunity for a rearrangement, as an intermediate is formed that can react further in more than one way (clue 2). Finally, the anchimeric assistance provided by an intramolecular nucleophile often reveals itself in the form of a rate increase (clue 3).

PROBLEM 21.6 Go back and find examples of these three clues in the preceding reactions.

Now let's look at more heteroatoms with nonbonding electrons that might participate in neighboring group reactions. For example, nitrogen bears a pair of electrons, and so can act as an internal nucleophile. Let's look at the reactions of the two bicyclic amines in Figure 21.16. Treatment of the endo chloride with ethyl alcohol leads to the endo ether, but the exo chloride gives neither the exo nor endo ether under the reaction conditions. Now we'll see why.

FIGURE 21.16 Treatment of only one isomer of this bicyclic amine gives the related ether. Note the retention of configuration.

Direct displacement of the chloride by ethyl alcohol cannot be the answer, as this potential reaction leads to a product that is not observed, the exo ether. The ether *is* formed, but it is formed with *retention* of stereochemistry, not the *inversion* demanded by the simple S_N2 reaction (Fig. 21.17).

FIGURE 21.17 A simple intermolecular S$_N$2 displacement predicts inversion of configuration.

At least one of the clues to the operation of a neighboring group effect is present in this reaction. The stereochemistry of the product is the opposite to what one would expect from a simple displacement (clue 1).

What's happening? By now you should be alert to the possibility of intramolecular displacement. The nitrogen with its pair of nonbonding electrons is an obvious possible neighboring group, and it is located in the proper position for a rear side displacement of the chloride (Fig. 21.18). The result is a cyclic ammonium ion, a ring containing a positively charged nitrogen. If this ring is opened by ethyl alcohol, the product ether must be formed with retention of stereochemistry. The intramolecular S$_N$2 displacement by nitrogen is the first inversion reaction, and the second intermolecular displacement, the opening of the ammonium ion by ethyl alcohol, provides the second. Two inversions produce net retention (Fig. 21.18).

FIGURE 21.18 If nitrogen acts as a neighboring group and displaces the leaving group, a cyclic ammonium ion is formed. The cyclic ion can be opened to give the observed product of retention of configuration.

Why does the exo compound not react in the same way? In this molecule, displacement of the leaving chloride would require a frontside S_N2 reaction, and this process is strongly disfavored (Chapter 7, p. 279; Fig. 21.19). So, the exo chloride reacts more slowly (and in a different way) from the endo isomer.

No
frontside
$S_N2!$

the cyclic ammonium
ion cannot be formed
this way

FIGURE 21.19 The exo chloride cannot be displaced by
nitrogen, as this reaction would require a frontside S_N2 reaction.

PROBLEM 21.7 The exo chloride gives the product shown below. Can you suggest a mechanism for this reaction?

CH_3CH_2OH
H_2O
50 °C

It is often difficult to see these reactions in bicyclic systems. The molecules are not easy to draw, and the extra atoms involved in the cage are distracting. There is nothing *fundamentally* difficult about these reactions, however. In fact, one could argue that once the drawing problems are mastered, the rigidity of the cage is a help. One need not worry about rotations about single bonds, or about finding the proper orientation for the reacting molecules. The stereochemical relationships are defined and preserved by the rigid cages. Here are some problems for practice. It should help to use models.

WORKED PROBLEM 21.8 Write mechanisms for the following reactions:

(a)

$$CH_3COO^- \ ^+Na$$
$$CH_3CH_2OH$$

(b)

$$CH_3CH_2O^- \ ^+Na$$
$$CH_3CH_2OH$$

(c)

$$HO^- \ ^+K$$
$$H_2O$$

ANSWER (a) The hardest part of this problem is (a). The internal nucleophile is the carbonyl oxygen. Notice that the trans stereochemistry of the starting material allows for a rearside S_N2 displacement of the good leaving group, tosylate. Displacement yields a cyclic cation that is nicely stabilized by resonance. Acetate ion adds to the carbon sharing the positive charge, and gives the product.

intramolecular | S_N2

The halogens are another set of heteroatoms with nonbonding electrons. Neighboring group reactions of halogens are almost entirely restricted to iodine, bromine, and chlorine, and very often reveal themselves through the stereochemistry of the resulting products.

As an example we will work through the conversion of *cis*-2-butene into 2,3-dibromobutane through the synthetic scheme shown in Figure 21.20.

FIGURE 21.20 A possible synthesis of 2,3-dibromobutane (no stereochemistry implied).

PROBLEM 21.9 What's wrong with fluorine? Why are neighboring group effects involving fluorine very rare?

PROBLEM 21.10 What is a much easier way of making the 2,3-dibromobutane of Figure 21.20?

In the bromination of *cis*-2-butene, the initial intermediate is the cyclic bromonium ion (Chapter 10, p. 442). It is opened by water to give the racemic pair of enantiomers shown in Figure 21.21.

FIGURE 21.21 Formation of a bromonium ion, followed by opening by water, gives a pair of enantiomeric bromohydrins.

An enantiomeric pair of bromohydrins

Treatment of a primary or secondary alcohol with hydrogen bromide gives the corresponding bromide through the following mechanism (Chapter 7, p. 293; Fig. 21.22).

FIGURE 21.22 The mechanism for formation of a bromide from alcohols and hydrogen bromide.

The reaction of Figure 21.22 would give the *meso*-2,3-dibromobutane from the bromohydrins of Figure 21.21. This compound is *not* the product that is really formed. The product *is* 2,3-dibromobutane, but it is the racemic form, not the meso compound that is produced (Fig. 21.23).

This is the product that is actually formed; note retention!

meso-2,3-Dibromobutane
not formed

Staggered-energy minimum

FIGURE 21.23 Reaction of the bromohydrin of Figure 21.20 by S$_N$2 displacement should give *meso*-2,3-dibromobutane. This compound is *not* the actual product.

Be absolutely certain you see this difference. If you are lost, reread Chapter 4, which works through the differences between meso compounds and racemic mixtures.

CONVENTION ALERT

Two conventions are being used here. First, the figures often carry through with only one enantiomer of a pair. Be sure to do Problem 21.11. Second, for clarity, eclipsed forms are often drawn. Be sure you remember that these are not really energy minima, but energy maxima.

PROBLEM 21.11 Figure 21.23 works the conventional reaction mechanism through with one of the enantiomers of Figure 21.21. You work it through with the other enantiomer.

The formation of the "wrong" stereoisomer is the clue to the operation of a neighboring group effect in this reaction. The product is formed not with the inversion required by the mechanism of Figure 21.22, but with retention (clue 1). There must be a neighboring group effect operating. After the hydroxyl group is protonated, it is displaced not by an external bromide ion, but by the bromine atom lurking within the same molecule. It is most important at this point to keep the stereochemical relationships clear. Displacement must be from the rear, as this is an intramolecular S_N2 reaction (Fig. 21.24).

FIGURE 21.24 If bromine acts as a neighboring group and displaces water, a bromonium ion is formed. Opening by bromide ion gives the observed product, a pair of enantiomeric dibromides.

An enantiomeric pair of dibromides

Notice that we have formed the same bromonium ion as in Figure 21.21, though by a different route. Like all bromonium ions, this one is subject to opening by nucleophilic attack. In this case, bromide ion can open the cyclic ion at two points (paths a and b) leading to the pair of enantiomers shown in Figure 21.24. The product shows no optical activity not because it is a meso compound, but because it is a racemic mixture of enantiomers.

PROBLEM 21.12 Work through the stereochemical relationships in the following reactions of *trans*-2-butene in great detail. Follow the reactions of all pairs of enantiomers. Contrast them with what you would expect from a reaction not involving a neighboring group effect.

meso-2,3-Dibromobutane

WORKED PROBLEM 21.13 There is a fundamental difference in the formation of a cyclic ion using a pair of nonbonding electrons or a pair of σ electrons. Carefully draw out the cyclic species for both cases and explain the difference. *Hint*: Count electrons.

ANSWER Problem 21.12 contains many examples of formation of a three-membered ring using a pair of nonbonding electrons. The result is a normal compound in which all bonds are two-electron bonds.

When participation is by a σ bond, everything changes. No longer is it possible to make what we could call a normal three-membered ring; there are not enough electrons. The two electrons originally in the σ bond must serve to bind three nuclei.

21.3   NEIGHBORING π SYSTEMS

21.3a Aromatic Neighboring Groups.
We have progressed from participation reactions of very strong nucleophiles such as alkoxides and sulfur, to those of less powerful displacing agents such as the halogens. Now we examine even weaker nucleophiles, the π systems of aromatic molecules and alkenes. We will finally move on to very weak nucleophiles indeed, and see σ bonds acting as intramolecular displacing agents. The π systems of aromatic rings can act as neighboring groups. The reactions of the 3-phenyl-2-butyl tosylates are good examples. D. J. Cram [1919–2001, awarded the Nobel prize in 1987 for work on crown ethers (Chapter 6, p. 264)] and his co-workers showed that in acetic acid, the optically active 3-phenyl-2-butyl tosylate shown in Figure 21.25 gives a racemic mixture of the corresponding 3-phenyl-2-butyl acetates.

FIGURE 21.25 This optically active isomer of 3-phenyl-2-butyl tosylate gives a racemic mixture of the corresponding acetates.

But this result is not what we would expect from a simple analysis of the possible displacement reactions of the starting material by either S_N2 or S_N1 substitution. The S_N2 reaction must give inversion in the substitution step, and so must form the optically active acetate shown in Figure 21.26.

FIGURE 21.26 Simple S_N2 displacement by acetic acid fails to predict the correct product. Note the eclipsed forms drawn in this figure! They are there for clarity only and should be replaced with the staggered forms.

Ionization in an S_N1 process gives an intermediate open carbocation that can be attacked at either face by acetic acid to give the two different optically active products of Figure 21.27.

FIGURE 21.27 An S_N1 ionization also fails to predict the product with the correct stereochemistry. A mixture of two optically active acetates should result.

PROBLEM 21.14 What is the stereochemical relationship (names) of the two products at the bottom right of Figure 21.27?

As neither of these reactions produces the observed product, some other mechanism must be operating in this reaction. One of the clues for a neighboring group participation is present—an unexpected stereochemical result, clue 1—so let's look around for possible internal nucleophiles.

Aromatic rings act as nucleophiles toward all manner of Lewis acids in the aromatic substitution reaction (Chapter 13), so perhaps we can use the π system of the phenyl ring present in 3-phenyl-2-butyl tosylate as a nucleophile. If the π system displaces the tosylate, it must do so from the rear. The result is the **phenonium ion** shown in Figure 21.28.

FIGURE 21.28 If the benzene ring acts as a neighboring group, a cyclic phenonium ion results. Notice that the phenonium ion has a plane of symmetry and is achiral.

PROBLEM 21.15 Draw resonance forms for the phenonium ion of Figure 21.28.

Note carefully the geometry of the phenonium ion. The two rings are perpendicular, not coplanar, and the ion is achiral, a meso compound. All chance of optical activity is lost at this point. Note also the close correspondence between this intermediate and that formed in an electrophilic aromatic substitution reaction (Chapter 14, p. 679; Fig. 21.29).

FIGURE 21.29 The phenonium ion is similar to the intermediate ion produced in electrophilic aromatic substitution.

Aromatic substitution

These two species are very similar

There is nothing particularly odd about this cyclic species. If it is opened in a second, intermolecular S_N2 reaction, there are two possible sites for attack. The two products are mirror images, and the racemic mixture of enantiomers of Figure 21.30 is exactly what was found experimentally.

FIGURE 21.30 Opening of the phenonium ion must occur in two equivalent ways (a and b) to give a pair of enantiomers—a racemic mixture.

PROBLEM 21.16 Work through the reaction of the other possible optically active isomer of 3-phenyl-2-butyl tosylate in acetic acid (see below). This time the product is optically active acetate.

21.3b Carbon–Carbon Double Bonds as Neighboring Groups. Carbon–carbon double bonds are also nucleophiles and can act as neighboring groups in intramolecular displacement reactions. Among the examples you already know are the sequence of intramolecular carbocation additions leading to steroids (Chapter 12, p. 604) and intramolecular carbene additions (Chapter 10, p. 460). In each of these reactions, a carbon–carbon double bond acts as a nucleophile toward an empty carbon $2p$ orbital as Lewis acid (Fig. 21.31).

FIGURE 21.31 Two examples of carbon–carbon double bonds acting as intramolecular nucleophiles.

Neighboring carbon–carbon π bonds can influence the rate and control the stereochemistry of a reaction just as a heteroatom can. For example, the tosylate of *anti*-7-hydroxybicyclo[2.2.1]hept-2-ene (7-hydroxynorbornene)* reacts in acetic acid 10^7 times faster than the corresponding syn compound, and a staggering 10^{11} times faster than the tosylate of the saturated alcohol, 7-hydroxynorbornane. The product is exclusively the anti acetate (Fig. 21.32).

FIGURE 21.32 The anti isomer of 7-norbornenyl tosylates gives the anti acetic acid. The anti tosylate reacts 10^7 faster than the syn isomer and 10^{11} faster than the saturated compound.

*The bicyclo[2.2.1]heptane cage is commonly known as the **norbornyl system**. This obscure name comes from a natural product, borneol, which has the structure shown in Figure 21.33, and which comes from the camphor tree, common in Borneo. The prefixes syn and anti locate the direction of the OH or OTs group with respect to the double bond and **nor** means "no methyl groups." Thus, 7-hydroxybicyclo[2.2.1]hept-2-ene is 7-hydroxynorborn-2-ene or, as there is only one possible place for the double bond, 7-hydroxynorbornene.

PROBLEM 21.17 There is only one possible place for the double bond in the bicyclo[2.2.1]heptyl ring system. Explain why.

Borneol   **Bornane**   **Norbornane**

FIGURE 21.33 Norborane (bicyclo[2.2.1]heptane) is related to the natural product borneol, *endo*-1,7,7-trimethylbicyclo[2.2.1]heptan-2-ol.

A consistent mechanistic explanation notes that only in the anti compound is the π system of the double bond in the correct position to assist in ionization of the tosylate. For the syn double bond to assist in ionization would require a frontside S_N2 displacement, which, as we know, is an unknown process.

The rate enhancement is attributed to anchimeric assistance by the double bond. Intramolecular displacement of the leaving group is faster than the possible intermolecular reactions and is therefore the dominant process. The stereochemistry of the reaction is explained by noting that addition of acetic acid to the intermediate ion is also an S_N2 reaction and must occur from the rear of the departing leaving group. The net result of the two S_N2 reactions (one intramolecular, one intermolecular) is retention (Fig. 21.34).

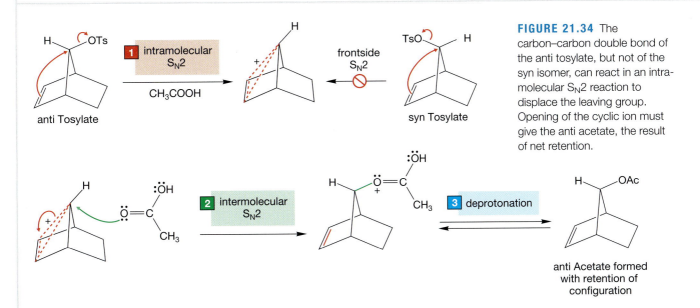

FIGURE 21.34 The carbon–carbon double bond of the anti tosylate, but not of the syn isomer, can react in an intramolecular S_N2 reaction to displace the leaving group. Opening of the cyclic ion must give the anti acetate, the result of net retention.

Think a bit now about the nature of the leaving group in the second, intermolecular S_N2 displacement. Just what is the leaving group, and what do all those dashed bonds in Figure 21.34 mean? To answer these questions well, we must first look closely at the structure of the postulated intermediate in this reaction. When we formed three-membered ring intermediates from heteroatoms, there was always a pair of nonbonding electrons available to do the intramolecular displacement. An example is the formation of a bromonium ion by displacement of water by neighboring bromine (Fig. 21.35).

FIGURE 21.35 In a bromonium ion, all bonds are normal two-electron bonds.

Contrast the bromonium ion of Figure 21.35 with the species formed by intramolecular displacement of tosylate using the π bond. In the bromonium ion, there are sufficient electrons to form the new σ bond in the three-membered ring. In the all-carbon species there are not. In this cyclic ion, two electrons must suffice to bind three atoms. The new ion contains an example of **three-center, two-electron bonding**, and the partial bonds are represented with dashed lines. A resonance formulation shows most clearly how the charge is shared by the three atoms in the ring (Fig. 21.36).

FIGURE 21.36 Two views of the cyclic ion as well as of the resonance forms contributing to the ion.

The two species are *very* different. When the bromonium ion is opened, it is a normal, two-electron σ bond that is broken. When the all-carbon three-membered ring is attacked, it is the two electrons binding the three carbons that act as the leaving group (Fig. 21.37).

FIGURE 21.37 When a nucleophile opens a bromonium ion, a pair of electrons is displaced. The same is true for the opening of the bridged, cyclic ion.

The process is still an S_N2 reaction, however, and there should be no reason for softening the requirement for rearside displacement.

Acceptance of the carbon–carbon double bond as a neighboring group requires you to expand your notions of bonding, and to accept the partial bonds making up the dashed three-center, two-electron system as a valid bonding arrangement. There are those who resisted this notion mightily, and although I believe they were ultimately shown to be incorrect, their resistance was anything but foolish. The postulated intermediate in this reaction is not a normal species and should be scrutinized with the greatest care.

PROBLEM 21.18 If three atoms share the positive charge, why is it only the 7-position that is attacked by the nucleophile? Explain the lack of another product. *Hint:* Look at the structure of the product formed by addition at the other position in the following reaction:

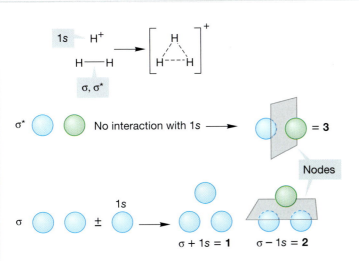

A molecular orbital approach shows how the two electrons can be stabilized through a symmetrical interaction with an empty carbon $2p$ orbital. The intellectual antecedents for this kind of bonding go back to the cyclic H_3^+ molecule in which two electrons are stabilized in a bonding molecular orbital through a cyclic overlap of hydrogen $1s$ orbitals. We can construct this system by allowing a $1s$ orbital to interact with σ and σ* of H_2 as in Figure 21.38 (recall our discussion of the allyl and cyclopropenyl π systems in Chapter 12).

FIGURE 21.38 The construction of the molecular orbitals of cyclic H_3^+.

The result is a system of three molecular orbitals: one bonding, two antibonding. Two, but only two, electrons can be accommodated in the bonding orbital. The same orbital pattern will emerge for any triangular array of orbitals.

PROBLEM 21.19 Work out the π molecular orbitals for the cyclopropenium ion below.

Cyclopropenium ion

Those resisting this notion (the "classicists") suggested that there was no need for this new kind of bonding, and that normal, two-electron models would do as well. Let's follow their argument. Remember as we go that they must explain both a rate acceleration for the anti tosylate as well as the overall retention of stereochemistry in the reaction.

The two compounds in question here, the syn and anti tosylates, are *different*, and there *must* be some rate difference between them. One can scarcely argue with this point, and one is reduced to relying on the magnitude and direction of the rate acceleration as evidence for the bridged ion. A classicist might argue that in cases such as this, the rate increase could easily come from the formation of a normal, unbridged or "classical," carbocationic intermediate. The anti tosylate has the leaving group in an excellent orientation to be attacked by the intramolecular nucleophile, the filled π orbital of the carbon–carbon double bond. The syn tosylate does not (Fig. 21.39).

FIGURE 21.39 A "classical" displacement reaction by the π bond of the anti isomer would give an open ion that could not be formed from the syn isomer.

As in the symmetrical displacement to form the bridged ion, the required attack on the leaving group from the rear could occur only in the anti compound. Of course, there are two equivalent intermediates possible (**A** and **B**), and it was suggested that they were in equilibrium. The compound with the charge at the 7-position (**C**) might also be a partner in the equilibrium. Be sure you see the difference between the *three* equilibrating structures of Figure 21.40 and the three resonance forms contributing to the *single* structure of Figure 21.36 (Fig. 21.40).

FIGURE 21.40 The first-formed ion **A** equilibrates with **B** and, perhaps **C**.

Addition of acetic acid to the new ions **A** and **B** must give the anti acetate, as the opening is an S$_N$2 reaction. Attack on ion **C** is more problematic, as it is difficult to explain the stereospecificity of the reaction. Still, it must be admitted that the syn and anti faces of the ion are different, and in principle, this difference could lead to a stereospecific reaction (Fig. 21.41).

FIGURE 21.41 Addition of acetic acid to the equilibrating ions **A** and **B** must lead to anti acetate. Attack at the 7-position (ion **C**) to give only the anti acetate presents some difficulties.

Stop right now and compare the two mechanistic hypotheses. In one, there is a symmetrical displacement of the leaving group to give a resonance-stabilized bridged intermediate. In the other, either the leaving group is displaced to give a carbocationic intermediate containing a three-membered ring (**A** and **B**) in equilibrium with at least one other ion, **C**, or a simple S$_N$1 ionization to **C** takes place (Fig. 21.42).

FIGURE 21.42 A comparison of the symmetrical, bridged intermediate with the open ions **A**, **B**, and **C**.

The whole difference between the two mechanisms is the formulation of the structure of the intermediate! The structures in the two versions of the mechanism for this reaction are only very subtly different. In one, the constituent structures are contributing resonance forms to a single intermediate bridged species; in the other they are equilibrating molecules, each with its own separate existence (Fig. 21.43).

FIGURE 21.43 The bridged ion is described with three resonance forms that resemble the open ions of the set of equilibrating structures.

Are we now down to questions of chemical trivia? Are we counting angels on pin heads? Many thought so, but I maintain that they were quite misguided in their criticisms of those engaged in the argument. All questions of mechanism come down to the existence and structures of intermediates, and to the number of barriers separating energy minima.

This argument has been recast many times in related reactions. Not all questions have been resolved, but those that have been have generally involved two kinds of evidence. Sometimes a direct observation of the ion in question was made, usually under **superacid** conditions in which a powerfully acidic, but nonnucleophilic medium is provided in which the ion is stable, at least at low temperature. In other reactions, an experiment like the following one provides evidence for symmetrical participation to produce a bridged intermediate.

The reactions in acetic acid of the monomethyl and dimethyl derivatives of the *p*-nitrobenzoate of *anti*-bicyclo[2.2.1]hepten-7-ol have been examined. The monomethyl compound reacts faster than the parent, as predicted by either mechanistic scheme. A bridged ion would be stabilized because one of its contributing resonance forms is tertiary. Another way of putting this is to note that the methyl group makes the double bond a stronger nucleophile (Fig. 21.44).

A more stable bridged ion —
one resonance form is a
tertiary carbocation

FIGURE 21.44 Formation of a symmetrical, bridged ion from the methyl-substituted molecule should be faster than from the parent.

PROBLEM 21.20 Draw resonance forms for the intermediate of Figure 21.44.

In the more traditional picture of the reaction, a tertiary carbocation can be formed and this would surely accelerate the rate of the reaction relative to the parent, which can produce only a less stable secondary carbocation (Fig. 21.45).

FIGURE 21.45 There should also be a rate increase in the formation of the unsymmetrical, open ion postulated by the classicists.

The observed rate increase of 13.3 is accommodated by both mechanisms.

Now consider the dimethyl compound. If a bridged intermediate is formed through a symmetrical displacement, both methyl groups exert their influence at the same time. If one methyl group induces a rate effect of 13.3, two should give an effect of 13.3 × 13.3 = 177 (Fig. 21.46).

FIGURE 21.46 Reaction of the dimethyl compound to give a bridged ion should be still faster than reaction of the monomethyl compound. The two methyl groups *simultaneously* stabilize the cation.

Relative rate = 1.0

Relative rate = 13.3

Predicted rate $(13.3)^2 = 177$
Both methyl groups stabilize the cation at the same time

On the other hand, if it is the formation of an open tertiary ion that is important, the second methyl increases the possibilities by a factor of 2, *but the two methyl groups never exert their influence at the same time.* The rate change should approximate a factor of 2 (Fig. 21.47).

FIGURE 21.47 In the open, unbridged ions, the two methyl groups can never stabilize the ion *at the same time.*

The fundamental difference between a mechanism involving a symmetrical bridged ion and the traditional mechanism, which postulates an open, less symmetrical ion, is that in the former case both methyl groups exert their stabilizing influence at the same time. In the pair of equilibrating open ions they cannot both act simultaneously.

The observed rate for the dimethyl compound is 148, much closer to that expected for the bridged intermediate than the open ion (Table 21.1).

TABLE 21.1 The Experimentally Determined Rates Favor a Symmetrical, Bridged Ion in Which Both Methyl Groups Act Simultaneously

Compound	Predicted by Bridged Symmetrical Ion	Predicted by Open Unsymmetrical Ion	Observed
	1	1	1
	n	n	13.3
	n^2 (177)	$2n$ (26.6)	148

SUMMARY

We have progressed from a discussion of obvious neighboring groups, heteroatoms with nonbonding electrons, to less obvious internal nucleophiles, the filled π orbitals of double bonds. In every case, a pair of internal electrons does an initial S_N2 displacement, which is followed by a second, external (intermolecular) S_N2 reaction. The result is overall retention of configuration. Now we turn to really obscure nucleophiles, filled σ orbitals.

21.4   SINGLE BONDS AS NEIGHBORING GROUPS

Surely, one of the weakest nucleophiles of all would be the electrons occupying the low-energy bonding σ orbitals. Yet there is persuasive evidence that even these tightly held electrons can assist in the ionization of a nearby leaving group. When the tosylate of 3-methyl-2-butanol is treated with acetic acid the result is formation of the rearranged tertiary acetate, not the "expected" secondary acetate (Fig. 21.48).

FIGURE 21.48 The reaction of the tosylate of 3-methyl-2-butanol in acetic acid involves a rearrangement.

This result is easily rationalized through a hydride shift to form the relatively stable tertiary carbocation. The interesting mechanistic question is whether the secondary carbocation is an intermediate, or whether the hydride moves *as the leaving group departs*. Thus, the timing of the steps becomes an issue. Is formation of the tertiary carbocation a two- or one-step process? This example recapitulates the classic mechanistic question: Is there a second intermediate—in this case the secondary carbocation—or is the reaction a concerted formation of the tertiary carbocation (Fig. 21.49)?

FIGURE 21.49 Two possible mechanisms for the acetolysis of 3-methyl-2-butyl tosylate.

PROBLEM 21.21 Draw Energy versus Reaction progress diagrams for the two possibilities of Figure 21.49. How many energy barriers separate starting material and the tertiary cation in the two mechanisms?

This question has been answered in this system by the observation of an **isotope effect**. The rates of reaction of the unlabeled tosylate and the deuterated tosylate of Figure 21.50 are measured. If the mechanism involves slow formation of the secondary carbocation, followed by a faster hydride shift to give the more stable tertiary carbocation, substitution of deuterium for hydrogen can have no great effect on the rate. Even though the carbon–deuterium bond is stronger than the carbon–hydrogen bond, *it is not broken in the rate-determining step of the reaction.* By contrast, if the hydride (or deuteride) migrates as the leaving group departs, the carbon–hydrogen or carbon–deuterium bond *is* broken in the rate-determining step and a substantial effect on the rate can be expected.

FIGURE 21.50 There can only be a substantial isotope effect if breaking the carbon–hydrogen or carbon–deuterium bond occurs in the rate-determining ionization step. The observation of a large isotope effect means that the carbon–hydrogen or carbon–deuterium bond must be breaking in the ionization step of the reaction.

The rate for the unlabeled compound was found to be more than twice as fast as that for the deuterated tosylate, which shows that the carbon–hydrogen or carbon–deuterium bond must be breaking in the rate-determining step of the reaction. This isotope effect ($k_H/k_D = 2.2$) reveals that the hydride must move *as the leaving group departs*. If the slow step in the reaction had involved a simple ionization to a secondary carbocation, *followed* by a hydride shift, only a very small effect of the label could have been observed (precedent suggests $k_H/k_D = 1.1 - 1.3$). The observation of a substantial isotope effect shows that the reaction mechanism must include a neighboring group effect, this time of neighboring hydride!

Now comes an even more subtle question. Given that even single bonds can participate in ionization reactions, are cyclic intermediates possible?

PROBLEM 21.22 There has been something of a controversy over the structure of the ethyl cation, $CH_3CH_2^+$. Contrast orbital pictures of the straightforward open ion with those of a cyclic version, which you might consider making through protonation at the center of the double bond of ethylene. Make analogies! Where have you seen a triangular array of molecular orbitals recently?

PROBLEM 21.23 Although it is hard to see why now, phenonium ions (p. 1192) were originally very controversial. Despite their evident similarity to the intermediates of aromatic substitution (Fig. 21.29), there was fierce opposition to the notion of their existence. Explain why a confusion between neighboring π participation and neighboring σ participation may have led to a misunderstanding.

The most famous argument of modern times in organic chemistry concerned the question of bridged intermediates produced by σ participation, and a fierce one it was—and still sometimes is. The reaction in question was the solvolysis (S_N1) reaction of the 2-norbornyl tosylates, the tosylates of *exo-* and *endo-*bicyclo[2.2.1]heptan-2-ol (Fig. 21.51).

exo-2-Norbornyl tosylate
Relative rate ~ 350

exo-2-Norbornyl acetate

endo-2-Norbornyl tosylate
Relative rate = 1

FIGURE 21.51 Both *endo-* and *exo-*2-norbornyl tosylate react in acetic acid to give *exo-*2-norbornyl acetate. The reaction of the exo isomer is faster.

The exo compound reacted at a faster rate than the endo compound, with the exact difference depending on reaction conditions. At first sight this rate increase would seem to provide evidence for bridging, but the opponents of participation quite rightly pointed out that the rate acceleration was very small (~ 350) and suggested that this effect could be the result of other differences between the exo and endo tosylates. This argument brings up a most important point. Whenever you are given a rate difference, you must ask yourself, Exactly what reaction does this rate difference refer to? In this case, the factor of 350 referred to the formation of acetate product. Is that what we are really interested in? Certainly not. When we speak of participation, we mean participation in *ionization*, ion formation, and that is not necessarily equivalent to product formation. In this case, the ion might be formed much faster than product. So we need a way to measure the rate acceleration in the ionization reaction itself, not the probably slower product-forming reaction (Fig. 21.52).

FIGURE 21.52 If the rate differential is determined by observing formation of acetate, we may be missing a fast, reversible ionization step.

Perhaps an Energy versus Reaction progress diagram makes the point more clearly. If we measure the rate of appearance of acetate (green), we miss the faster ionization reaction (red) (Fig. 21.53).

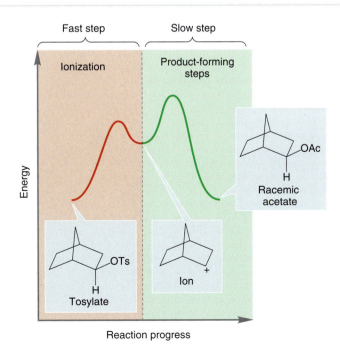

FIGURE 21.53 An Energy versus Reaction progress diagram for this reaction. A rapid ionization precedes a slower formation of acetate.

The stereochemistry of the reaction also points to participation. The product from the optically active exo tosylate is not that of a simple S_N2 reaction, which would be the inverted optically active endo acetate. Instead, it is the product of retention, the exo acetate. Moreover, optically active tosylate gives racemic acetate. We have been alerted to look for just this kind of stereochemical result as evidence of a neighboring group effect (Fig. 21.54)—it's clues 1 and 2.

FIGURE 21.54 A simple mechanism involving S_N2 displacement by acetic acid predicts optically active endo acetate, which is not what happens. The product is racemic exo acetate.

Proponents of participation asserted that these pieces of stereochemical data constituted powerful evidence for a bridged ion. First of all, it is only the exo tosylate that can form a bridged ion directly, and the exo compound did indeed react at a greater rate than the endo isomer (Fig. 21.55).

FIGURE 21.55 If the carbon–carbon σ bond assists in ionization, a cyclic, bridged ion can result. Opening of this ion by acetic acid, followed by deprotonation, would give the observed stereoisomer of the product.

PROBLEM 21.24 Draw resonance forms for the bridged ion of Figure 21.55.

It was also quite rightly pointed out that the bridged ion not only explained the observed retention of stereochemistry (exo tosylate gave exo acetate), but demanded a racemic product, as it itself was achiral. The bridged ion doesn't look symmetrical at first, but if you turn it over, it clearly is (Fig. 21.56).

Use this dotted carbon to orient yourself

rotate

Optically active

=

Achiral!

FIGURE 21.56 The observed formation of racemic product (as well as the racemization of recovered starting tosylate) can be explained by the achiral nature of the cyclic ion.

Addition of acetic acid must occur with equal probability at the two equivalent carbons of the bridged ion to give both enantiomers of *exo*-2-norbornyl acetate (Fig. 21.57).

deprotonate   add S_N2 add   deprotonate

S_N2

Enantiomers!

FIGURE 21.57 Addition of acetic acid to the bridged ion must give a pair of enantiomers in equal amounts—a racemic mixture.

This notion also gives a clue as to how to get a better measure of the rate of ionization than simply measuring the rate of acetate formation. *If the initial ionization is reversible*, then starting tosylate will be racemized as soon as the leaving group becomes symmetrically disposed with respect to the cation (Fig. 21.58).

FIGURE 21.58 If the leaving group (tosylate ion) re-adds to the cyclic ion to re-form starting material, racemization results.

So, we should be able to get a better measure of the rate of ionization by determining the rate of racemization of starting tosylate. This clever experiment has now been done and reveals a rate increase of about 1500 for the exo tosylate over the endo, a more impressive rate acceleration than the originally reported factor of 350.

Now we have the two traditional kinds of evidence for neighboring group participation: rate acceleration and retention of stereochemistry. It looks like a strong case can be made for participation. What could the opponents of bridging say?

First of all, they pointed out that the rate acceleration was small (they preferred to focus on the original factor of 350, but even 1500 isn't very large), and they were right. Moreover, one could invoke assistance by the σ bond in a simple way, without forming a cyclic ion. That would give some rate acceleration in the exo molecule, as there can be no assistance in the endo isomer (Fig. 21.59).*

FIGURE 21.59 Formation of two, classical, rapidly equilibrating ions can explain both the fast rate of the exo isomer and the racemization.

*The classicists actually argued most strongly that the rate of ionization of the endo compound was especially *slow* and that the exo compound was more or less normal.

But how would such a species explain the stereochemical results: racemization and retention of the exo configuration? Racemization was easy, one had only to permit the initially formed ion to equilibrate rapidly with the equivalent, mirror-image form (Fig. 21.59).

Always keep in mind the difference between the equilibrating open ions of Figure 21.59 and the bridged ion. The bridged ion is a single species. But be careful; the resonance forms of the bridged ion do look like the separate, equilibrating forms postulated by the classicists (Fig. 21.60).

Bridged ion

Resonance forms

Two equilibrating structures

FIGURE 21.60 The two mechanistic proposals compared.

WORKED PROBLEM 21.25 Explain the formation of *exo*-2-norbornyl acetate from the following reaction:

ANSWER This reaction is just another way of forming the 2-norbornyl cation. Symmetrical displacement of the leaving group by the double bond of the five-membered ring gives the ion.

(continued)

WORKED PROBLEM 21.25 *(continued)*

Opening by acetic acid in an intermolecular S_N2 reaction must produce the exo acetate.

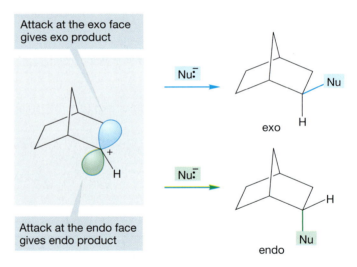

But how to explain retention: the exclusive formation of exo acetate? That was much harder for the classicists, and to my taste, never quite achieved. The two faces of the molecule are different, however, and one must admit that in principle, there might be sufficient difference between them to permit the exclusive formation of the exo acetate. Much chemistry of the norbornyl system reveals a real preference for exo reactions, and the question again became one of degree. How much exo addition could be tolerated by an open, unbridged ion (Fig. 21.61)?

Attack at the exo face gives exo product

Nu:⁻ → exo H

Attack at the endo face gives endo product

Nu:⁻ → endo Nu

FIGURE 21.61 The exo and endo faces of the open ion are different. Addition of a nucleophile could, in principle, occur from only the exo side.

In recent years, an increasing number of sophisticated spectroscopic techniques have been applied to this long-standing controversy. The 2-norbornyl cation has actually been detected by NMR spectroscopy at very low temperature in a highly polar, but nonnucleophilic solvent. Another spectroscopic technique, pioneered by Martin Saunders of Yale University (b. 1931) focuses on the general differences in the ^{13}C NMR spectra between equilibrating structures and bridged molecules. For example, the deuteriolabeled cyclohexenyl cation in Figure 21.62, certainly a resonance-stabilized species, shows essentially no difference between the ^{13}C NMR signals for the two carbons that share the positive charge.

FIGURE 21.62 In this resonance-stabilized carbocation, the two "red" carbons sharing the positive charge are not strongly differentiated in the ^{13}C NMR spectrum by deuterium substitution.

Saunders demonstrated that in equilibrating structures very great differences, on the order of 100 ppm, appear between such carbons. The Saunders technique gets right to the heart of the matter—to the center of the dispute over the 2-norbornyl cation—is it a single, resonance-stabilized bridged structure, or an equilibrating set of independent ions? Use of Saunders' technique in this case is complicated by the presence of hydride shifts in the 2-norbornyl cation, but nonetheless the weight of the evidence seems to favor strongly the bridged structure (Fig. 21.63).

FIGURE 21.63 The ^{13}C NMR spectrum shows only a tiny difference between the deuterated and undeuterated positively charged carbons. The 2-norbornyl cation is bridged.

Are all bicyclic cations bridged? For a while it certainly seemed so. At least they were drawn that way during what might be called the rococo phase of this theory. But there is no reason why they should *all* be bridged. Bridging is one form of stabilization, nothing more, and if other forms of stabilization are available, ions will be open, classical species. For example, the classicists were right when they argued that certain tertiary carbocations were unbridged. Tertiary carbocations are more stable than secondary ones, and so there is no need for the special stabilization of three-center, two-electron bridging. The proponents of bridging should not have been upset, but they certainly were (Fig. 21.64).*

FIGURE 21.64 But not all ions are bridged. This related, but tertiary carbocation is not bridged.

21.5   COATES' CATION

Robert M. Coates (b. 1938) of the University of Illinois has found an ion for which there is unequivocal evidence for a bridged structure. The molecule cannot be represented by a rapidly equilibrating set of open ions. The reaction of Coates' molecule in acetic acid to give an acetate is shown in Figure 21.65.

FIGURE 21.65 The solvolysis of this sulfonate in acetic acid gives the related acetate.

*The argument over the nature of the 2-norbornyl cation and related molecules was/is remarkable not only for its longevity, but for its intensity as well. I think the argument became very nasty and both sides definitely lost their cool. Despite the common notion of scientists as Dispassionate Seekers After Truth, we are all-too-prone to the human vices, in this case overdefending our ideas (our intellectual children). In one corner were the proponents of bridging, chief among them Saul Winstein and Donald Cram. In the other was H. C. Brown, who passionately maintained that all could be explained through conventional, unbridged intermediates. This argument was no battle of lightweights! Brown and Cram have since won the Nobel prize (for other work), and Winstein should have been so honored—and would have been, had he lived long enough. The dispute was raucous and featured many a shouting match at meetings and seminars. It is not easy to be insulting in the body of a scientific paper, because the referees (two or more readers who judge whether the work is publishable) will not stand for it. However, it is much easier to be nasty in the footnotes! There are some really juicy footnotes in the literature on this subject. In my view, both sides went overboard. Although there is much virtue in avoiding new, unwarranted explanations, I think the classicists were much too resistant to change. There is nothing inherently crazy about a bridged ion! Three-center, two-electron bonding is a fine way to hold molecules together and is quite common in inorganic chemistry, and even parts of organic chemistry. On the other hand, the bridgers overdefended their ground. Just because some ions are bridged doesn't mean that all are!

At low temperature in superacid, ionization with participation of the cyclopropane σ bond gives a bridged cation that must show three signals in its ^{13}C NMR spectrum (Fig. 21.66).

FIGURE 21.66 Participation of the carbon–carbon σ bond to give a symmetrical bridged ion. Notice that three different kinds of carbon exist. The ^{13}C NMR spectrum will consist of three lines.

In this molecule, an equilibrating set of ions must give quite a different ^{13}C NMR spectrum.

PROBLEM 21.26 How many signals would a rapidly equilibrating "Coates' cation" give in the ^{13}C NMR spectrum?

Equilibrating open ions

Happily, the sulfonate ionizes to give a stable ion at low temperature in superacid and the case is settled, as the spectrum consists of exactly the three signals required for a bridged structure: no more and no less. Coates' cation is bridged, there is no doubt about it.

21.6 SUMMARY

New Concepts

One concept dominates this chapter. The chemistry of many organic molecules containing leaving groups is strongly influenced by the presence of an internal nucleophile, a neighboring group. Internal nucleophiles can increase the rate of ionization of such compounds by providing anchimeric assistance in the ionization step. Moreover, the structures of the products derived from the ion produced through intramolecular displacement can be subtly different from those expected of simple intermolecular displacement. The three clues to the operation of a neighboring group effect are (1) an unusual stereochemical result, generally retention of configuration

where inversion might have been expected; (2) a rearrangement, often of a label, which is usually the result of formation of a cyclic ion through *intra*molecular displacement of the leaving group; and (3) an unexpectedly fast rate.

The cyclic ions formed by neighboring group displacement are of two kinds. There are normal (often called "classical") species in which all bonds are two-electron bonds, and more complex nonclassical or bridged species in which three-center, two-electron bonding is the rule. Figure 21.43 illustrates the difference.

Key Terms

anchimeric assistance (p. 1184)

episulfonium ion (p. 1183)

isotope effect (p. 1207)

α-lactone (p. 1179)

neighboring group effect (p. 1175)

nor (p. 1196)

norbornyl system (p. 1196)

phenonium ion (p. 1194)

superacid (p. 1203)

three-center, two-electron bonding (p. 1198)

Reactions, Mechanisms, and Tools

This chapter features intramolecular displacement using a variety of internal nucleophiles. Some, such as the heteroatoms bearing lone pairs of electrons, are easy to understand, and really just recapitulate intermolecular chemistry (Fig. 21.2).

Other nucleophiles, such as the electrons in π or σ bonds, are less obvious extensions of old chemistry. Here, the nucleophiles involved are very weak and there often is not a clear analogy to chemistry you already know. One case in which there is a good connection is formation of phenonium ions through the action of a benzene ring as a neighboring group.

The analogy is to the intermolecular reaction of benzenes with electrophiles in the aromatic substitution reaction (Figs. 21.28 and 21.29).

Displacement by carbon–carbon π or σ bonds can lead to bridged ions about which there has been much controversy. It now appears that such intermediates are involved in some ionizations, and that the three-center, two-electron bonding required is an energetically favorable situation, at least for highly electron-deficient systems such as carbocations (Fig. 21.60).

Common Errors

The problem with this material is mostly one of recognition. Once it is clear that it is a neighboring group problem, essentially every difficulty is resolved by a simple pair of S_N2 reactions: the first intramolecular, the second intermolecular. Search for the internal nucleophile, and use it to displace the leaving group.

There can be no denying that many of the structures encountered in neighboring group problems are complex and sometimes hard to untangle. A useful technique is to draw the result of the arrow formalism without moving any atoms, then

to relax the structure, probably containing long or even bent bonds, to a more realistic picture. Trying to do both steps at once is dangerous.

A minor problem appears in recognizing that in some bridged ions, the web of dashes representing three-center, two-electron bonding can operate as a leaving group. It is easier to see a pair of electrons in a normal carbon–leaving group σ bond (C–L) than it is to recognize the dashes representing partial bonds as a leaving group.

21.7 ADDITIONAL PROBLEMS

PROBLEM 21.27   Here comes a problem for rabbits. Provide a mechanism for this simple change.

$$BrCH_2CH_2CH_2CH_2NH_2 \quad \xrightarrow{\substack{1. \ H_2O, \ (CH_3)_2CHOH \\ 2. \ KOH/H_2O \\ (neutralize)}}$$

PROBLEM 21.28   Similarly, tetrahydrofuran is the product of this next reaction. Provide a mechanism and explain why this process is much faster than the similar reaction of 3-chloro-1-propanol, a molecule that gives the corresponding diol as product.

$$ClCH_2CH_2CH_2CH_2OH \quad \xrightarrow{H_2O}$$

$$ClCH_2CH_2CH_2OH \quad \xrightarrow{H_2O} \quad HOCH_2CH_2CH_2OH$$

PROBLEM 21.29   Now for something a little more interesting. Explain why the regiochemical results of the following two reactions are so different.

Et = CH₃CH₂
Bz = PhCH₂

PROBLEM 21.30   Explain why compound **1** reacts much faster than **2** (140,000 times) and **3** (190,000 times) in acetic acid. Does this information about rates let you decide whether the reaction proceeds through a bridged ion?

$Bs = SO_2 \!-\!\!\langle \ \rangle\!\!-\! Br$

PROBLEM 21.31   Compound **1** reacts 10^5–10^6 times faster than its cis isomer **2**. The product is **3**. Explain.

PROBLEM 21.32   When Z = OCH₃, about 93% of the products of solvolysis of **1** can be attributed to neighboring phenyl group participation. When Z = NO₂, essentially none of the products come from neighboring participation by phenyl. Explain.

PROBLEM 21.33   In Figure 21.32, we neglected to tell you about the product of reaction of *syn*-7-norbornenyl tosylate (**1**) with acetic acid. The product is compound **2**. Provide an arrow formalism mechanism and an explanation for the rate acceleration of 10^4 relative to **3**.

1
Rate = 10^4

2
(both stereoisomers)

3
Rate = 1

PROBLEM 21.34   Provide a mechanism for the following reaction of **1**. Be sure to explain the observed stereochemistry and the fact that the corresponding trans compound (**2**) does not react in this way.

1

2

PROBLEM 21.35   Reaction of **1** with bromine leads to a neutral product of the formula $C_{18}H_{17}BrO_2$. Propose a structure for the product and a mechanism for its formation.

1

PROBLEM 21.36   Provide an arrow formalism for the following reaction:

PROBLEM 21.37   Propose a mechanism for the following change. Be sure to explain the observed stereochemistry.

PROBLEM 21.38   What do the relative rate data suggest about the nature of the carbocationic intermediates in the following solvolysis reactions? *Hint*: See Figures 21.46, 21.47, and Table 21.1.

$HOAc =$

$Ns = SO_2$ —⟨ ⟩— NO_2

R	R	Relative Rate
H	H	1.0
CH_3	H	7.0
CH_3	CH_3	38.5

PROBLEM 21.39   Provide arrow formalism mechanisms for the following reactions. *Hint*: See Ch. 19, p. 1094.

(a)

(continued)

PROBLEM 21.39 *(continued)*

(b)

PROBLEM 21.40 Compound **1** reacts about 15 times faster than neopentyl brosylate (**2**). The product is isobutyraldehyde. Provide a mechanism and explain the rate difference.

1
Rate = 15

2
Rate = 1

PROBLEM 21.41 Compound **1** is the only product of the reaction of compound **2** under the conditions shown. Provide an arrow formalism mechanism, and deduce the stereochemistry of **2** from the stereochemistry of **1** shown.

2   **1**

PROBLEM 21.42 Treatment of the following two compounds with HCl leads to the same chloride. Explain, mechanistically. *Hint*: Watch the sulfur atom. What can sulfur do after an initial protonation of the oxygen atom?

PROBLEM 21.43 Heating β-hydroxyethylamide (**1**) with thionyl chloride affords β-chloroethylamide (**2**). In turn, amide **2** yields oxazoline **3** upon treatment with NaOH.

(a) Write arrow formalism mechanisms for the formations of **2** and **3**. Things are not quite as simple as they first appear. For example, if **1** is treated with thionyl chloride below 25 °C, compound **4** is obtained instead of **2**. Compound **4** has the same elemental composition as **2**, but unlike **2**, is soluble in water. Upon heating, **4** gives **2**, and compound **4** yields **3** upon treatment with aqueous sodium carbonate.

(b) Propose a structure for **4** and show how **4** is formed and converted into **2** and **3**.

PROBLEM 21.44   Outline the mechanistic steps for the following transformation:

1. NaOH/H_2O
2. Br_2
3. HCl (neutralize)

Now explain this more complicated reaction:

1. NaOH/H_2O
2. Br_2
3. HCl (neutralize)

PROBLEM 21.45   The 50 MHz ^{13}C NMR spectrum of the 2-norbornyl cation in $SbF_5/SO_2ClF/SO_2F_2$ solution exhibits an interesting temperature dependence. At −159 °C the spectrum consists of five lines at δ 20.4 [C(5)], 21.2 [C(6)],

36.3 [C(3) and C(7)], 37.7 [C(4)], and 124.5 [C(1) and C(2)]. However, at −80 °C the spectrum consists of three lines at δ 30.8 [C(3), C(5), and C(7)], 37.7 [C(4)], and 91.7 [C(1), C(2), and C(6)] ppm. Explain. *Hint*: Start with the low-temperature spectrum.

PROBLEM 21.46   2-Amino alcohols can be converted into aziridines if the hydroxyl group is first changed into a better leaving group. A related approach to aziridines might involve treating a 2-amino alcohol with thionyl chloride in the presence of base.

However, when 2-amino alcohol **1** was treated with thionyl chloride in the presence of triethylamine, isomeric compounds **3** and **4** ($C_{12}H_{17}NO_2S$) were obtained rather than the anticipated **2**. Propose structures for isomers **3** and **4** and explain their formation.

Special Topic: Carbohydrates

22

There exists no better thing in the world than beer to blur class and social differences and to make men equal.

—Emil Fischer*

22.1 PREVIEW

In the last chapter, we encountered neighboring group effects, the influence of one functional group upon the chemistry of another. In this chapter, we find molecules that bristle with functionality, and for which intramolecular neighboring group effects play most important roles, influencing both structure and reactivity. These are the **sugars**, also called **carbohydrates** or **saccharides**.

Sugars serve as the storage depots for the energy harvested from the sun through photosynthesis by green plants. Moreover, sugars are present in nucleic acids (Chapter 23) and are thus vital in controlling protein synthesis and in transmission of genetic information. More than 100 million tons of sugar are produced yearly, and each American consumes about 64 lb of sugar per year (down from over 100 lb just a few years ago). Sugars are vital to human survival, and it is somewhat surprising that their study took so long to emerge as a glamorous area of research. Perhaps the development of sugar chemistry was delayed by the simple technical difficulties of working with goopy syrups rather than nice, well-behaved solids. A full scale renaissance is underway now, however, and sugar chemistry is very much a hot topic.[†]

---

*Emil Fischer (1852–1919) was perhaps the greatest of all organic chemists. He won the Nobel prize of 1902 and made extraordinary contributions to many fields, including carbohydrates and proteins.

[†]Although commercial sugars are mostly harvested from plants (sugar beets, or sugar cane), there are some marvelously exotic laboratory syntheses. My favorite is that of Philip Shevlin (b. 1939) of Auburn University, who fired bare carbon atoms into water and obtained carbohydrates. Shevlin has suggested that this reaction may well have occurred on the icy surfaces of comets, transforming them into giant orbiting sugar cones.

The name carbohydrate comes from the formulas of these compounds, which can be factored into $C_x(H_2O)_y$. These molecules appear to be hydrates of carbon. Of course they are not literal hydrates, in which water molecules cluster around a carbon—but the classic demonstration experiment in which a sugar cube is treated with sulfuric acid does yield quite a spectacular sight as exothermic dehydration reactions produce steam and a cone of carbon.*

The sugar chemistry we will see involves only a few new reactions, but the polyfunctionality of these compounds can make for some complications. To be a successful sugar chemist you will have to be able to apply old reactions in new settings, and to deal with a good deal of stereochemical complexity as well. It's not a bad idea to review the concepts in Chapter 4 quickly before you start on this chapter. That's a hard thing to do—time is probably pressing, and it's always more interesting to move on to new stuff, but at least reread the summary sections of Chapter 4 and be prepared to look back if things get confusing. Dust off your models. There will be moments when they will be essential.

In this chapter, we will also meet Emil Fischer (1852–1919), surely one of the greatest of organic chemists, and a person of heroic theoretical and experimental achievements. Sugars can be hideously frustrating molecules to work with experimentally. They form syrups and are notoriously difficult to crystallize. These properties were especially important in Emil Fischer's time, as there were no fancy chromatographs and spectrometers in his basement, and separation was routinely achieved through crystallization. One of Fischer's experimental triumphs was the discovery that phenylhydrazone derivatives of many sugars were rather easily crystallized. That may seem a simple point now, but it certainly was not in 1875! One of the great logical triumphs of organic chemistry was the unraveling of the structures of the aldohexoses (we'll see just what this means later, but glucose, a vital source of human energy, is one of the aldohexoses), and Emil Fischer was at the heart of this work for which he won the Nobel prize for chemistry in 1902.

ESSENTIAL SKILLS

1. Sugar chemistry uses an important structural convention, the Fischer projection. To understand this chapter, you have to be able to move easily from Fischer projections to conventional structures. The directions are right there in Section 22.2.

2. Understanding the logic of structure determination in sugar chemistry is an essential skill. For example, Emil Fischer's determination of the structure of glucose is an exercise in logic. I think you should hone your skills and knowledge of the reactions of sugars so that you are able to follow that logic.

IMPORTANT DETAILS

1. The Fischer proof of the structure of glucose is not really a "detail," as it is one of the supreme intellectual achievements of human history! It is something one should know, at least in general.

2. Understanding the differences between the anomeric hydroxyl group in sugars, the one attached to C(1) in the cyclic forms, and all the others is critical to "getting" sugar chemistry.

*DO NOT try this experiment unsupervised! The reaction is very exothermic and sometimes violent. Sulfuric acid can be thrown about, and sulfuric acid in the eyes is no picnic!

22.2 NOMENCLATURE AND STRUCTURE

Sugars could all be named systematically, but their history and complexity has led to the retention of common names throughout the discipline. There is also a special stereochemical convention, named after Emil Fischer, that is used to simplify representations of these molecules that usually contain several stereogenic carbons.

The simplest sugar is the molecule called **glyceraldehyde**. It contains a single stereogenic carbon and thus can exist in two enantiomeric forms (Fig. 22.1).

It is fairly easy (I hope) to determine which of these molecules is (*R*) and which is (*S*), so no new convention is really necessary here (Fig. 22.1). Glyceraldehyde is an **aldotriose**. The "tri" tells you it is a three-carbon molecule, and the "aldo" tells you that the molecule contains an aldehyde. The "ose" means that the pattern of an aldehyde (CHO) at one end attached to a primary alcohol (CH_2OH) at the other end through a series of H—C—OH groups (in this case, just one) is present (Fig. 22.2).

(R)-Glyceraldehyde   **(S)-Glyceraldehyde**

FIGURE 22.1 The two enantiomers of the simplest sugar, glyceraldehyde.

WORKED PROBLEM 22.1 Generalize from the definition of a triose to generate the structures of an aldotetrose and an aldopentose.

ANSWER "Aldo" means that an aldehyde is present. A tetrose will have four carbons and a pentose will have five. All sugars (all ose's) have both a carbonyl group, here an aldehyde, and a primary alcohol (CH_2OH) group attached through a series of H—C—OH groups.

Three carbons = triose

Aldehyde group = aldotriose

Glyceraldehyde

FIGURE 22.2 Glyceraldehyde is an aldotriose because it has a three-carbon backbone and an aldehyde group.

CHO	CHO	CHO
H—C—OH	H—C—OH	H—C—OH
CH₂OH	H—C—OH	H—C—OH
	CH₂OH	H—C—OH
		CH₂OH

If glyceraldehyde is a triose, …   …then this compound is a tetrose…   …and this one is a pentose

The stereochemical convention called a **Fischer projection** translates the drawings of Figure 22.1 and Problem 22.1 into three-dimensional pictures. Fischer projections do not attempt to represent energy minimum conformations. Instead, they are designed to simplify the specifying of single stereoisomers. An analogy would be the representation of substituted ethanes as eclipsed "sawhorse" conformations (Chapter 2, p. 63). Drawings of these energy maxima are often used because they are easier to visualize than the real energy minima—the staggered conformations. Fischer projections are also eclipsed molecules, and are used because they are simple to draw and to read. We will see soon how to translate the Fischer projections into more accurate representations of the real molecules. In a Fischer projection, all horizontal lines are taken as coming forward, and vertical lines as retreating. The carbons in the frame are not written in. Aldo sugars are almost always written with the aldehyde at the top and the primary alcohol at the bottom (Fig. 22.3).

CONVENTION ALERT

FIGURE 22.3 Aldo sugars are written with the aldehyde group at the top and the primary alcohol at the bottom. In this scheme, called a Fischer projection, horizontal bonds are taken as coming toward the viewer and vertical bonds as retreating. If the OH adjacent to the primary alcohol at the bottom is on the right, the sugar is a member of the D series. If it is on the left, it is in the L series.

D-Glyceraldehyde　　=　　(R)-Glyceraldehyde

L-Glyceraldehyde　　=　　(S)-Glyceraldehyde

If the OH group attached to the carbon adjacent to the bottom CH_2OH group (that is, the stereogenic carbon most remote from the aldehyde group) is on the right, the molecule is designated as a "D" sugar; if this OH is on the left, the sugar is a member of the "L" series. *Notice that to get the designation right, the sugar must be drawn with the aldehyde at the top and the primary alcohol at the bottom.* There is absolutely no correlation between D and L and the sign of optical rotation (the direction of the rotation of the plane of plane-polarized light). A D sugar can rotate the plane of plane-polarized light either clockwise or counterclockwise. Clockwise rotation is indicated by prefacing the name of the sugar with (+), and counterclockwise rotation with (−). So we might write D-(+)-glyceraldehyde or D-(−)-erythrose. Both sugars are members of the D series, but one is dextrorotatory (rotates the plane of plane-polarized light in a clockwise sense) and the other is levorotatory (rotates the plane of plane-polarized light in a counterclockwise sense). Of course, if the sense of rotation is in one direction for the D stereoisomer, it is the opposite for the enantiomeric L molecule (Fig. 22.4).

FIGURE 22.4 There is no relation between D and L and the sign of optical rotation, (+) and (−).

D-(+)-Glycer-aldehyde

L-(−)-Glycer-aldehyde

Mirror

D-(−)-Erythrose

L-(+)-Erythrose

Mirror

PROBLEM 22.2 Write three-dimensional representations for the following molecules in their energy-minimum forms. They are shown in Fischer projection in the figure below. Assign each stereogenic carbon of the molecules in the following figure as either (R) or (S).

CHO
H——OH
H——OH
CH₂OH

CHO
HO——H
H——OH
CH₂OH

PROBLEM 22.3 Write Fischer projections for the following molecules:

(a)

H OH
OHC—C—C—CH₂OH
H OH H OH

(b)

H OH
OHC—C—C—CH₂OH
H OH HO H

(c)

HO H
OHC—C—C—CH₂OH
H OH H OH

(d)

H OH
OHC—C—C—CH₂OH
HO H H OH

There are also **keto sugars**, or **ketoses**, which contain no aldehyde group, but rather a ketone. Figure 22.5 shows 1,3-dihydroxyacetone, a simple ketotriose.

Four-carbon sugars are called **tetroses**. There are four **aldotetroses**, the enantiomeric pairs D- and L-erythrose and D- and L-threose. Remember, for a sugar to be D, the OH next to the bottom CH_2OH must be on the right in a Fischer projection (Fig. 22.6).

FIGURE 22.5
1,3-Dihydroxyacetone, a ketotriose.

CHO
H——OH
H——OH
CH₂OH
D-Erythrose

CHO
HO——H
HO——H
CH₂OH
L-Erythrose

CHO
HO——H
H——OH
CH₂OH
D-Threose

CHO
H——OH
HO——H
CH₂OH
L-Threose

FIGURE 22.6 The four aldotetroses.

Aldopentoses contain three stereogenic carbons, and so there must be a total of eight stereoisomers ($2^3 = 8$). Figure 22.7 shows the four members of the D series of aldopentoses and their names.

CHO
H——OH
H——OH
H——OH
CH₂OH
D-Ribose

CHO
HO——H
H——OH
H——OH
CH₂OH
D-Arabinose

CHO
H——OH
HO——H
H——OH
CH₂OH
D-Xylose

CHO
HO——H
HO——H
H——OH
CH₂OH
D-Lyxose

FIGURE 22.7 The four D-aldopentoses.

In the figures, the internal carbons in the sugar chains are not written in. Watch out for this convention.

Aldohexoses contain four stereogenic carbons, and so there must be a total of $2^4 = 16$ stereoisomers, 8 in the D series and 8 enantiomeric molecules in the L series (Fig. 22.8).

FIGURE 22.8 The eight D-aldohexoses.

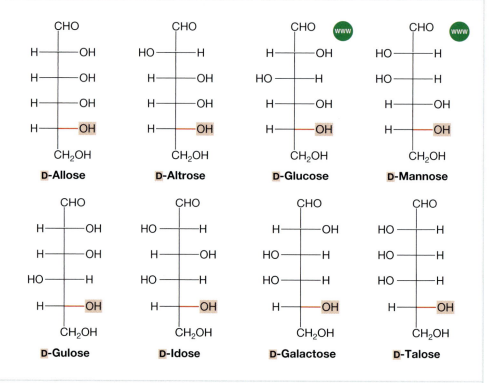

Figures 22.7 and 22.8 show only aldohexoses, but there are also many keto sugars among the **pentoses** and **hexoses**. Figure 22.9 shows D-fructose, a common ketohexose.

FIGURE 22.9 D-Fructose, a common ketohexose.

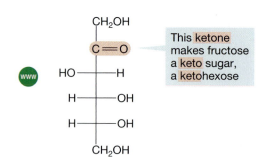

PROBLEM 22.4 Draw Fischer projections for the L-aldopentoses.

PROBLEM 22.5 How many D-aldoheptoses are possible?

PROBLEM 22.6 Treatment of D-glucose with sodium borohydride (NaBH₄) gives D-glucitol (sorbitol), $C_6H_{14}O_6$. Show the structure of D-glucitol and write a brief mechanism for this simple reaction.

One would expect these aldoses to combine the chemistries of alcohols and aldehydes, and that expectation would be correct. However, the proximity of these functional groups leads to complicated chemistry, and here is an important example. Although reduction with sodium borohydride, followed by hydrolysis, induces the expected change (see Problem 22.6), as the aldehyde is reduced to an alcohol, modern spectroscopic methods such as NMR and IR do *not* reveal the presence of substantial amounts of an aldehyde in the starting hexose (Fig. 22.10). Our first task is to understand this result, which is common in pentose and hexose chemistry. There are many typical reactions of carbonyl groups that take place in the apparent absence of an appropriate NMR or IR signal for the H—C=O functional group.

D-Glucose → (NaBH₄, H₂O) → **D-Glucitol** (~ 65%)

But
1. No large peak in IR for C=O
2. No peak in ¹H NMR for H—C=O

FIGURE 22.10 Although reduction with sodium borohydride, followed by hydrolysis, proceeds normally to give an alcohol, neither NMR nor IR spectroscopy reveals large amounts of an aldehyde in the starting material.

You know already (Chapter 16, p. 832) that carbonyl groups react with all manner of nucleophiles. Hydration and hemiacetal formation are typical examples (Fig. 22.11). A hemiacetal could certainly be formed in an intramolecular reaction in which a hydroxyl group at one point in a molecule reacted with a carbonyl group in another. Indeed, 4- and 5-hydroxy aldehydes exist mostly in the cyclic hemiacetal forms (Chapter 21, p. 1178). Figure 22.11 shows a typical reaction of this kind and makes the analogy to both hydration and intermolecular hemiacetal formation.

Hydration

Hemiacetal formation

Intramolecular hemiacetal formation

5-Hydroxyhexanal
(6%)

**2-Hydroxy-6-methyl-
tetrahydropyran**
A cyclic hemiacetal
(94%)

FIGURE 22.11 Intramolecular hemiacetal formation is analogous to hydration and intermolecular hemiacetal formation. Five- and six-membered ring hemiacetals are easily made and are often more stable than their open forms.

The key to the solution of the "missing carbonyl" problem is that for hexoses, the hemiacetal form is more stable than the open-chain isomer. We can even venture a reasonable guess as to which hydroxyl group participates in the ring-forming reaction. For the cyclic hemiacetal to be more stable than the acyclic hydroxy aldehyde, the ring formed surely must not introduce a lot of strain. That eliminates three of the five OH groups and points to either the OH at C(4), which leads to a five-membered ring, or that at C(5), which would produce a six-membered ring. These cyclic forms are named as **furanoses** (five-membered ring) or **pyranoses** (six-membered ring) by analogy to the five- and six-membered cyclic ethers: tetrahydrofuran and tetrahydropyran (Fig. 22.12).

FIGURE 22.12 In an aldohexose, intramolecular hemiacetal formation results in a furanose (five-membered ring) or pyranose (six-membered ring).

Cyclic forms are indicated in the name by combining the simple name of the sugar with "furanose" or "pyranose" to indicate the size of the ring. Figure 22.13 shows D-glucofuranose and D-glucopyranose in Fischer projection.

FIGURE 22.13 Fischer projections for D-glucofuranose and D-glucopyranose.

PROBLEM 22.7 Draw Fischer projections for D-mannopyranose, L-gulopyranose, and D-galactofuranose (see Fig. 22.8).

Both cyclic molecules, the five-membered furanose and the six-membered pyranose, are formed from most aldohexoses, but it is the six-membered ring that usually predominates. We'll prove this point a little later in this chapter, but for now accept that your guess was a good one, and let's go on to deal with the question of why typical carbonyl reactions appear for this cyclic hemiacetal. There is a simple answer to this puzzling question. The molecule isn't 100% in the cyclic form, it just exists predominantly in that form. It is in equilibrium with a small amount of the acyclic isomer, which *does* contain a carbonyl group and *does* react with reagents such as sodium borohydride. As the small amount of the reactive partner in the equilibrium is used up, more is generated (Fig. 22.14). Sugars that contain a hemiacetal, and are therefore in equilibrium with some open form, are called **reducing sugars**.

FIGURE 22.14 If two compounds are in equilibrium, irreversible reaction of the minor partner can result in complete conversion into a product. As long as the equilibrium exists, the small amount of the reactive molecule will be replenished as it is used up.

As long as an equilibrium exists, more open form will be regenerated as the reduction takes place

Open form
(< 0.05%)

Pyranose form
(> 99.95%)

TABLE 22.1 Pyranose and Furanose Forms of Aldohexoses at 25 °C

Aldohexose	Pyranose Form	Furanose Form
Allose	92	8
Altrose	70	30
Glucose	~100	<1
Mannose	~100	<1
Gulose	97	3
Idose	75	25
Galactose	93	7
Talose	69	31

Table 22.1 lists some common carbohydrates and gives the relative amounts found in pyranose and furanose forms. There is very little of the open form present at equilibrium.

However, because there is always a small amount of the reducible open form present, all of the material can eventually be reduced by sodium borohydride. If we return to our spectrometers, crank up the sensitivity, and look very hard for the aldehyde group present in the open-chain form, there it is, in the tiny amount required by our explanation.

Let's turn our attention to the structures of these cyclic, six-membered ring hemiacetals (pyranoses). The first thing to notice is that the intramolecular cyclization reaction creates a new stereogenic carbon that can be either (R) or (S). These two stereoisomers, called **anomers**, result from addition of the nucleophilic OH to the Lewis acid carbonyl group from the two possible sides. Anomers of aldoses differ only in the stereochemistry at C(1). If the new OH is on the same side in Fischer projection as the oxygen involved in the hemiacetal link, we have the α-anomer. If it is on the opposite side, it is the β-anomer. The stereoisomers are shown in Fischer projection in Figure 22.15, which also shows the schematic "squiggly bond" used to indicate that the new OH group at the anomeric carbon can be on the right or left in Fischer projection.

FIGURE 22.15 Intramolecular hemiacetal formation results in two C(1) stereoisomers called anomers.

PROBLEM 22.8 Identify the new stereogenic carbon in the α- and β-anomers of Figure 22.15 as either (R) or (S).

The last few figures have shown molecules with some very peculiar bonds in them. Notice in particular the "around the corner" bonds generally used to show attachment to the ring oxygen. The squiggly bond is used to show that both anomers can be present (Fig. 22.16).

CONVENTION ALERT

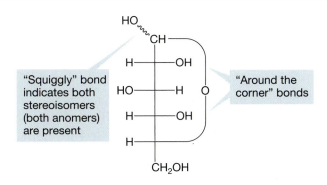

FIGURE 22.16 In these Fischer projections, we have drawn "around the corner" bonds.

Both α- and β-pyranose and furanose forms exist. Table 22.2 shows the relative amounts for the aldohexoses.

TABLE 22.2   α- and β-Pyranose and Furanose Forms of the Aldohexoses at 25 °C

Aldohexose	α-Pyranose Form	β-Pyranose Form	α-Furanose Form	β-Furanose Form
Allose	16	76	3	5
Altrose	27	43	17	13
Glucose	36	64	<1	<1
Mannose	66	34	<1	<1
Gulose	16	81	<1	3
Idose	39	36	11	14
Galactose	29	64	3	4
Talose	37	32	17	14

We know that bonds do not go around corners, so let's try to turn these schematics into something more sensible. There are many ways to do this task, and any way you devise will do as well as the following method. Just be certain that your method works! We will continue to use D-glucose as an example and work through a method of arriving at a three-dimensional structure. First, redraw the molecule so that both bonds to the ring oxygen are vertical; so that the ring is retreating from us into the page at both the top and bottom of the molecule. This transformation requires a simple rotation about a carbon–carbon bond. In the α-anomer, the OH at C(1) is on the side opposite the CH₂OH group, and in the β-anomer it is on the same side (Fig. 22.17).

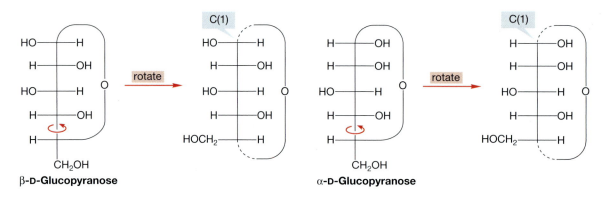

FIGURE 22.17 The first step in creating a three-dimensional drawing is rotation around the indicated carbon–carbon bond. This motion generates a new Fischer projection.

Now tip the molecule over in clockwise fashion to generate a flat six-membered ring. This kind of planar representation is called a **Haworth form**, after its inventor, Walter N. Haworth (1883–1950). Follow the CH₂OH group, and you can see that it is now pointing up. The remainder of the substituent OH groups can now be easily put in by using the CH₂OH as a guide. Figure 22.18 uses the β-anomer as an example.

A flat, Haworth, form

β-D-Glucopyranose

FIGURE 22.18 Next, tip the molecule over in clockwise fashion to produce a flat Haworth form.

PROBLEM 22.9 Draw the flat, Haworth form of the α-anomer.

Six-membered rings exist in energy minimum chair forms, and are not flat. The final step in this exercise is to allow the flat Haworth form to relax to a chair. There are two chairs possible for any six-membered ring. They are interconverted by the rotations around carbon–carbon bonds we called a ring flip (Chapter 5, p. 203). Be sure to check both forms to see which is lower in energy (Fig. 22.19). Sometimes this evaluation will be difficult, but for β-D-glucopyranose, it is easy. The form with all groups equatorial will be far more stable than that with all groups axial (Chapter 5, p. 209).

β-D-Glucopyranose

FIGURE 22.19 Now let the flat, Haworth form relax to a chair. Don't forget that there are always two possible chair forms.

WORKED PROBLEM 22.10 Follow this same procedure for the α-anomer.

ANSWER In the α-anomer, the hydroxyl group at C(1) will be axial, not equatorial as in the β-anomer. Otherwise all the substituent groups remain equatorial.

(more stable)   (less stable)

Of course, there are two chair forms, but the one with four equatorial groups will be the more stable one.

PROBLEM 22.11 Transform the Fischer projection into a three-dimensional picture of D-mannopyranose.

22.3   REACTIONS OF SUGARS

22.3a Mutarotation of Sugars. The following behavior is typical for aldohexoses and aldopentoses. Pure, crystallized α-D-glucopyranose has a specific rotation of +112°. The pure β-anomer has a specific rotation of +18.7°. However, an aqueous solution of *either* anomer steadily changes rotation until the value of +52.7° is reached. This phenomenon is called **mutarotation** and is common for sugars. Our task is to explain it. The answer comes from the realization that both α- and β-D-glucopyranoses are hemiacetals and exist in solution in equilibrium with a small amount of the open-chain aldohexose. Once the acyclic form is produced, it can re-form the hemiacetal to make *both* the α- and β-pyranose. Over time, an equilibrium mixture is formed and it is that equilibrated pair of anomers that gives the rotation of 52.7° (Fig. 22.20).

FIGURE 22.20 The α- and β-anomers can equilibrate through the small amount of the open form present at equilibrium.

β-Anomer
(64%)

Open chain
(trace)

α-Anomer
(36%)

PROBLEM 22.12 Write a mechanism for the acid-catalyzed mutarotation of D-glucopyranose.

22.3b Isomerization of Sugars in Base. In base, sugars rapidly equilibrate with other sugars, which is a more profound change than the equilibration of anomers we saw in the preceding section. Again, let's use D-glucose as an example. As Figure 22.21 shows, when treated with base, D-glucose equilibrates with another D-aldohexose, D-mannose, and a keto sugar, D-fructose (Fig. 22.21).

FIGURE 22.21 In base, D-glucose equilibrates with D-mannose and D-fructose, a keto sugar.

D-Glucopyranose
(63.4%)

$Ca(OH)_2$
H_2O
10 days

D-Mannopyranose
(2.4%)

$Ca(OH)_2$
H_2O
10 days

D-Fructose
(30.9%)

This process goes by the absolutely delightful name of the **Lobry de Bruijn–Alberda van Ekenstein reaction.***

PROBLEM 22.13 Although D-fructose is shown in Figure 22.21 in the open chain form, it exists mostly (~ 67%) in a pyranose form and partly (~ 31%) in a furanose form in which the hydroxyl group at C(5) is tied up in hemiacetal formation with the ketone. Draw Fischer projections for the pyranose and furanose forms of D-fructose.

The mechanism of this reaction is simpler than its provenance. As we have pointed out several times now, the predominant cyclic forms of these sugars are in equilibrium with small amounts of the open-chain isomers. In base, the α-hydrogen in the open, aldo form can be removed to produce a resonance-stabilized enolate. Reprotonation can regenerate either D-glucose, if reattachment occurs from the same side as proton loss, or D-mannose if reattachment is from the opposite side (Fig. 22.22).

FIGURE 22.22 A mechanism for the equilibration of D-glucose and D-mannose involves formation of an enolate followed by reprotonation.

PROBLEM 22.14 Make a three-dimensional drawing of the enolate in Figure 22.22 to show how both D-glucose and D-mannose can be formed on reprotonation.

If reprotonation occurs at oxygen, a "double enol" is formed. This compound can re-form a carbon–oxygen double bond in two ways. An aldehyde can be

*This "name reaction" is even better than it sounds. One might be forgiven for assuming that someone meeting these two chemists on a Dutch street in 1880 might say, "Goedemorgen, Lobry; hoe gaat het, Alberda?," should he or she be so presumptuously familiar as to address them by their first names. But that would be wrong! The reaction title, alas, incorporates only parts of their names. The reaction is named for Cornelius Adriaan van Troostenbery Lobry de Bruijn (1857–1904) and the slightly less spectacularly named Willem Alberda van Ekenstein (1858–1907).

generated, and this event simply re-forms the two aldohexoses, D-glucose and D-mannose. If a ketone is produced, it is D-fructose that is the product (Fig. 22.23).

FIGURE 22.23 Protonation on oxygen generates a double enol, which can lead to D-fructose (or the D-aldohexoses, D-glucose, and D-mannose).

22.3c Reduction. We have already seen one aldose reaction, the reduction of an aldohexose with sodium borohydride to give a 1,2,3,4,5,6-hexanehexaol. Other reducing agents also work (Fig. 22.24).

FIGURE 22.24 Reduction of D-glucose proceeds through the small amount of the open, aldo form present at equilibrium. As the open form is used up, it is regenerated through equilibration with the pyranose form.

WORKED PROBLEM 22.15 Reduction of D-altrose with sodium borohydride in water gives an optically active molecule, D-altritol. However, the same procedure applied to D-allose gives an optically inactive, meso hexa-alcohol. Explain.

ANSWER This problem is easy, but critical for the material following in the Fischer proof of the stereochemistry of glucose. The important point is to see that reduction of the aldehyde to the alcohol makes the molecule more symmetrical by making the two ends of the chain *both* CH_2OH groups. D-Allose is reduced to a meso compound but reduction of D-altrose gives an alcohol that is still optically active.

Open form of D-allose
(optically active)

Two superimposable mirror images —
a meso compound
(optically inactive)

Open form of D-altrose
(optically active)

Two nonsuperimposable mirror images
(this alcohol is still optically active)

22.3d Oxidation. Sugars contain numerous oxidizable groups, and methods have been developed in which one or more of them may be oxidized in the presence of the others. Mild oxidation converts only the aldehyde into a carboxylic acid; the primary and secondary hydroxyls are not touched. The reagent of choice is bromine in water, and the product is called an **aldonic acid** (Fig. 22.25).

FIGURE 22.25 Oxidation of an aldohexose with bromine in water gives an aldonic acid in which the end groups are still different.

D-Glucose
(open form)

D-Gluconic acid
(~ 96%)

In an aldonic acid, the end groups are still different

PROBLEM 22.16 Write a mechanism for this oxidation. *Hint for the first step*: What reaction is likely between an aldehyde and water?

PROBLEM 22.17 Examination of the NMR and IR spectra of typical aldonic acids often shows little evidence of the carboxylic acid group. Explain this odd behavior.

More vigorous oxidation with nitrous or nitric acid creates an **aldaric acid** in which both the aldehyde and primary alcohol have been oxidized to carboxylic acids. Notice that this reaction destroys the difference between the two ends of the molecule. In the aldaric acid, there is an acid group at both the top and bottom of the Fischer projection (Fig. 22.26).

FIGURE 22.26 Oxidation with nitric acid generates an aldaric acid in which the end groups are both carboxylic acids.

D-Galactose
(open form)

D-Galactaric acid
(89%)

In even more vigorous oxidations, the secondary hydroxyl groups are attacked and the six-carbon framework can be ruptured. You already know that vicinal diols (1,2-diols) are cleaved on treatment with periodate (Chapter 16, p. 872). When sugars are treated with periodate, mixtures of products are obtained, as the various 1,2-diols are cleaved. Treatment with periodate can chop the carbon frame into small fragments, and the structures of the fragments can often be used in structure determination.

22.3e Ether and Ester Formation.

22.3e Ether and Ester Formation. Alcohols can be made into ethers (Chapter 7, p. 331). One classic method in carbohydrate chemistry involves the treatment of the sugar with methyl iodide and silver oxide which leads to methylation at every free hydroxyl group in the molecule (Fig. 22.27).

FIGURE 22.27 Treatment of a sugar with methyl iodide and silver oxide—ether formation.

(~ 55%)

D-Glucopyranose   1,2,3,4,6-Pentamethoxy-D-glucose

One of the classic S_N2 reactions, the formation of ethers from alkoxides and alkyl halides, is known as the Williamson ether synthesis (Chapter 7, p. 331). Many variations on this theme are employed. Treatment of an aldohexose with benzyl chloride and potassium hydroxide results in formation of a tetra-ether, monoacetal through a series of Williamson ether syntheses. Be careful to notice the difference between the alkoxide group at C(1) and the others. The former is part of an acetal, whereas the others are simple ethers. Neither the existing ether at C(1) nor the pyranose ring connection is disturbed in the benzylation reaction (Fig. 22.28).

(~ 95%)

FIGURE 22.28 Benzylation through a series of Williamson ether syntheses.

Similarly, reaction with acetic anhydride or acetyl chloride will produce a polyacetylated compound in which all of the free hydroxyl groups are esterified (Fig. 22.29).

D-Glucopyranose → [Ac₂O, 98 °C, 8.5 h, acid catalyst] → A penta-ester (58%)

$$Ac_2O = \underset{H_3C}{\overset{O}{\underset{}{C}}}\text{-O-}\underset{}{\overset{O}{\underset{CH_3}{C}}}$$

FIGURE 22.29 The free hydroxyl groups are esterified with acetic anhydride.

Notice that in both these reactions the hydroxyl group tied up in hemiacetal formation remains unaffected by the reaction. It can form neither an ether nor an ester.

The presence of the hemiacetal function allows for selective ether formation when a sugar is treated with dilute acidic alcohol. The products are full acetals, and their generic name is **glycoside**. Six-membered acetals are **pyranosides** and five-membered acetals are **furanosides** (Fig. 22.30).

FIGURE 22.30 Treatment with dilute HCl and alcohol converts only the OH at the anomeric position [C(1)] into an acetal called a glycoside.

D-Glucopyranose → [HCl, CH₃OH, 64 °C, 72 h] → An acetal [*only* the OH at C(1) is converted into a CH₃O group] (~ 50%)

A "glycoside" (in this case, a "pyranoside")

Why is it that *only* the acetal hydroxyl group becomes a methoxyl group in this reaction? The first step is protonation of one of the many OH groups. All the OH groups in the molecule are reversibly protonated, but only one protonation leads to a resonance-stabilized cation on loss of water, and it will be this OH that is lost most easily. Addition of alcohol, followed by deprotonation, gives the acetal in which the hemiacetal OH has been converted into OCH_3 (Fig. 22.31).

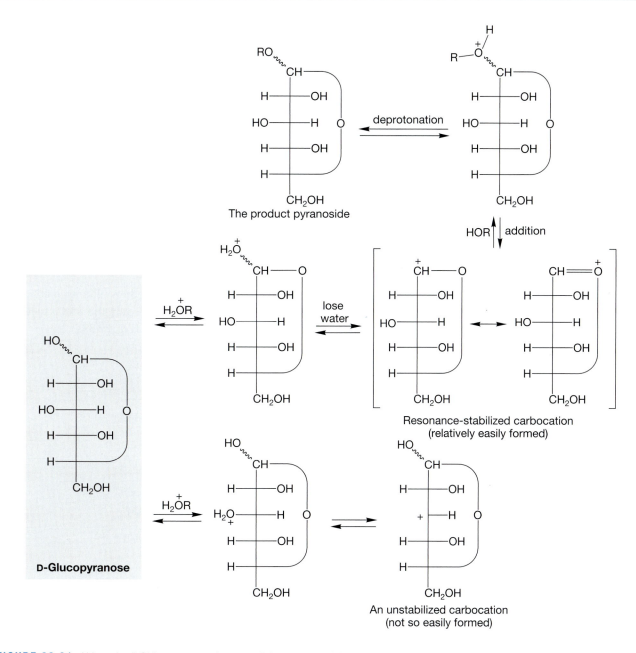

FIGURE 22.31 Although all OH groups can be reversibly protonated, loss of only the anomeric OH leads to a resonance-stabilized cation. Addition of alcohol at this position gives the glycoside.

Specific glycosides are named by replacing the "ose" of the simple sugar's name with "oside" and using the name of the group used to make the ether link. For example, if methyl alcohol were used to form a full acetal with the hemiacetal D-glucopyranose, both methyl α- and methyl β-D-glucopyranosides would be produced (Fig. 22.32).

FIGURE 22.32 Methyl α- and β-D-glucopyranosides.

In a similar vein, the tetra-ether, monoacetal formed in the Williamson ether synthesis can be hydrolyzed in acid to a compound in which only the acetal methoxyl group at C(1) has been transformed into a hydroxyl group (Fig. 22.33).

FIGURE 22.33 Hydrolysis of the fully methylated compound leads to a hemiacetal in which only the methoxyl group at the anomeric position [C(1)] has been converted into an OH.

A tetra-ether, monoacetal   A tetra-ether, hemiacetal

WORKED PROBLEM 22.18 Explain carefully why it is only the acetal methoxyl group that is converted into a hydroxyl group (Fig. 22.33).

ANSWER This kind of reaction takes place by protonation of an ether oxygen, loss of alcohol to give a carbocation, and addition of water followed by deprotonation.

(continued)

PROBLEM 22.18 *(continued)*

All of the ether oxygens can be protonated, but only one can lose an alcohol to give a resonance-stabilized carbocation. Addition of water to the cation, followed by deprotonation, leads to the tetra-ether shown in Figure 22.33 (shown in the α-form; both anomers can be formed here).

Resonance-stabilized carbocation

22.3f Osazone Formation. Treatment of aldo sugars with phenylhydrazine under acidic conditions initiates an odd, and very useful reaction. What would we expect? We know that the cyclic forms of aldohexoses are in equilibrium with the open-chain isomers that contain a free aldehyde group. Aldehydes react with substituted hydrazines to generate hydrazones (Chapter 16, p. 852), and it would be no surprise to see the reaction of Figure 22.34 in which the carbonyl group at C(1) has formed a phenylhydrazone derivative.

Open form   A phenylhydrazone

FIGURE 22.34 The small amount of free aldehyde present at equilibrium accounts for phenylhydrazone formation at C(1).

PROBLEM 22.19 Write a mechanism for the reaction of Figure 22.34.

What *is* a surprise, is to find that not only C(1) but C(2) as well has been transformed into a phenylhydrazone. The product is a 1,2-diphenylhydrazone, called an **osazone**, or a phenylosazone. Here are two clues to the mechanism of the reaction. First, it takes three equivalents of phenylhydrazine to complete the reaction and second, aniline and ammonia are byproducts (Fig. 22.35).

FIGURE 22.35 Osazone formation involves conversion of C(2) as well as C(1) into phenylhydrazones.

We will start our analysis with the reaction at C(1) to form a phenylhydrazone, a substituted imine. Imines (in this case a phenylhydrazone) are in equilibrium with enamines, just as ketones are in equilibrium with enols (Chapter 17, p. 903; Fig. 22.36).

FIGURE 22.36 The C(1) phenylhydrazone is an imine and therefore in equilibrium with an enamine. This enamine is also an enol.

In this case, the enamine is also an enol, and is therefore in equilibrium with a ketone that can react with a second mole of phenylhydrazine. Now aniline is eliminated to give a new imine, which reacts with the third mole of phenylhydrazine to give the final osazone product (Fig. 22.37).

FIGURE 22.37 Reaction of the ketone with phenylhydrazine leads to a new phenylhydrazone that can eliminate aniline to give a new imine. Reaction with a third equivalent of phenylhydrazine leads to the osazone.

Notice that osazone formation destroys the stereogenic carbon at C(2). A given osazone can be formed from *two* sugars, epimeric at C(2). Figure 22.38 makes this point using D-glucose and D-mannose, two aldohexoses that must give the same osazone. When we work through the Fischer proof of the structure of glucose and the other aldohexoses, we will make use of this point a number of times, so remember it.

D-Mannose   The same osazone!   **D-Glucose**

FIGURE 22.38 An osazone can be formed from two different aldo sugars that are stereoisomeric at C(2). The stereochemistry at C(2) is destroyed in the reaction.

SUMMARY

Many of the usual reactions of alcohols are encountered in sugar chemistry. What you have to watch out for are the extra complexities introduced by the polyalcohol nature of sugars, and the fact that one hydroxyl group is part of a hemiacetal. The chemistry of that OH is different from that of all the others.

22.3g Methods of Lengthening and Shortening Chains in Carbohydrates. Treatment of D-ribose with cyanide, followed by catalytic hydrogenation and hydrolysis, generates *two* new sugars: D-allose and D-altrose. This reaction is an example of a modern variation of the **Kiliani–Fischer synthesis** (Emil Fischer again and Heinrich Kiliani, 1855–1945). In the first step of this reaction, sodium cyanide (NaCN) adds to the carbonyl group to make a cyanohydrin (Chapter 16, p. 842). The key point is that a new stereogenic carbon is generated in this process. The new OH group can be on either the right or left in Fischer projection. The second step of the Kiliani–Fischer synthesis, catalytic reduction, gives a pair of imines that are hydrolyzed under the reaction conditions to aldehydes. The net result is the formation of two new sugars, each one carbon longer than the starting sugar. The new sugars differ *only* in their stereochemistry at the newly formed C(2) [they are C(2) epimers; Fig. 22.39].

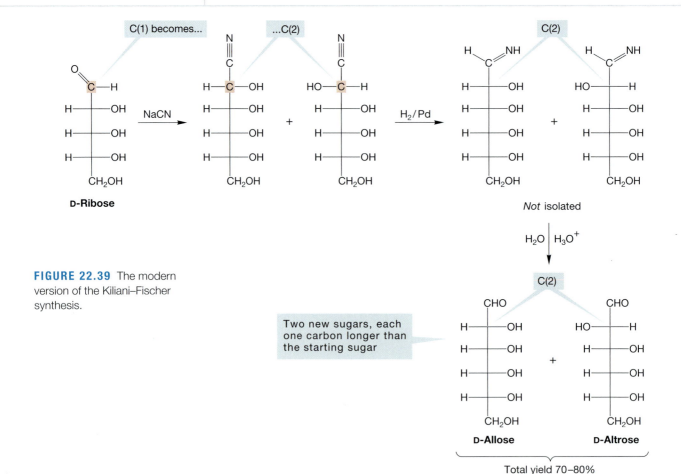

FIGURE 22.39 The modern version of the Kiliani–Fischer synthesis.

PROBLEM 22.20 Apply the Kiliani–Fischer synthesis to D-glyceraldehyde. What new sugars are formed? It is not necessary to write mechanisms for the reactions.

PROBLEM 22.21 The following two sugars are produced by Kiliani–Fischer synthesis from an unknown sugar. What is the structure of that unknown sugar?

D-Gulose D-Idose

The **Ruff degradation** (Otto Ruff, 1871–1939) is a method of shortening the backbone of a sugar by removal of the aldehyde at C(1) and creation of a new aldehyde at the old C(2). In this example, the calcium salt of D-galactonic acid is treated with ferric ion and hydrogen peroxide to give D-lyxose (Fig. 22.40).

D-Galactose Ca Salt of D-galactonic acid D-Lyxose (41%)

FIGURE 22.40 The Ruff degradation shortens the starting sugar by one carbon. It is the original aldehyde carbon that is lost.

The mechanism of this reaction remains obscure (chemistry is a living science—not everything is fully understood) but one imagines that radicals are involved. Metal ions can act as efficient electron-transfer agents, and a guess at the mechanism might involve formation of the carboxyl radical, decarboxylation, capture of the hydroxyl radical formed, and a final transformation of the hydrate into the aldehyde. Whatever the details of the mechanism, the stereochemistry at the original C(2) is lost, but there can be no changes at the old C(3), C(4), or C(5) (Fig. 22.41).

FIGURE 22.41 A possible mechanism for the Ruff degradation.

A similar process, if only in the outcome, is called the **Wohl degradation** (Alfred Wohl, 1863–1933). In a sense, this process reverses the Kiliani–Fischer synthesis. An oxime (a hydroxy imine) is formed first, and then transformed into the nitrile. Treatment with base hydrolyzes five acetoxy groups to hydroxyls. The resulting cyanohydrin then eliminates cyanide and generates the chain-shortened aldehyde (Fig. 22.42).

FIGURE 22.42 The Wohl degradation also shortens sugars through loss of the aldehyde carbon.

SUMMARY

The Kiliani–Fischer procedure lengthens the chain by one carbon atom. Both the Ruff and Wohl degradations shorten the chain by one carbon without disturbing the remaining stereogenic carbons.

22.4 THE RING SIZE IN GLUCOSE

We can take advantage of one of these reactions of sugars, combined with an old reaction, to determine the size of the ring in glucose and other aldohexoses. It is really very simple. D-Glucose is first converted into its methyl glycoside by treatment with methyl alcohol in acid. Figure 22.43 shows formation of methyl β-D-glucopyranoside. Now an old reaction is used. Recall that periodate cleaves 1,2-diols to dicarbonyl compounds (Chapter 16, p. 872). Therefore, treatment of methyl β-D-glucopyranoside with HIO_4 leads to the dialdehyde shown in Figure 22.43.

FIGURE 22.43 Formation of methyl β-D-glucopyranose and further oxidation with periodate.

If D-glucose had been mainly in the furanoside form, quite another structure would have been formed. Figure 22.44 shows the reaction of HIO_4 with methyl β-D-glucofuranoside.

FIGURE 22.44 Formation of methyl β-D-glucofuranose and further oxidation with periodate.

This procedure is one way in which the predominant ring structure in glucose, a six-membered pyranose, was determined.

22.5   THE FISCHER DETERMINATION OF THE STRUCTURE OF D-GLUCOSE (AND THE 15 OTHER ALDOHEXOSES)

In the following pages, we will approximate the reasoning that led Emil Fischer to the structure of D-glucose. The structures of the other 15 aldohexoses can be determined as we go along. As a bonus, we will also be able to derive the structures of the eight aldopentoses. A risk in this kind of discussion, which straightens out the real, highly twisted history of the Fischer determination, is that the work will seem too easy, too straightforward. Fischer's papers appeared in 1891, less than two decades after van't Hoff and LeBel's independent hypothesis of the tetrahedral carbon atom. Determining stereochemistry was no simple matter, and Fischer's work depended critically on a close analysis of the stereochemical relationships of complex molecules. Moreover, the experimental work was extraordinarily difficult and slow.* Fischer's work was monumental, and remains an example of the best we humans are able to accomplish.

Emil Fischer recognized that it was impossible for him to solve one of the central problems in this endeavor, the separation of the D series from the enantiomeric L series. In the 1880s there was no way to determine absolute configuration, which is hardly surprising, as the structural hypothesis of the tetrahedral carbon atom was barely 15 years old at the time. However, Fischer recognized that the 16 aldohexoses must be composed of eight pairs of mirror images, and that if he knew the structures of one set, he automatically could draw the structures of the mirror-image set. The problem was to tell which was which, and this he knew he could not do. So, he did the next best thing, he was arbitrary; one might say that he guessed. The odds aren't bad, 50:50, and there are times when it is best to accept what is possible and not give up because a perfect solution to the problem at hand isn't available. So Fischer simply *defined* one set of eight isomers as the D series, knowing that it might be necessary for history to correct him. In the event, fortune smiled and he made the right guess. I think it is nice that it came out that way. Be sure that you understand that the correctness of Fischer's guess wasn't due to cleverness—his success was not the result of a shrewd guess or the product of a well-developed intuition; there really was no way to know at the time. He was just lucky.

The Fischer proof starts with arabinose, arbitrarily assumed to be the D enantiomer. The first critical observation was that the Kiliani–Fischer synthesis applied to the pentose, D-arabinose, led to a pair of D-aldohexoses (as it must). These were D-glucose and D-mannose. Because the Kiliani–Fischer synthesis must produce a pair of sugars that have different configurations *only* at the newly generated stereogenic carbon, D-glucose and D-mannose must share the following partial structures (Fig. 22.45).

---

*Here is what Fischer had to say to his mentor, Adolf von Baeyer, about this matter in 1889. "The investigations on sugars are proceeding very gradually. . . . Unfortunately, the experimental difficulties in this group are so great, that a single experiment takes more time in weeks than other classes of compounds take in hours, so only very rarely a student is found who can be used for this work."

FIGURE 22.45 The first step in Emil Fischer's determination of the structure of glucose. The Kiliani–Fischer synthesis applied to D-arabinose leads to D-glucose and D-mannose, which must share the partial structures shown.

Let's take a moment at the beginning of this section to look carefully at Figure 22.45. What do we know, and how do we know it? Emil Fischer has a flask containing one of the enantiomers of the pentose called arabinose. He really doesn't know whether this enantiomer is D or L, that is, whether the OH at C(4) of arabinose is on the right or left. He "solves" this problem by guessing, by assuming that the sugar he has in the flask is a member of the D series defined as having the OH at C(4) on the right. We now know that he guessed correctly. He has no idea of the relative arrangement of the OH groups at C(3) and C(2). Kiliani–Fischer synthesis applied to D-arabinose gives D-glucose and D-mannose, one of which has the newly created OH on the carbon adjacent to the aldehyde on the right, the other of which has it on the left.

The second step in the Fischer proof determines the configuration of C(2) in D-arabinose [this carbon becomes C(3) in the two D-aldohexoses: D-glucose and D-mannose]. Fischer observed that the oxidation of D-arabinose with nitric acid gives an *optically active diacid.* The possibilities are shown in Figure 22.46. There are only three possible diacids that can be produced from a D-aldopentose, and two of them are meso. In the other, optically active diacid, the OH at C(2) of the diacid [which corresponds to C(3) in the aldohexoses] is on the left (Fig. 22.46).

These are the three possible diacids

FIGURE 22.46 Oxidation of D-arabinose leads to an optically active diacid, which shows that the OH at C(2) in D-arabinose is on the left.

Now we know a bit more about the partial structures of D-arabinose, D-glucose, and D-mannose. We can specify all but one OH in D-arabinose and in the two structures shared by the two D-aldohexoses. We do not yet know the position of the OH at C(3) of D-arabinose, which becomes C(4) in the D-aldohexoses, and we do not know which partial structure belongs to D-glucose and which to D-mannose (Fig. 22.47).

FIGURE 22.47 What we now know about the structures of D-arabinose, D-glucose, and D-mannose. Only the configuration at C(3) of D-arabinose, which becomes C(4) in D-glucose and D-mannose, is left to be determined.

In the third step, the position of the last OH is fixed by the observation that both D-glucose and D-mannose give *optically active diacids* on oxidation with nitric acid. There are two possibilities. The OH at C(4) is either on the left or the right. If it is on the right, both sugars do give an optically active diacid, in agreement with the experimental data (Fig. 22.48).

FIGURE 22.48 If the last unknown OH is on the right, two optically active diacids are produced on oxidation with nitric acid. This result matches the experimental results.

On the other hand, if the OH at C(4) is on the left, one of the diacids is meso, not optically active, which does not match the experimental data (Fig. 22.49).

If the OH at C(4) had been on the left

meso Diacid!

Optically active
diacid

FIGURE 22.49 However, if the unknown OH had been on the left, one possible diacid is meso, not optically active, which does not match the experimental results.

So, the correct position of the OH at C(3) in D-arabinose and at C(4) in the two D-aldohexoses, D-glucose and D-mannose, must be on the right. D-Glucose and D-mannose share the two structures in Figure 22.50. The next problem is to tell which is which.

Known from
Figure 22.46

Known from
Figures 22.48
and 22.49

Mirror

L-Arabinose   **D-Arabinose**

By definition,
a D sugar

Kiliani–Fischer
synthesis

A   **B**

These two structures are shared by
D-glucose and D-mannose

FIGURE 22.50 Now we know the structure of D-arabinose, L-arabinose, and the two structures (**A** and **B**) shared by D-glucose and D-mannose. We do not know which structure belongs to D-glucose and which to D-mannose.

We now know the full structure of D-arabinose as the positions of all three OH groups have been determined. Of course, if we know the structure of the D-enantiomer, we know that of its mirror image, the L-enantiomer (Fig. 22.50).

We return to the question of the structures of D-glucose and D-mannose. Which of the two structures in Figure 22.50 is that of D-glucose and which is that of D-mannose? The final necessary observation is that D-glucose *and another sugar,* L-gulose (notice that it's an L sugar!), gave the same optically active diacid when oxidized with nitric acid. We can find what the structure of the other sugar must be in the following way. Treatment with nitric acid destroys the identity of the groups at the ends of the chain, as an aldaric acid is formed (p. 1240). If we simply invert the sugar, placing the CH$_2$OH at the top and the CHO at the bottom, we will still get an aldaric acid on oxidation (Fig. 22.51).

FIGURE 22.51 Oxidation with nitric acid renders the ends of a sugar equivalent. Both the aldehyde end and the primary alcohol end are converted into the same group, a carboxylic acid.

An aldaric acid

There are two possibilities, **A** and **B**. If D-glucose has structure **A** of Figures 22.50 and 22.52, then L-gulose, the sugar that gives the same aldaric acid on oxidation, must have the structure shown at the right in Figure 22.52.

FIGURE 22.52 If D-glucose has the structure **A**, oxidation of another sugar, L-gulose, can give the same aldaric acid. Note in this case that the L-gulose is drawn with the CH$_2$OH at the top and the CHO at the bottom.

Let's see what happens if D-glucose has the other possible structure, **B**. Again, we invert the sugar **B** in an attempt to generate another sugar that can give the same aldaric acid. This time, however, the attempt fails, as the "new" sugar is identical with the first. Both are **B** (Fig. 22.53)! There is only one sugar that can give this diacid. There can be no second sugar. D-Glucose must have structure **A**, and the other, **B**, must therefore be D-mannose.

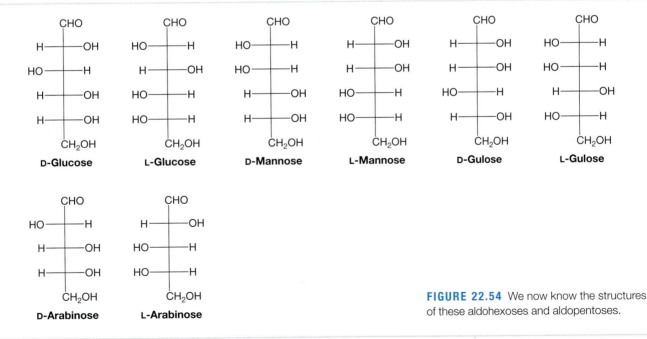

FIGURE 22.53 However, there is no sugar other than **B** that can give this aldaric acid. As this fact does not match the experimental results, D-glucose must have structure **A**.

Now let's take stock of what we know, which is rather more than we might think at first. We have the structures of 6 of the 16 possible aldohexoses, as we know D-glucose, D-mannose, and L-gulose and their enantiomers L-glucose, L-mannose, and D-gulose. In addition, we know the structures of D- and L-arabinose (Fig. 22.54).

FIGURE 22.54 We now know the structures of these aldohexoses and aldopentoses.

There are many ways to fill out the structures of the remaining aldopentoses and aldohexoses. Here is just one possible sequence. Let's first deal with the three remaining D-aldopentoses, given that we know the structure of D-arabinose. D-Ribose gives the same osazone as D-arabinose (Fig. 22.55), which means that the two sugars must differ only at the configuration at C(2). Therefore D-ribose must have the structure shown in Figure 22.55.

FIGURE 22.55 D-Arabinose and D-ribose give the same osazone, and therefore can differ only at C(2). The structure of D-ribose is therefore known. L-Ribose is simply the mirror image of the D isomer.

There are only two remaining D-aldopentoses. One of them, D-xylose, gives a meso diacid on oxidation, whereas the other, D-lyxose, gives an optically active diacid. The structures must be as shown in Figure 22.56.

FIGURE 22.56 D-Xylose gives a meso diacid and D-lyxose gives an optically active diacid, which fixes the structures of the remaining D-aldopentoses.

Now we know all four stereoisomers of the D-aldopentoses, and we can construct the L series by drawing the mirror images of the D series. That gives us all eight aldopentoses.

Let's complete the determination of the structures of the D-aldohexoses. D-Gulose and D-idose give the same osazone and must differ only in their configurations at C(2) (Fig. 22.57).

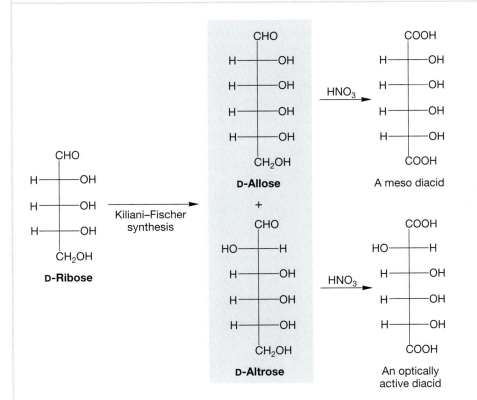

FIGURE 22.57 D-Gulose and D-idose give the same osazone, which determines the structure of D-idose and its mirror image L-idose.

The Kiliani–Fischer synthesis applied to D-ribose gives D-allose and D-altrose, which must share the two structures of Figure 22.58. The two structures can be assigned by noting that D-allose gives an optically inactive, meso diacid on treatment with nitric acid, whereas D-altrose gives an optically active diacid.

FIGURE 22.58 The Kiliani–Fischer synthesis applied to D-ribose gives D-allose and D-altrose. D-Allose gives a meso diacid and D-altrose gives an optically active diacid on oxidation. These observations determine the structures of D-allose, D-altrose, and their mirror-image L isomers.

There are only two remaining D-aldohexoses: D-talose and D-galactose. D-Galactose gives a meso diacid on oxidation with nitric acid, but D-talose gives an optically active diacid (Fig. 22.59).

FIGURE 22.59 D-Galactose gives a meso diacid and D-talose gives an optically active diacid. Now we know all the structures of all the aldohexoses.

D-Galactose → (HNO₃) → A meso diacid   D-Talose → (HNO₃) → An optically active diacid

The structures of the eight L isomers are simply the mirror images of those of the D series, and we have at last completed our task.

22.6   AN INTRODUCTION TO DI- AND POLYSACCHARIDES

A great many sugars containing 12, 18, and other multiples of 6 carbons are known. Some of the common names for sugars actually refer to dimeric species, not the monomeric, relatively simple D-aldohexoses or pentoses. For example, sucrose, lactose, and maltose are all C_{12} sugars. Starch and cellulose are polymers of simple repeating monosaccharide units, less some molecules of water. We will use lactose as an example of how the structures of such molecules can be unraveled.

The first important observation is that (+)-lactose, a disaccharide of the formula $C_{12}H_{22}O_{11}$, can be hydrolyzed in acid to a pair of simple aldohexoses: D-glucose and D-galactose (Fig. 22.60).

FIGURE 22.60 In acid, (+)-lactose is hydrolyzed to D-glucose and D-galactose.

(+)-Lactose $C_{12}H_{22}O_{11}$ → (H_3O^+, H_2O) → D-Glucose + D-Galactose

That observation identifies the two sugars making up (+)-lactose, but this simple experiment gives us another piece of less obvious information. Remember that it is only glycosidic linkages that are cleaved in acid—simple ether connections are not touched. We saw this first in the hydrolysis of *only* the glycosidic methoxyl group of fully methylated D-glucose (p. 1242). D-Glucose and D-galactose must be attached using the hydroxyl group at C(1) of one of the sugars. If the attachment were at any other position, dilute acid would not suffice to break up the dimeric structure (Fig. 22.61).

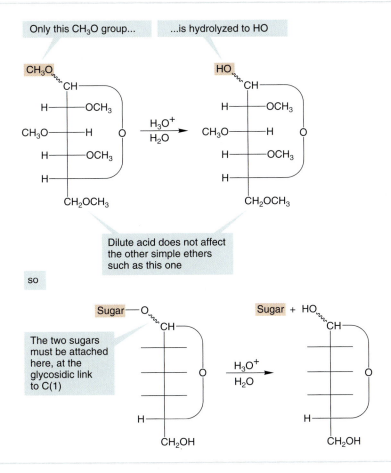

FIGURE 22.61 Hydrolysis in dilute acid means that there must be a glycosidic linkage in (+)-lactose.

We now need to know three things about the dimeric structure of lactose (Fig. 22.62).

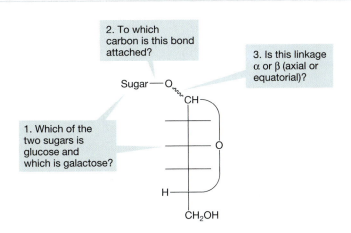

FIGURE 22.62 The remaining questions about the structure of (+)-lactose.

1. Which sugar is the "attacher" and which is the "attachee"?

2. Which hydroxyl group of one sugar is used to make the linkage to the anomeric position of the other?

3. Are the two sugars linked in an α or β way?

(+)-Lactose can be oxidized by bromine in water to the improbably named lactobionic acid. Bromine in water oxidizes only the aldehydic position of sugars to the acid (p. 1239). In one of the partners in the disaccharide there is no hemiacetal, and therefore no small amount of free aldehyde to be oxidized by bromine/water. Acid hydrolysis of lactobionic acid gives one molecule of D-galactose and one of D-gluconic acid. The remaining aldehyde group in (+)-lactose must belong to glucose—it is the galactose half that is attached to glucose through the OH at C(1) (Fig. 22.63).

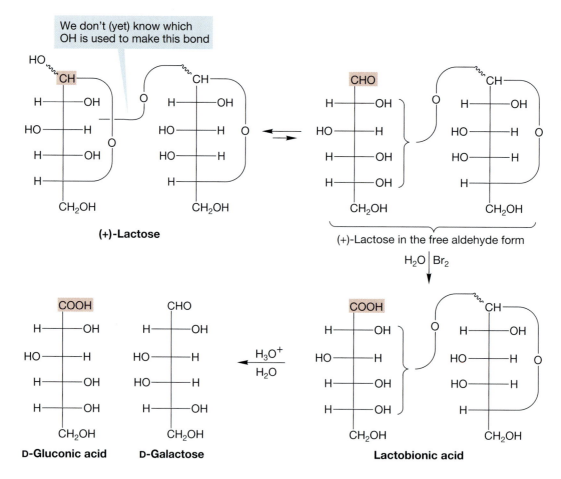

FIGURE 22.63 The acid group in lactobionic acid marks the position of the aldehyde in (+)-lactose. As hydrolysis of lactobionic acid gives a gluconic acid (*not* a galactonic acid), it is glucose that has the free aldehyde in (+)-lactose.

WORKED PROBLEM 22.22 If the two sugars had been attached in the other way, using C(1) of glucose as the full acetal, what would the products of the experiment shown in Figure 22.63 have been?

ANSWER If the two sugars making up lactose had been attached in the other sense, it would have been the galactose that had the aldehyde, and glucose that was attached through the OH group at C(1). Thus, it would have been the galactose aldehyde group that was oxidized to an acid in this hypothetical lactobionic acid. Hydrolysis would have given D-galactonic acid and D-glucose.

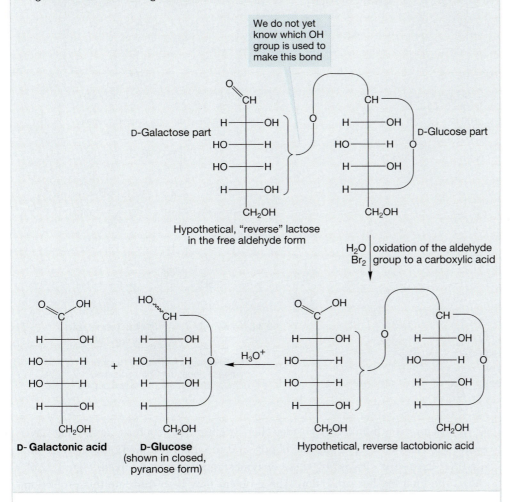

Now, which carbon of glucose is attached to C(1) of galactose? This question is answered by first using a Williamson ether synthesis to methylate all free hydroxyl groups in (+)-lactose. (+)-Lactose is then hydrolyzed in acid to give 2,3,4,6-tetramethylgalactose and 2,3,6-trimethylglucose. Hydrolysis liberates only the OH groups that were tied up in (+)-lactose in hemiacetals or full acetals. It leaves simple ether groups unchanged. It is the OH at C(5) that is used to form the pyranose ring in both sugars, and that accounts for its appearance as an unmethylated OH in the two partially methylated sugars of Figure 22.64. The remaining free OH—at C(4) of glucose—must mark the position attached to C(1) of galactose. It was tied up in an acetal linkage, not as a simple methyl ether (Fig. 22.64).

FIGURE 22.64 The position of the OH used to attach glucose to C(1) of galactose can be determined through a series of methylation and hydrolysis experiments.

(+)-Lactose is a β-linked glycoside. There are enzymes such as lactase that have evolved to cleave only β-glycosidic bonds, and to which α-linked disaccharides are inert. (+)-Lactose is cleaved by these enzymes and must be β-linked. Figure 22.65 shows the final Fischer projection. The problem now is to translate this projection into a three-dimensional representation.

FIGURE 22.65 A Fischer projection for (+)-lactose.

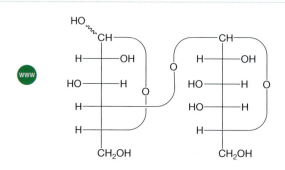

PROBLEM 22.23 Make a good three-dimensional drawing of (+)-lactose.

PROBLEM 22.24 Cellobiose is a disaccharide cleaved by hydrolysis into two molecules of D-glucose. Treatment of cellobiose with dimethyl sulfate in base leads to an octamethyl derivative. Hydrolysis with acid converts the octamethyl derivative into one molecule of 2,3,6-trimethyl-D-glucose and one of 2,3,4,6-tetramethyl-D-glucose. Cellobiose can be cleaved enzymatically with lactase. Provide a structure for cellobiose.

PROBLEM 22.25 Maltose is a disaccharide that differs very little from cellobiose. It has the same properties as cellobiose described in Problem 22.24, but is not cleaved by lactase (it is cleaved by another enzyme, maltase). Provide a three-dimensional structure for maltose. *Hint*: How else can the two sugars be linked?

Sometimes two sugars are linked in a way that connects the anomeric carbons [C(1)] of both sugars. These molecules are full acetals and oxidation to an aldonic acid is not possible because there is not even a small amount of free aldehyde. In contrast to other sugars, in which at least a small amount of an oxidizable free aldehyde remains (reducing sugars), these are **nonreducing sugars**. An example is sucrose, common table sugar, in which a molecule of D-glucose and a molecule of D-fructose are connected in this way (Fig. 22.66).

FIGURE 22.66 In nonreducing sugars, there is no free aldehyde group. Attachment must be between both C(1) atoms. Sucrose is an example.

There is also no reason that more than two sugars cannot be connected by glycosidic bonds, and larger polysaccharides are common. The limit of such bonding is a polymer. **Cellulose** is a polymer of glucose in which repeating units are connected linearly by β-glycosidic bonds between C(4) of one sugar and the anomeric C(1) of the other. **Starch** is a different, nonlinear polymer of glucose from which a linear molecule called amylose can be isolated. Amylose has the same structure as cellulose except that the sugars are connected in an α way (Fig. 22.67).

FIGURE 22.67 Cellulose and amylose.

Cellulose is poly-D-glucose linked β

Starch ⟶ amylose, which is poly-D-glucose linked α

22.7 SUMMARY

New Concepts

This chapter contains few new concepts, which does not mean that the chemistry is uncomplicated, however. As I have said many times, simple reactions taken in series can become very difficult indeed. The polyfunctionality and stereochemical complexity of sugars make for an intricate chemistry even though there are few really new reactions involved.

The stereochemical complications lead to the simplifying notion of a Fischer projection in which a stylized drawing is used to reduce complexity. The sugar is drawn vertically with the aldehyde group at the top. Horizontal bonds are taken as coming toward you and vertical bonds as retreating. If the OH of the bottom H—C—OH (the one adjacent to the CH₂OH) is

on the right, it is a D sugar, if it is on the left, it is an L sugar (Fig. 22.3).

When Emil Fischer worked out the structure of glucose in the late 1800s there was no way to tell absolute configuration. Fischer could not tell the D series from the enantiomeric L series, so he assumed that the OH at C(5) was on the right and worked out the relative configurations of the other groups. As it turned out, he guessed correctly.

A number of times in this chapter we encounter reactions in which a minor partner in an equilibrium reacts to give the product. As long as equilibrium is reestablished, removal of the minor partner can drive the reaction to completion (Fig. 22.14).

Key Terms

Reactions, Mechanisms, and Tools

Most of the reactions in this chapter can be found in the chemistry of carbonyl groups (Chapters 16 and 17) and alcohols (Chapters 6 and 7); only a few new reactions appear.

One of these is intramolecular hemiacetal formation. Carbonyl groups react with nucleophiles in the addition reaction (Chapter 16). When the nucleophile is an alcohol, hemiacetals are formed, but generally they are not favored at equilibrium. However, when a relatively strain-free ring can be formed in an intramolecular hemiacetal formation, the cyclic form can be favored. Such is the case for aldohexoses (and many other sugars), which exist mainly in the six-membered pyranose forms (Fig. 22.12).

The equilibration of the cyclic hemiacetal with a small amount of the open aldohexose form allows for either the α- or

β-anomer to produce an equilibrium mixture of the two anomers in a phenomenon called mutarotation (Fig. 22.20).

The reactivity of the hemiacetal OH allows for a differentiation of this OH from the others in the molecule. For example, protonation of the C(1) hemiacetal OH leads to a molecule that can form a resonance-stabilized carbocation on loss of water. Nucleophiles such as alcohols can add to this carbocation to give the full acetals called glycosides. Only the OH at C(1) can react in this way (Fig. 22.30).

Reaction with three equivalents of phenylhydrazine leads to the 1,2-phenylhydrazones called osazones (Fig. 22.35).

Methods have been developed both to lengthen (Kiliani–Fischer synthesis) and to shorten (Ruff and Wohl degradations) the backbone carbon chains of sugars.

Syntheses

Chain lengthening. The Kiliani–Fischer synthesis adds a new C(1) aldehyde and generates a pair of C(2) stereoisomers (Fig. 22.39).

Chain shortening. In the Ruff and Wohl degradations, the original aldehyde group is lost and a new aldehyde created at the old C(2) (Figs. 22.40 and 22.42).

Osazone formation (Fig. 22.35).

Monoacetal and tetra-ether, monoacetal derivatives of carbohydrates (Figs. 22.28 and 22.30).

Aldaric and aldonic acids can be made through oxidation of aldo sugars (Figs. 22.25 and 22.26).

Common Errors

At the heart of sugar chemistry is the difference between C(1), the anomeric center, and the other carbons. If you lose track of this difference, you lose the ability to follow the labeling studies establishing the ring size of simple sugars and the points of attachment of disaccharides.

It is also sometimes difficult to keep track of the open and closed (hemiacetal) forms of the sugars and sugar derivatives in

this chapter. We typically write what is convenient, not always good representations of structure. For example, open forms of sugars are often written even though most of the material is really tied up in the cyclic hemiacetal form. It is the small amount of the open form that is reactive. This concept can be confusing if you don't have the equilibrium between open and cyclic forms solidly in mind.

22.8 ADDITIONAL PROBLEMS

PROBLEM 22.26   Draw the following molecules in Fischer projection:
(a) Any D-aldoheptose.
(b) Any L-ketopentose.
(c) A D-aldotetrose that gives a meso diacid on oxidation with dilute nitric acid.
(d) Methyl α-D-galactopyranoside.
(e) The osazone of D-allose.
(f) Phenyl β-D-ribofuranoside.

PROBLEM 22.27   Make perfect three-dimensional drawings of all the pyranose forms of D-altrose. Before you start, decide how many will there be.

PROBLEM 22.28   (a) What sugar or sugars would result from the Ruff degradation applied to D-gulose?
(b) What sugar or sugars would result from the Kiliani–Fischer synthesis applied to D-lyxose?

PROBLEM 22.29   (a) What other sugar, if any, would give the same aldaric acid as D-talose when oxidized with nitric acid?
(b) What other sugar, if any, would give the same aldaric acid as D-xylose when oxidized with nitric acid?
(c) What other sugar, if any, would give the same aldaric acid as D-idose when oxidized with nitric acid?

PROBLEM 22.30   What other sugar would give the same osazone as L-talose?

PROBLEM 22.31   What are the principal organic products when D-lyxose is treated under the following conditions? Mechanisms are not necessary. D-Lyxose is drawn in the open form, but of course it is largely in the cyclic, furanose form. Keep this fact in mind.

(a) $NaBH_4$, H_2O
(b) Br_2, H_2O
(c) HNO_3
(d) CH_3I, Ag_2O

D-Lyxose

PROBLEM 22.32   What are the principal organic products when D-ribose is treated under the following conditions? Mechanisms are not necessary. D-Ribose is drawn in the open form, but of course is largely a cyclic furanose.

(a) Ac_2O, H_3O^+
(b) CH_3OH, H_3O^+
(c) $PhNHNH_2$
(d) 1. NaCN
    2. H_2/Pd
    3. H_2O
(e) 1. NH_2OH, CH_3OH, CH_3ONa
    2. Ac_2O, 100 °C
    3. CH_3OH, CH_3ONa
    4. H_2O, H_3O^+
(f) 1. Br_2/H_2O
    2. $Ca(OH)_2$
    3. $Fe_2(SO_4)_3$
    4. H_2O_2

D-Ribose

PROBLEM 22.33   Glucose is the most abundant aldohexose. Use a stereochemical analysis to speculate upon the reason for this prevalence.

PROBLEM 22.34   Your task is to distinguish between D-talose and D-galactose. The following sequence is suggested. First, oxidize to aldaric acids. Second, make the methyl esters. Then . . . ? Fill in the last step. Oh, we forgot to tell you that at the end of the second step your town's elected officials passed an ordinance forbidding any laboratory work as harmful to their chances for re-election. You are restricted to spectroscopy.

PROBLEM 22.35   Use Table 22.2 (p. 1234) to calculate the energy difference between the α- and β-pyranose forms of D-allose at 25 °C.

PROBLEM 22.36   Calculate the equilibrium ratio of α- and β-D-glucopyranose from the specific rotations of the pure anomers (α = +112°, β = +18.7°; equilibrium value = +52.7°). You might compare your calculation with the values given in Table 22.2.

PROBLEM 22.37   The disaccharide maltose ($C_{12}H_{22}O_{11}$) can be hydrolyzed in acid to two molecules of D-glucose. It is also hydrolyzed by the enzyme maltase, a molecule known to cleave only α-glycosidic linkages. Maltose can be oxidized by bromine in water to maltobionic acid (MBA, $C_{12}H_{22}O_{12}$). When MBA is treated with methyl iodide and base, followed by acid hydrolysis, the products are 2,3,4,6-tetramethyl-D-glucose and 2,3,5,6-tetramethyl-D-gluconic acid. Provide a three-dimensional structure for maltose and explain your reasoning.

PROBLEM 22.38 Draw in Fischer projection the nonreducing disaccharides composed of one molecule of D-allose and one molecule of D-altrose. First determine how many are possible, in principle, and then adopt some scheme for indicating what they are. Don't attempt to draw them all out in three dimensions.

PROBLEM 22.39 Write a mechanism for the acid-induced transformation of a D-aldopentose into furfural.

PROBLEM 22.40 Figures 22.22 and 22.23 give one mechanism for the Lobry de Bruijn–Alberda van Ekenstein reaction. Use the example given, in which D-glucose equilibrates with D-mannose and D-fructose, and provide another mechanism in which a hydride shift occurs.

PROBLEM 22.41 When α-D-xylopyranose is treated with hydrochloric acid in methyl alcohol, the major products are two isomers of methyl D-xylopyranoside.
(a) Draw the starting material and the two major products and explain how the two major products are formed from the single starting material.
(b) There are two minor products formed as well. Both are methyl furanosides. What are these two products and how are they formed?

PROBLEM 22.42 Outline a mechanism for the following reaction:

PROBLEM 22.43 Outline a mechanism for the following reaction:

PROBLEM 22.44 When D-glucose, or any polysaccharide containing D-glucose units such as cellulose, is pyrolyzed with an acid catalyst, a new compound, levoglucosan (**1**), is formed. Write a mechanism.

PROBLEM 22.45 Treatment of α-D-galactopyranose with acetone and sulfuric acid catalyst leads to a compound $C_{12}H_{20}O_6$ that still contains a primary alcohol. Draw a structure for this compound and explain its formation.

PROBLEM 22.46 Provide structures for **A–D**, and rationalize the following reactions of the starting material, the product formed in Problem 22.45:

PROBLEM 22.47   With many carbonyl compounds the primary OH participates in product formation. For example, many saccharides react with benzaldehyde in acid to give six-membered rings in which the primary alcohol is incorporated.

A schematic view
of a sugar

However, as shown in Problem 22.45, acetone leads to five-membered rings, not six-membered rings shown above.

First, draw the detailed three-dimensional structures of the six-membered rings formed by reaction between benz-aldehyde and acetone with the partial sugar shown above, then explain this odd observation.

Special Topic: Introduction to Amino Acids and Polyamino Acids (Peptides and Proteins)

23

When I returned from the physical shock of Nagasaki . . . , I tried to persuade my colleagues in governments and the United Nations that Nagasaki should be preserved exactly as it was then. I wanted all future conferences on disarmament . . . to be held in that ashy, clinical sea of rubble. I still think as I did then, that only in this forbidding context could statesmen make realistic judgements of the problems which they handle on our behalf.

Alas, my official colleagues thought nothing of my scheme. On the contrary, they pointed out to me that delegates would be uncomfortable in Nagasaki.

—Jacob Bronowski*

23.1   PREVIEW

For many people, fascination with science depends on the relationship between the material they study and biological processes. This connection seems natural as no one can deny the excitement of today's incredible pace of discovery in biological chemistry and molecular biology. It's quite impossible to be a decent molecular biologist without being first a chemist. Organic chemistry is particularly, if not uniquely, important. Molecular biology is organic chemistry (and inorganic and physical chemistry) applied to molecules of a scale not generally encountered in traditional, "small molecule" organic chemistry.†

---

*Jacob Bronowski (1908–1974) was a mathematician, poet, and playwright. He was the author and compelling narrator of the brilliant television series, *The Ascent of Man*. Don't miss it if it is replayed.

†Scales change depending on your point of view. To a theoretician, a molecule with 10–20 non-hydrogen "heavy atoms" is often too large to be easily approached. Daunting problems are presented, if not of intellectual content, certainly of cost. Even with the decrease in price of computer time, calculations on molecules of "biological" size often simply cost too much. To a molecular biologist, a molecule of only 10 heavy atoms would be laughably tiny—hardly visible on the scale of proteins and enzymes.

Size, and the attendant complexity of structure, generates genuinely new chemistry. We have already seen how important molecular shape is in small molecules, and now, in molecules of biological size, stereochemistry becomes more complex and even more vital to understanding mechanism. To a great extent, biological action depends on shape. Much of the immense architecture of amino acid polymers serves "only" to bring a reactive group to an appropriate point on a substrate molecule, which is itself held by the enzyme in just the right position for reaction.

This chapter can only begin to introduce this subject and attempt to show you how important organic chemistry is to it. The choice of subjects is certainly arbitrary—there is a vast number of possible topics. This chapter is held together, somewhat tenuously it must be admitted, by nitrogen atoms. Nitrogen atoms are always components of the alkaloids we saw in Chapter 6, and of amino acids, the subjects of this chapter.

ESSENTIAL SKILLS

1. In this chapter, we pass from small molecules to big ones. It is necessary to be able to apply old "small molecule" chemistry in these new larger and more complex molecules.

2. Manipulation of what are called blocking and activating groups is necessary for successful peptide synthesis, so you have to know which to use, and when and how to attach and detach them.

IMPORTANT DETAILS

1. It is only introduced here, but surely the general mechanism for polynucleotide replication is important. If you go on in biology or biochemistry, you'll learn much more about it.

23.2 AMINO ACIDS

23.2a Nomenclature. Of course there is a systematic nomenclature for amino acids, but as we have so often seen, the real world eschews the system and goes right on using familiar old common names. The generic term "amino acid" itself is shorthand for α-**amino acid.** It is the α that locates the amino group on the carbon adjacent to the carboxylate carbon and distinguishes the α-amino acids from the isomeric, but less important β and γ compounds (Fig. 23.1).

FIGURE 23.1 The amino group can be located on the hydrocarbon chain with a Greek letter, α, β, γ, or δ.

An α-amino acid   A β-amino acid

In another naming system, we use a "2" instead of the "α" and name the α-amino acids as members of a series of 2-amino-substituted carboxylic acids (Fig. 23.2).

Aminoacetic acid
(glycine)

α-Aminopropionic acid
2-aminopropionic acid
(alanine)

α-Amino-β-methylbutyric acid
2-amino-3-methylbutyric acid
(valine)

α-Amino-β-phenylpropionic acid
2-amino-3-phenylpropionic acid
(phenylalanine)

FIGURE 23.2 An α-amino acid is just an "R-substituted" 2-amino acetic acid.

In practice, the systematic name is rarely used, and common names are retained for most of these molecules. Biochemists tend to view amino acids as substituted α-aminoacetic acids. Note that as long as R doesn't equal H, the amino acids contain a stereogenic carbon atom (Fig. 23.3).

FIGURE 23.3 Most α-amino acids contain a stereogenic carbon.

Stereogenic carbon

Basic unit

R-Substituted α-amino acid

In addition, three-letter codes have been developed for the amino acids and are very commonly used, especially when describing the polymeric collections of amino acids known as polypeptides or in especially large examples, as **proteins**. The boundary between the smaller **peptides**, or "polypeptides," and the larger proteins is neither fixed nor always clear, and the terms are sometimes used interchangeably. More recently, an even shorter set of one-letter abbreviations has come into use. Table 23.1 collects the 20 α-amino acids, gives their common names and abbreviations, and collates some of their physical properties. Notice that R groups of many kinds appear, from simple hydrocarbon chains to more complicated acidic and basic groups.

TABLE 23.1   Properties of 20 Common Amino Acids

R	Common[a] Name	Abbreviations	mp (°C) L-Form[b]	pK_a (COOH)	pK_a ($^+NH_3$)	pK_a (Other)	Isoelectric Point pH
Simple alkane groups							
H	Glycine	Gly, G	229 (dec)	2.3	9.6		6.0
CH_3	Alanine	Ala, A	297 (dec)	2.3	9.7		6.0
$CH(CH_3)_2$	Valine*	Val, V	315 (dec)	2.3	9.7		6.0
$CH_2CH(CH_3)_2$	Leucine*	Leu, L	337 (dec)	2.3	9.6		6.0
$CHCH_3(CH_2CH_3)$	Isoleucine*	Ile , I	284 (dec)	2.3	9.6		6.0
Aromatic groups							
H_2C—⟨benzene ring⟩	Phenylalanine*	Phe, F	283 (dec)	1.8	9.2		5.9
H_2C—⟨benzene ring⟩—OH	Tyrosine	Tyr, Y	316 (dec)	2.2	9.1	10.1	5.7
H_2C—⟨imidazole ring⟩	Histidine*	His, H	288 (dec)	1.8	9.0	6.0	7.6
H_2C—⟨indole ring⟩	Tryptophan*	Trp, W	290 (dec)	2.4	9.4		5.9
Groups containing an OH (see also Tyr)							
CH_2OH	Serine	Ser, S	228 (dec)	2.2	9.2		5.7
$CH(CH_3)OH$	Threonine*	Thr, T	226 (dec)	2.2	9.2		5.6
Groups containing S							
CH_2SH	Cysteine	Cys, C		1.7	10.8	8.3	5.0
$CH_2CH_2SCH_3$	Methionine*	Met, M	283 (dec)	2.3	9.2		5.8
Groups containing a carbonyl group							
CH_2COOH	Aspartic Acid	Asp, D	270 (dec)	1.9	9.6	3.7	2.9
CH_2CH_2COOH	Glutamic Acid	Glu, E	211 (dec)	2.2	9.7	4.3	3.2
CH_2CONH_2	Asparagine	Asn, N	214 (dec)	2.0	8.8		5.4
$CH_2CH_2CONH_2$	Glutamine	Gln, Q	185 (dec)	2.2	9.1		5.7

(continued)

TABLE 23.1 Properties of 20 Common Amino Acids (*continued*)

R	Common[a] Name	Abbreviations	mp (°C) L-Form[b]	pK_a (COOH)	pK_a ($^+NH_3$)	pK_a (Other)	Isoelectric Point pH
Groups containing an amino group (see also Trp and His)							
$(CH_2)_4NH_2$	Lysine*	Lys, K	224 (dec)	2.2	8.9	10.3	9.7
$(CH_2)_3NH-C=NH(NH_2)$	Arginine*	Arg, R	207 (dec)	2.2	9.1	12.5	10.8
A secondary amine							
	Proline	Pro, P	221 (dec)	2.0	10.6		6.1

[a] Essential amino acids are marked by an asterisk (*).
[b] Decomposition is abbreviated as dec.

Why pick out this particular set of 20 from the vast number conceivable? These are the 20 amino acids encountered in peptides and proteins. Your body can synthesize one-half of these 20 from starting materials provided by the food you eat. One might think that a reasonable way for your insides to proceed would be to break down ingested amino acid polymers to their constituent amino acids, and then use these directly. But this is not how you do it. Instead, amino acids are metabolized (broken down chemically) into much simpler molecules from which the amino acids you need are resynthesized. But even your body is not a perfect synthetic chemist—unless you are a superior mutant, there are 10 of these 20 amino acids it is unable to make. These you must ingest directly, and if your diet fails to include one or more of these **essential amino acids** you will not last long. The essential amino acids are marked with an asterisk in Table 23.1.

23.2b Structure of Amino Acids. All but 1 of the 20 common amino acids in Table 23.1 is a primary amine. The only exception is proline, an amino acid containing a secondary amine (Fig. 23.4). All biologically important amino acids but the achiral glycine and one other are found in the (*S*) configuration. Many (*R*) amino acids are known in nature, but most play no important role in human biochemistry. Why? That's a good question and no certain answer exists. Did the first resolution produce an (*S*) amino acid, which then determined the sense of chirality of all the amino acids formed thereafter? Is the preference for (*S*) no more than the result of some primeval accident? No one knows. If it is, we may one day encounter (*R*) civilizations somewhere in the universe, provided we are sufficiently lucky and clever to survive long enough.

Basic unit
(primary amines)

Proline
(a secondary amine)

FIGURE 23.4 Of the 20 common α-amino acids, only proline is not a primary amine.

WORKED PROBLEM 23.1 Draw tetrahedral three-dimensional structures for (R) and (S) valine and aspartic acid.

ANSWER Draw the tetrahedron, assign priorities (Chapter 4, p. 157), and draw the arrow 1→2→3. If it runs clockwise, the molecule is (R); if counterclockwise, it is (S).

(S) Enantiomer   (R) Enantiomer
Valine

(S) Enantiomer   (R) Enantiomer
Aspartic acid

Amino acids are usually not described as (S) or (R). Instead, the Fischer projection system is used (Chapter 22, p. 1225). The acid group is drawn at the top, and, as with the sugars, vertical bonds are retreating from you and horizontal bonds are coming toward you. The position, left or right, of the amino group determines whether the amino acid is D (right) or L (left). As shown in Figure 23.5, a simple (S) amino acid is also L in the Fischer terminology used for carbohydrates.

FIGURE 23.5 An (S) amino acid is L in the Fischer terminology.

Fischer projections of L-alanine = 2-(S)-alanine

WORKED PROBLEM 23.2 Draw the L-isomers of valine and cysteine in Fischer projection.

ANSWER Remember the convention for Fischer projections. Horizontal bonds are coming toward you and vertical bonds are retreating from you.

L-Valine is the (S) enantiomer

L-Cysteine is the (R) enantiomer

As you can see from Table 23.1, amino acids are high-melting solids. Does this not seem curious? Common amino acids have molecular weights between 90 and 150, and this molecular weight does not seem sufficient to produce solids. This conundrum leads directly to Section 23.2c.

23.2c Acid–Base Properties of Amino Acids.

Why are amino acids solids? To answer this question, let's ask another. What reaction do you expect between the generic amine and the generic acid of Figure 23.6?

A simple organic acid has a pK_a of about 4.5, and an ammonium ion has a pK_a of about 8–10. An amine and an acid must react to give the ammonium salt of Figure 23.7, as the carboxylic acid is a much stronger acid than an ammonium ion. Exactly the same reaction must take place within a single amino acid to form the "inner salt" or **zwitterion**, a molecule in which full positive and negative charges coexist (Fig. 23.7).

FIGURE 23.6 What reaction must occur between these two molecules?

FIGURE 23.7 An amine and a carboxylic acid must undergo acid–base chemistry to give an ammonium salt of the acid. In amino acids, an intramolecular version of the same reaction gives a zwitterion.

Amino acids are ionic compounds under normal conditions, and that accounts for their high melting points. Other properties are in accord with this notion. They are miscible with water, but quite insoluble in typical nonpolar organic solvents. They have high dipole moments. Of course, the charge state of an amino acid depends on the pH of the medium in which it finds itself. Under neutral, or nearly neutral conditions they will exist as the zwitterions of Figure 23.7. If we lower the pH, eventually the carboxylate anion will protonate, and the amino acid will exist primarily in the ammonium ion form. Conversely, if we raise the pH, eventually the ammonium ion site will be deprotonated, and the molecule will exist as the carboxylate anion (Fig. 23.8).

FIGURE 23.8 At low pH, the zwitterion will protonate to give a net positively charged ion. At high pH, a carboxylate anion, net negatively charged, will be formed.

PROBLEM 23.3 The pK_a for a simple organic acid is 4–5. Why are amino acids so much more acidic?

Amino acids can be described by a series of pK_a values. The first refers to the deprotonation of the acid function, and, as you can see from Table 23.1, is in the range 1.7–2.6 for the 20 commonly encountered molecules.

The second pK_a value is for deprotonation of the ammonium ion ($^+NH_3CHRCOOH$), and is in the range 9–10.8 (Table 23.1).

The pH at which the concentration of the ammonium ion ($^+NH_3CHRCOOH$) equals that of the carboxylate ion ($NH_2CHRCOO^-$) is called the **isoelectric point**.

PROBLEM 23.4 Calculate the pH of the isoelectric point for alanine ($pK_a = 2.3$ for COOH and $pK_a = 9.7$ for $^+NH_3$). *Hint*: Write expressions for the K_a for each of the two equilibria involved, then multiply them. Remember that at the isoelectric point the concentrations of $^+NH_3-CHR-COOH$ and $H_2N-CHR-COO^-$ are equal.

At this pH the amino acid is primarily in the zwitterionic form, and the number of positively charged species exactly equals the number of negatively charged molecules.

As you can see from Table 23.1, many amino acids bear acidic or basic side chains, which complicates the picture somewhat and can make the determination of the isoelectric point a bit more difficult. These side groups are also involved in acid–base equilibria and there will be pK_a values associated with them. For example, arginine will be doubly protonated at very low pH (Fig. 23.9).

FIGURE 23.9 Arginine is twice protonated at low pH.

Arginine (Arg)

PROBLEM 23.5 In the guanidino group of arginine, where is the site of initial protonation?

The most acidic group is the carboxylic acid, $pK_a = 2.2$. Second is the ammonium ion, $pK_a = 9.1$. The guanidinium ion has a pK_a of 12.5 and will be the last position deprotonated as the pH is raised.

PROBLEM 23.6 Which ring nitrogen in histidine will be protonated first?

These acid–base properties give rise to an especially powerful separation method widely used in analysis of amino acid mixtures. This method is called **electrophoresis**.

It works this way. A mixture of amino acids is placed on moistened paper. At a given pH, different amino acids will exist as different mixtures of positive and negative ions. When a current is applied to the paper, the positively charged ions will migrate to the cathode (negative pole) and the negatively charged ions to the anode (positive pole) at rates depending on the pH and the strength of the field, which are controllable variables, and the size and net electric charge of the amino acid, which are individual properties of the individual amino acids. The net result is a separation, as the amino acids migrate at different rates. For example, at pH 5.5, the amino acids Lys, Phe, and Glu exist in the forms shown in Figure 23.10.

FIGURE 23.10 The charge states of Lys, Phe, and Glu at pH 5.5.

When a current is applied, Phe will not move, Lys will migrate to the cathode and Glu to the anode (Fig. 23.11).

FIGURE 23.11 Separation of three amino acids by electrophoresis.

23.2d Syntheses of Amino Acids. One very simple method is to alkylate ammonia using the S_N2 reaction of ammonia with an α-halo acid (Fig. 23.12).

FIGURE 23.12 Simple alkylation of an α-halo acid will give an α-amino acid.

THE GENERAL CASE

SPECIFIC EXAMPLES

Glycine
(65%)

Norleucine
(65%)

In Chapter 7 (p. 334), a closely related alkylation method appeared as a means of producing alkylamines. A problem for this synthetic procedure was the ease of overalkylation. The simple alkylation reaction is not generally a practical method for producing alkylamines. It is rather more useful in making α-amino acids. The α-bromo acids are easily available, and a vast excess of ammonia can be used to minimize overalkylation. Moreover, the first formed α-amino acids are not especially effective competitors in the S_N2 reaction. They are sterically congested and less basic than simple amines. This technique is an effective way of making both "natural" and "unnatural" amino acids.

PROBLEM 23.7 Devise simple syntheses of the following α-bromo acids starting from alcohols containing no more than four carbon atoms and inorganic reagents of your choice (p. 1012).

(a)   (b)

PROBLEM 23.8 Why are α-bromo acids (or their carboxylate salts) more easily alkylated than simple alkyl bromides in the S_N2 reaction (Chapter 12, p. 585)?

A successful variation of the simple alkylation synthesis involves a protected, disguised amine, and is called the Gabriel synthesis, or, sometimes, the **Gabriel malonic ester synthesis**. In this procedure, bromomalonic ester is first treated with potassium phthalimide in an S_N2 displacement of bromide ion. The nitrogen atom destined to become the amine nitrogen of the amino acid is introduced at this point, and rendered nonnucleophilic by the flanking carbonyl groups. In the next step, the malonic ester product is treated with base and an active alkyl halide to introduce the R group of the amino acid. Finally, treatment with acid or base, followed by neutralization, uncovers the amine. The phthalimide group is hydrolyzed to phthalic acid, and the malonic ester is hydrolyzed to a malonic acid. Gentle heating induces decarboxylation to give the amino acid. There are many variations on this general theme (Fig. 23.13).

FIGURE 23.13 The Gabriel malonic ester synthesis.

PROBLEM 23.9 Provide detailed mechanisms for the reactions of Figure 23.13.

WORKED PROBLEM 23.10 In a simpler version of this reaction, acetamidomalonic ester is used. Sketch out the steps in this related amino acid synthesis.

Acetamidomalonic ester   **Phenylalanine**

ANSWER It is basically the same process as that of Figure 23.13 and Problem 23.9. Advantage is first taken of the acidity of the doubly α hydrogen to introduce an R group.

Now acid (or base) is used to convert the esters into carboxylic acids, and to hydrolyze the amide. Heating decarboxylates the malonic acid and liberates the amino acid.

In the **Strecker synthesis** (Adolph Strecker, 1822–1871), a cyanide is used as the source of the acid portion of an amino acid. An aldehyde is treated with ammonia, the ultimate source of the amino group, and cyanide ion. In the first step of the reaction, a carbinolamine is formed (Chapter 16, p. 852) and water is lost to give an imine. The imine is then attacked by cyanide to give an α-amino cyanide. Acid hydrolysis of the nitrile gives the acid (Chapter 19, p. 1059; Fig. 23.14).

THE GENERAL CASE

Carbinolamine   Imine

A SPECIFIC EXAMPLE

Overall yield
(36%)

FIGURE 23.14 The Strecker amino acid synthesis.

The first step in this reaction, imine formation, is analogous to the first step in the Mannich condensation in which the role of the amine is also to produce a reactive imine (p. 949). This imine can next be attacked by the base cyanide to give the α-amino cyanide (Fig. 23.15).

proton transfers

α-Amino cyanide

Acid hydrolysis
of the cyanide

FIGURE 23.15 The mechanism of the Strecker synthesis.

As we have seen, amino acids in nature are almost all encountered in optically active (S), or L forms (save glycine, which is achiral). None of the syntheses outlined in this section automatically produces the natural (S) enantiomer specifically. So we either need a method for resolution (separation) of the racemic pairs of enantiomers formed in these syntheses or a method of selectively producing the (S) enantiomers.

23.2e Resolution of Amino Acids. The standard method is the one we first discussed in our chapter on chirality (Chapter 4, p. 176) and involves transformation of the pair of enantiomers, which have identical physical properties, into a pair of diastereomers, which do not, and can be separated by physical means (most often crystallization or chromatography).

PROBLEM 23.11 There actually is one physical property not shared by enantiomers. What is it?

Optically active amines are used to transform the enantiomeric amino acids into diastereomeric ammonium salts that can be separated by crystallization. The chiral amines of choice are alkaloids, which are easily available optically pure from sources in nature (Chapter 6, p. 266). The process of separation is outlined in Figure 23.16.

FIGURE 23.16 Separation, or resolution, of a pair of enantiomeric amino acids.

However, it seems that there must be a way to avoid the tedium of fractional crystallization and make the optically active compounds directly. After all, Nature does this somehow, and we should be clever enough to find out how this happens and imitate the process. One method nature uses to make optically active amino acids involves a reductive amination. Ammonia and an α-keto acid react in the presence of an enzyme and a source of hydride, usually NADH (Chapter 16, p. 879), to produce an optically active chiral amino acid (Fig. 23.17).

FIGURE 23.17 Synthesis of an optically active L-amino acid through reductive amination.

There are two basic imitative procedures. In one, actual microorganisms are used in fermentation reactions that can directly produce some L-amino acids.

In a more subtle imitative (biomimetic) procedure, enzymes are used to react with one enantiomer of a racemic pair of mirror-image molecules. As you might guess, it is usually the L isomer that is "eaten" by the enzyme, because enzymes have evolved in an environment of L isomers. So it is usually the D isomers that are ignored by enzymes. In an extraordinarily clever procedure, this preference for natural, L, enantiomers can be used to achieve a **kinetic resolution**. The mixture of enantiomers is first acetylated to create amide linkages. An enzyme, hog-kidney acylase, hydrolyzes the amide linkages *of only L-amino acids*. So, treatment of the pair of acetylated amino acids leads to formation of the free L-amino acid. The acylated D enantiomer is unaffected by the enzyme, which has evolved to react only with natural L-amino acids. The free L-amino acid can then easily be separated from the residual D acetylated material. The process is shown in Figure 23.18.

FIGURE 23.18 A kinetic resolution of a pair of enantiomeric amino acids.

23.3   REACTIONS OF AMINO ACIDS

The simple expectation that the chemical reactions of the amino acids would be a combination of acid (Chapters 18 and 19) and amine (Chapter 6) chemistry would be right. Indeed, we have already seen some of this in the acid–base chemistry described in Section 23.2c. A few more important and useful examples follow.

23.3a Acylation Reactions. The amino group of amino acids can be acylated using all manner of **acylating agents**. Typical examples of effective acylating agents are acetic anhydride, benzoyl chloride, and acetyl chloride (Fig. 23.19).

FIGURE 23.19 The amino group of an amino acid can be acylated.

23.3b Esterification Reactions. The acid function of amino acids can be esterified using typical esterification reactions. Fischer esterification is the simplest process (Fig. 23.20).

FIGURE 23.20 Fischer esterification of the acid group of an amino acid.

PROBLEM 23.12 Write a mechanism for the acylation reactions of Figure 23.19.

PROBLEM 23.13 If amino acids are typically in their zwitterionic forms, how are acylation and esterification reactions, which depend on the presence of free amine or acid groups, possible?

23.3c Reaction with Ninhydrin. Amino acids react with **ninhydrin** to form a deep purple molecule. It is this purple color that is used to detect amino acids. Although it may seem curious at first, all amino acids except proline, regardless of the identity of the R group, form the *same* purple molecule on reaction with ninhydrin. How can this be?

Ninhydrin is the hydrate of indan-1,2,3-trione, and like all hydrates, is in equilibrium with the related carbonyl compound (Fig. 23.21).

Indan-1,2,3-trione   Ninhydrin

FIGURE 23.21 Ninhydrin is the hydrate of indan-1,2,3-trione.

PROBLEM 23.14 Why is the trione of Figure 23.21 hydrated at the middle, not the side carbonyl group?

The amino acid first reacts with the trione to form an imine through the small amount of the free amine present at equilibrium. We have seen this kind of reaction many times (Chapter 16, p. 852). Decarboxylation produces a second imine that is hydrolyzed under the aqueous conditions to give an amine. It is in this hydrolysis step that the R group of the original amino acid is lost. Still another imine, the third in this sequence, is then formed through reaction of the amine with another molecule of ninhydrin (Fig. 23.22). The extensive conjugation in the product leads to a very low energy absorption in the visible region (Chapter 12, p. 506), and the evident purple color.

Ninhydrin

First imine

decarboxylation

hydrolysis
H_2O:

Second imine

$+ \ CO_2$

R Group of original amino acid lost here

Purple!

FIGURE 23.22 Formation of a purple molecule through reaction of an amino acid with ninhydrin.

By far, the most important reaction of amino acids is their polymerization to peptides and proteins through the formation of new amide bonds (Fig. 23.23). The following sections will be devoted to this reaction and some of its consequences.

FIGURE 23.23 Polypeptides are polyamino acids linked through amide bonds.

Amide bonds   Amide bond

SUMMARY

Many of the reactions of acids and amines appear in the chemistry of amino acids. Acylation and Fischer esterification are two examples. There are some new reactions, however, that involve both the amino and the acid groups. Reaction with ninhydrin is an example.

23.4   PEPTIDE CHEMISTRY

23.4a Nomenclature and Structure. Figure 23.24 shows the schematic construction of a three amino acid peptide (a tripeptide) made from alanine, serine, and valine. Don't confuse this picture with real synthetic procedures—we will get to them soon enough. Peptides are always written and named with the **amino terminus** on the left, with the structure, and name, proceeding toward the **carboxy terminus** on the right. So the tripeptide in Figure 23.24 is alanylserinylvaline, or Ala·Ser·Val, or A·S·V. The amino acid terminus always starts the name, and the carboxy terminus ends it.

Alanine   Serine   Valine   A tripeptide: Alanylserinylvaline
Ala · Ser · Val
A · S · V

Amide links to be made

FIGURE 23.24 The tripeptide, alanylserinylvaline, Ala-Ser-Val. The amino acid terminus starts the name, and the carboxy terminus ends it.

PROBLEM 23.15 Write structures for the peptides named in the box below.

Pro · Gly · Tyr

Asp · His · Cys

Glu · Thr · Phe

PROBLEM 23.16 Name the peptides drawn in Problem 23.15.

The sequence of amino acids in a peptide or protein constitutes its **primary structure**. Often these complex series of amino acids are linked in another, intermolecular way. Connections between chains are often formed through **disulfide bridges**. Cysteine is usually the amino acid involved in these bridges, which are made through oxidation of the thiol groups (Fig. 23.25). For the chemical oxidation of thiols to disulfides, see Chapter 16, page 874.

FIGURE 23.25 Formation of disulfide bonds can link one polyamino acid chain to another.

Cys

oxidation

Disulfide

Cys

A cross-linked pair of chains— the link is a disulfide (S–S) bridge

There are other important structural features of polyamino acids. The amide groups prefer planarity and the internal hydrogen bonds between the NH hydrogens and carboxy carbonyl groups, along with the structural constraints imposed by the disulfide links, lead to regions of order in most large polypeptides.

WORKED PROBLEM 23.17 Explain carefully why amides prefer planarity.

ANSWER Amides are stabilized by resonance.

This resonance stabilization is maximized when the 2*p* orbitals on N, C, and O overlap as much as possible, which requires sp^2 hybridization and planarity.

Especially common are right-handed, or α-helical regions, and repeating folded sections known as **β-pleated sheets**. Both of these arrangements allow regular patterns of hydrogen bonding to develop, thereby imposing order on the formally linear sequence of amino acids. In the α-helical regions the hydrogen bonds are between the coils of the helix, whereas in the pleated sheets they hold lengths of the chains in roughly parallel lines (Fig. 23.26). Other regions consist of completely disordered, **random coil**, series of amino acids. The combination of these structural features determines the folding pattern, or **secondary structure** of the polypeptide.

FIGURE 23.26 Two examples of ordered secondary structure of polypeptides, the α-helix and β-pleated sheet.

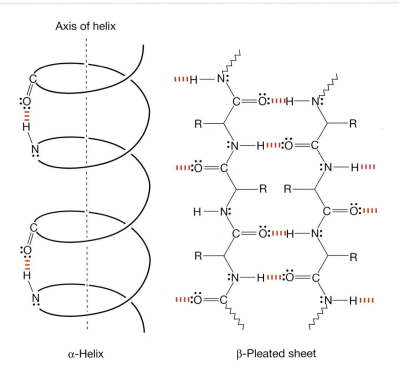

Axis of helix

α-Helix   β-Pleated sheet

There are even higher structural orders for these molecules, which is not an obvious point. It might have been that polypeptides were best described as ordered regions of secondary structure (α-helix or β-pleated sheets) connected by sections of random coil. In such a case, a protein would have no fixed overall structure beyond the sections defined by secondary structure. But this has been shown not to be the case. X-ray structure determinations of crystallized proteins have revealed these species to have *specific*, three-dimensional structures.* This **tertiary structure** of the protein is determined partly by hydrogen bonding, but also by van der Waals and electrostatic forces. These giant molecules are composed of a carbon backbone to which side chains, or R groups, are attached. These R groups are either hydrocarbon-like, nonpolar groups such as methyl, isopropyl, or benzyl, or highly polar amino, hydroxyl, or sulfhydryl groups. In the natural environment of a protein, water, it should be no surprise to find that the protein adjusts its shape so that the polar, **hydrophilic** groups are aimed outward toward the polar solvent, whereas the nonpolar, "greasy" **hydrophobic** hydrocarbon portions cluster inside the molecule, maximally protected from the hostile aqueous environment (Fig. 23.27).

Many proteins adopt these globular shapes, which maximize interior hydrophilic and exterior hydrophobic interactions, but others are fibrous, taking the form of a superhelix composed of rope-like coils of α-helices. Figure 23.28 shows a typical polyamino acid, the enzyme lactate dehydrogenase which incorporates five α-helices and six β-sheets.

R = Hydrocarbon-like, nonpolar side chains
R = Polar side chains

FIGURE 23.27 A globular protein, ordered so as to put the nonpolar side chains in the inside of the "glob" and the polar side chains outside in the polar solvent medium.

Lactate dehydrogenase

FIGURE 23.28 Lactate dehydrogenase.

*It is a general problem that until very recently the structures of proteins could only be determined by X-ray diffraction studies on crystallized materials. Work was limited by the availability of crystals, and crystallization of these molecules is usually difficult. Successful formation of X-ray quality crystals of a protein is cause for celebration (and publication) even before the work of structure determination begins. But how do we know that the structure of the molecule in solution is the same as that in the solid state crystal? Remember, biological activity is tied intimately to the detailed structure of these molecules—might we not be led astray by a structure that owed its shape to crystal packing forces in the solid state, and that was quite different in solution? Indeed we might. These days, it is becoming possible to use very high field NMR spectrometers, along with increasingly sophisticated NMR pulse techniques, to derive structures directly in solution. Remember that high-field spectrometers are anything but cheap; the cost is about 10^3 per megahertz of your tax dollars.

This tertiary structure of a protein is extraordinarily important for it determines the shape of these huge molecules and biological activity depends intimately on these shapes. Evolution has produced proteins in which pockets, or bays, exist into which certain molecules, or substrates, fit and can be bound in the proper orientation for reaction. The reaction could be complicated, or as simple as the hydrolysis of an ester. In other molecules, substrates are bound only for transportation purposes. These **active sites** are a direct consequence of the primary structure that determines the secondary and tertiary structures. Figure 23.29 gives a schematic representation of this process.

FIGURE 23.29 A binding site in a protein captures (binds) a small molecule for transport, or reaction with an appropriately located reactive site.

How do we know that the *same* tertiary structure is adopted by every protein molecule of a given amino acid sequence or primary structure? One clue is that synthetic proteins have the same biological activity as the natural materials. If we reproduce the primary sequence of a protein correctly, the proper biological activity appears. As this biological activity depends *directly* on the details of shape, the molecule must be folding properly into the appropriate tertiary structure.

Another clue comes from experiments in which the tertiary structure of a protein is deliberately disrupted in a reversible way, which is called **denaturing** the protein. Many ways of denaturing a protein are irreversible. When an egg is cooked, the thermal energy supplied denatures the proteins in the egg white in an irreversible way. Once the egg is fried, there is no unfrying it. Similar chemistry can be induced by pH change. The curdling of milk products is an example. But some denaturing reactions are reversible. For example, disulfide bonds can be broken through treatment with thiols. This disruption of the secondary structure transforms the ordered sections of the protein into the random coil arrangements. Similar reversible disruptions of order can be achieved through treatment with urea (in a reaction whose precise mechanism remains unknown), or, sometimes, through heating (Fig. 23.30).

FIGURE 23.30 One method of denaturing proteins, of destroying the secondary and tertiary structure, is to break the disulfide bonds. If air oxidization re-forms the disulfides, the higher order structures are regenerated and biological activity is reestablished.

Active globular protein   Denatured (random coil)

If these denatured proteins are allowed to stand, the thiol groups are reoxidized by air to the appropriate disulfide linkages and the original tertiary structure with its biological activity returns!

There is even higher order to some proteins. Some proteins are collections of smaller polyamino acid units held together by intramolecular attractions—van der Waals and electrostatic forces—into super structures with **quaternary structure**. The classic example of quaternary structure is **hemoglobin**, the protein responsible for oxygen absorption and transport in humans. To accomplish these functions it incorporates four heme units, each consisting of an iron atom surrounded by heterocyclic rings positioned so as to stabilize the metal. Figure 23.31 shows the heme unit in blue, before and after oxygen absorption.

Coordinated O_2

HOOC

HOOC

Heme

Imidazole

Protein

Bound heme unit

FIGURE 23.31 The unoxidized and oxidized heme structures of hemoglobin.

The protein surrounding the heme is called globin, and in each of the subunits of globin an imidazole is poised to hold the heme in position. There are four subunits in hemoglobin, two each of two slightly different peptide chains. These are held together by electrostatic and van der Waals forces as well as hydrogen bonding, and take the shape of a giant tetrahedron (Fig. 23.32).

FIGURE 23.32 The quaternary structure of hemoglobin.

Given all this general information about structure, we still face two difficult problems. First, how do we determine the primary sequence of an unknown protein, and second, once we know that structure, how can we synthesize it?

23.4b Determination of Protein Structure.

If suitable crystals can be obtained, and if the X-ray diffraction pattern can be solved in a reasonable amount of time, the structure of the protein is apparent. However, this process is not yet a general procedure. Crystallization of these molecules remains difficult, and although the determination of structure of small molecules by X-ray is now routine and rapid, the determination of the structures of proteins remains difficult and time consuming.

In principle, we can get an overall picture of the composition of a protein by first destroying all disulfide links in the molecule, and then hydrolyzing all the peptide bonds. *Remember*: These peptide bonds are amides and simple acid hydrolysis should convert them into acids (Chapter 19, p. 1054). This procedure regenerates all the amino acids of which the molecule is built. Figure 23.33 shows a hypothetical example. Of course, this technique does nothing to reveal primary, or higher structures.

FIGURE 23.33 Cleavage of a hexapeptide to its constituent amino acids. The disulfide bonds are first broken to produce two fragments, which are hydrolyzed to break the amide bonds.

In practice, the pieces formed when the disulfide bonds are broken are first separated and then hydrolyzed. This separation is no trivial task, but several methods are now available. Electrophoresis is a technique we saw as useful for separating simple amino acids (p. 1278) and it works for polyamino acids as well. Various kinds of chromatography are also effective. In **gel-filtration chromatography** the mixture is passed over a column of polymer beads which, on the molecular level, contain holes into which the smaller peptide fragments fit more easily than the larger pieces. Accordingly, the larger fragments pass more rapidly down the column, and a separation is achieved. **Ion-exchange chromatography** uses electrostatic attractions to hold more highly charged pieces on the column longer than more nearly neutral molecules. As in any chromatographic technique, the more tightly held molecules move more slowly than those less tightly held by the medium (Fig. 23.34).

FIGURE 23.34 A schematic description of chromatographic separation of polyamino acids.

Once the peptide pieces are separated, each can be hydrolyzed to its constituent amino acids. Now we need methods for determining how much of each amino acid is present. Again chromatography is used. The amino acids are separated on an ion-exchange column. As each amino acid is eluted from the column, it emerges into a chamber in which it reacts with ninhydrin (p. 1286) to form a purple dye. The intensity of the purple color is proportional to the amount of amino acid present, and can be plotted against time or amount of solvent used for elution from the column. This procedure produces a **chromatogram** consisting of a series of peaks of various sizes. How do we tell which peak corresponds to which amino acid? This identification is done by running samples of known amino acids through the column, determining their **retention times**, and matching these with those of the unknown amino acids (Fig. 23.35).

FIGURE 23.35 Polypeptides can be hydrolyzed to constituent amino acids, which can be separated by chromatography. Reaction of each amino acid with ninhydrin gives a purple color that can be quantitatively analyzed spectroscopically. The intensity of the purple color is proportional to the amount of amino acid formed.

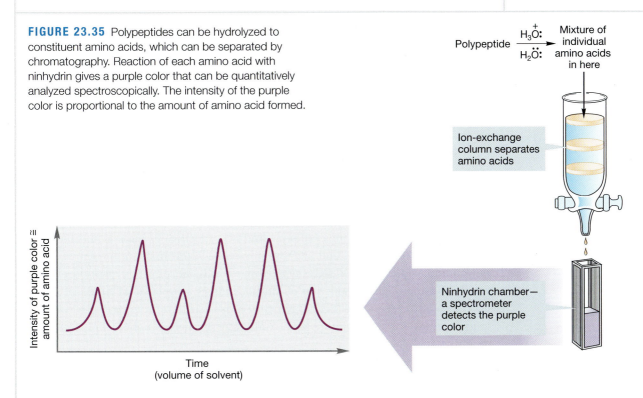

But this technique only gives us the coarsest picture of the structure of the protein: We know the constituent amino acids and their relative amounts, but we have no idea of their order, of the primary structure. An important task is to find a better way to get the actual sequence of amino acids in a protein. The first step is again to destroy the disulfide links and separate the constituent polypeptides. A simple method of determining the amino terminus is called the **Sanger degradation**, after Frederick Sanger (b. 1918). The peptide is allowed to react with 2,4-dinitrofluorobenzene. An adduct is formed that labels the terminal amino group with a 2,4-dinitrobenzene group (Fig. 23.36).

FIGURE 23.36 The amino acid at the amino terminus can be identified by reaction with 2,4-dinitrofluorobenzene, followed by hydrolysis. Only the amino acid at the amino end of the chain will be labeled.

Hydrolysis now generates a number of amino acids, but only the amino acid terminus is labeled! Unfortunately, we have had to destroy the entire peptide to make this determination!

It is also possible to discover the amino acid at the other end of a peptide chain, the carboxy terminus. The method of choice uses an enzymatic reaction in which one of a number of carboxypeptidases specifically cleaves the terminal amino acid at the carboxy end of the chain. As each terminal amino acid reacts, a new amino acid is revealed to be cleaved in turn by more carboxypeptidase. Careful monitoring of the production of amino acids as a function of time can give a good idea of the sequence (Fig. 23.37). Still, it is clear that things will get messy as long peptides are turned into a soup containing an increasingly complex mixture of amino acids.

FIGURE 23.37
Carboxypeptidase is an enzyme that cleaves only at the carboxy end of a polypeptide.

PROBLEM 23.18 What is the mechanism of adduct formation in Figure 23.36? What is the function of the nitro groups (Chapter 14, p. 726)?

A much better method would be one that included no overall hydrolysis step and that could be more controlled than the cleavage reactions set in motion by treatment with a carboxypeptidase. We need a method that unzips the protein from one end or the other, revealing at each step the terminal amino acid.

Methods have now been developed that allow just such a step-by-step sequencing of peptides. One effective way is called the **Edman degradation**, after Pehr Edman (1916–1977), and uses the molecule **phenylisothiocyanate**, Ph–N=C=S (Fig. 23.38). Isocyanates have appeared before (Chapter 19, p. 1098; Fig. 23.38), and an isothiocyanate is just the sulfur analogue of an isocyanate.

FIGURE 23.38 An isothiocyanate is just the sulfur version of an isocyanate, the product of a Curtius rearrangement.

PROBLEM 23.19 What is the mechanism of the isocyanate formation in Figure 23.38 (Chapter 19, p. 1098)?

Like isocyanates, the isothiocyanates react rapidly with nucleophiles. Reactions of isocyanates with ammonia or amines yield **ureas**. Isothiocyanates give a **thiourea** as the first formed intermediate (Fig. 23.39).

FIGURE 23.39 Isothiocyanates react with amines to give thioureas. Phenylisothiocyanate can be used to label the amino terminus of a polypeptide.

The thiourea can be cleaved in acid *without breaking the other amide bonds in the peptide.* As we saw in Chapter 21 (p. 1183), sulfur is an excellent neighboring group, and the introduction of the thiourea has the effect of placing a sulfur atom in just the right position to aid in cleavage of the terminal amino acid. The adjacent carbonyl is protonated, and then it is the neighboring sulfur that adds to it, triggering the cleavage reaction to give a molecule called a thiazolinone (Fig. 23.40).

FIGURE 23.40 The Edman procedure for sequencing polyamino acids uses phenylisothiocyanate. The amino terminus is identified by the structure of the phenylthiohydantoin formed. Notice that this method does not destroy the peptide chain as does the Sanger procedure.

Further acid treatment converts the thiazolinone into a phenylthiohydantoin containing the R group of the terminal amino acid, which can now be identified (Fig. 23.40). This process is indirect, as the initial product of reaction is not the phenylthiohydantoin, but a thiazolinone. The critical point is that the rest of the peptide remains intact and the Edman procedure can now be repeated, in principle, as many times as are needed to sequence the entire protein (Fig. 23.41).

FIGURE 23.41 The structures of polyamino acids can be determined through repeated applications of the Edman procedure, which is one method of sequencing.

PROBLEM 23.20 Provide a mechanism for the formation of the phenylthiohydantoin from the thiazolinone (Fig. 23.40).

But the practical world intrudes, and even the Edman procedure, good as it is, becomes ineffective for polypeptides longer than about 20–30 amino acids. Impurities build up at each stage in the sequence of determinations and eventually the reaction mixture becomes too complex to yield unequivocal results in phenylthiohydantoin formation. Accordingly, methods have been developed to cut proteins into smaller pieces at known positions in the amino acid sequence. These methods take advantage of both biological reactions in which enzymes break peptides only at certain amino acids, as well as a very few specific cleavages induced by small laboratory reagents.

Perhaps the best of these "chemical" (enzymes are chemicals, too) cleavages uses **cyanogen bromide (BrCN)** and takes advantage of the nucleophilicity of sulfur and the neighboring group effect to induce cleavage at the carboxy side of each methionine residue in a polypeptide (Fig. 23.42).

FIGURE 23.42 The carboxy end of each methionine is cleaved on treatment of a polypeptide with cyanogen bromide (BrCN).

The sulfur atom of methionine first displaces bromide from cyanogen bromide to form a sulfonium ion. Attack by the proximate carbonyl group of methionine displaces the excellent leaving group, methylthiocyanate, and forms a five-membered ring called a homoserine lactone. Hydrolysis steps follow to produce the end products, two smaller peptides. The full mechanism is shown in Figure 23.43.

FIGURE 23.43 The mechanism of cleavage using BrCN.

Most methods used to chop large peptides into smaller sequences of amino acids take advantage of enzymatic reactions. **Chymotrypsin**, for example, is an enzyme that cleaves peptides at the carboxyl groups of amino acids containing aromatic side chains (phenylalanine, tryptophan, or tyrosine). **Trypsin** cleaves only at the carboxy end of lysine and arginine. There are several other examples of this kind of specificity. Long peptide chains can be degraded into shorter molecules that can be effectively sequenced by the Edman procedure. A little detective work suffices to put together the entire sequence. Let's look at a simple example and then do a real problem.

Consider the decapeptide in Figure 23.44. We can determine the amino acid terminus through the Sanger procedure to be Phe. Next, we use the enzyme trypsin to cleave the molecule into three fragments. As we know the peptide must start with Phe, the first four amino acids are Phe·Tyr·Trp·Lys. The only other cleavage point is at Arg, so the Asp·Ile·Arg fragment must be the middle piece. If this tripeptide were at the end, with Arg as the carboxy terminus, no cleavage would be possible. The sequence must be Phe·Tyr·Trp·Lys·Asp·Ile·Arg·Glu·Leu·Val.

FIGURE 23.44 An example of peptide sequencing using Sanger and Edman procedures as well as enzymatic cleavage reactions.

PROBLEM 23.21 The nonapeptide bradykinin can be completely hydrolyzed in acid to give three molecules of Pro, two molecules each of Arg and Phe, and single molecules of Ser and Gly. Treatment with chymotrypsin gives the pentapeptide Arg·Pro·Pro·Gly·Phe, the tripeptide Ser·Pro·Phe, and Arg. End-group analysis shows that both the amino and carboxy termini are the same. Provide the sequence for bradykinin.

23.4c Synthesis of Peptides.
Now that we know how to sequence proteins, how about the reverse process—how can we put amino acids together in any sequence we want? How can we synthesize proteins? It is simply a matter of forming amide bonds in the proper order, and this task might seem easy. However, two difficulties surface as soon as we begin to think hard about the problem. First, there's a lot of work to do in order to make even a small protein. Imagine that we only want to create the nine amide linkages of the decapeptide in Figure 23.44. Even if we can make each amide bond in 95% yield, we are in trouble. A series of nine reactions

each proceeding in 95% yield, which probably seems quite good to anyone who has spent some time in an organic lab, gives an overall yield of only $(0.95)^9 = 63\%$. Obviously, to get anywhere in the real world of proteins, in which chain lengths of hundreds of amino acids are common, we are going to have to do much better than this.

Second, there is a problem of specificity. Suppose we want to make the trivial dipeptide Ala·Leu. Even if we can avoid simple acid–base chemistry, random reaction between these two amino acids will give us at least the four possible dimeric molecules, and in practice, other larger peptides will be produced as well (Fig. 23.45).

FIGURE 23.45 The possible combinations of two unprotected amino acids.

One simple strategy for avoiding random reaction would be to activate the carbonyl group of alanine by transforming the acid group of alanine into the acid chloride, and then treating it with leucine (Fig. 23.46).

FIGURE 23.46 Some specificity might be obtained by allowing the acid chloride of one amino acid (activated) to react with the other amino acid.

However, even this is hopeless! What is to prevent the original amino group of alanine from reacting with the acid chloride made from alanine? Nothing. Again, a mixture is certain to result (Fig. 23.47). Remember, we ultimately will have to make dozens, perhaps hundreds of amide bonds. We can tolerate no mediocre yields if we hope to make useful amounts of product. We need to find most efficient ways of making the amide bond specifically at the point we want it, and of avoiding side reactions.

FIGURE 23.47 Here, too, more than one product is inevitable.

In this trivial case, we need not only to activate the carboxyl group of alanine, but to block reaction at the amino group of alanine and the carboxyl group of leucine. Three positions must be modified (the carboxyl groups of alanine and leucine and the amino group of alanine). Left free for reaction are the two positions we hope to join, the activated carboxyl group of alanine and the unmodified amino group of leucine. Remember also that eventually the blocking, or protecting, groups will have to be removed (Fig. 23.48).

FIGURE 23.48 In any successful strategy, we must activate the carboxy end and block the amino end of one amino acid while blocking reaction at the carboxy end of the other amino acid.

Amino groups can be deactivated by converting them into carbamates through simple addition–elimination reactions (Chapter 19, p. 1099). Two popular methods involve the transformation of the free amine into a carbamate by reaction with di-*tert*-butyl dicarbonate or benzyl chloroformate (benzyloxycarbonyl chloride).Biochemists are even more addicted to acronyms than are organic chemists, and these protecting groups are called **tBoc** and **Cbz**, respectively (Fig. 23.49).

FIGURE 23.49 Two protecting, or blocking groups for amine ends of amino acids.

So now we have two ways of deactivating one amino group, in this case the amino group of alanine, so it cannot participate in the formation of the peptide amide bond. Now we need to deactivate the carboxy end of the other amino acid, leucine. This deactivation (protection) can be accomplished by simply transforming the acid into an ester (Fig. 23.50).

FIGURE 23.50 A carboxy end can be protected as a simple ester.

Now we need to do two more things. First, we still have to do the actual joining of the unprotected amino group of leucine to the unprotected carboxyl group of alanine, and we need to do this very efficiently and under the mildest possible conditions. Second, we need to deprotect the tied-up amino and carboxyl groups after the amide linkage has been formed.

Dicyclohexylcarbodiimide (DCC) is the reagent of choice for peptide bond making. Formation of the amide bond is a dehydration reaction (water is the other product), and DCC is a powerful dehydrating agent. At the end of the reaction it appears as dicyclohexylurea (DCU), the hydrated form of DCC (Fig. 23.51).

FIGURE 23.51 Coupling can be achieved using dicyclohexylcarbodiimide (DCC).

What is the mechanism of this coupling reaction? Remember that other cumulated double bonds as in ketenes (Chapter 19, p. 1061), isocyanates (Chapter 19, p. 1098), and isothiocyanates (Chapter 23, p. 1297) all react rapidly with nucleophiles. The related DCC is no exception, and it is attacked by amines to give molecules called guanidines. Carboxylates are nucleophiles too, and will also add to DCC (Fig. 23.52).

FIGURE 23.52 Cumulated double bonds react with all manner of nucleophiles, including amines and carboxylate anions.

The unprotected carboxylic acid group of alanine adds to DCC to give an intermediate that can react with a second carboxylic acid to give the anhydride. The amino acid, leucine, then adds through an addition–elimination mechanism to give the dipeptide. Note that the dipeptide is still protected in two places. The mechanism of amino acid coupling using DCC is shown in Figure 23.53.

FIGURE 23.53 The mechanism of amino acid coupling using DCC.

Now all we have to do is remove these protecting groups and, at last, we will have made our dipeptide, Ala·Leu. The Cbz blocking group is easily removed by catalytic hydrogenation, and tBoc by treatment with very mild acid, which does not cleave amide bonds. The acid can be regenerated by treatment of the ester with base. The ester will hydrolyze faster than the amide (Fig. 23.54).

Deprotect tBoc

1. mild acid
CF₃COOH
at 25 °C

2. neutralize

Ala·Leu

Deprotect Cbz

H₂
PtO₂

Ala·Leu

Deprotect ester

1. NaOH
H₂O

2. neutralize

Ala·Leu

What about the problem of yield? Remember, this process, with all its protection and deprotection steps, must be repeated many, many times in the synthesis of a large peptide or protein. These steps are efficient—they proceed in high yield as essentially no side pathways are followed. The problem of tedious repetition has been solved in an old fashioned way—by automation. Polypeptides can be synthesized by a machine invented by R. Bruce Merrifield (b. 1921). In Merrifield's procedure, the carboxy end of a tBoc-protected amino acid is first anchored to a polymeric material constructed of polystyrene in which some phenyl rings have been substituted with chloromethyl groups. The attachment of the amino acid involves a displacement of chlorine through an S_N2 reaction (Fig. 23.55).

Polystyrene

FIGURE 23.55 Polystyrene can be chloromethylated in a Friedel–Crafts procedure, and the chlorines replaced through S_N2 reaction with the unprotected carboxy end of an amino acid. This technique anchors an amino acid to the polymer chain.

PROBLEM 23.22 What is the function of the benzene ring of polystyrene? Could any ring-containing polymer have been used? Would cyclohexane have done as well, for example?

The tBoc group is then removed with mild acid hydrolysis, and a new, tBoc-protected amino acid is attached to the bound amino acid through DCC-mediated coupling. The steps of amino acid synthesis can now be carried out on this immobilized amino acid (Fig. 23.56).

FIGURE 23.56 The steps of amino acid synthesis can now be carried out on this immobilized amino acid.

This procedure can be repeated as many times as necessary. The growing chain cannot escape, as it is securely bound to the polystyrene; reagents can be added and byproducts can be washed away after each reaction. The final step is detachment of the peptide from the polymeric resin, usually through reaction with hydrogen fluoride (HF) or another acid (Fig. 23.57).

FIGURE 23.57 When synthesis is complete, the polypeptide can be detached from the polymer backbone by reaction with HF.

PROBLEM 23.23 Devise a synthesis of the dipeptide Leu·Ala starting from the amino acids leucine and alanine themselves and other reagents described in this chapter. Mechanisms are not necessary.

The first relatively primitive homemade Merrifield machine managed to produce a chain of 125 amino acids in an overall 17% yield, a staggering accomplishment. Merrifield was quite rightly rewarded with the Nobel prize for 1984.

23.5   NUCLEOSIDES, NUCLEOTIDES, AND NUCLEIC ACIDS

In Section 23.3c, the synthesis of proteins carried out by humans was described. In the natural world* quite another technique is used, and our uncovering of the outlines of how it works is one of the great discoveries of human history.

The **nucleic acids, deoxyribonucleic acid (DNA)**, and **ribonucleic acid (RNA)**, are polymers of **nucleotides** (phosphorylated nucleosides). A **nucleoside** is a β-glycoside formed between a sugar and a heterocyclic molecule (a base). In a ribonucleoside, the sugar ribose (Chapter 22, p. 1227) is attached to a base through a β-glycosidic linkage. In a deoxyribonucleoside, the sugar deoxyribose replaces ribose. 5-Phosphorylated nucleosides are called nucleotides (Fig. 23.58).

*It's an artificial distinction, almost as foolish as labels describing natural as opposed to artificial vitamin C. The Merrifield procedure is no less natural than what we are about to describe. It's not unnatural, it's just human-mediated.

Sugars

Ribose

2-Deoxyribose
(no oxygen at the 2-position)

Nucleosides

A ribonucleoside

A deoxyribonucleoside

Nucleotides

FIGURE 23.58 The sugars ribose and deoxyribose, and the corresponding nucleosides and nucleotides, phosphorylated nucleosides.

In the polymers (the nucleic acids DNA and RNA), the sugars, ribose in RNA and deoxyribose in DNA, are linked to each other through phosphoric acid groups attached to the 5′ position of one sugar and the 3′ position of the other. Each sugar still carries a heterocyclic base attached to the 1′ carbon. We saw these heterocyclic bases first in (Chapter 13, p. 662; Fig. 23.59).

FIGURE 23.59 Nucleic acids (DNA and RNA) are polymeric collections of nucleotides.

These bases can be different in RNA and DNA

This OH at the 2-position is present in RNA, but absent in DNA

One of the striking features of this replication machine, for that is exactly what we have begun to describe, is the economy with which it operates. So far, we have phosphoric acid and two sugars, ribose and deoxyribose, held together in polymeric fashion. One might have been pardoned for imagining that all manner of bases would have been attached to this backbone, and that the diversity of the world of living things would come from the introduction of detail in this way. But that would be wrong: There are only five bases attached to C(1) in DNA and RNA, and, in fact, three of them (cytosine, adenine, and guanine), are common to *both* kinds of nucleic acid and the other two (thymine and uracil) differ only in the presence or absence of a single methyl group. These bases are written in shorthand by using a single letter, C, A, G, T, or U (Fig. 23.60). The introduction of detailed information must come in another way.

RNA Bases **DNA Bases**

Cytosine (C)

Adenine (A)

Guanine (G)

Uracil (U) Thymine (T)

FIGURE 23.60 The five bases present in RNA and DNA. Three are common to both kinds of nucleic acid.

Notice the correspondence between these polymers and the other polymeric species we have seen, the proteins. In the polyamino acids, the units are linked by amide bonds. In DNA and RNA they are attached by phosphates. The information in proteins is carried by the side chains, the R groups whose steric and electronic demands determine the higher order structures—secondary, tertiary, and quaternary—

of these molecules. In DNA and RNA, it is the bases that perform the function of the R groups of proteins. There are only 20 common R groups, but there are even fewer bases—a mere four for each nucleic acid. Nevertheless, higher ordered structures are also found in the nucleic acids. Our next task is to see how this comes about.

A crucial observation was made by Erwin Chargaff (1905–2002) who noticed in 1950 that in many different DNAs the amounts of the bases adenine (A) and thymine (T) were always equal, as were the amounts of guanine (G) and cytosine (C). The four bases of DNA seemed to be found in pairs. This observation led to the reasonable notion that the members of the pairs must be associated in some way, and to the detailed picture of hydrogen-bonded dimeric structures of **base pairs** in which A was always associated with T and G with C. These combinations are the result of especially facile hydrogen bonding; the molecules fit well together (Fig. 23.61).

FIGURE 23.61 Adenine-thymine and guanine-cytosine form base pairs through effective hydrogen-bond formation.

The existence of base pairs shows how the order of bases on one polymeric strand of DNA can determine the structure of another strand, hydrogen bonded to the first. For a sequence of bases C C A T G C T A, the complementary sequence must be G G T A C G A T in order that the optimal set of hydrogen-bonded base pairs can be formed (Fig. 23.62).

FIGURE 23.62 The sequence of nucleotides on one polymer determines the sequence in another through formation of hydrogen-bonded base pairs, C-G and A-T.

Hydrogen-bonded base pairs as in Figure 23.61

This notion was also instrumental in the realization by James D. Watson (b. 1928) and Francis H. C. Crick (b. 1916), with immense help from X-ray diffraction patterns determined by Rosalind Franklin (1920–1958) and Maurice

Wilkins (b. 1916), that the structure of DNA was that of a double-stranded helix in which the two DNA chains were held together in the helix by A-T and G-C hydrogen-bonded base pairs. The diameter of the helix is about 20 Å, and there are 10 residues per turn of the helix. Each turn is about 34 Å long (Fig. 23.63).

Now we still have two outstanding problems: (1) Where does the density of information come from? How is the seemingly inadequate pool of five bases translated into the wealth of information that must be transferred in the synthesis of proteins containing specific sequences made from the 20 amino acids? (2) What is the actual mechanism of the information transferal? The second of these two questions is probably the easier to tackle.

The replication of DNA involves partial unwinding of the helix to yield two templates for the development of new DNA chains. Note that it is the order of bases in each strand that determines the growth of the next strand. Where there is a T, an A must appear, where there is a G, a C must attach, and so on. *Each strand of the original DNA must induce the formation of an exact duplicate of the strand to which it was originally attached* (Fig. 23.64).

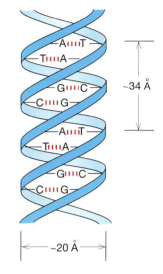

FIGURE 23.63 The famous double helix is held together by hydrogen bonds between base pairs.

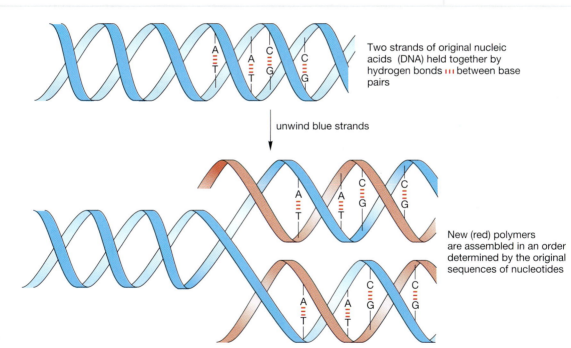

Two strands of original nucleic acids (DNA) held together by hydrogen bonds ··· between base pairs

unwind blue strands

New (red) polymers are assembled in an order determined by the original sequences of nucleotides

FIGURE 23.64 The replication machine works by first unraveling to form single-stranded sections. Assembly of new polymers is determined by the original base sequence. Where there is an A, a T must be added; a G requires a C.

Protein synthesis starts with a similar, but not quite identical process. A DNA strand can also act as a template for the production of an RNA polymer. The differences are that in RNA it is the sugar ribose that helps construct the sugar–phosphate backbone, not deoxyribose, and that the base uracil replaces thymine. The RNA produced is called **messenger RNA (mRNA)** (Fig. 23.65).

DNA   mRNA Molecules

Now replication uses a ribose, not deoxyribose, based
nucleotide. In RNA, the bases are A,C,G, and U *not* T

FIGURE 23.65 Synthesis of mRNA.

This new RNA molecule acts to direct protein synthesis in the following spectacularly simple way: Synthesis of each of the 20 common amino acids is directed by one or
more sequences of three bases, called **codons**. How many sequences of three bases can
be constructed from the available four RNA bases? The answer is $4^3 = 64$, more than
enough to code for the 20 common amino acids. Indeed, some amino acids are coded
by more than one codon. Some extra codons are needed to deal with the problems of
when to start amino acid production and when to stop. Table 23.2 gives three-base
codons for the 20 common amino acids and some frequently used commands.

TABLE 23.2   Codons for Amino Acids and SImple Commands

First Position (5′)	Second Position				Third Position (3′)
	U	C	A	G	
U	Phe	Ser	Tyr	Cys	U
	Phe	Ser	Tyr	Cys	C
	Leu	Ser	Stop	Stop	A
	Leu	Ser	Stop	Trp	G
C	Leu	Pro	His	Arg	U
	Leu	Pro	His	Arg	C
	Leu	Pro	Gln	Arg	A
	Leu	Pro	Gln	Arg	G
A	Ile	Thr	Asn	Ser	U
	Ile	Thr	Asn	Ser	C
	Ile	Thr	Lys	Arg	A
	Met	Thr	Lys	Arg	G
	Start	Thr	Lys	Arg	G
G	Val	Ala	Asp	Gly	U
	Val	Ala	Asp	Gly	C
	Val	Ala	Glu	Gly	A
	Val	Ala	Glu	Gly	G

As an example of how this process works, let's look at a strand of DNA with the sequence: T A C C G A A G C A C G A T T. Base pairing must produce a mRNA fragment: A U G G C U U C G U G C U A A (remember that in RNA the base U replaces T). A look at Table 23.2 shows that this translates into the following message: "Start - add Ala - add Ser - add Cys - Stop."

WORKED PROBLEM 23.24   Why would 20 amino acids be coded for by a total of 64 three-base sequences? Why is there so much overlap? Consider the alternative method of having two-base sequences coding for amino acid production. How many possible two-base sequences are there for a total of four bases?

ANSWER   There are only 2^4 possible combinations of four objects taken two at a time. So there are only 16 things that could be coded by two-base sequences of four different bases, which is not sufficient to code for the 20 common amino acids. Three-base codons turns out to be the most economical way to code for a system of 20 amino acids.

One further problem remains. How does mRNA direct amino acid synthesis? Other RNAs are used, called transfer RNA (tRNAs), which are relatively small and designed to interact with an enzyme, aminoacyl tRNA synthetase, to acquire one specific amino acid and carry it to the mRNA where it is added to the growing chain at the position of the correct codon.

23.6   SUMMARY

New Concepts

By far the major new concept in this chapter is that the assembly of α-amino acids into long polymeric chains leads to structures of higher order. The secondary structure of these peptides or proteins involves the regions of order within the polymer. Most often these are in the α-helical or β-pleated sheet arrangements, which are separated from each other by disordered sections of the chains, random coils. Disulfide bonds as well as electrostatic forces, van der Waals forces, and hydrogen bonding twist these molecules into shapes characteristic of individual proteins, the tertiary structure. Secondary and tertiary structure can be destroyed, sometimes only temporarily, by any of a number of denaturing processes. Finally, intermolecular forces can hold a number of these molecules together to form super molecules. We call this the quaternary structure.

These higher ordered structures are affected by the identities of the R groups on the constituent amino acids. It is the electronic and steric properties of the R groups that generate the particular secondary, tertiary, and quaternary structures of proteins.

These new structures require new analytical and synthetic techniques. Methods for determining the sequence of peptides and proteins involve physical techniques such as X-ray diffraction, as well as chemical techniques for revealing the terminal amino acids of the chains, and even for sequencing entire polymers.

Other biopolymers, the nucleic acids, are constructed not of amino acids, but of nucleotides, sugars bearing heterocyclic base groups and joined through phosphoric acid linkages. These molecules also have higher order, the most famous being the double-helical arrangement of DNA. Most remarkable of all is the ability of these molecules to carry the genetic code, to uncoil and direct the assembly not only of replicas of themselves, but also, through the mediation of RNA, of the polyamino acids called proteins. In the nucleic acids, the information is not carried in a diverse supply of R groups as it is in peptides and proteins, but in a small number of base groups. Hydrogen bonding in base pairs allows for the replication reactions, and three-base units called codons direct the assembly of amino acids into proteins.

Key Terms

Reactions, Mechanisms, and Tools

Reactions important in determining the sequence of a polymer of amino acids include several techniques for end-group analysis. There are enzymes (proteins themselves) that will cleave the amino acid marking the carboxy terminus of the polymer. The amino terminus can be found through the Sanger reaction.

In principle, an entire sequence can be unraveled using the Edman degradation in which phenylisothiocyanate is used in a procedure that cleaves the amino acid at the amino terminus and labels it as a phenylthiohydantoin.

Other agents are available for inducing specific cleavages in the polymers, which can yield smaller fragments for sequencing through the Edman procedure. These include various enzymes and small chemical reagents such as cyanogen bromide.

Synthesis of these polymeric molecules requires nothing more than the sequential formation of a large number of amide bonds. The DCC coupling procedure is effective, but it requires that the amino acid monomers be supplied in a form in which the ends to be attached are activated, or at least available, and other reactive points blocked, or protected. Those protecting and activating groups must be added in appropriate places and removed after the reaction is over (Fig. 23.66).

FIGURE 23.66 The DCC-mediated amino acid coupling. Protecting and activating groups must be added in appropriate places and removed after the reaction is over.

In turn, we need sources of the amino acids themselves. As with simple amines, a straightforward S_N2 alkylation procedure is usually inadequate. The best known syntheses of amino acids are the Gabriel and Strecker procedures.

Naturally occurring amino acids are optically active (except for glycine, which is achiral) and occur in the L form. Res-olution can be achieved through the classical technique of diastereomer formation followed by crystallization, or through other methods. The most clever of these involves a kinetic resolution in which advantage is taken of the ability of enzymes to attack only the naturally occurring (S) enantiomers, while the (R) isomer is left alone (Fig. 23.18).

Syntheses

1. AMINO ACIDS

Simple alkylation of ammonia using an α-bromo acid; overalkylation is a problem

Strecker synthesis

Gabriel synthesis. The R—X must be S_N2 active

2. DISULFIDE BRIDGES

thiols or air (oxygen)

Oxidation can convert the SH groups of cysteines into disulfide bonds between or within chains

3. GUANIDINES

Reaction of an amine (ammonia) with a carbodiimide

4. OPTICALLY ACTIVE AMINO ACIDS

1. optically active alkaloid
2. separate diastereomers
3. regenerate amino acids

Racemic mixture of (R) and (S)

(R) (S)

Classical resolution

(continued)

4. OPTICALLY ACTIVE AMINO ACIDS (continued)

A typical enzymatic synthesis of only the (S) enantiomer of an amino acid through reductive amination

The free (S) amino acid is produced in this kinetic resolution, along with acylated (R) amino acid

5. PROTECTED AMINO ACIDS

A tBoc-protected amino acid

A Cbz-protected amino acid

P = tBoc or Cbz   DCC-Mediated coupling; notice that the starting amino acids are prepared for reaction with protecting groups that are removed in later steps

6. THIOHYDANTOIN

The Edman degradation

7. THIOUREAS

Addition of an amine to an isothiocyanate

8. UREAS

Addition of an amine to an isocyanate

Hydration of a carbodiimide

23.7 ADDITIONAL PROBLEMS

PROBLEM 23.25

(a) Amides protonate on oxygen rather than nitrogen. Why?

(b) Amides ($pK_b \sim 14$) are weaker bases than amines ($pK_b \sim 5$). Why?

PROBLEM 23.26 Write structures for the following tripeptides and name them:

(a) Ala·Ser·Cys   (b) Met·Phe·Pro   (c) Val·Asp·His

PROBLEM 23.27 Name the following dipeptides:

(a)

(b)

PROBLEM 23.28 Draw (S)-serine and (S)-proline as tetrahedra.

PROBLEM 23.29 Write Fischer projections for L-serine and L-proline.

PROBLEM 23.30 Give the structures for the amino acids His, Arg, and Phe at pH = 3 and 12.

PROBLEM 23.31 Toward which electrode (anode or cathode) will the amino acids Asp and Lys migrate at pH 7?

PROBLEM 23.32 All of the amino acids listed in Table 23.1, except the achiral glycine, have the L configuration. Of the 19 amino acids with the L configuration, 18 are (S) amino acids. Which amino acid has an (R) configuration? Carefully explain why the configuration of this L amino acid is (R) rather than (S).

PROBLEM 23.33 In Section 23.3c, we saw that all the amino acids in Table 23.1, except proline, react with ninhydrin to form the same purple compound. In contrast, proline reacts with ninhydrin to give a yellow compound thought to have structure **1**. Propose a mechanism for the formation of **1**.

1

PROBLEM 23.34

(a) A peptide of unknown structure is treated with the Sanger reagent, 2,4-dinitrofluorobenzene, followed by acid hydrolysis. The result is formation of a compound containing signals integrating for eight aromatic hydrogens in its ^1H NMR spectrum. What information can you derive about the structure of this peptide?

(b) A second peptide, similarly treated, gives a compound in which a single hydrogen appears in the high resolution ^1H NMR spectrum as an eight-line signal in the region δ 3.5 ppm. What information can you derive about the structure of this peptide?

PROBLEM 23.35 A tripeptide of unknown structure is hydrolyzed in acid to one molecule of Ala, one of Cys, and one of Met. The tripeptide reacts with phenyl isothiocyanate, followed by treatment with acid, to give a phenylthiohydantoin that shows only a three-hydrogen doublet above δ 3 ppm in the ^1H NMR spectrum. The tripeptide is not cleaved by BrCN. Deduce the structure from these data.

PROBLEM 23.36 A dodecapeptide **1** is hydrolyzed in acid to give 2 Arg, Asp, Cys, His, 2 Leu, Lys, 2 Phe, and 2 Val. The Edman procedure yields a phenylthiohydantoin that shows two 3H doublets and a multiplet integrating for 1H at $\delta > 4.0$ ppm in the ^1H NMR spectrum. Treatment of **1** with trypsin leads to Val·His·Phe·Leu·Arg, Asp·Cys·Leu·Phe·Lys, and Val·Arg. Treatment of **1** with chymotrypsin leads to Val·His·Phe, Leu·Arg·Asp·Cys·Leu·Phe, and Lys·Val·Arg. Deduce the structure of **1** and explain your reasoning.

PROBLEM 23.37   In Section 23.4, we saw that the tBoc group can be used to protect (or block) the amino group of an amino acid during peptide synthesis. This group can be introduced by allowing the amino acid to react with di-*tert*-butyl dicarbonate (**1**). Once the amide linkage has been formed, the tBoc protecting group can be removed by mild acid "hydrolysis". Both of these reactions are illustrated below. Provide mechanisms for the introduction of the tBoc group and its removal. *Hint*: One of the products of the hydrolysis is isobutene.

PROBLEM 23.38   In Section 23.4c (Fig. 23.53), we saw that the acylating agent in DCC-mediated peptide bond formation is the anhydride. There is another intermediate that could act as the acylating agent. What is it, and how can it function as the acylating agent? *Hint*: This intermediate appears in Figure 23.53.

PROBLEM 23.39   Reaction of *N*-phenylalanine (**1**) with nitrous acid affords *N*-nitroso amino acid (**2**), which upon treatment with acetic anhydride yields "sydnone" (**3**). Sydnones belong to a class of compounds known as mesoionic compounds. These compounds cannot be satisfactorily represented by Lewis structures not involving charge separation. The name sydnone derives from the University of Sydney where the first examples were prepared in 1935. Reaction of **3** and dimethyl acetylenedicarboxylate (DMAD) gives pyrazole (**4**). Propose mechanisms for the formations of **2**, **3**, and **4**. *Hint*: For the formation of **4**, it may be helpful to consider the other possible Lewis structures for sydnone **3**.

PROBLEM 23.40   In Problem 23.12, you were asked to write a mechanism for the acylation of glycine. The mechanism presented in the answer is actually incomplete, at least when the reaction is run in the presence of excess acetic anhydride. For example, what is the purpose of the water in the second step? There is actually an intermediate involved in this reaction, the 2-oxazolin-5-one (**1**), colloquially referred to as an azlactone. Propose mechanisms for the formation of azlactone (**1**) and its hydrolysis to *N*-acetylglycine.

PROBLEM 23.41 In Problem 23.40, we saw that the azlactone (**1**) is an intermediate in the acetylation of glycine in the presence of excess acetic anhydride. The generation of azlactone (**1**) in this reaction lies at the heart of one of the oldest known amino acid syntheses. For example, if glycine (or *N*-acetylglycine) is treated with acetic anhydride in the presence of benzaldehyde and sodium acetate, the benzylidene azlactone (**2**) is formed. Azlactone (**2**) can then be converted into racemic phenylalanine (**3**) by the following sequence. Propose a mechanism for the formation of (**2**). *Hint*: Azlactone (**1**) is still an intermediate in this reaction.

PROBLEM 23.42 When optically active *N*-methyl-L-alanine (**1**) is heated with an excess of acetic anhydride, followed by treatment with water, the resulting *N*-acetyl-*N*-methylalanine (**2**) is racemic. If the acetylation reaction is run in the presence of dimethyl acetylenedicarboxylate (DMAD), the pyrrole **3** can be isolated in good yield. Account for the racemization of **2** and the formation of **3**. *Hint*: There is a common intermediate involved in these two reactions. Also, see Problem 23.39.

PROBLEM 23.43
(a) What mRNA sequence does the following DNA sequence produce? T A C G G G T T T A T C?
(b) What message does that mRNA sequence produce?

Key Terms

Absolute configuration   The arrangement in space of the atoms in an enantiomer, its stereochemical description as (R) or (S).

Acetal   The final product in the acid-catalyzed reaction of an aldehyde with an alcohol.

Acetoacetic ester   The product of the Claisen condensation of an acetate, such as

Acetone (dimethyl ketone)   The simplest ketone.

Acetylenes   Hydrocarbons of the general formula C_nH_{2n-2}. These molecules, also called alkynes, contain carbon–carbon triple bonds. The parent compound, $HC{\equiv}CH$, is called acetylene, or ethyne.

Acetylide   The anion formed by removal by base of a terminal hydrogen from an acetylene.

Achiral   Not chiral.

Acid chloride   A compound of the structure

Acid halide   A compound of the structure

$$R = F, Cl, Br, or I$$

Activation energy $(\Delta G^{\ddagger})$ The difference in free energy between the starting material and the transition state in a reaction. It is this amount of energy that is required for a molecule of starting material to be transformed into product.

Active site   The region in a protein in which a substrate molecule is bound and in which chemical reaction often takes place.

Acyl compound   A compound of the structure

Acylating agent   Any of a number of reagents capable of transferring an acyl group $(RC{=}O)$ to a nucleophilic site, usually N or O. Examples are acetyl chloride and acetic anhydide.

Acyl group   The group $CH_3{-}C{=}O$.

Acylium ion $(RC^+{=}O)$.

Addition–elimination reaction   A carbonyl group can be attacked to give a tetrahedral intermediate. If the carbon atom of the carbonyl was originally attached to a good leaving group, it can now be lost with regeneration of the carbonyl.

Alcohol $(R{-}OH)$ A molecule containing a simple hydroxyl group.

Aldaric acid   A diacid derived from an aldohexose by oxidation with nitric acid. It has the overall structure $HOOC{-}(CHOH)_4{-}COOH$. In an aldaric acid, the old aldehyde and primary alcohol ends of the sugar have become identical.

Aldehydes Compounds containing a monosubstituted carbon–oxygen double bond.

$$\underset{H}{\overset{R}{C}}=O$$

Aldohexose A hexose of the structure $OCH-(CHOH)_4-CH_2OH$.

Aldol condensation The acid- or base-catalyzed conversion of a ketone or aldehyde into a β-hydroxy aldehyde or β-hydroxy ketone. In acid, the enol is an intermediate; in base, it is the enolate anion.

Aldonic acid A monoacid derived from an aldohexose through oxidation with bromine in water. Only the aldehyde group is oxidized to the acid state. It has the structure $HOOC-(CHOH)_4-CH_2OH$

Aldopentose A pentose of the structure $OCH-(CHOH)_3-CH_2OH$.

Aldotetrose A tetrose of the structure $OCH-(CHOH)_2-CH_2OH$.

Aldotriose A triose of the structure $OCH-CHOH-CH_2OH$.

Alkaloid A nitrogen-containing compound, often polycyclic and generally of plant origin. The term is more loosely applied to other naturally occurring amines.

Alkanes The series of saturated hydrocarbons of the general formula C_nH_{2n+2}.

Alkenes Hydrocarbons of the general formula, C_nH_{2n}. These molecules, also called "olefins," contain carbon–carbon double bonds.

Alkoxide ion The conjugate base of an alcohol, RO^-.

Alkyl compounds Substituted alkanes. One or more hydrogens is replaced by another atom or group of atoms.

Alkyl halides Compounds of the formula $C_nH_{2n+1}X$, where X = F, Cl, Br, or I.

Alkynes Hydrocarbons of the general formula C_nH_{2n-2}. These molecules, also called acetylenes, contain carbon–carbon triple bonds.

Allene A 1,2-diene. A compound containing a carbon atom that is part of two double bonds.

Allyl group The group $H_2C=CH-CH_2$.

Allylic halogenation Specific formation of a carbon–halogen bond at the position adjacent to a carbon–carbon double bond.

Amide A compound of the structure

$$\underset{RO \qquad NR_2}{\overset{O}{\overset{\|}{C}}}$$

R can be H

Remember that this term also refers to the ions H_2N^-, RHN^-, R_2N^-, $RR'N^-$.

Amine A compound of the structure $R_3N:$, where R can be H or another group. Cyclic and aromatic amines are common.

Amine inversion The conversion of one pyramidal form of an amine into the other through a planar, sp^2 hybridized transition state.

Amine oxide A compound of the structure $R_3N^+-O^-$. Amine oxides are produced from the reaction of tertiary amines with hydrogen peroxide. See also *N*-**Oxide**.

α-Amino acids 2-Amino acetic acids, the monomeric constituents of the polymeric peptides and proteins.

Amino terminus The amino acid at the free amine end of the peptide polymer.

Ammonia ($H_3N:$) The simplest of all amines.

Ammonium ion R_4N^+, R = alkyl, aryl, or H.

Anchimeric assistance The increase in rate of a reaction that proceeds through intramolecular displacement over that expected of an intermolecular displacement.

Angle strain The increase in energy caused by the deviation of an angle from the ideal demanded by the hybridization.

Anhydride A compound formed by formal loss of water from two molecules of a carboxylic acid. The structure is

$$\underset{R \qquad\quad O \qquad\quad R}{\overset{O\quad\qquad O}{\overset{\|\quad\qquad \|}{C \qquad\quad C}}}$$

Aniline Aminobenzene.

Anion A negatively charged atom or molecule.

Anisole Methoxybenzene, $PhOCH_3$.

Annulene A cyclic polyene that is at least formally fully conjugated.

Anomers For aldoses, these are sugars differing only in the stereochemistry at C(1). They are C(1) stereoisomers.

Antarafacial motion Migration of a group from one side of a π system to the other in a sigmatropic shift.

anti Elimination An elimination reaction in which the dihedral angle between the breaking bonds, usually C—H and C—L, is 180°.

Antibonding molecular orbital Occupation of an antibonding molecular orbital by an electron destabilizes a molecule.

Aprotic An aprotic solvent is not a proton donor. An aprotic molecule does not bear a transferable proton.

Arene An aromatic compound containing a benzene ring or rings.

Arndt–Eistert reaction The use of the Wolff rearrangement to elongate the chain of a carboxylic acid by one carbon.

Aromatic character See **Aromaticity**.

Aromaticity The special stability of planar, cyclic, fully conjugated molecules with $4n+2$ π electrons. Such molecules will have molecular orbital systems with all bonding molecular orbitals completely filled and all antibonding and nonbonding molecular orbitals empty.

Arrow formalism A mapping device for chemical reactions. The electron pairs (lone pairs or bond pairs) are "pushed" using curved arrows that show the bonds that are forming and breaking in the reaction.

Atom A neutral atom consists of a nucleus, or core of protons and neutrons, orbited by a number of electrons equal to the number of protons.

Atomic orbital A three-dimensional representation of the solution of Schrödinger's equation describing the motion of an electron in the vicinity of a nucleus. Atomic orbitals have different shapes, which are determined by quantum numbers. The s orbitals are spherically symmetric, p orbitals roughly dumbbell shaped, and the d and f orbitals are even more complicated.

Aufbau principle When adding electrons to a system of orbitals, first fill the lowest energy orbital available before filling any higher energy orbitals. Electron-electron repulsion is minimized by filling systems of equi-energetic orbitals by singly occupying all orbitals with electrons of the same spin before doubly occupying any of them. See **Hund's rule.**

Axial hydrogens The set of six, straight up and down hydrogens in chair cyclohexane. Ring flip interconverts these hydrogens with the set of equatorial hydrogens.

Azetidine (azacyclobutane) A saturated four-membered ring containing one nitrogen atom.

Azides Compounds of the structure

$$R-\overset{..}{\underset{..}{N}}-\overset{+}{N}\equiv N\!:$$

Aziridine (azacyclopropane) A saturated three-membered ring containing one nitrogen atom.

Azo compounds Compounds of the structure, R—N=N—R′.

Bartlett–Condon–Schneider reaction An acid-catalyzed deuterium exchange reaction of alkanes. In the classic example, a trace of alkene catalyzes the exchange of all nine primary hydrogens of isobutane, but leaves the tertiary position unexchanged.

Base In biochemistry, one of the five heterocyclic molecules, adenine (A), thymine (T), uracil (U), cytosine (C), and guanine (G), attached to the 1′ position of the sugar (ribose or deoxyribose) in a nucleotide or nucleoside.

Base pair A hydrogen-bonded pair of bases, always A-T in DNA or A-U in RNA, and C-G in both DNA and RNA.

Base peak The largest peak in a mass spectrum, to which all other peaks are referred.

Benzaldehyde The common (and always used) name for the simplest aromatic aldehyde, "benzenecarboxaldehyde."

Benz[*a*]pyrene A powerfully carcinogenic polynuclear aromatic compound composed of five fused benzene rings.

Benzene The archetypal aromatic compound; a planar, regular hexagon of sp^2 hybridized carbons. The six $2p$ orbitals overlap to form a six-electron cycle above and below the plane of the ring. The molecular orbital system has three fully occupied bonding molecular orbitals and three unoccupied antibonding orbitals.

Benzhydryl group The group Ph_2CH.

Benzofuran A compound incorporating the structure

Benzoic acid (benzenecarboxylic acid) Ph—COOH.

Benzothiophene A compound incorporating the structure

Benzyl The $PhCH_2$ group.

Benzyne 1,2-Dehydrobenzene, C_6H_4.

Birch reduction The conversion of aromatic compounds into 1,4-cyclohexadienes through treatment with sodium in liquid ammonia–ethyl alcohol. Radical anions are the first formed intermediates.

tBoc A protecting group for the amino end of an amino acid that works by transforming the amine into a less basic carbamate.

Boltzmann distribution The range of energies of a set of molecules at a given temperature.

Bond dissociation energy (BDE) The amount of energy that must be applied to break a bond into two neutral species. See **Homolytic bond cleavage.**

Bonding molecular orbital Occupation of a bonding orbital by an electron stabilizes a molecule.

Bredt's rule Bredt noticed that there were no examples of bicyclic molecules with double bonds at the bridgehead position.

Bridged rings In a bridged bicyclic molecule, two rings share more than two atoms.

Bridgehead position The bridgehead positions are shared by the rings in a bicyclic molecule. In a bicyclic molecule, the three bridges emanate from the bridgehead positions.

Bromonium ion A three-membered ring containing bromine that is formed by the reaction of an alkene with Br_2. The bromine atom in the ring is positively charged.

***N*-Bromosuccinimide (NBS)** An effective agent for allylic bromination.

Brønsted acid A donor of a proton.

Brønsted base An acceptor of a proton.

Bullvalene   Bullvalene is a molecule with a fluxional structure. Every carbon of this $(CH)_{10}$ compound is bonded on time average to each of the other nine carbons.

Butyl group   The group $CH_3CH_2CH_2CH_2$.

sec-**Butyl group**   The group $CH_3CH_2CH(CH_3)$.

tert-**Butyl group**   The group $(CH_3)_3C$.

Cahn–Ingold–Prelog priority system   An arbitrary system for naming stereoisomers. It determines a priority system for ordering groups.

Cannizzaro reaction   The redox reaction of an aldehyde containing no α-hydrogens with hydroxide ion. Addition of hydroxide to the aldehyde is followed by hydride transfer to another aldehyde. Protonation generates a molecule of the carboxylic acid and the alcohol related to the original aldehyde.

Carbamates   Esters of carbamic acid. These molecules do not decarboxylate (lose CO_2) easily.

Carbamic acid   A compound of the structure.

These acids easily decarboxylate to give amines.

Carbanion   A compound containing a negatively charged carbon atom. A carbon-based anion.

Carbene   A short-lived neutral intermediate containing a divalent carbon atom. See also **Singlet** and **Triplet carbene**.

Carbenium ion   A name suggested for a molecule containing a trivalent, positively charged carbon. It is not widely used. See **carbocation**.

Carbinolamine   The initial product in the reaction between a carbonyl-containing molecule, $R_2C=O$, and an amine. It is analogous to a hemiacetal.

Carbocation   The compromise and currently widely used name for a molecule containing a trivalent, positively charged carbon atom.

Carbohydrate   A molecule whose formula can be factored into $C_x(H_2O)_y$. A "sugar" or "saccharide."

Carbonic acid   An unstable acid, H_2CO_3.

Carbonium ion   The traditional name for a molecule containing a trivalent, positively charged carbon. It has fallen into disuse.

Carbonyl compound   A compound containing a carbon–oxygen double bond.

Carbonyl group   The carbon–oxygen double bond.

$$C=O$$

Carboxylate anion   The resonance-stabilized anion formed on deprotonation of a carboxylic acid.

Carboxylic acid   A compound of the structure

Carboxy terminus   The amino acid at the free carboxylic acid end of the peptide polymer.

Catalyst   A catalyst functions to increase the rate of a chemical reaction. It is ultimately unchanged by the reaction and functions not by changing the energy of the starting material or product but by providing a lower energy pathway between them. Thus it operates to lower the energies of the transition states involved in the reaction.

Cation   A positively charged atom or molecule.

Cationic polymerization   A reaction in which an initially formed carbocation adds to an alkene that in turn adds to another alkene. Repeated additions can lead to polymer formation.

Cbz   A protecting group for the amino end of an amino acid that works by transforming the amine into a less basic carbamate.

Cellulose   A polymer of glucose in which C(4) of one glucose is linked in β fashion to C(1) of another.

Cephalosporin   A powerful antibiotic related to penicillin incorporating a β-lactam ring fused to a six-membered ring containing sulfur.

Chain reaction   A cycling reaction in which the species necessary for the first step of the reaction is produced in the last step. This intermediate then recycles and starts the process over again.

Chemical shift (δ)   The position, on the ppm scale, of a peak in an NMR spectrum. The chemical shift for ^{1}H and ^{13}C is given relative to a standard, TMS, and is determined by the chemical environment surrounding the nucleus.

Chichibabin reaction   Nucleophilic addition of an amide ion to pyridine (or a related heteroaromatic compound) leading to an aminopyridine. A key step involves hydride transfer.

Chiral   A chiral molecule is not superimposable on its mirror image.

Chirality   The ability of a molecule to exist in two nonsuperimposable mirror-image forms; "handedness."

Chromatogram   A plot of amount of component versus time for any chromatographic technique.

Chymotrypsin   An enzyme that cleaves sequences of amino acids after any amino acid containing an aromatic side chain.

cis   Hydrogens "on the same side." Applied to specify stereochemical (spatial) relationships in ring compounds and alkenes.

s-cis   The less stable, coiled form of a 1,3-diene.

Claisen condensation   A condensation reaction of esters in which an ester enolate adds to the carbonyl group of another ester. The result of this addition–elimination process is a β-keto ester.

Claisen–Schmidt condensation   A crossed aldol condensation of an aldehyde without α-hydrogens with a ketone that does have at least one α-hydrogen.

β-Cleavage   The fragmentation of a radical into a new radical and an alkene through breaking of the α–β bond.

Codon   A three-base sequence in a polynucleotide that directs the addition of a particular amino acid to a growing chain of amino acids. Some codons also give directions: "start assembly" "stop assembly," and so on.

Coenzyme   A molecule able to carry out a chemical reaction with another molecule only in cooperation with an enzyme. The enzyme's function is often to bring the substrate and the coenzyme together.

Concerted reaction   A single-barrier process. In a concerted reaction, starting material is converted into product with no intermediate structures.

Conformation   The three-dimensional structure of a molecule. Conformational isomers are interconverted by rotations about bonds.

Conformational analysis   The study of the relative energies of conformational isomers.

Conformational enantiomers   Enantiomers interconvertable through (generally easy) rotations around bonds within the molecule.

Conjugate acid   Some molecule plus a proton. Conjugate acids and bases are related by the gain and loss of a proton.

Conjugate base   Some molecule less a proton. Conjugate acids and bases are related by the gain and loss of a proton.

Conjugated double bonds   Double bonds in a 1,3–relationship are conjugated.

Conrotation   In a conrotatory process, the end p orbitals of a polyene rotate in the same sense (both clockwise or both counterclockwise).

Cope elimination   The thermal formation of alkenes through the pyrolysis of *N*-oxides.

Cope rearrangement   This [3,3] sigmatropic shift converts one 1,5-diene into another.

Coupling constant (*J*)   The magnitude (in hertz) of *J*, the measure of the spin–spin interaction between two nuclei.

Covalent bond   A bond formed by the sharing of electrons through the overlap of atomic or molecular orbitals.

Crossed aldol condensation   An aldol condensation between two different carbonyl compounds. This reaction is not very useful unless strategies are employed to limit the number of possible products.

Crown ether   A cyclic polyether often capable of forming complexes with metal ions. The ease of complexation depends on the size of the ring and the number of heteroatoms in the ring.

Cryptand   A three-dimensional, bicyclic, counterpart of a crown ether. Various heteroatoms (O, N, S) act to complex metal ions that fit into the cavity.

Cumulene   Any molecule containing at least three consecutive double bonds, $R_2C{=}C{=}C{=}CR_2$. The parent cumulene is butatriene.

Curtius rearrangement   The thermal or photochemical decomposition of an acyl azide to give an isocyanate.

Curved arrow formalism   The important bookkeeping device that keeps track of the motion of electrons in a chemical reaction.

Cyanides   Compounds of the structure RCN. These compounds are also commonly called nitriles.

Cyanogen bromide (BrCN)   A reagent able to cleave peptide chains after a methionine residue.

Cyanohydrin   The product of addition of hydrogen cyanide to a carbonyl compound.

Cycloaddition reaction   A reaction in which two π systems are converted into a ring. The Diels–Alder reaction and the 2 + 2 reaction of a pair of ethylenes to give a cyclobutane are examples.

Cycloalkenes   Ring compounds containing a double bond within the ring.

Cycloheptatrienylium ion   See **tropylium ion**.

Cyclopentadienyl anion   A five-carbon aromatic anion containing 6 π electrons (4n + 2, n = 1).

Daughter ion   In MS, an ion formed by the fragmentation of the first-formed parent ion.

Decarboxylation   The loss of carbon dioxide, a common reaction of 1,1-diacids and β-keto acids.

Decoupling   The removal of coupling between hydrogens or other nuclei (see **Coupling constant**) through either chemical exchange or electronic means.

Degenerate reaction   In a degenerate reaction, the starting material and product have the same structure.

Degree of unsaturation (Ω)   In a hydrocarbon, this is the total number of π bonds and rings.

Dehydrating agent   A reagent such as DCC or P_2O_5 that can promote the loss of water in a chemical reaction.

Dehydrobenzene   See **benzyne**.

Delocalization energy   The energy lowering conferred by the delocalization of electrons. In benzene, this is the amount by which benzene is more stable than the hypothetical 1,3,5-cyclohexatriene containing three localized double bonds. See **resonance energy**.

Denaturing   The destruction of the higher order structures of a protein, sometimes reversible, sometimes not.

Deoxyribonucleic acid (DNA)   A polymer of nucleotides made up of deoxyribose units connected by phosphoric acid links. Each sugar is attached to one of the bases, A, T, G, or C.

DEPT   Distortionless Enhancement with Polarization Transfer, a spectroscopic method of determining the number of hydrogens attached to a carbon.

Detergent   A long-chain alkyl sulfonic acid salt.

Dewar forms   Resonance forms for benzene in which overlap between 2p orbitals on two para carbons is emphasized. These forms superficially resemble Dewar benzene (bicyclo[2.2.0]hexa- 2,5-diene).

Dextrorotatory   The rotation of the plane of plane-polarized light in the clockwise direction.

Dial   A molecule containing two aldehyde groups, a dialdehyde.

Diastereomers   Stereoisomers that are not mirror images.

Diastereotopic   Diastereotopic hydrogens (or groups) are different both chemically and spectroscopically under all circumstances.

Diazirine   A three-membered ring containing both a double bond and two nitrogen atoms.

Diazo compounds   Compounds of the structure $R_2C=N_2$.

Diazo ketone   A compound of the structure

Diazonium ion   The group N_2^+ as in RN_2^+.

Diazotic acid   An intermediate in the nitrosation reactions of amines, R—N=N—OH. The enol form of a nitroso compound.

Dicyclohexylcarbodiimide (DCC)   A dehydrating agent effective in the coupling of amino acids through the formation of amide bonds.

Dieckmann condensation   An intramolecular, or cyclic, Claisen condensation.

Diels–Alder reaction   The concerted reaction of an alkene or alkyne with a 1,3-diene to form a six-membered ring.

Diene   Any molecule containing two double bonds.

Dihedral angle   The torsional, or twisting angle between two bonds. In an X—C—C—X system, the dihedral angle is the angle between the X—C—C and C—C—X planes.

Diimide (HN=NH)   A nonmetallic hydrogenating agent.

Diol   A molecule containing two OH groups. Also called a glycol.

Dione   A compound containing two ketone groups, a diketone.

Dioxirane   A three-membered ring containing two oxygen atoms. A versatile oxidizing agent.

Dipole moment   A dipole moment in a molecule results when two opposite charges or partial charges are separated.

1,3-Dipolar reagents   Molecules for which a good neutral structure cannot be written. Ozone is a typical example. These species undergo addition to π systems to give five-membered rings.

Diradical   A species containing two unpaired electrons, usually on different atoms.

Disproportionation   The reaction of a pair of radicals to give a saturated and unsaturated molecule by abstraction of a hydrogen by one radical from the position adjacent to the free electron of the other radical.

Disrotation In a disrotatory process, the end *p* orbitals of a polyene rotate in opposite senses (one clockwise and the other counterclockwise).

Disulfide bridges The attachment of amino acids through sulfur–sulfur bonds formed from the oxidation of cysteine CH_2SH side chains. Disulfide bridges can be formed within a single peptide or between two peptides.

Dithianes Six-membered rings containing a pair of sulfur atoms, usually in the 1,3 positions.

Double bond Two atoms can be attached by a double bond composed of one σ bond and one π bond.

E1cB Reaction An elimination reaction in which the first step is loss of a proton to give an anion. The anion then internally displaces the leaving group in a second step.

E1 Reaction The unimolecular elimination reaction. The ionization of the starting material is followed by the loss of a proton to base.

E2 Reaction The bimolecular elimination reaction. The proton and leaving group are lost in a single, base-induced step.

Eclipsed ethane The conformation of ethane in which all carbon–hydrogen bonds are as close as possible. This conformation is not an energy minimum, but the top of the barrier separating two molecules of the stable, staggered conformation of ethane.

Edman degradation The phenyl isothiocyanate-induced cleavage of the amino acid at the amino terminus of a peptide. Successive applications of the Edman technique can determine the sequence of a peptide.

Electrocyclic reaction The interconversion of a polyene and a ring compound. The end *p* orbitals of the polyene rotate so as to form the new σ bond of the ring compound.

Electron A particle of tiny mass (1/1845 of a proton) and a single negative charge.

Electron affinity A measure of the tendency for an atom or molecule to accept an electron.

Electronegativity The tendency for an atom to attract electrons.

Electronic spectroscopy The measurement of the absorption of energy when electromagnetic radiation of the proper energy is provided. An electron is promoted from the HOMO to the LUMO.

Electrophile A lover of electrons, a Lewis acid.

Electrophilic aromatic substitution The classic substitution reaction of aromatic compounds with Lewis acids. A hydrogen attached to the benzene ring is replaced by the Lewis acid and the aromatic ring is retained in the overall reaction.

Electrophoresis A technique for separating amino acids or chains of amino acids that takes advantage of the different charge states of different amino acids (or their polymers) at a given pH.

Enamine The nitrogen analogue of an enol, a vinyl amine. These compounds are nucleophilic and useful in alkylation reactions.

Enantiomers Nonsuperimposable mirror images.

Enantiotopic Enantiotopic hydrogens are chemically and spectroscopically equivalent except in the presence of optically active (single enantiomer) reagents.

Endergonic In an endergonic reaction the products are less stable than the starting materials.

endo Aimed "inside" the cage in a bicyclic molecule. In a Diels–Alder reaction, the endo product generally has the substituents aimed toward the newly produced double bond.

Endothermic reaction A reaction in which the bonds in the products are higher energy than those in the starting materials. In an endothermic reaction the product is less stable than the starting material.

Energy level See **Atomic orbital** and **Wave function**.

Enol A vinyl alcohol. These compounds usually equilibrate with the more stable keto forms.

Enolate The resonance-stabilized anion formed on treatment of an aldehyde or ketone containing an α-hydrogen with base.

Enthalpy change ($\Delta H°$) The difference in total bond energies between starting material and product in their standard states.

Entropy change ($\Delta S°$) The difference in disorder between the starting material and product in their standard states.

Episulfide A saturated three-membered ring containing a single sulfur atom. See **Thiirane**.

Episulfonium ion A three-membered ring containing a trivalent, positively charged sulfur atom.

Epoxide A saturated three-membered ring containing a single oxygen atom. See **Oxirane**.

Equatorial hydrogens The set of six hydrogens, more or less in the plane of the ring in chair cyclohexane. These hydrogens are interconverted with the set of axial hydrogens through ring "flipping" of the chair.

Equilibrium constant (K) The equilibrium constant is related to the difference in energy between starting material and products ($\Delta G°$) in the following way: $K = e^{-\Delta G°/RT}$.

Essential amino acid Any of the 10 amino acids that cannot be synthesized by humans and must be ingested directly.

Ester A compound of the structure

$$\begin{matrix} & O \\ & \| \\ R & C & OR \end{matrix}$$

Ester hydrolysis The conversion of an ester into an acid through treatment with an acid catalyst in excess water. The reverse of Fischer esterification. This reaction also occurs in base, and is called "saponification."

Ethene The simplest alkene is $H_2C{=}CH_2$. It is usually known as ethylene.

Ether A compound of the general structure, ROR, or ROR′.

Ethyl compounds Substituted ethanes; $CH_3CH_2{-}X$ compounds.

Ethylene The simplest alkene is $H_2C{=}CH_2$. It is more properly known as ethene, a name that is rarely used.

Exergonic In an exergonic reaction, the products are more stable than the starting materials.

exo Aimed "outside" the cage in a bicyclic molecule. In a Diels–Alder reaction, the exo product generally has the substituents aimed away from the newly produced double bond.

Exothermic reaction A reaction in which the bonds in the products are lower energy than those in the starting materials. In an exothermic reaction, the product is more stable than the starting material.

Extinction coefficient The proportionality constant ε in Beer's law, $A = \log I_0/I = \varepsilon lc$.

Fatty acids Long-chain carboxylic acids generated by the hydrolysis of fats. Fatty acids are derived from acetic acid and always contain an even number of carbons.

First-order reaction A reaction for which the rate depends on the product of a rate constant and the concentration of a single reagent.

Fischer esterification The conversion of a carboxylic acid into an ester by treatment with an acid catalyst in excess alcohol. The reverse of ester hydrolysis.

Fischer indole synthesis A synthesis of indoles starting from phenylhydrazones.

Fischer projection A schematic stereochemical representation. In sugars, the aldehyde group is placed at the top and the primary alcohol at the bottom. Horizontal bonds are taken as coming toward the viewer and vertical bonds as retreating. If the OH of the $H{-}C{-}OH$ adjacent to the CH_2OH group is on the right, the molecule is a D sugar; if it is on the left, it is an L sugar.

Fluxional structure In a molecule with a fluxional structure, a given atom does not have fixed nearest neighbors. Instead, the framework atoms move about over time, each being bonded on time average to all the others.

Force constant A property of a bond related to the bond strength: to the stiffness of the bond. Bonds with high force constants absorb at high frequency in the IR.

Formaldehyde The simplest possible aldehyde.

$$\begin{matrix} H \\ \ \ \ \diagdown \\ \ \ \ \ C{=}O \\ \ \ \ \diagup \\ H \end{matrix}$$

Four-center transition state The chair-like transition state for the Cope rearrangement in which carbons 1 and 1′ and 3 and 3′ are within bonding distance (eclipsed) but carbons 2 and 2′ are as far apart as possible.

Fragmentation pattern The characteristic spectrum of ions formed by decomposition of a parent ion produced in a mass spectrometer when a molecule is bombarded by high-energy electrons.

Free radical A neutral molecule containing an odd, unpaired electron. Also simply called "radical."

Friedel–Crafts acylation The electrophilic substitution of aromatic molecules with acyl chlorides facilitated by strong Lewis acids, usually $AlCl_3$.

Friedel–Crafts alkylation The electrophilic substitution of aromatic molecules with alkyl chlorides catalyzed by strong Lewis acids, usually $AlCl_3$.

Frost circle A device used to find the relative energies of the molecular orbitals of planar, cyclic, fully conjugated molecules. A polygon corresponding to the ring size of the molecule is inscribed in a circle, vertex down. The intersections of the polygon with the circle give the relative positions of the molecular orbitals.

Fully conjugated In a fully conjugated molecule, every carbon has a p orbital that overlaps effectively with the p orbitals on the adjacent atoms.

Fulvene A 5-methylene-1,3-cyclopentadiene. These are often synthesized through a Knoevenagel condensation using the cyclopentadienide anion and an aldehyde or ketone.

Functional group An atom or group of atoms that generally reacts the same way no matter what molecule it is in.

Furan An aromatic five-membered ring compound containing four CH units and one O atom.

Furanose   A sugar containing a five-membered cyclic ether.

Furanoside   A furanose in which the anomeric OH has been converted into an acetal.

Fused rings   Two rings sharing only two carbons.

Gabriel amine synthesis   A method of forming primary amines without overalkylation to give more substituted amines. Phthalimide is used as a masked amine to introduce the nitrogen atom.

Gabriel malonic ester synthesis   A synthesis of amino acids that uses phthalimide as a source of the amine nitrogen. Overalkylation is avoided by decreasing the nucleophilicity of the nitrogen in this way.

Gas chromatogaphy (GC)   A method of separation in which molecules are forced to equilibrate between the moving gas phase and a stationary phase packed in a column. The less easily a molecule is adsorbed in the stationary phase, the faster it moves through the column.

GC/IR   The combination of gas chromatography and infrared spectroscopy in which the molecules separated by a gas chromatograph are led directly into an infrared spectrometer for analysis.

GC/MS   The combination of gas chromatography and mass spectrometry in which the molecules separated by a gas chromatograph are led directly into a mass spectrometer for analysis.

Gel-filtration chromatography   A chromatographic technique that relies on polymeric beads containing molecule-sized holes. Molecules that fit easily into the holes pass more slowly down the column than larger molecules, which fit less well into the holes.

Geminal   1,1-Disubstituted.

Gibbs free energy change ($\Delta G°$)   The difference in free energy during a reaction. The parameter $\Delta G°$ is composed of an enthalpy ($\Delta H°$) term and an entropy ($\Delta S°$) term. $\Delta G° = \Delta H° - T\Delta S°$.

Glyceraldehyde   The aldotriose,
$$OCH-CHOH-CH_2OH$$

1,2-Glycol   A vicinal dialcohol. A molecule bearing hydroxyl groups on adjacent carbons. 1,2-Glycols come from the treatment of alkenes with osmium tetroxide or potassium permanganate, followed by hydrolysis, or by the acid- or base-catalyzed opening of epoxides.

Glycol   A dialcohol. See **diol**.

Glycoside   A sugar in which the anomeric OH at C(1) has been converted into an OR group.

Grignard reagent (RMgX)   A strongly basic organometallic reagent formed from a halide and magnesium in an ether solvent. An important and characteristic reaction is the addition to carbonyl groups.

Haloform   A compound of the structure HCX_3, where X is F, Cl, Br, or I.

Haloform reaction   The conversion of a methyl ketone into a molecule of a carboxylic acid and a molecule of haloform. The trihalo carbonyl compound is formed, base adds to the carbonyl group, and the trihalomethyl anion ($^{-}:CX_3$) is eliminated. The reaction works for X = Cl, Br, or I.

Halogenation   In aromatic systems, electrophilic substitution using chlorine or bromine and Lewis acid catalysts, usually $FeCl_3$ or $FeBr_3$.

Hammond postulate   The transition state for an endothermic reaction will resemble the product. It can be equivalently stated as, The transition state for an exothermic reaction will resemble the starting material.

Hantzsch synthesis   A synthesis of pyridines that uses multiple condensation reactions of β-keto esters, ammonia, and an aldehyde. A dihydropyridine is an intermediate.

Haworth form   The representation of sugars in which they are shown as planar rings.

Heat of formation ($\Delta H_f°$)   The heat evolved or required for the formation of a molecule from its constituent elements in their standard states. The more negative the heat of formation, the more stable the molecule.

Heisenberg uncertainty principle   For an electron, the uncertainty in position times the uncertainty in momentum (or speed) is a constant. We cannot know the exact position and momentum (speed) of an electron at the same time.

α-Helix   A right-handed coiled form adopted by many proteins in their secondary structures.

Hell–Volhard–Zelinsky (HVZ) reaction   The conversion of a carboxylic acid into either the α-bromo acid or the α-bromo acid bromide through reaction with PBr_3/Br_2.

Hemiacetal   The initial product when an alcohol adds to an aldehyde or ketone.

Hemoglobin   The protein in human blood responsible for oxygen transport.

Heteroaromatic compound   See **Heterobenzene**.

Heterobenzene   A benzene ring in which one (or more) ring carbons is replaced with another heavy (non-hydrogen) atom.

Heterocyclic compound   A ring compound containing at least one heteroatom in the ring.

Heterogeneous catalysis   A catalytic process in which the catalyst is insoluble.

Heterolytic bond cleavage The breaking of a bond to produce a pair of oppositely charged ions.

$$X \overset{\frown}{-} Y \longrightarrow X^+ + \overset{..}{:}Y$$

Hexose A six-carbon sugar.

High-performance liquid chromatography (HPLC) An especially effective version of column chromatography in which the stationary phase consists of many tiny spheres that provide an immense surface area for absorption.

Hofmann elimination The formation of the less substituted alkene in an elimination reaction.

Hofmann rearrangement The formation of amines through the treatment of amides with bromine and base. An intermediate isocyanate is hydrolyzed to a carbamic acid that decarboxylates.

HOMO Highest occupied molecular orbital.

Homogeneous catalysis A catalytic process in which the catalyst is soluble.

Homolytic bond cleavage The breaking of a bond to form two neutral species.

$$X \overset{\frown\frown}{-} Y \longrightarrow X\cdot + \cdot Y$$

Homotopic Homotopic hydrogens are identical, both chemically and spectroscopically, under all circumstances.

Hückel's rule All planar, cyclic, fully conjugated molecules with $4n + 2$ π electrons will be aromatic (especially stable). The rule works because such molecules will have molecular orbital systems in which all bonding molecular orbitals are completely full and in which no antibonding or nonbonding molecular orbitals are occupied.

Hund's rule For a given electronic configuration, the state with the greatest number of parallel spins is the lowest in energy. That is,

is lower in energy than

Hunsdiecker reaction The conversion of silver salts of carboxylic acids into alkyl halides, usually bromides.

Hybridization A mathematical model in which atomic orbital wave functions are combined to produce new, combination, or hybrid orbitals. The new orbitals are made up of fractions of the pure atomic orbital wave functions. Thus, an sp^3 hybrid is made of three parts p wave function and one part s wave function.

Hydrate The product of the reaction of a carbonyl compound with water.

Hydration The addition of water to a molecule. This reaction is generally acid catalyzed.

Hydride A negatively charged hydrogen ion ($H:^-$) bearing a pair of electrons.

Hydride shift The migration of hydrogen with a pair of electrons ($H:^-$).

Hydroboration The addition of BH_3 (in equilibrium with the dimer, B_2H_6) to π systems to form alkylboranes. These alkylboranes can be further converted into alcohols. Addition to an unsymmetrical alkene proceeds so as to give the less substituted alcohol.

Hydrocarbon cracking The thermal treatment of high molecular weight hydrocarbons to give lower molecular weight fragments. Bonds are broken to give radicals that abstract hydrogen to give alkanes, undergo β-cleavage, and disproportionate to give alkenes and alkanes.

Hydrocarbons Molecules containing only carbon and hydrogen.

Hydrogenation Addition of hydrogen (H_2) to the π bond of an alkene to give an alkane. A soluble or insoluble metallic catalyst is necessary. Alkynes also undergo hydrogenation to give alkanes, or under special conditions, alkenes.

Hydrogen bonding A low-energy bond between a pair of electrons, usually on oxygen, and a hydrogen.

Hydroperoxide (ROOH) The prefix *per* means "extra oxygen."

Hydrophilic "Water-loving," hence a polar group soluble in water.

Hydrophobic "Water-hating," hence a nonpolar group insoluble in water.

Hyperconjugation The stabilization of carbocations through the overlap of a filled σ orbital with the empty $2p$ orbital on carbon.

Imide A compound containing the structure O=C−NH−C=O, such as

Imine The nitrogen analogue of a ketone or aldehyde.

Immonium ion A compound of the structure

Indole　Any compound containing the following ring system:

Inductive effects　Electronic effects transmitted through the σ bonds. All bonds between different atoms are polar, and thus many molecules contain dipoles. These dipoles can affect reactions through induction.

Infrared spectroscopy (IR)　In IR, absorption of IR energy causes bonds to vibrate and rotate. The characteristic vibrational frequencies can be used to determine the functional groups present in a molecule.

Inhibitors　A species that can react with a radical, thus destroying it. Inhibitors interrupt chain reactions.

Initiation step　The first step in a radical chain reaction in which a free radical is produced that can start the chain-carrying or propagation steps.

Integration　The determination of the relative numbers of hydrogens corresponding to the signals in an NMR spectrum.

Inversion of configuration　The change in the handedness of a molecule during certain reactions. Generally, inversion of stereochemistry means that an (R) starting material would be transformed into an (S) product.

Ion　A charged atom or molecule.

Ion exchange chromatography　A separation technique that relies on differences in the affinity of molecules for a charged substrate.

Ionic bond　The electrostatic attraction between a positively charged atom or group of atoms and a negatively charged atom or group of atoms.

Ionization potential　A measure of the tendency of an atom or molecule to lose an electron.

Ipso attack　Addition to an aromatic ring at a position already occupied by a non-hydrogen substituent.

Isobutyl group　The group $(CH_3)_2CHCH_2$.

Isocyanate　A compound of the structure

Isoelectric point　The pH at which the amount of an amino acid present as the ammonium ion exactly equals the amount present as the carboxylate anion.

Isomers　Molecules of the same formula but different structures.

Isoprene　2-Methyl-1,3-butadiene.

Isoprene rule　Most terpenes are composed of isoprene units combined in "head-to-tail" fashion.

Isoprenoids　Compounds whose carbon skeleton is composed of isoprene units. Terpenes.

Isopropyl group　The group $(CH_3)_2CH$.

Isoquinoline　Any compound containing the following ring system:

Isotope effect　The effect on the rate of a reaction of the replacement of an atom with an isotope. The replacement of hydrogen with deuterium is especially common.

Karplus curve　The dependence between the coupling constant of two hydrogens on adjacent atoms and the dihedral angle between the carbon–hydrogen bonds.

Kekulé forms　Resonance forms for benzene in which overlap between adjacent carbons is emphasized. These forms superficially resemble 1,3,5-cyclohexatriene.

Ketene　A compound of the structure

Keto sugar　A sugar containing not the usual aldehyde group, but a ketone.

Ketone　A compound containing a carbon–oxygen double bond in which the carbon is attached to two other carbons (not hydrogen).

Ketose　A sugar containing not the usual aldehyde group, but a ketone.

Kiliani–Fischer synthesis　A method of lengthening the chain of an aldose by one carbon atom. A pair of sugars, epimeric at the new C(1) is produced.

Kinetic control　A reaction in which the product distribution is determined by the heights of the different transition states leading to products.

Kinetic enol　The most easily formed enol. It may or may not be the same as the most stable possible enol, the thermodynamic enol.

Kinetic resolution　A technique for separating a pair of enantiomers based on the selective transformation of one of them, often by an enzyme.

Kinetics　The study of the rates of reactions.

Knoevenagel condensation　Any of a number of condensation reactions related to the crossed aldol condensation. A stabilized anion, often an enolate, is first formed and then adds to the carbonyl group of another molecule. Dehydration often leads to the formation of the final product.

Kolbe electrolysis　The electrochemical conversion of carboxylate anions into hydrocarbons. The carboxyl radical produced loses CO_2 to give an alkyl radical that dimerizes.

Lactam   A cyclic amide.

Lactone   A cyclic ester.

α-Lactone   A three-membered lactone (cyclic ester).

Le Chatelier's principle   A system at equilibrium adjusts so as to relieve any stress upon it.

Leaving group   The departing group in a substitution or elimination reaction.

Levorotatory   The rotation of the plane of plane-polarized light in the counterclockwise direction.

Lewis acid   Any species that reacts with a Lewis base.

Lewis base   Any species with a reactive pair of electrons.

Lewis structure   In a Lewis structure, every electron (except the $1s$ electrons for atoms other than hydrogen or helium) is shown as a dot. In slightly more abstract Lewis structures, electrons in bonds are shown as lines connecting atoms.

Lithium diisopropylamide (LDA)   An effective base for alkylation reactions of carbonyl compounds. This compound is a strong base, but a poor nucleophile, and thus is effective at removing α-hydrogens but not at additions to carbonyl compounds.

Lobry de Bruijn–Alberda van Ekenstein reaction   The base-catalyzed interconversion of aldo and keto sugars. The key intermediate is the double enol formed by protonation of an enolate.

Lone-pair electrons   Electrons in an orbital that is not involved in binding atoms. See **nonbonding electrons**.

Long-range coupling   Any coupling between nuclei separated by more than three bonds. It is not usually observed in ^{1}H NMR spectroscopy.

LUMO   Lowest unoccupied molecular orbital

McLafferty rearrangement   A common process in MS through which a hydrogen in the γ-position of a carbonyl compound is transferred to oxygen to give an enol radical cation and an alkene.

Magid's Second Rule   "Always try the Michael reaction first."

Magid's Third Rule   "In times of desperation and/or despair, when all attempts at solution seem to have failed, try a hydride shift."

Malonic ester synthesis   The alkylation of malonic esters to produce substituted 1,1-diesters, followed by hydrolysis and decarboxylation of the malonic acids to give substituted acetic acids.

Malonic ester   The 1,1-diester, $ROOC-CH_2-COOR$.

Markovnikov's rule   In additions of Lewis acids to alkenes, the Lewis acid will add to the less substituted end of the alkene. The rule works because additions of Lewis acids produce the more substituted, more stable carbocation.

Mass spectrometry (MS)   The analysis of the ions formed by bombardment of a molecule with high-energy electrons. High-resolution MS can give the molecular formula of an ion, and an analysis of the fragmentation pattern can give information about the structure.

Meerwein–Ponndorf–Verley–Oppenauer (MPVO) equilibration   A method of oxidizing alcohols to carbonyl compounds, or, equivalently, of reducing carbonyls to alcohols, that uses an aluminum atom to clamp the partners in the reaction together. A hydride shift is involved in the crucial step.

Meisenheimer complex   The intermediate in nucleophilic aromatic substitution formed by the addition of a nucleophile to an aromatic compound activated by electron-withdrawing groups, often NO_2.

Mercaptan (RSH)   A thiol. The sulfur counterpart of an alcohol.

Mercaptide (RS⁻)   The sulfur counterpart of an alkoxide.

Meso compound   An achiral compound containing stereogenic atoms.

Messenger RNA (mRNA)   A polynucleotide produced from DNA whose base sequence codes for amino acid assembly.

Meta substitution   1,3-Substitution on a benzene ring.

Methane   The simplest stable hydrocarbon, CH_4.

Methine group   The group CH.

Methyl anion $^-$:CH_3.

Methyl cation $^+CH_3$.

Methyl compounds   Substituted methanes; CH_3-X compounds.

Methyl radical $\cdot CH_3$.

Methylene group   The group CH_2.

Micelle   An aggregated group of fatty acid salts (or other molecules) in which a hydrophobic center consisting of the hydrocarbon chains is protected from an aqueous environment by a spherical skin of polar hydrophilic groups.

Michael reaction   The addition of a nucleophile, often an enolate, to the β-position of a double bond of an α,β-unsaturated carbonyl group.

Microscopic reversibility   The notion that the lowest energy path for a reaction in one direction will also be the lowest energy path in the other direction.

Molecular orbital   An orbital not restricted to the region of space surrounding an atom, but extending over several atoms in a molecule. Molecular orbitals are formed through the overlap of atomic or molecular orbitals. Molecular orbitals can be bonding, nonbonding, or antibonding.

Molozonide   The initial product of the reaction of ozone and an alkene. This molecule contains a five-membered ring with three oxygen atoms in a row. See **Primary ozonide**.

Morphine   An important medicinal alkaloid isolated from the opium poppy.

Mutarotation   The interconversion of anomeric sugars in which an equilibrium mixture of α- and β-forms is reached.

NAD$^+$ Nicotinamide adenine dinucleotide, a biological oxidizing agent.

Neighboring group effect A general term for the influence of an internal nucleophile (very broadly defined) on the rate of a reaction, or on the structures of the products produced.

Newman projection A convention used to draw what one would see if one could look down a bond. The groups attached to the front atom are drawn in as three lines. The back atom is represented as a circle to which its attached bonds are drawn. Newman projections are enormously useful in seeing spatial (stereochemical) relationships in molecules.

Ninhydrin The hydrate of indan-1,2,3-trione, a molecule that reacts with amino acids to give the purple dye used in quantitative analysis of amino acids.

Nitration The electrophilic substitution of aromatic compounds by nitro groups using a source of the nitronium ion, $^+NO_2$.

Nitrene A reactive intermediate containing a neutral, monovalent nitrogen atom. The nitrogen counterpart of a carbene.

Nitrile A compound of the structure RCN, also commonly called cyanides.

Nitroso compound A compound containing the N=O group.

Node The region of zero electron density separating regions of opposite sign in an orbital. At a node the sign of the wave function is zero.

Nonbonding electrons Electrons in an orbital that is not involved in binding atoms. See **lone pair.**

Nonbonding orbital An electron in a nonbonding orbital neither helps to hold the molecule together nor tears it apart.

Nonreducing sugar A carbohydrate containing no amount of an oxidizable aldehyde group.

Nor A prefix meaning "no methyl groups."

Norbornyl system The bicyclo[2.2.1]heptyl system.

Nuclear magnetic resonance spectroscopy (NMR) Spectroscopy that detects the absorption of energy as the lower energy nuclear spin state in which the nuclear spin is aligned with an applied magnetic field flips to the higher energy spin state in which it is aligned against the field. See **Chemical shift** and **Coupling constant.**

Nucleic acid The polynucleotides DNA and RNA.

Nucleophile Generally, a Lewis base. A strong nucleophile has a high affinity for a carbon $2p$ orbital.

Nucleoside A sugar, either ribose or deoxyribose, bonded to a heterocyclic base at its 1' position.

Nucleotide A phosphorylated nucleoside; one of the monomers of which DNA and RNA are composed.

Nucleus The positively charged core of an atom containing the protons and neutrons.

Octet rule The notion that special stability attends the filling of the $2s$ and $2p$ atomic orbitals to achieve the electronic configuration of neon, a noble gas.

Off-resonance decoupling A technique used in ^{13}C NMR in which coupling between ^{13}C and ^1H is restricted to the hydrogens directly attached to the carbon. The number of directly attached hydrogens can be determined from the multiplicity of the observed signal for the carbon.

Olefins Hydrocarbons of the general formula, C_nH_{2n}. These molecules, also called alkenes, contain carbon–carbon double bonds.

Optical activity The rotation by a molecule of the plane of plane-polarized light.

Orbital A three-dimensional representation of the solution to the Schrödinger equation. See **Wave function.**

π Orbitals Molecular orbitals made from the overlap of p orbitals.

Organocuprates An organometallic reagent (R_2CuLi) notable for its ability to react with primary or secondary halides (R'—X) to generate hydrocarbons, R—R', and to add to the β-position of α,β-unsaturated carbonyl compounds.

Organolithium reagent (R—Li) A strongly basic organometallic reagent formed from a halide and lithium. A characteristic reaction is addition to carbonyl groups.

Organometallic reagents Molecules that contain both carbon and a metal. Usually carbon is at least partially covalently bonded to the metal. Examples are Grignard reagents, organolithium reagents, and lithium organocuprates.

Ortho 1,2-Disubstituted on a benzene ring.

Ortho esters Compounds of the structure

$$
\begin{array}{c}
\text{OR} \\
| \\
\text{R—C—OR} \\
| \\
\text{OR}
\end{array}
$$

Orthogonal orbitals Two noninteracting orbitals. The bonding interactions are exactly balanced by antibonding interactions

Ortho substitution 1,2-Substitution on a benzene ring.

Osazone A 1,2-phenylhydrazone formed by treatment of a sugar with three equivalents of phenylhydrazine.

Oxacyclopropane A three-membered ring containing oxygen, also called an epoxide or oxirane.

Oxaphosphetane The four-membered intermediate in the Wittig reaction that contains two carbons, a phosphorus, and an oxygen.

Oxetane A saturated, four-membered ring containing a single oxygen atom.

N-**Oxide**   A compound of the structure $R_3N^+-O^-$. They are produced from the reaction of tertiary amines with hydrogen peroxide. See also **Amine oxide**.

Oxirane   A three-membered ring containing oxygen, also called an epoxide or oxacyclopropane.

Oxonium ion   A molecule containing a trivalent, positively charged, oxygen atom (R_3O^+).

Oxymercuration   The mercury-catalyzed conversion of alkenes into alcohols. Addition is in the Markovnikov sense, and there are no rearrangements. A three-membered ring containing mercury is an intermediate in the reaction. Alkynes also undergo oxymercuration to give enols that are rapidly converted into carbonyl compounds under the reaction conditions.

Ozonide   The product of rearrangement of the primary ozonide formed on reaction of ozone with an alkene.

Ozonolysis   The reaction of ozone with π systems. The intermediate products are ozonides that can be transformed into carbonyl-containing compounds of various kinds.

p + 1 peak   An ion with a mass one greater than that of the parent ion. It appears because molecules contain atoms with isotopes of mass one greater than that of the most commonly observed isotope.

Para   1,4-Disubstituted on a benzene ring.

[*n*]paracyclophane   A benzene ring bridged 1,4 (para) with a chain of atoms.

Para substitution   1,4-Substitution on a benzene ring.

Parent ion   The ion directly formed in the mass spectrometer by ejection of an electron from the sample molecule. The parent ion commonly undergoes many fragmentation reactions before detection.

Pauli principle   No two electrons in an atom or molecule may have the same values of the four quantum numbers.

Penicillin   A powerful antibiotic. The key feature is the reactive β-lactam ring.

Pentose   A five-carbon sugar.

Peptide   A polyamino acid in which the constituent amino acids are linked through amide bonds. They are distinguished from proteins only by the length of the polymer.

Pericyclic reaction   A reaction in which the maintenance of bonding overlap between the lobes of the orbitals involved in bond making and breaking is controlling. All single-barrier (concerted) reactions can be regarded as pericyclic reactions.

Peroxides   Compounds of the structure $R-O-O-R$.

Phase-transfer catalysis   A synthetic method that often takes advantage of the ability of alkylammonium ions to transport ions from aqueous to organic media.

Phenol

Phenonium ion   The benzenonium ion produced through intramolecular displacement of a leaving group by the π system (not the σ system) of an aromatic ring.

Phenyl group   A common name for the benzene ring as substituent, C_6H_5-X.

Phenylisothiocyanate   The active ingredient, PhNCS, in the Edman degradation that converts the amino terminus of a peptide into a phenyl thiohydantoin.

Photohalogenation   The formation of alkyl halides through the irradiation of molecular X_2 (X = Br or Cl) in an alkane.

Phthalimide   The cyclic imide of phthalic acid.

Pinacol rearrangement   An acid-catalyzed rearrangement reaction leading from 1,2-diols to carbonyl compounds. The rearrangement is driven by the conversion of a localized carbocation into a resonance-stabilized carbocation.

Piperidine (azacyclohexane)   A six-membered ring amine.

pK_a   The negative logarithm of the acidity constant. A strong acid has a low pK_a, a weak acid has a high pK_a.

β-Pleated sheet   One of the common forms of secondary peptide structure in which hydrogen bonding holds chains of amino acids roughly in parallel lines.

Polar covalent bond   Any shared electron bond between two different atoms must be polar. Two nonidentical atoms must have different electronegativities and will attract the shared electrons to different extents, creating a dipole.

Polarimeter   A device for measuring the amount of rotation of the plane of plane-polarized light.

Polynuclear aromatic compound   An aromatic molecule composed of two or more fused aromatic rings.

ppm scale   The chemical shift scale commonly used in NMR spectroscopy in which the positions of absorptions are quoted as parts per million (ppm) of applied magnetic field. On the ppm scale, the chemical shift is independent of the magnetic field.

Primary amine   An amine bearing only one R group and two hydrogens.

Primary carbon   A carbon atom attached to one other carbon.

Primary ozonide   The initial product of the reaction of ozone and an alkene. This molecule contains a five-membered ring with three oxygen atoms in a row. See **Molozonide**.

Primary structure   The sequence of amino acids in a protein.

Product-determining step   The step in a multistep reaction that determines the structure or structures of the products.

Propagation step   The product-producing steps of a chain reaction. In the last propagation step, a molecule of product is formed along with a new radical that can carry the chain.

Propargyl group   The H$-$C$\equiv$C$-$CH$_2$ group.

Propyl group   The CH$_3$CH$_2$CH$_2$ group.

Protecting group   Sensitive functional groups in a molecule can be protected by a reaction that converts them into a less reactive group, called a protecting group. A protecting group must be removable to regenerate the original functionality.

Protein   A polyamino acid in which the constituent amino acids are linked through amide bonds. They are distinguished from peptides only through the length of the polymer.

Protic solvent   A solvent containing a hydrogen easily lost as a proton. Examples are water and most alcohols.

Pseudo-first-order reaction   A bimolecular reaction in which the concentration of one reagent (usually the solvent) does not change appreciably.

Pseudohalogen   Any of a variety of X$-$Y compounds that add to alkenes as do the halogens. Examples include I$-$NCO and I$-$N$_3$.

Pyranose   A sugar containing a six-membered cyclic ether.

Pyranoside   A pyranose in which the anomeric OH at C(1) has been converted into an OR group.

Pyridine   Any compound containing the following "azabenzene" ring system:

Pyridine *N*-oxide   A pyridine in which the nitrogen has been oxidized to the oxide stage.

Pyridinium ion   A quaternary ammonium ion formed from a pyridine.

Pyrolysis   The process of inducing chemical change by supplying heat energy. See **Thermolysis**.

Pyrophosphate   The group

O   O
‖   ‖
O$-$P$-$O$-$P$-$OH
| |
OH   OH

Pyrrole   Any compound containing the following ring structure:

Pyrrolidine (azocyclopentane) A saturated five-membered ring amine.

Quantum numbers These numbers evolve from the Schrödinger equation and characterize the various solutions to it. They may have only certain values, and these values determine the distance of an electron from the nucleus (n), the shape (l), and orientation (m_l), of the orbitals, and the electron spin (s).

Quaternary ammonium ions A compound of the structure $R_4N^+ {}^-X$.

Quaternary carbon A carbon attached to four other carbons.

Quaternary structure The structure of a self-assembled aggregate of two or more protein units.

Quinine An important antimalarial alkaloid containing the quinoline ring structure.

Quinoline Any compound containing the following ring structure:

Racemic mixture (racemate) A mixture containing equal amounts of two enantiomeric forms of a chiral molecule.

Radical A neutral molecule containing an odd, unpaired electron. Also called a "free radical."

Radical anion A negatively charged molecule containing both a pair of electrons and an odd, unpaired electron.

Radical cation A species that is positively charged yet contains a single unpaired electron. These are commonly formed by ejection of an electron when a molecule is bombarded with high-energy electrons in a mass spectrometer.

Random coil Disordered portions of a chain of amino acids.

Raney nickel A good reducing agent composed of finely divided nickel on which hydrogen has been adsorbed.

Rate constant (k) A fundamental property of a reaction that depends on the temperature, pressure, and solvent, but not on the concentrations of the reactants.

Rate-determining step The step in a reaction with the highest activation energy.

Reaction mechanism Loosely speaking, How does the reaction occur? How do the reactants come together? Are there any intermediates? What do the transition states look like? More precisely, a determination in terms of structure and energy of the stable molecules, reaction intermediates, and transition states involved in the reaction, along with a consideration of how the energy changes as the reaction progresses.

Reactive intermediates Molecules of great instability, and hence fleeting existence under normal conditions. Most carbon-centered anions, cations, and radicals are examples.

Rearrangement The migration of an atom or group of atoms from one place to another in a molecule. Rearrangements are exceptionally common in reactions involving carbocations.

Reducing sugar A sugar containing some amount of an oxidizable free aldehyde group.

Reductive amination A synthetic method in which an imine or enamine is generated from a carbonyl compound and an amine only to be reduced in situ to give the substituted amine.

Regiochemistry The orientation of a reaction taking place on an unsymmetrical substrate.

Regioselectivity (regiospecificity) If a reaction may produce one or more isomers and one predominates, the reaction is called regioselective. If one product strongly predominates, the reaction is regiospecific.

Reissert synthesis A synthesis of indoles starting from a nitrated alkylbenzene and an ester of oxalic acid.

Resolution The separation of a racemic mixture into its constituent enantiomeric molecules.

Resonance energy The energy lowering conferred by the delocalization of electrons. In benzene, this is the amount by which benzene is more stable than the hypothetical 1,3,5-cyclohexatriene containing three localized double bonds. See **Delocalization energy**.

Resonance forms Many molecules cannot be represented adequately by a single Lewis form. Instead, two or more different electronic representations must often be combined to give a good description of the molecule. These different representations are called resonance forms.

Retention of configuration The preservation of the handedness of a molecule in a reaction. Generally, retention of configuration means that an (R) starting material would be transformed into an (R) product.

Retention time The time necessary for elution of a component of a mixture in chromatography.

Ribonucleic acid (RNA) A polymer of nucleotides made up of ribose units connected by phosphoric acid links. Each sugar is attached to one of the bases, A, U, G, or C.

Robinson annulation A classic method of construction for six-membered rings using a ketone and methyl vinyl ketone. It involves a two-step sequence of a Michael reaction between the enolate of the original ketone and methyl vinyl ketone, followed by an intramolecular aldol condensation.

Rosenmund reduction The reduction of acid chlorides to aldehydes using a poisoned catalyst.

Ruff degradation A method for shortening the carbon backbone of a sugar by one carbon. The aldehyde carbon at C(1) is lost and a new aldehyde created at the old C(2).

Saccharide   A molecule whose formula can be factored into $C_x(H_2O)_y$. A "sugar" or "carbohydrate."

Sandmeyer reaction   The reaction of an aromatic diazonium ion with cuprous salts to form substituted aromatic compounds.

Sanger degradation   A method for determining the amino acid at the amino terminus of a chain using 2,4-dinitrofluorobenzene. Unfortunately, the Sanger degradation hydrolyzes, and thus destroys, the entire peptide.

Saponification   Base-induced hydrolysis of an ester, usually a fatty acid.

Saturated alkanes   Alkanes of the molecular formula C_nH_{2n+2}.

Saytzeff elimination   Formation of the more substituted alkene in an elimination reaction.

Schiff base   The nitrogen analogue of a ketone or aldehyde. See **imine.**

Schmidt reaction   The formation of amines through the reaction of carboxylic acids with hydrazoic acid, HN_3. The key step is an intramolecular rearrangement to generate an isocyanate that is then converted into an unstable carbamic acid under the reaction conditions. Decarboxylation gives the amine.

Second-order reaction   A reaction for which the rate depends on the product of a rate constant and the concentrations of two reagents.

Secondary amine   An amine bearing one hydrogen and two R groups.

Secondary carbon   A carbon attached to two other carbons.

Secondary structure   Ordered regions of a protein chain. The two most common types of secondary structure are the α-helix and the β-pleated sheet.

Selenoxide   A compound of the structure

Sigma bond (σ)   Any bond with cylindrical symmetry.

Sigmatropic shift   The migration of an atom or group, under orbital symmetry control, along a π system.

Singlet carbene   A singlet carbene contains only paired electrons. In a singlet carbene, the two nonbonding electrons have opposite spins and occupy the same orbital.

Six-center transition state   The boat-like transition state for the Cope rearrangement in which carbons 1 and 1′, 2 and 2′, and 3 and 3′ are all eclipsed.

Skraup synthesis   A synthesis of quinolines starting from aniline and glycerol (1,2,3-trihydroxypropane).

S_N1 Reaction   Substitution, nucleophilic, unimolecular. An initial ionization is followed by attack of the nucleophilic solvent.

S_N2 Reaction   Substitution, nucleophilic, bimolecular. In this reaction, the nucleophile dispaces the leaving group by attack from the rear. The stereochemistry of the starting material is inverted.

Soap   The salt of a fatty acid.

Sodium bicarbonate

Sodium carbonate

Solvated electron   The species formed when sodium is dissolved in ammonia. The product is a sodium ion and an electron surrounded by ammonia molecules. This electron can add to alkynes (and some other π systems) in a reduction step.

Solvation   The stabilizing effects of a solvent, typically the stabilization of ions by a polar solvent.

Solvolysis reaction   In such S_N1 and S_N2 reactions, the solvent acts as the nucleophile.

***sp* Hybrid**   A hybrid orbital made by the combination of one *s* and one *p* atomic orbital.

***sp²* Hybrid**   A hybrid orbital made by the combination of one *s* orbital and two *p* orbitals.

***sp³* Hybrid**   A hybrid orbital made by the combination of one *s* orbital and three *p* orbitals.

Spectroscopy   The study of the interactions between electromagnetic radiation and atoms and molecules.

Spiro rings   Two rings that share a single carbon.

Staggered ethane   The energy minimum conformation of ethane in which the carbon–hydrogen bonds (and the electrons in them) are as far apart as possible.

Starch   A nonlinear polymer of glucose containing amylose, an α-linked linear polymer of glucose.

Stereochemistry   The physical and chemical consequences of the arrangement in space of the atoms in molecules.

Stereogenic center   An atom, usually carbon, of such nature and bearing groups of such nature that it can have two non-equivalent configurations.

Steric   A generic word referring to the arrangement in space of atoms and groups of atoms.

Steroid   A class of four-ring compounds, always containing three six-membered rings and one five-membered ring. The ring system can be substituted in many ways, but the rings are always connected in the following fashion:

Strecker synthesis   A synthesis of amino acids using an aldehyde, cyanide, and ammonia, followed by hydrolysis.

Structural isomers   Molecules of the same formula but with different connectivities among the constituent atoms.

Substitution reaction   The replacement of a leaving group in R—L by a nucleophile, Nu:$^-$. Typically,

$$\text{Nu:}^- \ + \ \text{R—L} \longrightarrow \text{Nu—R} \ + \ \text{L:}^-$$

Sugar   A molecule whose formula can be factored into $C_x(H_2O)_y$. A "saccharide" or "carbohydrate."

Sulfide (RSR)   The sulfur counterpart of an ether. See **Thioether**.

Sulfonation   The electrophilic substitution of aromatic compounds by $SO_3/HOSO_2OH$ to give benzenesulfonic acids.

Sulfone (R—SO$_2$—R)   The final product of oxidation of a sulfide with hydrogen peroxide.

Sulfonic acid   A compound of the general structure RSO_2OH.

Sulfonium ion (R$_3$S$^+$)   A positively charged sulfur ion.

Sulfoxide   A compound of the structure

Superacid   A number of highly polar, highly acidic, but weakly or nonnucleophilic solvent systems. Superacids are often useful in stabilizing carbocations, at least at low temperature.

Suprafacial motion   Migration of a group in a sigmatropic shift in which bond breaking and bond making take place on the same side of the π system.

syn Elimination   An elimination reaction in which the dihedral angle between the breaking bonds, usually C—H and C—L, is 0°.

Termination step   The mutual annihilation of two radicals. Termination steps destroy chain-carrying radicals and end chains through bond formation.

Terpenes   Compounds whose carbon skeleton is composed of isoprene units.

Tertiary amine   An amine bearing no hydrogens and three R groups.

Tertiary carbon   A carbon attached to three other carbons.

Tertiary structure   The structure of a protein induced by its folding pattern.

Tetrahydropyranyl (THP) ethers   Protecting groups for alcohols of the structure

Tetramethylsilane (TMS)   The standard "zero point" on the ppm scale in NMR spectroscopy. The chemical shift of a nucleus is quoted in parts per million of applied magnetic field relative to the position of TMS.

Tetrose   A four-carbon sugar.

Thermal eliminations   A group of intramolecular, syn β-elimination reactions in which an ester or related group acts both as base and leaving group.

Thermodynamic control   In a reaction under thermodynamic control, the product distribution depends on the energy differences between the products.

Thermodynamic enol   The most stable enol. It may or may not be the same as the most rapidly formed enolate, the "kinetic enol."

Thermodynamics   The study of energetic relationships.

Thermolysis   The process of inducing chemical change by supplying heat energy. See **Pyrolysis**.

Thietane   A saturated four-membered ring containing a single sulfur atom.

Thiirane   A saturated three-membered ring containing a single, unoxidized sulfur atom. See **Episulfide**.

Thioacetal   The sulfur counterpart of an acetal.

Thioether (RSR)   The sulfur counterpart of an ether. A sulfide.

Thiol   The sulfur counterpart of an alcohol, RSH.

Thionyl chloride (SOCl$_2$)   An effective reagent for converting alcohols into chlorides and carboxylic acids into acid chlorides.

Thiophene   An aromatic five-membered ring compound containing four CH units and one S atom.

Thiourea   A compound of the structure

$$\overset{\displaystyle :\!S\!:}{\underset{\displaystyle RHN \qquad NHR}{\|\!\!\!C}}$$

Three-center, two-electron bonding   A bonding system in which only two electrons bind three atoms. This arrangement is common in electron-deficient molecules such as boranes and carbocations.

Torsional strain   Destabilization caused by the proximity of bonds (usually eclipsing) and the electrons in them.

trans   "Hydrogens on opposite sides." Used to specify stereochemical (spatial) relationships in ring compounds and alkenes.

s-trans   The more stable, extended form of a 1,3-diene,

Transamination   The conversion of one amine into another.

Transesterification   The formation of one ester from another. The reaction can be catalyzed by either acid or base.

Transition state   The high point in energy between starting material and product. The transition state is an energy maximum and not an isolable compound.

Triazoline   Any compound containing the following ring system, usually formed by the 1,3-dipolar cycloaddition of an azide to an alkene:

Triple bond   Two atoms can be attached by a triple bond composed of one σ bond and two π bonds.

Triplet carbene   In a triplet carbene, the two nonbonding electrons have the same spin and must occupy different orbitals.

Trityl cation   The triphenylmethyl cation.

Tropane alkaloids   A family of alkaloids, all of which contain the tropane ring system.

Tropylium ion   The 1,3,5-cycloheptatrienylium ion. This ion has $4n + 2$ π electrons ($n = 1$) and is aromatic.

Trypsin   An enzyme capable of cleaving a peptide after lysine or arginine.

Ultraviolet (UV) spectroscopy   Electronic spectroscopy using light of wavelength 200–400 nm.

Unsaturated alkanes   Alkanes of the formula C_nH_{2n}.

Unsaturated   Containing less than the maximum number of hydrogens. Alkenes and alkynes (as well as ring compounds) are unsaturated molecules.

Urea

Valence electrons   The outermost, or most loosely held, electrons.

van der Waals forces   Attractive or repulsive intermolecular forces in or between molecules caused by induced dipole-induced dipole interactions.

van der Waals strain   Repulsive forces in or between molecules caused by induced dipole–dipole interactions.

Vicinal   1,2-Disubstituted.

Vinyl group   The H$_2$C=CH group.

Visible (vis) spectroscopy   Electronic spectroscopy using light of wavelength 400–800 nm.

Wagner–Meerwein rearrangement   The 1,2-migration of an alkyl group in a carbocation.

Wave function (ψ)   A solution to the Schrödinger equation, another word for "orbital."

Wavenumber   The wavenumber ($\bar{\nu}$) equals $1/\lambda$ or ν/c.

Williamson ether synthesis   The S_N2 reaction of an alkoxide with R—L to give an ether.

Wittig reaction   The reaction of an ylide with a carbonyl group to give, ultimately, an alkene.

Wohl degradation   A method for shortening the carbon backbone of a sugar by one carbon. The aldehyde carbon at C(1) is lost and a new aldehyde created at the old C(2).

Wolff rearrangement   The formation of a ketene through the thermal or photochemical decomposition of a diazo ketone.

Woodward–Hoffmann theory   The notion that one must take account of the phase relationships between orbitals in order to understand reaction mechanisms. In a concerted reaction, bonding relationships must be maintained at all times.

Xanthate ester   A compound of the structure

Ylide   A compound containing opposite charges on adjacent atoms.

Zwitterion   A dipolar species in which full plus and minus charges coexist within the same molecule.

Index

compound	pK_a (water)	compound	pK_a (water)
alkanes	50–60	(acetone, H₃C–CO–CH₃)	19.3
(CH₂=CH–H)	45–50	(Ph–CO–CH₃)	18.3
(CH₂=CH–CH₂–H)	43	(cyclohexanone)	18.1
(benzene, C₆H₅–H)	43	(H₃C–CO–NH₂)	18
(toluene, C₆H₅–CH₃)	41	$(CH_3)_3COH$	17
NH_3	38	(H₃C–CHO)	16.7
$CHCl_3$	29	$(CH_3)_2CHOH$	16.5
$R-C{\equiv}C-H$	~25	CH_3CH_2OH	15.9
(H₃C–CO–NH₂)	25	H_2O	15.7
CH_3CN	24	((CH₃)₂CH–CHO)	15.5
(H₃C–CO–OCH₂CH₃)	24	CH_3OH	15.2
(CH₃CH₂–CO–CH₂CH₃)	19.9		